PERIODIC TABLE OF

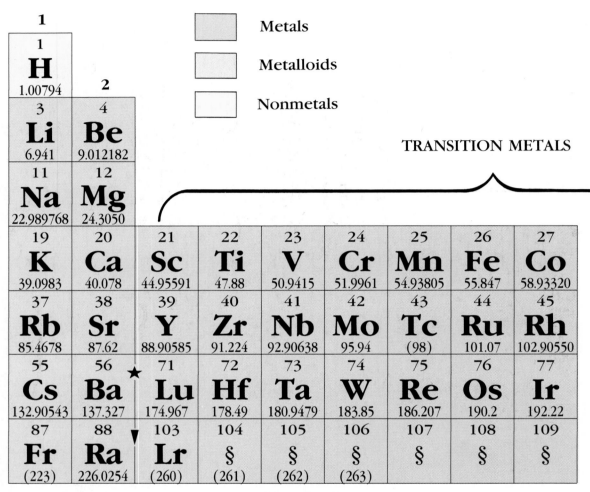

	Metals
	Metalloids
	Nonmetals

TRANSITION METALS

1		
1 **H** 1.00794	**2**	
3 **Li** 6.941	**4** **Be** 9.012182	
11 **Na** 22.989768	**12** **Mg** 24.3050	

19 **K** 39.0983	20 **Ca** 40.078	21 **Sc** 44.95591	22 **Ti** 47.88	23 **V** 50.9415	24 **Cr** 51.9961	25 **Mn** 54.93805	26 **Fe** 55.847	27 **Co** 58.93320
37 **Rb** 85.4678	38 **Sr** 87.62	39 **Y** 88.90585	40 **Zr** 91.224	41 **Nb** 92.90638	42 **Mo** 95.94	43 **Tc** (98)	44 **Ru** 101.07	45 **Rh** 102.90550
55 **Cs** 132.90543	56 **Ba** 137.327	71 **Lu** 174.967	72 **Hf** 178.49	73 **Ta** 180.9479	74 **W** 183.85	75 **Re** 186.207	76 **Os** 190.2	77 **Ir** 192.22
87 **Fr** (223)	88 **Ra** 226.0254	103 **Lr** (260)	104 **§** (261)	105 **§** (262)	106 **§** (263)	107 **§**	108 **§**	109 **§**

§ The International Union for Pure and Applied Chemistry has not adopted official names or symbols for these elements.

Note: Atomic masses shown here are 1985 IUPAC values.

★ Lanthanides

★ 57 **La** 138.9055	58 **Ce** 140.115	59 **Pr** 140.90765	60 **Nd** 144.24	61 **Pm** (145)

▼ Actinides

▼ 89 **Ac** 227.0278	90 **Th** 232.0381	91 **Pa** 231.03588	92 **U** 238.0289	93 **Np** 237.0482

THE ELEMENTS

				7	8
				1 **H** 1.00794	2 **He** 4.002602

3	4	5	6		
5 **B** 10.811	6 **C** 12.011	7 **N** 14.00674	8 **O** 15.9994	9 **F** 18.998403	10 **Ne** 20.1797
13 **Al** 26.981539	14 **Si** 28.0855	15 **P** 30.973762	16 **S** 32.066	17 **Cl** 35.4527	18 **Ar** 39.948

28 **Ni** 58.69	29 **Cu** 63.546	30 **Zn** 65.39	31 **Ga** 69.723	32 **Ge** 72.61	33 **As** 74.92159	34 **Se** 78.96	35 **Br** 79.904	36 **Kr** 83.80
46 **Pd** 106.42	47 **Ag** 107.8682	48 **Cd** 112.411	49 **In** 114.82	50 **Sn** 118.710	51 **Sb** 121.75	52 **Te** 127.60	53 **I** 126.90447	54 **Xe** 131.29
78 **Pt** 195.08	79 **Au** 196.96654	80 **Hg** 200.59	81 **Tl** 204.3833	82 **Pb** 207.2	83 **Bi** 208.98037	84 **Po** (209)	85 **At** (210)	86 **Rn** (222)

62 **Sm** 150.36	63 **Eu** 151.965	64 **Gd** 157.25	65 **Tb** 158.92534	66 **Dy** 162.50	67 **Ho** 164.93032	68 **Er** 167.26	69 **Tm** 168.93421	70 **Yb** 173.04

94 **Pu** (244)	95 **Am** (243)	96 **Cm** (247)	97 **Bk** (247)	98 **Cf** (251)	99 **Es** (252)	100 **Fm** (257)	101 **Md** (258)	102 **No** (259)

CHEMISTRY

Principles & Reactions

William L. Masterton
University of Connecticut

Cecile N. Hurley
University of Connecticut

SAUNDERS GOLDEN SUNBURST SERIES

SAUNDERS COLLEGE PUBLISHING

Harcourt Brace Jovanovich College Publishers

Fort Worth   Philadelphia   San Diego
New York   Orlando   Austin   San Antonio
Toronto   Montreal   London   Sydney   Tokyo

Text typeface: Baskerville
Compositor: General Graphic Services
Acquisitions Editor: John Vondeling
Developmental Editor: Sandra Kiselica
Project Editors: Becca Gruliow, Charlotte Hyland
Copy Editor: Joni Fraser
Art Director: Carol Bleistine
Art Assistant: Doris Bruey
Text Designer: Edward A. Butler
Cover Designer: Larry Didona
Text Artwork: J&R Technical Services
Layout Artist: Dorothy Chattin
Production Manager: Harry Dean

Cover Credit: Liquid, Solid, Gas by Dominique Sarraute © THE IMAGE BANK

Printed in the United States of America

CHEMISTRY: PRINCIPLES AND REACTIONS

ISBN 0-03-053028-8

Library of Congress Catalog Card Number: 88-043417

123456   071   9876543

DEDICATION

To Loris and Jim

*For their understanding when entropy was high
and their encouragement when energy was low.*

PREFACE

It is difficult for authors to praise the virtues of their own book. We could tell you it is so inspiring that students will be turned on to chemistry with no effort on your part, but we doubt you would believe that. We could tell you that the text is so clearly written that your students will learn chemistry with little or no effort on their part; certainly you wouldn't believe that. We can tell you that our two goals in writing this book have been to make it as clear and as interesting as possible. We hope you will believe that, because it is true.

As the title, "Chemistry: Principles and Reactions" implies, this text blends theory with practice and calculations with descriptive chemistry. Three major topics are emphasized in the first half of the book. These are,

Stoichiometry Molarity is introduced in Chapter 2, immediately following the mole concept. Chapter 3 deals with formulas and mass relations in reactions. It also includes sections on the formulas and names of ionic and simple molecular compounds. Our objective is to establish a firm foundation for the chemical principles to follow.

Atomic and Molecular Structure Chapter 5 describes the electronic structures of atoms and ions. The Periodic Table, introduced in

Chapter 1, is analyzed in some detail in Chapter 6. That chapter concludes with a discussion of the chemistry of the main-group metals (the alkalis, alkaline earths, and aluminum). Two chapters are devoted to covalent bonding: Chapter 8 emphasizes Lewis structures and Chapter 9 the VSEPR model of molecular structure.

States of Matter   The gas laws are covered in Chapter 4; liquids and solids are deferred to in Chapter 10, following chemical bonding. Chapter 11 emphasizes the physical properties of aqueous solutions, laying a foundation for later discussions of solution chemistry.

The second half of the book emphasizes two general areas:

Chemical Equilibrium   Chapter 13 applies equilibrium principles to gaseous systems; K_p is treated as well as K_c. Solution equilibria are first discussed in Chapter 14, where K_w, K_a, and K_b are introduced. In Chapter 15, these equilibrium constants are applied to explain the characteristics of buffers and acid-base titrations. The chapter concludes with a discussion of the solubility product constant, K_{sp}. Formation constants for complex ions are included in Chapter 16 (coordination chemistry). Once the principles of equilibrium have been introduced, they are referred to repeatedly in subsequent chapters.

Descriptive Chemistry   Two chapters (16 and 22) are devoted to the transition metals, two (23 and 24) to the nonmetals, and two more (26 and 27) to an introduction to organic chemistry and biochemistry. We have tried to avoid the dreary recitation of facts that has put a generation of students and their instructors to sleep. Instead of endlessly plodding through the chemistry of one element after another, we have organized descriptive chemistry around chemical principles. Perhaps the best example of this approach is seen in Chapter 22. There the chemistry of the transition metals is tied to the types of reactions undergone, first by the metals themselves, then by transition metal cations, and finally by their oxyanions (e.g., CrO_4^{2-} and MnO_4^{-}).

Thermodynamics is divided between two chapters. Thermochemistry, ΔH, and the First Law are presented in Chapter 7, after stoichiometry and the gas laws. This chapter serves as a quantitative interlude between qualitative discussions of atomic structure (Chapter 5 and 6) and chemical bonding (Chapters 8 and 9); it directly precedes the discussion of bond energies in Chapter 8. The more subtle aspects of thermodynamics, ΔG, ΔS, and the Second Law, are deferred to Chapter 20, immediately preceding the treatment of electrochemistry in Chapter 21.

We believe that there are three unique chapters found in this text:

Reactions in Aqueous Solution (Chapter 12)   This covers the writing and balancing of net ionic equations for precipitation, acid-base, and redox reactions. It also deals with the stoichiometry of these reactions and their application to volumetric analysis. Too often, this important material gets lost in discussions of more esoteric topics.

Qualitative Analysis (Chapter 17)   This deals with group separations in cation analysis and spot tests for a variety of anions. This material has an obvious application to the general chemistry laboratory. Beyond that, it offers a splendid opportunity to review the principles of solution chemistry presented in previous chapters.

Chemistry of the Atmosphere (Chapter 19)   Included here are such topics as nitrogen fixation, weather modification and the greenhouse effect, depletion of the ozone layer, and the ravages of acid rain. Throughout, the principles of chemical kinetics (Chapter 18) and equilibrium are applied. We hope you will find time to cover this material, because we believe it is both interesting and relevant.

More information about content can be found in the introductory discussions at the beginning of each chapter. Alternatively, you may wish to read the chapter summaries that are included in Appendix 5. Many pedagogical features of the text are described in the section addressed "To the Student," which follows this preface. We call your attention in particular to the Summary Problems and the Perspectives, two innovations in this text.

ANCILLARY PACKAGE

A complete resource package has been prepared to enhance the student's learning and the professor's teaching of *Chemistry: Principles and Reactions*.

Lecture Outline by Ronald O. Ragsdale (University of Utah). It helps students to organize the class lectures and frees them of extensive note taking.

Student Solutions Manual by John E. Bauman (University of Missouri, Columbia). It contains detailed solutions to half of the end-of-chapter problems and all the challenging problems.

Study Guide/Workbook by Cecile N. Hurley. It contains worked examples and problem-solving techniques to help the student's understanding of general chemistry. It also has fill-in-the-blanks, exercises, and self-tests to help the student gauge his mastery of the chapter.

Instructor's Manual by William L. Masterton. Included are basic skills, lecture outlines, classroom demonstrations, quizzes, and solutions to half of the end-of-chapter questions.

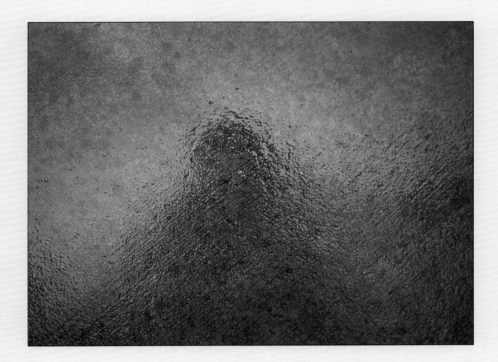

Chemical Principles in the Laboratory, 5th edition by Emil J. Slo-winski, Wayne C. Wolsey, and William L. Masterton. A revision of the best-selling general chemistry laboratory manual. It includes 43 experiments, each with a pre-lab study assignment. An instructor's manual is also available.

Overhead Transparencies   One-hundred four-color figures from the text.

Computerized Test Bank by Gordon Eggleton (Southeastern Oklahoma State University). A multiple-choice test bank with over 1000 test items for IBM PC or compatibles.

Printed Test Bank   Tests generated from the computerized test bank.

Audio-Tape Lessons by B. Shakhashiri, R. Schreiner, and P. A. Meyer (all of University of Wisconsin, Madison). Enables students to study and learn chemistry at their own pace. Students listen to the instructions on the tape and follow the examples in the workbook.

Tutorial Software   Wilkie Computerized Chemistry, for Apple and IBM. COMPress, for Apple II + and Apple IIe.

ACKNOWLEDGMENTS

We wish to express our deepest appreciation to two individuals who contributed a great deal to this text although their names do not appear on it. Emil Slowinski worked with us from the beginning, reviewing our material with candor and, above all, compassion. Beyond that, he wrote the marginal notes in his own inimitable style. Ruven Smith supervised and set up all the photography; somehow he managed to turn an onerous task into a delightful experience.

Conrad Stanitski supplied us with a thoughtful, detailed analysis of our manuscript. Other reviewers, to whom we are indebted, include:

John E. Bauman, University of Missouri, Columbia

Robert Conley, New Jersey Institute of Technology

Marsha C. Davies, Creighton University

Gordon Eggleton, Southeastern Oklahoma State University

Elizabeth S. Friedman, Los Angeles Valley College

Wyman K. Grindstaff, Southwest Mississippi State University

Anthony W. Harmon, University of Tennessee at Martin

Douglas W. Hensley, Louisiana Technical University

David L. Keeling, California Polytechnic State University

James Long, University of Oregon

James McClure, Southern Illinois University at Edwardsville

William E. Ohnesorge, Lehigh University

Ronald O. Ragsdale, University of Utah

Henry D. Schreiber, Virginia Military Institute

Thomas W. Sottery, University of Southern Maine

Robert S. Sprague, Lehigh University

Paul Walter, Skidmore College

Most of the photographs that appear in this book were taken by either Marna Clarke or Charles Winters. We express our appreciation to them and to Ed Kostiner, who allowed us to use the chemistry laboratories of the University of Connecticut for this project. Art Dimock, Lew Butler, and John Reynolds (see Figure 26.6) obtained the chemicals and equipment we needed, often on a few moments' notice. Jim Bobbitt, from the vast resources of his vineyard, supplied us with the condiments for Figure 26.7. He also gave us some advice as to the contents of Chapters 26 and 27.

Finally, we express our appreciation to all the people at Saunders who helped us with this book. In particular, we are grateful to Becca Gruliow, Charlotte Hyland, and Kate Pachuta for their thoughtfulness, good humor, and insightful suggestions. Special thanks go to Sandi Kiselica and John Vondeling, who deflected with equanimity the protestations showered upon them by two strong-willed authors.

William L. Masterton
Cecile N. Hurley

TO THE STUDENT

Over the next several months, you will probably receive a lot of advice from your instructor, teaching assistant, and fellow students about how to study chemistry. We hesitate to add our advice; experience as teachers and parents has taught us that students tend to do surprisingly well without it. We would, however, like to acquaint you with some of the learning tools in this text. They are briefly described and illustrated in the sample pages that follow.

Chapter-Opening Photographs
These illustrate, in a somewhat abstract way, chemical reactions of one type or another. Each photograph is accompanied by a chemical equation and a description of the reaction involved. All reactions shown are discussed in this text.

Included with the photograph at the beginning of each chapter is a poem or quotation. Be sure to read the one associated with Chapter 27, even if that chapter is not assigned.

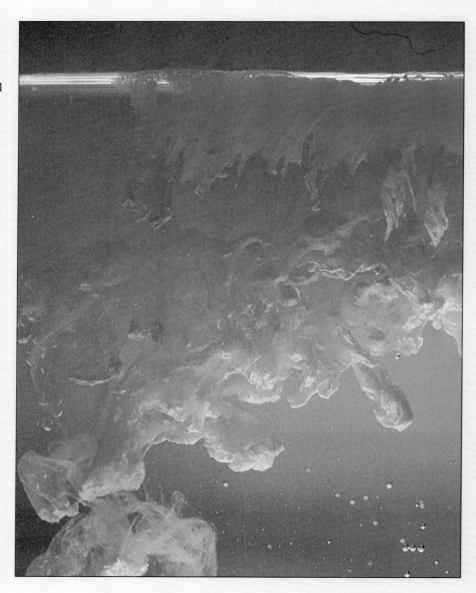

Flow Charts

Scattered throughout the book are several diagrams that suggest a general approach to solving a specific type of problem. This particular flow chart describes conversions between moles, grams, and number of particles. In using a flow chart, you should not attempt to memorize the various steps. Instead you should try to understand the process involved; once you do that, the sequence of steps will seem to be a logical one.

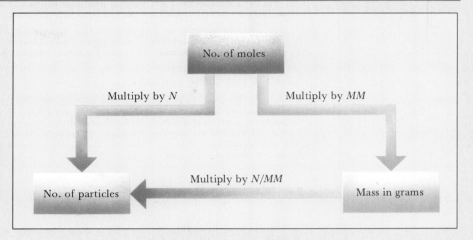

■ EXAMPLE 2.9

The bottle labeled "concentrated hydrochloric acid" in the lab contains 12.0 mol of the compound HCl per liter of solution, that is, M = 12.0 mol/L.

a. How many moles of HCl are there in 25.0 mL of this solution?
b. What volume of concentrated hydrochloric acid must be taken to contain 1.00 mol HCl?

Solution

The molarity of the HCl relates the number of moles of HCl to the number of liters of solution. Since there are 12.0 mol HCl per liter, we can say that

$$12.0 \text{ mol} \simeq 1 \text{ L solution}$$

This relation gives us the conversion factors we need to go from liters of solution to moles or vice versa.

a. We first convert 25.0 mL to liters:

$$\text{number liters solution} = 25.0 \text{ mL} \times \frac{1 \text{ L}}{1000 \text{ mL}} = 0.0250 \text{ L}$$

Now we convert to moles of HCl:

$$\text{no. moles HCl} = 0.0250 \text{ L} \times \frac{12.0 \text{ mol HCl}}{1 \text{ L}} = \boxed{0.300 \text{ mol HCl}}$$

b. To find the volume in liters, we start with the number of moles of HCl and use the conversion factor 1 L/12.0 mol HCl:

$$\text{number liters solution} = 1.00 \text{ mol HCl} \times \frac{1 \text{ L}}{12.0 \text{ mol HCl}}$$

$$= \boxed{0.0833 \text{ L}} \quad (83.3 \text{ mL})$$

Exercise

How many grams of HCl (molar mass = 36.46 g/mol) are there in 0.100 L of this solution? Answer: 43.8 g.

Examples and Exercises

In a typical chapter, you will find 10 or more examples, each designed to illustrate a particular principle. These have answers, screened in yellow to save you the trouble of highlighting them. More important, the step-wise solutions are worked out in some detail. Following each example is an exercise, for which only the answer is given. Try the exercise; if you do not get the right answer, the chances are you did not completely understand the principle illustrated. In that case, try again, getting help if you need it.

In Section 9.1, we pointed out that insofar as geometry is concerned, a multiple bond acts as if it were a single bond. In other words, the "extra" electron pairs in a double or triple bond have no effect upon the geometry of the molecule. This behavior is explained in terms of hybridization.

The extra electron pairs in a multiple bond (one pair in a double bond, two pairs in a triple bond) are not located in hybrid orbitals.

From this point of view, the geometry of a molecule is fixed by the electron pairs in hybrid orbitals about a central atom. These orbitals are directed to be as far apart as possible. The hybrid orbitals contain

— all unshared electron pairs.
— electron pairs forming single bonds.
— one and only one of the electron pairs in a double or triple bond.

To illustrate this rule, consider the ethylene (C_2H_4) and acetylene (C_2H_2) molecules. You will recall that the bond angles in these molecules are 120° for ethylene and 180° for acetylene. This implies sp^2 hybridization in C_2H_4 and sp hybridization in C_2H_2 (see Table 9.3). Using colored lines to represent hybridized electron pairs,

$$\begin{matrix} H & & H \\ & \diagdown\ \diagup & \\ & C=C & \\ & \diagup\ \diagdown & \\ H & & H \end{matrix} \qquad H—C≡C—H$$
ethylene   acetylene

C_2H_4—3 electron pairs in hybrid, so it is sp^2
C_2H_2—2 electron pairs in hybrid, so it is sp

In both cases, only one of the electron pairs in the multiple bond is hybridized.

Marginal Notes
Sprinkled throughout the text are a number of short notes that have been placed in the margin. Many of these are of the "now, hear this" variety; a few make points that we forgot to emphasize in the body of the text. Some, probably fewer than we think, are humorous.

Summary Problems
At the end of each chapter is a several-part problem covering all or nearly all of the main ideas in the chapter. You can test your understanding of the chapter by working this problem; you may wish to do this as part of your preparation for intraterm exams. A major advantage of a summary problem is that it ties together many different concepts, showing how they correlate with one another.

SUMMARY PROBLEM
Consider the thermochemical equation:

$$C_2H_6(g) + \tfrac{7}{2} O_2(g) \rightarrow 2\ CO_2(g) + 3\ H_2O(l); \qquad \Delta H = -1559.7\ \text{kJ}$$

a. Calculate ΔH for the thermochemical equations:

$$2\ C_2H_6(g) + 7\ O_2(g) \rightarrow 4\ CO_2(g) + 6\ H_2O(l)$$

$$2\ CO_2(g) + 3\ H_2O(l) \rightarrow C_2H_6(g) + \tfrac{7}{2} O_2(g)$$

b. The heat of vaporization of $H_2O(l)$ is $+44.0$ kJ/mol. Calculate ΔH for the equation:

$$C_2H_6(g) + \tfrac{7}{2} O_2(g) \rightarrow 2\ CO_2(g) + 3\ H_2O(g)$$

c. The heats of formation of $CO_2(g)$ and $H_2O(l)$ are -393.5 and -285.8 kJ/mol, respectively. Calculate ΔH_f^0 of $C_2H_6(g)$.
d. How much heat is evolved when 1.00 g of $C_2H_6(g)$ is burned to give $CO_2(g)$ and $H_2O(l)$ in an open container?
e. When 1.00 g of C_2H_6 is burned in a bomb calorimeter, the temperature rises from 22.00°C to 33.13°C. In a separate experiment, it is found that absorption of 10.0 kJ of heat raises the temperature of the calorimeter by 2.15°C. Calculate $C_{calorimeter}$ and $q_{reaction}$.
f. Calculate ΔE at 25°C for the equation:

$$C_2H_6(g) + \tfrac{7}{2} O_2(g) \rightarrow 2\ CO_2(g) + 3\ H_2O(l)$$

Answers
a. -3119.4 kJ; $+1559.7$ kJ   **b.** -1427.7 kJ   **c.** -84.7 kJ
d. 51.9 kJ   **e.** 4.65 kJ/°C, -51.8 kJ   **f.** -1553.5 kJ

QUESTIONS AND PROBLEMS

The questions and problems listed here are typical of those at the end of each chapter. Some involve discussion and most require calculations, writing equations, or other quantitative work. The topic emphasized in each question or problem is indicated in the heading, such as "Symbols and Formulas" or "Significant Figures." Those in the "Unclassified" category may involve more than one concept, including, perhaps, topics from a preceding chapter. "Challenge Problems," listed at the end of the set, require extra skill and/or effort.

The "classified" questions and problems (Problems 1–50 in this set) are arranged in matched pairs, one below the other, and illustrate the same concept. For example, Questions 1 and 2 below are nearly identical in nature; the same is true of Questions 3 and 4, and so on. Problems numbered in color are answered in Appendix 4.

Symbols and Formulas

1. Give the color, state of matter at room temperature, and symbol of the following elements. Use Tables 1.4 and 1.5.
  a. bromine   b. iron

2. Use Tables 1.4 and 1.5 to write the name and symbol for the element that is
  a. a green-yellow gas at 25°C and 1 atm.
  b. a red solid at 25°C and 1 atm.

3. Give the symbols for
  a. potassium   b. cadmium   c. gold
  d. antimony   e. rubidium

4. Name the elements whose symbols are
  a. Mn   b. Na   c. As   d. W   e. P

5. How many elements are there in the following groups?
  a. Group 2   b. Group 3   c. Group 8
  d. the subgroup headed by Cr

6. How many elements are there in the following periods?
  a. Period 1   b. Period 2   c. Period 3
  d. Period 4   e. Period 5

Measurements

7. Classify each of the following as units of mass, volume, length, density, energy, or pressure.
  a. mg   b. mL   c. cm^3 d. mm
  e. kg/m^3 f. mm Hg   g. kJ

8. Classify each of the following units as units of mass, volume, length, density, energy, or pressure.
  a. nm   b. kg   c. J   d. m^3
  e. g/cm^3 f. atm   g. kcal

9. Select the smaller member of each pair.

10. Select the smaller member of each pair.
  a. 27.12 g or 27.12 kg
  b. 35 cm^3 or 0.035 m^3
  c. 2.87 g/L or 2.87 g/cm^3
  d. 525 mm or 5.25×10^{-3} km

11. Most laboratory experiments are done at 25°C. Express this in
  a. °F   b. K

12. A child has a temperature of 104°F. What is his temperature in
  a. °C   b. K

13. Carbon dioxide, CO_2, at room temperature (70°F) is a gas. It can be frozen at -69.7°F and 5 atm pressure to solid carbon dioxide, popularly known as dry ice. What is the freezing point of carbon dioxide in °C?

14. Superconductors use liquid nitrogen as a coolant. Liquid nitrogen boils at -195.8°C. What is its boiling point in
  a. °F   b. K

Significant Figures

15. How many significant figures are there in each of the following?
  a. 1.92 mm   b. 0.032100 g
  c. 6.022×10^{23} atoms   d. 460.00 L
  e. 0.00036 cm^3 f. 2×10^9 nm

16. How many significant figures are there in each of the following?
  a. 23.437 m   b. 0.002017 g   c. 30.0×10^{20} nm
  d. 50.010 L   e. 2.30790 atm   f. 350 miles

17. A student prepares a salt solution by dissolving 85.638 g of sodium chloride (NaCl) in enough water to form 237 cm^3 of solution. Calculate the number of grams of salt per cubic centimeter of solution.

18. Calculate the volume of a sodium atom, which has a radius of 0.186 nm. Assume the atom is spherical. The volume of a sphere is given by the expression $V = 4\pi r^3/3$.

19. Calculate the following to the correct number of significant figures.

  a. $x = \dfrac{1.27 \text{ g}}{5.296 \text{ cm}^3}$ b. $x = \dfrac{12.235 \text{ g}}{1.01 \text{ L}}$

  c. $x = 12.2 \text{ g} + 0.38 \text{ g}$ d. $x = \dfrac{17.3 \text{ g} + 2.785 \text{ g}}{30.20 \text{ cm}^3}$

20. How many significant figures are there in the values of x obtained from

  a. $x = \dfrac{34.0300 \text{ g}}{12.09 \text{ cm}^3}$

  b. $x = 32.647 g   32.327 g$

End-of-Chapter Problems

Most of these are "classified," that is, grouped by type under a particular heading, such as symbols and formulas and significant figures. These classified problems are in matched pairs. The second member of each pair illustrates the same principle as the first; it is numbered in color and answered in Appendix 4. Your instructor may assign unanswered problems as homework. After these problems have been discussed, you should work the corresponding answered problems to make sure you know what's going on. Each chapter also contains a smaller number of "unclassified" and "challenge" problems. These are not paired; all of them are answered in Appendix 4.

The compounds dealt with in this chapter have relatively simple formulas. At most, they contain three elements; the atom ratios involve small numbers such as 1, 2, or 3. Many compounds have more complex formulas; among these are a class of mixed oxides that contain three or four different metals in addition to oxygen. Certain compounds of this type have been in the news lately because they have a very unusual property. Upon cooling to about 100 K, their electrical resistance drops to zero; for that reason they are referred to as superconductors. A typical formula is $YBa_2Cu_3O_7$*, but many other compositions are possible. Superconductivity is the most active research area in the physical sciences today, with knowledge of the subject increasing rapidly.

Potentially, superconductors have a host of practical applications. One of the simplest is in the transmission of electrical energy. A power line made of a superconducting material could carry electricity over hundreds of kilometers with virtually no energy loss. This would make it possible, for instance, to build nuclear power plants in remote areas far from population centers. On a more exotic level, it may be practical to design electric trains that travel at 500 km/h over superconducting tracks. The train would be literally suspended in the air, 20 cm off the ground (Fig. 3.7). This effect arises because the magnetic field induced when current flows through the superconductor repels that of a magnet above it. The extent of levitation depends upon a balance between gravitational and magnetic forces.

---

*Actually, these compounds are nonstoichiometric (Chapter 2). The number of oxygen atoms in the formula varies from 6.5 to 7.2, depending upon the method of preparation. For a discussion of the structure of superconductors, see *Chemical and Engineering News*, May 11, 1987, pp. 7–16.

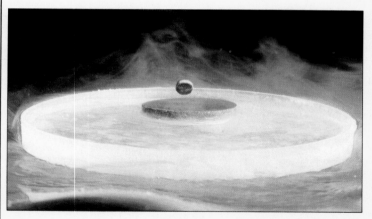

Figure 3.7 A pellet of superconducting material, previously cooled to 77 K with liquid nitrogen, floats above a magnet. (Courtesy of Edmund Scientific)

End-of-Chapter Perspectives

At the end of each chapter is a brief essay or "perspective," based directly upon a topic covered in the chapter. Some of these are historical, consisting mainly of biographical sketches of men and women referred to in the chapter. More commonly, the perspective deals with the applications of chemical principles or reactions in today's world. In several cases, the application of chemistry to nutrition is emphasized. The "modern perspective" shown here deals with superconductivity, a rapidly evolving and exciting area of research.

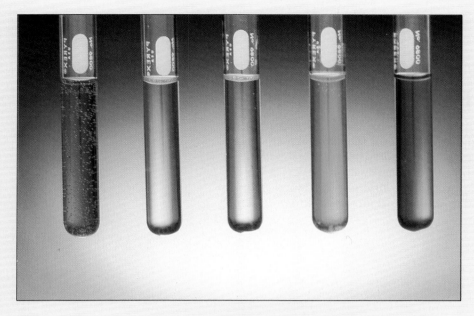

Figure 17.4 Several of the Group III cations are colorless in solution. Among these are CO^{2+} (red), Fe^{2+} (pale green), Mn^{2+} (pale pink), Ni^{2+} (green), and Cr^{3+} (purple).

Other features include:

Photography and Full-Color Presentation

Color photographs are used to make chemistry come alive. They depict chemical reactions, solutions, minerals, and applications of chemistry to daily life. A color design has also enabled us to develop a color scheme, for example, in the acid/base chapter acids are red and bases are blue.

Chapter Summaries

In Appendix 5 there is a summary of each chapter, along with a list of key words and equations. You may find these summaries useful, especially when studying for a test.

Mathematics Review

Appendix 3 reviews several mathematical concepts that are essential to your understanding of the text. These include exponential notation and the use of logarithms.

Index/Glossary

From time to time you will come across a technical word or phrase in the test that you don't understand. If that happens, look it up in the index, which has a glossary incorporated into it. For example, you will find the word ''mole'' defined and illustrated as well as referenced by page.

A NOTE ON SAFETY

The photographs that appear in this book were chosen to illustrate the discussion in the text and to show how chemical reactions actually occur. Some of the reactions illustrated involve very corrosive chemicals and occur very rapidly and with the evolution of a great deal of energy. They are unsafe unless carried out under appropriate conditions by trained chemists wearing protective clothing. Under *no* circumstances should you attempt these reactions yourself.

CONTENTS OVERVIEW

CONTENTS

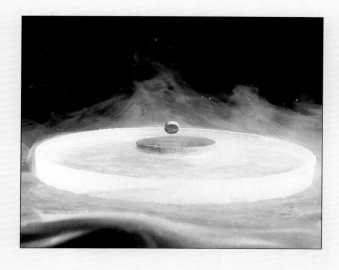

CHEMISTRY

Principles & Reactions

$$(NH_4)_2Cr_2O_7(s) \rightarrow Cr_2O_3(s) + N_2(g) + 4H_2O(g)$$

When ammonium dichromate is ignited it decomposes to give chromium(III) oxide, nitrogen gas, and water vapor.

CHAPTER
1

MATTER AND
MEASUREMENTS

Almost certainly, this is your first college course in chemistry; perhaps it is your first exposure to chemistry at any level. Unless you are a chemistry major (only about 5% of general chemistry students are in that category), you may wonder why you are taking this course and what you can expect to gain from it. To address that question, it will be helpful to look at some of the ways in which chemistry contributes to other disciplines.

If you're planning to be an engineer, you can be sure that many of the materials you work with have been synthesized by chemists. Some of these materials are organic (carbon containing); familiar plastics such as polyethylene (Chapter 27) fall in this category. Others are inorganic, like the new semiconductors made of gallium arsenide (Chapter 7). A few years down the road, you may find yourself working with the mixed metal oxides that have made superconductivity (Chapter 3) the hottest research topic of the 1980s.

Perhaps you are a health science major, looking forward to a career in medicine or pharmacy. If so, you will want to become familiar with the many life-saving products made by chemists over the last two decades. These range from drugs used in chemotherapy (Chapter 16) to synthetic polymers used in valves for artificial hearts (Chapter 27). Beyond this, you will find that the instruments used for diagnosis and testing have, for the most part, been developed and refined by chemists. These instruments relate to such modern techniques as atomic absorption spectroscopy (Chapter 5) as well as more specialized methods such as positron emission tomography (Chapter 25), used to study brain disorders.

Beyond career preparation, one of the objectives of a college education is to stimulate your curiosity about the world around you or, to abuse a cliché, to make you "a better informed citizen." In this text, we will look at some of the chemistry-related issues that face society today, including:

— the pros and cons of nuclear power (Chapter 25)

— the phenomenon of acid rain; its causes and prevention (Chapter 19)

— the role of trace metals in nutrition; their use and misuse (Chapter 22)

— carcinogens in the laboratory, the home, and the work place (Chapters 19 and 24)

We hope that when you complete this course you too will be convinced of the importance of chemistry in today's world. We should, however, caution you upon one point. Although we will talk about many of the applications of chemistry, *our main concern will be with the principles that govern chemical reactions.* Only by mastering these principles will you understand the basis of the applications referred to in the preceding paragraphs.

We begin our study of chemical principles by looking at the nature of measurements (Section 1.1), the uncertainties associated with them (Section 1.2), and the method used to convert measured quantities from one set of units to another (Section 1.3). Our approach here will be quantitative, emphasizing the simple mathematical relationships that lie at the heart of general chemistry. The last half of this chapter is more descriptive. We will deal with the chemical nature of different kinds of substances (Section 1.4) and with their properties (Section 1.5). The chapter closes with a "modern perspective" describing one of the most versatile techniques used to separate and identify substances—chromatography.

1.1 MEASUREMENTS

Chemistry is an experimental science. It involves measurements, most of them quantitative in nature. Consider, for example, the following directions for the laboratory preparation of aspirin:

> Add *2.0 g* of salicylic acid, *5.0 mL* of acetic anhydride, and 5 drops of 85% H_3PO_4 to a 50-mL Erlenmeyer flask. Heat in a water bath at *75°C* for *15 minutes*. Add *cautiously 20 mL* of water and transfer to an ice bath at *0°C*. Scratch the inside of the flask with a stirring rod to initiate crystallization. Separate aspirin from the solid-liquid mixture by filtering through a Buchner funnel *10 cm* in diameter.

Included in this description are the measured values of several quantities:

In chemistry, if you can't measure it, you don't know much about it

length ("a Buchner funnel 10 cm in diameter")

volume ("5.0 mL of acetic anhydride; 20 mL of water")

mass ("2.0 g of salicylic acid")

temperature ("a water bath at 75°C; an ice bath at 0°C")

time ("heat for 15 minutes")

In this section, we will consider how these quantities are measured and the nature of the units in which they are expressed.

LENGTH

Most of us are familiar with a simple measuring device found in every general chemistry laboratory. This is the meter stick, which reproduces, as accurately as possible, the basic unit of length in the metric system, the **meter** (m). The length of a normal stride is about 1 m; a typical room has a height of 2 to 3 m.

A meter stick is divided into 100 equal parts, each one *centimeter* (cm) in length (1 cm = 10^{-2} m). A centimeter, in turn, is divided into ten equal parts, each one millimeter (mm) long (1 mm = 10^{-3} m). A much larger unit, familiar to runners, is the kilometer (1 km = 10^3 m). The prefixes *kilo-, centi-,* and *milli-* are used in the metric system to designate units obtained by multiplying by 1000, 0.01, and 0.001, respectively. Another unit used to express the dimensions of tiny particles such as atoms is the *nanometer* (nm):

$$1 \text{ nm} = 10^{-9} \text{ m}$$

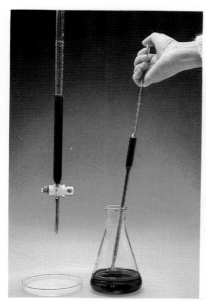

Figure 1.1 A buret (*left*) is used to deliver a variable volume of liquid, accurately measured. A pipet (*right*) is used to deliver a fixed volume (e.g., 25.00 mL) of liquid. (Charles D. Winters)

VOLUME

Units of volume in the metric system are simply related to those of length. The *cubic centimeter* (cm³) represents the volume of a cube one centimeter on an edge. A larger unit is the *liter* (L), which is exactly 1000 cm³:

$$1 \text{ L} = 1000 \text{ cm}^3 \tag{1.1}$$

A softball has a volume of about ½ L, a basketball a volume of somewhat more than 7 L. A *milliliter* (mL), 1/1000 of a liter, has the same volume as a cubic centimeter:

$$1 \text{ mL} = 1 \text{ cm}^3$$

The device most commonly used to measure volumes in general chemistry is the graduated cylinder. With this, we can measure out a known volume of a liquid, accurate to perhaps 0.1 mL. When greater accuracy is required, we use a pipet or buret (Fig. 1.1). A pipet is calibrated to deliver a fixed volume of liquid (for example, 25.00 ± 0.01 mL) when filled to the mark and allowed to drain normally. Variable volumes can be delivered with about the same accuracy from a buret. With a buret, final and initial volumes must be read carefully to calculate the volume of liquid withdrawn.

Written: 25 mL
Spoken: "25 em ell"

MASS

The mass of a sample is a measure of the amount of matter it contains. In the metric system, mass may be expressed in grams (g), kilograms (kg), or milligrams (mg):

$$1 \text{ kg} = 10^3 \text{ g}; \qquad 1 \text{ mg} = 10^{-3} \text{ g}$$

This book weighs about 2 kg (2 × 10^3 g).

1 mL of water weighs just about 1 g

Figure 1.2 Single pan analytical balance with digital readout; the solid sample and container weigh 46.289 g. (Marna G. Clarke)

Chemists "weigh" objects by measuring their mass on a balance (Fig. 1.2). Modern balances such as the one shown read out the mass of a sample to ±0.001 g or better almost instantaneously.

TEMPERATURE

The concept of temperature is familiar to all of us. This is because our bodies are so sensitive to temperature differences. When we pick up a piece of ice, we feel cold because its temperature is lower than that of our hand. After drinking a cup of coffee, we may refer to it as "hot," "lukewarm," or "atrocious." In the first two cases, at least, we are describing the extent to which its temperature exceeds ours. From a slightly different viewpoint, temperature is the factor that determines the direction of heat flow. When two objects at different temperatures are placed in contact with one another, heat flows from the one at the higher temperature to the one at the lower temperature.

To measure temperature we can use a mercury-in-glass thermometer. Here, we take advantage of the fact that mercury, like other substances, expands as temperature increases. When the temperature rises, the mercury in the thermometer expands up a narrow tube. The total volume of the tube is only about 2% of that of the bulb at the base. In this way, a rather small change in volume is made readily visible.

Thermometers used in chemistry are marked in degrees *Celsius* (centigrade), named after the Swedish astronomer Anders Celsius (1701–1744). On this scale, the freezing point of water is taken to be 0°C. The boiling point of water at one atmosphere pressure is 100°C. Household thermometers in the United States are commonly marked in Fahrenheit degrees. Daniel Fahrenheit (1686–1736) was a German instrument maker who was the first to use the mercury-in-glass thermometer. On this scale, the normal freezing and boiling points of water are taken to be 32° and 212°, respectively. That is:

$$32°F = 0°C \quad \text{and} \quad 212°F = 100°C$$

These days about the only country using the Fahrenheit scale is the USA

As you can see from Figure 1.3, the Fahrenheit degree is smaller than the Celsius degree. The distance between the freezing and boiling points of water is 100° on the Celsius scale and 180° on the Fahrenheit scale. The relation between temperatures expressed on the two scales is

$$°F = 1.8(°C) + 32 \tag{1.2}$$

Another temperature scale that is particularly useful in science is the absolute or Kelvin scale. The relationship between temperatures in K and °C is

$$K = °C + 273.15 \tag{1.3}$$

Lord Kelvin was 28 years old when he found this out

This scale is named after Lord Kelvin, an English physicist. He showed that it is impossible to reach a temperature lower than 0 K (−273.15°C).

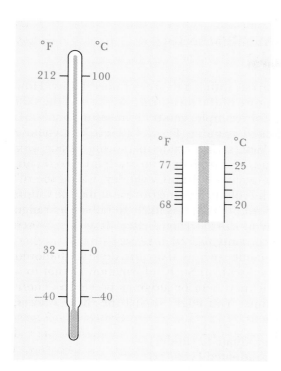

Figure 1.3 On the Celsius scale, the distance between the freezing and boiling points of water is 100°; on the Fahrenheit scale, it is 180°. This means that the Celsius degree is $\frac{9}{5}$ as large as the Fahrenheit degree, as is evident from the magnified section of the thermometer at the right.

■ EXAMPLE 1.1 _____

Express normal body temperature, 98.6°F, in °C and K.

Solution

Substituting in Equation 1.2, we obtain

$$98.6 = 1.8(°C) + 32$$

Solving:

$$1.8(°C) = 98.6 - 32 = 66.6$$

$$°C = 66.6/1.8 = \boxed{37.0}$$

Applying Equation 1.3:

$$K = 37.0 + 273.15 = \boxed{310.2}$$

Exercise

In a certain laboratory experiment, the temperature increases by 35°C. Express this temperature change (Δ) in K; in °F. Answer: Δ K = 35; Δ°F = 63°.

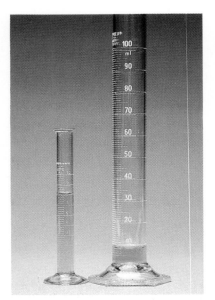

Figure 1.4 The uncertainty associated with a measurement depends upon the nature of the measuring device. To measure out small volumes of liquids precisely, a 10-mL graduated cylinder is much more effective than one with a volume of 100 mL. (Marna G. Clarke)

1.2 UNCERTAINTIES IN MEASUREMENTS; SIGNIFICANT FIGURES

Every measurement you make carries with it a degree of uncertainty. How large this is depends upon the nature of the measuring device and the skill with which you use it. Suppose, for example, you try to measure out 8 mL of liquid using a 100-mL graduated cylinder (Fig. 1.4). Here, the volume is uncertain to at least ± 1 mL. With such a crude measuring device, you would be lucky to obtain a volume closer to 8 than to 7 or 9 mL. To obtain greater precision, you might use a narrow 10-mL cylinder, where the divisions are much farther apart. The volume you measure now may be within 0.1 mL of the desired value of 8 mL; that is, it is likely to fall in the range 7.9 to 8.1 mL. Using a buret, you can do even better. If you are very careful, you may reduce the uncertainty to 0.01 mL.

Anyone making a measurement such as that just described should indicate the uncertainty associated with it. Such information is vital to a person who wants to repeat the experiment or judge its precision. There are many ways to show uncertainty. You might report the three volume measurements referred to above as

8 ± 1 mL   (large graduated cylinder)
8.0 ± 0.1 mL   (small graduated cylinder)
8.00 ± 0.01 mL   (buret)

In this text, we will drop the $\pm$ notation and simply write

8 mL;   8.0 mL;   8.00 mL

When we do this, it is understood that there is an *uncertainty of at least one unit in the last digit,* that is, 1 mL, 0.1 mL, 0.01 mL, respectively.

This method of citing the degree of confidence in a measurement is often described in terms of **significant figures**. We say that in 8.00 mL there are three significant figures. Each of the three digits in 8.00 has experimental meaning. Similarly, there are two significant figures in 8.0 mL and one significant figure in 8 mL.

COUNTING SIGNIFICANT FIGURES

Frequently, we need to know the number of significant figures in a measurement that someone else has reported. We do this by applying the following common-sense rules.

1. *All nonzero digits are significant.* There are three significant figures in 5.37 cm and four significant figures in 4.293 cm.
2. *Zeros between nonzero digits are significant.* There are three significant figures in 106 g or in 1.02 g.
3. *Zeros beyond the decimal point at the end of a number are significant.* When we say that the volume of a liquid is 8.00 mL, we imply that the two zeros are experimentally meaningful. The quantity 8.00 mL carries the same

degree of precision as 8.13 mL or 7.98 g; all of these quantities have three significant figures.

4. *Zeros preceding the first nonzero digit in a number are not significant.* In a mass measurement of 0.002 g, there is only one significant figure—the "2" at the end. The zeros serve only to fix the position of the decimal point. This becomes obvious if we express the mass in exponential (scientific) notation (Appendix 3). In that case, we would write 0.002 g as

$$2 \times 10^{-3} \text{ g}$$

This is a good way to find the number of significant figures

Now, clearly, there is only one significant figure. The uncertainty is $\pm 1 \times 10^{-3}$ g.

Sometimes the number of significant figures in a reported measurement is ambiguous. Suppose you are told that a coin weighs "50 g." You cannot be sure how many of these digits are meaningful. Perhaps the coin was weighed to the nearest gram (50 ± 1 g). If so, both the "5" and the "0" are known; there are two significant figures. Then again, the coin might have been weighed only to the nearest ten grams (50 ± 10 g). In this case, only the "5" is known accurately; there is only one significant figure. About all we can do in situations like this is to wish that the person who carried out the weighing had used exponential notation. The mass should have been reported as either

Here a decimal point would help clarify things

$$5.0 \times 10^1 \text{ g} \quad \text{(2 significant figures)}$$

or

$$5 \times 10^1 \text{ g} \quad \text{(1 significant figure)}$$

In general, *any ambiguity concerning the number of significant figures in a measurement can be resolved by using exponential notation.*

■ **EXAMPLE 1.2** ⎯⎯⎯⎯⎯⎯⎯⎯⎯⎯⎯⎯⎯⎯⎯⎯⎯⎯⎯⎯⎯⎯

Three different students weigh the same object, using different balances. They report the following masses:

a. 15.02 g   b. 15.0 g   c. 0.01502 kg

How many significant figures are there in each value?

Solution

a. 4

b. 3. The zero after the decimal point is significant. It indicates that the object was weighed to the nearest 0.1 g.

c. 4. The zeros at the left are not significant. They are there only because the mass was expressed in kilograms rather than grams. Note that "15.02 g" and "0.01502 kg" represent the same mass.

The same answers could have been obtained, perhaps with more confidence, by expressing the masses in exponential notation:

$$15.02 \text{ g} = 1.502 \times 10^1 \text{ g} \qquad \text{(4 significant figures)}$$
$$15.0 \text{ g} = 1.50 \times 10^1 \text{ g} \qquad \text{(3 significant figures)}$$
$$0.01502 \text{ kg} = 1.502 \times 10^{-2} \text{ kg} \qquad \text{(4 significant figures)}$$

Exercise
Give the number of significant figures in $2.6 \times 10^2 \text{ cm}^3$; $2.40 \times 10^{-3} \text{ cm}^3$.
Answer: 2; 3.

SIGNIFICANT FIGURES IN MULTIPLICATION AND DIVISION

Most of the quantities that we measure are not end results in themselves. Instead, they are used to calculate other quantities, often by multiplication or division. The precision of any such derived result is limited by those of the measurements on which it is based. **When experimental quantities are multiplied or divided, the number of significant figures in the result is the same as that in the quantity with the smallest number of significant figures.**

You can't increase the number of significant figures by multiplying numbers together

■ **EXAMPLE 1.3**
An overseas flight leaves New York in the late afternoon and arrives in London 8.50 hours later. The airline distance from New York to London is about 5.6×10^3 km, depending to some extent upon the flight path followed. What is the average speed of the plane, in kilometers per hour?

Solution
The average speed is given by the quotient:

$$\text{speed} = \frac{\text{distance traveled}}{\text{time elapsed}} = \frac{5.6 \times 10^3 \text{ km}}{8.50 \text{ h}}$$

Reporting all those numbers doesn't make any sense, so don't do it

If you carry out this division on your calculator, the number that appears is 658.82353. Noting that there are three significant figures in the time but only two in the distance, we conclude that the speed should be expressed to two significant figures. The average speed is 6.6×10^2 km/h.

Exercise
How long would it take a plane, traveling at an average speed of 634 km/h, to cover a distance of 3912 km? Answer: 6.17 h.

The rule is actually somewhat approximate, but good enough for our purposes

Expanding upon the calculation in Example 1.3, we can see the reason behind the rule for the number of significant figures retained in multiplication or division. When we say that the distance is "5.6×10^3 km," we mean that it lies between 5.5×10^3 and 5.7×10^3 km. Similarly, if the time is quoted to ± 0.01 h, its true value should lie between 8.49 and 8.51 h. We see then that the average speed might be as large as:

$$\frac{5.7 \times 10^3 \text{ km}}{8.49 \text{ h}} = 6.7 \times 10^2 \text{ km/h}$$

On the other hand, the average speed could be as little as:

$$\frac{5.5 \times 10^3 \text{ km}}{8.51 \text{ h}} = 6.5 \times 10^2 \text{ km/h}$$

Looking at the results of these calculations, we see that it is entirely reasonable to report the average speed to be 6.6×10^2 km/h, implying an uncertainty of ± 0.1 km/h. Conversely, it would be absurd to report an average speed of 658.82353 km/h, implying a precision far greater than that actually justified.

UNCERTAINTIES IN ADDITION AND SUBTRACTION

When measured quantities are added or subtracted, the uncertainty in the result is found in a quite different way than in multiplication or division. It is determined by the absolute *uncertainty* (rather than the number of significant figures) in the least precise measurement. Suppose, for example, we wish to calculate the total mass of a solution containing 10.21 g of instant coffee, a "pinch" (0.2 g) of sugar, and 256 g of water. The uncertainties in these masses are

	MASS (g)		UNCERTAINTY
Instant coffee	10.21	$\pm$	0.01 g
Sugar	0.2	$\pm$	0.1 g
Water	256	$\pm$	1 g
Total mass	266	$\pm$	1 g

The sum of the masses cannot be more precise than that of the water, which has the largest uncertainty, ± 1 g. The total mass should be reported as 266 g rather than 266.4 g or 266.41 g. The general rule illustrated by this example is as follows:

When experimental quantities are added or subtracted, the number of digits beyond the decimal point in the result is the same as that in the quantity with the smallest number of digits beyond the decimal point.

EXACT NUMBERS

In applying the rules that we have cited, you should keep in mind one important point. Some numbers involved in calculations are exact rather than approximate, because they represent defined rather than measured quantities. If you were asked to express in liters a measured volume of 536 cm^3, using the relation:

If you count the number of students in a room you should be able to get an exact number

$$1 \text{ L } = 1000 \text{ cm}^3$$

your answer should be given to three significant figures. This is the number of significant figures in the measured quantity, 536 cm³. The "1" and the "1000" in the equation are defined quantities. There are exactly 1000 cubic centimeters in exactly one liter. A similar situation applies with the equation relating Fahrenheit and Celsius temperatures:

$$°\text{F } = 1.8(°\text{C}) + 32$$

The numbers 1.8 and 32 are exact. Hence, they do not affect the precision of any calculation involving a temperature conversion.

A different type of exact number arises in certain types of calculations. Suppose we ask the question, "How much heat is evolved when *one kilogram* of coal is burned?" The question implies that we want to know how much heat is evolved when *exactly* one kilogram of coal burns. We would like the uncertainty in the answer to be independent of the mass of coal. By spelling out the mass as *one* kilogram, we imply an exact quantity.

1.3
CONVERSION OF UNITS

Often we need to convert measurements expressed in one unit (e.g., grams) to another unit (milligrams or kilograms). To do this, we follow what is known as a **conversion factor** approach. Suppose, for example, we want to convert a volume of 536 cm³ to liters. We know that

$$1 \text{ L } = 1000 \text{ cm}^3$$

Dividing both sides of this equation by 1000 cm³ gives a quotient equal to one:

$$\frac{1 \text{ L}}{1000 \text{ cm}^3} = \frac{1000 \text{ cm}^3}{1000 \text{ cm}^3} = 1$$

We multiply 536 cm³ by the quotient 1 L/1000 cm³, which is called a *conversion factor*. Since the conversion factor equals one, this does not change the value of the volume. However, it does accomplish the desired conversion of units:

$$536 \text{ cm}^3 \times \frac{1 \text{ L}}{1000 \text{ cm}^3} = 0.536 \text{ L}$$

The relation 1 L = 1000 cm³ can be used equally well to convert a volume in liters, say 1.28 L, to cubic centimeters. In this case, we obtain the necessary conversion factor by dividing both sides of the equation by 1 L:

$$\frac{1000 \text{ cm}^3}{1 \text{ L}} = \frac{1 \text{ L}}{1 \text{ L}} = 1$$

Multiplying 1.28 L by the quotient 1000 cm³/1 L converts the volume from liters to cubic centimeters:

$$1.28 \, \cancel{L} \times \frac{1000 \text{ cm}^3}{1 \, \cancel{L}} = 1280 \text{ cm}^3 = 1.28 \times 10^3 \text{ cm}^3$$

Notice that a single relation (1 L = 1000 cm³) gives us two conversion factors:

$$\frac{1 \text{ L}}{1000 \text{ cm}^3} \quad \text{and} \quad \frac{1000 \text{ cm}^3}{1 \text{ L}}$$

Both of these are equal numerically to one. In making a conversion, we choose the factor that cancels out the unit we want to get rid of:

initial quantity × conversion factor(s) = desired quantity

Conversions between English and metric units are made in a similar way (Example 1.4). The relations required can be obtained from Table 1.1.

TABLE 1.1 Relations Between Length, Volume, and Mass Units

METRIC		ENGLISH		METRIC-ENGLIGH	
Length					
1 km	$= 10^3$ m	1 ft	= 12 in	1 in	= 2.54 cm*
1 cm	$= 10^{-2}$ m	1 yd	= 3 ft	1 m	= 39.37 in
1 mm	$= 10^{-3}$ m	1 mile	= 5280 ft	1 mile	= 1.609 km
1 nm	$= 10^{-9}$ m $= 10$Å				
Volume					
1 m³	$= 10^6$ cm³ $= 10^3$ L	1 gallon	= 4 qt = 8 pt	1 ft³	= 28.32 L
1 cm³	= 1 mL $= 10^{-3}$ L	1 qt (Can.)	= 69.35 in³	1 L	= 0.8799 qt (Can.)
		1 qt (U.S. liq.)	= 57.75 in³	1 L	= 1.057 qt (U.S. liq.)
Mass					
1 kg	$= 10^3$ g	1 lb	= 16 oz	1 lb	= 453.6 g
1 mg	$= 10^{-3}$ g	1 short ton	= 2000 lb	1 g	= 0.03527 oz
1 metric ton	$= 10^3$ kg			1 metric ton	= 1.102 short ton

*This conversion factor is exact; the inch is defined to be exactly 2.54 cm. The other factors listed in this column are approximate, quoted to four significant figures. Additional digits are available if needed for very accurate calculations. For example, the pound is defined to be 453.59237 g.

■ **EXAMPLE 1.4** _____

According to a highway sign, the distance from St. Louis to Chicago is 295 miles. Express this distance in kilometers.

Solution

From Table 1.1, we note that the required relation is

1 mile = 1.609 km

The 1 is exact; the 1.609 is not (4 sig. fig.)

Since we want to convert from miles to kilometers, we want a conversion factor with kilometers in the numerator and miles in the denominator. The required conversion factor is 1.609 km/1 mile:

$$295 \text{ miles} \times \frac{1.609 \text{ km}}{1 \text{ mile}} = \boxed{475 \text{ km}}$$

Exercise
Determine the mass in grams of a hamburger that weighs 8.0 oz. Answer: 2.3×10^2 g.

Frequently, you will need to carry out more than one conversion to work a problem. This can be done by setting up successive conversion factors (Example 1.5).

■ **EXAMPLE 1.5**
A certain U.S. car has a fuel efficiency rating of 36.2 miles per gallon. Convert this to kilometers per liter.

Solution
The distance conversion can be made by using the relation

$$1 \text{ mile} = 1.609 \text{ km} \quad (1)$$

Using Table 1.1, we find that the volume conversion involves using two relations:

$$1 \text{ gallon} = 4 \text{ qt} \quad (2)$$

$$1 \text{ L} = 1.057 \text{ qt} \quad (3)$$

We set up the arithmetic in a single expression, writing conversion factors in such a way as to cancel units:

When converting from one quantity to another, conversion factors are the way to go

$$36.2 \frac{\text{miles}}{\text{gallon}} \times \frac{1.609 \text{ km}}{1 \text{ mile}} \times \frac{1 \text{ gallon}}{4 \text{ qt}} \times \frac{1.057 \text{ qt}}{1 \text{ L}} = \boxed{15.4 \text{ km/L}}$$

Notice that three consecutive conversions were required. We first converted miles to kilometers, obtaining the fuel efficiency in kilometers per gallon. Then we converted gallons to quarts, and, finally, quarts to liters. Our final answer was in the desired units, kilometers per liter.

Exercise
Convert a density of 3.50 g/cm³ to kilograms per cubic meter. Answer: 3.50×10^3 kg/m³.

Experiments in general chemistry often involve measured quantities other than those given in Table 1.1 (length, volume, mass). You may, for example, measure the time required to carry out a reaction. This might be expressed in days (d), hours (h), minutes (min), or seconds (s).

$$1 \text{ d} = 24 \text{ h}; \quad 1 \text{ h} = 60 \text{ min}; \quad 1 \text{ min} = 60 \text{ s}$$

In experiments involving gases, it is important to measure the pressure. Many different units are used to express pressure (Table 1.2). In this text, we will most commonly use the two units

TABLE 1.2 Relations Between Pressure and Energy Units

PRESSURE
1 atm = 760 mm Hg = 1.013×10^5 Pa = 14.70 lb/in^2
1 torr = 1 mm Hg
1 bar = 10^5 Pa

ENERGY
1 cal = 4.184 J = 4.129×10^{-2} L·atm
1 J = 10^7 ergs

— *millimeter of mercury* (mm Hg). This is the pressure exerted by a column of mercury 1 millimeter in height.

— *atmosphere* (atm). An atmosphere (760 mm Hg) is approximately the pressure of the atmosphere on a "normal" day at sea level.

Pascals (Pa) and kilopascals (1 kPa = 10^3 Pa) will be used less frequently.

In every reaction, there is an energy change whose magnitude can be measured. Chemists use a variety of energy units (Table 1.2). You may be most familiar with the calorie,* which is the amount of heat required to raise the temperature of 1 gram of water by 1°C. Increasingly, nowadays, scientists are using a different unit, the **joule**. This is the preferred unit of energy in the International System of Units. We will use the joule extensively throughout this text. The basic definition of this energy unit is given in Appendix 1. Notice from Table 1.2 that 1 calorie is about 4 joules; more exactly, 1 cal = 4.184 J. This means that an energy change expressed in joules will have a magnitude about four times that for the same energy change in calories. For example, it takes about 4 joules to raise the temperature of 1 gram of water 1°C.

The joule is a rather small energy unit. One joule of electrical energy would keep a 10-watt light bulb burning for only a tenth of a second! Most often, in chemical reactions, we will express energy changes in **kilojoules**:

1 watt = 1 J/s

1 kJ = 10^3 J

Conversions involving these units are carried out in the ordinary way, using the relations given in Table 1.2. (A more extensive list of relations between units of all types is given in Appendix 1.) Example 1.6 illustrates the conversion from calories to joules.

■ **EXAMPLE 1.6** _____

When one gram of gasoline burns in an automobile engine, the amount of energy given off is about 1.03×10^4 cal. Express this in joules.

---

*The "calorie" referred to by nutritionists is actually a kilocalorie (1 kcal = 10^3 cal). On a "2000-calorie" per day diet, you eat food capable of producing 2000 kcal = 2×10^3 kcal = 2×10^6 cal of energy.

Solution

From Table 1.2, we see that 1 cal = 4.184 J. To convert calories to joules, we use the conversion factor 4.184 J/1 cal.

$$1.03 \times 10^4 \text{ cal} \times \frac{4.184 \text{ J}}{1 \text{ cal}} = \boxed{4.31 \times 10^4 \text{ J}} \qquad (3 \text{ sig. fig.})$$

Exercise

Convert 836.8 J to calories. Answer: 2.000×10^2 cal.

The conversion factor approach shown in Examples 1.4 to 1.6 will be used throughout this text. If this is your first contact with it, it may seem awkward or artificial. You will find, however, that it is the best way to solve a wide variety of problems in chemistry. It is particularly useful when multiple conversions are required (Example 1.5) or when the units may be unfamiliar to you (Example 1.6).

SI UNITS

As indicated by the large number of entries in Tables 1.1 and 1.2, many different units are often used to express a single quantity. Pressure may be expressed in atmospheres, millimeters of mercury, pascals, etc. This proliferation of units has long been of concern to scientists. In 1960, the General Conference of Weights and Measures recommended a self-consistent set of units based upon the metric system. In the International System of Units (SI), a single base unit is used for each measured quantity. An extended discussion of this system is presented in Appendix 1. Table 1.3 lists the recommended SI units for each of the quantities referred to in this chapter, along with the SI prefixes most commonly used in chemistry.

The prefixes can be used to obtain larger or smaller multiples of the units listed. Thus we have:

$$1 \text{ dm} = 10^{-1} \text{ m} = 10 \text{ cm}$$

$$(1 \text{ dm})^3 = (10^{-1} \text{ m})^3 = (10 \text{ cm})^3$$

$$1 \text{ dm}^3 = 10^{-3} \text{ m}^3 = 10^3 \text{ cm}^3 = 1 \text{ L}$$

We will often use liters to express volumes. The m^3 is just too big

The cubic decimeter, which represents the volume of a cube 0.1 m on a side, is more commonly referred to as the *liter*.

TABLE 1.3   SI Units and Prefixes

QUANTITY	UNIT	SYMBOL	FACTOR	PREFIX	SYMBOL
Length	meter	m	10^6	mega	M
Volume	cubic meter	m^3	10^3	kilo	k
Mass	kilogram	kg	10^{-1}	deci	d
Temperature	kelvin	K	10^{-2}	centi	c
Time	second	s	10^{-3}	milli	m
Pressure	pascal	Pa	10^{-6}	micro	μ
Energy	joule	J	10^{-9}	nano	n

For the most part, we will use SI units in this text. For example, we will express atomic dimensions in nanometers rather than angstroms (1 nm $= 10^{-9}$ m $= 10$ Å). Energy changes will be expressed in joules rather than calories (1 cal $= 4.184$ J). In dealing with pressures, however, we will more commonly use the atmosphere or the millimeter of mercury as opposed to the pascal or kilopascal. If desired, non-SI units such as the millimeter of mercury can readily be converted to SI units.

1.4
TYPES OF MATTER

When a chemist carries out a reaction in the laboratory, he or she is seldom fortunate enough to obtain a single pure substance as a product. Instead, the product is most often a **mixture** of two or more substances. In this section, we will consider the composition of mixtures and a couple of simple techniques by which they can be resolved into the pure components.

Many thousands of different substances have been isolated from mixtures. These substances can be divided into two classes (Fig. 1.5). Some of them, called elements, cannot be broken down by chemical means into two or more pure substances. All other pure substances are compounds. A compound, by definition, is a pure substance that can be broken down into two or more elements.

Most material one encounters is a mixture

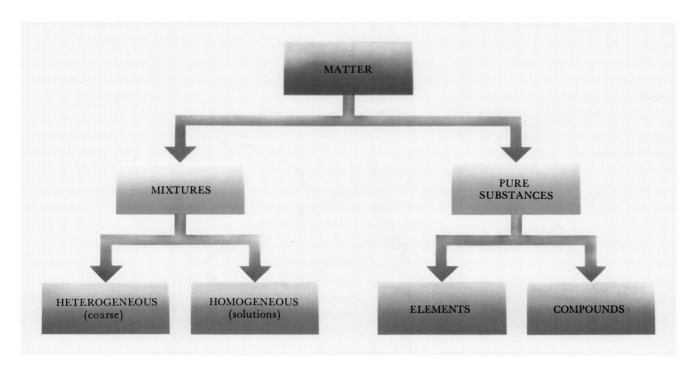

Figure 1.5 Classification of matter.

Figure 1.6 Mixtures can be *homogeneous*, as with brass, which is a solid solution of copper and zinc. Alternatively, they can be *heterogeneous*, as with granite, which contains discrete regions of different minerals (feldspar, mica, and quartz). (Charles D. Winters)

Distillation can also be used to separate two liquid substances, and is, in many industrial processes, including gasoline manufacture

Figure 1.7 By simple distillation, the two components of a water solution of potassium chromate can be separated from each other. Potassium chromate is yellow; water is colorless.

MIXTURES

We can distinguish between two different types of mixtures.

1. A *homogeneous* (uniform) mixture, called a *solution.* In discussing solutions, we often distinguish between the components by using the words "solute" and "solvent." Most commonly, the solvent is a liquid; the solute may be a solid, liquid, or gas. Soda water is a solution of carbon dioxide (solute) in water (solvent). Seawater is a more complex solution in which there are several solid solutes, including sodium chloride; the solvent is water. It is also possible to have solutions in the solid state. Brass (Fig. 1.6) is a solid solution containing the two metals copper (67–90%) and zinc (10–33%).

2. A *heterogeneous* (nonuniform) mixture, sometimes referred to as a coarse mixture. Most of the rocks and minerals in the earth's crust fall into this category. If you look at a piece of granite (Fig. 1.6), you can distinguish several components differing from one another in color.

Chemists, in carrying out reactions, ordinarily work with pure substances. To obtain a pure substance, it is often necessary to separate it from a mixture containing impurities. Such separations are based on differences in properties between the components of a mixture. Sometimes the process is a very simple one. Filtration, shown in Figure 1.10 (p. 23), can be used to separate the components of a heterogeneous solid-liquid mixture. A homogeneous solution of a solid in a liquid can be separated by distillation. A simple distillation apparatus that can be used to resolve a solution of potassium chromate in water is shown in Figure 1.7. When the solution is heated, the water vaporizes, leaving a yellow residue of potassium chromate in the distilling flask. The water may be collected by passing the vapor

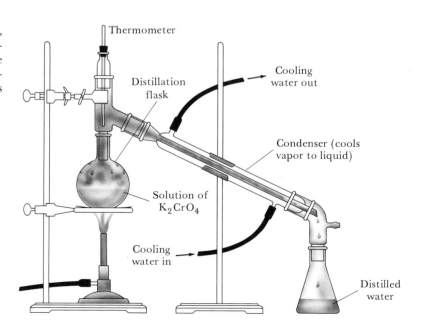

down a cold tube, thus condensing it as a liquid. In many arid areas of the world, distillation is used to obtain fresh water from seawater.

ELEMENTS AND THE PERIODIC TABLE

There are, at latest count, 109 known elements. Of these, 91 occur naturally. Many elements are familiar to all of us. The charcoal used in outdoor grills is nearly pure carbon. Electrical wiring, jewelry, and water pipes are often made from copper, a metallic element. Another such element, aluminum, is used in many household utensils. The shiny liquid in the thermometers you use is still another metallic element, mercury.

In chemistry, an element is identified by its symbol. This consists of one or two letters, usually derived from the name of the element. Thus the symbol for carbon is C; that for aluminum is Al. Sometimes the symbol comes from the Latin name of the element or one of its compounds. The two elements copper and mercury, which were known in ancient times, have the symbols Cu (*cuprum*) and Hg (*hydrargyrum*). Table 1.4 lists the names and symbols of the elements that we will be most concerned with in this text. Those shown in boldface are ones whose properties we will discuss in some detail.

You will need to learn the symbols for the elements, if you don't know them already

You are probably familiar with a device used to organize the properties of the elements. This is the Periodic Table, shown on the inside cover of the text. Elements that are similar chemically fall directly beneath one another in the table. In later chapters, we will consider the rationale for this and see how the table is used for a variety of purposes. At the moment, we need only be concerned with two general features of the Periodic Table.

1. The horizontal rows are referred to as **periods**. Thus, the first period consists of the two elements, hydrogen (H) and helium (He). The second period starts with lithium (Li) and ends with neon (Ne), and so on. Each period ends with an unreactive element called a noble gas (He, Ne, Ar, Kr, Xe, Rn).

2. The vertical groups are known as **groups** or **families**. Those at the far left and the right of the table are numbered: Groups 1 and 2 are at the

TABLE 1.4 Names and Symbols of Some of the More Familiar Elements

Aluminum	**Al**	**Chlorine**	**Cl**	**Lithium**	**Li**	Rubidium	Rb
Antimony	Sb	**Chromium**	**Cr**	**Magnesium**	**Mg**	Selenium	Se
Argon	**Ar**	**Cobalt**	**Co**	**Manganese**	**Mn**	Silicon	Si
Barium	**Ba**	**Copper**	**Cu**	**Mercury**	**Hg**	**Silver**	**Ag**
Beryllium	Be	**Fluorine**	**F**	Neon	Ne	**Sodium**	**Na**
Bismuth	Bi	Gold	Au	**Nickel**	**Ni**	**Strontium**	**Sr**
Boron	B	**Helium**	**He**	**Nitrogen**	**N**	**Sulfur**	**S**
Bromine	**Br**	**Hydrogen**	**H**	**Oxygen**	**O**	Tin	**Sn**
Cadmium	Cd	**Iodine**	**I**	**Phosphorus**	**P**	Uranium	U
Calcium	**Ca**	**Iron**	**Fe**	Platinum	Pt	Xenon	Xe
Carbon	**C**	Krypton	Kr	Plutonium	Pu	**Zinc**	**Zn**
Cesium	Cs	**Lead**	**Pb**	**Potassium**	**K**		

left, Groups 3 to 8 at the right. Elements that fall in these groups are often referred to as *main-group* elements. The elements in the center of Periods 4 through 6 are called *transition* elements (for example, Sc through Zn in Period 4).

Certain main groups are given special names. The elements in Group 1 are called *alkali metals;* those in Group 2 are referred to as *alkaline earth metals.* The Group 7 elements are called *halogens;* the *noble gases* constitute Group 8. Elements in the same main group show very similar chemical reactions. For example, sodium (Na) and potassium (K) in Group 1 both react violently with water to produce hydrogen gas.

COMPOUNDS

Most of these compounds contain the elements carbon and hydrogen, among others

As you might suppose, there are a great many more compounds than elements. The elements can combine with one another in many different ways. Millions of different compounds have been prepared in the laboratory or extracted from natural sources. Each year thousands of new compounds are reported.

Every compound contains two or more elements in fixed proportion by mass. Water, by far our most abundant compound, contains the two elements hydrogen and oxygen. In a 100-g sample of pure water, regardless of where it comes from, there are 11.19 g of hydrogen and 88.81 g of oxygen. In other words, the elemental composition of water is fixed; all samples of this pure substance contain 11.19% by mass of hydrogen and 88.81% by mass of oxygen. In contrast, a solution such as seawater can have a variable composition. The percentages of the various elements (H, O, Na, Cl, etc.) will vary at least slightly from one sample to another, depending upon the part of the world from which the sample is obtained.

The properties of compounds are very different from those of the elements they contain. Ordinary table salt, sodium chloride, is a white, unreactive solid. As you can guess from its name, it contains the two elements sodium and chlorine. Sodium (Na) is a shiny, extremely reactive metal. Chlorine (Cl) is a poisonous, greenish-yellow gas. Clearly, when these two elements combine to form sodium chloride, a profound change takes place.

Sodium is a shiny, highly reactive metal, ordinarily stored under toluene (bottle) to prevent reaction with air and water. Chlorine is a greenish-yellow gas, shown here in a high-pressure cylinder. Sodium chloride is a white, nontoxic solid (table salt). (Charles D. Winters)

Many different methods can be used to resolve compounds into their elements. Sometimes, but not often, heat alone is sufficient. Mercury(II) oxide, a compound of mercury and oxygen, decomposes to its elements when heated to 600°C. Joseph Priestley, an English chemist, discovered oxygen 200 years ago when he carried out this reaction by exposing a sample of mercury(II) oxide to an intense beam of sunlight focused through a powerful lens. A more common method of resolving compounds into elements is called electrolysis. This involves passing an electric current through a compound, usually in the liquid state. Through the process of electrolysis, it is possible to separate water into the two elements hydrogen and oxygen.

1.5
PROPERTIES OF SUBSTANCES

Every pure substance has its own unique set of properties that serve to distinguish it from all other substances. A chemist most often identifies an unknown substance by measuring its properties and comparing them to the properties recorded in the chemical literature for known substances. In this section we will consider a few of these properties, some of which you will probably measure in the general chemistry laboratory.

Elements and compounds are sometimes identified on the basis of their *chemical properties*. These are properties observed when a substance undergoes a chemical change, converting it to one or more other substances. You might, for example, show that a red solid is mercury(II) oxide by heating it in air. At about 600°C, this compound decomposes to mercury, a silvery liquid, and oxygen, a colorless gas.

More commonly, we measure the *physical properties* of elements or compounds. These are properties that can be measured without changing the chemical identity of a substance. Table 1.5 lists several physical properties of a variety of substances, both elements and compounds.

A substance with all the properties of benzene *is* benzene

DENSITY

The density of a substance is the ratio of its mass to its volume:

$$\text{density} = \frac{\text{mass}}{\text{volume}} \tag{1.4}$$

For liquids or gases, density can be found in a straightforward way by measuring, independently, the mass and volume of a sample (Example 1.7).

TABLE 1.5   Physical Properties of Substances

SUBSTANCE	DENSITY (g/cm³)*	mp (°C)	bp (°C)	SOLUBILITY (g/100 g water)*	COLOR
Elements					
Bromine(l)	3.12	−7	59	3.3	red
Chlorine(g)	0.00292	−101	−34	0.59	green-yellow
Copper(s)	8.94	1083	2567	~0	red
Iron(s)	7.87	1535	2750	~0	gray
Magnesium(s)	1.74	650	1120	~0	gray
Compounds					
Benzene(l)	0.879	5	80	0.13	colorless
Ethyl alcohol(l)	0.785	−112	78	infinite	colorless
Potassium nitrate(s)	2.11	334	—	40	white
Sodium chloride(s)	2.16	808	1473	36	white
Water(l)	1.00	0	100	—	colorless

*At 25°C, 1 atm.

■ **EXAMPLE 1.7**

To determine the density of ethyl alcohol, a student pipets a 5.00 cm³ sample into an empty flask weighing 15.246 g. He finds that the mass of the flask + ethyl alcohol = 19.171 g. Calculate the density of ethyl alcohol.

Solution

Mass ethyl alcohol = 19.171 g − 15.246 g = 3.925 g

Volume ethyl alcohol = 5.00 cm³

Density = 3.925 g/5.00 cm³ = 0.785 g/cm³

The density of water is just about 1 g/cm³

Exercise

What must be the volume of a flask if, when filled with ethyl alcohol, it contains 5.000 g of that substance? Answer: 6.37 cm³.

For solids, density is a bit more difficult to determine. A common approach is shown in Figure 1.8. The mass of the solid sample is determined in the usual way. Its volume is found indirectly (Example 1.8).

Figure 1.8   To determine the density of a solid, we need to know its volume as well as its mass, m. To find its volume, we add the solid to a flask of known volume, V. We then determine the volume of water, V_w, required to fill the flask. The density of the solid is $m/(V − V_w)$. (Marna G. Clarke)

■ EXAMPLE 1.8

Referring to Figure 1.8, a student finds that 24.960 g of water (density = 0.9971 g/cm^3) is required to fill the empty flask. The water is removed and the flask is carefully dried; pieces of copper weighing 51.236 g are then added. With the copper present, it is found that 19.244 g of water is required to fill the flask. Determine the:

a. volume of the flask   b. volume of the copper
c. density of copper

Solution

a. Since 24.960 g of water is required to fill the empty flask:

     Volume of flask = Volume of 24.960 g water

Treating the density of water, 0.9971 g/cm^3, as a conversion factor:

$$\text{Volume of flask} = 24.960 \text{ g} \times \frac{1 \text{ cm}^3}{0.9971 \text{ g}} = \boxed{25.03 \text{ cm}^3}$$

b. For the flask containing copper and water:

     Volume of flask = Volume of Cu + Volume of 19.244 g water

     Volume of Cu = Volume of flask − Volume of 19.244 g water

$$\text{Volume of 19.244 g water} = 19.244 \text{ g} \times \frac{1 \text{ cm}^3}{0.9971 \text{ g}} = 19.30 \text{ cm}^3$$

> Volume here is an additive quantity

     Volume of Cu = 25.03 cm^3 − 19.30 cm^3 = $\boxed{5.73 \text{ cm}^3}$

c.   Density of Cu = $\dfrac{51.236 \text{ g}}{5.73 \text{ cm}^3}$ = $\boxed{8.94 \text{ g/cm}^3}$

Exercise

If this flask could be completely filled with solid copper, what would be the mass of the copper? Answer: 224 g.

MELTING POINT AND BOILING POINT

The melting point is the temperature at which a substance changes from the solid to the liquid state. If the substance is pure, the temperature stays constant during melting. This means that, for a pure substance, the melting point is identical with the freezing point. Ice melts at 0°C; pure water freezes at that same temperature.

> We would say that melting is a reversible process

     The boiling point of a liquid is the temperature at which bubbles filled with vapor form within the liquid. For reasons to be discussed in Chapter 10, boiling point depends upon the pressure above the liquid. The normal boiling point (Table 1.5) is the temperature at which a liquid boils when the pressure above it is 1 atm (760 mm Hg). For a pure liquid, temperature remains constant during the boiling process.

Figure 1.9 The solubility of potassium nitrate (KNO_3) increases much more rapidly with temperature than does that of potassium chromate, K_2CrO_4.

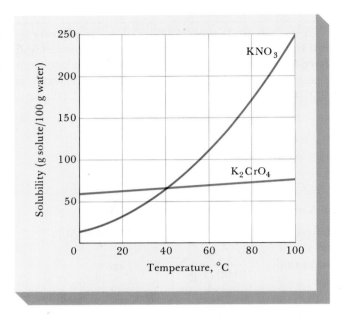

SOLUBILITY

The extent to which a substance dissolves in a particular solvent can be expressed in various ways. A common method is to state the number of grams of the substance that dissolves in 100 g of solvent at a given temperature. At 20°C, about 32 g of potassium nitrate dissolves in 100 g of water. At 100°C, the solubility of this solid is considerably greater, about 246 g/100 g of water (Figure 1.9).

■ **EXAMPLE 1.9** _____

Taking the solubility of potassium nitrate, KNO_3, to be 246 g/100 g water at 100°C and 32 g/100 g water at 20°C, calculate:

a. the mass of water required to dissolve one hundred grams of KNO_3 at 100°C.
b. The amount of KNO_3 that remains in solution when the mixture in (a) is cooled to 20°C.

Solution

We can solve this problem by the conversion factor approach, noting that:

at 100°C: 246 g KNO_3 ≏ 100 g water*

at 20°C: 32 g KNO_3 ≏ 100 g water

We treat the equivalences as if they were equalities.

_____

*The equivalence symbol, ≏, indicates that these two quantities are mathematically equivalent to one another. We can use this relationship to obtain the conversion factors necessary to work the problem.

a. Mass water required $= 100 \text{ g KNO}_3 \times \dfrac{100 \text{ g water}}{246 \text{ g KNO}_3} = \boxed{40.7 \text{ g water}}$

b. Since the solution contains 40.7 g of water:

mass KNO_3 in solution $= 40.7 \text{ g water} \times \dfrac{32 \text{ g KNO}_3}{100 \text{ g water}} = \boxed{13 \text{ g KNO}_3}$

Exercise

What happens to the rest of the KNO_3? Answer: 100 g − 13 g = 87 g of KNO_3 crystallizes out of solution when the temperature drops from 100°C to 20°C.

The general tendency for the solubility of solids to decrease when the temperature is lowered is the basis of a commonly used separation technique called **fractional crystallization**. In this process, we start with an impure solid, A, which we wish to purify. It contains a relatively small amount of impurity, B. The mixture of A and B is dissolved in the minimum amount of hot solvent, often boiling water (Figure 1.10). The solution is cooled, either to room temperature or to 0°C in an ice-water bath. If all goes well, the solid that separates out on cooling will be pure A. This can

(a)

(b)

(c)

(d)

(e)

Figure 1.10 Potassium dichromate, $K_2Cr_2O_7$, an orange-red solid, can be purified by fractional crystallization. The solid, containing a small amount of impurity (a), is dissolved in the minimum quantity of hot water (b). Upon cooling in an ice-water bath, $K_2Cr_2O_7$, which is less soluble at low temperatures, comes out of solution (c). It is separated by filtration and dried (d), (e). (Marna G. Clarke)

be filtered off and dried. The solution that remains (the *filtrate*) should contain all of the impurity along with a small amount of A. This filtrate is ordinarily discarded.

To obtain a pure solid by fractional crystallization, two conditions must apply:

Most pure inorganic substances are prepared by fractional crystallization

1. *The major component must be much more soluble at high temperatures than at low temperatures.* Otherwise, much of it will be lost; that is, it will remain in the solution upon cooling. Referring back to Example 1.9, you can see that it should be relatively easy to purify potassium nitrate by fractional crystallization. The solubility of KNO_3 at 20°C is only about 13% of that at 100°C. This means that 87% of the KNO_3 dissolved at 100°C should come out of solution upon cooling to 20°C. The situation is quite different with potassium chromate (Figure 1.9); here the solubility drops by only 18% when the temperature decreases from 100°C to 20°C. If we attempt to purify K_2CrO_4 by fractional crystallization, we will "lose" 82% of it; i.e., 82% will stay in solution and hence will not be recovered.

2. *The amount of impurity, B, must be rather small.* Otherwise, it will tend to crystallize out on cooling, contaminating the desired product, A. This is likely to happen with a mixture containing 10% to 20% B. Several crystallizations might be required to obtain pure A. With a 50-50 mixture, the technique is likely to be completely ineffective.

■ SUMMARY PROBLEM

Potassium dichromate is a reddish-orange compound containing the three elements potassium (26.58%), chromium (35.35%), and oxygen (38.07%). It has a density of 2.68 g/cm^3; its melting point is 398°C. At 20°C, its solubility is 12 g/100 g water; at 100°C, the solubility is 80 g/100 g water.

a. What are the symbols of the three elements present in potassium dichromate?
b. What is the volume of a sample of potassium dichromate weighing 32.349 g?
c. Express the density in pounds per cubic foot.
d. How many grams of potassium dichromate must be taken to contain one gram of the element chromium?
e. A solid mixture contains 88 g of potassium dichromate and 12 g of sodium dichromate. How much water at 100°C is required for this mixture? Assume the amount of water required to dissolve the potassium dichromate is sufficient to dissolve the impurity as well.
f. When the solution in (e) is cooled to 20°C, how much potassium dichromate remains in solution? How much crystallizes out?

Express all your answers to the correct number of significant figures; use the conversion factor approach throughout.

Answers
a. K, Cr, O   **b.** 12.1 cm^3 **c.** 167 lb/ft^3 **d.** 2.829 g
e. 1.1×10^2 g   **f.** 13 g, 75 g

A major task of chemistry is to separate the components of mixtures. Such separations are most often based upon differences in physical properties between the components. *Distillation* takes advantage of the difference in volatility between a liquid such as water and a solid such as sodium chloride. Two solids can be separated from one another by *fractional crystallization* because they differ in solubility in an appropriate solvent.

Perhaps the most widely used separation technique in chemistry today is **chromatography**, which can be applied to liquid or gaseous solutions. The word "chromatography, derived from the Greek *chroma* (color) is somewhat of a misnomer. This technique was originally applied (80 years ago) to separate colored plant pigments from one another. Today, it is routinely used to separate all kinds of substances, most of which are colorless. The only requirement is that the components differ in solubility and/or extent of adsorption on a solid or liquid surface.

To understand the principle behind chromatographic separations, consider the process of **column chromatography**, Figure 1.11, p. 26. The column, which may be a simple buret, is packed with a finely divided solid such as silica gel that tends to adsorb other solids on its surface. A solution containing the sample is poured into the column, where the sample is retained by the packing at the top (stage 1). A solvent in which the compounds A and B have different solubilities is then added. The sample is gradually washed down the column, continuously dissolving in the solvent and readsorbing on the solid packing. Component A, which is more soluble (less strongly adsorbed), moves more rapidly. The other component, B, which is more strongly attracted to the solid than to the solvent, moves very slowly. Gradually, the mixture is separated into zones or bands (stage 2). Finally, as more solvent is added, components A and B pass successively from the bottom of the column (stages 3 and 4).

Ordinary column chromatography is seldom used today, because it is a relatively inefficient separation technique. However, by packing a column with very fine particles of the order of 10^{-3} mm in diameter, it is possible to achieve much better separation. Unfortunately, with tightly packed particles of this size, it takes essentially forever for a liquid solution to pass down the column. To surmount this problem, liquid is pumped through the column under pressures of the order of 200 atm (3000 lb/in²) or more. This technique, known as **high-pressure liquid chromatography (HPLC)**, is a popular tool used for the separation of complex mixtures.

A quite different technique, known as **paper chromatography**, is often used in the general chemistry laboratory. The solid adsorbent is a strip of filter paper suspended vertically in a stoppered flask or cylinder (Figure 1.12). The sample is introduced as a spot near the bottom of the paper. Enough solvent is added to almost but not quite reach the sample spot. As the solvent climbs the paper by capillary action, the components of the sample gradually separate.

Thin layer chromatography (TLC) resembles paper chromatography except that a glass plate covered with a thin layer of an adsorbent such

MODERN
PERSPECTIVE:

CHROMATOGRAPHY

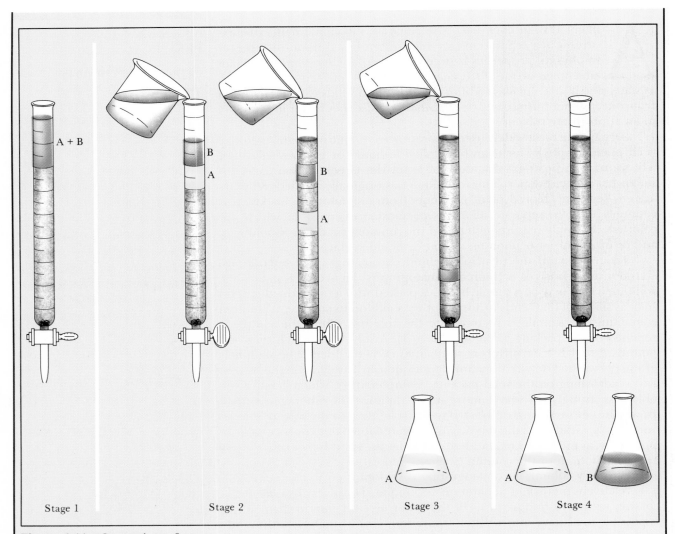

Stage 1   Stage 2   Stage 3   Stage 4

Figure 1.11 Separation of two components, A and B, by column chromatography. Component A is more soluble and/or less strongly adsorbed.

as silica gel is substituted for the filter paper. This technique is routinely used by organic chemists and biochemists to separate complex mixtures of natural products. It has also found application in forensic chemistry to compare fiber samples to one another. Dyes on these fibers are separated by TLC to give characteristic patterns (Figure 1.13) that may or may not match one another.

One of the most widely used chromatographic technique is **gas-liquid chromatography** (**GLC**). Here, the components, which may be gases or volatile liquids, are introduced as a gaseous mixture into one end of a heated glass tube. As little as one microliter (10^{-6} L) of sample may be used. The tube is packed with an inert solid whose surface is coated with a high-boiling, viscous liquid. An unreactive "carrier gas," often helium,

Figure 1.12 Apparatus used to carry out an experiment in paper chromatography in the general chemistry laboratory. The solvent in the bottom of the beaker moves up the paper and separates the components of an ink. (Charles Steele)

Figure 1.13 Separation of colored dyes by chromatography. The original spot is shown near the bottom of the figure; the solvent has advanced to the top.

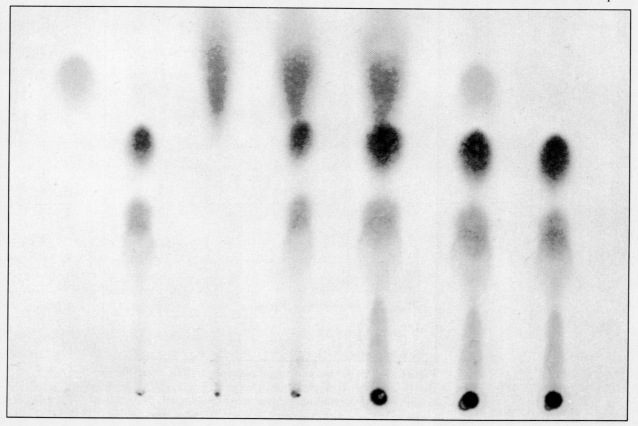

is passed through the tube. The components of the sample gradually separate as they vaporize into the helium or condense into the viscous liquid. Usually the more volatile fractions move faster and come out of the column first. As successive fractions leave the column, they activate a detector and recorder. The end result is a plot such as that shown in Figure 1.14.

Gas-liquid chromatography finds many applications outside the chemistry laboratory. If you've ever had an emissions test on the exhaust system of your car, GLC was almost certainly the analytical method used. Pollutants such as carbon monoxide and unburned hydrocarbons appear as peaks on a graph such as that shown in Figure 1.14. A computer sums the areas under these peaks, which are proportional to the concentrations of pollutants, and prints out a series of numbers that tells you whether your car passed or failed the test. Many of the techniques used to test people for drugs (marijuana, cocaine, etc.) or alcohol also make use of gas-liquid chromatography.

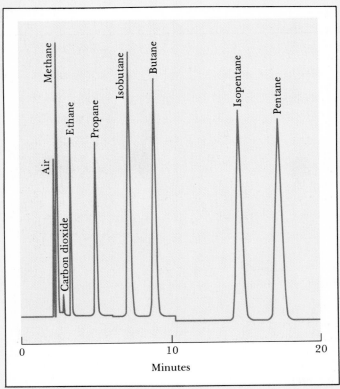

Figure 1.14 The components of natural gas (mostly methane) can be separated by gas-liquid chromatography. With some volatile mixtures, the sample can be as small as 10^{-6} L.

QUESTIONS AND PROBLEMS

The questions and problems listed here are typical of those at the end of each chapter. Some involve discussion and most require calculations, writing equations, or other quantitative work. The topic emphasized in each question or problem is indicated in the heading, such as "Symbols and Formulas" or "Significant Figures." Those in the "Unclassified" category may involve more than one concept, including, perhaps, topics from a preceding chapter. "Challenge Problems," listed at the end of the set, require extra skill and/or effort.

The "classified" questions and problems (Problems 1–50 in this set) are arranged in matched pairs, one below the other, and illustrate the same concept. For example, Questions 1 and 2 below are nearly identical in nature; the same is true of Questions 3 and 4, and so on. Problems numbered in color are answered in Appendix 4.

Symbols and Formulas

1. Give the color, state of matter at room temperature, and symbol of the following elements. Use Tables 1.4 and 1.5.
  a. bromine   b. iron

2. Use Tables 1.4 and 1.5 to write the name and symbol for the element that is
  a. a green-yellow gas at 25°C and 1 atm.
  b. a red solid at 25°C and 1 atm.

3. Give the symbols for
  a. potassium   b. cadmium   c. gold
  d. antimony   e. rubidium

4. Name the elements whose symbols are
  a. Mn   b. Na   c. As   d. W   e. P

5. How many elements are there in the following groups?
  a. Group 2   b. Group 3   c. Group 8
  d. the subgroup headed by Cr

6. How many elements are there in the following periods?
  a. Period 1   b. Period 2   c. Period 3
  d. Period 4   e. Period 5

Measurements

7. Classify each of the following as units of mass, volume, length, density, energy, or pressure.
  a. mg   b. mL   c. cm^3 d. mm
  e. kg/m^3 f. mm Hg   g. kJ

8. Classify each of the following units as units of mass, volume, length, density, energy, or pressure.
  a. nm   b. kg   c. J   d. m^3
  e. g/cm^3 f. atm   g. kcal

9. Select the smaller member of each pair.
  a. 303 m or 0.300 km
  b. 500 kg or 0.0500 g
  c. 1.50 cm^3 or 1.50×10^3 nm^3
  d. 25.0 g/cm^3 or 2.50×10^{-3} kg/m^3

10. Select the smaller member of each pair.
  a. 27.12 g or 27.12 kg
  b. 35 cm^3 or 0.035 m^3
  c. 2.87 g/L or 2.87 g/cm^3
  d. 525 mm or 5.25×10^{-3} km

11. Most laboratory experiments are done at 25°C. Express this in
  a. °F   b. K

12. A child has a temperature of 104°F. What is his temperature in
  a. °C   b. K

13. Carbon dioxide, CO_2, at room temperature (70°F) is a gas. It can be frozen at -69.7°F and 5 atm pressure to solid carbon dioxide, popularly known as dry ice. What is the freezing point of carbon dioxide in °C?

14. Superconductors use liquid nitrogen as a coolant. Liquid nitrogen boils at -195.8°C. What is its boiling point in
  a. °F   b. K

Significant Figures

15. How many significant figures are there in each of the following?
  a. 1.92 mm   b. 0.032100 g
  c. 6.022×10^{23} atoms   d. 460.00 L
  e. 0.00036 cm^3 f. 2×10^9 nm

16. How many significant figures are there in each of the following?
  a. 23.437 m   b. 0.002017 g   c. 30.0×10^{20} nm
  d. 50.010 L   e. 2.30790 atm   f. 350 miles

17. A student prepares a salt solution by dissolving 85.638 g of sodium chloride (NaCl) in enough water to form 237 cm^3 of solution. Calculate the number of grams of salt per cubic centimeter of solution.

18. Calculate the volume of a sodium atom, which has a radius of 0.186 nm. Assume the atom is spherical. The volume of a sphere is given by the expression $V = 4\pi r^3/3$.

19. Calculate the following to the correct number of significant figures.
  a. $x = \dfrac{1.27 \text{ g}}{5.296 \text{ cm}^3}$ b. $x = \dfrac{12.235 \text{ g}}{1.01 \text{ L}}$

  c. $x = 12.2 \text{ g} + 0.38 \text{ g}$ d. $x = \dfrac{17.3 \text{ g} + 2.785 \text{ g}}{30.20 \text{ cm}^3}$

20. How many significant figures are there in the values of x obtained from
  a. $x = \dfrac{34.0300 \text{ g}}{12.09 \text{ cm}^3}$

  b. $x = 32.647 \text{ g} - 32.327 \text{ g}$

  c. $x = (0.00630 \text{ cm})(2.003 \text{ cm})(200.0 \text{ cm})$

  d. $x = \dfrac{236.45 \text{ g} - 1.3 \text{ g}}{(3.4561 \text{ cm})(32.675 \text{ cm}^2)}$

Conversion Factors

21. Using Table 1.1, convert 17.5 quarts to
   a. liters   b. cubic meters   c. cubic feet
22. Using Table 1.1, convert 38.75 ft² to
   a. square miles   b. km²   c. m²
23. Using Table 1.2, convert 3.57 atm to
   a. mm Hg   b. pascals   c. lb/in²
24. Using Table 1.2, convert one kilojoule to
   a. J   b. calories   c. liter atmospheres
25. The unit of land measure in the metric system is the hectare; in the English system it is the acre. A square exactly 100 m on a side has an area of one hectare; if it is 208.7 ft on a side, the area is one acre. How many acres are there in one hectare?
26. In the United States, cigarettes are smoked at the rate of 2.0×10^4 cigarettes/s. At this rate, how many are smoked in one day?
27. A day on Mars is 8.864×10^4 seconds long and a year is 5.935×10^7 seconds long.
   a. How many earth days are there in one Mars day?
   b. How many earth days are there in one Mars year?
28. A typical aerobic walker on a track covers a lap (0.25 mile) in about 3.4 minutes. A champion marathon runner takes about five minutes to cover a mile. If an aerobic walker competed in the Boston marathon (42.16 km), how many minutes would you expect him to finish behind the winner?
29. During earlier times in England, land was measured in units such as fardells, nookes, yards, and kides:

   2 fardells = 1 nooke; 4 nookes = 1 yard;
   4 yards = 1 kide

Thus,
   a. 6.00 kides = _____ fardells
   b. 15 nookes = _____ kides
              = _____ fardells
30. When the Pharmacopoeia of London was compiled in 1618, the troy system of measure was used to prepare medicine. Among the units used were:

   20 grains = 1 scruple; 3 scruples = 1 drachm;
   8 drachms = 1 ounce; 12 ounces = 1 pound.

Thus,
   a. 12.05 pounds = _____ drachms
   b. 25.0 drachms = _____ pounds;
              = _____ ounces
31. In Germany, nutritional information is given in kilojoules instead of nutritional calories (1 nutritional calorie = 1 kcal). A packet of Rindfleisch-Suppe (meat soup) has the following information:

   250 mL of prepared soup = 235 kJ

A packet of the same soup sold in the United States would have the same nutritional information in kilocalories per cup. There are two cups to a pint. How many nutritional calories would be quoted per cup of prepared soup?
32. White gold is an alloy that typically contains 60.0% by mass of gold; the rest is platinum. If 175 g of gold are available, how many grams of platinum are required to combine with the gold to form this alloy?
33. Because of the rapid rise in the price of copper, the Bureau of the Mint in 1982 changed the composition of the penny. The new penny is now an alloy of 97.6% Zn and 2.4% Cu. Pre-1982 pennies were made up of 95% Cu and 5% Zn. If a new penny has a mass of 2.507 g, how many grams of each metal are there in the new penny?
34. One type of antiperspirant uses aluminum chlorohydrate as its active ingredient. This compound is made up of 30.93% Al, 45.86% O, 2.89% H, and 20.32% Cl. How many grams of each element are there in one ounce of this compound?

Physical and Chemical Properties

35. The following data refer to the element carbon. Classify each as a physical or chemical property.
   a. It is virtually insoluble in water.
   b. It exists in several forms, e.g., diamond, graphite.
   c. It is a solid at 25°C and 1 atm.
   d. It reacts with oxygen to form carbon dioxide.
36. The following data refer to the element iodine. Classify each as a physical or chemical property.
   a. Its density at 20°C and 1 atm is 4.93 g/cm³.
   b. Iodine has a purple color.
   c. It reacts with chlorine.
   d. Its normal melting point is 113.5°C.
37. The mass and volume of a piece of pure graphite are 0.600 g and 0.270 mL. What is the density of graphite?
38. A sample of mercury has a volume of 0.250 L and weighs 3.40 kg. Calculate the density of mercury in g/cm³.
39. A piece of metal weighing 20.32 g is added to a flask with a volume of 24.5 cm³. It is found that 18.52 g of water ($d = 1.00$ g/cm³) must be added to the metal to fill the flask. What is the density of the metal?
40. A solid with an irregular shape weighing 13.56 g is added to a flask with a volume of 20.35 cm³. It is found that 10.35 g of benzene ($d = 0.879$ g/cm³) must be added to the metal to fill the flask. What is the density of the metal?
41. A water bed filled with water has the dimensions 8.0 ft × 7.0 ft × 0.75 ft. Taking the density of water to be 1.00 g/cm³, determine the number of kilograms of water required to fill the water bed.
42. Blood plasma volume for adults is about 3.1 L. Its density is 1.020 g/cm³. How many pounds of blood plasma are there in your body?
43. Air is 21% oxygen by volume. Oxygen has a density of 1.31 g/L. What is the volume in liters of a room that holds enough air to contain 75 kg of oxygen?

44. A water solution contains 10.0% ethyl alcohol by mass and has a density of 0.983 g/cm³. What mass in grams of ethyl alcohol is present in 7.50 L of this solution?

45. The solubility of potassium chloride is 37.0 g/100 g water at 30°C. Calculate at 30°C:
a. the mass of potassium chloride that dissolves in 34.5 g of water.
b. the mass of water required to dissolve 34.5 g of potassium chloride.

46. At 25°C and 1 atm, 0.1449 g of carbon dioxide dissolves in 100.0 grams of water. Calculate at 25°C and 1 atm:
a. the mass of carbon dioxide that dissolves in 5.00 mL of water ($d = 1.00$ g/cm³).
b. the mass of water required to dissolve 5.00 g of carbon dioxide.

47. Use Figure 1.9 to estimate
a. the solubility of KNO_3 at 50°C.
b. the mass of water required to dissolve 20.0 g of KNO_3 at 50°C.
c. the mass of KNO_3 that dissolves in one liter of water ($d = 1.00$ g/cm³) at 50°C.

48. Use Figure 1.9 to estimate
a. the solubility of KNO_3 at 15°C.
b. the mass of KNO_3 that dissolves in 5.0 kg of water at 15°C.
c. the mass of water needed to dissolve 1.00 g of KNO_3 at 15°C.

49. You are given a mixture of 88 g of KNO_3 and 22 g of K_2CrO_4. You dissolve the mixture in 50.0 g of water at 100°C and cool the solution to 10°C. Using Figure 1.9, calculate
a. how much KNO_3 crystallizes out of solution.
b. how much K_2CrO_4 crystallizes out of solution.
c. what percentage of the KNO_3 you started out with remains in solution.

50. Repeat the calculations called for in Problem 49, this time for a mixture containing 95 g of KNO_3 and 15 g of K_2CrO_4.

Unclassified

51. How do you distinguish
a. chemical properties from physical properties?
b. a solute from a solution?
c. boiling point from normal boiling point?
d. a compound from a mixture?

52. How do you distinguish
a. an element from a compound?
b. an element from a mixture?
c. a solution from a heterogeneous mixture?
d. distillation from filtration?

53. An intensive property is independent of the mass of the sample. Which of the following properties are intensive?

a. heat required to raise the temperature 1°C
b. boiling point
c. solubility per 100 g of water

54. An extensive property is one that depends on the amount of the sample. Which of the following properties are extensive?
a. volume   b. density   c. temperature
d. energy   e. melting point

55. Diamonds ($d = 3.51$ g/cm³) are commonly measured in carats (1 carat = 200 mg). What is the volume of a 2.0-carat diamond?

56. A swimming pool that is 4.05 m wide and 6.10 m long has an average depth of 8.0 ft. The water solution in the pool ($d = 1.00$ g/cm³) contains 0.70% chlorine by mass as a disinfectant. What is the mass of chlorine in the pool?

57. Titanium is used in airplane bodies because it is both strong and light. It has a density of 4.55 g/cm³. If a cylinder of titanium is 4.75 cm long and has a mass of 104.2 g, calculate the diameter of the cylinder. ($V = \pi r^2 l$, where V is the volume of the cylinder, r is its radius, and l is the length.)

58. Take the price of gold to be $410 an ounce. Its density is 19.3 g/cm³. What is the value of a bar of gold two inches thick, a foot long and 6.50 inches wide?

59. A pycnometer is a device used to determine density. It weighs 20.455 g empty and 31.486 g when filled with water ($d = 1.000$ g/cm³). Pieces of an alloy are put into the empty, dry pycnometer. The mass of the alloy and pycnometer is 28.695 g. Water is added to the alloy to exactly fill the pycnometer. The mass of the pycnometer, water, and alloy is 38.689 g. What is the density of the alloy?

Challenge Problems

60. At what point is the temperature in °C exactly twice that in °F?

61. Oil spreads on water to form a film about 120 nm thick (two significant figures). How many square kilometers of ocean will be covered by the slick formed when two barrels of oil are spilled (1 barrel = 31.5 U.S. gallons)?

62. A laboratory experiment requires 0.750 g of aluminum wire ($d = 2.70$ g/cm³). The diameter of the wire is 0.0179 inches. Determine the length of the wire, in centimeters, to be used for this experiment. The volume of a cylinder is $\pi r^2 l$, where $r =$ radius and $l =$ length.

63. An average human male breathes about 8.50×10^3 L of air per day. The concentration of lead (Pb) in highly polluted urban air is 7.0×10^{-6} g Pb/m³ of air. Assume that 75% of the lead is present as particles less than 1.0×10^{-6} m in diameter and that 50% of the particles below that size are retained in the lungs. Calculate the mass of lead absorbed in this manner in one year by an average male living in this environment.

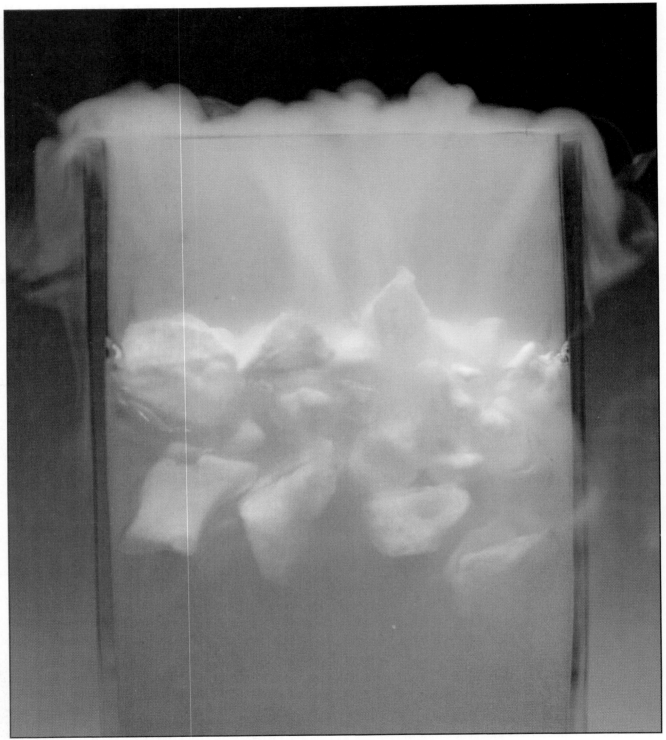

$$Ca^{2+}(aq) + 2OH^-(aq) + CO_2(g) \rightarrow CaCO_3(s) + H_2O$$

Precipitation of calcium carbonate by the addition of dry ice to a solution of calcium hydroxide.

CHAPTER
2

ATOMS, MOLECULES, AND IONS

*Atom from atom yawns as far
As moon from earth, or star from
star.*

RALPH WALDO EMERSON
ATOMS

I n Chapter 1, we described measurements carried out on macroscopic samples of matter. To measure properties such as melting point, boiling point, density, or solubility, we typically work with samples weighing between 10^{-2} and 10^2 grams. Quantities of matter in this range are easily visible and can be weighed accurately on an analytical balance.

In this chapter, we change our perspective to consider the submicroscopic building blocks of matter. Our primary concern will be with atoms (Section 2.1), those tiny particles characteristic of elements, which range in mass from 10^{-24} to 10^{-22} grams. We will also consider briefly two other particles derived from atoms: molecules and ions (Section 2.3).

Our discussion of atoms will focus upon their composition (Section 2.2) and their masses (Section 2.4), emphasizing the ways in which atoms of elements differ from one another insofar as structure and mass are concerned. Finally, in Sections 2.5 and 2.6, we will consider a basic counting unit used with atoms and other submicroscopic particles, the mole.

2.1
ATOMIC THEORY

The notion that matter consists of discrete particles is an old one. About 400 B.C. this idea appeared in the writings of Democritus, a Greek philosopher. He had been introduced to it by his teacher, Leucippus. The idea was rejected by Plato and Aristotle, who had a great deal more influence in the development of ideas than Democritus. For nearly two thousand years, the concept of the atom remained just that, an abstract idea divorced from experimental reality.

None of these men did any experiments, as far as we know

33

In 1808, an English schoolteacher and amateur meteorologist, John Dalton, developed an atomic model of matter that offered a simple explanation of several of the laws of chemistry. Some of Dalton's ideas had to be discarded as chemists learned more about the structure of matter. However, the essentials of his theory have withstood the test of time. Three of the main postulates of modern atomic theory, all of which came from Dalton, are given below with examples that illustrate their meaning.

1. *An element is composed of tiny particles called atoms. All atoms of a given element show the same chemical properties.* The element oxygen is made up of oxygen atoms. These atoms are much too small to be seen or weighed directly. All oxygen atoms behave chemically in the same way.

2. *Atoms of different elements have different properties. In an ordinary chemical reaction, no atom of any element disappears or is changed into an atom of another element.* The chemical behavior of oxygen atoms is different from that of hydrogen atoms or atoms of any other element. When hydrogen and oxygen react, all the hydrogen and oxygen atoms that react are present in the water formed. No atoms of any other element are formed.

3. *Compounds are formed when atoms of two or more elements combine. In a given compound, the relative numbers of atoms of each kind are definite and constant. In general, these relative numbers can be expressed as integers or simple fractions.* In the compound water, hydrogen atoms and oxygen atoms are combined with each other. For every oxygen atom present, there are always two hydrogen atoms.

The atomic theory explains three of the basic laws of chemistry.

1. The **Law of Conservation of Mass**. This law was first stated by the French chemist Antoine Lavoisier in 1789. In modern form, it says that *there is no detectable change in mass in an ordinary chemical reaction.* If atoms are "conserved" in a reaction (Postulate 2 above), mass will also be conserved.

> How could you check this experimentally?

Lavoisier has often been referred to as the "father of modern chemistry." He was among the first to carry out quantitative chemical experiments and draw logical conclusions from them. His experiments demonstrating the conservation of mass were described in the first modern textbook of chemistry, which he published in 1789. Five years later, at the age of 50, Lavoisier died on the guillotine, a victim of the hysteria that accompanied the French Revolution.

> This seldom happens to textbook authors nowadays

2. The **Law of Constant Composition**. This tells us that *a compound always contains the same elements in the same proportions by mass.* If the atom ratio of the elements in a compound is fixed (Postulate 3), their proportions by mass must also be fixed.

The validity of this law became generally recognized at about the same time that Dalton's theory appeared. Prior to 1808 many people agreed with the French chemist Berthollet. He believed that the composition of a compound could vary over wide limits, depending on how it was prepared. Joseph Proust, a Frenchman working in Madrid, refuted Berthollet by showing that the "compounds" Berthollet had cited were actually mixtures.

It is now known that in certain compounds, particularly metal oxides and sulfides, the atom ratio may vary slightly from a whole number ratio. Careful analyses of samples of nickel oxide prepared by heating nickel with oxygen at high temperatures give an atom ratio of 0.97 : 1.00 rather than the expected 1 : 1 ratio. Compounds of this type are sometimes referred to as "Berthollides" or, more frequently, *nonstoichiometric* compounds. The deviations from constant composition arise because of defects in the crystal structure. Certain nonstoichiometric compounds exhibit the property of superconductivity (see "Modern Perspective," Chapter 3).

3. The third postulate of the atomic theory is in many ways the most important. Among other things, it led Dalton to formulate the **Law of Multiple Proportions**. This law applies to the situation in which two elements form more than one compound. It states that, in these compounds *the masses of one element that combine with a fixed mass of the second element are in a ratio of small whole numbers* (for example, 2 : 1). To derive the Law of Multiple Proportions, Dalton reasoned along the following lines: Suppose elements A and B form two different compounds. In one of these (AB), one atom of A might be combined with one atom of B. The second compound (AB_2) might contain two atoms of B per atom of A. If this is true, the mass of B combined with a fixed mass (such as one gram) of A would be twice as great in the second compound. In other words, the masses of B *per gram of A* in the two compounds would be in a 2 : 1 ratio. The meaning of the Law of Multiple Proportions is further illustrated in Figure 2.1.

Despite the successes of Dalton's atomic theory, it was not immediately accepted by all scientists. In particular, many chemists felt it was a waste of time to speculate on the particulate nature of matter. As late as 1900, the well-known German chemist Wilhelm Ostwald, in writing a general chemistry textbook that was used for a generation, deliberately avoided all mention of atoms or other elementary particles.

"Be not the first by whom the new is tried, nor yet the last to cast the old aside"

Figure 2.1 Chromium forms two different oxides, Cr_2O_3 (green) and CrO_3 (red). In Cr_2O_3, 2.167 g of chromium is combined with one gram of oxygen; in CrO_3, there is 1.083 g of chromium per gram of oxygen. The ratio 2.167 : 1.083 is that of two small numbers,

$$2.167 : 1.083 = 2 : 1$$

illustrating the Law of Multiple Proportions. (Marna G. Clarke)

2.2
COMPONENTS OF THE ATOM

Like any useful scientific theory, the atomic theory raised more questions than it answered. Scientists wondered whether atoms, tiny as they are, could be broken down into still smaller particles. Nearly 100 years passed before the existence of subatomic particles was confirmed in the laboratory. Some of the experiments that led to the discovery of these particles, along with a few of the scientists who did pioneer work in this area, are described in an "Historical Perspective" at the end of this chapter. Here we will simply describe the properties of the three subatomic particles that are of particular interest to chemists and see how they combine to form atoms of different elements.

PROTONS, NEUTRONS, AND ELECTRONS

Some of the properties of these fundamental particles are listed in Table 2.1. Note that protons and neutrons are located in the **nucleus** of an atom. The nucleus contains more than 99.9% of the mass of the atom, even though its diameter is only about 1/10,000 of that of the atom. If an atom could be expanded to cover this page, its nucleus would be barely visible as a tiny dot one-tenth the size of the period at the end of this sentence.

An atom is mostly empty space

The proton has a mass nearly equal to that of the lightest atom, the hydrogen atom. Reflecting that fact, we assign the proton a *mass number* of 1. The proton carries a unit positive charge ($+1$). All atoms contain an integral number of protons in the nucleus; this number ranges from 1 to over 100. The neutron is an uncharged particle with a mass about equal to that of the proton. The neutron, like the proton, has a mass number of 1.

The electron carries a unit negative charge (-1), equal in magnitude but opposite in sign to that of the proton. The electron is much lighter than the proton or neutron; its mass is only about 1/2000 that of a hydrogen atom. In round number terms, we assign the electron a mass number of 0. The number of electrons in an atom, like the number of protons, can vary from 1 to over 100. We will have more to say in Chapter 5 about how electrons are arranged relative to one another. At this time, we need only point out that electrons are found in the outer regions of atoms, where they constitute what amounts to a cloud of negative charge about the nucleus.

TABLE 2.1   Properties of Subatomic Particles

PARTICLE	LOCATION	RELATIVE CHARGE	RELATIVE MASS	ACTUAL CHARGE (C = coulombs)	ACTUAL MASS
Proton	nucleus	$+1$	1.0	$+1.6022 \times 10^{-19}$ C	1.6727×10^{-24} g
Neutron	nucleus	0	1.0	0	1.6750×10^{-24} g
Electron	outside nucleus	-1	0.00055	-1.6022×10^{-19} C	9.1095×10^{-28} g

ATOMIC NUMBER

All the atoms of a particular element have the same number of protons in the nucleus. This number is a basic property of an element, called its **atomic number**:

atomic number = number of protons   (2.1)

In a neutral atom the number of protons in the nucleus is exactly equal to the number of electrons outside the nucleus. Consider, for example, the elements hydrogen (at. no. = 1) and helium (at. no. = 2). All hydrogen atoms have one proton in the nucleus; all helium atoms have two. In a neutral hydrogen atom there is one electron outside the nucleus; in a helium atom there are two.

The number of protons equals the number of electrons because the atom is electrically neutral

H atom:	1 proton,	1 electron,	atomic number = 1
He atom:	2 protons,	2 electrons,	atomic number = 2
U atom:	92 protons,	92 electrons,	atomic number = 92

Atomic numbers of the elements are given in the Periodic Table (inside front cover). They are printed in black directly above the symbol of the element. Notice that atomic number increases steadily as we move across the table. Indeed, the location of an element in the table is determined by its atomic number.

MASS NUMBER; ISOTOPES

As with protons and neutrons, we can assign mass numbers to atoms. Recall that the mass number is 1 for both a proton and a neutron. Hence, we can find the mass number of an atom by adding up the number of protons and neutrons in the nucleus:

mass number = number of protons + number of neutrons   (2.2)

For an atom with 17 protons and 20 neutrons in the nucleus,

mass number = 17 + 20 = 37

All atoms have integral mass numbers

As we have noted, all atoms of a given element have the same atomic number (number of protons). They may, however, differ from one another in mass and hence in mass number. This can happen because, although the number of protons in the nucleus of a given kind of atom is fixed, the number of neutrons is not. It may vary and often does. Consider the element hydrogen (at. no. = 1). There are three different kinds of hydrogen atoms. They all have one proton in the nucleus. A "light" hydrogen atom (the most common type) has no neutrons in the nucleus (mass no. = 1). Another type of hydrogen atom (deuterium) has one neutron (mass no. = 2). Still a third type (tritium) has two neutrons (mass no. = 3).

Atoms that contain the same number of protons but a different number of neutrons are called **isotopes**. The three kinds of hydrogen atoms just described are isotopes of that element. They have masses that are very nearly in the ratio $1:2:3$. Among the isotopes of the element uranium are the following:

ISOTOPE	ATOMIC NUMBER	MASS NUMBER	NUMBER OF PROTONS	NUMBER OF NEUTRONS
Uranium-235	92	235	92	143
Uranium-238	92	238	92	146

The composition of a nucleus is shown by its nuclear symbol. Here, the atomic number appears as a subscript at the lower left of the symbol of the element. The mass number is written as a superscript at the upper left. The nuclear symbols of the isotopes of hydrogen are

Each H atom has one electron

$$^1_1H, \ ^2_1H, \ ^3_1H$$

For the isotopes of uranium, we write the nuclear symbols

Each U atom has 92 electrons

$$^{235}_{92}U, \ ^{238}_{92}U$$

Quite often isotopes of an element are distinguished from one another simply by writing the mass number after the symbol of the element. In discussing the isotopes of uranium, we might refer to U-235 and U-238 rather than writing out the nuclear symbols.

■ **EXAMPLE 2.1**

a. Write nuclear symbols for three isotopes of oxygen (at. no. = 8) in which there are 8, 9, and 10 neutrons, respectively.
b. One of the most harmful species in nuclear waste is a radioactive isotope of strontium, $^{90}_{38}Sr$. How many protons are there in this nucleus? How many neutrons?

Solution

a. The mass numbers must be as follows: $8 + 8 = 16$; $8 + 9 = 17$; $8 + 10 = 18$. Thus we have

$$^{16}_8O, \ ^{17}_8O, \ ^{18}_8O$$

To do nuclear structure in this course, you have to be able to add and subtract

b. The number of protons is given by the atomic number, $\boxed{38}$. To obtain the number of neutrons we subtract the number of protons from the mass number:

$$\text{number of neutrons} = 90 - 38 = \boxed{52}$$

Exercise
Write the nuclear symbol for an atom that contains 32 protons and 38 neutrons (use the Periodic Table to find the symbol of the element). Answer: $^{70}_{32}Ge$.

2.3
MOLECULES AND IONS

Isolated atoms rarely occur in nature. For the most part, they are too reactive to be found by themselves. Instead, atoms tend to combine with one another in various ways. As a result, the structural units in most elements and all compounds are more complex than simple atoms. Two of the most important "building blocks" of matter are molecules and ions, both of which are formed from atoms.

MOLECULES

The basic structural unit in most volatile (easily vaporized) substances is the molecule. *A molecule is a group of two or more atoms held together by strong forces called chemical bonds.* Many elements consist of diatomic molecules. The hydrogen molecule is typical; its structure may be shown as

An H—H molecule is much more stable than two separate H atoms

H—H

where the dash is used to represent the bond joining the two hydrogen atoms. The molecule of the gaseous compound hydrogen chloride is also diatomic. It has the structure

H—Cl

Most often, molecular substances are represented by molecular formulas. In a molecular formula, the number of atoms of each element is indicated by a subscript written after the symbol of the element. The molecular formulas of the two substances just described are

hydrogen: H_2 (2 H atoms per molecule)
hydrogen chloride: HCl (1 H atom, 1 Cl atom per molecule)

(Note that no subscript is used when there is only one atom of a particular type present.)

Most molecular substances are composed of molecules more complex than those just cited. The water molecule, H_2O, consists of a central oxygen atom bonded to two hydrogen atoms. In the ammonia molecule, NH_3, a central nitrogen atom is bonded to three hydrogen atoms. Methane, CH_4, the major component of natural gas, has as a structural unit a molecule in which a carbon atom is bonded to four hydrogen atoms. The structures of these molecules are shown in compact and expanded forms in Figure 2.2.

Many simple molecules have a lot of symmetry

IONS

If enough energy is applied, one or more electrons can be removed from a neutral atom. This leaves a positively charged particle somewhat smaller than the original atom. Electrons may also be added to certain atoms to form negatively charged particles larger than the original atom. Charged particles are called **ions**. An example of a positive ion (**cation**) is the Na^+

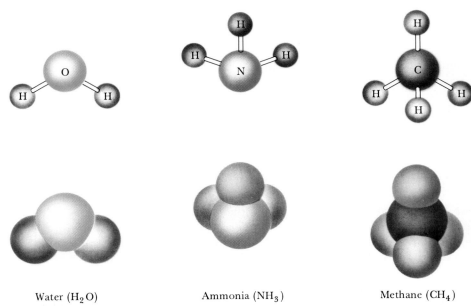

Water (H_2O)   Ammonia (NH_3)   Methane (CH_4)

Figure 2.2   Ball-and-stick and space-filling models of the molecules H_2O, NH_3, and CH_4. In each case, the hydrogen atoms are bonded to a central, nonmetal atom (O, N, C).

ion, formed from a sodium atom by the loss of a single electron. Another cation is Ca^{2+}, derived from a calcium atom by removing two electrons.

$$Na\ atom \quad \rightarrow \quad Na^+\ ion \quad + e^-$$
$$(11\ p^+,\ 11\ e^-) \quad (11\ p^+,\ 10\ e^-)$$

$$Ca\ atom \quad \rightarrow \quad Ca^{2+}\ ion \quad + 2\ e^-$$
$$(20\ p^+,\ 20\ e^-) \quad (20\ p^+,\ 18\ e^-)$$

Two common negative ions (**anions**) are the chloride ion, Cl^-, and the oxide ion, O^{2-}. These are formed when atoms of chlorine or oxygen acquire electrons:

$$Cl\ atom \quad + \quad e^- \rightarrow \quad Cl^-\ ion$$
$$(17\ p^+,\ 17\ e^-) \qquad\qquad (17\ p^+,\ 18\ e^-)$$

$$O\ atom \quad + 2e^- \rightarrow \quad O^{2-}\ ion$$
$$(8\ p^+,\ 8\ e^-) \qquad\qquad (8\ p^+,\ 10\ e^-)$$

Notice that when an ion is formed the number of protons in the nucleus is unchanged. It is the number of electrons that increases or decreases. Negative ions contain more electrons than protons; positive ions contain fewer electrons than protons. The charge of the ion is indicated by a superscript at the upper right. The O^{2-} ion has a -2 charge; it has two more electrons than the oxygen atom. The Na^+ ion has a $+1$ charge; it has one fewer electron than the sodium atom.

Usually atoms gain or lose electrons by reacting with atoms of a different kind:

$$Na + Cl \rightarrow Na^+ + Cl^-\ (in\ NaCl)$$

■ EXAMPLE 2.2

Give the number of protons and electrons in the Sc^{3+} ion.

Solution

Using the Periodic Table, we find that Sc (scandium) has an atomic number of 21. Hence, the Sc^{3+} ion contains $\boxed{21\ protons}$ and

$$21 - 3 = \boxed{18\ electrons}$$

Exercise

Give the symbol of an ion that has $10\ e^-$ and $7\ p^+$; an ion that has $10\ e^-$ and $12\ p^+$. Answers: N^{3-}; Mg^{2+}.

Many compounds are made up of ions. They are called ionic compounds. Since a bulk sample of matter must be electrically neutral, ionic compounds always contain both cations ($+$ charge) and anions ($-$ charge). Ordinary table salt, sodium chloride, is made up of an equal number of Na^+ and Cl^- ions. The structure of a sodium chloride crystal is shown in Figure 2.3. Notice that there are no discrete molecules. Positive and negative ions are bonded together in a continuous network. Calcium oxide (quicklime), which is made up of an equal number of Ca^{2+} and O^{2-} ions, has a similar structure.

In the ionic compounds sodium oxide and calcium chloride there are unequal numbers of cations and anions. To maintain electrical neutrality, the amount of positive charge in the compound must equal the amount of negative charge. This means that in sodium oxide there must be two Na^+ ions for every O^{2-} ion. A similar situation applies with calcium chloride: two Cl^- ions are required to balance one Ca^{2+} ion.

You can't buy a bottle of Na^+ ions

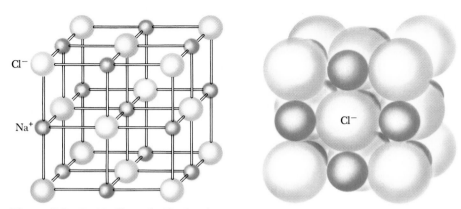

Figure 2.3 In NaCl, each Cl^- ion is surrounded by six Na^+ ions and vice versa. A single crystal of NaCl contains many billions of ions arranged in this pattern.

The composition of an ionic compound is indicated by writing its formula. Here, subscripts are used to indicate the relative numbers of ions of each type. For the four compounds just discussed,

COMPOUND	FORMULA	INTERPRETATION
Sodium chloride	NaCl	1 Na$^+$ ion for 1 Cl$^-$ ion
Calcium oxide	CaO	1 Ca^{2+} ion for 1 O^{2-} ion
Sodium oxide	Na$_2$O	2 Na$^+$ ions for 1 O^{2-} ion
Calcium chloride	CaCl$_2$	1 Ca^{2+} ion for 2 Cl$^-$ ions

2.4
MASSES OF ATOMS

As pointed out earlier, individual atoms are too small to be seen, let alone weighed. It is possible, however, to determine the relative masses of different atoms; that is, we can find out how heavy an atom is compared to an atom of a different element. In this section, we will consider how this is done, using quantities called **atomic masses**. Later, we will see how it is possible to calculate the masses of individual atoms, using a quantity referred to as **Avogadro's number**.

ATOMIC MASSES; THE CARBON-12 SCALE

Relative masses of atoms of different elements are expressed in terms of their atomic masses. The atomic mass of an element tells us how heavy, on the average, an atom of that element is compared to an atom of another element. The atomic mass scale is based on the most common isotope of carbon, $^{12}_{6}$C, which is assigned a mass of exactly 12 atomic mass units (amu):

$$\text{mass C-12 atom} = 12 \text{ amu (exactly)} \tag{2.3}$$

When we say that hydrogen has an atomic mass of 1.00794 amu, we imply that a hydrogen atom is about $\frac{1}{12}$ as heavy as a C-12 atom. More exactly:

$$\frac{\text{mass H atom}}{\text{mass C-12 atom}} = \frac{1.00794 \text{ amu}}{12 \text{ amu}} = 0.0839950$$

Again, since helium has an atomic mass of 4.002602 amu, an He atom weighs about $\frac{1}{3}$ as much as a C-12 atom, or about four times as much as an H atom:

$$\frac{\text{mass He atom}}{\text{mass C-12 atom}} = \frac{4.002602 \text{ amu}}{12 \text{ amu}} = 0.3335502;$$

$$\frac{\text{mass He atom}}{\text{mass H atom}} = \frac{4.002602 \text{ amu}}{1.00794 \text{ amu}} = 3.97107$$

In the general case, for two elements X and Y:

$$\frac{\text{mass of an atom of Y}}{\text{mass of an atom of X}} = \frac{\text{atomic mass of Y}}{\text{atomic mass of X}} \qquad (2.4)$$

This is really a definition of atomic mass

Atomic masses are listed in the Periodic Table (inside front cover). They are given in black below the symbol of the element (H = 1.00794, He = 4.002602, and so on). Notice that, with a few exceptions, atomic mass increases in the same order as atomic number. Another table of atomic masses is given on the inside back cover of this text. Here the elements are arranged alphabetically.

MEASUREMENT OF ATOMIC MASSES; THE MASS SPECTROMETER

Atomic masses can be determined with great accuracy using a mass spectrometer (Fig. 2.4). Here, a gas such as helium, at very low pressures, is bombarded by high-energy electrons. A few helium atoms are converted to He$^+$ ions. These ions are focused into a narrow beam and accelerated by a voltage of 500 to 2000 V toward a magnetic field. The field deflects the ion beam out of its straight-line path, toward the collector.

For ions of a given charge, let us say +1, the extent of deflection depends upon the mass of the ion. The heavier the ion, the less it will be deflected. At a given magnetic field strength and accelerating voltage (V$_1$), a $^{12}_{6}$C$^+$ ion might arrive at point A (Fig. 2.4). Under the same conditions, $^4_2$He$^+$ ions might be deflected to point B. By changing the accelerating

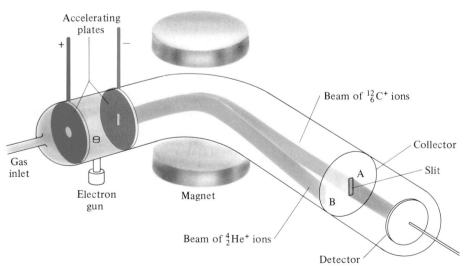

Figure 2.4 The mass spectrometer. A beam of gaseous ions is deflected in the magnetic field toward the collector plate. Light ions are deflected more than heavy ones. By comparing the accelerating voltages required to bring the two ions to the same point, it is possible to determine the relative masses of the ions.

voltage to V_2, it is possible to bring $^4_2\text{He}^+$ ions to point A instead. The accelerating voltages needed to bring the two ions to the same point are related by the equation

$$\frac{\text{mass } ^4_2\text{He}^+}{\text{mass } ^{12}_6\text{C}^+} = \frac{V_1}{V_2} \tag{2.5}$$

By carefully measuring V_1 and V_2, we can calculate the relative masses of ^4_2He and $^{12}_6\text{C}$. This way, we find that ^4_2He has a mass 0.3336 times that of $^{12}_6\text{C}$. Hence, on the carbon-12 scale, the mass of ^4_2He is

0.3336(12.00 amu) = 4.003 amu

This isn't the only way to find an atomic mass, but it is the most accurate way

Naturally occurring helium consists almost entirely of this isotope; therefore, the atomic mass of the element helium is about 4.003 amu.

When the element being studied in the mass spectrometer has more than one isotope, the situation is a bit more complex. Consider neon, which has three isotopes, $^{20}_{10}\text{Ne}$, $^{21}_{10}\text{Ne}$, and $^{22}_{10}\text{Ne}$. A beam of neon ions is split into three parts when it passes through the magnetic field of the mass spectrometer. This produces three peaks on a recorder attached to the detecting unit of the mass spectrometer (Fig. 2.5). The atomic masses of the three isotopes are obtained in the usual way. From the heights of the peaks or, more accurately, from the areas under the peaks, we can determine the relative *abundances* of the isotopes. Thus we find that:

	ATOMIC MASS	ABUNDANCE
Ne-20	20.00 amu	90.92%
Ne-21	21.00 amu	0.26%
Ne-22	22.00 amu	8.82%

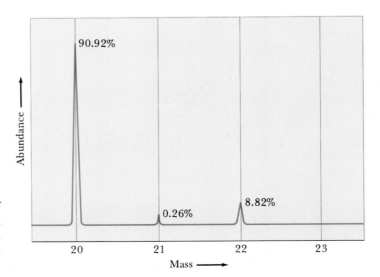

Figure 2.5 The mass spectrum of neon, which contains isotopes with atomic masses of 20.00 amu (90.92%), 21.00 amu (0.26%) and 22.00 amu (8.82%).

We interpret these data to mean that, out of every 10,000 Ne atoms, 9092 are Ne-20 isotopes, 26 are Ne-21 isotopes, and 882 are Ne-22 isotopes. Given this information, along with the atomic masses of the individual isotopes, it is possible to calculate the atomic mass of the element neon, as we will see in Example 2.3.

The mass spectrometer is one of the most widely used instruments in a modern chemistry research laboratory. It is used primarily to detect and analyze for "molecular ions," i.e., cations derived from molecules by the loss of electrons and/or atoms. If we put CH_4 molecules into a mass spectrometer, such species as CH_4^+, CH_3^+, and CH_2^+ are produced. Each of these ions appears as a distinct peak on a recorder, serving to identify the parent molecule, CH_4, and indicate its relative concentration in a mixture. Mass spectroscopy has many advantages as an analytical method. For one thing, it can be applied to samples weighing as little as 10^{-10} g. Moreover, it can easily distinguish between molecular fragments differing in mass by less than one part per thousand, such as $^{13}_{6}C^+$ (13.003 amu) and $(^{12}_{6}C^1_1H)^+$ (13.007 amu).

ATOMIC MASSES AND ISOTOPIC ABUNDANCES

The atomic mass of an element is a weighted average of the atomic masses of its individual isotopes, taking into account their relative abundances. For an element Y that consists of isotopes Y_1, Y_2, and so on:

$$\text{atomic mass Y} = (\text{atomic mass } Y_1) \times (\% \text{ of } Y_1)/100$$
$$+ (\text{atomic mass } Y_2) \times (\% \text{ of } Y_2)/100$$
$$+ \cdots \tag{2.6}$$

The percentages of the several isotopes must, of course, add up to 100:

$$\% \text{ of } Y_1 + \% \text{ of } Y_2 + \cdots = 100\%$$

■ **EXAMPLE 2.3** _____

Using the information given earlier for the isotopes of neon, calculate the atomic mass of that element.

Solution
Applying Equation 2.6 to neon:

$$\text{atomic mass} = 20.00 \text{ amu} \times \frac{90.92}{100.00} + 21.00 \text{ amu} \times \frac{0.26}{100.00}$$
$$+ 22.00 \text{ amu} \times \frac{8.82}{100.00}$$

$$= 20.18 \text{ amu}$$

Exercise
Carbon consists essentially of two isotopes with atomic masses of 12.00 amu and 13.00 amu. Taking the abundance of the heavier isotope to be 1.1%, calculate the atomic mass of the element. Answer: 12.01 amu.

Calculations of the type shown in Example 2.3, using data obtained with a mass spectrometer, can give answers precise to seven or eight significant figures. The accuracy of tabulated atomic masses is limited mostly by variations in natural abundances. Sulfur is an interesting case in point. It consists largely of two isotopes, $^{32}_{16}S$ and $^{34}_{16}S$. The abundance of sulfur-34 varies from about 4.18% in sulfur deposits in Texas and Louisiana to 4.34% in volcanic sulfur from Italy. This leads to an uncertainty of 0.006 amu in the atomic mass of sulfur (32.066 ± 0.006 amu).

The atomic mass of fluorine, F, with one isotope, is 18.998403

MASSES OF INDIVIDUAL ATOMS; AVOGADRO'S NUMBER

For most purposes in chemistry, it is sufficient to know the relative masses of different atoms. We would like, however, to go one step further and calculate the mass in grams of individual atoms. We will now consider how this can be done.

To start with, let us consider the elements helium and hydrogen. We have seen that a helium atom is about four times as heavy as a hydrogen atom. It follows that a sample containing 100 helium atoms weighs four times as much as 100 hydrogen atoms. Again, if we compare samples of the two elements containing a million atoms each, the masses will be in a 4:1 ratio (helium to hydrogen). Turning this argument around, we conclude that a sample of helium weighing four grams must contain the same number of atoms as a sample of hydrogen weighing one gram. More exactly:

If a nickel weighs twice as much as a dime, there are equal numbers of coins in 1000 g of nickels and 500 g of dimes

> no. of He atoms in 4.003 g helium = no. of H atoms in 1.008 g hydrogen

This reasoning is readily extended to other elements, such as those shown in Figure 2.6. A sample of an element with a mass in grams equal to its atomic mass contains a certain definite number of atoms, N, regardless of the identity of the element.

The question now arises as to the numerical value of N; that is, how many atoms are there in 4.003 g of helium, 1.008 g of hydrogen, 32.07 g of sulfur, etc.? As it happens, this problem is one that has been studied for at least a century. Several ingenious experiments have been designed to determine this number, known as Avogadro's number (see Problem 70, end of chapter). As you can imagine, it is huge. (Remember that atoms are tiny. There must be a lot of them in 4.003 g He, 1.008 g H, etc.) To four significant figures, Avogadro's number is

$$6.022 \times 10^{23}$$

To get some idea of how large this number is, suppose the entire population of the world were assigned to counting the atoms in 4.003 g of helium. If each person counted one atom per second and worked a 48-hour week, the task would take more than ten million years.

The importance of Avogadro's number in chemistry should be clear. *It represents the number of atoms in a sample of an element with a mass in grams numerically equal to its atomic mass.* Thus there are

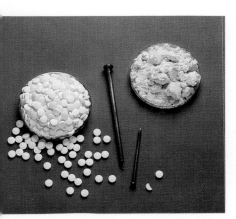

Figure 2.6 One mole of various elements and compounds. There is 55.85 g of iron in the iron nails, 32.07 g of sulfur in the pile of yellow sulfur, and 180.2 g of the compound $C_9H_8O_4$ in the aspirin tablets. (Marna G. Clarke)

6.022×10^{23} H atoms in 1.008 g H   atomic mass H = 1.008 amu

6.022×10^{23} He atoms in 4.003 g He   atomic mass He = 4.003 amu

6.022×10^{23} S atoms in 32.07 g S   atomic mass S = 32.07 amu

Knowing Avogadro's number and the atomic mass of an element, it is possible to calculate the mass of an individual atom (Example 2.4a). We can also determine the number of atoms in a weighed sample of any element (Example 2.4b).

■ **EXAMPLE 2.4** ——————————————————————————

Taking Avogadro's number to be 6.022×10^{23}, calculate

a. the mass of a sulfur atom.

b. the number of sulfur atoms in a 1.000-g sample of the element.

Solution

a. We know that 6.022×10^{23} S atoms weigh 32.07 g:

$$6.022 \times 10^{23} \text{ S atoms} = 32.07 \text{ g}$$

This is really an equivalence equation

This relation gives us a factor to "convert" a sulfur atom to grams:

$$\text{mass S atom} = 1 \text{ S atom} \times \frac{32.07 \text{ g}}{6.022 \times 10^{23} \text{ S atoms}}$$

$$= \boxed{5.325 \times 10^{-23} \text{ g}}$$

b. Here we want to go in the opposite direction, from grams to number of atoms. The appropriate conversion factor is

6.022×10^{23} atoms/32.07 g

$$\text{number of S atoms} = 1.000 \text{ g} \times \frac{6.022 \times 10^{23} \text{ S atoms}}{32.07 \text{ g}}$$

$$= \boxed{1.878 \times 10^{22} \text{ atoms}}$$

Exercise

Suppose the element referred to in the example had been copper instead of sulfur. What would have been the answers in (a) and (b)? Answers: 1.055×10^{-22} g; 9.476×10^{21} atoms.

Calculations such as those just carried out confirm what we have been saying all along: Atoms have very, very small masses ranging from 2×10^{-24} g (H atom) to 4×10^{-22} g (U atom).

2.5
THE MOLE

People in different professions often use special counting units. You and I eat eggs one at a time, but farmers sell them by the dozen. We spend dollar bills one at a time, but Congress distributes them by the billion.

Chemists have their own counting unit—Avogadro's number. Since atoms and molecules are so small, it is convenient to talk about groups of 6.022×10^{23} of them. This counting unit is used so often in chemistry that it is given a special name—the **mole** (abbreviated as **mol**). To a chemist, **a mole means Avogadro's number of items**. Thus,

$$1 \text{ mol H atoms} = 6.022 \times 10^{23} \text{ H atoms}$$
$$1 \text{ mol O atoms} = 6.022 \times 10^{23} \text{ O atoms}$$
$$1 \text{ mol H}_2 \text{ molecules} = 6.022 \times 10^{23} \text{ H}_2 \text{ molecules}$$
$$1 \text{ mol H}_2\text{O molecules} = 6.022 \times 10^{23} \text{ H}_2\text{O molecules}$$
$$1 \text{ mol electrons} = 6.022 \times 10^{23} \text{ electrons}$$
$$1 \text{ mol pennies} = 6.022 \times 10^{23} \text{ pennies}$$

(One mole of pennies is a lot of money. It's enough to pay all the expenses of the United States for the next billion years or so.)

This definition allows us to convert directly between moles and numbers of particles. For example,

$$\text{no. particles in 1.24 mol} = 1.24 \text{ mol} \times \frac{6.022 \times 10^{23} \text{ particles}}{1 \text{ mol}}$$
$$= 7.47 \times 10^{23} \text{ particles}$$
$$\text{no. moles in } 3.24 \times 10^{22} \text{ particles} = 3.24 \times 10^{22} \text{ particles} \times \frac{1 \text{ mol}}{6.022 \times 10^{23} \text{ particles}}$$
$$= 0.0538 \text{ mol}$$

MOLAR MASSES FROM FORMULA MASSES

We can readily find the mass in grams of one mole of atoms. Recall that there are 6.022×10^{23} atoms in 1.01 g of hydrogen (atomic mass H = 1.01 amu) or 16.00 g of oxygen (atomic mass O = 16.00 amu). Hence:

1 mol H weighs 1.01 g;   1 mol O weighs 16.00 g

In general, one mole of atoms of any element has a mass in grams numerically equal to the atomic mass of the element.

We can readily extend this idea to substances that consist of molecules or ions. For the substances represented by the formulas H_2, H_2O, and NaCl,

We set up the mole so that these relations would be true

1 mol H_2 contains 2 mol H atoms and weighs 2(1.01 g) = 2.02 g
1 mol H_2O contains 2 mol H atoms and 1 mol O atoms and weighs 2(1.01 g) + 16.00 g = 18.02 g
1 mol NaCl contains 1 mol Na atoms (22.99 g) and 1 mol Cl atoms (35.45 g) and weighs 22.99 g + 35.45 g = 58.44 g

In general, we can say that, for any substance, the mass of one mole in grams is given by the *formula mass,* defined to be the sum of the atomic masses in the formula. Thus:

FORMULA	FORMULA MASS	MOLAR MASS
He	4.003 amu	4.003 g/mol
H_2	2(1.008 amu) = 2.016 amu	2.016 g/mol
H_2O	2(1.008 amu) + 16.00 amu = 18.02 amu	18.02 g/mol
NaCl	22.99 amu + 35.45 amu = 58.44 amu	58.44 g/mol

In other words, **the molar mass of a substance, in grams per mole, is numerically equal to its formula mass.** Notice that:

1. The formula mass of a substance can be found by simply adding the atomic masses of the atoms in the formula. If the "formula" consists of a single atom, as with helium, the formula mass is the atomic mass. If the formula corresponds to a molecule like H_2 or H_2O, we often refer to the formula mass as a *molecular mass*. Thus we would say that the molecular masses of H_2 and H_2O are 2.016 amu and 18.02 amu, respectively.

2. In order to find the molar mass of a substance, *we must know its formula.* It would be ambiguous, to say the least, to refer to the "molar mass of hydrogen." One mole of hydrogen atoms, represented by the symbol H, weighs 1.008 g; the molar mass of H is 1.008 g/mol. One mole of hydrogen molecules, represented by the formula H_2, weighs 2.016 g; the molar mass of H_2 is 2.016 g/mol.

■ **EXAMPLE 2.5** _____

Calculate the molar masses (g/mol) of

a. potassium chromate, K_2CrO_4
b. sucrose, $C_{12}H_{22}O_{11}$

Solution

a. We start by calculating the formula mass of K_2CrO_4:

$$\text{formula mass } K_2CrO_4 = 2(\text{atomic mass K}) + \text{atomic mass Cr}$$
$$+ 4(\text{atomic mass O})$$
$$= 2(39.10 \text{ amu}) + 52.00 \text{ amu} + 4(16.00 \text{ amu})$$
$$= 194.20 \text{ amu}$$

$$\text{molar mass } K_2CrO_4 = \boxed{194.20 \text{ g/mol}}$$

The main use of formula mass is to find molar mass

b.
$$\text{formula mass } C_{12}H_{22}O_{11} = 12(\text{atomic mass C}) + 22(\text{atomic mass H})$$
$$+ 11(\text{atomic mass O})$$
$$= 12(12.01 \text{ amu}) + 22(1.01 \text{ amu})$$
$$+ 11(16.00 \text{ amu})$$
$$= 342.34 \text{ amu}$$

$$\text{molar mass } C_{12}H_{22}O_{11} = \boxed{342.34 \text{ g/mol}}$$

Exercise

What is the molar mass of H_2SO_4, sulfuric acid? Answer: 98.08 g/mol.

MOLE-GRAM CONVERSIONS

Very often in chemistry we have to convert from moles to grams or vice versa. Such conversions are readily made by knowing the molar masses of the substances involved. They are illustrated in Examples 2.6 and 2.7.

■ **EXAMPLE 2.6** ────────────────────────────────
Determine the number of moles in 212 g of

a. K_2CrO_4 b. $C_{12}H_{22}O_{11}$

Solution

a. Recall from Example 2.5a that the molar mass of K_2CrO_4 is 194.20 g/mol:

$$1 \text{ mol } K_2CrO_4 = 194.20 \text{ g } K_2CrO_4$$

$$\text{no. moles } K_2CrO_4 = 212 \text{ g} \times \frac{1 \text{ mol}}{194.20 \text{ g}} = \boxed{1.09 \text{ mol}}$$

b. From Example 2.5b,

$$1 \text{ mol } C_{12}H_{22}O_{11} = 342.34 \text{ g } C_{12}H_{22}O_{11}$$

$$\text{no. moles } C_{12}H_{22}O_{11} = 212 \text{ g} \times \frac{1 \text{ mol}}{342.34 \text{ g}} = \boxed{0.619 \text{ mol}}$$

Exercise
How many moles are there in 212 g of H_2SO_4? Answer: 2.16 mol.

■ **EXAMPLE 2.7** ────────────────────────────────
Find the mass in grams of 1.69 mol of phosphoric acid, H_3PO_4.

Solution
First, we need to know the molar mass of H_3PO_4:

$$\begin{aligned}\text{formula mass} &= 3(\text{atomic mass H}) + \text{atomic mass P} \\ &\quad + 4(\text{atomic mass O}) \\ &= 3(1.01 \text{ amu}) + 30.97 \text{ amu} + 4(16.00 \text{ amu}) \\ &= 98.00 \text{ amu} \\ \text{molar mass } H_3PO_4 &= 98.00 \text{ g/mol}\end{aligned}$$

Now, we can make the required conversion:

$$\text{mass } H_3PO_4 = 1.69 \text{ mol} \times \frac{98.00 \text{ g}}{1 \text{ mol}} = \boxed{166 \text{ g}}$$

Exercise
What is the mass in grams of 1.69 mol H_2O? Answer: 30.5 g.

Moles are important because they give us a handle on the masses of substances that react chemically

Conversions of the type we have just carried out come up over and over again in chemistry. They will be required in nearly every chapter of this text. Clearly, you must know what is meant by a mole. Remember, a mole always represents a certain number of items, 6.022×10^{23}. Its mass, however, differs with the substance involved: A mole of H_2O, 18.02 g, weighs considerably more than a mole of H_2, 2.02 g, even though they both contain the same number of molecules. In the same way, a dozen bowling balls weigh a lot more than a dozen eggs, even though each involves the same number of items.

We have now considered several different types of conversions involving numbers of particles, moles, and grams. These are summarized in the flow chart shown in Figure 2.7.

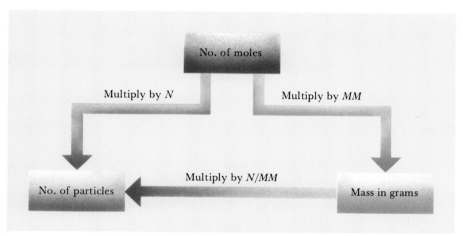

Figure 2.7   Conversions between numbers of particles, numbers of moles, and masses in grams can be made using

N = Avogadro's number = 6.022×10^{23} particles/mole
MM = molar mass (g/mol)

To convert in the direction of the arrows, you multiply by the indicated conversion factor. To go in the reverse direction, you would invert the indicated conversion factor.

2.6
MOLES IN SOLUTION; MOLARITY

To obtain a given amount of a pure solid in the laboratory, you would weigh it out on a balance. Suppose, however, the solid is present as a solute dissolved in a solvent such as water. In this case, you ordinarily measure out a given volume of the water solution, perhaps using a graduated cylinder. The amount of solute you obtain in this way depends not only upon the volume of solution but also upon the *concentration* of solute, i.e., the amount of solute in a given amount of solution.

Perhaps the most useful way to express solute concentration is in terms of **molarity (M)**. The molarity of a solution represents the number of moles

Most chemical reactions occur in solution

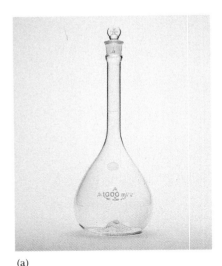

(a)

(b)

(c)

Figure 2.8 To prepare one liter of 0.100 M K_2CrO_4, you would start by weighing out 19.4 g of potassium chromate (Example 2.8). The solid is transferred to a 1000-mL volumetric flask. Enough water is added to dissolve, by swirling, all of the potassium chromate. More water is then added to bring the level up to the mark on the neck. Finally, the flask is shaken repeatedly until a homogeneous solution is obtained. (Marna G. Clarke)

of solute per liter of solution. Specifically, molarity is defined by the equation

$$\text{molarity} = \frac{\text{number moles solute}}{\text{number liters solution}} \tag{2.7}$$

Thus a solution containing 1.20 mol of solute in 2.50 L of solution would have a molarity of

$$\frac{1.20 \text{ mol}}{2.50 \text{ L}} = 0.480 \text{ mol/L}$$

One liter of such a solution would contain 0.480 mol of solute; 100 mL would contain 0.0480 mol of solute, and so on.

To prepare a solution to a desired molarity, we follow the procedure shown in Figure 2.8. If less accuracy is required, we might substitute a graduated cylinder or even a beaker for the volumetric flask. The calculations involved are illustrated in Example 2.8.

■ **EXAMPLE 2.8** _____

Consider the solution of potassium chromate shown in Figure 2.8.

a. How many grams of K_2CrO_4 were required to form one liter of the solution?
b. If, by error, 20.0 g of K_2CrO_4 were used, what would be the molarity of the solution?

Solution
Recall from Example 2.5a that the molar mass of K_2CrO_4 is 194.20 g/mol.

a. Since we have one liter of solution:

$$\text{mass } K_2CrO_4 = 1.00 \text{ L} \times 0.100 \frac{\text{mol}}{\text{L}} \times 194.20 \frac{\text{g}}{\text{mol}} = \boxed{19.4 \text{ g}}$$

b. no. moles K_2CrO_4 = 20.0 g $\times$ $\dfrac{1 \text{ mol}}{194.2 \text{ g}}$ = 0.103 mol

$\quad$ molarity = $\dfrac{0.103 \text{ mol}}{1.00 \text{ L}}$ = $\boxed{0.103 \text{ mol/L}}$

Exercise
If 19.4 g of K_2CrO_4 were dissolved in enough water to fill a 0.250 L volumetric flask to the mark, what would be the molarity? Answer: 0.400 M.

We can use the molarity of a solution to calculate

— the number of moles of solute in a given volume of solution.
— the volume of solution containing a given number of moles of solute.

Here, as in so many other cases, we use a conversion factor approach (Example 2.9).

■ EXAMPLE 2.9

The bottle labeled "concentrated hydrochloric acid" in the lab contains 12.0 mol of the compound HCl per liter of solution, that is, M = 12.0 mol/L.

a. How many moles of HCl are there in 25.0 mL of this solution?
b. What volume of concentrated hydrochloric acid must be taken to contain 1.00 mol HCl?

Solution
The molarity of the HCl relates the number of moles of HCl to the number of liters of solution. Since there are 12.0 mol HCl per liter, we can say that

$\quad$ 12.0 mol $\stackrel{\frown}{=}$ 1 L solution

This relation gives us the conversion factors we need to go from liters of solution to moles or vice versa.

a. We first convert 25.0 mL to liters:

$\quad$ number liters solution = 25.0 mL $\times$ $\dfrac{1 \text{ L}}{1000 \text{ mL}}$ = 0.0250 L

Now we convert to moles of HCl:

$\quad$ no. moles HCl = 0.0250 L $\times$ $\dfrac{12.0 \text{ mol HCl}}{1 \text{ L}}$ = $\boxed{0.300 \text{ mol HCl}}$

b. To find the volume in liters, we start with the number of moles of HCl and use the conversion factor 1 L/12.0 mol HCl:

$\quad$ number liters solution = 1.00 mol HCl $\times$ $\dfrac{1 \text{ L}}{12.0 \text{ mol HCl}}$

$\quad\quad$ = $\boxed{0.0833 \text{ L}}$ (83.3 mL)

Exercise
How many grams of HCl (molar mass = 36.46 g/mol) are there in 0.100 L of this solution? Answer: 43.8 g.

With a little practice you will become comfortable with molecules, moles, and molarity. We try to minimize the jargon, but we do need some

Knowing the molarity of a solution, you can readily obtain a specified amount of solute. All you have to do is to calculate the required volume, as in Example 2.9b. Upon measuring out that volume, you should obtain the desired number of moles or grams of solute. Concentrations of reagents in the general chemistry laboratory are most often expressed as molarities. We will have more to say about molarity and other concentration units in Chapter 11.

■ **SUMMARY PROBLEM**

Hydrogen peroxide, which has the molecular formula H_2O_2, is widely used as a bleach, germicide, and disinfectant.

  a. How many protons and electrons are there in a molecule of H_2O_2?
  b. What is the formula mass of H_2O_2? The molar mass?
  c. What is the mass in grams of a molecule of H_2O_2? The number of molecules in 1.00 g of H_2O_2?
  d. How many moles are there in 10.0 g of H_2O_2? What is the mass in grams of 0.0235 mol H_2O_2?
  e. The solution of hydrogen peroxide sold in drug stores contains about 3.0 g of H_2O_2 per hundred milliliters of solution. What is the molarity of H_2O_2? What volume of solution must be taken to obtain one mole of H_2O_2?

Answers
a. 18; 18   **b.** 34.0 amu; 34.0 g/mol   **c.** 5.65×10^{-23} g; 1.77×10^{22}
d. 0.294 mol; 0.799 g   **e.** 0.88 mol/L; 1.1 L

John Dalton, the father of atomic theory, is reputed to have said, "Thou canst not split an atom." He was proven wrong by the combined efforts of three British physicists. These were J. J. Thomson (1856–1940), Ernest Rutherford (1871–1937), and James Chadwick (1891–1974), who discovered, in turn, the electron, the proton, and the neutron. The key member of this trio was Rutherford, certainly the greatest experimental physicist of his time, one of the greatest of all time. A student of Thomson, Rutherford became the mentor of Chadwick and a host of other distinguished scientists.

In the 1890s, J. J. Thomson, director of the Cavendish laboratory at Cambridge, studied the conduction of electricity through gases at low pressures. When the tube shown in Figure 2.9 is partially evacuated and connected to a spark coil, an electric current flows through it. Associated with this flow are colored rays of light spreading out from the negative electrode (cathode). From the nature of the deflection of these *cathode rays* in an electric field, Thomson showed that they consist of a stream of negatively charged particles that he called **electrons**. He deduced that the electron must be a basic particle common to all atoms and that its mass is a tiny fraction of that of the atom. The discovery of this first subatomic particle earned Thomson the Nobel Prize in physics in 1906. Curiously enough, Thomson never really followed up on this research; someone has said that he "opened the door to modern physics but never walked through it."

Rutherford, a native of New Zealand, walked through that door and never looked back. He was a student of Thomson at the time of the discovery of the electron. From Cambridge he went to McGill University in Montreal and began an investigation of radioactivity that was to continue, with occasional interruptions, for the rest of his life. In less than ten years, Rutherford established the nature of α, β, and γ radiation,

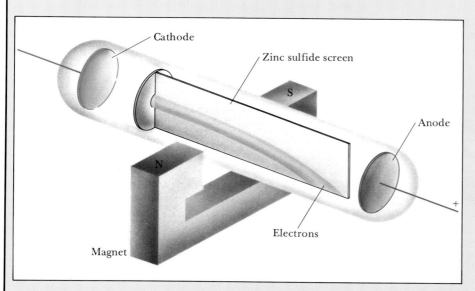

Cathode

Zinc sulfide screen

S

Anode

N

+

Electrons

Magnet

Figure 2.9 Cathode-ray tube. The ray, shown here as a yellow beam, is made up of fast-moving electrons. In an electric or magnetic field, the beam is deflected in such a way as to indicate that it carries a negative charge.

showed that radioactivity resulted in the transmutation of elements, and discovered the rate law for radioactive decay (Chapter 25). For this work he received the Nobel Prize in chemistry in 1908. There was a touch of irony in this award, in that many prominent chemists of that era were highly critical of Rutherford, considering him to be little better than a latter-day alchemist.

At the University of Manchester in England, Rutherford directed a series of experiments that changed forever our ideas about the structure of the atom. The experiments were carried out by Johannes Geiger, a German physicist, and Ernest Marsden, an undergraduate. They bombarded thin foils of gold and other metals (Fig. 2.10) with α-particles (He atoms minus their electrons). With a fluorescent screen, they observed the extent to which these particles were scattered. To no one's surprise, most of the α-particles passed through the foil unchanged in direction. A few, however, were reflected back at acute angles. This was a totally unexpected result, inconsistent with the model of the atom in vogue at that time. In Rutherford's words: "It was as though you had fired a 15-inch shell at a piece of tissue paper and it had bounced back and hit you." Eighteen months later, in 1911, Rutherford published an explanation. By an elegant mathematical analysis of the forces involved, he showed that the atom must contain a tiny, positively charged **nucleus** accounting for most of its mass.

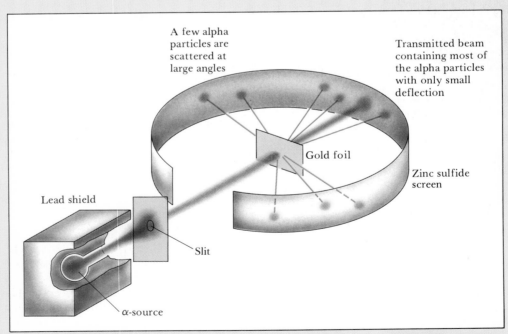

Figure 2.10 Rutherford's scattering experiment. Most of the α-particles are essentially undeflected, but a few are scattered at large angles. In order to cause the large deflections, atoms must contain heavy, positively charged nuclei.

Once the existence of the nucleus had been established, scientists began to speculate upon its composition. In 1919, Rutherford, who had just succeeded J. J. Thomson as director of the Cavendish laboratory at Cambridge, announced the discovery of the **proton**. He showed that it was ejected from the nucleus of a nitrogen atom when it collides with an α-particle. A year later, Rutherford suggested that the nucleus must contain an uncharged particle with the same mass as the proton. He and his close associate, James Chadwick, spent ten years searching for that particle. In Chadwick's words, "I did a lot of experiments, some of which were quite stupid. I suppose I got that from Rutherford. He would do some damn silly experiments at times, and we did some together. There was always the possibility of something turning up. But until 1932 it was a frustrating time."

In 1932 Chadwick discovered the **neutron**; three years later he won the Nobel Prize in physics. Two years after that, Rutherford died suddenly and an era ended in experimental physics.

We only write up the successful experiments

Figure 2.11 J. J. Thomson and Ernest Rutherford (*right*) talking, perhaps, about nuclear physics, more likely about yesterday's cricket match. (AIP Niels Bohr Library, Bainbridge Collection)

QUESTIONS AND PROBLEMS

Symbols, Formulas, and Equations

1. Using information given in this chapter, write the chemical formula for
a. ammonia   b. water
c. methane   d. hydrochloric acid

2. Write chemical formulas for the following compounds
a. sodium chloride   b. calcium chloride
c. calcium oxide   d. sodium oxide

3. Using the figures in this chapter, give the color and physical state at room temperature of
a. Cr_2O_3 b. water

4. Using the figures in this chapter, give the color and physical state at room temperature of
a. K_2CrO_4 b. CrO_3

Atomic Theory and Laws

5. State, in your own words, the three basic postulates of the atomic theory of matter.

6. State, in your own words, how the Law of Multiple Proportions is explained in terms of atomic theory.

7. Which of the three laws (if any) listed on pp. 34 and 35 is illustrated by each of the following statements?
a. Lavoisier found that when mercury(II) oxide, HgO, decomposes, the total mass of mercury and oxygen formed equals the mass of mercury(II) oxide decomposed.
b. Analysis of the calcium carbonate, $CaCO_3$, found in the marble of Carrara, Italy, and in the stalactites of the Carlsbad Caverns of New Mexico gives the same value for the percent calcium.
c. The atom ratio of oxygen to hydrogen is twice as large in one compound as it is in another compound made up of the two elements.
d. Hydrogen occurs as a mixture of two isotopes, one of which is twice as heavy as the other.

8. Which of the three laws (if any) listed on pp. 34 and 35 is illustrated by each of the following statements?

a. A cold pack has the same mass before and after the seal between the two compartments is broken.
b. It is highly improbable that the formula for carbon monoxide gas found in London, England, is $C_{1.1}O_{2.5}$.
c. The mass of phosphorus, P, combined with one gram of hydrogen, H, in the highly toxic gas phosphine, PH_3, is a little more than twice the mass of nitrogen, N, combined with one gram of hydrogen in ammonia gas, NH_3.

9. Sulfur and oxygen form two compounds. In the first compound, the mass percent of oxygen is 50.0; in the second, it is 60.0.
a. Calculate the mass of oxygen combined with one gram of sulfur in each compound.
b. Use the two values from (a) to calculate the ratio of the masses of oxygen per gram of sulfur in the two compounds.
c. Suggest reasonable formulas for the two compounds.

10. Ethane and ethene are two gases containing only hydrogen and carbon atoms. In a certain sample of ethane, 4.53 g of hydrogen is combined with 18.0 g of carbon. In a sample of ethene, 7.25 g of hydrogen is combined with 43.20 g of carbon.
a. Show how these data illustrate the Law of Multiple Proportions.
b. Suggest reasonable formulas for the two compounds.

11. When magnesium ribbon, Mg(s), is heated in oxygen gas, magnesium oxide, a white powder, is produced. In one experiment, 3.56 g of magnesium ribbon is completely consumed in reacting with 7.00 g of oxygen to produce 5.93 g of magnesium oxide; some oxygen remains unreacted. In a second experiment, 2.50 g of magnesium ribbon reacts with 1.10 g of oxygen gas. This time, all the oxygen is consumed; some unreacted magnesium remains and 2.75 g of magnesium oxide is produced. Show that these results are consistent with the Law of Constant Composition.

12. Mercury(II) oxide, a red powder, can be decomposed by heating to produce liquid mercury and oxygen gas. When a sample of this compound is decomposed, 3.87 g of oxygen and 48.43 g of mercury are produced. In a second experiment, 15.68 g of mercury is allowed to react with an excess of oxygen; 16.93 g of red mercury(II) oxide is produced. Show that these results are consistent with the Law of Constant Composition.

Nuclear Symbols and Isotopes

13. Studies show that there is an inverse relationship between the selenium content of the blood and the incidence of breast cancer in women. $^{80}_{34}Se$ is the most abundant form of naturally occurring selenium. How many protons are there in a Se-80 atom? How many neutrons?

14. The eruption of Mount St. Helens in Washington state produced a considerable amount of a radioactive gas, radon-222. Write the nuclear symbol for this isotope of radon (Rn).

15. a. Do the symbols $^{57}_{26}Fe$ and $_{26}Fe$ have the same meaning?
b. Do the symbols $^{57}_{26}Fe$ and ^{57}Fe convey the same information?

16. Explain how the two isotopes of copper, Cu-63 and Cu-65, differ from each other. Write nuclear symbols for each isotope.

17. An isotope of cobalt used in radiation therapy for certain cancers is Co-60. How many
a. protons are in its nucleus?
b. neutrons are in its nucleus?
c. electrons are in a cobalt atom?
d. neutrons, protons, and electrons are in the Co^{2+} ion formed from this isotope?

18. An isotope of iodine used in thyroid disorders is $^{131}_{53}I$. How many
a. protons are in its nucleus?
b. neutrons are in its nucleus?
c. electrons are in an iodine atom?
d. neutrons and protons are in the I^- ion formed from this isotope?

19. Complete the table below, using the Periodic Table when necessary:

SYMBOL	CHARGE	NUMBER OF PROTONS	NUMBER OF NEUTRONS	NUMBER OF ELECTRONS
——	0	9	10	——
P	0	——	16	——
——	+3	26	30	——
——	——	16	16	18

20. Complete the table below, using the Periodic Table when necessary:

NUCLEAR SYMBOL	CHARGE	NUMBER OF PROTONS	NUMBER OF NEUTRONS	NUMBER OF ELECTRONS
$^{79}_{35}Br$	0	——	——	——
——	-3	7	7	——
——	+5	33	42	——
$^{90}_{40}Zr$	0	——	——	——

Atomic Masses

21. Arrange the following in order of increasing mass:
a. 2 Na atoms   b. 3 Mg atoms
c. a K atom   d. a K^+ ion

22. Calculate the mass ratio of an arsenic (As) atom to an atom of
a. Mg   b. Sn   c. Hg

23. Suppose the atomic mass of C-12 were taken to be 1.000 amu instead of 12.000 amu. On that basis, what would be the atomic mass of the following?

   a. O   b. Cl   c. Ni

24. If the atomic mass of fluorine, F, were taken to be exactly 19 amu, what would be the atomic mass of the following?

   a. Br   b. O   c. Au

25. Strontium consists of four isotopes with masses of 83.9134 amu (0.5%), 85.9094 amu (9.9%), 86.9089 amu (7.0%), and 87.9056 amu (82.6%). Calculate the average atomic mass of strontium.

26. Naturally occurring europium (Eu) consists of two isotopes with masses 150.9199 amu (48.03%) and 152.9212 amu (51.97%). Calculate the atomic mass of Eu.

27. Naturally occurring boron consists of two isotopes with atomic masses 10.013 amu and 11.009 amu. Estimate the abundances of these isotopes. The atomic mass of naturally occurring boron is 10.811 amu.

28. Nitrogen has two isotopes, N-14 and N-15, with atomic masses of 14.0031 amu and 15.001 amu, respectively. What is the abundance of N-15?

29. Magnesium (atomic mass = 24.305 amu) consists of three isotopes with masses 23.98 amu, 24.98 amu, and 25.98 amu. The abundance of the middle isotope is 10.1%. Estimate the abundances of the other two isotopes.

30. Chromium (atomic mass = 51.9961 amu) has four isotopes. Their masses are 49.9461 amu, 51.9405 amu, 52.9407 amu, and 53.9389 amu. The last two isotopes have abundances of 9.50% and 2.36%, respectively. Estimate the other two abundances.

31. Silicon consists of three isotopes with masses 27.977 amu, 28.977 amu, and 29.974 amu. Their abundances are 92.34%, 4.70%, and 2.96%, respectively. Sketch the mass spectrum for silicon.

32. Chlorine has two isotopes, Cl-35 and Cl-37. Their abundances are 75.53% and 24.47%, respectively. Assuming that the only hydrogen isotope present is H-1,

   a. How many different HCl molecules are possible?

   b. What is the sum of the mass numbers of the two atoms in each molecule?

   c. Sketch the mass spectrum for HCl if all the positive ions are obtained by removing a single electron from an HCl molecule.

Molecules and Ions

33. Give the number of protons and electrons in

   a. an S^{2-} ion   b. an S_8 molecule

   c. an H_2S molecule   d. an H^+ ion

34. Give the number of protons and electrons in

   a. a P_4 molecule   b. a PCl_5 molecule

   c. a P^{3-} ion   d. a P^{5+} ion

35. Complete the following table:

SPECIES	NUMBER OF NEUTRONS	NUMBER OF PROTONS	NUMBER OF ELECTRONS
$^{55}_{25}Mn^{2+}$	___	___	___
	46	34	36
$^{194}_{78}Pt^{3+}$	___	___	___

36. Complete the following table:

SPECIES	NUMBER OF PROTONS	NUMBER OF ELECTRONS
As^{5+}	___	___
N_2O_4	___	___
I^-	___	___
CH_4	___	___

37. Give the formulas of all the compounds containing no ions other than K^+, Sr^{2+}, Br^-, or O^{2-}.

38. Give the formulas of compounds in which

   a. the cation is Li^+; the anion is S^{2-} or N^{3-}.

   b. the anion is O^{2-}; the cation is Fe^{2+} or Fe^{3+}.

Avogadro's Number

39. The atomic mass of tungsten, W, is 183.85 amu. Calculate the

   a. mass in grams of one tungsten atom.

   b. number of atoms in one milligram of tungsten.

40. Mercury has atomic mass 200.59 amu. Calculate the

   a. mass of 2.0×10^{15} atoms.

   b. number of atoms in one nanogram of mercury.

41. Determine the

   a. mass of one billion gold (Au) atoms.

   b. number of atoms in one ounce of gold.

42. Determine the

   a. mass of a trillion (one million million) lead (Pb) atoms.

   b. number of atoms in one ounce of lead.

43. The national debt in 1987 was 2.565 trillion dollars. How many moles of pennies would be required to pay off the national debt?

44. Calculate the number of molecules in an eight-ounce glass of water. (1 lb = 16 oz = 453.6 g)

45. How many electrons are there in

   a. an aluminum atom?

   b. a mole of aluminum atoms?

   c. 0.2843 mol aluminum?

   d. 0.2843 g aluminum?

46. How many protons are there in

   a. one neon (Ne) atom?

   b. a mole of neon atoms?

   c. 20.02 mol of neon?

   d. 20.02 g neon?

Molar Masses and Mole-Gram Conversions

47. Calculate the molar masses (g/mol) of
   a. gallium, Ga, a metal that literally melts in your hands.
   b. laughing gas, N_2O, one of the first anesthetics used.
   c. cane sugar, $C_{12}H_{22}O_{11}$.

48. Calculate the molar masses (g/mol) of
   a. ammonia (NH_3).
   b. baking soda ($NaHCO_3$).
   c. osmium (Os) metal.

49. Convert the following to moles:
   a. 0.830 g of strychnine ($C_{21}H_{22}N_2O_2$), present in rat poison.
   b. two hundred fifty milligrams of aspirin ($C_9H_8O_4$).
   c. a gram of Vitamin C ($C_6H_8O_6$).

50. Convert the following to moles:
   a. 3.86 g of carbon dioxide, CO_2.
   b. 0.485 g of ethyl alcohol (C_2H_5OH).
   c. 6.00×10^3 g of hydrazine, N_2H_4, a rocket propellant.

51. Calculate the mass in grams of 5.75 mol of
   a. nitrogen atoms   b. nitrogen molecules (N_2)
   c. ammonia molecules

52. Calculate the mass in grams of 6.83 mol of
   a. vinyl chloride, C_2H_3Cl, the starting material for a plastic.
   b. stannous fluoride, SnF_2, the active ingredient in toothpastes for cavity prevention.
   c. nitrogen dioxide, NO_2, the brown gas responsible for the color of smog.

53. Complete the following table for ethylene glycol, $C_2H_6O_2$, an antifreeze used in cars.

NUMBER OF GRAMS	NUMBER OF MOLES	NUMBER OF MOLECULES	NUMBER OF C ATOMS
0.1245	——	——	——
——	0.0375	——	——
——	——	2.0×10^{25}	——
——	——	——	3.6×10^{12}

54. Complete the following table for acetone, C_3H_6O, the main component of nail polish remover.

NUMBER OF GRAMS	NUMBER OF MOLES	NUMBER OF MOLECULES	NUMBER OF H ATOMS
0.2500	——	——	——
——	0.2000	——	——
——	——	2.5×10^{10}	——
——	——	——	8.0×10^{15}

Molarity

55. How would you prepare 425 mL of 0.628 M
   a. K_2CrO_4 b. NaI   c. $C_6H_{12}O_6$

56. Given the pure solid and water, how would you prepare
   a. 0.500 L of 1.25 M KOH?
   b. 0.750 L of 3.50 M $CuSO_4$?

57. You are given a bottle labeled "3.00 M HNO_3."
   a. How many moles of HNO_3 are there in 12.45 mL of this solution?
   b. What volume of this solution contains 0.800 mol HNO_3?

58. On a shelf are bottles marked 0.125 M NaCl and 6.00 M acetic acid, $HC_2H_3O_2$.
   a. How many grams of solute are there in 35.0 mL of each solution?
   b. What volume of each solution must be taken to obtain 0.0854 mol of solute?

59. Complete the table below for aqueous solutions.

SOLUTE	MASS OF SOLUTE	VOLUME	MOLARITY
Na_2CO_3	3.58 g	0.250 L	——
CH_3OH	——	0.500 L	6.00
$Ba(NO_3)_2$	5.89 g	——	1.21
$Al_2(SO_4)_3$	——	0.455 L	0.105

60. Complete the table below for aqueous solutions.

SOLUTE	MASS OF SOLUTE	VOLUME	MOLARITY
$CaCl_2$	45.0 g	0.400 L	——
H_2O_2	——	0.300 L	0.155
K_2CO_3	5.98 g	——	2.50
ZnI_2	——	0.150 L	0.0536

Unclassified

61. The density of ethyl alcohol, C_2H_5OH, is 0.789 g/cm³ at 25°C. Calculate
   a. the number of moles in 355 mL of ethyl alcohol.
   b. the mass of 2.87 mol of ethyl alcohol.

62. The molecular formula of morphine, a pain-killing narcotic, is $C_{17}H_{19}NO_3$.
   a. What is its molar mass?
   b. What fraction of the atoms in morphine is accounted for by carbon?
   c. Which element contributes least to the molar mass?

63. The hormone adrenaline has the molecular formula $C_9H_{13}NO_3$. Its normal concentration in blood plasma is 6.0×10^{-8} g/L. How many adrenaline molecules are there in one liter of plasma?

64. Arrange the following in order of increasing numbers of atoms.
   a. 10 Cl_2 molecules   b. 1×10^{-10} mol Cl
   c. 1×10^{-12} mol Cl_2 d. 1×10^{-10} g Cl
   e. 1×10^{-12} g Cl_2

65. Lithium consists of two isotopes with masses 6.015 and 7.016 amu. What is the percentage of Li-6 (6.015 amu) in a lithium sample with atomic mass 6.999 amu?

66. Carbon tetrachloride, CCl_4, was a popular dry-cleaning agent until it was shown to be carcinogenic. It has a density of 1.594 g/cm³. What volume of carbon tetrachloride will contain a total of 5.00×10^{24} molecules of CCl_4?

67. Some brands of salami contain 0.090% sodium benzoate ($NaC_7H_5O_2$) by mass as a preservative. If you eat 2.52 oz of this salami, how many atoms of sodium will you consume, assuming salami contains no other source of that element?

Challenge Problems

68. Taking the mass of an electron to be 9.11×10^{-28} g, determine the mass, to eight significant figures, of one mole of
 a. Na^+ ions b. I^- ions c. NaI

69. Chlorophyll, the substance responsible for the green color of leaves, has one magnesium atom per chlorophyll molecule and contains 2.72% magnesium by mass. What is the molar mass of chlorophyll?

70. By x-ray diffraction, it is possible to determine the geometric pattern in which atoms are arranged in a crystal and the distances between atoms. In a crystal of silver, four atoms effectively occupy the volume of a cube 0.409 nm on an edge. Taking the density of silver to be 10.5 g/cm³, calculate the number of atoms in one mole of Ag.

71. Suppose you arranged a mole of moles, head to tail, in a straight line stretching from the earth to the moon. Assume your average mole is 6 inches long; the distance to the moon is 250,000 miles, more or less. How many columns of moles would there be between the earth and the moon?

72. Each time you inhale, you take in about 500 mL of air; each milliliter of air contains about 2.5×10^{19} molecules. It has been estimated that Abraham Lincoln, in delivering the Gettysburg Address, inhaled about 200 times.
 a. How many molecules did Lincoln take in?
 b. In the entire atmosphere, there are about 1.8×10^{20} mol of air. What fraction of the molecules in the earth's atmosphere was inhaled by Lincoln at Gettysburg?
 c. In the next breath that you take, estimate the number of molecules that were inhaled by Lincoln at Gettysburg.

73. Show by calculation whether the Law of Conservation of Mass, within experimental error, is obeyed in the following experiment: Eighteen grams of aluminum metal are added to 25.00 mL of hydrochloric acid ($d = 1.025$ g/cm³). After the experiment, there are 12.00 g of aluminum left, and 30.95 g of solution made up of aluminum ions, water, and chloride ions. There are 8.16 L of hydrogen gas ($d = 0.0824$ g/L) produced.

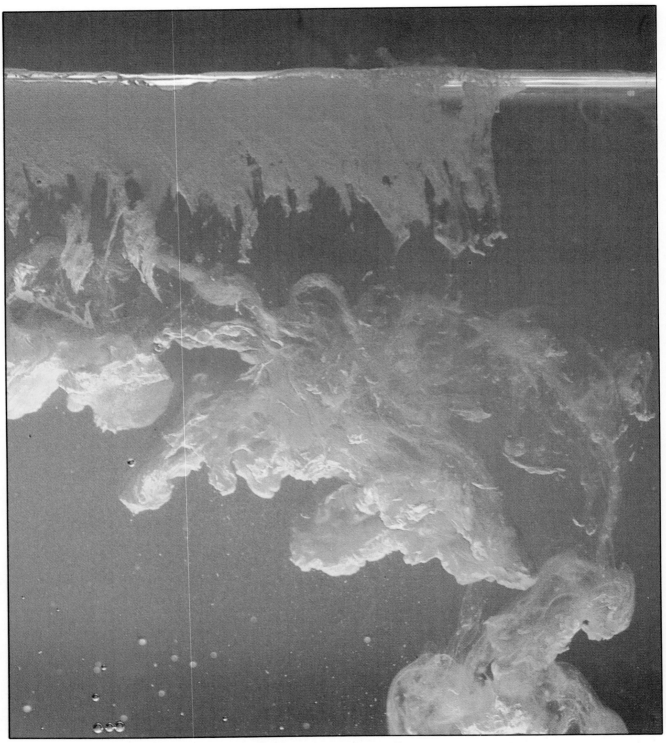

$$Pb^{2+}(aq) \ + \ 2I^-(aq) \ \rightarrow \ PbI_2(s)$$

Precipitation of lead iodide by mixing solutions containing Pb^{2+} and I^- ions.

CHAPTER 3

CHEMICAL FORMULAS AND EQUATIONS

Chemical formulas were touched on briefly in Chapter 2. In this chapter we will examine them more closely, considering their meaning (Section 3.1) and how they can be used to calculate the relative masses of different elements in a compound (Section 3.2). We will also see how the formula of a compound can be determined experimentally (Section 3.3) or, for ionic compounds, predicted with pencil and paper (Section 3.4).

Compounds can be identified by names as well as formulas. In Section 3.5, we will see how certain simple compounds are named in a systematic way. With that background, we will go on to discuss how chemical reactions are represented by equations (Section 3.6).

Chemical equations serve many purposes. In the most general sense, they describe what happens in reactions. Here we will be particularly interested in how to use them to relate amounts of reactants and products (Sections 3.7 and 3.8). Such relationships are readily obtained by using the mole concept introduced in Chapter 2.

3.1
TYPES OF FORMULAS

The **simplest** (*empirical*) formula of a compound gives the smallest whole-number ratio between the numbers of atoms of different elements in the compound. An example is the simplest formula of water, H_2O. This tells us that there are twice as many hydrogen atoms as oxygen atoms in water. In the compound potassium chlorate, there are three elements: potassium, chlorine, and oxygen. These are present in an atom ratio of 1 K : 1 Cl : 3 O. Hence, the simplest formula of potassium chlorate is $KClO_3$.

The **molecular** formula indicates the actual number of atoms of each type in a molecule. The molecular formula may be the same as the simplest formula. This is the case with water, H_2O: there are two atoms of hydrogen and one oxygen atom in a water molecule. In other cases, the molecular formula is a whole-number multiple of the simplest formula. Consider, for example, the compound of hydrogen and oxygen known as hydrogen peroxide. Here we write the molecular formula H_2O_2 to indicate that two hydrogen atoms are combined with two oxygen atoms in the hydrogen peroxide molecule. The simplest formula of this compound would be HO.

Sometimes, to represent a compound, we go beyond the simplest or molecular formula. We write its formula in such a way as to suggest the structure of the compound. Consider, for example, methyl alcohol, which has the molecular structure:

$$H - \overset{\overset{\displaystyle H}{|}}{\underset{\underset{\displaystyle H}{|}}{C}} - O - H$$

Reflecting the bonding pattern in this molecule, we customarily write the formula of methyl alcohol as CH_3OH rather than CH_4O.

We also use a special type of formula for compounds called *hydrates*. These are solids, usually ionic in nature, that contain water molecules in their crystal lattice (Figure 3.1). An example is hydrated barium chloride. This compound contains two moles of water, H_2O, for every mole of barium chloride, $BaCl_2$. Its formula is:

$$BaCl_2 \cdot 2H_2O$$

(A dot is used to separate the formulas of the two compounds, $BaCl_2$ and H_2O.) We interpret the formula $CuSO_4 \cdot 5H_2O$ in a similar way. This crystalline hydrate contains five moles of water for every mole of copper sulfate, $CuSO_4$.

The molecular formula tells us more than the simplest formula, so we use it when we can

Figure 3.1 $CuSO_4 \cdot 5H_2O$ (*upper left*) is blue; the anhydrous solid (*lower left*) is white. $CoCl_2 \cdot 6H_2O$ (*upper right*) is pink; lower hydrates such as $CoCl_2 \cdot 4H_2O$ are purple (*lower right*) or blue. (Marna G. Clarke)

3.2
PERCENT COMPOSITION FROM FORMULA

As we have seen, the formula of a compound tells us the relative number of each kind of atom present. It can also be used to determine the mass percents of the elements in the compound. To do this, it is convenient to work with one mole of the compound (Example 3.1).

■ **EXAMPLE 3.1** _____
Sodium hydrogen carbonate, commonly called "bicarbonate of soda," is used in many commercial products to relieve an upset stomach. It has the simplest formula $NaHCO_3$. What are the mass percents of Na, H, C, and O in sodium hydrogen carbonate (at. mass Na = 22.99 amu, H = 1.01 amu, C = 12.01 amu, O = 16.00 amu)?

Solution

In one mole of $NaHCO_3$, there are

$$22.99 \text{ g Na (1 mol)} \qquad 1.01 \text{ g H (1 mol)}$$
$$12.01 \text{ g C (1 mol)} \qquad 48.00 \text{ g O (3 mol)}$$

The mass of one mole of $NaHCO_3$ is

$$22.99 \text{ g} + 1.01 \text{ g} + 12.01 \text{ g} + 48.00 \text{ g} = 84.01 \text{ g}$$

Since 84.01 g of $NaHCO_3$ contains 22.99 g of Na, it follows that:

$$\text{mass percent Na} = \frac{\text{mass Na}}{\text{mass NaHCO}_3} \times 100 = \frac{22.99 \text{ g}}{84.01 \text{ g}} \times 100 = \boxed{27.36}$$

Similarly, for hydrogen, carbon, and oxygen:

$$\text{mass percent H} = \frac{1.01 \text{ g}}{84.01 \text{ g}} \times 100 = \boxed{1.20}$$

$$\text{mass percent C} = \frac{12.01 \text{ g}}{84.01 \text{ g}} \times 100 = \boxed{14.30}$$

$$\text{mass percent O} = \frac{48.00 \text{ g}}{84.01 \text{ g}} \times 100 = \boxed{57.14}$$

The percentages add up to 100, as they should:

$$27.36 + 1.20 + 14.30 + 57.14 = 100.00$$

Exercise

What are the mass percents of carbon and oxygen in CO_2? Answer: 27.29% C; 72.71% O.

The calculations in Example 3.1 illustrate an important characteristic of formulas. In one mole of $NaHCO_3$, there is 1 mol Na (22.99 g), 1 mol H (1.01 g), 1 mol C (12.01 g), and 3 mol O (48.00 g). In other words, the mole ratio is 1 mol Na : 1 mol H : 1 mol C : 3 mol O. This is the same as the atom ratio in $NaHCO_3$, 1 atom Na : 1 atom H : 1 atom C : 3 atoms O. In general, we can say that **the subscripts in a formula represent not only the atom ratio in which the different elements are combined, but also the mole ratio.** For example,

The atom ratio equals the mole ratio. Can you see why?

FORMULA	ATOM RATIO	MOLE RATIO
H_2O	2 atoms H : 1 atom O	2 mol H : 1 mol O
KNO_3	1 atom K : 1 atom N : 3 atoms O	1 mol K : 1 mol N : 3 mol O
$C_{12}H_{22}O_{11}$	12 atoms C : 22 atoms H : 11 atoms O	12 mol C : 22 mol H : 11 mol O

The formula of a compound can also be used in a straightforward way to find the mass of an element in a weighed sample of the compound (Example 3.2).

■ **EXAMPLE 3.2**

An iron-containing ore responsible for the red color of soils in many parts of the country is limonite, which has the formula $Fe_2O_3 \cdot \frac{3}{2} H_2O$. What mass of iron can be obtained from a metric ton (10^3 kg = 10^6 g) of limonite?

Solution

In one mole of limonite, there are:

2 mol Fe, 3 mol O, $\frac{3}{2}$ mol H_2O

The mass of one mole of limonite is

$$2(55.85 \text{ g}) + 3(16.00 \text{ g}) + 1.5(18.02 \text{ g}) = 186.73 \text{ g}$$

It follows that 2(55.85 g) = 111.70 g of Fe can be obtained from 186.73 g of limonite. We might say that 111.70 g of iron is *chemically equivalent* to 186.73 g of limonite.

111.70 g Fe $\backsimeq$ 186.73 g limonite

Treating this "chemical equivalence" as an equality, we obtain the conversion factor 111.70 g Fe/186.73 g limonite, which can be used to find the mass of iron that can be extracted from a metric ton of limonite.

$$1.0000 \times 10^6 \text{ g limonite} \times \frac{111.70 \text{ g Fe}}{186.73 \text{ g limonite}} = \boxed{5.9819 \times 10^5 \text{ g Fe}}$$

This corresponds to 5.9819×10^2 kg or 0.59819 metric ton of iron.

Exercise

What mass of limonite is required to produce one metric ton of Fe? Answer: 1.6717 metric ton.

> The mole concept is really helpful when dealing with problems of this sort

3.3
FORMULAS FROM EXPERIMENT

We saw in Section 3.2 that given the formula of a compound, we can calculate the mass percents of the elements present. As you might suppose, this process can be reversed. If we know the mass percents of the elements from experiment, we can obtain the formula of a compound. The formula found in this way is the simplest formula. To go one step further and determine the molecular formula, it is necessary to know one other quantity, the molar mass.

SIMPLEST FORMULA FROM PERCENT COMPOSITION

If we know the mass percents of the elements in a compound, we can deduce its simplest formula. Consider the mineral cassiterite, a compound of tin and oxygen. Chemical analysis shows that this compound contains 78.8% tin by mass and 21.2% oxygen. For simplicity, let us work with a 100-g sample of cassiterite in which there are:

$$0.788 \times 100 \text{ g} = 78.8 \text{ g Sn}; \quad 0.212 \times 100 \text{ g} = 21.2 \text{ g O}$$

The numbers of moles of tin and oxygen in this 100-g sample are

$$\text{no. moles Sn} = 78.8 \text{ g Sn} \times \frac{1 \text{ mol Sn}}{118.7 \text{ g Sn}} = 0.664 \text{ mol Sn}$$

Using 100 g simplifies the math

$$\text{no. moles O} = 21.2 \text{ g O} \times \frac{1 \text{ mol O}}{16.00 \text{ g O}} = 1.33 \text{ mol O}$$

Since we have 1.33 mol of oxygen and 0.664 mol of tin in this sample, it follows that the mole ratio of oxygen to tin in this or any other sample of cassiterite is:

$$\frac{1.33 \text{ mol O}}{0.664 \text{ mol Sn}} = 2.00 \frac{\text{mol O}}{\text{mol Sn}}$$

In other words, there are two moles of oxygen for every mole of tin. As pointed out earlier, *the mole ratio is equal to the atom ratio.* Hence, the simplest formula must be SnO_2.

The same approach can be used to determine the simplest formula of a compound containing more than two elements. The arithmetic is a bit longer but the reasoning is the same (Example 3.3).

Cassiterite, an ore of tin with the formula SnO_2. (George Whiteley, Photoresearchers)

■ **EXAMPLE 3.3** _____

Potassium dichromate, an orange solid, contains the three elements potassium, chromium, and oxygen. Analysis of a sample of potassium dichromate gives the following mass percents:

$$K = 26.6; \quad Cr = 35.4; \quad O = 38.0$$

From these data, determine the simplest formula.

Solution

Again, for convenience, we work with a 100-g sample. In this sample, there are 26.6 g K, 35.4 g Cr, and 38.0 g O. Noting that the molar masses of K, Cr, and O are 39.10, 52.00, and 16.00 g/mol, respectively, we have, in the 100-g sample:

$$\text{no. moles K} = 26.6 \text{ g K} \times \frac{1 \text{ mol K}}{39.10 \text{ g K}} = 0.680 \text{ mol K}$$

$$\text{no. moles Cr} = 35.4 \text{ g Cr} \times \frac{1 \text{ mol Cr}}{52.00 \text{ g Cr}} = 0.681 \text{ mol Cr}$$

$$\text{no. moles O} = 38.0 \text{ g O} \times \frac{1 \text{ mol O}}{16.00 \text{ g O}} = 2.38 \text{ mol O}$$

To find the simplest mole ratio in the compound, we divide by the smallest number, 0.680:

$$\frac{0.681 \text{ mol Cr}}{0.680 \text{ mol K}} = 1.00 \frac{\text{mol Cr}}{\text{mol K}}; \quad \frac{2.38 \text{ mol O}}{0.680 \text{ mol K}} = 3.50 \frac{\text{mol O}}{\text{mol K}}$$

We conclude that for every mole of potassium, there is 1.00 mol of chromium and 3.50 mol of oxygen. The mole ratio and hence the atom ratio could be expressed as

1 K : 1.00 Cr : 3.50 O

Multiplying by two to obtain the simplest whole-number ratio, we arrive at

2 K : 2 Cr : 7 O

The simplest formula contains the smallest possible set of integers

The simplest formula of potassium dichromate is $K_2Cr_2O_7$.

Exercise
Hexane, a colorless organic liquid, contains 83.6% C and 16.4% H by mass. What is its simplest formula? Answer: C_3H_7.

The calculations in Example 3.3 are typical of those involved in determining simplest formulas from percent composition. Frequently, the mole ratio that you calculate directly involves one or more fractional numbers (for example, 3.50, 2.33). When this happens, multiply through by the smallest integer (2, 3, etc.) that will give a whole-number ratio. Thus,

Since there is experimental error, you may need to round off to obtain integers

$$\frac{3.50}{1.00} \times \frac{2}{2} = \frac{7.00}{2.00} = 7:2$$

$$\frac{2.33}{1.00} \times \frac{3}{3} = \frac{6.99}{3.00} = 7:3$$

DETERMINATION OF PERCENT COMPOSITION AND SIMPLEST FORMULA FROM EXPERIMENT

Once the mass percents of the elements in a compound are known, the calculation of the simplest formula is straightforward. Finding these percentages, however, is not a simple matter. They must be determined by experiment. Many different methods are possible. To illustrate the principle involved, consider the oxide of tin referred to earlier. To *analyze* this compound—that is, to determine the mass percents of tin and oxygen—we might heat a weighed sample of the oxide, say 0.800 g, with hydrogen gas. At high temperatures a reaction occurs: the oxygen in the compound is converted to water vapor, which escapes with the excess hydrogen. The solid residue that remains is pure tin; its mass is 0.630 g. The mass percent of tin in the oxide must then be

$$\text{mass percent Sn} = \frac{0.630 \text{ g}}{0.800 \text{ g}} \times 100 = 78.8$$

The mass percent of oxygen is found by subtracting from 100:

$$\text{mass percent O} = 100.0 - 78.8 = 21.2$$

Many simple compounds containing only two elements can be analyzed by a procedure similar to this. The general approach is to carry out a reaction in which one of the elements is produced in the pure state.

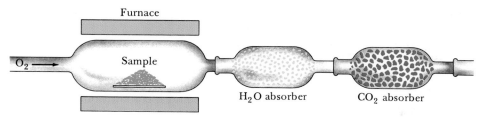

Figure 3.2 Combustion train used for carbon-hydrogen analysis. The absorbent for water is magnesium perchlorate, $Mg(ClO_4)_2$. Carbon dioxide is absorbed by finely divided sodium hydroxide, NaOH. Alternatively, H_2O and CO_2 can be determined by vapor phase chromatography. Only a few milligrams of sample are needed for analysis.

Simple organic compounds such as hexane (containing C and H only) or ethyl alcohol (containing C, H, and O) can be analyzed using the apparatus shown in Figure 3.2. A weighed sample of the compound, often only a few milligrams, is burned in oxygen, a process called combustion. The carbon present in the sample is converted to carbon dioxide; the hydrogen present is converted to water. The amounts of CO_2 and H_2O produced can be determined by measuring the increases in mass of the two absorption tubes or by vapor phase chromatography (Ch. 1). From these changes in mass, the masses of carbon and hydrogen can be calculated. If oxygen was originally present in the sample, its mass is determined by difference; that is,

These days VPC is usually the method of choice, since it is fast

$$\text{mass O} = \text{mass sample} - (\text{mass C} + \text{mass H})$$

Once the masses of the elements are known, we can calculate the percent composition of the compound and/or its simplest formula (Example 3.4).

■ **EXAMPLE 3.4** _____

Ethyl alcohol contains the three elements carbon, hydrogen, and oxygen. Combustion of a 5.00-g sample of ethyl alcohol gives 9.55 g CO_2 and 5.87 g H_2O. Calculate

a. the masses of C, H, and O in the 5.00-g sample, assuming all the carbon is converted to CO_2 and all the hydrogen to H_2O.
b. the percent composition of ethyl alcohol.
c. the simplest formula of ethyl alcohol.

Solution

a. Let us first calculate the mass of carbon in 9.55 g CO_2. Since one mole of CO_2, 44.01 g, contains one mole of C, 12.01 g, these two quantities are equivalent to one another:

$$12.01 \text{ g C} \simeq 44.01 \text{ g CO}_2$$

They are equivalent because they contain the same amount of carbon

This relation gives us the conversion factor we need to convert from grams of CO_2 to grams of carbon:

$$\text{mass C} = 9.55 \text{ g CO}_2 \times \frac{12.01 \text{ g C}}{44.01 \text{ g CO}_2} = \boxed{2.61 \text{ g C}}$$

One mole of H_2O (18.02 g) contains 2 mol H (2.02 g). Hence,

$$2.02 \text{ g H} \simeq 18.02 \text{ g } H_2O$$

$$\text{mass H} = 5.87 \text{ g } H_2O \times \frac{2.02 \text{ g H}}{18.02 \text{ g } H_2O} = \boxed{0.658 \text{ g H}}$$

Since the total mass of the sample is 5.00 g, and it contains only C, H, and O,

$$\text{mass O} = 5.00 \text{ g} - (2.61 \text{ g} + 0.658 \text{ g}) = \boxed{1.73 \text{ g O}}$$

b. $\text{mass percent C} = \dfrac{\text{mass C}}{\text{mass sample}} \times 100 = \dfrac{2.61 \text{ g}}{5.00 \text{ g}} \times 100 = \boxed{52.2}$

Similarly,

$$\text{mass percent H} = \frac{0.658 \text{ g}}{5.00 \text{ g}} \times 100 = \boxed{13.2}$$

$$\text{mass percent O} = \frac{1.73 \text{ g}}{5.00 \text{ g}} \times 100 = \boxed{34.6}$$

c. We could now determine the simplest formula as in Example 3.3, using the percent composition data calculated in (b). Another approach, however,

You don't have to use percent composition

is to work directly with the masses of C, H, and O in the 5.00-g sample. We found in (a) that the sample contained

$$2.61 \text{ g C}; \qquad 0.658 \text{ g H}; \qquad 1.73 \text{ g O}$$

Let us find the number of moles of each element in the sample:

$$\text{no. moles C} = 2.61 \text{ g C} \times \frac{1 \text{ mol C}}{12.01 \text{ g C}} = 0.217 \text{ mol C}$$

$$\text{no. moles H} = 0.658 \text{ g H} \times \frac{1 \text{ mol H}}{1.01 \text{ g H}} = 0.651 \text{ mol H}$$

$$\text{no. moles O} = 1.73 \text{ g O} \times \frac{1 \text{ mol O}}{16.00 \text{ g O}} = 0.108 \text{ mol O}$$

Now, we find the mole ratio:

$$\frac{0.217 \text{ mol C}}{0.108 \text{ mol O}} = 2.01 \frac{\text{mol C}}{\text{mol O}}; \qquad \frac{0.651 \text{ mol H}}{0.108 \text{ mol O}} = 6.03 \frac{\text{mol H}}{\text{mol O}}$$

Rounding off to whole numbers, we have

$$2 \text{ mol C} : 6 \text{ mol H} : 1 \text{ mol O}$$

The simplest formula of ethyl alcohol is $\boxed{C_2H_6O}$.

Exercise

Using the percentages obtained in (b), find the number of moles of each element in a 100-g sample of ethyl alcohol and its simplest formula. Answer: 4.35 mol C, 13.1 mol H, 2.16 mol O; C_2H_6O.

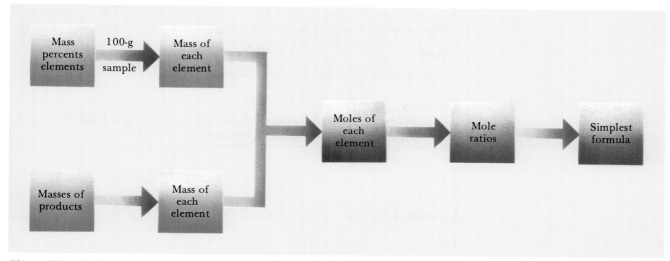

Figure 3.3 Outline of the procedure used to determine the simplest formula of a compound from experimental data.

As we have just seen in Example 3.4, it is not necessary to work with the percent composition of a compound to find its simplest formula. All we need to know are the masses of each element in a sample of the compound. The sample may weigh 5.00 g, 100 g, or have any other mass.

To complete our discussion, it may be helpful to review the general approach used to obtain simplest formulas from experimental data (Fig. 3.3). Regardless of the nature of the data given, we first find the mass and then the number of moles of each element. That information leads directly to the mole ratio. This in turn is identical with the atom ratio given by the simplest formula.

MOLECULAR FORMULA FROM SIMPLEST FORMULA

As pointed out in Section 3.1, the molecular formula is a whole-number multiple of the simplest formula. That multiple may be 1 as in H_2O, 2 as in H_2O_2, or 3 as in C_3H_6. To find out what the multiple is, we need to know one more piece of data, the molar mass. By comparing the observed molar mass to that corresponding to the simplest formula, we can determine whether the multiple is 1, 2, 3, etc.

■ **EXAMPLE 3.5** _____

The simplest formula of vitamin C is found by analysis to be $C_3H_4O_3$. From another experiment, the molar mass is found to be about 180 g/mol. What is the molecular formula of vitamin C?

Molar masses can often be found by mass spectroscopy

Solution

Let us first calculate the formula mass of $C_3H_4O_3$:

$$3(12.0 \text{ amu}) + 4(1.01 \text{ amu}) + 3(16.0 \text{ amu}) = 88.0 \text{ amu}$$

Thus, the molar mass of $C_3H_4O_3$ must be 88.0 g/mol. The approximate molar mass, 180 g/mol, is twice this (180/88 = 2.0). Hence, the simplest formula is multiplied by 2 to obtain the molecular formula:

$$\text{molecular formula of vitamin C} = 2 \times C_3H_4O_3 = \boxed{C_6H_8O_6}$$

Exercise
The simplest formula of hexane is C_3H_7. Its molar mass is about 86 g/mol. What is the molecular formula of hexane? Answer: C_6H_{14}.

3.4
FORMULAS OF IONIC COMPOUNDS

We have just seen how formulas can be determined by experiment. It is also possible to predict "on paper" the formulas of certain substances, notably, simple ionic compounds. This is done using the principle of **electrical neutrality**, referred to briefly in Chapter 2. Consider, for example, the ionic compound calcium chloride. The ions present are Ca^{2+} and Cl^-. It follows that if the compound is to be electrically neutral, the formula of calcium chloride must be $CaCl_2$; two Cl^- ions are required to balance one Ca^{2+} ion.

If you know the ionic charges, you can find the formula very easily

To apply this principle, you must know the charges of the ions involved. In the remainder of this section, we will consider the charges of two types of ions:

— *monatomic* ions, formed when a single atom gains or loses electrons.

— *polyatomic* ions, which are charged particles containing more than one atom.

MONATOMIC IONS

Figure 3.4 shows the charges of the more common monatomic ions, superimposed upon the Periodic Table. Notice that the diagonal line or stairway that runs from the upper left to the lower right of the table separates positive ions (cations) from negative ions (anions). In particular,

— *metals*, located below and to the left of this line, form *cations* by losing electrons. For example,

$$\text{Na atom } (11\ p^+,\ 11\ e^-) \rightarrow \text{Na}^+ \text{ ion } (11\ p^+,\ 10\ e^-) + e^-$$

$$\text{Mg atom } (12\ p^+,\ 12\ e^-) \rightarrow \text{Mg}^{2+} \text{ ion } (12\ p^+,\ 10\ e^-) + 2\ e^-$$

— *nonmetals*, located above and to the right of the line, form *anions* by gaining electrons. For example,

$$\text{F atom } (9\ p^+,\ 9\ e^-) + e^- \rightarrow \text{F}^- \text{ ion } (9\ p^+,\ 10\ e^-)$$

$$\text{O atom } (8\ p^+,\ 8\ e^-) + 2\ e^- \rightarrow \text{O}^{2-} \text{ ion } (8\ p^+,\ 10\ e^-)$$

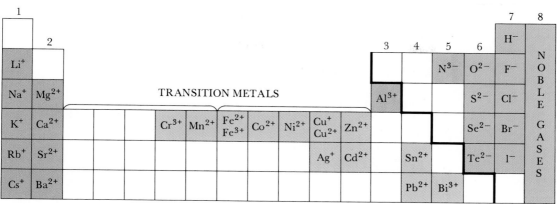

Figure 3.4 Charges of ions found in solid ionic compounds. The step-like diagonal line separates metals from nonmetals and cations from anions. Ions shown in blue have the same number of electrons as the neighboring noble gas atom. For example, S^{2-}, Cl^-, K^+, and Ca^{2+} all have 18 electrons, as does the Ar atom. The transition and post-transition metal cations shown in red do not have a noble gas structure.

Notice that each of the ions just described (Na^+, Mg^{2+}, F^-, and O^{2-}) contains the same number of electrons, 10. This is the number of electrons found in an atom of the noble gas neon ($10\ p^+$, $10\ e^-$). In general, we find that **elements that are close to a noble gas (Group 8) in the Periodic Table tend to form ions that have the same number of electrons as the noble gas atom.**

Ions that have the same number of electrons as the nearest noble gas atom are shown in blue in Figure 3.4. Notice that

The ions of the Group 1 metals have $+1$ charges.
The ions of the Group 2 metals have $+2$ charges.
The ions of the Group 6 nonmetals have -2 charges.
The ions of the Group 7 nonmetals have -1 charges.

The Periodic Table is useful here

Several metals that are farther removed from the noble gases in the Periodic Table form positive ions. The structures of these ions are not related in any direct way to those of noble gas atoms. Indeed, there is no simple way to predict their charges. Some of the more important of these ions are listed in Figure 3.4. They are derived from the **transition metals** (those in the groups near the center of the Periodic Table) and the post-transition metals in Groups 4 and 5. The most common charge among these ions, you will note, is $+2$. Charges of $+1$ (Ag^+) and $+3$ (Cr^{3+}, Bi^{3+}) are less common. Several of these metals form more than one cation. Thus, we have

Fe^{2+} and Fe^{3+} Cu^+ and Cu^{2+}

Using Figure 3.4 and the principle of electrical neutrality, it is possible to predict the formulas of a large number of ionic compounds (Example 3.6).

■ **EXAMPLE 3.6** _____

Predict the formulas of the ionic compounds formed by

a. lithium and oxygen (Li and O)
b. magnesium and chlorine (Mg and Cl)
c. nickel and sulfur (Ni and S)
d. bismuth and fluorine (Bi and F)

Solution

a. Li_2O; 2 Li^+ ions required to balance 1 O^{2-} ion

b. $MgCl_2$; 1 Mg^{2+} ion required to balance 2 Cl^- ions

c. NiS; 1 Ni^{2+} ion required to balance 1 S^{2-} ion

d. BiF_3; 1 Bi^{3+} ion required to balance 3 F^- ions

Exercise

Give the formulas of zinc iodide, silver sulfide, and aluminum oxide. Answer: ZnI_2; Ag_2S; Al_2O_3.

Notice that, *in writing the formula of an ionic compound, the positive ion is always placed first.* Thus, we write $CaCl_2$ not Cl_2Ca, Li_2O not OLi_2, and so on.

POLYATOMIC IONS

Many familiar compounds contain polyatomic ions. Sodium hydroxide (lye), NaOH, contains the hydroxide ion, OH^-. Calcium carbonate (limestone), $CaCO_3$, contains the carbonate ion, CO_3^{2-}. Table 3.1 lists some of the more common polyatomic ions, along with their names and charges. Notice that, with the exception of the ammonium ion, NH_4^+, all of these polyatomic ions are anions.

■ **EXAMPLE 3.7** _____

Using Figure 3.4 and Table 3.1, predict the formulas of

TABLE 3.1   Some Common Polyatomic Ions

+1	−1		−2		−3	
NH_4^+ (ammonium)	OH^-	(hydroxide)	CO_3^{2-}	(carbonate)	PO_4^{3-}	(phosphate)
	NO_3^-	(nitrate)	SO_4^{2-}	(sulfate)		
	ClO_3^-	(chlorate)	CrO_4^{2-}	(chromate)		
	ClO_4^-	(perchlorate)	$Cr_2O_7^{2-}$	(dichromate)		
	MnO_4^-	(permanganate)				
	HCO_3^-	(hydrogen carbonate)				

a. barium hydroxide
b. potassium sulfate
c. ammonium phosphate

Solution

a. One Ba^{2+} ion is required to balance two OH^- ions. The formula is written as $Ba(OH)_2$ to indicate that relationship.

b. K_2SO_4; 2 K^+ ions required to balance 1 SO_4^{2-} ion

c. $(NH_4)_3PO_4$; 3 NH_4^+ ions required to balance 1 PO_4^{3-} ion

You will need to know the formulas and charges of the ions in Table 3.1

Exercise
What is the formula of potassium hydroxide? barium phosphate? ammonium sulfate? Answer: KOH; $Ba_3(PO_4)_2$; $(NH_4)_2SO_4$.

3.5
NAMES OF COMPOUNDS

There are several million known compounds, all of which have systematic names of one type or another. Most of these are carbon compounds, which fall into the area of organic chemistry (Chap. 26). At this time, we need only be concerned with naming inorganic compounds, a simpler task. Moreover, we will restrict our attention to two types of inorganic compounds:

— simple ionic compounds of the type described in Section 3.4. Nearly all of these contain cations derived from metals, and either monatomic or polyatomic anions.
— binary molecular compounds formed from two different nonmetals.

IONIC COMPOUNDS

The name of an ionic compound consists of two words: the first word gives the name of the positive ion (cation), the second that of the negative ion (anion). Thus, we have

COMPOUND	CATION	ANION	NAME
NaCl	Na^+	Cl^-	sodium chloride
K_2SO_4	K^+	SO_4^{2-}	potassium sulfate
$Zn(NO_3)_2$	Zn^{2+}	NO_3^-	zinc nitrate

The system for naming compounds is almost, but not quite, obvious

To assign names to individual ions, note the following:

1. Monatomic positive ions take the name of the metal from which they are derived. Examples include

Na^+ *sodium* K^+ *potassium* Zn^{2+} *zinc*

There is one complication: certain metals form more than one type of positive ion. An example is iron, where we have the Fe^{2+} and Fe^{3+} ions. To distinguish between these ions, the charge must be indicated in the name. This is done by giving the charge as a Roman numeral in parentheses after the name of the metal:

Fe^{2+} *iron(II)* Fe^{3+} *iron(III)*

Similarly, for the two cations of copper, we write

Cu^{+} *copper(I)* Cu^{2+} *copper(II)*

This system is still fairly common

(An older system used the suffixes *-ic* for the ion of higher charge and *-ous* for the ion of lower charge. These were added to the stem of the Latin name of the metal, so that the Fe^{3+} ion was referred to as ferric, the Fe^{2+} ion as ferrous.)

2. Monatomic negative ions are named by adding the suffix *-ide* to the stem of the name of the nonmetal from which they are derived. Thus, we have

			H^-	hydride	
N^{3-}	nitride	O^{2-}	oxide	F^-	fluoride
		S^{2-}	sulfide	Cl^-	chloride
		Se^{2-}	selenide	Br^-	bromide
		Te^{2-}	telluride	I^-	iodide

3. Polyatomic ions are given special names (Table 3.1, p. 74).

■ **EXAMPLE 3.8** _____

Name the following ionic compounds:

a. CaS   b. $Al(NO_3)_3$ c. $FeCl_2$

Solution

a.   Calcium sulfide.

b. The   ions   present   are Al^{3+} and NO_3^-;   the   name   is   aluminum nitrate.

c. Recall that the chloride ion has a -1 charge (Cl^-). The positive ion must then be Fe^{2+}. The name is   iron(II) chloride.

Exercise
Name Al_2O_3, $Fe(NO_3)_3$, and Ag_2Se. Answer: aluminum oxide; iron(III) nitrate; silver selenide.

BINARY MOLECULAR COMPOUNDS

When a pair of nonmetals forms only one compound, that compound is named very simply. The name of the element whose symbol appears first in the formula is written first. The second part of the name is formed by adding the suffix *-ide* to the stem of the name of the second nonmetal. Examples include

HCl   hydrogen chloride

H$_2$S   hydrogen sulfide

Same system as with the analogous ionic compounds

More often, a pair of nonmetals forms more than one compound. In those cases, the names used are a bit more complex. The most common system of naming such compounds uses the Greek prefixes *di* = two, *tri* = three, *tetra* = four, *penta* = five, *hexa* = six, *hepta* = seven, and *octa* = eight to show the number of atoms of each element. The several oxides of nitrogen have the names

N$_2$O$_5$ dinitrogen pentoxide

N$_2$O$_4$ dinitrogen tetroxide

NO$_2$ nitrogen dioxide

N$_2$O$_3$ dinitrogen trioxide

NO   nitrogen oxide

N$_2$O   dinitrogen oxide

In rare cases, the prefix *mono* is used:

CO$_2$ carbon dioxide

CO   carbon monoxide

We don't ever want to confuse CO$_2$ and CO

■ **EXAMPLE 3.9** _____

Give the names of

a. SO$_2$ b. SO$_3$ c. PCl$_3$ d. PCl$_5$

Solution

a.   sulfur dioxide   b.   sulfur trioxide

c.   phosphorus trichloride   d.   phosphorus pentachloride

Exercise

Give the names of SF$_4$ and SF$_6$. Answer: sulfur tetrafluoride; sulfur hexafluoride.

_____

Many of the best known binary compounds of the nonmetals have acquired common names. These are widely and, in some cases, exclusively used. Examples include

H$_2$O	water	PH$_3$	phosphine
H$_2$O$_2$	hydrogen peroxide	AsH$_3$	arsine
NH$_3$	ammonia	NO	nitric oxide
N$_2$H$_4$	hydrazine	N$_2$O	nitrous oxide

Nobody calls water "dihydrogen oxide"

Certain binary molecular compounds react with water to produce an *acidic* solution, containing H$^+$ ions. One such compound is hydrogen chloride, HCl; in water solution it exists as aqueous H$^+$ and Cl$^-$ ions. The water solution of hydrogen chloride is given a special name; it is referred to as hydrochloric acid. A similar situation applies with HBr and HI:

PURE SUBSTANCE		WATER SOLUTION	
$HCl(g)$	hydrogen chloride	$H^+(aq), Cl^-(aq)$	hydrochloric acid
$HBr(g)$	hydrogen bromide	$H^+(aq), Br^-(aq)$	hydrobromic acid
$HI(g)$	hydrogen iodide	$H^+(aq), I^-(aq)$	hydriodic acid

Two other acids that you will find in the general chemistry laboratory are *nitric* and *sulfuric* acids, water solutions of HNO_3 and H_2SO_4, respectively.

3.6
WRITING AND BALANCING CHEMICAL EQUATIONS

You start with reactants and end up with products

When a chemical reaction occurs, starting materials, which we call *reactants*, are converted to other substances, called *products*. A reaction can be described in words, but it is more useful to represent it by a chemical equation. Formulas of reactants appear on the left side of a chemical equation. They are separated by an arrow from the formulas of the products, written on the right side of the equation. In a balanced chemical equation, there is the same number of atoms of a given element on the two sides.

Beginning students are sometimes led to believe that writing a chemical equation is a simple, mechanical process. Nothing could be further from the truth. One point that seems obvious is often overlooked. *You cannot write an equation unless you know what happens in the reaction that it represents.* All the reactants and all the products must be identified. Moreover, you must know their formulas and physical states.

To illustrate how we arrive at a balanced equation, consider a reaction that was used in the Titan rocket motor to launch the Gemini spacecraft. The reactants were two liquids, hydrazine and dinitrogen tetroxide. As we saw in Section 3.5, these compounds have the molecular formulas N_2H_4 and N_2O_4, respectively. The products of the reaction are gaseous nitrogen, N_2, and water vapor, H_2O. To write a balanced equation for this reaction, we proceed as follows:

1. Write a "skeleton" equation in which the formulas of the reactants appear on the left and those of the products on the right. In this case,

$$N_2H_4 + N_2O_4 \rightarrow N_2 + H_2O$$

2. Indicate the physical state of each reactant and product, after the formula, by writing

(*g*)   for a gaseous substance
(*l*)   for a pure liquid
(*s*)   for a solid
(*aq*) for an ion or molecule in water (aqueous) solution

In this case, we have:

$$N_2H_4(l) + N_2O_4(l) \rightarrow N_2(g) + H_2O(g)$$

Titan rocket motor. (NASA)

3. Balance the equation by taking into account the Law of Conservation of Mass. This requires that there be the same number of atoms of each element on both sides of the equation. To accomplish this, start by writing a coefficient of 4 for H_2O, thus obtaining 4 oxygen atoms on both sides:

$$N_2H_4(l) + N_2O_4(l) \rightarrow N_2(g) + 4 H_2O(g)$$

Now consider the hydrogen atoms. There are $4 \times 2 = 8$ H atoms on the right. To obtain 8 H atoms on the left, write a coefficient of 2 for N_2H_4:

$$2 N_2H_4(l) + N_2O_4(l) \rightarrow N_2(g) + 4 H_2O(g)$$

Finally, consider nitrogen. There are a total of $(2 \times 2) + 2 = 6$ nitrogen atoms on the left. To balance, nitrogen, write a coefficient of 3 for N_2:

$$2 N_2H_4(l) + N_2O_4(l) \rightarrow 3 N_2(g) + 4 H_2O(g) \qquad (3.1)$$

This is the final balanced equation for the reaction of hydrazine with dinitrogen tetroxide.

Two points concerning the balancing process are worth emphasizing:

1. Equations are balanced by adjusting coefficients in front of formulas, never by changing subscripts within formulas. In arriving at Equation 3.1, we balanced nitrogen by writing $3 N_2$, which indicates three N_2 molecules. On paper, we could have obtained six nitrogen atoms on the right by writing N_6, but that would have been absurd. Elemental nitrogen exists as diatomic molecules, N_2; there is no such thing as an N_6 molecule.

6N is also no good; the formula must be that of the species actually present

2. In balancing an equation, it is best to start with an element that appears in only one species on each side of the equation. In this case, we could have started with either oxygen or hydrogen. Nitrogen would have been a poor choice, however, since there are nitrogen atoms in both reactant molecules, N_2H_4 and N_2O_4.

Ordinarily, if you are asked to balance a chemical equation, you will be given the formulas of reactants and products. However, for one type of reaction you should be able to come up with this information on your own. We refer to the reaction of a metal with a nonmetal to form an ionic compound containing monatomic ions listed in Figure 3.4. Recall that in Example 3.6 we showed how to predict the formulas of ionic compounds of this type, *all of which are solids*. The vast majority of reacting elements can be shown as monatomic solids, e.g., Na(s), Al(s), etc. The principal exceptions to this rule are shown in Figure 3.5. In a chemical equation oxygen is represented as $O_2(g)$, iodine as $I_2(s)$, and so on.

■ **EXAMPLE 3.10** _____

Write a balanced equation for the reaction between

a. lithium and oxygen   b. bismuth and fluorine

Solution

a. Recall from Example 3.6 that the formula of lithium oxide is Li_2O. The skeleton equation is

		GROUP 7	GROUP 8
		$H_2(g)$	
GROUP 5	GROUP 6		
$N_2(g)$	$O_2(g)$	$F_2(g)$	
$P_4(s)$	$S_8(s)$	$Cl_2(g)$	
		$Br_2(l)$	
		$I_2(s)$	

Figure 3.5 Molecular elements and their physical states in the Periodic Table. In writing equations, sulfur is usually shown as monatomic for simplicity.

$$Li + O_2 \rightarrow Li_2O$$

(Note that oxygen is diatomic.) To balance oxygen atoms, we write a coefficient of 2 for Li_2O. This requires a coefficient of 4 for Li:

$$4\,Li + O_2 \rightarrow 2\,Li_2O$$

Inserting the proper physical states, we obtain

$$4\,Li(s) + O_2(g) \rightarrow 2\,Li_2O(s)$$

b. The formula of bismuth fluoride is BiF_3, as deduced in Example 3.6. The skeleton equation is

$$Bi + F_2 \rightarrow BiF_3$$

The final, balanced equation is

$$2\,Bi(s) + 3\,F_2(g) \rightarrow 2\,BiF_3(s)$$

Exercise
Write a balanced equation for the reaction of aluminum with iodine. Answer: $2\,Al(s) + 3\,I_2(s) \rightarrow 2\,AlI_3(s)$.

3.7
MASS RELATIONS IN REACTIONS

One of the main uses of balanced equations is in relating the masses of reactants and products in a reaction. We can use a balanced equation to determine, for a given amount of one reactant, how much of another reactant is required or how much product is formed. Calculations of this sort are based upon a very important principle:

The coefficients of a balanced equation represent numbers of moles of reactants and products.

To show that this statement is valid, recall the equation:

$$2\,N_2H_4(l) + N_2O_4(l) \rightarrow 3\,N_2(g) + 4\,H_2O(g)$$

The coefficients in this equation represent numbers of molecules; that is,

One possible interpretation of the equation

$$2 \text{ molecules } N_2H_4 + 1 \text{ molecule } N_2O_4 \rightarrow$$
$$3 \text{ molecules } N_2 + 4 \text{ molecules } H_2O$$

A balanced equation remains valid if each coefficient is multiplied by the same number, including *Avogadro's number, N:*

Another, usually more practical, interpretation

$$2N \text{ molecules } N_2H_4 + N \text{ molecules } N_2O_4 \rightarrow$$
$$3N \text{ molecules } N_2 + 4N \text{ molecules } H_2O$$

As we saw in Chapter 2, however, a mole represents Avogadro's number of items, N. Thus, we can write

$$2 \text{ mol } N_2H_4 + 1 \text{ mol } N_2O_4 \rightarrow 3 \text{ mol } N_2 + 4 \text{ mol } H_2O$$

which is the relation we set out to demonstrate. From a slightly different standpoint, we might say that, in this reaction

$$2 \text{ mol } N_2H_4 \simeq 1 \text{ mol } N_2O_4 \simeq 3 \text{ mol } N_2 \simeq 4 \text{ mol } H_2O$$

where the symbol $\simeq$ indicates that the quantities listed are chemically equivalent to one another in this reaction.

■ **EXAMPLE 3.11** _____

For the reaction

$$2 \text{ N}_2\text{H}_4(l) + \text{N}_2\text{O}_4(l) \rightarrow 3 \text{ N}_2(g) + 4 \text{ H}_2\text{O}(g)$$

determine

a. the number of moles of N_2O_4 required to react with 2.72 mol N_2H_4.
b. the number of moles of N_2 produced from 2.72 mol N_2H_4.

Solution

a. The conversion factor required follows directly from the coefficients of the balanced equation:

$$2 \text{ mol } N_2H_4 \simeq 1 \text{ mol } N_2O_4$$

$$\text{no. moles } N_2O_4 = 2.72 \text{ mol } N_2H_4 \times \frac{1 \text{ mol } N_2O_4}{2 \text{ mol } N_2H_4}$$

$$= \boxed{1.36 \text{ mol } N_2O_4}$$

b. In this case,

$$2 \text{ mol } N_2H_4 \simeq 3 \text{ mol } N_2$$

$$\text{no. moles } N_2 = 2.72 \text{ mol } N_2H_4 \times \frac{3 \text{ mol } N_2}{2 \text{ mol } N_2H_4} = \boxed{4.08 \text{ mol } N_2}$$

Exercise
In the reaction between aluminum and iodine (Example 3.10) how many moles of AlI_3 are formed from 1.68 mol I_2, reacting with excess Al? Answer: 1.12 mol AlI_3.

_____

The approach followed in Example 3.11 is readily extended to reactions involving masses in grams of reactants and products. To do this, we convert moles to grams, knowing the molar masses of the substances involved. In this way, it is possible to use the coefficients of a balanced equation to relate

— moles of one substance to grams of another (Example 3.12a).
— grams of one substance to grams of another (Examples 3.12b and c).

■ **EXAMPLE 3.12**

The ammonia used to make fertilizers for lawns and gardens is made by reacting nitrogen of the air with hydrogen. The balanced equation for the reaction is

$$3 H_2(g) + N_2(g) \rightarrow 2 NH_3(g)$$

Determine

a. the mass in grams of ammonia, NH_3, formed when 1.34 mol N_2 reacts.
b. the mass in grams of N_2 required to form 1.00 kg NH_3.
c. the mass in grams of H_2 required to react with 6.00 g N_2.

The chemical equation makes these calculations possible

Solution

a. The balanced equation tells us that

$$1 \text{ mol } N_2 \simeq 2 \text{ mol } NH_3$$

The formula mass of NH_3 is 14.01 amu + 3(1.01) amu = 17.04 amu. Hence the molar mass of NH_3 is 17.04 g/mol and

$$1 \text{ mol } NH_3 = 17.04 \text{ g } NH_3$$

These two relations give us the conversion factors we need to find the mass in grams of NH_3 formed from 1.34 mol N_2:

$$\text{mass } NH_3 = 1.34 \text{ mol } N_2 \times \frac{2 \text{ mol } NH_3}{1 \text{ mol } N_2} \times \frac{17.04 \text{ g } NH_3}{1 \text{ mol } NH_3}$$

$$= \boxed{45.7 \text{ g } NH_3}$$

b. We carry out a three-step conversion:
  1. grams $NH_3 \rightarrow$ moles NH_3 (1 mol NH_3 = 17.04 g NH_3)
  2. moles $NH_3 \rightarrow$ moles N_2 (1 mol $N_2 \simeq$ 2 mol NH_3)
  3. moles $N_2 \rightarrow$ grams N_2 (1 mol N_2 = 28.02 g N_2)

Here's a case where conversion factors come into their own

$$\text{mass } N_2 = 1000 \text{ g } NH_3 \times \underset{(1)}{\frac{1 \text{ mol } NH_3}{17.04 \text{ g } NH_3}} \times \underset{(2)}{\frac{1 \text{ mol } N_2}{2 \text{ mol } NH_3}} \times \underset{(3)}{\frac{28.02 \text{ g } N_2}{1 \text{ mol } N_2}}$$

$$= \boxed{822 \text{ g } N_2}$$

c. As in (b), we first convert grams of nitrogen to moles (1 mol N_2 = 28.02 g N_2). Then we convert moles of N_2 to moles of H_2, using the coefficients of the balanced equation (3 mol $H_2 \simeq$ 1 mol N_2). Finally, we convert moles of H_2 to grams (1 mol H_2 = 2.02 g H_2):

$$\text{mass } H_2 = 6.00 \text{ g } N_2 \times \frac{1 \text{ mol } N_2}{28.02 \text{ g } N_2} \times \frac{3 \text{ mol } H_2}{1 \text{ mol } N_2} \times \frac{2.02 \text{ g } H_2}{1 \text{ mol } H_2}$$

$$= \boxed{1.30 \text{ g } H_2}$$

Exercise

For this reaction, calculate the mass in grams of H_2 required to form 1.0 g NH_3. Answer: 0.18 g H_2.

3.8
LIMITING REACTANT AND THEORETICAL YIELD

When the two elements antimony and iodine are heated in contact with one another (Fig. 3.6), they react to form antimony triiodide.

$$2\ Sb(s)\ +\ 3\ I_2(s)\ \rightarrow\ 2\ SbI_3(s) \tag{3.2}$$

The coefficients in this equation tell us that two moles of Sb (243.5 g) react exactly with three moles of I_2 (761.4 g) to form two moles of SbI_3 (1004.9 g). If we mix antimony and iodine in a 2:3 mole ratio, we expect both reactants to be completely consumed, forming pure antimony triiodide.

Ordinarily, in the laboratory, reactants are not mixed in exactly the ratio required for reaction. Instead, we use an excess of one reactant, usually the cheaper one. We might, for example, mix 3.00 mol Sb with 3.00 mol I_2. In that case, the antimony would be in excess, since only 2.00 mol Sb is required to react with 3.00 mol I_2. After the reaction is over, we expect to have 1.00 mol Sb remaining:

$$\begin{aligned} \text{excess Sb} &= 3.00 \text{ mol originally} - 2.00 \text{ mol consumed} \\ &= 1.00 \text{ mol} \end{aligned}$$

The 3.00 mol I_2 should be completely consumed in forming the 2.00 mol SbI_3:

$$\text{moles SbI}_3 \text{ formed} = 3.00 \text{ mol I}_2 \times \frac{2 \text{ mol SbI}_3}{3 \text{ mol I}_2} = 2.00 \text{ mol SbI}_3$$

After the reaction is over, the solid obtained would be a mixture of product, 2.00 mol SbI_3 (1004.9 g), together with 1.00 mol unreacted Sb (121.8 g).

In situations such as this, we distinguish between the reactant in excess (Sb) and the other reactant (I_2), called the **limiting reactant**. The amount of product formed is determined (limited) by the amount of limiting reactant. With 3.00 mol I_2, we cannot get more than 2.00 mol SbI_3, regardless of how large an excess of Sb we use. The amount of product that would be formed if all the limiting reactant were consumed is referred to as the **theoretical yield** of product. If we mix 3.00 mol I_2 with excess Sb, the coefficients of Equation 3.2 tell us that the theoretical yield of SbI_3 is 2.00 mol.

Often, you will be told the amounts of two different reactants and asked to determine which is the limiting reactant and calculate the theoretical yield of product. To do this, it helps to follow a systematic procedure. The one we will use involves three steps:

1. Calculate the amount of product that would be formed if the first reactant were completely consumed.
2. Repeat this calculation for the second reactant; that is, calculate how much product would be formed if all of that reactant were consumed.
3. Choose the smaller of the two amounts calculated in (1) and (2). This is the theoretical yield of product; the reactant that produces the smaller

Figure 3.6 When antimony (gray powder) comes in contact with iodine (violet vapor), an exothermic reaction takes place (*middle picture*). The product is a red solid, SbI_3. (Charles D. Winters)

amount is the limiting reactant. The other reactant is in excess; only part of it is consumed.

To illustrate this procedure, it may help to cite an example far removed from chemistry. Let us suppose that a fast-food restaurant is making "double cheeseburgers" by putting a hamburger between two pieces of cheese:

2 slices cheese + 1 hamburger patty → 1 double cheeseburger

Suppose further that the restaurant has 250 slices of cheese and 150 hamburger patties. How many double cheeseburgers can they make?

If all the cheese is consumed,

$$250 \text{ slices cheese} \times \frac{1 \text{ double cheeseburger}}{2 \text{ slices cheese}} = 125 \text{ double cheeseburgers}$$

If, on the other hand, all the hamburger patties are used up,

$$150 \text{ hamburger patties} \times \frac{1 \text{ double cheeseburger}}{1 \text{ hamburger patty}} = 150 \text{ double cheeseburgers}$$

Here the "limiting reactant" is the cheese. The "theoretical yield" of double cheeseburgers is 125. When all the cheese is consumed, there will be $150 - 125 = 25$ hamburger patties left over. If we tried to make 150 double cheeseburgers, 25 of them would contain no cheese and we would have 25 unhappy customers.

■ **EXAMPLE 3.13** _____

Consider the reaction

$$2 \text{ Sb}(s) + 3 \text{ I}_2(s) \rightarrow 2 \text{ SbI}_3(s)$$

Determine the limiting reactant and the theoretical yield of product if we start with

a. 1.20 mol Sb and 2.40 mol I_2
b. 1.20 g Sb and 2.40 g I_2

Solution

a. We calculate the amount of product formed from the two reactants, using conversion factors obtained directly from the coefficients of the balanced equation:

$$2 \text{ mol Sb} \simeq 2 \text{ mol SbI}_3; \qquad 3 \text{ mol I}_2 \simeq 2 \text{ mol SbI}_3$$

1. If antimony is the limiting reactant,

$$\text{no. moles SbI}_3 = 1.20 \text{ mol Sb} \times \frac{2 \text{ mol SbI}_3}{2 \text{ mol Sb}} = 1.20 \text{ mol SbI}_3$$

2. If iodine is the limiting reactant,

$$\text{no. moles SbI}_3 = 2.40 \text{ mol I}_2 \times \frac{2 \text{ mol SbI}_3}{3 \text{ mol I}_2} = 1.60 \text{ mol SbI}_3$$

We can only use up one reactant. The limiting reactant is the one that gives the smaller amount of product. Once that reactant is gone, the show is over

3. The theoretical yield of SbI_3 is the *smaller* quantity, 1.20 mol; the limiting reactant is therefore Sb , and there is an excess of I_2.

b. To calculate the amount of product, we follow the three-step path used in Example 3.12b. Note that the molar masses of Sb, I_2, and SbI_3 are 121.8 g/mol, 253.8 g/mol, and 502.5 g/mol respectively.

1. Mass SbI_3 formed if all the Sb is consumed:

$$1.20 \text{ g Sb} \times \frac{1 \text{ mol Sb}}{121.8 \text{ g Sb}} \times \frac{2 \text{ mol SbI}_3}{2 \text{ mol Sb}} \times \frac{502.5 \text{ g SbI}_3}{1 \text{ mol SbI}_3} = 4.95 \text{ g SbI}_3$$

2. Mass SbI_3 formed if all the I_2 is consumed:

$$2.40 \text{ g I}_2 \times \frac{1 \text{ mol I}_2}{253.8 \text{ g I}_2} \times \frac{2 \text{ mol SbI}_3}{3 \text{ mol I}_2} \times \frac{502.5 \text{ g SbI}_3}{1 \text{ mol SbI}_3} = 3.17 \text{ g SbI}_3$$

3. The theoretical yield of SbI_3 is 3.17 g; the limiting reactant is I_2.

Exercise

How many grams of Sb are required to react with the I_2 in (b)? How many grams of Sb are left over? Answer: 0.768 g; 0.43 g.

Remember that, in deciding upon the theoretical yield of product, you *choose the smaller of the two calculated amounts.* To see why this must be the case, let us refer back to Example 3.13b. There we started with 1.20 g Sb and 2.40 g I_2. We decided that the theoretical yield of SbI_3 was 3.17 g and 0.43 g Sb was left over. Thus:

$$1.20 \text{ g Sb} + 2.40 \text{ g I}_2 \rightarrow 3.17 \text{ g SbI}_3 + 0.43 \text{ g Sb}$$

This makes sense: 3.60 g of reactants yields a total of 3.60 g of "products," including the unreacted antimony. Suppose, however, we had chosen 4.95 g SbI_3 as the theoretical yield. We would then have the nonsensical situation:

$$1.20 \text{ g Sb} + 2.40 \text{ g I}_2 \rightarrow 4.95 \text{ g SbI}_3$$

This violates the Law of Conservation of Mass; we cannot get 4.95 g of product from 3.60 g of reactants.

The theoretical yield is the maximum amount of product that we can hope to obtain. In calculating the theoretical yield, we assume that the limiting reactant is completely (that is, 100%) converted to product. In the real world, this is unlikely to happen. Some of the limiting reactant may be consumed in competing reactions. Some of the product may be lost in separating it from the reaction mixture. For these and other reasons, the **actual yield** in a reaction is ordinarily less than the theoretical yield. Putting it another way, the **percent yield** is expected to be less than 100.

The actual yield is what you get. The theoretical yield is what you would get if everything went perfectly

$$\text{percent yield} = \frac{\text{actual yield}}{\text{theoretical yield}} \times 100 \qquad (3.3)$$

To illustrate how percent yield is calculated, consider again the reaction described in Example 3.13b. Let us suppose that the actual yield of SbI_3 was 2.42 g. Since the theoretical yield was 3.17 g, we have:

$$\text{percent yield} = \frac{2.42\,\text{g}}{3.17\,\text{g}} \times 100 = 76.3$$

■ **SUMMARY PROBLEM**

When bismuth reacts with fluorine, an ionic compound is formed.

a. What is the name of the compound? What is its formula?
b. Calculate the mass percent of fluorine in the compound. Calculate the mass of fluorine required to form 16.5 g of the compound.
c. Write a balanced equation for the reaction of bismuth with fluorine.
d. How many moles of F_2 are required to react with 0.240 mol Bi? With 1.60 g Bi?
e. If 5.00 g of bismuth reacts with excess fluorine, what mass of product is formed?
f. If 5.00 g of bismuth reacts with 5.00 g of fluorine, what is the theoretical yield of product?
g. When the compound referred to in (a) and (b) reacts with water, one of the products is a compound containing 85.65% Bi, 6.56% O, and 7.79% F. What is the simplest formula of this compound?

Answers
a. bismuth(III) fluoride; BiF_3 b. 21.43; 3.54 g
c. $2\,Bi(s) + 3\,F_2(g) \rightarrow 2\,BiF_3(s)$ d. 0.360 mol; 0.0115 mol
e. 6.36 g   f. 6.36 g   g. BiOF

The compounds dealt with in this chapter have relatively simple formulas. At most, they contain three elements; the atom ratios involve small numbers such as 1, 2, or 3. Many compounds have more complex formulas; among these are a class of mixed oxides that contain three or four different metals in addition to oxygen. Certain compounds of this type have been in the news lately because they have a very unusual property. Upon cooling to about 100 K, their electrical resistance drops to zero; for that reason they are referred to as superconductors. A typical formula is $YBa_2Cu_3O_7$*, but many other compositions are possible. Superconductivity is the most active research area in the physical sciences today, with knowledge of the subject increasing rapidly.

Potentially, superconductors have a host of practical applications. One of the simplest is in the transmission of electrical energy. A power line made of a superconducting material could carry electricity over hundreds of kilometers with virtually no energy loss. This would make it possible, for instance, to build nuclear power plants in remote areas far from population centers. On a more exotic level, it may be practical to design electric trains that travel at 500 km/h over superconducting tracks. The train would be literally suspended in the air, 20 cm off the ground (Fig. 3.7). This effect arises because the magnetic field induced when current flows through the superconductor repels that of a magnet above it. The extent of levitation depends upon a balance between gravitational and magnetic forces.

---

*Actually, these compounds are nonstoichiometric (Chapter 2). The number of oxygen atoms in the formula varies from 6.5 to 7.2, depending upon the method of preparation. For a discussion of the structure of superconductors, see *Chemical and Engineering News*, May 11, 1987, pp. 7–16.

MODERN PERSPECTIVE

SUPERCONDUCTORS

Figure 3.7 A pellet of superconducting material, previously cooled to 77 K with liquid nitrogen, floats above a magnet. (Courtesy of Edmund Scientific)

The phenomenon of superconductivity was discovered by a Dutch physicist Kammerlingh Onnes in 1911. For many years, all known superconductors were metals or metal alloys which had to be cooled to very low temperatures, approaching absolute zero, before they became superconducting. This required the use of liquid helium (bp = 4 K), whose high cost severely limited the practical applications of superconductors.

In 1975, a group at Du Pont led by Arthur Sleight made the first mixed-oxide superconductor, a solid containing the metals Ba, Pb, and Bi, in addition to oxygen. A little more than a decade later, Bednorz and Muller at the IBM laboratories in Zurich showed that a La–Ba–Cu oxide became superconducting at 35 K, a temperature considerably higher than any previously achieved. For their work, Bednorz and Muller received the 1987 Nobel Prize in physics.

Early in 1987, a research group at the University of Houston, working with Y–Ba–Cu oxides, raised the critical temperature for superconductivity to 90 K. This was an important breakthrough, because it now became possible to replace liquid helium with liquid nitrogen (bp = 77 K), reducing the cooling cost by a factor of 1000. There have been many reports since then of materials showing superconductivity near or even above room temperature. At this time (July, 1988), none of these claims have been substantiated.

Mixed-oxide superconductors can be made by heating together the oxides or carbonates of the appropriate metals. A typical reaction, taking place in a furnace at 1000°C might be:

$$\tfrac{1}{2}\,Y_2O_3(s) + 2\,BaCO_3(s) + 3\,CuO(s) + \tfrac{1}{2}\,O_2(g) \rightarrow YBa_2Cu_3O_7(s) + 2\,CO_2(g)$$

In May 1988; superconductors containing no Cu ions were prepared

The solid product must be annealed for several hours in an oxygen atmosphere at 500°C. It is relatively expensive, slowly deteriorates upon exposure to moisture, and is difficult to fabricate into useful shapes (e.g., wires or ribbons). Very recently (February, 1988), a new type of mixed-oxide superconductor, considerably easier to make and with superior physical properties, has been reported. This material contains four different metals; it has the approximate composition $Bi_2CaSr_2Cu_2O_x$, where x is between 8 and 9.

QUESTIONS AND PROBLEMS

Symbols, Formulas, and Equations

1. Using the figures in the chapter, give the color, formula, and physical state of
a. anhydrous copper sulfate
b. cobalt chloride hydrate with six molecules of water
c. iodine

2. Using the figures in the chapter, give the color, formula, and physical state of
a. antimony   b. copper sulfate hydrate
c. cobalt chloride hydrate with 4 molecules of water
3. Write the formulas of the following ionic compounds. Use the Periodic Table and, if necessary, Table 3.1 and Figure 3.4.

a. iron(III) sulfate   b. potassium nitrate
c. lead(II) oxide   d. barium chlorate
e. copper(I) selenide   f. manganese(II) carbonate
g. sulfuric acid   h. sodium hydrogen carbonate

4. Follow the directions of Problem 3 for the following compounds:

a. sodium dichromate   b. tin(II) sulfide
c. aluminum phosphate   d. gold(III) chloride
e. chromium(III) oxide f. nitric acid
g. calcium phosphate   h. calcium perchlorate

5. Complete the following table of ionic compounds:

NAME	FORMULA
barium nitride	————
copper(II) hydroxide	————
————	Ag_2Te
————	$Fe_2(CO_3)_3$
strontium chromate	————
————	$NaHCO_3$
ammonium perchlorate	————

6. Complete the following table of ionic compounds:

NAME	FORMULA
————	$CsOH$
sodium permanganate	————
lithium dichromate	————
————	NH_4Cl
aluminum sulfate	————
————	$Ba(NO_3)_2$

7. Complete the following table of molecular compounds:

NAME	FORMULA
————	NH_3
ditellurium dichloride	————
————	XeO_3
————	$BrCl_3$
————	P_2O_5
dinitrogen tetroxide	————

8. Complete the following table of molecular compounds.

NAME	FORMULA
————	N_2H_4
hydrogen peroxide	————
————	XeF_4
————	S_4N_4
nitrogen trifluoride	————
carbon tetrachloride	————

Percent Composition from Formula

9. Turquoise has the following chemical formula: $CuAl_6(PO_4)_4(OH)_8 \cdot 4H_2O$. Calculate the mass percent of each element in turquoise.

10. The active ingredient of some antiperspirants is aluminum chlorohydrate, $Al_2(OH)_5Cl$. Calculate the mass percent of each element in this ingredient.

11. The hormone thyroxine secreted by the thyroid gland has the formula $C_{15}H_{11}NO_4I_4$. How many milligrams of iodine can be extracted from 5.00 g of thyroxine?

12. One of the most common rocks on earth is feldspar. One type of feldspar has the formula $CaAl_2Si_2O_8$. How much aluminum can be obtained from mining and smelting one metric ton (10^3 kg) of feldspar?

13. A 250.0-mg sample of a commercial headache remedy contains 152 mg of aspirin, $C_9H_8O_4$. Assume all the carbon in the sample is in aspirin.

a. What is the mass percent of aspirin in the product?
b. How many grams of carbon are there in the aspirin contained in a tablet of this product weighing 0.611 g?

14. The active ingredient in Pepto-Bismol (an over-the-counter remedy for upset stomach) is bismuth subsalicylate, $C_7H_5BiO_4$. Analysis of a 1.50 g sample of Pepto-Bismol yields 346 mg of bismuth. What percent by mass of the sample is bismuth subsalicylate? (Assume that there is no other bismuth-containing compound in Pepto-Bismol.)

15. Toluene is now widely used instead of benzene, since benzene has been proven to be carcinogenic. Combustion of 1.000 g of toluene, a compound of carbon and hydrogen atoms, gives 3.348 g carbon dioxide. What are the mass percents of carbon and hydrogen in toluene?

16. Hexachlorophene, a compound made up of atoms of carbon, hydrogen, chlorine, and oxygen, is an ingredient in germicidal soaps. Combustion of a 1.000-g sample yields 1.407 g of carbon dioxide, 0.134 g of water, and 0.5228 g of chlorine gas. What are the mass percents of carbon, hydrogen, chlorine, and oxygen in hexachlorophene?

Simplest Formula from Analysis

17. Arsenic reacts with chlorine to form a chloride. If 1.587 g of arsenic reacts with 3.755 g of chlorine, what is the simplest formula of the chloride?

18. If 7.35 g of chromium reacted directly with oxygen to form 10.74 g of a metal oxide, what is the simplest formula of the oxide?

19. Determine the simplest formulas of the following three compounds:

a. citric acid, which has the composition: 37.51% C, 4.20% H, and 58.29% O.
b. tetraethyl lead, the gasoline additive, which has the composition: 29.71% C, 6.234% H, and 64.07% Pb.

c. saccharine, the artificial sweetener, which has the composition: 45.90% C, 2.75% H, 26.20% O, 17.50% S, and 7.65% N.

20. Determine the simplest formulas of the following compounds:

a. rubbing alcohol, a compound made up of carbon, hydrogen, and oxygen atoms, which has the composition: 59.96% C and 13.42% H.

b. the food enhancer, monosodium glutamate (MSG), which has the composition: 35.51% C, 4.77% H, 37.85% O, 8.29% N, and 13.60% Na.

c. the sedative chloral hydrate, which has the composition: 14.52% C, 1.828% H, 64.30% Cl, and 19.35% O.

21. Vanillin, a flavoring agent, is made up of carbon, hydrogen, and oxygen atoms. When a sample of vanillin weighing 2.500 g burns in oxygen, 5.79 g of carbon dioxide and 1.18 g water are obtained. What is the empirical formula of vanillin?

22. Methyl salicylate is a common "active ingredient" in liniments such as Ben-Gay. It is also known as oil of wintergreen. It is made up of carbon, hydrogen, and oxygen atoms. When a sample of methyl salicylate weighing 5.287 g is burned in excess oxygen, 12.24 g of carbon dioxide and 2.522 g of water are formed. What is the empirical formula for oil of wintergreen?

23. The hormone norepinephrine is an antidote to the allergic reaction resulting from a bee sting. It is made up of hydrogen, carbon, oxygen, and nitrogen atoms. A 2.587 g sample of norepinephrine burned in oxygen yields 5.387 g of carbon dioxide and 1.526 g of water. Another experiment determines that the hormone contains 8.281% nitrogen by mass. What is the empirical formula of norepinephrine?

24. Dimethyl nitrosamine is a known carcinogen. It can be formed in the intestinal tract when digestive juices react with the nitrite ion in preserved and smoked meats. It is made up of carbon, hydrogen, nitrogen, and oxygen atoms. A 4.319-g sample of dimethyl nitrosamine burned in oxygen yields 5.134 g of carbon dioxide and 3.173 g of water. The compound contains 37.82% by mass of nitrogen. What is the empirical formula of dimethyl nitrosamine?

25. Dimethylhydrazine, the fuel used in the Apollo lunar descent module, has a molar mass of 60.10 g/mol. It is made up of carbon, hydrogen, and nitrogen atoms. The combustion of 2.859 g of the fuel in excess oxygen yields 4.190 g carbon dioxide and 3.428 g of water. What are the empirical and molecular formulas for dimethylhydrazine?

26. Hexamethylenediamine (MM = 116.2 g/mol), a compound made up of carbon, hydrogen, and nitrogen atoms, is used in the production of nylon. When 6.315 g of hexamethylenediamine are burned in oxygen, 14.36 g

of carbon dioxide and 7.832 g of water are obtained. What are the empirical and molecular formulas of the compound?

27. A sample of a compound of chlorine and oxygen reacts with excess hydrogen to give 0.3059 g HCl and 0.5287 g of water. What is the simplest formula of the compound?

28. A 2.103-g sample of a copper oxide, when heated in a stream of hydrogen gas, yields 0.476 g of water. What is the formula of the copper oxide?

29. Washing soda is a hydrate of sodium carbonate. Its formula is $Na_2CO_3 \cdot xH_2O$. A 2.714-g sample of washing soda is heated until a constant mass of 1.006 g of Na_2CO_3 is reached. What is x?

30. A certain hydrate of potassium aluminum sulfate (alum) has the formula $KAl(SO_4)_2 \cdot xH_2O$. When a hydrate sample weighing 5.459 g is heated to remove all the water, 2.583 g of $KAl(SO_4)_2$ remains. What is the mass percent of water in the hydrate? What is x?

Balancing Equations

31. Balance the following equations.

a. $Au_2S_3(s) + H_2(g) \rightarrow Au(s) + H_2S(g)$

b. $C_3H_8(g) + O_2(g) \rightarrow CO_2(g) + H_2O(g)$

c. $SiO_2(s) + C(s) \rightarrow Si(s) + CO(g)$

32. Balance the following equations.

a. $UO_2(s) + HF(l) \rightarrow UF_4(s) + H_2O(l)$

b. $PH_3(g) + O_2(g) \rightarrow P_4O_{10}(s) + H_2O(g)$

c. $C_2H_3Cl(l) + O_2(g) \rightarrow CO_2(g) + H_2O(g) + HCl(g)$

33. Write balanced equations for the reactions of sodium with the following nonmetals to form ionic solids.

a. nitrogen   b. oxygen, forming O^{2-} ions
c. sulfur   d. bromine   e. iodine

34. Write balanced equations for the reaction of chlorine with the following metals to form solids that you can take to be ionic.

a. aluminum   b. strontium   c. lithium
d. chromium   e. silver

35. Write a balanced equation for the

a. reaction of boron trifluoride gas with water to give liquid hydrogen fluoride and solid boric acid (H_3BO_3).

b. reaction of magnesium oxide with iron to form iron(III) oxide and magnesium.

c. the decomposition of dinitrogen oxide gas to its elements.

d. the reaction of calcium carbide (CaC_2) solid with water to form calcium hydroxide and acetylene (C_2H_2) gas.

e. the reaction of solid calcium cyanamide ($CaCN_2$) with water to form calcium carbonate and ammonia gas.

36. Write a balanced equation for the

a. decomposition of ammonium nitrate to dinitrogen oxide gas and steam.

b. reaction of ammonia gas with oxygen to form nitrogen and steam.

c. decomposition of potassium chlorate to potassium chloride and oxygen.

d. reaction of solid diborontrioxide with carbon and chlorine gas to produce two gases, boron trichloride and carbon monoxide.

e. combustion of liquid benzene (C_6H_6) to carbon dioxide and water.

Mole-Mass Relations in Reactions

37. One way to remove nitrogen oxide (NO) from smoke-stack emissions is to react it with ammonia:

$$4NH_3(g) + 6NO(g) \rightarrow 5N_2(g) + 6H_2O(l)$$

Fill in the blanks below.

a. 12.3 mol NO reacts with _____ mol ammonia.

b. 5.87 mol NO yields _____ mol nitrogen.

c. 0.2384 mol nitrogen requires _____ mol nitrogen oxide.

d. 13.9 mol NH_3 produces _____ mol water.

38. The reaction of the mineral fluorapatite with sulfuric acid follows the equation

$$Ca_{10}F_2(PO_4)_6(s) + 7H_2SO_4(l) \rightarrow$$
$$2HF(g) + 3Ca(H_2PO_4)_2(s) + 7CaSO_4(s)$$

Fill in the blanks below.

a. 7.38 mol fluorapatite yields _____ mol calcium sulfate.

b. 3.98 mol calcium dihydrogen phosphate requires _____ mol H_2SO_4.

c. 0.379 mol sulfuric acid reacts with _____ mol fluorapatite.

d. 4.983 mol fluorapatite produces _____ mol $Ca(H_2PO_4)_2$.

39. Using the equation given in Problem 37, calculate

a. the mass of nitrogen produced from 2.93 mol of nitrogen oxide.

b. the mass of ammonia required to form 7.65 mol of water.

c. the mass of nitrogen oxide that yields 0.356 g water.

d. the mass of ammonia required to react with 20.0 g nitrogen oxide.

40. Using the equation given in Problem 38, calculate

a. the mass of calcium sulfate formed from 0.349 mol of fluorapatite.

b. the number of moles of fluorapatite required to form one gram of hydrogen fluoride.

c. the mass of fluorapatite required to yield 3.79 g of calcium sulfate.

d. the mass of hydrogen fluoride formed from 13.98 g of sulfuric acid.

41. Iron ore consists mainly of iron(III) oxide. When iron(III) oxide is heated with an excess of coke (carbon), iron metal and carbon monoxide are produced.

a. Write a balanced equation for the reaction.

b. How many moles of iron(III) oxide are required to form 12.79 mol of iron?

c. How many grams of carbon monoxide are formed from 13.68 g of coke?

42. Nitrogen trichloride gas reacts with water to form ammonia and hypochlorous acid, $HClO(aq)$, the main component of bleach.

a. Write a balanced equation for this reaction.

b. How many moles of ammonia are produced from 275 mL of water ($d = 1.00$ g/cm^3)?

c. How many grams of nitrogen trichloride are required to produce 14.8 g of ammonia?

43. A wine cooler contains about 4.5% ethyl alcohol, C_2H_5OH, by mass. Assume that the alcohol in 2.85 kg of wine cooler is produced by the fermentation of the glucose in grapes in the reaction

$$C_6H_{12}O_6(aq) \rightarrow 2C_2H_5OH(l) + 2CO_2(g)$$

a. How many grams of glucose are needed to produce the ethyl alcohol in the wine?

b. What volume of carbon dioxide gas ($d = 1.80$ g/L) is produced at the same time?

44. When corn is allowed to ferment, ethyl alcohol is produced from the glucose in corn according to the fermentation reaction given in Problem 43.

a. What volume of ethyl alcohol ($d = 0.789$ g/cm^3) is produced from one pound of glucose?

b. Gasohol is a mixture of 10 cm^3 of ethyl alcohol per 90 cm^3 of gasoline. How many grams of glucose are required to produce the ethyl alcohol in one gallon of gasohol?

45. Oxygen masks for producing O_2 in emergency situations contain potassium superoxide, KO_2. It reacts with CO_2 and H_2O in exhaled air to produce oxygen:

$$4KO_2(s) + 2H_2O(g) + 4CO_2(g) \rightarrow$$
$$4KHCO_3(s) + 3O_2(g)$$

If a person wearing such a mask exhales 0.702 g CO_2/min, how many grams of KO_2 are consumed in ten minutes?

46. A crude oil burned in electrical generating plants contains about 1.2% sulfur by mass. When the oil burns, the sulfur forms sulfur dioxide gas:

$$S(s) + O_2(g) \rightarrow SO_2(g)$$

How many liters of SO_2 ($d = 2.60$ g/L) are produced when one metric ton (10^3 kg) of oil burns?

Limiting Reactant; Theoretical Yield

47. A tool set contains 5 wrenches, 4 screwdrivers, and 3 pairs of pliers. The manufacturer has in stock 1000 pairs of pliers, 2000 screwdrivers, and 3000 wrenches. Can an order for 500 tool sets be filled? Explain.

48. A textbook chapter contains 6 figures, 2 tables, and 3 photographs. The publisher has on hand 1200 sets of figures, 250 sets of tables, and 800 photograph sets. How many copies of the chapter can be assembled?

49. Chlorine and fluorine react to form gaseous chlorine trifluoride. You start with 1.75 mol chlorine and 3.68 mol fluorine.
   a. Write a balanced equation for the reaction.
   b. What is the limiting reactant?
   c. What is the theoretical yield of chlorine trifluoride in moles?
   d. How many moles of the excess reactant remain unreacted?

50. A gaseous mixture containing 7.50 mol hydrogen gas and 9.00 mol chlorine gas reacts to form hydrogen chloride gas.
   a. Write a balanced equation for the reaction.
   b. Which reactant is limiting?
   c. If all the limiting reactant is consumed, how many moles of hydrogen chloride are formed?
   d. How many moles of the excess reactant remain unreacted?

51. The space shuttle uses aluminum metal and ammonium perchlorate, NH_4ClO_4, in its reusable booster rockets. The products of the reaction are aluminum oxide, aluminum chloride, NO, and steam. The reaction mixture contains 5.75 g of Al and 7.32 g of NH_4ClO_4.
   a. What is the theoretical yield of aluminum chloride, $AlCl_3$? Note that all of the chlorine comes from NH_4ClO_4.
   b. If 1.87 g of aluminum chloride is formed, what is the percent yield?

52. Oxyacetylene torches used for welding reach temperatures near 2000°C. The reaction involved is the combustion of acetylene, C_2H_2:

$$2\ C_2H_2(g) + 5\ O_2(g) \rightarrow 4\ CO_2(g) + 2\ H_2O(g)$$

Starting with 175 g of both acetylene and oxygen, what is the theoretical yield, in grams, of carbon dioxide? If 52.5 L of carbon dioxide ($d = 1.80$ g/L) are produced, what is the percent yield?

53. In the Ostwald process, nitric acid, HNO_3, is produced from ammonia by a three-step process:

$$4NH_3(g) + 5O_2(g) \rightarrow 4NO(g) + 6H_2O(g)$$

$$2NO(g) + O_2(g) \rightarrow 2NO_2(g)$$

$$3NO_2(g) + H_2O(g) \rightarrow 2HNO_3(aq) + NO(g)$$

Assuming a 75% yield in each step, how many grams of nitric acid can be made from one hundred liters of ammonia ($d = 0.695$ g/L)?

54. A century ago, $NaHCO_3$ was prepared from Na_2SO_4 by a three-step process:

$$Na_2SO_4(s) + 4\ C(s) \rightarrow Na_2S(s) + 4\ CO(g)$$

$$Na_2S(s) + CaCO_3(s) \rightarrow CaS(s) + Na_2CO_3(s)$$

$$Na_2CO_3(s) + H_2O(l) + CO_2(g) \rightarrow 2\ NaHCO_3(s)$$

How many kilograms of $NaHCO_3$ could be formed from one kilogram of Na_2SO_4, assuming an 82% yield in each step?

55. Aspirin, $C_9H_8O_4$, is prepared by reacting salicylic acid, $C_7H_6O_3$, with acetic anhydride, $C_4H_6O_3$, in the reaction

$$2C_7H_6O_3(s) + C_4H_6O_3(l) \rightarrow 2C_9H_8O_4(s) + H_2O(l)$$

A student is told to prepare 25.0 g of aspirin. He is also told that he should use a 50.0% excess of acetic anhydride and expect to get a 65.0% yield in the reaction. How many grams of each reactant should he use?

56. A student prepares phosphorous acid, H_3PO_3, by reacting solid phosphorus triiodide with water:

$$PI_3(s) + 3\ H_2O(l) \rightarrow H_3PO_3(s) + 3\ HI(g)$$

The student needs to obtain 0.100 L of H_3PO_3 ($d = 1.651$ g/cm³). The procedure calls for a 40.0% excess of water and a yield of 55.0%. How much phosphorus triiodide should she weigh out? What volume of water ($d = 1.00$ g/cm³) should she use?

Unclassified

57. Criticize each of the following statements.
   a. In every reaction, one reactant is limiting and the other is in excess.
   b. The compound $Co_4(CO)_{12}$ has the same percent composition as $Co_{12}(CO)_4$.
   c. In an ionic compound, the number of cations is always twice the number of anions.
   d. The molecular formula of calcium chloride is $CaCl_2$.

58. Which of the following statements are always true? Never true? Not always true?
   a. A compound with the molecular formula C_6H_6 has the same simplest formula.
   b. The mass percent of copper in CuO is less than in Cu_2O.
   c. The limiting reactant is the one present in the smallest amount.
   d. Since $C_3H_6O_3$ and $C_6H_{12}O_6$ reduce to the same formula, they represent the same compound.

59. How many moles of ions are there in
   a. 0.700 mol of sodium oxide?
   b. 0.338 mol of strontium hydroxide?
   c. 12.89 g of aluminum sulfate?

60. How many moles of
   a. cations are there in 1.78 mol of ammonium phosphate?

b. anions are there in 0.833 mol of chromium(III) oxide?

c. ions are there in 1.837 g sodium dichromate?

61. Write a balanced equation for the combustion in air of a gaseous hydrocarbon containing 85.56% carbon, which has a molar mass of 42 g/mol. (Hydrocarbons are compounds containing only the elements carbon and hydrogen.)

62. Determine the percent composition of the compound formed when barium is burned in chlorine gas.

63. All the fertilizers listed below contribute nitrogen to the soil. If all these fertilizers are sold for the same price per gram of nitrogen, which will cost the least per 50-lb bag?

urea $(NH_2)_2CO$

ammonia NH_3

ammonium nitrate NH_4NO_3

guanidine $HNC(NH_2)_2$

64. Most wine is prepared by the fermentation of the glucose in grape juice by yeast:

$$C_6H_{12}O_6(aq) \rightarrow 2C_2H_5OH(aq) + 2CO_2(g)$$

How many grams of glucose should there be in grape juice to produce 750.0 mL of wine that is 11.0% ethyl alcohol, C_2H_5OH ($d = 0.789$ g/cm³), by volume?

Challenge Problems

65. Limestone is almost completely made up of calcium carbonate. When calcium carbonate reacts with hydrochloric acid, the following reaction occurs:

$$CaCO_3(s) + 2 H^+(aq) \rightarrow Ca^{2+}(aq) + CO_2(g) + H_2O$$

If a 25.0-g sample of calcium carbonate is added to 75.5 mL HCl ($d = 1.125$ g/cm³, 26.0% HCl by mass), calculate the molarity of the HCl remaining in solution (if any) after the reaction is complete. Assume no volume change.

66. A 5.025-g sample of calcium is burned in air to produce a mixture of two ionic compounds, calcium oxide and calcium nitride. Water is added to this mixture. It reacts with calcium oxide to form 4.832 g of calcium hydroxide. How many grams of calcium oxide are formed? How many grams of calcium nitride?

67. A mixture of potassium chloride and potassium bromide weighing 3.595 g is heated with chlorine, which converts the mixture completely to potassium chloride. The total mass of potassium chloride after the reaction is 3.129 g. What percent of the original mixture is potassium bromide?

68. A sample of an oxide of vanadium weighing 4.589 g was heated with hydrogen gas to form water and another oxide of vanadium weighing 3.782 g. The second oxide was treated further with hydrogen until only 2.573 g of vanadium metal remained.

a. What are the simplest formulas of the two oxides?

b. What is the total mass of water formed in the successive reactions?

69. A sample of cocaine, $C_{17}H_{21}O_4N$, is diluted with sugar, $C_{12}H_{22}O_{11}$. When a 1.00-mg sample of this mixture is burned, 1.00 mL of carbon dioxide ($d = 1.80$ g/L) is formed. What is the percentage of cocaine in the mixture?

$$CaH_2(s) + 2H_2O \rightarrow 2H_2(g) + Ca^{2+}(aq) + 2OH^-(aq)$$

Calcium hydride reacts vigorously with water to yield hydrogen gas and aqueous calcium hydroxide.

CHAPTER 4

GASES

Religious faith is a most filling vapor.
It swirls occluded in us under tight Compression to uplift us out of weight—
As in those buoyant bird bones thin as paper,
To give them still more buoyancy in flight.
Some gas like helium must be innate.

ROBERT FROST
INNATE HELIUM

In the gas phase, all substances show remarkably similar physical behavior. Consider, for example, their molar volumes. At 0°C and 1 atm, 1 mol of every gas occupies almost exactly the same volume, about 22.4 L. This is true of $O_2(g)$, $N_2(g)$, $CH_4(g)$, and any other gas you care to mention. Moreover, the volumes of different gases respond in almost exactly the same way to changes in amount, changes in pressure, or changes in temperature. In particular, we find that it is possible to write a single equation that relates the volume of any gas to amount, temperature, and pressure. This equation is known as the **Ideal Gas Law**; it is central to almost everything we will have to say about gases in this chapter. After a brief review of the measurement of volume, amount, temperature, and pressure (Section 4.1), we will look at the nature and applications of the Ideal Gas Law in Sections 4.2 and 4.3. We will apply this law to gas mixtures in Section 4.4 and consider deviations from it in Section 4.6.

The physical behavior of gases can be explained quite simply in terms of a model of particle motion known as the kinetic theory (Section 4.5). This model emphasizes the fact that gases differ fundamentally from liquids and solids in one important respect. The atoms or molecules in a gas are relatively far removed from one another. We can treat these atoms or molecules as completely independent particles, ignoring interactions between them in the gas state.

In the air you breathe, the molecules are about ten diameters apart

4.1
MEASUREMENTS ON GASES

To specify the state of a gaseous substance completely we must cite the values of four variables: volume, amount, temperature, and pressure. The first three of these quantities have been discussed in some detail earlier in

95

this text; we need only review them briefly here. Pressure is a somewhat more abstract quantity; we will need to consider what gas pressure means, how it may be measured, and the units in which it is expressed.

VOLUME, AMOUNT, AND TEMPERATURE

A gas expands uniformly to fill any container in which it is placed. This means that the volume of a gas is the volume of its container. Volumes of gases can be expressed in liters, cubic centimeters, or cubic meters:

$$1 \text{ L} = 10^3 \text{ cm}^3 = 10^{-3} \text{ m}^3 \tag{4.1}$$

Most commonly, the amount of matter in a gaseous sample is expressed in terms of the number of moles. In some cases, we may quote the mass in grams. These two quantities are simply related:

$$n = \text{g/MM} \tag{4.2}$$

where n is the number of moles, g is the number of grams, and MM stands for molar mass.

We ordinarily measure the temperature of a gas using a thermometer marked in degrees Celsius. However, *in any calculation involving gases, temperatures must be expressed on the Kelvin scale.* To convert between °C and K, we use the relation introduced in Chapter 1:

$$T(\text{K}) = t(°\text{C}) + 273.15 \tag{4.3}$$

Typically, we express temperatures only to the nearest degree. In that case, the Kelvin temperature can be found by simply adding 273 to the Celsius temperature.

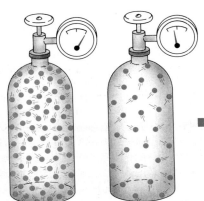

Figure 4.1 When the valve on a cylinder of compressed oxygen is opened, molecules of O_2 escape into the atmosphere. The pressure, as indicated by the gauge on the cylinder, drops; the number of molecules per unit volume inside the cylinder also decreases. In general, for an ideal gas, pressure is directly proportional to molar concentration: $P = \text{constant} \times \dfrac{n}{V}$ (at constant T).

■ **EXAMPLE 4.1**

A balloon with a volume of 2.80×10^4 m³ contains 4.68×10^6 g of helium at 18°C. Express the volume of the balloon in liters, the amount of helium in moles, and the temperature in K.

Solution

$$V = 2.80 \times 10^4 \text{ m}^3 \times \frac{1 \text{ L}}{10^{-3} \text{ m}^3} = \boxed{2.80 \times 10^7 \text{ L}}$$

$$n = 4.68 \times 10^6 \text{ g} \times \frac{1 \text{ mol}}{4.00 \text{ g}} = \boxed{1.17 \times 10^6 \text{ mol}}$$

$$T = 18 + 273 = \boxed{291 \text{ K}}$$

Exercise

Suppose the balloon contained 3.16×10^7 g of hot air at 41°C. What would be the values of n and T in that case? Assume air contains, in mole percent, 78% N_2, 21% O_2, and 1% Ar and hence has a molar mass of 0.78(28.0 g/mol) + 0.21(32.0 g/mol) + 0.01(39.9 g/mol) = 29.0 g/mol. Answer: $n = 1.09 \times 10^6$; $T = 314$ K.

PRESSURE

Pressure is force per unit area. We can interpret gas pressure in terms of the force exerted by the gas on a surface with which it is in contact. Consider, for example, a cylinder of compressed oxygen at a pressure of "1000 pounds per square inch." On one square inch of the inside surface of this cylinder, the oxygen exerts a force equivalent to that of a 1000-lb weight working against gravity.

From a different point of view, we can think of pressure as a direct measure of the concentration of gas particles, i.e., the number of atoms or molecules per unit volume (Fig. 4.1). In a cylinder of compressed oxygen, the concentration of O_2 molecules is relatively high. When the valve on the cylinder is opened, molecules stream out of it. Their concentration drops sharply, and gas pressure falls off accordingly.

The most familiar gas, and the only one known until about 1750, is the air we breathe. Although the facts of atmospheric pressure are really quite simple, they were not clearly understood until about 1640. Evangelista Torricelli, an Italian mathematician and a student of Galileo, was the first person to measure the pressure of the atmosphere accurately. The device he built to do this is still used today. It is called the mercury barometer (Fig. 4.2). This consists of a closed glass tube filled with mercury and inverted over a pool of mercury. When the tube is first inverted, mercury flows into the reservoir, leaving a nearly perfect vacuum above the mercury in the tube. After a few seconds, the mercury reaches a constant level. As shown in Figure 4.2, the pressure exerted by the mercury column exactly balances that of the atmosphere. Hence, the height of the column is a measure of the atmospheric pressure. At or near sea level, it typically varies from 740 to 760 mm, depending upon weather conditions.

Because of the way in which gas pressure is measured, it is often expressed in **millimeters of mercury (mm Hg)***. Thus, we might say that the atmospheric pressure on a certain day is 752 mm Hg. This means that the pressure of the air is equal to that exerted by a column of mercury 752 mm high.

Another unit commonly used to express gas pressure is the standard atmosphere, or simply **atmosphere (atm)**. This is the pressure exerted by a column of mercury 760 mm high with the mercury at 0°C. If we say that a gas has a pressure of 0.98 atm, we mean that the pressure is 98% of that exerted by a mercury column 760 mm high. Other pressure units include

— pounds per square inch. A mass of 1 lb resting on a surface 1 in² in area exerts a pressure of 1 lb/in².

— *kilopascal* (a metric pressure unit). A mass of 10 g resting on a surface 1 cm² in area exerts a pressure of approximately 1 kPa. Atmospheric pressure is ordinarily close to 100 kPa.

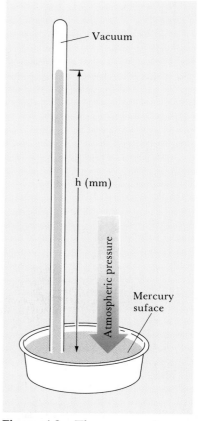

Figure 4.2 The mercury barometer. At the lower mercury surface, the pressure both inside and outside the tube must be that of the atmosphere. Inside the tube, the pressure is exerted by the mercury column h mm high. Hence, the atmospheric pressure must equal h mm Hg.

100 kPa is sometimes called 1 bar

---

*The pressure exerted by a column of mercury one millimeter high under certain specified conditions (0°C at sea level) is defined as 1 *torr*. More commonly, the unit torr, introduced to honor Torricelli, is used as a synonym for millimeter of mercury. Throughout this text, we will use millimeters of mercury rather than torr because the former has a clearer physical meaning.

To convert between these pressure units, we use the relations

$$1 \text{ atm} = 760 \text{ mm Hg} = 14.70 \text{ lb/in}^2 = 101.3 \text{ kPa} \tag{4.4}$$

■ **EXAMPLE 4.2** ───────────────────────────────

An announcer on a radio station in Montreal reports the atmospheric pressure to be 99.6 kPa. What is the pressure in

a. atmospheres   b. millimeters of mercury

Solution

Using the conversion factor approach,

a. $99.6 \text{ kPa} \times \dfrac{1 \text{ atm}}{101.3 \text{ kPa}} = \boxed{0.983 \text{ atm}}$

b. $0.983 \text{ atm} \times \dfrac{760 \text{ mm Hg}}{1 \text{ atm}} = \boxed{747 \text{ mm Hg}}$

Exercise

Express a pressure of 729 mm Hg in atmospheres and in inches of mercury, a unit often used by weather forecasters in the United States. Answer: 0.959 atm, 28.7 in Hg.

4.2
THE IDEAL GAS LAW

As pointed out at the beginning of this chapter, all gases closely resemble each other in one important aspect: the dependence of volume upon amount, temperature, and pressure. In particular,

1. *Volume is directly proportional to amount.* Figure 4.3A shows a typical plot of volume (V) versus number of moles (n) for a gas. Notice that the graph is a straight line passing through the origin. The general equation for such a plot is

$$V = k_1 n \quad \text{(constant } T, P) \tag{4.5}$$

where k_1 is a "constant," that is, it is independent of individual values of V and n and of the nature of the gas. This is the equation of a direct proportionality.

Equation 4.5 can be interpreted in a slightly different way to mean that equal volumes of different gases, measured at the same temperature and pressure, contain equal numbers of moles. This can be regarded as a modern form of *Avogadro's Law*, first proposed in 1811; Avogadro referred to "molecules" rather than moles. We will have more to say, in the historical perspective at the end of this chapter, about Avogadro and other scientists who did pioneering work with gases.

2. *Volume is directly proportional to absolute temperature.* The dependence of volume (V) on the Kelvin temperature (T) is shown in Figure 4.3B. Here

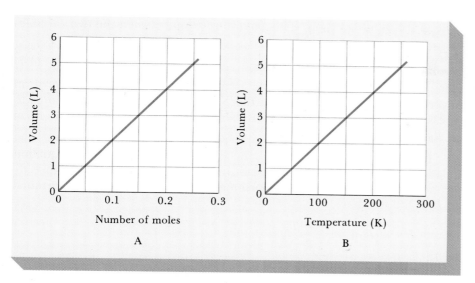

Figure 4.3 At constant pressure, the volume of a gas is directly proportional to the number of moles (A) and to the absolute temperature (B).

again, the graph is a straight line through the origin. The equation of the line is

$$V = k_2T \qquad \text{(constant } n, P)$$ (4.6)

where k_2 is a constant independent of the values of V or T. This relationship was first suggested, in a different form, by two French scientists, Charles and Gay-Lussac. It is often referred to as the *Law of Charles and Gay-Lussac*.

3. *Volume is inversely proportional to pressure.* Figure 4.4 shows a typical plot of volume (V) versus pressure (P). Notice that V decreases as P increases.

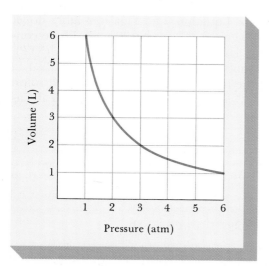

Figure 4.4 At constant temperature, the volume of a gas sample is inversely proportional to pressure. In this case, the volume decreases from 6 L to 1 L when the pressure increases from 1 atm to 6 atm.

The graph is a hyperbola. The general relation between the two variables is

$$V = k_3/P \qquad \text{(constant } n, T)\tag{4.7}$$

The quantity k_3, like k_1 and k_2, is a constant. This is the equation of an inverse proportionality. The fact that volume is inversely proportional to pressure was first established by Robert Boyle. Equation 4.7 is one form of *Boyle's Law*.

This was one of the first physical laws to be discovered

Equations 4.5 through 4.7 can be combined into a single equation relating volume to amount, temperature, and pressure. Since V is directly proportional to n, directly proportional to T, and inversely proportional to P, it follows that

$$V = \text{constant} \times \frac{n \times T}{P}$$

This equation is ordinarily written in a different form. We represent the constant by the symbol R and multiply both sides of the equation by P. This leads to the **Ideal Gas Law**:

This is one of the few relations in science that contains four variables

$$PV = nRT \tag{4.8}$$

where P is the pressure, V the volume, n the number of moles, and T the Kelvin temperature. The quantity R appearing in the Ideal Gas Law is a true constant, independent of P, V, n, or T.

The Ideal Gas Law can be used to solve a wide variety of problems involving the physical behavior of gases. Later in this section, we will see how this is done. First, let us consider the value of R.

EVALUATION OF R, THE GAS CONSTANT

To determine the value of R, we need only establish by experiment one set of values for P, V, n, and T in Equation 4.8. This is readily done. Consider, for example, gaseous oxygen at 0°C and 1.00 atm. These conditions are often referred to as standard temperature and pressure (STP). We find that at STP, 32.0 g (1.00 mol) of O_2 occupies a volume of 22.4 L. The same is true of other gases; the molar volumes of N_2, H_2, and so on are found to be about 22.4 L at 0°C and 1.00 atm. Solving the Ideal Gas Law for R,

$$R = \frac{PV}{nT}$$

Substituting $P = 1.00$ atm, $V = 22.4$ L, $n = 1.00$ mol, and $T = 0 + 273 = 273$ K,

$$R = \frac{1.00 \text{ atm} \times 22.4 \text{ L}}{1.00 \text{ mol} \times 273 \text{ K}} = 0.0821 \text{ L·atm/(mol·K)}$$

The value of R obtained under the most precise conditions, using oxygen at low pressures, is 0.082056 L·atm/(mol·K). Note that R involves the units

TABLE 4.1   Values of R in Different Units

VALUE	WHERE USED	HOW OBTAINED
$0.0821 \dfrac{\text{L·atm}}{\text{mol·K}}$	Gas Law problems with V in liters, P in atm	From known values of P, V, T, n
$8.31 \dfrac{\text{L·kPa}}{\text{mol·K}}$	Gas Law problems with V in liters, P in kPa	$1\ \text{atm} = 101.3\ \text{kPa}$
$8.31 \dfrac{\text{J}}{\text{mol·K}}$	Equations involving energy in joules	$1\ \text{L·atm} = 101.3\ \text{J}$
$8.31 \dfrac{\text{kg·m}^2}{\text{s}^2\text{·mol·K}}$	Calculation of molecular velocity (Eqn. 4.26)	$1\ \text{J} = 1\ \text{kg·m}^2/\text{s}^2$

of atmospheres, liters, moles, and K. These units must be used for pressure, volume, amount, and temperature in any problem where this value of R is employed.

In most of our work in this chapter, we will use 0.0821 L·atm/(mol·K) as the value of R. For certain purposes, however, we will need R in other units. Table 4.1 lists values of R in various sets of units.

FINAL AND INITIAL STATE PROBLEMS

In a common type of problem, a gas undergoes a change from an "initial" to a "final" state. In this process, two or more of the four variables (V, n, T, P) change. You are asked to determine the effect of this change upon a particular variable, perhaps the volume V. For example, starting with a sample of gas with a volume of 1.00 L at STP, you might be asked to determine the final volume when the pressure is increased to 2.50 atm, holding temperature and amount constant.

The Ideal Gas Law is readily applied to problems of this type. What you do is to use the law to derive a relationship between the variables involved, in this case pressure and volume. Since n and T remain constant, we can write:

Initial state: $P_1V_1 = nRT$

Final state: $P_2V_2 = nRT$

Since P_2V_2 and P_1V_1 are both equal to nRT, they must be equal to each other. That is:

$$P_2V_2 = P_1V_1 \qquad \text{(constant } n, T\text{)} \tag{4.9}$$

This is the most common form for Boyle's Law

Applying this general relation to the problem cited above, we have

$$V_2 = V_1 \times \frac{P_1}{P_2} = 1.00\ \text{L} \times \frac{1.00\ \text{atm}}{2.50\ \text{atm}} = 0.400\ \text{L}$$

In a very similar way, we can use the Ideal Gas Law to obtain a "two-point" relation between volume and temperature when pressure and amount are held constant. The Law tells us that $V/T = nR/P$, so:

Initial state: $\dfrac{V_1}{T_1} = nR/P$

Final state: $\dfrac{V_2}{T_2} = nR/P$

Hence:

This is the common form for Charles' Law

$$\frac{V_2}{T_2} = \frac{V_1}{T_1} \qquad \text{(constant } n, P) \tag{4.10}$$

Using this general relation, we can find the final volume when a 1.00-L gas sample, originally at STP, is heated to 25°C at constant pressure:

$$V_2 = V_1 \times \frac{T_2}{T_1} = 1.00 \text{ L} \times \frac{298 \text{ K}}{273 \text{ K}} = 1.09 \text{ L}$$

Relations 4.9 and 4.10, which we derived from the Ideal Gas Law, are ones that you may well have seen before. They are forms of Boyle's Law and Charles' Law, respectively, equivalent to Equations 4.7 and 4.6. The approach we used is a general one that can be applied to obtain relations between any two or three of the four variables that govern the physical behavior of gases. It is particularly useful for dealing with "final and initial state" problems that do not fit into a familiar pattern (Example 4.3).

■ **EXAMPLE 4.3**

A 250-mL flask, open to the atmosphere, contains 0.0110 mol of air at 0°C (Fig. 4.5). How much air remains in the flask after heating to 100°C?

Solution

It should be clear from Figure 4.5 that in this problem, *the volume and pressure remain constant.* The volume is 250 mL throughout; the pressure is that of the atmosphere. What we need then is a relation between the number of moles n and the temperature T, holding P and V constant.

You need to think about this for a bit

Figure 4.5 At constant P and V, the number of moles of gas, n, is inversely proportional to the temperature, T. When an open flask, originally at 0°C in an ice-water bath, is transferred to a boiling water bath at 100°C, about 27% of the air is expelled.

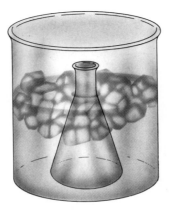

Initial state: $n_1 T_1 = PV/R$

Final state: $n_2 T_2 = PV/R$

It follows that:

$$n_2 T_2 = n_1 T_1; \qquad n_2 = n_1 \times \frac{T_1}{T_2}$$

Since $n_1 = 0.0110$ mol; $T_1 = 0 + 273 = 273$ K; and $T_2 = 100 + 273 = 373$ K,

$$n_2 = 0.0110 \text{ mol} \times \frac{273 \text{ K}}{373 \text{ K}} = \boxed{0.00805 \text{ mol}}$$

We conclude that 0.00805/0.0110, or about 73%, the air remains in the flask at 100°C. About 27% of the air escapes when the temperature is raised.

Exercise
At what temperature is half of the air driven out of the flask? Answer: 273°C.

CALCULATION OF *P, V, n,* OR *T*

In Example 4.3 the value of the gas constant R was not needed, since R was canceled from the calculations. In a different type of problem, we use R to calculate one of the four quantities, P, V, n, or T, when we know the values of the other three. Example 4.4 illustrates the kind of calculation involved. Later in this chapter and in succeeding chapters we will deal with other examples of this type.

■ **EXAMPLE 4.4** _____
If 3.00 g SF_6 gas is introduced into an evacuated 5.00-L container at 92°C, what is the pressure in atmospheres in the container?

Solution
In this problem, only one state is involved and, for it, $PV = nRT$:

$P = ?;$ $V = 5.00$ L; $T = 273 + 92 = 365$ K

$R = 0.0821$ L·atm/(mol·K)

$$n = 3.00 \text{ g } SF_6 \times \frac{1 \text{ mol}}{146.1 \text{ g } SF_6} = 0.0205 \text{ mol}$$

Substituting,

$$P = \frac{nRT}{V} = \frac{0.0205 \text{ mol} \times 0.0821 \frac{\text{L·atm}}{\text{mol·K}} \times 365 \text{ K}}{5.00 \text{ L}} = \boxed{0.123 \text{ atm}}$$

Here all the quantities in the Ideal Gas Law enter the calculation directly. If we use 0.0821 L·atm/(mol·K) for R, the units for P, V, n, and T must be those that appear in R. Any quantity that is not given in these units must be converted to them before substituting in the Ideal Gas Law. Here, for example, we converted grams of SF_6 to moles by dividing by the molar mass, 146.1 g/mol.

Exercise
What is the pressure exerted by 2.0 mol O_2 in a 10.0-L flask at 27°C?
Answer: 4.9 atm.

CALCULATION OF MOLAR MASS AND DENSITY

For certain problems, it is convenient to rewrite the Ideal Gas Law in a slightly different form, expressing amount in grams rather than moles. To do this, we use Equation 4.2:

$$n = g/MM$$

where g is the mass of the gas in grams and MM is the molar mass in g/mol. Substituting for n in the Ideal Gas Law, we have

$$PV = \frac{gRT}{MM} \tag{4.11}$$

The Ideal Gas Law in this form is useful for calculating the following:

1. The molar mass (MM) of a gas, knowing the mass (g) of a given volume (V) at a certain temperature (T) and pressure (P). Solving Equation 4.11 for MM,

$$MM = \frac{gRT}{PV} \tag{4.12}$$

2. The density (d) of a gas of known molar mass (MM) at a given temperature (T) and pressure (P). From Equation 4.11, on solving for the density, we obtain,

$$d = \frac{g}{V} = \frac{P \times MM}{RT} \tag{4.13}$$

■ **EXAMPLE 4.5**

A sample of xenon tetrafluoride is collected in a flask with a volume of 226 mL at a pressure of 749 mm Hg and a temperature of 12°C. The mass of the gas is found to be 1.973 g. Calculate the molar mass of xenon tetrafluoride.

Solution
We can calculate the molar mass directly, using Equation 4.12:

$$g = 1.973\ g; \quad R = 0.0821\ L·atm/(mol·K); \quad T = 12 + 273 = 285\ K$$

$$P = 749 \text{ mm Hg} \times \frac{1 \text{ atm}}{760 \text{ mm Hg}} = 0.986 \text{ atm}$$

$$V = 226 \text{ mL} \times \frac{1 \text{ L}}{1000 \text{ mL}} = 0.226 \text{ L}$$

$$MM = \frac{1.973 \text{ g} \times 0.0821 \dfrac{\text{L·atm}}{\text{mol·K}} \times 285 \text{ K}}{0.986 \text{ atm} \times 0.226 \text{ L}} = \boxed{207 \text{ g/mol}}$$

It's rather remarkable that we can get information about the molecules in a gas from density data

The calculated value compares closely with the theoretical value for XeF_4, 207.3 g/mol.

Exercise
What is the molar mass of a gas whose density is 5.00 g/L at 25°C and 1.00 atm? Answer: 122 g/mol.

■ **EXAMPLE 4.6**
Taking the molar mass of dry air to be 29.0 g/mol, calculate the density of air at 27°C and 1.00 atm.

Solution
We use Equation 4.13 to calculate the density:

$$d = \frac{P \times MM}{RT} = \frac{1.00 \text{ atm} \times 29.0 \text{ g/mol}}{0.0821 \dfrac{\text{L·atm}}{\text{mol·K}} \times 300 \text{ K}} = \boxed{1.18 \text{ g/L}}$$

Exercise
What is the density of $O_2(g)$ at 1.00 atm and 27°C? Answer: 1.30 g/L.

Equation 4.13 tells us that gas density is directly proportional to molar mass and inversely proportional to temperature. Balloons, which must contain a gas lighter than the surrounding air, are designed to take advantage of one or the other of these effects. "Hot air" balloons are filled with air at a temperature higher than that of the atmosphere, making the air inside less dense than that outside. First used in France in the 18th century, they are now seen in balloon races and other sporting events. Heat is supplied on demand using a propane burner.

The other type of balloon uses a gas with a molar mass less than that of air. Hydrogen (MM = 2.02 g/mol) has the greatest lifting power, because it has the lowest density of all gases. However, it has not been used in manned balloons since 1937, when the Hindenberg, a hydrogen-filled airship, exploded and burned. Helium (MM = 4.00 g/mol) is slightly less effective than hydrogen but a lot safer to work with, since it is nonflammable. It is used in a variety of balloons, ranging from the small ones used at parties to meteorological balloons with volumes of a billion liters.

There are no hydrogen gas balloons these days

4.3
VOLUMES OF GASES INVOLVED IN REACTIONS

We saw in Chapter 3 that a balanced equation can be used to relate moles or masses of substances taking part in a reaction. Where gases are involved, we can extend these relations to include volumes. To do this, we use the Ideal Gas Law and the conversion factor approach described in Chapter 3. The calculations required are illustrated in Examples 4.7 and 4.8.

■ EXAMPLE 4.7 _____

What mass in grams of hydrogen peroxide must be used to produce 1.00 L $O_2(g)$, measured at 25°C and 1.00 atm? The reaction is

$$2 \ H_2O_2(aq) \rightarrow O_2(g) + 2 \ H_2O$$

Solution

We follow a three-step procedure:

CONVERSION	REQUIRED RELATION
1. volume $O_2 \rightarrow$ moles O_2	$n = PV/RT$
2. moles $O_2 \rightarrow$ moles H_2O_2	2 mol $H_2O_2 \simeq$ 1 mol O_2
3. moles $H_2O_2 \rightarrow$ grams H_2O_2	1 mol H_2O_2 = 34.0 g H_2O_2

Substituting into the Ideal Gas Law,

$$n \ O_2 = \frac{1.00 \ \text{atm} \times 1.00 \ \text{L}}{0.0821 \ \text{L·atm/(mol·K)} \times 298 \ \text{K}} = 0.0409 \ \text{mol} \ O_2$$

The last two steps can be carried out in a single-line calculation (Chapter 3):

$$\text{grams } H_2O_2 = 0.0409 \ \text{mol } O_2 \times \frac{2 \ \text{mol } H_2O_2}{1 \ \text{mol } O_2} \times \frac{34.0 \ \text{g } H_2O_2}{1 \ \text{mol } H_2O_2}$$

$$= \quad 2.78 \ \text{g } H_2O_2$$

We conclude that it takes somewhat less than 3 g of hydrogen peroxide to form 1 L of oxygen gas at room temperature and atmospheric pressure.

Exercise

What mass of H_2O is formed as a by-product in the reaction described in this example? Answer: 1.47 g H_2O.

_____

■ EXAMPLE 4.8 _____

How many liters of oxygen, measured at 740 mm Hg and 24°C, are required to burn 1.00 g of octane, $C_8H_{18}(l)$, to carbon dioxide and water?

Solution

We first write the balanced equation for the reaction, using the approach described in Chapter 3:

$$2 \ C_8H_{18}(l) \ + \ 25 \ O_2(g) \ \rightarrow \ 16 \ CO_2(g) \ + \ 18 \ H_2O(l)$$

Now we again follow a three-step approach:
1. Convert 1.00 g of octane to moles (1 mol C_8H_{18} = 114 g C_8H_{18}).
2. Convert moles of octane to moles of oxygen (2 mol $C_8H_{18} \simeq$ 25 mol O_2).
3. Convert moles of oxygen to volume, using the Ideal Gas Law.

Sometimes it seems that general chemistry is just one conversion factor problem after another

Combining the first two steps,

$$\text{no. moles } O_2 = 1.00 \text{ g } C_8H_{18} \times \frac{1 \text{ mol } C_8H_{18}}{114 \text{ g } C_8H_{18}} \times \frac{25 \text{ mol } O_2}{2 \text{ mol } C_8H_{18}}$$
$$= 0.110 \text{ mol } O_2$$

We find the volume of oxygen by substituting into the Ideal Gas Law:

$$V = \frac{nRT}{P} = \frac{0.110 \text{ mol} \times 0.0821 \text{ L·atm/(mol·K)} \times 297 \text{ K}}{(740/760) \text{ atm}}$$
$$= \boxed{2.75 \text{ L}}$$

This reaction is very similar to that occurring in an automobile engine. Clearly, a large volume of air (which is only 21% O_2 by volume) must pass through the engine during combustion.

In going a mile a typical engine will bring in about 1000 liters of air

Exercise

A typical cylinder in an automobile engine has a volume of 500 mL. Suppose the cylinder is filled with air (21% O_2 by volume) at 50°C and 1.00 atm. What mass of octane must be injected to react with the oxygen in the cylinder? Answer: 0.0361 g.

The two examples just worked illustrate what is perhaps the most important application of the gas laws to chemistry. Here your objective is to relate the volume of a gas A to the mass of another species, B, involved in a reaction with A. The general path that you follow is shown in the flow diagram in Figure 4.6.

LAW OF COMBINING VOLUMES

When two different gases are involved in a reaction, there is a simple relationship between their volumes, measured at the same temperature and

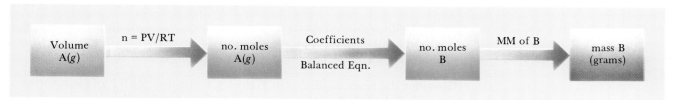

Figure 4.6 To convert between volume of a gaseous species, A, and mass of another species, B, involved in a reaction, follow a three-step path.

Figure 4.7   When water is electrolyzed, the volume of hydrogen gas formed in the tube at the left is twice that of oxygen (right tube), in accordance with the equation: $2H_2O(l) \rightarrow 2H_2(g) + O_2(g)$. (Charles D. Winters)

pressure. To find this relationship, consider two gases, 1 and 2, both at T and P. Applying the Ideal Gas Law to both gases,

$$V_2 = n_2RT/P$$
$$V_1 = n_1RT/P$$

Dividing the first equation by the second, RT/P cancels and we obtain

$$V_2/V_1 = n_2/n_1 \qquad (4.14)$$

This tells us that the volume ratio in which two gases react at T and P is the same as the reacting mole ratio.

To see what Equation 4.14 implies, consider the reaction

$$N_2(g) + 3\ H_2(g) \rightarrow 2\ NH_3(g) \qquad (4.15)$$

We found in Chapter 3 that the coefficients of a balanced equation can be interpreted in terms of moles. In this case, 1 mol N_2 reacts with 3 mol H_2 to form 2 mol NH_3:

$$1\ \text{mol}\ N_2 \simeq 3\ \text{mol}\ H_2 \simeq 2\ \text{mol}\ NH_3$$

However, since the reacting volume ratio, at a given temperature and pressure, is the same as the mole ratio, 1 L N_2 must react with 3 L H_2 to form 2 L NH_3:

$$1\ \text{L}\ N_2 \simeq 3\ \text{L}\ H_2 \simeq 2\ \text{L}\ NH_3 \qquad \text{(same}\ T, P)$$

We see that the reacting volume ratio is given by the coefficients of the balanced equation. Another example of this relationship is shown in Figure 4.7. In general, we can say that:

The volumes of different gases involved in a reaction, if measured at the same temperature and pressure, are in the same ratio as the coefficients in the balanced equation.

This relation, known as the Law of Combining Volumes, was first proposed in a somewhat different form by Gay-Lussac in 1808.

■ **EXAMPLE 4.9**
In Reaction 4.15, what volume of H_2 at 22°C and 719 mm Hg is required to form 12.0 L NH_3, measured at the same temperature and pressure?

Solution
From the coefficients of the balanced equation,

$$3\ \text{L}\ H_2 \simeq 2\ \text{L}\ NH_3$$

$$\text{volume}\ H_2 = 12.0\ \text{L}\ NH_3 \times \frac{3\ \text{L}\ H_2}{2\ \text{L}\ NH_3} = \boxed{18.0\ \text{L}\ H_2}$$

Note that it makes no difference what the temperature and pressure are, so long as they are the same for both gases.

Exercise
Consider the reaction shown in Figure 4.7. Suppose 24.5 L of H_2 is formed at 25°C and 1.00 atm. What volume of O_2 is produced at the same time? What volume of water is consumed? Answer: 12.2 L of $O_2(g)$; approximately 18.0 mL of $H_2O(l)$.

4.4
GAS MIXTURES; PARTIAL PRESSURES, MOLE FRACTIONS

So far we have concentrated upon the behavior of pure gases. Frequently, however, we deal with gaseous mixtures where more than one substance is present. Here, a relation discovered by John Dalton in 1801 is very helpful. Dalton's Law states the following:

The total pressure of a gas mixture is the sum of the partial pressures of the components of the mixture.

For a mixture of two gases A and B,

$$P_{tot} = P_A + P_B \qquad (4.16)$$

Here, P_A is the partial pressure of gas A; P_B is the partial pressure of gas B. The partial pressure of a gas is the pressure it would exert if it were alone in the container at the same temperature as the mixture.

Dalton's Law is perhaps applied most often in calculations involving gases collected over water (Fig. 4.8). Here, the gas being collected is mixed with water vapor; molecules of H_2O escape from the liquid and are carried along with the gas. The only pressure that can be measured directly is the total pressure of the mixture. The partial pressure of the water vapor is readily obtained. It has a fixed value at a given temperature, shown in a table of water vapor pressures (Appendix 1). By subtracting the partial pressure of $H_2O(g)$ from the total pressure of the mixture, we obtain the partial pressure of the gas under study. This in turn can be used to determine the number of moles of that gas present (Example 4.10).

Dalton had quite a few good ideas, and some that weren't so good

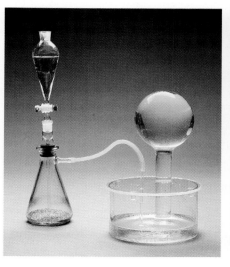
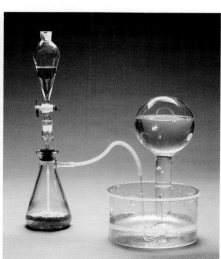

Figure 4.8 When a gas is collected by displacing water, it becomes saturated with water vapor; the partial pressure of $H_2O(g)$ is equal to the vapor pressure of liquid water at the temperature of the system. (Marna G. Clarke)

■ **EXAMPLE 4.10** ───────────────────────────

A student prepares a sample of hydrogen gas by electrolyzing water at 25°C. She collects 152 mL H_2 at a total pressure of 758 mm Hg. Using Appendix 1 to find the vapor pressure of water, calculate

a. the partial pressure of hydrogen gas.
b. the number of moles of hydrogen collected.

Solution

a. The collected gas is a mixture of hydrogen and water vapor. Using Dalton's Law,

$$P_{tot} = P_{H_2} + P_{H_2O} \qquad P_{H_2} = P_{tot} - P_{H_2O}$$

The total pressure is given as 758 mm Hg; from Appendix 1 we find that the vapor pressure of water at 25°C is 23.76 mm Hg. Hence,

$$P_{H_2} = 758 \text{ mm Hg} - 23.76 \text{ mm Hg} = \boxed{734 \text{ mm Hg}}$$

b. We use the Ideal Gas Law, where P is the partial pressure of hydrogen:

$$n = \frac{PV}{RT} = \frac{(734/760 \text{ atm})(0.152 \text{ L})}{\left(0.0821 \dfrac{\text{L·atm}}{\text{mol·K}} \right) (298 \text{ K})} = \boxed{0.00600 \text{ mol}}$$

Exercise
Hydrogen gas is collected over water at a total pressure of 744 mm Hg at 20°C. Using the table of water vapor pressures in Appendix 1, calculate the partial pressure of the $H_2(g)$. Answer: 726 mm Hg.

The validity of Dalton's Law is readily shown using the Ideal Gas Law. For a mixture of two gases A and B, we can write

$$P_{tot} = n_{tot} \times \frac{RT}{V} = (n_A + n_B) \times \frac{RT}{V}$$

Separating terms on the right side of this equation,

$$P_{tot} = n_A \frac{RT}{V} + n_B \frac{RT}{V}$$

Applying the Ideal Gas Law again, this time to obtain the partial pressures of the individual gases,

$$P_A = n_A \frac{RT}{V} ; \qquad P_B = n_B \frac{RT}{V}$$

Hence, $P_{tot} = P_A + P_B$, which is Equation 4.16.

Does a molecule of A exert the same pressure as a molecule of B? Answer: Yes!

PARTIAL PRESSURE AND MOLE FRACTION

As we have just seen, for a gas mixture,

$$P_A = n_A \frac{RT}{V} \qquad \text{and} \qquad P_{tot} = n_{tot} \frac{RT}{V}$$

If we divide P_A by P_{tot}, we obtain

$$\frac{P_A}{P_{tot}} = \frac{n_A}{n_{tot}} \qquad \text{or} \qquad P_A = \frac{n_A}{n_{tot}} \times P_{tot} \qquad\qquad (4.17)$$

The fraction n_A/n_{tot} is referred to as the **mole fraction** of A in the mixture. It is the fraction of the total number of moles that is accounted for by gas A. Using X to represent the mole fraction,

$$X_A = \frac{n_A}{n_{tot}} \qquad\qquad (4.18)$$

Substituting for n_A/n_{tot} in Equation 4.17, we have

$$P_A = X_A P_{tot} \qquad\qquad (4.19)$$

In other words, *the partial pressure of a gas in a mixture is equal to its mole fraction multiplied by the total pressure.*

If half of the molecules in a gas are O_2, the partial pressure of O_2 will equal half the total pressure

■ **EXAMPLE 4.11** _____

Calculate the mole fractions of H_2 and H_2O in the gas mixture referred to in Example 4.10.

Solution

In Example 4.10a, we found that the partial pressure of H_2 was 734 mm Hg; the total pressure is 758 mm Hg. Solving Equation 4.19 for mole fractions, we obtain

$$X_{H_2} = \frac{P_{H_2}}{P_{tot}} = \frac{734 \text{ mm Hg}}{758 \text{ mm Hg}} = \boxed{0.968}$$

The mole fraction of water is found similarly, taking its partial pressure to be 23.76 mm Hg:

$$X_{H_2O} = \frac{P_{H_2O}}{P_{tot}} = \frac{23.76 \text{ mm Hg}}{758 \text{ mm Hg}} = \boxed{0.0313}$$

Exercise

The mole fraction of nitrogen in air is about 0.78. What is the partial pressure of N_2 in air at a total pressure of 740 mm Hg? Answer: 5.8×10^2 mm Hg.

4.5
KINETIC THEORY OF GASES

The fact that the Ideal Gas Law applies to all gases indicates that the gaseous state is a relatively simple one to treat from a theoretical point of view. Gases must have certain properties in common that cause them to follow the same natural law. Between about 1850 and 1880, Maxwell, Boltzmann, Clausius, and others developed the kinetic theory of gases to explain these similarities in the behavior of gases. They based it on the idea that all gases behave similarly insofar as particle motion is concerned. Since that time, the kinetic theory has had to be modified only slightly. In its present form it is one of the most successful of scientific theories. It ranks in stature with the atomic theory of matter.

POSTULATES OF THE KINETIC THEORY

1. Gases consist of particles (atoms or molecules) in continuous, random motion. These particles undergo frequent collisions with one another and with the container walls. The pressure exerted by a gas is due to the forces associated with wall collisions (Fig. 4.9).

2. Collisions between gas particles are elastic. When two particles collide, their individual energies may change. Typically, one speeds up and the other slows down. The total energy, however, remains the same. No kinetic energy is converted to heat. As a result, the temperature of a gas insulated from its surroundings does not change.

3. The volume occupied by the particles is negligibly small, compared to that of the vessel in which they are confined. Typically, at room temperature and atmospheric pressure, about 99.8% of the total volume of a gas is "empty space." The volume of the particles becomes more significant at high pressures and low temperatures, but this effect is ordinarily ignored in derivations based on the kinetic theory.

4. Attractive forces between particles have a negligible effect on their behavior. The atoms or molecules in a gas are treated as independent particles, ignoring interactions between neighboring particles. This assumption is justified at ordinary pressures and temperatures, where the particles are far removed from one another on the average.

5. The average energy of translational motion of a gas particle is directly proportional to temperature. At a given temperature, this average energy is the same for all gases. The energy associated with the motion of a particle from one place to another is related to its speed by the equation

$$E_t = \frac{mu^2}{2}$$

Here, E_t is the average kinetic energy of translation, m is the mass of the particle, and u is the corresponding velocity, which we will call the average speed. Kinetic theory tells us that E_t, and hence $mu^2/2$, are directly proportional to temperature:

$$E_t = \frac{mu^2}{2} = cT \tag{4.20}$$

In this equation, T is the absolute temperature in K. The quantity c is a constant that has the same value for all gases.

　　The postulates of the kinetic theory can be used to derive many of the properties of gases, including the Ideal Gas Law. Such derivations are generally not simple, reflecting the fact that the motion of atoms or molecules in gases is relatively complex. Their velocities are constantly changing in both magnitude and direction as they collide with each other and with the walls of their container. To treat this kind of motion in a rigorous way

Figure 4.9 The pressure exerted by a gas on its container is the same in all directions and is caused by collisions of gas molecules with the container walls.

The energy of a rocket is mostly translational; that of a spinning top is rotational requires mathematics of a high level of sophistication. We will not attempt to present such a mathematical development. Instead, we will concentrate upon some of the simpler applications of kinetic theory to the experimental behavior of gases.

GAS PRESSURE

As noted in Section 4.2, the pressure of a gas (P) is dependent upon container volume (V), number of moles (n), and temperature (T). We can readily understand the effect of these factors in terms of kinetic theory, which tells us that gas pressure is due to collisions with the walls of the container.

1. P increases as n increases (constant V, T) because, with more particles present, there are more collisions per unit time.
2. P increases as V decreases (constant n, T) because, in a smaller volume, gas particles strike the walls more often.
3. P increases as T increases (constant n, V) because, at a higher temperature, particles move more rapidly and collisions occur more often and with greater force.

Putting these effects together we conclude that, at constant temperature, gas pressure should increase directly with concentration, n/V. This effect, referred to earlier in this chapter, is an obvious consequence of the Ideal Gas Law, which tells us that $P = nRT/V$. Among other things, it accounts for the decrease of atmospheric pressure with altitude. This pressure is proportional to the mass of the column of air above a given point. Gravitational effects cause air to become "thinner" (fewer molecules per unit volume) as height increases. Pressure drops accordingly, from 1 atm at sea level to 0.1 atm at an altitude of 16 km, to 0.01 atm at 32 km, and so on.

At 15,000 ft the air pressure is about ¾ atm. Until you are acclimated, breathing at that altitude is difficult

DIFFUSION AND EFFUSION OF GASES; GRAHAM'S LAW

Kinetic theory can be applied in a straightforward way to explain relative rates of diffusion and effusion of different gases. *Diffusion* refers to the movement of gas particles through space, from a region of high concentration to one of low concentration. If your instructor momentarily opens a cylinder of chlorine gas at the lecture table, you will soon recognize the sharp odor of Cl_2, particularly if you have a front-row seat in the classroom. On a more pleasant note, the odor associated with a freshly baked apple pie also reaches you via gaseous diffusion.

Diffusion is a relatively slow process; a sample of gas introduced at one location may take an hour or more to distribute itself uniformly throughout a room. At first glance, this seems surprising, since gas particles are moving very rapidly. One can calculate from kinetic theory that at room temperature, O_2 molecules on the average have a velocity in the range 400–500 m/s, corresponding to about 1000 miles per hour. However, molecules are constantly colliding with one another; at 25°C and 1 atm, an O_2

A diffusing molecule has roughly the same problem as a commuter trying to get off a Tokyo subway at rush hour

Figure 4.10 Bromine diffuses much more slowly in liquid water (*left*) than in air (*right*) because the molecules are much closer together in the liquid state.

molecule undergoes more than a billion collisions per second with its neighbors. This slows down the net movement of gas molecules in any given direction. Even so, diffusion occurs more rapidly in gases than in liquids, where the molecules are much closer together (Fig. 4.10). Bromine molecules diffuse through air perhaps 10–100 times as fast as they do through water.

As we have implied, diffusion is a rather complex process so far as molecular motion is concerned. *Effusion*, the flow of gas particles through tiny pores or pinholes, is easier to analyze mathematically. The relative rates of effusion of different gases through a small opening depend upon two factors: the concentrations and relative velocities of their particles. If we compare two different gases A and B at the same pressure, their concentrations are equal and we need only be concerned with velocity. Thus, we can say that:

$$\frac{\text{rate of effusion of A}}{\text{rate of effusion of B}} = \frac{\text{average speed of A}}{\text{average speed of B}} = \frac{u_A}{u_B} \tag{4.21}$$

By Equation 4.20, if the two gases are at the same temperature,

$$\frac{m_A u_A^2}{2} = \frac{m_B u_B^2}{2} = cT$$

or

$$\frac{u_A^2}{u_B^2} = \frac{m_B}{m_A} = \frac{MM_B}{MM_A} \tag{4.22}$$

where MM_A and MM_B are the molar masses of A and B. If Equation 4.22 is solved for the ratio u_A/u_B, which is then substituted into 4.21, we obtain

$$\frac{\text{rate of effusion of A}}{\text{rate of effusion of B}} = \left(\frac{MM_B}{MM_A}\right)^{1/2} \tag{4.23}$$

Among other things, Graham discovered dialysis

This relation in a somewhat different form was discovered experimentally by the Scottish chemist Thomas Graham in 1829. Graham was interested in a wide variety of chemical and physical problems, among them the separation of the components of air. Graham's Law can be stated as:

At a given temperature and pressure, the rate of effusion of a gas is inversely proportional to the square root of its molar mass.

Somewhat by coincidence, this law applies to gaseous diffusion as well as effusion.

Graham's Law gives us a way of determining molar masses of gases. All we need do is to compare the rate of effusion (diffusion) of the gas in question to that of another gas of known molar mass. Usually, we measure either the distances moved by the two gases in the same time (Fig. 4.11) or the time required for equal amounts of the two gases to effuse at the same temperature and pressure. Time is inversely related to rate:

$$\text{rate} = \frac{\text{distance}}{\text{time}}$$

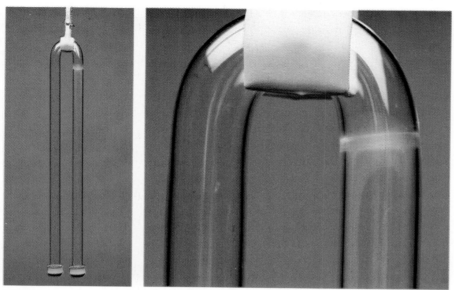

Figure 4.11   When ammonia gas, injected into the left arm of the U-tube, comes in contact with hydrogen chloride (*right-arm*), they react to form a white deposit of ammonium chloride: $NH_3(g) + HCl(g) \rightarrow NH_4Cl(s)$. Since NH_3 (MM = 17 g/mol) diffuses faster than HCl (MM = 36.5 g/mol), the deposit forms closer to the HCl end of the tube. (Marna G. Clarke)

The faster the gas effuses, the less time is required for a given amount to effuse. Hence:

$$\frac{\textbf{time}_B}{\textbf{time}_A} = \frac{\textbf{rate}_A}{\textbf{rate}_B} = \left(\frac{\textbf{MM}_B}{\textbf{MM}_A}\right)^{1/2} \tag{4.24}$$

In other words, the time required for effusion increases with molar mass: heavy molecules take longer to effuse.

■ **EXAMPLE 4.12** _____

In an effusion experiment, it required 45 s for a certain number of moles of an unknown gas, X, to pass through a small opening into a vacuum. Under the same conditions, it took 28 s for the same number of moles of Ar to effuse. Find the molar mass of the unknown gas.

Solution

We use the relation between time of effusion and molar mass, Equation 4.24:

$$\frac{\text{time for Ar}}{\text{time for X}} = \frac{28}{45} = \left(\frac{\text{MM of Ar}}{\text{MM of X}}\right)^{1/2}$$

To solve this equation, we square both sides:

This method is not as accurate as the
gas density method

$$\frac{\text{MM of Ar}}{\text{MM of X}} = \left(\frac{28}{45}\right)^2 = 0.39$$

Solving for the molar mass of the unknown gas,

$$\text{MM of X} = \frac{\text{MM of Ar}}{0.39} = \frac{39.9 \text{ g/mol}}{0.39} = \boxed{1.0 \times 10^2 \text{ g/mol}}$$

Exercise
A certain gas takes only one fourth as long to effuse as O_2. What is its molar mass? Answer: 2.0 g/mol.

A practical application of Graham's Law arose during World War II, when scientists were studying the fission of uranium atoms as a source of energy in the atomic bomb. It became necessary to separate $^{235}_{92}\text{U}$, which is fissionable, from the more abundant isotope of uranium, $^{238}_{92}\text{U}$. Since the two isotopes have almost identical chemical properties, chemical separation was not feasible. Instead, an effusion process was worked out using uranium hexafluoride, UF_6. This compound is a gas at room temperature. Preliminary experiments indicated that $^{235}_{92}\text{UF}_6$ could indeed be separated from $^{238}_{92}\text{UF}_6$ by effusion. The separation factor is very small, since the rates of effusion of these two species are nearly equal:

$$\frac{\text{rate of effusion of } ^{235}_{92}\text{UF}_6}{\text{rate of effusion of } ^{238}_{92}\text{UF}_6} = \left(\frac{352}{349}\right)^{1/2} = 1.004$$

so a great many repetitive separations are necessary. An enormous plant was built for this purpose in Oak Ridge, Tennessee. In this process, UF_6 effuses many thousands of times through porous barriers. The lighter fractions move on to the next stage while heavier fractions are recycled through earlier stages. Eventually, a nearly complete separation of the two isotopes is achieved.

AVERAGE SPEEDS OF GAS PARTICLES

We have seen that Equation 4.20 can be used in a straightforward way to calculate relative speeds of two different kinds of particles. For many purposes in chemistry, that calculation is sufficient. It is possible, however, to go one step further and calculate the average speed of a given kind of gas particle. Here again we use Equation 4.20,

$$\frac{mu^2}{2} = cT$$

This time, we need an expression for the constant c in this equation. By a development that we will not attempt to go through here, it can be shown that

$$c = \frac{3R}{2N} \tag{4.25}$$

where R is the gas constant and N is Avogadro's number. Substituting in Equation 4.20, we obtain

$$mu^2 = \frac{3RT}{N} \quad \text{or} \quad u^2 = \frac{3RT}{mN}$$

However, the product of the mass of a particle, m, times the number of particles in a mole, N, is simply the molar mass, MM. Therefore,

$$u^2 = \frac{3RT}{MM}$$

or:

$$u = \left(\frac{3RT}{MM}\right)^{1/2} \tag{4.26}$$

■ **EXAMPLE 4.13** _____

Find the average speed of an oxygen molecule in air at room temperature (25°C).

Solution

In Equation 4.26, $T = 25 + 273 = 298$ K, $R = 8.31$ kg·m²/(s²·mol·K), and MM = 0.0320 kg/mol. (Note that molar mass must be expressed in *kilograms* per mole if units are to cancel properly.)

$$
\begin{aligned}
u &= \left(\frac{3RT}{MM}\right)^{1/2} \\
&= \left(\frac{3 \times 8.31 \text{ kg·m}^2/(\text{s}^2\text{·mol·K}) \times 298 \text{ K}}{0.0320 \text{ kg/mol}}\right)^{1/2} \\
&= \boxed{482 \text{ m/s}}
\end{aligned}
$$

By a simple conversion (1 m/s = 2.24 miles/h) we can show that, in more familiar units, the average speed is about 1000 miles/h!

Molecules move pretty fast in a gas, but don't go very far between collisions, about 10^{-5} cm

Exercise

What is the average speed of an H_2 molecule at 25°C? Answer: 1.92×10^3 m/s.

DISTRIBUTION OF MOLECULAR SPEEDS AND ENERGIES

As we have seen, gas molecules are moving very rapidly at ordinary temperatures. The average velocity of an O_2 molecule at 25°C is 482 m/s, while that of H_2 is even higher, 1920 m/s. We must remember, however, that not all molecules in these gases have these speeds. The motion of particles in a gas is utterly chaotic. In the course of a second, a particle undergoes millions of collisions with other particles. As a result, the speed and direction of motion of a particle is constantly changing. Over a period of time, the

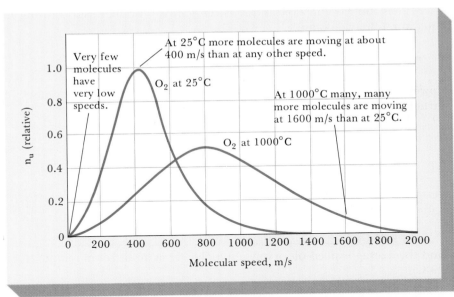

Figure 4.12 The distribution of molecular velocities in oxygen gas at two different temperatures, 25°C and 1000°C. At the higher temperature, the fraction of molecules moving at very high speeds is much greater.

speed will vary from almost zero to some very high value, considerably above the average.

In 1860, James Clerk Maxwell showed that different possible speeds are distributed among particles in a definite way. Indeed, he developed a mathematical expression for this distribution. His results are shown graphically in Figure 4.12 for O_2 at 25°C and at 1000°C. On the graph, we plot the relative number of molecules having a certain speed, u, against that speed. We see that at 25°C this number increases rapidly with the speed, up to a maximum of about 400 m/s. This is the most probable speed of an oxygen molecule at 25°C. Above about 400 m/s, the number of molecules moving at any particular speed decreases. A molecule is only about one-fifth as likely to be moving at 800 m/s as it is to be moving at 400 m/s. For speeds in excess of about 1200 m/s, the fraction of molecules drops off to nearly zero. In general, most molecules have speeds rather close to the average value.

As temperature increases, the speed of the molecules increases. The distribution curve for molecular speeds (Fig. 4.12) shifts to the right and becomes broader. The chance of a molecule having a very high speed is much greater at 1000°C than it is at 25°C. Note, for example, that a large number of molecules have speeds greater than 1200 m/s at 1000°C.

Since translational energy is simply related to speed (Equation 4.20), we can talk about a distribution of energies as well as speeds. A plot of the relative number of particles having a certain energy, E, against that energy would resemble Figure 4.12. Here again, *the fraction of particles having very*

Sort of like people; most of them go along with the crowd

high energies increases sharply with temperature. As we will see in Chapter 18 it is these high-energy particles that take part in chemical reactions.

4.6
REAL GASES

In this chapter, we have used the Ideal Gas Law in all our calculations, assuming it applies exactly. Under ordinary conditions (and in nearly all problems in this text), this assumption is a good one; however, all real gases deviate at least slightly from the Ideal Gas Law. Table 4.2 shows the extent to which two gases, O_2 and CO_2, deviate from ideality at different temperatures and pressures. The data compare the experimentally observed molar volume, $\overline{V}_{obs}$:

$$\text{molar volume} = \overline{V}_{obs} = V_{obs}/n$$

with the molar volume calculated from the Ideal Gas Law:

$$\overline{V}_{ideal} = RT/P$$

It should be obvious from the Table that deviations from ideality become larger at *high pressures* and *low temperatures*. Moreover, the deviations are larger for CO_2 than for O_2. All of these effects can be correlated in terms of a simple, commonsense observation:

In general, the closer a gas is to the liquid state, the more it will deviate from the Ideal Gas Law.

High pressures and low temperatures are, of course, the conditions required to liquefy any gas. Moreover, as you can see from Table 4.2, carbon dioxide is much easier to liquefy than oxygen.

From a molecular standpoint, deviations from the Ideal Gas Law arise because it neglects two factors (recall postulates 3 and 4 of the kinetic theory):

1. The finite volume of gas particles
2. Attractive forces between gas particles

TABLE 4.2   Real vs. Ideal Gases; Percent Deviation* in Molar Volume

P(atm)	O₂ 50°C	O₂ 0°C	O₂ −50°C	CO₂ 50°C	CO₂ 0°C	CO₂ −50°C
1	−0.0%	−0.1%	− 0.2%	− 0.4%	−0.7%	−1.4%
10	−0.4%	−1.0%	− 2.1%	− 4.0%	−7.1%	
40	−1.4%	−3.7%	− 8.5%	−17.9%		
70	−2.2%	−6.0%	−14.4%	−34.2%	Condenses to Liquid	
100	−2.8%	−7.7%	−19.1%	−59.0%		

*% dev. $= \dfrac{(\overline{V}_{obs} - \overline{V}_{ideal})}{\overline{V}_{ideal}} \times 100$

We will now consider in turn the effect of these two factors on the molar volumes of real gases.

ATTRACTIVE FORCES

Notice that in Table 4.2 all the deviations are negative; the observed molar volume is less than that predicted by the Ideal Gas Law. This effect can be attributed to attractive forces between gas particles. These forces tend to pull the particles toward one another, reducing the space between them. As a result, the particles are crowded into a smaller volume, just as if an additional external pressure were applied. The observed molar volume, $\overline{V}_{obs}$, becomes less than $\overline{V}_{ideal}$, and the deviation from ideality is *negative:*

Due to molecular attractions, a given pressure will compress a gas to a smaller volume than the ideal value

$$\frac{\overline{V}_{obs} - \overline{V}_{ideal}}{\overline{V}_{ideal}} < 0$$

The magnitude of this effect depends upon the strength of the attractive forces and hence upon the nature of the gas. Intermolecular attractive forces are stronger in CO_2 than they are in O_2, which explains why the deviation from ideality of $\overline{V}_{obs}$ is greater with carbon dioxide. Attractive forces are ultimately responsible for liquefaction, which explains why carbon dioxide is more readily condensed to a liquid than is oxygen.

PARTICLE VOLUME

Figure 4.13 shows a plot of $\overline{V}_{obs}/\overline{V}_{ideal}$ versus pressure for methane at 25°C. Up to about 150 atm, methane shows a steadily increasing negative deviation from ideality, as might be expected on the basis of attractive forces. At 150 atm, $\overline{V}_{obs}$ is only about 70% of $\overline{V}_{ideal}$.

Figure 4.13 Below about 350 atm, attractive forces between CH_4 molecules cause the observed molar volume of methane gas at 25°C to be less than that calculated from the Ideal Gas Law. At 350 atm, the effect of the attractive forces is just balanced by that of the finite volume of CH_4 molecules, and the gas appears to behave ideally. Above 350 atm, the effect of finite molecular volume predominates and $\overline{V}_{obs} > \overline{V}_{ideal}$.

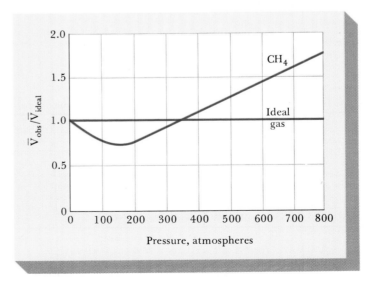

At very high pressures, methane behaves quite differently. Above 150 atm, the ratio $\overline{V}_{obs}/\overline{V}_{ideal}$ increases, becoming 1 at about 350 atm. Above that pressure, methane shows a *positive* deviation from the Ideal Gas Law:

$$\frac{\overline{V}_{obs} - \overline{V}_{ideal}}{\overline{V}_{ideal}} > 0$$

This effect is by no means unique to methane; it is observed with all gases. If the data in Table 4.2 is extended to very high pressures, we find that oxygen and carbon dioxide behave like methane; $\overline{V}_{obs}$ becomes larger than $\overline{V}_{ideal}$.

An increase in molar volume above that predicted by the Ideal Gas Law is related to the finite volume of gas particles. These particles contribute to the observed volume, making $\overline{V}_{obs}$ greater than $\overline{V}_{ideal}$. Ordinarily, this effect becomes evident only at high pressures, where the particles are quite close to one another.

VAN DER WAALS EQUATION

It is possible to write equations involving P, V, and T for gases that take into account attractions between particles and finite particle volumes. One of these is the van der Waals equation:

$$(P + a/\overline{V}^2)(\overline{V} - b) = RT \tag{4.27}$$

where $\overline{V}$ is the molar volume of the gas. In this equation, a and b are constants, independent of P, V, and T. They do, however, vary from one gas to another (Table 4.3). The term $a/\overline{V}^2$ reflects the attractive forces between particles, while the constant b corrects for the effect of particle volume; b is roughly equal to the molar volume of the liquid (for H_2O, b = 0.030 L; molar volume $H_2O(l)$ = 0.018 L). The van der Waals equation is much better than the Ideal Gas Law for predicting the behavior of gases at moderate to high pressures.

TABLE 4.3   van der Waals Constants

GAS	a $\left(\dfrac{L^2 \cdot atm}{mol^2}\right)$	b (L/mol)
H_2	0.244	0.027
O_2	1.360	0.032
N_2	1.390	0.039
CH_4	2.253	0.043
CO_2	3.592	0.043
SO_2	6.714	0.056
Cl_2	6.493	0.056
H_2O	5.464	0.030

■ SUMMARY PROBLEM

When aluminum foil is added to a solution of hydrochloric acid, the following reaction occurs:

$$2 \text{ Al}(s) + 6 \text{ H}^+(aq) \rightarrow 2 \text{ Al}^{3+}(aq) + 3 \text{ H}_2(g)$$

a. If 255 mL of H_2 is produced, measured at 22°C and 740 mm Hg, what is the yield of hydrogen in moles? In grams?

b. What pressure would be required to compress the sample of H_2 referred to in (a) to one-half of its volume at 22°C? At what temperature would the volume be doubled, if the pressure is held constant at 740 mm Hg?

c. What is the density of $H_2(g)$ at 22°C and 740 mm Hg?

d. Suppose the hydrogen is collected over water at 25°C and a total pressure of 756 mm Hg. What is the partial pressure of $H_2(g)$? If the volume of the sample is 235 mL, what is the mass of H_2?

e. What is the mole fraction of hydrogen in the wet gas in (d)?

f. What mass of aluminum is required to generate 1.00 L of $H_2(g)$ at 25°C and 1.00 atm?

g. What volume of H_2 at STP could be generated from 250 mL of "dilute" hydrochloric acid, in which the molarity of H^+ is 6.0 mol/L?

h. Compare the rate of effusion of H_2 to that of He; compare the time required for equal numbers of moles of H_2 and CO_2 to effuse.

Answers

a. 0.0103 mol; 0.0207 g   b. 1480 mm Hg; 317°C   c. 0.0810 g/L

d. 732 mm Hg; 0.0187 g   e. 0.968   f. 0.735 g   g. 17 L

h. 1.41; 0.214

For more than 300 years, gases have been involved, in one way or another, with the evolution of chemistry and physics. Many famous scientists have made their reputations through experiments or theoretical calculations involving gases. Among these are several individuals mentioned in this or other chapters: Boyle, Charles, Gay-Lussac, Avogadro, Cannizzaro, and Maxwell.

Robert Boyle (1627–1691)

Boyle was born in Ireland, the 14th child of the Earl of Cork. He was one of the first experimental scientists. Among other things, Boyle invented a vacuum pump and used it to show that air is necessary for combustion, respiration, and the transmission of sound. His discovery, in 1660, of the law that bears his name ($P_2V_2 = P_1V_1$) grew out of his work on the properties of air.

Boyle was a prolific author on a wide variety of topics ranging from religious philosophy to the structure of matter. The best known of his publications is a book entitled *The Sceptical Chemist*. Here he attacked Aristotle's theory of four elements (earth, air, fire, and water) and the three principles of Paracelsus (salt, sulfur, and mercury). Instead, Boyle proposed that matter was composed of particles of various sorts that could arrange themselves into groups. Groups of one kind constituted a chemical substance. In this sense, he used concepts of atomic and molecular theory similar to those we have today.

Jacques Charles (1746–1823) and Joseph Gay-Lussac (1778–1850)

Charles started his career as a minor bureaucrat in the government of France. When he was dismissed from that post in an economy move, he turned to the study of gases. On December 1, 1783, Charles was on the second balloon ever to lift a human being off the surface of the earth. This accomplishment so impressed Louis XVI that Charles was given a laboratory at the Sorbonne.

In 1787, Charles discovered that the temperature dependence of gas volume could be expressed by the equation

$$V = V_0 (1 + \alpha t)$$

where V_0 is the volume at 0°C, t is the Celsius temperature, and α is a constant. The quantity α is the same for all gases and is approximately equal to 1/273; that is, when the temperature increases 1°C, the volume of a gas increases by 1/273 of its value at 0°C.

Charles never published his work; Gay-Lussac found out about it by accident. He repeated Charles' work and published his results in 1802. Gay-Lussac, like Charles, was fascinated by balloons. In 1804, he ascended to a height of 7 km in a hydrogen-filled balloon. This altitude record remained unbroken for 50 years. Unlike Charles, Gay-Lussac had a keen interest in chemistry. Among his accomplishments were the isolation of

Jacques Charles' ascent in a hydrogen balloon at Paris, December 1, 1783. (The Granger Collection, New York)

the element boron, the preparation of hydrogen fluoride, and the identification of prussic acid, an extremely toxic gas, as HCN.

Gay-Lussac's most important contribution was the Law of Combining Volumes, published in 1808. This law was based on studies of several reactions involving gases, including that between hydrogen and oxygen. Gay-Lussac found that

2 volumes hydrogen + 1 volume oxygen → 2 volumes water vapor

Gay-Lussac seems not to have grasped the implications of his work. However, John Dalton saw where the Law of Combining Volumes led, and he didn't like what he saw. He realized that this simple, integral relation between volumes implied an equally simple relation between particles:

2 particles hydrogen + 1 particle oxygen → 2 particles water

At this point, Dalton, equating particles with atoms, was in trouble. One atom of oxygen could hardly yield two particles of water, both of which must contain at least one atom of oxygen (in Dalton's words, "Thou canst not split an atom"). Faced with what he took to be a direct challenge to the atomic theory, Dalton attempted to discredit Gay-Lussac's results, citing contradictory experiments of his own. His argument accomplished little except to prove that Dalton was a better theoretician than experimentalist.

Avogadro (1776–1856) and Cannizzaro (1826–1910)

In 1811, an Italian physicist at the University of Turin with the improbable name of Lorenzo Romano Amedeo Carlo Avogadro di Quarequa e di Cerreto resolved the dispute between Gay-Lussac and Dalton. He pointed out that Dalton had confused the concepts of atoms and molecules. If an oxygen molecule is *diatomic*, two molecules of water, each containing one atom of oxygen, can be formed from one oxygen molecule. Avogadro interpreted Gay-Lussac's work with hydrogen, oxygen, and water to mean that

2 H_2 molecules + 1 O_2 molecule → 2 H_2O molecules

Going one step further, Avogadro suggested that *equal volumes of all gases at the same temperature and pressure contain the same number of molecules.*

Avogadro's Law, which we have just stated, offers a simple method of determining molar masses. All one has to do is to determine the masses of a fixed volume, say 1 L, of different gases at the same temperature and pressure. These masses must be in the same ratio as the molar masses of the different gases. Unfortunately, this argument, which seems so obvious today, made little or no impact on Avogadro's contemporaries. For one thing, scientists of that time refused to believe that an element could form a diatomic molecule.

Avogadro's ideas lay fallow for nearly half a century. They were revived in 1860 by a fellow countryman, Stanislao Cannizzaro, Professor of Chemistry at Genoa. He showed that Avogadro's Law could be used to determine not only molar masses, but also, indirectly, atomic masses. Cannizzaro had more impact than Avogadro, perhaps because he presented his ideas more clearly. He made a major contribution to the development of the atomic mass scale discussed in Chapter 2.

James Clerk Maxwell (1831–1879)

Maxwell was born in 1831 in Scotland. After his education at the University of Edinburgh and at Cambridge he became a professor of natural philosophy, first in Scotland and later at Cambridge. His mathematical abilities became apparent at an early age and were applied in many areas. In addition to his accomplishments with gases, which laid the foundation of the science now known as statistical mechanics, Maxwell worked extensively in thermodynamics, developing several fundamental equations that bear his name. His greatest successes, however, were in connection with the theory of light and electricity: he discovered and formulated the general equations of the electromagnetic field. He was the first to recognize that light is a form of electromagnetic radiation and anticipated the development of what we now call radio waves. Maxwell was forced to be inactive for several years of his life because of illness; he was 48 years old when he died, having completed much of his work by the time he was 30.

The number of truly outstanding theoreticians the world has known is very small, certainly numbering less than 100. James Clerk Maxwell belongs among the elite of this group. He was truly an intellectual giant, to be ranked with Newton, Einstein, and J. Willard Gibbs (Chapter 20).

QUESTIONS AND PROBLEMS

Measurements on Gases

1. A bedroom 11 ft × 12 ft × 8.0 ft contains 35.41 kg of air at 25°C. Express the volume of the room in liters, the amount of air in moles (MM of air = 29.0 g/mol), and the temperature in Kelvin.

2. A three-gallon tank of oxygen contains 0.461 mol O_2 gas at 27°C. Express the volume of the tank in liters, the amount of O_2 in the tank in grams, and the temperature of the tank in Kelvin.

3. Complete the following table of pressure conversions.

mm Hg	ATMOSPHERES	KILOPASCALS
728	——	——
——	1.28	——
——	——	99.8

4. Carry out the indicated conversions between pressure units.

mm Hg	ATMOSPHERES	KILOPASCALS
158	——	——
——	0.795	
——	——	128

Ideal Gas Law: Initial and Final States

5. A sample of carbon dioxide gas occupies a volume of 5.75 L at 0.890 atm. If the temperature and the number of moles remain constant, calculate the volume when the pressure is
  a. increased to 1.25 atm
  b. decreased to 0.350 atm

6. The pressure of a 5.00-L sample of xenon gas is 725 mm Hg. Assuming that the temperature and moles of

xenon are unchanged, calculate the new pressure when the volume becomes
  a. 8.75 L   b. 1.35 L

7. A nitrogen sample at 30°C has a volume of 1.75 L. If the pressure and the amount of gas remain unchanged, determine the volume when
  a. the Celsius temperature is doubled.
  b. the Celsius temperature is halved.

8. A gas is originally at a temperature of 50°C. To what temperature must it be heated to triple its volume (with n and P constant)?

9. A basketball is inflated in a garage at 20°C to a gauge pressure of 8.0 psi. Gauge pressure is the pressure above atmospheric pressure, which is 14.7 psi. The ball is used in the driveway at a temperature of -5°C and feels "flat." What is the actual pressure of the air in the ball? What is the gauge pressure?

10. A tire is inflated to a gauge pressure of 28.0 psi at 67°F. After several hours of driving, the gauge pressure in the tire is 34.0 psi. What is the temperature of the air in the tire in °F? Assume volume changes are negligible.

11. A 2.90-cm³ air bubble forms in a deep lake at a depth where the temperature is 8°C at a total pressure of 1.98 atm. The bubble rises to a depth where the temperature and pressure are 15°C and 1.50 atm, respectively. Assuming that the amount of air in the bubble has not changed, calculate its new volume.

12. On a cold day, a person takes in a breath of 450 mL of air at 756 mm Hg and -10°C. What is the volume of this air in the lungs at 37°C and 758 mm Hg?

13. An open flask contains 0.200 mol of air. Atmospheric pressure is 745 mm Hg and room temperature is 68°F. How many moles are present in the flask when the pressure is 1.10 atm and the temperature is 33°C?

14. A closed syringe contains 0.01765 mol of the foul-smelling gas hydrogen sulfide, H_2S. The pressure in the container is 725 mm Hg and the temperature is 23°C. An additional 0.00125 mol of H_2S is injected into the syringe. The temperature rises to 35°C. Assuming constant volume, calculate the new pressure in atmospheres.

15. Houses with well-water systems have ballast tanks to hold a supply of water. Typically, water flows into the tank from the well until the gauge pressure (pressure in excess of 15 lb/in²) of the air in the tank reaches 50 lb/in². As water is drawn from the tank, the air above it expands and its pressure drops. When the gauge pressure reaches 20 lb/in², the pump delivers more water to the tank. Suppose a 1.50-m³ tank is 83% full of water when the gauge pressure is 50 lb/in². How much water can be withdrawn before the pump turns on at 20 lb/in²?

16. Frequently, ballast tanks of the type described in Problem 15 lose most of their air and become "water-logged." When this happens, the pump operates more frequently.

Suppose the 1.50-m³ tank referred to in Problem 15 has to be 92% full of water before the gauge pressure reaches 50 lb/in². How much water can be withdrawn from the tank before the pump turns on at 20 lb/in²?

Ideal Gas Law: Calculation of One Variable

17. On a warm day, an amusement park balloon is filled with 47.8 g of helium. The temperature is 33°C and the pressure in the balloon is 2.25 atm. Calculate the volume of the balloon.

18. A ten-liter gas cylinder contains 3.8×10^2 g of nitrogen. What pressure is exerted by the nitrogen at 25°C?

19. A drum used to transport crude oil has a volume of 162 L. How many water molecules, as steam, are required to fill the drum at 1.00 atm and 100°C? What volume of liquid water (d $H_2O(l) = 1.00$ g/cm³) is required to produce that amount of steam?

20. How many moles of air are there in a 125-mL Erlenmeyer flask if the pressure is 755 mm Hg and the temperature is 20°C?

21. Use the Ideal Gas Law to complete the following table for ammonia gas.

PRESSURE	VOLUME	TEMPERATURE	MOLES	GRAMS
2.50 atm	___	0°C	___	32.0
___	75.0 mL	30°C	___	0.385
768 mm Hg	6.0 L	100°C	___	___
195 kPa	58.7 L	___	___	19.8

22. Complete the following table for dinitrogen tetroxide gas, assuming ideal gas behavior.

PRESSURE	VOLUME	TEMPERATURE	MOLES	GRAMS
735 mm Hg	___	15°C	___	18.9
___	489 mL	38°C	___	27.8
1.49 atm	0.885 L	45°C	___	___
239 kPa	2.75 L	___	___	45.0

Ideal Gas Law: Density and Molar Mass

23. Calculate the densities of the following gases at 27°C and 763 mm Hg.
  a. uranium hexafluoride   b. carbon monoxide
  c. chlorine

24. Calculate the densities of the following gases at 68°F and 115 kPa.
  a. nitrogen   b. argon   c. iodine trichloride

25. Assuming water vapor behaves ideally at 100°C and 1.00 atm, what is the ratio of the density of $H_2O(l)$ to $H_2O(g)$ under these conditions? The density of liquid water at 100°C is 0.958 g/cm³.

26. Recent measurements show that the atmosphere on Venus consists mostly of carbon dioxide. The temperature at the surface of Venus is 460°C; the pressure is 75 atm.

Compare the density of CO_2 on Venus's surface to that on the earth's surface at 25°C and one atmosphere.

27. Cyclopropane mixed in the proper ratio with oxygen can be used as an anesthetic. At 755 mm Hg, and 25°C, it has a density of 1.71 g/L.
a. What is the molar mass of cyclopropane?
b. Cyclopropane is made up of 85.7% C and 14.3% H. What is its molecular formula?

28. Methyl isocyanate is the volatile, highly toxic gas that killed thousands of people in Bhopal, India, in 1984. It is made up of 42.1% C, 5.26% H, 24.6% N, and oxygen. A sample of the volatile liquid is vaporized completely in a 125-ml flask at 99°C and 745 mm Hg. The vapor is condensed and weighs 0.232 g. What is the molecular formula of methyl isocyanate?

29. To prevent a condition called the "bends," deep-sea divers breathe a mixture containing, in mole percent, 10.0% O_2, 10.0% N_2, and 80.0% He.
a. Calculate the molar mass of this mixture.
b. What is the ratio of the density of this gas to that of pure oxygen?

30. Exhaled air contains 74.5% N_2, 15.7% O_2, 3.6% CO_2, and 6.2% H_2O (mole percent).
a. Calculate the molar mass of exhaled air.
b. Calculate the density at 22°C and 750 mm Hg and compare the value obtained to that for ordinary air (MM = 29.0 g/mol).

31. A 2.00-g sample of $SX_6(g)$ has a volume of 329.5 cm^3 at 1.00 atm and 20°C. Identify the element X. Name the compound.

32. A 1.58-g sample of $C_2H_3X_3(g)$ has a volume of 297 mL at 769 mm Hg and 35°C. Identify the element X.

Gases in Reactions

33. When hydrogen sulfide gas, H_2S, reacts with oxygen, sulfur dioxide gas and steam are produced.
a. Write a balanced equation for the reaction.
b. How many liters of sulfur dioxide would be produced from 4.0 L of oxygen? Assume 100% yield and that all gases are measured at the same temperature and pressure.

34. Nitrogen trifluoride gas reacts with steam to form the gases HF, NO, and NO_2.
a. Write a balanced equation for the reaction.
b. What volume of nitrogen oxide, NO, is formed when six liters of nitrogen trifluoride are made to react with 6.00 L of steam? Assume 100% yield and a constant temperature and pressure throughout the reaction.

35. Hydrogen cyanide, HCN, is a poisonous gas that was used in the gas chambers of Hitler's concentration camps. It can be formed by the reaction

$$NaCN(s) + H^+(aq) \rightarrow HCN(g) + Na^+(aq)$$

What mass of sodium cyanide, NaCN, is required to make 8.53 L of hydrogen cyanide at 22°C and 751 mm Hg?

36. Calcium reacts with water, yielding hydrogen gas and calcium hydroxide. How many mL of water (d = 1.00 g/cm^3) are required to produce 7.00 L of dry hydrogen at 1.05 atm and 30°C?

37. Oxygen can be made by the decomposition of hydrogen peroxide:

$$2H_2O_2(aq) \rightarrow 2H_2O + O_2(g)$$

a. What volume of oxygen gas at 27°C and 745 mm Hg can be made from 5.50 mL of a solution containing 3.00 mass percent hydrogen peroxide, and having a density of 1.01 g/cm^3?
b. The label on the bottle of hydrogen peroxide claims that the solution produces ten times its own volume of oxygen gas. Does it?

38. Diborane, $B_2H_6(s)$, is a highly explosive compound formed by the reaction

$$3NaBH_4(s) + 4BF_3(g) \rightarrow 2B_2H_6(g) + 3NaBF_4(s)$$

a. What mass of sodium borohydride, $NaBH_4$, is required to form one liter of diborane at STP (0°C and 1.00 atm)?
b. What volume of boron trifluoride at 20°C and 742 mm Hg is required to produce 6.00 g of sodium borofluoride, $NaBF_4$?

Dalton's Law

39. Suppose exhaled air (Problem 30) is at a pressure of 751 mm Hg. Calculate the partial pressures of nitrogen, oxygen, carbon dioxide, and water.

40. A gaseous mixture contains 5.78 g of methane, 2.15 g of neon, and 6.80 g of sulfur dioxide. What pressure is exerted by the mixture inside a 75.0-L cylinder at 85°C? Which gas contributes the greatest pressure?

41. A student collects 355 cm^3 of oxygen saturated with water vapor at 27°C. The mixture exerts a total pressure of 775 mm Hg. At 27°C, the vapor pressure of $H_2O(l)$ = 26.7 mm Hg.
a. What is the partial pressure of oxygen in the sample?
b. How many grams of oxygen does the sample contain?

42. To prepare a sample of hydrogen gas, a student reacts zinc with hydrochloric acid. The overall reaction is

$$Zn(s) + 2H^+(aq) \rightarrow Zn^{2+}(aq) + H_2(g)$$

The hydrogen is collected over water at 24°C and the total pressure is 758 mm Hg (vp $H_2O(l)$ = 22.4 mm Hg).
a. What is the partial pressure of hydrogen?
b. How many grams of hydrogen are there in a 2.00-L sample of wet gas?

43. A sample of oxygen gas is collected over water at 25°C (vp $H_2O(l)$ = 23.8 mm Hg). The wet gas occupies a volume of 12.83 L at a total pressure of 745 mm Hg. If all the water is removed, what volume will the dry oxygen occupy at a pressure of 762 mm Hg and a temperature of 50°C?

44. A sample of gas collected over water at 42°C occupies a volume of one liter. The wet gas has a pressure of 0.986 atm. The gas is dried and the dry gas occupies 1.04 L with a pressure of 1.00 atm at 90°C. What is the vapor pressure of water at 42°C?

Kinetic Theory

45. A gas effuses 1.25 times faster than nitrogen trifluoride at the same temperature and pressure.
   a. Is the gas heavier or lighter than nitrogen trifluoride?
   b. Calculate the ratio of the molar mass of nitrogen trifluoride to that of the unknown gas.

46. What is the ratio of the rate of effusion of the most dense gas known, uranium hexafluoride, to that of the most abundant gas, nitrogen?

47. There is a tiny leak in a system containing neon gas at 22°C and 760 mm Hg. After one minute, the pressure of Ne drops to 749 mm Hg. The neon is replaced by helium, again at 22°C and 760 mm Hg. What would you expect the pressure of helium to be after one minute?

48. It takes 32.0 s for ammonia to effuse down a capillary tube. How long will it take hydrogen chloride to effuse down an identical capillary tube at the same conditions of temperature and pressure?

49. For a sulfur tetrafluoride gas molecule,
   a. at what temperature will its average kinetic energy be twice that at 32°C?
   b. at what temperature will it have half the average speed it has at 25°C?

50. At what temperature will a xenon tetrafluoride molecule have the same
   a. average kinetic energy as an F_2 molecule at 15°C?
   b. average speed as an F_2 molecule at 15°C?

51. Calculate the average speed of
   a. a Cl_2 molecule at 25°C.
   b. an argon atom at −25°C.

52. A professional tennis player can serve a tennis ball traveling at 45 m/s. At what temperature will a nitrogen molecule have the same average speed?

Real Gases

53. The normal boiling points of CO and SO_2 are −192°C and −10°C, respectively.
   a. At 25°C and 1 atm, which gas would you expect to have a molar volume closest to the ideal value?
   b. If you wanted to reduce the deviation from ideal gas behavior, in what direction would you change the temperature? The pressure?

54. A sample of $CH_4(g)$ is at 50°C and 20 atm. Would you expect it to behave more ideally or less ideally if

   a. the pressure were reduced to 1 atm?
   b. the temperature were reduced to −50°C?

55. Using Figure 4.13, estimate the density of methane gas at 200 atm and 25°C and compare to the value calculated from the Ideal Gas Law.

56. Calculate the pressure of one mole of oxygen gas in a 125-mL container at 0°C using the
   a. Ideal Gas Law.
   b. van der Waals equation (see Table 4.3).

Unclassified

57. The pressure exerted by a column of liquid is directly proportional to its density and to its height. If you wanted to fill a barometer with ethyl alcohol (C_2H_5OH, $d = 0.789$ g/cm^3), how long a tube should you use (d Hg = 13.6 g/cm^3)?

58. For an ideal gas, sketch graphs of
   a. V vs. P at constant T, n.
   b. P vs. n at constant V, T.
   c. u vs. T at constant P, V.
   d. E_{trans} vs. P at constant T, n.

59. For an ideal gas, sketch graphs of
   a. V vs. T at constant P, n.
   b. P vs. T at constant V, n.
   c. n vs. T at constant P, V.
   d. E_{trans} vs. T at constant P, n.

60. A mixture of 3.5 mol neon and 3.9 mol chlorine gas occupies a 5.00-L container at 27°C. Which gas has the larger
   a. average translational energy?
   b. partial pressure?
   c. mole fraction?
   d. effusion rate?

61. Given that 1.00 mol nitrogen and 1.00 mol fluorine gas are in separate containers at the same temperature and pressure, calculate each of the following ratios:
   a. volume F_2/volume N_2.
   b. density F_2/density N_2.
   c. average translational energy F_2/average translational energy N_2.
   d. number of atoms F/number of atoms N.

62. A Porsche 928 S4 engine has a cylinder volume of 618 cm^3. The cylinder is full of air at 75°C and 1.00 atm.
   a. How many moles of oxygen are in the cylinder? (Mole percent of oxygen in air = 21.)
   b. Assume that the hydrocarbons in gasoline have an average molar mass of 100 g/mol and react with oxygen in a 1:12 mole ratio. How many grams of gasoline should be injected into the cylinder to react with the oxygen?

63. Gasoline is a mixture of hydrocarbons of which octane, C_8H_{18}, is typical. The combustion of octane is represented by the equation:

$$2 \; C_8H_{18}(l) + 25 \; O_2(g) \rightarrow 16 \; CO_2(g) + 18 \; H_2O(l)$$

What volume of air (21% by volume oxygen) is required at 25°C and 1.00 atm for the combustion of one gallon of octane ($d = 0.692 \; g/cm^3$)?

64. An intermediate reaction used in the production of nitrogen-containing fertilizers is that between ammonia and oxygen:

$$4 \; NH_3(g) + 5 \; O_2(g) \rightarrow 4 \; NO(g) + 6 \; H_2O(g)$$

A 150.0-L reaction chamber is charged with reactants to the following partial pressures at 500°C: $P_{NH_3} = 1.3$ atm; $P_{O_2} = 0.80$ atm. What is the theoretical yield of NO in grams?

65. Glycine is an amino acid made up of carbon, hydrogen, oxygen, and nitrogen atoms. Combustion of a 0.2036-g sample gives 132.9 mL of CO_2 at 25°C and 1.00 atm and 0.122 g of water. What are the percentages of carbon and hydrogen in glycine? Another sample of glycine weighing 0.2500 g is treated in such a way that all the nitrogen atoms are converted to $N_2(g)$; this gas has a volume of 40.8 mL at 25°C and 1.00 atm. What is the percentage of nitrogen in glycine? The percentage of oxygen? What is the empirical formula of glycine?

66. At 25°C and 380 mm Hg, the density of sulfur dioxide is 1.31 g/L. The rate of effusion of sulfur dioxide through an orifice is 4.48 mL/s. What is the density of a sample of gas that effuses through an identical orifice at the rate of 6.78 mL/s under the same conditions? What is the molar mass of the gas?

67. A research laboratory has two steel cylinders of equal volume; they are at the same temperature. Cylinder I contains carbon dioxide gas, while cylinder II contains nitrogen.

a. If each cylinder contains one kilogram of gas, which cylinder has the larger pressure?

b. Suppose cylinder I has a pressure of 7.0 atm at 50°C, while cylinder II has a pressure of 5.5 atm at 10°C. Which cylinder contains the larger number of molecules?

Challenge Problems

68. A tube 3.0 ft long is originally filled with air. Samples of NH_3 and HCl, at the same temperature and pressure, are introduced simultaneously at opposite ends of the tube.

When the two gases meet, a white ring of $NH_4Cl(s)$ forms. How far from the end at which ammonia was introduced will the ring form?

69. The Rankine temperature scale resembles the Kelvin scale in that 0° is taken to be the lowest attainable temperature (0°R = 0 K). However, the Rankine degree is the same size as the Fahrenheit degree, whereas the Kelvin degree is the same size as the Celsius degree. What is the value of the gas law constant in L·atm/(mol·°R)?

70. A 0.2500-g sample of an Al–Zn alloy reacts with HCl to form hydrogen gas:

$$Al(s) + 3H^+(aq) \rightarrow Al^{3+}(aq) + \tfrac{3}{2} H_2(g)$$

$$Zn(s) + 2H^+(aq) \rightarrow Zn^{2+}(aq) + H_2(g)$$

The hydrogen produced has a volume of 0.153 L at 25°C and 1.00 atm. What is the mass percent of aluminum in the alloy?

71. The buoyant force on a balloon is equal to the mass of air it displaces. The gravitational force on the balloon is equal to the sum of the masses of the balloon, the gas it contains, and the balloonist. If the balloon and balloonist together weigh 175 kg, what would the diameter of a spherical hydrogen-filled balloon have to be in meters if the rig is to get off the ground at 22°C and 752 mm Hg? (Take MM air = 29.0 g/mol.)

72. A mixture in which the mole ratio of hydrogen to oxygen is 2:1 is used to prepare water by the reaction

$$2H_2(g) + O_2(g) \rightarrow 2H_2O(g)$$

The total pressure in the container is 0.820 atm at 25°C before the reaction. What is the final pressure in the container at 125°C after reaction, assuming a 75.0% yield and no volume change?

73. The volume percent of a gas, A, in a mixture is defined by the equation

$$\text{volume percent A} = \frac{100 \times V_a}{V}$$

where V is the total volume of the mixture and V_a is the volume that gas A would occupy alone at the same temperature and pressure. Show that, assuming ideal gas behavior, the volume percent of a gas in a mixture is the same as its mole percent. Explain why the volume percent differs from the mass percent.

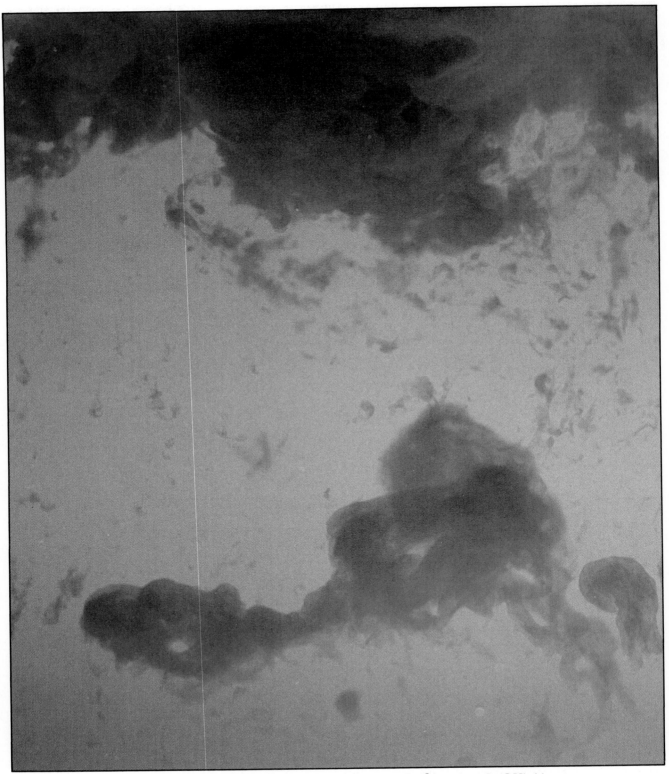

$$2Co^{2+}(aq) + 6NH_3(aq) + 2OH^-(aq) \rightarrow Co(NH_3)_6{}^{2+}(aq) + Co(OH)_2(s)$$

Precipitation of cobalt(II) hydroxide by adding aqueous ammonia to a solution of cobalt(II) chloride.

CHAPTER 5

ELECTRONIC STRUCTURE

In Chapter 2, we considered the structure of the atom briefly. You will recall that every atom has a tiny, positively charged nucleus, made up of protons and neutrons. The number of protons in the nucleus is characteristic of the atoms of a particular element. It is referred to as the atomic number of the element.

The nucleus is surrounded by electrons that carry a negative charge. The charge of an electron is equal in magnitude but opposite in sign to that of a proton. In a neutral atom, the number of electrons is equal to the number of protons and, hence, to the atomic number of the element. A cation (positive ion) is formed by the loss of one or more electrons. By gaining one or more electrons, an atom is converted to an anion (negative ion).

In this chapter, we will focus upon the electron arrangements in atoms (Section 5.5) and monatomic ions (Section 5.6). In preparation for that, it will be helpful to examine the energies of these electrons, starting with the simplest atom, hydrogen (Section 5.2) and proceeding to multielectron atoms (Sections 5.3 and 5.4). Energies of electrons in atoms are derived from the quantum theory (Section 5.1), which was developed during the early part of this century.

Chemical properties, of atoms and molecules, are determined by electron arrangements.

5.1
THE QUANTUM THEORY

The quantum theory was proposed by the German physicist Max Planck in 1900 to explain the properties of radiation emitted by hot bodies. A few years later, in 1905, it was used by Albert Einstein in describing the emission of electrons by metals exposed to light. Still later, in 1913, Niels Bohr used the quantum theory to develop a mathematical model of the hydrogen atom (Section 5.2). We now know that the quantum theory is a general one that applies to all the interactions of matter with energy. Here, we will discuss the postulates of the theory as they apply to electrons in atoms.

POSTULATES OF QUANTUM THEORY

1. Atoms can only exist in certain states, characterized by definite amounts of energy. When an atom changes its state, it absorbs or emits an amount of energy exactly equal to the energy difference between the states.

Atoms can have various forms of energy. One form of particular importance arises from the motion of electrons about the atomic nucleus and from electrical interactions between the electrons, carrying a negative charge, and the positively charged nucleus. This kind of energy is called *electronic energy*. Only certain values of electronic energy are allowed for an atom. The energy of systems that can exist only in discrete states is said to be *quantized*. Because the electronic energy of an atom is quantized, a change in the electronic energy level (state) of an atom involves the absorption or emission of a definite amount, or *quantum*, of energy.

When an atom goes from one allowed energy state to another, it must absorb or emit just enough energy to bring its own energy to that of the final state. The lowest electronic energy state for an atom is called the **ground state**. An electronic **excited state** is any state with energy greater than that of the ground state.

2. When atoms absorb or emit light in moving from one energy state to another, the wavelength λ of the light is related to the energies of the two states by the equation

$$E_{\text{hi}} - E_{\text{lo}} = \frac{hc}{\lambda} \tag{5.1}$$

In this equation, E_{hi} is the energy of the higher state (the one with more energy). The quantity E_{lo} is the energy of the lower state (the one with less energy). The symbols h and c represent physical constants; h is called Planck's constant, and c is the speed of light.

A ray of light can be considered to consist of particles called photons. Each photon of wavelength λ has an energy of hc/λ. An atom can move from one electronic energy state to another by absorbing or emitting a photon. If it absorbs a photon of energy hc/λ, it moves from a lower to a higher energy state and its energy increases by hc/λ; that is,

$$\Delta E = E_{\text{final}} - E_{\text{initial}} = E_{\text{hi}} - E_{\text{lo}} = hc/\lambda$$

When an atom loses energy, all of that energy goes into the photon that is emitted. Total energy is conserved

If, on the other hand, the atom emits a photon of energy hc/λ, it moves from a higher to a lower energy state. In that case, its energy *decreases* by hc/λ:

$$\Delta E = E_{\text{final}} - E_{\text{initial}} = E_{\text{lo}} - E_{\text{hi}} = -hc/\lambda$$

3. The allowed energy states of atoms can be described by sets of numbers called quantum numbers.

It is possible to set up mathematical equations to describe the energies of electrons in atoms. Such equations usually have several solutions, each of which involves one or more quantum numbers. These numbers distinguish a particular energy state from all the others. Quantum numbers are ordinarily associated with individual electrons in an atom. Each electron is assigned a set of quantum numbers according to a set of rules (Section 5.4).

ATOMIC SPECTRA

Equation 5.1 is particularly useful in dealing with *atomic spectra,* the patterns of light given off by excited atoms of an element. Excitation can be accomplished by a flame, by a spark, or by electrons that have been accelerated by falling through a few volts of potential. The effect is to promote an electron in an atom from its ground state to a higher-energy, excited state. Almost immediately, the electron drops back to a lower state, emitting energy in the form of light. In agreement with quantum theory, the light is emitted at a discrete, sharply defined wavelength. The atomic spectrum of an element consists of a series of "lines" produced in this way, each at a different wavelength, each corresponding to a different transition between energy states in the atom.

Each element has a characteristic spectrum which can be used to identify it. Figure 5.1 shows spectra for several different elements. You have

Figure 5.1 *Absorption* ~~Emission~~ *spectra.* The continuous spectrum at the top contains all visible wavelengths and is typical of what is obtained when a solid such as iron is heated. The atomic spectra of gaseous Na, H, Ca, Hg, and Ne are quite different. They consist of discrete lines at certain definite wavelengths.

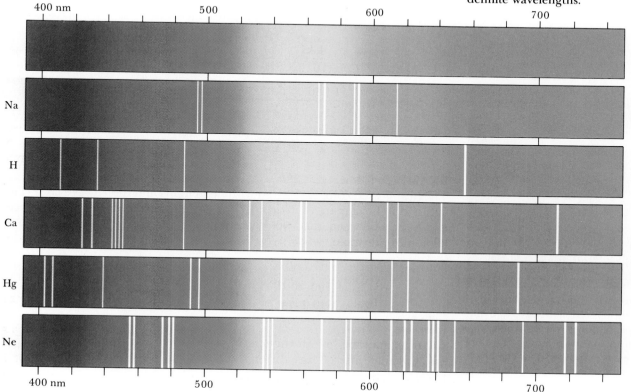

almost certainly observed at least one of these, perhaps without realizing it. The intense yellow light given off when soft glass tubing is heated is due to sodium atoms in the glass. There are two strong lines in the yellow region of the atomic spectrum of sodium at 589.0 and 589.6 nm. These two lines are also responsible for the yellow color of sodium vapor lamps used to illuminate highways.

By accurately measuring the wavelength of a line in the atomic spectrum of an element, we can find the difference in energy between the two states involved. Equation 5.1 is used to obtain $E_{hi} - E_{lo}$ from λ. If we substitute

$$h = 6.626 \times 10^{-34} \frac{\text{J·s}}{\text{particle}}; \qquad c = 2.998 \times 10^{8} \frac{\text{m}}{\text{s}}$$

we obtain $E_{hi} - E_{lo}$ in *joules per particle* when λ is expressed in *meters*:

$$E_{hi} - E_{lo} = \frac{hc}{\lambda} = \frac{(6.626 \times 10^{-34})(2.998 \times 10^{8})}{\lambda} \frac{\text{J·m}}{\text{particle}}$$

$$= \frac{1.986 \times 10^{-25}}{\lambda} \frac{\text{J·m}}{\text{particle}}$$

Ordinarily, we express energy differences in **kilojoules per mole** rather than joules per particle. Since wavelengths for electron transitions are very short, it is convenient to express them in **nanometers** rather than meters. To obtain an equation with the desired units, we use the relations

$$1 \text{ mol} = 6.022 \times 10^{23} \text{ particles}; \qquad 1 \text{ kJ} = 10^{3} \text{ J}; \qquad 1 \text{ nm} = 10^{-9} \text{ m}$$

These relations give us the conversion factors we need:

$$E_{hi} - E_{lo} = \frac{1.986 \times 10^{-25}}{\lambda} \frac{\text{J·m}}{\text{particle}} \times \frac{6.022 \times 10^{23} \text{ particles}}{1 \text{ mol}}$$

$$\times \frac{1 \text{ kJ}}{10^{3} \text{ J}} \times \frac{1 \text{ nm}}{10^{-9} \text{ m}}$$

$$E_{hi} - E_{lo} = \frac{(1.986 \times 6.022 \times 10^{4})}{\lambda} \frac{\text{kJ·nm}}{\text{mol}}$$

$$E_{hi} - E_{lo} = \frac{1.196 \times 10^{5}}{\lambda} \frac{\text{kJ·nm}}{\text{mol}} \tag{5.2}$$

Figure 5.2 The He–Ne laser gives off orange-red light at 640.2 nm, corresponding to an electronic transition in the neon atom. The same process is responsible for the bright red color of neon lights. (Marna G. Clarke)

EXAMPLE 5.1

The orange-red color associated with neon lights and the He–Ne laser (Fig. 5.2) is due in part to a line at 640.2 nm. Calculate the energy difference, $E_{hi} - E_{lo}$, between the two states that are responsible for this line, in kilojoules per mole.

Solution

$$E_{hi} - E_{lo} = \frac{1.196 \times 10^{5}}{640.2 \text{ nm}} \frac{\text{kJ·nm}}{\text{mol}} = \boxed{186.8 \text{ kJ/mol}}$$

Exercise

There is another line in the spectrum of neon at a wavelength of 585.2 nm. Calculate the energy difference in kilojoules per mole; in joules per particle. Answer: 204.4 kJ/mol; 3.394×10^{-19} J/particle.

From Example 5.1 or, more generally, from the form of Equation 5.2, it should be clear that *wavelength, λ, is inversely related to energy difference, $E_{hi} - E_{lo}$.* The shorter the wavelength of the light, the larger the difference in energy. This point is further illustrated in Figure 5.3, which shows the wavelengths corresponding to various regions of the electromagnetic spectrum. Notice in particular that in the visible region (400–800 nm), wavelength *increases* and energy *decreases* as we move from violet through yellow to red light. At wavelengths below the visible spectrum, we have high-energy ultraviolet radiation. At the other limit, above 800 nm, is infrared radiation, which has a relatively low energy.

Ultraviolet light has more energy per photon than infrared light

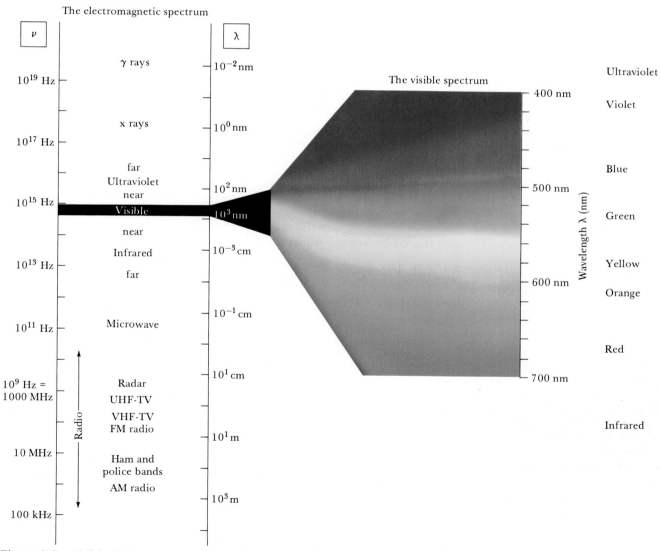

Figure 5.3 Visible light comprises only a tiny portion of the entire spectrum of electromagnetic radiation.

WAVELENGTH, FREQUENCY AND AMPLITUDE

Instead of specifying the wavelength, λ, of light, we may cite its *frequency*, ν. The frequency tells us how many wave cycles (successive crests or troughs) pass a given point in unit time. If we are told that, for a certain kind of wave, 10^8 cycles pass a given point in one second, we would say that*

$$\text{frequency} = 10^8 \text{ cycles per second}$$
$$= 10^8/s = 10^8 \text{ s}^{-1}$$

Notice that the units of frequency are those of "reciprocal time", e.g., "per second" or (seconds)$^{-1}$. This is always the case, whether we are talking about light or any other phenomenon. If you have three lectures per week in this course, we might say that the frequency of lectures is 3/week or 3 (week)$^{-1}$.

The velocity at which a wave moves can be found by multiplying the length of a wave cycle (λ) by the number of cycles passing a point in unit time (ν). For light waves,

$$\nu\lambda = c$$

where c, the speed of light, has a constant value, 2.998×10^8 m/s. This means that the frequency of light is related to its wavelength by the equation

$$\nu = \frac{c}{\lambda} = \frac{2.998 \times 10^8 \text{ m/s}}{\lambda} \tag{5.3}$$

■ **EXAMPLE 5.2** _____

In the sodium atom, there are two states that differ in energy by 203.1 kJ/mol. When an electron transition occurs from the higher of these states to the lower, energy is given off as yellow light. For this light, calculate

a. λ in nanometers, using Equation 5.2.
b. ν in (seconds)$^{-1}$, using Equation 5.3.

Solution

a. $\lambda = \dfrac{1.196 \times 10^5 \text{ kJ·nm}}{E_{hi} - E_{lo} \quad \text{mol}} = \dfrac{1.196 \times 10^5}{203.1} \text{ nm} = \boxed{588.9 \text{ nm}}$

b. In order to cancel units properly, we must express λ in meters (1 nm = 10^{-9} m):

$$\nu = \frac{2.998 \times 10^8}{588.9 \times 10^{-9}} \frac{\text{m/s}}{\text{m}} = \boxed{5.091 \times 10^{14}/s}$$

(Frequencies of visible light are typically of the order of 10^{14} cycles per second; a lot of waves pass a given point in 1 s!)

One hundred thousand billion of them

---

*The unit "cycle per second" has been given a special name, hertz (Hz). Frequencies are often cited in hertz or megahertz (1 MHz = 10^6 Hz). We might say that for the radiation referred to here, $\nu = 10^8$ Hz = 10^2 MHz.

Exercise

An FM radio station broadcasts at a frequency of 106.5 MHz = 106.5 × 10^6/s. What is the wavelength in meters of these radio waves? Answer: 2.815 m.

Another characteristic of wave motion is the **amplitude** (Fig. 5.4). The amplitude measures the maximum displacement of the wave from its midline. As it happens, the two waves shown in Figure 5.4 have the same amplitude. This need not be true, as you well know if you've ever gone surfing or deep-sea fishing.

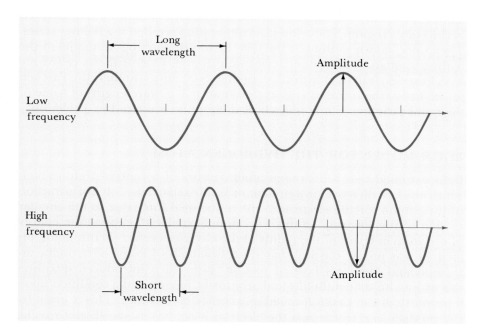

Figure 5.4 Three characteristics of a wave are its amplitude, wavelength, and frequency. The *amplitude* (ψ) is the height of a crest, or the depth of a trough. The *wavelength* (λ) is the distance between successive crests or troughs. The *frequency* (ν) is the number of wave cycles (successive crests or troughs) that pass a given point in a given time.

5.2
THE ATOMIC SPECTRUM OF HYDROGEN AND THE BOHR MODEL

By 1900, scientists had begun to speculate on the way in which positive and negative charges are arranged in atoms. Part of the problem was solved in 1911 when Rutherford showed the existence of atomic nuclei (Chap. 2). Two years later, Niels Bohr, a young Danish physicist, took the next step. He developed a mathematical model for the behavior of the electron in the hydrogen atom.

Bohr based his approach on the Rutherford atom. He also utilized the quantum theory of Planck. Most important, he had accurate information concerning energy differences between electronic states in the hydrogen atom. This came from earlier experimental studies, in the 1880s, of the

TABLE 5.1   Wavelengths (nm) of Lines in the Atomic Spectrum of Hydrogen

ULTRAVIOLET (LYMAN SERIES)	VISIBLE (BALMER SERIES)	INFRARED (PASCHEN SERIES)
121.53	656.28	1875.09
102.54	486.13	1281.80
97.23	434.05	1093.80
94.95	410.18	1004.93
93.75	397.01	
93.05		

atomic spectrum of hydrogen. At that time, a series of lines was discovered in the visible region. These lines constitute what is called the Balmer series. They were known to Bohr at the time he was developing his model of the hydrogen atom. Later, other series were found. One of these, called the Lyman series, is in the ultraviolet region (<400 nm). Another, the Paschen series, lies in the infrared region (>800 nm). The wavelengths of some of the more prominent lines in each of these series are listed in Table 5.1.

BOHR MODEL FOR THE HYDROGEN ATOM

Bohr assumed that a hydrogen atom consists of a central proton about which an electron moves in a circular orbit. He related the force of attraction of the proton for the electron to the centrifugal force due to the circular motion of the electron. In this way, Bohr was able to express the energy of the atom in terms of the radius of the electron's orbit. To this point, his analysis was purely classical, based on Coulomb's Law of electrostatic attraction and Newton's laws of motion. Bohr then introduced quantum theory into his model. Boldly and arbitrarily, he imposed a condition upon a property of the electron called its angular momentum. This is given by the product mvr, where m is the electron mass, v is its speed, and r is the radius of its orbit about the nucleus. Bohr proposed that the angular momentum of an electron in a hydrogen atom is limited to values given by the equation

Much like the earth moves around the sun

$$mvr = \mathbf{n}h/2\pi \qquad\qquad (5.4)$$

where h is Planck's constant and $\mathbf{n}$ is a quantum number that can have any positive integral value (1, 2, 3, . . .). According to the Bohr model of the hydrogen atom, the angular momentum of the electron is quantized. It cannot have just any value; the angular momentum mvr is restricted to values for which $\mathbf{n}$ is a positive integer (h, 2, and π are constants). Thus, the angular momentum can change only by discrete amounts, i.e., integral multiples of $h/2\pi$.

Bohr found that his quantum condition also restricted the allowed energies of the electron in the hydrogen atom. It could have only those values given by the equation

$$E = \frac{-B}{\mathbf{n}^2} \qquad (\mathbf{n} = 1, 2, 3, . . .)$$

where B is a constant that can be calculated from theory. The value of B is 2.179×10^{-18} J/particle. Substituting,

$$E = \frac{-2.179 \times 10^{-18}}{n^2} \frac{J}{\text{particle}}$$

Here the particle is an electron in an H atom

To find E in the common units of kilojoules per mole, we carry out two successive conversions:

$$E = \frac{-2.179 \times 10^{-18}}{n^2} \frac{J}{\text{particle}} \times \frac{6.022 \times 10^{23} \text{ particles}}{1 \text{ mol}} \times \frac{1 \text{ kJ}}{10^3 \text{ J}}$$

$$= \frac{-1312}{n^2} \frac{\text{kJ}}{\text{mol}} \qquad (5.5)$$

Before going further with the Bohr model, let us make three points:

1. In setting up his model, Bohr designated zero energy as the point at which the proton and electron are completely separated. Energy has to be absorbed to reach that point. This means that the electron, in all its allowed energy states within the atom, must have an energy below zero, i.e., must be negative. Hence, the minus sign in Equation 5.5.

2. In the normal hydrogen atom, the electron is in its **ground state**, for which $n = 1$. When an electron absorbs energy, it moves to a higher, **excited state**. In a hydrogen atom, the first excited state has $n = 2$; for the second excited state, $n = 3$, and so on.

3. When an excited electron gives off energy in the form of light, it drops back to a lower energy state. Some of these transitions are shown in Figure 5.5. Notice that the electron may return to

— *the ground state* ($n = 1$). Electrons returning to this state produce the Lyman lines in the hydrogen spectrum. The transition between $n = 2$ and $n = 1$ yields one such line, that between $n = 3$ and $n = 1$ gives another, and so on.

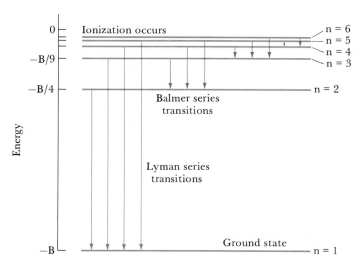

Figure 5.5 Some energy levels and transitions in the hydrogen atom. Lines in the Balmer series arise from transitions from upper levels ($n > 2$) to the $n = 2$ level. In the Lyman series, the lower level is $n = 1$.

— *an excited state.* Electrons returning to **n** = 2 from **n** = 3, 4, . . . are responsible for lines in the Balmer series. Transitions back to **n** = 3 give the Paschen series.

Bohr used his model to calculate the wavelengths of the Balmer lines in the spectrum of hydrogen. He found excellent agreement, 0.1% or better, between theory and experiment. We can repeat his calculations by first using Equation 5.5 to obtain the energies of the individual states. Then, by subtraction, we get the energy difference between two states. Finally, using Equation 5.2, we obtain a predicted value for the wavelength.

■ **EXAMPLE 5.3**
Find the wavelength, in nanometers, of the line in the Balmer series that results from the transition from **n** = 3 to **n** = 2.

Solution
Using the Bohr relation (Equation 5.5), we can calculate the energies of the two levels and hence the energy difference between the two states. Then, using the basic relation between energy difference and wavelength (Equation 5.2), we can find λ for this line:

$$E_3 = \frac{-1312}{9} \text{ kJ/mol} = -145.8 \text{ kJ/mol}$$

$$E_2 = \frac{-1312}{4} \text{ kJ/mol} = -328.0 \text{ kJ/mol}$$

$$E_{hi} - E_{lo} = E_3 - E_2 = -145.8 \text{ kJ/mol} - (-328.0 \text{ kJ/mol})$$
$$= 182.2 \text{ kJ/mol}$$

Solving Equation 5.2 for λ,

$$\lambda = \frac{1.196 \times 10^5}{E_{hi} - E_{lo}} \frac{\text{kJ·nm}}{\text{mol}} = \frac{1.196 \times 10^5}{182.2} \text{ nm} = \boxed{656.4 \text{ nm}}$$

Exercise
Repeat this calculation for the first line in the Lyman series (**n** = 2 to **n** = 1). Answer: 121.5 nm.

Another quantity that can be calculated from the Bohr model is the *ionization energy* of the hydrogen atom. This is the energy that must be absorbed to remove an electron from the gaseous atom, starting from the ground state:

$$\text{H}(g) \rightarrow \text{H}^+(g) + e^-; \quad \Delta E = \text{ionization energy}$$

From Equation 5.5, it is possible to calculate the ionization energy of hydrogen. To do this, we calculate the energy change, ΔE, when an electron moves from the ground state (**n** = 1, E = −1312 kJ/mol) to the state where it is completely removed from the atom (**n** = ∞, E = 0).

$$\Delta E = 0 - (-1312 \text{ kJ/mol}) = 1312 \text{ kJ/mol}$$

The value calculated in this way by Bohr was in excellent agreement with the experimental value for the ionization energy of the hydrogen atom.

■ **EXAMPLE 5.4** _____

Calculate ΔE, in kilojoules per mole, when a hydrogen electron in the first excited state ($\mathbf{n} = 2$) is completely removed from the atom.

Solution

When the electron is completely removed from the atom, $E = 0$. Hence:

$$\Delta E = 0 - E_2$$

But, from Equation 5.5:

$$E_2 = \frac{-1312 \text{ kJ}}{4 \text{ mol}} = -328.0 \text{ kJ/mol}$$

$$\Delta E = 0 - (-328.0 \text{ kJ/mol}) = \boxed{328.0 \text{ kJ/mol}}$$

Exercise

What is the ionization energy in joules per electron from the ground state of hydrogen? Answer: 2.179×10^{-18} J.

Niels Bohr (1885–1962) was one of the giants of 20th-century physics. His contributions went far beyond the model of the hydrogen atom, with which his name is associated. As director of the Institute of Theoretical Physics at Copenhagen, he encouraged and guided the development of the new science of quantum mechanics (Section 5.3) in the 1920s. A gentle, modest man, Bohr was respected and admired by scientists and politicians alike.

In the 1930s, Bohr's research turned to the structure of the nucleus and nuclear reactions. He visited the United States for the first time in 1939, on the eve of World War II, bringing word of the discovery of nuclear fission in Germany. Later he served as an adviser to American physicists working on the atomic bomb, the use of which he strongly opposed.

Bohr, like all the other individuals mentioned in this chapter, was not a chemist. His only real contact with chemistry came as an undergraduate at the University of Copenhagen. His chemistry teacher, Niels Bjerrum, who later became his close friend and sailing companion, recalled that Bohr set a record for broken glassware that lasted half a century.

Nobody does everything right

5.3
THE QUANTUM MECHANICAL ATOM

Bohr's theory for the structure of the hydrogen atom was highly successful. Scientists of the day must have thought they were on the verge of being able to predict the allowed energy levels of all atoms. However, the extension of Bohr's ideas to atoms with two or more electrons gave, at best, only qualitative agreement with experiment. Consider, for example, what happens when Bohr's theory is applied to the helium atom. Here, the errors in calculated energies and wavelengths are of the order of 5% instead of the 0.1% error with hydrogen. There appeared to be no way the theory could be modified to make it work well with helium or other atoms. Indeed, it soon became apparent that there was a fundamental problem with the Bohr model. The idea of an electron moving about the nucleus in a well-defined orbit at a fixed distance from the nucleus with a definite energy had to be abandoned.

WAVE NATURE OF THE ELECTRON; THE DE BROGLIE RELATION

Prior to 1900, it was supposed that light was wavelike in nature. However, the work of Planck and Einstein suggested that, in many processes, light behaves as if it consists of particles, called photons. Within 20 years, the dualistic nature of light became generally accepted. Then, in 1924, a young Frenchman, Louis de Broglie (1892–1987), in his doctoral thesis at the Sorbonne, made a revolutionary suggestion about the nature of matter. Reasoning as physicists do, often with striking success, de Broglie suggested that if light can have a particle nature, then matter might have wave properties. He showed that the wavelength, λ, associated with a particle of mass m moving at speed v would be

$$\lambda = h/mv \tag{5.6}$$

where h is Planck's constant. Within a few years, Davisson and Germer at Bell Telephone Laboratories and G. P. Thomson (son of J. J. Thomson) in England showed that a beam of electrons does indeed have wave properties. Moreover, the observed wavelength of the electron was exactly that predicted by de Broglie.

Using Equation 5.6, it is possible to show how the Bohr quantum number $\mathbf{n}$ arises in a natural way. To do this, consider Figure 5.6. Here,

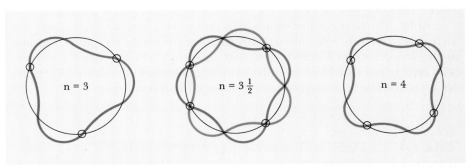

Figure 5.6 Imagine the electron of the hydrogen atom in the form of a wave moving in a circular path about the nucleus. If the wave is to be stable, it must trace the same path in successive orbits. This is the case for the $\mathbf{n} = 3$ wave, where three wavelengths bring the electron exactly back to its starting point. The condition also applies to the $\mathbf{n} = 4$ wave; four wavelengths correspond exactly to a trip around the circle. The $\mathbf{n} = 3\frac{1}{2}$ wave is unstable; the electron does not trace the same path in successive orbits (the second orbit is shown by a red path). The general condition for stability is: $2\pi r = \mathbf{n}\lambda$, where $2\pi r$ is the circumference of the orbit, λ is the wavelength of the electron, and $\mathbf{n}$ is an integer.

we imagine an electron in the form of a wave moving about the nucleus along the circumference of a circle. Under these conditions, there is a restriction on the wavelengths that the electron can have. In successive revolutions, the waves must be exactly in phase with each other; that is, they must have exactly the same height (*amplitude*) at any given point. This means that a wave must fit into the circumference of the circle, $2\pi r$, an integral number of times, **n**. In other words,

$$2\pi r = \mathbf{n}\lambda \tag{5.7}$$

where λ is the wavelength and **n** is a whole number—that is, 1, 2, 3, . . . , but *not* 1.5, 2.1, etc. Combining Equations 5.6 and 5.7, we obtain

$$2\pi r = \frac{\mathbf{n}h}{mv}$$

or

$$mvr = \frac{\mathbf{n}h}{2\pi}$$

This is the condition that Bohr imposed arbitrarily on the momentum of the electron in the hydrogen atom (Equation 5.4). Using the de Broglie relation, this condition becomes physically reasonable.

WAVE FUNCTIONS AND ELECTRON CLOUDS

If an electron behaves like a wave, there is a fundamental problem with the Bohr atom. How does one specify the "position" of a wave at a particular instant? We can hope to determine its wavelength, its energy, and even its amplitude, but there is no obvious way to tell precisely where the electron is. Indeed, since a wave extends over space, the very idea of the position of the electron within an atom becomes nebulous, to say the least.

It's hard to nail down a wave

Scientists in the 1920s, speculating on this problem, became convinced that an entirely new approach was required to treat electrons in atoms, ions, and molecules. A new discipline, called *quantum mechanics,* was developed to describe the motion of small particles confined to tiny regions of space. In the quantum mechanical atom, no attempt is made to specify the position of an electron at a given instant; nor does quantum mechanics concern itself with the path that an electron takes about the nucleus. (After all, if we can't say where the electron is, we certainly don't know how it got there). Instead, quantum mechanics deals only with the *probability* of finding a particle within a given region of space.

In 1926, Erwin Schrödinger, an Austrian physicist working in Zurich, made a major contribution to quantum mechanics. Expanding upon the ideas of de Broglie, Schrödinger went a step further. He wrote down an equation from which one could calculate the amplitude (height) ψ, of the electron wave at various points in space. Although we will not use the Schrödinger wave equation in any calculations, we write it down so you will have some idea of its nature. For a single particle of mass m moving in only one dimension (x), the equation has the form:

Schrödinger may have derived his equation from a similar one governing vibrating violin strings

$$\frac{d^2\psi}{dx^2} + \frac{8\pi^2 m}{h^2}(E - V)\psi = 0 \qquad (5.8)$$

where E and V are the total and potential energies respectively of the particle, and h is Planck's constant. The quantity ψ, commonly called a *wave function*, can be interpreted as the amplitude of the wave associated with the particle. The first term in the equation is the second derivative of ψ with respect to x; for motion in three dimensions, similar terms in y and z must be added.

As you can see, the wave equation is a rather complex one, involving the notation of calculus. It can, however, be solved exactly for the hydrogen atom. It turns out that there are many expressions for ψ that will satisfy this equation. Each of these solutions is associated with a set of quantum numbers (Section 5.4). Using these quantum numbers, it is possible to calculate the allowable energies for an electron in a hydrogen atom.

For the hydrogen electron, the square of the wave function, ψ^2, is directly proportional to the probability of finding the electron at a particular point. If ψ^2 at point A is twice as large as at point B, then we are twice as likely to find the electron at A as at B. Putting it another way, over time the electron will turn up at A twice as often as at B.

Figure 5.7 shows, in two different ways, how ψ^2 for the hydrogen electron in its ground state ($\mathbf{n} = 1$) varies as we move out from the nucleus. In Figure 5.7a, the depth of color is directly proportional to ψ^2 and hence to the probability of finding the electron at a point. As you can see, the color fades as we move out from the nucleus in any given direction; the value of ψ^2 drops accordingly. This feature is emphasized in Figure 5.7b, where ψ_x^2 is plotted against distance along the x axis. Notice that ψ_x^2 has its greatest value at the nucleus ($x = 0$), decreasing smoothly and rapidly as x increases.

The H atom in its ground state is spherical

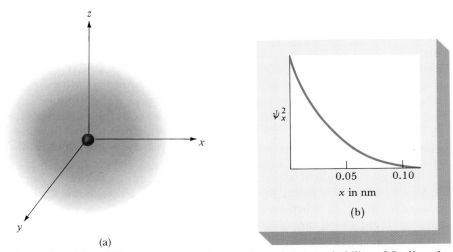

(a)

(b)

Figure 5.7   Two different ways of showing the relative probability of finding the hydrogen electron in its ground state at various distances from the nucleus. In (a), the depth of color is proportional to ψ^2. In (b), the probability, ψ_x^2, of finding the electron at a given distance along the x axis is plotted.

5.4
QUANTUM NUMBERS, ENERGY LEVELS, AND ORBITALS

Schrödinger found that the electron in the hydrogen atom could be described by three quantum numbers. These are now given the symbols $\mathbf{n}$, ℓ, and $\mathbf{m}_\ell$. There are three such numbers because the electron requires three coordinates to describe its motion.

For reasons we will discuss later in this section, we find that it takes four, rather than three, quantum numbers to completely describe the state of a particular electron in an atom. The fourth quantum number is given the symbol $\mathbf{m}_s$. Each electron in an atom has a set of four quantum numbers, $\mathbf{n}$, ℓ, $\mathbf{m}_\ell$, and $\mathbf{m}_s$, which fix its energy and the size, shape, and orientation of its electron cloud. We will now discuss the quantum numbers of electrons as they are used in atoms beyond hydrogen.

FIRST QUANTUM NUMBER, n; PRINCIPAL ENERGY LEVELS

The first quantum number, given the symbol $\mathbf{n}$, is of primary importance in determining the energy of an electron. For the hydrogen atom, the energy depends only upon $\mathbf{n}$ (recall Equation 5.5). In other atoms, the energy of each electron depends mainly, but not completely, upon the value of $\mathbf{n}$. As $\mathbf{n}$ increases, the energy of the electron increases and, on the average, it is found farther out from the nucleus. The quantum number $\mathbf{n}$ can take on only integral values, starting with 1:

The quantum numbers serve to classify the electron

$$\mathbf{n} = 1, 2, 3, 4, \ldots \tag{5.9}$$

In an atom, the value of $\mathbf{n}$ corresponds to what we call a **principal energy level**. Thus, an electron for which $\mathbf{n} = 1$ is said to be in the first principal level. If $\mathbf{n} = 2$, we are dealing with the second principal level, and so on.

SECOND QUANTUM NUMBER, ℓ; SUBLEVELS (s, p, d, f)

Each principal energy level includes one or more **sublevels**. The sublevels are denoted by the second quantum number, ℓ. As we shall see later, the general shape of the electron cloud associated with an electron is determined by ℓ. Larger values of ℓ produce more complex shapes.

The quantum numbers $\mathbf{n}$ and ℓ are related. We find that ℓ can take on any integral value starting with 0 and going up to a maximum of $(\mathbf{n} - 1)$; that is,

The conditions on n, ℓ and m_ℓ come from the solution of the Schrödinger equation for the H atom

$$\ell = 0, 1, 2, \ldots, (\mathbf{n} - 1) \tag{5.10}$$

If $\mathbf{n} = 1$, there is only one possible value of ℓ, namely 0. This means that, in the first principal level, there is only one sublevel, for which $\ell = 0$. If $\mathbf{n} = 2$, two values of ℓ are possible, 0 and 1. In other words, there are two sublevels ($\ell = 0$ and $\ell = 1$) within the second principal energy level. Similarly,

if $\mathbf{n} = 3$: $\ell = 0, 1,$ or 2   (three sublevels)
if $\mathbf{n} = 4$: $\ell = 0, 1, 2,$ or 3   (four sublevels)

In general, **in the nth principal level, there are n different sublevels.**

Another method is commonly used to designate sublevels. Instead of giving the quantum number ℓ, we use a letter (s, p, d, or f*) to indicate the sublevel. A sublevel for which $\ell = 0$ is referred to as an **s sublevel**. If $\ell = 1$, we are dealing with a **p sublevel**. A **d sublevel** is one for which $\ell = 2$; in an **f sublevel**, $\ell = 3$:

Quantum number ℓ 0   1   2   3
Type of sublevel   s   p   d   f

Usually, in designating a sublevel, we include a number to indicate the principal level as well. Thus, we refer to a 1s sublevel ($\mathbf{n} = 1, \ell = 0$), a 2s sublevel ($\mathbf{n} = 2, \ell = 0$), and a 2p sublevel ($\mathbf{n} = 2, \ell = 1$). The sublevel designations for the first four principal levels are shown in Table 5.2.

TABLE 5.2   Sublevel Designations for the First Four Principal Levels

n	1	2		3			4			
ℓ	0	0	1	0	1	2	0	1	2	3
Sublevel	1s	2s	2p	3s	3p	3d	4s	4p	4d	4f

Within a given principal level (same value of **n**), sublevels increase in energy in the order

$$\mathbf{ns} < \mathbf{np} < \mathbf{nd} < \mathbf{nf} \tag{5.11}$$

Thus a 2p sublevel has a slightly higher energy than a 2s sublevel. By the same token, when $\mathbf{n} = 3$, the 3s sublevel has the lowest energy, the 3p is intermediate, and the 3d has the highest energy.

THIRD QUANTUM NUMBER, $\mathbf{m}_\ell$; ORBITALS

The terminology seems complicated at first, but you need to learn it

Each sublevel contains one or more **orbitals**, designated by the third quantum number, $\mathbf{m}_\ell$. This quantum number tells us how the electron cloud surrounding the nucleus is directed in space. The value of $\mathbf{m}_\ell$ is related to that of ℓ. For a given value of ℓ, $\mathbf{m}_\ell$ can have any integral value, including 0, between ℓ and $-\ell$; that is,

$$\mathbf{m}_\ell = \ell, \ldots, +1, 0, -1, \ldots, -\ell \tag{5.12}$$

To illustrate how this rule works, consider an s sublevel ($\ell = 0$). Here $\mathbf{m}_\ell$ can have only one value, 0. This means that an s sublevel contains only one orbital, referred to as an *s orbital*. The electron cloud associated with

---

*These letters come from the adjectives used by spectroscopists to describe spectral lines; *s*harp, *p*rincipal, *d*iffuse, and *f*undamental.

Figure 5.8 All s orbitals are spherically symmetrical; the density of the electron cloud is independent of direction from the nucleus, located at the center of the orbital (part a). The 2s orbital is considerably larger than the 1s orbital (part b). This is another way of saying that, on the average, the electron is farther from the nucleus when it is in the 2s as opposed to the 1s orbital.

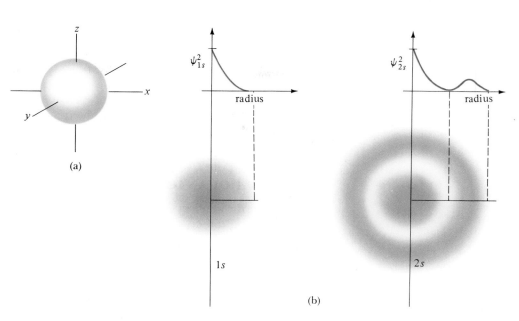

an s orbital is spherically symmetrical; the density of the cloud varies with distance from the nucleus but is independent of direction. Most commonly, an s orbital is shown as a simple sphere (Fig. 5.8a). The radius of the sphere indicates the region within which there is a specified probability of finding the electron. If we agree that the sphere should contain 90% of the density of the electron cloud, then a 1s orbital in a hydrogen atom has a radius of 0.14 nm.

As the value of **n** increases, the s orbital gets larger. That is, the electron is more likely to be found farther out from the nucleus. This effect is shown in Figure 5.8b, where we plot the square of the wave function vs. distance from the nucleus for the 1s and 2s orbitals.

For a p sublevel ($\ell = 1$), we can have:

$$\mathbf{m}_\ell = 1, 0, \text{ or } -1$$

This means that within a given p sublevel, there are three different orbitals. They are described by the quantum numbers $\mathbf{m}_\ell = 1, 0$ and -1; more commonly they are referred to as p_x, p_y, and p_z orbitals. Figure 5.9 shows the shapes and orientations of these orbitals.

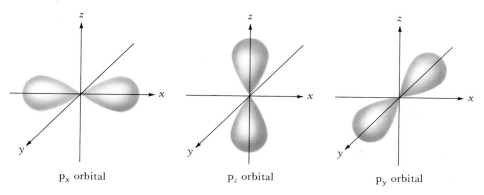

p_x orbital p_z orbital p_y orbital

Figure 5.9 Electron clouds corresponding to the three p orbitals. The electron density in one of these orbitals is symmetrical about the x axis (p_x orbital). In another orbital, it is symmetrical about the z axis (p_z orbital), and in the third it is symmetrical about the y axis (p_y orbital). We describe this situation by saying that the three orbitals are directed at 90° angles to each other.

These orbitals have a fairly complex geometry

We can extend these ideas to d and f sublevels:

d sublevel; $\ell = 2$; $m_\ell = 2, 1, 0, -1, -2$;   5 orbitals

f sublevel; $\ell = 3$; $m_\ell = 3, 2, 1, 0, -1, -2, -3$;   7 orbitals

In general, for a sublevel of quantum number ℓ, there are a total of $2\ell + 1$ orbitals. All of the orbitals in a given sublevel (e.g., $2p_x$, $2p_y$, $2p_z$) have essentially the same energy.

FOURTH QUANTUM NUMBER, m_s; ELECTRON SPIN

To describe an electron in an atom completely, we need to specify a fourth quantum number, m_s. This quantum number is associated with the spin of the electron. An electron has magnetic properties that correspond to those of a charged particle spinning on its axis (Fig. 5.10). Either of two spins are possible, clockwise or counterclockwise.

The yellow sodium doublet cannot be explained with only three quantum numbers

The quantum number m_s was introduced to make theory consistent with experiment. In that sense, it differs from the first three quantum numbers, which came from the solution to the Schrödinger wave equation for the hydrogen atom. This quantum number is not related to n, ℓ, or m_ℓ. It can have either of two possible values:

$$m_s = +1/2 \text{ or } -1/2 \tag{5.13}$$

Electrons that have the same value of m_s (i.e., both $+\frac{1}{2}$ or both $-\frac{1}{2}$) are said to have *parallel* spins. Electrons that have different m_s values (i.e., one $+\frac{1}{2}$ and the other $-\frac{1}{2}$) are said to have *opposed* spins.

Figure 5.10 In some respects, an electron behaves as if it were a spherical particle spinning about its axis. There is an analogy between the alignment of electron spins and the alignment of bar magnets (*top*). Within an orbital, the more stable arrangement is the one in which the two electrons have opposed spins (*lower right*).

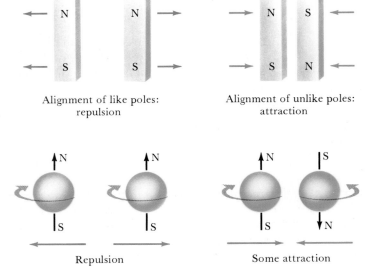

Alignment of like poles: repulsion   Alignment of unlike poles: attraction

Repulsion   Some attraction

PAULI EXCLUSION PRINCIPLE

We have now considered the four quantum numbers that characterize an electron in an atom. There is an important rule, called the Pauli exclusion principle, that relates to these numbers. It tells us that **no two electrons in an atom can have the same set of four quantum numbers**. This principle was first stated in 1925 by Wolfgang Pauli, a colleague of Bohr, again to make theory consistent with the properties of atoms.

Our model for electronic structure is a mix of theory and experiment

The Pauli exclusion principle has an implication that may not be obvious at first glance. It requires that no more than two electrons can fit into an orbital. Moreover, if two electrons occupy the same orbital they must have opposed spins. To see that this is the case, consider the 2s orbital. Any electron in this orbital must have

$$\mathbf{n} = 2, \qquad \ell = 0, \qquad \mathbf{m}_\ell = 0$$

To satisfy the Pauli exclusion principle, the electrons in this orbital must have different $\mathbf{m}_s$ values. But there are only two possible values of $\mathbf{m}_s$. Hence only two electrons can enter that orbital. If the orbital is filled, one electron must have $\mathbf{m}_s = +\frac{1}{2}$, and the other $\mathbf{m}_s = -\frac{1}{2}$. In other words, the two electrons must have opposed spins. Their quantum numbers are

$$\mathbf{n} = 2, \qquad \ell = 0, \qquad \mathbf{m}_\ell = 0, \qquad \mathbf{m}_s = +\tfrac{1}{2}$$
$$\mathbf{n} = 2, \qquad \ell = 0, \qquad \mathbf{m}_\ell = 0, \qquad \mathbf{m}_s = -\tfrac{1}{2}$$

The same argument can be applied equally well to any other orbital.

CAPACITIES OF PRINCIPAL LEVELS, SUBLEVELS, AND ORBITALS

The rules that we have given for quantum numbers fix the capacities of principal levels, sublevels, and orbitals. In summary,

1. Each principal level of quantum number $\mathbf{n}$ contains a total of $\mathbf{n}$ sublevels.
2. Each sublevel of quantum number ℓ contains a total of $2\ell + 1$ orbitals; that is,

an s sublevel ($\ell = 0$) contains 1 orbital

a p sublevel ($\ell = 1$) contains 3 orbitals

a d sublevel ($\ell = 2$) contains 5 orbitals

an f sublevel ($\ell = 3$) contains 7 orbitals

3. Each orbital can hold two electrons, which must have opposed spins.

Two electrons in the same orbital are said to be paired

Applying these rules to the first three principal energy levels, we obtain Table 5.3, p. 150. At the bottom of the table, each arrow indicates an electron. In each orbital, there are two electrons with opposed spins. The number of electrons in a sublevel is found by adding up the electrons in the orbitals within that sublevel. For example, in a p sublevel ($\ell = 1$), there

TABLE 5.3   Allowed Sets of Quantum Numbers for Electrons in Atoms

Level n	1	2				3								
Sublevel ℓ	0	0	1			0	1			2				
Orbital m_ℓ	0	0	1	0	−1	0	1	0	−1	2	1	0	−1	−2
Spin m_s ↑ $= +\frac{1}{2}$ ↓ $= -\frac{1}{2}$	⇅	⇅	⇅	⇅	⇅	⇅	⇅	⇅	⇅	⇅	⇅	⇅	⇅	⇅

are six electrons, two in each of three orbitals. To find the total number of electrons in a principal level, we add the electrons in the sublevels within that principal level.

■ **EXAMPLE 5.5** _____

a. How many electrons can fit into the principal level for which $n = 2$?
b. What is the capacity for electrons of the 3d sublevel?

Solution

a. In the $n = 2$ level, there are two sublevels, the 2s and the 2p. Of these, the 2s contains one orbital with a capacity of $2e^-$. The 2p contains three orbitals, each of which can hold $2e^-$:

total capacity $= 1(2) + 3(2) =$ 8

Note that under the $n = 2$ level in Table 5.3 there are eight electrons, in four pairs.

b. Like any d sublevel, the 3d contains five orbitals. Each orbital can hold $2e^-$. This result is confirmed in Table 5.3, which shows five electron pairs $(10e^-)$ in the $\ell = 2$ sublevel.

Exercise
What is the total capacity for electrons of the fourth principal energy level?
Answer: 32.

Using the approach in Example 5.5, we can readily obtain the total electron capacity of any principal level or sublevel. In Table 5.4, these

TABLE 5.4   Capacities of Electronic Levels and Sublevels in Atoms

LEVEL n	TOTAL NUMBER OF ELECTRONS IN LEVEL, $2n^2$	MAXIMUM NUMBER OF ELECTRONS IN SUBLEVELS, $2(2\ell + 1)$			
		s	p	d	f
1	2	2	—	—	—
2	8	2	6	—	—
3	18	2	6	10	—
4	32	2	6	10	14

capacities are listed through $n = 4$. Notice that the capacity of a principal level is $2n^2$, where n is the principal quantum number.

5.5
ELECTRON ARRANGEMENTS IN ATOMS

Given the rules referred to in Section 5.4, it is possible to assign quantum numbers to each electron in an atom. Beyond that, electrons can be assigned to specific principal levels, sublevels, and orbitals. There are several ways to do this. Perhaps the simplest way to describe the arrangement of electrons in an atom is to give its **electron configuration**. This tells us the number of electrons in each principal level and sublevel. With an **orbital diagram**, we go one step further and indicate the arrangement of electrons within orbitals. Finally, we can specify the set of four quantum numbers for each electron.

Table 5.5 gives the electron configuration, orbital diagram, and quantum numbers of the electrons in hydrogen ($1\ e^-$) and helium ($2\ e^-$). Note that

— in electron configurations, a superscript is used to indicate the number of electrons in a given sublevel.

— in an orbital diagram, arrows are used to indicate electron spins. Thus ↑ indicates an electron with $m_s = +\frac{1}{2}$; ↓ shows an electron with the opposite spin, $m_s = -\frac{1}{2}$.

— quantum numbers are specified in the order n, ℓ, m_ℓ, m_s.

Throughout the remainder of this section, we will consider the electron configurations, orbital diagrams, and electron quantum numbers in multielectron atoms. We should point out that throughout this discussion, we will be dealing with **isolated, gaseous atoms in their ground states**.

ELECTRON CONFIGURATIONS

To arrive at the electron configurations of atoms, we must know the order in which different sublevels are filled. Electrons enter the available sublevels in order of their increasing energy. Usually, a sublevel is filled to capacity before the next one is entered. The relative energies of different sublevels

TABLE 5.5   Electron Arrangements in the Hydrogen and Helium Atoms

	$_1$H (ONE ELECTRON)	$_2$HE (TWO ELECTRONS)
Electron configuration	$1s^1$	$1s^2$
Orbital diagram	1s (↑)	1s (↑↓)
Quantum numbers (n, ℓ, m_ℓ, m_s)	$1, 0, 0, +\frac{1}{2}$	↑ : $1, 0, 0, +\frac{1}{2}$ ↓ : $1, 0, 0, -\frac{1}{2}$

can be obtained from experiment. Figure 5.11 is a plot of these energies for atoms through the **n** = 4 principal level.

From Figure 5.11, it is possible to predict the electron configurations of atoms of elements with atomic numbers 1 through 36. Since an s sublevel can hold only two electrons, the 1s is filled at helium ($1s^2$). With lithium (at. no. = 3), the third electron has to enter a new sublevel: this is the 2s, the lowest sublevel of the second principal energy level; lithium has one electron in this sublevel ($1s^2 2s^1$). With beryllium (at. no. = 4), the 2s sublevel is filled ($1s^2 2s^2$). The next six elements (at. no. 5 through 10) fill the 2p sublevel. Their electron configurations are

$_5$B	$1s^2 2s^2 2p^1$	$_8$O	$1s^2 2s^2 2p^4$
$_6$C	$1s^2 2s^2 2p^2$	$_9$F	$1s^2 2s^2 2p^5$
$_7$N	$1s^2 2s^2 2p^3$	$_{10}$Ne	$1s^2 2s^2 2p^6$

Beyond neon, we enter the third principal level. The 3s sublevel is filled at magnesium:

$$_{12}\text{Mg}\quad 1s^2 2s^2 2p^6 3s^2$$

Six more electrons are required to fill the 3p sublevel. This is filled to capacity at argon:

$$_{18}\text{Ar}\quad 1s^2 2s^2 2p^6 3s^2 3p^6$$

After argon, we observe an "overlap" of principal energy levels. The next electron enters the *lowest* sublevel of the fourth principal level (4s) instead of the *highest* sublevel of the third principal level (3d). Potassium (at. no. = 19) has one electron in the 4s sublevel; calcium (at. no. = 20) fills it with two electrons:

$$_{20}\text{Ca}\quad 1s^2 2s^2 2p^6 3s^2 3p^6 4s^2$$

Now, the 3d sublevel starts to fill with scandium (at. no. = 21). Recall that a d sublevel has a capacity of ten electrons. Hence the 3d sublevel becomes filled at zinc (at. no. = 30):

$$_{30}\text{Zn}\quad 1s^2 2s^2 2p^6 3s^2 3p^6 4s^2 3d^{10}$$

The next sublevel, 4p, is filled at krypton (at. no. = 36):

$$_{36}\text{Kr}\quad 1s^2 2s^2 2p^6 3s^2 3p^6 4s^2 3d^{10} 4p^6$$

Why can't the electron configuration of Li be 1s³?

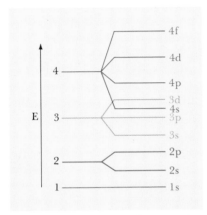

Figure 5.11   In general, energy increases with the principle quantum number, **n**. It is possible, however, for the lowest sublevel of **n** = 4 (i.e., 4s) to be below the highest sublevel of **n** = 3 (i.e., 3d). This appears to be the situation in the potassium and calcium atoms, where successive electrons enter the 4s rather than the 3d sublevel.

■ **EXAMPLE 5.6** _____

Find the electron configuration of the sulfur atom and the nickel atom.

Solution
S atom: at. no. = 16; 16 electrons. The sublevels fill to capacity in order of increasing energy. There are two 1s electrons, two 2s, six 2p, two 3s, and four 3p electrons, making a total of 16. The configuration of the S atom is therefore $1s^2 2s^2 2p^6 3s^2 3p^4$.

Ni atom: at. no. = 28; 28 electrons. Proceeding as before, this time until we have added 28 electrons, we obtain as the electron configuration of the Ni atom: $1s^2 2s^2 2p^6 3s^2 3p^6 4s^2 3d^8$. (Note that the 4s sublevel fills before the 3d.)

Exercise
Give the electron configuration of Mn (at. no. = 25). Answer: $1s^2 2s^2 2p^6 3s^2 3p^6 4s^2 3d^5$.

We can use this general approach to find the electron configuration of any atom. The method, as we have seen, involves adding electrons one by one as atomic number increases. This is sometimes referred to as the Aufbau (building-up) process. To follow it, all we need to know is the order in which sublevels are filled. For the first 36 electrons, this is readily deduced from the preceding discussion or from Figure 5.11. Beyond that, the order of filling can be deduced from the Periodic Table, which we will discuss in Chapter 6.

A rule-of-thumb that can be helpful in writing electron configurations of atoms is the $\mathbf{n} + \ell$ rule. It states that for any pair of sublevels, the one with the lower sum of $\mathbf{n} + \ell$ fills first. Thus, a 4s sublevel ($\mathbf{n} = 4$, $\ell = 0$; $4 + 0 = 4$) is occupied in an atom before a 3d sublevel ($\mathbf{n} = 3$, $\ell = 2$; $3 + 2 = 5$). When the sum of $\mathbf{n} + \ell$ is the same for a pair of sublevels, the one with the lower $\mathbf{n}$ value fills first; for example, a 3d sublevel ($3 + 2 = 5$) is filled before a 4p sublevel ($4 + 1 = 5$).

Table 5.6 shows the order in which various sublevels are filled, following the $\mathbf{n} + \ell$ rule. Note, for example, that for the four sublevels in which $\mathbf{n} + \ell = 7$, the filling order starts with the sublevel of lowest $\mathbf{n}$: $4f < 5d < 6p < 7s$.

Up to atomic number 36, the order and populations are:

$$1s^2 2s^2 2p^6 3s^2 3p^6 4s^2 3d^{10} 4p^6$$

TABLE 5.6   Order of Filling of Atomic Orbitals (left to right)

ORBITAL	1s	2s	2p	3s	3p	4s	3d	4p	5s	4d	5p	6s	4f	5d	6p	7s	5f	6d
e^- in atom	2	4	10	12	18	20	30	36	38	48	54	56	70	80	86	88	102	112
$\mathbf{n}$	1	2	2	3	3	4	3	4	5	4	5	6	4	5	6	7	5	6
ℓ	0	0	1	0	1	0	2	1	0	2	1	0	3	2	1	0	3	2
$\mathbf{n} + \ell$	1	2	3	3	4	4	5	5	5	6	6	6	7	7	7	7	8	8

Table 5.7, p. 154, lists the *abbreviated electron configurations* of elements with atomic numbers 1 through 106. Here, to save space, we start from the preceding noble gas. Thus for the element nickel we write:

[Ar]$4s^2 3d^8$

The symbol [Ar] indicates that the first 18 electrons in the Ni atom have the argon configuration, $1s^2 2s^2 2p^6 3s^2 3p^6$.

You will notice from Table 5.7 that the elements generally follow the rules we have discussed. At most, a single electron deviates from the expected order. In the first transition series, this happens with chromium and copper, where one of the 4s electrons is "promoted" to the 3d level. In actual fact, these two levels are very close to one another in energy throughout this transition series, so anomalies of this type are hardly surprising.

We don't know why this is so, but then there are many things we don't know

TABLE 5.7   Ground State Electron Configurations of Gaseous Atoms

1	H	$1s^1$	37	Rb	[Kr] $5s^1$	71	Lu	[Xe] $6s^24f^{14}5d^1$
2	He	$1s^2$	38	Sr	[Kr] $5s^2$	72	Hf	[Xe] $6s^24f^{14}5d^2$
3	Li	[He] $2s^1$	39	Y	[Kr] $5s^24d^1$	73	Ta	[Xe] $6s^24f^{14}5d^3$
4	Be	[He] $2s^2$	40	Zr	[Kr] $5s^24d^2$	74	W	[Xe] $6s^24f^{14}5d^4$
5	B	[He] $2s^22p^1$	41	Nb	[Kr] $5s^14d^4$	75	Re	[Xe] $6s^24f^{14}5d^5$
6	C	[He] $2s^22p^2$	42	Mo	[Kr] $5s^14d^5$	76	Os	[Xe] $6s^24f^{14}5d^6$
7	N	[He] $2s^22p^3$	43	Tc	[Kr] $5s^24d^5$	77	Ir	[Xe] $6s^24f^{14}5d^7$
8	O	[He] $2s^22p^4$	44	Ru	[Kr] $5s^14d^7$	78	Pt	[Xe] $6s^14f^{14}5d^9$
9	F	[He] $2s^22p^5$	45	Rh	[Kr] $5s^14d^8$	79	Au	[Xe] $6s^14f^{14}5d^{10}$
10	Ne	[He] $2s^22p^6$	46	Pd	[Kr] $4d^{10}$	80	Hg	[Xe] $6s^24f^{14}5d^{10}$
11	Na	[Ne] $3s^1$	47	Ag	[Kr] $5s^14d^{10}$	81	Tl	[Xe] $6s^24f^{14}5d^{10}6p^1$
12	Mg	[Ne] $3s^2$	48	Cd	[Kr] $5s^24d^{10}$	82	Pb	[Xe] $6s^24f^{14}5d^{10}6p^2$
13	Al	[Ne] $3s^23p^1$	49	In	[Kr] $5s^24d^{10}5p^1$	83	Bi	[Xe] $6s^24f^{14}5d^{10}6p^3$
14	Si	[Ne] $3s^23p^2$	50	Sn	[Kr] $5s^24d^{10}5p^2$	84	Po	[Xe] $6s^24f^{14}5d^{10}6p^4$
15	P	[Ne] $3s^23p^3$	51	Sb	[Kr] $5s^24d^{10}5p^3$	85	At	[Xe] $6s^24f^{14}5d^{10}6p^5$
16	S	[Ne] $3s^23p^4$	52	Te	[Kr] $5s^24d^{10}5p^4$	86	Rn	[Xe] $6s^24f^{14}5d^{10}6p^6$
17	Cl	[Ne] $3s^23p^5$	53	I	[Kr] $5s^24d^{10}5p^5$	87	Fr	[Rn] $7s^1$
18	Ar	[Ne] $3s^23p^6$	54	Xe	[Kr] $5s^24d^{10}5p^6$	88	Ra	[Rn] $7s^2$
19	K	[Ar] $4s^1$	55	Cs	[Xe] $6s^1$	89	Ac	[Rn] $7s^2$ $6d^1$
20	Ca	[Ar] $4s^2$	56	Ba	[Xe] $6s^2$	90	Th	[Rn] $7s^2$ $6d^2$
21	Sc	[Ar] $4s^23d^1$	57	La	[Xe] $6s^2$ $5d^1$	91	Pa	[Rn] $7s^25f^26d^1$
22	Ti	[Ar] $4s^23d^2$	58	Ce	[Xe] $6s^24f^15d^1$	92	U	[Rn] $7s^25f^36d^1$
23	V	[Ar] $4s^23d^3$	59	Pr	[Xe] $6s^24f^3$	93	Np	[Rn] $7s^25f^46d^1$
24	Cr	[Ar] $4s^13d^5$	60	Nd	[Xe] $6s^24f^4$	94	Pu	[Rn] $7s^25f^6$
25	Mn	[Ar] $4s^23d^5$	61	Pm	[Xe] $6s^24f^5$	95	Am	[Rn] $7s^25f^7$
26	Fe	[Ar] $4s^23d^6$	62	Sm	[Xe] $6s^24f^6$	96	Cm	[Rn] $7s^25f^76d^1$
27	Co	[Ar] $4s^23d^7$	63	Eu	[Xe] $6s^24f^7$	97	Bk	[Rn] $7s^25f^9$
28	Ni	[Ar] $4s^23d^8$	64	Gd	[Xe] $6s^24f^75d^1$	98	Cf	[Rn] $7s^25f^{10}$
29	Cu	[Ar] $4s^13d^{10}$	65	Tb	[Xe] $6s^24f^9$	99	Es	[Rn] $7s^25f^{11}$
30	Zn	[Ar] $4s^23d^{10}$	66	Dy	[Xe] $6s^24f^{10}$	100	Fm	[Rn] $7s^25f^{12}$
31	Ga	[Ar] $4s^23d^{10}4p^1$	67	Ho	[Xe] $6s^24f^{11}$	101	Md	[Rn] $7s^25f^{13}$
32	Ge	[Ar] $4s^23d^{10}4p^2$	68	Er	[Xe] $6s^24f^{12}$	102	No	[Rn] $7s^25f^{14}$
33	As	[Ar] $4s^23d^{10}4p^3$	69	Tm	[Xe] $6s^24f^{13}$	103	Lr	[Rn] $7s^25f^{14}6d^1$
34	Se	[Ar] $4s^23d^{10}4p^4$	70	Yb	[Xe] $6s^24f^{14}$	104		[Rn] $7s^25f^{14}6d^2$
35	Br	[Ar] $4s^23d^{10}4p^5$				105		[Rn] $7s^25f^{14}6d^3$
36	Kr	[Ar] $4s^23d^{10}4p^6$				106		[Rn] $7s^25f^{14}6d^4$

ORBITAL DIAGRAMS; HUND'S RULE

For many purposes, electron configurations are sufficient to describe the arrangement of electrons in atoms. The energy of an electron is determined primarily by the principal level and sublevel in which it is located. Sometimes, however, we want to indicate how electrons are distributed within orbitals. To do this, we use orbital diagrams such as those shown in Table 5.5, p. 151, for hydrogen and helium.

To show how orbital diagrams are obtained from electron configurations, consider the boron atom (at. no. = 5). Its electron configuration is $1s^22s^22p^1$. We know that the pair of electrons in the 1s orbital must have opposed spins. The same is true for the two electrons in the 2s orbital. There are three orbitals in the 2p sublevel. The single 2p electron in boron could be in any one of these orbitals. Its spin could be either "up" ($m_s = +\frac{1}{2}$) or "down" ($m_s = -\frac{1}{2}$). The orbital diagram is ordinarily written

$$\begin{array}{cccc} & 1s & 2s & 2p \\ {}_5B & (\uparrow\downarrow) & (\uparrow\downarrow) & (\uparrow)(\ )(\ ) \end{array}$$

with the first electron in an orbital arbitrarily designated by an "up" arrow, $\uparrow$.

With the next element, carbon, a complication arises. Where should we put the sixth electron? We could put it in the same orbital as the other 2p electron, in which case it would have to have the opposite spin, $\downarrow$. It could go into one of the other two orbitals, either with a parallel spin, $\uparrow$, or an opposed spin, $\downarrow$. Experiment shows that there is an energy difference between these arrangements. The most stable is the one in which the two electrons are in different orbitals with parallel spins. The orbital diagram of the carbon atom is

$$\begin{array}{cccc} & 1s & 2s & 2p \\ {}_6C & (\uparrow\downarrow) & (\uparrow\downarrow) & (\uparrow)(\uparrow)(\ ) \end{array}$$

This situation arises frequently with other atoms. There is a general principle that applies in all such cases. Known as Hund's rule, it can be stated as follows:

Within a given sublevel, the order of filling is such that there is the maximum number of half-filled orbitals. The single electrons in these half-filled orbitals have parallel spins.

Following this principle, we show the orbital diagrams for the elements boron through neon in Figure 5.12. Notice that

— in all filled orbitals, the two electrons have opposed spins. Such electrons are often referred to as being *paired*. There are four paired electrons

Atom	Orbital diagram			Electron configuration
B	$(\uparrow\downarrow)$	$(\uparrow\downarrow)$	$(\uparrow\ )(\ \ )(\ \ )$	$1s^2\,2s^2\,2p^1$
C	$(\uparrow\downarrow)$	$(\uparrow\downarrow)$	$(\uparrow\ )(\uparrow\ )(\ \ )$	$1s^2\,2s^2\,2p^2$
N	$(\uparrow\downarrow)$	$(\uparrow\downarrow)$	$(\uparrow\ )(\uparrow\ )(\uparrow\ )$	$1s^2\,2s^2\,2p^3$
O	$(\uparrow\downarrow)$	$(\uparrow\downarrow)$	$(\uparrow\downarrow)(\uparrow\ )(\uparrow\ )$	$1s^2\,2s^2\,2p^4$
F	$(\uparrow\downarrow)$	$(\uparrow\downarrow)$	$(\uparrow\downarrow)(\uparrow\downarrow)(\uparrow\ )$	$1s^2\,2s^2\,2p^5$
Ne	$(\uparrow\downarrow)$	$(\uparrow\downarrow)$	$(\uparrow\downarrow)(\uparrow\downarrow)(\uparrow\downarrow)$	$1s^2\,2s^2\,2p^6$
	1s	2s	2p	

Figure 5.12 Orbital diagrams showing electron arrangements for atoms with five to ten electrons. Electrons entering orbitals of equal energy remain unpaired as long as possible.

in the B, C, and N atoms, six in the oxygen atom, eight in the fluorine atom, and ten in the neon atom.

— in accordance with Hund's rule, within a given sublevel there are as many half-filled orbitals as possible. Electrons in such orbitals are said to be *unpaired*. There is one unpaired electron in atoms of B and F, two unpaired electrons in C and O atoms, and three unpaired electrons in the N atom. When there are two or more unpaired electrons, as in C, N, and O, those electrons have parallel spins.

Hund's rule keeps the electrons as far apart as possible

■ **EXAMPLE 5.7** _____

Construct orbital diagrams for atoms of chlorine and iron.

Solution

We first need to know the electron configurations of the atoms. For a Cl atom with 17 electrons, the electron configuration is $1s^2 2s^2 2p^6 3s^2 3p^5$. For an Fe atom with 26 electrons, the configuration is $1s^2 2s^2 2p^6 3s^2 3p^6 4s^2 3d^6$.

In writing an orbital diagram, we deal first with orbitals in completed sublevels. Each such orbital is filled with two electrons of opposed spins (↑ and ↓). Then we turn our attention to partially filled sublevels. Here we add electrons one by one to the available orbitals, keeping spins parallel (↑ and ↑) as much as possible, in accordance with Hund's rule:

	1s	2s	2p	3s	3p
$_{17}$Cl	(↑↓)	(↑↓)	(↑↓)(↑↓)(↑↓)	(↑↓)	(↑↓)(↑↓)(↑)

	1s	2s	2p	3s	3p	4s	3d
$_{26}$Fe	(↑↓)	(↑↓)	(↑↓)(↑↓)(↑↓)	(↑↓)	(↑↓)(↑↓)(↑↓)	(↑↓)	(↑↓)(↑)(↑)(↑)(↑)

Orbital diagrams are most useful when we deal with electron spins

Notice that there is one unpaired electron in the Cl atom; in the Fe atom, there are four unpaired electrons.

Exercise

Show the orbital diagram for the 3d sublevel for Ni (at. no. = 28). Answer: (↑↓)(↑↓)(↑↓)(↑)(↑).

Hund's rule, like the Pauli exclusion principle, is based upon experiment. It is possible to determine the number of unpaired electrons in an atom. With solids, this is done by studying their behavior in a magnetic field. If there are unpaired electrons present, the solid will be attracted into the field. Such a substance is said to be *paramagnetic*. If the atoms in the solid contain only paired electrons, it is slightly repelled by the field. Substances of this type are called *diamagnetic*. With gaseous atoms, the atomic spectrum can also be used to establish the presence and number of unpaired electrons.

QUANTUM NUMBERS

Within limits, it is possible to state a complete set of quantum numbers for electrons in atoms. This was done in Table 5.5 for hydrogen and helium. Ordinarily, to specify the quantum numbers, it is simplest to start with the orbital diagram. Consider, for example, the boron atom, whose orbital diagram is given in Figure 5.12. For the first four electrons, we readily arrive at the sets of quantum numbers:

$$1, 0, 0, +\tfrac{1}{2}; \quad 1, 0, 0, -\tfrac{1}{2}; \quad 2, 0, 0, +\tfrac{1}{2}; \quad 2, 0, 0, -\tfrac{1}{2}$$

We note that the fifth electron is in a 2p orbital. Its $\mathbf{m}_\ell$ value could be 1, 0, or −1. Since all these orbitals have the same energy, it makes no difference which one we choose. If we arbitrarily use the highest $\mathbf{m}_\ell$ value, 1, for the first orbital to be filled, the quantum numbers for the fifth electron in boron become:

$$2, 1, 1, +\tfrac{1}{2}$$

Following the same procedure, the set of quantum numbers for the sixth electron in carbon would be

$$2, 1, 0, +\tfrac{1}{2}$$

Notice that the two 2p electrons in carbon have parallel spins: $\mathbf{m}_s = +\tfrac{1}{2}$ in both cases. For the seventh electron in nitrogen, the set of quantum numbers is:

$$2, 1, -1, +\tfrac{1}{2}$$

Summarizing the quantum numbers of each electron in the nitrogen atom,

	1s		2s		2p		
	(↑	↓)	(↑	↓)	(↑)	(↑)	(↑)
n	1	1	2	2	2	2	2
ℓ	0	0	0	0	1	1	1
$\mathbf{m}_\ell$	0	0	0	0	1	0	−1
$\mathbf{m}_s$	$+\tfrac{1}{2}$	$-\tfrac{1}{2}$	$+\tfrac{1}{2}$	$-\tfrac{1}{2}$	$+\tfrac{1}{2}$	$+\tfrac{1}{2}$	$+\tfrac{1}{2}$

■ **EXAMPLE 5.8**

Write a complete set of quantum numbers for the four unpaired 3d electrons in an iron atom. Its orbital diagram is shown in Example 5.7.

Solution

For all these electrons, $\mathbf{n} = 3$ (third principal level), and $\ell = 2$ (d sublevel). The values of $\mathbf{m}_\ell$ run from 2 to −2. If, in the orbital diagram, we start with $\mathbf{m}_\ell = 2$ at the left, we have

	(⇅)	(↑)	(↑)	(↑)	(↑)
$\mathbf{m}_\ell$	2	1	0	−1	−2

The unpaired electrons must all have the same spin. Since they are drawn as ↑, we arbitrarily assign them an $\mathbf{m}_s$ value of $+\tfrac{1}{2}$. The complete set of quantum numbers for the four unpaired electrons is

$$3, 2, 1, +\tfrac{1}{2}; \quad 3, 2, 0, +\tfrac{1}{2}; \quad 3, 2, -1, +\tfrac{1}{2}; \quad 3, 2, -2, +\tfrac{1}{2}$$

This is one set. Are there any others that would be OK?

Exercise

Give a complete set of quantum numbers for each paired 3d electron in Fe. Answer: $3, 2, 2, +\tfrac{1}{2}; 3, 2, 2, -\tfrac{1}{2}$.

5.6
ELECTRON ARRANGEMENTS IN MONATOMIC IONS

The electronic structures of cations and anions are closely related to those of the atoms from which they are derived. In this section, we will emphasize the electron configurations of these ions. You should, however, be able to derive orbital diagrams and quantum numbers of electrons in ions, following the principles discussed in Section 5.5.

IONS WITH NOBLE GAS STRUCTURES

As pointed out in Chapter 3, *elements close to a noble gas in the Periodic Table form ions that have the same number of electrons as the noble gas atom.* This means that these ions have noble gas electron configurations ($1s^2$ for He; $\mathbf{ns^2np^6}$ for the other noble gases). Thus the three elements preceding neon (N, O, and F) and the three elements following neon (Na, Mg, and Al) all form ions with the neon configuration, $1s^2 2s^2 2p^6$. The three nonmetal atoms achieve this structure by gaining electrons to form anions:

The noble gas electron configurations are remarkably stable

$$_7N\ (1s^2 2s^2 2p^3) + 3e^- \rightarrow\ _7N^{3-}\ (1s^2 2s^2 2p^6)$$

$$_8O\ (1s^2 2s^2 2p^4) + 2e^- \rightarrow\ _8O^{2-}\ (1s^2 2s^2 2p^6)$$

$$_9F\ (1s^2 2s^2 2p^5) + e^- \rightarrow\ _9F^-\ (1s^2 2s^2 2p^6)$$

The three metal atoms acquire the neon structure by losing electrons to form cations:

$$_{11}Na\ (1s^2 2s^2 2p^6 3s^1) \rightarrow\ _{11}Na^+\ (1s^2 2s^2 2p^6) + e^-$$

$$_{12}Mg\ (1s^2 2s^2 2p^6 3s^2) \rightarrow\ _{12}Mg^{2+}\ (1s^2 2s^2 2p^6) + 2e^-$$

$$_{13}Al\ (1s^2 2s^2 2p^6 3s^2 3p^1) \rightarrow\ _{13}Al^{3+}\ (1s^2 2s^2 2p^6) + 3e^-$$

The species N^{3-}, O^{2-}, F^-, Ne, Na^+, Mg^{2+} and Al^{3+} are said to be *isoelectronic;* they have the same electron configuration.

There are a great many ions that have noble gas configurations, suggesting that this electron arrangement is a particularly stable one. Figure 5.13 shows 24 ions of this type. Notice from the figure that

Figure 5.13 Species (cations, anions, or atoms) with noble gas structures. The color coding indicates species that are *isoelectronic* (same electronic structure).

							H^-	He
Li^+	Be^{2+}				N^{3-}	O^{2-}	F^-	Ne
Na^+	Mg^{2+}	Al^{3+}				S^{2-}	Cl^-	Ar
K^+	Ca^{2+}	Sc^{3+}				Se^{2-}	Br^-	Kr
Rb^+	Sr^{2+}	Y^{3+}				Te^{2-}	I^-	Xe
Cs^+	Ba^{2+}	La^{3+}						

1. The anions are formed from neutral atoms having one, two, or three *less* electrons than the *next* noble gas. Those in Group 7 (outer configuration $\mathbf{ns^2np^5}$) *gain one* electron per atom to form -1 ions. Group 6 nonmetals (outer configuration $\mathbf{ns^2np^4}$) *gain two* electrons per atom to form -2 ions. Nitrogen (outer configuration $2s^22p^3$) forms -3 ions when it reacts with very reactive metals such as lithium or magnesium. Other Group 5 elements rarely, if ever, form -3 ions.

2. The cations are formed by metals whose neutral atoms have one, two, or three electrons *more* than the preceding noble gas. Atoms of Group 1 ($\mathbf{ns^1}$) and Group 2 ($\mathbf{ns^2}$) achieve noble gas configurations by losing one and two electrons, respectively. Accordingly, the corresponding cations have $+1$ and $+2$ charges (Group 1 = $+1$; Group 2 = $+2$). Atoms of aluminum in Group 3 and the metals in the scandium subgroup all have three electrons more than the preceding noble gas. They form $+3$ ions: Al^{3+}, Sc^{3+}, ... (Boron, the first member of Group 3, does not form ions.)

> If an atom can attain a noble gas structure by gaining or losing up to three electrons, it will usually form an ion with that structure

TRANSITION METAL IONS

The transition metals to the right of the scandium subgroup do not form ions with noble gas configurations. To do so, they would have to lose four or more electrons. The energy requirement is too high for that to happen. However, as pointed out in Chapter 3, these metals do form cations with charges of $+1$, $+2$, or $+3$. It is important to note that **when transition metal atoms form positive ions, the outer s electrons are lost first.** Consider, for example, the formation of the Mn^{2+} ion from the Mn atom:

$$_{25}Mn\ (Ar\ 4s^23d^5) \rightarrow {}_{25}Mn^{2+}\ (Ar\ 3d^5) + 2e^-$$

Notice that it is the 4s electrons that are lost rather than the 3d electrons. We know this is the case because the Mn^{2+} ion has been shown to have five unpaired electrons (the five 3d electrons). If two 3d electrons had been lost, the Mn^{2+} ion would have had only three unpaired electrons.

All the transition metals form cations by a similar process, i.e., loss of outer s electrons. Only after those electrons are lost are electrons removed from the inner d sublevel. Consider, for example, what happens with iron, which, you will recall, forms two different cations. First the 4s electrons are lost to give the Fe^{2+} ion:

$$_{26}Fe\ (Ar\ 4s^23d^6) \rightarrow {}_{26}Fe^{2+}\ (Ar\ 3d^6) + 2e^-$$

Then an electron is removed from the 3d level to form the Fe^{3+} ion:

$$_{26}Fe^{2+}\ (Ar\ 3d^6) \rightarrow {}_{26}Fe^{3+}\ (Ar\ 3d^5) + e^-$$

In the Fe^{2+} and Fe^{3+} ions, as in all transition metal ions, there are no outer s electrons.

The behavior just described suggests that in transition metal cations (as opposed to the atoms), the inner d sublevels are lower in energy than the outer s sublevels; that is,

Fourth period cations: 3d lower in energy than 4s

Fifth period cations: 4d lower in energy than 5s

Sixth period cations: 5d lower in energy than 6s

■ **EXAMPLE 5.9** _____

Give the abbreviated electron configurations of

a. a Sc atom and a Sc^{3+} ion.
b. a Zn atom and a Zn^{2+} ion.
c. a Se atom and a Se^{2-} ion.

Solution

In cation formation, electrons are lost. The ion formed has a positive charge equal to the number of electrons lost. In anion formation, electron gain yields an ion with a negative charge equal to the number of electrons gained.

a. To form a $+3$ ion, scandium (at. no. $= 21$) loses three electrons. The Sc^{3+} ion has 18 electrons and the argon structure:

$$_{21}Sc\ (Ar\ 4s^2 3d^1) \rightarrow\ _{21}Sc^{3+}\ (Ar) + 3e^-$$

b. A Zn atom loses two electrons to form the Zn^{2+} ion:

$$_{30}Zn\ (Ar\ 4s^2 3d^{10}) \rightarrow\ _{30}Zn^{2+}\ (Ar\ 3d^{10}) + 2e^-$$

Note the loss of 4s rather than 3d electrons.

c. Two electrons are gained to give the electron configuration of the noble gas krypton:

$$_{34}Se\ (Ar\ 4s^2 3d^{10} 4p^4) \rightarrow\ _{34}Se^{2-}\ (Kr)$$

Exercise

What is the abbreviated electron configuration of $_{44}Ru^{3+}$? Answer: $(Kr\ 4d^5)$.

_____

_____

■ **SUMMARY PROBLEM**

Consider the element cobalt (at. no. $= 27$).

a. Give the electron configuration of the Co atom; the Co^{3+} ion.
b. Give the orbital diagram (beyond Ar) for Co and Co^{3+}.
c. Give the quantum numbers for all the d electrons in the Co atom; In the Co^{3+} ion.
d. How many unpaired electrons are there in the Co atom? In the Co^{3+} ion?
e. How many p electrons are there in the Co atom? In the Co^{3+} ion?
f. There is a line in the cobalt spectrum at 345.35 nm. In what region of the spectrum (ultraviolet, visible, or infrared) is this line found? What is the frequency of the line? What is the energy difference between the two levels responsible for this line, in kilojoules per mole?

g. The ionization energy of cobalt from the ground state is 758 kJ/mol. Assume that the transition in (f) is from the ground state to an excited state. If that is the case, calculate the ionization energy from the excited state.

Answers

a. $1s^22s^22p^63s^23p^64s^23d^7$; $1s^22s^22p^63s^23p^63d^6$

b.   4s   3d

Co   (↑↓) (↑↓)(↑↓)(↑)(↑)(↑)
Co^{3+} (   ) (↑↓)(↑)(↑)(↑)(↑)

c. Co: $3,2,2,\frac{1}{2}$; $3,2,2,-\frac{1}{2}$; $3,2,1,\frac{1}{2}$; $3,2,1,-\frac{1}{2}$; $3,2,0,\frac{1}{2}$; $3,2,-1,\frac{1}{2}$; $3,2,-2,\frac{1}{2}$ (other combinations are possible)

Co^{3+}: $3,2,2,\frac{1}{2}$; $3,2,2,-\frac{1}{2}$; $3,2,1,\frac{1}{2}$; $3,2,0,\frac{1}{2}$; $3,2,-1,\frac{1}{2}$; $3,2,-2,\frac{1}{2}$ (other combinations are possible)

d. 3; 4   **e.** 12; 12   **f.** UV; 8.681×10^{14}/s; 346.3 kJ/mol

g. 412 kJ

MODERN
PERSPECTIVE

SPECTROSCOPY

Earlier in this chapter, we showed how the atomic spectrum of hydrogen served to establish the energy levels of the hydrogen electron. Similar deductions can be made with multielectron atoms, although the spectra are considerably more complex. Much of the evidence for electron configurations comes from atomic spectra. For example, the nature of the lines in the spectrum of atomic chromium indicates that the gaseous atom contains one rather than two 4s electrons.

More generally, **atomic spectroscopy** can be used to analyze for the presence of different metals in a sample (qualitative analysis) and to estimate how much of a particular metal is present (quantitative analysis). The method is fast and reasonably accurate, and can be applied to a wide variety of samples. Most important, it can be used to detect metals at concentrations as low as 1×10^{-7} mol/L.

In *atomic emission spectroscopy,* the sample is most often introduced directly into a high-temperature flame (2000°C–3000°C). Under these conditions, any molecules or polyatomic ions are broken down into individual atoms. Electrons in these atoms are promoted to excited energy levels; when they return to lower levels, they give off light at discrete wavelengths. Typically, the strongest emissions are in the ultraviolet, except for the Group 1 and Group 2 metals, where they are in the visible region of the spectrum. The emitted light passes to a detector, which activates a recorder that produces the type of spectrum shown in Figure 5.14. The location (i.e., wavelength) of various lines in the spectrum is characteristic of a particular metal and serves to identify it. The height or, more accurately, the total area of a particular peak in the spectrum is directly proportional to the amount of metal present and so can be used to establish its concentration.

Atomic absorption spectroscopy can be used to determine metals at very low concentrations. Once again, the sample is introduced into a high-temperature flame to assure dissociation into atoms. This time, though, ultraviolet or visible radiation is passed through the flame; the quantity measured is the wavelength at which light is absorbed. These wavelengths may or may not be the same as those observed in emission. The important point is that the spectrum produced is characteristic of a particular metal. These two techniques, atomic emission spectroscopy and atomic absorption spectroscopy, are used in a variety of fields outside of chemistry. People doing research in nutrition use atomic spectroscopy routinely (along with chromatography and mass spectroscopy) to establish the presence of essential trace metals in human and animal tissue.

In the general chemistry laboratory, you are most likely to encounter spectroscopy in the form of **molecular absorption spectroscopy**. Here the sample being analyzed is at room temperature; the species that absorbs radiation is a molecule or polyatomic ion. Figure 5.15, p. 164, shows the absorption spectrum of the MnO_4^- ion. Notice the broad band in the green-yellow region of the spectrum (~500–570 nm), midway through the visible range. Light transmitted through a solution of $KMnO_4$ is rich in the blue (shorter wavelength) and red (longer wavelength) components. Hence it appears purple, a mixture of blue and red.

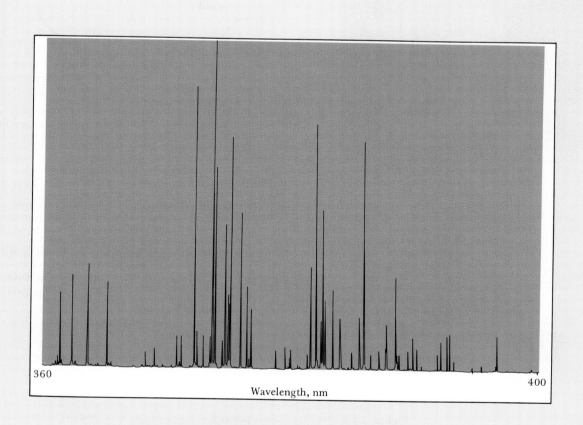

Wavelength, nm

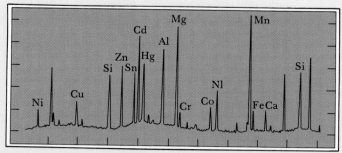

Figure 5.14 Atomic emission spectroscopy. The diagram at the top shows the iron spectrum between 360 and 400 nm; note the large number of lines, each corresponding to a different electronic transition in the iron atom. The lower spectrum was obtained with a sample containing 14 elements; the horizontal axis covers a total distance of only about 2 nm, indicating the high resolution characteristic of modern spectrometers.

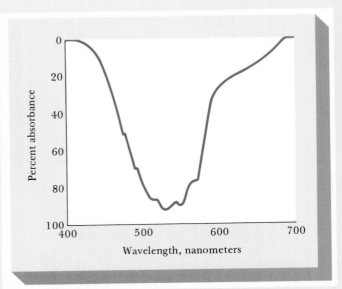

Figure 5.15 The purple color of a solution of KMnO₄ is caused by strong absorption in the yellow-green region of the spectrum. (Charles D. Winters)

Figure 5.16 Infrared absorption spectrum of ascorbic acid (Vitamin C). Note the strong absorptions (80% or more) at 1 (6100 nm), 2 (7600 nm), 3 (8800 nm), and 4 (9600 nm). These serve to identify this particular compound.

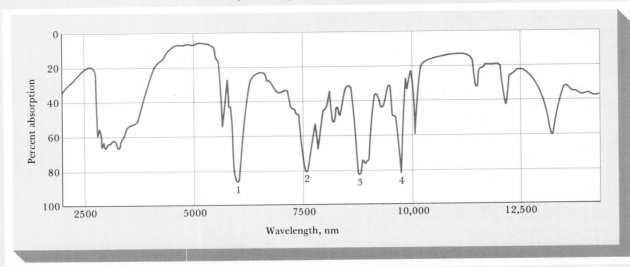

Many molecules absorb radiation outside the visible region. In particular, all organic molecules absorb in the infrared, above 800 nm. By exposing a sample to infrared radiation covering a range of wavelengths and measuring absorption as a function of wavelength, a spectrum of the type shown in Figure 5.16 is obtained. Such an infrared (IR) spectrum serves as a "fingerprint" to identify a substance. Organic chemists use this technique routinely. One of its major advantages is that it can be applied to samples weighing less than a milligram.

Infrared absorption spectroscopy can also be used to deduce the structure of molecules. In simple cases, like HCl and H_2O, it is possible from a study of the IR spectrum to determine the distance between atoms in the molecule, and the strength of the chemical bonds holding the molecule together. With more complex molecules, it is even possible to deduce from the spectrum the geometric pattern in which the atoms are arranged (Chap. 9).

QUESTIONS AND PROBLEMS

Energy, Wavelength and Frequency

1. Compare the energies and frequencies of two photons, one with a short wavelength, the other with a long wavelength.

2. Compare the energy of an electron in the ground state to that in an excited state. In which state does the electron have the higher energy?

3. A photon of yellow light has a wavelength of 585 nm. Calculate the
  a. frequency.
  b. energy difference in joules per particle.
  c. energy difference in kilojoules per mole.

4. An infrared ray has a frequency of 5.71×10^{12}/s. Calculate the
  a. wavelength.
  b. energy difference in joules per particle.
  c. energy difference in kilojoules per mole.

5. The first ionization energy of potassium is 419 kJ/mol. Do x-rays with a wavelength of 80 nm have sufficient energy to ionize potassium?

6. Carbon dioxide absorbs energy at a frequency of 2.001×10^{13}/s.
  a. Calculate the wavelength of this absorption in nanometers.
  b. In what spectral range does this absorption occur?
  c. What is the energy difference in kilojoules per mole?

7. Two states differ in energy by 5.00×10^2 kJ/mol. When an electron moves from the higher to the lower of these states, calculate

  a. λ in nanometers
  b. ν in $(\text{seconds})^{-1}$

8. Two states differ in energy by 4.08×10^{-18} J/particle. When an electron moves from the higher to the lower state, calculate
  a. λ in meters
  b. λ in nanometers
  c. ν in $(\text{seconds})^{-1}$

9. When sunlight is scattered by air molecules, the sky appears blue. Blue light has a frequency of 6.6×10^{14} Hz. Calculate
  a. the wavelength in nanometers associated with this radiation.
  b. the energy in J/particle associated with this frequency.

10. Compact disc players use lasers with light at a wavelength of 785 nm.
  a. What is the frequency of this light?
  b. What is the energy associated with one photon of this light?

Bohr Model of the Hydrogen Atom

11. Consider the following transitions

  (1) $n = 1$ to $n = 3$
  (2) $n = 3$ to $n = 4$
  (3) $n = 3$ to $n = 2$
  (4) $n = 5$ to $n = 3$

  a. Which one involves the ground state?
  b. Which absorbs the most energy?
  c. Which emits the most energy?

12. Consider the transitions in Question 11. For which of the transitions is energy absorbed? For which is energy emitted?

13. According to the Bohr model, the radius of a circular orbit is given by the expression

$$r \text{ (in nm)} = 0.0529 \, n^2$$

Draw successive orbits for the hydrogen electron at $n = 1, 2, 3,$ and 4. Indicate by arrows transitions between orbits that lead to lines in the
a. Lyman series
b. Balmer series

14. Using Equation 5.5, calculate E for $n = 1, 2, 3,$ and 4. Make a one-dimensional graph showing energy, at different values of n, increasing vertically. On this graph, indicate by vertical arrows transitions in the
a. Lyman series
b. Balmer series

15. For the Brackett series, $n_{lo} = 4$. Calculate the wavelength in nanometers, of a transition from $n = 6$ to $n = 4$.

16. The Paschen series lines in the atomic spectrum of hydrogen result from electronic transitions from $n > 3$ to $n = 3$. Calculate the wavelength, in nanometers, of a line in this series resulting from the $n = 7$ to $n = 3$ transition.

17. A line in the Balmer series occurs at 434.05 nm. Calculate the n_{hi} for the transition associated with this line.

18. In the Pfund series $n_{lo} = 5$. Calculate the longest wavelength possible for a transition in this series.

19. Calculate the ionization energy (kJ/mol) for the hydrogen electron from the $n = 5$ state.

20. Calculate ΔE, in kilojoules per mole, when a hydrogen electron in the second excited state ($n = 3$) is completely removed from the atom.

de Broglie Relation

21. Compare the velocity of an electron (mass $= 9.11 \times 10^{-28}$ g) and a neutron (mass $= 1.67 \times 10^{-24}$ g), both with de Broglie wavelengths of 0.100 nm (1 J $= 1$ kg·m²/s²).

22. Compare the de Broglie wavelengths (in nm) of an electron (mass $= 9.11 \times 10^{-28}$ g) and a proton (mass $= 1.67 \times 10^{-24}$ g), each moving at 1.00% the speed of light ($c = 2.998 \times 10^8$ m/s; 1 J $= 1$ kg·m²/s²).

23. A golf ball weighs about 0.100 lb. Calculate the de Broglie wavelength of a golf ball traveling at 1.00×10^2 miles per hour.

24. Calculate the de Broglie wavelength of a shotgun pellet weighing 1.2 g moving with a velocity of 150 m/s.

Energy Levels, Sublevels, and Quantum Numbers

25. What are the possible values for m_ℓ for
a. $\ell = 2$ b. $\ell = 4$ c. $n = 4$ (all sublevels)

26. What are the possible values of m_ℓ for
a. $\ell = 0$ b. $\ell = 3$ c. $n = 3$ (all sublevels)

27. For the following pairs of orbitals, indicate which is higher in energy in a many-electron atom.
a. 3p or 4p   b. 4s or 4d
c. 2s or 3d   d. 5s or 4f

28. Which orbital in each of the following pairs is lower in energy in a many-electron atom?
a. 3s or 3p   b. 4p or 4d
c. 1s or 2s   d. 4d or 5f

29. What type of electron orbital (i.e., s, p, d, or f) is designated:
a. $n = 2, \ell = 1, m_\ell = 1$?
b. $n = 4, \ell = 3, m_\ell = 0$?
c. $n = 1, \ell = 0, m_\ell = 0$?

30. What type of electron orbital (i.e., s, p, d, or f) is designated by:
a. $n = 4, \ell = 1, m_\ell = 0$?
b. $n = 3, \ell = 2, m_\ell = 0$?
c. $n = 2, \ell = 0, m_\ell = 0$?

31. Give the number of orbitals in
a. the principal level $n = 3$.
b. a 4d sublevel
c. an f sublevel

32. State the total capacity for electrons in
a. the principal level $n = 4$
b. a 4d sublevel
c. a 4d orbital

33. State the relationship, if any, between the following pairs of quantum numbers:
a. n and ℓ b. ℓ and m_ℓ c. m_ℓ and m_s

34. What is the
a. minimum n value for $\ell = 2$?
b. letter used to designate the sublevel with $\ell = 2$?
c. number of orbitals in a sublevel with $\ell = 3$?
d. number of different sublevels when $n = 3$?

35. Given the following sets of electron quantum numbers, indicate those that could not occur and explain your answer.
a. $1, 1, 0, +\frac{1}{2}$
b. $2, 1, 0, +\frac{1}{2}$
c. $2, 0, 1, -\frac{1}{2}$
d. $2, 1, 0, 0$
e. $3, 2, 0, -\frac{1}{2}$

36. Arrange the following sets of quantum numbers in order of increasing energy. If they have the same energy, place them together.
a. $4, 2, -1, +\frac{1}{2}$
b. $1, 0, 0, -\frac{1}{2}$
c. $3, 1, 1, -\frac{1}{2}$
d. $2, 0, 0, +\frac{1}{2}$
e. $2, 1, 0, +\frac{1}{2}$
f. $3, 1, 1, +\frac{1}{2}$

Electron Configurations

37. Write the electron configuration (ground state) for
a. Cl   b. Cr   c. Se   d. Si

38. Write the electron configuration (ground state) for
a. Mg   b. Co   c. K   d. In

39. Write the abbreviated electron configuration for
a. Br   b. Ti   c. Cd   d. F

40. Write the abbreviated electron configuration for
a. O   b. Ba   c. Ni   d. Te

41. Give the symbol of the element of lowest atomic number whose ground state has
a. a d electron
b. three f electrons
c. a completed s sublevel
d. six p electrons

42. Give the symbol of the element of lowest atomic number whose ground state has
a. a completed d sublevel
b. three 4d electrons
c. five 3p electrons
d. one s electron

43. What fraction of the total number of electrons is in s sublevels in
a. Ar?   b. Cd?   c. Ne?

44. What fraction of the total number of electrons is in p sublevels in
a. Ca?   b. Cs?   c. Ru?

45. Which of the following electron configurations are for atoms in the ground state? In excited states? Which are impossible?
a. $1s^1 1p^1$
b. $2s^1 2p^1$
c. $1s^2 2s^2 2p^5$
d. $[_{10}Ne] 3s^2 3p^3 3d^1$
e. $[_{10}Ne] 3s^2 3p^7 4s^2 5d^1$
f. $1s^2 2s^1 2p^3 3s^2$

46. Which of the following electron configurations are for atoms in the ground state? In excited states? Which are not possible?
a. $1s^2 2s^2$
b. $1s^2 2s^2 4s^1$
c. $[_{10}Ne] 3s^2 3p^6 4s^1$
d. $[_{18}Ar] 3s^1$
e. $[_2He] 2s^2 2p^6 2d^3$
f. $[_{10}Ne] 3p^1$

Orbital Diagrams; Hund's Rule

47. Give the orbital diagram of
a. C   b. Fe   c. P   d. Ar

48. Give the orbital diagram of
a. Si   b. Nb   c. Sr   d. Sb

49. Give the symbol of the atom with the following orbital diagram beyond argon.

	4s	3d	4p
a.	(↑↓)	(↑↓)(↑↓)(↑↓)(↓)(↓)	( )( )( )
b.	(↑↓)	(↑↓)(↑↓)(↓)(↓)(↓)	( )( )( )
c.	(↑↓)	(↑↓)(↑↓)(↑↓)(↑↓)(↑↓)	(↓)(↓)( )

50. Give the symbol of the atom that has the following orbital diagram:

	1s	2s	2p	3s	3p
a.	(↑↓)	(↑↓)	(↑↓)(↑↓)(↑↓)	(↑↓)	( )( )( )
b.	(↑↓)	(↑↓)	(↑↓)(↑↓)(↑↓)	(↑↓)	(↓)(↓)(↓)
c.	(↑↓)	(↑↓)	(↑↓)(↓)(↓)	( )	( )( )( )

51. Give the symbols of
a. all the elements that have filled 3p sublevels.
b. all the metals in the 5th period that have no unpaired electrons.
c. all the nonmetals in the 3rd period that have one unpaired electron.
d. all the elements in the 3rd period in which all sublevels are half-full.

52. Give the symbols of
a. all the elements in which all the 4d orbitals are half-full.
b. all the metals in the 5th period that have one unpaired electron.
c. all the non-metals in the 4th period that have no unpaired electrons.
d. all the elements in the 5th period in which the 5p sublevel is half-full.

53. Give the number of unpaired electrons in an atom of
a. O   b. Sr   c. V

54. How many unpaired electrons are there in an atom of
a. S?   b. Tc?   c. B?

55. Give the symbol of the element(s) in the second transition series with the following number of unpaired electrons per atom:
a. 0   b. 1   c. 2
d. 3   e. 4   f. 5

56. In what main group(s) of the Periodic Table do element(s) have the following number of filled p orbitals in the outermost principal level?
a. 0   b. 1   c. 2   d. 3

Quantum Numbers in Atoms

57. Assign a set of four quantum numbers to each electron in the nitrogen atom.

58. Assign a set of four quantum numbers to each electron in the fluorine atom.

59. Assign a set of four quantum numbers to
a. the 5s electron in Rb.
b. all the 4d electrons in Tc.
c. all the p electrons in P.

60. Assign a set of four quantum numbers to
   a. the 4s electrons in Ca.
   b. all the 3d electrons in Cr.
   c. all the 3p electrons in Cd.
61. For how many electrons in strontium does
   a. $n = 5$, $\ell = 1$, $m_\ell = 0$, $m_s = +\frac{1}{2}$?
   b. $n = 4$, $\ell = 1$?
   c. $n = 3$, $\ell = 2$, $m_\ell = 1$?
62. For how many electrons in arsenic, As, does
   a. $n = 4$, $\ell = 1$, $m_\ell = 0$?
   b. $n = 3$, $\ell = 2$?
   c. $n = 3$, $\ell = 2$, $m_\ell = -1$?

Electron Arrangement in Ions

63. Write the electron configuration for
   a. a calcium atom; a Ca^{2+} ion.
   b. a selenium atom; an Se^{2-} ion.
   c. a zirconium atom; a Zr^{2+} ion.
   d. a Co^{2+} ion; a Co^{3+} ion.
64. Give the electron configuration of
   a. a nitrogen atom; an N^{3-} ion.
   b. a potassium atom; a K^+ ion.
   c. a scandium atom; an Sc^{3+} ion.
   d. an Fe^{2+} ion; an Fe^{3+} ion.
65. How many unpaired electrons are there in each of the following ions?
   a. Cr^{3+} b. Cr^{2+} c. Br^- d. Cu^+
66. Give the number of unpaired electrons in
   a. Sn   b. Sn^{2+} c. Sn^{4+} d. Cd^{2+} e. Na^+

Unclassified

67. A light bulb radiates 7.0% of the energy supplied to it as visible light. If the wavelength of visible light is assumed to be 565 nm, how many photons per second are emitted by a 75-W bulb (1 W = 1J/s)?
68. A carbon dioxide laser produces radiation of wavelength 10.6 μm (1 μm = 10^{-6} m). If the laser produces about one joule of energy/pulse, how many photons are produced per pulse?
69. Explain the difference between
   a. the Bohr model of the atom and the quantum mechanical model.
   b. wavelength and frequency.
   c. paramagnetism and diamagnetism.
   d. the geometries of the three different p orbitals.
70. Explain in your own words what is meant by
   a. the Pauli exclusion principle.
   b. Hund's rule.
   c. a line in an atomic spectrum.
   d. the principal quantum number.
71. Indicate whether each of the following statements is true or false. If false, correct the statement.
   a. An electron transition from $n = 3$ to $n = 1$ absorbs energy.
   b. Light emitted by an $n = 4$ to $n = 2$ transition will have a shorter wavelength than that from an $n = 5$ to $n = 2$ transition.
   c. A sublevel of $\ell = 4$ has a capacity of 18 electrons.
   d. An atom of a Group 6 element has two unpaired electrons.
72. Criticize the following statements:
   a. The energy of a photon is inversely proportional to frequency.
   b. The energy of the hydrogen electron is inversely proportional to the quantum number n.
   c. Electrons start to enter the fourth principal level as soon as the third is full.
73. No currently known elements contain electrons in g ($\ell = 4$) orbitals in the ground state. If an element is discovered that has electrons in the g orbital, what is the lowest value for n in which these g orbitals could exist? What are the possible values of m_ℓ? How many electrons could a set of g orbitals hold?

Challenge Problems

74. The energy of any one-electron species in its nth state is given by $E = -BZ^2/n^2$, where Z is the charge on the nucleus and B is 2.179×10^{-18} J/particle. Find the ionization energy of the Li^{2+} ion in its first excited state in kilojoules per mole.
75. In 1885, Johann Balmer, a numerologist, derived the following relation for the wavelength of lines in the visible spectrum of hydrogen:

$$\lambda = 364.6\ n^2/(n^2 - 4)$$

where λ is in nanometers and n is an integer that can be 3, 4, 5, ... Show that this relation follows from the Bohr equation and Equation 5.2. Note that, for the Balmer series, the electron is returning to the $n = 2$ level.
76. Suppose the rules for assigning quantum numbers were as follows:

$$n = 1, 2, 3, \ldots$$
$$\ell = 0, 1, 2, \ldots, n$$
$$m_\ell = 0, 1, 2, \ldots, \ell + 1$$
$$m_s = +\tfrac{1}{2} \text{ or } -\tfrac{1}{2}$$

Prepare a table similar to Table 5.3, based on these rules, for $n = 1$ and $n = 2$. Give the electron configuration for an atom with eight electrons.
77. Suppose that the spin quantum number could have the values $\frac{1}{2}$, 0, and $-\frac{1}{2}$. Assuming that the rules governing the values of the other quantum numbers and the order of filling sublevels were unchanged,
   a. what would be the electron capacity of an s sublevel? A p sublevel? A d sublevel?

b. how many electrons could fit in the **n** = 3 level?

c. what would be the electron configuration of the element with atomic number 8? 17?

78. In the photoelectric effect, electrons are ejected from a metal surface when light strikes it. A certain minimum energy, E_{min}, is required to eject an electron. Any energy absorbed beyond that minimum gives kinetic energy to the electron. It is found that when light at a wavelength of 540 nm falls on a cesium surface, an electron is ejected with a kinetic energy of 2.60×10^{-20} J. When the wavelength is 400 nm, the kinetic energy is 1.54×10^{-19} J.

a. Calculate E_{min} for cesium, in joules.

b. Calculate the longest wavelength, in nanometers, that will eject electrons from cesium.

$$2K(s) \; + \; 2H_2O \rightarrow H_2(g) \; + \; 2K^+(aq) \; + \; 2OH^-(aq)$$

Potassium reacts vigorously with water to yield hydrogen gas and aqueous potassium hydroxide.

CHAPTER 6

THE PERIODIC TABLE AND THE MAIN-GROUP METALS

*Not chaos-like crush'd and bruis'd,
But, as the world, harmoniously
confus'd,
Where order in variety we see,
And where, though all things differ,
all agree.*

ALEXANDER POPE
WINDSOR FOREST

Scientists constantly seek ways to organize factual material so that similarities, differences, and trends become more apparent. The most useful device for this purpose in chemistry is the Periodic Table of the elements, referred to several times in previous chapters. The Periodic Table, you will recall, organizes elements in horizontal rows, called *periods*, and vertical columns, called *groups* or *families*. As we move across a period or down a group, the physical properties of elements change in a smooth, regular fashion. Within a given group, the elements show very similar chemical properties.

In this chapter, we will examine the Periodic Table in some detail, with emphasis upon its use in correlating the properties of elements. After a brief historical discussion of the evolution of the Periodic Table (Section 6.1), we will consider how the electron configuration of an element is related to its position in the table (Section 6.2). With that background, we will examine how the properties of elements change as we move across or down the table (Section 6.3).

In the second half of this chapter, we will focus on the properties of metals, which account for more than 80 of the 109 elements known today. We will consider, in Section 6.4, the general characteristics of metals, emphasizing their physical properties. Finally, in Sections 6.5 and 6.6, we will look at the reactivity and properties of the metals in Groups 1, 2, and 3 of the Periodic Table.

6.1
DEVELOPMENT OF THE PERIODIC TABLE

In 1864, J. A. R. Newlands, an English chemist, proposed the first version of the Periodic Table. He organized his table by arranging the known elements in order of increasing atomic mass in horizontal rows seven elements long (Fig. 6.1). He pointed out that the eighth element in a sequence had chemical properties very similar to those of the starting one. Newlands referred to this principle as the Law of Octaves. Unfortunately, Newlands' periodic table met only with ridicule and scorn from members of the Chemical Society in London. Indeed, they refused to publish his paper.

It's rumored they suggested he take up the cello

Shortly thereafter, a much more successful proposal for a periodic table was made in Russia by Dmitri Mendeleev, Professor of Chemistry at the University of Saint Petersburg. In writing a textbook of chemistry, Mendeleev devoted separate chapters to families of elements with similar chemical properties, including the alkali metals (Group 1 in the modern Periodic Table), the alkaline earth metals (Group 2), and the halogens (Group 7). Reflecting on the behavior of these elements and others, Mendeleev was struck by the fact that their chemical properties vary periodically (in cycles) with their atomic mass. The Periodic Table of Mendeleev first appeared in a paper presented at a meeting of the Russian Chemical Society on March 6, 1869. The acclaim that greeted this and successive papers lifted Mendeleev from obscurity to fame. His textbook, *Principles of Chemistry*, first appeared in 1870 and was widely adopted through eight editions.

Why did the Mendeleev version meet with such success while that of Newlands failed? We can see reasons for this by comparing the two versions as shown in Figure 6.1. For the elements H through Ca, Newlands' ar-

Figure 6.1 Two early versions of the Periodic Table. Both are in the condensed form used by early chemists, with transition metals placed with main-group elements. The elements shown in blue were out of place in Newlands' table.

Newlands (1864)

						H
Li	Be	B	C	N	O	F
Na	Mg	Al	Si	P	S	Cl
K	Ca	Cr	Ti	Mn	Fe	Co, Ni
Cu	Zn	Y	In	As	Se	Br
Rb	Sr	La, Ce	Zr	Nb, Mo	Ru, Rh	Pd
Ag	Cd	U	Sn	Sb	Te	I
Cs	Ba, V					

Mendeleev (as revised, 1871)

I	II	III	IV	V	VI	VII	VIII
R_2O	RO	R_2O_3	RO_2	R_2O_5	RO_3	R_2O_7	RO_4
H							
Li	Be	B	C	N	O	F	
Na	Mg	Al	Si	P	S	Cl	
K	Ca	—	Ti	V	Cr	Mn	Fe, Co, Ni
Cu	Zn	—	—	As	Se	Br	Ru, Rh, Pd
Ag	Cd	In	Sn	Sb	Te	I	
Cs	Ba						

rangement worked well; beyond calcium he was in trouble. Although chromium and aluminum both form oxides of the formula R_2O_3 (R is the metal), placing chromium below aluminum requires that chromium (at. mass = 52 amu) precede titanium (at. mass = 48 amu). Additionally, iron certainly does not resemble oxygen chemically, and neither cobalt nor nickel even vaguely belong with the halogens.

In contrast to Newlands, Mendeleev was creative enough to realize that the elements beyond calcium would align properly only if he left some empty spaces in the Periodic Table. He boldly suggested that new elements would be discovered to occupy the gaps he had left for them. Going one step further, Mendeleev predicted detailed physical and chemical properties for three such elements: "ekaboron" (scandium), "ekaaluminum" (gallium), and "ekasilicon" (germanium). Mendeleev's predictions were based on the known properties of other elements in the same group of the Periodic Table. To estimate the atomic mass of germanium, for example, he took an average of those of Si (at. mass = 28 amu) and Sn (at. mass = 118 amu):

Mendeleev liked to live dangerously

$$\text{predicted atomic mass Ge} \approx \frac{28 \text{ amu } + 118 \text{ amu}}{2} = 73 \text{ amu}$$

Since the oxides of silicon and tin were known to have the formulas SiO_2 and SnO_2, respectively, Mendeleev could predict with some confidence that the formula of the oxide of germanium would be GeO_2. Compounds of elements in the same group of the Periodic Table tend to have similar formulas.

By 1886, all of the elements predicted by Mendeleev had been isolated. They were shown to have properties very close to those he predicted (Table 6.1). The spectacular agreement between prediction and experiment removed any doubts about the validity and value of Mendeleev's Periodic Table.

Another person who made a major contribution to the development of the Periodic Table was the German chemist Lothar Meyer. In July 1868, in the process of revising his highly successful text, *Modern Theories of Chemistry,* Meyer compiled a Periodic Table containing 56 elements. He also prepared extensive graphs showing that such properties as molar vol-

TABLE 6.1   Predicted and Observed Properties of Germanium (Ekasilicon)

PROPERTY	PREDICTED BY MENDELEEV (1871)	OBSERVED (1886)
Atomic mass	73 amu	72.3 amu
Density	5.5 g/cm³	5.47 g/cm³
Specific heat	0.31 J/(g·°C)	0.32 J/(g·°C)
Melting point	very high	960°C
Formula of oxide	RO_2	GeO_2
Formula of chloride	RCl_4	$GeCl_4$
Density of oxide	4.7 g/cm³	4.70 g/cm³
Boiling point of chloride	100°C	86°C

ume were a periodic function of atomic mass. Unaware of Mendeleev's work, Meyer published his results in 1870. In 1882, the two men jointly were awarded the Davy Medal, the highest honor of the Royal Society. Five years later, the Society belatedly awarded the same medal to its own member, J. A. R. Newlands.

The Periodic Table of Mendeleev was based on atomic masses; elements were arranged in order of increasing atomic mass. If you look at a modern Periodic Table, you will find three cases in which elements are out of order insofar as atomic mass is concerned: argon (at. mass = 39.95 amu) comes before potassium (at. mass = 39.10 amu), cobalt (at. mass = 58.93 amu) before nickel (at. mass = 58.69 amu), and tellurium (at. mass = 127.60 amu) before iodine (at. mass = 126.90 amu). Chemically, however, the positions of these elements make sense. Argon, for example, is clearly a noble gas, not an alkali metal.

These anomalies were resolved in 1914 by a young Englishman, Henry Moseley, a student of Rutherford. Moseley was studying the properties of the radiation (x-rays) given off when elements were bombarded by high-energy electrons. He discovered that a plot of the square root of the x-ray frequency, $\nu^{1/2}$, versus atomic mass gave a nearly straight line. There were, however, three pairs of elements that fell off the line. You guessed it: the elements that were out of order were Ar and K, Co and Ni, and Te and I. When Moseley plotted order number in the Periodic Table rather than atomic mass, these elements fell neatly in line. A graph of $\nu^{1/2}$ versus order number was a perfect straight line.

Moseley's work showed that order number in the Periodic Table had a significance that went beyond atomic mass. Moseley suggested a simple explanation*: "There is every reason to suppose that the integer that controls the x-ray spectrum is the charge on the nucleus." Today we relate position in the Periodic Table to atomic number. The Periodic Law is stated in these terms: *The properties of the chemical elements are a periodic (cyclic) function of atomic number.* By the same token, we can say the following:

The Periodic Table is an arrangement of elements in order of increasing atomic number in horizontal rows of such a length that elements with similar chemical properties fall directly beneath one another.

6.2
ELECTRON ARRANGEMENTS AND THE PERIODIC TABLE

If you examine the Periodic Table on the inside front cover of this text, you will note that there are seven horizontal rows, called *periods*. Each period, except the first, starts with an alkali metal (Li, Na, K, Rb, Cs, Fr). Each period, except the last, which is incomplete, ends with a noble gas (He, Ne, Ar, Kr, Xe, Rn). The length of successive periods varies from 2 to 32 elements, as shown in Table 6.2.

---

*Moseley's paper was published in 1913. Two years later, he died at the age of 27 in the senseless slaughter of British troops at Gallipoli.

TABLE 6.2   Structure of the Periodic Table

PERIOD	NUMBER OF ELEMENTS	BEGINS WITH	ENDS WITH
1	2	$_1$H	$_2$He
2	8	$_3$Li	$_{10}$Ne
3	8	$_{11}$Na	$_{18}$Ar
4	18	$_{19}$K	$_{36}$Kr
5	18	$_{37}$Rb	$_{54}$Xe
6	32	$_{55}$Cs	$_{86}$Rn
7	32	$_{87}$Fr	at. no. 118

The vertical columns of elements in the Periodic Table are referred to as *groups*. Each group is assigned a number, written at the top of the vertical column. Unfortunately, from the time of Mendeleev, there has been no general agreement as to the numbering system to be used. Quite recently, the International Union of Pure and Applied Chemistry (IUPAC) has recommended that the 18 groups be numbered consecutively from left to right, 1 to 18. As of this writing, it is not clear how widely this system will be accepted. The numbering system used in this text is one that is in common use at this time. We will refer to the **main-group elements**, those in the two groups at the far left and the six groups at the right, as Groups 1 through 8. The **transition elements**, in the center of the Periodic Table, are not assigned group numbers. The same is true of the **lanthanides** (at. no. = 57 to 70) and **actinides** (at. no. = 89 to 102), listed separately at the bottom of the table.

The structure of the Periodic Table was established by experiment, based on the properties of elements. The fact that successive elements in a group resemble each other chemically suggests that there must be a basic similarity in the structure of their atoms. The nature of this similarity became apparent when electron configurations were established in the first two decades of this century.

To understand how position in the Periodic Table relates to electron configuration, consider the metals in the first two groups. Atoms of the Group 1 elements all have one s electron in the outermost principal energy level (Table 6.3). In each Group 2 atom, there are two s electrons in the outermost level. A similar relationship applies to the elements in any group:

The atoms of elements in a group of the Periodic Table have the same outer electron configuration. Elements within a group show similar

TABLE 6.3   Electron Configurations of the Group 1 and 2 Elements

GROUP 1		GROUP 2	
$_3$Li	$[_2$He$]$**2s^1**	$_4$Be	$[_2$He$]$**2s^2**
$_{11}$Na	$[_{10}$Ne$]$**3s^1**	$_{12}$Mg	$[_{10}$Ne$]$**3s^2**
$_{19}$K	$[_{18}$Ar$]$**4s^1**	$_{20}$Ca	$[_{18}$Ar$]$**4s^2**
$_{37}$Rb	$[_{36}$Kr$]$**5s^1**	$_{38}$Sr	$[_{36}$Kr$]$**5s^2**
$_{55}$Cs	$[_{54}$Xe$]$**6s^1**	$_{56}$Ba	$[_{54}$Xe$]$**6s^2**

chemical behavior because it is the outer electrons that are involved in chemical reactions.

In the remainder of this section, we will examine some of the implications of this general principle.

FILLING OF ELECTRON SUBLEVELS IN THE PERIODIC TABLE

As we have just seen, the group in which an element is located depends upon its outer electron configuration. This means that the order in which electron sublevels are filled establishes the structure of the Periodic Table. Figure 6.2 shows how this applies for all the elements. Notice the following points:

1. *The elements in Groups 1 and 2 are filling an s sublevel.* Thus, Li and Be in the second period fill the 2s sublevel. Na and Mg in the third period fill the 3s sublevel, and so on.

Figure 6.2 The Periodic Table can be used to deduce the electron configurations of atoms. Elements in Groups 1 and 2 fill an **ns** sublevel, where **n** is the number of the period. Elements in Groups 3 through 8 (except for H and He) fill an **np** sublevel. The transition metals fill an (**n** − 1)d sublevel. For example, the elements Sc through Zn in the fourth period fill the 3d sublevel. The lanthanides fill the 4f sublevel, while the actinides fill the 5f.

Group / Period	1	2											3	4	5	6	7	8
																	1 H	2 He
1																	1s	
	3 Li	4 Be											5 B	6 C	7 N	8 O	9 F	10 Ne
2	2s														2p			
	11 Na	12 Mg											13 Al	14 Si	15 P	16 S	17 Cl	18 Ar
3	3s														3p			
	19 K	20 Ca	21 Sc	22 Ti	23 V	24 Cr	25 Mn	26 Fe	27 Co	28 Ni	29 Cu	30 Zn	31 Ga	32 Ge	33 As	34 Se	35 Br	36 Kr
4	4s						3d								4p			
	37 Rb	38 Sr	39 Y	40 Zr	41 Nb	42 Mo	43 Tc	44 Ru	45 Rh	46 Pd	47 Ag	48 Cd	49 In	50 Sn	51 Sb	52 Te	53 I	54 Xe
5	5s						4d								5p			
	55 Cs	56 Ba	71 Lu	72 Hf	73 Ta	74 W	75 Re	76 Os	77 Ir	78 Pt	79 Au	80 Hg	81 Tl	82 Pb	83 Bi	84 Po	85 At	86 Rn
6	6s						5d								6p			
	87 Fr	88 Ra	103 Lr	104	105	106												
7	7s																	

	57 La	58 Ce	59 Pr	60 Nd	61 Pm	62 Sm	63 Eu	64 Gd	65 Tb	66 Dy	67 Ho	68 Er	69 Tm	70 Yb
6							4f							
	89 Ac	90 Th	91 Pa	92 U	93 Np	94 Pu	95 Am	96 Cm	97 Bk	98 Cf	99 Es	100 Fm	101 Md	102 No
7							5f							

2. *The elements in Groups 3 through 8 (six elements in each period) fill p sublevels,* which have a capacity of six electrons. In the second period, the 2p sublevel starts to fill with B (at. no. = 5) and is completed with Ne (at. no. = 10). In the third period, the elements Al (at. no. = 13) through Ar (at. no. = 18) fill the 3p sublevel.

3. *The transition metals, in the center of the Periodic Table, fill d sublevels.* Remember that a d sublevel can hold ten electrons. In the fourth period, the ten elements Sc (at. no. = 21) through Zn (at. no. = 30) fill the 3d sublevel. In the fifth period, the 4d sublevel is filled by the elements Y (at. no. = 39) through Cd (at. no. = 48). The ten transition metals in the sixth period fill the 5d sublevel. Note that transition metals fill sublevels in which the principal quantum number is one less than the period number, for example, 3d in the fourth period. From a slightly different point of view, we might say that the transition metals fill *inner* d sublevels.

The most outstanding characteristic of the transition metals is their ability to form colored compounds (Fig. 6.3). Nearly all brightly colored inorganic compounds contain a transition metal such as chromium, manganese, nickel, or copper. Qualitatively, we can explain the colors of transition metal compounds in terms of electron configurations. In most transition metal ions there are unfilled or half-filled inner d orbitals. These differ by only small amounts of energy from orbitals occupied by electrons. The energy difference is often comparable to that of visible light. By absorbing light of a particular color, an electron can move from a lower to a higher orbital. This absorption of light accounts for the color we see when we look at a transition metal compound or its water solution. We will have more to say about the colors of transition metal compounds when we discuss complex ions in Chapter 16.

4. *The two sets of 14 elements listed separately at the bottom of the table are filling f sublevels* with a principal quantum number two less than the period number. That is:

— 14 elements in the sixth period (at. no. = 57 to 70) are filling the 4f sublevel. These elements are sometimes called rare earths or, more commonly nowadays, **lanthanides**, after the name of the first element in the series, lanthanum (La).

— 14 elements in the seventh period (at. no. = 89 to 102) are filling the 5f sublevel. The first element in this series is actinium (Ac); collectively, these elements are referred to as **actinides**.

Because the lanthanides have similar properties, their compounds are difficult to separate from one another. Until quite recently, samples of pure compounds of these elements were not available except for those of cerium, the most abundant member of the series. Chromatographic processes are now used to separate compounds of the lanthanide metals. The availability of these compounds in highly pure form has led to several commercial applications. A brilliant red phosphor now used in color TV receivers contains a small amount of europium oxide, Eu_2O_3. This is added to a base of yttrium oxide, Y_2O_3, or gadolinium oxide, Gd_2O_3.

The actinide metals are all radioactive. Only two of these elements, uranium and thorium, are found in appreciable amounts in nature. The

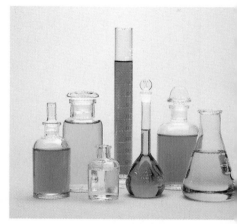

Figure 6.3 Water solutions of transition metal compounds (except for those of Zn^{2+}) tend to be brightly colored. The cations in these solutions are, from left to right: Cu^{2+}(blue), Fe^{3+}(yellow), Zn^{2+} (colorless), Cr^{3+}(purple), Co^{2+} (pink), Ni^{2+}(green), and Mn^{2+}(light pink). (Marna G. Clarke)

These elements are similar because their electron configurations differ only in the number of inner f electrons

other elements were first observed in the products of controlled nuclear reactions. In most cases, they have been produced in only very small amounts. Isotopes of uranium and plutonium are used as the fuel elements in nuclear fission (Chap. 25).

OUTER ELECTRON CONFIGURATIONS FROM THE PERIODIC TABLE

We could, if we had to, use Figure 6.2 to deduce the complete electron configuration of any element. More commonly, we use the Periodic Table itself for a less ambitious purpose. Specifically, we use it to obtain the outer electron configurations of the main-group elements:

Group	1	2	3	4	5	6	7	8
Outer configuration	ns^1	ns^2	ns^2np^1	ns^2np^2	ns^2np^3	ns^2np^4	ns^2np^5	ns^2np^6

where **n** *is the number of the period in which the element is located;* that is, $n = 1$ for the first period (H, He), $n = 2$ for the second period (Li → Ne), and so on. The value for **n** is also the principal quantum number of the highest occupied energy level in that atom.

■ **EXAMPLE 6.1** _____

Using the Periodic Table and principles of electronic structure discussed in Chapter 5, give

a. the outer electron configurations of radium (Ra) and bromine (Br).
b. the orbital diagram for the outer electrons in these two atoms.
c. a set of quantum numbers for each of the outer electrons in radium.

Solution

Ra has two outer electrons; Br has seven

a. Radium (at. no. = 88) is in the seventh period, in Group 2; its outer electron configuration is $7s^2$. Bromine (at. no. = 35) is in the fourth period in Group 7. Its outer electron configuration is $4s^2 4p^5$.

b. The orbital diagrams are

<pre>
          7s   4s   4p
    Ra (↑↓)   Br (↑↓) (↑↓)(↑↓)(↑)
</pre>

c. For the two 7s electrons in radium, the four quantum numbers, in the usual order, are $7, 0, 0, +\frac{1}{2}$ and $7, 0, 0, -\frac{1}{2}$.

Exercise
Give the outer electron configuration of tin, Sn. Answer: $5s^2 5p^2$.

6.3
SOME TRENDS IN THE PERIODIC TABLE

The Periodic Table can be used for a variety of purposes. In particular, it is useful in correlating properties on an atomic scale. In this section, we will consider trends in four such properties.

ATOMIC RADIUS

Strictly speaking, the "size" of an atom is a rather nebulous concept. The electron cloud surrounding the nucleus does not have a sharp boundary. We can, however, define and measure a quantity known as the atomic radius, assuming a spherical atom. Ordinarily, the atomic radius is taken to be one-half the distance of closest approach between atoms in the elemental substance (Fig. 6.4).

The atomic radii of the main-group elements are shown at the top of Figure 6.5, p. 180. Notice that, in general, **atomic radius**

—decreases as we move across a period from left to right in the Periodic Table.
—increases as we move down a group in the Periodic Table.

These trends can be rationalized in terms of the *effective nuclear charge* experienced by an electron at the outer edge of an atom. For any electron, the effective nuclear charge, Z_{eff}, is given by the expression

$$Z_{eff} \approx Z - S \qquad (6.1)$$

where Z is the actual nuclear charge, i.e., the atomic number, and S is the number of electrons in inner, complete energy levels. These electrons shield the outer electron from the nucleus, in effect reducing the positive charge attracting that electron toward the nucleus. For the three atoms Na, Mg, and Al:

ELECTRON CONFIGURATION	Z	NUMBER OF INNER ELECTRONS = S	APPROXIMATE VALUE OF Z_{eff}
$_{11}$Na $1s^2 2s^2 2p^6 3s^1$	11	2 + 2 + 6 = 10	11 − 10 = 1
$_{12}$Mg $1s^2 2s^2 2p^6 3s^2$	12	2 + 2 + 6 = 10	12 − 10 = 2
$_{13}$Al $1s^2 2s^2 2p^6 3s^2 3p^1$	13	2 + 2 + 6 = 10	13 − 10 = 3

According to the argument we have just made, effective nuclear charge should increase as we move across a period from left to right in the Periodic Table. Thus, Z_{eff} should be about 1 for Na, 2 for Mg, 3 for Al, and so on. The greater the effective nuclear charge, the stronger will be the attractive force pulling an outer electron in toward the nucleus. As this force increases, the electron is drawn in closer to the nucleus and atomic radius decreases. As a result, atoms get smaller as we move across from left to right in the Periodic Table.

As we move down in the Periodic Table, the effective nuclear charge experienced by an outer electron should remain essentially constant. For example, for the elements Li, Na, K, Rb, and Cs in Group 1, the value of

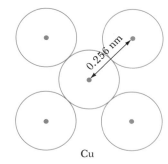

Cu

Atomic radius = $\dfrac{0.256 \text{ nm}}{2}$ = 0.128 nm

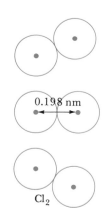

0.198 nm

Cl$_2$

Atomic radius = $\dfrac{0.198 \text{ nm}}{2}$ = 0.099 nm

Figure 6.4 Atomic radii are determined by assuming that atoms in closest contact in an element touch one another. The atomic radius is taken to be one-half of the closest internuclear distance.

180

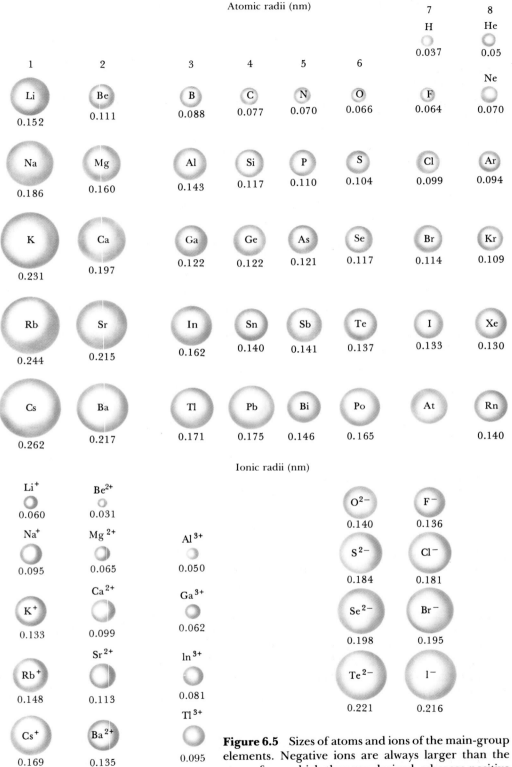

Atomic radii (nm)

Ionic radii (nm)

Figure 6.5 Sizes of atoms and ions of the main-group elements. Negative ions are always larger than the atoms from which they are derived, whereas positive ions are smaller.

Z_{eff} should be about 1. In that sense, the outer s electron in an alkali metal atom behaves like the electron in a hydrogen atom, where $Z = 1$. However, you will recall from our discussion in Chapter 5 that the distance of the electron from the nucleus increases with the principal quantum number. Hence, we expect atomic radius to increase as we move from Li (2s electron) to Na (3s electron), to K (4s electron), and so on. A similar argument can be applied to other groups, explaining why atoms get larger as we move down a group.

IONIC RADIUS

The radii of cations and anions derived from atoms of the main-group elements are shown at the bottom of Figure 6.5. The trends referred to above for atomic radii are clearly visible here as well. Notice, for example, that ionic radius increases as we move down a group in the Periodic Table (e.g., $Li^+ < Na^+ < K^+$; $F^- < Cl^- < Br^-$). Moreover, if we compare cations in the same period with one another, their radii decrease from left to right ($Na^+ > Mg^{2+} > Al^{3+}$). The same effect is shown by anions ($S^{2-} > Cl^-$; $Se^{2-} > Br^-$).

It is of particular interest to compare the radii of cations and anions to those of the atoms from which they are derived. It should be clear from examination of the two halves of Figure 6.5 that

— *positive ions are smaller than the metal atoms from which they are formed.* The Na^+ ion has a radius, 0.095 nm, only a little more than half that of the Na atom, 0.186 nm.

— *negative ions are larger than the nonmetal atoms from which they are formed.* The radius of the Cl^- ion, 0.181 nm, is nearly twice that of the Cl atom, 0.099 nm.

These effects combine to make anions, on the whole, larger than cations. Compare, for example, the Cl^- ion (radius = 0.181 nm) to the Na^+ ion (radius = 0.095 nm). This means that in sodium chloride, and indeed in the vast majority of all ionic compounds, most of the space in the crystal lattice is taken up by anions.

The differences in radii between atoms and ions can be explained quite simply. A cation is smaller than the corresponding metal atom because the excess of protons in the ion draws the outer electrons in closer to the nucleus. In contrast, an extra electron in an anion adds to the repulsion between outer electrons. This makes a negative ion larger than the corresponding nonmetal atom.

■ EXAMPLE 6.2 _____

Using only the Periodic Table, arrange each of the following sets of atoms and ions in order of increasing size.

a. Mg, Al, Ca   b. S, Cl, S^{2-} c. Fe, Fe^{2+}, Fe^{3+}

Solution

a. Since calcium lies below magnesium in the Periodic Table, the Ca atom should be larger than the Mg atom. Since aluminum lies to the right of magnesium, the Al atom should be smaller than the Mg atom. Hence the predicted order is: Al < Mg < Ca. A glance at Figure 6.5 should convince you that this order is correct.

b. The Cl atom should be smaller than the S atom, taking into account the positions of the two elements in the Periodic Table. Since anions are larger than the corresponding atoms, the S^{2-} ion should be larger than the S atom: Cl < S < S^{2-}. The observed radii are 0.099 nm for Cl, 0.104 nm for S, and 0.184 nm for S^{2-}.

c. The two cations (Fe^{2+}, Fe^{3+}) should be smaller than the atom from which they are derived, Fe. Of the two, the Fe^{3+} ion should be the smaller; its greater positive charge draws the outer electrons in closer to the nucleus.

Thus we have: Fe^{3+} < Fe^{2+} < Fe.

Exercise
Arrange the species: Na, K, Na^+ in order of increasing radius. Answer: Na^+ < Na < K.

IONIZATION ENERGY

It takes energy to knock an electron out of an atom

The ionization energy is a measure of how difficult it is to remove an electron from an atom. It is always true that energy must be *absorbed* to bring about ionization; ionization energies are always *positive* quantities. Ordinarily, they are quoted in kilojoules per mole.

The first ionization energy is the energy change for the removal of the outermost electron from a gaseous atom to form a +1 ion:

$$M(g) \rightarrow M^+(g) + e^-; \qquad \Delta E_1 = \text{first ionization energy} \qquad (6.2)$$

We may also refer to second, third, etc., ionization energies. For example,

$$M^+(g) \rightarrow M^{2+}(g) + e^-; \qquad \Delta E_2 = \text{second ionization energy}$$

In general, the larger the ionization energy, the more difficult it is to remove an electron.

First ionization energies of the main-group elements are listed in Figure 6.6. Notice that this **ionization energy**

—increases as we move across the Periodic Table from left to right.
—decreases as we move down in the Periodic Table.

Comparing Figure 6.6 with 6.5, we see an inverse correlation between ionization energy and atomic radius. The smaller the atom, the more tightly

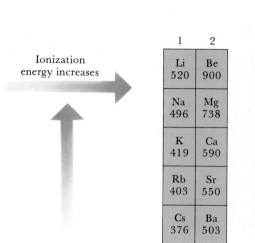

Ionization energy increases

its electrons are held to the nucleus and the more difficult they are to remove. This follows from Coulomb's Law:

$$E = \frac{\text{constant} \times Q_1 \times Q_2}{d}$$

$Q_1 = Z_{\text{eff}}$; Q_2 = charge on electron

which tells us that the electrostatic attractive energy, E, between two particles of opposite charge (Q_1, Q_2) is inversely related to the distance d between the charges. Large atoms, such as those of the heavier Group 1 metals, have rather low ionization energies. The outer electrons are far away from the nucleus and hence are relatively easy to remove.

■ **EXAMPLE 6.3** _____

Consider the three elements B, C, and Al. Using only the Periodic Table, predict which of the three elements has

a. the largest atomic radius; the smallest atomic radius.
b. the largest ionization energy; the smallest ionization energy.

Solution
The three elements form a block in the table:

    B   C
    Al

a. Since atomic radius increases as we go down in the table and decreases as we go across, Al should be the largest and C the smallest atom.

The observed atomic radii are as follows: B = 0.088 nm; C = 0.077 nm; Al = 0.143 nm.

b. Since ionization energy increases as we go across in the table and decreases as we go down,   C should have the highest value and Al the smallest.   The observed values are as follows: B = 801 kJ/mol; C = 1086 kJ/mol; Al = 578 kJ/mol.

Exercise
Compare the two elements Ca and Rb with regard to atomic radius and ionization energy. Answer: Rb has the larger radius and the smaller ionization energy.

As we remove successive electrons from an atom, ionization energy increases. In the case of potassium, we have

$$K(g) \rightarrow K^+(g) + e^-; \quad \Delta E_1 = 419 \text{ kJ/mol}$$
$$K^+(g) \rightarrow K^{2+}(g) + e^-; \quad \Delta E_2 = 3051 \text{ kJ/mol}$$
$$K^{2+}(g) \rightarrow K^{3+}(g) + e^-; \quad \Delta E_3 = 4411 \text{ kJ/mol}$$

This trend is readily explained. As the positive charge on the species increases (0 in the K atom, $+1$ in the K^+ ion, $+2$ in the K^{2+} ion), electrons are attracted more and more strongly toward the nucleus. Hence, it becomes more and more difficult to remove these electrons.

Notice, however, that ionization energy does not increase smoothly as successive electrons are removed. For potassium, there is a big jump between the first and second ionization energies; ΔE_2 is seven times ΔE_1. In contrast, ΔE_3 is less than twice ΔE_2. This makes sense in terms of the electronic structures of the species from which electrons are being removed. Recall that the potassium atom has the electron configuration $[Ar] 4s^1$. The first ionization corresponds to the removal of the 4s electron, which is far from the nucleus. This electron should leave rather easily, so it is not surprising that ΔE_1 is comparatively small. Once this electron is gone, we are dealing with a K^+ ion, which has the electron configuration of the noble gas argon. We would expect it to be very difficult to remove an electron from such a stable species; that electron would have to come from a completed inner level ($\mathbf{n} = 3$), relatively close to the nucleus. Small wonder then that ΔE_2 is much, much larger than ΔE_1. The next electron to be removed (from the K^{2+} ion) comes from that same level, so we would expect ΔE_3 to be not too much larger than ΔE_2.

The argument we have just gone through explains, in a sense, why potassium, in its reactions, forms the K^+ cation rather than K^{2+}, K^{3+}.... There is not enough energy available in chemical reactions to remove inner electrons. An analogous situation applies with calcium (Table 6.4). Here the large jump comes between the 2nd and 3rd ionization energies; the Ca^{2+} ion is a very stable species with a noble gas electronic structure. With transition metals such as iron and copper, the situation is quite different (Table 6.4). Here successive ionization energies increase smoothly, since noble gas structures are not involved. This accounts qualitatively for the tendency of transition metals to form more than one stable cation, depending upon the nature of the reaction in which they participate. It is comparatively easy to convert Cu^+ to Cu^{2+} or Fe^{2+} to Fe^{3+}.

In chemical reactions, only the outer electrons are affected

TABLE 6.4 Successive Ionization Energies of K, Ca, Fe, and Cu*

	ΔE_1 (kJ/mol)	ΔE_2 (kJ/mol)	ΔE_3 (kJ/mol)
$_{19}$K	419	3051	4411
$_{20}$Ca	590	1145	4912
$_{26}$Fe	759	1561	2957
$_{29}$Cu	746	1958	3554

*Ionization energies shown in color require removing electrons from a noble-gas structure, that of argon.

ELECTRON AFFINITY

The (first) electron affinity is the energy change when a gaseous atom acquires an electron to form a -1 ion.

$$X(g) + e^- \rightarrow X^-(g); \quad \Delta E = \text{electron affinity} \qquad (6.3)$$

The electron affinities of several nonmetal atoms are listed in Figure 6.7. When the -1 anion formed is a stable species, energy is given off in its formation and ΔE has a negative sign. This is the case with the halogen atoms, all of which form stable -1 ions with noble gas structures. For example:

$$F(g) + e^- \rightarrow F^-(g); \quad \Delta E = -322 \text{ kJ/mol} \qquad (6.4)$$

An F atom wants to pick up an electron and become an F$^-$ ion

The situation is quite different when an electron is added to a noble gas such as neon. The Ne$^-$ ion (11 electrons) is a highly unstable species; energy is absorbed in its formation and ΔE is a positive quantity.

$$Ne(g) + e^- \rightarrow Ne^-(g); \quad \Delta E = +29 \text{ kJ/mol} \qquad (6.5)$$

In general, we expect atoms that form -1 ions readily to have a large, negative value of the electron affinity. Atoms that form less stable -1 ions may have electron affinity values that are small negative numbers or even, like neon, positive quantities.

From what was said earlier, you might expect to find a correlation between electron affinity and atomic radius. Small atoms should have a relatively strong attraction for electrons and hence should have large negative electron affinities. The larger the atom, the less negative its electron

3	4	5	6	7	8
B −23	C −123	N 0	O −142	F −322	Ne +29
	Si −120	P −74	S −200	Cl −348	Ar +35
	Ge −116	As −77	Se −195	Br −324	Kr +39
		Sb −101	Te −190	I −295	Xe +41

Figure 6.7 Electron affinities (kilojoules per mole) of some nonmetals.

affinity should be. From Figure 6.7, we see that this trend is only roughly followed; there are many exceptions. For example, nitrogen has less affinity for an electron than carbon, even though the nitrogen atom is smaller than the carbon atom. This effect has been attributed to the stability of the half-filled p sublevel in the nitrogen atom, which makes the addition of another electron comparatively difficult.

$$\text{N }(1s^2 2s^2 2p^3) + e^- \rightarrow \text{N}^-(1s^2 2s^2 2p^4); \quad \Delta E = 0 \quad (6.6)$$

6.4
METALS AND THE PERIODIC TABLE

In a broad sense, we can classify elements as metals or nonmetals. In this section we will consider how metallic character depends upon position in the Periodic Table. We will then consider some of the general properties that distinguish metals from nonmetals. Finally, we will see how these properties are explained by a simple model of the structure of metals.

METALS, NONMETALS, AND METALLOIDS

The diagonal line or stairway that runs from the left to the lower right of the Periodic Table separates metals from nonmetals. *Elements below and to the left of this line are metals.* Included among the metals are

— all of the elements in Group 1 (except H) and Group 2; also Al (but not B) in Group 3

— all of the transition elements

— the elements to the right of the transition series in Groups 3 (Ga, In, Tl), 4 (Sn, Pb), and 5 (Bi)—collectively, these elements are often referred to as *post-transition metals*

— the lanthanides and actinides

Most of the elements are metals

With one exception (mercury, which melts at $-40°C$), all of the metals are solids at 25°C and 1 atm.

Elements above and to the right of the diagonal line are classified as nonmetals. At 25°C and 1 atm, about half of these elements are gases, one (bromine) is a liquid, and the rest are solids. There are relatively few nonmetals, about 20, including the noble gases.

Along the diagonal line in the Periodic Table are several elements that are difficult to classify exclusively as metals or nonmetals. They have properties in between those of elements in the two classes. In particular, their electrical conductivities are intermediate between those of metals and nonmetals. The six elements

B	Si	Ge	As	Sb	Te
boron	silicon	germanium	arsenic	antimony	tellurium

are often called **metalloids**. Certain metalloids, notably silicon and germanium, are used in semiconductor devices such as the "chips" found in calculators, computers, and solar cells.

As this discussion implies, metallic character varies in a systematic way with position in the Periodic Table. Metallic character *decreases* as one moves *across* from left to right within a given period. For example, the third period starts with three elements that are distinctly metallic (Na, Mg, Al), moves on to a metalloid (Si), and is completed by four nonmetals (P, S, Cl, Ar). Metallic character *increases* as one moves *down* a given group. Consider, for example, Group 4. C is a nonmetal; Si and Ge are metalloids, while Sn and Pb have the properties we associate with metals.

PHYSICAL PROPERTIES OF METALS

The characteristic properties that we associate with metals include the following:

1. *High electrical conductivity.* Metals typically have electrical conductivities several hundred times greater than those of typical nonmetals. Silver is the best electrical conductor but is too expensive for general use. Copper, with a conductivity close to that of silver, is the metal most commonly used for electrical wiring. Although a much poorer conductor than copper, mercury is used in many electrical devices, such as "silent" light switches, where a liquid conductor is required.

Glass is a poor conductor of heat and electricity

2. *High thermal conductivity.* Of all solids, metals are the best conductors of heat. Saucepans used for cooking commonly contain aluminum, copper, or stainless steel; their handles are made of a nonmetallic material that is a good thermal insulator.

3. *Ductility, malleability.* Most metals are ductile (capable of being drawn out into a wire) and malleable (capable of being hammered into thin sheets). Gold, for example, can be hammered into sheets so thin that they are transparent. A wire 160 m long and only 1×10^{-5} m in radius can be drawn out from 1 g of gold. In a metal, the electrons act somewhat like a glue holding the atomic nuclei together. As a result, metal crystals can be deformed without shattering. A crystal of a nonmetal breaks into small pieces if you hammer it or try to draw it into a wire.

4. *Luster.* Polished metal surfaces reflect light. Most metals have a silvery white "metallic" color because they reflect light of all wavelengths. Gold and copper absorb some light in the blue region of the visible spectrum and so appear yellow (gold) or red (copper).

In contrast to metals, nonmetals are ordinarily nonconductors of electricity (the graphite form of carbon is a notable exception). Crystals of nonmetals such as iodine or sulfur shatter under stress; they are neither malleable nor ductile. As one would expect, nonmetal crystals reflect light poorly and are poor conductors of heat.

METALLIC BONDING

The unique properties of metals suggest a structure in which electrons are relatively mobile. Only if this is true can we explain why metals are good conductors of electricity and form positive ions readily.

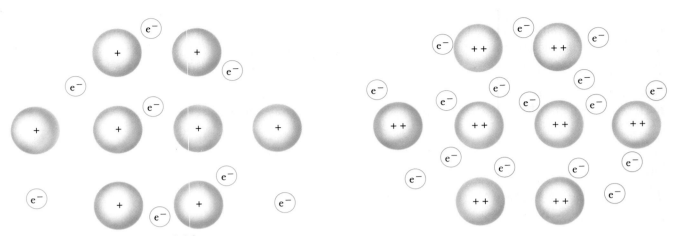

Figure 6.8 Electron-sea model for metallic bonding in sodium (Na^+, e^-) and magnesium (Mg^{2+}, $2 e^-$).

Electrons in metals have no permanent address

Electron mobility can be explained by a simple model of bonding in metals, known as the **electron-sea** model. The metallic crystal is pictured as an array of positive ions, e.g., Na^+, Mg^{2+}. These are anchored in position, like buoys in a mobile "sea" of electrons. These electrons are not attached to any particular positive ion, but rather can wander through the crystal. Figure 6.8 shows what a tiny portion of a metal crystal looks like according to the electron-sea model.

This simple picture of metallic structure offers an obvious explanation of the high electrical conductivity of metals. It can also be used to explain many of the other properties of metals. High thermal conductivity is explained by assuming that heat is carried through the metal by collisions between electrons, which occur frequently. Since electrons are not restricted to a particular bond, they can absorb and re-emit light over a wide wavelength range. This explains why metal surfaces are excellent reflectors.

6.5
MAIN-GROUP METALS: GROUP 1, GROUP 2, AND ALUMINUM

These elements, particularly the alkali metals in Group 1 and the alkaline earth metals in Group 2, show to an unusual extent the properties we associate with metals. All of these elements are excellent thermal and electrical conductors. Their surfaces, when freshly exposed, show a typical metallic luster (Fig. 6.9). In their reactions, these metals form cations with noble gas structures and charges of $+1$ (e.g., Li^+, Na^+, and K^+), $+2$ (e.g., Mg^{2+}, Ca^{2+}, and Sr^{2+}), or $+3$ (Al^{3+}).

PHYSICAL PROPERTIES

Table 6.5 summarizes some of the physical properties of the alkali and alkaline earth metals. Looking at the table, we note some of the trends referred to earlier in this chapter. In particular, atomic radius decreases

Figure 6.9 When sodium is freshly cut with a knife, the exposed surface has a metallic luster. (Marna G. Clarke)

as we move from Group 1 to Group 2 (compare Na, 0.186 nm, to Mg, 0.160 nm). As we move down a group, atomic radius increases (compare Mg, 0.160 nm, to Ca, 0.197 nm).

As we would expect, ionization energy is inversely related to atomic radius. The larger Group 1 atoms have smaller ionization energies than their neighbors in Group 2 (compare Na, 496 kJ/mol, to Mg, 738 kJ/mol).

TABLE 6.5   Physical Properties of the Group 1 and Group 2 Metals

	GROUP 1				
PROPERTY	$_3Li$	$_{11}Na$	$_{19}K$	$_{37}Rb$	$_{55}Cs$
Atomic mass (amu)	6.94	22.99	39.10	85.47	132.91
Outer electron configuration	$2s^1$	$3s^1$	$4s^1$	$5s^1$	$6s^1$
Atomic radius (nm)	0.152	0.186	0.231	0.244	0.262
Ionization energy (kJ/mol)	520	496	419	403	376
Density (g/cm³)	0.534	0.971	0.862	1.53	1.87
Melting point (°C)	186	98	64	39	28
Boiling point (°C)	1326	889	774	688	690
Flame color	red	yellow	violet	purple	blue
Atomic spectrum (nm)	670.8	589.6	766.5	780.0	852.1
(strong lines)	610.4	589.0	404.4	420.2	455.5

	GROUP 2				
PROPERTY	$_4Be$	$_{12}Mg$	$_{20}Ca$	$_{38}Sr$	$_{56}Ba$
Atomic mass (amu)	9.01	24.30	40.08	87.62	137.33
Outer electron configuration	$2s^2$	$3s^2$	$4s^2$	$5s^2$	$6s^2$
Atomic radius (nm)	0.111	0.160	0.197	0.215	0.217
Ionization energy (kJ/mol)	900	738	590	550	503
Density (g/cm³)	1.85	1.74	1.55	2.60	3.51
Melting point (°C)	1283	650	845	770	725
Boiling point (°C)	2970	1120	1420	1380	1640
Flame color	—	—	red	crimson	green

In both groups, ionization energy decreases as we move down in the Periodic Table; the larger the atom, the smaller the ionization energy.

The horizontal trend in atomic radius explains in part why the Group 2 metals have higher densities than do the Group 1 metals. An alkaline earth atom is smaller than that of an alkali metal in the same period. It is also slightly heavier (at. mass Na = 23 amu, Mg = 24 amu). This means that atoms of the Group 2 metals have the larger density, as do the metals themselves (d Na = 0.971 g/cm^3, Mg = 1.74 g/cm^3).

When a compound of an alkali metal is heated in a Bunsen burner flame, an electron is excited to a higher energy state. When this electron returns to a lower state, energy is emitted as visible light. This process also occurs with compounds of the heavier metals in Group 2: Ca, Sr, and Ba. The flame colors listed in Table 6.5 and shown in Figure 6.10 can be used to identify many of these metals. Fireworks displays are vivid, large-scale "flame tests," using compounds of sodium (yellow), strontium (red), and barium (green). Flash powder used at rock concerts is a mixture of magnesium and potassium chlorate, $KClO_3$.

■ **EXAMPLE 6.4** _____
Consider the element aluminum.

a. How would you expect the atomic radius and density of Al to compare to those of Mg?

b. Of the strong lines in the spectrum of Al, the one with the longest wavelength is at 396.15 nm. Would you expect Al to show a flame color? What is the energy difference, in kilojoules per mole, associated with this line?

Figure 6.10 Flame tests for the alkali metals: Li(red), Na(yellow), and K(violet).

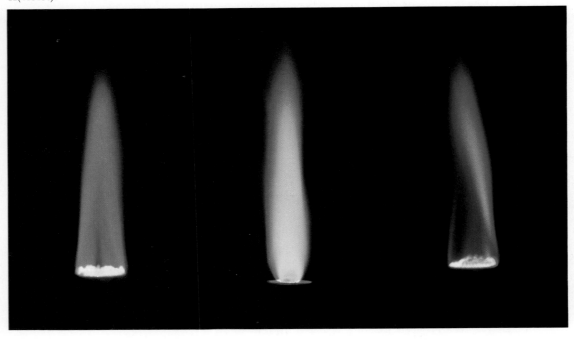

Solution

a. Since Al lies to the right of Mg in the Periodic Table, it is expected to have a smaller radius. With a smaller radius and a slightly larger atomic mass, Al is expected to have a larger density than Mg and it does (2.71 g/cm³ vs. 1.74 g/cm³).

b. This line is in the ultraviolet (λ <400 nm); all the other strong lines must be even further into the UV. Aluminum does not show a characteristic flame color. To calculate the energy difference, we use the basic equation referred to in Chapter 5:

If you can't see the light, it means that it doesn't have any color, even though it is there

$$E_{hi} - E_{lo} = \frac{1.196 \times 10^5}{\lambda} \frac{kJ \cdot nm}{mol}$$

Substituting λ = 396.15 nm:

$$E_{hi} - E_{lo} = \frac{1.196 \times 10^5}{396.15} \; kJ/mol = \boxed{301.9 \; kJ/mol}$$

Exercise

The first ionization energy of Al is 578 kJ/mol. Can you suggest why this value is smaller than that for B? Answer: See discussion, p. 183.

CHEMICAL PROPERTIES

The chemical properties of the Group 1 and Group 2 metals are summarized in Table 6.6. Anyone who has ever worked with the metals knows that they are highly reactive. The alkali metals and the heavier alkaline

TABLE 6.6 Reactions of Alkali Metals and Alkaline Earth Metals

METAL*	COMBINING SUBSTANCE	REACTION
		Group 1
All	Hydrogen	$2\ M(s) + H_2(g) \rightarrow 2\ MH(s)$
All	Halogens	$2\ M(s) + X_2 \rightarrow 2\ MX(s)$
Li	Nitrogen	$6\ M(s) + N_2(g) \rightarrow 2\ M_3N(s)$
All	Sulfur	$2\ M(s) + S(s) \rightarrow M_2S(s)$
Li	Oxygen	$4\ M(s) + O_2(g) \rightarrow 2\ M_2O(s)$
Na	Oxygen	$2\ M(s) + O_2(g) \rightarrow M_2O_2(s)$
K, Rb, Cs	Oxygen	$M(s) + O_2(g) \rightarrow MO_2(s)$
All	Water	$2\ M(s) + 2\ H_2O \rightarrow 2\ M^+(aq) + 2\ OH^-(aq) + H_2(g)$
		Group 2
Ca, Sr, Ba	Hydrogen	$M(s) + H_2(g) \rightarrow MH_2(s)$
All	Halogens	$M(s) + X_2 \rightarrow MX_2(s)$
Mg, Ca, Sr, Ba	Nitrogen	$3\ M(s) + N_2(g) \rightarrow M_3N_2(s)$
Mg, Ca, Sr, Ba	Sulfur	$M(s) + S(s) \rightarrow MS(s)$
Be, Mg, Ca, Sr, Ba	Oxygen	$2\ M(s) + O_2(g) \rightarrow 2\ MO(s)$
Ba	Oxygen	$M(s) + O_2(g) \rightarrow MO_2(s)$
Ca, Sr, Ba	Water	$M(s) + 2\ H_2O \rightarrow M^{2+}(aq) + 2\ OH^-(aq) + H_2(g)$
Mg	Water	$M(s) + H_2O(g) \rightarrow MO(s) + H_2(g)$

*Abbreviated as "M" in reactions.

earth metals (Ca, Sr, Ba) are commonly stored under dry mineral oil or kerosene to prevent them from reacting with oxygen or water vapor in the air. Magnesium is less reactive; it is commonly available in the form of ribbon or powder. Beryllium, as one would expect from its position in the Periodic Table, is the least metallic element in these two groups. It is also the least reactive toward water, oxygen, or other nonmetals.

■ EXAMPLE 6.5

Write balanced equations for the reaction of

a. sodium with hydrogen.
b. barium with oxygen.
c. aluminum with nitrogen, assuming the reaction is similar to that of nitrogen with the Group 1 and Group 2 metals.

Solution

a. $2\,Na(s) + H_2(g) \rightarrow 2\,NaH(s)$

b. $2\,Ba(s) + O_2(g) \rightarrow 2\,BaO(s);$ $Ba(s) + O_2(g) \rightarrow BaO_2(s)$

c. In the compounds Li_3N and Mg_3N_2, nitrogen is in the form of the N^{3-} ion. Since aluminum forms the Al^{3+} ion, we would expect the formula of aluminum nitride to be AlN. The equation is:

$2\,Al(s) + N_2(g) \rightarrow 2\,AlN(s)$

Exercise
Write the balanced equation for the reaction of lithium with a. nitrogen; b. water. Answers:

a. $6\,Li(s) + N_2(g) \rightarrow 2\,Li_3N(s)$
b. $2\,Li(s) + 2\,H_2O \rightarrow 2\,Li^+(aq) + 2\,OH^-(aq) + H_2(g)$

When potassium metal comes in contact with chlorine gas, it bursts into flame; the reaction is: $2K(s) +$ $Cl_2(g) \rightarrow 2KCl(s)$. (Charles Steele)

On electrolysis of molten NaH, the H^- ions go to the positive (+) electrode

A tank of compressed H_2 gas would not be practical

The compounds formed by the reaction of hydrogen with the alkali and alkaline earth metals contain H^- ions; for example, sodium hydride consists of Na^+ and H^- ions. These white crystalline solids are often referred to as saline hydrides because of their physical resemblance to NaCl. Chemically, they behave quite differently from sodium chloride; for example, they react with water to produce hydrogen gas. Typical reactions are:

$$NaH(s) + H_2O \rightarrow H_2(g) + Na^+(aq) + OH^-(aq) \tag{6.7}$$

$$CaH_2(s) + 2\,H_2O \rightarrow 2\,H_2(g) + Ca^{2+}(aq) + 2\,OH^-(aq) \tag{6.8}$$

In this way, saline hydrides can serve as compact, portable sources of hydrogen gas for inflating life rafts and balloons.

The alkali metals react vigorously with water (Figure 6.11) to evolve hydrogen and form a water solution of the alkali hydroxide. The reaction of sodium is typical:

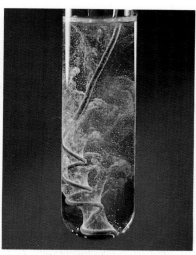

Figure 6.11 When a *very small* piece of sodium is added to water, it reacts violently: $Na(s) + H_2O \rightarrow Na^+(aq) + OH^-(aq) + \frac{1}{2}H_2(g)$. The OH^- ions formed turn the color of the organic dye phenolphthalein from colorless to pink. (Marna G. Clarke)

When aluminum wire is treated with dilute hydrochloric acid, a reation occurs: $Al(s) + 3H^+(aq) \rightarrow Al^{3+}(aq) + 3/2H_2(g)$. (Charles Steele)

$$2 \, Na(s) + 2 \, H_2O \rightarrow 2 \, Na^+(aq) + 2 \, OH^-(aq) + H_2(g) \qquad (6.9)$$

This reaction evolves a considerable amount of heat; the hydrogen produced often catches fire.

Among the Group 2 metals, Ca, Sr, and Ba react with water in much the same way as the alkali metals. The reaction with calcium is:

$$Ca(s) + 2 \, H_2O \rightarrow Ca^{2+}(aq) + 2 \, OH^-(aq) + H_2(g) \qquad (6.10)$$

Beryllium does not react with water at all. Magnesium reacts very slowly with boiling water, but reacts more readily with steam at high temperatures:

$$Mg(s) + H_2O(g) \rightarrow MgO(s) + H_2(g); \qquad (6.11)$$

This reaction, like Reaction 6.9 or 6.10, produces enough heat to ignite the hydrogen. Firefighters who try to put out a magnesium fire by spraying water on it have discovered this reaction, often with tragic results. The best way to extinguish burning magnesium is to dump dry sand on it.

Aluminum does not react directly with water (which is a reassuring thought if you own an automobile with an aluminum engine block). It does, however, dissolve in water solutions that are highly acidic (H^+ ions) or highly basic (OH^- ions). The reactions are:

$$Al(s) + 3 \, H^+(aq) \rightarrow Al^{3+}(aq) + \tfrac{3}{2} \, H_2(g) \qquad (6.12)$$

$$Al(s) + 3 \, H_2O + OH^-(aq) \rightarrow Al(OH)_4^-(aq) + \tfrac{3}{2} \, H_2(g) \qquad (6.13)$$

Commercial drain cleaners nowadays often contain a mixture of powdered aluminum and sodium hydroxide. The OH^- ions dissolve substances such as grease, hair, and paper; the hydrogen gas produced by Reaction 6.13 breaks up solid plugs in pipes.

EXAMPLE 6.6
What volume of hydrogen, measured at 20°C and 750 mm Hg, is evolved by the reaction with water of

a. 1.00 g of Ca?   b. 1.00 g of CaH_2?

Solution

a. The reaction is given by Equation 6.10. We first find the number of moles of hydrogen and then apply the Ideal Gas Law:

$$\text{no. moles } H_2 = 1.00 \text{ g Ca} \times \frac{1 \text{ mol Ca}}{40.08 \text{ g Ca}} \times \frac{1 \text{ mol } H_2}{1 \text{ mol Ca}}$$

$$= 0.0250 \text{ mol } H_2$$

$$V = \frac{(0.0250 \text{ mol})(0.0821 \text{ L·atm/mol·K})(293 \text{ K})}{(750/760 \text{ atm})} = \boxed{0.609 \text{ L}}$$

b. Here we use Equation 6.8:

$$\text{no. moles } H_2 = 1.00 \text{ g } CaH_2 \times \frac{1 \text{ mol } CaH_2}{42.10 \text{ g } CaH_2} \times \frac{2 \text{ mol } H_2}{1 \text{ mol } CaH_2}$$

$$= 0.0475 \text{ mol } H_2$$

Calculating the volume as in (a), we obtain a value of $\boxed{1.16 \text{ L,}}$ nearly twice as much $H_2(g)$ as in (a).

Exercise
How much Al is required to produce 1.00 L of $H_2(g)$ at 25°C and 1.00 atm by Reaction 6.13? Answer: 0.735 g.

You will note from Table 6.6 that several products are possible when a main-group metal reacts with oxygen. The product may be a normal oxide, a peroxide, or a superoxide.

Oxides   The normal oxide contains the oxide anion, O^{2-}, and the metal ion:

Group 1 oxides: M_2O (2 M^+ ions, 1 O^{2-} ion)
Group 2 oxides: MO (M^{2+} ion, O^{2-} ion)
Aluminum oxide: Al_2O_3 (Al^{3+} ion, O^{2-} ion)

Lithium is the only Group 1 metal that forms the normal oxide in good yield by direct reaction with oxygen. The other Group 1 oxides (Na_2O, K_2O, Rb_2O, Cs_2O) must be prepared by other means. In contrast, the Group 2 metals usually react with oxygen to give the normal oxide. Beryllium and magnesium must be heated strongly to give BeO and MgO (Fig. 6.12). Calcium and strontium react more readily to give CaO and SrO. Barium, the most reactive of the Group 2 metals, catches fire when exposed to moist air. The product is a mixture of the normal oxide, BaO, and the peroxide, BaO_2. A freshly cut piece of aluminum, upon exposure to air, quickly

To understand Chapter 6, you need to know what was in the earlier chapters

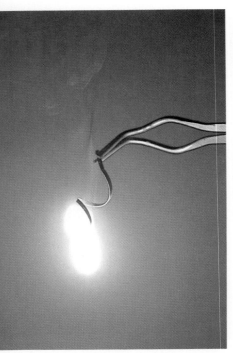

Figure 6.12 A piece of magnesium ribbon bursts into flame when heated in air; the product of the reaction is MgO(s). (Charles D. Winters)

acquires a thin coating of Al_2O_3. This oxide adheres tightly to the surface of the aluminum, protecting it from further reaction. The presence of the oxide film explains why aluminum pots and pans do not react with air or water solutions.

Thank goodness for the oxide film

Peroxides These compounds contain the peroxide ion, O_2^{2-}. The most important peroxides are those of sodium and barium, Na_2O_2 and BaO_2. Upon addition to water, they react vigorously to form hydrogen peroxide, H_2O_2:

$$Na_2O_2(s) + 2\ H_2O \rightarrow 2\ Na^+(aq) + 2\ OH^-(aq) + H_2O_2(aq) \quad (6.14)$$

$$BaO_2(s) + 2\ H_2O \rightarrow Ba^{2+}(aq) + 2\ OH^-(aq) + H_2O_2(aq) \quad (6.15)$$

These reactions can be used to produce small amounts of hydrogen peroxide in the laboratory. Commercially, both Na_2O_2 and BaO_2 are used as bleaching agents.

Superoxides The superoxide ion, O_2^-, is present in these unusual compounds, the most important of which is KO_2. Virtually all the potassium metal produced today is used to make potassium superoxide, KO_2. The reaction

$$K(s) + O_2(g) \rightarrow KO_2(s) \quad (6.16)$$

is carried out by spraying the metal into air. This serves not only to furnish the oxygen required but also to cool the product. Potassium superoxide is used in self-contained breathing apparatus for firefighters and miners. It reacts with the moisture in exhaled air to generate oxygen:

$$4\ KO_2(s) + 2\ H_2O(g) \rightarrow 3\ O_2(g) + 4\ KOH(s) \quad (6.17)$$

The carbon dioxide in the exhaled air is removed by reaction with the KOH formed in Reaction 6.17:

KO_2 is perfect for this job

$$KOH(s) + CO_2(g) \rightarrow KHCO_3(s) \quad (6.18)$$

A person using a mask charged with KO_2 can rebreathe the same air for an extended period of time. This allows that person to enter an area where there are poisonous gases or oxygen-deficient air.

■ **EXAMPLE 6.7** ⎯⎯⎯⎯⎯⎯⎯⎯⎯⎯⎯⎯⎯⎯⎯⎯⎯⎯⎯⎯⎯

Consider the compounds strontium hydride, radium peroxide, and cesium superoxide.

a. Give the formulas of these compounds.
b. Write equations for the formation of these compounds from the elements.
c. Write equations for the reactions of strontium hydride and radium peroxide with water.

Solution

a. SrH_2 (Sr^{2+}, H^- ions); RaO_2 (Ra^{2+}, O_2^{2-} ions); CsO_2 (Cs^+, O_2^- ions)

b. $Sr(s) + H_2(g) \rightarrow SrH_2(s)$

$Ra(s) + O_2(g) \rightarrow RaO_2(g)$

$Cs(s) + O_2(g) \rightarrow CsO_2(s)$

c. $SrH_2(s) + 2 H_2O \rightarrow 2 H_2(g) + Sr^{2+}(aq) + 2 OH^-(aq);$
(compare Eqn. 6.8)

$RaO_2(s) + 2 H_2O \rightarrow Ra^{2+}(aq) + 2 OH^-(aq) + H_2O_2(aq);$
(compare Eqn. 6.15)

Exercise
Write an equation for the reaction of cesium superoxide with water vapor.
Answer: $4 CsO_2(s) + 2 H_2O(g) \rightarrow 3 O_2(g) + 4 CsOH(s)$.

PREPARATION OF SODIUM AND ALUMINUM

Because the main-group metals are so reactive, they are very difficult to prepare chemically. Instead, these metals are ordinarily made from their compounds by a process called *electrolysis,** in which electrical energy is used to bring about a chemical reaction. A simple electrolytic cell used to prepare sodium metal from molten sodium chloride is shown in Figure 6.13. Electrons from a storage battery or other source of direct electric current enter the cell through the iron *electrode* at the left. Here, they convert Na^+ ions to sodium atoms:

$2 Na^+ + 2 e^- \rightarrow 2 Na$

Electric current is carried through the cell by the movement of ions. Cations (Na^+ ions) move in one direction; anions (Cl^- ions) move in the opposite direction. At the graphite rod near the center of the cell, Cl^- ions lose electrons to form Cl_2 molecules:

$2 Cl^- \rightarrow Cl_2 + 2 e^-$

The electrons given off by this half-reaction return to the storage battery. The overall cell reaction is simply:

$$2 NaCl(l) \rightarrow 2 Na(l) + Cl_2(g) \tag{6.19}$$

The cell must be operated at a high temperature, about 600°C, to keep the sodium chloride melted. (Solid NaCl does not conduct a current, since the ions are not free to move.) About 14 kJ of electrical energy must be supplied

---

*Electrolysis is discussed in greater detail in Chapter 21.

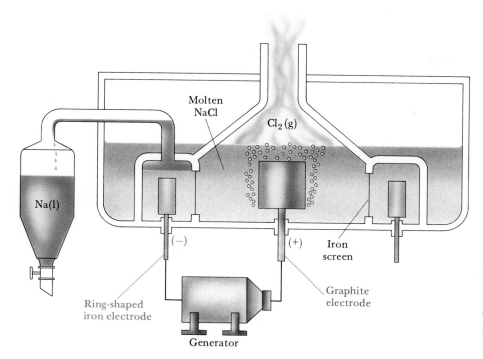

Molten NaCl

$Cl_2(g)$

Na(l)

(−)

(+) Iron screen

Ring-shaped iron electrode

Graphite electrode

Generator

Figure 6.13 Electrolysis of molten sodium chloride, containing some $CaCl_2$ to lower the melting point. The iron screen is used to prevent sodium and chlorine from coming in contact with each other.

for every gram of sodium produced. The sodium metal (mp = 98°C) is drawn off as a liquid. The chlorine gas formed at the other electrode is a valuable by-product.

Aluminum metal, like sodium, is obtained by electrolysis (Fig. 6.14, p. 198). Here the reactant is the oxide, rather than the chloride. The overall cell reaction is:

$$2\ Al_2O_3(l) \rightarrow 4\ Al(l) + 3\ O_2(g) \qquad (6.20)$$

Cryolite, Na_3AlF_6, is added to Al_2O_3 to produce a mixture melting at about 1000°C. (A mixture of AlF_3, NaF, and CaF_2 may be substituted for cryolite.) The cell is heated electrically to keep the mixture molten so that ions can move through it, carrying the electric current. About 30 kJ of electrical energy is consumed per gram of aluminum formed. The high energy requirement explains in large part the value of recycling aluminum cans.

The process for obtaining aluminum from bauxite was worked out in 1886 by Charles Hall, a chemistry student at Oberlin College. The problem that Hall faced was to find a way to electrolyze Al_2O_3 at a temperature below its melting point of 2000°C. His general approach was to look for ionic compounds in which Al_2O_3 would dissolve at a reasonable temperature. After several unsuccessful attempts, Hall found that cryolite was the ideal "solvent." Curiously enough, the same electrolytic process was worked out by Heroult in France, also in 1886. Each young man (both were 22 years old) was entirely unaware of the other's work!

Some graduate students are more productive than others

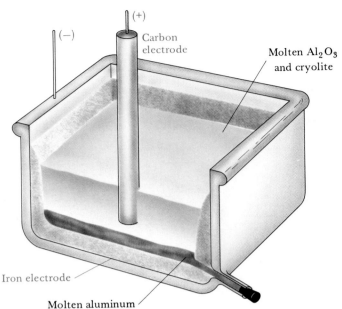

Figure 6.14 Electrolytic preparation of aluminum. Aluminum, being more dense than cryolite collects at the bottom of the cell and so is protected from oxidation by the air.

■ **EXAMPLE 6.8**

Magnesium metal is ordinarily obtained from seawater, where its concentration is 0.0520 mol/L. It is precipitated as $Mg(OH)_2$ by treating the seawater with limewater, a solution of $Ca(OH)_2$. The precipitate is reacted with hydrochloric acid to convert it to the chloride, $MgCl_2$. Finally, the molten chloride is electrolyzed to form magnesium metal.

a. Write a balanced equation for the electrolysis reaction.
b. What volume of seawater must be treated to give one kilogram of magnesium?

Solution

a. $MgCl_2(l) \rightarrow Mg(l) + Cl_2(g)$

b.   Volume of seawater = $1000 \text{ g Mg} \times \dfrac{1 \text{ mol Mg}}{24.30 \text{ g Mg}} \times \dfrac{1 \text{ L seawater}}{0.0520 \text{ mol Mg}}$

     =   791 L seawater

Exercise

What mass of $Ca(OH)_2$ is required to treat enough seawater to obtain a kilogram of magnesium? Answer: 3.05 kg.

6.6
SOME IMPORTANT COMPOUNDS OF THE MAIN-GROUP METALS

Many of the compounds that you work with in the general chemistry laboratory contain a main-group metal. Beyond that, compounds of sodium, calcium, and aluminum play a major role in many industrial processes. In this section, we will look at a few of the more important compounds of these elements.

SODIUM COMPOUNDS

NaCl Sodium chloride may be obtained by evaporation of seawater or from huge underground deposits where it occurs in impure form ("rock salt"). The largest of these deposits, more than 100 m thick, underlies parts of Oklahoma, Texas, and Kansas. The salt is obtained by forming a water solution that is pumped to the surface. Upon evaporation, crystals of NaCl form.

All sodium compounds are soluble in water

The NaCl produced in this way is contaminated with small amounts of $MgCl_2$ and $CaCl_2$. These compounds are present in ordinary table salt. They tend to pick up water from the air, causing the salt to "cake" and refuse to pour. Small amounts of chemical drying agents are added to prevent this. About 0.01% KI is also added to table salt. This prevents a condition called goiter, an enlargement of the thyroid gland caused by iodine deficiency.

NaOH About 10^7 metric tons of sodium hydroxide are produced annually in the United States. It is used in making soap, textiles, and paper as well as in the metallurgy of aluminum. In the laboratory, you are likely to encounter NaOH either as a solid (flakes or pellets) or in water solution. Concentrated sodium hydroxide solutions are very corrosive to skin and clothing, so you should be very careful in working with them.

Sodium hydroxide gives off a considerable amount of heat when it is added to water. The solution process can be represented simply as:

$$NaOH(s) \rightarrow Na^+(aq) + OH^-(aq) \tag{6.21}$$

although the first step appears to involve the formation of a hydrate, $NaOH \cdot H_2O$. Often, enough heat is released to raise the temperature to near the boiling point. The solution is strongly basic because of the high concentration of OH^- ions. Over a period of time, concentrated NaOH solutions attack glass, becoming cloudy in the process. They also pick up CO_2 from the air:

$$2\ OH^-(aq) + CO_2(g) \rightarrow H_2O + CO_3^{2-}(aq) \tag{6.22}$$

This contaminates the NaOH solution with sodium carbonate, Na_2CO_3.

Na_2CO_3 and $NaHCO_3$ These two compounds, sodium carbonate and sodium hydrogen carbonate, occur as the mineral trona, a hydrate that has the composition $Na_2CO_3 \cdot NaHCO_3 \cdot 2H_2O$. Large deposits of trona are found near Green River, Wyoming. They are the major source of the 7×10^6 metric tons of Na_2CO_3 and smaller amounts of $NaHCO_3$ produced annually in the United States.

Sodium carbonate, Na_2CO_3, is commonly called "soda ash"; its deca-hydrate $Na_2CO_3 \cdot 10H_2O$ is called "washing soda." Soda ash is used to make glass, soap, paper, and many chemicals. Currently, $Na_2CO_3 \cdot 10H_2O$ is used in detergents as a substitute for phosphates. Washing soda is not a serious pollutant, since carbonates occur naturally in surface waters.

Sodium hydrogen carbonate, $NaHCO_3$, is commonly called bicarbonate of soda or baking soda. It is an ingredient of many commercial products used to relieve indigestion. The HCO_3^- ions react with H^+ ions, thus relieving "excess stomach acidity":

$$HCO_3^-(aq) + H^+(aq) \rightarrow CO_2(g) + H_2O \qquad (6.23)$$

A similar reaction accounts for the use of $NaHCO_3$ in baking powders. An acidic component of the baking powder provides the H^+ ions required for Reaction 6.23. The carbon dioxide is formed as tiny bubbles that expand upon warming, causing bread or pastries to "rise."

CALCIUM COMPOUNDS

By far the most abundant and widely used compound of calcium is the carbonate, $CaCO_3$. Calcium carbonate is found in many different forms in nature. The purest of these is the transparent mineral calcite. In less pure form, $CaCO_3$ is found as marble, limestone, and dolomite ($CaCO_3 \cdot MgCO_3$).

Calcium carbonate is very insoluble in pure water. However, it dissolves to an appreciable extent in ground water, which contains dissolved carbon dioxide from the atmosphere:

$$CaCO_3(s) + H_2O + CO_2(g) \rightarrow Ca^{2+}(aq) + 2\,HCO_3^-(aq) \qquad (6.24)$$

This reaction is responsible for the formation of limestone caves. Such caves often contain icicle-like formations of calcium carbonate, called stalactites and stalagmites (Fig. 6.15). These are formed when Reaction 6.24 reverses within the cave.

When limestone is heated to about 800°C, it decomposes:

$$CaCO_3(s) \rightarrow CaO(s) + CO_2(g) \qquad (6.25)$$

The solid product, calcium oxide, is often referred to as quicklime. It reacts with water to form calcium hydroxide:

$$CaO(s) + H_2O(l) \rightarrow Ca(OH)_2(s) \qquad (6.26)$$

Calcium hydroxide is sometimes called slaked lime or, more simply, lime. It is used to "sweeten" acidic soils and in many processes where a cheap, strong base is required. Mortar is made by mixing $Ca(OH)_2$ with sand and water. The mortar "sets" by picking up CO_2 from the air:

$$Ca(OH)_2(s) + CO_2(g) \rightarrow CaCO_3(s) + H_2O(l) \qquad (6.27)$$

The calcium carbonate formed binds the particles of sand together.

Figure 6.15 Stalactites (*upper*) and stalagmites (*lower*) consist of calcium carbonate. They are formed when a water solution containing Ca^{2+} and HCO_3^- ions enters a cave. Carbon dioxide is released and calcium carbonate precipitates: $Ca^{2+}(aq) + 2HCO_3^-(aq) \rightarrow CaCO_3(s) + H_2O + CO_2(g)$. (Dick George, Tom Stack & Associates)

Another important calcium compound is the mineral called gypsum. This is a hydrate of calcium sulfate, $CaSO_4\cdot2H_2O$. Gypsum is used in making cement, wallboard, and pottery. Other uses depend upon a reaction that takes place when gypsum is heated, losing three-fourths of its water of crystallization:

$$CaSO_4\cdot2H_2O(s) \rightarrow CaSO_4\cdot\tfrac{1}{2}H_2O(s) + \tfrac{3}{2}H_2O(g) \qquad (6.28)$$

The product is called plaster of Paris, or sometimes simply plaster. When it is ground to a fine powder and mixed with water, plaster is converted back to gypsum:

$$CaSO_4\cdot\tfrac{1}{2}H_2O(s) + \tfrac{3}{2}H_2O(l) \rightarrow CaSO_4\cdot2H_2O(s) \qquad (6.29)$$

This process takes place with an increase in volume. Hence, the material expands to fill completely any space to which it is confined. This explains its use in making models of statues, patching holes in walls, and forming casts for broken bones.

Gypsum is the filling in a wallboard sandwich

ALUMINUM COMPOUNDS

Aluminum oxide, Al_2O_3, occurs in nature in nearly pure form as corundum, which is used as an abrasive in "sandpaper." Several precious stones consist of aluminum oxide with small amounts of impurities that give the stones their color. Among these are rubies and sapphires. These can be made synthetically by melting Al_2O_3 with the appropriate impurity (Cr_2O_3 for rubies, FeO and TiO_2 for sapphires). The synthetic stones can be distinguished from the natural gems by the presence of tiny air bubbles, visible only under the microscope.

Aluminum forms a series of "double salts" containing other metals in addition to aluminum. These salts, commonly called *alums,* have the general formula $MAl(SO_4)_2\cdot12H_2O$, where M is a +1 cation. The best known of these is potassium alum, $KAl(SO_4)_2\cdot12H_2O$. It crystallizes when a water solution containing K_2SO_4 and $Al_2(SO_4)_3$ is slowly evaporated. Large amounts of potassium alum and aluminum sulfate are used to purify water supplies. The addition of lime, $Ca(OH)_2$, brings about the reaction:

$$Al^{3+}(aq) + 3\,OH^-(aq) \rightarrow Al(OH)_3(s) \qquad (6.30)$$

As the aluminum hydroxide precipitates, it adsorbs on its surface suspended particles and bacteria.

It's easy to grow large alum crystals

■ SUMMARY PROBLEM

Consider the element strontium (Sr).

a. What is the outer electron configuration of Sr?
b. Compare Sr to Ca and Rb in atomic radius and first ionization energy.
c. How does the Sr^{2+} ion compare in size to the Sr atom? To the Ba^{2+} ion?
d. As successive electrons are removed from the Sr atom, where does the largest jump in ionization energy occur?

e. How does strontium compare to sulfur in electrical conductivity? Malleability? Luster?

f. Write a balanced equation for the reaction of strontium with oxygen; with water.

g. Give the formulas of strontium hydride; strontium nitride; strontium sulfide.

h. Electrolysis of strontium chloride gives strontium metal and chlorine gas. What mass of the chloride must be electrolyzed to form one kilogram of strontium? One liter of $Cl_2(g)$ at STP?

Answers

a. $5s^2$

b. atomic radius: Rb > Sr > Ca; ionization energy: Ca > Sr > Rb

c. Sr^{2+} is smaller than either Sr or Ba^{2+} **d.** between ΔE_2 and ΔE_3

e. better electrical conductor, more malleable, higher luster

f. $2\ Sr(s)\ +\ O_2(g)\ \rightarrow\ 2\ SrO(s)$

$Sr(s)\ +\ 2\ H_2O\ \rightarrow\ Sr^{2+}(aq)\ +\ 2\ OH^-(aq)\ +\ H_2(g)$

g. SrH_2; Sr_3N_2; SrS   **h.** 1.809 kg; 7.08 g

Four of the main-group cations, Na^+, K^+, Mg^{2+}, and Ca^{2+}, are essential in human nutrition (Table 6.7). In addition, small amounts of lithium compounds have been found to be effective in treating mental disorders, including manic depression. On the other hand, traces of airborne beryllium compounds are highly toxic. Long-term exposure to Be^{2+} at a concentration in air of 2×10^{-9} g/m³ can cause lung damage.

TABLE 6.7 Nutritional Properties of the Main-Group Cations

ION	APPROX. MASS, 70-kg ADULT	DAILY REQUIREMENT	FUNCTION IN BODY	MAJOR FOOD SOURCES
Na^+	63 g	~2 g	Principal cation *outside* cell fluid	Table salt, drinking water, bread, many processed and preserved foods, ham, bacon
K^+	150 g	~2 g	Principal cation *within* cell fluid	Fruits, nuts, fish, instant coffee, wheat bran
Mg^{2+}	21 g	0.35 g	Activates enzymes for body processes	Chocolate, nuts, green vegetables, instant coffee, wheat bran
Ca^{2+}	1160 g	0.8 g	Bone and tooth formation	Milk, cheeses, other dairy products, broccoli, canned salmon (with bones)

Na^+ Ions

The requirement for Na^+ in the diet has been recognized for centuries. Table salt, NaCl, was a major item of commerce in ancient Greece and Rome. On the American frontier, it was the only "food" that could not be prepared in the farm kitchen. Instead, it had to be purchased at the nearest store, which might be many miles away, and transported to the farm, often on the back of the farmer.

Today no one need worry about sodium deficiency. On the average, Americans take in about 4 to 8 g of Na^+ per day; 1 to 3 g is adequate for nutritional purposes. For many people with high blood pressure, it is imperative that the intake of sodium be reduced. One obvious way to do this is to use less table salt. Substitutes, containing K^+ or NH_4^+ ions as a partial or total replacement for Na^+, are available.

Processed foods often contain surprisingly large amounts of Na^+ ions, since NaCl is commonly added to these foods as a flavoring agent or preservative. A single serving of canned peas may contain as much as 200 mg of sodium; the same quantity of fresh peas contain essentially no sodium. Other processed foods that are high in Na^+ include cottage cheese, cornflakes, pickles, and sardines. Some seasonings, including soy sauce, catsup, and mustard, are rich in sodium; others, such as curry powder, thyme, and pepper, are virtually sodium-free.

Ca^{2+} Ions

About 90% of the calcium in the body is found in bones and teeth, largely in the form of hydroxyapatite, $Ca(OH)_2 \cdot 3Ca_3(PO_4)_2$. Calcium ions in bones exchange readily with those in the blood; about 0.6 g of Ca^{2+} enters and leaves your bones every day. In a normal adult this exchange is in balance, but in elderly people, particularly women, there is sometimes a net loss of bone calcium, leading to the disease known as osteoporosis. It is generally supposed that osteoporosis is related to Ca^{2+} deficiency in early adult years, although the evidence is far from conclusive.

Frankly, ...

The richest sources of calcium by far are milk and milk products such as yogurt, cheese, or ice cream. A single cup of milk supplies about 40% of your daily requirement of Ca^{2+}. If you don't like milk, alternatives include broccoli and blackstrap molasses, both of which are high in calcium. Another possibility is to take a calcium supplement, such as calcium carbonate, in the form of tablets. If you choose this approach, make sure you read the label. A 100-mg $CaCO_3$ tablet (40% Ca) supplies only about 5% of your daily requirement of calcium.

QUESTIONS AND PROBLEMS

Symbols, Formulas, and Equations

1. Give the formulas of the following compounds:
a. table salt
b. washing soda
c. hydrogen peroxide
d. lime

2. Give the formulas of the following compounds:
a. potassium superoxide
b. baking soda
c. corundum
d. limestone

3. Write the equation for
a. the electrolysis of molten sodium chloride.
b. the conversion of gypsum to plaster of Paris.
c. the relief of "excess stomach acidity" with bicarbonate of soda.

4. Write the equation for
a. the electrolysis of molten magnesium chloride.
b. the conversion of plaster of Paris to gypsum.
c. the formation of lime from quicklime.

Periodic Table

5. Define or describe the following terms related to the Periodic Table:
a. alkaline earth metal
b. group
c. alkali metal
d. transition metal
e. noble gas

6. Describe or define the following terms related to the Periodic Table:
a. period
b. halogen
c. lanthanide
d. actinide
e. metalloid

7. In the sixth period, what is the symbol and name for the element that
a. completes the 5d level?
b. completes the 4f level?

8. In the seventh period, what would be the atomic number of the element that
a. completes the period?
b. completes the 6d sublevel?

9. Give the name and symbol for the element that
a. starts to fill the 5p sublevel.
b. completes the 5p sublevel.
c. completes the 6s sublevel.
d. is the last member of the lanthanide series.

10. Give the name and symbol for the element that
a. starts the 7s sublevel.
b. starts the 6p sublevel.
c. has a half-filled 6p sublevel.
d. starts the actinide series.

11. State the outer electron configuration for
a. radium (Ra)
b. astatine (At)
c. lead (Pb)
d. antimony (Sb)

12. State the outer electron configuration for
a. radon (Rn)
b. francium (Fr)
c. tellurium (Te)
d. bismuth (Bi)

13. Arrange the elements Rb, Te, and I in order of
a. increasing atomic radius
b. increasing ionization energy
c. decreasing metallic character

14. Arrange the elements Mg, S, and Cl in order of
a. increasing atomic radius
b. increasing ionization energy
c. increasing metallic character

15. Which of the four atoms Na, Mg, S, or K has the
a. smallest atomic radius?
b. lowest ionization energy?
c. most metallic character?

16. Which of the following atoms, Ca, As, Br, and Rb, has the
a. largest atomic radius?
b. highest ionization energy?
c. least metallic character?

17. Classify each of the following as a metal, nonmetal, or metalloid:

a. sulfur   b. silicon   c. scandium
d. antimony   e. tin

18. Classify each of the following as metal, nonmetal, or metalloid:
a. boron   b. beryllium   c. bromine
d. barium   e. bismuth

19. Give the symbols and names of all elements
a. with atomic number greater than 30 that are chemically similar to magnesium.
b. that have the outer electron configuration $ns^2(n - 1)d^3$.
c. that are actinides with atomic number greater than 100.

20. Give the symbols and names of all elements
a. that are nonmetals with the outer configuration ns^2np^4.
b. that are metals chemically similar to cesium but have atomic numbers less than 50.
c. that just complete an s sublevel.
d. that have a single outer p electron.

21. Predict the melting point of strontium given those of calcium (845°C) and barium (725°C). (The observed value is 770°C.)

22. Mendeleev predicted the properties of gallium by averaging the values for the elements above and below gallium in the Periodic Table. Use this method to estimate the density of gallium, given that the densities of Al and In are 2.70 and 7.31 g/cm³, respectively (measured density of Ga = 5.91 g/cm³).

23. Select the larger member of each pair.
a. K and K^+ b. O and O^{2-}
c. Tl and Tl^{3+} d. Ni^{2+} and Ni^{3+}

24. Select the larger member of each pair.
a. N and N^{3-} b. Ba and Ba^{2+}
c. Cr and Cr^{3+} d. Cu^+ and Cu^{2+}

25. List the following species in order of decreasing radius:
a. K, Ca, Ca^{2+}, Rb   b. S, Te^{2-}, Se, Te

26. List the following species in order of decreasing radius:
a. Co, Co^{2+}, Co^{3+} b. Cl, Cl^-, Br^-

Groups 1, 2, and Aluminum

27. Give the formula and name of the compound formed when strontium reacts with
a. nitrogen   b. bromine
c. water   d. oxygen

28. Give the formula and name of the compound formed when potassium reacts with
a. nitrogen   b. iodine   c. water
d. hydrogen   e. sulfur

29. Write a balanced equation and give the names of the products for the reaction of
a. magnesium with chlorine.
b. barium peroxide with water.

c. lithium with sulfur.
d. sodium with water.

30. Write a balanced equation and give the names of the products for the reaction of
a. sodium peroxide and water.
b. calcium and oxygen.
c. rubidium and oxygen.
d. strontium hydride and water.

31. Write balanced equations for the reaction of aluminum with
a. $O_2(g)$ b. $F_2(g)$ c. $N_2(g)$ d. $Cl_2(g)$

32. Give the formula of
a. aluminum sulfate   b. aluminum bromide
c. aluminum hydroxide   d. potassium alum

33. When sodium is exposed to air, it forms a peroxide. This compound, upon addition to water, gives a solution of sodium hydroxide. The same solution can be obtained by adding sodium to water. Write balanced equations for the reactions just described.

34. When calcium is exposed to air, it forms an oxide. This compound reacts with water to form calcium hydroxide, which can also be made by adding calcium to water. Write a balanced equation for each reaction just described.

35. Identify the Group 2 element that
a. does not give a flame test.
b. forms a peroxide readily.
c. is least metallic.
d. is found in limestone.

36. Which Group 1 metal
a. is found in bicarbonate of soda?
b. has the lowest melting point?
c. gives a yellow flame test?
d. has an atomic mass less than that of the preceding noble gas?

37. Consider the elements in the second period. Indicate where the largest jump in ionization energy is expected to occur, following the example given:

Li: between ΔE_1 and ΔE_2

a. C: between _____ and _____
b. B: between _____ and _____
c. F: between _____ and _____

38. The first three ionization energies for calcium are 590, 1145, and 4912 kJ/mol. The corresponding values for argon are 1520, 2666, and 3931 kJ/mol. Explain why
a. the first two ionization energies are much lower for calcium than for argon.
b. the third ionization energy for calcium is higher than that for argon.

39. In the electrolysis of NaCl, what volume of Cl_2 at STP is formed at the same time that 1.00 g of sodium is produced at the other electrode?

40. In the electrolysis of Al_2O_3, the product at one electrode is 2.00 L of O_2 at 25°C and 751 mm Hg. What mass

of aluminum is formed at the same time at the other electrode?

41. What is the energy difference in kilojoules per mole between the two levels associated with the strong line in the lithium spectrum at 670.8 nm?

42. A certain alkali metal has two energy levels separated by 153.3 kJ/mol. When an electron moves from the higher to the lower level, what is the wavelength of the radiation emitted? Referring to Table 6.5, identify the metal.

Unclassified

43. Explain why
a. negative ions are larger than their corresponding atoms.
b. scandium, a transition metal, forms an ion with a noble gas structure.
c. it is easier to make Na^+ than Na^{2+}.
d. the electron affinity of chlorine is a negative quantity but that of helium is positive.

44. Explain why
a. transition metals form many colored compounds.
b. atomic radius decreases going across a period in the Periodic Table.
c. Group 2 metals form +2 ions.
d. electron affinity is expected to become a larger negative number when atomic radius decreases.

45. Criticize the following statements:
a. Energy is given off when an electron is removed from an atom.
b. Elements are located in the Periodic Table in sequence of increasing atomic mass.
c. The alkali metals are stored under water to protect them from air.

46. Indicate whether each of the following is true or false. If false, correct the statement.
a. Effective nuclear charge stays about the same going across a period.
b. Group 7 elements have an np^7 outer electron configuration.
c. All alkali metals form oxides when reacting with oxygen.

47. What role did each of the following have in developing the Periodic Table?
a. Lothar Meyer   b. J. A. R. Newlands

48. What role did each of the following have in developing the Periodic Table?
a. Henry Moseley   b. John Dalton
c. D. Mendeleev

49. Name and give the symbol for the element with the characteristic given below:
a. Electron configuration of $1s^22s^22p^63s^23p^3$.
b. Lowest ionization energy in Group 7.
c. Its +2 ion has the configuration $[_{18}Ar]\ 3d^2$.
d. Alkali metal with the largest atomic radius.
e. Largest ionization energy in the third period.

50. Given the following elements with their respective outer electron configurations, answer the following questions without using the Periodic Table.

$$X = 5s^2 \qquad Y = 4s^24p^4$$

a. Characterize each element as a metal, nonmetal, or metalloid.
b. Which has the larger ionization energy?
c. Which is more metallic?
d. Which has the weaker electron affinity?
e. Which has the larger atomic radius?
f. What are the possible ions for each element?

51. At 700°C, barium peroxide decomposes to barium oxide. Oxygen is given off in the process. If the oxygen liberated by heating 22.5 g of barium peroxide is collected in a 500-cm³ flask at 25°C, what is the pressure in the flask?

52. A sample of sodium liberates 2.73 L hydrogen at 752 mm Hg and 22°C when it is added to a large amount of water. How much sodium is used?

53. A self-contained breathing apparatus contains 248 g of potassium superoxide. A firefighter exhales 116 L of air at 37°C and 748 mm Hg. The volume percent of water in exhaled air is 6.2. What mass of potassium superoxide is left after the water in the exhaled air reacts with it?

54. Consider the electrolysis of 3.50 kg of aluminum oxide. How many grams of aluminum are produced? If 15.0% of the oxygen formed reacts with the carbon electrode to form carbon dioxide, what volume of CO_2 at 25°C and one atmosphere pressure is formed?

55. A sample of limestone is partially converted to calcium oxide by heating. When the residue is treated with concentrated hydrochloric acid, the following reactions occur:

$$CaCO_3(s) + 2H^+(aq) \rightarrow Ca^{2+}(aq) + H_2O + CO_2(g)$$

$$CaO(s) + 2H^+(aq) \rightarrow Ca^{2+}(aq) + H_2O$$

It is found that 2.83 L of 12.0 M HCl is required to react with one kilogram of the residue. What is the percent by mass of $CaCO_3$ in the residue?

56. Plot the melting points and atomic numbers of the alkali metals using Table 6.5. Deduce the physical state of francium at room temperature (25°C). How many other elements do you know that have the same physical state at the same conditions?

57. Barium atoms impart a green color to a flame due to an electronic transition that gives off light at 554 nm. What is the energy difference, $E_{hi} - E_{lo}$, for this transition?

58. Magnesium does not give a flame test. The strongest line in the Mg spectrum arises from a transition between two states that differ in energy by 419.3 kJ/mol. What is the wavelength of this line? In what region of the spectrum does it fall?

59. Predict some of the properties of element 116, which we will call Z.

a. Give its abbreviated electron configuration.
b. What element will it most resemble chemically?

Challenge Problems

60. Astatine is a synthetic element, first made by nuclear bombardment in 1940. On graph paper, plot the quantities listed below versus atomic number for the more common halogens. (Data can be obtained from Appendix 2.) Draw smooth curves through your points and extrapolate to estimate the atomic radius, melting point, and boiling point of astatine.
   a. atomic radius   b. melting point
   c. boiling point

61. A sample of 20.00 g of barium reacts with oxygen to form 22.38 g of a mixture of barium oxide and barium peroxide. Determine the composition of the mixture.

62. Using data in Table 6.5, calculate the densities of (spherical) atoms of potassium and calcium and compare to the listed densities of these metals.

63. Sodium hydrogen carbonate is a home remedy for "excess stomach acid." The relieving reaction is

$$HCO_3^-(aq) + H^+(aq) \rightarrow CO_2(g) + H_2O$$

To relieve 0.250 L of excess stomach acid (0.12 M HCl), how many teaspoonsful (1 teaspoonful = 7.8 g) of $NaHCO_3$ should you take? What volume of carbon dioxide will be produced at body temperature, 98.6°F, and 1.00 atm?

$$Fe_2O_3(s) + 2Al(s) \rightarrow 2Fe(s) + Al_2O_3(s)$$

Reaction of aluminum powder with iron(III) oxide to give aluminum oxide and iron.

CHAPTER
7

THERMOCHEMISTRY

*Some say the world will end in fire,
Some say in ice.
From what I've tasted of desire
I hold with those who favour fire.*

ROBERT FROST
FIRE AND ICE

Chemical reactions are accompanied by energy changes, which may take various forms. Consider, for example, the reactions

$$CH_4(g) + 2\ O_2(g) \rightarrow CO_2(g) + 2\ H_2O(l)$$

$$2\ N_2H_4(l) + N_2O_4(l) \rightarrow 3\ N_2(g) + 4\ H_2O(g)$$

$$2\ H_2O(l) \rightarrow 2\ H_2(g) + O_2(g)$$

The first reaction takes place when natural gas, which is mostly methane, burns. It supplies the heat required to cook a steak on a gas range or heat water with a Bunsen burner. The reaction between hydrazine, N_2H_4, and dinitrogen tetroxide furnishes the mechanical energy to lift a rocket and its payload from the surface of the earth. In contrast, electrical energy must be supplied to decompose water to its elements, hydrogen and oxygen.

    In this chapter, our main concern will be with one type of energy change: the heat flow associated with a chemical reaction or physical change. Studies of heat flow constitute the science of *thermochemistry*. Our discussion will focus upon:

— the direction (or sign) of the heat flow (Section 7.1)

— the experimental determination of the sign and magnitude of the heat flow (Section 7.2)

— the concept of enthalpy (heat content) and the enthalpy change, ΔH (Section 7.3)

— the calculation of ΔH, using thermochemical equations (Sections 7.4, 7.5)

    In the last two sections of this chapter we will look at energy changes from a broader standpoint. We will examine the relation between heat flow

Heat is a form of energy

209

and other forms of energy in Section 7.6. In Section 7.7 we will review the natural sources of energy that are available today and speculate about energy sources of the future.

7.1
SOME BASIC CONCEPTS

To understand the principles of thermochemistry, you need to develop a basic vocabulary, involving a dozen or so terms. To begin with, in any discussion of the direction of heat flow, it is important to distinguish carefully between **system** and **surroundings**. The system is, quite simply, the sample of matter upon which we focus attention. It might be a beaker of water, a reaction mixture, or a gas undergoing expansion. For all practical purposes, the surroundings ordinarily include only those materials in close contact with the system. In principle, though, they include all of the universe outside the system. The system shown in Figure 7.1 consists of a 50-g sample of water contained within a 100-mL beaker. The surroundings consist of a hot plate and the air around the beaker.

System + Surroundings = Universe

STATE PROPERTIES

To specify the **state** of a system, we cite its composition, temperature, and pressure. When we say that the system in Figure 7.1a consists of

50.00 g of $H_2O(l)$ at 50.0°C and 1 atm

we know precisely what we are dealing with. When the water in Figure 7.1 is heated, its state changes, perhaps to one described as

50.00 g $H_2O(l)$ at 80.0°C and 1 atm

Figure 7.1   When 50.00 g of water is heated from 50.0°C to 80.0°C, 6280 J of heat is absorbed. It follows that:
—the heat capacity of the sample is 6280 J/30.0°C = 209 J/°C
—the specific heat of water is 6280 J/(30.0°C × 50.0 g) = 4.18 J/g·°C

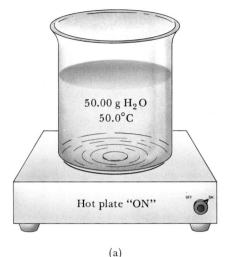

50.00 g H_2O
50.0°C

Hot plate "ON"

50.00 g H_2O
80.0°C

Hot plate "OFF"

(a)

(b)

Certain quantities, called **state properties**, depend only upon the state of a system, not upon the way it reached that state. Volume is such a property. If the mass, temperature, and pressure of a sample of water are specified, we can immediately calculate its volume. Noting that the densities of water at 50.0°C and 80.0°C are 0.9881 g/mL and 0.9718 g/mL, respectively, we have:

$$V \text{ of } 50.00 \text{ g of } H_2O(l) \text{ at } 50.0°C, 1 \text{ atm} = \frac{50.00 \text{ g}}{0.9881 \text{ g/mL}} = 50.60 \text{ mL}$$

$$V \text{ of } 50.00 \text{ g of } H_2O(l) \text{ at } 80.0°C, 1 \text{ atm} = \frac{50.00 \text{ g}}{0.9718 \text{ g/mL}} = 51.45 \text{ mL}$$

When a system moves from an initial to a final state, its properties change. For a state property, like volume, the magnitude of the change depends only upon the nature of the two states. It is independent of the path by which the change occurs. For example, if a 50.00-g sample of water goes from an initial state of 50.0°C and 1 atm to a final state of 80.0°C and 1 atm:

$$\Delta V = V_{final} - V_{initial} = 51.45 \text{ mL} - 50.60 \text{ mL} = 0.85 \text{ mL}$$

The volume increases by 0.85 mL regardless of whether heating takes place directly or by an indirect path, perhaps by first heating to 100°C and then cooling to 80°C. In general, the change in a state property is independent of the number of steps by which the change occurs.

The cost of water would not be a state property

SIGN OF HEAT FLOW

Let us look again at Figure 7.1, this time concentrating upon the direction in which heat is flowing. In Figure 7.1a, while the hot plate is "on," the water is absorbing heat; there is a flow of heat from the surroundings (the hot plate) to the system (the 50-g sample of water). We describe this situation by saying that the heat flow, q, for the system is a positive quantity:

Heat flows from a sample at a higher temperature into one at a lower

q is + when heat flows into the system from the surroundings.

In Figure 7.1b, with the hot plate turned off, the water cools, giving off heat to the air around it. When heat flows out of the system to the surroundings, q for the system is a negative quantity:

q is − when heat flows out of the system to the surroundings.

HEAT CAPACITY AND SPECIFIC HEAT

The magnitude of the heat flow when a system undergoes a temperature change is determined by its **heat capacity**, which is defined by the equation:

$$q = C \times \Delta t \tag{7.1}$$

where q is the heat flow, C is the heat capacity, and Δt is the temperature change:

$$\Delta t = t_{final} - t_{initial} \tag{7.2}$$

From Equation 7.1, we see that

$$C = q/\Delta t$$

We can think of heat capacity as the amount of heat required to raise the temperature of a system one degree Celsius. If we find that 6280 J must be absorbed to raise the temperature of the sample of water shown in Figure 7.1 from 50.0°C to 80.0°C, the heat capacity of the sample must be

$$C = 6280 \text{ J}/30.0°C = 209 \text{ J}/°C$$

For a pure substance, the heat capacity is the product of mass and **specific heat**:

$$C = m \times S.H. \tag{7.3}$$

where C is the heat capacity, m is the mass in grams, and $S.H.$ represents specific heat. Combining Equations 7.1 and 7.3, we have

$$q = S.H. \times m \times \Delta t \tag{7.4}$$

which relates heat flow to specific heat, mass, and temperature change. From Equation 7.4 we see that

$$S.H. = q/(m \times \Delta t)$$

The specific heat of a substance is the amount of heat required to raise the temperature of one gram of the substance one degree Celsius. Noting that for the sample of water shown in Figure 7.1:

$$q = 6280 \text{ J}; \qquad m = 50.0 \text{ g}; \qquad \Delta t = 30.0°C$$

we conclude that the **specific heat of water** must be:

Amounts of heat can also be expressed in calories: 1 cal = 4.18 J, so the specific heat of water is 1 cal/g°C

$$S.H. = \frac{6280 \text{ J}}{50.0 \text{ g} \times 30.0°C} = 4.18 \text{ J}/g \cdot °C$$

Specific heat, like density or melting point, is a characteristic property of a substance. Most substances have specific heats considerably smaller than that of water (Table 7.1).

■ **EXAMPLE 7.1** _____
How much heat is given off by a 50.0 g sample of copper when it cools from 80.0°C to 50.0°C?

Solution
Using Equation 7.4 and the specific heat of copper given in Table 7.1:

$$q = 0.382 \frac{\text{J}}{g \cdot °C} \times 50.0 \text{ g} \times (50.0 - 80.0)°C = \boxed{-573 \text{ J}}$$

The negative sign indicates that heat flows from the copper to the surroundings.

Exercise
What is the final temperature when a 10.0 g sample of ethyl alcohol, C_2H_5OH, originally at 25.0°C, absorbs one kilojoule of heat? Answer: 66.2°C.

TABLE 7.1 Specific Heats of Some Elements and Compounds

	S.H. (J/g·°C)		S.H. (J/g·°C)
$Br_2(l)$	0.474	$CO_2(g)$	0.843
$Cl_2(g)$	0.478	$C_2H_5OH(l)$	2.43
$Cu(s)$	0.382	$C_6H_6(l)$	1.72
$Mg(s)$	1.01	$NaCl(s)$	0.866

HEAT FLOW IN REACTIONS

We can apply the ideas we have developed concerning heat flow to chemical reactions. Here, the chemicals are the reaction system and change from reactants to products as the reaction proceeds. We distinguish between **exothermic** and **endothermic** reactions. An exothermic reaction is one in which heat flows out of the reaction system into the surroundings. This means that q for the reaction system is a negative quantity:

Exothermic reaction: $q < 0$ exo: out of system

In an endothermic reaction, heat flows in the opposite direction, from the surroundings into the reaction system. Since the system absorbs heat, q is a positive quantity:

Endothermic reaction: $q > 0$ endo: into system

Most of the reactions that you carry out in the laboratory are exothermic. A familiar example is the combustion of methane gas:

$$CH_4(g) + 2\ O_2(g) \rightarrow CO_2(g) + 2\ H_2O(l);\qquad q < 0$$

This reaction gives off heat to the surroundings, which might be the air around a Bunsen burner in the laboratory or a potato being baked in a gas oven in the kitchen. In either case, the effect of the heat transfer is to raise the temperature of the surroundings.

A familiar example of an endothermic reaction is the melting of ice. We can represent this process by the equation:

$$H_2O(s) \rightarrow H_2O(l);\qquad q > 0$$

The melting of ice absorbs heat from the surroundings, which might be the water in a glass of iced tea. Here, as is usually the case, the temperature of the surroundings drops when they transfer heat to the system, in this case the ice, to make an endothermic reaction take place.

7.2
MEASUREMENT OF HEAT FLOW; CALORIMETRY

To make accurate thermochemical measurements, we use a device known as a *calorimeter*. This apparatus contains water and/or other materials whose heat capacities are readily determined. The walls of the calorimeter are insulated to minimize exchange of heat with the air.

COFFEE-CUP CALORIMETER

Figure 7.2 shows a simple calorimeter often used in the general chemistry laboratory. It consists of a polystyrene foam cup partially filled with water. The cup has a tightly fitting cover through which an accurate thermometer is inserted. Since polystyrene foam is a good insulator, there is very little heat flow through the walls of the cup. Essentially all the heat evolved by a reaction taking place within the calorimeter is absorbed by the water. We ordinarily neglect the small amount of heat absorbed by the polystyrene foam and the glass thermometer.

To use this apparatus, we start by adding a weighed amount of water to the cup. We then put on the cover, making sure that the thermometer bulb is below the surface of the water. The initial temperature is recorded. The reaction under study is then carried out within the calorimeter. After the reaction is over, the final temperature is measured. Knowing the two temperatures and the mass of the water, we can calculate q for the reaction system. To illustrate the procedure involved, let us consider the reaction:

$$NH_4NO_3(s) \rightarrow NH_4^+(aq) + NO_3^-(aq)$$

This is the process taking place within an instant cold pack (Fig. 7.3), used to treat muscle sprains, minor burns, and other injuries.

If it's hurt, cool it. That may help, and won't make it worse

■ EXAMPLE 7.2

When 1.00 g of ammonium nitrate, NH_4NO_3, is added to 50.0 g of water in a coffee-cup calorimeter, the solid dissolves via the above equation. The temperature of the water drops from 25.00°C to 23.32°C. Assume there is no heat flow through the walls of the calorimeter. Taking the specific heat of water to be 4.18 J/g·°C, calculate:

a. q for the water in the calorimeter.
b. q for the reaction system, i.e., the sample of ammonium nitrate.

Solution

a. Applying Equation 7.4:

$$q_{water} = 4.18 \frac{J}{g \cdot °C} \times 50.0 \text{ g} \times (23.32 - 25.00)°C = \boxed{-351 \text{ J}}$$

The water in the calorimeter gives up 351 J of heat when the endothermic reaction takes place.

b. Since no heat flows into or through the insulating walls of the calorimeter, the following relation must hold:

$$q_{reaction} + q_{water} = 0$$

$$q_{reaction} = -q_{water} = \boxed{+351 \text{ J}}$$

We see that the heat flow for the reaction system is equal in magnitude but opposite in sign to that of the water, which we can consider to be the surroundings.

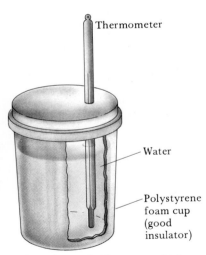

Thermometer

Water

Polystyrene foam cup (good insulator)

Figure 7.2 Coffee-cup calorimeter. The heat given off by a reaction is absorbed by the water. Knowing the mass of the water, its specific heat (4.18 J/g·°C), and the temperature change as read on the thermometer, we can calculate the heat flow, q, for the reaction.

Exercise

Suppose 40.0 g of water were used instead of 50.0 g. What would the temperature change be when 1.00 g of NH_4NO_3 dissolves? Answer: $-2.10°C$.

The argument we went through in part (b) of Example 7.2 can be generalized to any reaction occurring in a well-insulated calorimeter. *The heat flow for the reaction system will be equal in magnitude but opposite in sign to that for the calorimeter and its contents:*

$$q_{reaction} = -q_{calorimeter} \qquad (7.5)$$

This is the key equation for the calorimeter

In the coffee-cup calorimeter, the term on the right of Equation 7.5 is simply the heat flow for the water, q_{water}, because the styrofoam cup does not absorb or evolve heat. Hence we have:

$$q_{reaction} = -q_{water}$$

or:

$$q_{reaction} = -(4.18 \text{ J/g·°C}) \times m_{water} \times \Delta t \qquad \text{(coffee cup)} \qquad (7.6)$$

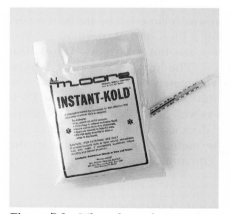

Figure 7.3 When the seal separating the compartments in a 'cold pack' is broken, the following endothermic reaction occurs: $NH_4NO_3(s) \rightarrow NH_4^+(aq) + NO_3^-(aq)$; $\Delta H = +28.1$ kJ, and the temperature, as read on the thermometer, drops. (Marna G. Clarke)

BOMB CALORIMETER

A coffee-cup calorimeter is suitable for measuring heat flows for reactions in solution. However, it cannot be used for reactions involving gases, which would escape from the cup; nor would it be appropriate for reactions in which the products reach high temperatures. The bomb calorimeter, shown in Figure 7.4, p. 216, is more versatile. To use it, we add a weighed sample of the reactant(s) to the heavy-walled steel container, called a "bomb." This is then sealed and lowered into a metal vessel that fits snugly within the insulating walls of the calorimeter. An amount of water sufficient to cover

Figure 7.4 Bomb calorimeter. The heat flow, q, for the reaction is calculated by multiplying the temperature change by the heat capacity of the calorimeter, which is determined in a preliminary experiment.

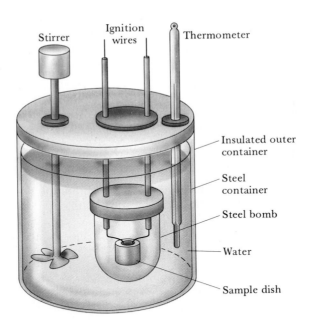

Stirrer

Ignition wires

Thermometer

Insulated outer container

Steel container

Steel bomb

Water

Sample dish

the bomb is added, and the entire apparatus is closed. The initial temperature is measured precisely. The reaction is then started, perhaps by electrical ignition. In an exothermic reaction, the hot products give off heat to the walls of the bomb and to the water. The final temperature is taken to be the highest value read on the thermometer.

To analyze the heat flow here, we start by noting that Equation 7.5 is a general one that applies to the bomb as well as to the coffee-cup calorimeter. On the other hand, Equation 7.6 does *not* apply to the bomb calorimeter, because heat is absorbed by the metal bomb in addition to the water. To evaluate $q_{calorimeter}$, we apply Equation 7.1:

$$q_{calorimeter} = C_{calorimeter} \times \Delta t$$

where $C_{calorimeter}$ is the total heat capacity of the calorimeter (water + bomb) and Δt is the temperature change for the reaction. Substituting into Equation 7.5, we obtain the simple operating equation for the bomb calorimeter:

Most heats of reaction are determined experimentally through this equation

$$q_{reaction} = -(C_{calorimeter}) \times \Delta t \quad \text{(bomb calorimeter)} \quad (7.7)$$

■ **EXAMPLE 7.3**

The reaction between hydrogen and chlorine: $H_2(g) + Cl_2(g) \rightarrow 2\ HCl(g)$ can be studied in a bomb calorimeter. It is found that when a 1.00 g sample of H_2 reacts completely, the temperature rises from 20.00°C to 29.82°C. Taking the heat capacity of the calorimeter to be 9.33 kJ/°C, calculate the amount of heat evolved in the reaction.

Solution
Applying Equation 7.7:

$$q_{reaction} = -9.33\ \frac{kJ}{°C} \times (29.82 - 20.00)°C = \boxed{-91.6\ kJ}$$

We conclude that 91.6 kJ is evolved when 1.00 g of H_2 combines with Cl_2 to form HCl. In other words, 91.6 kJ of heat flows from the reaction system to the surroundings, which consist of the bomb and the water around it.

Exercise

What would you expect $q_{reaction}$ to be if one mole of H_2, i.e., 2.016 g, reacts instead of one gram? Answer: -185 kJ.

In order to use Equation 7.7, we must know the total heat capacity of the bomb calorimeter, i.e., the amount of heat required to raise the temperature of the bomb and water by one degree Celsius. This is ordinarily determined in a preliminary experiment in which a known amount of heat is added and the increase in temperature measured. Suppose, for example, we find that with the system referred to in Example 7.3, the temperature rises 5.24°C when 48.9 kJ of heat is supplied to the calorimeter. Thus we have:

We can add the heat electrically

$$C_{calorimeter} = \frac{48.9 \text{ kJ}}{5.24°C} = 9.33 \text{ kJ/°C}$$

7.3 ENTHALPY

We have referred several times to the "heat flow for the reaction system," symbolized as $q_{reaction}$. At this point, you may well find this concept a bit nebulous and wonder if it could be made more concrete by relating $q_{reaction}$ to some property of reactants and products. This can indeed be done; the situation is particularly simple for reactions taking place at constant pressure. Under that condition, the heat flow for the reaction system is equal to the difference in **enthalpy** (H) between products and reactants. That is:

$$q_{reaction} \text{ at constant pressure } = \Delta H = H_{products} - H_{reactants} \quad (7.8)$$

Enthalpy is a type of chemical energy, sometimes referred to as "heat content." Reactions that occur in the laboratory in an open container or in the world around us take place at a constant pressure, that of the atmosphere. For such reactions, Equation 7.8 is valid, making enthalpy a very useful quantity.

From Equation 7.8 and our earlier discussion (Section 7.1) of the sign of heat flow, q, it follows that, at constant pressure:

Exothermic reaction: $\Delta H = (H_{products} - H_{reactants}) < 0 \quad (7.9)$

Endothermic reaction: $\Delta H = (H_{products} - H_{reactants}) > 0 \quad (7.10)$

These relationships are shown graphically in Figure 7.5. Figure 7.5a shows the situation for an exothermic reaction such as

$$CH_4(g) + 2 O_2(g) \rightarrow CO_2(g) + 2 H_2O(l); \quad \Delta H < 0$$

Here, the products, 1 mol of $CO_2(g)$ and 2 mol $H_2O(l)$, have a lower enthalpy than the reactants, 1 mol of $CH_4(g)$ and 2 mol $O_2(g)$. The decrease

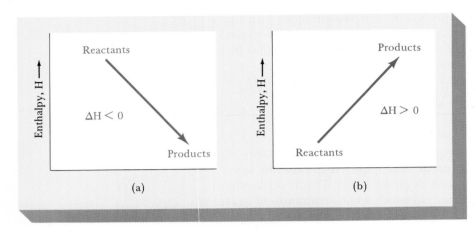

Figure 7.5 In an exothermic reaction (a), the products have a lower enthalpy than the reactants; thus, ΔH is negative, and heat is given off to the surroundings. In an endothermic reaction (b), the products have a higher enthalpy than the reactants, so ΔH is positive, and heat is absorbed from the surroundings.

in enthalpy is the source of the heat evolved to the surroundings. In Figure 7.5b, we have an endothermic reaction such as

$$H_2O(s) \rightarrow H_2O(l); \qquad \Delta H > 0$$

Since liquid water has a higher enthalpy than ice, heat must be transferred from the surroundings to melt the ice.

The development that we have just gone through is based upon the **Law of Conservation of Energy**. This law states that *energy is neither created nor destroyed in an ordinary chemical or physical change*. Here we say that the heat flow observed in the surroundings is exactly compensated for by the change in enthalpy of the reaction mixture. In an exothermic reaction, enthalpy is converted into heat, which flows into the surroundings. In an endothermic reaction, heat from the surroundings is converted into enthalpy of the reaction system.

The enthalpy of a substance, like its volume, is a state property. A sample of one gram of liquid water at 25°C and 1 atm has a fixed enthalpy, H. The value of H includes contributions from many sources, among them:

— the kinetic energy (energy of motion) of the H_2O molecule

— the electronic energies of the hydrogen and oxygen atoms

— the energies associated with the chemical bonds holding the molecule together

All we can do with energy is convert it from one form to another. We cannot change its amount

As you might guess even from this abbreviated list, the enthalpy of a substance is not readily evaluated. In practice, we make no attempt to obtain absolute values for H. Instead, we concentrate upon the enthalpy change, ΔH, when a system undergoes a change in state. The value of ΔH is relatively easy to establish in the laboratory. When we find that, at 1 atm pressure, 4.18 J of heat must be absorbed to raise the temperature of one gram of water from 24.00°C to 25.00°C, we conclude that, for the process:

$$1.00 \text{ g } H_2O(l, 24.00°C, 1 \text{ atm}) \rightarrow 1.00 \text{ g } H_2O(l, 25.00°C, 1 \text{ atm}); \qquad \Delta H = +4.18 \text{ J}$$

In other words, the enthalpy of one gram of liquid water at 25°C is 4.18 J larger than it is at 24°C.

Since enthalpy is a state property, ΔH, like ΔV, is independent of path. This is illustrated in Figure 7.6, which shows three different ways of raising the temperature of one gram of water from 24°C to 25°C. Regardless of how we bring about this change in state, ΔH remains constant at $+4.18$ J.

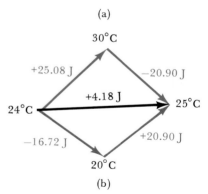

Figure 7.6 When 1.00 g of water is heated directly from 24.00°C to 25.00°C, ΔH is $+4.18$ J. The change in enthalpy is also $+4.18$ J when the indirect paths (a) and (b) are followed.

7.4
THERMOCHEMICAL EQUATIONS

A thermochemical equation shows the sign and magnitude of the heat flow in a reaction. This is done by indicating, at the right of the chemical equation, the appropriate value of ΔH. To show where a thermochemical equation comes from, consider the process by which ammonium nitrate dissolves in water:

$$NH_4NO_3(s) \rightarrow NH_4^+(aq) + NO_3^-(aq)$$

You will recall (Example 7.2) that a simple experiment with a coffee-cup calorimeter shows that when one gram of NH_4NO_3 dissolves, $q_{reaction} = +351$ J. Since the calorimeter is open to the atmosphere, the pressure is constant and we can say that

$$\Delta H \text{ for } 1.00 \text{ g } NH_4NO_3 = +351 \text{ J} = +0.351 \text{ kJ}$$

When one mole (80.05 g) of NH_4NO_3 dissolves, we expect ΔH to be eighty times as great:

$$\Delta H \text{ for } 1.00 \text{ mol } NH_4NO_3 = +0.351 \text{ kJ} \times 80.05 = +28.1 \text{ kJ}$$

We can now write the thermochemical equation:

$$NH_4NO_3(s) \rightarrow NH_4^+(aq) + NO_3^-(aq); \quad \Delta H = +28.1 \text{ kJ} \quad (7.11)$$

This equation tells us that when one mole of NH_4NO_3 dissolves in water to form one mole of NH_4^+ ions and one mole of NO_3^- ions:

$$\Delta H = [H \text{ of } 1 \text{ mol } NH_4^+(aq) + H \text{ of } 1 \text{ mol } NO_3^-(aq)]$$
$$- H \text{ of } 1 \text{ mol } NH_4NO_3(s)$$
$$= +28.1 \text{ kJ}$$

The sum of the molar enthalpies of the ions is 28.1 kJ higher than the molar enthalpy of the salt

By an entirely analogous procedure, using the calorimetric data referred to in Example 7.3, we arrive at the following thermochemical equation for the formation of HCl from the elements*:

$$H_2(g) + Cl_2(g) \rightarrow 2 HCl(g); \quad \Delta H = -185 \text{ kJ} \quad (7.12)$$

The thermochemical equations 7.11 and 7.12 are typical of those we will use throughout the rest of this text. It is important to realize that:

---

*Since a bomb calorimeter is sealed, the pressure will change if the number of moles of gas changes when reaction takes place. If that happens, $q_{reaction}$ is not exactly equal to ΔH (Section 7.6). In this reaction, the number of moles of gas does not change, so the final and initial pressures are the same and q is equal to ΔH.

— the sign of ΔH indicates whether the reaction, when carried out at constant pressure is endothermic (positive ΔH) or exothermic (negative ΔH).

— in interpreting a thermochemical equation, the coefficients represent numbers of moles (ΔH is -185 kJ when **1 mol** H_2 + **1 mol** $Cl_2 \rightarrow$ **2 mol** HCl).

— the phases (physical states) of all species must be specified, using the symbols (s), (l), (g), or (aq). The enthalpy of one mole of $H_2O(g)$ at 25°C is 44 kJ larger than that of one mole of $H_2O(l)$; the difference, which represents the heat of vaporization of water, is clearly significant.

— the value quoted for ΔH applies when products and reactants are at the same temperature, ordinarily taken to be 25°C unless specified otherwise.

LAWS OF THERMOCHEMISTRY

To make effective use of thermochemical equations, we apply three basic laws of thermochemistry.

1. *The magnitude of ΔH is directly proportional to the amount of reactant or product.* In effect, we assumed the validity of this rule in obtaining the thermochemical equations 7.11 and 7.12 from calorimetric data. For example, to calculate ΔH for dissolving one mole of NH_4NO_3, we multiplied the heat flow per gram by the molar mass, 80.05 g/mol. The rule is a common sense one, consistent with experience. Fuel oil is sold by the gallon because the amount of heat obtained from its combustion is directly proportional to the number of gallons burned.

This rule allows us to relate ΔH for a thermochemical equation to the amount of product or reactant. We do this by a simple extension of the conversion factor approach used in Chapter 3 with ordinary chemical equations. Consider, for example, the equation:

$$H_2(g) + Cl_2(g) \rightarrow 2\ HCl(g);\quad \Delta H = -185\ kJ$$

where we can say that

$$1\ mol\ H_2 \stackrel{\frown}{=} 1\ mol\ Cl_2 \stackrel{\frown}{=} 2\ mol\ HCl \stackrel{\frown}{=} -185\ kJ$$

Half a mole of H_2 would yield 92.5 kJ

The first two equivalences relate numbers of moles of reactants and products; the last one tells us that 185 kJ of heat is evolved for every mole of H_2 that reacts, *or* for every mole of Cl_2 reacting, *or* for every two moles of HCl formed.

■ **EXAMPLE 7.4** _____

Consider the thermochemical equation for the formation of HCl from the elements. Calculate ΔH when:

a. 1.00 g of Cl_2 reacts   b. 1.00 g of HCl is formed

Solution

a. Since 1 mol Cl_2 = 70.91 g Cl_2 and, from the thermochemical equation, 1 mol $Cl_2 \stackrel{\frown}{=} -185$ kJ,

$$\Delta H = 1.00 \text{ g Cl}_2 \times \frac{1 \text{ mol Cl}_2}{70.91 \text{ g Cl}_2} \times \frac{-185 \text{ kJ}}{1 \text{ mol Cl}_2} = \boxed{-2.61 \text{ kJ}}$$

b. Again we carry out a two-step conversion using the relations:

$$1 \text{ mol HCl} = 36.46 \text{ g}; \quad 2 \text{ mol HCl} \doteq -185 \text{ kJ}$$

$$\Delta H = 1.00 \text{ g HCl} \times \frac{1 \text{ mol HCl}}{36.46 \text{ g HCl}} \times \frac{-185 \text{ kJ}}{2 \text{ mol HCl}} = \boxed{-2.54 \text{ kJ}}$$

Exercise
What is ΔH for the thermochemical equation: $\frac{1}{2} H_2(g) + \frac{1}{2} Cl_2(g) \rightarrow HCl(g)$?
Answer: -92.5 kJ.

This relation between ΔH and amounts of substances is equally useful in dealing with chemical reactions or phase changes.

■ **EXAMPLE 7.5** _____
Using a coffee-cup calorimeter, it is found that when an ice cube weighing 24.6 g melts, it absorbs 8.19 kJ of heat. Calculate ΔH for the thermochemical equation: $H_2O(s) \rightarrow H_2O(l)$.

Solution
In effect, we are asked to calculate ΔH for the melting of one mole of ice, 18.02 g. From the statement of the problem:

$$24.6 \text{ g ice} \doteq 8.19 \text{ kJ}$$

Hence we have:

$$\Delta H \text{ per mole ice} = 1 \text{ mol ice} \times \frac{18.02 \text{ g ice}}{1 \text{ mol ice}} \times \frac{8.19 \text{ kJ}}{24.6 \text{ g ice}} = +6.00 \text{ kJ}$$

The thermochemical equation is: $H_2O(s) \rightarrow H_2O(l); \quad \Delta H = +6.00 \text{ kJ}$

Exercise
How much liquid water is cooled from 25.0 to 0.0°C when one mole of ice melts? Answer: 57.4 g.

The heat absorbed when a solid melts ($s \rightarrow l$) is referred to as the **heat of fusion**; that absorbed when a liquid vaporizes ($l \rightarrow g$) is called the **heat of vaporization**. Heats of fusion (ΔH_{fus}) and vaporization (ΔH_{vap}) are most often expressed in kilojoules per mole (kJ/mol). Values for several different substances are given in Table 7.2, p. 222.

For any substance,
$\Delta H_{fus} > 0$
$\Delta H_{vap} > 0$

2. *ΔH for a reaction is equal in magnitude but opposite in sign to ΔH for the reverse reaction.* Another way to state this rule is to say that the amount of heat evolved in a reaction is exactly equal to the amount of heat absorbed in the reverse reaction. Using this rule, we conclude that since the heat of fusion of ice at 0°C is 6.00 kJ/mol:

$$0°C: \quad H_2O(s) \rightarrow H_2O(l); \quad \Delta H = +6.00 \text{ kJ}$$

TABLE 7.2 ΔH(kJ/mol) for Phase Changes

SUBSTANCE		mp(°C)	ΔH_{fus}*	bp(°C)	ΔH_{vap}*
Benzene	C_6H_6	5	9.84	80	30.8
Bromine	Br_2	−7	10.8	59	29.6
Mercury	Hg	−39	2.33	357	59.4
Naphthalene	$C_{10}H_8$	80	19.3	218	43.3
Water	H_2O	0	6.00	100	40.7

*Values of ΔH_{fus} are given at the melting point, values of ΔH_{vap} at the boiling point. The heat of vaporization of water decreases from 44.9 kJ/mol at 0°C to 44.0 kJ/mol at 25°C to 40.7 kJ/mol at 100°C.

ΔH for the freezing of one mole of liquid water at the same temperature must be:

$$0°C:\quad H_2O(l) \rightarrow H_2O(s);\quad \Delta H = -6.00 \text{ kJ}$$

The validity of this rule should be apparent from Figure 7.7. The fact that ΔH for the fusion of ice is 6.00 kJ implies that the molar enthalpy of liquid water is 6.00 kJ greater than the molar enthalpy of ice. It follows that when we go in the opposite direction, ΔH must be −6.00 kJ.

It's as far up from the lobby to the roof as it is down from the roof to the lobby

■ **EXAMPLE 7.6**

For the reaction $H_2(g) + \frac{1}{2} O_2(g) \rightarrow H_2O(l)$, $\Delta H = -285.8$ kJ. Calculate ΔH for the reaction $2 H_2O(l) \rightarrow 2 H_2(g) + O_2(g)$

Solution

Applying Rules 1 and 2 in succession,

1. $2 H_2(g) + O_2(g) \rightarrow 2 H_2O(l);\quad \Delta H = 2(-285.8 \text{ kJ}) = -571.6 \text{ kJ}$

2. $2 H_2O(l) \rightarrow 2 H_2(g) + O_2(g);\quad \Delta H = -(-571.6 \text{ kJ}) = +571.6 \text{ kJ}$

We conclude that 571.6 kJ must be absorbed to decompose 2 mol of liquid water to its elements.

Exercise

Given $2 Al_2O_3(s) \rightarrow 4 Al(s) + 3 O_2(g)$, $\Delta H = +3351.4$ kJ, obtain ΔH for the formation of one mole of Al_2O_3 from the elements. Answer: −1675.7 kJ.

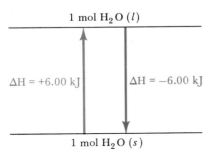

1 mol H_2O (l)

$\Delta H = +6.00$ kJ $\Delta H = -6.00$ kJ

1 mol H_2O (s)

Figure 7.7 The magnitude of ΔH for freezing water or melting ice is the same, 6.00 kJ/mol. Only the sign differs for the two processes.

3. *The value of ΔH for a reaction is the same whether it occurs directly or in a series of steps.* This means that if a thermochemical equation can be expressed as the sum of two or more other equations,

Equation (3) = Equation (1) + Equation (2) + · · ·

then ΔH for the overall equation is the sum of the ΔH's for the individual equations:

$$\Delta H_3 = \Delta H_1 + \Delta H_2 + \cdots$$

This relationship is referred to as **Hess's Law.** It is a direct consequence of the fact that H is a state function, making ΔH independent of path.

To illustrate Hess's Law, consider the reaction between tin and chlorine:

$$Sn(s) + 2\ Cl_2(g) \rightarrow SnCl_4(l) \tag{7.13}$$

We can imagine that this reaction takes place in two steps. In the first step, 1 mol Sn reacts with 1 mol Cl_2 to form 1 mol $SnCl_2$. Then, in a second step, the $SnCl_2$ reacts with another mole of Cl_2 to form $SnCl_4$:

Step 1: $Sn(s) + Cl_2(g) \rightarrow SnCl_2(s);$ $\Delta H_1 = -325.1$ kJ
Step 2: $SnCl_2(s) + Cl_2(g) \rightarrow SnCl_4(l);$ $\Delta H_2 = -186.2$ kJ
            $Sn(s) + 2\ Cl_2(g) \rightarrow SnCl_4(l)$

Since these two equations add to give Equation 7.13:

$$\Delta H \text{ for Reaction 7.13} = \Delta H_1 + \Delta H_2$$
$$= -325.1 \text{ kJ} - 186.2 \text{ kJ} = -511.3 \text{ kJ}$$

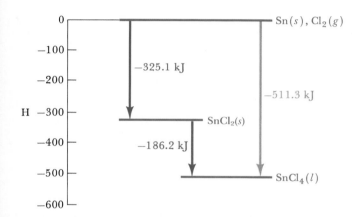

Figure 7.8 ΔH for the reaction: $Sn(s) + 2Cl_2(g) \rightarrow SnCl_4(l)$ is the same whether it is carried out in one step (red arrow) or two steps (green arrows), going through $SnCl_2(s)$ as an intermediate.

This situation is shown graphically in Figure 7.8.

Hess's Law can be applied to physical changes (Example 7.7) as well as chemical reactions.

■ **EXAMPLE 7.7** _____

Calculate ΔH for the process:

$$H_2O(s, 0.0°C) \rightarrow H_2O(l, 25.0°C)$$

That is, calculate the amount of heat that must be absorbed to convert one mole of ice at 0°C to one mole of liquid water at 25°C. Take the heat of fusion of ice to be 6.00 kJ/mol and the specific heat of liquid water to be 4.18 J/g·°C.

Solution
We can consider this process to be the sum of two other processes:

1. melting of ice at 0°C: $H_2O(s, 0.0°C) \rightarrow H_2O(l, 0.0°C)$
2. heating of water from $0 - 25°C$: $H_2O(l, 0.0°C) \rightarrow H_2O(l, 25.0°C)$

Melting occurs at a constant temperature, the melting point

For the melting of ice: $\Delta H_1 = +6.00$ kJ.
For heating one mole (18.0 g) of water:

$$\Delta H_2 = 4.18 \ \frac{J}{g \cdot {}^\circ C} \times 18.0 \text{ g} \times 25.0^\circ C = 1880 \text{ J} = 1.88 \text{ kJ}$$

Applying Hess's Law, we conclude that, for the process specified:

$$\Delta H = \Delta H_1 + \Delta H_2 = \boxed{+7.88 \text{ kJ}}$$

Exercise
If 9.12 kJ of heat were added to one mole of ice at 0.0°C, what would be the final temperature of the water formed? Answer: 41.5°C.

We will consider several other applications of Hess's Law later in this chapter and in succeeding chapters. In some cases, such as the one just cited, we will use Hess's Law to obtain ΔH for a process, knowing the individual values of ΔH for each step. In other cases, we will use the law to calculate ΔH for a particular step, knowing the values of ΔH for every other step and for the overall process.

7.5
HEATS OF FORMATION

We have now written several thermochemical equations. In each case, we have cited the corresponding value of ΔH. Literally thousands of such equations would be needed to list the ΔH values for all the reactions that have been studied. Clearly, we need some more concise way of recording data of this sort. These data should be in a form that can easily be used to calculate ΔH for a reaction that interests us. It turns out that there is a simple way to do this, using quantities known as heats of formation. Let us first consider how these quantities are defined. Then we will show how they can be used to calculate ΔH for reactions.

MEANING OF ΔH_f^0

The standard molar heat of formation of a compound, ΔH_f^0, is equal to the enthalpy change when one mole of the compound is formed at a constant pressure of 1 atm and a fixed temperature, ordinarily 25°C, from the elements in their stable states at that pressure and temperature. From the equations:

$$Ag(s, 25^\circ C) + \tfrac{1}{2} Cl_2(g, 25^\circ C, 1 \text{ atm}) \rightarrow AgCl(s, 25^\circ C); \quad \Delta H = -127.1 \text{ kJ}$$

$$\tfrac{1}{2} N_2(g, 1 \text{ atm}, 25^\circ C) + O_2(g, 1 \text{ atm}, 25^\circ C) \rightarrow NO_2(g, 1 \text{ atm}, 25^\circ C); \quad \Delta H = +33.2 \text{ kJ}$$

we conclude that

$$\Delta H_f^0 \ AgCl(s) = -127.1 \text{ kJ/mol}; \quad \Delta H_f^0 \ NO_2(g) = +33.2 \text{ kJ/mol}$$

Heats of formation are listed for a variety of compounds in Table 7.3. Notice that, with a few exceptions, heats of formation are usually negative

TABLE 7.3 Heats of Formation, ΔH_f^0 (kJ/mol), of Compounds at 25°C, 1 atm

AgBr(s)	−100.4	CaCl₂(s)	−795.8	H₂O(g)	−241.8	NH₄NO₃(s)	−365.6
AgCl(s)	−127.1	CaCO₃(s)	−1206.9	H₂O(l)	−285.8	NO(g)	+90.2
AgI(s)	−61.8	CaO(s)	−635.1	H₂O₂(l)	−187.8	NO₂(g)	+33.2
AgNO₃(s)	−124.4	Ca(OH)₂(s)	−986.1	H₂S(g)	−20.6	N₂O₄(g)	+9.2
Ag₂O(s)	−31.0	CaSO₄(s)	−1434.1	H₂SO₄(l)	−814.0	NaCl(s)	−411.2
Al₂O₃(s)	−1675.7	CdCl₂(s)	−391.5	HgO(s)	−90.8	NaF(s)	−573.6
BaCl₂(s)	−858.6	CdO(s)	−258.2	KBr(s)	−393.8	NaOH(s)	−425.6
BaCO₃(s)	−1216.3	Cr₂O₃(s)	−1139.7	KCl(s)	−436.7	NiO(s)	−239.7
BaO(s)	−553.5	CuO(s)	−157.3	KClO₃(s)	−397.7	PbBr₂(s)	−278.7
BaSO₄(s)	−1473.2	Cu₂O(s)	−168.6	KClO₄(s)	−432.8	PbCl₂(s)	−359.4
CCl₄(l)	−135.4	CuS(s)	−53.1	KNO₃(s)	−369.8	PbO(s)	−219.0
CHCl₃(l)	−134.5	Cu₂S(s)	−79.5	MgCl₂(s)	−641.3	PbO₂(s)	−277.4
CH₄(g)	−74.8	CuSO₄(s)	−771.4	MgCO₃(s)	−1095.8	PCl₃(g)	−287.0
C₂H₂(g)	+226.7	Fe(OH)₃(s)	−823.0	MgO(s)	−601.7	PCl₅(g)	−374.9
C₂H₄(g)	+52.3	Fe₂O₃(s)	−824.2	Mg(OH)₂(s)	−924.5	SiO₂(s)	−910.9
C₂H₆(g)	−84.7	Fe₃O₄(s)	−1118.4	MgSO₄(s)	−1284.9	SnO₂(s)	−580.7
C₃H₈(g)	−103.8	HBr(g)	−36.4	MnO(s)	−385.2	SO₂(g)	−296.8
CH₃OH(l)	−238.7	HCl(g)	−92.3	MnO₂(s)	−520.0	SO₃(g)	−395.7
C₂H₅OH(l)	−277.7	HF(g)	−271.1	NH₃(g)	−46.1	ZnI₂(s)	−208.0
CO(g)	−110.5	HI(g)	+26.5	N₂H₄(l)	+50.6	ZnO(s)	−348.3
CO₂(g)	−393.5	HNO₃(l)	−174.1	NH₄Cl(s)	−314.4	ZnS(s)	−206.0

quantities. This means that the formation of a compound from the elements is ordinarily exothermic. Conversely, when a compound decomposes to the elements, heat usually must be absorbed.

You will note from Table 7.3 that there are no entries for elemental species such as $Br_2(l)$ and $O_2(g)$. This is a consequence of the way in which heats of formation are defined. In effect, the elements in their stable states at 25°C and 1 atm are taken to have zero enthalpy in establishing values for heats of formation of other species. That is:

If you make the element from the element, $\Delta H = 0 = \Delta H_f^0$

$$\Delta H_f^0 \; Br_2(l) \;=\; \Delta H_f^0 \; O_2(g) \;=\; 0$$

CALCULATION OF ΔH^0

Heats of formation can be used to calculate the enthalpy change for a thermochemical equation. To do this, we apply the general rule:

The standard enthalpy change, ΔH^0, corresponding to a thermochemical equation is equal to the sum of the standard heats of formation of the product compounds minus the sum of the standard heats of formation of the reactant compounds. Using the symbol Σ to represent "the sum of":

$$\Delta H^0 = \Sigma \, \Delta H_f^0 \text{ products } - \Sigma \, \Delta H_f^0 \text{ reactants} \qquad (7.14)$$

Strictly speaking, ΔH^0 calculated from Equation 7.14 represents the enthalpy change at 25°C and 1 atm. Actually, ΔH is essentially independent of pressure and varies relatively little with temperature, changing by perhaps 1 to 10 kJ per 100°C.

■ EXAMPLE 7.8

Using Table 7.3, calculate ΔH^0 for the combustion of one mole of propane, C_3H_8, according to the equation

$$C_3H_8(g) + 5\ O_2(g) \rightarrow 3\ CO_2(g) + 4\ H_2O(l)$$

Solution

Expressing ΔH^0 in terms of heats of formation of products and reactants,

$$\Delta H^0 = 3\ \Delta H^0_f\ CO_2(g) + 4\ \Delta H^0_f\ H_2O(l) - [\Delta H^0_f\ C_3H_8(g) + 5\ \Delta H^0_f\ O_2(g)]$$

Taking the heat of formation of $O_2(g)$ to be zero and substituting values for the other substances from Table 7.3.

$$\Delta H^0 = 3\ \text{mol}\left(-393.5\ \frac{kJ}{mol}\right) + 4\ \text{mol}\left(-285.8\ \frac{kJ}{mol}\right) - 1\ \text{mol}\left(-103.8\ \frac{kJ}{mol}\right)$$

$$= -2219.9\ kJ$$

Exercise

The reaction between aluminum and iron(III) oxide, Fe_2O_3, is strongly exothermic. Referred to as the thermite reaction (Fig. 7.9), it was once used to weld steel rails. Using Table 7.3, calculate ΔH^0 for the reaction: $2\ Al(s) + Fe_2O_3(s) \rightarrow 2\ Fe(s) + Al_2O_3(s)$. Answer: -851.5 kJ.

Figure 7.9 When a piece of burning magnesium ribbon is used as a fuse, a finely divided mixture of aluminum powder and iron(III) oxide undergoes an exothermic reaction: $2Al(s) + Fe_2O_3(s) \rightarrow 2Fe(s) + Al_2O_3(s)$; $\Delta H° = -851.5$ kJ. Enough heat is generated to produce molten iron. (Charles D. Winters)

Equation 7.14 can also be used to calculate the heat of formation of one substance if we know ΔH^0 for the reaction and the heats of formation of all other substances involved in the reaction. Example 7.9 illustrates the calculations involved.

■ **EXAMPLE 7.9** _____

The thermochemical equation for the combustion of benzene, C_6H_6, is

$$C_6H_6(l) + \tfrac{15}{2} O_2(g) \rightarrow 6 CO_2(g) + 3 H_2O(l); \Delta H^0 = -3267.4 \text{ kJ}$$

Using Table 7.3 to obtain heats of formation for CO_2 and H_2O, calculate the heat of formation of benzene.

Solution

Taking the heat of formation of $O_2(g)$ to be zero, the relation for ΔH^0 is

$$\Delta H^0 = 6 \Delta H_f^0\, CO_2(g) + 3 \Delta H_f^0\, H_2O(l) - \Delta H_f^0\, C_6H_6(l)$$

Substituting values for ΔH^0, $\Delta H_f^0\, CO_2$, and $\Delta H_f^0\, H_2O(l)$, we have

$$-3267.4 \text{ kJ} = 6 \text{ mol} \left(-393.5 \frac{\text{kJ}}{\text{mol}}\right) + 3 \text{ mol} \left(-285.8 \frac{\text{kJ}}{\text{mol}}\right)$$
$$- 1 \text{ mol} [\Delta H_f^0\, C_6H_6(l)]$$

Solving,

$$\Delta H_f^0\, C_6H_6(l) = +49.0 \text{ kJ/mol}$$

Exercise

Given the following thermochemical equation: $3 O_2(g) \rightarrow 2 O_3(g)$; $\Delta H^0 = +285.4$ kJ, calculate the heat of formation of ozone, $O_3(g)$. Answer: $+142.7$ kJ/mol.

Many of the heats of formation in Table 7.3 were determined by combustion reactions like this one

The relation between ΔH^0 and heats of formation given by Equation 7.14 is perhaps the most useful in all of thermochemistry. To show its validity, consider the reaction,

$$CO(g) + \tfrac{1}{2} O_2(g) \rightarrow CO_2(g)$$

This equation is the sum of two other chemical equations:

$C(s) + O_2(g) \rightarrow CO_2(g)$	(7.15a)
$CO(g) \rightarrow C(s) + \tfrac{1}{2} O_2(g)$	(7.15b)
$CO(g) + \tfrac{1}{2} O_2(g) \rightarrow CO_2(g)$	(7.15)

Applying Hess's Law,

$$\Delta H^0 \text{ for React. 7.15} = \Delta H^0 \text{ for React. 7.15a} + \Delta H^0 \text{ for React. 7.15b}$$

By definition, ΔH^0 for Reaction 7.15a is the heat of formation of carbon dioxide, $\Delta H_f^0\, CO_2$. Since Reaction 7.15b is the reverse of that for the formation of CO from the elements, its ΔH^0 is $- \Delta H_f^0\, CO$. Hence,

$$\Delta H^0 \text{ for Reaction 7.15} = \Delta H_f^0\, CO_2 - \Delta H_f^0\, CO$$

This is the relation we would have obtained directly by applying Equation 7.14 to this reaction.

The analysis we have just gone through can be applied to any reaction. Equation 7.14 is a special case of Hess's Law. It is simpler to apply the equation directly, as we did in Example 7.8, rather than go through a Hess's Law calculation.

HEATS OF FORMATION OF IONS IN SOLUTION

It is possible to set up a table, very much like Table 7.3, for heats of formation of ions in water solution. There is, however, one problem. We cannot measure the heat of formation of an individual ion. In any reaction involving ions, at least two of them are present, as required by the principle of electrical neutrality. Consider, for example, the reaction that occurs when HCl is added to water:

$$HCl(g) \rightarrow H^+(aq) + Cl^-(aq)$$

Here, two different ions are formed, H^+ and Cl^- ions. The same situation applies in all other reactions. We cannot carry out a reaction to form only the H^+ ion, with no other ions involved.

To get around this dilemma, **the heat of formation of the H^+ ion is arbitrarily taken to be zero:**

$$\Delta H_f^0 \; H^+(aq) = 0 \tag{7.16}$$

We need to fix ΔH_f^0 for one ion, so we choose H^+, for no really good reason

Having done this, we can establish heats of formation for other ions. Take, for instance, the Cl^- ion. We can measure ΔH^0 for the reaction that occurs when HCl is added to water; it is -74.9 kJ/mol; that is,

$$HCl(g) \rightarrow H^+(aq) + Cl^-(aq); \qquad \Delta H^0 = -74.9 \text{ kJ} \tag{7.17}$$

Applying Equation 7.14,

$$-74.9 \text{ kJ} = \Delta H_f^0 \; H^+(aq) + \Delta H_f^0 \; Cl^-(aq) - \Delta H_f^0 \; HCl(g)$$

Taking the heat of formation of $H^+(aq)$ to be zero and that of HCl(g) to be -92.3 kJ (Table 7.3), we have

$$-74.9 \text{ kJ} = 0 + \Delta H_f^0 \; Cl^-(aq) + 92.3 \text{ kJ}$$

or

$$\Delta H_f^0 \; Cl^-(aq) = -74.9 \text{ kJ} - 92.3 \text{ kJ} = -167.2 \text{ kJ}$$

Heats of formation for other ions in water solution can be established in a similar way. Table 7.4 lists values for several common ions. This table can be used, much like Table 7.3, to calculate ΔH for reactions in solution. To do that, we use Equation 7.14:

$$\Delta H^0 = \Sigma \, \Delta H_f^0 \text{ products} - \Sigma \, \Delta H_f^0 \text{ reactants}$$

realizing that ΔH_f^0 for $H^+(aq)$ is zero as is ΔH_f^0 for an element in its stable state.

TABLE 7.4 Heats of Formation, ΔH_f^0 (kJ/mol), of Aqueous Ions at 25°C, 1 M

CATIONS				ANIONS			
$Ag^+(aq)$	+105.6	$Hg^{2+}(aq)$	+171.1	$Br^-(aq)$	−121.6	$HPO_4^{2-}(aq)$	−1292.1
$Al^{3+}(aq)$	−531.0	$K^+(aq)$	−252.4	$CO_3^{2-}(aq)$	−677.1	$HSO_4^-(aq)$	−887.3
$Ba^{2+}(aq)$	−537.6	$Mg^{2+}(aq)$	−466.8	$Cl^-(aq)$	−167.2	$I^-(aq)$	−55.2
$Ca^{2+}(aq)$	−542.8	$Mn^{2+}(aq)$	−220.8	$ClO_3^-(aq)$	−104.0	$MnO_4^-(aq)$	−541.4
$Cd^{2+}(aq)$	−75.9	$Na^+(aq)$	−240.1	$ClO_4^-(aq)$	−129.3	$NO_2^-(aq)$	−104.6
$Cu^+(aq)$	+71.7	$NH_4^+(aq)$	−132.5	$CrO_4^{2-}(aq)$	−881.2	$NO_3^-(aq)$	−205.0
$Cu^{2+}(aq)$	+64.8	$Ni^{2+}(aq)$	−54.0	$Cr_2O_7^{2-}(aq)$	−1490.3	$OH^-(aq)$	−230.0
$Fe^{2+}(aq)$	−89.1	$Pb^{2+}(aq)$	−1.7	$F^-(aq)$	−332.6	$PO_4^{3-}(aq)$	−1277.4
$Fe^{3+}(aq)$	−48.5	$Sn^{2+}(aq)$	−8.8	$HCO_3^-(aq)$	−692.0	$S^{2-}(aq)$	+33.1
$H^+(aq)$	0.0	$Zn^{2+}(aq)$	−153.9	$H_2PO_4^-(aq)$	−1296.3	$SO_4^{2-}(aq)$	−909.3

■ **EXAMPLE 7.10**

When hydrochloric acid is added to a solution of sodium carbonate, carbon dioxide gas is formed (Fig. 7.10). The equation for the reaction is

$$2\ H^+(aq)\ +\ CO_3^{2-}(aq)\ \rightarrow\ CO_2(g)\ +\ H_2O(l)$$

Using heats of formation from Tables 7.3 and 7.4, calculate ΔH^0 for this reaction.

Solution

Applying the general relation given above,

$$\Delta H^0 = \Delta H_f^0\ CO_2(g) + \Delta H_f^0\ H_2O(l) - 2\ \Delta H_f^0\ H^+(aq) - \Delta H_f^0\ CO_3^{2-}(aq)$$

Taking $\Delta H_f^0\ H^+(aq)$ to be zero and obtaining the other heats of formation from Tables 7.3 and 7.4,

$$\Delta H^0 = 1\ mol\left(-393.5\ \frac{kJ}{mol}\right) + 1\ mol\left(-285.8\ \frac{kJ}{mol}\right)$$
$$- 1\ mol\left(-677.1\ \frac{kJ}{mol}\right) = \boxed{-2.2\ kJ}$$

Exercise

Determine ΔH^0 for the reaction: $Zn(s) + 2\ H^+(aq) \rightarrow Zn^{2+}(aq) + H_2(g)$.
Answer: −153.9 kJ.

7.6
ENTHALPY, ENERGY, AND THE FIRST LAW OF THERMODYNAMICS

In this chapter, we have thus far emphasized one type of energy change—the change in enthalpy, ΔH. You will recall that ΔH represents the heat flow in a process carried out at constant pressure. For our purposes, the "process" is usually a chemical reaction and the "constant pressure" is that of the atmosphere. In general, we can say that

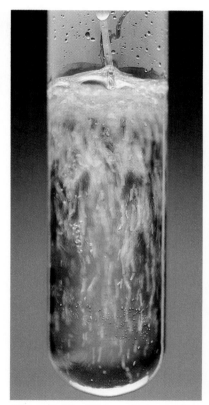

Figure 7.10 When solutions of hydrochloric acid and sodium carbonate are mixed, the following reaction takes place: $2H^+(aq) + CO_3^{2-}(aq) \rightarrow CO_2(g) + H_2O(l)$; $\Delta H = -2.2$ kJ. Bubbles of carbon dioxide gas are formed. (Marna G. Clarke)

$$\Delta H = q_{\text{reaction}} \qquad \text{(constant } P) \qquad\qquad (7.18)$$

It is possible to develop equations similar to 7.18 for all types of energy changes in all types of processes. To do this, we turn to the science of *thermodynamics*.

FIRST LAW OF THERMODYNAMICS

Thermodynamics distinguishes between two types of energy. One of these is **heat**, referred to earlier in this chapter, given the symbol q. The other type of energy is **work**, represented by the symbol w. The thermodynamic definition of work is quite different from its colloquial meaning. Quite simply, *work includes all forms of energy except heat.*

In chemistry, we ordinarily deal with only two types of work. One of these is electrical work, such as that supplied by a storage battery. We will have more to say about electrical work in later chapters. Here our interest is in mechanical work, such as that done by a gas expanding against a restraining pressure (Fig. 7.11). This is the most common type of work in the chemistry laboratory (except, perhaps, for the work done by students). Typically, the gas is one that is produced (or consumed) in a reaction. The "restraining pressure" is simply the constant pressure of the atmosphere.

The quantities q and w have direction as well as magnitude. As pointed out earlier, heat may flow into a system, raising its temperature. Conversely, heat can flow out of a system, lowering its temperature. When a gas expands (Fig. 7.11), it does work upon the surroundings, pushing back the atmosphere. Under these conditions, we would say that energy flows out of the gaseous system in the form of work. When a gas is compressed, work is done on the gas by the surroundings and energy flows into the gaseous system.

In thermodynamics, the direction of energy flow is indicated by specifying the signs of q and w. The conventions we will follow are

q is $+$ when the system absorbs energy in the form of heat from the surroundings

q is $-$ when the system evolves energy as heat to the surroundings

Sometimes even the profs work there too

Figure 7.11 The work done by a gas expanding against a force f is $f \times \Delta d$, where Δd is the distance through which the force is displaced. But, since pressure, P, is force per unit area, a, and volume, V, is the product of area and distance, the work done is: $f \times \Delta d = P \times a \times \Delta d = P\Delta V$. Here, the metal weights exert a force, f, or a pressure, P, on the piston of cross-sectional area a. Expansion occurs as gases are formed by reaction.

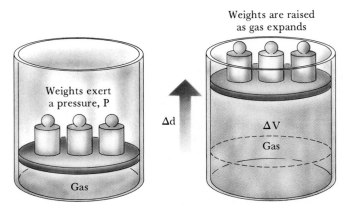

Weights are raised as gas expands

Weights exert a pressure, P

Δd

ΔV

Gas

Gas

w is + when the system absorbs energy by having work done on it by the surroundings

w is − when the system evolves energy by doing work on the surroundings

With this background, we can state the **First Law of Thermodynamics**:

In any process, the total change in energy of the system, ΔE, is equal to the sum of the heat absorbed, q, and the work, w, done on the system.

$$\Delta E = q + w$$

(7.19)   If you do work on the system, $w > 0$, and the energy of the system goes up

The quantity E in this equation is called the *internal energy,* or simply, the energy.

■ EXAMPLE 7.11

Calculate ΔE of a gas for a process in which the gas

a. absorbs 20 J of heat and does 12 J of work by expanding.
b. evolves 30 J of heat and has 52 J of work done on it as it contracts.

Solution

a. $q = +20$ J; $w = -12$ J, since the gas does work on the surroundings

$$\Delta E = +20\ J - 12\ J = \boxed{+8\ J}$$

b. $q = -30$ J; $w = +52$ J, since work is done on the gas by the surroundings

$$\Delta E = -30\ J + 52\ J = \boxed{+22\ J}$$

Exercise

In a certain process, a gas absorbs 25 J of heat; its volume remains constant so that no work is done. What is ΔE? Answer: +25 J.

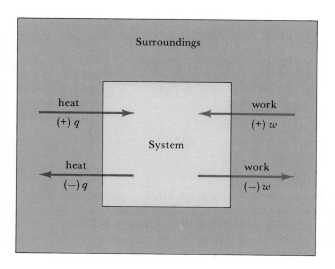

E AND ΔE

Energy, like enthalpy, is a state property. Its value is fixed when the state of the system is specified. A sample of one mole of $H_2O(l)$ at 25°C and 1 atm has a certain definite energy, just as it has a fixed enthalpy. In thermodynamics, we do not ordinarily worry about absolute energies of substances. Instead, we concentrate upon determining the value of ΔE for chemical or physical changes. This situation is, of course, similar to that which obtains with enthalpy, where we focus attention on changes in *H* rather than on *H* itself.

For a reaction system, there is a simple way to determine ΔE. This is to carry out the reaction in a bomb calorimeter and measure the heat flow, *q*. Since there is no change in volume, no mechanical work is done, and *w* is zero. Hence, using Equation 7.19 we conclude that for a reaction taking place in a bomb calorimeter,

In a bomb calorimeter, we measure ΔE

$$\Delta E = q_{\text{reaction}} \qquad (\text{constant } V) \tag{7.20}$$

where q_{reaction} is the heat flow for the reaction carried out at constant volume and ΔE is the difference in energy between products and reactants.

■ **EXAMPLE 7.12**

When a 1.00-g sample of methane is burned in a bomb calorimeter with a heat capacity of 5.96 kJ/°C, the temperature rises from 20.00°C to 29.26°C.

a. Determine q_{reaction} for the combustion of the 1.00-g sample.
b. Find ΔE for the thermochemical equation:

$$CH_4(g) + 2\,O_2(g) \rightarrow CO_2(g) + 2\,H_2O(l)$$

Solution

a. We use Equation 7.7; the calculation is similar to that in Example 7.3:

$$q_{\text{reaction}} = -5.96 \text{ kJ/°C} \times 9.26°C = \boxed{-55.2 \text{ kJ}}$$

b. Here, we need to calculate the heat flow when one mole of CH_4, 16.04 g, burns:

$$q \text{ (one mole } CH_4) = 1 \text{ mol } CH_4 \times \frac{16.04 \text{ g } CH_4}{1 \text{ mol } CH_4} \times \frac{-55.2 \text{ kJ}}{1.00 \text{ g } CH_4}$$

$$= -885 \text{ kJ}$$

Since the volume is constant, $\Delta E = q = \boxed{-885 \text{ kJ}}$.

Exercise

Suppose the pressure before reaction, when we have a mixture of $CH_4(g)$ and $2\,O_2(g)$, is 1.00 atm. If the temperature and volume remain constant, what is the final pressure exerted by $CO_2(g)$ and $2\,H_2O(l)$? Answer: Approx. $\frac{1}{3}$ atm.

ΔH AND ΔE

The enthalpy of a substance is related in a simple way to its energy. Thermodynamics defines enthalpy, H, to be:

$$H = E + PV \tag{7.21}$$

Here, E is energy, P is pressure, and V is volume. Ordinarily, the difference between H and E for a substance is quite small. Consider, for example, one mole of $CH_4(g)$, or indeed any other gas, at 25°C. Using the Ideal Gas Law with the gas constant in the appropriate units:

$$PV \text{ of } CH_4(g) \text{ at } 25°C = nRT = 1.00 \text{ mol} \times 8.31 \frac{J}{mol \cdot K} \times 298 \text{ K}$$
$$= 2.48 \times 10^3 \text{ J} = 2.48 \text{ kJ}$$

For a liquid or solid, the PV term is much, much smaller. For water at 25°C and 1 atm, where the molar volume is 18.0 mL, we find, using the energy conversion factor: 0.1013 kJ = 1 L·atm

$$PV \text{ of } H_2O(l) = (1.00 \text{ atm})(0.0180 \text{ L}) \times \frac{0.1013 \text{ kJ}}{1 \text{ L·atm}} = 0.0018 \text{ kJ}$$

This amounts to less than two joules, less than the experimental error in most thermochemical measurements.

Using Equation 7.21, we can readily obtain an expression relating ΔH and ΔE in a chemical reaction:

$$\Delta H = \Delta E + \Delta(PV)$$
$$= \Delta E + (PV)_{\text{products}} - (PV)_{\text{reactants}}$$

To simplify this equation, we assume that all gases follow the Ideal Gas Law, $PV = nRT$, and that the quantity PV is negligibly small for all species except gases. In that case, we have:

$$\Delta H = \Delta E + RT \times n_{\text{gaseous products}} - RT \times n_{\text{gaseous reactants}}$$
$$\Delta H = \Delta E + \Delta n_{\text{gas}}(RT) \tag{7.22}$$

where Δn_{gas} is the change in the number of moles of gas in the reaction. Ordinarily, the difference between ΔH and ΔE is very small, often within experimental error.

■ **EXAMPLE 7.13** _____

For the reaction:

$$CH_4(g) + 2 O_2(g) \rightarrow CO_2(g) + 2 H_2O(l)$$

calculate

a. the difference between ΔH and ΔE at 25°C, using Equation 7.22.
b. ΔH, using the results of Example 7.12.

Solution

a. There are three moles of gaseous reactants and only one mole of gaseous product, $CO_2(g)$. So: $\Delta n_{\text{gas}} = 1 \text{ mol} - 3 \text{ mol} = -2 \text{ mol}$

$$\Delta H - \Delta E = -2 \text{ mol} \times 8.31 \frac{J}{\text{mol·K}} \times 298 \text{ K} = \boxed{-4.95 \text{ kJ}}$$

b. In Example 7.12b, we found that $\Delta E = -885$ kJ

$$\Delta H = -885 \text{ kJ} - 4.96 \text{ kJ} = \boxed{-890 \text{ kJ}}$$

Exercise
Calculate the difference between ΔH and ΔE for the reaction: $H_2(g) + Cl_2(g) \rightarrow 2 \text{ HCl}(g)$ Answer: 0.

You will recall from our earlier discussion that the heat flow for a reaction at constant pressure is equal to ΔH; we have seen in this section that the heat flow at constant volume is equal to ΔE. This means that Equation 7.22 could be written in the form:

$$q_P = q_V + \Delta n_{gas}RT \tag{7.23}$$

where $q_P = q_{reaction}$ at constant pressure and $q_V = q_{reaction}$ at constant volume. From Equation 7.23, we see that these two types of heat flow differ from one another. More generally, we find that the heat flow, q, is dependent upon the conditions under which a reaction is carried out. The same is true of the work, w. For an expansion against a constant pressure, the mechanical work done by the system is $P\Delta V$ (recall Figure 7.11); at constant volume, $w = 0$. We see then that q and w, unlike ΔE and ΔH, are dependent upon path; in other words, *heat and work are not state properties*.

Thermodynamicists like to work with state properties when they can

7.7
SOURCES OF ENERGY

The United States today uses vast amounts of energy, about 8×10^{16} kJ/yr. With only about 6% of the world's population, it accounts for about 30% of worldwide energy usage. In the past 30 years, energy usage in the United States has more than doubled. Figure 7.12 shows where this energy comes

Figure 7.12 The major sources of energy in the United States today are petroleum (42%), natural gas (26%), and coal (21%). Nuclear fission and water power make smaller contributions, although they account for nearly one-fourth of electrical energy. The "biomass" contribution comes almost entirely from the combustion of wood and wood wastes. In 1860, wood supplied 90% of our energy.

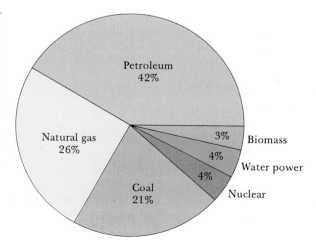

from. As you can see, nearly 90% is obtained from the combustion of the fossil fuels petroleum, natural gas, and coal.

When we "use" energy, we take the available energy from the fuel

FOSSIL FUELS

These materials consist largely of hydrocarbons, organic compounds containing the two elements carbon and hydrogen. Natural gas is mostly (70% to 95%) methane, CH_4. Smaller amounts of ethane, C_2H_6, and propane, C_3H_8, are also present, along with nitrogen and, in some cases, helium. Petroleum is a complex mixture of liquid hydrocarbons with 5 to 20 carbon atoms per molecule. Dissolved in the liquid are some hydrocarbon gases (C_3H_8, C_4H_{10}) and high-molar-mass solids of the type found in paraffin.

Of the fossil fuels, coal has the most complex composition. It is generally believed to consist of a mixture of elemental carbon and solid hydrocarbons of high molar mass, up to 3000 g/mol. The atom ratio of hydrogen to carbon in coal is quite low, about 0.8 compared to 2.2 in petroleum. Coal also contains 1% to 3% sulfur, most of it in the form of metal sulfides such as FeS_2.

When hydrocarbons burn in air, they release energy as heat. Typical reactions include:

$$CH_4(g) + 2\ O_2(g) \rightarrow CO_2(g) + 2\ H_2O(l); \qquad \Delta H = -890 \text{ kJ}$$

$$C_8H_{18}(l) + \tfrac{25}{2}\ O_2(g) \rightarrow 8\ CO_2(g) + 9\ H_2O(l); \qquad \Delta H = -5440 \text{ kJ}$$

$$C_{10}H_8(s) + 12\ O_2(g) \rightarrow 10\ CO_2(g) + 4\ H_2O(l); \qquad \Delta H = -5155 \text{ kJ}$$

About 40–50 kJ of heat is evolved per gram of fuel; with octane, C_8H_{18}, we have:

$$\Delta H = \frac{-5440 \text{ kJ}}{1 \text{ mol}} \times \frac{1 \text{ mol}}{114 \text{ g}} = -47.7 \text{ kJ/g}$$

The heat evolved may be used directly to warm our homes and places of business. More commonly, it is converted to mechanical or electrical energy by machines.

Until about 1950, all of the petroleum used in the United States came from domestic wells. Today, upwards of one-third of our petroleum is imported. For economic and political reasons, our dependence upon foreign oil is disturbing. Put quite simply, the United States is running out of petroleum and the natural gas associated with it. Each year, known reserves of petroleum decrease; more is used than is discovered. The production of oil in the United States peaked in 1970 and has decreased considerably since then.

Reserves of coal in the United States have an energy value 30 times that of reserves of petroleum. In principle, coal deposits could probably fulfill our energy needs for a century or more. For some purposes, coal is at least as suitable as petroleum. In the 1970s, when the price of oil skyrocketed, many electrical power plants converted from oil or natural gas to coal. On the other hand, it would be inconvenient, to say the least, to use a solid fuel such as coal as an energy source in automobiles.

Someday we will have to make our gasoline from coal

Another problem with coal is the adverse effect that its combustion has on both people and the environment. Most of the sulfur dioxide that pollutes our air comes from burning coal, which typically contains 1% to 3% sulfur. Sulfur dioxide is the major factor in the formation of acid rain (Chap. 19). It is also responsible each year for the deaths of several thousand people who suffer from chronic bronchitis, asthma, or emphysema. Moreover, coal mining is a dirty, hazardous occupation. About 1 of every 1000 underground miners is killed each year by cave-ins or explosions. At least 30% of those coal miners who have worked for many years suffer from "black lung" or other lung disease.

NUCLEAR ENERGY

At the present time, about 12% of our electrical energy comes from a process called *nuclear fission* (Chap. 25). Here, a heavy atom splits into smaller fragments when struck by a neutron. Energy is evolved as heat, which is then converted into electrical energy. In nuclear plants now in operation, the fuel used is the relatively rare isotope of uranium, U-235. A typical fission reaction is

$$^{235}_{92}\text{U} + ^{1}_{0}n \rightarrow ^{90}_{38}\text{Sr} + ^{144}_{54}\text{Xe} + 2\,^{1}_{0}n; \qquad \Delta H = -2 \times 10^{10}\text{ kJ} \qquad (7.24)$$

You can get an idea of the enormous amount of energy available from fission by calculating the amount of heat evolved per gram of uranium:

$$\frac{-2 \times 10^{10}\text{ kJ}}{235\text{ g}} = -1 \times 10^{8} \text{ (100 million) kJ/g}$$

This compares to about 50 kJ/g for petroleum and natural gas.

Balanced against this advantage of nuclear fuels is the problem of storing the fission products. These products, such as strontium-90, are dangerously radioactive and will remain so for many years to come. Moreover, there is always the possibility of a nuclear accident that could release radioactive material to the surroundings. The near miss at Three Mile Island in Pennsylvania in 1979 and the disaster at Chernobyl in the Ukraine in 1986 has influenced public opinion against fission reactors. For these reasons, among others, the development of nuclear energy has been much slower than predicted, and its future is uncertain.

In many ways, the ideal energy source of the future would be *nuclear fusion* (Chap. 25). Here, light nuclei such as deuterium, $^{2}_{1}\text{H}$, and tritium, $^{3}_{1}\text{H}$, combine to form heavier nuclei. A typical fusion reaction is

$$^{2}_{1}\text{H} + ^{3}_{1}\text{H} \rightarrow ^{4}_{2}\text{He} + ^{1}_{0}n; \qquad \Delta H = -1.7 \times 10^{9}\text{ kJ} \qquad (7.25)$$

This reaction evolves about three to four times as much energy as fission, per gram of fuel:

$$\frac{-1.7 \times 10^{9}\text{ kJ}}{5.0\text{ g}} = -3.4 \times 10^{8}\text{ kJ/g}$$

Fusion, unlike fission, does not produce long-lived radioactive products; in that sense, a fusion reactor would be much less hazardous. Moreover, the light isotopes required for fusion are quite common; they could supply all our energy needs for thousands of years.

Unfortunately, no one up to now has been able to use Reaction 7.25 to generate a sustained flow of energy. For this to happen, a basic problem must be solved. There is a huge electrical repulsion between two small, positively charged hydrogen nuclei (H^+ ions). To overcome this repulsion, the nuclei must be accelerated to enormous velocities, corresponding to temperatures of millions of degrees. Maintaining such temperatures requires the development of a whole new technology. It seems unlikely that fusion reactors will contribute to meeting our energy needs before the year 2000, if then.

SOLAR ENERGY

The problem of trapping and using solar energy is one that has intrigued scientists for generations (and federal granting agencies from 1974 to 1980). Each year the earth receives the enormous total of 5×10^{21} kJ of energy from the sun. A tiny fraction of this, through evaporation and condensation of water, supplies us with hydroelectric power. If this fraction could be increased to 0.0001 (1/100 of 1%), it would meet all the world's energy needs.

There is room for some new ideas in this area

At present, the most practical application of solar energy is to heat water. Solar water heaters are common in Israel, Japan, and Australia. Solar energy is also being used to heat water and some homes in the United States. So-called "active" solar heating systems use a roof collector of the type shown in Figure 7.13. Here, the base of a shallow metal tray is painted

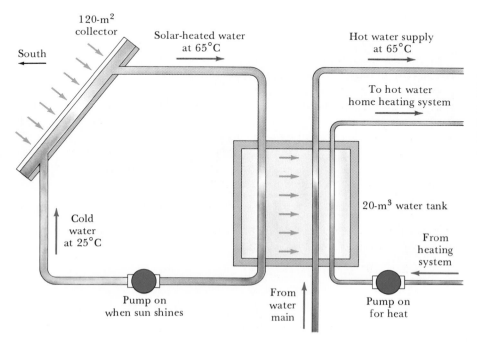

Figure 7.13 An "active" solar heating system. The large insulated storage tank acts as a heat exchanger, transferring heat from the roof collector to hot water faucets and radiators in the house.

black to absorb as much sunlight as possible. The glass cover exerts a "greenhouse" effect, allowing sunlight to reach the collector but preventing radiant heat from escaping. Water, passing through the collector several times, is heated to about 65°C. This water passes into a storage tank, from which heat is transferred throughout the house as in an ordinary hot water heating system.

There are several problems with solar heating systems of this type. For one thing, the components tend to break down frequently. Leaks, freeze-ups, and failures due to corrosion have been all too common. Moreover, the supply of heat is cut off in periods of cold, cloudy weather and during long winter nights. These conditions prevail during the heating season in most of the United States. Hence, an auxiliary heating system of the ordinary type must be available. This factor makes the economics of "active" solar heating systems unattractive at the present time.

Recently, emphasis has shifted to "passive" solar heating systems that involve no machinery or circulating water. Instead, a new house is designed, or an existing house remodeled, to absorb as much of the sun's rays as possible. A common technique is to cover most of the south wall of the house with double-glazed windows. During the day, heat passing through the windows is absorbed within the house, perhaps by a concrete floor next to the windows. When the sun sets, the window area is closed off by insulating drapes that prevent heat from escaping.

Another way to store solar energy, on a much larger scale, is to use salt ponds (Fig. 7.14). In such a pond, the concentration of sodium chloride is kept at a much higher level at the bottom than at the surface. Absorption of solar energy in the lower layers of the pond may be enhanced by using a black, heat-absorbing liner. As heat is absorbed, the temperature of the

You still need a backup heating system, at least in Minnesota

Figure 7.14 A 2000-m³ solar pond in Australia that produces enough energy to power a nearby winery. (Solar Energy Research Institute)

water at the bottom of the pond rises. In an ordinary pond, temperatures would rapidly equilibrate because the hot water, with a lower density, would rise to the surface. Here, this does not occur because the concentrated salt solution has a considerably higher density than the water above it. As a result, the temperature near the bottom of the pond can continue to rise, sometimes reaching as high as 100°C. Solar ponds are used for heating in Israel, adjacent to the Dead Sea, and in the Salton Sea area of California.

Within the next decade, solar energy may become competitive with fossil fuels for purposes other than heating. This can come about through the conversion of solar to electrical energy using *photochemical cells*. These cells contain semiconductor materials, such as silicon, cadmium telluride (CdTe), and gallium arsenide, (GaAs). When sunlight strikes the semiconductor, spread out in the form of a very thin film, electrons are set free. Movement of these charged particles creates an electric current. At present, the cost of electricity from photovoltaic cells is perhaps five times that from ordinary sources. Research and development is concentrated upon cost reduction, achieved by increasing the efficiency of energy conversion.

If these cells could be made cheaply, they could change our whole energy picture

■ SUMMARY PROBLEM

Consider the thermochemical equation:

$$C_2H_6(g) + \tfrac{7}{2} O_2(g) \rightarrow 2\ CO_2(g) + 3\ H_2O(l); \qquad \Delta H = -1559.7 \text{ kJ}$$

a. Calculate ΔH for the thermochemical equations:

$$2\ C_2H_6(g) + 7\ O_2(g) \rightarrow 4\ CO_2(g) + 6\ H_2O(l)$$

$$2\ CO_2(g) + 3\ H_2O(l) \rightarrow C_2H_6(g) + \tfrac{7}{2} O_2(g)$$

b. The heat of vaporization of $H_2O(l)$ is $+44.0$ kJ/mol. Calculate ΔH for the equation:

$$C_2H_6(g) + \tfrac{7}{2} O_2(g) \rightarrow 2\ CO_2(g) + 3\ H_2O(g)$$

c. The heats of formation of $CO_2(g)$ and $H_2O(l)$ are -393.5 and -285.8 kJ/mol, respectively. Calculate ΔH_f^0 of $C_2H_6(g)$.

d. How much heat is evolved when 1.00 g of $C_2H_6(g)$ is burned to give $CO_2(g)$ and $H_2O(l)$ in an open container?

e. When 1.00 g of C_2H_6 is burned in a bomb calorimeter, the temperature rises from 22.00°C to 33.13°C. In a separate experiment, it is found that absorption of 10.0 kJ of heat raises the temperature of the calorimeter by 2.15°C. Calculate $C_{calorimeter}$ and $q_{reaction}$.

f. Calculate ΔE at 25°C for the equation:

$$C_2H_6(g) + \tfrac{7}{2} O_2(g) \rightarrow 2\ CO_2(g) + 3\ H_2O(l)$$

Answers
a. -3119.4 kJ; $+1559.7$ kJ   **b.** -1427.7 kJ   **c.** -84.7 kJ
d. 51.9 kJ   **e.** 4.65 kJ/°C, -51.8 kJ   **f.** -1553.5 kJ

MODERN
PERSPECTIVE

ENERGY BALANCE IN THE
HUMAN BODY

Allllll of us require energy to maintain life processes and do the muscular work associated with such activities as studying, writing, walking, or jogging. The fuel used to produce that energy is the food we eat. In this discussion, we will focus upon "energy input" and "energy output" and the balance between them.

Energy Input

In principle, the energy values of foods can be measured in a bomb calorimeter, a device discussed earlier in this chapter. More commonly, energy values are estimated on the basis of their content of carbohydrate, protein, and fat:

Carbohydrate: 17 kJ (4.0 kcal) of energy per gram*
Protein: 17 kJ (4.0 kcal) of energy per gram
Fat: 38 kJ (9.0 kcal) of energy per gram

Alcohol, which doesn't fit into any of these categories, furnishes about 29 kJ (7.0 kcal) per gram.

Using these factors, you can readily calculate energy input for a meal or snack. Suppose, for example, you eat a quarter-pound hamburger (about 100 g when cooked) inside a bun weighing 25 g. Hamburger contains about 24% protein and 20% fat; the rest is water. Bread has about 50% carbohydrate, 10% protein, and 5% fat.

ENERGY INPUT

Hamburger:	0.24×100 g $\times$ 17 kJ/g	=	410 kJ
	0.20×100 g $\times$ 38 kJ/g	=	760 kJ
			1170 kJ
Bun:	0.50×25 g $\times$ 17 kJ/g	=	210 kJ
	0.10×25 g $\times$ 17 kJ/g	=	40 kJ
	0.05×25 g $\times$ 38 kJ/g	=	50 kJ
			300 kJ

Total = 1170 kJ + 300 kJ = 1470 kJ (350 kcal)

This approach to estimating energy input is tedious, at least if you eat a well-balanced meal with many different foods. A simpler alternative is to use a table such as Table 7.5, which lists approximate energy values for standard portions of different types of foods. With a little practice, you can use the table to estimate your energy input within ± 10%. That's probably as close as you'll come with any method of calculation.

Energy Output

Energy output can be divided, more or less arbitrarily, into two categories.

---

*In most nutrition texts, energy is expressed in kilocalories rather than kilojoules, although that situation is changing. Remember that: 1 kcal = 4.18 kJ; 1 kg = 2.20 lb.

TABLE 7.5 Energy Values of Food Portions*

	kJ	kcal
8 oz glass (250 g) skim milk	330	80
whole milk	710	170
beer	420	100
1 portion vegetable	100	25
1 portion fruit	170	40
1 portion bread or starchy vegetable	300	70
1 oz (30 g) lean meat	230	55
medium-fat meat	320	77
high-fat meat	420	100
1 teaspoon (5 g) fat or oil	190	45
1 teaspoon (5 g) sugar	80	20

*Adapted from E.N. Whitney and E.M. Hamilton, *Understanding Nutrition,* 3rd ed., West Publishing Co., St. Paul, 1984.

1. *Metabolic energy.* This is the largest item in most people's energy budget; it consists of the energy necessary to maintain life. Included is the work required to keep your heart beating, your lungs inhaling and exhaling air, and your nervous system transmitting signals to the brain. We'll also include here the energy consumed in the digestion and absorption of food.

It's possible to measure metabolic rates in the laboratory by determining the amount of carbon dioxide exhaled in a given time. More commonly, metabolic rates are estimated based on a person's sex, weight, and age. For a young adult female, the metabolic rate is about 100 kJ per day per kilogram of body weight. For a young adult male, the figure is about 10% higher. Thus, for a 20-year-old man weighing 150 lb:

$$\text{Metabolic rate} = 150 \text{ lb} \times \frac{1 \text{ kg}}{2.20 \text{ lb}} \times 110 \frac{\text{kJ}}{\text{d·kg}} = 7500 \frac{\text{kJ}}{\text{d}}$$
$$(1800 \text{ kcal per day})$$

For a woman of the same age weighing 120 lb:

$$\text{Metabolic rate} = 120 \text{ lb} \times \frac{1 \text{ kg}}{2.20 \text{ lb}} \times 100 \frac{\text{kJ}}{\text{d·kg}} = 5500 \frac{\text{kJ}}{\text{d}}$$
$$(1300 \text{ kcal per day})$$

2. *Muscular activity.* Energy is consumed whenever your muscles contract, whether in writing, walking, playing tennis, or even sitting in class. Generally speaking, energy expenditure for muscular activity ranges from 50% to 100% of metabolic energy, depending upon life style. If the 20-year-old man referred to earlier is a student who does nothing more strenuous than lift textbooks (and that not too often), the energy spent on muscular activity might be:

Studying doesn't consume much energy, believe it or not

$$0.50 \times 7500 \frac{\text{kJ}}{\text{d}} = 3800 \frac{\text{kJ}}{\text{d}}$$

His total energy output per day would be 7500 kJ + 3800 kJ ≈ 11300 kJ (2700 kcal). If the woman referred to lives a moderately active life, her total energy output might be:

$$0.75 \times 5500 \frac{kJ}{d} + 5500 \frac{kJ}{d} \approx 9600 \frac{kJ}{d} \left(2300 \frac{kcal}{d}\right)$$

Energy Balance

To maintain constant weight, your daily energy input, as calculated from the foods you eat, should be about 700 kJ (170 kcal) greater than output. The difference allows for the consumption of about 40 g of protein to maintain body tissues and fluids. If the excess of input over output is greater than 700 kJ/day, the unused food, whether it be carbohydrate, protein, or fat, is converted to fatty tissue and stored as such in the body.

Fatty tissue consists of about 85% fat and 15% water; its energy value is:

$$0.85 \times 38 \frac{kJ}{g} \times \frac{1000 \text{ g}}{1 \text{ kg}} = 32,000 \text{ kJ/kg}$$

or:

$$0.85 \times 9.0 \frac{kcal}{g} \times \frac{454 \text{ g}}{1 \text{ lb}} = 3500 \text{ kcal/lb}$$

This means that to lose one pound of fat, energy input must be decreased by about 3500 kcal. To lose weight at a sensible rate of one pound per week, it is necessary to reduce energy intake by 500 kcal (2100 kJ) per day.

It's also possible, of course, to lose weight by increasing muscular activity. Table 7.6 shows the amount of energy consumed per hour with various types of exercise. In principle, you can lose a pound a week by climbing mountains for an hour each day, provided you're not already doing that (most people aren't). Alternatively, you could spend 1½ hours a day ice skating or water skiing, depending on the weather.

On his rowing machine, the author of these marginal notes, EJS, can burn off 100 kcal in about 10 min.

TABLE 7.6   Energy Consumed by Various Types of Exercise*

kJ/hour	kcal/hour	
1000–1250	240–300	Walking (3 mph), bowling, golf (pulling cart)
1250–1500	300–360	Volleyball, calisthenics, golf (carrying clubs)
1500–1750	360–420	Ice skating, roller skating
1750–2000	420–480	Tennis (singles), water skiing
2000–2500	480–600	Jogging (5 mph), downhill skiing, mountain climbing
>2500	>600	Running (6 mph), handball, squash

*Adapted from Jane Brody: *Jane Brody's Nutrition Book: A Lifetime Guide to Good Eating for Better Health and Weight Control by the Personal Health Columnist for the New York Times.* Norton, New York, 1981.

QUESTIONS AND PROBLEMS

Formulas, Symbols, and Equations

1. Using Table 7.2, write thermochemical equations for the following reactions:
   a. benzene boiling   b. naphthalene melting
   c. bromine freezing

2. Using Table 7.2, write thermochemical equations for the following reactions:
   a. mercury freezing   b. naphthalene boiling
   c. benzene melting

3. Write chemical equations for the formation of one mole of the following compounds from the elements in their stable state at 25°C and 1 atm.
   a. aluminum oxide (*s*)   b. carbon tetrachloride (*l*)
   c. sodium sulfate (*s*)   d. potassium dichromate (*s*)

4. Write chemical equations for the formation of one mole of the following compounds from the elements in their stable state at 25°C and 1 atm.
   a. potassium superoxide (*s*)   b. hydrazine (*l*)
   c. ammonium carbonate (*s*)
   d. dinitrogen tetroxide (*g*)

Specific Heat and Calorimetry

5. The specific heat of solid aluminum is 0.902 J/g·°C. Calculate the amount of heat that must be absorbed to raise the temperature of a 4.75-g aluminum sample from 25.0°C to 26.8°C.

6. The specific heat of solid platinum is 0.133 J/g·°C. Calculate the temperature change in °C if a 37.0-g sample of platinum absorbs 125 J.

7. It is found that 252 J of heat must be absorbed to raise the temperature of 50.0 g of nickel from 20.0°C to 31.4°C. What is the specific heat of nickel?

8. The specific heat of iron is 0.449 J/g·°C. A sample of iron absorbs 175 J of energy when the temperature increases from 24.00°C to 29.75°C. What is the mass of the sample?

9. When 2.80 g of calcium chloride dissolves in 100.0 g of water, the temperature of the water rises from 20.50°C to 25.40°C. Assume that all the heat is absorbed by the water (specific heat = 4.18 J/g·°C).
   a. Write a balanced equation for the solution process.
   b. What is q for the process described above?
   c. Is the solution process exothermic or endothermic?
   d. How much heat is absorbed by the water if one mole of calcium chloride is dissolved?

10. When 1.34 g of potassium bromide dissolves in 74.0 g of water in a coffee-cup calorimeter, the temperature drops from 18.000°C to 17.279°C. Assume all the heat absorbed in the solution process comes from the water (specific heat = 4.18 J/g·°C).
   a. Write a balanced equation for the solution process.
   b. Is the process exothermic or endothermic?

   c. What is q when 1.34 g of potassium bromide dissolves?
   d. What is q when one mole of potassium bromide dissolves?

11. A sample of sucrose, $C_{12}H_{22}O_{11}$, weighing 4.50 g is burned in a bomb calorimeter. The heat capacity of the calorimeter is 2.411×10^4 J/°C. The temperature rises from 22.15°C to 25.22°C. Calculate q for the combustion of one mole of sucrose.

12. When 5.00 mL of ethyl alcohol, $C_2H_5OH(l)$ ($d = 0.789$ g/cm³), is burned in a bomb calorimeter, the temperature in the bomb rises from 19.75°C to 28.67°C. The calorimeter heat capacity is 13.24 kJ/°C. Calculate q for the combustion of one mole of ethyl alcohol.

13. When 1.750 g of methane, CH_4, burns in a bomb calorimeter, the temperature in the bomb rises 3.293°C. Under these conditions, 885.3 kJ of heat is evolved per mole of methane burned. Calculate the calorimeter heat capacity in J/°C.

14. A 350.0-mg sample of benzoic acid gave off 9.258 kJ of heat when burned in a bomb calorimeter. The temperature rose 2.152°C. Calculate the calorimeter heat capacity in joules per degree Celsius.

15. A 0.2500-g sample of naphthalene, $C_{10}H_8$, is burned in a bomb calorimeter, heat capacity = 4999 J/°C. If the bomb is initially at 20.00°C, what is the final temperature of the system? (When one mole of naphthalene burns, 5.15×10^3 kJ of heat are evolved.)

16. A 4.00-g sample of salicylic acid, $C_7H_6O_3$, is burned in a bomb calorimeter with a heat capacity of 2.046×10^4 J/°C. The combustion produces 87.7 kJ of heat, and the temperature after combustion is 26.10°C. Calculate the initial temperature.

17. When 1.00 g of acetylene, $C_2H_2(g)$, burns in a bomb calorimeter, the temperature rises 10.94°C. In a separate experiment, it is found that when 8.16 kJ of heat is added to the calorimeter, the temperature increases by 1.79°C. Calculate
   a. the heat capacity of the calorimeter.
   b. q for the combustion of one gram of acetylene.
   c. q for the combustion of one mole of acetylene.

18. When 1.00 g of ethylene, $C_2H_4(g)$, burns in a bomb calorimeter, the temperature rises 12.92°C. In a separate experiment, it is found that the addition of 6.23 kJ of heat to the calorimeter raises its temperature by 1.61°C. What is
   a. the heat capacity of the calorimeter?
   b. q for the combustion of one gram of ethylene?
   c. q for the combustion of one mole of ethylene?

Laws of Thermochemistry

19. The combustion of one mole of benzene, $C_6H_6(l)$, in oxygen liberates 3.268×10^3 kJ of heat. The products of the reaction are carbon dioxide and water.

a. Write the thermochemical equation for the combustion of benzene.
b. Is the reaction exothermic or endothermic?
c. Draw a diagram, similar to Figure 7.5, for this reaction.
d. Calculate ΔH when 10.00 g of benzene are burned.
e. How many grams of benzene must be burned to evolve one kilojoule of heat?

20. One mole of nickel carbonyl gas, $Ni(CO)_4$, decomposes upon heating to nickel and carbon monoxide. In addition, 160.7 kJ of heat is absorbed.
a. Write a thermochemical equation for this reaction.
b. Is this reaction exothermic or endothermic?
c. Draw a diagram, similar to Figure 7.5, for this reaction.
d. Calculate ΔH when 1.000 g of nickel carbonyl decomposes.
e. How many grams of nickel carbonyl decompose when 10.00 kJ of heat is absorbed?

21. Consider the dissociation of water into ions.

$$H_2O(l) \rightarrow H^+(aq) + OH^-(aq) \qquad \Delta H = 55.8 \text{ kJ}$$

a. Calculate ΔH when one mole of water is formed from the ions.
b. What is ΔH when 1.00 g of water is formed?

22. Consider the dissolving of silver bromide,

$$AgBr(s) \rightarrow Ag^+(aq) + Br^-(aq) \qquad \Delta H = 84.4 \text{ kJ}$$

a. Calculate ΔH when one mole of silver bromide solid precipitates from solution.
b. What is ΔH when one gram of silver bromide is precipitated?

23. When ignited, ammonium dichromate decomposes in a fiery display. This is the reaction for the "laboratory volcanoes" often demonstrated in science fair projects. The decomposition of 1.000 g of ammonium dichromate produces 1.19 kJ of energy, nitrogen gas, steam, and chromium(III) oxide.
a. Write a balanced thermochemical equation for the reaction.
b. What is ΔH for the formation of 11.2 L of nitrogen at 0°C and 1.00 atm?

24. When magnesium and carbon dioxide react, 16.7 kJ is evolved per gram of magnesium. The products are solid carbon and magnesium oxide.
a. Write a balanced thermochemical equation for the reaction.
b. How much heat is evolved if one gram of products is formed from this reaction?

25. A typical fat in the body is glyceryl trioleate ($C_{57}H_{104}O_6$). When it is metabolized in the body, it combines with oxygen to produce carbon dioxide, water, and 3.022×10^4 kJ of heat per mole of fat.

a. Write a balanced thermochemical equation for the metabolism of fat.
b. How many grams of fat would have to be burned to heat one liter of water ($d = 1.00$ g/mL) from 25.00°C to 30.00°C? The specific heat of water is 4.18 J/g·°C.

26. Using the metabolism of fat reaction in Problem 25, how many kilojoules of energy must be evolved in the form of heat if you want to get rid of one pound of this fat by combustion? How many calories is this? How many nutritional calories (1 nutritional calorie = 1×10^3 calories)?

27. Which requires the absorption of a greater amount of heat, melting 100.0 g of naphthalene or boiling 100.0 g of water? Use Table 7.2.

28. Which evolves more heat, the freezing of 25.0 g of mercury or the condensation of 100.0 g of benzene vapor? Use Table 7.2.

29. Calculate the amount of heat involved in the conversion of 25.00 g of water at 25°C to steam at 100°C.

30. Calculate the amount of heat involved in freezing 10.00 g of water at 37°C to ice at 0°C.

Hess's Law

31. A lead ore, galena, consisting mainly of lead(II) sulfide, is the principal source of lead. To obtain the lead, the ore is first heated in the air to form lead oxide.

$$PbS(s) + \tfrac{3}{2}O_2(g) \rightarrow PbO(s) + SO_2(g) \qquad \Delta H = -415.4 \text{ kJ}$$

The oxide is then reduced to metal with carbon.

$$PbO(s) + C(s) \rightarrow Pb(s) + CO(g) \qquad \Delta H = +108.5 \text{ kJ}$$

Calculate ΔH for the reaction of one mole of lead(II) sulfide with oxygen and carbon, forming lead, sulfur dioxide, and carbon monoxide.

32. To produce silicon, used for semiconductors, from sand (SiO_2), a reaction is used which can be broken down into three steps:

$$SiO_2(s) + 2C(s) \rightarrow Si(s) + 2CO(g) \qquad \Delta H = 689.9 \text{ kJ}$$

$$Si(s) + 2Cl_2(g) \rightarrow SiCl_4(g) \qquad \Delta H = -657.0 \text{ kJ}$$

$$SiCl_4(g) + 2Mg(s) \rightarrow 2MgCl_2(s) + Si(s) \qquad \Delta H = -625.6 \text{ kJ}$$

Write the thermochemical equation for the overall reaction for the formation of silicon from silicon dioxide. What is ΔH for the formation of one mole of silicon? Is the reaction endothermic or exothermic?

33. Given the following thermochemical equations:

$$2H_2(g) + O_2(g) \rightarrow 2H_2O(l) \qquad \Delta H = -571.6 \text{ kJ}$$

$$N_2O_5(g) + H_2O(l) \rightarrow 2HNO_3(l) \qquad \Delta H = -73.7 \text{ kJ}$$

$$\tfrac{1}{2}N_2(g) + \tfrac{3}{2}O_2(g) + \tfrac{1}{2}H_2(g) \rightarrow HNO_3(l) \qquad \Delta H = -174.1 \text{ kJ}$$

Calculate ΔH for the formation of one mole of dinitrogen pentoxide from its elements in their stable state at 25°C and 1 atm.

34. Given the following thermochemical equations:

$$C_2H_2(g) + \tfrac{5}{2}O_2(g) \rightarrow 2CO_2(g) + H_2O(l) \quad \Delta H = -1299.5 \text{ kJ}$$

$$C(s) + O_2(g) \rightarrow CO_2(g) \quad \Delta H = -393.5 \text{ kJ}$$

$$H_2(g) + \tfrac{1}{2}O_2(g) \rightarrow H_2O(l) \quad \Delta H = -285.8 \text{ kJ}$$

Calculate ΔH for the decomposition of one mole of acetylene, $C_2H_2(g)$, to its elements in their stable state at 25°C and 1 atm.

ΔH^0 and Heats of Formation

35. Given:

$$2 \text{ CuO}(s) \rightarrow 2 \text{ Cu}(s) + O_2(g) \quad \Delta H^0 = 314.6 \text{ kJ}$$

a. Determine the heat of formation of CuO.
b. Calculate ΔH^0 for the formation of 13.58 g of CuO.

36. Given

$$2Al_2O_3(s) \rightarrow 4Al(s) + 3O_2(g) \quad \Delta H^0 = 3351.4 \text{ kJ}$$

a. What is the heat of formation of aluminum oxide?
b. What is ΔH^0 for the formation of 12.50 g of aluminum oxide?

37. Limestone, $CaCO_3$, when subjected to a temperature of 900°C in a kiln, decomposes to calcium oxide and carbon dioxide. How much heat is evolved or absorbed when one gram of limestone decomposes? (Use Table 7.3.)

38. A first-aid hot pack is made up of two compartments: one has solid calcium chloride, the other has water. When heat is required, you are instructed to twist the hot pack. This breaks the diaphragm separating both compartments. Calcium chloride dissolves in water, giving calcium ions, chloride ions, and heat. How much heat is evolved if the hot pack contains five grams of calcium chloride? (Use Tables 7.3 and 7.4.)

39. Use Tables 7.3 and 7.4 to obtain ΔH^0 for the following reactions:

a. $2Br^-(aq) + I_2(s) \rightarrow Br_2(l) + 2I^-(aq)$
b. $Cr(OH)_3(s) + 5OH^-(aq) + 3Fe^{3+}(aq) \rightarrow 3Fe^{2+}(aq) + 4H_2O(l) + CrO_4^{2-}(aq)$ (ΔH_f^0 for $Cr(OH)_3 = -1064.0$ kJ/mol)
c. $SO_4^{2-}(aq) + 4H^+(aq) + Ni(s) \rightarrow Ni^{2+}(aq) + SO_2(g) + 2H_2O(l)$

40. Use Tables 7.3 and 7.4 to obtain ΔH^0 for the following reactions.

a. $2Br^-(aq) + 2H_2O(l) \rightarrow Br_2(l) + H_2(g) + 2OH^-(aq)$
b. $Pb^{2+}(aq) + Sn(s) \rightarrow Pb(s) + Sn^{2+}(aq)$
c. $3Mn^{2+}(aq) + 2NO_3^-(aq) + 2H_2O(l) \rightarrow 2NO(g) + 3MnO_2(s) + 4H^+(aq)$

41. Use Table 7.3 to calculate ΔH^0

a. for the combustion of one mole of ethyl alcohol, C_2H_5OH, to form carbon dioxide and water (l).

b. for the combustion of one mole of ethyl alcohol, C_2H_5OH, to form carbon dioxide and steam.

42. Use Table 7.3 to calculate ΔH^0 for

a. the reaction of ammonia and oxygen gas to form one mole of nitrogen and steam.
b. the reaction of ammonia and oxygen gas to form two moles of nitrogen and $H_2O(l)$.

43. When one mole of calcium carbonate reacts with ammonia, solid calcium cyanamide, $CaCN_2$, and liquid water are formed. The reaction absorbs 90.1 kJ of heat.

a. Write a balanced thermochemical equation for the reaction.
b. Using Table 7.3, calculate ΔH_f^0 for calcium cyanamide.

44. When one mole of ethylene gas, C_2H_4, is burned in oxygen and hydrogen chloride, the products are liquid ethylene chloride, $C_2H_4Cl_2$, and liquid water; 318.7 kJ of heat is evolved.

a. Write a thermochemical equation for the reaction.
b. Using Table 7.3, calculate the standard heat of formation of ethylene chloride.

45. Use Tables 7.3 and 7.4 to determine the heat of formation of the Co^{2+} ion given $\Delta H^0 = -1936.2$ kJ for the reaction

$$2MnO_4^-(aq) + 16H^+(aq) + 5Co(s) \rightarrow$$
$$2Mn^{2+}(aq) + 5Co^{2+}(aq) + 8H_2O(l)$$

46. Use Tables 7.3 and 7.4 to determine the heat of formation of lead sulfate from the following equation:

$$PbO_2(s) + SO_4^{2-}(aq) + 4H^+(aq) + 2I^-(aq) \rightarrow$$
$$PbSO_4(s) + 2H_2O(l) + I_2(s) \quad \Delta H^0 = -194.4 \text{ kJ}$$

47. Glucose, $C_6H_{12}O_6(s)$ ($\Delta H_f^0 = -1275.2$ kJ/mol) is converted to ethyl alcohol, $C_2H_5OH(l)$, and carbon dioxide in the fermentation of grape juice. What quantity of heat is liberated when a liter of wine containing 12.0% ethyl alcohol by volume ($d = 0.789$ g/cm³) is produced by the fermentation of grape juice?

48. When ammonia reacts with dinitrogen oxide gas ($\Delta H_f^0 = 82.05$ kJ/mol), liquid water and nitrogen gas are formed. How much heat is liberated or absorbed by the reaction that produces 275 mL of nitrogen gas at 25°C and 777 mm Hg?

First Law of Thermodynamics

49. Calculate

a. q when a system does 72 J of work and its internal energy decreases by 90 J.
b. ΔE for a gas that releases 35 J of heat and has 128 J of work done on it.

50. Find

a. ΔE when a gas absorbs 45 J of heat and has 32 J of work done on it.
b. q when 62 J of work are done on a system and its internal energy is increased by 84 J.

51. Calculate ΔE at 25°C for the reaction

$$Ca(s) + O_2(g) + H_2(g) \rightarrow Ca(OH)_2(s)$$

52. Using Table 7.2, calculate ΔE for the vaporization of one mole of water at 100°C.

53. Consider the combustion of one mole of acetylene, C_2H_2, which yields carbon dioxide and liquid water.
   a. Using Table 7.3, calculate ΔH.
   b. Calculate ΔE at 25°C.

54. Consider the combustion of propane, C_3H_8, the fuel that is commonly used in portable gas barbecue grills. The products of combustion are carbon dioxide and liquid water.
   a. Write a thermochemical equation for the combustion of one mole of propane.
   b. How much heat would be evolved if 1.00 g of propane were burned in a bomb calorimeter (25°C)?

Sources of Energy

55. In one type of solar heating system, the sun's energy is used to dissolve sodium sulfate decahydrate ($Na_2SO_4 \cdot 10H_2O$), also known as Glauber's salt. The reaction is

$$Na_2SO_4 \cdot 10H_2O(s) \rightarrow 2\,Na^+(aq) + SO_4^{2-}(aq)$$
$$+ 10H_2O(l) \qquad \Delta H = 79.8 \text{ kJ}$$

On cooling, the reverse reaction occurs, giving off heat to raise the temperature of water in a storage tank. How much water can be warmed from 20.0°C to 68.0°C (the temperature of water that is normally in hot water tanks) by the heat evolved when 100.0 g of Glauber's salt is formed?

56. If the solar heating system using Glauber's salt (Problem 55) is used to heat water for a hundred-gallon water heater, what is the minimum amount of Glauber's salt needed so that on a sunny day, enough energy is absorbed to increase the temperature in the tank from 22.0°C to 65.0°C?

57. It is estimated that the United States burns about 7.0×10^{11} cubic meters of natural gas a year. The heat of combustion of natural gas is about 43 kJ/g and its density is about 0.73 g/L.
   a. Calculate how much of the energy consumed in the United States annually comes from natural gas.
   b. If the United States uses 8.0×10^{16} kJ of energy annually, what percent of energy consumption comes from natural gas?

58. The United States has large coal reserves. The heat of combustion of coal is about 32 kJ/g. If the United States uses 0.50 billion metric tons of coal, how much energy comes from coal? What percent of energy comes from coal? (See Problem 57.)

59. An average home electric dishwasher uses 50 L of hot water per load. The owner reduces the temperature of the wash water from 65°C to 55°C.
   a. How much energy is saved per month (30 days) if the dishwasher is run once a day?

   b. If the water is heated electrically at 6.0¢ per kilowatt hour, how much money is saved in a month ($1 \text{ kW} \cdot \text{h} = 3.60 \times 10^3$ kJ)?

60. To conserve energy, a home owner resets the hot water heater temperature control from 64°C to 52°C. The heater tank holds 150 L of water ($d = 1.00 \text{ g/cm}^3$, S.H. = 4.18 J/g·°C).
   a. How much energy can be saved per tankful by this change?
   b. If the water is heated electrically at a rate of 6.0¢ per kilowatt hour, how much money is saved per tankful ($1 \text{ kW} \cdot \text{h} = 3.60 \times 10^3$ kJ)?

Unclassified

61. Given the following reactions:

$$N_2H_4(l) + O_2(g) \rightarrow N_2(g) + 2H_2O(g) \quad \Delta H^0 = -534.2 \text{ kJ}$$

$$H_2(g) + \tfrac{1}{2}O_2(g) \rightarrow H_2O(g) \qquad \Delta H^0 = +241.8 \text{ kJ}$$

Calculate the heat of formation of hydrazine.

62. In World War II, the Germans made use of unusable airplane parts by grinding them up into powdered aluminum. This was made to react with ammonium nitrate to give powerful bombs. The products of this reaction were nitrogen gas, steam, and aluminum oxide. If 10.00 kg of ammonium nitrate is mixed with 10.00 kg of powdered aluminum, how much heat is generated?

63. One way to lose weight is to exercise. Walking briskly at 4.0 mph for an hour consumes 1700 kJ of energy. How many miles would you have to walk to lose one pound of body fat? (One gram of body fat is equivalent to about 32 kJ of energy.)

64. Brass has a density of 8.25 g/cm³ and a specific heat of 0.362 J/g·°C. A cube of brass 7.50 mm on an edge is heated on a Bunsen burner flame to a temperature of 95.0°C. It is then immersed into 20.0 mL of water ($d = 1.00 \text{ g/cm}^3$) at 22.0°C in an insulated container. Assuming no heat loss, what is the final temperature of the water?

65. Some solutes have large heats of solution, and care should be taken in preparing solutions of these substances. The heat evolved when sodium hydroxide dissolves is 44.5 kJ/mol. What is the final temperature of the water, originally at 20.0°C, used to prepare 500.0 cm³ of 6.00 M NaOH solution? Assume all the heat is absorbed by 500 cm³ of water, specific heat = 4.18 J/g·°C.

66. A 1.00-L sample of a gaseous mixture at 0°C and 1 atm (STP) evolves, upon complete combustion at constant pressure, 82.58 kJ of heat. If the gas is a mixture of ethane (C_2H_6) and propane (C_3H_8), what is the mole % of ethane in the mixture?

Challenge Problems

67. Biking at 13 mph (moderate pace) uses about 2000 kJ of energy per hour. This comes from the oxidation of foods, which is about 30% efficient. How much energy do you "save" by biking one mile instead of driving a car that gets 20.0 mi/gal of gasoline (d gasoline = 0.68 g/cm^3; heat of combustion = 48 kJ/g)?

68. On a hot day, you take a six-pack of beer on a picnic, cooling it with ice. Each (aluminum) can weighs 38.5 g and contains 12.0 oz of beer. The specific heat of aluminum is 0.902 J/g·°C; take that of beer to be 4.10 J/g·°C.

  a. How much heat must be absorbed from the six-pack to lower the temperature from 25.0°C to 5.0°C?

  b. How much ice must be melted to absorb this amount of heat (ΔH_{fus} of ice is given in Table 7.2)?

69. A cafeteria sets out glasses of tea at room temperature; the customer adds ice. Assuming the customer wants to have some ice left when the tea cools to 0°C, what fraction of the total volume of the glass should be left empty for adding ice? Make any reasonable assumptions needed to work this problem.

70. The thermite reaction was once used to weld rails:

$$2 \ Al(s) + Fe_2O_3(s) \rightarrow Al_2O_3(s) + 2 \ Fe(s)$$

  a. Calculate ΔH for this reaction using heat of formation data.

  b. Take the specific heats of Al_2O_3 and Fe to be 0.77 and 0.45 J/g·°C, respectively. Calculate the temperature to which the products of this reaction will be raised, starting at room temperature, by the heat given off in the reaction.

  c. Will the reaction produce molten iron (mp Fe = 1535°C, ΔH_{fus} = 270 J/g)?

$$Cu(s) \ + \ 2Ag^+(aq) \rightarrow Cu^{2+}(aq) \ + \ 2Ag(s)$$

The "Christmas tree" reaction between copper metal and Ag^+ ions in solution.

CHAPTER

8

COVALENT BONDING

Might not things external to ourselves
Vibrate in union with our moods and passions,
Atom to atom in love of strange affinity?

OSCAR WILDE

Throughout earlier chapters, we have referred frequently to ionic compounds such as NaCl and ZnO. In these compounds, oppositely charged ions (Na^+ and Cl^-; Zn^{2+} and O^{2-}) are held to one another by strong electrostatic forces called *ionic bonds.* The forces that hold nonmetal atoms to one another in molecules such as H_2 or polyatomic ions such as OH^- are quite different from ionic bonds. They are referred to as **covalent bonds**.

A covalent bond consists of an electron pair shared between two atoms. Perhaps the simplest bond of this type is that which joins H atoms in the H_2 molecule. When two hydrogen atoms, each with one electron, come together, they form a bond. This may be shown as

H : H

where the dots represent electrons. More commonly, we show the structure of the H_2 molecule as

H—H

The understanding is that the dash represents a covalent bond, an electron pair shared by the two atoms.

Most of this chapter will be devoted to the covalent bond and the electronic structures of species containing covalent bonds. We will consider

— the nature of the forces responsible for covalent bonding (Section 8.1)

— the distribution of outer level electrons in molecules and polyatomic ions where atoms are joined by covalent bonds (Section 8.2). These distributions are most simply described in terms of *Lewis structures;* a major objective of this chapter is to make you proficient in writing these structures.

— the properties of the covalent bond (Section 8.3), including bond polarity, bond distance, and bond energy.

Figure 8.1 Electron density in H_2. The depth of color is proportional to the probability of finding an electron in a particular region.

All stable chemical bonds have an energy minimum of this sort

8.1
NATURE OF THE COVALENT BOND

The structures

$$H:H \quad \text{or} \quad H\text{—}H$$

used to represent the covalent bond in the H_2 molecule can be misleading if they are taken to mean that the two electrons are fixed in position between the two nuclei. A more accurate picture of the electron density in H_2 is shown in Figure 8.1. At a given instant, the two electrons may be located at any of various points about the two nuclei. However, they are more likely to be between the nuclei than at the far ends of the molecule.

A question that has long intrigued chemists is: Why should the sharing of two electrons between two nuclei result in increased stability? Why, for example should the H_2 molecule be more stable, by about 400 kJ/mol, than two isolated H atoms? The first plausible answer to this question was put forth in 1927 by two physicists, W. A. Heitler and T. London. They used quantum mechanics to calculate the interaction energy of two hydrogen atoms as a function of the distance between them.

At large distances of separation (far right, Fig. 8.2), there is no interaction between two hydrogen atoms. As they come closer together (moving to the left in Fig. 8.2), the two atoms experience an attraction. This leads gradually to an energy minimum, at an internuclear distance of 0.074 nm. At this distance of separation of the two atoms, the molecule is in its most stable state. It takes 436 kJ of energy to separate a mole of H_2 molecules at this energy minimum into isolated atoms. At distances less than 0.074 nm, repulsive forces become increasingly important and the energy curve rises steeply.

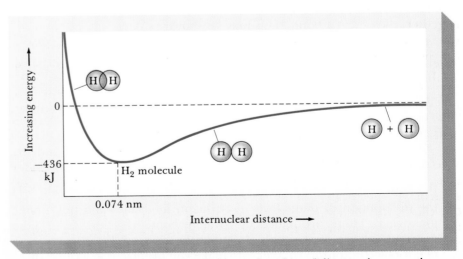

Figure 8.2 Energy of the H_2 molecule as a function of distance between the two nuclei. The minimum in the curve occurs at the observed internuclear distance, 0.074 nm. Energy is compared to that of two separated hydrogen atoms.

The existence of the energy minimum shown in Figure 8.2 accounts for the stability of the H_2 molecule. The question remains, however: What causes this energy minimum? There are two major factors involved:

1. When two hydrogen atoms approach each other, the electron on one atom is attracted to the nucleus of the other. This lowers the electrostatic energy of the system and so stabilizes it. (The kinetic energy of motion of the electrons actually increases, but the change in electrostatic energy dominates.) At close distances, the repulsion between particles of like charge (electron-electron and nucleus-nucleus) becomes significant. The minimum total energy occurs at 0.074 nm.

2. In the H_2 molecule, the electrons interact with both nuclei equally. It turns out that setting up the Schrödinger wave equation (Ch. 5) under these conditions requires that we don't distinguish between electrons, but allow them to be in the orbital of either hydrogen atom. In the molecule the atomic orbitals overlap; the electrons exchange between these orbitals, oriented at the distance (0.074 nm) that minimizes total energy.

The Heitler-London model explains the stability of the H_2 molecule reasonably well. Unfortunately, the equations upon which it is based cannot be solved for any but the simplest of molecules. Later, in Chapter 9, we will describe an extension of this model that explains qualitatively the stability of more complex molecules. In the remainder of this chapter, our goal is more modest. We will look at how electron pairs are distributed between two atoms joined by one or more covalent bonds.

Computers are getting better and better at dealing with this kind of problem

8.2
LEWIS STRUCTURES; THE OCTET RULE

The idea of the covalent bond was first suggested by the American physical chemist G. N. Lewis in 1916. He pointed out that the electron configuration of the noble gases appears to be a particularly stable one. Noble-gas atoms are themselves extremely unreactive. Lewis suggested that *atoms, by sharing electrons to form an electron-pair bond, can acquire a stable, noble-gas structure.* Consider, for example, two hydrogen atoms, each with one electron. The process by which they combine to form an H_2 molecule can be shown as

As with atoms and ions, noble gas structures tend to be stable in molecules

$$H \cdot \; + \; H \cdot \; \rightarrow \; \left(H \; : \; H \right)$$

using dots to represent electrons; the circles emphasize that the pair of electrons in the covalent bond can be considered to occupy the 1s orbital of either hydrogen atom. In that sense, each atom in the H_2 molecule has the electronic structure of the noble gas helium, with the electron configuration $1s^2$.

This idea is readily extended to other simple molecules containing nonmetal atoms. An example is the F_2 molecule. You will recall that a fluorine atom has the electron configuration $1s^2 2s^2 2p^5$. It has seven electrons in its outermost principal energy level ($n = 2$). These electrons are referred to as **valence electrons**. Showing them as dots about the symbol of the element, we can represent the fluorine atom as

$$: \ddot{\underset{..}{F}} \cdot$$

When two of these atoms combine, we have:

$$: \ddot{\underset{..}{F}} \cdot \ + \ \cdot \ddot{\underset{..}{F}} : \ \rightarrow \ \left(: \ddot{\underset{..}{F}} \mathbin{\vdots} \ddot{\underset{..}{F}} : \right)$$

From our drawing of the F_2 molecule, you can see that each atom owns six valence electrons outright and shares two others. Putting it another way, each F atom is surrounded by eight valence electrons. By this model, both F atoms achieve the electron configuration $1s^2 2s^2 2p^6$, which is that of the noble gas neon. This, according to Lewis, explains why the F_2 molecule is stable and why F atoms combine to form F_2 rather than F_3, F_4, ...

The structures just written (without the circles) are referred to as Lewis structures or sometimes, irreverently, as "flyspeck formulas." In writing the Lewis structure for a species, we include only those electrons in the outermost energy level. These so-called *valence electrons* are the ones that take part in covalent bonding. For the main-group elements, the only ones that we will be concerned with here, **the number of valence electrons is equal to the group number in the Periodic Table** (Table 8.1).

TABLE 8.1   Lewis Structures of Atoms

GROUP	OUTER ELECTRON CONFIGURATION	NUMBER OF VALENCE ELECTRONS	LEWIS STRUCTURE EXAMPLE
1	ns^1	1	$H \cdot$
2	ns^2	2	$\cdot Be \cdot$
3	$ns^2 np^1$	3	$\cdot \overset{.}{B} \cdot$
4	$ns^2 np^2$	4	$\cdot \overset{.}{C} \cdot$
5	$ns^2 np^3$	5	$\cdot \overset{..}{N} \cdot$
6	$ns^2 np^4$	6	$\cdot \overset{..}{O} \cdot$
7	$ns^2 np^5$	7	$: \overset{..}{F} \cdot$
8	$ns^2 np^6$	8	$: \overset{..}{\underset{..}{Ne}} :$

In the Lewis structure of a molecule, a covalent bond between atoms is ordinarily shown as a straight line between bonded atoms. Unshared electron pairs, belonging entirely to one atom, are shown as dots. The Lewis structures for the molecules formed by hydrogen with C, N, O, and F are

$$CH_4 \qquad NH_3 \qquad H_2O \qquad HF$$

Notice that in each case the central atom (C, N, O, or F) is surrounded by eight valence electrons. In each of these molecules, a single electron pair is shared between two bonded atoms. These bonds are called **single bonds**. There are four single bonds in CH_4, three in NH_3, two in H_2O, and one in HF. Pairs of electrons not used in bonding are called *unshared pairs*. There are none in CH_4, one unshared pair in NH_3, two unshared pairs in H_2O, and three unshared pairs in HF.

Unshared pairs are very common in molecules

Bonded atoms can share more than one electron pair. When this happens, we say that multiple bonds join the atoms. A **double bond** occurs when bonded atoms share two electron pairs; in a **triple bond**, three pairs of electrons are shared. In ethylene (C_2H_4) and acetylene (C_2H_2), the carbon atoms are linked by a double bond and a triple bond, respectively. Using two parallel lines to represent a double bond and three for a triple bond, we write the structures of these molecules as

ethylene, C_2H_4 acetylene, C_2H_2

Note that each carbon is surrounded by eight valence electrons and each hydrogen by two. **The atoms most commonly joined by multiple bonds are C, O, N, and S.**

Many polyatomic ions can be assigned simple Lewis structures. For example, the OH^- and NH_4^+ ions can be shown as

In both these ions, hydrogen atoms are joined by covalent bonds to non-metal atoms (O, N). In both ions there are eight valence electrons. With the OH^- ion, this is one more than the number contributed by the neutral atoms ($6 + 1 = 7$). The extra electron is accounted for by the -1 charge of the ion. With the NH_4^+ ion, four hydrogen atoms and a nitrogen atom supply nine valence electrons ($4 + 5 = 9$). One of these is missing in the NH_4^+ ion, accounting for its $+1$ charge.

These examples illustrate the principle that atoms in covalently bonded species tend to have noble-gas structures. This rule is often referred to as the **octet rule**. Nonmetals, except for hydrogen, achieve a noble-gas struc-

Most molecules obey the octet rule

ture by sharing in an "octet" of electrons (eight). Hydrogen atoms, in stable molecules or polyatomic ions, are surrounded by two electrons.

WRITING LEWIS STRUCTURES

For very simple molecules, Lewis structures can often be written by inspection. Usually, though, you will save time and avoid confusion by following these steps:

1. Count the number of valence electrons. For a molecule, simply sum up the valence electrons of the atoms present. For a polyatomic anion, one electron is added for each unit of negative charge. For a polyatomic cation, a number of electrons equal to the positive charge must be subtracted.

2. Draw a skeleton structure for the species, joining atoms by single bonds. In some cases, only one arrangement of atoms is possible; in others, experimental evidence must be used to decide between two or more alternative structures.

Most of the molecules and polyatomic ions that we will deal with consist of a *central atom* bonded to two or more *terminal atoms*, located at the outer edges of the molecule or ion. For such species (e.g., NH_4^+, SO_2, CCl_4), it is relatively easy to derive the skeleton structure. **The central atom is usually the one written first in the formula** (N in NH_4^+, S in SO_2, C in CCl_4); **put this in the center of the molecule or ion. Terminal atoms are most often hydrogen, oxygen, or a halogen; bond these atoms to the central atom.**

3. Deduct two valence electrons for each single bond written in step 2. Distribute the remaining electrons as unshared pairs so as to give each atom eight electrons, if possible.

The application of these steps and some further guiding principles are shown in Example 8.1.

■ **EXAMPLE 8.1** _____
Draw Lewis structures for

a. the hypochlorite ion, OCl^-
b. methyl alcohol, CH_4O
c. the silicate ion, SiO_4^{4-}

Solution

a. To obtain the number of valence electrons, note that oxygen is in Group 6 (6 valence e^-), chlorine in Group 7 (7 valence e^-), and the polyatomic ion has a charge of -1:

total valence electrons $= 6 + 7 + 1 = 14$

With two atoms, only one skeleton is possible:

O—Cl

The single bond in the skeleton consumes 2 electrons, leaving 12. These are distributed as unshared pairs, 6 around each atom:

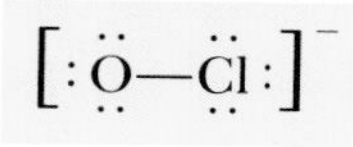

$$\left[\,:\ddot{\text{O}}-\ddot{\text{C}}\text{l}\,:\right]^{-}$$

This is indeed a reasonable structure for the OCl^{-} ion. Each atom has an octet of electrons, 6 unshared (dots) and 2 shared (single bond).

b. The number of valence electrons in CH_4O is readily obtained when we note that carbon is in Group 4, hydrogen in Group 1, and oxygen in Group 6:

total valence electrons $= 4 + 4(1) + 6 = 14$

To derive the skeleton, we start by noting that *hydrogen atoms must be located at the outer edges of the molecule,* since a hydrogen atom can only form one covalent bond, thereby acquiring a share in two electrons. The central atom is *carbon,* which typically *forms four bonds with no unshared pairs.* The skeleton is:

There are almost no exceptions to this rule

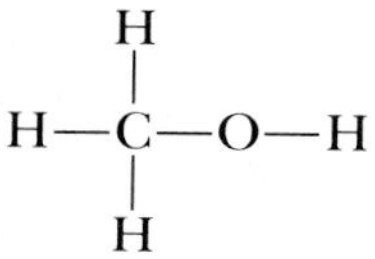

```
    H
    |
H — C — O — H
    |
    H
```

There are five single bonds in the skeleton; count them! That means that there are four electrons left: $14 - 10 = 4$ valence electrons. This is just enough to complete the octet of oxygen. The Lewis structure is

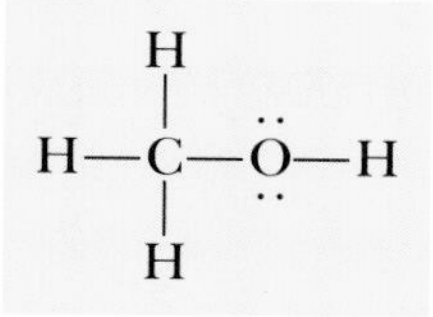

```
    H
    |    ..
H — C — Ö — H
    |    ..
    H
```

c. Total valence e^{-} in $SiO_4^{4-} = 4 + 4(6) + 4 = 32$. The silicate ion is typical of a general class of polyatomic ions known as *oxyanions;* others include NO_3^{-}, CO_3^{2-}, and PO_4^{3-}. In all of these ions the oxygens are terminal atoms, located at the outer edges. The central atom is the one written first in the formula, Si in the SiO_4^{4-} ion. The skeleton is:

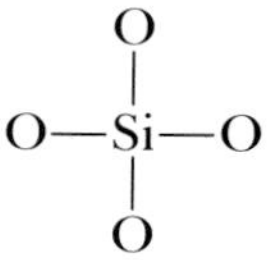

```
      O
      |
O — Si — O
      |
      O
```

Using rule 3, we deduct eight electrons for the four covalent bonds in the skeleton structure, leaving 24. Putting six electrons around each oxygen atom gives us a plausible Lewis structure for the silicate ion:

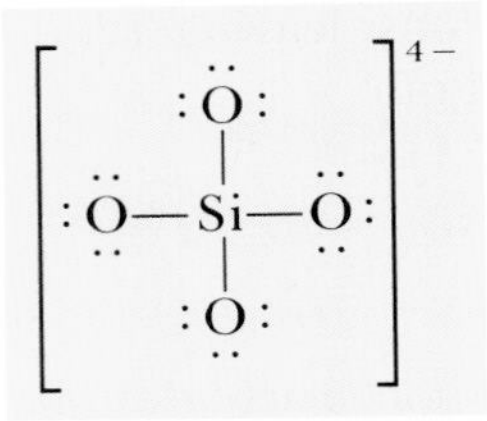

$$\left[\begin{array}{c} \ddot{O} \\ | \\ \ddot{O}-Si-\ddot{O} \\ | \\ \ddot{O} \end{array}\right]^{4-}$$

Exercise

Draw the Lewis structure of the arsenate ion, AsO_4^{3-}. Answer:

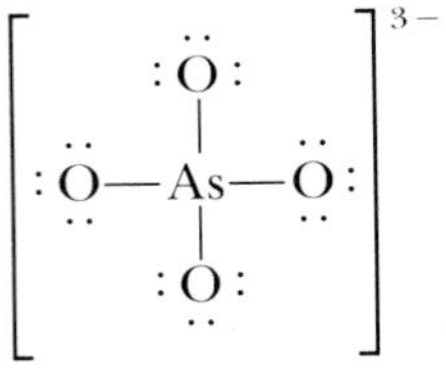

$$\left[\begin{array}{c} \ddot{O} \\ | \\ \ddot{O}-As-\ddot{O} \\ | \\ \ddot{O} \end{array}\right]^{3-}$$

Multiple bonds occur when there is a shortage of electrons

Sometimes, when you follow these rules, you find in the last step that there are "too few electrons to go around." That is, there are not enough electrons left to give each atom an octet. This dilemma can be avoided by making electron pairs do "double duty." By moving an unshared pair to a position between bonded atoms, it is counted in the octet of both atoms. The result of such a shift is to form a multiple bond from a single bond. *Forming a double bond by moving one electron pair corrects a deficiency of two electrons.* If you are four electrons shy, you must form a triple bond (or two double bonds). The process involved is shown in Example 8.2.

■ **EXAMPLE 8.2**

Draw Lewis structures for

a. SO_2 b. H_2CO c. N_2

Solution

a. Since all three atoms in SO_2 are in Group 6, we have $3(6) = 18$ valence electrons. The central atom is sulfur; the oxygen atoms are at the ends of the molecule. The skeleton is:

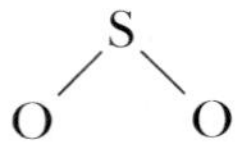

$$\begin{array}{c} S \\ / \ \ \backslash \\ O \qquad O \end{array}$$

Subtracting four electrons for the two single bonds in the skeleton leaves $18 - 4 = 14$. These electrons could be consumed by filling out the oxygen octets, which accounts for 12, and putting two around the sulfur:

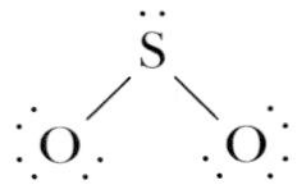

$$\begin{array}{c} \ddot{S} \\ / \ \ \backslash \\ \ddot{O} \qquad \ddot{O} \end{array}$$

This leaves sulfur with only six valence electrons, two less than an octet. To correct this, shift an unshared pair from one of the oxygen atoms to form a double bond with sulfur. The final Lewis structure is:

$$\ddot{S}\diagup\kern-0.5em\diagdown$$
$$\ddot{O}.\qquad.\ddot{O}:$$

There is no way you can figure this out if you don't count electrons properly

b. Total valence e^- = 2 + 4 + 6 = 12. Carbon is the central atom, bonded to terminal oxygen and hydrogen atoms. The skeleton is:

$$\begin{array}{c} H \\ \diagdown \\ C-O \\ \diagup \\ H \end{array}$$

Valence e^- left = 12 − 6 = 6. We might put these as unshared pairs around the oxygen atom:

$$\begin{array}{c} H \\ \diagdown \\ C-\ddot{O}: \\ \diagup \\ H \end{array}$$

This, however, leaves carbon with only six valence electrons. To supply the two electrons to complete the octet, an unshared pair from the oxygen is used to form a double bond with carbon. The correct Lewis structure is

$$\begin{array}{c} H \\ \diagdown \\ C=\ddot{O} \\ \diagup \\ H \end{array}$$

Note that carbon here, as in almost all cases, forms four bonds. In this case, it participates in two single bonds and one double bond.

c. The skeleton is N—N. Since nitrogen is in Group 5, the total number of valence electrons is 2(5) = 10. The single bond in the skeleton consumes two electrons, so there are eight left. Distributing these as unshared pairs gives the structure

$$:\ddot{N}-\ddot{N}:$$

Note that both N atoms are two electrons shy of an octet. In other words, there is a deficiency of four electrons. Move two unshared pairs, one from each nitrogen, to give a triple bond:

$$:N\equiv N:$$

Experimentally, we find that the N≡N bond is very strong

Exercise

Draw the Lewis structure of HOCN. Answer: $H-\ddot{O}-C\equiv N:$

RESONANCE FORMS

In certain cases, the Lewis structure does not adequately describe the properties of the ion or molecule that it represents. Consider, for example, the SO_2 structure derived in Example 8.2. This structure implies that there are two different kinds of sulfur-to-oxygen bonds in SO_2. One of these appears to be a single bond, the other a double bond. Yet experiment tells us that there is only one kind of bond in the molecule. In particular, we find that the two sulfur-to-oxygen distances are identical, 0.143 nm.

One way to explain this situation is to assume that each of the bonds in SO_2 is intermediate between a single and double bond. The fact that the observed bond distance is half-way between those expected for a single and double bond lends support to this idea. To express this concept within the framework of Lewis structures, we write two structures

It's the average of the two

with the understanding that the true structure is intermediate between them. These are referred to as *resonance forms*. The concept of resonance is invoked whenever a single Lewis structure does not adequately describe the properties of a substance.

Another species for which it is necessary to invoke the idea of resonance is the nitrate ion. Here three equivalent structures can be written:

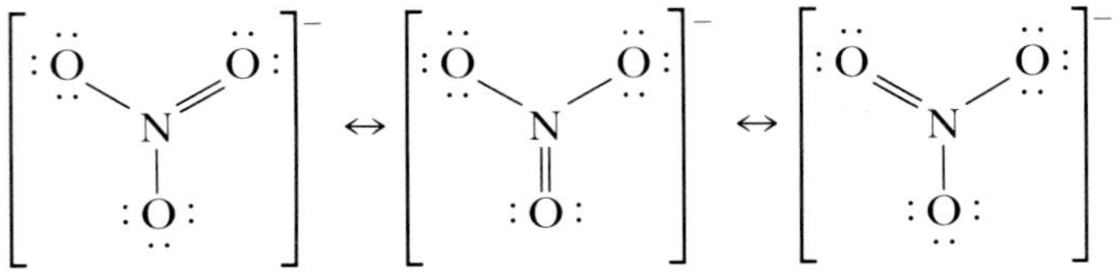

to explain the experimental observation that the three nitrogen to oxygen bonds in the NO_3^- ion are identical in all respects.

We will encounter other examples of molecules and ions whose properties can be interpreted in terms of resonance. In this connection it may be well to point out the following:

1. Resonance forms do not imply different kinds of molecules with electrons shifting eternally between them. There is only one type of SO_2 molecule; its structure is intermediate between those of the two resonance forms drawn for sulfur dioxide.

2. Resonance can be anticipated when it is possible to write two or more Lewis structures that are about equally plausible. In the case of the nitrate ion, the three structures we have written are equivalent. One could, in principle, write many other structures, but none of them would put eight electrons around each atom.

3. Resonance forms differ only in the distribution of electrons, not in the arrangement of atoms. The structure

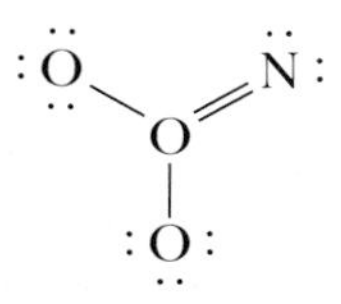

$$:\ddot{O}\diagdown_{O}\diagup N: \\ \quad\ \ | \\ \quad :\ddot{O}:$$

For all resonance forms, the atoms stay fixed, only the electrons move around

could not be a resonance form of the nitrate ion, since the atoms are arranged in a quite different way.

■ **EXAMPLE 8.3** _____

Write three resonance forms for SO_3.

Solution

SO_3 has the same number of atoms (4) and valence electrons (24) as NO_3^-. We would expect the Lewis structures to be similar. The resonance forms are

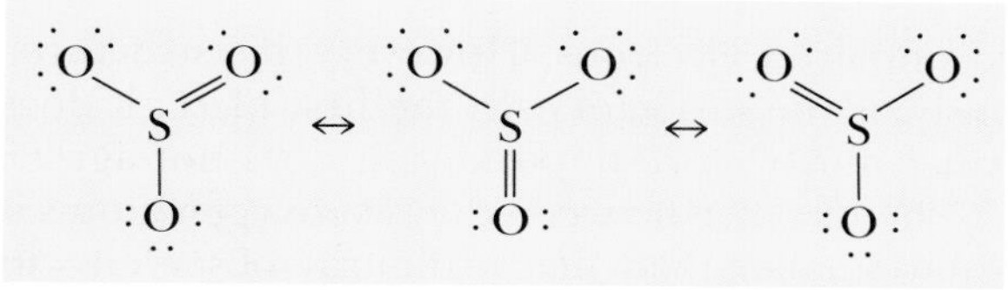

$$:\ddot{O}\diagdown_{S}\diagup\ddot{O}: \quad \leftrightarrow \quad :\ddot{O}\diagdown_{S}\diagup\ddot{O}: \quad \leftrightarrow \quad :\ddot{O}\diagdown_{S}\diagup\ddot{O}: \\ \quad\ \ | \qquad\qquad\qquad | \qquad\qquad\qquad | \\ \quad :\ddot{O}: \qquad\qquad\ :\ddot{O}: \qquad\qquad\ :\ddot{O}:$$

Exercise

One Lewis structure for CO_2 is $:\ddot{O}=C=\ddot{O}:$ Suggest other Lewis structures for CO_2 that would obey the octet rule. Answer: $:\ddot{O}-C\equiv O:$ and $:O\equiv C-\ddot{O}:$ These are ordinarily considered to be resonance forms of CO_2.

_____

EXCEPTIONS TO THE OCTET RULE

Although most of the molecules and polyatomic ions that we talk about in general chemistry follow the octet rule, there are some familiar species that do not. Among these are molecules containing an odd number of valence electrons. Nitric oxide, NO, and nitrogen dioxide, NO_2, fall in this category:

     NO   no. valence electrons = 5 + 6 = 11

     NO_2 no. valence electrons = 5 + 6(2) = 17

For such *odd electron* species (sometimes called free radicals) it is impossible to write Lewis structures in which each atom obeys the octet rule. The NO molecule is considered to be a resonance hybrid with the two contributing structures

There is no way all the electrons can be paired if there is an odd number of them

$$\cdot\ddot{N}=\ddot{O}: \quad \leftrightarrow \quad :\ddot{N}=\ddot{O}\cdot$$

Several different resonance structures can be written for NO_2, of which the more plausible are of the type

Species such as NO and NO_2, in which there are unpaired electrons, are **paramagnetic**; they show a weak attraction toward a magnetic field. Elementary oxygen is also paramagnetic, which suggests that the conventional Lewis structure is incorrect (see Fig. 8.3).

$$:\overset{..}{O}\!=\!\overset{..}{O}:$$

since it requires that all the electrons be paired. The paramagnetism of oxygen could be explained by the structure

$$:\overset{..}{O}\!-\!\overset{..}{O}:$$

in which there are two unpaired electrons. However, this structure, like the one written previously, is unsatisfactory. In the first place, it does not conform to the octet rule; much more important, it does not agree with experimental evidence. The distance between the two oxygen atoms in O_2 (0.121 nm) is considerably smaller than that ordinarily observed with an O—O single bond (0.148 nm). In summary, it appears that the O_2 molecule contains a double bond *and* two unpaired electrons.

These properties of oxygen are difficult to explain in terms of simple Lewis structures. As we shall see in Chapter 9, the molecular orbital approach leads to a more satisfactory picture of the electron distribution in the O_2 molecule.

We never said that the octet rule was perfect

Figure 8.3 Oxygen is attracted into a magnetic field and can actually be suspended between the poles of an electromagnet. Both the paramagnetism and the blue color are due to the unpaired electrons in the O_2 molecule. (S. Ruren Smith)

paramagnetism unpaired electron

There are other species for which Lewis structures written to conform to the octet rule are unsatisfactory. Examples include the fluorides of beryllium and boron, which exist in the vapor as molecules of BeF_2 and BF_3, respectively. Although one could write multiple bonded structures for these molecules in accordance with the octet rule, experimental evidence suggests the structures

$$: \ddot{F} - Be - \ddot{F}: \quad \text{and} \quad \begin{array}{c} : \ddot{F} \qquad \ddot{F}: \\ \diagdown \qquad \diagup \\ B \\ | \\ : \ddot{F}: \end{array}$$

in which the central atom is surrounded by four and six valence electrons, respectively, rather than eight.

EXPANDED OCTETS

The largest class of molecules to "violate" the octet rule consists of species in which the central atom is surrounded by more than four pairs of valence electrons. Typical molecules of this type are phosphorus pentachloride, PCl_5, and sulfur hexafluoride, SF_6. The Lewis structures of these molecules are:

$$\begin{array}{c} : \ddot{C}l: \\ | \\ : \ddot{C}l \diagdown \, | \diagup \ddot{C}l: \\ P \\ : \ddot{C}l \diagup \, \diagdown \ddot{C}l: \end{array} \qquad \begin{array}{c} : \ddot{F}: \\ : \ddot{F} \diagdown \, | \diagup \ddot{F}: \\ S \\ : \ddot{F} \diagup \, | \diagdown \ddot{F}: \\ : \ddot{F}: \end{array}$$

As you can see, the central atoms in these molecules have "expanded octets." In PCl_5, the phosphorus atom is surrounded by 10 valence electrons (5 shared pairs); in SF_6, there are 12 valence electrons (6 shared pairs) around the sulfur atom.

In molecules of this type, the terminal atoms are most often halogens (F, Cl, Br, I); in a few molecules, one or more oxygen atoms are at the end of the molecule. The central atom is a nonmetal in the 3rd, 4th, or 5th period of the Periodic Table. Most frequently, it is one of the following elements:

There aren't a great many molecules with expanded octets, but they do exist

	GROUP 5	GROUP 6	GROUP 7	GROUP 8
3rd period	P	S	Cl	
4th period	As	Se	Br	Kr
5th period	Sb	Te	I	Xe

Notice that all of these atoms have d orbitals available for bonding (3d, 4d, 5d). These are the orbitals in which the "extra" pairs of electrons are located in such species as PCl_5 and SF_6.

Sometimes, as with PCl_5 and SF_6, it is obvious from the formula that we are dealing with a molecule in which the central atom has an expanded octet. Often, however, it is by no means obvious that this is the case; at first glance formulas such as ClF_3 or XeF_4 look completely straightforward. It turns out, though, that when you attempt to draw Lewis structures for these molecules, it quickly becomes apparent that something unusual is involved. You find that you have a surplus of electrons. That is, you have more valence electrons than are needed to give each atom a noble-gas structure. When this happens, the *extra electrons should be distributed around the central atom.*

■ **EXAMPLE 8.4** _____

Draw the Lewis structure of XeF_4.

Solution

We follow the usual sequence. Since Xe is in Group 8 and fluorine is in Group 7:

$$\text{total valence } e^- = 8 + 4(7) = 36$$

The skeleton is:

$$\begin{array}{c} F \\ | \\ F-Xe-F \\ | \\ F \end{array}$$

With four bonds in the skeleton, we are left with $36 - 8 = 28$ valence electrons. To give each fluorine a noble-gas structure, we put 6 valence electrons around it:

$$\begin{array}{c} :\!\ddot{F}\!: \\ | \\ :\!\ddot{F}-Xe-\ddot{F}\!: \\ | \\ :\!\ddot{F}\!: \end{array}$$

We still have $28 - 24 = 4$ valence electrons left, even though each atom has a noble-gas structure. We distribute these as unshared pairs around the xenon atom, giving:

$$\begin{array}{c} :\!\ddot{F}\!: \\ | \\ :\!\ddot{F}-\ddot{Xe}-\ddot{F}\!: \\ | \\ :\!\ddot{F}\!: \end{array}$$

We conclude that in this molecule there are *six* pairs of electrons around the xenon atom.

Exercise

How many pairs of valence electrons are there around the central Cl atom in the Lewis structure for ClF_3? Answer: 5.

8.3
COVALENT BOND PROPERTIES

Three important properties of a covalent bond are its polarity, length (distance), and strength (bond energy). The polarity describes how the bonding electrons are distributed between the two bonded atoms. The bond length is the distance between the centers of the two bonded atoms. The strength of the bond is reflected by its bond energy, which tells us how much energy is required to break the bond.

POLAR AND NONPOLAR COVALENT BONDS

As we might expect, the two electrons in the H_2 molecule are shared equally by the two nuclei. Stated another way, a bonding electron is as likely to be found in the vicinity of one nucleus as another. Bonds of this type are described as **nonpolar**. We find nonpolar bonds whenever the two atoms joined are identical, as in H_2 or F_2.

In the HF molecule, the distribution of the bonding electrons is somewhat different from that found in H_2 or F_2. Here, the density of the electron cloud is greatest about the fluorine atom. The bonding electrons, on the average, are shifted toward fluorine and away from the hydrogen (atom Y in Figure 8.4). Bonds in which the electron density is unsymmetrical are referred to as **polar bonds**.

Atoms of two different elements always differ at least slightly in their affinity for electrons. Hence, covalent bonds between unlike atoms are always polar. Consider, for example, the H—F bond. Since fluorine has a stronger attraction for electrons than does hydrogen, bonding electrons are displaced toward the fluorine atom. The H—F bond is polar, with a partial negative charge at the fluorine atom and a partial positive charge at the hydrogen atom.

The ability of an atom to attract to itself the electrons in a covalent bond is described in terms of **electronegativity**. The greater the electronegativity of an atom, the greater its affinity for bonding electrons. Electronegativities can be estimated in various ways. One method, based on bond energies, leads to Table 8.2, p. 264. Here each element is assigned a number ranging from 4.0 for the most electronegative element, fluorine, to 0.7 for cesium, the least electronegative. Among the main-group elements, electronegativity increases moving from left to right in the Periodic Table. Ordinarily, it decreases as we move down a given group.

The extent of polarity of a covalent bond is related to the difference in electronegativities (EN) of the bonded atoms. If this difference is large, as in HF (EN H = 2.1, F = 4.0), the bond will be strongly polar. Where the difference is small, as in H—C (EN H = 2.1, C = 2.5), the bond will be only slightly polar. Thus, in the carbon-hydrogen bond, the bonding electrons are only slightly displaced toward the carbon atom.

In a pure (nonpolar) covalent bond, the electrons are equally shared. In a pure ionic bond, there has been a complete transfer of electrons from one atom to another. We can think of a polar covalent bond as being

Figure 8.4 If two atoms, X and Y, differ in electronegativity, the bond between them is polar. The electron cloud associated with the bonding electrons is concentrated around the more electronegative atom, in this case, X.

Electronegativity is always a positive number

TABLE 8.2   Electronegativity Values

H 2.1						
Li 1.0	Be 1.5	B 2.0	C 2.5	N 3.0	O 3.5	F 4.0
Na 0.9	Mg 1.2	Al 1.5	Si 1.8	P 2.1	S 2.5	Cl 3.0
K 0.8	Ca 1.0	Sc 1.3	Ge 1.8	As 2.0	Se 2.4	Br 2.8
Rb 0.8	Sr 1.0	Y 1.2	Sn 1.8	Sb 1.9	Te 2.1	I 2.5
Cs 0.7	Ba 0.9	La–Lu 1.0–1.2	Pb 1.9	Bi 1.9	Po 2.0	At 2.2

intermediate between these two extremes. In this sense, we can relate bond polarity to *partial ionic character*. The greater the difference in electronegativity between two elements, the more ionic will be the bond between them. The relation between these two variables is shown in Figure 8.5. A difference of 1.7 units corresponds to a bond with about 50% ionic character. Such a bond might be described as being halfway between a pure covalent (nonpolar) and a pure ionic bond.

It is clearly an oversimplification to refer to a bond between two elements as being "ionic" or "covalent." Consider, for example, the bonding in compounds formed by a Group 1 or 2 metal with a nonmetal in Group

However, we usually refer to them that way

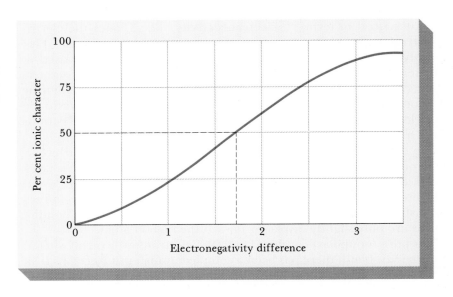

Figure 8.5 Relation between ionic character of a bond and the difference in electronegativity of the bonded atoms.

6 or 7. The difference in electronegativity ranges from a minimum of 0.6 for Be—Te to a maximum of 3.3 for Cs—F. The percentage of ionic character varies similarly, from about 10% for BeTe to 95% for CsF.

■ **EXAMPLE 8.5** —————————————————————————————
Using only the Periodic Table, arrange the following bonds in order of increasing polarity (ionic character): F—Cl, F—F, F—Na.

Solution
The F—F bond must come first; since the atoms are identical, the bond is nonpolar. To decide which bond comes next, recall that electronegativity increases as we move across in the Table and decreases as we move down. Chlorine lies below fluorine so fluorine should be somewhat more electronegative. On the other hand, sodium lies far to the left of fluorine and is below it. Hence, sodium should have a much lower electronegativity than fluorine. The proper sequence is

$$F—F < F—Cl < F—Na$$

We progress from a nonpolar to a slightly polar to a strongly polar, essentially ionic, bond.

Exercise
Which of the three bonds, B—C, C—N, or B—Si would you expect to be least polar? Answer: B—Si.

BOND DISTANCES

It is possible to estimate the distance between atoms joined by a covalent bond by adding their atomic radii (Appendix 2). For a nonpolar bond, such as that in F_2, we calculate a value just twice the atomic radius:

F—F distance = 2(radius F) = 0.064 nm + 0.064 nm = 0.128 nm

This is exactly the observed internuclear distance, since the atomic radius of a nonmetal, often referred to as the covalent radius, is taken to be one-half the distance between centers of bonded atoms in the element. In effect, we assume that the atoms in a molecule like F_2 or Cl_2 are touching.

When two unlike atoms are joined by a covalent bond, the observed bond distance is usually smaller than the sum of the atomic radii. Thus, for HF, we would calculate the following, using atomic radii:

H—F distance = radius H + radius F = 0.037 nm + 0.064 nm
= 0.101 nm

The observed distance is only 0.092 nm. There is a simple way to explain the bond shortening in HF and other polar molecules. It appears that the introduction of partial ionic character into a covalent bond strengthens it and, hence, tends to pull the bonded atoms closer together.

The more electrons, the shorter the bond

When we compare bond distance in a multiple bond to that of a single bond between the same two atoms, we find that the multiple bond distance is smaller. Compare, for example, the three types of carbon-carbon bonds:

C—C (in C_2H_6)   bond distance = 0.154 nm

C=C (in C_2H_4)   bond distance = 0.133 nm

C≡C (in C_2H_2)   bond distance = 0.120 nm

BOND ENERGIES

The strength of the bond between two atoms can be described using a quantity called bond energy. (More properly, but less commonly, it is called bond enthalpy.) **The bond energy is defined as ΔH when one mole of bonds is broken in the gaseous state.** From the equations (see also Fig. 8.6).

$$H_2(g) \rightarrow 2\,H(g); \quad \Delta H = +436 \text{ kJ} \tag{8.1}$$

$$Cl_2(g) \rightarrow 2\,Cl(g); \quad \Delta H = +243 \text{ kJ} \tag{8.2}$$

we conclude that the H—H bond energy is +436 kJ/mol while that for Cl—Cl is +243 kJ/mol. In Reaction 8.1, 1 mol of H—H bonds in H_2 is broken; in Reaction 8.2, 1 mol of Cl—Cl bonds is broken.

Bond energies for a variety of single bonds are listed in Table 8.3. You will note that bond energy is always a positive quantity. Energy is always absorbed when chemical bonds are broken. Conversely, heat is given off when bonds are formed from gaseous atoms. Thus, we have

$$H(g) + Cl(g) \rightarrow HCl(g); \quad \Delta H = -(\text{B.E. H—Cl}) = -431 \text{ kJ}$$

$$H(g) + F(g) \rightarrow HF(g); \quad \Delta H = -(\text{B.E. H—F}) = -565 \text{ kJ}$$

Here the abbreviation B.E. is used to stand for bond energy (kJ/mol of bonds).

The single-bond energies listed in Table 8.3 cover a wide range, from 138 kJ/mol for the O—O bond to 565 kJ/mol for the H—F bond. One

Figure 8.6 Bond energies in H_2 and Cl_2. The H—H bond energy is greater than the Cl—Cl bond energy (+436 kJ/mol vs. +243 kJ/mol). This means that the bond in H_2 is stronger than that in Cl_2.

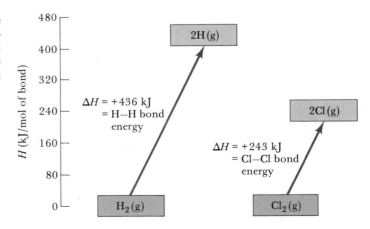

TABLE 8.3 Single Bond Energies (kJ/mol) at 25°C

	H	C	N	O	S	F	Cl	Br	I
H	436	414	389	464	339	565	431	368	297
C		347	293	351	259	485	331	276	218
N			159	222	—	272	201	243	—
O				138	—	184	205	201	201
S					226	285	255	213	—
F						153	255	255	277
Cl							243	218	209
Br								193	180
I									151

factor that contributes to bond strength is polarity. In general, a polar bond is stronger than might be expected if the bonding electrons were equally shared. To illustrate this effect, consider the H—F bond. If the bonding electrons were equally shared, the bond energy should be an average of those for the nonpolar bonds H—H and F—F:

$$\frac{436 \text{ kJ/mol} + 153 \text{ kJ/mol}}{2} = 295 \text{ kJ/mol}$$

The actual bond energy in H—F is nearly twice as great, 565 kJ/mol. We might say that the "extra" bond energy due to polarity is

$$565 \text{ kJ/mol} - 295 \text{ kJ/mol} = 270 \text{ kJ/mol}$$

The extent of this effect depends upon the difference in electronegativity between the bonded atoms. Consider the H—I bond, where the difference in electronegativity is only 0.4 (EN I = 2.5, H = 2.1) as compared to 1.9 for H—F (EN F = 4.0, H = 2.1). The H—I bond energy from Table 8.3 is 297 kJ/mol. The average of the H—H and I—I bond energies is

$$\frac{436 \text{ kJ/mol} + 151 \text{ kJ/mol}}{2} = 293 \text{ kJ/mol}$$

In this case, the "extra" bond energy is only 4 kJ/mol:

$$297 \text{ kJ/mol} - 293 \text{ kJ/mol} = 4 \text{ kJ/mol}$$

In general, the greater the difference in electronegativity between bonded atoms, the more strongly polar the bond, and the greater the "extra" bond energy.

The electronegativity scale was set up with this in mind

Another factor that affects bond energy is the number of electron pairs between the bonded atoms. The bond energy is larger for a multiple bond than for a single bond between the same two atoms. In other words, the multiple bond is stronger. This effect is shown in Table 8.4, p. 268, where the bond energies of single, double, and triple bonds are compared. It appears that the extra electron pairs strengthen the bond, making it more difficult to separate the bonded atoms from each other.

Notice that the factors that increase bond energy are those that shorten bond distance as well. Polar bonds tend to be both stronger and shorter

TABLE 8.4   Comparison of Bond Energies (kJ/mol)

SINGLE BOND	BOND ENERGY	DOUBLE BOND	BOND ENERGY	TRIPLE BOND	BOND ENERGY
C—C	347	C=C	612	C≡C	820
C—N	293	C=N	615	C≡N	890
C—O	351	C=O	715	C≡O	1075
C—S	259	C=S	477	—	—
N—N	159	N=N	418	N≡N	941
N—O	222	N=O	607	—	—
O—O	138	O=O	498	—	—
S—O	347	S=O	498	—	—

than pure covalent bonds between like atoms. Again, multiple bonding, involving C, N, O, or S atoms, brings the atoms closer to one another and increases the forces between them.

ESTIMATION OF ΔH FROM BOND ENERGIES

In Chapter 7, we showed how heats of formation can be used to calculate ΔH for a chemical reaction. In principle, a table of bond energies can be used for this same purpose, at least for reactions involving only gaseous, molecular species. To show how this is done, consider the reaction:

$$H_2(g) + Cl_2(g) \rightarrow 2\ HCl(g) \tag{8.3}$$

We can consider (Fig. 8.7) that this reaction occurs in two steps.

1. The bonds in the H_2 and Cl_2 molecules are broken, giving gaseous atoms:

The bond energy is the amount of energy required to break a mole of bonds

$$\left. \begin{array}{l} H_2(g) \rightarrow 2\ H(g) \\ Cl_2(g) \rightarrow 2\ Cl(g) \end{array} \right\} \begin{array}{l} \Delta H_1 = \text{B.E. H—H} + \text{B.E. Cl—Cl} \\ = 436\ kJ \quad + 243\ kJ = \quad +679\ kJ \end{array}$$

This step is endothermic; energy must be absorbed to break the bonds in H_2 and Cl_2.

2. The gaseous atoms formed in step 1 combine with one another, forming new bonds in HCl:

$$2\ H(g) + 2\ Cl(g) \rightarrow 2\ HCl(g);$$

$$\Delta H_2 = -2(\text{B.E. H—Cl}) = -862\ kJ$$

Energy is evolved in this step because two moles of H—Cl bonds are *formed*. By Hess's Law, the overall energy change is the sum of ΔH_1 and ΔH_2:

$$\Delta H = (\text{B.E. H—H} + \text{B.E. Cl—Cl}) - 2\ (\text{B.E. H—Cl})$$
$$= +679\ kJ - 862\ kJ = -183\ kJ$$

The argument we have just gone through can be applied to any reaction. We can estimate ΔH for the reaction by finding the difference between the energy that must be absorbed to break the old bonds in the

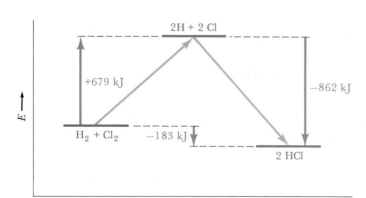

Figure 8.7 For the reaction: $H_2(g) + Cl_2(g) \rightarrow 2HCl(g)$; $\Delta H = -183$ kJ. This value for ΔH is the difference between the heat absorbed when the bonds in H_2 and Cl_2 are broken ($+679$ kJ), and the heat evolved when two moles of bonds are formed in HCl (-862 kJ).

reactants and the energy that is evolved when new bonds are formed in the products:

$$\Delta H = \Sigma \text{ B.E. bonds broken} - \Sigma \text{ B.E. bonds formed} \qquad (8.4)$$

Qualitatively, Equation 8.4 tells us that a reaction is *endothermic* ($\Delta H > 0$) if the bonds in the reactant molecules are stronger than those in the products. Conversely, a reaction is *exothermic* ($\Delta H < 0$) if relatively weak bonds in the reactants give way to stronger bonds in the products.

> This is the physical explanation for the energy effects in chemical reactions

■ **EXAMPLE 8.6**
Using bond energies, estimate ΔH for the reaction of ethylene, C_2H_4, with chlorine:

$$\underset{\text{H}}{\overset{\text{H}}{}}\quad$$

H—C=C—H(g) + Cl—Cl(g) → H—C—C—H(g)
with H H above each carbon and Cl Cl below each carbon

Solution
We start by taking an inventory of bonds:

    reactants: 4 C—H bonds, 1 C=C bond, 1 Cl—Cl bond
    products: 4 C—H bonds, 1 C—C bond, 2 C—Cl bonds

Note that the four C—H bonds are present in both reactants and products; they are neither broken nor formed.

The bonds broken are those in the reactants, one C=C and one Cl—Cl bond. So the first term in Equation 8.4 becomes:

    B.E. C=C + B.E. Cl—Cl = 612 kJ + 243 kJ = 855 kJ

The bonds formed are those in the products, one C—C and two C—Cl bonds. So the second term in Equation 8.4 is:

    B.E. C—C + 2 (B.E. C—Cl) = 347 kJ + 2(331 kJ) = 1009 kJ

Thus we have: $\Delta H = 855$ kJ $-$ 1009 kJ $= \boxed{-154 \text{ kJ}}$

Exercise
Calculate ΔH for the reaction of ethylene with fluorine to give $C_2H_4F_2$.
Answer: -552 kJ.

We should point out one difficulty with calculating ΔH for reactions from bond energies. In most cases it is not possible to assign an exact value to a bond energy. The bond energy varies to some extent with the species in which the bond is found. Consider, for example, the enthalpy changes for the two reactions

$$H-O-H(g) \rightarrow H(g) + O-H(g); \qquad \Delta H = +499 \text{ kJ}$$

$$O-H(g) \rightarrow H(g) + O(g); \qquad \Delta H = +428 \text{ kJ}$$

The ΔH values are quite different, even though both involve breaking an O—H bond. The bond energy quoted in Table 8.3, $+464$ kJ/mol, is an average, calculated from ΔH values for many different reactions in which O—H bonds are broken.

A similar situation applies with other bond energies; the amount of energy required to break a given type of bond varies from one molecule to another. For that reason, the value of ΔH calculated from tables of bond energies is likely to be in error, often by 10 kJ or more. That is why, in Example 8.6, we asked you to "estimate" the value of ΔH. The accurate value for ΔH of this reaction, calculated from heats of formation as in Chapter 7, is -182 kJ/mol.

The discrepancy between the experimental value of ΔH and that calculated from bond energies is particularly large when molecules that show resonance are involved. To illustrate this effect, consider the reaction:

$$CO(g) + \tfrac{1}{2} O_2(g) \rightarrow CO_2(g) \qquad\qquad (8.5)$$

The true value of ΔH is found from heat of formation data in Chapter 7:

$$\Delta H = \Delta H_f^0 \, CO_2(g) - \Delta H_f^0 \, CO(g) = -283 \text{ kJ}$$

To estimate ΔH from bond energies, we apply Equation 8.4, realizing that there is a $C\equiv O$ bond in CO, an $O=O$ bond in O_2, and 2 $C=O$ bonds in a CO_2 molecule:

$$\Delta H_{est} = \text{B.E. } C\equiv O + \tfrac{1}{2}(\text{B.E. } O=O) - 2\,(\text{B.E. } C=O) = -106 \text{ kJ}$$

We see that the true value of ΔH is some 177 kJ more negative than that calculated from bond energies. This large difference is related to the fact that CO_2 is a resonance hybrid:

$$:O\equiv C-\ddot{O}: \;\leftrightarrow\; :\ddot{O}=C=\ddot{O}: \;\leftrightarrow\; :\ddot{O}-C\equiv O:$$

The "extra" 177 kJ of energy evolved in Reaction 8.5 is often referred to as the *resonance energy* of carbon dioxide. In general, molecules that show resonance are unusually stable; their energies are considerably lower than would be predicted on the basis of bond energies.

The values of ΔH as found from bond energies must be taken with a grain of salt

Sometimes several grains

■ SUMMARY PROBLEM

There are several oxides of nitrogen; among the more common are N_2O, NO, NO_2, N_2O_4.

a. Write the Lewis structures of these molecules (in each case, the oxygens are terminal atoms).

b. Which of these molecules "violate" the octet rule?

c. Draw resonance structures for N_2O.

d. Describe each of the bonds in N_2O_4 as being polar or nonpolar. In the polar bonds, which atom acts as the positive pole?

e. How would you expect the nitrogen-to-oxygen bond distance in NO to compare to the average bond distance in NO_2?

f. Estimate ΔH for the following reactions, using bond energies (take the bond in O_2 to be a double bond):

$$N_2(g) + O_2(g) \rightarrow 2\ NO(g)$$

$$2\ NO_2(g) \rightarrow N_2O_4(g)$$

Answers

a. 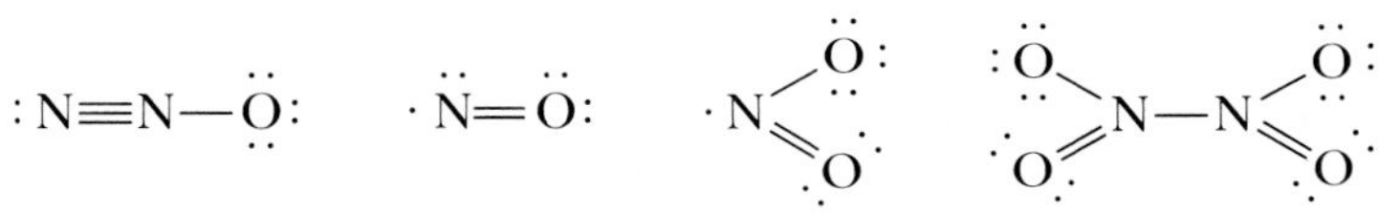 $:N\equiv N-\ddot{O}:$ $\cdot\ddot{N}=\ddot{O}:$

b. NO, NO_2 **c.** $:N\equiv N-\ddot{O}: \leftrightarrow :\ddot{N}=N=\ddot{O}: \leftrightarrow :\ddot{N}-N\equiv O:$

d. one nonpolar bond; four polar N to O bonds; N is + pole

e. shorter

f. $+225$ kJ, -159 kJ (the true values are $+181$ kJ and -58 kJ)

HISTORICAL PERSPECTIVE

GILBERT NEWTON LEWIS (1875–1946)

Figure 8.8 G. N. Lewis in a typical setting. When he traveled, Lewis took a supply of cigars with him, even if it meant leaving clothing, chemistry journals, or other items behind. (Dr. Glenn T. Seaburg, University of California, Lawrence Berkeley Laboratory)

Lewis certainly deserved a Nobel prize, but *he* never received one

The Lewis structures discussed in this chapter were the product of the American physical chemist, G. N. Lewis. Born in Massachusetts, Lewis grew up in Nebraska, then came back East to obtain his B.S. (1896) and Ph.D. (1899) at Harvard. Although he stayed on for a few years as an instructor, Lewis seems never to have been happy at Harvard. A precocious student and an intellectual rebel, he was repelled by the highly traditional atmosphere that prevailed in the chemistry department there in his time. Many years later, he refused an honorary degree from his alma mater.

After leaving Harvard, Lewis made his reputation at M.I.T., where he was promoted to full professor in only four years. In 1912, he moved across the country to the University of California at Berkeley as Dean of the College of Chemistry and department chairman. He remained there for the rest of his life. Under his guidance, the chemistry department at Berkeley became perhaps the most prestigious in the country. Among the faculty and graduate students that he attracted were five future Nobel Prize winners: Harold Urey in 1934, William Giauque in 1949, Glenn Seaborg in 1951, Willard Libby in 1960, and Melvin Calvin in 1961.

In administering the chemistry department at Berkeley, Lewis demanded excellence in both research and teaching. Virtually the entire staff was involved in the general chemistry program; at one time eight full professors carried freshman sections. Several department members became leaders of chemical education in America. Among them was Joel Hildebrand (1881–1983), who came to California in 1913 and was still active in teaching and research 60 years later.

Lewis's approach to training graduate students was unconventional but effective. He allowed them to circulate freely through the storeroom, picking up any chemicals or equipment needed for their research. Moreover, he encouraged graduate students to participate actively in the research seminars he conducted with the faculty at Berkeley. The story is told that at one such meeting a student interrupted a discussion by Lewis, challenging the validity of his ideas. After a moment's silence, Lewis turned to the student and said, "A very impertinent remark, young man, but very pertinent."

Lewis's interest in chemical bonding and structure dates from 1902. In attempting to explain "valence" to a class at Harvard, he devised an atomic model to rationalize the octet rule. His model was deficient in many respects; for one thing, Lewis visualized cubic atoms with electrons located at the corners. Perhaps this explains why his ideas of atomic structure were not published until 1916. In that year, Lewis conceived of the electron-pair bond, perhaps his greatest single contribution to chemistry. At that time, it was widely believed that all bonds were ionic; Lewis's ideas were rejected by many well-known organic chemists.

In 1923, Lewis published a classic book (recently reprinted by Dover Publications) entitled *Valence and the Structure of Atoms and Molecules*. Here, in Lewis's characteristically lucid style, we find many of the basic principles of covalent bonding discussed in this chapter. Included are electron-dot structures, the octet rule, and the concept of electronegativity. Here too is the Lewis definition of acids and bases (Chap. 14). That same year, Lewis published with Merle Randall a text called *Thermodynamics and the Free Energy of Chemical Substances*. Sixty years later, a revised edition of that text is still widely used in graduate courses in chemistry.

The years from 1923 to 1938 were relatively unproductive for G. N. Lewis insofar as his own research was concerned. The applications of the electron-pair bond came largely in the areas of organic and quantum chemistry; in neither of these fields did Lewis feel at home. In the early 1930s, he published a series of relatively minor papers dealing with the properties of deuterium. Then, in 1939, he began to publish in the field of photochemistry. Of approximately 20 papers in this area, several were of fundamental importance, comparable in quality to the best work of his early years. Retired officially in 1945, Lewis died a year later while carrying out an experiment on fluorescence.

QUESTIONS AND PROBLEMS

Lewis Structures

1. Write the Lewis structure for the following molecules and polyatomic ions. In each case, the first atom listed is the central atom.
   a. $POCl_3$ b. NF_3 c. ClO_4^- d. $GeCl_4$

2. Follow the directions of Question 1 for
   a. PH_3 b. SiF_4 c. NHO_2 d. NH_4^+

3. Follow the direction of Question 1 for
   a. PCl_4^+ b. N_3^- c. CN^- d. ClO_2^-

4. Follow the directions of Question 1 for
   a. IO_3^- b. ClF_2^+ c. SO_3^{2-} d. AsO_3^{3-}

5. Follow the directions of Question 1 for
   a. SeF_6 b. PCl_5 c. $TeBr_4$ d. KrF_2

6. Follow the directions of Question 1 for
   a. BrF_3 b. XeO_2F_2 c. I_3^- d. IF_5

7. Draw Lewis structures for the following species (the skeleton is indicated by the way the molecule is written):
   a. $H_2N—OH$ b. $H_2C—CO$ c. $H_2C—N—N$

8. Follow the directions of Question 7 for the following species:
   a. $(HO—CO_2)^-$ b. $H_3C—CO_2H$ c. $(HO)_2—S—O$

9. Radioastronomers have detected the isoformyl ion, HOC^+, in outer space. Write the Lewis structure for this ion.

10. Formation of dioxirane, H_2CO_2, has been suggested as a factor in smog formation. The molecule contains an oxygen-oxygen bond. Draw its Lewis structure.

11. Formic acid is the irritating substance that gets on your skin when an ant bites. Its formula is HCOOH. Carbon is the central atom and it has no O—O bonds. Write its Lewis structure.

12. Peroxyacetylnitrate (PAN) is the substance in smog that makes your eyes water. Its skeletal structure is

   $H_3C—CO_3—NO_2$

It has one O—O bond and three O—N bonds. Draw its Lewis structure.

13. Two different molecules have the formula $C_2H_4Br_2$. Draw a Lewis structure for each molecule. (All the H and Br atoms are bonded to carbon.)

14. There are two compounds with the molecular formula C_2H_6O. Draw a Lewis structure for each compound.

15. Give the formula of a polyatomic ion which you would expect to have the same Lewis structure as
   a. HCl   b. F_2 c. N_2 d. CCl_4

16. Give the formula of a molecule which you would expect to have the same Lewis structure as
   a. ClO^- b. $H_2PO_4^-$ c. PH_4^+ d. SiO_4^{4-}

17. Write a Lewis structure for
   a. HSO_3^- b. SCN^- c. $NFCl_2$
   d. $P_2O_7^{4-}$ (no O—O bonds)

18. Write Lewis structures for the following species:
   a. BCl_4^- b. $S_2O_3^{2-}$ c. $H_2PO_4^-$ d. ClO_3^-

19. Write reasonable Lewis structures for the following species, none of which follow the octet rule.
   a. CH_3 b. CO^- c. NO   d. BCl_3

20. Write reasonable Lewis structures for the following species, none of which obey the octet rule.
   a. NO_2 b. BeH_2 c. SO_2^- d. CO^+

Resonance Forms

21. Draw resonance structures for
a. CO_3^{2-} b. SeO_3 c. CS_3^{2-}

22. Draw resonance structures for
a. SCN^- b. HCO_2^- c. $HONO_2$

23. The oxalate ion, $C_2O_4^{2-}$, has the skeleton structure

a. Complete the Lewis structure of this ion.
b. Draw three resonance forms for $C_2O_4^{2-}$, equivalent to the Lewis structure drawn in (a).
c. Is

another resonance form of the oxalate ion? Explain.

24. The Lewis structure for hydrazoic acid may be written as

a. Draw two other resonance forms for this molecule.
b. Is

another resonance form of hydrazoic acid? Explain.

25. The skeletal structure for disulfur dinitride, S_2N_2, is

Draw the resonance forms of this molecule.

26. Borazine, $B_3N_3H_6$, has a structure similar to benzene:

Draw the resonance forms of the molecule.

Bond Properties

27. Arrange the following bonds in order of increasing polarity using only the Periodic Table:

C—N   C—C   C—O   C—F

28. Using the electronegativity trends in the Periodic Table, list the following bonds in order of decreasing polarity.

N—S   N—F   N—Se   N—N

29. Which of the following bonds would you expect to be the most polar?

O—H   O—F   N—H   N—O

30. Which of the following bonds would you expect to be the least polar?

K—O   Ca—O   C—O   Cl—O

31. For each of the following pairs, determine which bond is more polar. Use Table 8.2.
a. N—P or N—I   b. N—H or P—H
c. N—H or N—F

32. Follow the directions for Question 31 for the following pairs:
a. N—H or N—Cl   b. N—S or P—S
c. N—O or P—O

33. On which atom is the partial negative charge located in the following polar bonds?
a. N—O   b. F—Br   c. H—O   d. N—C

34. On which atom is the partial positive charge located in the following bonds?
a. C—Cl   b. O—S   c. H—F   d. Cl—I

35. Compare the lengths of the C—N bonds in
a. H_3C—N—H   b. $(O{=}C{=}N)^-$ c. Cl—$C{\equiv}N$
      |
      H

In which species is the bond longest? Shortest?

36. Compare the lengths of the carbon–carbon bonds in
a. H_3C—CH_3 b. $H_2C{=}CH_2$ c. $HC{\equiv}CH$

37. Which of the bonds referred to in Question 35 is the weakest? Strongest?

38. Which of the bonds referred to in Question 36 is the weakest? Strongest?

ΔH and Bond Energies

39. Calculate the amount of heat that must be absorbed (kJ/mol) to dissociate into atoms:
a. Cl_2 b. C_2H_6 c. CCl_4

40. Calculate ΔH per mole for the formation of the following molecules from gaseous atoms.
a. N_2 b. CH_4 c. NH_3

41. Using bond energies, estimate ΔH for the reaction between hydrogen gas and bromine gas to form one mole of hydrogen bromide gas.

42. Using bond energies, estimate ΔH for the reaction between hydrogen and nitrogen to form one mole of hydrazine gas, N_2H_4.

43. Using Tables 8.3 and 8.4, estimate ΔH for the reaction between carbon monoxide and chlorine gas to form one mole of phosgene, Cl—C—Cl.

$$\underset{\text{O}}{\overset{\|}{}}$$

44. Isopropyl alcohol, sold as rubbing alcohol, can be prepared by the reaction between acetone, the primary ingredient in nail polish remover, and hydrogen:

$$\underset{\text{O}}{\overset{\|}{H_3C—C—CH_3}}(g) + H_2(g) \rightarrow \underset{\text{OH}}{\overset{\text{H}}{H_3C—C—CH_3}}(g)$$

Calculate ΔH for this reaction.

45. The first step in the manufacture of acrylonitrile, a chemical widely used in the plastics industry, is

$$\underset{\text{O}}{H_2C—CH_2}(g) + HCN(g) \rightarrow$$

$$HO—CH_2—CH_2—CN(g)$$

Calculate ΔH for this reaction.

46. The calculation of ΔH using bond energies works only for gaseous molecules. In the reaction for the formation of one mole of hydrogen cyanide, HCN, from its elements, nitrogen and hydrogen gases react with carbon in its solid state. Calculate ΔH for this reaction using bond energies and adding the reaction

$$C(\text{graphite}) \rightarrow C(g) \qquad \Delta H = 717 \text{ kJ}$$

into your calculations.

Unclassified

47. Complete the following statements:
a. Bond energy _____ as the number of bonds between two atoms increases.
b. Atoms in a double bond are _____ than those joined by a single bond.
c. Bonds between atoms in molecules consist of _____ _____.
d. Electronegativity _____ going down in the Periodic Table.

48. Explain why
a. molecules with an odd number of valence electrons do not follow the octet rule.
b. there is a strong attractive force between H atoms at short distances.

c. energy is evolved in the formation of HF from the elements.

49. Explain the meaning of the following terms:
a. valence electrons   b. unshared electron pairs
c. expanded octets   d. multiple bonds

50. Explain the following terms:
a. octet   b. resonance
c. covalent bond   d. odd-electron species

51. Write Lewis structures for
a. the sulfate ion   b. the phosphate ion
c. the chlorate ion

52. Consider the dichromate ion. It has no metal-metal nor oxygen-oxygen bonds. Write a Lewis structure for the dichromate ion. Consider chromium to have 6 valence electrons.

53. Solid sulfur in its native state is written S_8. Write a Lewis structure for S_8 given that S_8 is an eight-membered ring.

54. Use bond energies to calculate ΔH for the reaction between ammonia and oxygen to form steam and one mole of nitrogen. (Assume a double bond in O_2.) Compare with ΔH^0 obtained from heats of formation.

55. Using the values given in Appendix 2, estimate the C—H and C—Cl bond lengths in methylene chloride, CH_2Cl_2.

56. In which of the following molecules does the sulfur atom have an expanded octet?
a. SO_2 b. SF_4 c. SO_2Cl_2 d. SF_6

57. What experimental evidence leads us to believe that there is a double bond in the O_2 molecule? unpaired electrons? Can you write a reasonable Lewis structure for O_2?

Challenge Problems

58. Benzyne, C_6H_4, is a cyclic molecule. Write a plausible Lewis structure for it and indicate possible resonance structures.

59. Write as many Lewis structures as you can for N_2F_2, following the octet rule.

60. A certain element reacts with chlorine to form a compound which is a gas at 85°C and 1.00 atm with a density of 4.66 g/L.
a. What is the molar mass of the gas?
b. Identify the element involved.
c. Draw the Lewis structure of the molecule.

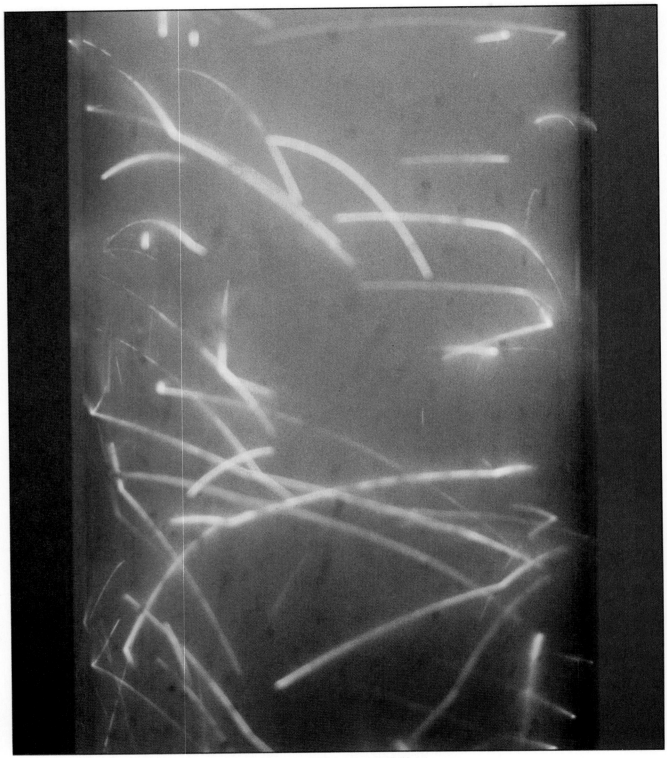

$$2Sb(s) \; + \; 3Cl_2(g) \; \rightarrow \; 2SbCl_3(s)$$

Antimony powder reacts with chlorine gas to give antimony(III) chloride

CHAPTER 9

MOLECULAR STRUCTURE

What immortal hand or eye
Could frame thy fearful symmetry?

WILLIAM BLAKE
THE TIGER

In Chapter 8, we considered how to draw Lewis structures to describe the arrangement of electrons in molecules. In this chapter, we will build upon our knowledge of Lewis structures to obtain answers to important questions related to molecular structure:

1. What is the "shape" of a molecule? That is, how are the atoms located in space with respect to one another (Section 9.1)? As we will see, starting with the Lewis structure of a molecule, we can use a simple model to predict

— the angles between bonds formed by a central atom.

— the orientation of other atoms around a central atom.

A knowledge of molecular geometry is also helpful in predicting whether a molecule will be polar or nonpolar (Section 9.2).

Molecular geometry can be determined from Lewis structures

2. How are valence electrons in a molecule distributed among electron orbitals? What are the shapes of these orbitals and in what order are they occupied? Here we will consider two different approaches:

— *the atomic orbital, or valence bond, model* (Section 9.3). In this approach, the valence electrons are distributed among orbitals characteristic of the unbonded atoms. These may be individual s and p orbitals of the type described in Chapter 5. More commonly, they are hybrid orbitals, formed by mixing s, p, and, in somes cases, d orbitals with one another.

— *the molecular orbital model* (Section 9.5). In this model, valence electrons are distributed among orbitals characteristic of the molecule as a whole. Molecular orbital theory has been applied with striking success to a wide variety of substances; however, our discussion will be limited to simple diatomic molecules such as N_2, O_2, and F_2.

9.1
MOLECULAR GEOMETRY

The geometry of a diatomic molecule such as Cl_2 or HCl can be described very simply. Since two points define a straight line, the molecule must be linear:

Cl—Cl   H—Cl

No question about it, Cl_2 and HCl are linear

With molecules containing three or more atoms, the geometry is not so obvious. Here we must be concerned with the angles between bonded atoms, called *bond angles*. Consider, for example, a molecule of the type XY_2, where X represents the central atom and Y an atom bonded to it. Two geometries are possible. The atoms might be in a straight line, giving a linear molecule with a bond angle ($\angle$ Y—X—Y) of 180° (Fig. 9.1). On the other hand, they could be arranged in a nonlinear pattern. In that case, XY_2 would be a bent molecule with a bond angle less than 180°.

Experimentally, the geometry of a molecule in the gas phase may be determined by the technique of *absorption spectroscopy*, referred to in Chapter 5. Most often, absorption is measured in the infrared (800–10^6 nm) or microwave (10^6–10^9 nm) regions of the spectrum, where molecules absorb energy through vibration or rotation. From the wavelengths at which absorption occurs, bond distances and bond angles can be calculated. Another technique that is used to determine molecular geometry in the gas state is *electron diffraction*. Here one measures the angles through which a high energy beam of electrons ($\sim 10^6$ kJ/mol) is scattered in passing through a gaseous sample. Still another technique, applicable to solid samples, is x-ray diffraction (Chap. 10).

The major features of molecular geometry can be predicted on the basis of a quite simple principle—electron pair repulsion. This principle is the essence of the *valence-shell electron-pair repulsion* (VSEPR) model, first suggested by N. V. Sidgwick and H. M. Powell in 1940. It was developed and expanded later by R. J. Gillespie and R. S. Nyholm. According to VSEPR theory, **the "density clouds" associated with electron pairs surrounding an atom repel one another and are oriented to be as far apart as possible.**

This is a very useful rule, one of the best

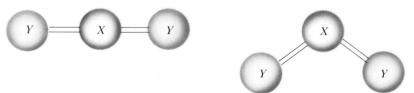

Figure 9.1 Two possible geometries for a molecule of general formula XY_2. Certain molecules of this type, including BeF_2 and CO_2, are linear, with a 180° bond angle. Others, including H_2O and SO_2, are bent, with a bond angle less than 180°. Knowing the Lewis structure, it is possible to predict whether the molecule will be linear or bent (see text discussion).

In this section we will apply this model to predict the geometry of some rather simple molecules and polyatomic ions. In all of these species, a central atom is surrounded by from two to six pairs of electrons.

IDEAL GEOMETRIES FOR TWO TO SIX ELECTRON PAIRS

We begin our study of molecular geometry by considering species in which a central atom, A, is surrounded by from two to six electron pairs, all of which are used to form single covalent bonds with terminal atoms, X. These species are described by the general formulas AX_2, AX_3, ... AX_6; it is understood that there are no unshared pairs around atom A.

To find out how the electron pair clouds surrounding the central atom are oriented with respect to one another, it is instructive to carry out the "experiment" shown in Figure 9.2. Here we show the positions taken naturally by two to six balloons tied together at the center. The balloons, like the electron pair clouds they represent, arrange themselves so as to be as far from one another as possible. The geometries they adopt are shown in a somewhat different way in Figure 9.3 on page 280, where the bond angles are indicated for molecules of the types AX_2 to AX_6.

The geometries shown in the figures for two and three electron pairs are those associated with species in which the central atom has less than an octet of electrons. Molecules of this type include BeF_2 and BF_3, which have the Lewis structures

$$\ddot{\underset{..}{F}}-Be-\ddot{\underset{..}{F}}\colon \qquad 180° \qquad\qquad \ddot{\underset{..}{F}}\diagdown\ \diagup\ddot{F}\colon$$
$$B\qquad 120°$$
$$\underset{..}{\overset{..}{F}}\colon$$

Two electron pairs are as far apart as possible when they are directed at 180° to one another. This gives BeF_2 a **linear** structure. The three electron pairs around the boron atom in BF_3 are directed toward the corners of an **equilateral triangle**; the bond angles are 120°. The BF_3 molecule is *planar*; all four atoms are in the same plane.

In species that follow the octet rule, the central atom is surrounded by four electron pairs. If each of these pairs forms a single bond with a terminal atom, a molecule of the type AX_4 results. The four bonds are directed toward the corners of a regular **tetrahedron**. All the bond angles are 109.5°, the tetrahedral angle. This geometry is found in many poly-

These geometries allow the electron pairs to get as far apart as possible

Figure 9.2 Geometries to be expected from the VSEPR model for molecules with two to six electron pairs, represented by balloons, surrounding a central atom.

Figure 9.3 Geometries to be expected from the VSEPR model for molecules with 2 to 6 electron pairs around a central atom; see also Figure 9.2.

Species type	Orientation of electron pairs	Predicted bond angles	Example	Ball and stick model
AX_2	Linear	180°	BeF_2	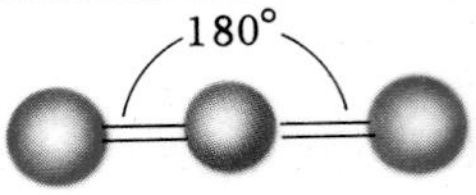
AX_3	Equilateral triangle	120°	BF_3	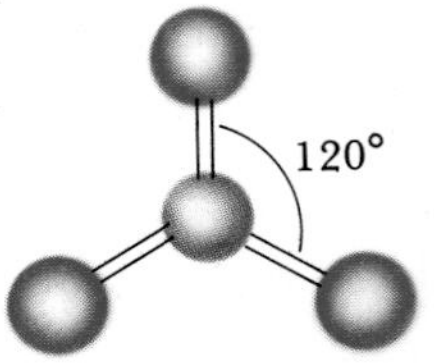
AX_4	Tetrahedron	109.5°	CH_4	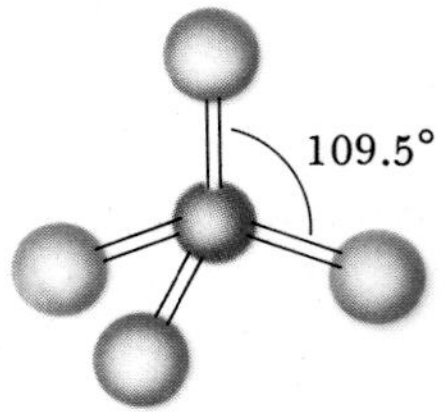
AX_5	Triangular Bipyramid	90° 120° 180°	PCl_5	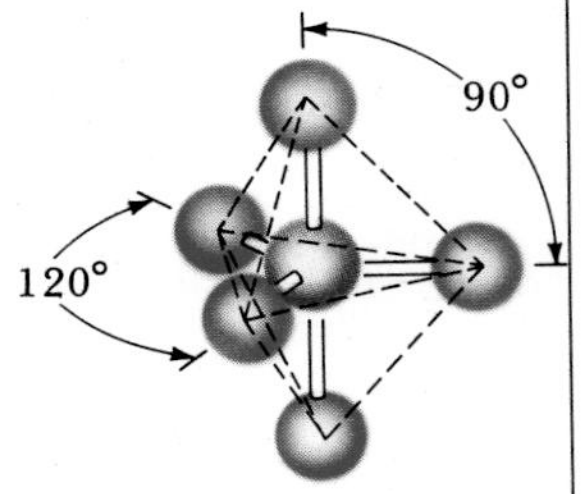
AX_6	Octahedron	90° 180°	SF_6	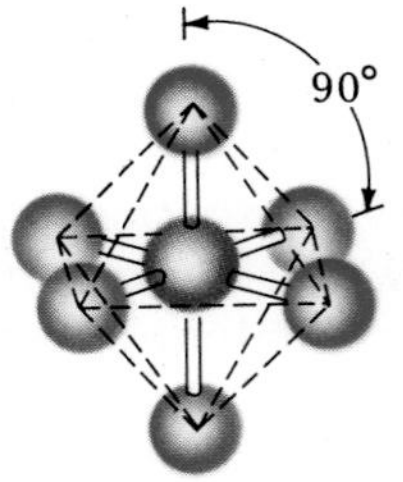

atomic ions (NH_4^+, SO_4^{2-}, etc.) and in a wide variety of organic molecules, the simplest of which is methane, CH_4.

Molecules of the type AX_5 and AX_6 require that the central atom have an expanded octet. The geometries of these molecules are shown at the bottom of Figure 9.3. In PF_5, the five bonding pairs are directed toward the corners of a **triangular bipyramid**, which we can visualize as two triangular pyramids fused together. Three of the five fluorine atoms are located at the corners of an equilateral triangle with the phosphorus atom at the center; the other two fluorine atoms are directly above and below the central P atom. In SF_6, the six bonds are directed toward the corners of a regular **octahedron**. We can think of an octahedron as being formed by fusing two square pyramids through the base. Four of the fluorine atoms in SF_6 are located at the corners of a square at the center of which is the sulfur atom; one fluorine atom is directly above the S atom, another directly below it. We will have more to say about octahedral geometry later (Chap. 16), when we discuss compounds of the transition metals.

EFFECT OF UNSHARED PAIRS ON GEOMETRY

In many molecules and polyatomic ions, one or more of the electron pairs around the central atom is unshared. The VSEPR model is readily extended to predict the geometries of these species. Here we should stress two points:

1. The electron-pair geometry is approximately the same as that observed when only single bonds are involved. The bond angles are either equal to the ideal values listed in Figure 9.3, or a little less than these values.
2. The *molecular geometry* is quite different when one or more unshared pairs are present. In describing molecular geometry, we refer only to the positions of the bonded atoms. These positions can be determined experimentally; positions of unshared pairs cannot be established by experiment. Hence the locations of unshared pairs are not specified in describing molecular geometry.

With these principles in mind, let us consider the NH_3 molecule. You will recall from Chapter 8 that the nitrogen atom in this molecule has an octet structure. It is surrounded by three single bonds and one unshared pair:

$$\overset{\displaystyle ..}{\underset{\displaystyle H}{\overset{\displaystyle N}{\diagup \, | \, \diagdown}}} \quad$$

H   H

At the left of Figure 9.4, we show the apparent orientation of the four electron pairs around the N atom in NH_3. Notice that, as in CH_4, the four pairs are directed toward the corners of a tetrahedron. At the right of Figure 9.4, we show the positions of the atoms in NH_3. The nitrogen atom is located above the center of an equilateral triangle formed by the three

The four electron pairs are tetrahedral
The atoms lie in a triangular pyramid
NH_3 is *not* tetrahedral

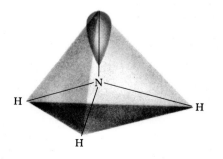

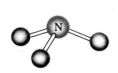

Figure 9.4 Two ways of showing the geometry of the NH_3 molecule. The orientation of the electron pairs, including the unshared pair (blue lobe), is shown at the left. The orientation of the atoms is shown at the right. The nitrogen atom is located directly above the center of the equilateral triangle formed by the three hydrogen atoms. The NH_3 molecule is described as a triangular pyramid.

hydrogen atoms. We describe the molecular geometry of NH_3 by saying that it is a **triangular pyramid**. The nitrogen atom is at the apex of the pyramid, while the three hydrogen atoms form its triangular base. The molecule is three-dimensional, as the word "pyramid" implies.

The development we have just gone through for NH_3 is readily extended to the water molecule, H_2O. Here, the Lewis structure shows that the central oxygen atom is surrounded by two single bonds and two unshared pairs:

$$ H \diagdown \overset{\displaystyle \overset{..}{O}}{} \diagdown H $$

The diagram at the left of Figure 9.5 emphasizes that the four electron pairs are oriented tetrahedrally. At the right, the positions of the atoms are shown. Clearly they are not in a straight line; we describe the H_2O molecule as being **bent**.

Experimentally, we find that the bond angles in NH_3 and H_2O are slightly less than the ideal value of 109.5°. In NH_3 (three single bonds, one unshared pair around N), the bond angle is 107°. In H_2O (two single bonds, two unshared pairs around O), the bond angle is about 105°.

These effects can be explained in a rather simple way. An unshared pair is attracted by one nucleus, that of the atom to which it belongs. In contrast, a bonding pair is attracted by two nuclei, those of the two atoms it joins. Hence we might expect the electron cloud of an unshared pair to spread out over a larger volume than that of a bonding pair. In NH_3, this tends to force the bonding pairs closer to one another, thereby reducing the bond angle. Where there are two unshared pairs, as in H_2O, this effect is more pronounced. In general, the VSEPR model predicts that unshared electron pairs will occupy slightly more space than bonding pairs.

The line of reasoning we have just gone through for NH_3 and H_2O can be readily extended to a molecule like GeF_2:

$$:\overset{..}{\underset{..}{F}} \diagup \overset{\displaystyle \overset{..}{Ge}}{} \diagdown \overset{..}{\underset{..}{F}}: $$

Here, a central germanium atom is surrounded by three electron pairs, one of which is unshared. The geometry of GeF_2 can be derived from that of BF_3 (Fig. 9.3). We would predict that GeF_2 should be a bent molecule with a bond angle somewhat less than the ideal value of 120°. This is indeed the case.

Unshared pairs have bigger electron clouds than bonding pairs

Figure 9.5 Two ways of showing the geometry of the H_2O molecule. At the left, the two unshared pairs are shown. As you can see from the drawing at the right, H_2O is a bent molecule. The bond angle, 105°, is a little smaller than the tetrahedral angle, 109.5°. The unshared pairs spread out over a larger volume than that occupied by the bonding pairs.

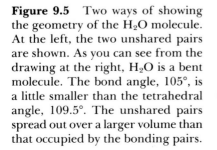
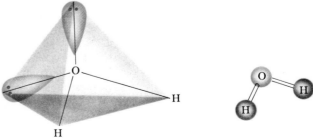

TABLE 9.1 Geometries with 2, 3, or 4 Electron Pairs Around a Central Atom

NO. OF TERMINAL ATOMS (X) + UNSHARED PAIRS (E)	SPECIES TYPE	IDEAL BOND ANGLES	MOLECULAR GEOMETRY	EXAMPLES
2	AX_2	180°	linear	BeF_2, CO_2
3	AX_3	120°	equilateral triangle	BF_3, SO_3
	AX_2E	120°*	bent	GeF_2, SO_2
4	AX_4	109.5°	tetrahedron	CH_4
	AX_3E	109.5°*	triangular pyramid	NH_3
	AX_2E_2	109.5°*	bent	H_2O

*In these species, the observed bond angle is ordinarily somewhat less than the ideal value.

Table 9.1 summarizes our discussion of the molecular geometries of species in which a central atom is surrounded by two, three, or four electron pairs. The table is organized in terms of the number of terminal atoms, X, and unshared pairs, E, surrounding the central atom, A.

■ **EXAMPLE 9.1** _____

Predict the geometries of the following molecules:

a. BeH_2 b. OF_2 c. PF_3

Solution

a. In the BeH_2 molecule, there are only four valence electrons (two from the Be atom, one from each of the two H atoms). The Lewis structure has to be

H—Be—H

There are no unshared pairs around the beryllium atom, so this is a type AX_2 molecule (Table 9.1). The molecule is linear, with a bond angle of 180°.

b. In OF_2 there are: $6 + 2(7) = 20$ valence electrons. The Lewis structure is:

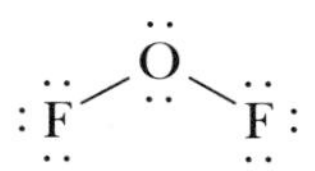

The central oxygen atom is bonded to two fluorine atoms and also has two unshared pairs. This is an AX_2E_2 molecule, analogous to water. The OF_2 molecule is bent, with a bond angle somewhat less than the ideal tetrahedral value of 109.5° (the observed value is 103°).

c. In PF_3 there are: $5 + 3(7) = 26$ valence electrons. You should be able to show that the Lewis structure is:

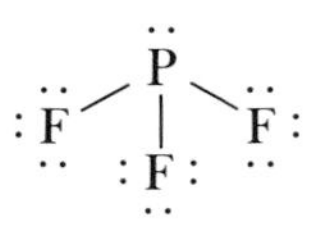

There are three bonded atoms and one unshared pair around P; PF_3 is a type AX_3E molecule, similar to NH_3. It is a triangular pyramid with a bond angle somewhat less than $109.5°$ (the observed angle is $104°$).

Exercise
Draw the Lewis structure for HOCl, predict the ideal bond angle, and describe the geometry. Answer: H—Ö—Cl̈:, $109.5°$, bent.

Example 9.1 illustrates a couple of points concerning molecular geometry.

1. You must know or be able to derive the Lewis structure of a species before you can predict its geometry.

2. The way a Lewis structure is written does not necessarily imply its geometry. In writing the Lewis structures of OF_2 and PF_3, we arranged the atoms in such a way as to suggest the proper geometry. More commonly, they are written:

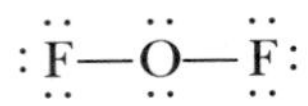

:F̈—Ö—F̈:   and   :F̈—P̈—F̈:
                                          |
                                        :F̈:

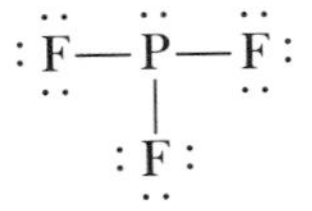

with no indication of the bond angles.

In many expanded-octet molecules, one or more of the electron pairs around the central atom is unshared. Recall, for example, the xenon tetrafluoride molecule, whose Lewis structure was derived in Chapter 8:

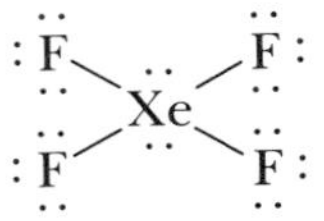

:F̈   ..   F̈:
    \   ..   /
      >Xe< AX_4E_2
    /   \
:F̈   ..   F̈:

Here there are six electron pairs around the xenon atom; four of these are covalent bonds to fluorine and the other two pairs are unshared. We classify this molecule as AX_4E_2. Other examples include the molecules XeF_2 and ClF_3:

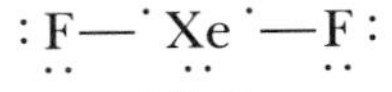

:F̈— Xe —F̈:   :F̈— Cl —F̈:
        AX_2E_3 |
                                    :F̈:
                                  AX_3E_2

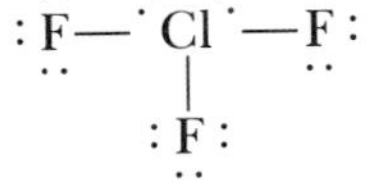

Both of these are expanded-octet molecules; the central atom is surrounded by five electron pairs rather than four.

To deduce the geometries of molecules such as these, we need to know what happens when successive corners are removed from an octahedron (6 electron pairs) or a triangular bipyramid (5 electron pairs). In many cases, more than one geometry is possible. The observed structure is one which assigns more space to unshared pairs than to bonding pairs. Consider, for example, the XeF_4 molecule, where two geometries are possible.

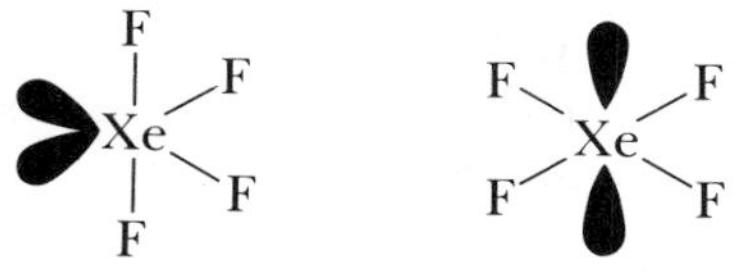

In the structure at the left, the lone pairs are close to one another and occupy relatively little space. In the structure at the right, they are far apart, occupying the maximum amount of space. VSEPR theory predicts and experiment confirms that the structure at the right is the correct one. The xenon atom is at the center of a square with fluorine atoms at each corner of the square.

XeF_4 is square planar

Following this principle, the VSEPR model predicts the geometries shown in Fig. 9.6, p. 286. The structures shown there include those of all molecules having 5 or 6 electron pairs around the central atom, one or more of which may be unshared. Notice that with 5 pairs, the unshared pairs always occupy *equatorial* positions, i.e., positions at the vertices of the equilateral triangle surrounding the central atom. The *axial* positions, at the top and bottom of the triangular bipyramid, are always occupied by bonded atoms.

■ **EXAMPLE 9.2**

Consider the following species

a. $SbCl_5$ b. IF_5 c. I_3^-

Draw the Lewis structures and predict the geometries, using Figure 9.6.

Solution

a. No. of valence $e^- = 5 + 7(5) = 40$. The Lewis structure is:

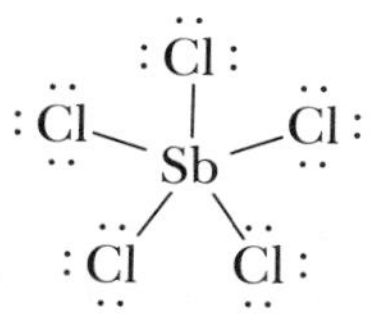

There are no unpaired electrons; this is an AX_5 molecule, like PF_5. The molecule is a triangular bipyramid.

b. No. of valence $e^- = 7 + 7(5) = 42$, two more than in (a). The Lewis structure is:

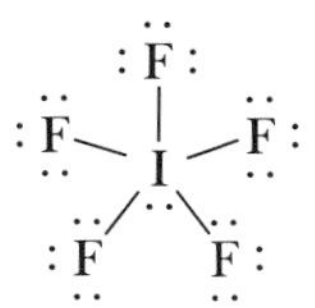

IF_5 is an AX_5E molecule; there are a total of six electron pairs around the iodine atom. The molecule is a square pyramid with the iodine atom in the center of the square.

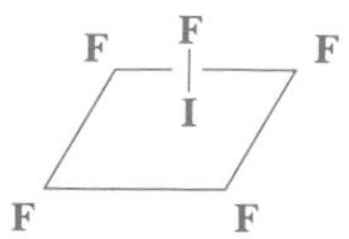

Figure 9.6 Geometries of molecules with expanded octets. The red spheres represent terminal atoms. The open ellipses represent unshared electron pairs.

VSEPR

5 ELECTRON PAIRS

Species type	Structure	Description	Example	Bond angles
AX_5	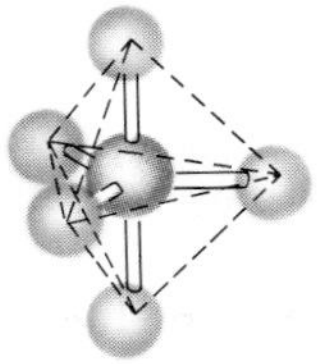	Triangular bipyramidal	PF_5	90°, 120°, 180°
AX_4E	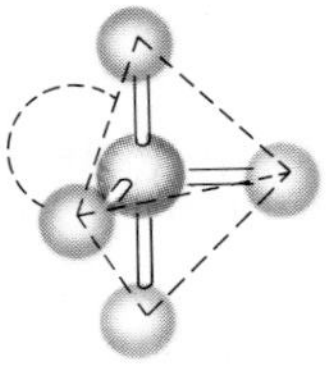	See-saw	SF_4	90°, 120°, 180°
AX_3E_2	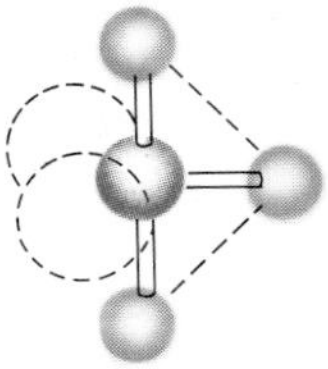	T-shaped	ClF_3	90°, 180°
AX_2E_3	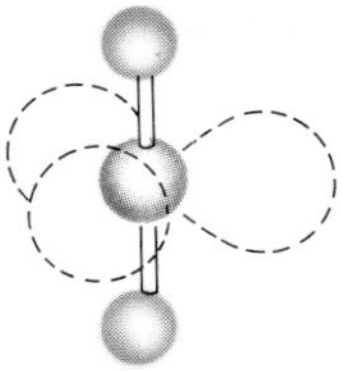	Linear	XeF_2	180°

6 ELECTRON PAIRS

AX_6	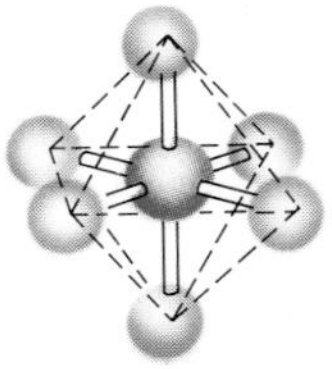	Octahedral	SF_6	90°, 180°
AX_5E	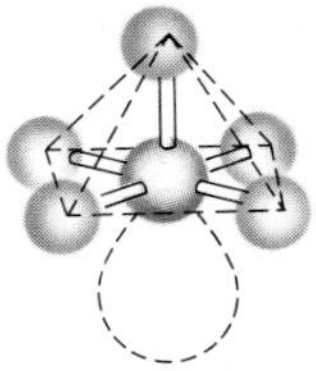	Square pyramidal	ClF_5	90°, 180°
AX_4E_2	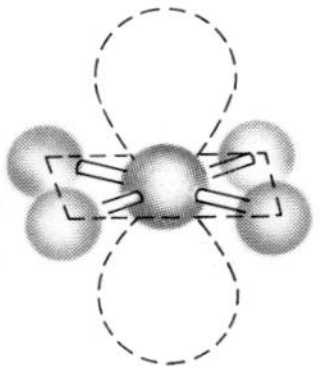	Square planar	XeF_4	90°, 180°

c. There are $3(7) + 1 = 22$ valence e^-. The Lewis structure is:

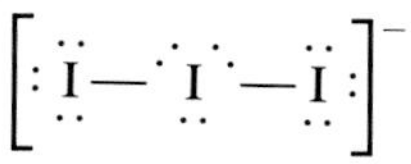

$$\left[:\ddot{I} - \dot{\ddot{I}} - \ddot{I}: \right]^-$$

Type: AX_2E_3. The ion is linear with a 180° bond angle.

Exercise
Which one of the following species has a geometry that *cannot* be predicted from Figure 9.6: SeF_4, SeF_6, SeF_6^{2-}? Answer: SeF_6^{2-}.

MULTIPLE BONDS

The VSEPR model is readily extended to species in which double or triple bonds are present. Here, a simple principle applies: *insofar as molecular geometry is concerned, a multiple bond behaves as if it were a single electron pair.* We can readily understand why this should be the case. The four electrons in a double bond, or the six electrons in a triple bond, must be located between the two atoms, as are the two electrons in a single bond. This means that the electron pairs in a multiple bond must occupy the same region of space as those in a single bond. Hence, the "extra" electron pairs in a multiple bond have no effect upon geometry.

To illustrate this principle, consider the CO_2 molecule. Its Lewis structure is

$$:\ddot{O}\!=\!C\!=\!\ddot{O}:$$

The central atom, carbon, has two double bonds and no unshared pairs. For purposes of determining molecular geometry, we pretend that the double bonds are single bonds, ignoring the "extra" bonding pairs. The bonds are directed to be as far apart as possible, giving a 180° O—C—O bond angle. The CO_2 molecule, like BeF_2, is linear:

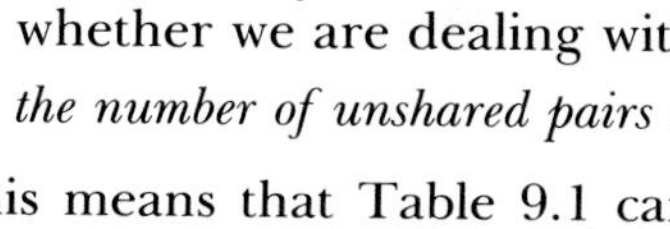

$$\underset{180°}{F\!-\!Be\!-\!F} \qquad \underset{180°}{O\!=\!C\!=\!O}$$

This principle can be restated in a somewhat different way for molecules in which there is a single central atom. The geometry of such a molecule depends only upon:

— *the number of terminal atoms bonded to the central atom, X,* irrespective of whether we are dealing with single, double, or triple bonds.

— *the number of unshared pairs around the central atom, E.*

With CO_2, $X = 2$
$E = 0$

$X + E = 2$ (linear)

This means that Table 9.1 can be used in the usual way to predict the geometry of a species containing multiple bonds.

■ **EXAMPLE 9.3** _____

Predict the geometries of the ClO_3^- ion, the NO_3^- ion, and N_2O molecule, which have the Lewis structures

a. $\left[:\ddot{O}-\underset{\underset{\displaystyle :\ddot{O}:}{|}}{\ddot{Cl}}-\ddot{O}: \right]^{-}$ b. $\left[:\ddot{O}-\underset{\underset{\displaystyle :\ddot{O}:}{\|}}{N}-\ddot{O}: \right]^{-}$ c. $:N{=}N{=}\ddot{O}:$

Solution

a. The central atom, chlorine, is bonded to three oxygen atoms; it has one unshared pair. The ClO_3^- ion is of the type AX_3E, similar to NH_3. It is a triangular pyramid; the ideal bond angle is 109.5°.

b. The central atom, nitrogen, is bonded to three oxygen atoms; it has no unshared pairs. The NO_3^- ion is of the type AX_3, similar to BF_3 and SO_3. It has the geometry of an equilateral triangle, with the nitrogen atom at the center. The bond angle is 120°.

c. The central nitrogen atom is bonded to two other atoms with no unshared pairs. The type is AX_2, analogous to BeF_2 and CO_2. The molecule is linear, with a bond angle of 180°.

Exercise
What is the ideal bond angle in SO_2, whose Lewis structure was derived in Chapter 8? Answer: 120°.

We can readily extend these ideas to molecules in which there is no single central atom. Consider, for example, the acetylene molecule, C_2H_2. You will recall that here the two carbon atoms are joined by a triple bond:

$$H{-}C{\equiv}C{-}H$$

Each carbon atom behaves as if it were surrounded by two electron pairs. Both of the bond angles (H—C≡C and C≡C—H) are 180°. The molecule is linear; the four atoms are in a straight line. The two "extra" electron pairs in the triple bond do not affect the geometry of the molecule.

In ethylene, C_2H_4, there is a double bond between the two carbon atoms. The molecule has the geometry to be expected if each carbon atom had only three pairs of electrons around it.

$$\underset{H}{\overset{H}{\diagdown}}C{=}C\underset{H}{\overset{H}{\diagup}}$$

The six atoms are located in a plane, with bond angles of 120°.

SUMMARY

It is possible, using the VSEPR model, to predict the molecular geometry and ideal bond angle for any species in which a central atom is surrounded by from two to six electron pairs. To do this:

1. **Draw the Lewis structure of the species.**
2. **Count the number of bonded atoms (X) and unshared pairs (E) around the central atom.**

3. Using the results of (2), classify the species in one of the categories listed in Table 9.1 (2 to 4 electron pairs) or Figure 9.6 (5 or 6 electron pairs). Use the information from these sources to predict the molecular geometry and ideal bond angle.

9.2
POLARITY OF MOLECULES

A polar molecule is one that contains positive and negative poles. There is a partial positive charge (positive pole) at one point in the molecule and a partial negative charge (negative pole) of equal magnitude at another point. As shown in Figure 9.7, polar molecules orient themselves in the presence of an electric field. The positive pole in the molecule aligns with the external negative charge, and the negative pole with the external positive charge. In contrast, there is no charge separation in a nonpolar molecule. In an electric field, nonpolar molecules, such as H_2, show no preferred orientation; they are oriented randomly.

The extent to which a molecule lines up in an electric field depends upon a quantity known as dipole moment. The *dipole moment*, μ (mu), is defined as the product of the partial positive charge, Q, times the distance between the two charges, d

$$\mu = Q \times d \qquad (9.1)$$

A molecule in which two partial charges of magnitude 1×10^{-20} coulomb are separated by 0.1 nm (10^{-10} m) would have a dipole moment of

$$\mu = (10^{-20} \text{ C}) \times (10^{-10} \text{ m}) = 10^{-30} \text{ C·m}$$

Dipole moments of molecules can be determined experimentally; some typical values are listed in Table 9.2. Nonpolar molecules have a dipole moment of zero.

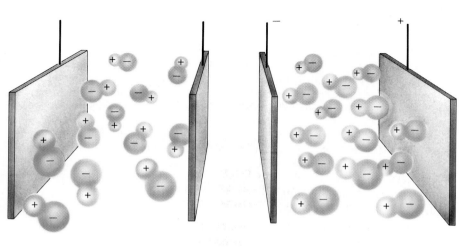

Field off   Field on

Figure 9.7 Orientation of polar molecules in an electric field. In the absence of the field, polar molecules are randomly oriented. In an electric field, polar molecules such as HF line up as shown ("Field on"). Nonpolar molecules such as H_2 do not line up.

TABLE 9.2   Dipole Moments (in units of 10^{-30} C·m)

H_2	0	$BeCl_2$	0	CH_4	0
F_2	0	CO_2	0	CH_3Cl	6.24
O_2	0	SO_2	5.42	CH_2Cl_2	5.40
HF	6.36	H_2O	6.17	$CHCl_3$	3.34
HCl	3.43	H_2S	3.12	CCl_4	0
HI	1.47	BCl_3	0		
		NH_3	4.88		

If a molecule is diatomic, we can readily decide whether it is polar or nonpolar. We need only be concerned with bond polarity, as discussed in Chapter 8. If the two atoms are the same, as in H_2 or F_2, the bond is nonpolar; so is the molecule. Suppose, on the other hand, that the two atoms differ, as in HF. In this case, the bond is polar; the bonding electrons are shifted toward the more electronegative F atom. There is a negative pole at the F atom and a positive pole at the H atom. This is sometimes indicated by writing

$$H \overset{+}{\longrightarrow} F$$

If a molecule has a nonzero dipole moment, the molecule is polar

The arrow points toward the negative end of the polar bond (F atom); the + sign is at the positive end (H atom). The HF bond and hence the HF molecule is a *dipole;* it contains positive and negative poles. From Table 9.2, we see that the dipole moment of HF is 6.36×10^{-30} C·m; it is known from spectroscopic measurements that the internuclear distance in HF is 0.0918 nm. It follows that the partial charge in the HF dipole is

$$Q = \frac{\mu}{d} = \frac{6.36 \times 10^{-30} \text{ C·m}}{0.0918 \times 10^{-9} \text{ m}} = 6.93 \times 10^{-20} \text{ C}$$

The charge on the electron is 1.60×10^{-19} C. We see that the partial charge in HF is a little more than 4/10 of the full electron charge:

$$(6.93 \times 10^{-20})/(1.60 \times 10^{-19}) = 0.43$$

In that sense, we might say that the bond in the HF molecule has about 43% ionic character.

If a molecule contains more than two atoms, it is not so easy to decide whether it is polar or nonpolar. In this case, we must be concerned not only with bond polarity but also with molecular geometry. To illustrate what is involved, consider the molecules shown in Figure 9.8.

1. In BeF_2 there are two polar Be—F bonds; in both bonds, the electron density is concentrated around the more electronegative fluorine atom. However, since the BeF_2 molecule is linear, the two Be $\overset{+}{\longrightarrow}$ F dipoles are in opposite directions and cancel one another. The molecule has no net dipole and hence is nonpolar. From a slightly different point of view, we might say that in BeF_2 the centers of positive and negative charge coincide with each other, at the Be atom. There is no way that a BeF_2 molecule can line up in an electric field.

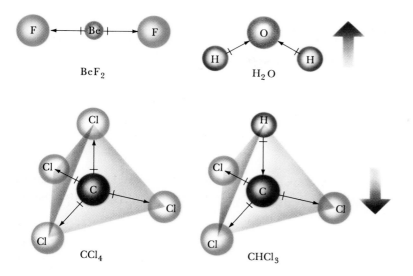

Figure 9.8 Polarity of molecules. In all these molecules, the bonds are polar, as indicated by the ↔ notation. However, in BeF_2 and CCl_4, the bond dipoles cancel and the molecule is nonpolar. In H_2O and $CHCl_3$, on the other hand, there is a net dipole and the molecule is polar. (The broad arrow beside the molecule points to the negative pole.)

2. Since oxygen is more electronegative than hydrogen, an O—H bond is polar, with the electron density highest around the oxygen atom. In the bent H_2O molecule, the two H ↔ O dipoles do not cancel each other. Instead, they add to give the H_2O molecule a net dipole. The center of negative charge is located at the O atom; this is the negative pole of the molecule. The center of positive charge is located midway between the two H atoms; the positive pole of the molecule is at that point. The H_2O molecule is polar (see Table 9.2). It lines up in an electric field with the oxygen atom oriented toward the positive electrode.

3. Carbon tetrachloride, CCl_4, is another molecule which, like BeF_2, is nonpolar despite the presence of polar bonds. Each of its four bonds is a dipole, C ↔ Cl. However, because the four bonds are arranged symmetrically around the carbon atom, they cancel. As a result, the molecule has no net dipole; it is nonpolar. If one of the Cl atoms in CCl_4 is replaced by hydrogen, the situation changes. The $CHCl_3$ molecule is not symmetrical; the H ↔ C dipole does not cancel with the three C ↔ Cl dipoles. Hence the $CHCl_3$ molecule is polar.

Nonpolar species:

X_2
linear XY_2
planar XY_3
tetrahedral XY_4

■ **EXAMPLE 9.4** _____

Determine whether each of the following is polar or nonpolar:

a. SO_2 b. BF_3 c. CO_2

Solution

a. In Example 9.3, we decided that the SO_2 molecule is bent, with a bond angle of 120°. We would expect it to be   polar.   There should be a positive pole (partial + charge) at the less electronegative sulfur atom and a negative pole (partial − charge) midway between the two oxygens.

b. BF_3, as noted in Figure 9.3, is a planar, triangular molecule. Because of this symmetrical geometry, the B $\longleftrightarrow$ F dipoles cancel.

The molecule is   nonpolar.

c. As pointed out earlier, CO_2 is linear, analogous to BeF_2. The C $\longleftrightarrow$ O dipoles cancel one another; the molecule is   nonpolar   (see Table 9.2).

Exercise
A certain molecule XY_4 is found to be polar. How many pairs of electrons are there around the central atom: 4, 5, or 6? Answer: 5.

9.3
ATOMIC ORBITALS; HYBRIDIZATION

To this point, we have used Lewis structures to describe the arrangement of atoms in molecules. These can be very useful. Lewis structures allow us to predict molecular geometries and polarities. However, they do not give us information about the energies of electrons in molecules. Neither do they tell us anything about the orbitals in which the bonding electrons are located.

In the 1930's a theoretical treatment of the covalent bond was developed by Linus Pauling and J. C. Slater, among others. It is referred to as the *atomic orbital*, or **valence bond**, model. According to this model, a covalent bond consists of a pair of electrons of opposed spin within an atomic orbital. For example, a hydrogen atom forms a covalent bond by accepting an electron from another atom to complete its 1s orbital. Using orbital diagrams, we could write

1s

Isolated H atom   (↑)
H atom in a stable molecule   (↑↓)

The second electron, shown in color, is contributed by another atom. This could be another H atom in H_2, a F atom in HF, a C atom in CH_4, and so on.

This simple model is readily extended to other atoms. The fluorine atom (electron configuration $1s^2 2s^2 2p^5$) has a half-filled p orbital:

1s   2s   2p

Isolated F atom   (↑↓)   (↑↓)   (↑↓) (↑↓) (↑)

By accepting an electron from another atom, F can complete this 2p orbital:

	1s	2s	2p
F atom in HF, F_2, . . .	(⇅)	(⇅)	(⇅) (⇅) (⇅)

This structure ties in nicely with the octet rule

According to this model, it would seem that in order for an atom to form a covalent bond, it must have an unpaired electron. Indeed, the number of bonds formed by an atom should be determined by its number of unpaired electrons. Since hydrogen has an unpaired electron, a H atom should form one covalent bond, as indeed it does. The same holds for the F atom, which ordinarily forms only one bond. Noble gas atoms, such as He and Ne, which have no unpaired electrons, should not form bonds at all. This is the case for He and Ne.

When we try to extend this simple idea beyond hydrogen, the halogens, and the noble gases, problems arise. Consider, for example, the three atoms Be (at. no. = 4), B (at. no. = 5), and C (at. no. = 6):

	1s	2s	2p
Be atom	(⇅)	(⇅)	( ) ( ) ( )
B atom	(⇅)	(⇅)	(↑) ( ) ( )
C atom	(⇅)	(⇅)	(↑) (↑) ( )

Notice that the beryllium atom has no unpaired electrons, the boron atom has one, and the carbon atom two. Simple valence bond theory would predict that Be, like He, should not form covalent bonds. A boron atom should form one bond, carbon two. Experience tells us that these predictions are wrong. Beryllium forms two bonds in BeF_2; B forms three bonds in BF_3. Carbon, in all its stable compounds, forms four bonds, not two.

To explain these and other discrepancies, simple valence bond theory must be modified. It is necessary to invoke a new kind of atomic orbital, called a hybrid orbital.

HYBRID ORBITALS: sp, sp², sp³, sp³d, sp³d²

The formation of two bonds by beryllium can be explained if we assume that, prior to reaction, one of the 2s electrons is promoted to the 2p level. Thus we could write

	1s	2s	2p			1s	2s	2p
ground state Be atom	(⇅)	(⇅)	( ) ( ) ( )	→	excited Be atom	(⇅)	(↑)	(↑) ( ) ( )

Electron promotion allows Be to form two bonds

With two unpaired electrons, the Be atom can form two covalent bonds, as in BeF_2:

	1s	2s	2p
Be in BeF_2	(⇅)	(⇅)	(⇅) ( ) ( )

(The colored arrows indicate the two electrons supplied by the fluorine atoms. The horizontal lines enclose the orbitals involved in bond formation.) Forming two Be—F bonds releases enough energy to more than compensate for that absorbed in promoting the 2s electron of beryllium.

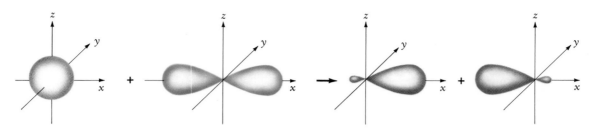

s orbital   p orbital   Two sp orbitals

Figure 9.9 Formation of sp hybrid orbitals. The mixing of an s orbital with a p orbital gives two new orbitals known as sp hybrids.

There is one basic objection to this model. It implies that two different kinds of bonds are formed. One would be an "s" bond, since the two electrons are filling an s orbital. The other would be a "p" bond, with a pair of electrons in a p orbital. Experimentally, we find that the properties of the two bonds in BeF_2 are identical. This suggests that the two orbitals used for bond formation by beryllium must be equivalent. In atomic orbital terminology, we say that an s and a p orbital have been mixed, or *hybridized* to give two new equivalent orbitals. These new orbitals are referred to as **sp hybrid orbitals**:

The sp hybrid orbitals are linear

one s atomic orbital + one p atomic orbital → *two* sp hybrid orbitals

Notice (Fig. 9.9) that *the number of hybrid orbitals formed is equal to the number of atomic orbitals mixed.* This is always true in hybridization of orbitals.

A similar argument can be used to explain why boron forms three bonds in BF_3. We assume that one of the 2s electrons in boron is promoted to the 2p level:

	1s	2s	2p
Excited B atom	(↑↓)	(↑)	(↑)(↑)( )

Each fluorine atom puts an electron into a half-filled orbital to give

	1s	2s	2p
B in BF_3	(↑↓)	(↑↓)	(↑↓)(↑↓)( )

The orbitals in which the bonding electrons are located, enclosed by horizontal lines, are referred to as **sp² hybrid orbitals**. They are formed by mixing an s orbital with two p orbitals:

one s atomic orbital + two p atomic orbitals → *three* sp^2 hybrid orbitals

To explain why carbon forms four bonds, we again invoke promotion of an electron:

	1s	2s	2p
Excited C atom	(↑↓)	(↑)	(↑)(↑)(↑)

followed by hybridization. In CH_4 or any other molecule in which carbon forms four single bonds, the carbon atom has the orbital diagram

		1s	2s	2p
C in CH_4		(⇅)	(⇅)	(⇅)(⇅)(⇅)

The four orbitals in which the bonding electrons are located are called **sp³ hybrid orbitals**. They are formed by mixing an s orbital with three p orbitals:

one s atomic orbital + three p atomic orbitals → *four* sp³ hybrid orbitals

You will recall that the bond angles in NH_3 and H_2O are very close to that in CH_4. This suggests that the four electron pairs surrounding the central atom in NH_3 and H_2O, like those in CH_4, occupy sp³ hybrid orbitals. In NH_3, three of these orbitals are filled by bonding electrons, the other by the unshared pair on the nitrogen atom. In H_2O, two of the sp³ orbitals of the oxygen atom contain bonding electron pairs; the other two contain unshared pairs. The situation in NH_3 and H_2O is not unique. In general, we find that *unshared as well as shared electron pairs can be located in hybrid orbitals.*

The "extra" electron pairs in an expanded octet are accommodated by using d orbitals. The phosphorus atom in PCl_5 and the sulfur atom in SF_6 make use of 3d as well as 3s and 3p orbitals:

		3s	3p	3d
P atom in PCl_5	[₁₀Ne]	(⇅)	(⇅)(⇅)(⇅)	(⇅)( )( )( )( )
S atom in SF_6	[₁₀Ne]	(⇅)	(⇅)(⇅)(⇅)	(⇅)(⇅)( )( )( )

The orbitals used by the five pairs of bonding electrons surrounding the phosphorus atom in PCl_5 are **sp³d hybrid orbitals**.

one s orbital + three p orbitals + one d orbital → *five* sp³d hybrid orbitals

Similarly, in SF_6, the six pairs of bonding electrons are located in **sp³d² hybrid orbitals**:

one s orbital + three p orbitals + two d orbitals → *six* sp³d² hybrid orbitals

In all species where a central atom is surrounded by more than eight electrons, d orbitals are used. With 10 valence electrons, one d orbital is involved; with 12 valence electrons, two d orbitals must be used. This explains why nonmetal atoms in the second period (N, O, F) do not form expanded octets. There are no 2d orbitals; d orbitals first become available when **n** = 3; that is, in the third period.

We should emphasize that hybrid orbitals have their own unique properties, quite different from those of the orbitals from which they are formed. Table 9.3 gives the orientation in space of hybrid orbitals. The geometries, as found mathematically by quantum mechanics, are exactly as predicted by VSEPR theory. In each case, the several hybrid orbitals are directed to be as far apart as possible.

TABLE 9.3   Hybrid Orbitals and Their Geometries

NUMBER OF ELECTRON PAIRS	ATOMIC ORBITALS	HYBRID ORBITALS	ORIENTATION	EXAMPLES
2	s, p	sp	linear	BeF_2, CO_2
3	s, two p	sp^2	equilateral triangle	BF_3, SO_3
4	s, three p	sp^3	tetrahedron	CH_4, NH_3, H_2O
5	s, three p, d	sp^3d	triangular bipyramid	PCl_5, SF_4, ClF_3
6	s, three p, two d	sp^3d^2	octahedron	SF_6, ClF_5, XeF_4

■ **EXAMPLE 9.5**

Give the hybridization of

a. carbon in CF_4 b. phosphorus in PF_3 c. sulfur in SF_4

Solution

The Lewis structures are needed to apply the rules cited above.

a. For CF_4, the Lewis structure is

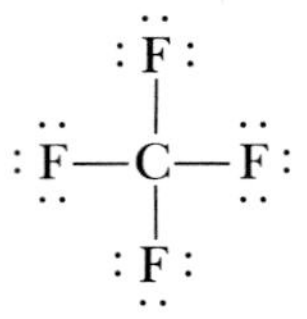

$$
\begin{array}{c}
\ddot{:}\!\ddot{F}\!\ddot{:} \\
| \\
:\!\ddot{F}\!-\!C\!-\!\ddot{F}\!: \\
| \\
:\!\ddot{F}\!: \\
\end{array}
$$

Carbon forms four single bonds, as in CH_4. The hybridization is sp^3.

b. The Lewis structure of PF_3 is

$$
\begin{array}{c}
:\!\ddot{F}\!-\!\ddot{P}\!-\!\ddot{F}\!: \\
| \\
:\!\ddot{F}\!: \\
\end{array}
$$

The nature of the hybrid is determined by the number of electron pairs it contains

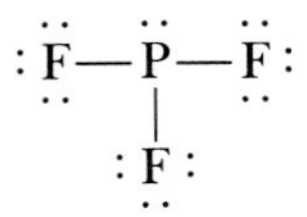

The phosphorus atom is surrounded by four electron pairs; the hybridization is sp^3,   as in NH_3.

c. When you write the Lewis structure of SF_4,

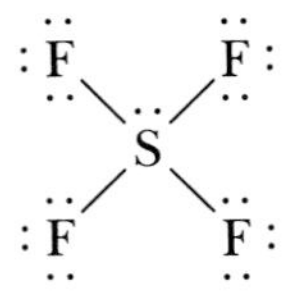

$$
\begin{array}{c}
:\!\ddot{F} \qquad \ddot{F}\!: \\
\diagdown\; \ddot{}\; \diagup \\
S \\
\diagup \qquad \diagdown \\
:\!\ddot{F} \qquad \ddot{F}\!: \\
\end{array}
$$

it becomes obvious that the sulfur atom is surrounded by five electron pairs; its hybridization is sp^3d.

Exercise

What is the hybridization of Ge in GeF_2, whose Lewis structure was shown earlier in this chapter? Answer: sp^2.

HYBRIDIZATION IN MOLECULES CONTAINING MULTIPLE BONDS

In Section 9.1, we pointed out that insofar as geometry is concerned, a multiple bond acts as if it were a single bond. In other words, the "extra" electron pairs in a double or triple bond have no effect upon the geometry of the molecule. This behavior is explained in terms of hybridization.

The extra electron pairs in a multiple bond (one pair in a double bond, two pairs in a triple bond) are not located in hybrid orbitals.

From this point of view, the geometry of a molecule is fixed by the electron pairs in hybrid orbitals about a central atom. These orbitals are directed to be as far apart as possible. The hybrid orbitals contain

— all unshared electron pairs.
— electron pairs forming single bonds.
— one and only one of the electron pairs in a double or triple bond.

To illustrate this rule, consider the ethylene (C_2H_4) and acetylene (C_2H_2) molecules. You will recall that the bond angles in these molecules are 120° for ethylene and 180° for acetylene. This implies sp^2 hybridization in C_2H_4 and sp hybridization in C_2H_2 (see Table 9.3). Using colored lines to represent hybridized electron pairs,

$$\begin{array}{cc} H & H \\ \diagdown & \diagup \\ & C{=}C \\ \diagup & \diagdown \\ H & H \end{array} \qquad H{-}C{\equiv}C{-}H$$

ethylene   acetylene

C_2H_4—3 electron pairs in hybrid, so it is sp^2
C_2H_2—2 electron pairs in hybrid, so it is sp

In both cases, only one of the electron pairs in the multiple bond is hybridized.

■ **EXAMPLE 9.6** _____

Describe the hybridization of

a. nitrogen in N_2 b. nitrogen in the NO_3^- ion   c. carbon in CO_2

Solution

a. The Lewis structure of N_2 is

: N≡N :

The hybridization is sp; the unshared pairs of electrons and one of the three shared pairs are hybridized.

b. The structure of the NO_3^- ion is

$$\left[\ddot{\underset{\cdot\cdot}{O}}{-}N{-}\overset{\cdot\cdot}{\underset{\cdot\cdot}{O}}\text{:} \right]^-$$
$$\underset{\overset{\|}{\underset{\cdot\cdot}{\ddot{O}}\text{:}}}{}$$

Three electron pairs are in hybrid orbitals; two single bonds and one of the pairs in the double bond. Three hybrid orbitals are required; the hybridization is sp^2.

c. The carbon atom in CO_2 forms two double bonds:

$$: \overset{..}{O} = C = \overset{..}{O} :$$

One electron pair in each double bond is hybridized. This requires two hybrid orbitals; carbon shows sp hybridization.

Exercise

What is the hybridization of the oxygen atoms in CO_2? Answer: sp^2.

9.4
SIGMA AND PI BONDS

We have noted that the extra electron pairs in a multiple bond are not hybridized and have no effect upon molecular geometry. At this point, you may well wonder what happened to those electrons. Where are they in molecules like C_2H_4 and C_2H_2?

To answer this question, it is necessary to consider the *shape* or spatial distribution, of the orbitals filled by bonding electrons in molecules. From this point of view, we can distinguish between two types of bonding orbitals. The first of these, and by far the more common, is called a **sigma (σ) bonding orbital**. A sigma bond is formed when two orbitals, each with a single electron, overlap "head-to-head" (Fig. 9.10 a,b). When this happens, the orbital formed is symmetrical about the bond axis. The electron density in a sigma bond is concentrated in the region directly between the two bonded atoms.

All single bonds are σ bonds

Sigma bonds are formed when two s orbitals overlap, as in H_2, or when two p orbitals overlap head-to-head, as in F_2. Most commonly, they are formed when hybrid orbitals around a central atom combine with other orbitals. For example, the four bonds in CH_4 (Fig. 9.10 c), formed by the overlap of the sp^3 orbitals of carbon with the 1s orbitals of hydrogen, are sigma bonds. Indeed, in all the molecules we will consider, *hybrid orbitals are associated with sigma bonds.*

A quite different type of bonding orbital is formed when two parallel p orbitals overlap side-to-side (Fig. 9.11). This orbital consists of two lobes, one above the bond axis, the other below it. Along the bond axis itself, the electron density is zero. A bond that consists of two electrons located in an orbital of this type is referred to as a **pi (π) bond**.

With this background, let us consider the bonding in ethylene, C_2H_4. As we have seen, each carbon atom is sp^2 hybridized. The orbital diagram of either carbon atom is:

	1s	2s	2p
C in C_2H_4	($\uparrow\downarrow$)	($\uparrow\downarrow$)	($\uparrow\downarrow$) ($\uparrow\downarrow$) ($\uparrow$)

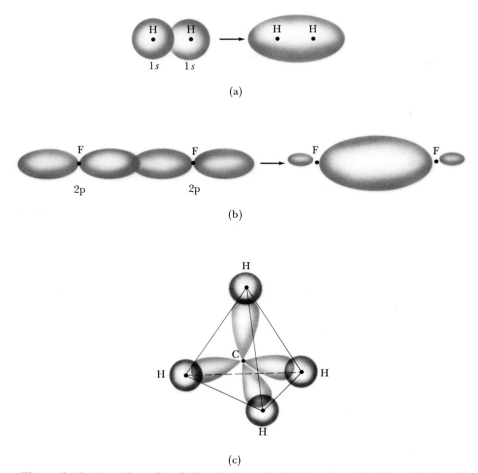

Figure 9.10 In a sigma bond, the electron density is symmetrical about the bond axis. Sigma bonds are formed when two s orbitals overlap (a), when two p orbitals overlap head-to-head (b), or when a hybrid orbital overlaps an s orbital (c).

Three of the four valence electrons of carbon are located in three sp^2 hybrid orbitals. The fourth valence electron is located by itself in an unhybridized p orbital.

This situation is shown at the top of Figure 9.12, p. 300. Let us focus attention on the carbon atom at the left. Each of the three sp^2 hybrid orbitals, shown in blue, contains a pair of bonding electrons. Two of these pairs bond the carbon atom to hydrogen atoms; the third pair joins the carbon atom at the left to the other carbon atom, shown at the right. The two lobes of the p orbital shown in red at the left contain the fourth valence electron. The carbon atom at the right has exactly the same structure; there are two electrons in each of three sp^2 hybrid orbitals and one electron in an unhybridized p orbital.

The electron pairs in the sp^2 orbitals of the carbon atoms in ethylene form sigma bonds, shown in blue at the bottom of Figure 9.12. The two

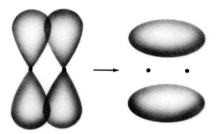

Figure 9.11 In a pi bond, the electron density is concentrated in two lobes, one above the bond axis, the other below it. Pi bonds are formed when two p orbitals overlap side-to-side.

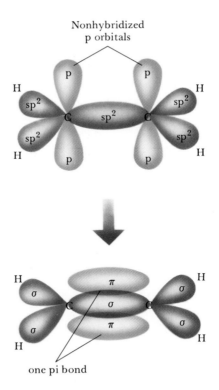

Figure 9.12 Bonding in ethylene.

parallel p orbitals overlap sideways to form a pi bonding orbital; the two lobes of this orbital are shown in red at the bottom of Figure 9.12. We would say that the double bond in ethylene consists of one sigma bond and one pi bond.

The bonding in the acetylene molecule

$$H{-}C{\equiv}C{-}H$$

is explained in a similar way (Fig. 9.13). The triple bond in acetylene consists of a sigma bond and two pi bonds. The lobes of the pi orbitals are wrapped around the sigma orbital like a roll around a hot dog.

As a general rule, we can say that in molecules and polyatomic ions:

— *each single covalent bond is a sigma bond.*

— *each double covalent bond consists of a sigma bond and a pi bond.*

— *each triple covalent bond consists of one sigma bond and two pi bonds.*

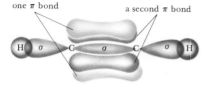

Figure 9.13 Bonding in acetylene, C_2H_2.

■ **EXAMPLE 9.7** _____

Give the number of sigma and pi bonds in molecules with the following Lewis structures:

a. H—C≡C—C̈l: :Ö═N̈—C̈l:

Solution

a. In C_2HCl, there are   three sigma bonds,   the two single bonds and one of the bonds in the triple bond between the carbon atoms. There are   two pi bonds,   the remaining bonds in the triple bond.

b. There are three bonds in the molecule, a double bond and a single bond. Of these,   two are sigma bonds   (the single bond and one of the double bonds). There is   one pi bond,   formed by the "extra" electron pair in the double bond.

Exercise
Give the number of sigma and pi bonds in CO_2. Answer: two of each.

9.5
MOLECULAR ORBITALS

In Section 9.3, we used valence bond theory to explain bonding in molecules. It accounts, at least qualitatively, for the stability of the covalent bond in terms of the overlap of atomic orbitals. By invoking hybridization, valence bond theory can account for the molecular geometries predicted by electron-pair repulsion. Where Lewis structures are inadequate, as in SO_2, the concept of resonance allows us to explain the observed properties.

A major weakness of valence bond theory has been its inability to predict the magnetic properties of molecules. We discussed this problem in Chapter 8 with regard to the O_2 molecule. The same problem arises with the B_2 molecule found in boron vapor at high temperatures. This molecule, like O_2, is paramagnetic but has an even number (six) of valence electrons. The octet rule, or valence bond theory, would predict that all electrons are paired in these molecules. This is inconsistent with their paramagnetism, which requires two unpaired electrons.

The deficiencies of valence bond theory arise from an inherent weakness. It assumes that the electrons in a molecule occupy atomic orbitals of the individual atoms. In the CH_4 molecule, the bonding is described in terms of the 1s orbitals of the H atom and the four sp^3 orbitals of the C atom. Clearly, this is an approximation. Each bonding electron in CH_4 must really be in an orbital characteristic of the molecule as a whole.

Following this idea, molecular orbital theory tries to treat bonds in terms of orbitals involving an entire molecule. The molecular orbital approach involves three basic operations:

1. The atomic orbitals of atoms are combined to give a new set of molecular orbitals characteristic of the molecule as a whole. Here, *the number of molecular orbitals formed is equal to the number of atomic orbitals combined.* When two H atoms combine to form H_2, two s orbitals, one from each atom, yield two molecular orbitals. In another case, six p orbitals, three from each atom, give a total of six molecular orbitals. In general, the atomic orbitals involved must be of comparable energy. We expect a 2s orbital of one

An MO is a mix of two atomic orbitals, one from each atom in the bond

lithium atom to combine with the 2s orbital of another Li atom, but not with a 1s orbital, which has a much lower energy.

2. The molecular orbitals are arranged in order of increasing energy. In principle these energies can be obtained by solving the Schrödinger equation (Ch. 5). In practice, we cannot make precise calculations for any but the simplest of molecules. Instead, the relative energies of molecular orbitals are deduced from experimental observations. Spectra and magnetic properties of molecules are used.

3. The valence electrons in a molecule are distributed among the available molecular orbitals. The process followed is much like that used with electrons in atoms. In particular, we find the following:

a. *Each molecular orbital can hold a maximum of two electrons.*
b. *Electrons go into the lowest energy molecular orbital available.* A higher orbital starts to fill only when each orbital below it has its quota of two electrons.
c. *Hund's rule is obeyed.* When two orbitals of equal energy are available to two electrons, one electron goes into each, giving two half-filled orbitals.

To illustrate molecular orbital theory, we will apply it to the diatomic molecules of the elements in the first two periods of the Periodic Table.

We find MO structures in much the same way that we obtain electron configurations

HYDROGEN AND HELIUM; COMBINATION OF 1s ORBITALS

Molecular orbital (MO) theory predicts that two 1s orbitals will combine to give two molecular orbitals. One of these has an energy lower than that of the atomic orbitals from which it is formed (Fig. 9.14). Placing electrons in this orbital gives a species that is more stable than the isolated atoms. For that reason the lower molecular orbital in Figure 9.14 is called a **bonding orbital**. The other molecular orbital has a higher energy than the corresponding atomic orbitals. Electrons entering it are in an unstable, higher energy state. It is referred to as an **antibonding orbital**. In energy, this orbital lies about as far above the individual atomic orbitals as the bonding orbital lies below them.

The electron density in these molecular orbitals is shown at the right of Figure 9.14. Notice that the bonding orbital has a high density between

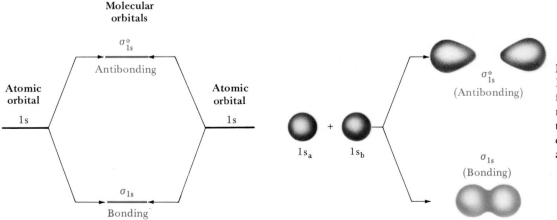

Figure 9.14 Molecular orbital formation. Two molecular orbitals are formed by combining two 1s atomic orbitals.

the nuclei. This accounts for its stability. In the antibonding orbital, the chance of finding the electron between the nuclei is very small. The electron density is concentrated at the far ends of the "molecule." This means that the nuclei are less shielded from each other than they are in the isolated atoms. Small wonder that the antibonding orbital is higher in energy than the atomic orbitals from which it is formed.

The electron density in both molecular orbitals is symmetrical about the axis between the two nuclei. This means that both of these are sigma orbitals. In MO notation, the 1s bonding orbital is designated as σ_{1s}. The antibonding orbital is given the symbol σ_{1s}^*. An asterisk designates an antibonding orbital.

In the H_2 molecule, there are two 1s electrons. They fill the σ_{1s} orbital, giving a single bond. In the He_2 molecule, there would be four electrons, two from each atom. These would fill the bonding and antibonding orbitals. As a result, the number of bonds (the *bond order*) in He_2 is zero. The general relation is

$$\text{no. of bonds} = \text{bond order} = \frac{B - AB}{2} \tag{9.2}$$

where B is the number of electrons in bonding orbitals and AB is the number of electrons in antibonding orbitals. In H_2, B = 2 and AB = 0, so we have one bond. In He_2, B = AB = 2, so the number of bonds is zero. The stability of He_2 would be no greater than that of two isolated He atoms. The He_2 molecule should not and does not exist.

Electrons in antibonding orbitals counteract the effect of those in bonding orbitals

SECOND PERIOD ELEMENTS; COMBINATION OF 2s AND 2p ORBITALS

Among the diatomic molecules of the second period elements are three familiar ones, N_2, O_2, and F_2. The molecules Li_2, B_2, and C_2 are less common but have been observed and studied in the gas phase. In contrast, the molecules Be_2 and Ne_2 are either highly unstable or nonexistent. Let us see what molecular orbital theory predicts about the structure and stability of these molecules. We start by considering how the atomic orbitals containing the valence electrons (2s and 2p) are used to form molecular orbitals.

Combining two 2s atomic orbitals, one from each atom, gives two molecular orbitals. These are very similar to the ones discussed above. They are designated as σ_{2s}(sigma, bonding, 2s) and σ_{2s}^*(sigma, antibonding, 2s).

Consider now what happens to the 2p orbitals. In an isolated atom, there are three such orbitals, oriented at right angles to each other. We call these atomic orbitals p_x, p_y, and p_z (top of Figure 9.15, p. 304). When two p_x atomic orbitals, one from each atom, overlap head-to-head, they form two sigma orbitals. These are a bonding orbital, σ_{2p}, and an antibonding orbital, σ_{2p}^*. The situation is quite different when the p_z orbitals overlap. Since they are oriented parallel to one another, they overlap side-to-side (Fig. 9.15c). The two molecular orbitals formed in this case are pi orbitals; one is a bonding orbital, π_{2p}, the other a nonbonding orbital, π_{2p}^*. In an entirely similar way, the p_y orbitals of the two atoms interact to form another pair of pi molecular orbitals, π_{2p} and π_{2p}^* (these orbitals are not shown in Fig. 9.15).

Figure 9.15 When p orbitals from two different atoms (a) overlap, there are two quite different possibilities. If they overlap head-to-head (b), two sigma molecular orbitals are produced. If, on the other hand, they overlap side-to-side (c), two pi molecular orbitals result.

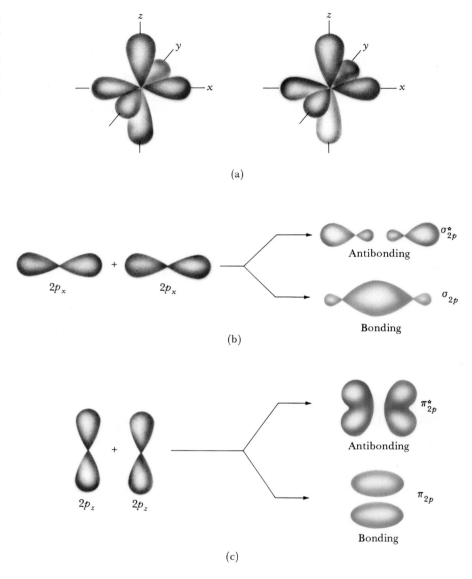

(a)

(b)

(c)

Summarizing the molecular orbitals available to the valence electrons of the second period elements, we have

From 2s atomic orbitals   one σ_{2s} and one σ_{2s}^* orbital.

From 2p$_x$ atomic orbitals   one σ_{2p} and one σ_{2p}^* orbital.

From 2p$_y$ and 2p$_z$ atomic orbitals   two π_{2p} and two π_{2p}^* orbitals.

The relative energies of these orbitals are shown in Figure 9.16. This order applies in the molecules of the second period elements, at least through N_2.*

---

*It appears that beyond N_2, the σ_{2p} orbital lies below the two π_{2p} orbitals. This does not affect the filling order because, in O_2, all three of these orbitals are filled with electrons.

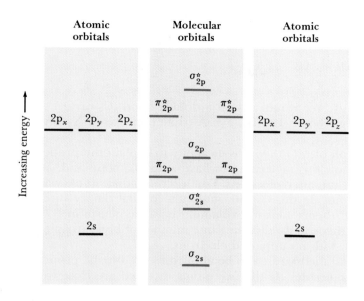

Figure 9.16 Relative energies, so far as filling order is concerned, for the molecular orbitals formed by combining 2s and 2p atomic orbitals.

To obtain the MO structure of the diatomic molecules of the elements in the second period, we fill the available molecular orbitals in order of increasing energy. The results are shown in Table 9.4. Note the agreement between MO theory and the properties of these molecules. In particular,

TABLE 9.4 Predicted and Observed Properties of Diatomic Molecules of Second Period Elements

	OCCUPANCY OF ORBITALS							
	σ_{2s}	σ_{2s}^*	π_{2p}	π_{2p}	σ_{2p}	π_{2p}^*	π_{2p}^*	σ_{2p}^*
Li_2	(⇅)	( )	( )	( )	( )	( )	( )	( )
Be_2	(⇅)	(⇅)	( )	( )	( )	( )	( )	( )
B_2	(⇅)	(⇅)	(↑)	(↑)	( )	( )	( )	( )
C_2	(⇅)	(⇅)	(⇅)	(⇅)	( )	( )	( )	( )
N_2	(⇅)	(⇅)	(⇅)	(⇅)	(⇅)	( )	( )	( )
O_2	(⇅)	(⇅)	(⇅)	(⇅)	(⇅)	(↑)	(↑)	( )
F_2	(⇅)	(⇅)	(⇅)	(⇅)	(⇅)	(⇅)	(⇅)	( )
Ne_2	(⇅)	(⇅)	(⇅)	(⇅)	(⇅)	(⇅)	(⇅)	(⇅)

	PREDICTED PROPERTIES		OBSERVED PROPERTIES	
	Number of Unpaired e^-	*Bond Order*	*Number of Unpaired* e^-	*Bond Energy (kJ/mol)*
Li_2	0	1	0	105
Be_2	0	0	0	unstable
B_2	2	1	2	289
C_2	0	2	0	628
N_2	0	3	0	941
O_2	2	2	2	494
F_2	0	1	0	153
Ne_2	0	0	0	nonexistent

the number of unpaired electrons predicted agrees with experiment. There is also a general correlation between the predicted bond order:

$$\text{bond order} = \frac{B - AB}{2}$$

and the bond energy. We would expect a double bond (C_2 or O_2) to be stronger than a single bond (Li_2, B_2, F_2). A triple bond, as in N_2, should be still stronger.

A major triumph of MO theory is its ability to explain the properties of O_2. It explains how this molecule can have a double bond and, at the same time, have two unpaired electrons. You will recall that simple valence bond theory could not do this. Again, molecular orbital theory succeeds where valence bond theory fails in explaining paramagnetism in B_2. For N_2 and F_2, the two theories predict the same number of bonds (triple and single, respectively), with no unpaired electrons.

Molecular orbital theory can also be applied to predict the properties of polar diatomic molecules. If the two atoms involved are close to one another in the same period of the Periodic Table, their orbital energies are similar and we can use the approach described above to find the MO structure.

■ **EXAMPLE 9.8** ————————————————————————
Using MO theory, predict the electronic structure, number of bonds, and number of unpaired electrons in the NO molecule.

Solution
We have 11 valence electrons to account for (five from nitrogen, six from oxygen). Placing these in molecular orbitals in order of increasing energy,

$$\sigma_{2s} \quad \sigma_{2s}^* \quad \pi_{2p} \quad \pi_{2p} \quad \sigma_{2p} \quad \pi_{2p}^* \quad \pi_{2p}^* \quad \sigma_{2p}^*$$

NO   (↑↓) (↑↓) (↑↓) (↑↓) (↑↓) (↑) ( ) ( )

There are eight electrons in bonding orbitals (two in σ_{2s}, four in π_{2p}, two in σ_{2p}) and three in antibonding orbitals (two in σ_{2s}^*, one in π_{2p}^*). Hence,

$$\text{no. bonds} = \frac{8 - 3}{2} = \frac{5}{2}$$

There is one unpaired electron (in the π_{2p}^* orbital).

Exercise
How many bonds and unpaired electrons are there in NO^+? Answer: 3, 0.

We would predict the NO bond to be stronger than the O_2 bond, and it is

The bonding in molecules containing more than two atoms can also be described in terms of molecular orbitals. We will not attempt to do this here; the energy level structure is considerably more complex than the one we have considered. However, one point is worth mentioning. In polyatomic species, a pi molecular orbital can be considered to be spread over the entire molecule rather than being concentrated between two atoms.

The electrons in such an orbital are shared by all the atoms in the molecule rather than being "localized" between two specific atoms.

To illustrate what is meant by "delocalized" molecular orbitals, consider the benzene molecule, C_6H_6. It is known that the molecule is a planar hexagon; the bond angles are 120° and all the carbon-carbon bond distances are 0.140 nm, intermediate between the values to be expected for single and double bonds. In the valence bond model, this evidence is interpreted to mean that there are two resonance forms:

These are most often abbreviated as:

Molecular orbital theory gives a more plausible picture of the carbon-carbon bonding in C_6H_6. It tells us that the hexagonal framework is established by six C—C sigma bonds, shown as straight lines in Figure 9.17. That leaves three electron pairs that are spread symmetrically over the entire molecule to form delocalized pi bonds. These are shown as doughnut-like lobes in Figure 9.17. This model for benzene is often represented as:

By MO theory, the pi electron density is constant all around the ring

with the understanding that:

— there is a carbon atom at each corner of the hexagon.
— there is a H atom bonded to each carbon atom.
— the circle in the center of the molecule represents the three pairs of delocalized electrons.

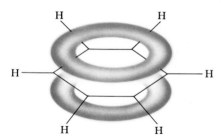

Figure 9.17 In benzene, three electron pairs are not localized on particular carbon atoms. Instead, they are spread out over two lobes of the shape shown, one above the plane of the benzene ring and the other below it.

■ SUMMARY PROBLEM

Consider the molecules: SO_3, XeO_3, XeO_4, and XeF_2

a. Draw the Lewis structures of these molecules.
b. Describe their geometries, including ideal bond angles.
c. Classify each molecule as being polar or nonpolar.
d. Give the hybridization of the central atom in each molecule.
e. State the number of pi and sigma bonds in each molecule.

Answers

a.

b. SO_3: equilateral triangle, 120° XeO_3: triangular pyramid, 109.5°
   XeO_4: tetrahedron, 109.5° XeF_2: linear, 180°
c. All nonpolar except XeO_3 **d.** sp^2 in SO_3; sp^3 in XeO_3 and XeO_4;
sp^3d in XeF_2 **e.** SO_3: 3 σ, 1 π; XeO_3: 3 σ; XeO_4: 4 σ; XeF_2: 2 σ

Among the molecular structures described in this chapter were those of several compounds of the noble gas xenon. Until about 25 years ago, the noble gases were referred to as "inert" gases. They were believed to be completely unreactive. The first noble gas compound was discovered at the University of British Columbia in 1962 by Neil Bartlett, a 29-year-old chemist. In the course of his research on platinum-fluorine compounds, he isolated a reddish solid that he showed to be $O_2^+(PtF_6^-)$. Bartlett realized that the ionization energy of Xe (1170 kJ/mol) is virtually identical to that of the O_2 molecule (1165 kJ/mol). This encouraged him to attempt to make the analogous compound $XePtF_6$. His success opened up a new era in noble gas chemistry.

The most stable binary compounds of xenon are the three fluorides, XeF_2, XeF_4, and XeF_6. Xenon difluoride can be prepared quite simply by exposing a 1:1 mol mixture of xenon and fluorine to ultraviolet light; colorless crystals of XeF_2 (mp = 129°C) form slowly.

$$Xe(g) + F_2(g) \rightarrow XeF_2(s)$$

The higher fluorides are formed using excess fluorine. For example, XeF_4 (Fig. 9.18) is formed when a 1:5 mol mixture of the elements is heated to 400°C:

$$Xe(g) + 2\ F_2(g) \rightarrow XeF_4(s)$$

The hexafluoride can be prepared by heating a 1:20 mol mixture of xenon and fluorine at high pressures:

$$Xe(g) + 3\ F_2(g) \rightarrow XeF_6(s)$$

The fluorides of xenon are stable in dry air at room temperature. However, they react with water to form compounds in which one or more of the fluorine atoms has been replaced by oxygen. Thus xenon hexafluoride reacts rapidly with water to give the trioxide:

$$XeF_6(s) + 3\ H_2O(l) \rightarrow XeO_3(s) + 6\ HF(g)$$

Xenon trioxide is highly unstable; it detonates if warmed above room temperature. It has a positive heat of formation of about +400 kJ/mol, which helps to explain why XeO_3 cannot be prepared by reacting the elements with one another.

Most of the compounds of xenon have expanded octet structures (XeO_3 is an exception). The VSEPR model has been strikingly successful in explaining the geometries of such molecules as XeF_2 (linear), XeF_4 (square planar), and XeO_3 (triangular pyramid). The XeF_6 molecule is unique among hexafluorides in that it does not have a regular octahedral structure. This is predictable by VSEPR theory. There are seven pairs of electrons around the xenon atom in XeF_6, one of which is unshared.

The chemistry of xenon is much more extensive than that of any other noble gas. Only one binary compound of krypton, KrF_2, has been prepared. It is a colorless solid with a positive heat of formation and decomposes at room temperature. The chemistry of radon is difficult to

Figure 9.18 Crystals of xenon tetrafluoride, XeF_4. (Argonne National Laboratory)

study because all of its isotopes are radioactive. Indeed, the radiation given off is so intense that it decomposes any reagent added to radon in an attempt to bring about a reaction.

There has been a great deal of speculation about the possibility of forming compounds of argon. Like krypton and xenon, argon has d orbitals available for bonding and so should be able to form expanded-octet molecules such as ArF_2 or ArF_4. However, all attempts to prepare such molecules have failed. It is now generally agreed that the Ar—F bond is weak, with a bond energy close to zero. In contrast, calculations suggest that the Ar—O bond might be strong enough to permit formation of the compound ArO_3, analogous to XeO_3. Unfortunately, no one knows how to make a noble gas oxide without going through the fluoride.

QUESTIONS AND PROBLEMS

Molecular Geometry

1. Describe the geometry of each of the following molecules. (The way the Lewis structure is written does not necessarily imply the geometry.)

a. :Ö—S̈=O: b. :N≡N—Ö:

c. :S̈=C=O: d. :C̈l—N̈—H
                            |
                            H

2. Describe the geometry of each of the following molecules. (The way the Lewis structure is written does not necessarily imply the geometry.)

a. :C̈l—Be—C̈l: b. :C̈l—G̈e—C̈l:

c. :O=N̈—C̈l: d. H—C̈—B̈r:
                            |
                          :B̈r:

3. Describe the geometry of the following polyatomic ions:

a. $\left[:\ddot{O}—\ddot{N}=\ddot{O}\right]^{-}$ b. $\left[\begin{array}{c} H \\ | \\ H—N—H \\ | \\ H \end{array}\right]^{+}$

c. $\left[\begin{array}{c} :\ddot{O}—\ddot{N}—\ddot{O}: \\ \| \\ :O: \end{array}\right]^{-}$ d. $\left[:\ddot{S}=C=\ddot{N}:\right]^{-}$

4. Describe the geometry of the following polyatomic ions:

a. $\left[\begin{array}{c} :\ddot{O}: \\ | \\ :\ddot{O}—Mn—\ddot{O}: \\ | \\ :\ddot{O}: \end{array}\right]^{-}$ b. $\left[:\ddot{O}—\ddot{O}—\ddot{O}:\right]^{2-}$

c. $\left[\begin{array}{c} :\ddot{O}—C—\ddot{O}: \\ \| \\ :O: \end{array}\right]^{2-}$ d. $\left[\begin{array}{c} :\ddot{O}—\ddot{C}l—\ddot{O}: \\ | \\ :\ddot{O}: \end{array}\right]^{-}$

5. Predict the geometry of the following molecules:
a. RnF_4 b. PCl_5 c. SeF_6 d. $TeBr_4$
6. Predict the geometry of the following molecules:
a. $SeCl_4$ b. SF_5Cl c. KrF_2 d. IF_3
7. Give all the ideal bond angles (109.5°, 120°, or 180°) in the following molecules and ions.

a. $\left[:\ddot{O}—\ddot{N}=\ddot{O}\right]^{-}$ b. $H—\overset{\overset{H}{|}}{\underset{\underset{H}{|}}{C}}—C≡C—H$

c. $\left[\begin{array}{c} :\ddot{O}: \\ | \\ :\ddot{O}—Si—\ddot{O}: \\ | \\ :\ddot{O}: \end{array}\right]^{4-}$

8. Give all the ideal bond angles (109.5°, 120°, or 180°) in the following molecules and ions.

a. $\left[\begin{array}{c} :\ddot{O}—C—\ddot{O}: \\ \| \\ :O: \end{array}\right]^{2-}$ b. $Cl—\overset{\overset{H}{|}}{C}=\overset{\overset{H}{|}}{C}—H$

c. 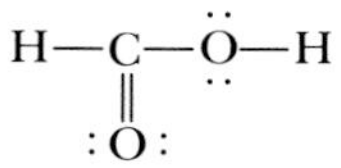 H—C—O—H
‖
:O:

9. Draw the Lewis structure and describe the geometry of
   a. HOCl   b. OCN⁻   c. NF₃   d. N₃⁻
10. Draw the Lewis structure and describe the geometry of
   a. C₂HCl   b. HCO₂⁻   c. CH₃Cl   d. PCl₄⁺
11. Describe the geometry of the species in which there are, around the central atom,
   a. three bonds and two unshared pairs.
   b. six bonds.
   c. four bonds and one unshared pair.
12. Describe the geometry of the species in which there are, around the central atom,
   a. four bonds and two unshared pairs.
   b. five bonds and one unshared pair.
   c. five bonds.
13. Draw the Lewis structure and describe the geometry of
   a. SiF₄   b. BrO₃⁻   c. HOClO   d. AsCl₃
14. Draw the Lewis structure and describe the geometry of
   a. Cl₂CO   b. NI₃   c. PO₃³⁻   d. O₃
15. For the structures shown in Question 3, determine
   a. the number of unshared pairs around the central atom.
   b. the number of atoms bonded to the central atom.
   c. the bond angle, using Table 9.1.
16. For the structures shown in Question 4, follow the directions of Question 15.
17. Peroxypropionyl nitrate (PPN) is an eye irritant found in smog. Its skeleton structure is

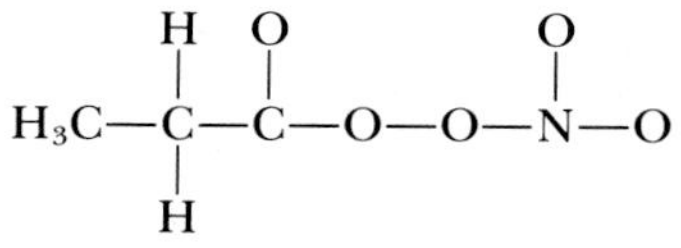

H   O   O
| | |
H₃C—C—C—O—O—N—O
|
H

Write the Lewis structure for PPN and give all the bond angles.
18. An objectionable component of smoggy air is acetyl-peroxide, which has the skeleton structure

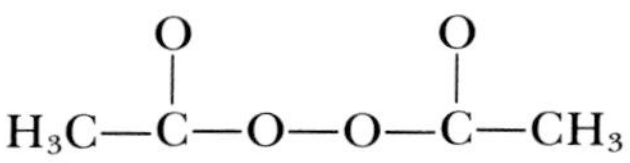

O   O
| |
H₃C—C—O—O—C—CH₃

   a. Draw the Lewis structure of this compound.
   b. Indicate all the bond angles.
19. List all the bond angles in the following hydrocarbons.
   a. 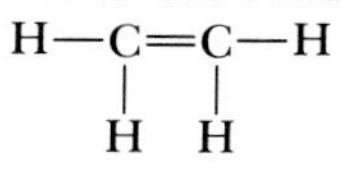 H—C=C—H   b. C₂H₆
        | |
        H   H
   c. H—C≡C—H
20. List all the bond angles in the following hydrocarbons.
   a. CH₄   b. H₃C—C≡C—H

c. 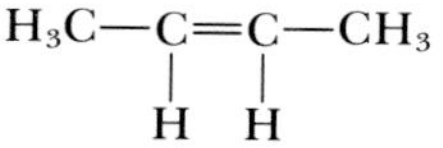 H₃C—C=C—CH₃
        | |
        H   H

21. In each of the following molecules, a central atom is surrounded by a total of three atoms or unshared pairs: BCl₃, SnCl₂, SO₂. In which of these molecules would you expect the bond angle to be less than 120°? Explain your reasoning.
22. Consider the following molecules: SiH₄, PH₃, H₂S. In each case, a central atom is surrounded by four electron pairs. In which of these molecules would you expect the bond angle to be less than 109.5°? Explain your reasoning.

Molecular Polarity
23. Which of the species in Question 1 are dipoles?
24. Which of the species in Question 2 are dipoles?
25. Which of the species in Question 3 are dipoles?
26. Which of the species in Question 4 are dipoles?
27. There are three compounds with the formula C₂H₂Cl₂:

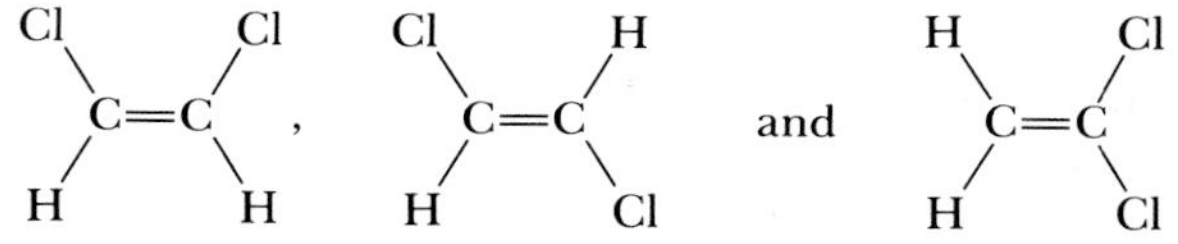

Cl   Cl   Cl   H   H   Cl
  \\   /   \\   /   \\   /
    C=C   ,   C=C   and   C=C
  /   \\   /   \\   /   \\
H   H   H   Cl   H   Cl

Which of these molecules are polar? Nonpolar?
28. There are two different molecules with the formula N₂F₂:

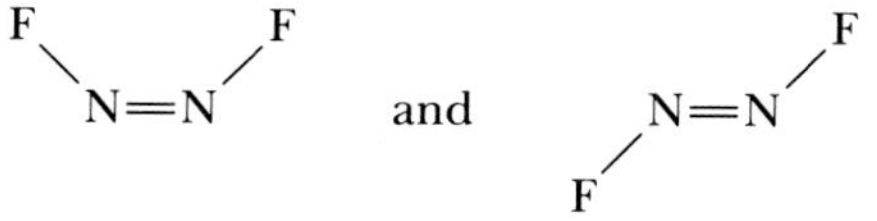

F   F   F
  \\   /   /
    N=N   and   N=N
                                /
                              F

Is either molecule polar? Explain.
29. Taking the dipole moment of HCl to be 3.43×10^{-30} C·m, the internuclear distance in HCl to be 0.127 nm, and the charge of the electron to be 1.60×10^{-19} C, estimate the percent ionic character in the H—Cl bond. Follow the text discussion for HF.
30. Follow the directions of Question 29 for the HI molecule. The dipole moment can be found in Table 9.2; the internuclear distance is 0.161 nm.

Hybridization
31. Give the hybridization of the central atom in each molecule in Question 1.
32. Give the hybridization of the central atom in each molecule in Question 2.
33. Give the hybridization of the central atom in each molecule in Question 3.
34. Give the hybridization of the central atom in each molecule in Question 4.
35. Consider the species in Question 5.
   a. How many electron pairs are there around the central atom in each species?
   b. What is the hybridization of the central atom?

36. Consider the species in Question 6. Answer the questions in Question 35.

37. In each of the following polyatomic ions, the central atom has an expanded octet. Determine the number of electron pairs around the central atom and the hybridization in

a. ClF_2^- b. $GeCl_4^{2-}$ c. ICl_4^-

38. Follow the directions of Question 37 for the following polyatomic ions:

a. XeF_3^- b. $SiCl_6^{2-}$ c. PCl_4^-

39. Give the hybridization of each C, N, and O atom in nitrobenzene (unshared electron pairs are not shown):

40. Give the hybridization of each C, N, and O atom in 3-pyridine carboxylic acid (unshared electron pairs are not shown):

41. What is the hybridization of nitrogen in

a. $\left[O-N-O \atop \quad\; \| \atop \quad\; O \right]^-$ b. $H-\overset{\cdot\cdot}{N}-H \atop \quad\;\; |\atop \quad\;\; Cl$

c. $:N\equiv N:$ d. $:N\equiv N-O$

42. What is the hybridization of carbon in

a. CH_3Cl b. $\left[O-C-O \atop \quad\;\; \| \atop \quad\;\; O \right]^{2-}$

c. $O=C=O$ d. $H-C-OH \atop \quad\;\; \| \atop \quad\;\; O$

43. What is the hybridization of the central atom (underlined) in

a. $F_2\underline{C}O$ b. $\underline{O}F_2$ c. $O\underline{P}Cl_3$

44. What is the hybridization of the central atom (underlined) in

a. $HO\underline{I}O_2$ b. $(H_2N)_2\underline{C}O$ c. $\underline{N}HF_2$

Sigma and Pi Bonds

45. Give the number of sigma and pi bonds in each species listed in Question 41.

46. Give the number of sigma and pi bonds in each species listed in Question 42.

47. Give the formula of an ion or molecule in which an atom of

a. B forms four bonds using sp^3 hybrid orbitals.

b. C forms two pi bonds.

c. N forms a pi and two sigma bonds.

d. S forms a pi bond.

48. Give the formula of an ion or molecule in which an atom of

a. B forms three bonds using sp^2 hybrid orbitals.

b. O forms a pi bond.

c. N forms four bonds using sp^3 hybrid orbitals.

d. C forms four bonds, in three of which it uses sp^2 hybrid orbitals.

Molecular Orbitals

49. Using MO theory, list the number of electrons in each of the 2s and 2p molecular orbitals, the number of bonds, and the number of unpaired electrons in

a. O_2^- b. CO   c. F_2^-

50. Follow the directions of Problem 49 for

a. O_2 b. C_2^- c. O_2^{2-}

51. Suppose in building up molecular orbitals the σ_{2p} were placed below the π_{2p}. Prepare a diagram similar to Table 9.4 based on this assignment. For which species would this change in relative energies affect the prediction of number of bonds? Number of unpaired electrons?

52. Consider the +1 ions formed by removing an electron from each of the diatomic molecules of the elements of atomic numbers 3 through 9. Using the MO approach, determine the number of bonds and unpaired electrons in each of these ions.

53. Using MO theory, predict the number of bonds in

a. OF^+ b. H_2^+ c. CN^-

54. Using MO theory, predict the number of bonds in

a. OF^- b. H_2^- c. CN^+ d. F_2

Unclassified

55. Describe how you predict

a. the geometry around a central atom surrounded by five electron pairs.

b. the approximate bond angles in a molecule.

c. the hybridization of an atom in a molecule.

56. Describe how you predict

a. the number of sigma and pi bonds in a molecule.

b. whether or not a molecule is polar.

c. the number of valence electrons around the central atom in a nonmetal halide.

57. Define or describe the meaning of each of the following terms:

a. dipole   b. dipole moment

c. hybrid orbital   d. sigma bond

58. Use molecular orbital theory to explain why diatomic beryllium is not stable.

59. Consider the $S_2O_3^{2-}$ ion, which has a skeleton similar to that of SO_4^{2-}.

a. Draw the Lewis structure.

b. Describe the geometry, including the approximate bond angles.

c. Is this species a dipole?

d. What is the hybridization for each atom?

60. Consider the SF_6 molecule.

a. Draw the Lewis structure.

b. Describe the geometry around the sulfur atom.

c. Is this species a dipole?

d. What is the hybridization about the sulfur?

61. Consider acetyl salicylic acid, better known as aspirin. Its structure is

a. How many sigma and pi bonds are there in aspirin?

b. What are the approximate values of the angles marked (in blue) A, B, and C?

c. What is the hybridization of each atom marked (in red) 1, 2, and 3?

62. Complete the following table

Challenge Problems

63. A compound of chlorine and fluorine, ClF_x, reacts at about 75°C with uranium to produce uranium hexafluoride and chlorine fluoride, ClF. A certain amount of uranium produced 5.63 g of uranium hexafluoride and 457 mL of chlorine fluoride at 75°C and 3.00 atm. What is x? Using valence bond theory, describe the geometry, polarity, and bond angle of the compound and the hybridization of chlorine. How many sigma and pi bonds are there?

64. Draw the Lewis structure and describe the geometry of the hydrazine molecule, N_2H_4. Would you expect this molecule to be polar?

65. The geometries shown in Figure 9.6 are not the only ones possible for species in which the central atom has an expanded octet and one or more pairs of unshared electrons. Suggest other possible geometries for ClF_3 and XeF_2. Can you explain, using the VSEPR model, why the geometries in Figure 9.6 are preferred?

66. Consider the polyatomic ion IO_6^{5-}. How many pairs of electrons are there around the central iodine atom? What is its hybridization? Describe the geometry of the ion.

67. There are two compounds with the molecular formula C_3H_6.

a. Write the Lewis structure for each one.

b. What is the hybridization of each carbon atom in each molecule?

c. Describe the geometry and ideal bond angles in the two molecules.

68. The molecule XeF_6 does not have an octahedral structure even though there are six atoms bonded to xenon. Explain this and suggest a possible geometry for the molecule.

SPECIES	ATOMS AROUND CENTRAL ATOM A	UNSHARED PAIRS AROUND A	GEOMETRY	HYBRIDIZATION	POLARITY (Assume all X atoms the same)
AX_2E_2	___	___	___	___	___
___	3	0	___	___	___
AX_4E_2	___	___	___	___	___
___	___	___	triangular bipyramid	___	___

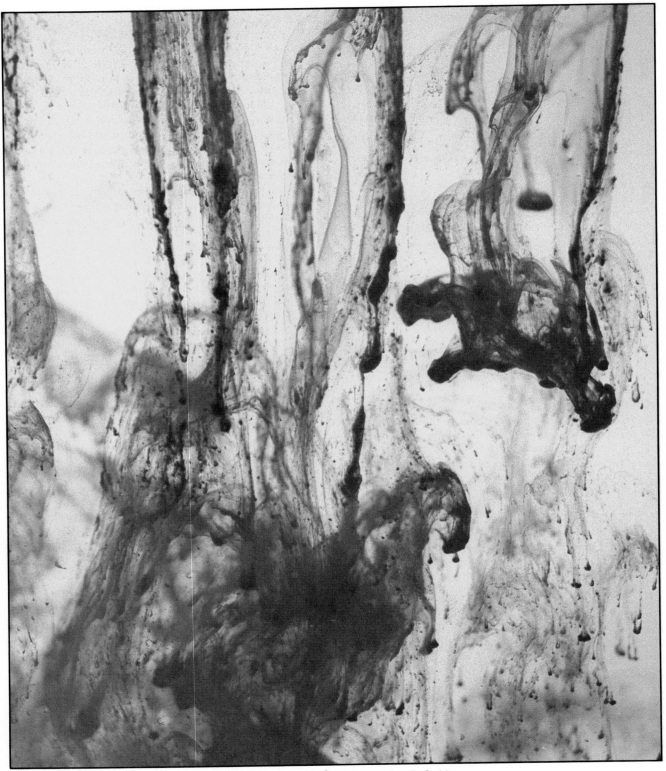

$2Ag^+(aq) + CrO_4^{2-}(aq) \rightarrow Ag_2CrO_4(s)$
Precipitation of silver chromate (red)

CHAPTER
10

LIQUIDS AND
SOLIDS

*See plastic nature working to this end,
The single atoms each to other tend.
Attract, attracted to, the next in place
Form'd and impell'd its neighbor to embrace.*

ALEXANDER POPE

C hapter 4 was devoted to a study of the physical behavior of gases. We found that, at ordinary temperatures and pressures, all gases behave similarly in that they follow the Ideal Gas Law. In this chapter we will examine liquids and solids. Unfortunately, there are no simple relations analogous to the Ideal Gas Law that can be used to correlate the properties of liquids and solids. In these two states of matter, each substance has its own unique properties, quite different from those of other substances.

There is a simple explanation for this difference in behavior between gases on the one hand and liquids or solids on the other. In a liquid or solid, the particles are much closer together than they are in a gas. Indeed, the particles—which may be atoms, ions, or molecules—touch each other. As a result, attractive forces become important; they play a major role in determining the physical properties of liquids and solids. Since attractive forces between particles vary in nature and magnitude from one substance to another, different liquids and solids have quite different properties.

In studying liquids and solids, we will be concerned with two major topics:

1. The equilibria between different phases of a pure substance. We start by examining the conditions under which a liquid is in equilibrium with its vapor (Section 10.1). Then, in Section 10.2, we will consider the principles that relate to all three types of phase equilibria (liquid-vapor, solid-vapor, liquid-solid).

2. The relation between particle structure and physical properties. We will look at the four structural types of substances: *molecular, network covalent, ionic,* and *metallic* (Sections 10.3–10.6). Substances in each of these categories have characteristic properties that are directly related to the strength of the forces between particles.

We know a lot more about gases than we do about liquids

315

Crystalline solids differ from liquids in one very important way. The particles in a crystal, whether they are atoms, molecules, or ions, cannot move past one another. Instead, their motion is restricted to small vibrations about a fixed position. In Section 10.7, we will look at the crystal structures of some representative metallic and ionic solids.

10.1
LIQUID-VAPOR EQUILIBRIUM

A lake is an open container

All of us are familiar with the process of vaporization, in which a liquid is converted to a vapor. In an open container, *evaporation* continues until all the liquid is gone. If the container is closed, the situation is quite different. At first, the movement of molecules is primarily in one direction, from liquid to vapor. Here, however, the vapor molecules cannot escape from the container. Some of them collide with the surface and re-enter the liquid. As time passes, and the concentration of molecules in the vapor increases, so does the rate of condensation. Eventually, it becomes equal to the rate of vaporization. When this happens, we say that the liquid and vapor are in a state of *dynamic equilibrium:*

$$\text{liquid} \rightleftharpoons \text{vapor}$$

The double arrow implies that the forward process (vaporization) and the reverse process (condensation) are occurring at the same rate.

VAPOR PRESSURE

Once equilibrium is reached between liquid and vapor, the number of molecules per unit volume in the vapor does not change with time. This means that *the pressure exerted by the vapor over the liquid remains constant.* The pressure of vapor in equilibrium with a liquid is called the **vapor pressure**. This quantity is a characteristic property of a given liquid at a particular temperature. It varies from one liquid to another, depending upon the strength of the intermolecular forces. At 25°C, the vapor pressure of water is 24 mm Hg; that of benzene at the same temperature is 92 mm Hg.

It is important to realize that

as long as both liquid and vapor are present, the pressure exerted by the vapor is independent of the volume of the container

To see what this statement means, consider Figure 10.1. This shows the vaporization of a liquid at constant temperature in a container whose volume can be varied by raising the piston. We start by placing a small amount of liquid in the container with the piston at the level shown in Figure 10.1A. Vaporization occurs until the equilibrium vapor pressure, say, 50 mm Hg, is established. This equilibrium is then disturbed by raising the piston to the level shown in Figure 10.1B. The pressure drops for a moment. However, liquid quickly evaporates to establish the original pressure, 50 mm Hg. This process is repeated each time the piston is raised to create a larger volume. Eventually, the container volume becomes large enough so that

Even if you double the volume, the pressure holds constant, as long as some liquid is present

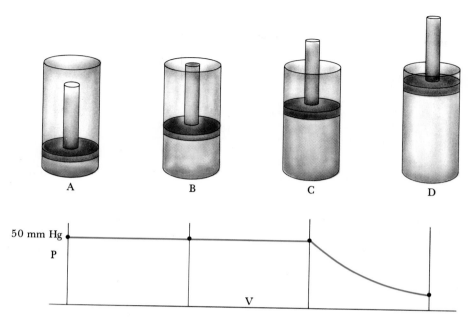

Figure 10.1   The pressure of a vapor in equilibrium with a liquid is independent of the volume of the container (A and B). When all the liquid is vaporized (C), a further increase in volume (D) decreases the pressure in accordance with Boyle's Law. Since system C consists only of gas, its volume can be calculated from the Ideal Gas Law if the temperature and amount of sample are known.

all the liquid vaporizes (Fig. 10.1C). From that point on, with no liquid present, the system behaves as an ideal gas and follows Boyle's Law (Fig. 10.1D).

The process just described can be reversed. If the vapor is compressed to the point where its pressure becomes equal to the vapor pressure of the liquid (point C in Fig. 10.1), liquid begins to condense. Further decrease in volume does not cause an increase in pressure. Instead, condensation continues, keeping the pressure of the vapor constant (points A and B in Fig. 10.1).

■ **EXAMPLE 10.1** _____

Consider a sample of 1.00 g of water, which has an equilibrium vapor pressure of 24 mm Hg at 25°C.

a. If this sample is introduced into a 10.0-L flask, how much liquid water remains when equilibrium is established?
b. What would be the minimum volume of a flask in which the 1.00-g sample would be completely vaporized?

Solution

a. Let us use the Ideal Gas Law to calculate the mass of $H_2O(g)$ at equilibrium.

$$g = \frac{P \times V \times MM}{RT} = \frac{(24/760 \text{ atm})(10.0 \text{ L})(18.0 \text{ g/mol})}{\left(0.0821 \dfrac{\text{L·atm}}{\text{mol·K}}\right)(298 \text{ K})} = 0.23 \text{ g}$$

The amount of liquid water left would be 1.00 g − 0.23 g = $\boxed{0.77 \text{ g}}$

b. Again, we use the Ideal Gas Law, this time calculating the volume.

$$V = \frac{gRT}{P \times MM} = \frac{(1.00 \text{ g}) \left(0.0821 \frac{\text{L·atm}}{\text{mol·K}}\right) (298 \text{ K})}{(24/760 \text{ atm})(18.0 \text{ g/mol})} = \boxed{43 \text{ L}}$$

Exercise

If the volume of the container were 60.0 L, how much water would be required to reach equilibrium? Answer: 1.4 g.

At $V = 60$ L, $P = 17$ mm Hg
All gas, no equilibrium

VAPOR PRESSURE VS. TEMPERATURE

The vapor pressure of a liquid always increases as temperature rises. Water evaporates more readily on a hot, dry day. Stoppers in bottles of volatile liquids such as ether or gasoline may pop out when the temperature rises.

The vapor pressure of water, which is 24 mm Hg at 25°C, becomes 92 mm Hg at 50°C and 1 atm (760 mm Hg) at 100°C. The data for water are plotted at the left of Figure 10.2. As you can see, the graph of vapor pressure vs. temperature is not a straight line, as would be the case if we plotted pressure vs. temperature for an ideal gas. Instead, the slope increases steadily. This reflects the fact that as temperature increases, a larger and larger fraction of molecules have sufficient energy to escape from the liquid. This increases the concentration of molecules in the vapor and hence increases the pressure they exert. At 100°C, the number of H_2O molecules in the vapor in equilibrium with liquid is 25 times as great as at 25°C.

In working with the relationship between two variables, such as vapor pressure and temperature, scientists prefer to deal with linear (straight-line) functions. Straight-line graphs are easier to construct and to interpret. In this case, it is possible to obtain a linear function by making a simple shift in variables. Instead of plotting vapor pressure (P) vs. temperature

Figure 10.2 A plot of vapor pressure vs. temperature for water (or any other liquid) is a curve with a steadily increasing slope. On the other hand, a plot of the logarithm of vapor pressure vs. the reciprocal of the absolute temperature is a straight line.

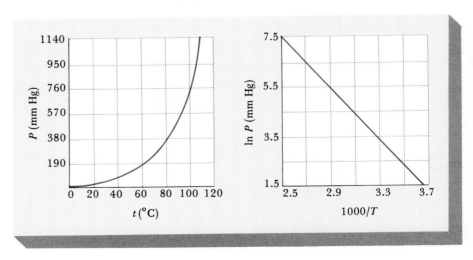

(T), we plot the *natural logarithm** of the vapor pressure (ln P) vs. the reciprocal of the absolute temperature (1/T). Such a plot for water is shown at the right of Figure 10.2; as with all other liquids, a plot of ln P vs. 1/T is a straight line.

The general equation of the straight line in Figure 10.2 is

$$\ln P = A - \frac{\Delta H_{vap}}{RT} \tag{10.1}$$

Here, ΔH_{vap} is the heat of vaporization of the liquid, in joules per mole, R is the gas constant and, in the units needed here, has a value of 8.31 J/mol·K. The quantity A is a constant for a particular liquid; its value need not concern us here.

For many purposes it is convenient to have a two-point equation relating the vapor pressure (P_2) at one temperature (T_2) to that (P_1) at another temperature (T_1). We obtain such an equation by applying Equation 10.1 at the two temperatures:

at T_2: $\ln P_2 = A - \dfrac{\Delta H_{vap}}{RT_2}$

at T_1: $\ln P_1 = A - \dfrac{\Delta H_{vap}}{RT_1}$

By subtracting, we eliminate the constant A:

$$\ln P_2 - \ln P_1 = -\frac{\Delta H_{vap}}{R}\left[\frac{1}{T_2} - \frac{1}{T_1}\right] = \frac{\Delta H_{vap}}{R}\left[\frac{1}{T_1} - \frac{1}{T_2}\right]$$

Rearranging to a somewhat more convenient form:

$$\ln \frac{P_2}{P_1} = \frac{\Delta H_{vap}}{R}\left[\frac{T_2 - T_1}{T_2 T_1}\right] \tag{10.2}$$

Equation 10.2 is known as the Clausius-Clapeyron equation, honoring Rudolph Clausius and B. P. E. Clapeyron. Clausius was a prestigious nineteenth-century German scientist. Clapeyron was a French engineer who first proposed a modified version of the equation in 1834. The use of the equation is illustrated in Example 10.2.

■ **EXAMPLE 10.2** _____
A certain organic compound has a vapor pressure of 91 mm Hg at 40°C. Calculate its vapor pressure at 25°C, taking the heat of vaporization to be 5.31×10^4 J/mol.

Solution

It is most convenient to use the subscript 2 for the *higher* temperature and pressure. With that choice, we have

_____

*Many natural laws are most simply expressed in terms of natural logarithms, which are based upon the number e = 2.71828 ... If $y = e^x$, then the natural logarithm of y, ln y, is equal to x.

$$T_2 = 40 + 273 = 313 \text{ K}; \qquad T_1 = 25 + 273 = 298 \text{ K};$$
$$P_2 = 91 \text{ mm Hg}; \qquad\qquad P_1 = ?$$

Substituting in Equation 10.2 with $R = 8.31$ J/mol·K:

$$\ln \frac{P_2}{P_1} = \ln \frac{91 \text{ mm Hg}}{P_1} = \frac{5.31 \times 10^4 \text{ J/mol}}{8.31 \text{ J/mol·K}} \left[\frac{313 \text{ K} - 298 \text{ K}}{313 \text{ K} \times 298 \text{ K}} \right]$$
$$= 1.03$$

Taking antilogs:

$$\frac{91 \text{ mm Hg}}{P_1} = e^{1.03} = 2.8$$

(e is the base of natural logarithms; on many calculators, to find the number whose natural logarithm is 1.03, you first enter 1.03, then press in succession the "INV" and "ln x" keys).

Solving for P_1,

$$P_1 = \frac{91 \text{ mm Hg}}{2.8} = \boxed{32 \text{ mm Hg}}$$

This value is reasonable in the sense that lowering the temperature should reduce the vapor pressure.

Exercise
Calculate the vapor pressure of this compound at 50°C. Answer: 170 mm Hg.

BOILING POINT

When we heat a liquid in an open container bubbles form, usually at the bottom, where heat is applied. The first small bubbles we see are air, driven out of solution by the increase in temperature. Eventually, at a certain temperature, large vapor bubbles form throughout the liquid. These vapor bubbles rise to the surface and break. When this happens, we say that the liquid is boiling.

The temperature at which a liquid boils depends upon the pressure above it. To understand why this is the case, consider Figure 10.3. This shows vapor bubbles rising in a boiling liquid. For a vapor bubble to form, the pressure within it, P_1, must be at least equal to the pressure above it, P_2. Since P_1 is simply the vapor pressure of the liquid, we conclude that **a liquid boils at a temperature at which its vapor pressure becomes equal to the pressure above its surface**. If this pressure is 1 atm (760 mm Hg), we refer to the temperature as the **normal boiling point**. (When the term "boiling point" is used without qualification, normal boiling point is implied.) The normal boiling point of water is 100°C; its vapor pressure is 760 mm Hg at that temperature.

As you might expect, the boiling point of a liquid can be reduced by lowering the pressure above it. Water can be made to boil at 25°C by evacuating the space above it. When a pressure of 24 mm Hg, the equilib-

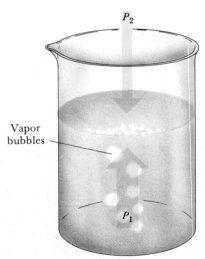

Figure 10.3 A liquid boils when its vapor pressurre (P_1) slightly exceeds the pressure above it (P_2).

Vapor bubbles

rium vapor pressure at 25°C, is reached, the water starts to boil. Chemists often take advantage of this effect in purifying a high-boiling compound that might decompose or oxidize at its normal boiling point. They boil it at a reduced temperature under vacuum and condense the vapor.

If you have been fortunate enough to camp in the high Sierras or the Rockies, you may have noticed that it takes longer at high altitudes to cook foods in boiling water. The reduced pressure lowers the temperature at which water boils in an open container. This slows down the physical and chemical changes that take place when foods like potatoes or eggs are cooked. In principle, this problem can be solved by using a pressure cooker. In that device, the pressure that develops is high enough to raise the boiling point of water above 100°C. Pressure cookers are indeed used in places like Cheyenne, Wyoming (elevation 1848 m), but not by mountain climbers, who have to carry all their equipment on their backs.

The boiling point of a liquid depends on how you boil it

They have enough trouble as it is

CRITICAL TEMPERATURE AND PRESSURE

Consider an experiment in which liquid carbon dioxide is introduced into an otherwise evacuated glass tube, which is then sealed (Fig. 10.4). At 0°C, the pressure above the liquid is 34 atm, the equilibrium vapor pressure of $CO_2(l)$ at that temperature. As the tube is heated, some of the liquid is converted to vapor and the pressure rises, to 44 atm at 10°C and 56 atm at 20°C. Nothing spectacular happens (unless there happens to be a weak spot in the tube) until we reach 31°C, at which point the vapor pressure is 73 atm. Suddenly, as we pass this temperature, the meniscus between the liquid and vapor disappears! The tube now contains only one phase.

It is impossible to have liquid carbon dioxide at temperatures above 31°C, no matter how much pressure is applied. Even at pressures as high as 1000 atm, carbon dioxide gas does not liquefy at 35°C or 40°C. This behavior is typical of all substances. There is a temperature, called the

Figure 10.4 When liquid carbon dioxide under pressure is heated, some of it vaporizes (a); vapor bubbles form in the liquid (b). Suddenly, at 31°C, the critical temperature of CO_2, the meniscus disappears (c). Above that temperature, only one phase is present, regardless of the applied pressure. (Marna G. Clarke)

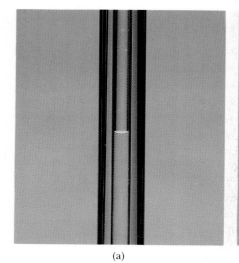
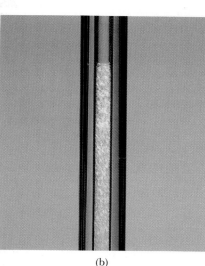
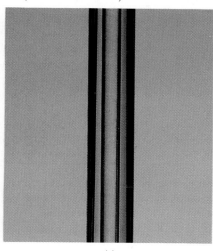

(a) (b) (c)

TABLE 10.1   Critical Temperatures (°C)

"PERMANENT GASES"		"CONDENSABLE GASES"		"LIQUIDS"	
Helium	−268	Carbon dioxide	31	Ethyl ether	194
Hydrogen	−240	Ethane	32	Ethyl alcohol	243
Nitrogen	−147	Propane	97	Benzene	289
Argon	−122	Ammonia	132	Bromine	311
Oxygen	−119	Chlorine	144	Water	374
Methane	−82	Sulfur dioxide	158		

critical temperature, above which the liquid phase of a pure substance cannot exist. The pressure that must be applied to cause condensation at that temperature is called the **critical pressure**. Quite simply, the critical pressure is the vapor pressure of the liquid at the critical temperature.

That is the maximum possible vapor pressure for that liquid

Table 10.1 lists the critical temperatures of several common substances. The species in the column at the left all have critical temperatures below 25°C. They are often referred to as "permanent gases." Applying pressure at room temperature will not condense a permanent gas. It must be cooled as well. When you see a truck labeled "liquid nitrogen" on the highway, you can be sure that the cargo trailer is refrigerated to at least −147°C, the critical temperature of N_2.

"Permanent gases" are most often stored and sold in steel cylinders under high pressures, often 150 atm or greater. When the valve on a cylinder of N_2 or O_2 is opened, gas escapes and the pressure drops in accordance with the Ideal Gas Law. The substances listed in the center column of Table 10.1, all of which have critical temperatures above 25°C, are handled quite differently. They are available commercially as liquids in high-pressure cylinders. When the valve on a cylinder of propane is opened, the gas that escapes is replaced by vaporization of liquid. The pressure quickly returns to its original value. Only when the liquid is completely vaporized does the pressure drop as gas is withdrawn (recall Fig. 10.1). This indicates that almost all of the propane is gone, and it's time to order a new tank.

Above the critical temperature, a substance is perhaps best described as a "fluid" rather than a gas. Supercritical fluids have unusual solvent properties that have led to many practical applications. Above its critical temperature, 374°C, water is a good solvent for organic compounds, capable of extracting hydrocarbons from coal. Supercritical CO_2 is being used to extract caffeine from coffee and nicotine from tobacco.

10.2
PHASE DIAGRAMS

In Section 10.1, we discussed several features of the equilibrium between a liquid and its vapor. For a pure substance, at least two other types of phase equilibria need be considered: the equilibrium between a solid and its vapor, which is similar in general to liquid-vapor equilibrium, and that between solid and liquid at the melting (freezing) point. Many of the important relations in all these equilibria can be shown in a **phase diagram**. A phase diagram is a graph that shows the pressures and temperatures at which different phases are in equilibrium with each other. The phase diagram of water is shown in Figure 10.5. This figure, which covers a wide range of temperatures and pressures, is obviously not drawn to scale.

To understand what a phase diagram implies, consider first the three lines AB, AC, and AD in Figure 10.5. Each of these lines tells us the pressures and temperatures at which two adjacent phases are in equilibrium.

1. Line AB is a portion of the vapor pressure–temperature curve of liquid water. At any temperature and pressure along this line, liquid water is in equilibrium with water vapor. From the curve we see that at point A, these two phases are in equilibrium at 0°C and about 5 mm Hg (more exactly, 0.01°C and 4.56 mm Hg). At B, corresponding to 100°C, the pressure exerted by the vapor in equilibrium with liquid water is 1 atm. The extension of line AB beyond point B gives the equilibrium vapor pressure of the liquid above the normal boiling point. The line ends at 374°C, the critical temperature of water, where the pressure is 218 atm.

Only if the state of the system lies on a line can we have two phases in equilibrium

2. Line AC represents the vapor pressure curve of ice. At any point along this line such as point A (0°C, 5 mm Hg) or point C, which might represent −3°C and 3 mm Hg, ice and vapor are in equilibrium with each other.

3. Line AD gives the temperatures and pressures at which liquid water is in equilibrium with ice.

Point A on the phase diagram is the only one at which all three phases, liquid, solid, and vapor, are in equilibrium with each other. It is called the **triple point**. For water, the triple point temperature is 0.01°C. At this temperature liquid water and ice have the same vapor pressure, 4.56 mm Hg.

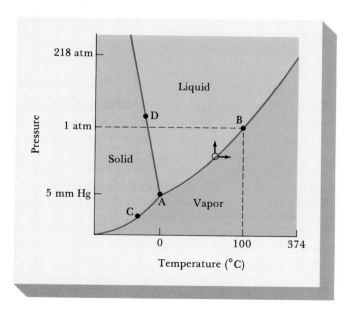

Figure 10.5 Phase diagram for water. The triple point is at 0°C, 5 mm Hg; the critical point is at 374°C, 218 atm.

In the three areas of the phase diagram labeled "solid," "liquid," and "vapor," only one phase is present. To understand this, consider what happens to an equilibrium mixture of two phases when the pressure or temperature is changed. Suppose we start at the point on AB indicated by an open circle. Here liquid water and vapor are in equilibrium with each other, let us say at 50°C and 92 mm Hg. If we increase the pressure on this mixture, condensation occurs. The phase diagram confirms this; by increasing the pressure at 50°C (*vertical arrow*), we move up into the liquid region. In another experiment, we might hold the pressure constant but increase the temperature. This change should cause the liquid to vaporize. The phase diagram tells us that this is indeed what happens: an increase in temperature (*horizontal arrow*) shifts us into the vapor region.

The vapor all condenses

■ **EXAMPLE 10.3** _____

Consider a sample of H_2O at point D in Figure 10.5.

a. What phase(s) is (are) present?
b. If the temperature of the sample were reduced at constant pressure, what would happen?
c. How would you convert the sample to vapor without changing the temperature?

Solution

a. Point D is on the solid-liquid equilibrium line. Ice and liquid water are present.
b. This corresponds to moving horizontally to the left from point D. The sample freezes completely to ice.
c. Reduce the pressure to below the triple point value, perhaps to 4 mm Hg.

Exercise
What phase(s) is (are) present at 5 mm Hg and 5°C? Answer: only $H_2O(g)$.

SUBLIMATION

The process by which a solid changes directly to vapor without passing through the liquid phase is called *sublimation*. A solid can sublime only at temperatures below the triple point; above that temperature it will melt to liquid (Fig. 10.5). At temperatures below the triple point, a solid can be made to sublime by reducing the pressure of the vapor above it to less than the equilibrium value. To illustrate what this means, consider the conditions under which ice sublimes. This happens on a cold, dry, winter day when the temperature is below 0°C and the pressure of water vapor in the air is less than the equilibrium value (4.56 mm Hg at 0°C). The rate of sublimation can be increased by evacuating the space above the ice. This is how foods are freeze-dried. The food is frozen, put into a vacuum chamber, and evacuated to a pressure of 1 mm Hg or less. The ice crystals formed upon freezing sublime, which leaves a product whose mass is only a fraction of that of the original food.

A lot of snow in Minnesota is lost by sublimation

Iodine sublimes more readily than ice because its triple point pressure, 90 mm Hg, is much higher. Iodine sublimes upon heating in an open container (Fig. 10.6) below the triple point temperature, 114°C. If we exceed the triple point, the solid melts. No such problem arises with solid carbon dioxide (dry ice). It has a triple point pressure above 1 atm (5.2 atm at −57°C). Liquid carbon dioxide cannot be made by heating dry ice in an open container. No matter what we do, solid CO_2 passes directly to the vapor at 1 atm pressure.

FUSION

For a pure substance, the **melting point** is identical to the **freezing point**. It represents the temperature at which solid and liquid phases are in equilibrium. Melting points are usually measured in an open container, i.e., at atmospheric pressure. For most substances, the melting point at 1 atm (the "normal" melting point) is virtually identical with the triple point temperature. For water, the difference is only 0.01°C.

Although the effect of pressure upon melting point is very small, we are often interested in its direction. To decide whether the melting point will be increased or decreased by compression, we apply a simple principle: **an increase in pressure favors the formation of the more dense phase**. We can distinguish between two types of behavior (Fig. 10.7):

1. *The solid is the more dense phase* (Fig. 10.7A). The solid-liquid equilibrium line is inclined to the right, shifting away from the y-axis as it rises. At higher pressures, the solid becomes stable at temperatures above the normal melting point. In other words, the melting point is raised by an increase in pressure. This behavior is shown by most substances.

2. *The liquid is the more dense phase* (Fig. 10.7B). The liquid-solid line is inclined to the left, toward the y-axis. An increase in pressure favors the formation of liquid; that is, the melting point is decreased by raising the pressure. Water is one of the few substances that behaves this way; ice is

Figure 10.6 When solid iodine is heated at temperatures below its triple point, 114°C, it sublimes; the vapor condenses to solid on a cool surface (upper tube). (Marna G. Clarke)

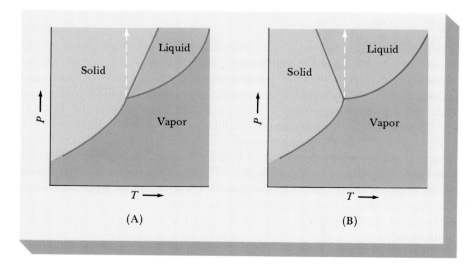

(A)

(B)

Figure 10.7 Effect of pressure on the melting point of a solid. (A) When the solid is the more dense phase, an increase in pressure converts liquid to solid; the melting point increases. (B) If the liquid is the more dense phase, an increase in pressure converts solid to liquid, and the melting point decreases.

If you cool a water pipe below 0°C, the pipe often breaks. Why?

less dense than liquid water. The effect is exaggerated for emphasis in Figure 10.7B. Actually, an increase in pressure of 134 atm is required to lower the melting point of ice by 1°C.

10.3
MOLECULAR SUBSTANCES

All of the substances that we dealt with in Sections 10.1 and 10.2 are molecular in nature. They are built up of small, discrete molecules such as H_2O, CO_2, and I_2. In this section we will look at some of the general properties of molecular substances and the nature of the forces between molecules.

As a class, molecular substances tend to be:

1. *Nonconductors of electricity when pure.* Since molecules are uncharged, they cannot carry an electric current. In most cases, including iodine, I_2, and ethyl alcohol, C_2H_5OH, water solutions of molecular substances are nonconductors. A few polar molecules, including HCl, react with water to form ions:

$$HCl(g) \rightarrow H^+(aq) + Cl^-(aq)$$

and hence produce a conducting water solution.

2. *Insoluble in water but soluble in nonpolar solvents such as CCl_4 or benzene.* Iodine is typical of most molecular substances; it is only slightly soluble in water (0.0013 mol/L at 25°C), much more soluble in benzene (0.48 mol/L). A few molecular substances, including ethyl alcohol, are very soluble in water. As we will see later in this section, such substances have intermolecular forces similar to those in water.

3. *Volatile, with appreciable vapor pressures at room temperature.* This accounts for the odors we associate with such molecular substances as H_2S (rotten eggs) and ethyl butyrate (oil of pineapple). It also explains why many molecular solids, notably iodine, can be purified by sublimation (recall Fig. 10.6).

4. *Low melting and boiling points.* Many molecular substances are gases at 25°C and 1 atm (for example, N_2, O_2, and CO_2), which means that they have boiling points below 25°C. Others (such as H_2O and benzene) are liquids with melting (freezing) points below room temperature. Of the molecular substances that are solids at ordinary temperatures, most melt around 100°C (mp I_2 = 114°C, mp naphthalene = 80°C). The upper limit for melting and boiling points of most molecular substances is about 300°C.

The generally low melting and boiling points of molecular substances reflect the fact that the forces between molecules (intermolecular forces) are weak. All we need do to melt or boil a molecular substance is to set the molecules free from one another (Fig. 10.8a). This requires only that we supply enough energy to overcome the attractive forces between molecules. The forces within a molecule (intramolecular forces) are strong, since they are covalent bonds. These bonds remain intact when a molecular substance melts or boils.

It's a lot easier to separate two molecules than to separate two atoms within a molecule

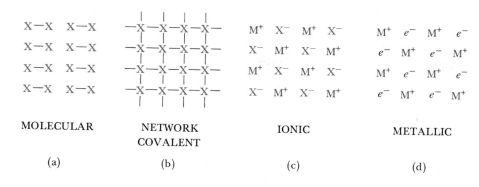

MOLECULAR	NETWORK COVALENT	IONIC	METALLIC
(a)	(b)	(c)	(d)

Figure 10.8 Four different types of lattices (see text discussion). M^+ represents a cation, X^- an anion, X a nonmetal atom, and e^- an electron. Covalent bonds are shown by straight lines joining atoms.

The boiling points of different molecular substances are directly related to the strength of the intermolecular forces involved. **The stronger the intermolecular forces, the higher the boiling point of the substance.** In the remainder of this section, we will examine the nature of the three different types of intermolecular forces: *dispersion forces, dipole forces,* and *hydrogen bonds.*

DISPERSION FORCES

The most common type of intermolecular force, found in all molecular substances, is referred to as a dispersion force. It is basically electrical in nature and involves an attraction between temporary dipoles in adjacent molecules. To understand the origin of dispersion forces, consider Figure 10.9, where we deal with the charge distribution in the H_2 molecule.

On the average, electrons in a nonpolar molecule such as H_2 are as close to one nucleus as the other. However, at a given instant, the electron cloud may be concentrated at one end of the molecule (position 1A in Fig. 10.9). A fraction of a second later it may be at the opposite end of the molecule (position 1B). The situation is similar to that of a person watching a tennis match from a position directly in line with the net. At one instant his eyes are focused on the player to his left. A moment later, they shift to the player on his right. Over a period of time, he looks to one side as often as the other. The "average" position of focus of his eyes is straight ahead.

It's hard work watching the U.S. Open, but somebody's got to do it

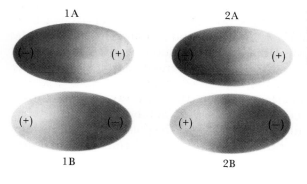

Figure 10.9 Temporary dipoles in adjacent H_2 molecules line up to create an electrical attractive force known as the dispersion force. Deeply shaded areas indicate regions where the electron cloud is momentarily concentrated.

The momentary concentration of the electron cloud on one side or the other creates a temporary dipole in the H_2 molecule. One side of the molecule, shown in deeper color in Figure 10.9, acquires a partial negative charge; the other side has a partial positive charge of equal magnitude. This temporary dipole induces a similar dipole in an adjacent molecule. When the electron cloud in the first molecule is at 1A, the electrons in the second molecule are attracted to 2A. As the first electron cloud shifts to 1B, the electrons of the second molecule are pulled back to 2B. These temporary dipoles, both in the same direction, lead to an attractive force between the molecules. This is the dispersion force.

A similar argument can be made to explain the existence of dispersion forces in all molecules. The strength of these forces depends upon how many electrons there are in a molecule and how readily these electrons are dispersed or *polarized*. Both the number of electrons and the ease of polarization increase with molecular size. Large molecules, where the electrons are relatively far from the nucleus, are easier to polarize than small molecules. In general, molecular size and molar mass parallel one another. As a result, we ordinarily find that **as molar mass increases, dispersion forces become stronger and the boiling point of nonpolar molecular substances increases** (Table 10.2).

TABLE 10.2   Effect of Molar Mass on Boiling Points of Molecular Substances

NOBLE GASES*			HALOGENS			HYDROCARBONS		
	MM (g/mol)	bp (°C)		MM (g/mol)	bp (°C)		MM (g/mol)	bp (°C)
He	4	−269	F_2	38	−188	CH_4	16	−161
Ne	10	−246	Cl_2	71	−34	C_2H_6	30	−88
Ar	40	−186	Br_2	160	59	C_3H_8	44	−42
Kr	84	−152	I_2	254	184	$n\text{-}C_4H_{10}$	58	0

*Strictly speaking, the noble gases are "atomic" rather than molecular. However, like molecules, the noble gas atoms are attracted to one another by dispersion forces.

DIPOLE FORCES

Polar molecules, in addition to dispersion forces, show another type of intermolecular force, called a dipole force. The origin of this force is shown in Figure 10.10, which shows the orientation of polar molecules, such as ICl, in a crystal. Adjacent molecules line up so that the negative pole of one molecule (small Cl atom) is as close as possible to the positive pole (large I atom) of its neighbor. Under these conditions, there is an electrical attractive force, referred to as a dipole force, between adjacent polar molecules.

When iodine chloride is heated to 27°C, the rather weak dipole forces are unable to keep the molecules rigidly aligned and the solid melts. Dipole forces are still important in the liquid state, since the polar molecules remain close to one another. Only in the gas, where the molecules are far apart, do dipole forces become negligible. Hence, boiling points as well as melting

Br_2, with the same molar mass, melts at −7°C. No dipole forces at all

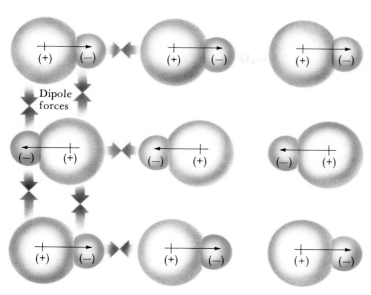

Figure 10.10 Dipole forces in the ICl crystal. The (+) and (−) indicate fractional charges on the I and Cl atoms in the polar molecules. The existence of these partial charges causess the molecules to line up in the pattern shown. Adjacent molecules are attracted to each other by dipole forces between the (+) pole of one molecule and the (−) pole of the other.

Dipole forces

points of polar compounds such as ICl are somewhat higher than those of nonpolar substances of comparable molar mass. This effect is shown in Table 10.3.

■ **EXAMPLE 10.4**

Explain, in terms of intermolecular forces, why

a. the boiling point of O_2 (−183°C) is higher than that of N_2 (−196°C).
b. the boiling point of NO (−151°C) is higher than that of either O_2 or N_2.

Solution

a. Only dispersion forces are involved with these nonpolar molecules. Since the molar mass of O_2 is greater (32.0 g/mol vs. 28.0 g/mol for N_2), its dispersion forces are stronger, making its boiling point higher.
b. Dispersion forces in NO (30 g/mol) are comparable in strength to those in O_2 and N_2. The polar NO molecule shows an additional type of intermolecular force not present in N_2 or O_2: the dipole force. As a result, its boiling point is the highest of the three substances.

TABLE 10.3 Boiling Points of Nonpolar vs. Polar Substances

NONPOLAR			POLAR		
Formula	MM (g/mol)	bp (°C)	Formula	MM (g/mol)	bp (°C)
N_2	28	−196	CO	28	−192
SiH_4	32	−112	PH_3	34	−88
GeH_4	77	−90	AsH_3	78	−62
Br_2	160	59	ICl	162	97

Exercise
Which would you expect to have the higher boiling point, Cl_2 or ICl?
Answer: ICl (polar, larger molar mass).

HYDROGEN BONDS

Ordinarily, polarity has a relatively small effect on the boiling points of molecular substances. For example, in the series HCl → HBr → HI, boiling point increases steadily with molar mass. This happens despite the fact that polarity decreases as we move from HCl to HI. We find, however, that when hydrogen is bonded to a small, highly electronegative atom (N, O, F), polarity has a much greater effect upon boiling point. Hydrogen fluoride, HF, despite its low molar mass (20 g/mol), has the highest boiling point of all the hydrogen halides. Water (MM = 18 g/mol) and ammonia (MM = 17 g/mol) also have abnormally high boiling points (Fig. 10.11). In these cases, the effect of polarity reverses the normal trend expected from molar mass alone.

The unusually high boiling points of HF, H_2O, and NH_3 result from an unusually strong type of dipole force called a hydrogen bond. The

Figure 10.11 Boiling points of hydrogen compounds of elements in Groups 4 to 7. Among the Group 4 hydrides, boiling point increases with molar mass. In Groups 5, 6, and 7, the first member (NH_3, H_2O, HF) has an abnormally high boiling point.

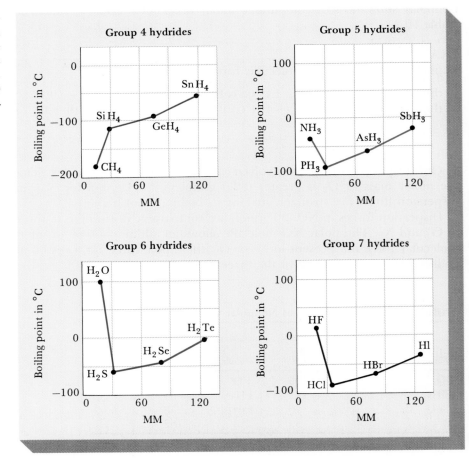

hydrogen bond is a force exerted between an H atom bonded to an F, O, or N atom in one molecule and an unshared pair on the F, O, or N atom of a neighboring molecule:

The hydrogen bond is the strongest intermolecular force

$$X—H \text{-----} :X—H \qquad X = N, O, \text{ or } F$$

↑
hydrogen bond

There are two reasons that hydrogen bonds are stronger than ordinary dipole forces:

1. The difference in electronegativity between hydrogen (2.1) and fluorine (4.0), oxygen (3.5), or nitrogen (3.0) is quite large. It causes the bonding electrons in molecules such as HF, H_2O, and NH_3 to be primarily associated with the more electronegative atom (F, O, or N). So, the hydrogen atom, insofar as its interaction with a neighboring molecule is concerned, behaves almost like a bare proton. The hydrogen bond is strongest in HF, where the difference in electronegativity is greatest. It is weakest in NH_3, where the difference in electronegativity is relatively small.

2. The small size of hydrogen allows the unshared electron pair of an F, O, or N atom of one molecule to approach the H atom in another very closely. It is significant that hydrogen bonding occurs only with these three elements. All of them have small atomic radii. The larger chlorine and sulfur atoms, with electronegativities (3.0, 2.8) similar to nitrogen, do not form hydrogen bonds in such compounds as HCl and H_2S.

Hydrogen bonds can exist in many molecules other than HF, H_2O, and NH_3. The basic requirement is simply that hydrogen be bonded to fluorine, oxygen, or nitrogen (Example 10.5).

■ **EXAMPLE 10.5** _____

Would you expect to find hydrogen bonds in

a. ethyl alcohol

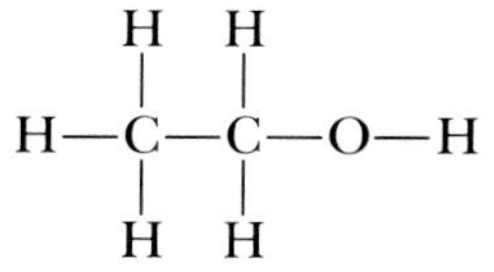

```
    H  H
    |  |
H—C—C—O—H
    |  |
    H  H
```

b. dimethyl ether

```
      H        H
      |        |
  H—C—O—C—H
      |        |
      H        H
```

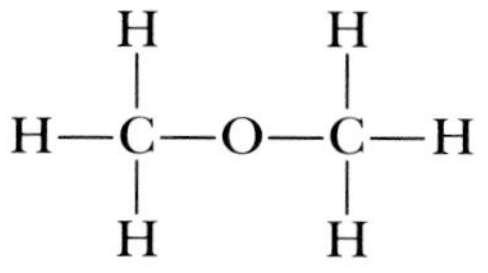

c. hydrazine, N_2H_4

Solution

a. There should be hydrogen bonds in ethyl alcohol, since an H atom is bonded to oxygen. Hydrogen bonding between two CH_3CH_2OH molecules might be shown as

H bonds stabilize the configurations of many large molecules, such as proteins

$$CH_3-CH_2-O \quad \overset{H}{\underset{H}{\diagdown}} \quad O-CH_2-CH_3$$

b. Since all the hydrogen atoms are bonded to carbon in dimethyl ether, there should be no hydrogen bonds. Reflecting this difference in behavior, dimethyl ether boils at $-25°C$, as compared to $78°C$ for ethyl alcohol.

c. The Lewis structure of hydrazine is:

$$H-\overset{..}{N}-\overset{..}{N}-H$$
$$\;\;\;\;\;\underset{H}{|}\;\;\underset{H}{|}$$

Since hydrogen is bonded to nitrogen, hydrogen bonding can occur between neighboring N_2H_4 molecules. The boiling point of hydrazine ($114°C$) is much higher than that of molecular O_2 ($-183°C$), which has the same molar mass.

Exercise

Would you predict hydrogen bonding in acetic acid, $CH_3-\overset{\overset{\textstyle O}{\|}}{C}-OH$? in acetone, $CH_3-\overset{\overset{\textstyle O}{\|}}{C}-CH_3$? Answer: In acetic acid but not in acetone.

Water has many unusual properties in addition to its high boiling point. In particular, water, in contrast to nearly all other substances, expands upon freezing. Ice is one of the very few solids that has a density less than that of the liquid from which it is formed (d ice at $0°C = 0.917$ g/cm³, d water at $0°C = 1.000$ g/cm³). This behavior is an indirect result of hydrogen bonding. When water freezes to ice, an open hexagonal pattern of molecules results (Fig. 10.12). Each oxygen atom in an ice crystal is bonded to four hydrogens. Two of these are attached by ordinary covalent bonds at a distance of 0.099 nm. The other two involve hydrogen bonds 0.177 nm in length. The large proportion of "empty space" in the ice structure explains why ice is less dense than water.

The hydrogen bonds in H_2O are about twice as long as the O—H bonds

■ **EXAMPLE 10.6**

What types of intermolecular forces are present in H_2? CCl_4? OCS? NH_3?

Solution

The H_2 and CCl_4 molecules are nonpolar (Chap. 9). Thus, only dispersion forces are present among neighboring molecules. Both OCS and NH_3 are polar molecules:

$$:\overset{..}{O}=C=\overset{..}{S}: \qquad\qquad H\overset{\overset{\textstyle \overset{..}{N}}{\diagup\;|\;\diagdown}}{\;}\underset{H}{\;}H$$

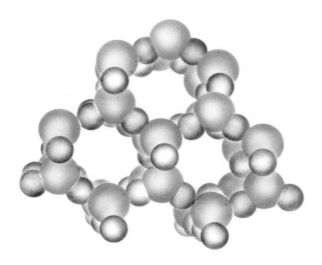

Figure 10.12 In ice, the water molecules are arranged in an open hexagonal pattern that gives ice its low density. Each oxygen atom is bonded covalently to two hydrogen atoms and forms hydrogen bonds with two other hydrogen atoms.

There are dipole forces as well as dispersion forces with OCS molecules. In NH_3, there are hydrogen bonds and dispersion forces.

Exercise
In which one of the above substances would you expect dispersion forces to be weakest? Strongest? Answer: H_2; CCl_4.

We have now discussed three types of intermolecular forces: dispersion forces, dipole forces, and hydrogen bonds. You should bear in mind that all of these forces are relatively weak compared to ordinary covalent bonds. Consider, for example, the situation in H_2O. The total intermolecular attractive force in ice is about 50 kJ/mol; the O—H bond energy is an order of magnitude greater, about 464 kJ/mol. This explains why it is a lot easier to melt ice or boil water than it is to decompose water into the elements. Even at a temperature of 1000°C and 1 atm, only about one H_2O molecule in a billion decomposes to hydrogen and oxygen atoms.

10.4
NETWORK COVALENT SUBSTANCES

Covalent bond formation need not lead to small, discrete molecules. Instead it can produce structures (Fig. 10.8b, p. 327) with nonmetal atoms joined by a continuous network of covalent bonds. Substances with this type of

TABLE 10.4   Melting Points of Network Covalent Substances

	mp (°C)		mp (°C)
C	3570	SiC	>2700
Si	1414	BN	>3000
Ge	937	SiO_2	>1400

structure are always solids at 25°C; they are referred to as *network covalent solids*. The entire crystal, in effect, consists of one huge molecule.

Network covalent solids have certain characteristic properties. They are

1. *High melting, often with melting points above 1000°C* (Table 10.4). To melt the solid, covalent bonds between atoms must be broken. In this respect, solids of this type differ markedly from molecular solids, which have much lower melting points.

2. *Insoluble in all common solvents.* For solution to occur, covalent bonds throughout the solid would have to be broken.

3. *Poor electrical conductors.* In a typical network covalent substance, there are no charged particles to carry a current.

GRAPHITE AND DIAMOND

The element carbon has two different crystalline forms, both of the network covalent type. Graphite and diamond are referred to as **allotropes** of carbon; the word allotropy is used to describe the situation where an element exists in two or more different structural forms in the same physical state. Both graphite and diamond have high melting points, above 3500°C. However, as you can see from Figure 10.13, the bonding patterns in the two crystals are quite different.

The graphite crystal is planar, with the carbon atoms arranged in a hexagonal pattern. Each carbon atom is bonded to three others, forming one double bond and two single bonds. The forces between the layers in graphite are of the dispersion type and are quite weak. Thus, the layers can readily slide past one another so that graphite is soft and slippery to the touch. When you write with a "lead" pencil, which is really made of graphite, thin layers of graphite rub off onto the paper.

In diamond, each carbon atom forms single bonds with four other carbon atoms arranged tetrahedrally around it. The bonds are strong enough

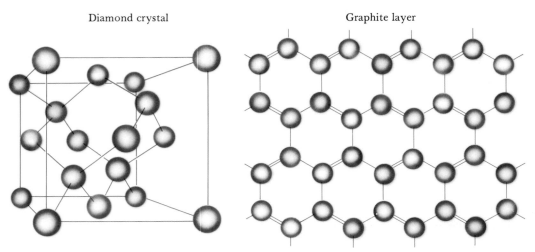

Diamond crystal   Graphite layer

Figure 10.13   Structure of diamond and graphite.

(bond energy C—C = 347 kJ/mol) to produce a rugged, three-dimensional lattice. Diamond is one of the hardest of substances, and is used in industry in cutting tools and quality grindstones.

At room temperature and atmospheric pressure, graphite is the stable form of carbon. Diamond, in principle, should slowly transform to graphite under ordinary conditions. Fortunately for the owners of diamond rings this transition occurs at zero rate unless the diamond is heated to about 1500°C, at which temperature the conversion occurs rapidly. For understandable reasons, no one has ever become very excited over the commercial possibilities of this process. The more difficult task of converting graphite to diamond has aroused much greater enthusiasm.

Since diamond has a higher density than graphite (3.51 vs. 2.26 g/cm³), its formation should be favored by high pressures. Theoretically, at 25°C and 15,000 atm graphite should turn to diamond. However, under those conditions the reaction has a negligible rate. At higher temperatures it goes faster, but the required pressure goes up too; at 2000°C, a pressure of about 100,000 atm is needed. In 1954, scientists at the General Electric laboratories were able to achieve these high temperatures and pressures and converted graphitic carbon to diamond. Nowadays, all industrial diamonds are synthetic.

Here pure scientific research paid off handsomely

SILICON DIOXIDE

Quartz, the most common form of SiO_2, is the main component of sea sand. In quartz, each silicon atom bonds tetrahedrally to four oxygen atoms. Each oxygen atom bonds to two silicon atoms, thus linking adjacent tetrahedra to one another (Fig. 10.14). Note that no discrete SiO_2 molecules are present; the network of covalent bonds extends throughout the entire

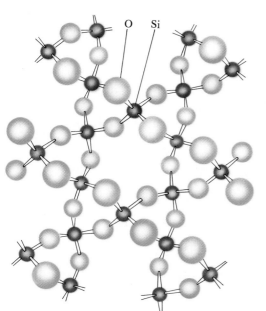

Figure 10.14 Crystal structure of quartz, SiO_2. Every silicon atom (small) is linked tetrahedrally to four oxygen atoms (large).

crystal. Unlike most solids, quartz does not melt sharply to a liquid. Instead, it turns to a viscous mass over a wide temperature range, first softening at about 1400°C. The viscous fluid probably contains long —Si—O—Si—O— chains, with enough bonds broken so that the material can flow.

Ordinary glass is made by heating a mixture of sand, limestone ($CaCO_3$), and soda ash (Na_2CO_3) to the melting point. Carbon dioxide is given off from the hot mass, which contains the oxides of silicon, sodium, and calcium in a mole ratio of about 7:1:1. The "soft" glass produced this way softens at about 600°C and has a wider range of high viscosity than quartz. Hard glass, called Pyrex or Kimax, is made from a melt of silicon, boron, aluminum, sodium, and potassium oxides. It is much superior to soft glass in withstanding chemical attack and thermal shock. Hard glass softens at about 800°C and is readily worked in the flame of a gas-oxygen torch. Nearly all the glassware used in chemical equipment nowadays is made of hard glass.

Small amounts of various substances are often added to glass to produce different colors. Chromium(III) oxide (Cr_2O_3) gives a green glass, CoO a blue glass, and MnO_2 a violet glass. Milky white glass contains calcium fluoride, CaF_2; SnO_2 produces an opaque glass. Eyeglasses that darken in the sunlight contain small amounts of white, finely dispersed silver chloride, $AgCl$. Exposure to sunlight converts some of the Ag^+ ions to metallic silver, which is black. The reaction is reversible; in a dark room, $AgCl$ is re-formed and the glass becomes clear again.

Pyrex, photogray glass, and pyroceram were invented by chemists at the Corning Glass Works

10.5
IONIC COMPOUNDS

A very important class of substances referred to frequently in earlier chapters consists of ionic compounds. As pointed out earlier, such compounds contain two different types of structural units: a cation, carrying a positive charge, and an anion, carrying a negative charge. The cation is ordinarily derived from a metal atom by the loss of electrons (e.g., Na^+, Ca^{2+}, Ni^{2+}, Fe^{3+}). The anion may be formed when a nonmetal atom gains electrons (F^-, O^{2-}) or it may be polyatomic (OH^-, NO_3^-, CrO_4^{2-}). Typical ionic compounds include:

$$NaCl \quad\quad CaF_2 \quad\quad Ni(OH)_2 \quad\quad Fe(NO_3)_3$$

With few exceptions, *compounds formed by metals are ionic*. Invariably, ionic compounds are solids at room temperature. In this section, we will look at some of the properties of ionic compounds and examine the factors affecting the strength of ionic bonds between cations and anions.

GENERAL PROPERTIES OF IONIC SUBSTANCES

An ionic solid consists of cations and anions, for example, Na^+, Cl^-. No simple, discrete molecules are present in $NaCl$ or other ionic compounds; rather, the ions are held in a regular, repeating arrangement (Fig. 10.8C) by strong **ionic bonds**, electrostatic interactions between oppositely charged ions. Because of this structure, ionic solids show the following properties:

1. *Ionic solids are nonvolatile and high-melting* (typically 600°C to 2000°C). Ionic bonds must be broken to melt the solid, separating oppositely charged ions from each other. Only at high temperatures do the ions acquire enough kinetic energy for this to happen.

2. *Ionic solids do not conduct electricity* because the charged ions are fixed in position. They become good conductors, however, when melted or dissolved in water. In both cases, the melt or solution, the ions (such as Na^+ and Cl^-) are free to move through the liquid and thus can conduct an electric current.

3. *Many, but not all, ionic compounds* (for example, NaCl but not $CaCO_3$) *are soluble in water,* a polar solvent. In contrast, ionic compounds are insoluble in nonpolar solvents such as benzene (C_6H_6) or carbon tetrachloride (CCl_4).

COULOMB'S LAW AND LATTICE ENERGY

The relative strengths of different ionic bonds can be estimated from **Coulomb's Law**, which gives the electrical energy of interaction between a cation and anion in contact with one another:

$$E = \frac{k \times Q_1 \times Q_2}{d} \qquad (10.3)$$

Here, Q_1 and Q_2 are the charges of anion and cation, and d, the distance between the centers of the two ions, is the sum of the ionic radii:

$$d = r \text{ cation} + r \text{ anion} \qquad (10.4)$$

The quantity k is a constant whose magnitude need not concern us here. Since the cation and anion have opposite charges, E is a negative quantity. This makes sense; energy is evolved when two oppositely charged ions, originally far apart with $E = 0$, approach one another closely. Conversely, energy has to be absorbed to separate the ions from each other.

From Equation 10.3, we conclude that the strength of an ionic bond should depend upon two factors:

1. *The charges of the ions.* The bond in CaO ($+2$, -2 ions) is considerably stronger than in NaCl ($+1$, -1 ions).

2. *The size of the ions.* The ionic bond in NaCl is somewhat stronger than that in KBr, because the internuclear distance is smaller in NaCl:

$$d \text{ NaCl} = r \text{ Na}^+ + r \text{ Cl}^- = 0.095 \text{ nm} + 0.181 \text{ nm} = 0.276 \text{ nm}$$
$$d \text{ KBr} = r \text{ K}^+ + r \text{ Br}^- = 0.133 \text{ nm} + 0.195 \text{ nm} = 0.328 \text{ nm}$$

For an ionic compound in which there are equal numbers of cations and anions, the ionic bond strength is directly related to the *lattice energy*. Lattice energy is defined as the enthalpy change per mole when an ionic solid is formed from gaseous ions:

$$M^+(g) + X^-(g) \rightarrow MX(s); \Delta H = \text{lattice energy MX} \qquad (10.5)$$

We see from Table 10.5 that the lattice energy of CaO is about four times that of NaCl, which in turn is somewhat larger than that of KBr.

TABLE 10.5   Lattice Energies of Ionic Solids

	CHARGES OF IONS	DISTANCE BETWEEN IONS IN CONTACT	LATTICE ENERGY (kJ/mol)
KBr	+1, −1	0.328 nm	−689
KCl	+1, −1	0.314 nm	−718
NaCl	+1, −1	0.276 nm	−787
CaO	+2, −2	0.239 nm	−3481

10.6 METALS

The unique properties of metals—malleability, ductility, and high thermal and electrical conductivities—were described in Chapter 6. These properties were rationalized in terms of the electron sea model of metallic bonding (Fig. 10.8d, p. 327). By Coulomb's Law, we expect the lattice energy and hence the melting point of a metal to increase with the charge of the cation (e.g., $Na^+ < Mg^{2+} < Al^{3+}$). In general, the melting points of metals cover a wide range, from −39°C for mercury to 3410°C for tungsten.

In addition to those we have mentioned, metals as a class have the following properties. They are

1. *Nonvolatile.* Mercury is the only metal whose volatility is of concern at ordinary temperatures. Its vapor pressure is very low (2×10^{-6} atm at 25°C, 4×10^{-4} atm at 100°C); however, mercury vapor is toxic at the parts per million level. That is why your instructor will insist that you clean up any mercury spilled in the laboratory.

2. *Insoluble in water and other common solvents.* No metals "dissolve" in water in the true sense. As we saw in Chapter 6, a few very active metals react chemically with water to form hydrogen gas. Liquid mercury dissolves many metals, forming solutions called amalgams. Perhaps the most familiar amalgams are those of silver and gold. These are formed when ores of these metals are extracted with mercury. An Ag-Sn-Hg amalgam is used in filling teeth.

Silver amalgam is still the best material for filling your cavities

Much of what we have said about the properties of the four structural types of substances discussed in Sections 10.3 to 10.6 is summarized in Table 10.6 and in Example 10.7.

■ **EXAMPLE 10.7**

A certain substance is a liquid at room temperature and is insoluble in water. Suggest which of the four types of basic structural units is present in the substance and list additional experiments that could be carried out to confirm your prediction.

Solution

The fact that the substance is a liquid suggests that it is probably molecular. The fact that it is insoluble in water agrees with this classification. To

$$\ln \frac{P_2}{P_1} = \frac{\Delta H_{vap}}{R}\left(\frac{1}{T_1} - \frac{1}{T_2}\right)$$

TABLE 10.6   Structures and Properties of Types of Substances

TYPE	STRUCTURAL UNITS	FORCES WITHIN UNITS	FORCES BETWEEN UNITS	PROPERTIES	EXAMPLES
Molecular	Molecules				
	a. Nonpolar	Covalent bond	Dispersion	Low mp, bp; often gas or liquid at 25°C. Nonconductors. Insoluble in water, soluble in organic solvents.	H_2 CCl_4
	b. Polar	Covalent bond	Dispersion, dipole, H bond	Similar to nonpolar but generally higher mp and bp, more likely to be water soluble.	HCl NH_3
Network Covalent	Atoms	Covalent bond	—	Hard, very high-melting solids. Nonconductors. Insoluble in common solvents.	C SiO_2
Ionic	Ions	—	Ionic bond	High mp. Conductors in molten state or water solution. Often soluble in water, insoluble in organic solvents.	NaCl MgO $CaCO_3$
Metallic	Cations, mobile electrons	—	Metallic bond	Variable mp. Good conductors in solid. Insoluble in common solvents.	Na Fe

confirm that it is indeed molecular, we might measure the conductivity of the liquid, which should be essentially zero. It should also be soluble in most organic solvents.

Exercise
A certain solid is high melting (above 1000°C) and is a nonconductor. Which type of substance is it likely to be? How could you make a final decision as to the type of substance? Answer: It could be ionic or network covalent; test the conductivity of the melt.

10.7
CRYSTAL STRUCTURES

Solids tend to crystallize in definite geometric forms (Fig. 10.15) that can often be seen by the naked eye. In ordinary table salt, we can distinguish small, cubic crystals of NaCl. Large, beautifully formed crystals of such minerals as fluorite, CaF_2, are found in nature. It is possible to observe distinct crystal forms of many metals under a microscope.

Crystals have definite geometric forms because the particles that form them are arranged in a definite, three-dimensional pattern. The nature of this pattern can be deduced by a technique known as x-ray diffraction (Fig. 10.16). A beam of x-rays of known wavelength is directed at a crystal; the angles and intensities of the diffracted beams are measured. From this information, it is possible to calculate the distances between successive layers

Figure 10.15   Crystals of fluorite (CaF_2), amethyst (SiO_2), and halite (NaCl). (Marna G. Clarke)

Figure 10.16 X-ray diffraction. If the sample is powdered, as shown in this schematic drawing, x-rays of a single wavelength are used. From the angles at which the rays are diffracted, we can calculate the distances between planes of atoms in the crystal.

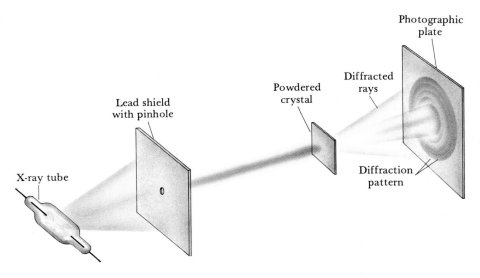

of atoms or ions in the crystal. This can be done using the Bragg equation, first suggested by W.H. Bragg and his son W.L. Bragg in 1913:

$$d = n \lambda/2 \sin \theta \qquad (10.6)$$

Here, θ is the angle between the diffracted beam and the layer of atoms, λ is the wavelength of the x-rays, d is the distance between successive layers, and n is an integer $(1, 2, 3, \ldots)$ known as the order number of the diffracted beam.

Knowing the value of d in Equation 10.6, it is a relatively simple matter to find atomic and ionic radii in crystals. It is much more difficult to deduce the geometric pattern in which the particles are arranged. The basic approach is to assume a particular pattern and calculate the angles and intensities of the diffracted beams expected for that structure. Comparison with experimental data can then suggest refinements which give better agreement. Prior to the computer age, this was a tedious process. It took Dorothy Hodgkin at Oxford eight years to work out the structure of Vitamin B_{12}, for which she received the Nobel Prize in Chemistry in 1964. Now, using computers, x-ray crystallographers can determine more quickly the structure of proteins, in which a single molecule may contain a thousand or more atoms.

Even proteins can be crystallized under the right conditions

UNIT CELLS; METALS

The basic information that comes out of x-ray diffraction studies concerns the dimensions and geometric form of the **unit cell**, the smallest structural unit that, repeated over and over again in three dimensions, generates the crystal. There are many different types of unit cells. We will consider three of the simpler cells found in metals (Fig. 10.17): the simple cubic cell, the face-centered cubic cell, and the body-centered cubic cell.

1. Simple cubic cell. This is a cube that consists of eight atoms whose centers are located at the corners of the cell. Atoms at adjacent corners of the cube touch one another.

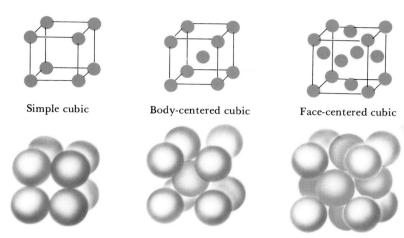

Simple cubic   Body-centered cubic   Face-centered cubic

Figure 10.17 Three types of cubic lattices. In each case, there is an atom at each of the eight corners of the cube. In the body-centered lattice, there is an additional atom in the center of the cube. In the face-centered lattice, there is an atom in the center of each of the six faces.

2. Face-centered cubic cell (FCC). Here, there is an atom at each corner of the cube and one in the center of each of the six faces of the cube. In this structure, atoms at the corners of the cube do not touch one another; they are forced slightly apart. Instead, contact occurs along a face diagonal. The atom at the center of each face touches atoms at opposite corners of the face.

3. Body-centered cubic cell (BCC). This is a cube with atoms at each corner and one in the center of the cube. Here again, corner atoms do not touch each other. Instead, contact occurs along the body diagonal; the atom at the center of the cube touches atoms at opposite corners.

If the nature and dimensions of the unit cell of a metallic crystal are known, we can readily calculate the atomic radius of the metal. It should be evident from Figure 10.18 that the following relationships apply:

$$\text{simple cubic cell:} \quad 2r = s \qquad (10.7)$$

$$\text{face-centered cubic cell:} \quad 4r = s\sqrt{2} \qquad (10.8)$$

$$\text{body-centered cubic cell:} \quad 4r = s\sqrt{3} \qquad (10.9)$$

Figure 10.18 Relation between atomic radius, r, and length of edge, s, for cubic cells.

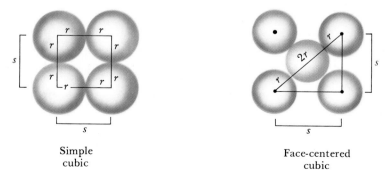

Simple
cubic

Face-centered
cubic

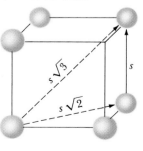

Body-centered
cubic

■ **EXAMPLE 10.8**

Silver crystallizes in a structure with a face-centered cubic (FCC) unit cell 0.407 nm on an edge. Calculate the atomic radius of silver.

Solution

The appropriate relation for a face-centered cubic cell is

$$4r = s\sqrt{2}$$

Substituting $s = 0.407$ nm and solving for r:

This is how we find the radii of metal atoms

$$r = \frac{0.407 \text{ nm} \times \sqrt{2}}{4} = \frac{0.407 \text{ nm} \times 1.41}{4} = \boxed{0.144 \text{ nm}}$$

Exercise

Vanadium crystallizes in a body-centered cubic (BCC) unit cell 0.303 nm on an edge. What is the atomic radius of vanadium? Answer: 0.131 nm.

In analyzing the properties of unit cells, it is important to keep in mind that many such cells are in contact with one another in a crystal, interlocking to form a three-dimensional network. As a result, many of the atoms in a unit cell do not belong exclusively to that cell. Specifically:

An atom at the center of a cube belongs only to that cube

— an atom at the center of the face of a cube is shared by another cube that touches that face. In effect then, only half of that atom can be assigned to a given cell.

— an atom at the corner of a cube forms a part of eight different cubes that touch at that point. In this sense, only one-eighth of a corner atom "belongs" to a particular cell.

Following this reasoning we conclude that the number of atoms to be assigned to each type of cell is:

simple cubic: 8 corner atoms $\times \frac{1}{8} = 1$ atom per cube

body-centered cubic: 8 corner atoms $\times \frac{1}{8} + 1$ center atom $= 2$ atoms per cube

face-centered cubic: 8 corner atoms $\times \frac{1}{8} + 6$ face atoms $\times \frac{1}{2} = 4$ atoms per cube

One way to see what these numbers mean is to imagine that you were asked to construct a crystal by fitting together a large number of unit cells, say, one million. If the crystal is simple cubic, it will contain 1 million atoms. If the structure is body-centered, you will need 2 million atoms to construct the crystal. If it is face-centered, 4 million atoms will be required.

■ **EXAMPLE 10.9**

Using the information given in Example 10.8 for silver and the additional fact that the density of silver (MM = 107.9 g/mol) is 10.6 g/cm³, estimate Avogadro's number.

Solution

One way to analyze this problem is to first calculate the volume of the unit cell, knowing from Example 10.8 that it has an edge 0.407 nm long. Then, knowing the density of the metal, we can calculate the mass of silver in the unit cell. Finally, knowing that there are four Ag atoms in an FCC cell, we can calculate the mass of a silver atom, from which Avogadro's number follows.

1. Length of cell $= 0.407$ nm $\times \dfrac{10^{-9} \text{ m}}{1 \text{ nm}} \times \dfrac{10^2 \text{ cm}}{1 \text{ m}} = 4.07 \times 10^{-8}$ cm

   Volume of cell $= (4.07 \times 10^{-8} \text{ cm})^3 = 6.74 \times 10^{-23} \text{ cm}^3$

2. Mass of Ag in cell $= 10.6 \dfrac{\text{g}}{\text{cm}^3} \times 6.74 \times 10^{-23} \text{ cm}^3 = 7.14 \times 10^{-22}$ g

3. Mass of Ag atom $= (7.14 \times 10^{-22} \text{ g})/4 = 1.79 \times 10^{-22}$ g

   Molar mass Ag $=$ Mass of one Ag atom $\times$ Avogadro's number

   $$107.9 \text{ g} = 1.79 \times 10^{-22} \text{ g} \times N$$

   $$N = \frac{107.9}{1.79 \times 10^{-22}} = 6.03 \times 10^{23}$$

This is a good way to determine Avogadro's number

Exercise
Barium crystallizes in a BCC structure and has a density of 3.51 g/cm³. Taking Avogadro's number to be 6.02×10^{23} and the molar mass of Ba to be 137.3 g/mol, estimate the length of the edge of the unit cell. Answer: 0.507 nm.

When the atoms in a crystal are all of the same size, as with metals, they tend to pack closely together. From this point of view, the simple cubic structure is very unstable, since it contains a large amount of empty space. A body-centered cubic structure has less waste space; about 20 metals have this type of unit cell. A still more efficient way of packing spheres of the same size is the face-centered cubic structure, where the amount of empty space is a minimum. About 40 different metals have a structure based on either the face-centered cubic cell or a close relative in which the packing is equally efficient (hexagonal closest-packed structure).

In the close-packed structures each atom touches 12 others

IONIC CRYSTALS

The geometry of ionic crystals, where there are two different kinds of ions, is more difficult to describe than that of metals. However, in many cases we can visualize the packing in terms of the unit cells discussed above. Lithium chloride, LiCl, is a case in point. Here, the larger Cl^- ions form a face-centered cubic lattice (Fig. 10.19). The smaller Li^+ ions fit into "holes" between the Cl^- ions. This puts an Li^+ ion at the center of each edge of the cube.

In the sodium chloride crystal, the Na^+ ion is slightly too large to fit into holes in a face-centered lattice of Cl^- ions (Fig. 10.19). As a result,

Figure 10.19 Three types of lattices in ionic crystals. In LiCl, the Cl^- ions are in contact with each other, forming a face-centered cubic lattice. In NaCl, the Cl^- ions are forced slightly apart by the larger Na^+ ions. In CsCl, the structure is quite different; the large Cs^+ ion at the center touches Cl^- ions at each corner of the cube.

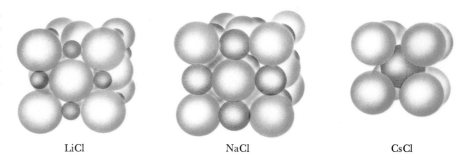

LiCl   NaCl   CsCl

the Cl^- ions are pushed slightly apart so that they are no longer touching and only Na^+ ions are in contact with Cl^- ions. However, the relative positions of positive and negative ions remain the same as in LiCl: each anion is surrounded by six cations and each cation by six anions.

The structures of LiCl and NaCl are typical of all the alkali halides (Group 1 cation, Group 7 anion) except those of cesium. In CsCl, the Cs^+ ion is much too large to fit into a face-centered cubic array of Cl^- ions. We find that CsCl crystallizes in a quite different structure (Fig. 10.19). Here, each Cs^+ ion is located at the center of a cube outlined by Cl^- ions. The Cs^+ ion at the center touches all the Cl^- ions at the corners; the Cl^- ions do not touch each other. As you can see, each Cs^+ ion is surrounded by eight Cl^- ions. It is also true that each Cl^- ion is surrounded by eight Cs^+ ions.

The CsCl crystal is simple cubic in both Cs^+ and Cl^- ions

■ **SUMMARY PROBLEM**

Consider methyl alcohol, an organic compound of molecular formula CH_3OH.

a. What kind of intermolecular forces would you expect to find in methyl alcohol?

b. How would you expect the normal boiling point of CH_3OH to compare to that of O_2? To C_2H_5OH? To NaCl? To SiO_2?

c. Would you expect CH_3OH to conduct electricity? How would its water solubility compare to that of O_2? To SiO_2?

d. Give the symbol of a solid nonmetal that is a better conductor than CH_3OH. Give the formula of a solid compound that, unlike CH_3OH, has a network structure.

e. The normal boiling point of CH_3OH is 65°C, the normal freezing point is −98°C, and the critical temperature is 239°C. Draw a rough sketch of the phase diagram of methyl alcohol, assuming the solid is more dense than the liquid.

f. If you wanted to purify methyl alcohol by sublimation, at what temperature would you operate?

g. A sample of methyl alcohol vapor at 250°C and 1 atm is cooled at constant pressure. At what temperature does liquid first appear?

CH_3OH structure. (Charles D. Winters)

h. Taking the normal boiling point of CH_3OH to be 65°C and its heat of vaporization to be 38.0 kJ/mol, estimate its normal vapor pressure at 25°C. Estimate the temperature at which its vapor pressure is 0.500 atm.

i. A sample of 1.00 mL of CH_3OH ($d = 0.792$ g/mL) is placed in a one-liter flask at 35°C, where its vapor pressure is 203 mm Hg. Show by calculation whether there is any liquid left in the flask when equilibrium is established.

Answers

a. Dispersion, H bonds   **b.** Higher than O_2, lower than the others
c. no; higher than O_2 or SiO_2 **d.** C(graphite) or Si; SiO_2, NaCl, and
so on   **e.** See Figure 10.7a   **f.** below about -98°C   **g.** 65°C
h. 123 mm Hg, 49°C   **i.** yes

Among the most important of the network covalent structures are those of the element silicon and the silicate minerals, in which silicon atoms are bonded to oxygen.

Elemental Silicon

A crystal of pure silicon has the diamond structure and is an electrical nonconductor. The valence electrons are used in the four covalent bonds that each atom forms with its neighbors. Conductivity increases dramatically when small amounts of arsenic (Group 5) or boron (Group 3) are introduced into the crystal (Fig. 10.20). As little as 0.0001% of these elements present in Si helps to produce the semiconductor devices used in transistors or solar cells.

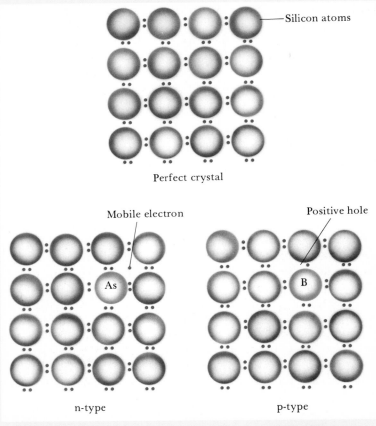

Figure 10.20 Semiconductors derived from silicon. In n-type semiconductors, the impurity atoms furnish mobile electrons to the crystal. In p-type semiconductors there is a deficiency of electrons, since the impurity atoms have three rather than four valence electrons.

To understand the effect of impurities on the conductivity of silicon or germanium, consider Figure 10.20. An atom of arsenic, with five valence electrons, can fit into the crystal lattice of Si. Its atomic radius (0.121 nm) is close to that of Si (0.117 nm). To do so, however, an arsenic atom must give up its fifth valence electron. This electron can move through the crystal under the influence of an electrical field. This gives an **n-type semiconductor** (current carried by the flow of negative charge). If an atom of boron or another element with three valence electrons is introduced into the lattice, a different situation arises. An electron deficiency is created at the site occupied by the foreign atom. It is surrounded by seven valence electrons rather than eight. In this sense, there is a "positive hole" in the lattice. In an electrical field, an electron moves from a neighboring atom to fill that hole. By so doing, it creates an electron deficiency around the atom which it leaves. The result is a **p-type semiconductor**. In effect, positive holes move through the lattice.

Within the past few years, research groups in the United States, Japan, and Western Europe have come up with a variety of semiconductor materials that are competitive with crystalline silicon. Among these is amorphous silicon, in which the atoms are not arranged in a fixed geometric pattern. If you have a solar-powered calculator, the chances are that the light-absorbing element is a very thin film of amorphous silicon.

Silicates

More than 90% of the rocks, minerals, and soils found in the earth's crust are classed as silicates; they contain silicon atoms bonded tetrahedrally to four oxygen atoms. Silicates are essentially ionic; the negative ion is an oxyanion in which silicon is the central atom. The anion may be a very simple species (e.g., SiO_4^{4-}) or exceedingly complex, depending on the geometric pattern in which Si and O atoms are linked. The negative charge of the silicon oxyanion is balanced by positive charges on the cations of one or more metals (e.g., Na^+, Mg^{2+}, Al^{3+}) that fit into the crystal lattice.

The simplest silicates are those containing discrete oxyanions, such as those in Fig. 10.21, p. 348. Of these, the simplest is the SiO_4^{4-} anion, found in the semiprecious stone zircon, $ZrSiO_4$, and the garnets, of which $Ca_3Cr_2(SiO_4)_3$ is typical. In the oxyanion at the right of Figure 10.21, six tetrahedra are linked through oxygen atoms at the corners to form a hexagonal ring structure. This is the anion present in beryl, $Be_3Al_2Si_6O_{18}$. The gemstone emerald is beryl with a small amount of chromium(III) as an impurity; the Cr^{3+} ions give emerald its characteristic green color.

Most silicate minerals have network covalent structures in which SiO_4^{4-} tetrahedra are linked together in one, two, or three dimensions. Two such structures are in Fig. 10.22, p. 349. In the structure at the left, tetrahedra are linked together to form an infinite chain; this occurs in the fibrous minerals such as diopside, $CaSiO_3 \cdot MgSiO_3$. The silicon-to-oxygen ratio in these minerals is readily derived by noting that two of the four oxygen

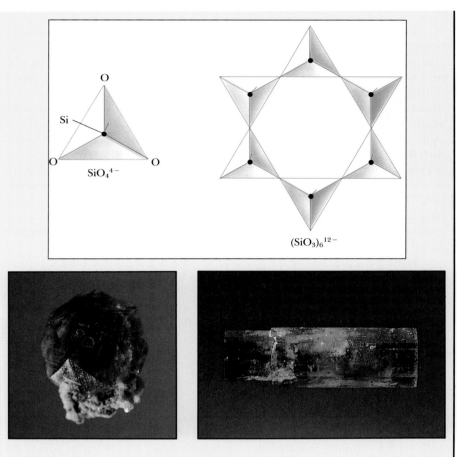

Figure 10.21 Two silicate anions. Garnets have the structure on the left, while emeralds crystallize in the hexagonal lattice at the right. (Allen B. Smith, Tom Stack and Associates)

atoms of each tetrahedron are shared with the next tetrahedron in the chain. So, for every silicon atom, the number of oxygens is:

$$2 + \frac{1}{2}(2) = 3$$

The best known of the fibrous silicates are the amphiboles or asbestos minerals, which have a structure similar to diopside except that two chains are linked together to form a double strand. Asbestos has been used since Roman times for its insulating and fire-resistant properties. Today, asbestos is being removed from private homes and public buildings because of the potential health hazards associated with exposure to the material. Workers in the asbestos industry have abnormally high rates of lung

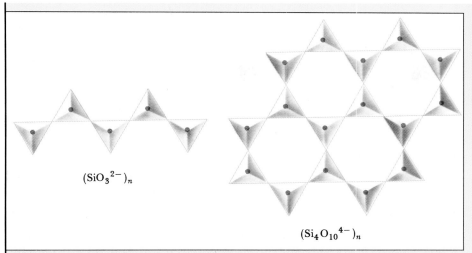

$(SiO_3{}^{2-})_n$

$(Si_4O_{10}{}^{4-})_n$

Figure 10.22 Two infinite silicate lattices. Asbestos contains a double anion chain similar to the one shown at the left. The structure of mica is that of the infinite sheet, of which a portion is shown at the right.

disease, including lung cancer. Presumably, asbestos fibers become embedded in lung tissue, causing an irritation that may lead to the formation of tumors many years later.

The silicate structure shown at the right of Figure 10.22 is typical of layer minerals such as talc, $Mg_3(OH)_2Si_4O_{10}$. Here, $SiO_4{}^{4-}$ tetrahedra are linked together in two dimensions to give an infinite layer. Three of four oxygen atoms are shared with adjacent tetrahedra, so the number of oxygens per silicon is:

$$1 + \frac{1}{2}(3) = \frac{5}{2}$$

The sheets are held together only by weak dispersion forces, so they easily slide past one another. As a result, talcum powder has a slippery feeling, much like graphite. Mica is also a layer silicate with a somewhat more complex structure in which cations are located between the layers. The attraction between oppositely charged ions makes sheets of mica more difficult to separate. The shiny luster of mica is exploited by adding small amounts of it to paint to give a silvery, metallic effect.

The simplest three-dimensional silicate network is that described previously for silicon dioxide, SiO_2. Here, all the oxygen atoms in a $SiO_4{}^{4-}$ tetrahedron are shared, and the oxygen-to-silicon ratio is

$$\frac{1}{2}(4) = 2$$

Many common minerals may be considered derivatives of SiO_2 in which some of the silicon atoms have been replaced by aluminum. An example is feldspar, in which an Al atom may be substituted for $\frac{1}{4}$ of the Si atoms, as in $KAlSi_3O_8$, or for $\frac{1}{2}$ of the Si atoms, as in $Ca(AlSiO_4)_2$. Granite is a heterogeneous mixture of three different silicates: mica, feldspar, and quartz.

Layer structure of mica. (Professor Carl Rettenmeyer, Museum of Natural History, University of Connecticut)

Another class of three-dimensional silicates are the materials known as zeolites (Fig. 10.23). These contain relatively large cavities or tunnels in which H_2O molecules or cations such as Na^+ or Ca^{2+} may be trapped. A variety of zeolites occur naturally. They can also be made in the laboratory, in which case the dimensions of the "holes" can be made to order.

Zeolites are widely used in home water softeners. When hard water, containing Ca^{2+} (or Mg^{2+}) ions, flows through a column packed with a zeolite, NaZ, the following reaction occurs:

This is called an ion exchange column

$$Ca^{2+}(aq) + 2\ NaZ(s) \rightarrow CaZ_2(s) + 2\ Na^+(aq)$$

Sodium ions migrate out of the cavities; Ca^{2+} ions move in. The effect is to replace Ca^{2+} ions in the water by Na^+ ions. In this way, the objectionable properties of hard water (precipitation of soap, formation of boiler scale, etc.) are removed.

Synthetic zeolites are also used as "molecular sieves" to trap small molecules such as H_2O. Anhydrous alcohol is made from 95% alcohol by treatment with a zeolite that quantitatively removes the water. Molecular sieves are widely used in the petroleum industry to separate small, compact hydrocarbon molecules from larger or more elongated molecules that will not fit into holes in the structure.

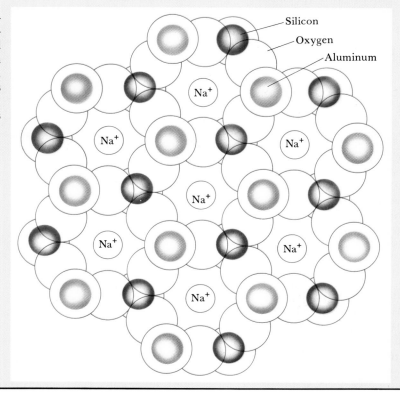

Figure 10.23 Structure of a natural zeolite with the simplest formula $NaAlSiO_4$. Na^+ ions are held loosely in the anionic lattice, which consists of Al, Si, and O atoms. When water containing Ca^{2+} ions passes through the zeolite, Ca^{2+} ions enter the lattice while Na^+ ions take their place in solution.

QUESTIONS AND PROBLEMS

Vapor Pressure

1. A sample of benzene vapor in a flask of constant volume exerts a pressure of 325 mm Hg at 80°C. The flask is slowly cooled.

a. Assuming no condensation, use the Ideal Gas Law to calculate the pressure of the vapor at 50°C; at 60°C.

b. Compare your answers in (a) to the equilibrium vapor pressures of benzene: 269 mm Hg at 50°C, 389 mm Hg at 60°C.

c. On the basis of your answers to (a) and (b), predict the pressure exerted by the benzene at 50°C; at 60°C.

2. The vapor pressure of solid iodine at 30°C is 0.466 mm Hg.

a. How many milligrams of iodine will sublime into an evacuated 1.00-L flask?

b. If 2.0 mg of I_2 are used, what will the final pressure be?

c. If 10 mg of I_2 are used, what will the final pressure be?

3. Ethyl ether, $C_4H_{10}O$, was a widely used anesthetic in the early days of surgery. It has a vapor pressure of 537 mm Hg at room temperature, 25°C. A ten-mL sample ($d = 0.708$ g/mL) is placed in a sealed 0.250-L flask.

a. Will there be any liquid left in the flask at equilibrium at 25°C?

b. What is the maximum volume the flask can have if equilibrium is to be established between liquid and vapor?

c. If the flask has a volume of 5.00 L, what is the pressure of ethyl ether in the flask?

4. A humidifier is used in a bedroom kept at 22°C. The bedroom's volume is 4.0×10^4 L. Assume that the air is originally dry and no moisture leaves the room while the humidifier is operating.

a. If the humidifier has a capacity of 3.00 gallons of water, will there be enough water to saturate the room with water vapor (vp of water at 22°C = 19.83 mm Hg; d water = 1.00 g/cm³)?

b. What is the final pressure of water vapor in the room when the humidifier has vaporized two-thirds of its water supply?

5. Carbon tetrachloride was once widely used in the dry cleaning industry. Its use was discouraged, however, after it was determined to be carcinogenic. It has a vapor pressure of 9.708 mm Hg at −20.0°C and 90.00 mm Hg at 20.0°C. Use the Clausius-Clapeyron equation to estimate the

a. heat of vaporization.

b. temperature at which the vapor pressure is 0.50 atm.

6. Octane, C_8H_{18}, is a principal component of gasoline. It has a vapor pressure of 145 mm Hg at 75.0°C and 20.0 mm Hg at 32.0°C. Use the Clausius-Clapeyron equation to estimate the

a. heat of vaporization of octane.

b. vapor pressure of octane on a warm day (85°F).

7. Mexico City has an elevation of 7400 ft above sea level. Its normal atmospheric pressure is 589 mm Hg. At what temperature does water boil in Mexico City? (ΔH_{vap} of water = 40.7 kJ/mol.)

8. Mount Kilimanjaro has an altitude of 19,350 ft. Water boils at about 81°C atop Mount Kilimanjaro. What is the normal atmospheric pressure on the mountain?

9. Mercury is used extensively in studying the effect of pressure on the volume of gases. It is fairly inert at the surface, and most gases are insoluble in it. Should we worry about the contribution of the vapor pressure of mercury to the pressure of gases above it? Use the following data to determine the vapor pressure of mercury at 0°C and 100°C: ΔH_{vap} of Hg = 59.4 kJ/mol; normal boiling point = 357°C.

10. When water boils inside a pressure cooker, its vapor pressure is 1.500×10^3 mm Hg. Use the Clausius-Clapeyron equation to calculate the temperature of the water. Take the heat of vaporization to be 40.7 kJ/mol.

11. The data below are for the vapor pressure of toluene, a common solvent.

vp (mm Hg)	40	60	80	100
t (°C)	32	40	47	52

Plot ln vp vs. $1/T$ (Equation 10.1) and use your graph to estimate the heat of vaporization of toluene.

12. Consider the following data for the vapor pressure of acetone.

vp (mm Hg)	39	116	283	613
t (°C)	−10	10	30	50

Follow the instructions of Problem 11 to estimate the heat of vaporization of acetone.

13. The critical point of CO is −139°C, 35 atm. Liquid CO has a vapor pressure of 6 atm at −171°C. Which of the following statements must be true?

a. CO is a gas at −171°C and 1 atm.

b. A tank of CO at 20°C can have a pressure of 35 atm.

c. CO gas cooled to −145°C and 40 atm pressure will condense.

d. The normal boiling point of CO lies above −171°C.

14. The normal boiling point of SO_2 is $-10°C$. At $32°C$ its vapor pressure is 5 atm. Which of the following statements concerning sulfur dioxide must be true?
    a. A tank of SO_2 at $32°C$ that has a pressure of 4 atm must contain liquid SO_2.
    b. A tank of SO_2 at $32°C$ that has a pressure of 1 atm cannot contain liquid SO_2.
    c. The critical temperature of SO_2 must be greater than $30°C$.

Phase Diagrams

15. Referring to Figure 10.5, state what phase(s) is (are) present at
    a. $-3°C$, 5 mm Hg   b. $25°C$, 1 atm
    c. $70°C$, 20 mm Hg
16. Referring to Figure 10.5, state what phase(s) is (are) present at
    a. 1 atm, $100°C$ b. 0.5 atm, $100°C$
    c. 0.5 atm, $0°C$
17. Iodine has a triple point at $114°C$, 90 mm Hg. Its critical temperature is $535°C$. The density of the solid is 4.93 g/cm^3, while that of the liquid is 4.00 g/cm^3. Sketch a phase diagram for iodine, and use it to fill in the blanks below, using either "liquid" or "solid."
    a. Iodine vapor at 80 mm Hg condenses to the _____ when cooled sufficiently.
    b. Iodine vapor at $125°C$ condenses to the _____ when enough pressure is applied.
    c. Iodine vapor at 700 mm Hg condenses to the _____ when cooled above the triple point temperature.
18. Argon gas has its triple point at $-189.3°C$ and 516 mm Hg. It has a critical point at $-122°C$ and 48 atm. The density of the solid is 1.65 g/cm^3, while that of the liquid is 1.40 g/cm^3. Sketch the phase diagram for argon and use it to fill in the blanks below with the words "boils," "melts," "sublimes," or "condenses."
    a. Solid argon at 500 mm Hg _____ when the temperature is increased.
    b. Solid argon at 2 atm _____ when the temperature is increased.
    c. Argon gas at $-150°C$ _____ when the pressure is increased.
    d. Argon gas at $-165°C$ _____ when the pressure is increased.
19. A pure substance A has a liquid vapor pressure of 320 mm Hg at $125°C$, 800 mm Hg at $150°C$, and 60 mm Hg at the triple point, $85°C$. The melting point of A decreases slightly as pressure increases.
    a. Sketch a phase diagram for A.
    b. From the phase diagram, estimate the normal boiling point.
    c. What changes occur when, at a constant pressure of 320 mm Hg, the temperature drops from 150 to $100°C$?

20. A pure substance X has the following properties: mp $= 90°C$, increasing slightly as pressure increases; normal bp $= 120°C$; liquid vp $= 65$ mm Hg at $100°C$, 20 mm Hg at the triple point.
    a. Draw a phase diagram for X.
    b. Label solid, liquid, and vapor regions of the diagram.
    c. What changes occur if, at a constant pressure of 100 mm Hg, the temperature is raised from $100°C$ to $150°C$?

Intermolecular Forces

21. Arrange the following in order of increasing boiling point:
    a. Ar   b. He   c. Ne   d. Xe
22. Arrange the following in order of decreasing boiling point:
    a. I_2 b. F_2 c. Cl_2 d. Br_2
23. Which of the following would you expect to show dispersion forces? Dipole forces?
    a. CH_4 b. CCl_4 c. CO   d. CO_2
24. Which of the following would you expect to show dispersion forces? Dipole forces?
    a. O_2 b. H_2O c. $CHCl_3$ d. CH_3Cl
25. Which of the following would show hydrogen bonding?
    a. $CH_3—CH_3$ b. $CH_3—OH$
    c. CH_3NH_2 d. $HO—OH$
26. Which of the following would show hydrogen bonding?
    a. CH_3F b. NH_3
    c. $H_3C—O—CH_3$ d. $[H—F—H]^+$
27. Explain in terms of intermolecular forces why
    a. ICl has a higher melting point than Br_2.
    b. C_2H_6 has a higher boiling point than CH_4.
    c. H_2O_2 has a higher boiling point than C_2H_6.
    d. C_2H_5OH has a lower boiling point than NaF.
28. Explain in terms of intermolecular forces why
    a. Br_2 is lower melting than NaBr.
    b. C_2H_5OH is higher boiling than butane, C_4H_{10}.
    c. Water is higher boiling than hydrogen telluride, H_2Te.
    d. Acetic acid, $H_3C—\overset{O}{\underset{\|}{C}}—OH$, is lower boiling than benzoic acid, $C_6H_5\overset{O}{\underset{\|}{C}}—OH$.

29. In which of the following processes are covalent bonds broken?
    a. Melting mothballs made up of naphthalene, $C_{10}H_8$.
    b. Boiling water.
    c. Boiling ethyl alcohol.
    d. Dissolving iodine in water.
30. In which of the following processes is it necessary to break covalent bonds as opposed to simply overcoming intermolecular forces?

a. Electrolysis of water.

b. Subliming dry ice.

c. Melting glass.

d. Decomposing dinitrogen tetroxide to nitrogen dioxide.

31. For each of the following pairs, choose the member with the lower boiling point. Explain your reason in each case.

a. SO_2 or CO_2 b. HCl or HBr

c. H_2 or O_2 d. F_2 or Cl_2

32. Follow the directions for Question 31 for the following substances:

a. AsH_3 or PH_3 b. C_6H_6 or $C_{10}H_8$

c. NH_3 or PH_3 d. LiCl or C_3H_8

33. What are the strongest attractive forces that must be overcome to

a. boil silicon hydride (SiH_4)?

b. melt iodine (I_2)?

c. dissolve chlorine in carbon tetrachloride?

d. vaporize calcium chloride?

34. What are the strongest attractive forces that must be overcome to

a. melt ice?

b. boil carbon tetrachloride (CCl_4)?

c. sublime bromine (Br_2)?

d. melt benzene (C_6H_6)?

Types of Solids

35. Classify as metallic, network covalent, ionic, or molecular, a solid that

a. melts below 100°C and is insoluble in water.

b. conducts electricity only when melted.

c. is insoluble in water and conducts electricity.

36. Classify as metallic, network covalent, ionic, or molecular, a solid that

a. is insoluble in water, melts above 500°C, and does not conduct electricity.

b. dissolves in water but is nonconducting either as an aqueous solution, a solid, or molten.

c. dissolves in water, melts above 100°C, and conducts electricity when present in an aqueous solution.

37. Of the four general types of solids, which one(s)

a. are generally low boiling?

b. are ductile and malleable?

c. are generally nonvolatile?

38. Of the four general types of solids, which one(s)

a. are generally insoluble in water?

b. are very high melting?

c. conduct electricity as solids?

39. Classify each of the following species as molecular, network covalent, ionic, or metallic.

a. K   b. $CaCO_3$ c. C_8H_{18}

d. C(diamond)   e. HCl(g)

40. Classify each of the following species as molecular, network covalent, ionic, or metallic.

a. mercury   b. iodine bromide

c. sand   d. chromium(III) sulfate

e. chlorine gas

41. Give the formula of a solid containing oxygen that is

a. polar molecular   b. ionic

c. network covalent   d. nonpolar molecular

42. Give the formula of a solid containing carbon that is

a. molecular   b. ionic

c. network covalent   d. metallic

43. Describe the nature of the structural units in

a. NaI   b. N_2 c. CO_2 d. W

44. Describe the nature of the structural units in

a. C(graphite)   b. SiC

c. $FeCl_2$ d. C_2H_2

Crystal Structure

45. Aluminum (at. radius = 0.143 nm) crystallizes with a face-centered cubic unit cell. What is the length of a side of the cell?

46. Xenon crystallizes with a face-centered cubic unit cell. The edge of the unit cell is 0.620 nm. Calculate the atomic radius of xenon.

47. Chromium crystallizes with a body-centered cubic unit cell. The volume of the unit cell is 2.376×10^{-2} nm³. What is the atomic radius of chromium?

48. Sodium crystallizes with a body-centered cubic unit cell. Its atomic radius is 0.186 nm. What is the volume of the cell?

49. Potassium iodide has a unit cell similar to that of NaCl (Fig. 10.19). The ionic radii of K^+ and I^- are 0.133 nm and 0.216 nm, respectively. What is the length of

a. one side of the cube?

b. the face diagonal of the cube?

50. In the LiCl structure shown in Figure 10.19, the chloride ions form a face-centered cubic unit cell 0.513 nm on an edge. The ionic radius of Cl^- is 0.181 nm.

a. Along a cell edge, how much space is there between the Cl^- ions?

b. Would an Na^+ ion ($r = 0.095$ nm) fit into this space? a K^+ ion ($r = 0.133$ nm)?

51. Consider the CsCl cell (Fig. 10.19). The ionic radii of Cs^+ and Cl^- are 0.169 and 0.181 nm, respectively. What is the length of

a. the body diagonal?

b. the side of the cell?

52. For a cell of the CsCl type (Fig. 10.19), how is the length of one side of the cell, s, related to the sum of the radii of the ions, $r_{cation} + r_{anion}$?

53. Using the result of Problem 45 and taking the density of aluminum to be 2.70 g/cm³, estimate Avogadro's number.

54. Using the data given in Problem 46 and assuming Avogadro's number to be 6.02×10^{23}, estimate the density of Xe(s).
55. Consider the CsCl unit cell shown in Figure 10.19. How many Cs^+ ions are there per unit cell? How many Cl^- ions?
56. Consider the NaCl unit cell shown in Figure 10.19. Looking only at the front face (5 large Cl^- ions, 4 small Na^+ ions),
   a. How many cubes share each of the Na^+ ions in this face?
   b. How many cubes share each of the Cl^- ions in this face?

Unclassified
57. Criticize each of the following statements.
   a. Vapor pressure is inversely proportional to volume.
   b. Vapor pressure is directly proportional to temperature.
   c. Boiling point increases with molar mass.
   d. The melting point of a molecular substance depends upon the strength of the covalent bonds within the molecule.
58. Which of the following statements are true?
   a. The critical pressure is the highest vapor pressure that a liquid can have.
   b. To sublime a solid, it must be heated above the triple point.
   c. NaF melts higher than F_2 because its molar mass is larger.
   d. One metal crystallizes in a BCC cell, another in a face-centered cubic cell of the same size. The two atomic radii must be equal.
59. Explain the difference between
   a. a face-centered cubic and a body-centered cubic unit cell.
   b. melting and sublimation.
   c. a boiling point and a normal boiling point.
   d. a phase diagram and a vapor pressure curve.
60. Differentiate between
   a. intramolecular forces and intermolecular forces.
   b. a molecular solid and a network covalent solid.
   c. a dipole force and a hydrogen bond.
   d. an ionic solid and a molecular solid.
61. How would you expect the lattice energy of BaO to compare to that of MgO? To that of CsF?
62. Using the data in Table 10.5, calculate k (Eqn. 10.3) for each compound. In principle, k should be about the same in each case.
63. The density of liquid mercury at 20°C is 13.6 g/cm³; its vapor pressure is 1.2×10^{-3} mm Hg.
   a. What volume (cm³) is occupied by one mole Hg(l) at 20°C?

   b. What volume (cm³) is occupied by one mole Hg(g) at 20°C and the equilibrium vapor pressure?
   c. The atomic radius of Hg is 0.155 nm. Calculate the volume (cm³) of one mole Hg atoms ($V = 4\pi r^3/3$).
   d. From your answers to (a), (b), and (c), calculate the percentage of the total volume occupied by the atoms in Hg(l) and Hg(g) at 20°C and 1.2×10^{-3} mm Hg.
64. Explain the difference in boiling points for each of the following pairs of substances:
   a. ethyl alcohol, C_2H_5OH (79°C) and ethyl ether, C_2H_5—O—C_2H_5 (34.6°C).
   b. HF (20°C) and HCl (−85°C).
   c. LiF (1717°C) and HF (20°C).
65. Elemental boron is almost as hard as diamond. It is insoluble in water, does not conduct electricity at room temperature, and melts at 2300°C. What type of solid is boron?
66. An experiment is performed to determine the vapor pressure of formic acid. A 30.0-L volume of helium gas at 20.0°C is passed through 10.00 g of liquid formic acid (HCOOH) at 20.0°C. After the experiment 7.50 g of liquid formic acid remains. Assume that the helium gas becomes saturated with formic acid vapor and that the total gas volume and temperature remain constant. What is the vapor pressure of formic acid at 20.0°C?
67. Which of the following substances can be liquified by applying pressure at 25°C?

SUBSTANCE	CRITICAL POINTS
Ethane	32°C, 48 atm
Carbon disulfide	273°C, 76 atm
Krypton	−63°C, 54 atm
Nitrogen oxide	−94°C, 65 atm

Challenge Problems
68. A flask with a volume of 10.0 L contains 0.400 g of hydrogen gas and 3.20 g of oxygen. The mixture is ignited and the reaction

$$2 H_2(g) + O_2(g) \rightarrow 2H_2O$$

goes to completion. The mixture is cooled to 27°C. Assuming 100% yield,
   a. what physical state(s) of water are present in the flask?
   b. what is the final pressure in the flask?
69. Chloroform (CHCl₃) is the liquid that usually saturates the handkerchief that is pressed over the nose of an unsuspecting victim in TV spy stories. Chloroform has a vapor pressure of 199 mm Hg at 25.0°C. If half a pint of chloroform ($d = 1.489$ g/mL) is spilled on the floor of a room that is 12 ft × 13 ft × 8.0 ft, will all the chloroform vaporize?
70. It has been suggested that the pressure exerted on a skate blade is sufficient to melt the ice beneath it and form

a thin film of water, which makes it easier for the blade to slide over the ice. Assume that a skater weighs 120 lb and the blade has an area of 0.10 in². Calculate the pressure exerted on the blade (1 atm = 15 lb/in²). From information in the text, calculate the decrease in melting point at this pressure. Comment on the plausibility of this explanation, and suggest another mechanism by which the water film might be formed.

71. As shown in Figure 10.19, Li^+ ions fit into a closely packed array of Cl^- ions, but Na^+ ions do not. What is the value of the r_{cation}/r_{anion} ratio at which a cation just fits into a structure of this type?

72. Calculate the percent of the total volume that is occupied by atoms in a simple cubic cell; by atoms in a face-centered cubic cell.

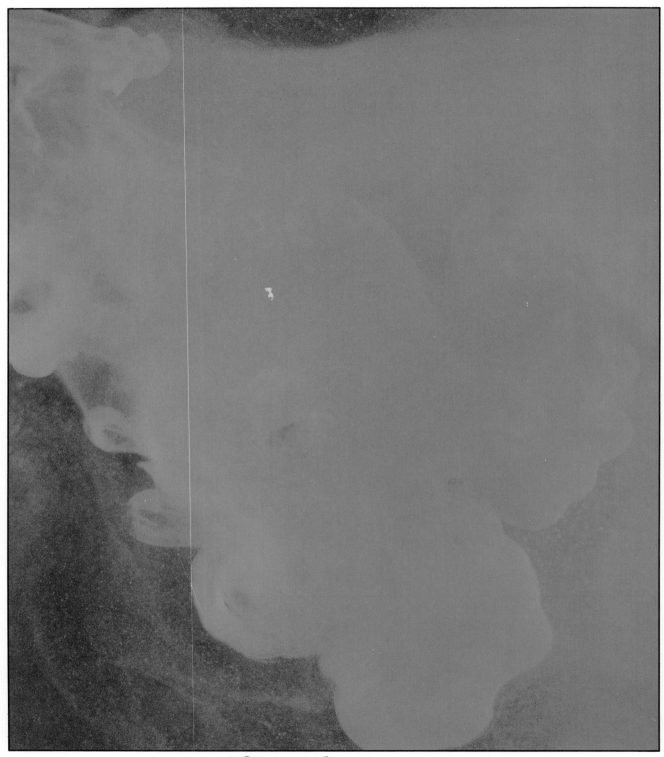

$$Ba^{2+}(aq) \ + \ SO_4{}^{2-}(aq) \ \rightarrow \ BaSO_4(s)$$
Precipitation of barium sulfate by mixing solutions containing Ba^{2+} and $SO_4{}^{2-}$ ions.

CHAPTER
11

SOLUTIONS

When water turns ice does it remember one time it was water?
When ice turns back into water does it remember it was ice?

CARL SANDBURG
METAMORPHOSES

I n the course of a day, you use or make solutions many times. Your morning cup of coffee is a solution of solids (sugar and coffee) in a liquid (water). The gasoline you fill your gas tank with is a solution of several different liquid hydrocarbons. The soda you drink at a study break is a solution of a gas (carbon dioxide) in a liquid (water).

A solution is a homogeneous mixture of a *solute* (substance being dissolved) distributed through a *solvent* (substance doing the dissolving). Solutions exist in any of the three physical states: gas, liquid, or solid. Air, the most common gaseous solution, is a mixture of nitrogen, oxygen, and lesser amounts of other gases. Many metal alloys are solid solutions. An example is the U.S. "nickel" coin (25% Ni, 75% Cu). The most familiar solutions are those in the liquid state, especially ones in which water is the solvent. Aqueous solutions are most important for our purposes in chemistry and will be emphasized in this chapter.

We will consider several aspects of solutions. These include:

— methods of expressing solution concentrations by specifying the relative amounts of solute and solvent (Section 11.2).

— factors affecting solubility, including the nature of the solute and the solvent, the temperature, and the pressure (Section 11.3).

— the effect of solutes upon such solvent properties as vapor pressure, freezing point, and boiling point (Section 11.4).

To start with, it will be helpful to review some of the terms used to describe the nature and concentrations of the components of a solution (Section 11.1).

Homogeneous means having constant composition down to the molecular or ionic level. Milk is not a true solution

357

This column of solid sodium acetate was formed by slowly pouring a supersaturated solution over a seed crystal. (Charles D. Winters)

11.1
SOLUTION TERMINOLOGY

Several different adjectives can be used to indicate the relative amounts of solute and solvent in a solution. We may describe a solution containing a small amount of solute as being "dilute." Another solution containing more solute in the same amount of solvent might be called "concentrated." In a few cases, these terms, through tradition, have taken on a quantitative meaning. Solutions of certain common reagents, labeled dilute or concentrated, have the compositions specified in Table 11.1.

Relative concentrations of solutions are often expressed in a different way by using the terms "saturated," "unsaturated," and "supersaturated." A **saturated** solution is one that is, or could be, in equilibrium with undissolved solute. An **unsaturated** solution contains a lower concentration of solute than the saturated solution. The unsaturated solution is not at equilibrium. If solute is added, it dissolves until saturation is reached. A **supersaturated** solution contains more than the equilibrium concentration of solute. It is unstable in the presence of excess solute.

To illustrate the meaning of these terms, consider sucrose (cane sugar), $C_{12}H_{22}O_{11}$ (Fig. 11.1). A saturated aqueous solution of sucrose at 20°C contains 203.9 g $C_{12}H_{22}O_{11}$ in 100 g of water. Any solution in which the concentration of sucrose at 20°C is less than this value is unsaturated. If more sucrose is added to such a solution, it will dissolve until the concentration reaches the saturation value.

A supersaturated solution contains more than 203.9 g $C_{12}H_{22}O_{11}$ in 100 g of water at 20°C. To prepare such a solution we take advantage of the fact that the solubility of sucrose increases with temperature. At 80°C, we can dissolve 362.1 g of sucrose in 100 g of water. If this solution is cooled carefully to 20°C, without shaking or stirring, the excess solute stays in solution. This produces a supersaturated solution. If a small seed crystal of sucrose is now added, crystallization quickly takes place. The excess solute (362.1 g − 203.9 g = 158.2 g) comes out of solution, establishing equilibrium between the saturated solution and the sucrose crystals.

The crystallization of excess solute is a common problem in the preparation of candies and in the storage of jam and honey. From these su-

TABLE 11.1   Concentrations of Laboratory Acid and Base Solutions

		SOLUTE	MOLES SOLUTE PER LITER	MASS PERCENT SOLUTE	DENSITY (g/cm^3)
Hydrochloric acid	conc.	HCl	12	36	1.18
	dilute		6	20	1.10
Nitric acid	conc.	HNO_3	16	72	1.42
	dilute		6	32	1.19
Sulfuric acid	conc.	H_2SO_4	18	96	1.84
	dilute		3	25	1.18
Ammonia	conc.	NH_3*	15	28	0.90
	dilute		6	11	0.96

*Often labeled "NH_4OH."

Figure 11.1 When sugar is added to the unsaturated solution (*top*), it dissolves. When added to the saturated solution (*bottom left*), it does not dissolve. With the supersaturated solution (*bottom right*), addition of sugar causes crystallization of the excess solute. (Marna G. Clarke)

persaturated solutions, sugar separates either as tiny crystals, causing the "graininess" in fudge, or as large crystals, which often appear in honey kept for a long time (Fig. 11.2).

NONELECTROLYTES VS. ELECTROLYTES

Pure water does not conduct electricity (Fig. 11.3a, p. 360). Solutes in water can be classified according to the conductivity of the solutions they form. We can distinguish between two types of solutes.

1. Nonelectrolytes form water solutions that do not conduct an electric current (Fig. 11.3B). Typically, these substances are molecular and dissolve as molecules. Since molecules are neutral, they do not migrate in an electric field. Hence, they do not conduct an electric current. The processes by which methyl alcohol, CH_3OH, and sugar, $C_{12}H_{22}O_{11}$, dissolve in water can be represented by the simple equations:

$$CH_3OH(l) \rightarrow CH_3OH(aq)$$

$$C_{12}H_{22}O_{11}(s) \rightarrow C_{12}H_{22}O_{11}(aq)$$

2. Electrolytes are solutes whose water solutions conduct an electric current (Fig. 11.3c). *These substances produce ions in solution.* The charged ions mi-

Figure 11.2 Rock candy is formed by the crystallization of sugar from a saturated solution that is slowly cooled. (Marna G. Clarke)

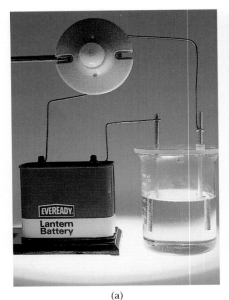
(a)

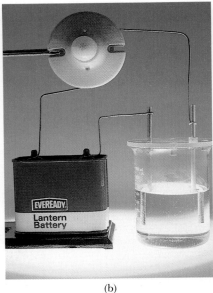
(b)

(c)

Figure 11.3 An apparatus for testing electrical conductivity. For an electric current to flow, the solution must contain ions, which carry electric charge. (Marna G. Clarke)

grate in an electric field, thereby carrying a current. Sodium chloride is a familiar example of an electrolyte. Solid sodium chloride consists of Na^+ and Cl^- ions. When sodium chloride dissolves in water, these ions are set free (Fig. 11.4). The solution process can be represented most simply as:

$$NaCl(s) \rightarrow Na^+(aq) + Cl^-(aq)$$

With some ionic solutes, discrete hydrated cations such as $Zn(H_2O)_4^{2+}$ and $Al(H_2O)_6^{3+}$ are formed. However, when we write equations for the dissolving of ionic solids, we ordinarily show the products as simple, nonhydrated ions in solution (Example 11.1).

■ **EXAMPLE 11.1** _____
Write equations for the dissociation in water of each of the following ionic solids:

a. KI   b. Li_2CO_3 c. $Fe(NO_3)_3$ d. $Ce_2(SO_4)_3$

Solution
In each case, the "reactant" is the ionic solid. The "products" are the ions released into water.

a. $KI(s) \rightarrow K^+(aq) + I^-(aq)$
b. $Li_2CO_3(s) \rightarrow 2\ Li^+(aq) + CO_3^{2-}(aq)$
c. $Fe(NO_3)_3(s) \rightarrow Fe^{3+}(aq) + 3\ NO_3^-(aq)$
d. $Ce_2(SO_4)_3(s) \rightarrow 2\ Ce^{3+}(aq) + 3\ SO_4^{2-}(aq)$

Dissolved salts are usually completely ionized

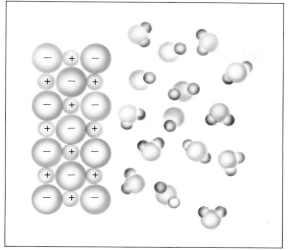 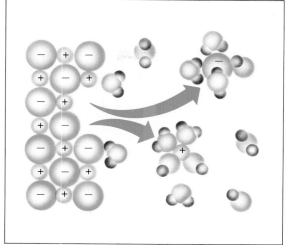

Figure 11.4 Dissolving NaCl in water. The attraction between the oxygen atoms in H_2O molecules and Na^+ ions in NaCl tends to bring the Na^+ ions into solution. At the same time, Cl^- ions, attracted to the hydrogen atoms of H_2O molecules, also enter the solution. The H_2O molecules that surround the ions tend to prevent oppositely charged ions from recombining.

Exercise
One mole of Na_2SO_4 and one mole of KCl are dissolved separately in water. Which solution contains the greater number of moles of ions? Answer: Na_2SO_4; 3 mol.

Electrolyte systems are important biologically. Blood plasma is an electrolyte containing many different ions, among them Na^+, K^+, Mg^{2+}, Ca^{2+}, Cl^-, HCO_3^-, SO_4^{2-}, and HPO_4^{2-}. The concentrations of these ions in blood must stay nearly constant. Slight variations, particularly in K^+, can be critical. Many anorexics and bulimics change the K^+ ion concentration in their blood quite rapidly when they fluctuate between starving and overeating. Fatalities from these eating disorders have been attributed to the rapid change in concentration of K^+.

As noted in Chapter 10, hydrogen chloride, HCl, is typical of a certain class of molecular compounds that act as electrolytes *in water solution*. Pure hydrogen chloride as a solid, liquid, or gas has the properties of a molecular substance. For example, liquid HCl does not conduct an electric current. However, when put into water, HCl dissolves to form a conducting solution typical of an electrolyte. A reaction occurs with water to produce aqueous H^+ ions and Cl^- ions:

$$HCl(aq) \rightarrow H^+(aq) + Cl^-(aq)$$

Table 11.2 (p. 362) summarizes the behavior of different kinds of solutes. Note from the table that a mole of a nonelectrolyte produces *one*

TABLE 11.2   **Solution Behavior of Some Aqueous Solutes**

SOLUTE	TYPE	SOLUTION EQUATION	MOLES OF SOLUTE PARTICLES PER MOLE SOLUTE
$I_2(s)$	Nonelectrolyte	$I_2(s) \rightarrow I_2(aq)$	1
$C_6H_{12}O_6(s)$ (glucose)	Nonelectrolyte	$C_6H_{12}O_6(s) \rightarrow C_6H_{12}O_6(aq)$	1
$C_{12}H_{22}O_{11}(s)$ (sucrose)	Nonelectrolyte	$C_{12}H_{22}O_{11}(s) \rightarrow C_{12}H_{22}O_{11}(aq)$	1
$HCl(g)$	Electrolyte	$HCl(g) \rightarrow H^+(aq) + Cl^-(aq)$	2 (1 H^+ and 1 Cl^-)
$LiBr(s)$	Electrolyte	$LiBr(s) \rightarrow Li^+(aq) + Br^-(aq)$	2 (1 Li^+ and 1 Br^-)
$K_2SO_4(s)$	Electrolyte	$K_2SO_4(s) \rightarrow 2 K^+(aq) + SO_4{}^{2-}(aq)$	3 (2 K^+ and 1 $SO_4{}^{2-}$)
$LaBr_3(s)$	Electrolyte	$LaBr_3(s) \rightarrow La^{3+}(aq) + 3 Br^-(aq)$	4 (1 La^{3+} and 3 Br^-)

mole of particles in solution. On the other hand, electrolytes provide *more than one mole* of particles per mole of solute dissolved.

11.2
CONCENTRATION UNITS

The properties of a solution depend strongly upon the relative amounts of solute and solvent present. These are described by citing the *concentration* of solute, which tells us how much solute is present for a given amount of solvent or solution. Concentrations can be expressed in various ways. Sometimes we state the mass percent (commonly called weight percent) of solute:

Usage is sometimes loose. A 5% NaCl solution is 5 mass percent NaCl

$$\text{mass percent solute} = \frac{\text{mass solute}}{\text{total mass solution}} \times 100 \qquad (11.1)$$

In a solution prepared by dissolving 24 g of NaCl in 152 g of water:

$$\text{mass percent NaCl} = \frac{24 \text{ g}}{24 \text{ g} + 152 \text{ g}} \times 100 = \frac{24}{176} \times 100 = 14$$

When the amount of solute is very small, as with trace impurities in water, concentration is often expressed in parts per million (ppm) or parts per billion (ppb). For liquid solutions:

$$\text{ppm solute} = \frac{\text{mass solute}}{\text{total mass solution}} \times 10^6 \qquad (11.2a)$$

$$\text{ppb solute} = \frac{\text{mass solute}}{\text{total mass solution}} \times 10^9 \qquad (11.2b)$$

At least it's not supposed to

In the United States and Canada, drinking water cannot contain more than 5×10^{-4} mg of mercury per gram of sample. In parts per million that would be

$$\text{ppm Hg} = \frac{5 \times 10^{-4} \text{ mg Hg}}{1 \times 10^3 \text{ mg}} \times 10^6 = 0.5$$

Note that it does not matter what the units of mass are as long as they are the same in numerator and denominator.

Comparing Equations 11.1 and 11.2, we see that there is a very simple relationship between mass percent of solute and parts per million or parts per billion:

$$\text{ppm} = \text{mass percent} \times 10^4$$

$$\text{ppb} = \text{mass percent} \times 10^7$$

More commonly, in expressing concentrations, we work with the number of moles of solute. In the remainder of this section, we will consider three concentration units of this type. Two of these, *mole fraction* and *molarity*, were referred to in previous chapters (Chapters 2 and 4). The third, *molality*, is discussed here for the first time.

MOLE FRACTION (X)

You may recall from Chapter 4 that the mole fraction (X) of a component, A, in a solution is given by the relation:

$$X_A = \frac{\text{no. moles } A}{\text{total no. moles all components}} \qquad (11.3)$$

Mole fraction is well named. Its meaning is implied by its name

If 1.20 mol of methyl alcohol, CH_3OH, is dissolved in 16.8 mol of water, we have

$$X_{CH_3OH} = \frac{1.20 \text{ mol}}{1.20 \text{ mol} + 16.8 \text{ mol}} = \frac{1.20}{18.0} = 0.0667$$

The sum of the mole fractions of all the components of a solution must be 1. That is:

$$X_A + X_B + \cdots = 1 \qquad (11.4)$$

For the methyl alcohol solution just referred to, there are only two components, CH_3OH and water. The mole fraction of water is readily found.

$$X_{H_2O} = 1 - X_{CH_3OH} = 1 - 0.0667 = 0.9333$$

■ EXAMPLE 11.2 ─────────────────────────

What are the mole fractions of CH_3OH and H_2O in a solution prepared by dissolving 1.20 g of methyl alcohol in 16.8 g of water?

Solution

We first find the numbers of moles of CH_3OH and H_2O. Their molar masses are 32.0 and 18.0 g/mol, in that order:

$$\text{no. moles } CH_3OH = 1.20 \text{ g} \times \frac{1 \text{ mol}}{32.0 \text{ g}} = 0.0375 \text{ mol } CH_3OH$$

$$\text{no. moles } H_2O = 16.8 \text{ g} \times \frac{1 \text{ mol}}{18.0 \text{ g}} = 0.933 \text{ mol } H_2O$$

To find the mole fraction of CH_3OH, we apply the defining relation, Equation 11.3:

$$X_{CH_3OH} = \frac{\text{no. moles } CH_3OH}{\text{no. moles } CH_3OH + \text{no. moles } H_2O} = \frac{0.0375}{0.0375 + 0.933}$$

$$= \boxed{0.0386}$$

We could carry out a similar calculation to obtain the mole fraction of water, but it is simpler to use Equation 11.4:

$$X_{H_2O} = 1 - X_{CH_3OH} = 1 - 0.0386 = \boxed{0.9614}$$

In the solution about 96% of the molecules are H_2O and 4% are CH_3OH

Exercise
Calculate the mole fraction of ethyl alcohol, C_2H_5OH, in a solution prepared by dissolving 1.20 g of C_2H_5OH in 16.8 g of water. Answer: 0.0272.

MOLALITY (m)

The concentration unit molality, given the symbol m, is defined as the number of moles of solute per kilogram (1000 g) of *solvent*.

$$\text{molality } (m) = \frac{\text{no. moles solute}}{\text{no. kilograms solvent}} \qquad (11.5)$$

The molality of a solution is readily calculated if the masses of solute and solvent are known (Example 11.3).

■ **EXAMPLE 11.3** —————————————————————————————
A solution contains 12.0 g of glucose, $C_6H_{12}O_6$, in 95.0 g of water. Calculate the molality of glucose.

Solution
To obtain the number of moles of solute, note that the molar mass of glucose is:

$$6(12.0 \text{ g/mol}) + 12(1.0 \text{ g/mol}) + 6(16.0 \text{ g/mol}) = 180.0 \text{ g/mol}$$

$$\text{no. moles solute} = 12.0 \text{ g} \times \frac{1 \text{ mol}}{180.0 \text{ g}} = 0.0667 \text{ mol}$$

$$\text{no. kg solvent} = 95.0 \text{ g water} \times \frac{1 \text{ kg water}}{1000 \text{ g water}} = 0.0950 \text{ kg water}$$

$$\text{molality} = \frac{0.0667 \text{ mol solute}}{0.0950 \text{ kg solvent}} = \boxed{0.702 \text{ } m}$$

Alternatively, we could find the molality by a conversion-factor approach:

$$\frac{12.0 \text{ g solute}}{95.0 \text{ g solvent}} \times \frac{1 \text{ mol solute}}{180.0 \text{ g solute}} \times \frac{10^3 \text{ g solvent}}{1 \text{ kg solvent}} = \frac{0.702 \text{ mol solute}}{1 \text{ kg solvent}}$$

Exercise
A solution contains 23.0 g of ethyl alcohol, C_2H_5OH, dissolved in 30.0 g of water. Calculate the molality of ethyl alcohol in this solution. Answer: 16.7 m.

MOLARITY (M)

In general chemistry, the most common concentration unit is molarity, which is the number of moles of solute per liter of solution.

Molality is used less often, morality not at all

$$molarity\ (M) = \frac{no.\ moles\ solute}{no.\ liters\ solution}$$ (11.6)

In Chapter 2, we used this defining equation to calculate one of the three quantities (molarity, moles solute, and volume solution), knowing the values of the other two quantities. Here we will consider some other types of calculations involving molarity.

In the laboratory you will frequently need to prepare a certain volume of a solution of a specified molarity. To do this, you will most often start with pure solute. In that case, you:

Molarity is best when volume measurements are involved

— calculate the mass of solute required, using the defining equation for molarity, Equation 11.6, and the molar mass of the solute.

— weigh out the calculated mass of solute and dissolve in enough solvent to give the desired volume of solution.

■ **EXAMPLE 11.4** ⎯⎯⎯⎯⎯⎯⎯⎯⎯⎯⎯⎯⎯⎯⎯⎯⎯⎯⎯⎯⎯⎯
How would you prepare 0.150 L of a 0.500 M NaOH solution, starting with solid sodium hydroxide and water?

Solution
We first calculate the number of moles of NaOH required. To do that, we treat molarity as a conversion factor:

0.500 mol NaOH ≏ 1 L solution

$$no.\ moles\ NaOH\ required = 0.150\ L \times \frac{0.500\ mol\ NaOH}{1\ L}$$

no. moles A = $V \times M_A$

$$= 0.0750\ mol\ NaOH$$

To find the mass of NaOH required, note that the molar mass is 40.0 g/mol:

$$mass\ NaOH\ required = 0.0750\ mol \times \frac{40.0\ g}{1\ mol} = 3.00\ g$$

You should weigh out 3.00 g of NaOH and dissolve in enough water to form 0.150 L (150 mL) of solution.

Exercise
What mass of NaOH would be obtained by evaporating to dryness 1.00 mL of this solution? Answer: 0.0200 g.

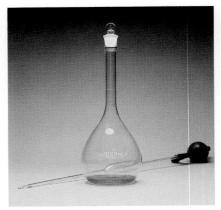 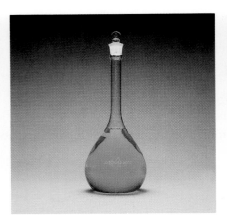 

Figure 11.5 To dilute a solution to a specified concentration, enough water is added to bring the final volume to the required value. (Marna G. Clarke)

Another way to prepare a known volume of a solution of a desired molarity is to start with a concentrated solution rather than pure solute. In this case you

— calculate the volume of the concentrated solution required.

— measure out that volume and add enough solvent to give the desired volume of the more dilute solution.

The calculations required with this method are readily made if you keep a simple point in mind. Adding solvent cannot change the number of moles of solute. In other words, the number of moles of solute is the same before and after dilution.

moles solute concentrated solution = moles solute dilute solution

In both solutions, the number of moles of solute can be found by multiplying the molarity (M) by the volume in liters (V). Hence:

$$M_c V_c = M_d V_d \tag{11.7}$$

where the subscripts c and d stand for concentrated and dilute solutions, respectively. The use of Equation 11.7 is illustrated in Example 11.5.

■ **EXAMPLE 11.5**

How would you prepare 0.150 L of 0.500 M NaOH, starting with a 6.00 M solution of NaOH?

Solution

This question might be restated as: What volume of 6.00 M NaOH should be diluted with water to yield 0.150 L of 0.500 M NaOH? That quantity can readily be calculated using Equation 11.7. We need to know the volume of concentrated solution, V_c. Solving Equation 11.7 for V_c,

$$V_c = V_d \times \frac{M_d}{M_c}$$

Here:

V_d = volume of dilute solution = 0.150 L

M_d = molarity of dilute solution = 0.500 mol/L

M_c = molarity of concentrated solution = 6.00 mol/L

$$V_c = 0.150 \text{ L} \times \frac{0.500}{6.00} = 0.0125 \text{ L, or } 12.5 \text{ mL}$$

You should measure out 12.5 mL of 6.00 M NaOH and dilute with water to give 0.150 L of solution.

Exercise
What mass of NaOH is present in 0.0125 L of 6.00 M NaOH? in 0.150 L of 0.500 M NaOH? Answer: 3.00 g.

Many water solutions in the general chemistry laboratory as well as in everyday life contain ionic solutes. Frequently, you will need to relate the molarity of such a solute to the molarities of its ions in solution. To do this, it is convenient to write dissociation equations of the type given in Example 11.1. Thus from the equation:

$$Li_2CO_3(s) \rightarrow 2 \text{ Li}^+(aq) + CO_3^{2-}(aq)$$

we see that

$$1 \text{ mol } Li_2CO_3 \rightarrow 2 \text{ mol Li}^+ + 1 \text{ mol } CO_3^{2-}$$

so a 1 M Li_2CO_3 solution is 2 M in Li^+ and 1 M in CO_3^{2-}.

But we label it as 1 M Li_2CO_3

■ EXAMPLE 11.6

What is the concentration, in moles per liter, of each ion in

a. 0.080 M K_2SO_4?   b. 0.40 M $LaBr_3$?

Solution

a. $K_2SO_4(s) \rightarrow 2 \text{ K}^+(aq) + SO_4^{2-}(aq)$

$1 \text{ mol } K_2SO_4 \rightarrow 2 \text{ mol K}^+ + 1 \text{ mol } SO_4^{2-}$

The concentrations of the ions are

$$\text{conc. K}^+ = \frac{0.080 \text{ mol } K_2SO_4}{1 \text{ L}} \times \frac{2 \text{ mol K}^+}{1 \text{ mol } K_2SO_4} = \boxed{0.16 \text{ } M}$$

$$\text{conc. SO}_4^{2-} = \frac{0.080 \text{ mol } K_2SO_4}{1 \text{ L}} \times \frac{1 \text{ mol } SO_4^{2-}}{1 \text{ mol } K_2SO_4} = \boxed{0.080 \text{ } M}$$

b. $LaBr_3(s) \rightarrow La^{3+}(aq) + 3 \text{ Br}^-(aq)$

$1 \text{ mol } LaBr_3 \rightarrow 1 \text{ mol } La^{3+} + 3 \text{ mol Br}^-$

$\text{conc. La}^{3+} = \text{conc. } LaBr_3 = \boxed{0.40 \text{ } M}$

$\text{conc. Br}^- = 3(\text{conc. } LaBr_3) = 3(0.40 \text{ } M) = \boxed{1.2 \text{ } M}$

Exercise
Calculate the number of moles of K^+ in 1.5 L of 0.080 M K_2SO_4. Answer: 0.24 mol.

CONVERSIONS BETWEEN CONCENTRATION UNITS

Often you will have occasion to convert from one concentration unit to another. This problem arises in making up solutions of such acids as HCl, HNO_3, and H_2SO_4. Typically, the analysis or assay that appears on the label (Fig. 11.6) does not give the molarity or molality of the acid. Instead, it gives the mass percent of solute and the density of the solution. Example 11.7 illustrates how the molarity can be calculated from this information.

■ EXAMPLE 11.7

Calculate the molarity of a concentrated solution of nitric acid (HNO_3) that is 70.8% by mass HNO_3; the solution has a density of 1.424 g/mL.

Solution
It is convenient to start arbitrarily with 100.0 g of solution. Thus we can express the mass percent as

$$70.8\% \ HNO_3 = \frac{70.8 \text{ g nitric acid}}{100.0 \text{ g solution}}$$

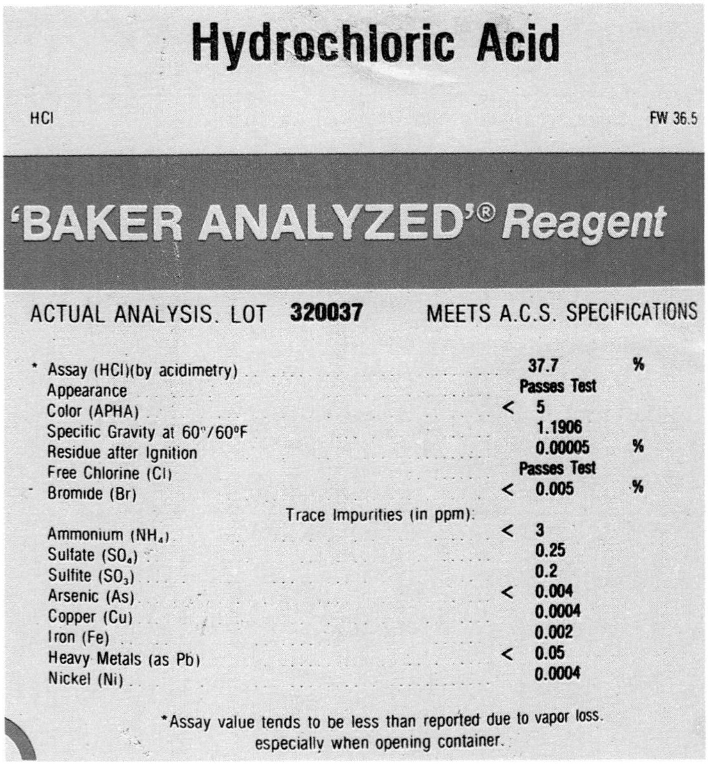

Figure 11.6 The label on a bottle of concentrated hydrochloric acid typically gives the mass percent of HCl ("assay") and the density (or "specific gravity") of the solution. Given that information, it is possible to calculate the molality, molarity, and mole fraction of HCl. (Marna G. Clarke)

To find the number of moles of HNO_3 in 70.8 g, we first determine the molar mass of HNO_3 to be 63.0 g/mol. Then

$$\text{moles } HNO_3 = 70.8 \text{ g } HNO_3 \times \frac{1 \text{ mol}}{63.0 \text{ g } HNO_3} = 1.12 \text{ mol}$$

We next determine the volume of 100.0 g (our assumed mass) of solution. Using the density as a conversion factor, we obtain

$$V = 100.0 \text{ g solution} \times \frac{1 \text{ mL of solution}}{1.424 \text{ g solution}} \times \frac{1 \text{ L}}{1000 \text{ mL}}$$
$$= 0.0702 \text{ L}$$

The molarity of the solution can now be calculated. Since molarity is moles solute/liter solution, we have:

$$M = \frac{1.12 \text{ mol solute}}{0.0702 \text{ L solution}} = \boxed{16.0 \text{ mol/L}}$$

Exercise
Calculate the mass percent of HCl in a 12.0 M solution, density 1.192 g/mL. Answer: 36.7%.

In this example, we were given the mass percent of HNO_3 and asked to calculate the molarity. In another case, we might want to determine the molality of a solution given its molarity. Such conversions are best carried out systematically. Figure 11.7 (p. 370) is a flow chart that shows how to convert between molarity, molality, and mole fraction. Notice that you have to know the density of the solution to go from molarity to any other concentration unit.

To illustrate how Figure 11.7 is used, let us consider how we might calculate the molality of HCl in dilute (6.0 M) hydrochloric acid.

■ **EXAMPLE 11.8** ————————————————————————
Dilute HCl has a molarity of 6.0 mol/L and a density of 1.10 g/cm³. Calculate the molality of HCl.

Solution
We follow the blue arrows in Figure 11.7. We assume a volume of one liter and use the density to calculate the total mass of solution. From this, we subtract the mass of solute (6.0 M HCl, MM = 36.5 g/mol). That gives us the mass in grams of solvent, which is readily converted to kilograms. Finally, knowing the moles of solute and kilograms of solvent, we find molality.

1. Total mass 1 L solution = $1.00 \text{ L} \times \dfrac{10^3 \text{ cm}^3}{1 \text{ L}} \times 1.10 \dfrac{\text{g}}{\text{cm}^3} = 1100 \text{ g}$

2. Mass HCl in 1 L solution = $6.0 \text{ mol} \times \dfrac{36.5 \text{ g}}{1 \text{ mol}} = 220 \text{ g}$

3. Mass water in 1 L solution = $1100 \text{ g} - 220 \text{ g} = 880 \text{ g} = 0.880 \text{ kg}$

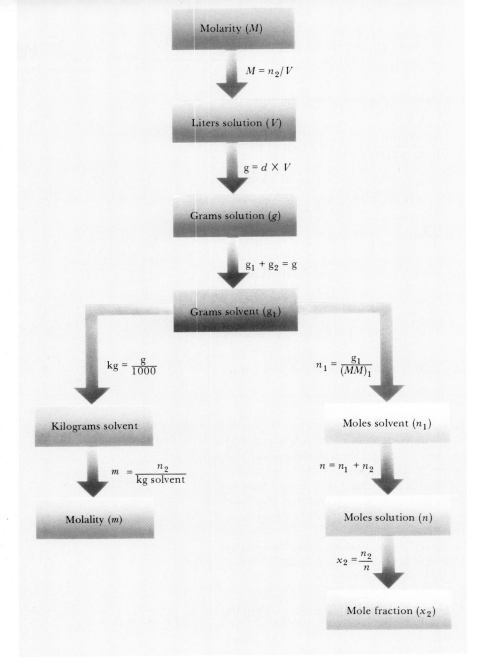

Figure 11.7 Flow sheet for conversions between concentration units (the subscripts 1 and 2 refer to solvent and solute, respectively). Molarity can be converted to molality or mole fraction of solute by following the arrows; use the relations given to go from one box to the next. To carry out other types of conversions (e.g., molality to molarity), simply reverse the arrows.

In making conversions, it is convenient to make the denominator of the starting concentration unit equal to unity. Thus:

Starting Unit	Take:	Then:
$M = n_2/V$	1 L solution	$n_2 = M$
$m = n_2/\text{kg solvent}$	1 kg solvent	$n_2 = m$
$X_2 = n_2/(n_1 + n_2)$	1 mol solution	$n_2 = X_2$

The advantage of this approach is that *the number of moles of solute becomes numerically equal to the concentration unit.*

4. molality $= \dfrac{6.0 \text{ mol solute}}{0.880 \text{ kg solvent}} = $ 6.8 m

Exercise

Taking the molality of concentrated HCl to be 15.4 m and its density to be 1.18 g/cm^3, find the molarity. Answer: 11.7 M. Start with one kilogram of solvent and reverse the path followed in the example. That is, find, in order, the grams of solvent (1000 g), the grams of solute (562 g), the grams of solution (1562 g), the volume of solution (1.32 L), and finally the molarity (15.4 mol/1.32 L = 11.7 M).

11.3
PRINCIPLES OF SOLUBILITY

The extent to which a solute dissolves in a particular solvent depends upon several factors. The most important of these are:

— the nature of solvent and solute particles and the interactions between them.

— the temperature at which the solution is formed.

— the pressure of a gaseous solute.

In this section we will consider in turn the effect of each of these factors upon solubility.

SOLUTE-SOLVENT INTERACTIONS

In discussing solubility, it is sometimes stated that "like dissolves like." A more meaningful way to express this idea is to say that two substances with intermolecular forces of about the same type and magnitude are likely to be very soluble in one another. To illustrate, consider the hydrocarbons pentane, C_5H_{12}, and hexane, C_6H_{14}, which are completely miscible with each other. Molecules of these nonpolar substances are held together by dispersion forces of about the same magnitude. A pentane molecule experiences little or no change in intermolecular forces when it goes into solution in hexane.

Many organic liquids are completely soluble in one another

Most nonpolar substances have very small water solubilities. Petroleum, a mixture of hydrocarbons, spreads out in a thin film on the surface of a body of water rather than dissolving. The mole fraction of pentane, C_5H_{12}, in a saturated water solution is only 0.0001. These low solubilities are readily understood in terms of the structure of liquid water. To dissolve appreciable amounts of pentane in water, it would be necessary to break the hydrogen bonds holding H_2O molecules together. There is no comparable attractive force between C_5H_{12} and H_2O to supply the energy required to break into the water structure.

Of the relatively few organic compounds that dissolve readily in water, most contain —OH groups. Three familiar examples are methyl alcohol (methanol), ethyl alcohol (ethanol), and ethylene glycol, all of which are infinitely soluble in water.

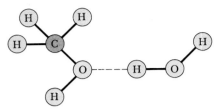

Figure 11.8 Methyl alcohol is a small molecule that readily forms hydrogen bonds (shown by the broken line) with water. This explains why CH_3OH is infinitely soluble in water.

H—C—OH   H—C—C—OH   H—C—C—H
methyl alcohol   ethyl alcohol   ethylene glycol

In these compounds, as in water, the principal intermolecular forces are hydrogen bonds. When a substance like methyl alcohol dissolves in water, it forms hydrogen bonds with H_2O molecules (Fig. 11.8). These hydrogen bonds, joining a CH_3OH molecule to an H_2O molecule, are about as strong as those in the pure substances.

Not all organic compounds that contain —OH groups are soluble in water (Table 11.3). As molar mass increases, the polar —OH group represents an increasingly smaller portion of the molecule. At the same time, the nonpolar hydrocarbon portion becomes larger. As a result, solubility decreases with molar mass. Butanol, $CH_3CH_2CH_2CH_2OH$ is much less soluble in water than methanol, CH_3OH. The hydrocarbon portion, shaded in red, is much larger in butanol than in methanol.

The solubility (or insolubility) of different vitamins is of concern in nutrition. Molecules of Vitamins B and C contain several —OH groups that can form hydrogen bonds with water (Fig. 11.9). As a result, they are water-soluble, readily excreted by the body, and must be consumed daily. In contrast, Vitamins A, D, E, and K, whose molecules are relatively nonpolar, are water-insoluble. These vitamins are not so readily excreted; they tend to stay behind in fatty tissues. This means that the body can draw on its reservoir of Vitamins A, D, E, and K to deal with sporadic deficiencies. Conversely, megadoses of these vitamins can lead to very high, possibly toxic, concentrations in the body.

It's good to stay away from megadoses of anything

EFFECT OF TEMPERATURE UPON SOLUBILITY

When an excess of a solid such as sodium chloride is shaken with water, it forms a saturated solution. An equilibrium is established between the solid and its ions in solution.

$$NaCl(s) \rightleftharpoons Na^+(aq) + Cl^-(aq)$$

TABLE 11.3 Solubilities of Alcohols in Water

SUBSTANCE	FORMULA	SOLUBILITY (g solute/L H_2O)
Methanol	CH_3OH	Completely soluble
Ethanol	CH_3CH_2OH	Completely soluble
Propanol	$CH_3CH_2CH_2OH$	Completely soluble
Butanol	$CH_3CH_2CH_2CH_2OH$	74
Pentanol	$CH_3CH_2CH_2CH_2CH_2OH$	27
Hexanol	$CH_3CH_2CH_2CH_2CH_2CH_2OH$	6.0
Heptanol	$CH_3CH_2CH_2CH_2CH_2CH_2CH_2OH$	1.7

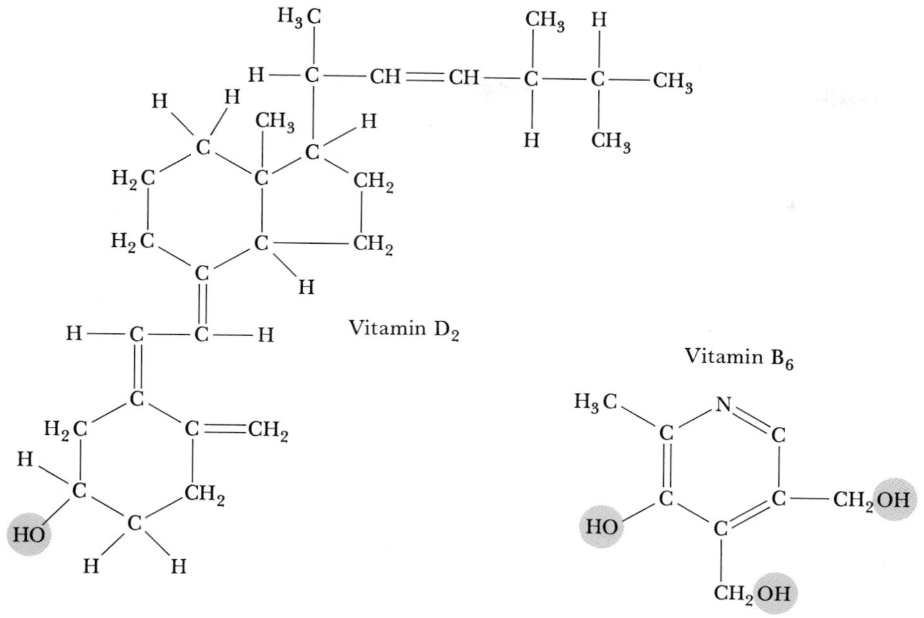

Figure 11.9 Molecular structures for Vitamins D_2 and B_6. Polar groups are in color. Vitamin D_2 is water-insoluble; Vitamin B_6 is water-soluble.

A similar type of equilibrium is established when a gas such as carbon dioxide is bubbled through water:

$$CO_2(g) \rightleftharpoons CO_2(aq)$$

We can predict the effect of a temperature change on solubility equilibria such as these by applying a simple principle. **An increase in temperature always favors an endothermic process.** This means that if the solution process absorbs heat ($\Delta H > 0$), an increase in temperature increases the solubility. In other words, more solute goes into solution at higher temperatures. Conversely, if the solution process is exothermic ($\Delta H < 0$), an increase in temperature will decrease the solubility; heating will drive the solute out of solution.

Dissolving a solid in a liquid is usually an endothermic process; heat must be absorbed to break down the crystal lattice.

$$\text{solid} + \text{liquid} \rightleftharpoons \text{solution}; \quad \Delta H > 0 \qquad (11.8)$$

Applying the principle referred to above, we expect the solubilities of solids to increase as the temperature rises. Experience confirms this prediction. More sugar dissolves in hot coffee than in cold coffee. Cooling a saturated solution usually causes a solid to crystallize out of solution, indicating that it is less soluble at the lower temperature.

Dissolving a gas in a liquid usually evolves heat ($\Delta H < 0$):

$$\text{gas} + \text{liquid} \rightleftharpoons \text{solution}; \quad \Delta H < 0$$

The gas in a sense condenses

This means that the reverse process (gas coming out of solution) is endothermic. Hence, it is favored by an increase in temperature; gases become less soluble as the temperature rises. This rule is followed by all gases in

Figure 11.10 The solubility of $O_2(g)$ in water decreases as temperature rises (A) and increases as pressure increases (B). In A, the pressure is held constant at 1 atm; in B, the temperature is held constant at 25°C.

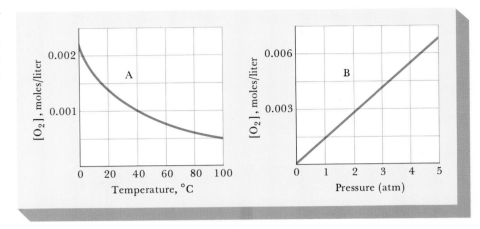

water (Fig. 11.10). You have probably noticed this effect when heating water in an open pan or beaker. Bubbles of air are driven out of the water by an increase in temperature. The characteristic flat taste of boiled water is due to the fact that it does not contain any dissolved air.

EFFECT OF PRESSURE UPON SOLUBILITY

Pressure has a major effect on solubility only for gas-liquid systems. At a given temperature, a rise in pressure increases the solubility of a gas. Indeed, at low to moderate pressures, gas solubility is directly proportional to pressure (Henry's Law; Figure 11.10B).

$$C_g = kP_g \tag{11.9}$$

If you double the pressure, you double the solubility

where P_g is the partial pressure of the gas over the solution, C_g is its molarity in the solution, and k is a constant characteristic of the particular gas-liquid system. This effect arises because increasing the pressure raises the concentration of molecules in the gas phase. To balance this change and maintain equilibrium, more gas molecules enter the solution, increasing their concentration in the liquid phase.

■ **EXAMPLE 11.9**
The solubility of pure nitrogen in blood at body temperature, 37°C, and one atmosphere, is $6.2 \times 10^{-4}\ M$. If a diver breathes air ($X\ N_2 = 0.78$) at a depth where the total pressure is 2.5 atm, calculate the concentration of nitrogen in his blood.

Solution
To solve this problem, we first use the solubility data for pure nitrogen to calculate k in Equation 11.9.

$$k = \frac{\text{concentration } N_2}{\text{pressure } N_2} = \frac{6.2 \times 10^{-4}\ M}{1.00\ \text{atm}} = 6.2 \times 10^{-4}\ M/\text{atm}$$

Next we calculate the pressure of nitrogen. Recall from Chapter 4 that partial pressure is the product of mole fraction times total pressure. Thus

$$P\ N_2 = 0.78 \times 2.5 \text{ atm} = 2.0 \text{ atm}$$

Knowing both k and $P\ N_2$, we can now calculate the concentration of nitrogen:

$$\text{conc. } N_2 = 6.2 \times 10^{-4} \frac{M}{\text{atm}} \times 2.0 \text{ atm} = \boxed{1.2 \times 10^{-3}\ M}$$

Exercise
The partial pressure of oxygen in the lungs at 37°C is 101 mm Hg. Calculate the solubility of oxygen (mol/L), taking k to be $1.06 \times 10^{-3}\ M/\text{atm}$. Answer: 0.000141 M.

The influence of partial pressure on gas solubility is used in bottling carbonated beverages such as beer, sparkling wines, and many soft drinks. These beverages are bottled under pressures of CO_2 as high as 4 atm. When the bottle or can is opened, the pressure above the liquid drops to 1 atm and the carbon dioxide bubbles rapidly out of solution. Pressurized containers for shaving cream, whipped cream, and cheese spreads work on a similar principle. Pressing a valve reduces the pressure on dissolved gas, causing it to rush from solution, carrying liquid with it as a foam.

Sometimes, particularly with beer, too rapidly

Another consequence of the effect of pressure on gas solubility is the painful, sometimes fatal, affliction known as the "bends." This occurs when a person goes rapidly from deep water (high pressure) to the surface (lower pressure). The rapid decompression causes air, dissolved in blood and other body fluids, to bubble out of solution. These bubbles impair blood circulation and affect nerve impulses. To minimize these effects, deep-sea divers and aquanauts breathe a helium-oxygen mixture rather than compressed air (nitrogen-oxygen). Helium is only about one-third as soluble as nitrogen, and hence, much less gas comes out of solution upon decompression.

11.4
COLLIGATIVE PROPERTIES OF SOLUTIONS

The properties of a solution differ considerably from those of the pure solvent. Those solution properties which depend primarily on the *concentration of solute particles* rather than their nature are called **colligative properties**. These properties include vapor pressure lowering, osmotic pressure, boiling point elevation, and freezing point depression. In this section, we will look at the relations between colligative properties and solute concentration, starting with nonelectrolytes and then going on (briefly) to electrolytes.

With colligative properties, an ion is as effective as a molecule

The relationships among colligative properties and solute concentration are best regarded as limiting laws. They are approached most closely when the solution is very dilute. In practice, the relationships we will discuss are valid, for nonelectrolytes, to within a few per cent at concentrations as high as 1 M.

VAPOR PRESSURE LOWERING (NONELECTROLYTES)

The rate at which water molecules escape from the surface is reduced in the presence of a nonvolatile solute. Concentrated aqueous solutions of nonelectrolytes such as glucose or sucrose evaporate more slowly than pure water. This reflects the fact that the vapor pressure of water in the solution is less than that of pure water. This decrease in vapor pressure is a true colligative property; that is, it is independent of the nature of the solute but directly proportional to its concentration. We find, for example, that the vapor pressure of water above a 0.10 M solution of either glucose or sucrose at 0°C is the same, about 0.008 mm Hg less than that of pure water. In 0.30 M solution, the vapor pressure lowering is almost exactly three times as great, 0.025 mm Hg.

The relationship between solvent vapor pressure and concentration is ordinarily expressed in the form of Raoult's Law:

$$P_1 = X_1 P_1{}^0 \tag{11.10}$$

In this equation, P_1 is the vapor pressure of solvent over the solution, $P_1{}^0$ is the vapor pressure of the pure solvent at the same temperature, and X_1 is the mole fraction of solvent. Note that since X_1 in a solution must be less than 1, P_1 must be less than $P_1{}^0$.

We can obtain a direct expression for the vapor pressure lowering by making the substitution $X_2 = 1 - X_1$, where X_2 is the mole fraction of solute in the two-component system.

$$P_1 = (1 - X_2) P_1{}^0$$

Rearranging,

$$P_1{}^0 - P_1 = X_2 P_1{}^0$$

The quantity $(P_1{}^0 - P_1)$ is the vapor pressure lowering (VPL). It is the difference between the solvent vapor pressure in the pure solvent and in solution.

VPL depends on the total mole fraction of solute particles, but not on their nature

$$VPL = X_2 P_1{}^0 \tag{11.11}$$

Equation 11.11 can be used to calculate the vapor pressure lowering in a solution (Example 11.10).

■ **EXAMPLE 11.10** ——————————————————————

A solution contains 102 g of sugar, $C_{12}H_{22}O_{11}$, in 375 g of water. Calculate

a. the mole fraction of sugar.
b. the vapor pressure lowering at 25°C (vp pure water = 23.76 mm Hg).

Solution

a. The molar mass of $C_{12}H_{22}O_{11}$ is 342 g/mol; that of H_2O is 18.0 g/mol; thus,

$$\text{no. moles } C_{12}H_{22}O_{11} = 102 \text{ g} \times \frac{1 \text{ mol}}{342 \text{ g}} = 0.298 \text{ mol}$$

$$\text{no. moles } H_2O = 375 \text{ g} \times \frac{1 \text{ mol}}{18.0 \text{ g}} = 20.8 \text{ mol}$$

$$X_{sugar} = \frac{0.298}{0.298 + 20.8} = \frac{0.298}{21.1} = \boxed{0.0141}$$

b. Applying Equation 11.11,

$$\text{VPL} = X_{sugar} \times P^0_{H_2O} = 0.0141 \times 23.76 \text{ mm Hg}$$

$$= \boxed{0.335 \text{ mm Hg}}$$

VPL tends to be a small effect

We conclude that the vapor pressure of water over this solution is:

23.76 mm Hg − 0.335 mm Hg = 23.42 mm Hg

Exercise
What is the vapor pressure of water over this solution at 100°C? Answer: 749.3 mm Hg.

One interesting effect of vapor pressure lowering is shown at the left of Figure 11.11. Here, we start with two beakers, one containing pure water and the other containing a sugar solution. These are placed next to each other, under a bell jar (Fig. 11.11A). As time passes, the liquid level in the beaker containing the solution rises. The level of pure water in the other beaker falls. Eventually, by evaporation and condensation, all the water is transferred to the solution (Fig. 11.11B). At the end of the experiment, the beaker that contained pure water is empty. The driving force behind this process is the difference in vapor pressure of water in the two beakers. *Water moves from a region in which its vapor pressure is high* (pure water) *to one*

That would take a long time

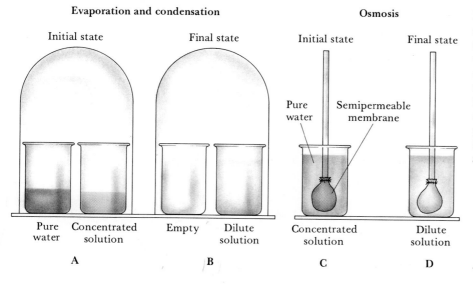

Evaporation and condensation

Initial state   Final state

Pure   Concentrated   Empty   Dilute
water   solution   solution

A   B

Osmosis

Initial state   Final state

Pure   Semipermeable
water   membrane

Concentrated   Dilute
solution   solution

C   D

Figure 11.11 Water tends to move spontaneously from a region where its vapor pressure is high to a region where it is low. In A → B, movement of water molecules occurs through the air trapped under the bell jar. In C → D, water molecules move by osmosis through a semipermeable membrane. The driving force is the same in the two cases, although the mechanism differs.

in which its vapor pressure is low (sugar solution). This is a general tendency, followed by all liquids, and is responsible for a variety of natural processes, including osmosis.

OSMOTIC PRESSURE (NONELECTROLYTES)

The apparatus shown in Figure 11.11C and D can be used to achieve a result similar to that found in the bell jar experiment. Here, a sugar solution is separated from water by a "semipermeable" membrane. This may be an animal bladder, a slice of vegetable tissue, or a piece of parchment. The membrane, by a mechanism that is not well understood, allows water molecules to pass through it, but not sugar molecules. Here, as before, water moves from a region where its vapor pressure is high (pure water) to a region where it is low (sugar solution). This process, taking place through a membrane permeable only to the solvent, is called **osmosis**. As a result of osmosis, the water level rises in the tube and drops in the beaker (Fig. 11.11D).

You might wonder why the process shown at the right of Figure 11.11 does not continue indefinitely. After all, no matter how dilute the solution inside the tube, its vapor pressure is less than that of the pure water outside. The process stops because the column of water in the tube exerts sufficient pressure to prevent water from moving into the tube. The height of this column and hence the pressure it exerts will depend upon several factors, the most important of which is the concentration of solute.

We define the **osmotic pressure**, π, as being equal to the external pressure, P, just sufficient to prevent osmosis (Fig. 11.12). If P is less than π, osmosis takes place in the normal way and water moves through the membrane into the solution (Fig. 11.12A). By making the external pressure large enough, it is possible to reverse this process (Fig. 11.12B). When $P > \pi$, water molecules move through the membrane from the solution to pure water. This process, called *reverse osmosis*, is used to obtain fresh water from seawater in arid regions of the world.

During the osmosis, only water goes through the membrane

Figure 11.12 Osmosis can be prevented by applying to the solution a pressure P that just balances the osmotic pressure, π. If $P < \pi$, normal osmosis occurs. If $P > \pi$, water flows in the opposite direction. This process, called reverse osmosis, can be used to obtain fresh water from seawater.

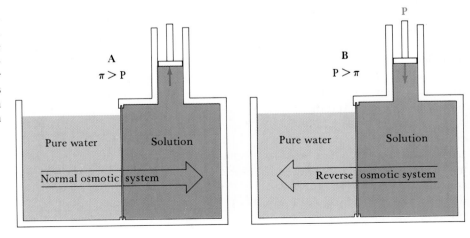

Figure 11.13 When a cucumber is pickled, water moves out of the cucumber by osmosis into the concentrated brine solution. A prune placed in pure water swells as water moves into the prune, again by osmosis. (Marna G. Clarke)

Osmotic pressure, like vapor pressure lowering, is a colligative property. For a nonelectrolyte, π is directly proportional to molarity, M. The equation relating the two quantities is very similar to the Ideal Gas Law:

$$\pi = \frac{nRT}{V} = MRT \qquad (11.12)$$

where R is the gas law constant, 0.0821 L·atm/(mol·K), and T is the Kelvin temperature. Even in dilute solution, the osmotic pressure is quite large. Suppose, for example, we are dealing with a 0.10 M solution at 25°C:

$$\pi = (0.10)(0.0821)(298) \text{ atm} = 2.4 \text{ atm}$$

A pressure of 2.4 atm is equivalent to a column of water 25 m (more than 80 ft) high.

If a cucumber is placed in a concentrated brine solution, it shrinks and assumes the wrinkled skin of a pickle. The skin of the cucumber acts as a semipermeable membrane. The water solution inside the cucumber is more dilute than the solution surrounding it. As a result, water flows out of the cucumber into the brine (Fig. 11.13).

When a prune is placed in water, the skin also acts as a semipermeable membrane. This time the solution inside the prune is more concentrated than the water, so that water flows into the prune, making the prune less wrinkled.

Dialysis, which occurs at cell walls in plants and animals, is similar to osmosis, except that small solute molecules and ions as well as solvent molecules pass through the membrane. This process is used to treat patients suffering from kidney failure (Fig. 11.14). In an artificial kidney, blood is circulated from the patient through cellophane tubes, which act as semipermeable membranes. The tubes are immersed in a solution that contains all of the essential ions and small molecules in blood at the appropriate concentrations. These solute species do not move; only waste products dialyze through the membrane. This purifies the blood, leaving its composition unchanged otherwise.

Figure 11.14 Artificial kidney (dialysis machine). (Dan McCoy, Black Star)

BOILING POINT ELEVATION AND FREEZING POINT DEPRESSION (NONELECTROLYTES)

We ordinarily find that a solution boils at a *higher* temperature and freezes at a *lower* temperature than the pure solvent. Consider, for example, a solution containing 18.0 g of glucose, $C_6H_{12}O_6$, in 100 g of water. This solution boils at 100.52°C at 1 atm pressure; the normal boiling point of pure water is 100.00°C. The glucose solution freezes at −1.86°C (fp pure water = 0.00°C).

In discussing the boiling points or freezing points of solutions, we use the terms boiling point elevation, ΔT_b, and freezing point depression, ΔT_f. These are defined so as to be positive quantities. Thus,

$$\Delta T_b = \text{bp solution} - \text{bp pure solvent} \qquad (11.13)$$

$$\Delta T_f = \text{fp pure solvent} - \text{fp solution} \qquad (11.14)$$

For a one-molal glucose solution just referred to:

$$\Delta T_b = 100.52°C - 100.00°C = 0.52°C$$

$$\Delta T_f = 0.00°C - (-1.86°C) = 1.86°C$$

Boiling point elevation is a direct result of vapor pressure lowering. At any given temperature, a solution of a nonvolatile solute* has a vapor pressure *lower* than that of the pure solvent. Hence, a *higher* temperature must be reached before the solution boils—that is, before its vapor pressure becomes equal to the external pressure. Figure 11.15 illustrates this reasoning graphically.

Figure 11.15 Since a nonvolatile solute lowers the vapor pressure of a solvent, the boiling point of a solution will be higher and the freezing point lower than the corresponding values for the pure solvent. Water solutions freeze *below* 0°C at point A and boil *above* 100°C at point B.

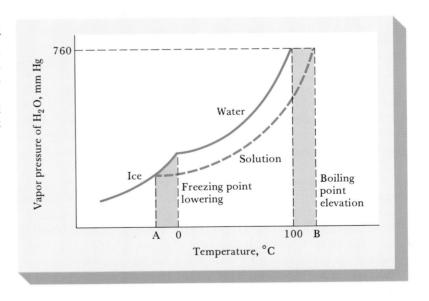

*Volatile solutes ordinarily lower the boiling point because they contribute to the total vapor pressure of the solution.

TABLE 11.4 Molal Freezing Point and Boiling Point Constants

SOLVENT	fp (°C)	k_f (°C/m)	bp (°C)	k_b (°C/m)
Water	0.00	1.86	100.00	0.52
Acetic acid	16.66	3.90	117.90	2.53
Benzene	5.50	5.10	80.10	2.53
Cyclohexane	6.50	20.2	80.72	2.75
Camphor	178.40	40.0	207.42	5.61
p-Dichlorobenzene	53.1	7.1	174.1	6.2
Naphthalene	80.29	6.94	217.96	5.80

The freezing point depression, like the boiling point elevation, is a direct result of the lowering of the solvent vapor pressure by the solute. Notice from Figure 11.15 that the freezing point of the solution is the temperature at which the solvent in solution has the same vapor pressure as the pure solid solvent. This implies that it is pure solvent (e.g., ice) that separates when the solution freezes.

Boiling point elevation and freezing point depression, like vapor pressure lowering, are colligative properties. They are directly proportional to solute concentration, generally expressed as molality, m.

The equations relating boiling point elevation and freezing point depression to molality are

$$\Delta T_b = k_b \times m \tag{11.15}$$

$$\Delta T_f = k_f \times m \tag{11.16}$$

The proportionality constants in these equations, k_b and k_f, are called the *molal boiling point constant* and the *molal freezing point constant*. Their magnitudes depend upon the nature of the solvent (Table 11.4). Note that when the solvent is water

$$k_b = 0.52°C/m; \qquad k_f = 1.86°C/m$$

The use of Equations 11.15 and 11.16 is illustrated in Example 11.11.

■ **EXAMPLE 11.11** _____

An antifreeze solution is prepared containing 50.0 cm³ of ethylene glycol, $C_2H_6O_2$ ($d = 1.12$ g/cm³), in 50.0 g of water. Calculate the freezing point of this "50-50" mixture.

Solution
We first calculate the mass of ethylene glycol in the solution, then the number of moles of ethylene glycol (molar mass $C_2H_6O_2 = 62.0$ g/mol), and then the molality. Finally, we use Equation 11.16 to calculate the freezing point lowering ($k_f = 1.86°C/m$).

$$\text{mass } C_2H_6O_2 = 50.0 \text{ cm}^3 \times 1.12 \frac{g}{\text{cm}^3} = 56.0 \text{ g}$$

Presumably an iceberg contains very little salt

That's what we use in Minnesota

$$\text{no. moles } C_2H_6O_2 = 56.0 \text{ g} \times \frac{1 \text{ mol}}{62.0 \text{ g}} = 0.903 \text{ mol}$$

$$\text{molality} = \frac{0.903 \text{ mol } C_2H_6O_2}{0.0500 \text{ kg water}} = 18.1 \; m$$

$$\Delta T_f = k_f \times m = 1.86°C/m \times 18.1 \; m = 33.7°C$$

We conclude that the freezing point of the solution should be 33.7°C below that of pure water (0°C). Hence, the solution should freeze at $-33.7°C$. Actually, the freezing point is somewhat lower, about $-37°C$ or $-35°F$. The deviation occurs because Equation 11.16 is a limiting law, strictly valid only at low concentrations.

Exercise
Estimate the boiling point of this solution at 1 atm. Answer: 109.4°C.

As you know, we take advantage of the freezing point depression when we add antifreeze to automobile radiators in winter. Ethylene glycol is the solute commonly used in so-called "permanent" antifreezes. It has a high boiling point (197°C), is virtually nonvolatile at 100°C and, as we have seen, raises the boiling point of water. Hence, antifreeze containing ethylene glycol does not boil away in summer driving.

DETERMINATION OF MOLAR MASSES OF NONELECTROLYTES FROM COLLIGATIVE PROPERTIES

The molar mass of a gas or volatile liquid can be obtained from gas density measurements, as described in Chapter 4. This method is useless for nonvolatile solids or for substances that decompose on heating. Such substances do not ordinarily exist as vapors. It turns out that colligative properties, particularly freezing point depression, can be used to determine molar masses of a wide variety of nonelectrolytes. The approach used is illustrated in Example 11.12.

■ **EXAMPLE 11.12** ——————————————————
A student dissolves 1.50 g of a newly prepared compound in 75.0 g of cyclohexane. She measures the freezing point of the solution to be 2.70°C; that of pure cyclohexane is 6.50°C. Cyclohexane has a k_f of 20.2°C/m. Using these data, calculate the molar mass of the compound.

Solution
We follow a three-step procedure. First, we obtain ΔT_f, using Equation 11.14. Then we use Equation 11.16 to calculate the molality. Finally, we obtain the molar mass, using the defining equation for molality (Equation 11.5):

1. $\Delta T_f = $ fp pure cyclohexane $-$ fp solution $= 6.50°C - 2.70°C = 3.80°C$

2. $\Delta T_f = k_f \times m$; $m = \dfrac{\Delta T_f}{k_f} = \dfrac{3.80°C}{20.2°C/m} = 0.188\ m$

For good results in this experiment, you need a sensitive thermometer and good technique

3. $m = \dfrac{\text{no. moles solute}}{\text{no. kilograms solvent}} = \dfrac{\text{no. grams solute/MM}}{\text{no. kilograms solvent}}$

where MM is the molar mass (g/mol). Solving the above equation for MM,

$$MM = \frac{\text{no. grams solute}}{(m)(\text{no. kilograms solvent})}$$

All the quantities on the right side of this equation are known. We know that 1.50 g of solute is dissolved in 75.0 g of solvent (0.0750 kg); we calculated m to be 0.188. Hence:

$$MM = \frac{1.50}{(0.188)(0.0750)}\ \frac{g}{mol} = \boxed{106\ g/mol}$$

Exercise
A solution of 5.00 g of a compound X in 60.0 g of water freezes at $-1.00°C$. What is the molar mass of X? Answer: 155 g/mol.

In carrying out a molar mass determination by freezing point depression, we must choose a solvent in which the solute is readily soluble. Usually, several such solvents are available. Of these, we tend to pick one that has the largest k_f. This makes ΔT_f large and thus reduces the percent error in the freezing point measurement. From this point of view, cyclohexane or other organic solvents are better choices than water, since their k_f values are larger.

Molar masses can also be detemined using other colligative properties. Osmotic pressure measurements are often used, particularly for solutes of high molar mass where the concentration is likely to be quite low. The advantage of using osmotic pressure is that the effect is relatively large.

■ **EXAMPLE 11.13** _____

A solution contains 1.0 g of hemoglobin dissolved in enough water to form 100 cm³ of solution. The osmotic pressure at 20°C is found to be 2.75 mm Hg. Calculate:

a. the molarity of hemoglobin.
b. the molar mass, MM, of hemoglobin.

Solution

a. Rearranging Equation 11.12 to solve for the molarity, M,

molarity $= \pi/RT$

But, $\pi = (2.75/760)$ atm; $R = 0.0821$ L·atm/(mol·K); $T = 293$ K. Hence:

$$\text{molarity} = \frac{2.75/760}{(0.0821)(293)}\ mol/L = \boxed{1.50 \times 10^{-4}\ mol/L}$$

b. From the defining equation for molarity:

$$\text{molarity} = \frac{\text{no. moles solute}}{\text{no. liters solution}} = \frac{\text{no. grams solute/MM}}{\text{no. liters solution}}$$

Solving for the molar mass, MM,

$$\text{MM} = \frac{\text{no. grams solute}}{(\text{molarity})(\text{no. liters solution})}$$

Recall that there is 1.0 g of hemoglobin in 100 cm³ of solution. Hence:

$$\text{MM} = \frac{1.0 \text{ g}}{(1.50 \times 10^{-4} \text{ mol/L})(0.100 \text{ L})} = \boxed{6.7 \times 10^4 \text{ g/mol}}$$

No way could you measure this molar mass by FP depression

Exercise
What would be the osmotic pressure at 20°C of a solution containing 1.0 g of hemoglobin (MM = 6.7×10^4 g/mol) per liter? Answer: 0.27 mm Hg, one-tenth of that above.

COLLIGATIVE PROPERTIES OF ELECTROLYTES

As noted earlier, colligative properties of dilute solutions are directly proportional to the concentration of solute *particles*. On this basis, we would predict that, at a given concentration, an electrolyte would have a greater effect upon these properties than a nonelectrolyte. When one mole of a nonelectrolyte such as glucose dissolves in water, 1 mol of solute molecules is obtained. On the other hand, one mole of the electrolyte NaCl yields 2 mol of ions. With calcium chloride, $CaCl_2$, 3 mol of ions are produced per mole of solute.

This reasoning is confirmed experimentally. Suppose, for example, we compare the vapor pressure of 1 *M* solutions of glucose, sodium chloride, and calcium chloride. We find that the vapor pressure lowering is smallest for glucose and largest for calcium chloride. In other words,

vp pure water > vp glucose soln. > vp NaCl soln. > vp $CaCl_2$ soln.

Many electrolytes form saturated aqueous solutions whose vapor pressures are so low that the solids pick up water (*deliquesce*) when exposed to moist air. This occurs with calcium chloride, whose saturated solution has a vapor pressure only 30% that of pure water. If dry $CaCl_2$ is exposed to air in which the relative humidity is greater than 30%, it absorbs water and forms a saturated solution. Deliquescence continues until the vapor pressure of the solution becomes equal to that of the water in the air.

Those salt solutions also cause a lot of car rust

The freezing points of electrolyte solutions, like their vapor pressures, are lower than those of nonelectrolytes at the same concentration. Sodium chloride and calcium chloride are used to lower the melting point of ice on highways; their aqueous solutions can have freezing points as low as −21°C and −55°C, respectively. The equation for the freezing point lowering of an electrolyte is similar to that for nonelectrolytes, except for the introduction of a multiplier, *i*. For aqueous solutions of electrolytes

$$\Delta T_f = 1.86°C/m \times m \times i \tag{11.17}$$

If we assume that the ions of an electrolyte behave independently, we would predict that i should be equal to *the number of moles of ions per mole of electrolyte.* Thus i should be 2 for NaCl and $MgSO_4$, 3 for $CaCl_2$ and Na_2SO_4, and so on.

■ **EXAMPLE 11.14** _____

Estimate the freezing points of 0.20 m solutions of

a. KNO_3 b. $MgSO_4$ c. $Cr(NO_3)_3$

Assume that i in Equation 11.17 is the number of moles of ions formed per mole of electrolyte.

Solution

a. One mole of KNO_3 forms two moles of ions:

$$KNO_3(s) \rightarrow K^+(aq) + NO_3^-(aq)$$

Hence, i should be 2 and we have

$$\Delta T_f = (1.86°C)(0.20)(2) = 0.74°C; \qquad \boxed{T_f = -0.74°C}$$

b. $MgSO_4$ behaves like KNO_3, that is, $i = 2$:

$$MgSO_4(s) \rightarrow Mg^{2+}(aq) + SO_4^{2-}(aq)$$

$$\Delta T_f = (1.86°C)(0.20)(2) = 0.74°C; \qquad \boxed{T_f = -0.74°C}$$

c. For $Cr(NO_3)_3$, $i = 4$:

$$Cr(NO_3)_3(s) \rightarrow Cr^{3+}(aq) + 3\ NO_3^-(aq)$$

$$\Delta T_f = (1.86°C)(0.20)(4) = 1.5°C; \qquad \boxed{T_f = -1.5°C}$$

Exercise

What is the estimated freezing point of 0.20 m Na_2CO_3? Answer: $-1.1°C$.

Looking at the data in Table 11.5, we see that the situation is not so simple as our discussion might imply. The observed freezing point lowerings of NaCl and $MgSO_4$ are smaller than we would predict from Equation 11.17 with $i = 2$. In other words, at any finite concentration, the multiplier i is less than 2. It approaches a limiting value of 2 as the solution becomes more and more dilute. This behavior is generally typical of electrolytes. Interactions occur between positive and negative ions in solution. As a result, they do not behave as completely independent particles. The effect is ordinarily to make the freezing point depression or other colligative property less than we would expect from the number of ions in solution.

But larger than if no dissociation occurred

TABLE 11.5   Freezing Point Lowerings of Solutions

MOLALITY	ΔT_f OBSERVED (°C)		i (CALC. FROM EQ. 11.17)	
	NaCl	MgSO$_4$	NaCl	MgSO$_4$
0.005	0.0182	0.0160	1.96	1.72
0.01	0.0360	0.0285	1.94	1.53
0.02	0.0714	0.0534	1.92	1.44
0.05	0.176	0.121	1.89	1.30
0.10	0.348	0.225	1.87	1.21
0.20	0.685	0.418	1.84	1.12
0.50	1.68	0.995	1.81	1.07

■ SUMMARY PROBLEM

Consider camphor, $C_{10}H_{16}O$ (MM = 152 g/mol), a substance obtained from the Formosa camphor tree. It has considerable use in the polymer and drug industries.

A solution of camphor is prepared by mixing 30.0 g of camphor with 1.25 L of ethanol, C_2H_5OH (d = 0.789 g/mL). Assume no change in volume when the solution is prepared.

1. What is the mass percent of camphor in the solution?
2. What is the concentration of camphor in parts per million?
3. What is the molarity of the solution?
4. What is the molality of the solution?
5. The vapor pressure of pure ethanol at 25°C is 59.0 mm Hg. What is the vapor pressure of ethanol in the solution at this temperature?
6. What is the osmotic pressure of the solution at 25°C?
7. What is the boiling point of the solution? The normal boiling point of ethanol is 78.26°C (k_b = 1.22°C/m).
8. Camphor can also be used as a solvent. It is used in the Rast method for determining molar mass. The molar mass of cortisone acetate is determined by dissolving 2.50 g in 50.0 g camphor (k_f = 40.0°C/m). The freezing point of the mixture is determined to be 173.44°C; that of pure camphor is 178.40°C. What is the molar mass of cortisone acetate?

Answers

1. 2.95%   2. 2.95 × 10⁴ ppm   3. 0.158 M 4. 0.200 m
5. 58.5 mm Hg   6. 3.87 atm   7. 78.50°C   8. 403 g/mol

P olluted water is, quite simply, water that is unfit for the purpose for which it is intended. It is entirely possible for water to be polluted for one purpose yet pure enough for other uses. Water containing 0.1 ppm of lead ion, Pb^{2+}, is considered to be unfit for drinking but is perfectly satisfactory for most industrial purposes. There are a great many different kinds of water pollutants (Table 11.6). Rather than attempting to discuss all these types of pollutants, we will focus on a few examples of current interest.

Organic Molecules

The species listed in Table 11.6 are typical of about 30 relatively simple organic molecules that contaminate groundwater in many areas of the United States. Chronic exposure to these chemicals, even at the part per million level, can have serious consequences. Carbon tetrachloride, CCl_4, can cause kidney and liver damage; trichloroethylene, $Cl_2C{=}CHCl$, attacks the central nervous system. Benzene, C_6H_6, can cause leukemia.

Organic molecules typically leach into soils and water supplies from landfills and industrial dumpsites. Many were deposited a generation ago, when most people were unconcerned and unaware of the dangers of pollution. In a few cases, chlorination of drinking water has caused problems. Chloroform, $CHCl_3$, is formed when chlorine reacts with certain naturally occurring organic molecules.

A more complex molecule that has received a great deal of attention in recent years is dioxin:

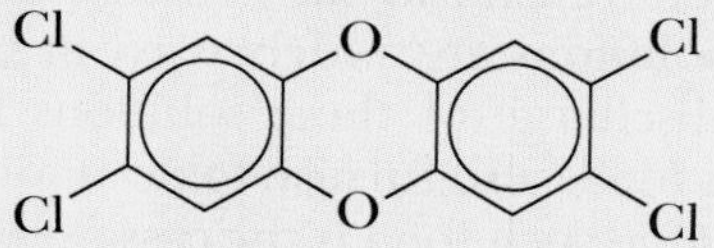

This compound is extremely toxic to certain laboratory animals; as little as 1×10^{-8} mol is lethal to guinea pigs. The effect on human beings of chronic dioxin exposure is debatable. There is evidence implicating dioxin in certain types of cancer.

Dioxin is a common contaminant of certain herbicides including Agent Orange, which was used as a defoliant in Vietnam. Several years ago, it was discovered that a contractor had sprayed several different areas in

TABLE 11.6 Classification of Water Pollutants

TYPE	EXAMPLES
Oxygen-demanding wastes	Human, animal wastes; decaying vegetation
Infectious agents	Bacteria and viruses
Organic molecules	CCl_4, $CHCl_3$, $Cl_2C{=}CHCl$, C_6H_6
Inorganic ions	$CrO_4{}^{2-}$, CN^-, Be^{2+}, Cd^{2+}, Hg^{2+}, Pb^{2+}, Tl^+
Heat	Water used for cooling in industry
Radioactive substances	Fallout products, radioactive waste

Missouri with an oil heavily contaminated with dioxin. The dioxin levels were so high in the town of Times Beach that the Environmental Protection Agency evacuated the residents, buying out and destroying their homes at a cost of $50 million.

Another organic molecule which has received a great deal of attention as an environmental contaminant is *dichlorodiphenyltrichloroethane* (DDT):

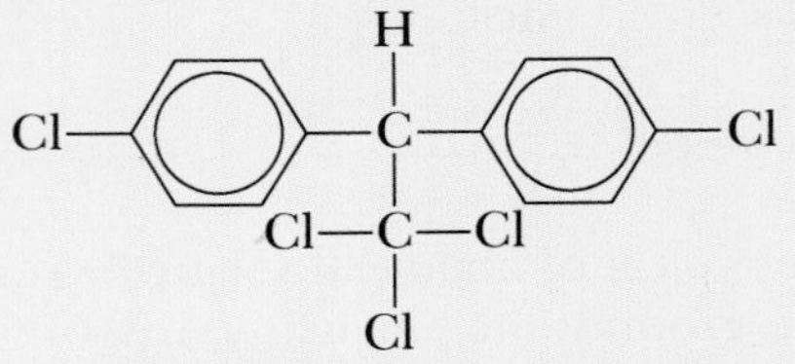

For at least 20 years, DDT was the most commonly used insecticide. Its use was severely restricted in 1973 because of its toxicity to wildlife. The high solubility of DDT in fatty tissue coupled with its long-term stability in the environment led to the near extinction of the golden eagle and the peregrine falcon, among other species.

It was heavily used for mosquito control

Inorganic Ions

The toxicity of the cyanide ion has been known for centuries. Sodium and potassium cyanide have long been used as poisons (particularly by authors in murder mysteries). The CN^- ion inactivates the enzyme cytochrome oxidase, which regulates oxygen transfer to body tissues; death occurs through asphyxiation. Cyanide solutions are widely used in metallurgy and electroplating; accidental discharge of these solutions into water supplies can be disastrous. Certain foods also furnish toxic amounts of CN^- if consumed in large quantities. One such food is the cassava root, a dietary staple in some parts of Africa.

Of the toxic cations listed in Table 11.6, most are associated primarily with air pollution rather than water pollution. An exception is mercury, which can enter water supplies through careless discharge of the element or its compounds. Certain bacteria living at the bottom of lakes and rivers can convert mercury into the extremely toxic compound dimethyl mercury, $Hg(CH_3)_2$. This is taken up by fish, which are then eaten by humans. Thirty years ago, 40 Japanese fishermen or members of their families died of brain damage caused by dimethyl mercury.

Thermal Pollution

The electrical power industry requires large quantities of cooling water to condense steam that has been used to generate electrical energy. In a conventional power plant, at least 2 kJ of heat are discharged to the water circulated through condensers for every kilojoule of electrical energy produced. In a nuclear power plant, this ratio is even higher, as high as three to one.

The discharge of hot water into a river or lake raises its temperature, typically by 5° to 10°C at the source. Occasionally temperatures as high as 60°C have been recorded. Even a relatively small temperature increase can have an adverse effect on aquatic life; salmon and trout cannot live in water much above 25°C. At higher temperatures, the rate of metabolism increases, creating a greater demand for oxygen. However, less oxygen is available, since its solubility decreases with increasing temperature (recall Fig. 11.10).

QUESTIONS AND PROBLEMS

Symbols, Formulas and Equations

1. Write the formulas of the following compounds:
   a. cane sugar   b. nitric acid
   c. calcium chloride

2. Write the formulas of the following compounds:
   a. glucose   b. sulfuric acid
   c. potassium chromate

3. Write an equation for the dissociation in water of each of the following electrolytes:
   a. chromium(III) nitrate   b. calcium phosphate
   c. magnesium iodide
   d. rubidium hydrogen carbonate

4. Write an equation for the dissociation in water of each of the following electrolytes:
   a. potassium perchlorate   b. scandium(III) sulfate
   c. calcium bromide   d. nickel(II) chlorate

5. Write an equation to represent the dissolving of the following species in water:
   a. cane sugar   b. hydrogen chloride gas
   c. aluminum nitrate   d. iodine

6. Write a reaction to represent the dissolving of the following species in water:
   a. bromine
   b. potassium permanganate
   c. ammonium phosphate   d. glucose

Concentrations of Solutions

7. How many moles of ions are present in water solutions prepared by dissolving 0.25 mol of
   a. calcium bromide?   b. magnesium sulfate?
   c. iron(III) nitrate?   d. nickel(II) sulfate?

8. How many moles of ions are present in water solutions prepared by dissolving 0.33 mol of
   a. cobalt(II) nitrate?   b. lithium carbonate?
   c. cesium sulfate?   d. aluminum sulfate?

9. A solution is prepared by dissolving 4.87 g of sodium dichromate in 25.0 mL of water ($d = 1.00$ g/mL). Calculate
   a. the mass percent of sodium dichromate in the solution.
   b. the mass percent of water in the solution.
   c. the mole fraction of sodium dichromate in the solution.

10. A solution is made by adding 3.50 mL of ethyl acetate, $C_4H_8O_2$ ($d = 0.901$ g/mL), to 25.00 mL of water ($d = 1.00$ g/mL). Assuming volumes are additive, calculate
   a. the mass percent of ethyl acetate in solution.
   b. the volume percent of water in solution.
   c. the mole fraction of ethyl acetate in solution.

11. The "proof" of an alcoholic beverage is twice the volume percent of ethyl alcohol, C_2H_5OH, in solution. For a 90-proof vodka, what is the molality of the ethyl alcohol? Take the densities of ethyl alcohol and water to be 0.789 g/mL and 1.00 g/mL, respectively.

12. Household chlorine bleach contains 5.00% by mass of sodium hypochlorite, NaClO. If the only other component were water, what would be the molality of ClO^-?

13. The Dead Sea contains 58 moles of bromide ion in 1.0×10^3 kg of water. Calculate the concentration, in ppm, of bromide ion in the Dead Sea.

14. The Salton Sea in California contains a relatively large amount of lithium ions. Its concentration of Li^+ is 1.9 ppm. How many moles of lithium ions are present in ten kilograms of water from the Salton Sea?

15. Complete the following table for water solutions of sodium permanganate, $NaMnO_4$.

	MASS SOLUTE	MOLES SOLUTE	V SOLUTION	MOLARITY
a.	10.3 g	_____	315 mL	_____
b.	_____	2.65	_____	0.832
c.	_____	_____	3.85 L	1.58

16. Complete the following table for water solutions of oxalic acid, $H_2C_2O_4$.

	MASS SOLUTE	MOLES SOLUTE	V SOLUTION	MOLARITY
a.	12.5 g	———	456 mL	———
b.	———	0.0375	———	0.138
c.	———	———	1.75 L	0.496

17. Complete the following table for water solutions of acetic acid, $HC_2H_3O_2$.

	MOLALITY	MASS PERCENT SOLUTE	PPM OF SOLUTE	MOLE FRACTION SOLUTE
a.	0.257	———	———	———
b.	———	5.00	———	———
c.	———	———	1542	———
d.	———	———	———	0.387

18. Complete the following table for water solutions of copper(II) sulfate.

	MOLALITY	MASS PERCENT SOLUTE	PPM OF SOLUTE	MOLE FRACTION SOLUTE
a.	———	12.00	———	———
b.	———	———	586	———
c.	0.389	———	———	———
d.	———	———	———	0.0534

19. Describe in detail how you would prepare 350.0 mL of 0.250 M potassium chromate solution starting with
  a. solid potassium chromate
  b. 1.50 M potassium chromate solution

20. Describe how you would prepare 5.00 L of 0.250 M formic acid, HCOOH, starting with
  a. 0.750 M formic acid   b. 0.350 M formic acid

21. A solution is made by diluting 175 mL of 0.238 M aluminum nitrate, $Al(NO_3)_3$, solution with water to a final volume of 5.00×10^2 mL. Calculate
  a. the molarity of aluminum nitrate, aluminum ion, and nitrate ion in the diluted solution.
  b. the number of moles of nitrate ion in the original solution.

22. A solution is made by diluting 125 mL of 0.230 M potassium phosphate solution with water to a final volume of 7.50×10^2 mL. Calculate
  a. the molarities of potassium phosphate, potassium ion, and phosphate ion in the diluted solution.
  b. the molarities of potassium ion and phosphate ion in the original solution.

23. The concentrated sulfuric acid available in the laboratory is 98.0% sulfuric acid by mass. Its density is 1.83 g/mL.
  a. Calculate the molarity of concentrated sulfuric acid.

b. How would you prepare 1.50 L of 3.00 M sulfuric acid solution from concentrated sulfuric acid?

24. A bottle of phosphoric acid is labeled "85.0% H_3PO_4 by mass; density = 1.689 g/cm³." Calculate the molarity, molality, and mole fraction of the phosphoric acid.

25. Complete the following table for potassium hydroxide solutions.

	DENSITY (g/cm³)	MOLARITY	MOLALITY	MASS PERCENT OF SOLUTE
a.	1.05	1.13	———	———
b.	1.29	———	———	30.0
c.	1.43	———	14.2	———

26. Complete the following table for ammonium sulfate solutions.

	DENSITY (g/cm³)	MOLARITY	MOLALITY	MASS PERCENT OF SOLUTE
a.	1.06	0.886	———	———
b.	1.15	———	———	26.0
c.	1.23	———	3.11	———

Solubilities

27. Choose the member of each set that you would expect to be more soluble in water. Explain your answer.
  a. ethane, CH_3CH_3, or methyl alcohol, CH_3OH.
  b. potassium chloride or carbon tetrachloride.
  c. methyl fluoride, CH_3F, or hydrogen fluoride.
  d. benzene, C_6H_6, or hydrogen peroxide,
    H—O—O—H.

28. Choose the member of each set that you would expect to be more soluble in water. Explain your answer.
  a. nitrogen trifluoride or sodium fluoride.
  b. ammonia or methane.
  c. carbon dioxide or silicon dioxide.
  d. methyl alcohol, CH_3OH, or methyl ether,
    H_3C—O—CH_3.

29. Consider the process by which ammonium chloride dissolves in water:

$$NH_4Cl(s) \rightarrow NH_4^+(aq) + Cl^-(aq)$$

  a. Using data from tables in Chapter 7, calculate ΔH for this reaction.
  b. Would you expect the solubility of NH_4Cl to increase or decrease if the temperature is lowered?

30. A certain gaseous solute dissolves in water, evolving 4.8 kJ/mol of heat. Its solubility at 25°C and 2.00 atm is 0.010 M. Would you expect the solubility to be greater or less than 0.010 M at

a. 0°C and 5 atm? b. 50°C and 1 atm?
c. 15°C and 2 atm? d. 25°C and 1 atm?

31. The concentration of hydrogen sulfide, H_2S, in hot springs is relatively high. This accounts for the rotten-egg smell around the "mud pots" in Yellowstone National Park. The solubility of hydrogen sulfide in water at 25°C is 0.0932 M at 1.00 atm. If the partial pressure of H_2S at Yellowstone is 0.12 atm, calculate the molarity of H_2S in the mud pots.

32. A carbonated beverage is made by saturating water with carbon dioxide at 0°C and a pressure of 3.0 atm. The bottle is then opened at room temperature (25°C), and comes to equilibrium with air in the room containing CO_2 ($P_{CO_2} = 3.4 \times 10^{-4}$ atm).

a. What is the concentration of carbon dioxide in the bottle before it is opened?

b. What is the concentration of carbon dioxide in the bottle after it has been opened and come to equilibrium with the air?

The Henry's Law constant for the solubility of CO_2 in water is 0.0769 M/atm at 0°C and 0.0313 M/atm at 25°C.

Colligative Properties

33. Calculate the vapor pressure lowering in an aqueous sucrose solution at 22°C (vp pure water = 19.83 mm Hg) if the mole fraction of sucrose is

a. 0.0100 b. 0.100 c. 0.120

What is the vapor pressure of water over each of these solutions?

34. Repeat the calculations called for in Problem 33 at 90°C (vp = 525.8 mm Hg).

35. What is the vapor pressure at 20°C of a solution of 3.50 g naphthalene, $C_{10}H_8$, in 28.75 g of benzene, C_6H_6 (vp pure C_6H_6 = 74.7 mm Hg). Assume the vapor pressure of naphthalene is negligible.

36. The vapor pressure of pure chloroform at 70.0°C is 1.34 atm. How much iodine should be dissolved in one liter of chloroform, $CHCl_3$, ($d = 1.49$ g/cm^3) to lower the vapor pressure by 1.00×10^2 mm Hg?

37. Calculate the osmotic pressure at 25°C in solutions of urea, $CO(NH_2)_2$, containing the following masses of solute per liter of solution:

a. 10.0 g b. 50.0 g c. 100.0 g

38. Lysozyme is an enzyme that cleaves bacterial cell walls. A sample of lysozyme extracted from egg white has a molar mass of 13,930 g/mol. If 15.0 mg of this enzyme is dissolved in 175 mL of water ($d = 1.00$ g/cm^3) at 25°C, what is the osmotic pressure of this solution?

39. Calculate the freezing point and normal boiling point of each of the following solutions:

a. 25.0 g of propylene glycol, $C_3H_8O_2$, in 250.0 mL of water ($d = 1.00$ g/cm^3).

b. 25.0 mL of methyl alcohol, CH_3OH ($d = 0.792$ g/cm^3), in 325 mL of water ($d = 1.00$ g/cm^3).

40. How many grams of the following nonelectrolyte solutes would have to be dissolved in 100.0 mL of water ($d = 1.00$ g/cm^3) to give a solution freezing at -1.50°C? What would be the normal boiling point of the solution?

a. glucose, $C_6H_{12}O_6$ b. citric acid, $C_6H_8O_7$

41. What is the freezing point and normal boiling point of a solution made by adding 20.0 mL of isopropyl alcohol, C_3H_7OH, to 80.0 mL of water? The density of isopropyl alcohol is 0.785 g/cm^3, while the density of water is 1.00 g/cm^3.

42. An automobile radiator is filled with an antifreeze solution prepared by mixing equal volumes of ethylene glycol, $C_2H_6O_2$ ($d = 1.12$ g/cm^3) and water ($d = 1.00$ g/cm^3). Estimate the freezing point of the mixture. Will this mixture protect automobile engines in Connecticut if the lowest temperature expected is -20°F?

43. Using Table 11.4, calculate the freezing point and the normal boiling point of solutions of 12.50 g of naphthalene, $C_{10}H_8$, in 100.0 g of

a. p-dichlorobenzene b. benzene c. cyclohexane

44. When 20.25 g of lactic acid, $C_3H_6O_3$, are dissolved in 250.0 mL of acetone ($d = 0.791$ g/mL), the resulting solution boils at 57.89°C. If the boiling point of pure acetone is 55.95°C, what is the boiling point constant for acetone?

45. A compound contains 42.9% C, 2.4% H, 16.6% N, and 38.1% O. The addition of 3.16 g of this compound to 75.0 mL of cyclohexane ($d = 0.779$ g/cm^3) gives a solution with a freezing point at 0.0°C. Using Table 11.4, determine the molecular formula of the compound.

46. Lauryl alcohol is obtained from the coconut and is an ingredient in many hair shampoos. Its empirical formula is $C_{12}H_{26}O$. A solution of 5.00 g of lauryl alcohol in 100.0 g of benzene boils at 80.78°C. Using Table 11.4, find the molecular formula of lauryl alcohol.

47. The freezing point of p-dichlorobenzene is 53.1°C; its k_f value is 7.10°C/m. A solution of 1.52 g of sulfanilamide (a sulfa drug) in 10.0 g of p-dichlorobenzene freezes at 46.7°C. What is the molar mass of sulfanilamide?

48. A student dissolved menthol ($C_{10}H_{19}OH$), a crystalline nonelectrolyte with the taste of peppermint, in 100.0 g of cyclohexane. The solution froze at -1.95°C. Using Table 11.4, calculate the percent by mass of menthol in the solution.

49. When aqueous solutions are introduced into the blood stream by injection, the solution must have the same osmotic pressure as blood. The solution is called "isotonic" with blood. At 25°C, the average osmotic pressure of blood is 7.7 atm. What is the molarity of a glucose solution isotonic with blood?

50. Refer to Problem 49 and determine the concentration of an isotonic saline solution (NaCl in H_2O). Recall that

NaCl is an electrolyte; assume complete dissociation and ideal behavior.

51. A biochemist isolated a new protein and determined its molar mass by osmotic pressure measurements. She used 0.270 g of the protein in 50.0 mL of solution and observed an osmotic pressure of 3.86 mm Hg for this solution at 25°C. What should she report as the molar mass of the new protein?

52. The molar mass of a type of hemoglobin was determined by osmotic pressure measurement. A student measured an osmotic pressure of 4.60 mm Hg for a solution at 20°C containing 3.27 g of hemoglobin in 0.200 L of solution. What is the molar mass of the hemoglobin?

53. Arrange 0.10 m solutions of the following solutes in order of decreasing freezing point:
a. $C_2H_6O_2$ b. $CrCl_3$ c. $Al_2(SO_4)_3$ d. Na_2CO_3

54. Estimate the freezing points of 0.10 m solutions of
a. K_2SO_4 b. $CsNO_3$ c. $Al(NO_3)_3$

55. The freezing point of 0.10 m $KHSO_3$ is $-0.38°C$. Which of the following equations best represents what happens when $KHSO_3$ dissolves in water?
a. $KHSO_3(s) \rightarrow KHSO_3(aq)$
b. $KHSO_3(s) \rightarrow K^+(aq) + HSO_3^-(aq)$
c. $KHSO_3(aq) \rightarrow K^+(aq) + H^+(aq) + SO_3^{2-}(aq)$

56. The freezing point of 0.20 m HF is $-0.38°C$. Is HF primarily nonionized in this solution (HF molecules), or is it dissociated to H^+ and F^- ions?

Unclassified

57. How would you prepare a saturated solution of $CO_2(g)$ in water? A supersaturated solution?

58. You are given a clear water solution containing KNO_3. How would you determine experimentally whether the solution is unsaturated, saturated, or supersaturated?

59. Explain why
a. the freezing point of 0.10 m $CaCl_2$ is lower than that of 0.10 m $MgSO_4$.
b. a water solution of NaCl conducts an electric current but NaCl(s) does not.
c. the solubility of solids in water usually increases as temperature increases.
d. pressure must be applied to cause reverse osmosis to occur.

60. Explain why
a. 0.10 M $CaCl_2$ has a higher electrical conductivity than 0.10 M NaCl.
b. the solubility of gases in water decreases as temperature increases.
c. a water solution of HCl conducts an electric current but HCl(l) does not.
d. molality and molarity are nearly the same in dilute water solution.

61. In your own words, explain
a. why the concentrations of solutions used for intravenous feeding must be controlled carefully.
b. why fish in a lake (and fishermen) seek deep, shaded places during summer afternoons.
c. why champagne "fizzes" in a glass.
d. the differences between molality and molarity.

62. In your own words, explain why
a. seawater has a lower freezing point than fresh water.
b. we believe that vapor pressure lowering is a colligative property.
c. one often obtains a "grainy" product when making fudge (a supersaturated sugar solution).
d. what causes the "bends" in divers.

63. Criticize the following statements:
a. A saturated solution is always a concentrated solution.
b. The water solubility of a solid always decreases with a drop in temperature.
c. For aqueous solutions, molarity and molality are equal.
d. The freezing point depression of a 0.10 m $CaCl_2$ solution is twice that of a 0.10 m KCl solution.
e. A 0.10 m sucrose solution and a 0.10 m NaCl solution have the same osmotic pressure.

64. Explain, in your own words,
a. how to determine experimentally whether a pure substance is an electrolyte or nonelectrolyte.
b. why a cold glass of beer goes "flat" upon warming.
c. why the molality of a solute is ordinarily larger than its mole fraction.
d. why the boiling point is raised by the presence of a solute.

65. A water solution containing 275.9 g of sugar, $C_{12}H_{22}O_{11}$, per liter has a density of 1.104 g/cm^3 at 20°C. Calculate the
a. molarity   b. molality
c. vapor pressure (vp H_2O at 20°C = 17.54 mm Hg)
d. freezing point

66. A water solution containing 28% by mass of iron(III) chloride has a density of 1.271 g/cm^3. Assuming ideal behavior, calculate the
a. molarity   b. molality
c. osmotic pressure at 25°C   d. boiling point

67. So far as its colligative properties are concerned, seawater behaves like a 0.60 M solution of NaCl, with i = 1.9. Calculate the minimum pressure (above atmospheric pressure) that must be applied to obtain pure water from seawater by reverse osmosis at 25°C.

Challenge Problems

68. A solution contains 158.2 g KOH per liter; its density is 1.13 g/cm^3. A lab technician wants to prepare 0.250 molal KOH, starting with 100.0 mL of this solution. How much water or solid KOH should be added to the 100.0-mL portion?

69. Show that the following relation is generally valid for any solution:

$$m = \frac{M}{d - \dfrac{M(MM)}{1000}}$$

where m is molality, M is molarity, d is solution density (g/mL), and MM is the molar mass of the solute. Using this equation, explain why molality approaches molarity in dilute solution when water is the solvent, but not with other solvents.

70. The water-soluble nonelectrolyte X has a molar mass of 410 g/mol. A 0.100-g mixture containing this substance and sugar (MM = 342 g/mol) is added to 1.00 g of water to give a solution freezing at $-0.500°C$. Estimate the mass percent of X in the mixture.

71. A martini, weighing about 5.0 oz (142 g), contains 30% by mass of alcohol. About 15% of the alcohol in the martini passes directly into the blood stream (7.0 L for an adult). Estimate the concentration of alcohol in the blood (g/cm³) of a person who drinks two martinis before dinner.

(A concentration of 0.0010 g/cm³ or more is frequently considered indicative of intoxication in a "normal" adult.)

72. When water is added to a mixture of aluminum metal and sodium hydroxide, hydrogen gas is produced; this reaction is used in commercial drain cleaners:

$$2\ Al(s) + 6\ H_2O(l) + 2\ OH^-(aq) \rightarrow$$
$$2\ Al(OH)_4^-(aq) + 3\ H_2(g)$$

A sufficient amount of water is added to 49.92 g of NaOH to make 0.600 L of solution; 41.28 g Al is added to this solution and hydrogen gas is formed.

a. Calculate the molarity of the initial NaOH solution.
b. How many moles of hydrogen were formed?
c. The hydrogen was collected over water at 25°C and 758.6 mm Hg. The vapor pressure of water at this temperature is 23.8 mm Hg. What volume of hydrogen was generated?

73. It is found experimentally, that the volume of a gas that dissolves in a given amount of water is independent of the pressure of the gas; that is, if 5 cm³ of a gas dissolves in 100 g of water at 1 atm pressure, 5 cm³ will dissolve at a pressure of 2 atm, 5 atm, 10 atm, ... Show that this relationship follows logically from Henry's Law and the Ideal Gas Law.

$$Ag^+(aq) \ + \ Cl^-(aq) \ \rightarrow \ AgCl(s)$$
Precipitation of silver chloride.

CHAPTER
12

REACTIONS IN AQUEOUS SOLUTION: AN OVERVIEW

Water, water, everywhere,
Nor any drop to drink.

SAMUEL TAYLOR COLERIDGE
THE RIME OF THE ANCIENT MARINER

So far, most of the chemical reactions we have studied are those in which pure substances react with each other. We have looked at reactions where liquids like gasoline or ethanol react with oxygen gas. We have considered reactions where gases like hydrogen and oxygen react to form a new compound, water. In this chapter, we will discuss for the first time a somewhat different type of reaction in which the reactants are ions or molecules in water solution.

Reactions between pure substances usually occur slowly. If iron(III) nitrate crystals are added to sodium hydroxide pellets, chemical change occurs slowly, at the interface where the two solids come in contact with each other. However, if an aqueous solution of iron(III) nitrate is added to an aqueous solution of sodium hydroxide, a red gelatinous precipitate results almost instantaneously. Most reactants need to be in aqueous solution for a chemical change to occur quickly at room temperature.

In this chapter, we will look at three types of reactions that take place in water: precipitation reactions (Section 12.1), acid-base reactions (Section 12.2), and oxidation-reduction reactions (Section 12.3). These reactions are described qualitatively by writing a particular type of chemical equation called a *net ionic equation*. Quantitatively, they are described in terms of their stoichiometry, i.e., the mass relationships between reactants and products (Section 12.4). Finally (Section 12.5), we will investigate how reactions in aqueous solution serve as a basis for quantitative chemical analysis.

Chemists ordinarily carry out inorganic reactions in solution

12.1
PRECIPITATION REACTIONS

When water solutions of two different electrolytes are mixed, we often find that an insoluble solid precipitates out of solution. To identify the solid, we must know which ionic compounds are soluble in water and which are

(a)

(b)

Figure 12.1 LiClO₃ (a) is extremely soluble. In contrast, HgS (b) is about as insoluble as it is possible for a compound to be. (Marna G. Clarke)

insoluble. Information on solubilities can be summarized by a set of solubility rules. With the help of these rules, we can write net ionic equations to represent precipitation reactions. These equations can be used in the usual way to relate amounts of reactants and products.

SOLUBILITIES OF IONIC SOLIDS

When an ionic solid dissolves in water, there is a strong interaction between polar H_2O molecules and the charged ions that make up the solid. The extent to which solution occurs depends upon a balance between two forces, both electrical in nature:

1. The force of attraction between H_2O molecules and the ions of the solid, which tends to bring the solid into solution. If this factor predominates, we expect the compound to be very soluble in water, as is the case with NaCl, NaOH, and many other ionic solids.

2. The force of attraction between oppositely charged ions, which tends to keep them in the solid state. If this is the major factor, we expect the water solubility to be very low. The fact that $CaCO_3$ and $BaSO_4$ are almost insoluble in water implies that interionic attractive forces predominate with these ionic solids.

Unfortunately, we cannot estimate from first principles the relative strengths of these two forces for a given solid. For this reason, among others, we cannot predict in advance the water solubilities of electrolytes. Ionic solids cover an enormous range of solubilities. At one extreme we have the white solid lithium chlorate, $LiClO_3$, which dissolves to the extent of 35 mol/L at room temperature. Mercury(II) sulfide, found in nature as the red mineral cinnabar, is at the other extreme. Its calculated solubility at 25°C is 10^{-26} mol/L. This means that, in principle at least, about 200 L of a saturated solution of HgS would be required to contain a single pair of Hg^{2+} and S^{2-} ions (Fig. 12.1).

Information on the solubility of common ionic solids, in the form of solubility rules, is given in Table 12.1. These rules are quite simple to

You must learn these rules to work with precipitation reactions

TABLE 12.1   Solubility Rules*

NO_3^-	All nitrates are soluble.
Cl^-	All chlorides are soluble except AgCl, Hg_2Cl_2, and $PbCl_2$.
SO_4^{2-}	Most sulfates are soluble; exceptions include $SrSO_4$, $BaSO_4$, and $PbSO_4$.
CO_3^{2-}	All carbonates are insoluble except those of the Group 1 elements and NH_4^+.
OH^-	All hydroxides are insoluble except those of the Group 1 elements, $Sr(OH)_2$, and $Ba(OH)_2$. ($Ca(OH)_2$ is slightly soluble.)
S^{2-}	All sulfides except those of Group 1 and 2 elements and NH_4^+ are insoluble.

*Insoluble compounds are those that precipitate when we mix equal volumes of 0.1 M solutions of the corresponding ions.

Figure 12.2 Of the compounds of Pb^{2+} (a), $PbCl_2$ and $PbSO_4$ are insoluble white solids; $Pb(NO_3)_2$ (center) dissolves readily to form a colorless solution. The Fe^{3+} cation (b) is yellow, as seen in the solutions of the soluble salts $Fe(NO_3)_3$ and $FeCl_3$; $Fe(OH)_3$ is an insoluble, red solid. Two insoluble compounds of Ni^{2+} (c) are NiS (black) and $Ni(OH)_2$ (green); $NiSO_4$ (far right) is soluble to give a bright green solution. All of these observations are consistent with the solubility rules. (Charles D. Winters)

interpret. For example, the following facts should be evident from the table:

— $Ni(NO_3)_2$ is soluble. (All nitrates are soluble.)
— $BaCl_2$ is soluble. ($BaCl_2$ is not one of the three insoluble chlorides listed.)
— PbS is insoluble. (Pb is not a Group 1 or Group 2 element.)

Figure 12.2 shows how these rules apply to some common ionic compounds.

■ **EXAMPLE 12.1** ———————————————————

Using the solubility rules in Table 12.1, predict what will happen when the following pairs of aqueous solutions are mixed.

a. $CuSO_4$ and $NaNO_3$ b. Na_2CO_3 and $CaCl_2$

Solution

a. An aqueous solution of $CuSO_4$ contains Cu^{2+} and SO_4^{2-} ions; a solution of $NaNO_3$ contains Na^+ and NO_3^- ions. In principle at least, four different ionic solids could form: $CuSO_4$, $Cu(NO_3)_2$, Na_2SO_4, and $NaNO_3$.

In solution, salts exist as ions

  (Remember, an ionic solid must contain one cation and one anion, and it must be electrically neutral.) Note from Table 12.1 that all four of these compounds are water-soluble. Thus when aqueous solutions of $CuSO_4$ and $NaNO_3$ are mixed, no precipitate will form.

b. Following the same reasoning as in part (a), we find that the ions present in the combined solutions before any reaction occurs are Na^+, CO_3^{2-}, Ca^{2+}, and Cl^-. The solids that could possibly form are: Na_2CO_3, NaCl, $CaCO_3$, and $CaCl_2$. All these are water-soluble except $CaCO_3$ (see

Figure 12.3 When solutions of sodium carbonate, Na_2CO_3, and calcium chloride, $CaCl_2$, are mixed, a white precipitate of $CaCO_3$ is obtained. (Charles D. Winters)

Table 12.1 and Fig. 12.3). When the solutions are mixed, $CaCO_3$ will precipitate.

Exercise
Predict what will happen when the following pairs of solutions are mixed:

a. Colorless solutions of saltpeter (KNO_3) and lithium chloride (LiCl).
b. Colorless solutions of sodium sulfate (Na_2SO_4) and lead nitrate ($Pb(NO_3)_2$).

Answer: a. no precipitate; b. a precipitate of $PbSO_4$ forms.

Sometimes insolubility can be useful. This is the case with barium sulfate, $BaSO_4$. Patients who need x-rays of their gastrointestinal tract are given a "barium sulfate cocktail," which consists of a thick, chalky suspension of $BaSO_4$ in water. This coats the gastrointestinal tract so it will appear opaque when the x-ray pictures are developed (Fig. 12.4). Since barium sulfate is insoluble, the toxic Ba^{2+} ion does not enter the blood stream.

EQUATIONS FOR PRECIPITATION REACTIONS

The precipitation reaction that occurs when solutions of Na_2CO_3 and $CaCl_2$ are mixed can be represented by a simple equation. We have seen that the product of the reaction is solid $CaCO_3$ (Example 12.1), formed by the reaction between Ca^{2+} and CO_3^{2-} ions in aqueous solution. The equation for the reaction is

$$Ca^{2+}(aq) + CO_3^{2-}(aq) \rightarrow CaCO_3(s) \tag{12.1}$$

Notice that we include in the equation only those ions that participate in the reaction. To be specific, Na^+ and Cl^- ions do not appear in Equation 12.1. They are "spectator ions," which are present in the solution before and after the precipitation of calcium carbonate. Equations such as this, which involve ions and exclude any species that do not take part in the reaction, are often referred to as *net ionic equations*. They do not differ in any essential way from the equations we have been writing all along. Other examples of net ionic equations for precipitation reactions include

— the precipitation of Ag_2CrO_4 when solutions of $AgNO_3$ and K_2CrO_4 are mixed:

$$2Ag^+(aq) + CrO_4^{2-}(aq) \rightarrow Ag_2CrO_4(s) \tag{12.2}$$
(spectator ions: NO_3^- and K^+)

— the precipitation of $Fe(OH)_3$ when solutions of $FeCl_3$ and NaOH are mixed:

$$Fe^{3+}(aq) + 3OH^-(aq) \rightarrow Fe(OH)_3(s) \tag{12.3}$$
(spectator ions: Cl^- and Na^+)

■ **EXAMPLE 12.2** _____
Write a net ionic equation for any precipitation reaction that occurs when 0.1 M solutions of the following ionic compounds are mixed:

a. NaOH and $Cu(NO_3)_2$ b. $Ba(OH)_2$ and $NiSO_4$
c. RbCl and LiOH

Solution

In each case, we first decide what ions are present when the solutions are mixed. Then we write down the formulas of the four solids that could precipitate. Using the solubility rules, we decide what compound, if any, precipitates. Finally, we write an equation for the precipitation reaction.

a. *Ions in solution:* Na^+, OH^-, NO_3^-, Cu^{2+}
   Possible products: NaOH, $NaNO_3$, $Cu(OH)_2$, $Cu(NO_3)_2$
   Identity of precipitate: The solubility rules indicate that only $Cu(OH)_2$ is insoluble.

   Equation: $\quad Cu^{2+}(aq) + 2OH^-(aq) \rightarrow Cu(OH)_2(s)$

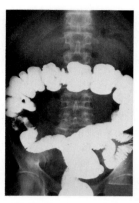

Figure 12.4 An x-ray photo of the gastrointestinal tract after a person drank a barium sulfate "cocktail."

b. The possible products are $Ba(OH)_2$, $BaSO_4$, $Ni(OH)_2$, $NiSO_4$. $BaSO_4$ is listed in Table 12.1 as being insoluble. Since Ni is not a Group 1 element, we deduce that $Ni(OH)_2$ is insoluble. Thus, we predict that both these compounds will precipitate. We write two precipitation equations, since two entirely different precipitation reactions take place:

$$Ba^{2+}(aq) + SO_4^{2-}(aq) \rightarrow BaSO_4(s)$$
$$Ni^{2+}(aq) + 2OH^-(aq) \rightarrow Ni(OH)_2(s)$$

Experiment confirms our deductions. If we look at the precipitate under a microscope, we see white $BaSO_4$ crystals and green particles of $Ni(OH)_2$.
c. All possible products—RbCl, RbOH, LiOH, and LiCl—are soluble. No precipitation reaction occurs, so no equation can be written.

All spectators, no players

Exercise

Write net ionic equations for the precipitation reactions that occur when solutions of $SrCl_2$ and Na_2SO_4 are mixed; NH_4Cl and $Pb(NO_3)_2$. Answer:

$$Sr^{2+}(aq) + SO_4^{2-}(aq) \rightarrow SrSO_4(s)$$

$$Pb^{2+}(aq) + 2Cl^-(aq) \rightarrow PbCl_2(s)$$

12.2
ACID-BASE REACTIONS

For our purposes in this chapter, we will use the following working definitions of acids and bases, first proposed by the Swedish chemist Svante Arrhenius:

An acid is a species that supplies H^+ ions to water

A base is a species that supplies OH^- ions to water

Later, in Chapter 14, we will present more general models of acids and bases, but these definitions will suffice for now.

There are two types of acids, strong and weak, which differ in the extent of their dissociation in water. Strong acids dissociate completely, forming H^+ ions and anions. A typical strong acid is HCl. It undergoes the following reaction upon addition to water:

$$HCl(aq) \rightarrow H^+(aq) + Cl^-(aq)$$

In a 0.1 M solution of hydrochloric acid, there is 0.1 mol/L of H^+ ions, 0.1 mol/L of Cl^- ions, and no HCl molecules. There are six common strong acids, whose names and molecular formulas are listed in Table 12.2.

A weak acid is only partially dissociated to H^+ ions in water. Many different species behave as weak acids; those considered in this chapter are all molecules with an ionizable hydrogen atom. If we represent the general formula of a weak acid as HB, we can show its dissociation as

$$HB(aq) \rightleftharpoons H^+(aq) + B^-(aq)$$

A solution of HB contains mainly HB molecules

The double arrow implies that this reaction is reversible. In other words, a mixture is formed, containing significant amounts of both products and reactants. With the weak acid hydrogen fluoride, a 0.1 M solution contains about 0.01 mol/L of H^+ ions, 0.01 mol/L of F^- ions, and 0.09 mol/L of HF molecules. We will have a great deal more to say about the dissociation of weak acids in Chapter 14.

Bases, like acids, are classified as strong or weak. A strong base dissociates completely in water to release OH^- ions and metal cations. As you can see from Table 12.2, the strong bases are hydroxides of Group 1 metals and the heavier Group 2 metals. The dissociation of sodium hydroxide is typical:

TABLE 12.2   Common Strong Acids and Bases

ACID	NAME	DISSOCIATION
HCl	Hydrochloric acid	$HCl(aq) \rightarrow H^+(aq) + Cl^-(aq)$
HBr	Hydrobromic acid	$HBr(aq) \rightarrow H^+(aq) + Br^-(aq)$
HI	Hydriodic acid	$HI(aq) \rightarrow H^+(aq) + I^-(aq)$
HNO_3	Nitric acid	$HNO_3(aq) \rightarrow H^+(aq) + NO_3^-(aq)$
$HClO_4$	Perchloric acid	$HClO_4(aq) \rightarrow H^+(aq) + ClO_4^-(aq)$
H_2SO_4	Sulfuric acid	$H_2SO_4(aq) \rightarrow H^+(aq) + HSO_4^-(aq)$*

BASE	NAME	DISSOCIATION
LiOH	Lithium hydroxide	$LiOH(s) \rightarrow Li^+(aq) + OH^-(aq)$
NaOH	Sodium hydroxide	$NaOH(s) \rightarrow Na^+(aq) + OH^-(aq)$
KOH	Potassium hydroxide	$KOH(s) \rightarrow K^+(aq) + OH^-(aq)$
RbOH	Rubidium hydroxide	$RbOH(s) \rightarrow Rb^+(aq) + OH^-(aq)$
CsOH	Cesium hydroxide	$CsOH(s) \rightarrow Cs^+(aq) + OH^-(aq)$
$Ca(OH)_2$	Calcium hydroxide	$Ca(OH)_2(s) \rightarrow Ca^{2+}(aq) + 2\,OH^-(aq)$
$Sr(OH)_2$	Strontium hydroxide	$Sr(OH)_2(s) \rightarrow Sr^{2+}(aq) + 2\,OH^-(aq)$
$Ba(OH)_2$	Barium hydroxide	$Ba(OH)_2(s) \rightarrow Ba^{2+}(aq) + 2\,OH^-(aq)$

*Sulfuric acid is a strong acid in the sense that its first dissociation, shown here, goes to completion. The HSO_4^- ion further dissociates to form H^+ and SO_4^{2-} ions. This second dissociation occurs to a sufficient extent that in its reactions with bases, one mole of H_2SO_4 effectively produces two moles of H^+.

$$NaOH(s) \rightarrow Na^+(aq) + OH^-(aq)$$

In a 0.1 M solution of sodium hydroxide, there is 0.1 mol/L of Na^+ ions, 0.1 mol/L of OH^- ions, and no NaOH molecules.

Weak bases are species that are only partially dissociated to OH^- ions in water. Typically, the OH^- ions are generated by the reaction of the weak base with H_2O molecules. This is the case with all of the molecular weak bases that we will consider in this chapter. Of these, the simplest is the ammonia molecule, NH_3; its reaction with water is represented by the equation:

$$NH_3(aq) + H_2O \rightleftharpoons NH_4^+(aq) + OH^-(aq)$$

The double arrow, as always, implies that the reaction does not go to completion. In a 0.1 M solution of ammonia, there is about 0.001 mol/L of NH_4^+, 0.001 mol/L of OH^-, and nearly 0.099 mol/L of NH_3.

A common class of weak bases consists of the organic molecules known as *amines*. An amine can be considered to be a derivative of ammonia in which one or more hydrogen atoms have been replaced by hydrocarbon groups. In the simplest case, methyl amine, a hydrogen atom is replaced by a —CH_3 group to give the CH_3NH_2 molecule, which reacts with water in a manner very similar to NH_3:

Such as $^-CH_3$, $^-C_2H_5$, and $^-C_6H_5$

$$CH_3NH_2(aq) + H_2O \rightleftharpoons CH_3NH_3^+(aq) + OH^-(aq)$$

When an acidic water solution is mixed with a solution containing a base, a reaction occurs. The nature of this reaction and thus the equation we write for it depend upon whether the acid and base are strong or weak. We will look at several different types of acid-base reactions. Those that we consider, for all practical purposes, go to completion; the reactants are essentially all converted to products.

REACTIONS OF STRONG ACIDS WITH STRONG BASES

Consider what happens when we add a solution of a strong acid such as HNO_3 to a solution of a strong base such as NaOH. Since HNO_3 is a strong acid, the reacting species is H^+. Similarly, since NaOH is a strong base, the reacting species is OH^-. The equation for the reaction is simply

$$H^+(aq) + OH^-(aq) \rightarrow H_2O \qquad (12.4)$$

This is the net reaction that occurs when *any strong acid reacts with any strong base*. Note that we do not include in the equation spectator ions such as Na^+ or NO_3^-, which do not take part in the reaction.

REACTIONS OF WEAK ACIDS WITH STRONG BASES

When a strong base such as NaOH is added to a weak acid solution, a reaction similar in many ways to Reaction 12.4 occurs. The equation for the reaction, however, is different. To see why, let us consider the nature of the **reacting species** in the two solutions. The strong base NaOH is

In an acid-base reaction, the principal acidic species reacts with the principal basic species

completely dissociated to Na^+ and OH^- ions, and the reacting species is OH^-. In a solution of a weak acid, HB, the situation is quite different. Here there are very few H^+ and B^- ions. The principal reacting species is the undissociated HB molecule. The acid-base reaction involves the HB molecule and the OH^- ion as reactants. The products are an H_2O molecule and a B^- ion in solution. The equation for the reaction is

$$HB(aq) + OH^-(aq) \rightarrow H_2O + B^-(aq) \tag{12.5}$$

where HB stands for any weak acid, such as HF, $HC_2H_3O_2$, and so on. By the same token, B^- is the anion derived from that acid (F^-, $C_2H_3O_2^-$, etc.). Thus, for the reaction between solutions of sodium hydroxide, NaOH, and hydrogen fluoride, HF, we write the equation:

$$HF(aq) + OH^-(aq) \rightarrow H_2O + F^-(aq)$$

In another case, for the reaction between solutions of calcium hydroxide, $Ca(OH)_2$, and acetic acid, $HC_2H_3O_2$,* we write:

$$HC_2H_3O_2(aq) + OH^-(aq) \rightarrow C_2H_3O_2^-(aq) + H_2O$$

Notice that here, as always, we do not include spectator ions such as Na^+ or Ca^{2+}. These equations, like all the ones written in this chapter, are net ionic equations; they contain only those ions and/or molecules that take part in the reaction.

REACTIONS OF STRONG ACIDS WITH WEAK BASES

As an example of this type of reaction, consider what happens when an aqueous solution of a strong acid like HCl is added to an aqueous solution of ammonia, NH_3. Here, the reacting species for the strong acid is H^+ (Table 12.3). In contrast, since NH_3 is a weak base, the reacting species is the ammonia molecule. The acid-base reaction is

$$H^+(aq) + NH_3(aq) \rightarrow NH_4^+(aq) \tag{12.6}$$

In another case, for the reaction of a strong acid such as HNO_3 with an amine of general formula R_3N, we would write the equation:

$$H^+(aq) + R_3N(aq) \rightarrow R_3NH^+(aq) \tag{12.7}$$

From our discussion thus far, it should be clear that the reactions between acids and bases differ from one another largely in the nature of the reacting species. These species are listed in Table 12.3. You should find this table helpful in writing the equations called for in Example 12.3.

---

*Acetic acid is an organic compound with the structure:

$$\begin{array}{c} \quad\;\; H \\ \quad\;\; | \\ H-C-C-O-H \\ \quad\;\; | \;\;\; \| \\ \quad\;\; H \;\;\; O \end{array}$$

In chemical equations it is often represented as CH_3COOH (or even as HAc).

TABLE 12.3   Reacting Species for Acids and Bases

REACTANT	SPECIES WRITTEN IN THE EQUATION
Strong acid	H^+
Strong base	OH^-
Weak acid	HB
Weak base	R_3N (R = H, CH_3—, etc.)

■ **EXAMPLE 12.3** _____

Write a net ionic equation for each of the following reactions in dilute water solution:

a. nitrous acid (HNO_2) with sodium hydroxide (NaOH)
b. methyl amine (CH_3NH_2) with perchloric acid ($HClO_4$)
c. hydrobromic acid (HBr) with potassium hydroxide (KOH)

Solution

a. Nitrous acid is a weak acid, while sodium hydroxide is a strong base. Hence the reacting species are

    solution of HNO_2: HNO_2

    solution of NaOH: OH^-

The acid-base reaction is analogous to Reaction 12.5:

$$HNO_2(aq) + OH^-(aq) \rightarrow H_2O + NO_2^-(aq)$$

b. Methyl amine is a weak base, so its reacting species is the molecule CH_3NH_2. Perchloric acid is a strong acid. Its reacting species is H^+. Thus the reacting species are:

    solution of methyl amine: CH_3NH_2

    solution of $HClO_4$: H^+

The acid-base reaction is

$$CH_3NH_2(aq) + H^+(aq) \rightarrow CH_3NH_3^+(aq)$$

c. Hydrobromic acid (HBr) is a strong acid. Potassium hydroxide (KOH) is a strong base. The reacting species are

    solution of HBr: H^+

    solution of KOH: OH^-

The acid-base reaction is

$$H^+(aq) + OH^-(aq) \rightarrow H_2O$$

Exercise
Write the equation for the reaction between acetic acid and calcium hydroxide; between sulfuric acid and ammonia. Answers:

$$HC_2H_3O_2(aq) + OH^-(aq) \rightarrow H_2O + C_2H_3O_2^-(aq)$$
$$NH_3(aq) + H^+(aq) \rightarrow NH_4^+(aq)$$

The procedure followed in Example 12.3 is a general one. If you are asked to write an equation for an acid-base reaction between two solutions, you follow what amounts to a three-step procedure:

1. Decide upon the nature of the reacting solutions. These may be strong acids, strong bases, weak acids, or weak bases.
2. Decide what the reacting species is in each solution (Table 12.3).
3. Write a balanced net ionic equation for the acid-base reaction. There are three possibilities.*

    a. strong acid—strong base: $H^+(aq) + OH^-(aq) \rightarrow H_2O$
    b. weak acid—strong base: $HB(aq) + OH^-(aq) \rightarrow H_2O + B^-(aq)$
    c. strong acid—weak base: $H^+(aq) + R_3N(aq) \rightarrow R_3NH^+(aq)$
                                                $(R = H, CH_3, etc.)$

12.3
OXIDATION-REDUCTION REACTIONS

An oxidation-reduction reaction, commonly called a **redox reaction**, involves an exchange of electrons. The species that *loses* electrons is said to be *oxidized*. The other species, which *gains* electrons, is *reduced*. A simple example of a redox reaction is that which occurs when zinc metal is added to hydrochloric acid. Bubbles of hydrogen gas are evolved and H^+ ions in the solution are replaced by Zn^{2+} ions. The net ionic equation for the reaction that occurs is

$$Zn(s) + 2H^+(aq) \rightarrow Zn^{2+}(aq) + H_2(g) \tag{12.8}$$

Here, zinc atoms are oxidized to Zn^{2+} ions:

    oxidation: $Zn(s) \rightarrow Zn^{2+}(aq) + 2e^- \tag{12.8a}$

while H^+ ions are reduced to H_2 molecules:

    reduction: $2H^+(aq) + 2e^- \rightarrow H_2(g) \tag{12.8b}$

From this example, it should be clear that

— *oxidation and reduction occur together*, in the same reaction; you can't have one without the other.

> The electrons don't appear in the overall reaction (12.8)

---

*In principle, we could have a fourth type of reaction, that between a weak acid and a weak base. Such reactions do not go to completion; they produce an equilibrium mixture of reactants and products. For that reason, among others, we will not consider such reactions in this chapter.

— *there is no net change in the number of electrons in a redox reaction.* Those given off in the oxidation half-reaction are taken on by another species in the reduction half-reaction.

— *electrons appear on the right side of an oxidation half-equation* (Equation 12.8a); *in a reduction half-equation* (Equation 12.8b), *they appear on the left side.*

The redox reaction between zinc and hydrochloric acid is a simple one; looking at Equation 12.8, we can readily see what is happening to the electrons. Many redox reactions in water solution are more complex. The exchange of electrons is not apparent in the final equation. To simplify the electron bookkeeping in such cases, it is helpful to introduce the concept of oxidation number.

OXIDATION NUMBER

Oxidation number was developed to treat, in a rather arbitrary way, the distribution of electrons in covalent bonds. Consider the HCl molecule. The electrons in the H—Cl covalent bond are displaced strongly toward the more electronegative chlorine. Insofar as "electron bookkeeping" is concerned, we might assign these electrons to the chlorine atom:

$$H \mid : \ddot{\underset{..}{Cl}} :$$

By assigning electrons in this way we have, in a sense, given a -1 charge to chlorine. It now has one more valence electron (eight) than an isolated chlorine atom (which has seven valence electrons). The hydrogen atom, stripped of its valence electron by this assignment, has in effect acquired a $+1$ charge.

The accounting system we have just gone through is widely used in inorganic chemistry. The concept of oxidation number is used to refer to

(a)

(b)

When a copper wire is immersed in a solution of silver nitrate, a redox reaction occurs, forming crystals of silver metal (white) and Cu^{2+} ions in solution (blue). The equation is: $Cu(s) + 2Ag^+(aq) \rightarrow Cu^{2+}(aq) + 2Ag(s)$. (Charles Steele) the charge an atom would have if the bonding electrons were assigned arbitrarily to the more electronegative element. In the HCl molecule, hydrogen is said to have an oxidation number of $+1$, and chlorine an oxidation number of -1. In water the bonding electrons are assigned to the more electronegative oxygen atom:

$$H \; \Big| \; :\!\overset{\cdot\cdot}{\underset{\cdot\cdot}{O}}\!: \; \Big| \; H$$

This gives oxygen an oxidation number of -2 (eight valence electrons vs. six in the neutral atom). A hydrogen atom in water has an oxidation number of $+1$ (zero valence electrons vs. one in the neutral atom). In a nonpolar covalent bond, the bonding electrons are split evenly between the two atoms:

$$:\!\overset{\cdot\cdot}{\underset{\cdot\cdot}{Cl}}\!\cdot \; \Big| \; \cdot\overset{\cdot\cdot}{\underset{\cdot\cdot}{Cl}}\!: \qquad \text{oxidation number Cl} = 0$$

We should emphasize that the oxidation number of an atom in a molecule is an artificial concept. Unlike the charge of an ion, oxidation number cannot ordinarily be determined by experiment. The hydrogen atom in the HCl or H_2O molecule does not carry a full positive charge. We might regard its oxidation number of $+1$ in these molecules as a "pseudocharge."

RULES FOR ASSIGNING OXIDATION NUMBERS

In principle, oxidation numbers could be determined in any species by assigning bonding electrons in the manner just described. Such a method would require, however, that we know the Lewis structure of the species. In practice, oxidation numbers are ordinarily obtained in a much simpler way, applying certain arbitrary rules. There are four such rules:

1. The oxidation number of an element in an elementary substance is 0. For example, the oxidation number of chlorine in Cl_2 or of phosphorus in P_4 is 0.

2. The oxidation number of an element in a monatomic ion is equal to the charge of that ion. In the ionic compound NaCl, sodium has an oxidation number of $+1$, chlorine an oxidation number of -1. The oxidation numbers of aluminum and oxygen in Al_2O_3 (Al^{3+}, O^{2-} ions) are $+3$ and -2, respectively.

3. Certain elements have the same oxidation number in all or almost all their compounds. The Group 1 metals always exist as $+1$ ions in their compounds and, hence, are assigned an oxidation number of $+1$. By the same token, Group 2 elements always have oxidation numbers of $+2$ in their compounds. Fluorine, the most electronegative of all elements, has an oxidation number of -1 in all of its compounds.

Oxygen is ordinarily assigned an oxidation number of -2 in its compounds. (An exception arises in compounds containing the peroxide ion, O_2^{2-}, where the oxidation number of oxygen is -1.)

Oxidation numbers make it easier to analyze redox reactions

These charges are given in Fig. 3.4, p. 73

Hydrogen in its compounds ordinarily has an oxidation number of $+1$. (The only exception is in metal hydrides such as NaH and CaH_2, where hydrogen is present as the H^- ion, and hence is assigned an oxidation number of -1.)

4. The sum of the oxidation numbers of all the atoms in a neutral species is 0; in an ion, it is equal to the charge of that ion. The application of this very useful principle is illustrated in Example 12.4.

■ **EXAMPLE 12.4** _____
What is the oxidation number of sulfur in Na_2SO_4? Of manganese in MnO_4^-?

Solution
For Na_2SO_4, taking the oxidation number of sodium to be $+1$ and that of oxygen to be -2, we have

$$2(+1) + \text{oxid. no. S} + 4(-2) = 0; \quad \text{oxid. no. S} = 8 - 2 = \boxed{+6}$$ *Simple algebra does it*

In the MnO_4^- ion, taking the oxidation number of oxygen to be -2 and realizing that the sum must be -1:

$$\text{oxid. no. Mn} + 4(-2) = -1; \quad \text{oxid. no. Mn} = \boxed{+7}$$

Exercise
What is the oxidation number of Cr in Na_2CrO_4? in $Cr_2O_7^{2-}$? Answer: $+6$.

_____

OXIDATION AND REDUCTION—A WORKING DEFINITION

The concept of oxidation number leads directly to a working definition of the terms oxidation and reduction. **Oxidation** is defined as **an increase in oxidation number**, and **reduction** as a **decrease in oxidation number**. Examples include

$$2Al(s) + 3Cl_2(g) \rightarrow 2AlCl_3(s) \quad \begin{matrix} \text{Al oxidized (oxid. no. } 0 \rightarrow +3) \\ \text{Cl reduced (oxid. no. } 0 \rightarrow -1) \end{matrix} \quad (12.9)$$

$$4As(s) + 5O_2(g) \rightarrow 2As_2O_5(s) \quad \begin{matrix} \text{As oxidized (oxid. no. } 0 \rightarrow +5) \\ \text{O reduced (oxid. no. } 0 \rightarrow -2) \end{matrix} \quad (12.10)$$

These definitions are compatible with our earlier interpretation of oxidation and reduction in terms of loss and gain of electrons. An element that loses electrons must increase in oxidation number. The gain of electrons always results in an decrease in oxidation number. However, defining oxidation and reduction in terms of changes in oxidation number has one distinct advantage. It greatly simplifies the electron bookkeeping in redox reactions. Consider, for example, the reaction

$$HCl(g) + HNO_3(l) \rightarrow NO_2(g) + \tfrac{1}{2}Cl_2(g) + H_2O(l) \quad (12.11)$$

Analysis in terms of oxidation number reveals that chlorine is oxidized (oxid. no. = −1 in HCl, 0 in Cl_2). Nitrogen is reduced (oxid. no. = +5 in HNO_3, +4 in NO_2). It is much more difficult to decide precisely which atoms are "losing" or "gaining" electrons in this reaction.

In a redox reaction, we usually have at least two reactants. One of these is referred to as the oxidizing agent, another as the reducing agent.

An oxidizing agent brings about the oxidation of another species. To do this, it must accept electrons from that species. Hence, the oxidizing agent is itself reduced in the reaction.

A reducing agent brings about the reduction of another species. To do this, it must donate electrons to that species. Hence, the reducing agent is itself oxidized in the reaction.

To illustrate what these statements mean, consider Reaction 12.9. Here, Cl_2 is the oxidizing agent. It oxidizes Al (0 → +3). The Cl_2 is itself reduced to Cl^- ions (0 → −1). The reducing agent is aluminum metal, Al. It reduces Cl_2 molecules to Cl^- ions. In the process, Al is oxidized to Al^{3+} ions.

BALANCING REDOX EQUATIONS

Frequently you will be working with redox reactions where the coefficients in the balanced equation are by no means obvious. In this section, we will discuss a general approach to balancing such equations. It is called the **half-equation** (or *ion-electron*) method and is perhaps the most straightforward way to balance a variety of redox equations for reactions in water solution.

The first step in this method involves breaking the overall equation down into two half-equations. One of these is an oxidation, the other is a reduction. The two half-equations are balanced separately. Finally, they are combined in such a way as to obtain an overall equation in which there is no net change in the number of electrons.

Reaction Between Cr^{3+} and Cl^- Ions

To illustrate the half-equation method, we start with a simple example. This involves the reaction that occurs when a direct electric current is passed through a water solution of chromium(III) chloride, $CrCl_3$, producing the two elements chromium and chlorine. The unbalanced equation for the reaction is

$$Cr^{3+}(aq) + Cl^-(aq) \rightarrow Cr(s) + Cl_2(g)$$

To balance this equation, we proceed as follows:

1. *Split the equation into two half-equations,* one oxidation and one reduction:

    reduction: $Cr^{3+}(aq) \rightarrow Cr(s)$ (1a)

    oxidation: $Cl^-(aq) \rightarrow Cl_2(g)$ (1b)

2. *Balance these half-equations, first with respect to atoms and then with respect to charge.* Equation 1a is balanced insofar as atoms are concerned, since there is one atom of Cr on both sides. The charges, however, are unbalanced;

the Cr atom on the right has 0 charge while the Cr^{3+} ion on the left has a charge of $+3$. To correct this, we add three electrons to the left of 1a, arriving at

$$Cr^{3+}(aq) + 3e^- \rightarrow Cr(s) \qquad (2a)$$

This is the reduction half-equation

Equation 1b must first be balanced with respect to atoms, by providing two Cl^- ions to give one molecule of Cl_2:

$$2Cl^-(aq) \rightarrow Cl_2(g)$$

To balance charges, two electrons must be added to the right, giving a charge of -2 on both sides:

$$2Cl^-(aq) \rightarrow Cl_2(g) + 2e^- \qquad (2b)$$

This is the oxidation half-equation

3. Having arrived at two balanced half-equations, *combine them so as to make the number of electrons gained in reduction equal to the number lost in oxidation.* In Equation 2a, three electrons are gained; in 2b, two electrons are given off. To arrive at a final equation in which no electrons appear, multiply 2a by 2, 2b by 3, and add the resulting half-equations:

$$
\begin{array}{lll}
2 \times 2a: & 2Cr^{3+}(aq) + 6e^- \rightarrow 2Cr(s) & (3a) \\
3 \times 2b: & 6Cl^-(aq) \rightarrow 3Cl_2(g) + 6e^- & (3b) \\
\hline
& 2Cr^{3+}(aq) + 6Cl^-(aq) \rightarrow 2Cr(s) + 3Cl_2(g) & (3)
\end{array}
$$

Reaction Between MnO_4^- and Fe^{2+} (Acidic Solution)

The equation just worked out was easy to balance because it involved only two species. One of these (Cr^{3+}) was reduced; the other (Cl^-) was oxidized. In many redox reactions, species other than those being reduced or oxidized take part in the reaction. Most commonly, such species contain hydrogen (oxid. no. $= +1$) or oxygen (oxid. no. $= -2$). The presence of ions or molecules of this type makes the redox equation more difficult to balance. However, the half-equation method can still be applied.

To illustrate the approach used, consider the reaction between Fe^{2+} ions and MnO_4^- ions in acidic solution (Fig. 12.5, p. 410):

$$MnO_4^-(aq) + Fe^{2+}(aq) \rightarrow Mn^{2+}(aq) + Fe^{3+}(aq) \qquad \text{(acidic solution)}$$

This reaction can be represented by the (unbalanced) equation:

$$MnO_4^-(aq) + H^+(aq) + Fe^{2+}(aq) \rightarrow Mn^{2+}(aq) + Fe^{3+}(aq) + H_2O$$

In acidic aqueous solution, H^+ and H_2O can be reactants or products

Note that the two elements that undergo a change in oxidation number are manganese ($+7 \rightarrow +2$) and iron ($+2 \rightarrow +3$). Neither hydrogen nor oxygen changes oxidation number, yet atoms of these elements participate in the reaction. The oxygen atoms originally in the MnO_4^- ion end up in H_2O molecules; the H^+ ions meet the same fate.

To balance this equation, we proceed as follows:

1. Recognize which species undergo oxidation and reduction. Represent these by appropriate half-equations:

$$
\begin{array}{lll}
\text{oxidation:} & Fe^{2+}(aq) \rightarrow Fe^{3+}(aq) & (1a) \\
\text{reduction:} & MnO_4^-(aq) + H^+(aq) \rightarrow Mn^{2+}(aq) + H_2O & (1b)
\end{array}
$$

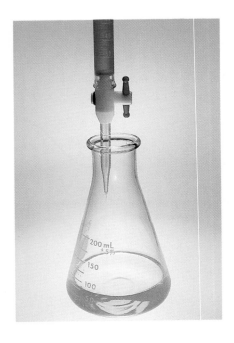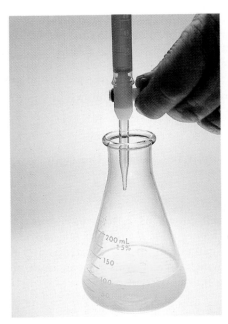

Figure 12.5 When potassium permanganate (buret) is added to an acidic solution containing Fe^{2+} ions (flask), a redox reaction occurs. The equation for the reaction is derived and balanced in the text. As reaction takes place, the purple color characteristic of the MnO_4^- ion fades. (Charles D. Winters)

2. Balance the two half-equations with respect to atoms and charge. Half-equation 1a is balanced by writing

$$Fe^{2+}(aq) \rightarrow Fe^{3+}(aq) + e^- \tag{2a}$$

To balance half-equation 1b, we first make sure that there are the same number of Mn atoms on both sides—one. Next, oxygen is balanced by using H_2O. In this case a coefficient of 4 in front of H_2O on the right accounts for the four oxygens in an MnO_4^- ion:

Balance oxygen before hydrogen

$$MnO_4^-(aq) + H^+(aq) \rightarrow Mn^{2+}(aq) + 4\ H_2O$$

To complete the atom balance, the number of hydrogen atoms on the two sides must be equalized. The four H_2O molecules on the right contain eight hydrogen atoms; there must then be eight H^+ ions on the left:

$$MnO_4^-(aq) + 8H^+(aq) \rightarrow Mn^{2+}(aq) + 4H_2O$$

This half-equation now has a charge of $+2$ on the right and $+7$ on the left $(-1 + 8)$. To balance each side with a $+2$ charge, five electrons are added to the left:

$$MnO_4^-(aq) + 8H^+(aq) + 5e^- \rightarrow Mn^{2+}(aq) + 4H_2O \tag{2b}$$

3. Finally, half-equations 2a and 2b are combined so as to eliminate electrons from the final equation. To do this, we multiply 2a by 5 and add to 2b:

No electrons appear in the final equation

$$5 \times 2a: \quad\quad\quad\quad 5Fe^{2+}(aq) \rightarrow 5Fe^{3+}(aq) + 5e^- \tag{3a}$$
$$2b: \quad MnO_4^-(aq) + 8H^+(aq) + 5e^- \rightarrow Mn^{2+}(aq) + 4H_2O \tag{3b}$$

$$\overline{MnO_4^-(aq) + 8H^+(aq) + 5Fe^{2+}(aq) \rightarrow Mn^{2+}(aq) + 4H_2O + 5Fe^{3+}(aq)}$$

It is helpful to compare the procedure we have just gone through for the $MnO_4^- - Fe^{2+}$ reaction to that discussed earlier for the simpler $Cr^{3+} - Cl^-$ reaction. They differ only in the balancing of the half-equations (step 2). This process is more tedious where MnO_4^- is involved, because of the extra elements present (H, O). In general, to balance a redox half-equation involving H^+ ions and H_2O molecules, you proceed as follows:

a. Balance the atoms of the element being oxidized or reduced.
b. Balance oxygen, using H_2O molecules.
c. Balance hydrogen, using H^+ ions.
d. Balance charge, using e^-.

Reaction Between MnO_4^- and I^- (Basic Solution)

Often we need to write balanced equations for redox reactions taking place in basic solution. Here, we should not have H^+ ions in the final equation; their concentration in basic solution is very small. Instead, hydrogen in such equations should be in the form of OH^- ions or H_2O molecules. A simple way to accomplish this is to eliminate any H^+ ions appearing in the half-equations, "neutralizing" them by adding an equal number of OH^- ions to both sides. Consider, for example, the oxidation, in basic solution, of iodide by permanganate ions:

$$I^-(aq) + MnO_4^-(aq) \rightarrow I_2(aq) + MnO_2(s) \quad \text{(basic solution)}$$

One can proceed exactly as in the foregoing example to obtain the half-equations

oxidation: $\qquad 2I^-(aq) \rightarrow I_2(aq) + 2e^- \qquad\qquad$ (2a)

reduction: $\qquad MnO_4^-(aq) + 4H^+(aq) + 3e^- \rightarrow MnO_2(s) + 2H_2O$ (2b)

Start out as though the solution were acidic

The H^+ ions appearing in the reduction half-equation must now be removed to obtain an equation valid in basic solution. To do this, four OH^- ions are added to both sides, and water is formed on the left by combining H^+ with OH^- ions:

$$MnO_4^-(aq) + 4H^+(aq) + 3e^- \rightarrow MnO_2(s) + 2H_2O$$
$$\underline{\qquad\quad + 4OH^-(aq) \qquad\qquad\qquad\qquad\qquad + 4OH^-(aq)}$$
$$MnO_4^-(aq) + 4H_2O + 3e^- \rightarrow MnO_2(s) + 2H_2O + 4OH^-(aq)$$

Then add OH^- to get rid of the H^+ ions

Eliminating two water molecules from each side, we arrive at

$$MnO_4^-(aq) + 2H_2O + 3e^- \rightarrow MnO_2(s) + 4OH^-(aq) \qquad (2b')$$

for the reduction half-reaction in basic solution. To obtain the overall equation, we proceed as before, combining 2a and 2b' in such a way as to make the electron gain equal the electron loss:

$3 \times 2a$: $\qquad\qquad\qquad\qquad 6I^-(aq) \rightarrow 3I_2(aq) + 6e^- \qquad\qquad$ (3a)

$\underline{2 \times 2b': \quad 2MnO_4^-(aq) + 4H_2O + 6e^- \rightarrow 2MnO_2(s) + 8OH^-(aq) \quad (3b)}$

$6I^-(aq) + 2MnO_4^-(aq) + 4H_2O \rightarrow 3I_2(aq) + 2MnO_2(s) + 8OH^-(aq)$ (3)

■ **EXAMPLE 12.5**

Balance the equation

$$Cl_2(g) \rightarrow Cl^-(aq) + ClO_3^-(aq) \qquad \text{(acidic solution)}$$

Solution

Don't panic just because the same species is both oxidized and reduced

In this case, the same species, Cl_2, is being both oxidized and reduced. Part of the chlorine is reduced to Cl^- (oxid. no. Cl: $0 \rightarrow -1$), while part of it is oxidized to ClO_3^- (oxid. no. Cl: $0 \rightarrow +5$).

1. reduction: $Cl_2(g) \rightarrow Cl^-(aq)$

    oxidation: $Cl_2(g) \rightarrow ClO_3^-(aq)$

We now balance these equations in acidic solution, using H^+ ions and H_2O molecules. The reduction half-equation is readily balanced. We need two Cl^- ions to balance chlorine atoms and two electrons on the left to balance charge:

$$Cl_2(g) + 2e^- \rightarrow 2Cl^-(aq)$$

To balance the oxidation half-equation, we start by balancing Cl atoms:

$$Cl_2(g) \rightarrow 2ClO_3^-(aq)$$

To balance oxygen, we add 6 H_2O molecules to the left:

$$Cl_2(g) + 6H_2O \rightarrow 2ClO_3^-(aq)$$

To balance hydrogen, we add 12 H^+ ions to the right:

$$Cl_2(g) + 6H_2O \rightarrow 2ClO_3^-(aq) + 12H^+(aq)$$

To balance charge (0 on the left, $+10$ on the right at this point), we add $10e^-$ to the right

$$Cl_2(g) + 6H_2O \rightarrow 2ClO_3^-(aq) + 12H^+(aq) + 10e^-$$

2. $Cl_2(g) + 2e^- \rightarrow 2Cl^-(aq)$

    $Cl_2(g) + 6H_2O \rightarrow 2ClO_3^-(aq) + 12H^+(aq) + 10e^-$

We now combine the half-equations in such a way as to eliminate electrons. To do this, we multiply the reduction half-equation by 5 and add it to the oxidation half-equation:

3. $5Cl_2(g) + 10e^- \rightarrow 10Cl^-(aq)$

    $\underline{Cl_2(g) + 6H_2O \rightarrow 2ClO_3^-(aq) + 12H^+(aq) + 10e^-}$

    $6Cl_2(g) + 6H_2O \rightarrow 10Cl^-(aq) + 2ClO_3^-(aq) + 12H^+(aq)$

This equation can be simplified by dividing all the coefficients by two. Doing this, we obtain

$$3Cl_2(g) + 3H_2O \rightarrow 5Cl^-(aq) + ClO_3^-(aq) + 6H^+(aq)$$

When asked to balance a redox equation, check to make sure that your final equation is the one with the smallest whole-number coefficients. Sometimes, as here, a simplification is possible.

Exercise

Write the balanced equation for this reaction in basic solution. Answer:
$3Cl_2(g) + 6OH^-(aq) \rightarrow 5Cl^-(aq) + ClO_3^-(aq) + 3H_2O$.

Reactions such as the one considered in Example 12.5, where the same species is both oxidized and reduced, are given a special name. They are called *disproportionation* reactions. In all disproportionations, part of the reactant increases in oxidation number while part decreases in oxidation number.

12.4
SOLUTION STOICHIOMETRY

We have looked at three different kinds of reactions involving ions in water solution. So far, our approach has been qualitative; we have concentrated upon writing and balancing equations to represent these reactions. Now we shift to a quantitative approach, looking at the stoichiometry (mass relations) associated with reactions in water solution. To start with, we will consider how to calculate the amount of product formed, knowing the volumes and molarities of the reacting solutions. The approach we will follow is, with minor modifications, that used in Chapter 3. It is outlined in some detail in the flow chart shown in Figure 12.6, p. 414.

■ **EXAMPLE 12.6** ────────────────────────────

When aqueous solutions of sodium hydroxide and iron(III) nitrate are mixed, a red gelatinous precipitate forms. Calculate the mass of precipitate formed when 50.00 mL of 0.200 M NaOH and 30.00 mL of 0.125 M $Fe(NO_3)_3$ are mixed.

Solution

1. Since the problem says that a precipitate is formed, this is a precipitation reaction involving the ions Na^+, OH^-, Fe^{3+}, and NO_3^-.

2. Possible products are $NaNO_3$, NaOH, $Fe(OH)_3$, $Fe(NO_3)_3$. All the possible products are soluble except $Fe(OH)_3$. The precipitation reaction is

$$Fe^{3+}(aq) + 3OH^-(aq) \rightarrow Fe(OH)_3(s)$$

3. We calculate the number of moles of each reactant:

$$\text{moles } Fe^{3+} = 0.03000 \text{ L } Fe(NO_3)_3 \times \frac{0.125 \text{ mol } Fe(NO_3)_3}{1 \text{ L } Fe(NO_3)_3} \times \frac{1 \text{ mol } Fe^{3+}}{1 \text{ mol } Fe(NO_3)_3}$$

$$= 3.75 \times 10^{-3} \text{ mol } Fe^{3+}$$

Note from the formula of iron(III) nitrate that there is one mole of Fe^{3+} per mole of $Fe(NO_3)_3$.

$$\text{moles } OH^- = 0.05000 \text{ L NaOH} \times \frac{0.200 \text{ mol NaOH}}{1 \text{ L NaOH}} \times \frac{1 \text{ mol } OH^-}{1 \text{ mol NaOH}}$$

$$= 1.00 \times 10^{-2} \text{ mol } OH^-$$

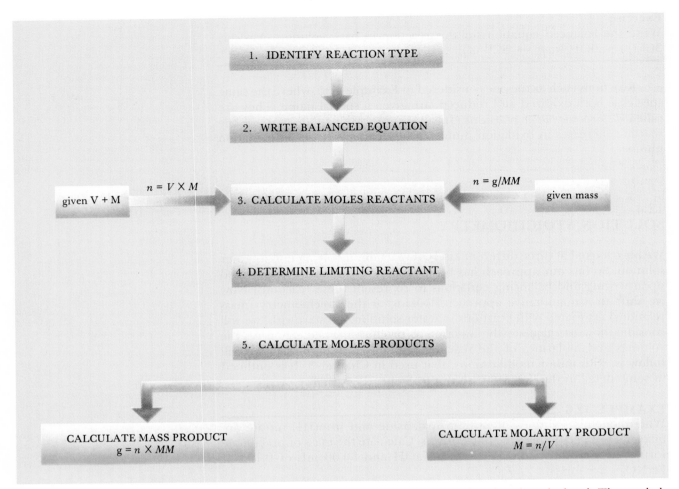

Figure 12.6 Stoichiometry flow chart for solution reactions where amount of product is to be calculated. The symbols have the following meanings: n = no. moles, M = molarity, V = volume in liters, g = mass in grams, MM = molar mass.

4. If Fe^{3+} is limiting:

$$\text{moles Fe(OH)}_3 = 3.75 \times 10^{-3} \text{ mol Fe}^{3+} \times \frac{1 \text{ mol Fe(OH)}_3}{1 \text{ mol Fe}^{3+}}$$

$$= 3.75 \times 10^{-3} \text{ mol Fe(OH)}_3$$

If OH^- is limiting:

$$\text{moles Fe(OH)}_3 = 1.00 \times 10^{-2} \text{ mol OH}^- \times \frac{1 \text{ mol Fe(OH)}_3}{3 \text{ mol OH}^-}$$

$$= 3.33 \times 10^{-3} \text{ mol Fe(OH)}_3$$

Since 3.33×10^{-3} is less than 3.75×10^{-3}, OH^- is the limiting reactant.

5. 3.33×10^{-3} mol of $Fe(OH)_3$ is formed.

6. We convert to the required unit, grams:

$$3.33 \times 10^{-3} \text{ mol Fe(OH)}_3 \times \frac{106.9 \text{ g Fe(OH)}_3}{1 \text{ mol Fe(OH)}_3} = \boxed{0.356 \text{ g Fe(OH)}_3}$$

Exercise

What mass of precipitate is formed if 100.0 mL of 0.150 M magnesium nitrate is added to 50.0 mL of 0.0150 M barium hydroxide? Answer: 0.0437 g.

■ **EXAMPLE 12.7**

Barium hydroxide, $Ba(OH)_2$, reacts with acetic acid, $HC_2H_3O_2$. Calculate the concentrations of all ions and molecules, except H_2O, after reaction if 75.00 mL of 0.0200 M $Ba(OH)_2$ is added to 22.68 mL of 0.500 M $HC_2H_3O_2$. Assume the final volume is the sum of the initial volumes.

Solution

1. This is a reaction between a strong base, $Ba(OH)_2$, and a weak acid, $HC_2H_3O_2$.

2. The equation for the reaction is

$$OH^-(aq) + HC_2H_3O_2(aq) \rightarrow H_2O + C_2H_3O_2^-(aq)$$

3. The moles of the reacting species are

$$\text{moles OH}^- = 0.07500 \text{ L Ba(OH)}_2 \times \frac{0.0200 \text{ mol Ba(OH)}_2}{1 \text{ L Ba(OH)}_2} \times \frac{2 \text{ mol OH}^-}{1 \text{ mol Ba(OH)}_2}$$

$$= 0.00300 \text{ mol OH}^-$$

$$\text{moles HC}_2\text{H}_3\text{O}_2 = 0.02268 \text{ L HC}_2\text{H}_3\text{O}_2 \times \frac{0.500 \text{ } M \text{ HC}_2\text{H}_3\text{O}_2}{1 \text{ L HC}_2\text{H}_3\text{O}_2}$$

$$= 0.0113 \text{ mol HC}_2\text{H}_3\text{O}_2$$

4. We look for the limiting reactant:

$$OH^-: \quad 0.00300 \text{ mol OH}^- \times \frac{1 \text{ mol C}_2\text{H}_3\text{O}_2^-}{1 \text{ mol OH}^-} = 0.00300 \text{ mol C}_2\text{H}_3\text{O}_2^-$$

$$HC_2H_3O_2: \quad 0.0113 \text{ mol HC}_2\text{H}_3\text{O}_2 \times \frac{1 \text{ mol C}_2\text{H}_3\text{O}_2^-}{1 \text{ mol HC}_2\text{H}_3\text{O}_2} = 0.0113 \text{ mol C}_2\text{H}_3\text{O}_2^-$$

Since $0.00300 < 0.0113$, the limiting reactant is OH^-.

5. We now calculate the number of moles of each species after reaction.

$C_2H_3O_2^-$: 0.00300 mol

OH^-: ~0 mol (It is the limiting reactant, so it is all used up.)

$HC_2H_3O_2$: 1 mol $HC_2H_3O_2$ is equivalent to 1 mol $C_2H_3O_2^-$. Thus 0.00300 mol $HC_2H_3O_2$ is used up to make 0.00300 mol $C_2H_3O_2^-$. Mol $HC_2H_3O_2$ left = $0.0113 - 0.00300 = 0.0083$ mol

Ba^{2+}: $0.0200 \dfrac{\text{mol}}{\text{L}} \times 0.07500 \text{ L} = 1.50 \times 10^{-3}$ mol. Note that since Ba^{2+}

is a spectator ion, the final amount of Ba^{2+} is equal to the original amount.

6. We convert to the desired unit, molarity, using the relation

$$\text{Molarity} = \frac{\text{no. of moles}}{V \text{ in liters}}$$

This example covers the waterfront of solution stoichiometry

The volume is: $75.00 \text{ mL} + 22.68 \text{ mL} = 97.68 \text{ mL}$

conc. OH^-: $0 \; M$

conc. $C_2H_3O_2^-$: $\dfrac{0.00300 \text{ mol}}{0.09768 \text{ L}} = \; 0.0307 \; M$

conc. $HC_2H_3O_2$: $\dfrac{0.0083 \text{ mol}}{0.09768 \text{ L}} = \; 0.085 \; M$

conc. Ba^{2+}: $\dfrac{0.00150 \text{ mol}}{0.09768 \text{ L}} = \; 0.0154 \; M$

Exercise

Magnesium hydroxide is a substance commonly used in antacids. If an antacid tablet contains 200.0 mg of $Mg(OH)_2$, what volume of $0.100 \; M$ HCl is required to completely react with the antacid? Answer: 68.6 mL.

In many reactions, we are concerned not with the amount of product formed, but rather with the amount of one reactant required to react with a known amount of the other reactant. Calculations of this type are really quite simple; all you need do is to follow the path outlined in the flow chart shown in Figure 12.7, p. 417.

■ EXAMPLE 12.8

A solution of permanganate ion, MnO_4^-, reacts with iron(II) in solution to give manganese(II) and iron(III) ions. How many milliliters of $0.684 \; M$ $KMnO_4$ solution will react completely with 27.50 mL of $0.250 \; M$ $Fe(NO_3)_2$?

Solution

1. The reaction is a redox reaction.
2. This equation was balanced in Section 12.3, where we obtained:

$$8H^+(aq) + MnO_4^-(aq) + 5Fe^{2+}(aq) \rightarrow 5Fe^{3+}(aq) + Mn^{2+}(aq) + 4H_2O$$

3. We have all the information we need to calculate the number of moles of Fe^{2+}

$$\text{moles } Fe^{2+} = 0.02750 \text{ L Fe(NO}_3)_2 \times \frac{0.250 \text{ mol Fe(NO}_3)_2}{1 \text{ L Fe(NO}_3)_2} \times \frac{1 \text{ mol Fe}^{2+}}{1 \text{ mol Fe(NO}_3)_2}$$

$$= 6.88 \times 10^{-3} \text{ mol Fe}^{2+}$$

4. From the coefficients of the balanced equation, we can readily calculate the number of moles of MnO_4^-:

$$\text{moles } MnO_4^- = 0.00688 \text{ mol } Fe^{2+} \times \frac{1 \text{ mol } MnO_4^-}{5 \text{ mol } Fe^{2+}} = 0.00138 \text{ mol } MnO_4^-$$

5. Finally, we convert to the desired unit, milliliters of $KMnO_4$

$$0.00138 \text{ mol } MnO_4^- \times \frac{1 \text{ mol } KMnO_4}{1 \text{ mol } MnO_4^-} = 0.00138 \text{ mol } KMnO_4$$

$$\text{Volume} = 0.00138 \text{ mol } KMnO_4 \times \frac{1 \text{ L}}{0.684 \text{ mol } KMnO_4}$$

$$= 0.00202 \text{ L} = \boxed{2.02 \text{ mL}}$$

Exercise

When household bleach, a solution of NaClO, is treated with excess iodide ions in acid, the hypochlorite ions, ClO^-, are reduced to chloride ions and the iodide ions are oxidized to I_3^-. Balance the redox equation involved and calculate the percent by mass of NaClO in bleach if 42.4 mL of 2.50 M KI is required to react completely with 50.0 g of bleach. Answer: 5.26%.

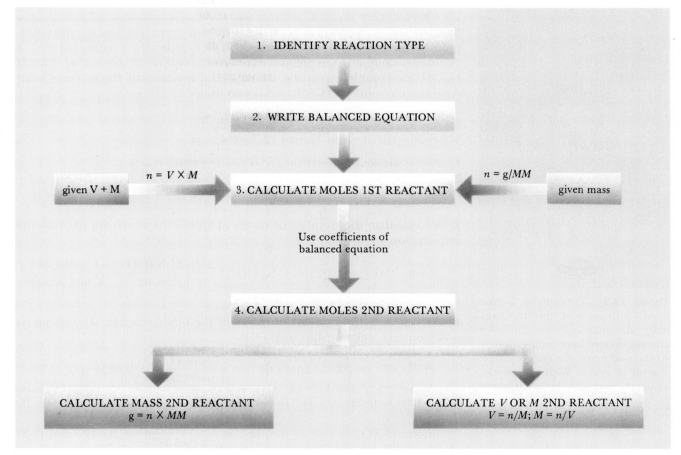

Figure 12.7 Stoichiometry flow chart for solution reactions where amount of reactant is to be determined.

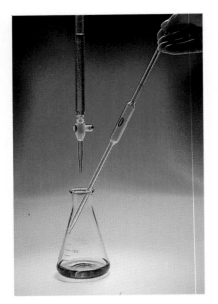

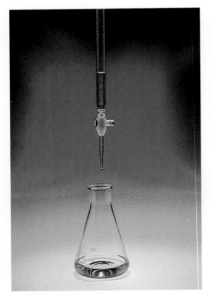

Figure 12.8 Titration of a base (flask) with an acid (buret). Originally, the indicator has the color characteristic of basic solution (blue). At the end point, the color changes sharply to green. With excess acid, we get the acid color, yellow. (Charles D. Winters)

12.5
SOLUTION REACTIONS IN VOLUMETRIC ANALYSIS

Reactions involving aqueous solutions are extremely useful in **quantitative analysis**, where we determine the concentration of a species in solution or its percent in a solid mixture. In this section, we will focus upon a branch of quantitative analysis called *volumetric analysis*, which depends primarily upon volume measurements.

A volumetric analysis involves a **titration**, where we measure the volume of a standard solution of known concentration required to react with a measured amount of a sample. The *titrant* is delivered from a buret, which accurately measures the volume delivered (Fig. 12.8). The point in the titration at which just enough titrant has been added to react completely with the substance being analyzed for is called the **equivalence point**. This point is usually detected using an indicator that changes color at the *end point*, telling us it's time to stop adding titrant.

Calculations for titrations are very similar to those described in Section 12.4; indeed, Figure 12.7 also applies very nicely here.

EXAMPLE 12.9

Sulfuric acid is used in car batteries. It is important to know its concentration, because a low concentration of H_2SO_4 indicates a dying battery. A 10.00-mL sample taken from a car battery requires 25.68 mL of 3.26 M NaOH for complete reaction. What is the molarity of the sulfuric acid? Assume one mole of H_2SO_4 yields two moles of H^+ ions.

Solution

We follow the steps in Figure 12.7.

1. This is an acid-base (strong acid–strong base) reaction.
2. The equation for the reaction is

$$H^+(aq) + OH^-(aq) \rightarrow H_2O$$

3. We calculate the number of moles of OH^-, for which we have all the information we need:

$$\text{moles } OH^- = 0.02568 \text{ L NaOH} \times \frac{3.26 \text{ } M \text{ NaOH}}{1 \text{ L NaOH}} \times \frac{1 \text{ mol } OH^-}{1 \text{ mol NaOH}}$$

$$= 0.0837 \text{ mol } OH^-$$

4. We calculate the number of moles of the other reactant (H^+) from its stoichiometric relationship with OH^-:

$$\text{moles } H^+ = 0.0837 \text{ mol } OH^- \times \frac{1 \text{ mol } H^+}{1 \text{ mol } OH^-} = 0.0837 \text{ mol } H^+$$

5. We now convert mol H^+ to the desired unit, M H_2SO_4:

$$\text{moles } H_2SO_4 = 0.0837 \text{ mol } H^+ \times \frac{1 \text{ mol } H_2SO_4}{2 \text{ mol } H^+} = 0.0419 \text{ mol } H_2SO_4$$

$$M = \frac{0.0419 \text{ mol } H_2SO_4}{0.010000 \text{ L}} = \boxed{4.19 \text{ mol/L}}$$

Exercise

Vinegar is an aqueous solution of acetic acid ($HC_2H_3O_2$). A 10.00-mL sample was titrated with KOH. It took 42.10 mL of 0.200 M KOH to reach the equivalence point. What is the molarity of the acetic acid solution? Answer: 0.842 M.

■ **EXAMPLE 12.10** _____

The percent of tin (Sn) in an alloy can be determined by treating the alloy with acid, which converts tin metal to Sn^{2+} ions in solution. These are titrated with Ce^{4+} ions, producing tin(IV) and Ce(III) ions. A 2.00-g sample of the alloy was dissolved and titrated with a solution of cerium nitrate, $Ce(NO_3)_4$. It required 38.46 mL of 0.2164 M $Ce(NO_3)_4$ to react completely with 45.00 mL of the solution of dissolved alloy. What is the percent tin in the alloy?

Solution

1. This is a redox reaction.
2. The balanced equation for this reaction is

$$2Ce^{4+}(aq) + Sn^{2+}(aq) \rightarrow Sn^{4+}(aq) + 2Ce^{3+}(aq)$$

3. We have enough information to calculate the number of moles of Ce^{4+}.

$$\text{moles } Ce^{4+} = 0.03846 \text{ L } Ce(NO_3)_4 \times \frac{0.2164 \text{ mol } Ce(NO_3)_4}{1 \text{ L } Ce(NO_3)_4} \times \frac{1 \text{ mol } Ce^{4+}}{1 \text{ mol } Ce(NO_3)_4}$$

$$= 0.008323 \text{ mol } Ce^{4+}$$

4. We calculate the number of moles of the other reactant based on the stoichiometric relationship from the balanced equation:

$$\text{moles } Sn^{2+} = 0.008323 \text{ mol } Ce^{4+} \times \frac{1 \text{ mol } Sn^{2+}}{2 \text{ mol } Ce^{4+}} = 0.004161 \text{ mol } Sn^{2+}$$

5. We convert moles of Sn^{2+} to the desired unit, % Sn. To find a mass percent, we need to know the mass of the tin and the mass of the sample. The mass of the sample is given. The mass of tin can be obtained by changing moles of Sn^{2+} to grams of Sn.

$$\text{g Sn} = 0.004161 \text{ mol } Sn^{2+} \times \frac{1 \text{ mol Sn}}{1 \text{ mol } Sn^{2+}} \times \frac{118.71 \text{g Sn}}{1 \text{ mol Sn}}$$

$$= 0.4940 \text{ g Sn}$$

$$\% \text{ Sn} = \frac{0.4940 \text{ g Sn}}{2.00\text{-g sample}} \times 100 = \boxed{24.7}$$

Exercise

Oxalic acid, $H_2C_2O_4$, is sometimes used as a bleach. It reacts with MnO_4^- ion, yielding Mn^{2+} and $CO_2(g)$. A solution of oxalic acid requires 24.86 mL of 0.1500 M $KMnO_4$ for complete reaction. How many grams of oxalic acid are there in the solution? Answer: 0.839 g.

■ SUMMARY PROBLEM

Nitric acid, HNO_3, and strontium hydroxide, $Sr(OH)_2$, are a strong acid and a strong base, respectively. They participate in a variety of acid-base reactions with other species. In addition, nitric acid is often involved in redox reactions because the NO_3^- ion is readily reduced. Finally, strontium hydroxide can form precipitates involving either the Sr^{2+} ion or the OH^- ion.

a. Write the equation for the reaction between solutions of
   (1) strontium hydroxide and aluminum nitrate.
   (2) nitric acid and ammonia.
   (3) strontium hydroxide and nitric acid.
   (4) strontium hydroxide and acetic acid ($HC_2H_3O_2$).
   (5) nitric acid and Cu(s) forming nitrogen oxide gas and copper(II) ion.
b. When 15.00 mL of 0.2050 M strontium hydroxide reacts with 22.00 mL of 0.1380 M cobalt(III) sulfate, two precipitates form. Calculate the mass of each precipitate and the concentrations of all ions after reaction, assuming the final volume is the sum of the initial volumes.
c. Nitric acid is used to analyze a sulfide ore; S^{2-} ions are oxidized to sulfur, while NO_3^- ions are reduced to NO(g). It is found that 64.5 mL of a solution 0.125 M in NO_3^- is required to react exactly with a 1.00-g sample of ore. Calculate the percent of sulfur in the ore.
d. To determine the molarity of a solution of hydrobromic acid, it is titrated with 0.2384 M strontium hydroxide. It is determined that 20.50 mL of strontium hydroxide is required to react with 33.20 mL of hydrobromic acid. What is the molarity of the hydrobromic acid?

Answers
a. (1) $Al^{3+}(aq) + 3OH^-(aq) \rightarrow Al(OH)_3(s)$
(2) $H^+(aq) + NH_3(aq) \rightarrow NH_4^+(aq)$ **(3)** $H^+(aq) + OH^-(aq) \rightarrow H_2O$
(4) $HC_2H_3O_2(aq) + OH^-(aq) \rightarrow H_2O + C_2H_3O_2^-(aq)$
(5) $2NO_3^-(aq) + 8H^+(aq) + 3Cu(s) \rightarrow 2NO(g) + 4H_2O + 3Cu^{2+}(aq)$
b. ~0 moles of Sr^{2+} and OH^-, 0.5646 g $SrSO_4$, 0.2254 g $Co(OH)_3$; 0.1631 M SO_4^{2-}; 0.1087 M Co^{3+} **c.** 38.8%   **d.** 0.2944 M

I f you've ever wandered into the general chemistry storeroom, you've probably seen rows upon rows of reagent bottles filled with white or colored solids, bearing such labels as "silver iodide, AgI", "copper(II) nitrate, $Cu(NO_3)_2$", or "magnesium chloride, $MgCl_2$." These compounds and indeed most ionic solids are made by the types of reactions discussed in this chapter.

Precipitation Reactions

Insoluble ionic solids are readily prepared by simple precipitation. For example, silver iodide, a compound used in photography, is made by mixing solutions of silver nitrate and potassium iodide (Fig. 12.9). The reaction is:

$$Ag^+(aq) + I^-(aq) \rightarrow AgI(s)$$

Filtration and drying of the precipitate gives high-purity silver iodide.

You may be surprised to learn that precipitation reactions can be used to prepare soluble ionic solids. To illustrate the principle involved, suppose you wish to convert a sample of rubidium chloride, RbCl, to rubidium nitrate, $RbNO_3$. One way to do this is to add an equivalent amount of silver nitrate solution. A precipitation reaction occurs; AgCl(s) forms, leaving Rb^+ and NO_3^- ions in solution:

$$Rb^+(aq) + Cl^-(aq) + Ag^+(aq) + NO_3^-(aq) \rightarrow AgCl(s) + Rb^+(aq) + NO_3^-(aq)*$$

The silver chloride is removed by filtration; evaporation of the solution remaining gives solid rubidium nitrate:

$$Rb^+(aq) + NO_3^-(aq) \rightarrow RbNO_3(s)$$

---

*This equation contains spectator ions, because they become participants in the next reaction.

Figure 12.9 Silver iodide can be made by a simple precipitation reaction, mixing solutions containing Ag^+ ions and I^- ions. (Charles D. Winters)

Acid-Base Reactions

This type of reaction offers a general method for preparing the class of inorganic solids known as *salts*. Indeed, historically, salts were defined to be the solids produced by acid-base reactions. Today, we define a salt to be any ionic compound in which the anion is *not* OH^- or O^{2-}. Examples of salts include $NaCl$, $Cu(NO_3)_2$, and $Al_2(SO_4)_3$ (but not $NaOH$ or CaO).

Let us suppose that you need to prepare calcium bromide, starting with calcium hydroxide. One way to accomplish this is to neutralize the hydroxide with an equivalent amount of hydrobromic acid:

$$Ca^{2+}(aq) + 2OH^-(aq) + 2H^+(aq) + 2Br^-(aq) \rightarrow Ca^{2+}(aq) + 2Br^-(aq) + 2H_2O$$

Evaporation of the resulting solution gives solid $CaBr_2$:

$$Ca^{2+}(aq) + 2Br^-(aq) \rightarrow CaBr_2(s)$$

Adding these two equations gives the overall reaction:

$$Ca^{2+}(aq) + 2OH^-(aq) + 2H^+(aq) + 2Br^-(aq) \rightarrow CaBr_2(s) + 2H_2O$$

Insoluble hydroxides or oxides can also be used. Thus we can prepare $Cu(NO_3)_2$ by treating either $Cu(OH)_2$ or CuO with dilute nitric acid and evaporating the resulting solution. The overall reactions are:

$$Cu(OH)_2(s) + 2H^+(aq) + 2NO_3^-(aq) \rightarrow Cu(NO_3)_2(s) + 2H_2O$$

$$CuO(s) + 2H^+(aq) + 2NO_3^-(aq) \rightarrow Cu(NO_3)_2(s) + H_2O$$

The experimental procedure followed is indicated in Figure 12.10.

Figure 12.10 To prepare $Cu(NO_3)_2$, we start by adding CuO (black) to nitric acid, HNO_3. Formation of a blue solution indicates that reaction has occurred to form hydrated Cu^{2+} ions. Evaporation of the resulting solution gives crystals of $Cu(NO_3)_2 \cdot 6H_2O$, which are blue. (Charles D. Winters)

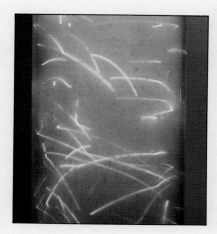

Figure 12.11 When powdered antimony is added to chlorine gas, a vigorous redox reaction occurs, as shown by the flashes of light. The equation for the reaction is: $2Sb(s) + 3Cl_2(g) \rightarrow 2SbCl_3(s)$. (Charles D. Winters)

Oxidation-Reduction Reactions

Redox reactions are more widely used than any other type to prepare inorganic compounds. In particular, they are readily applied to make binary ionic compounds (Fig. 12.11). Suppose, for example, you want to make magnesium chloride, $MgCl_2$, starting with magnesium metal. Two quite different redox reactions can be used.

1. React magnesium metal directly with chlorine gas.

$$Mg(s) + Cl_2(g) \rightarrow MgCl_2(s)$$

The white, water-soluble product, containing Mg^{2+} and Cl^- ions, is easily separated from any unreacted metal, which is insoluble in water.

2. React magnesium with hydrochloric acid

$$Mg(s) + 2H^+(aq) + 2Cl^-(aq) \rightarrow Mg^{2+}(aq) + 2Cl^-(aq) + H_2(g)$$

The solution formed is evaporated to obtain the solid salt.

$$Mg^{2+}(aq) + 2Cl^-(aq) \rightarrow MgCl_2(s)$$

QUESTIONS AND PROBLEMS

Precipitation Reactions and Solubility

1. Write the formulas of the following compounds. Using Table 12.1, decide which of them are soluble in water.
  a. calcium chloride   b. ammonium sulfide
  c. barium sulfate   d. iron(II) hydroxide

2. Follow the instructions for Question 1 for the following compounds:
  a. potassium carbonate   b. magnesium hydroxide
  c. cerium(III) nitrate   d. nickel(II) sulfate

3. Name the reagent that you would add to a solution of nickel(II) nitrate to precipitate
  a. nickel(II) carbonate   b. nickel(II) sulfide
  c. nickel(II) hydroxide

4. Describe how you would prepare

  a. copper(II) hydroxide from a solution of copper(II) chloride.
  b. strontium sulfate from sulfuric acid.
  c. mercury(I) chloride from a solution of mercury(I) nitrate, $Hg_2(NO_3)_2$.

5. Write net ionic equations for the formation of
  a. a green precipitate when solutions of nickel(II) nitrate and sodium hydroxide are mixed.
  b. a white precipitate (that is soluble in acid) when potassium hydroxide and magnesium chloride are mixed.

6. Write net ionic equations to explain the formation of
  a. a red precipitate when solutions of iron(III) chloride and sodium hydroxide are mixed.
  b. two different precipitates, one yellow and the other white, when solutions of cadmium(II) sulfate and strontium sulfide are mixed.

7. Using Table 12.1, decide whether a precipitate will form when the following solutions are mixed. If a precipitate forms, write a net ionic equation for the reaction.
a. aluminum sulfate and sodium chloride
b. lead(II) nitrate and ammonium chloride
c. barium nitrate and chromium(III) sulfate
d. potassium nitrate and sodium hydroxide
e. arsenic(III) chloride and sodium sulfide

8. Follow the directions of Question 7 for solutions of
a. copper(II) chloride and sodium hydroxide
b. copper(II) sulfate and potassium sulfide
c. mercury(I) nitrate, $Hg_2(NO_3)_2$, and hydrochloric acid
d. iron(III) sulfate and barium hydroxide
e. nickel(II) nitrate and ammonium carbonate

9. Write a net ionic equation for any precipitation reaction that occurs when 0.1 M solutions of the following are mixed:
a. sodium carbonate and barium chloride
b. zinc(II) sulfate and ammonium sulfide
c. calcium chloride and silver nitrate
d. strontium sulfide and potassium hydroxide

10. Follow the directions for Question 9 for the following pairs of solutions:
a. scandium(III) chloride and nickel(II) nitrate
b. sodium carbonate and calcium chloride
c. ammonium sulfide and lead(II) nitrate
d. iron(III) nitrate and potassium hydroxide

Acid-Base Reactions

11. For an acid-base reaction, what is the reacting species in the following acids?
a. nitric acid   b. hydrofluoric acid
c. perchloric acid   d. sulfurous acid (H_2SO_3)
e. propionic acid ($HC_3H_5O_2$)

12. For an acid-base reaction, what is the reacting species in the following acids?
a. nitrous acid (HNO_2)   b. sulfuric acid
c. acetic acid ($HC_2H_3O_2$)   d. hydrochloric acid
e. lactic acid ($HC_3H_5O_3$)

13. For an acid-base reaction, what is the reacting species in the following bases?
a. sodium hydroxide   b. ethylamine ($C_2H_5NH_2$)
c. ammonia   d. barium hydroxide

14. For an acid-base reaction, what is the reacting species in the following bases?
a. lithium hydroxide   b. dimethyl amine, $(CH_3)_2NH$
c. strontium hydroxide   d. pyridine, (C_5H_5N)

15. Classify the following compounds as acids or bases, weak or strong:
a. hydrobromic acid   b. cesium hydroxide
c. phosphoric acid   d. trimethyl amine, $(CH_3)_3N$

16. Follow the directions for Question 15 for
a. hypochlorous acid ($HClO$)
b. barium hydroxide

c. ammonia
d. hydrochloric acid

17. Write a net ionic equation for each of the following acid-base reactions in water:
a. butyric acid ($HC_4H_7O_2$) with lithium hydroxide
b. ammonia with hydriodic acid
c. hydrofluoric acid with strontium hydroxide

18. Write a balanced net ionic equation for each of the following acid-base reactions in water:
a. perchloric acid and potassium hydroxide
b. nitrous acid (HNO_2) and sodium hydroxide
c. aniline ($C_6H_5NH_2$) and sulfuric acid

19. Consider the equation: $H^+(aq) + OH^-(aq) \rightarrow H_2O$. For which of the following pairs would this be the correct equation for the acid-base reaction? If it is not correct, write the proper equation for the acid-base reaction between the pair.
a. hydrochloric acid and calcium hydroxide
b. HBr and CH_3NH_2
c. nitric acid and ammonia
d. H_2SO_4 and NaOH
e. perchloric acid and potassium hydroxide
f. HF and barium hydroxide

20. Follow the directions of Question 19 for the following pairs:
a. nitric acid and $C_2H_5NH_2$
b. perchloric acid and cesium hydroxide
c. $HC_2H_3O_2$ and LiOH
d. sulfuric acid and calcium hydroxide
e. barium hydroxide and hydriodic acid

Redox Reactions

21. Give the oxidation number of each atom in
a. hydrogen carbonate ion (HCO_3^-)
b. magnesium sulfate
c. sulfur hexafluoride
d. iodic acid (HIO_3)
e. sodium molybdate (Na_2MoO_4)

22. Give the oxidation number of each atom in
a. nitrite ion (NO_2^-)   b. TeF_8^{2-}
c. dinitrogen trioxide   d. thiosulfate ion ($S_2O_3^{2-}$)
e. perchlorate ion

23. Give the oxidation number of each atom in
a. Sb_4O_{10} b. HPO_3^{2-} c. RuF_5 d. $C_2H_6O_2$

24. Give the oxidation number of each atom in
a. CaC_2O_4 b. HSO_4^- c. $Na_2Fe_2O_4$ d. NOF
e. N_2H_4

25. Classify each of the following half-equations as oxidation or reduction:
a. $Ca(s) \rightarrow Ca^{2+}(aq)$ b. $OH^-(aq) \rightarrow O_2(g)$
c. $NO_3^-(aq) \rightarrow NO(g)$ d. $AuCl_4^-(aq) \rightarrow AuCl_2^-(aq)$

26. Classify each of the following half-equations as oxidation or reduction:
a. $Co^{3+}(aq) \rightarrow Co^{2+}(aq)$ b. $Cl_2(g) \rightarrow ClO_3^-(aq)$
c. $Fe^{3+}(aq) \rightarrow Fe^{2+}(aq)$ d. $Hg(l) \rightarrow Hg_2^{2+}(aq)$

27. For each unbalanced equation given below, identity the species oxidized and the species reduced; identify the oxidizing agent and the reducing agent.
  a. $Mg(s) + O_2(g) \rightarrow MgO(s)$
  b. $Cr_2O_7^{2-}(aq) + Sn^{2+}(aq) \rightarrow Cr^{3+}(aq) + Sn^{4+}(aq)$

28. For each unbalanced equation given below, identify the species oxidized and the species reduced; identify the oxidizing agent and the reducing agent.
  a. $FeS(s) + NO_3^-(aq) + H^+(aq) \rightarrow$
$$NO(g) + SO_4^{2-}(aq) + Fe^{2+}(aq) + H_2O$$
  b. $C_2H_4(g) + O_2(g) \rightarrow CO_2(g) + H_2O(l)$

29. Balance the equations in Question 27.

30. Balance the equations in Question 28.

31. Write balanced equations for the following reactions in acid solution.
  a. $Hg^{2+}(aq) + Cu(s) \rightarrow Hg(l) + Cu^{2+}(aq)$
  b. $Zn(s) + VO_3^-(aq) \rightarrow Zn^{2+}(aq) + V^{2+}(aq)$
  c. $H_2O_2(aq) + Cr_2O_7^{2-}(aq) \rightarrow Cr^{3+}(aq) + O_2(g)$
  d. $MnO_2(s) + Cl^-(aq) \rightarrow Mn^{2+}(aq) + Cl_2(g)$
  e. $IO_3^-(aq) + I^-(aq) \rightarrow I_3^-(aq)$

32. Write balanced equations for the following reactions in acid solution.
  a. $P_4(s) \rightarrow PH_3(g) + HPO_3^{2-}(aq)$
  b. $H_3AsO_3(aq) + BrO_3^-(aq) \rightarrow H_3AsO_4(aq) + Br^-(aq)$
  c. $MnO_4^-(aq) + HSO_3^-(aq) \rightarrow Mn^{2+}(aq) + SO_4^{2-}(aq)$
  d. $Sn^{2+}(aq) + O_2(g) \rightarrow Sn^{4+}(aq) + H_2O$
  e. $Pt(s) + NO_3^-(aq) + Cl^-(aq) \rightarrow PtCl_6^{2-}(aq)$
$$+ NO(g) + H_2O$$

33. Write balanced net ionic equations for the following reactions in acid solution:
  a. Solid phosphorus (P_4) reacts with hypochlorous acid (HClO) to form phosphoric acid, H_3PO_4, and chloride ion.
  b. Tellurium, Te, is oxidized by nitrate ion to form solid tellurium dioxide and $NO(g)$.
  c. An aqueous solution of bromine is reduced to Br^-; at the same time iodide ions are oxidized to iodate ions, IO_3^-.

34. Write balanced net ionic equations for the following reactions in acid solution:
  a. Silver is dissolved in nitric acid, forming aqueous silver nitrate and nitrogen dioxide gas.
  b. Solid copper(II) sulfide is dissolved in nitric acid, forming copper(II) nitrate (aq), sulfur, and $NO(g)$.
  c. Tin(II) ion reacts with periodate ion (IO_4^-), yielding iodide ion and tin(IV) ion.

35. Write balanced equations for the following reactions in basic solution:
  a. $ClO^-(aq) + CrO_2^-(aq) \rightarrow Cl^-(aq) + CrO_4^{2-}(aq)$
  b. $Al(s) + H_2O \rightarrow Al(OH)_4^-(aq) + H_2(g)$
  c. $Ni^{2+}(aq) + Br_2(l) \rightarrow NiO(OH)(s) + Br^-(aq)$

36. Write balanced equations for the following reactions in basic solution:
  a. $S_2O_3^{2-}(aq) + I_2(s) \rightarrow SO_4^{2-}(aq) + I^-(aq)$

  b. $CN^-(aq) + MnO_4^-(aq) \rightarrow CNO^-(aq) + MnO_2(s)$
  c. $Cr(OH)_3(s) + ClO_3^-(aq) \rightarrow CrO_4^{2-}(aq) + Cl^-(aq)$

Solution Stoichiometry

37. A precipitate forms when solutions of silver nitrate and scandium(III) chloride are mixed:
$$Ag^+(aq) + Cl^-(aq) \rightarrow AgCl(s)$$
  a. What volume of 0.0385 M scandium(III) chloride is required to completely react with 22.0 mL of 0.130 M silver nitrate?
  b. What mass in grams of silver chloride is formed?

38. Mixing solutions of sodium oxalate, $Na_2C_2O_4$, and lanthanum(III) chloride precipitates lanthanum(III) oxalate:
$$3C_2O_4^{2-}(aq) + 2La^{3+}(aq) \rightarrow La_2(C_2O_4)_3(s)$$
  a. What is the molarity of a solution of lanthanum(III) chloride if 25.0 mL is required to react with 13.85 mL of 0.0225 M sodium oxalate?
  b. How many grams of precipitate are formed?

39. What volume of 0.100 M lead nitrate is required to precipitate completely
  a. 25.0 mL of 0.0832 M nickel(II) sulfate?
  b. 55.8 mL of 0.222 M hydrochloric acid?
  c. 18.7 mL of 0.389 M potassium chromate?

40. What volume of 0.0750 M cobalt(II) sulfate is required to react completely with
  a. 28.7 mL of 0.183 M sodium hydroxide?
  b. 42.5 mL of 0.189 M barium nitrate?
  c. 68.4 mL of 0.273 M ammonium sulfide?

41. A student finds that 38.4 mL of 0.215 M hydrochloric acid is required to neutralize a 20.0-mL sample of barium hydroxide. What is the molarity of $Ba(OH)_2$?

42. What volume of 0.317 M potassium hydroxide is required to neutralize 32.0 mL of 0.164 M nitric acid?

43. Determine the volume of 0.250 M hydrochloric acid required to titrate
  a. 30.0 mL of 0.278 M lithium hydroxide.
  b. 17.6 mL of 0.0162 M strontium hydroxide.
  c. 15.0 mL of a solution ($d = 0.958$ g/cm^3) containing 10.0% by mass of NH_3.

44. Calculate the volume of 0.309 M sodium hydroxide required to titrate
  a. 15.9 mL of 0.190 M hydrofluoric acid.
  b. 22.9 mL of 0.296 M perchloric acid.
  c. 10.0 g of concentrated acetic acid ($HC_2H_3O_2$) that is 95.0% pure.

45. Iodine, I_2, reacts with the thiosulfate ion, $S_2O_3^{2-}$, to give the iodide ion and the tetrathionate ion, $S_4O_6^{2-}$.
  a. Write a balanced net ionic equation for the reaction.
  b. If 10.0 g of iodine is dissolved in enough water to make 3.00×10^{-1} L of solution, what volume of 0.125 M sodium thiosulfate will be needed for complete reaction?

46. An acidic solution of potassium permanganate reacts with oxalate ions, $C_2O_4^{2-}$, to form carbon dioxide and manganese(II) ions.
   a. Write a balanced equation for the reaction.
   b. If 38.4 mL of 0.150 M potassium permanganate is required to titrate 25.2 mL of sodium oxalate solution, what was the concentration of oxalate ion?

47. Consider the reaction between copper and nitric acid, for which the unbalanced equation is: $Cu(s) + H^+(aq) + NO_3^-(aq) \rightarrow Cu^{2+}(aq) + NO_2(g) + H_2O$
   a. Balance the equation.
   b. What volume of 16.0 M HNO_3 is needed to furnish enough H^+ ions to react with 5.87 g of copper?
   c. What volume of $NO_2(g)$ is formed at 25°C and 1.00 atm when 5.87 g of Cu reacts?

48. Consider the reaction of iron(II) hydroxide with oxygen, for which the unbalanced equation is: $Fe(OH)_2(s) + O_2(g) + H_2O \rightarrow Fe(OH)_3(s)$
   a. Balance the equation (basic solution).
   b. What mass of $Fe(OH)_3$ is formed from 4.00 g of $Fe(OH)_2$?
   c. What volume of $O_2(g)$ at 22°C and 1.00 atm is required to react with 4.00 g of iron(II) hydroxide?

49. Consider the acidic reaction between dichromate ions and iron(II) ions. The products are chromium(III) ions and iron(III) ions. It is found that 23.8 mL of a potassium dichromate solution reacts with 29.3 mL of a 0.0325 M iron(II) nitrate solution.
   a. What is the molarity of the potassium dichromate solution?
   b. If a 0.100 M solution of an acid furnishing one mole of H^+ per mole of acid is used to acidify the solution, what is the minimum volume of acid required?

50. Consider the acidic reaction between permanganate ions and manganese(II) ions to form manganese dioxide, $MnO_2(s)$.
   a. What volume of 0.0859 M potassium permanganate is required to react with 39.0 mL of 0.175 M manganese(II) nitrate?
   b. What is the mass of precipitate formed in (a)?

Chemical Analysis

51. A chemist analyzes an alloy for silver. A 2.50-g sample of the alloy is first dissolved in nitric acid. This brings the silver into solution as the silver(I) ion. The solution is titrated with 38.4 mL of 0.500 M potassium thiocyanate, KSCN, to completely precipitate the silver as silver thiocyanate, AgSCN(s). What is the mass percent of silver in the alloy?

52. The mass percent of barium in an alloy is determined by reacting the alloy with sulfuric acid, which precipitates the barium ion as barium sulfate. An alloy sample weighing 1.587 g requires 22.9 mL of 0.150 M sulfuric acid for the complete precipitation of barium ion. What is the mass percent of barium in the alloy?

53. A Vitamin C capsule is analyzed by titrating it with 0.250 M sodium hydroxide. It is found that 10.3 mL of base is required to react with a capsule weighing 0.518 g. What is the percentage of Vitamin C, $C_6H_8O_6$, in the capsule? (One mole of Vitamin C reacts with one mole of hydroxide ion.)

54. The percentage of sodium bicarbonate, $NaHCO_3$, in a powder used for stomach upsets is found by titrating with 0.187 M hydrochloric acid. If 20.5 mL of hydrochloric acid is required to react with 0.375 g of the powder, what is the percentage of sodium bicarbonate in the sample? The reaction is: $H^+(aq) + HCO_3^-(aq) \rightarrow CO_2(g) + H_2O$.

55. An artificial fruit beverage contains 12.0 g of tartaric acid, $H_2C_4H_4O_6$, to achieve tartness. It is titrated with a basic solution that has a density of 1.045 g/cm^3 and contains 5.00 mass % KOH. What volume of the basic solution is required? (One mole of tartaric acid reacts with two moles of hydroxide ion.)

56. Lactic acid, $C_3H_6O_3$, is the acid present in sour milk. A 0.100-g sample of pure lactic acid requires 12.95 mL of 0.0857 M sodium hydroxide for complete reaction. How many moles of hydroxide ion are required to neutralize one mole of lactic acid?

57. A wire weighing 0.100 g and containing 99.78% Fe is dissolved in HCl. The iron is completely oxidized to Fe^{3+} by bromine water. The solution is then treated with tin(II) chloride to bring about the reaction

$$Sn^{2+}(aq) + 2Fe^{3+}(aq) \rightarrow Sn^{4+}(aq) + 2Fe^{2+}(aq)$$

If 9.47 mL of tin(II) chloride solution is required for complete reaction, what is its molarity?

58. Limonite, an ore of iron, is brought into solution and titrated with potassium permanganate, $KMnO_4$. The unbalanced equation for the redox reaction (acid solution) is:

$$Fe^{2+}(aq) + MnO_4^-(aq) \rightarrow Fe^{3+}(aq) + Mn^{2+}(aq)$$

It is found that a 0.500-g sample of limonite requires 40.0 mL of 0.0187 M $KMnO_4$. What is the percent of iron in the limonite sample? If the iron in limonite is present as Fe_2O_3, what is the percent of Fe_2O_3 in the ore?

59. Laws passed in some states define a drunk driver as one who drives with a blood alcohol level of 0.1% by mass or higher. The level of alcohol can be determined by titrating blood plasma with potassium dichromate according to the unbalanced equation

$$H^+(aq) + Cr_2O_7^{2-}(aq) + C_2H_5OH(aq) \rightarrow Cr^{3+}(aq) + CO_2(g) + H_2O$$

Assuming that the only substance that reacts with dichromate in blood plasma is alcohol, is a person legally drunk if 45.02 mL of 0.05000 M potassium dichromate is required to titrate a fifty gram sample of blood plasma?

60. The molarity of iodine in solution can be determined by titration with H_3AsO_3; the unbalanced equation for the redox reaction (acid solution) is:

$$I_2(aq) + H_3AsO_3(aq) \rightarrow I^-(aq) + H_3AsO_4(aq)$$

What is the molarity of I_2 if 28.9 mL of the solution reacts exactly with 0.750 g of H_3AsO_3?

Unclassified

61. Classify each of the following as a precipitation, acid-base, or redox reaction:
  a. the reaction between solutions of sulfuric acid and barium nitrate.
  b. the reaction between solutions of sulfuric acid and calcium hydroxide.
  c. the reaction of hydrochloric acid with aluminum metal.
  d. the reaction of a solution of tin(II) chloride with air to form SnO_2.

62. Consider the reaction between permanganate ions and hydrogen peroxide, H_2O_2, in acid solution:

$$2MnO_4^-(aq) + 5H_2O_2(aq) + 6H^+(aq)$$
$$\rightarrow 2Mn^{2+}(aq) + 5O_2(g) + 8H_2O$$

What volume of oxygen gas at 25°C and 1.00 atm is obtained if 25.0 mL of 0.583 M $KMnO_4$ reacts with 5.00 mL of 0.200 M HCl and 1.55 g of H_2O_2?

63. Gold metal will dissolve only in *aqua regia,* a mixture of concentrated hydrochloric acid and concentrated nitric acid. The products of the reaction between gold and aqua regia are $AuCl_4^-(aq)$, $NO(g)$, and H_2O.
  a. Write a balanced net ionic equation for the redox reaction, treating HCl and HNO_3 as strong acids.
  b. What ratio of hydrochloric acid to nitric acid should be used?
  c. What volumes of 12 M HCl and 16 M HNO_3 are required to furnish the Cl^- and NO_3^- ions to react with 10.0 g of gold?

64. The iron content of hemoglobin is determined by destroying the hemoglobin molecule and producing small water-soluble ions and molecules. The iron in the aqueous solution is reduced to iron(II) ion and then titrated against potassium permanganate. In the titration, iron(II) is oxidized to iron(III) and permanganate is reduced to manganese(II) ion. A 5.00-g sample of hemoglobin requires 32.3 mL of a 0.002100 M solution of potassium permanganate. What is the mass percent of iron in hemoglobin?

65. A sample of limestone weighing 0.145 g is dissolved in 50.00 mL of 0.100 M hydrochloric acid. The following reaction occurs:

$$CaCO_3(s) + 2H^+(aq) \rightarrow Ca^{2+}(aq) + CO_2(g) + H_2O$$

It is found that 13.05 mL of 0.175 M NaOH is required to titrate the excess HCl left after reaction with the limestone. What is the mass percent of $CaCO_3$ in the limestone?

Challenge Problems

66. Calcium in blood or urine can be determined by precipitation as calcium oxalate, CaC_2O_4. The precipitate is dissolved in strong acid and titrated with potassium permanganate. The products of the reaction are carbon dioxide and manganese(II) ion. A 24-h urine sample is collected from an adult patient, reduced to a small volume, and titrated with 26.2 cm^3 of 0.0946 M $KMnO_4$. How many grams of calcium oxalate are in the sample? Normal range for Ca^{2+} output for an adult is 100–300 mg per 24 h. Is the sample within the normal range?

67. Stomach acid is approximately 0.020 M HCl. What volume of this acid is neutralized by an antacid tablet that weighs 330 mg and contains 41.0% $Mg(OH)_2$, 36.2% $NaHCO_3$, and 22.8% NaCl? The reactions involved are:

$$Mg(OH)_2(s) + 2H^+(aq) \rightarrow Mg^{2+}(aq) + 2H_2O$$

$$HCO_3^-(aq) + H^+(aq) \rightarrow CO_2(g) + H_2O$$

68. A few years ago, 20,000 gallons of concentrated nitric acid (72% HNO_3 by mass, d = 1.42 g/cm^3) spilled from a tank car in a Denver railyard. Sodium carbonate was spread on the acid to react with it to form carbon dioxide gas. The reaction is:

$$CO_3^{2-}(aq) + 2H^+(aq) \rightarrow CO_2(g) + H_2O$$

How many grams of sodium carbonate were required?

69. A solution contains both iron(II) and iron(III) ions. A 50.00-mL sample of the solution is titrated with 35.0 mL of 0.0280 M $KMnO_4$, which oxidizes Fe^{2+} to Fe^{3+}; the MnO_4^- ion is reduced to Mn^{2+}. Another 50.00-mL sample of the solution is treated with zinc, which reduces all the Fe^{3+} to Fe^{2+}. The resulting solution is again titrated with 0.0280 M $KMnO_4$; this time 48.0 mL is required. What are the concentrations of Fe^{2+} and Fe^{3+} in the solution?

$4Fe(s) + 3O_2(g) \rightarrow 2Fe_2O_3(s)$
Iron powder burning in a Bunsen flame.

CHAPTER
13

GASEOUS
EQUILIBRIUM

The cumbrous elements, Earth,
Flood, Aire, Fire
Flew upward, spirited various forms,
Each had his place appointed, each
his course
The rest in circuit walles this
Universe.

JOHN MILTON
PARADISE LOST

In Chapter 10, we described the equilibrium between liquid and gaseous water. When a sample of liquid water is placed in a closed container at constant temperature, part of it vaporizes. After a few minutes, if sufficient water is present, a dynamic equilibrium is established, that is, both forward and reverse reactions take place continually at the same rate.

$$H_2O(l) \rightleftharpoons H_2O(g)$$

The double arrow implies that the two processes, vaporization ($\rightarrow$) and condensation ($\leftarrow$), are occurring at the same rate. Once equilibrium is established, the relative amounts of liquid and vapor do not change with time.

The state of this equilibrium system at a given temperature can be described in a simple way. We can cite either the equilibrium concentration of the water vapor (0.0327 mol/L at 100°C) or its equilibrium pressure (1.00 atm at 100°C). The concentration or pressure of vapor in equilibrium with liquid is independent of the volume of the container or the amount of water we started with. It depends only upon the temperature of the system.

Chemical reactions carried out in closed containers resemble in many ways the system just discussed. They are reversible. Ordinarily, the reactants are not completely consumed. Instead, we obtain an equilibrium mixture containing both products and reactants. At equilibrium, forward and reverse reactions take place at the same rate. As a result, the concentrations of all species at equilibrium remain constant as time passes.

In this chapter, we will examine the properties of chemical systems that reach equilibrium in the gas state. A typical example is the decomposition of dinitrogen tetroxide (N_2O_4). Dinitrogen tetroxide in the liquid state was one of the fuels used on the lunar lander for the NASA Apollo missions. In the gas phase it decomposes to gaseous nitrogen dioxide:

They can be reversed by changing the conditions in the reaction mixture

429

$$N_2O_4(g) \rightleftharpoons 2NO_2(g)$$

The state of this equilibrium system cannot be described as simply as that of the $H_2O(l)-H_2O(g)$ system. In particular, the concentration of NO_2 at equilibrium is not fixed; it can take on any of a number of values, depending upon the concentration of N_2O_4. However, as we will see in Section 13.1, there is a simple relationship between the equilibrium concentrations of these two gases. This relationship is expressed in terms of a quantity called the **equilibrium constant** and given the symbol K_c. The equilibrium constant for a reaction is one of its characteristic properties. It has a fixed value at a given temperature, independent of such factors as initial concentration, container volume, or pressure. Our discussion throughout this chapter will focus upon the properties of this equilibrium constant. We will see how, among other things, one can

— write the expression for K_c corresponding to any chemical equilibrium involving gases (Section 13.2).

— use the value of K_c to predict the extent to which a reaction will take place (Section 13.3).

— use K_c along with Le Châtelier's principle to predict what will happen when an equilibrium system is disturbed in some way (Section 13.4).

It is possible to describe the chemical equilibrium between gaseous species in terms of their partial pressures rather than their concentrations. Here it is convenient to work with a different type of equilibrium constant given the symbol K_p. In Section 13.5, we will consider how K_p is expressed, how it is related to K_c, and how it is used in equilibrium calculations.

13.1
THE $N_2O_4-NO_2$ EQUILIBRIUM SYSTEM

Consider what happens when a sample of N_2O_4, a colorless gas, is placed in a closed, evacuated container at 100°C. Instantly, a reddish-brown color develops. This color is due to nitrogen dioxide, NO_2, formed by decomposition of part of the N_2O_4 (Fig. 13.1):

$$N_2O_4(g) \rightarrow 2NO_2(g) \tag{13.1a}$$

At first, this is the only reaction taking place. As soon as some NO_2 is formed, however, the reverse reaction can occur:

$$2NO_2(g) \rightarrow N_2O_4(g) \tag{13.1b}$$

As time passes, Reaction 13.1a slows down and 13.1b speeds up. Soon, their rates become equal. A dynamic equilibrium has been established:

$$N_2O_4(g) \rightleftharpoons 2NO_2(g) \tag{13.1}$$

At equilibrium, appreciable amounts of both gases are present. From that point on their concentrations are constant, as long as the volume of the container and the temperature remain unchanged.

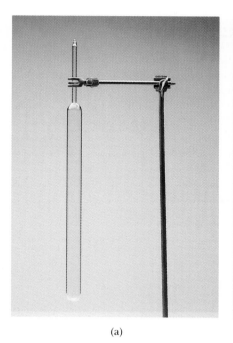
(a)

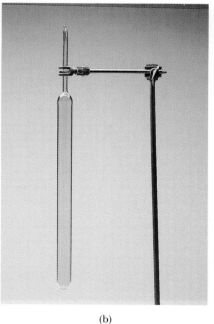
(b)

(c)

Figure 13.1 Dinitrogen tetroxide, N_2O_4 (a) is colorless; nitrogen dioxide, NO_2 (c) is a brownish-red gas. The concentration of NO_2 in an equilibrium mixture of the two gases (b) is directly related to the depth of the brown color. (Marna G. Clarke)

The approach to equilibrium in this system is illustrated by the data in Table 13.1 and Fig. 13.2, p. 432. We start with 0.100 mol N_2O_4 in a 1.00-L container at 100°C. The original concentration of N_2O_4 is thus 0.100 mol/L. Since there is no NO_2 present at the beginning of the experiment, its original concentration is zero. As equilibrium is approached, the *net reaction* taking place is that in the forward direction (13.1a). The concentration of N_2O_4 drops rapidly at first and then more slowly. The concentration of NO_2 increases. Finally, both concentrations level off and become constant. This indicates that we are at equilibrium. From Table 13.1 or Figure 13.2,

$$[N_2O_4] = 0.040 \text{ mol/L}$$

$$[NO_2] = 0.120 \text{ mol/L}$$

These concentrations don't change, since N_2O_4 is being formed at the same rate as it reacts

The square brackets, here and elsewhere throughout this text, represent equilibrium concentrations in moles per liter. It is important that these be distinguished from original or other nonequilibrium concentrations.

TABLE 13.1 Establishment of Equilibrium at 100°C in the System $N_2O_4(g) \rightleftharpoons 2NO_2(g)$

Time (s)	0	20	40	60	80	100
Conc. N_2O_4 (mol/L)	0.100	0.070	0.050	**0.040***	**0.040**	**0.040**
Conc. NO_2 (mol/L)	0.000	0.060	0.100	**0.120***	**0.120**	**0.120**

*Equilibrium concentrations are in bold type.

Figure 13.2 Approach to equilibrium in the N_2O_4–NO_2 system. The concentration of N_2O_4 starts off at 0.100 M, drops sharply at first, and finally levels off at its equilibrium value, 0.040 M. Meanwhile, the concentration of NO_2 rises rapidly at first, then increases more slowly, and finally becomes constant at its equilibrium value, 0.120 M.

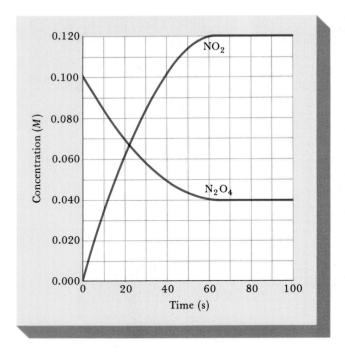

There are many ways to approach equilibrium in the N_2O_4–NO_2 system. Table 13.2 gives data for three experiments in which the original concentrations are quite different. Experiment 1 is that just described, in which we start with pure N_2O_4. In Experiment 2, we start from the other side of the equilibrium system, with pure NO_2 at a concentration of 0.100 mol/L. As we approach equilibrium, some of the NO_2 combines to form N_2O_4:

$$2NO_2(g) \rightarrow N_2O_4(g)$$

Finally, in Experiment 3, we start with a mixture of N_2O_4 and NO_2, both at a concentration of 0.100 mol/L. Some of the N_2O_4 decomposes in this

TABLE 13.2   Equilibrium Measurements in the N_2O_4–NO_2 System at 100°C

		ORIGINAL CONC. (mol/L)	EQUILIBRIUM CONC. (mol/L)
Expt. 1	N_2O_4	0.100	0.040
	NO_2	0.000	0.120
Expt. 2	N_2O_4	0.000	0.014
	NO_2	0.100	0.072
Expt. 3	N_2O_4	0.100	0.070
	NO_2	0.100	0.160

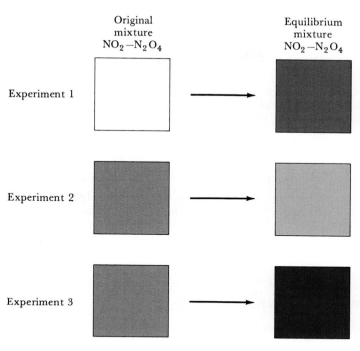

Original
mixture
NO_2–N_2O_4

Equilibrium
mixture
NO_2–N_2O_4

Experiment 1

Experiment 2

Experiment 3

Figure 13.3   When NO_2 (brown) is allowed to come to equilibrium with N_2O_4 (colorless), the composition of the equilibrium mixture can differ widely from one experiment to another. The intensity of the color shown here is directly proportional to the concentration of NO_2. The experiments shown are those outlined in Table 13.2. In all these experiments, the ratio $[NO_2]^2/[N_2O_4]$ is the same at equilibrium and is equal to the equilibrium constant, K_c.

experiment. At equilibrium, there is less N_2O_4 and more NO_2 than there was originally (see Fig. 13.3).

Looking at the data in Table 13.2, you might wonder whether these three experiments have anything in common. Specifically, is there any relationship between the equilibrium concentrations of NO_2 and N_2O_4 that is valid for all the experiments? It turns out that there is, although it is not an obvious one. The quotient $[NO_2]^2/[N_2O_4]$ is the same, about 0.36, in each case:

Expt. 1 $$\frac{[NO_2]^2}{[N_2O_4]} = \frac{(0.120)^2}{0.040} = 0.36$$

Expt. 2 $$\frac{[NO_2]^2}{[N_2O_4]} = \frac{(0.072)^2}{0.014} = 0.37$$ *Rather surprising, but true*

Expt. 3 $$\frac{[NO_2]^2}{[N_2O_4]} = \frac{(0.160)^2}{0.070} = 0.36$$

This relationship holds for any equilibrium mixture containing NO_2 and N_2O_4 at 100°C. Regardless of where we start, we find that at equilibrium,

$$\frac{[NO_2]^2}{[N_2O_4]} = 0.36 \text{ at } 100°C$$

Further experiments with this system at various temperatures lead to the following general conclusion:

At any given temperature, the quantity

$$\frac{[NO_2]^2}{[N_2O_4]}$$

is a constant, independent of the amounts of N_2O_4 and NO_2 that we start with, the volume of the container, or the total pressure. This constant is referred to as the **equilibrium constant**, K_c, for the reaction

$$N_2O_4(g) \rightleftharpoons 2NO_2(g)$$

At 100°C, K_c for this reaction is 0.36; at 150°C it has a different value, about 3.2. Any mixture of NO_2 and N_2O_4 at 100°C will react to form more NO_2 or more N_2O_4, until the ratio (conc. NO_2)^2/(conc. N_2O_4) becomes equal to 0.36. At 150°C, reaction will occur until this ratio becomes equal to 3.2.

13.2
THE EQUILIBRIUM CONSTANT, K_c

THE EQUILIBRIUM EXPRESSION

We have seen that for the system

$$N_2O_4(g) \rightleftharpoons 2NO_2(g); \qquad K_c = \frac{[NO_2]^2}{[N_2O_4]}$$

For every gaseous system, a similar expression can be written. The form of that expression can be obtained readily. For the general gas-phase system,

$$a\ A(g)\ +\ b\ B(g)\ \rightleftharpoons\ c\ C(g)\ +\ d\ D(g)$$

where A, B, C, D represent different substances and a, b, c, d are their coefficients in the balanced equation

$$K_c = \frac{[C]^c \times [D]^d}{[A]^a \times [B]^b} \qquad\qquad (13.2)$$

This is the general expression for K_c for a reaction involving only gases

Notice that in the expression for K_c,

— the equilibrium concentrations of **products** (right side of equation) appear in the **numerator**.

— the equilibrium concentrations of **reactants** (left side of equation) appear in the **denominator**.

■ **EXAMPLE 13.1** ─────────────────────────────

Consider the synthesis of ammonia from hydrogen gas and nitrogen gas. A German chemist, Fritz Haber, pioneered the large-scale production of ammonia using this reaction (the Haber process).

$$3H_2(g)\ +\ N_2(g)\ \rightleftharpoons\ 2NH_3(g)$$

Write the equilibrium expression for the Haber process.

Solution

Using Equation 13.2, we obtain

$$K_c = \frac{[NH_3]^2}{[H_2]^3 \times [N_2]}$$

Exercise

Write the equilibrium expression for the decomposition of two moles of hydrogen iodide gas into hydrogen gas and iodine gas. Answer: $K_c = [H_2] \times [I_2]/[HI]^2$.

CHANGING THE CHEMICAL EQUATION FOR AN EQUILIBRIUM SYSTEM

It is important to realize that *the expression for K_c depends upon the form of the chemical equation written to describe the equilibrium system.* To illustrate what this statement means, consider the N_2O_4–NO_2 system, for which we wrote

$$N_2O_4(g) \rightleftharpoons 2NO_2(g); \qquad K_c = \frac{[NO_2]^2}{[N_2O_4]}$$

There are many other equations that could be used to describe this system. We could have written

$$\tfrac{1}{2}N_2O_4(g) \rightleftharpoons NO_2(g)$$

In this case the expression for the equilibrium constant would have been

$$K_c' = \frac{[NO_2]}{[N_2O_4]^{1/2}}$$

Comparing the expressions for K_c and K_c', we see that K_c' is the square root of K_c. This illustrates a general rule, sometimes referred to as the **Coefficient Rule**, which states that:

If the coefficients in a balanced equation are multiplied by a factor, n, the equilibrium constant is raised to the nth power:

$$K_c' = (K_c)^n \tag{13.3}$$

In this particular case, $n = \tfrac{1}{2}$, because we divided the coefficients by 2. Hence, K_c' is the square root of K_c.

Another equation that might be written to describe the N_2O_4–NO_2 system is:

$$2NO_2(g) \rightleftharpoons N_2O_4(g); \qquad K_c'' = \frac{[N_2O_4]}{[NO_2]^2}$$

This chemical equation is simply the reverse of that written originally; N_2O_4 and NO_2 switch sides of the equation. Notice that K_c'' is the reciprocal of K_c; the numerator and denominator have been inverted. This illustrates the so-called **Reciprocal Rule**,

For a value of K_c to be useful, you need to know the equation for the associated reaction

The equilibrium constants for forward and reverse reactions are the reciprocals of each other.

$$K_c'' = 1/K_c \tag{13.4}$$

Applying these two rules to the N_2O_4–NO_2 system at 100°C, where, as pointed out earlier, $K_c = 0.36$:

$$K_c' = (0.36)^{1/2} = 0.60 \text{ at } 100°C$$

$$K_c'' = 1/0.36 = 2.8 \text{ at } 100°C$$

ADDING CHEMICAL EQUATIONS

A property of K_c that we will find very useful in this and succeeding chapters is expressed by the **Rule of Multiple Equilibria**, which states that:

If a reaction can be expressed as the sum of two or more reactions, K_c for the overall reaction is the product of the equilibrium constants of the individual reactions. That is, if

Reaction 3 = Reaction 1 + Reaction 2

Then

$$K_c(\text{Reaction 3}) = K_c(\text{Reaction 1}) \times K_c(\text{Reaction 2}) \tag{13.5}$$

To illustrate the application of this rule, consider the following reactions at 700°C:

$$SO_2(g) + \tfrac{1}{2}O_2(g) \rightleftharpoons SO_3(g); \qquad K_c = 20 \tag{13.6a}$$

$$NO_2(g) \rightleftharpoons NO(g) + \tfrac{1}{2}O_2(g); \qquad K_c = 0.45 \tag{13.6b}$$

If we add these two equations, the "$\tfrac{1}{2}O_2(g)$" cancels and we obtain

$$SO_2(g) + NO_2(g) \rightleftharpoons SO_3(g) + NO(g) \tag{13.6}$$

So it follows from the rule that

$$
\begin{aligned}
K_c(\text{Reaction 13.6}) &= K_c(\text{Reaction 13.6a}) \times K_c(\text{Reaction 13.6b}) \\
&= 20 \times 0.45 = 9.0
\end{aligned}
$$

The validity of this rule can be demonstrated by writing the expressions for K_c for Reactions 13.6a and 13.6b and multiplying:

$$
\begin{aligned}
K_c(13.6a) \times K_c(13.6b) &= \frac{[SO_3]}{[SO_2] \times [O_2]^{1/2}} \times \frac{[NO] \times [O_2]^{1/2}}{[NO_2]} \\
&= \frac{[SO_3] \times [NO]}{[SO_2] \times [NO_2]}
\end{aligned}
$$

Looking at Reaction 13.6, we see that the quotient we have just obtained is precisely equal to the K_c expression for that reaction.

HETEROGENEOUS EQUILIBRIA

In the reactions described so far, all the reactants and products have been gaseous; the equilibrium system is *homogeneous*. In certain reactions, one or more of the substances involved is a pure liquid or solid; the others are gases. We call such a system *heterogeneous*. Examples include:

$$CO_2(g) + H_2(g) \rightleftharpoons CO(g) + H_2O(l) \tag{13.7}$$

$$I_2(s) \rightleftharpoons I_2(g) \tag{13.8}$$

Experimentally, we find, that for heterogeneous systems such as these:

1. *The position of the equilibrium is independent of the amount of solid or liquid.* Adding or removing a liquid or solid has no effect on the equilibrium.

But there must be at least a little solid or liquid present

2. *We do not need to include terms for liquids or solids in the expression for K_c.*

To see why this is the case, consider Reaction 13.7 taking place at, say, 100°C. The equation tells us that liquid water is present. There must also be water vapor present, in equilibrium with the liquid. Indeed, it is actually $H_2O(g)$ that takes part in the gas-phase reaction. The pressure of the vapor is 1.00 atm, the vapor pressure of liquid water at 100°C. The concentration of water vapor, calculated from the Ideal Gas Law, is 0.0327 mol/L. This concentration remains constant as long as there is any liquid water present, regardless of what [CO], [H_2] or [CO_2] may be. It doesn't matter whether we have 0.1 g, 18 g, or 1 kg of liquid water; it will still be true that:

$$\text{conc. } H_2O(g) = \text{constant} \tag{13.9}$$

If we denote by K_c' the equilibrium constant that would apply if the system were homogeneous:

$$K_c' = \frac{[CO] \times [H_2O]}{[CO_2] \times [H_2]} = \frac{[CO] \times \text{constant}}{[CO_2] \times [H_2]}$$

$$K_c'/\text{constant} = \frac{[CO]}{[CO_2] \times [H_2]} = K_c$$

$K_c' \neq K_c$

The equilibrium constant that we work with for this heterogeneous system is K_c; in effect the constant concentration of $H_2O(g)$ has been incorporated into that constant.

A similar argument can be used to justify dropping the term for solid I_2 in the equilibrium constant expression for Reaction 13.8.

■ **EXAMPLE 13.2** _____

Write the expressions for K_c for:

a. The reduction of black solid copper(II) oxide (1 mol) with hydrogen to form copper metal and steam.

b. The reaction of 1 mol of steam with red hot coke (carbon) to form a mixture of hydrogen and carbon monoxide, called water gas.

Solution

a. The reaction is: $CuO(s) + H_2(g) \rightleftharpoons Cu(s) + H_2O(g)$

The equilibrium expression is: $K_c = [H_2O]/[H_2]$

b. The reaction is

$$C(s) + H_2O(g) \rightleftharpoons H_2(g) + CO(g)$$

The equilibrium expression is

$$K_c = \frac{[H_2] \times [CO]}{[H_2O]}$$

Exercise
Suppose that the reaction in (b) is carried out at a temperature low enough so that liquid water is present. What is the expression for K_c? Answer: $K_c = [H_2] \times [CO]$.

DETERMINATION OF K_c

Numerical values of equilibrium constants are determined in much the same way as we described for the N_2O_4–NO_2 system. Typical calculations are shown in Example 13.3 and 13.4.

■ **EXAMPLE 13.3**
Ammonium chloride is sometimes used as a flux in soldering because it decomposes upon heating:

$$NH_4Cl(s) \rightleftharpoons NH_3(g) + HCl(g)$$

The HCl formed removes oxide films from metals to be soldered. When this reaction system reaches equilibrium at 500°C in a 5.0-L container, there is 2.0 mol NH_3, 2.0 mol HCl, and 1.0 mol NH_4Cl. Calculate K_c for this system at 500°C.

To evaluate K_c we need to know the equilibrium concentrations of all reactants and products

Solution
The expression for K_c is

$$K_c = [NH_3] \times [HCl]$$

(There is no term for solid NH_4Cl.) The equilibrium concentrations are found by dividing the number of moles by the volume in liters:

$$[NH_3] = \frac{2.0 \text{ mol}}{5.0 \text{ L}} = 0.40 \text{ mol/L}; \quad [HCl] = \frac{2.0 \text{ mol}}{5.0 \text{ L}} = 0.40 \text{ mol/L}$$

$$K_c = 0.40 \times 0.40 = 0.16$$

Exercise

Suppose that for the system at equilibrium

$$2SO_3(g) \rightleftharpoons 2SO_2(g) + O_2(g)$$

$[SO_2] = [O_2] = 0.10\ M$ and $[SO_3] = 0.20\ M$. Calculate K_c. Answer: 0.025.

■ **EXAMPLE 13.4** _____

Consider the equilibrium system

$$2HI(g) \rightleftharpoons H_2(g) + I_2(g) \qquad\qquad\qquad (13.10)$$

Suppose we start with pure HI at a concentration of 0.100 mol/L at 520°C. The equilibrium concentration of H_2 is found to be 0.010 mol/L. Calculate

a. $[I_2]$ b. $[HI]$ c. K_c for Reaction 13.10

Solution

a. According to Equation 13.10, one mole of I_2 is formed for every mole of H_2. The concentrations of H_2 and I_2 must therefore increase by the same amount. Since they both start at zero, the two concentrations must be equal at equilibrium. The equilibrium concentration of I_2, like that of H_2, is 0.010 mol/L.

b. Two moles of HI are required to form one mole of H_2. The decrease in HI concentration must be twice the increase in the concentration of H_2:

> The H_2 and I_2 came from the HI

$$\text{decrease in HI conc.} = 2(0.010\ \text{mol/L}) = 0.020\ \text{mol/L}$$

The original concentration of HI is 0.100 mol/L; Hence,

$$\text{equilibrium conc. HI} = 0.100\ \text{mol/L} - 0.020\ \text{mol/L}$$

$$= \boxed{0.080\ \text{mol/L}}$$

It may be helpful to summarize the reasoning we have gone through by means of a table.

$$2HI(g) \rightleftharpoons H_2(g) + I_2(g)$$

	ORIG. CONC. (mol/L)	CHANGE IN CONC. (mol/L)	EQUIL. CONC. (mol/L)
HI	0.100	−0.020	0.080
H_2	0.000	+0.010	0.010
I_2	0.000	+0.010	0.010

> This kind of table can be very useful in analyzing equilibrium systems

c. $$K_c = \frac{[H_2] \times [I_2]}{[HI]^2} = \frac{0.010 \times 0.010}{(0.080)^2} = \boxed{0.016}$$

Exercise
Suppose that, at a different temperature, we start with pure HI at 0.100 mol/L and find that its equilibrium concentration is 0.074 mol/L. What is the value of K_c at this temperature? Answer: $K_c = 0.031$.

In dealing with chemical systems, it is important to distinguish between the *equilibrium constant*, K_c, and sets of *equilibrium concentrations* of products and reactants. At a given temperature, the equilibrium constant can have only one value, independent of original concentrations, container volume, or any other factor. In contrast, for a system containing two or more gases, there can be an infinite number of different sets of equilibrium concentrations. Table 13.2 lists a few of these for the N_2O_4–NO_2 system at 100°C. Recall that, for each of these sets, K_c is 0.36.

13.3
APPLICATIONS OF THE EQUILIBRIUM CONSTANT

A knowledge of the equilibrium constant for a reaction tells us a great deal, qualitatively and quantitatively, about the extent to which it will occur. Often, knowing the magnitude of the equilibrium constant, we can predict whether a reaction is likely to be feasible. Consider, for example, a possible method for "fixing" atmospheric nitrogen—converting it to a compound— by reaction with oxygen:

$$N_2(g) + O_2(g) \rightleftharpoons 2NO(g) \qquad (13.11)$$

$$K_c = \frac{[NO]^2}{[N_2] \times [O_2]} = 1 \times 10^{-30} \qquad \text{at } 25°C$$

If K_c is very small, the reaction won't go very far to the right

We see that the equilibrium constant for Reaction 13.11 is a very small number. This means that the equilibrium concentration of NO, which appears in the numerator, must be very small relative to the concentrations of N_2 and O_2, which are in the denominator. This tells us that a mixture of nitrogen and oxygen will react to a very small extent to produce NO. Clearly, this would not be a suitable way to fix nitrogen, at least at 25°C.

An alternative approach to nitrogen fixation involves reacting it with hydrogen:

$$N_2(g) + 3H_2(g) \rightleftharpoons 2NH_3(g)$$

$$K_c = \frac{[NH_3]^2}{[N_2] \times [H_2]^3} = 5 \times 10^8 \text{ at } 25°C$$

If K_c is very large, the reaction goes essentially to completion

Here the situation is quite different from that discussed above. Since the equilibrium constant is large, the equilibrium system must contain mostly NH_3, which appears in the numerator. We expect a mixture of N_2 and H_2 to be almost completely converted to NH_3 at equilibrium. (Unfortunately, it takes essentially forever for equilibrium to be reached in this system at 25°C, but that's a problem we'll worry about in Chapter 18.)

In general, we can say that if the equilibrium constant is very large, the forward reaction will tend to proceed far to the right. The equilibrium system will contain mostly products (right side of the chemical equation) with very little unreacted starting materials. Conversely, if K_c is very small, virtually no reaction will occur in the forward direction. The equilibrium system will consist almost entirely of unreacted starting materials, with very little in the way of products. Finally, if the equilibrium constant is neither "very large" nor "very small," we expect the equilibrium system to contain appreciable amounts of both products and reactants.

The type of prediction we have just made is qualitative. Knowing the magnitude of K_c, we can make a ballpark estimate of the composition of the equilibrium mixture. However, with a little work, we can use the equilibrium constant to make quantitative predictions. Specifically, we can use the value of K_c to make quantitative predictions about the *direction* or *extent* of reaction.

DIRECTION OF REACTION

We have seen (Section 13.2) that, for the general gas-phase reaction

$$a \, A(g) + b \, B(g) \rightleftharpoons c \, C(g) + d \, D(g)$$

the expression for the equilibrium constant K_c is

$$K_c = \frac{[C]^c \times [D]^d}{[A]^a \times [B]^b}$$

The square brackets are used to designate equilibrium concentrations in moles per liter.

When we carry out a reaction in the laboratory, the original concentration quotient Q, expressed as

$$Q = \frac{(\text{orig. conc. C})^c \times (\text{orig. conc. D})^d}{(\text{orig. conc. A})^a \times (\text{orig. conc. B})^b}$$

will seldom be equal numerically to K_c. Frequently, we start with pure reactants, A and B, and no products, C and D. In that case, $Q = 0$. If in another experiment, we started with C and D, but no A or B, we would have $Q = \infty$. Clearly, in either case, we are not at equilibrium.

If the concentration quotient, Q, at the beginning of an experiment is not equal to K_c, reaction will occur in one direction or the other so as to bring the concentrations of products and reactants to the ratio required at equilibrium. We can distinguish two possibilities:

1. If $\quad Q < K_c$

the reaction will proceed from left to right, i.e.,

$$a \, A(g) + b \, B(g) \rightarrow c \, C(g) + d \, D(g)$$

In this way, the concentrations of products increase and those of reactants decrease. As this happens, the concentration quotient, Q, increases until it becomes equal to K_c. When we reach that point, we are at equilibrium and there is no further change.

2. If $Q > K_c$

we conclude that the concentrations of products are "too high" and those of the reactants "too low" to meet the equilibrium condition. Reaction must proceed in the reverse direction, i.e.,

$$c\ C(g) + d\ D(g) \rightarrow a\ A(g) + b\ B(g)$$

increasing the concentrations of A and B while reducing those of C and D. This lowers the concentration quotient, Q, to its equilibrium value given by K_c.

■ **EXAMPLE 13.5** ───────────────────────────────

Consider the following system at 100°C:

$$N_2O_4(g) \rightleftharpoons 2NO_2(g); \qquad K_c = 0.36$$

Predict the direction in which the system will move to reach equilibrium if we start with 0.20 mol N_2O_4 and 0.20 mol NO_2 in a 4.0-L container.

Solution

orig. conc. N_2O_4 = 0.20 mol/4.0 L = 0.050 mol/L
orig. conc. NO_2 = 0.20 mol/4.0 L = 0.050 mol/L

$$Q = \frac{(\text{orig. conc. } NO_2)^2}{(\text{orig. conc. } N_2O_4)} = \frac{(0.050)^2}{0.050} = 0.050$$

Since $0.050 < K_c = 0.36$,   reaction must proceed in the forward direction

to produce NO_2. The concentration of NO_2 increases, that of N_2O_4 decreases, and eventually the concentration quotient becomes equal to K_c.

Exercise

Suppose orig. conc. N_2O_4 = orig. conc. NO_2 = 1.0 M. Which way does reaction occur to establish equilibrium? Answer: Reverse direction (right to left).

EQUILIBRIUM CONCENTRATIONS

The equilibrium constant for a chemical system can be used to calculate the concentrations of the species present at equilibrium. In the simplest case, we can use K_c to obtain the equilibrium concentration of one species when we know those of all the other species (Example 13.6).

■ **EXAMPLE 13.6** ───────────────────────────────

We pointed out earlier that for the equilibrium system

$$N_2(g) + O_2(g) \rightleftharpoons 2NO(g)$$

K_c is 1×10^{-30} at 25°C. Suppose that in a mixture at 25°C the equilibrium concentrations of N_2 and O_2 are 0.040 and 0.010 mol/L, respectively. Calculate the equilibrium concentration of NO.

Solution
The expression for K_c is

$$K_c = \frac{[NO]^2}{[N_2] \times [O_2]} = 1 \times 10^{-30}$$

Substituting for the concentrations of N_2 and O_2, we have

$$\frac{[NO]^2}{(0.040)(0.010)} = 1 \times 10^{-30}$$

$$[NO]^2 = (4 \times 10^{-2})(1 \times 10^{-2})(1 \times 10^{-30}) = 4 \times 10^{-34}$$

$$[NO] = \quad 2 \times 10^{-17} \text{ mol/L}$$

At 25°C you can't make NO from N_2 and O_2

The concentration of NO at equilibrium is extremely small relative to those of N_2 and O_2. This confirms the qualitative prediction that we made at the beginning of this section.

Exercise
At 2000°C, K_c for this system is much larger, about 0.10. Calculate [NO] at 2000°C if $[N_2] = 0.040\ M$, $[O_2] = 0.010\ M$. Answer: 0.0063 M.

More commonly, we use K_c to determine the equilibrium concentrations of all species, reactants and products, knowing their original concentrations. The reasoning here is somewhat more complex than that in Example 13.6. In general we follow a five-step process:

1. Write the balanced equation for the reaction, if it is not given.

2. Write the equilibrium expression for the reaction using Equation 13.2.

3. Express the equilibrium concentration of all species in terms of a single unknown, x. As you will see, there are many different ways in which you can choose the variable x. Once you have chosen x, all the equilibrium concentrations have to be related to x in a way that is consistent with the coefficients of the balanced equation. To do this, it helps to set up an equilibrium table of the sort used in Example 13.4.

4. Substitute the equilibrium terms into the expression for K_c. This gives an algebraic equation involving x. Simplify this equation if possible, then solve it for x.

5. Having found x, calculate the equilibrium concentrations of all species.

■ **EXAMPLE 13.7** _____
For the system

$$CO_2(g) + H_2(g) \rightleftharpoons CO(g) + H_2O(g)$$

K_c is 0.64 at 900 K. Suppose we start with CO_2 and H_2, both at a concentration of 0.100 mol/L. When the system reaches equilibrium, what are the concentrations of products and reactants?

Solution

1. The balanced equation is given with the problem.

2. The equilibrium expression is

$$K_c = \frac{[CO] \times [H_2O]}{[CO_2] \times [H_2]}$$

3. We must first express all the equilibrium concentrations in terms of a single unknown, x. Let us take

x = no. moles per liter CO formed

All the coefficients in the balanced equation are the same, 1. It follows that

no. moles per liter H_2O formed = no. moles per liter CO formed = x

no. moles per liter CO_2 consumed = no. moles per liter CO formed = x

no. moles per liter H_2 consumed = no. moles per liter CO formed = x

We conclude that, in reaching equilibrium, the concentrations of CO and H_2O *increase* by x, while those of CO_2 and H_2 *decrease* by the same amount, x. Putting this reasoning in the form of a table,

$$CO_2(g) + H_2(g) \rightleftharpoons CO(g) + H_2O(g)$$

	ORIG. CONC. (mol/L)	CHANGE IN CONC. (mol/L)	EQUIL. CONC. (mol/L)
CO_2	0.100	$-x$	$0.100 - x$
H_2	0.100	$-x$	$0.100 - x$
CO	0.000	$+x$	x
H_2O	0.000	$+x$	x

You need to think about this for a while

4. We are now ready to substitute into the expression for K_c:

$$K_c = 0.64 = \frac{[CO] \times [H_2O]}{[CO_2] \times [H_2]} = \frac{x^2}{(0.100 - x)^2}$$

This is a second-order equation in x. Such equations can always be solved using the quadratic formula (see Example 13.8). Here, however, we can simplify the arithmetic by noting that the right side of the equation is a perfect square. Taking the square root of both sides, we have

Keep the math simple when you can

$$(0.64)^{1/2} = 0.80 = \frac{x}{0.100 - x}$$

Solving for x: $x = 0.80(0.100 - x) = 0.080 - 0.80x$

$1.80x = 0.080;$ $x = 0.044$

5. Referring back to the equilibrium table,

$$[CO] = [H_2O] = x = \boxed{0.044 \text{ mol/L}}$$

$$[CO_2] = [H_2] = 0.100 - x = \boxed{0.056 \text{ mol/L}}$$

Exercise
Suppose the original concentrations of CO_2 and H_2 were 0.100 M and 0.200 M. In that case, what would be the algebraic equation obtained in step 4? Answer:

$$0.64 = \frac{x^2}{(0.100 - x)(0.200 - x)}$$

This equation is not so easy to solve

The arithmetic involved in equilibrium calculations can be relatively simple, as it was in Example 13.7. Sometimes it is more complex (Example 13.8). The reasoning involved, however, is the same. It is always helpful to set up an equilibrium table to summarize the analysis of the problem.

■ **EXAMPLE 13.8** _____
Consider the system

$$N_2O_4(g) \rightleftharpoons 2NO_2(g); \qquad K_c = 0.36 \text{ at } 100°C$$

Suppose we start with pure N_2O_4 at a concentration of 0.100 mol/L. What are the equilibrium concentrations of NO_2 and N_2O_4?

Solution

1. The balanced equation is given.
2. The equilibrium expression is

$$K_c = \frac{[NO_2]^2}{[N_2O_4]}$$

3. Some of the N_2O_4 will decompose to achieve equilibrium. Let x be the number of moles per liter of N_2O_4 that decomposes. The balanced equation tells us that two moles of NO_2 are formed for every mole of N_2O_4 that decomposes. Hence, if the concentration of N_2O_4 decreases by x, that of NO_2 must increase by $2x$. We set up an equilibrium table,

$$N_2O_4(g) \rightleftharpoons 2NO_2(g)$$

	ORIG. CONC. (mol/L)	CHANGE IN CONC. (mol/L)	EQUIL. CONC. (mol/L)
N_2O_4	0.100	$-x$	$0.100 - x$
NO_2	0.000	$+2x$	$2x$

The concentration of NO_2 changes twice as fast as that of N_2O_4 and in the opposite direction

4. $K_c = 0.36 = \dfrac{(2x)^2}{0.100 - x} = \dfrac{4x^2}{0.100 - x}$

This time, we cannot solve for x as simply as in Example 13.7. The denominator of the right side is not a perfect square. Instead, we use the general method of solving a quadratic equation. This involves rearranging to the form

$$ax^2 + bx + c = 0$$

and applying the quadratic formula

$$x = \frac{-b \pm \sqrt{b^2 - 4ac}}{2a}$$

To convert our expression to the desired form, we proceed as follows:

$$4x^2 = 0.36(0.100 - x) = 0.036 - 0.36x$$

$$4x^2 + 0.36x - 0.036 = 0$$

We can simplify a bit by dividing both sides of the equation by 4:

$$x^2 + 0.090x - 0.0090 = 0$$

Thus, $a = 1$, $b = 0.090$, and $c = -0.0090$.

$$x = \frac{-0.090 \pm \sqrt{(0.090)^2 + 4(0.0090)}}{2} = \frac{-0.090 \pm \sqrt{0.0441}}{2}$$

$$= \frac{-0.090 \pm 0.21}{2} = \frac{0.12}{2} \text{ or } \frac{-0.30}{2} \text{ (that is, } 0.060 \text{ or } -0.15\text{)}$$

Of the two answers, only 0.060 is plausible. A value of -0.15 for x is not physically significant, since it would give a negative concentration for NO_2.

Negative concentrations are very, very rare, indeed impossible

5. $[NO_2] = 2x = 2(0.060 \text{ mol/L}) = \boxed{0.120 \text{ mol/L}}$

$[N_2O_4] = 0.100 - x = \boxed{0.040 \text{ mol/L}}$

Compare these values with those listed in Experiment 1, Table 13.2. The calculations we have just carried out refer to that experiment.

Exercise
Suppose we had chosen our unknown x to be the number of moles per liter of NO_2 formed rather than the number of moles per liter of N_2O_4 that decompose. What would be the algebraic equation obtained in step 2? Answer: $0.36 = x^2/(0.100 - \frac{1}{2}x)$.

Example 13.8 and the exercise that follows illustrate an important point concerning equilibrium problems of this type. There are many different ways in which the unknown can be chosen. It doesn't matter what choice you make, provided you are consistent in relating equilibrium concentrations. If you solve the algebraic equation cited in the exercise:

$$\frac{x^2}{0.100 - \frac{1}{2}x} = 0.36$$

you should find that $x = 0.120$, so that $[NO_2] = 0.120$ mol/L, $[N_2O_4] = 0.100 - \frac{1}{2}x = 0.100 - 0.060 = 0.040$ mol/L. These are, of course, the same values obtained with a different choice of unknown in Example 13.8.

As indeed they should be

13.4
EFFECT OF CHANGES IN CONDITIONS UPON AN EQUILIBRIUM SYSTEM

Once a system has attained equilibrium, it is possible to change the ratio of products to reactants by changing the external conditions. We will consider three ways in which a chemical equilibrium can be disturbed:

1. Adding or removing a gaseous reactant or product.
2. Changing the volume of the system.
3. Changing the temperature.

We can deduce the direction in which an equilibrium will shift when one of these changes is made by applying Le Châtelier's principle. This states the following:

If a system at equilibrium is disturbed by some change, the system will shift so as to partially counteract the effect of the change, if possible.

There is a natural tendency, in people and reactions, to resist change

ADDING OR REMOVING A GASEOUS SPECIES

According to Le Châtelier's principle, **if we disturb a chemical system at equilibrium by adding a gaseous species (reactant or product), the reaction will proceed in such a direction as to consume part of the added species. Conversely, if we remove a gaseous species, the system will shift so as to restore part of that species.**

Let us apply this general rule to the equilibrium system

$$N_2O_4(g) \rightleftharpoons 2NO_2(g)$$

Suppose this system has reached equilibrium at a certain temperature. We might disturb the equilibrium by

— *adding N_2O_4.* If we do this, reaction will occur in the forward direction (left to right). In this way, part of the N_2O_4 will be consumed.

— *adding NO_2,* which causes the reverse reaction (right to left) to occur, using up part of the NO_2 added.

— *removing N_2O_4.* Here, reaction occurs in the reverse direction to restore part of the N_2O_4.

— *removing NO_2,* which causes the forward reaction to occur, restoring part of the NO_2 removed.

It is possible to use K_c to calculate the extent to which reaction occurs when an equilibrium is disturbed by adding or removing a product or reactant. To show how this is done, we consider the effect of adding hydrogen iodide to the HI—H$_2$—I$_2$ system (Example 13.9):

$$2HI(g) \rightleftharpoons H_2(g) + I_2(g); \qquad K_c = \frac{[H_2] \times [I_2]}{[HI]^2} = 0.016 \text{ at } 520°C$$

■ **EXAMPLE 13.9** _____

In Example 13.4 we saw that this system is at equilibrium at 520°C when [HI] = 0.080 mol/L and [H$_2$] = [I$_2$] = 0.010 mol/L. Suppose that to this mixture we add enough HI to raise its concentration temporarily to 0.096 mol/L. When equilibrium is restored, what will [HI], [H$_2$], and [I$_2$] be?

Solution

1. Here, the "original concentrations" are those that prevail immediately after the equilibrium is disturbed:

orig. conc. HI = 0.096 mol/L;

orig. conc. H$_2$ = orig. conc. I$_2$ = 0.010 mol/L

Since we added HI, some of it must decompose to bring the system back to equilibrium. When that happens, some H$_2$ and some I$_2$ are formed. Let x be the number of moles per liter of H$_2$ that is produced by this process. Since H$_2$ and I$_2$ are formed in a 1:1 mole ratio, the concentration of I$_2$ must also increase by x. Since two moles of HI are required to form one mole of H$_2$, the concentration of HI must decrease by $2x$. The equilibrium table is

$$2HI(g) \rightleftharpoons H_2(g) + I_2(g)$$

	ORIG. CONC. (mol/L)	CHANGE IN CONC. (mol/L)	EQUIL. CONC. (mol/L)
H$_2$	0.010	$+x$	$0.010 + x$
I$_2$	0.010	$+x$	$0.010 + x$
HI	0.096	$-2x$	$0.096 - 2x$

2. Substituting into the expression for K_c,

$$0.016 = \frac{(0.010 + x)(0.010 + x)}{(0.096 - 2x)^2}$$

To solve this equation, we first take the square root of both sides. Since $(0.016)^{1/2} = 0.13$, we have

$$0.13 = \frac{0.010 + x}{0.096 - 2x}$$

which solves to give $x = 0.002$.

3. Thus,

[H$_2$] = 0.010 + 0.002 = 0.012 mol/L
[I$_2$] = 0.010 + 0.002 = 0.012 mol/L
[HI] = 0.096 - 0.004 = 0.092 mol/L

Note that the equilibrium concentration of HI, 0.092 mol/L, is greater than it was originally (0.080 mol/L), but less than it was immediately after equilibrium was disturbed (0.096 mol/L).

When HI was added, reaction occurred to use some of it up

Exercise

Suppose that, to the original equilibrium mixture, we add enough HI to raise its concentration temporarily to 0.200 mol/L. Which of the following would be a reasonable value for the final concentration of HI: 0.060 M, 0.080 M, 0.175 M, 0.200 M, or 0.225 M? Answer: 0.175 M. (The other answers are impossible. Why?)

Notice that throughout this discussion, we have referred to the effect of adding or removing a gaseous species. If we add a pure solid or liquid to an equilibrium system, nothing happens! Consider, for example, the system

$$CaCO_3(s) \rightleftharpoons CaO(s) + CO_2(g)$$

We might establish equilibrium at 840°C with 0.100 mol $CaCO_3$, 0.100 mol CaO, and 0.011 mol CO_2 in a one-liter container. The equilibrium constant at this temperature is

$$K_c = [CO_2] = 0.011$$

Suppose now that we add some $CaCO_3$ to this equilibrium system. None of it reacts. If it did, it would produce CO_2, changing the concentration of that product. That is impossible; the equilibrium concentration of CO_2 at 840°C is fixed at 0.011 mol/L. The same argument can be used to show that the position of this equilibrium would not be affected by adding CaO, removing $CaCO_3$, or removing CaO. In general,

Adding or removing a species disturbs an equilibrium system only if the concentration of that species appears in the expression for the equilibrium constant.

CHANGES IN VOLUME

To understand how a change in container volume can change the position of an equilibrium, consider again the N_2O_4–NO_2 system

$$N_2O_4(g) \rightleftharpoons 2NO_2(g)$$

Suppose we decrease the volume of this system (Fig. 13.4, p. 460). The immediate effect is to increase the number of molecules per unit volume. According to Le Châtelier's Principle, the system will shift so as to partially counteract this change. There is a simple way that this can happen. Some of the NO_2 molecules combine with each other to form N_2O_4 (diagram at right of Fig. 13.4). That is, reaction occurs in the reverse direction. Since two molecules of NO_2 form only one molecule of N_2O_4, this reduces the number of molecules.

It is possible, using the value of K_c, to calculate the extent to which NO_2 is converted to N_2O_4 when the volume is decreased. The results of

Figure 13.4 Effect of a decrease in volume upon the $N_2O_4(g) \rightleftharpoons 2NO_2(g)$ system at equilibrium. The immediate effect (*middle cylinder*) is to crowd the same number of moles of gas into a smaller volume and so to increase the total pressure. This is partially compensated for by the conversion of some of the NO_2 to N_2O_4, thereby reducing the total number of moles of gas.

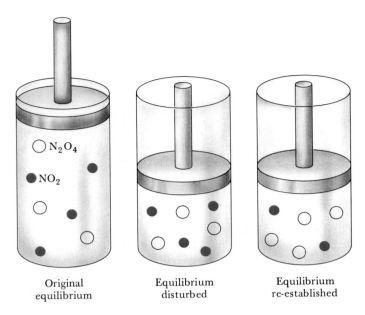

| Original equilibrium | Equilibrium disturbed | Equilibrium re-established |

such calculations are given in Table 13.3. As the volume is decreased from 10.0 to 1.0 L, more and more of the NO_2 is converted to N_2O_4. Notice that the total number of moles of gas decreases steadily as a result of this conversion.

The analysis we have gone through for the N_2O_4–NO_2 system can be applied to any chemical equilibrium involving gases. **When the volume of an equilibrium system is decreased, reaction takes place in the direction that decreases the total number of moles of gas. When the volume is increased, the reaction that increases the total number of moles of gas takes place.**

The application of this principle to several different systems is shown in Table 13.4. In System 2, the number of moles of gas decreases, from $\frac{3}{2}$ to 1, as the reaction goes to the right. Hence, decreasing the volume causes the forward reaction to occur; an increase in volume has the reverse effect. Notice that it is the change in the number of moles of *gas* that determines which way the equilibrium shifts (System 4). When there is no change in the number of moles of gas (System 5), a change in volume has no effect on the position of the equilibrium.

TABLE 13.3 Effect of Change in Volume on the Equilibrium System: $N_2O_4(g) \rightleftharpoons 2NO_2(g)$; $K_c = 0.36$ at 100°C

V (L)	MOLES NO_2	MOLES N_2O_4	MOLES TOTAL
10.0	1.20	0.40	1.60
5.0	0.96	0.52	1.48
2.0	0.68	0.66	1.34
1.0	0.52	0.74	1.26

TABLE 13.4 Effect of a Change in Volume upon the Position of Gaseous Equilibria

SYSTEM	V INCREASES	V DECREASES
1. $N_2O_4(g) \rightleftharpoons 2NO_2(g)$	$\rightarrow$	$\leftarrow$
2. $SO_2(g) + \frac{1}{2}O_2(g) \rightleftharpoons SO_3(g)$	$\leftarrow$	$\rightarrow$
3. $N_2(g) + 3H_2(g) \rightleftharpoons 2NH_3(g)$	$\leftarrow$	$\rightarrow$
4. $C(s) + H_2O(g) \rightleftharpoons CO(g) + H_2(g)$	$\rightarrow$	$\leftarrow$
5. $N_2(g) + O_2(g) \rightleftharpoons 2NO(g)$	0	0

Changes in volume of gaseous systems at equilibrium ordinarily result from changes in applied pressure. To cut the volume of the N_2O_4–NO_2 system in half, we would have to increase the pressure on the piston in Figure 13.4. Instead of saying that the shift in the position of the equilibrium comes about because of a change in volume, we might ascribe the shift to the pressure change that accompanies the volume decrease. Specifically, we could say that an *increase in pressure shifts the position of the equilibrium in such a way as to decrease the number of moles of gas* ($2NO_2 \rightarrow N_2O_4$).

In discussing the effect of pressure on an equilibrium system, we must be careful to specify how the pressure change comes about. There are many different ways in which we could change the total pressure in the N_2O_4–NO_2 system without changing the volume. We might, for example, add helium at constant volume. This increases the total number of moles and hence the total pressure. It has no effect, however, upon the position of the equilibrium. In general, we do not shift the position of an equilibrium by adding an unreactive gas, since that does not change the concentrations of reactants or products.

To disturb an equilibrium system, you need to change the concentration of at least one reactant or product

■ EXAMPLE 13.10

The pressure on each of the following systems is reduced so that the volume increases from 2 L to 10 L. Which way does the equilibrium shift?

a. $2CO_2(g) \rightleftharpoons 2CO(g) + O_2(g)$
b. $H_2(g) + I_2(g) \rightleftharpoons 2HI(g)$
c. $H_2(g) + I_2(s) \rightleftharpoons 2HI(g)$

Solution

a. $\rightarrow$ (2 mol gas $\rightarrow$ 3 mol gas). The immediate effect of the volume increase is to decrease the number of molecules per unit volume. This is partially compensated for by increasing the number of molecules.

b. No effect. Same number of moles of gas on both sides.

c. $\rightarrow$ (1 mol gas $\rightarrow$ 2 mol gas). Note that it is the number of moles of gas that is important; solids or liquids don't count.

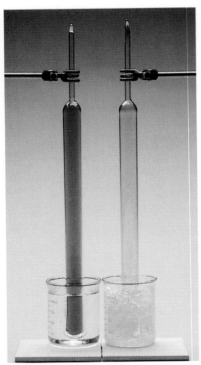

Figure 13.5 Effect of temperature on the N_2O_4–NO_2 system. At 0°C (*tube at right*) N_2O_4, which is colorless, predominates. At 50°C (*tube at left*), some of the N_2O_4 has dissociated to give the deep brown color of NO_2. (Marna G. Clarke)

The system shifts as T changes because K_c varies with T

Exercise

To increase the yield of NO_2 in the reaction

$$NO(g) + \tfrac{1}{2}O_2(g) \rightleftharpoons NO_2(g)$$

would you compress the system, expand the system, or add an inert gas? Answer: Compress it.

CHANGES IN TEMPERATURE

Let us suppose that we increase the temperature of an equilibrium mixture of N_2O_4 and NO_2. According to Le Châtelier's principle, the system will shift so as to partially counteract the temperature increase. This can be achieved if the forward reaction occurs:

$$N_2O_4(g) \rightarrow 2NO_2(g); \qquad \Delta H = +57.2 \text{ kJ}$$

Since the reaction is endothermic, it absorbs some of the heat used to raise the temperature. The result is to increase the concentration of NO_2 at the expense of N_2O_4. This effect brings about a color change. The reddish-brown color of NO_2 is much deeper at the higher temperature (Fig. 13.5).

In general, we can say that, for a system at equilibrium,

— **an increase in temperature causes the endothermic reaction to occur;**

$$N_2O_4(g) \rightarrow 2NO_2(g); \qquad \Delta H = +57.2 \text{ kJ}$$

— **a decrease in temperature causes the exothermic reaction to occur;**

$$2NO_2(g) \rightarrow N_2O_4(g); \qquad \Delta H = -57.2 \text{ kJ}$$

You will recall from our earlier discussion that the equilibrium constant of a system changes with temperature. The effect of a change in temperature upon an equilibrium system is often expressed in terms of its effect upon K_c:

If the forward reaction (left to right) is endothermic, K_c becomes larger as the temperature increases. If the forward reaction is exothermic, K_c becomes smaller as the temperature increases.

The effect of temperature upon the magnitude of K_c is often very large. For the synthesis of ammonia from the elements, K_c changes by a factor of 50,000 when the temperature is raised from 200°C to 600°C (Table 13.5).

TABLE 13.5   Effect of Temperature upon K_c

1. $N_2O_4(g) \rightleftharpoons 2NO_2(g)$;	$\Delta H = +57.2$ kJ	
t (°C) 0	50	100
K_c 0.0005	0.022	0.36

2. $N_2(g) + 3H_2(g) \rightleftharpoons 2NH_3(g)$;	$\Delta H = -92.2$ kJ	
t (°C) 200	400	600
K_c 650	0.50	0.014

■ **EXAMPLE 13.11** _____

Consider the system $I_2(g) \rightleftharpoons 2I(g)$; $\Delta H = +151$ kJ. Suppose the system is at equilibrium at 1000°C. In which direction will reaction occur if

a. I atoms are added?
b. the system is compressed?
c. the temperature is increased?

Solution

a. $2I(g) \rightarrow I_2(g)$

b. $2I(g) \rightarrow I_2(g)$ This decreases the number of particles per unit volume.

c. $I_2(g) \rightarrow 2I(g)$ This reaction absorbs heat.

Exercise

What effect, if any, will each of these changes (a, b, c) have on the magnitude of K_c? Answer: Only (c) changes K_c, making it larger.

13.5
EQUILIBRIUM EXPRESSIONS INVOLVING PARTIAL PRESSURES

So far, we have dealt with the type of equilibrium constant designated as K_c. As we have pointed out repeatedly, in the expression for K_c, equilibrium concentrations are expressed in moles per liter (molarity). There is another type of equilibrium constant in common use with gases. This constant is given the symbol K_p. In the expression for K_p, equilibrium concentrations are expressed as partial pressures in atmospheres.

The mathematical expression for K_p for a reaction is identical to that for K_c except that partial pressures replace molarities. Thus, we have

The partial pressures and concentrations of gases are directly related

$$N_2O_4(g) \rightleftharpoons 2NO_2(g); \qquad K_c = \frac{[NO_2]^2}{[N_2O_4]}; \qquad K_p = \frac{(P_{NO_2})^2}{P_{N_2O_4}}$$

$$N_2(g) + 3H_2(g) \rightleftharpoons 2NH_3(g); \qquad K_c = \frac{[NH_3]^2}{[N_2] \times [H_2]^3}; \qquad K_p = \frac{(P_{NH_3})^2}{P_{N_2} \times (P_{H_2})^3}$$

The symbol P in the expression for K_p represents an *equilibrium partial pressure in* **atmospheres**.

K_p, like K_c, is dependent only upon temperature. Its value is independent of the starting amounts of products or reactants, the volume of the container, or the applied pressure. In the expression for K_p, as with K_c, terms for solids and liquids are not included.

■ **EXAMPLE 13.12**

Write expressions for K_p for the following reactions:

a. The formation of 1 mol of phosphorus pentachloride gas from gaseous phosphorus trichloride and chlorine.
b. The thermal decomposition of 1 mol of solid calcium carbonate into lime (CaO) and carbon dioxide gas.

Solution

a. The reaction is

$$PCl_3(g) + Cl_2(g) \rightleftharpoons PCl_5(g)$$

The K_p expression is

If you can write the equation for K_c, you can write the one for K_p

$$K_p = \frac{P_{PCl_5}}{(P_{PCl_3}) \times (P_{Cl_2})}$$

b. The reaction is

$$CaCO_3(s) \rightleftharpoons CaO(s) + CO_2(g)$$

The K_p expression is

$$K_p = P_{CO_2}$$

Exercise

Write the K_p expression for the decomposition of one mole of solid ammonium nitrate into dinitrogen oxide gas and steam. Answer: $K_p = (P_{N_2O}) \times (P_{H_2O})^2$.

For most reactions, K_p and K_c have different numerical values at a given temperature. To relate K_p to K_c, we start by relating partial pressure, P, to concentration in moles per liter, n/V. For a gas A, we can express the partial pressure using the Ideal Gas Law:

P_A is in atmospheres

$$P_A = \frac{n_A RT}{V}$$

where R is the gas constant, 0.0821 L·atm/(mol·K), and T is the Kelvin temperature. At equilibrium, n_A/V becomes, by definition, [A], and we have

$$P_A = [A] \times RT \tag{13.12}$$

To show how Equation 13.12 can be used to relate K_p to K_c, consider the N_2O_4–NO_2 system:

$$N_2O_4(g) \rightleftharpoons 2NO_2(g); \qquad K_p = \frac{(P_{NO_2})^2}{P_{N_2O_4}}$$

Substituting for partial pressures, using Equation 13.12

$$K_p = \frac{[NO_2]^2 \times (RT)^2}{[N_2O_4] \times RT} = \frac{[NO_2]^2 \times RT}{[N_2O_4]}$$

However, $K_c = [NO_2]^2/[N_2O_4]$, so that for this system,

$$K_p = K_c \times RT$$

We could go through a similar argument to relate K_p to K_c for any gaseous system. It is simpler, however, to work with the following general equation, valid for all systems:

$$K_p = K_c \times (RT)^{\Delta n_g} \qquad (13.13)$$

K_p and K_c are always related this way

The exponent Δn_g in this equation is the change in the number of moles of gas in the equation written to represent the equilibrium system.

Δn_g = no. moles gaseous products − no. moles gaseous reactants (13.14)

Thus we have:

$$N_2O_4(g) \rightleftharpoons 2NO_2(g) \qquad \Delta n_g = +1$$

$$N_2(g) + 3H_2(g) \rightleftharpoons 2NH_3(g) \qquad \Delta n_g = -2$$

$$2HI(g) \rightleftharpoons H_2(g) + I_2(g) \qquad \Delta n_g = 0$$

$$CaCO_3(s) \rightleftharpoons CaO(s) + CO_2(g) \qquad \Delta n_g = +1$$

■ **EXAMPLE 13.13**

At 300°C, K_c for the system

$$N_2(g) + 3H_2(g) \rightleftharpoons 2NH_3(g)$$

is 9.5. Calculate K_p for this system at the same temperature.

Solution

$$\Delta n_g = -2; \qquad T = 573 \text{ K};$$

$$K_p = K_c \times (0.0821 \times 573)^{-2} = \frac{9.5}{(0.0821 \times 573)^2} = \boxed{4.3 \times 10^{-3}}$$

Exercise

At 520°C, K_c for the system $2HI(g) \rightleftharpoons H_2(g) + I_2(g)$ is 0.016. What is the value of K_p at 520°C? Answer: 0.016.

The equilibrium constant K_p can be used to calculate the direction or extent of reaction in much the same way that we used K_c earlier in the chapter. The only difference is that we work with partial pressures in atmospheres rather than concentrations in moles per liter. When we say that K_p for the reaction:

$$CaCO_3(s) \rightleftharpoons CaO(s) + CO_2(g)$$

is "1.0" at 840°C, we mean that the equilibrium pressure of CO_2 at this temperature is one atmosphere.

In working problems involving K_p, it is helpful to realize that, in an equilibrium system, partial pressure, like concentration, is directly proportional to number of moles. In a container of fixed volume at constant temperature

$$P = \frac{nRT}{V} = \text{constant} \times n$$

This means that changes in partial pressure of different species, like changes in concentration, are related through the coefficients of the balanced equation for the equilibrium system. Thus we have:

$$N_2O_4(g) \rightleftharpoons 2NO_2(g) \qquad \Delta \text{ conc. } NO_2 = -2\Delta \text{ conc. } N_2O_4$$

$$\Delta P_{NO_2} = -2\Delta P_{N_2O_4}$$

$$H_2(g) + I_2(g) \rightleftharpoons 2HI(g) \qquad \Delta \text{ conc. } HI = -2\Delta \text{ conc. } H_2$$

$$= -2\Delta \text{ conc. } I_2$$

$$\Delta P_{HI} = -2\Delta P_{H_2} = -2\Delta P_{I_2}$$

As a result, the analysis of an equilibrium problem using K_p is entirely analogous to that used with K_c. Indeed, we can follow the same five-step procedure referred to earlier (p. 443).

■ **EXAMPLE 13.14** _____

Consider the reaction between sulfur dioxide gas and nitrogen dioxide gas to produce nitrogen oxide and sulfur trioxide. When the equation is balanced using lowest whole number coefficients, $K_p = 85.0$ at 460°C. Suppose nitrogen dioxide and sulfur dioxide, both with partial pressures of 0.500 atm, are initially present in a 5.00-L flask. Calculate equilibrium partial pressures for all gases.

Solution

1. The balanced equation for the reaction is

$$SO_2(g) + NO_2(g) \rightleftharpoons NO(g) + SO_3(g)$$

Since pressures of gases are easily measured, a chemist doing gas equilibrium problems would usually work with K_p rather than K_c

2. The equilibrium expression is

$$K_p = 85.0 = \frac{(P_{NO}) \times (P_{SO_3})}{(P_{SO_2}) \times (P_{NO_2})}$$

3. We must first express all the equilibrium partial pressures in terms of a single unknown x. Let x be the equilibrium pressure (in atmospheres) of NO. Since there is no NO to start with, the equilibrium pressure x must be reached by increasing the pressure of NO by $+x$. Hence the change in pressure of NO is $+x$. Remember that changes in partial pressure are related through the coefficients of the balanced equation. In this case, each coefficient is 1. Thus the partial pressures of SO_2 and NO_2 *decrease* by x, while that of SO_3 *increases* by x. Putting this reasoning in the form of a table, we get

$$SO_2(g) + NO_2(g) \rightleftharpoons SO_3(g) + NO(g)$$

	INITIAL PRESSURE (atm)	CHANGE IN PRESSURE (atm)	EQUIL. PRESSURE (atm)
SO_2	0.500	$-x$	$0.500 - x$
NO_2	0.500	$-x$	$0.500 - x$
NO	0.000	$+x$	x
SO_3	0.000	$+x$	x

This table is very similar to the one with K_c

4. We now substitute into the equilibrium expression and obtain

$$K_p = 85.0 = \frac{(x)(x)}{(0.500 - x)(0.500 - x)} = \frac{x^2}{(0.500 - x)^2}$$

Simplifying the equation by taking the square root of both sides we get

$$(85.0)^{1/2} = 9.22 = \frac{x}{0.500 - x}$$

Solving for x:

$$x = 9.22(0.500 - x) = 4.61 - 9.22x$$

$$x = 0.451 \text{ atm}$$

5. Referring back to the equilibrium table

$$P_{NO} = P_{SO_3} = x = \boxed{0.451 \text{ atm}}$$

$$P_{NO_2} = P_{SO_3} = 0.500 - x = \boxed{0.049 \text{ atm}}$$

Exercise
For the system $BrF(g) + F_2(g) \rightleftharpoons BrF_3(g)$; $K_p = 1$ at 900°C. What is the algebraic expression obtained in Step 4 if initial pressures are $P_{BrF_3} = 0.0100$ atm, $P_{BrF} = 0.100$ atm, and $P_{F_2} = 0.200$ atm? Answer: $\frac{0.0100 + x}{(0.100 - x)(0.200 - x)}$.

■ **SUMMARY PROBLEM**

Consider bromine chloride, an interhalogen compound. It is formed by the reaction between red-orange bromine vapor and yellow chlorine gas; BrCl is itself a gas. The reaction is endothermic.

a. Write the chemical equation for the formation of BrCl, using simplest whole-number coefficients.
b. Write the equilibrium expression for K_c.
c. At 400°C, after the reaction reached equilibrium, the mixture contained 0.82 M BrCl, 0.20 M Br_2 and 0.48 M Cl_2. Calculate K_c for the reaction.
d. Write the equilibrium expression for K_p.

e. What is K_p at 400°C?
f. Initially, a 2.00-L flask contains Cl_2 with partial pressure 0.51 atm and Br_2 with partial pressure 0.34 atm. After equilibrium is established, the partial pressure of BrCl is 0.46 atm. Calculate the equilibrium pressures of Cl_2 and Br_2.
g. Initially, a 2.00-L reaction vessel contains 0.15 mol of each gas. In what direction will the reaction proceed? What are the equilibrium concentrations of each gas after equilibrium is established? If 0.050 mol BrCl are added, in what direction will equilibrium shift? What are the equilibrium concentrations after equilibrium has been re-established?
h. In what direction will the system shift if at equilibrium
    1. the volume is increased?
    2. helium gas is added?
    3. the temperature is increased?

Answers

a. $Br_2(g) + Cl_2(g) \rightleftharpoons 2BrCl(g)$ b. $K_c = \dfrac{[BrCl]^2}{[Br_2] \times [Cl_2]}$ c. 7.0

d. $K_p = \dfrac{(P_{BrCl})^2}{P_{Br_2} \times P_{Cl_2}}$ e. 7.0   f. 0.28 atm; 0.11 atm

g. $\rightarrow$; $[Br_2] = [Cl_2] = 0.048\ M$, $[BrCl] = 0.129\ M$; $\leftarrow$; $[Br_2] = [Cl_2] = 0.054\ M$, $[BrCl] = 0.142\ M$ h. (1) no change; (2) no change; (3) $\rightarrow$

Throughout this chapter, we have emphasized the principles of equilibrium in the gas state. It may be appropriate to end the chapter on a more descriptive note, dealing with the preparation and properties of hydrogen, an important gas mentioned in this and in earlier chapters.

Preparation

Small amounts of hydrogen are commonly made in the laboratory by reacting zinc with hydrochloric acid:

$$Zn(s) + 2H^+(aq) \rightarrow Zn^{2+}(aq) + H_2(g) \qquad (13.15)$$

or by the electrolysis of water:

$$H_2O(l) \rightarrow H_2(g) + \tfrac{1}{2}O_2(g) \qquad (13.16)$$

Both of these processes are relatively expensive—Reaction 13.15 because of the cost of the chemicals, Reaction 13.16 because of the large amount of electrical energy consumed (118 kJ per gram of hydrogen).

Commercially, hydrogen gas can be prepared by heating methane with steam. At 1100°C, the reaction

$$CH_4(g) + H_2O(g) \rightarrow CO(g) + 3H_2(g) \qquad (13.17)$$

goes essentially to completion ($K_p = 2400$). To remove the carbon monoxide from the product, the gas mixture is cooled to 400°C, where the exothermic reaction

$$CO(g) + H_2O(g) \rightarrow CO_2(g) + H_2(g) \qquad \Delta H = -41.2 \text{ kJ} \quad (13.18)$$

takes place ($K_p = 10$ at 400°C). The carbon dioxide is removed quite simply by bubbling the gas through water, in which CO_2 is much more soluble than hydrogen. The overall equation for the preparation of hydrogen by this process can be written:

$$CH_4(g) + 2H_2O(g) \rightarrow CO_2(g) + 4H_2(g) \qquad (13.19)$$

The mixture of carbon monoxide and hydrogen produced by Reaction 13.17 is commonly referred to as *synthesis gas*. It can also be made by passing steam through hot coke (carbon):

$$C(s) + H_2O(g) \rightarrow CO(g) + H_2(g) \qquad (13.20)$$

Synthesis gas can be converted to a liquid fuel by heating to 200–300°C in the presence of a catalyst, usually a compound of iron or cobalt. Under these conditions, CO and H_2 react to form useful hydrocarbons. The following reactions are typical:

$$9H_2(g) + 16CO(g) \rightarrow C_8H_{18}(l) + 8CO_2(g) \qquad (13.21)$$

$$17H_2(g) + 8CO(g) \rightarrow C_8H_{18}(l) + 8H_2O(l) \qquad (13.22)$$

A 60% yield of hydrocarbons in the gasoline range can be obtained. This process, known as the Fischer-Tropsch synthesis, was used in Germany

Hydrogen gas is commonly prepared in the laboratory by reacting zinc with acid. (Charles Steele)

during World War II to produce a low-grade gasoline. The only full-scale plant using this process today is located in South Africa, where coal is plentiful but all oil must be imported. Several small units have been built in the United States to study various modifications of the Fischer-Tropsch synthesis. At this point, the product is considerably more expensive than ordinary gasoline.

Isotopes of Hydrogen; Physical Properties

Of the three isotopes of hydrogen, the lightest one, ^1_1H, is by far the most common. Deuterium, ^2_1H, comprises 0.0156% of natural hydrogen; tritium, ^3_1H, which is highly radioactive, is found only in trace amounts in nature. The physical properties of these isotopes (Table 13.6) differ from one another considerably. Notice that volatility, as measured by boiling point or heat of vaporization, decreases as molar mass increases from H_2 to D_2 to T_2.

When water is electrolyzed, the hydrogen gas formed is slightly enriched in the more volatile, light isotope, ^1_1H. Conversely, the water remaining contains a slightly higher fraction of deuterium than it did originally. By repeated electrolyses, a small amount of nearly pure "heavy water," D_2O, can be prepared. This is used in certain types of nuclear reactors. It is also the starting material used by chemists to make deuterated compounds, which in turn are used to trace the path of hydrogen atoms through complex chemical reactions.

The bond energy of hydrogen, 436 kJ/mol, is unusually large for a nonpolar single bond. The "atomic hydrogen" torch takes advantage of this factor. Diatomic H_2 molecules are first dissociated to atoms in an electric arc; the atoms are then allowed to recombine on the surface of a metal. The large amount of heat given off in the reaction

$$H(g) + H(g) \rightarrow H_2(g); \qquad \Delta H = -436 \text{ kJ} \qquad (13.23)$$

produces a temperature close to 4000°C. This is high enough to melt just about any metal, including tantalum (mp = 2980°C) and tungsten (mp = 3410°C).

Chemical Properties

As we saw in Chapter 6, hydrogen reacts with Group 1 and Group 2 metals to form ionic hydrides containing the H^- ion. The reactions with sodium and calcium are typical:

TABLE 13.6 Properties of Hydrogen, Deuterium, and Tritium

	H_2	D_2	T_2
Molar mass (g/mol)	2.016	4.028	6.032
Melting point (°C)	−259.2	−254.4	−252.5
Boiling point (°C)	−252.8	−249.5	−248.1
ΔH_{vap} (kJ/mol)	0.90	1.23	1.39
Bond energy (kJ/mol)	435.9	443.4	446.9

$$2Na(s) + H_2(g) \rightarrow 2NaH(s) \qquad\qquad\qquad (13.24)$$

$$Ca(s) + H_2(g) \rightarrow CaH_2(g) \qquad\qquad\qquad (13.25)$$

Certain transition metals behave quite differently toward hydrogen. When water is electrolyzed using palladium electrodes, very little hydrogen gas is produced. The metal absorbs up to one thousand times its volume of hydrogen. The fact that the conductivity of palladium drops off sharply as it takes up hydrogen suggests a chemical interaction between the two elements. It appears that in this and other *interstitial hydrides*, hydrogen is present as H atoms, which fit into "holes" or interstices in the metal lattice.

It will actually swell up

Of the reactions of hydrogen with nonmetals, that with oxygen is perhaps the most familiar (Fig. 13.6).

$$H_2(g) + \tfrac{1}{2}O_2(g) \rightarrow H_2O(l); \qquad \Delta H = -285.8 \text{ kJ} \qquad (13.26)$$

The amount of heat given off per gram of fuel burned:

$$285.8 \text{ kJ}/2.016 \text{ g} = 141.8 \text{ kJ/g}$$

exceeds that for any other fuel, which explains why this reaction is used as a source of energy in the space program. Liquid hydrogen and liquid oxygen were used in the Saturn V rocket that put astronauts on the moon; they are also used to power the space shuttle (Fig. 13.7).

In organic chemistry, hydrogen gas is often used to convert carbon-carbon double bonds to single bonds:

$$\begin{matrix} \diagdown \\ \diagup \end{matrix} C{=}C \begin{matrix} \diagup \\ \diagdown \end{matrix} + \text{ H}{-}\text{H} \rightarrow {-} \underset{\underset{\text{H}}{|}}{\overset{|}{\text{C}}} {-} \underset{\underset{\text{H}}{|}}{\overset{|}{\text{C}}} {-}$$

 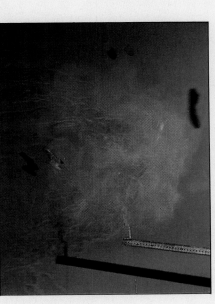

Figure 13.6 When a hydrogen-filled balloon is ignited from a distance (caution!), it explodes. The reaction is: $2H_2(g) + O_2(g) \rightarrow 2H_2O(l)$. (Charles D. Winters)

Commercially, this reaction is used to convert the multiple bonds in unsaturated fats to the single bonds characteristic of saturated fats. In this way, vegetable oils are converted to solid cooking fats and margarine. Removing the multiple bonds makes the fats much less susceptible to oxidation and rancidity. On the other hand, we now know that saturated fats raise the level of cholesterol in the blood. This in turn promotes the formation of fatty deposits on the inner walls of arteries, which can bring on a heart attack or stroke.

Figure 13.7 The reaction between liquid hydrogen and liquid oxygen is used to power the space shuttle. (NASA)

QUESTIONS AND PROBLEMS

Establishment of Equilibrium

1. The following data is for the system

$$A(g) \rightleftharpoons 2 B(g)$$

Time (s)	0	15	30	45	60	75	90
Conc. A (M)	0.500	0.450	0.400	0.360	0.340	0.325	0.325
Conc. B (M)	0.000	0.100	0.200	0.280	0.320	0.350	0.350

At what time (s) is the system in equilibrium?

2. For the system in Question 1,

a. What is the concentration of B after 2 minutes?
b. How does the rate of the forward reaction compare with the rate of the reverse reaction at 45 s? At 90 s?

3. Complete the table below for the reaction

$$A(g) + 2B(g) \rightleftharpoons C(g)$$

Time (s)	0	20	40	60	80	100
Conc. A (M)	0.200	0.150	___	___	0.100	___
Conc. B (M)	0.250	___	___	0.070	___	___
Conc. C (M)	0.000	___	0.075	___	___	0.100

4. The following data apply to the *unbalanced* equation

$$A(g) \rightleftharpoons B(g)$$

Time (min)	0	2	4	6	8
Conc. A (M)	0.600	0.480	0.390	0.345	0.315
Conc. B (M)	0.000	0.040	0.070	0.085	0.095

a. Based on the data, balance the equation.
b. Has the system reached equilibrium? Explain.

Expressions for K_c

5. Write equilibrium constant (K_c) expressions for the following reactions:
a. $CS_2(g) + 4H_2(g) \rightleftharpoons CH_4(g) + 2H_2S(g)$
b. $2H_2O_2(g) \rightleftharpoons 2H_2O(g) + O_2(g)$
c. $C_3H_8(g) + 5O_2(g) \rightleftharpoons 3CO_2(g) + 4H_2O(g)$

6. Write equilibrium constant expressions (K_c) for the following reactions:
a. $IF(g) \rightleftharpoons \frac{1}{2}I_2(g) + \frac{1}{2}F_2(g)$
b. $2C_5H_6(g) \rightleftharpoons C_{10}H_{12}(g)$
c. $P_4O_{10}(g) + 6PCl_5(g) \rightleftharpoons 10POCl_3(g)$

7. Write equilibrium constant expressions (K_c) for the following reactions:
a. $ZnO(s) + CO(g) \rightleftharpoons Zn(l) + CO_2(g)$
b. $2HgO(s) \rightleftharpoons 2Hg(l) + O_2(g)$
c. $2PbS(s) + 3O_2(g) \rightleftharpoons 2PbO(s) + 2SO_2(g)$

8. Write equilibrium constant expressions (K_c) for the following reactions:
a. $Ni(s) + 4CO(g) \rightleftharpoons Ni(CO)_4(g)$
b. $2Ag_2O(s) \rightleftharpoons 4Ag(s) + O_2(g)$
c. $4KO_2(s) + 2H_2O(g) \rightleftharpoons 4KOH(s) + 3O_2(g)$

9. Given the following descriptions of reversible reactions, write a balanced equation (simplest whole-number coefficients) and the equilibrium constant expression (K_c) for each.
a. Liquid acetone, C_3H_6O, is in equilibrium with its vapor.
b. Hydrogen gas reduces nitrogen dioxide to form ammonia and steam.
c. Chlorine gas reacts with liquid carbon disulfide to produce the liquids carbon tetrachloride and disulfur dichloride.

10. Given the following descriptions of reversible reactions, write a balanced equation (simplest whole-number coefficients) and the equilibrium expression (K_c) for each.
a. Nitrogen gas reacts with solid sodium carbonate and solid carbon to produce carbon monoxide gas and solid sodium cyanide.
b. Solid magnesium nitride reacts with water vapor to form magnesium hydroxide solid and ammonia gas.
c. Solid barium carbonate decomposes to solid barium oxide and carbon dioxide gas.

11. Write a chemical equation for an equilibrium system that would lead to the following expressions for K_c:

a. $\dfrac{[O_2]^3}{[O_3]^2}$ b. $\dfrac{[SO_3]^2}{[O_2] \times [SO_2]^2}$

c. $\dfrac{[NH_3]^2}{[N_2] \times [H_2]^3}$ d. $\dfrac{[H_2O]^2 \times [Cl_2]^2}{[O_2] \times [HCl]^4}$

12. Write a chemical equation for an equilibrium system that would lead to the following expressions for K_c:

a. $\dfrac{[H_2] \times [Br_2]}{[HBr]^2}$ b. $\dfrac{[CH_3OH]}{[CO] \times [H_2]^2}$

c. $\dfrac{[C_2H_4] \times [H_2]}{[C_2H_6]}$ d. $\dfrac{[SO_3]}{[SO_2] \times [O_2]^{1/2}}$

Calculation of K_c

13. At 25°C, $K_c = 2.2 \times 10^{-3}$ for the reaction

$$ICl(g) \rightleftharpoons \tfrac{1}{2}I_2(g) + \tfrac{1}{2}Cl_2(g)$$

Calculate K_c at 25°C for
a. $2ICl(g) \rightleftharpoons I_2(g) + Cl_2(g)$
b. $2I_2(g) + 2Cl_2(g) \rightleftharpoons 4ICl(g)$

14. At 627°C, $K_c = 56$ for the reaction

$$2SO_2(g) + O_2(g) \rightleftharpoons 2SO_3(g)$$

Calculate K_c at 627°C for
a. $2SO_3(g) \rightleftharpoons 2SO_2(g) + O_2(g)$
b. $4SO_2(g) + 2O_2(g) \rightleftharpoons 4SO_3(g)$

15. Given that, for the reactions

$$SnO_2(s) + 2H_2(g) \rightleftharpoons Sn(s) + 2H_2O(g)$$

$$CO(g) + H_2O(g) \rightleftharpoons CO_2(g) + H_2(g)$$

K_c is 21 and 0.034, respectively, calculate K_c for the reaction: $SnO_2(s) + 2CO(g) \rightleftharpoons Sn(s) + 2CO_2(g)$

16. Given that K_c for the reactions

$$H_2(g) + S(s) \rightleftharpoons H_2S(g)$$

$$S(s) + O_2(g) \rightleftharpoons SO_2(g)$$

is 1.0×10^{-3} and 5.0×10^6, respectively, calculate K_c for the reaction: $H_2(g) + SO_2(g) \rightleftharpoons H_2S(g) + O_2(g)$

17. Given the following data at 25°C

$$2NO(g) \rightleftharpoons N_2(g) + O_2(g) \qquad K_c = 1 \times 10^{30}$$

$$2NO(g) + Br_2(g) \rightleftharpoons 2NOBr(g) \qquad K_c = 2 \times 10^3$$

Calculate K_c for the formation of one mole of NOBr from its elements in the gaseous state at 25°C.

18. Given the following data at a certain temperature

$$2N_2(g) + O_2(g) \rightleftharpoons 2N_2O(g); \qquad K_c = 1.2 \times 10^{-35}$$

$$N_2O_4(g) \rightleftharpoons 2NO_2(g) \qquad K_c = 4.6 \times 10^{-3}$$

$$\tfrac{1}{2}N_2(g) + O_2(g) \rightleftharpoons NO_2(g) \qquad K_c = 4.1 \times 10^{-9}$$

Calculate K_c for the reaction between one mole of dinitrogen oxide gas and oxygen gas to give dinitrogen tetroxide gas.

19. Calculate K_c at a certain temperature for

$$2CO_2(g) \rightleftharpoons 2CO(g) + O_2(g)$$

given that the equilibrium concentrations of carbon monoxide, oxygen, and carbon dioxide at that temperature are 0.35 M, 0.18 M, and 0.029 M, respectively.

20. For the system

$$CO(g) + 3H_2(g) \rightleftharpoons CH_4(g) + H_2O(g)$$

analysis shows that at a certain temperature, the concentrations of methane, steam, hydrogen gas, and carbon monoxide gas are 0.150 M, 0.233 M, 0.259 M, and 0.513 M, respectively. Calculate K_c at that temperature.

21. Carbon dioxide reacts with solid carbon to give carbon monoxide. At 700°C and equilibrium, a 5.0-L flask contains 0.27 mol carbon monoxide, 0.58 mol of carbon dioxide, and 0.44 mol carbon.
 a. Write a balanced equation for the reaction of one mole of carbon dioxide with carbon.
 b. Calculate K_c for the reaction at 700°C.

22. At 227°C, carbon monoxide gas reacts with hydrogen gas to form one mole of methanol, $CH_3OH(g)$. At equilibrium and 227°C, a 10.0-L flask contains 0.114 moles of methanol gas, 0.718 moles of carbon monoxide, and 1.23 moles of hydrogen gas.
 a. Write a balanced equation for the reaction.
 b. Calculate K_c for the reaction at 227°C.

23. When 0.0930 mol nitrogen oxide gas and 0.0652 mol bromine gas are sealed into a one-liter container at 77°C, the following equilibrium is established:

$$2NO(g) + Br_2(g) \rightleftharpoons 2NOBr(g)$$

Given that [NOBr] = 0.0612 M, calculate K_c at 77°C.

24. One of the steps in the manufacture of sulfuric acid is the oxidation of sulfur dioxide gas with oxygen to produce sulfur trioxide gas. The reaction is

$$SO_2(g) + \tfrac{1}{2}O_2(g) \rightleftharpoons SO_3(g)$$

At 627°C, 0.0123 moles of oxygen and 0.0271 moles of sulfur dioxide are placed in a sealed one-liter container. When equilibrium is established, the equilibrium concentration of sulfur trioxide is found to be 0.0198 M. Calculate K_c for the reaction at 627°C.

25. Consider the decomposition of ammonia to its elements:

$$2NH_3(g) \rightleftharpoons N_2(g) + 3H_2(g)$$

In a sealed flask at 633°C, initial concentrations are 0.296 M for ammonia, 0.170 M for nitrogen, and 0.095 M for hydrogen gas. When equilibrium is established at the same temperature, the equilibrium concentration of ammonia is 0.268 M. Calculate K_c for the decomposition of ammonia at 633°C.

26. Consider the decomposition of nitrosyl chloride, $NOCl(g)$:

$$2NOCl(g) \rightleftharpoons 2NO(g) + Cl_2(g)$$

At 220°C in a sealed flask, initial concentrations are 0.520 M nitrosyl chloride, 0.010 M nitrogen oxide, and 0.053 M chlorine. When equilibrium is established at the same temperature, it is found that 4.23% of the nitrosyl chloride is decomposed. Calculate K_c for the decomposition at 220°C.

K_c; Direction and Extent of Reaction

27. A gaseous reaction mixture contains 0.30 mol SO_2, 0.16 mol Cl_2, and 0.50 mol SO_2Cl_2 in a 2.0-L container; $K_c = 0.011$ for $SO_2Cl_2(g) \rightleftharpoons SO_2(g) + Cl_2(g)$.
 a. Is the system at equilibrium? Explain.
 b. If it is not at equilibrium, in which direction will the system move to reach equilibrium?

28. K_c is 0.56 at 300°C for the system

$$PCl_5(g) \rightleftharpoons PCl_3(g) + Cl_2(g)$$

In a 5.0-L flask, a gaseous mixture consists of 0.45 mol Cl_2, 0.90 mol PCl_3, and 0.12 mol PCl_5.
 a. Is the mixture at equilibrium? Explain.
 b. If not at equilibrium, which way will the system shift to establish equilibrium?

29. For the reaction

$$2NO_2(g) \rightleftharpoons 2NO(g) + O_2(g)$$

K_c at a certain temperature is 0.50. Predict the direction in which the reaction will occur to establish equilibrium if one starts with
 a. Conc. NO_2 = conc. NO = conc. O_2 = 0.10 M.
 b. 1.23 mol NO_2 and 0.168 mol O_2 in a 14.5-L container.
 c. 2.0 mol NO_2, 0.40 mol NO, and 0.10 mol O_2 in a 10-L container.

30. The reversible reaction between hydrogen chloride gas and one mole of oxygen produces steam and chlorine gas. K_c at a certain temperature is 1.6. Predict the direction in which the system will move to reach equilibrium if one starts with
 a. Conc. HCl = conc. O_2 = conc. H_2O = 0.20 M.
 b. 1.20 mol hydrogen chloride gas, 0.60 mol oxygen, 1.40 mol steam, and 0.80 mol chlorine in a 4.0-L container.

31. For the reaction

$$Cl_2(g) \rightleftharpoons 2Cl(g)$$

$K_c = 1.2 \times 10^{-6}$ at 1100°C. If one starts with 1.0 mol Cl_2 in a 2.0-L container at 1100°C, will the equilibrium system contain
 a. mostly Cl_2?   b. mostly Cl atoms?
 c. about equal amounts of Cl_2 and Cl?

32. At 2400°C, K_c for the reaction in Problem 31 is 3.6×10^{-2}. At this temperature, will the equilibrium system contain mostly Cl_2, mostly Cl, or appreciable amounts of both species?

K_c; Equilibrium Concentrations

33. At a certain temperature, K_c is 1.3×10^{10} for the reaction

$$2H_2(g) + S_2(g) \rightleftharpoons 2H_2S(g)$$

What is the equilibrium concentration of hydrogen sulfide if those of hydrogen and sulfur gases are $0.0034\ M$ and $0.0017\ M$, respectively?

34. At a certain temperature, K_c is 0.040 for the decomposition of two moles of bromine chloride to its elements. An equilibrium mixture at this temperature contains equal concentrations of bromine and chlorine gases, $0.0325\ M$. What is the equilibrium concentration of bromine chloride?

35. For the system

$$PCl_5(g) \rightleftharpoons PCl_3(g) + Cl_2(g)$$

$K_c = 33.3$ at a certain temperature. In an equilibrium mixture at that temperature, $[PCl_3] = 4.0 \times [PCl_5]$. What is the equilibrium concentration of Cl_2?

36. For the system

$$2HI(g) \rightleftharpoons H_2(g) + I_2(g)$$

$K_c = 0.016$ at 800 K. If, at 800 K, $[HI] = 0.50\ M$ and $[H_2] = [I_2]$, calculate the equilibrium concentration of H_2.

37. At 460°C, the reaction

$$SO_2(g) + NO_2(g) \rightleftharpoons NO(g) + SO_3(g)$$

has $K_c = 85.0$. What will be the equilibrium concentrations of the four gases if a mixture of SO_2 and NO_2 is prepared in which they both have initial concentrations of $0.0750\ M$?

38. At 650°C, the reaction

$$H_2(g) + CO_2(g) \rightleftharpoons CO(g) + H_2O(g)$$

has a K_c value of 0.771. If 2.00 mol of both hydrogen and carbon dioxide are placed in a 4.00-L container and allowed to react, what will be the equilibrium concentrations of all four gases?

39. Hydrogen gas and cyanogen gas, C_2N_2, react at a certain temperature to form hydrogen cyanide. The equation for the reaction is

$$H_2(g) + C_2N_2(g) \rightleftharpoons 2HCN(g)$$

K_c for this system is 1.50. A mixture of 0.200 mol hydrogen, 0.150 mol cyanogen, and 0.0100 mol hydrogen cyanide is sealed in a 1.00-L flask. Calculate equilibrium concentrations for all gases.

40. Carbonyl bromide, $COBr_2$, can be formed by reacting carbon monoxide with bromine gas. The equation for the reaction is

$$CO(g) + Br_2(g) \rightleftharpoons COBr_2(g)$$

Take K_c for this system to be 5.26. A mixture of 0.400 mol carbon monoxide, 0.300 mol bromine, and 0.0200 mol carbonyl bromide is sealed in a 5.00-L flask. Calculate equilibrium concentrations for all gases.

41. For the reaction

$$CO(g) + H_2O(g) \rightleftharpoons CO_2(g) + H_2(g)$$

equilibrium is established at a certain temperature when the concentrations of CO, H_2O, CO_2, and H_2 are 0.010, 0.020, 0.012, and 0.012 mol/liter, respectively.
  a. Calculate K_c.
  b. If enough CO is added to raise its concentration temporarily to $0.020\ M$, what will be the concentrations of all species after equilibrium has been re-established?

42. Water gas, a commercial fuel, is made by reacting hot coke with steam:

$$C(s) + H_2O(g) \rightleftharpoons CO(g) + H_2(g)$$

When equilibrium is established at 900°C, the concentrations of CO, H_2, and H_2O are $0.040\ M$, $0.040\ M$, and $0.010\ M$, respectively.
  a. Calculate K_c.
  b. If enough steam is added to raise its concentration temporarily to $0.040\ M$, what will be the concentrations of CO and H_2 when equilibrium is re-established?

Le Châtelier's Principle

43. Consider the system

$$SO_3(g) \rightleftharpoons SO_2(g) + \tfrac{1}{2}O_2(g)$$

ΔH for the forward reaction is 98.9 kJ. Predict whether the forward or reverse reaction will occur when the equilibrium is disturbed by
  a. adding oxygen gas.
  b. compressing the system at constant temperature.
  c. adding argon gas.
  d. decreasing the temperature.
  e. removing sulfur dioxide gas.

44. Consider the system

$$4NH_3(g) + 3O_2(g) \rightleftharpoons 2N_2(g) + 6H_2O(l);$$
$$\Delta H = -1530.4\ kJ$$

How will the amount of ammonia at equilibrium be affected by
  a. removing oxygen gas?
  b. adding nitrogen gas?
  c. adding water?
  d. expanding the container at constant temperature?
  e. increasing the temperature?

45. Predict the direction in which each of the following equilibria will shift if the pressure on the system is decreased by expansion.

a. $Ni(s) + 4CO(g) \rightleftharpoons Ni(CO)_4(g)$
b. $ClF_5(g) \rightleftharpoons ClF_3(g) + F_2(g)$
c. $HBr(g) \rightleftharpoons \frac{1}{2}H_2(g) + \frac{1}{2}Br_2(g)$

46. Predict the direction in which each of the following equilibria will shift if the pressure on the system is increased by compression.

a. $H_2O(g) + C(s) \rightleftharpoons CO(g) + H_2(g)$
b. $SbCl_5(g) \rightleftharpoons SbCl_3(g) + Cl_2(g)$
c. $CO(g) + H_2O(g) \rightleftharpoons CO_2(g) + H_2(g)$

47. In the forward reaction of the Haber process (the commercial preparation of ammonia), two moles of ammonia are produced by reacting nitrogen and hydrogen gases. K_c is 5×10^8 at 25°C and 1 at 367°C. Is the forward reaction exothermic? Based only on equilibrium conditions, what are the optimal conditions (T and P) for the Haber process?

48. Consider the equilibrium system

$$Fe_3O_4(s) + 4H_2(g) \rightleftharpoons 3Fe(s) + 4H_2O(g)$$

At 900°C, K_c is 80. Using the appropriate tables in Appendix 1, determine whether K_c will increase or decrease with an increase in temperature. Explain.

49. The system

$$2\,X(g) + Z(g) \rightleftharpoons Q(g)$$

is at equilibrium when the concentration of X is 0.60 M. Sufficient X is added to raise its concentration temporarily to 1.0 M. When equilibrium is re-established, the concentration of X could be which of the following?

a. 0.80 M b. 0.60 M c. 0.90 M
d. 1.0 M e. 1.2 M

50. For the system in Problem 49, the initial equilibrium concentration of Z was 0.40 M. When equilibrium is re-established, the concentration of Z could be

a. 0.40 M b. 0.50 M c. 0.35 M
d. 0.30 M e. 0.00 M

K_p Calculations

51. Consider the reaction

$$H_2O(l) \rightleftharpoons H_2O(g)$$

a. Using the table of vapor pressures of water in Appendix 1, calculate K_p for pressures in atmospheres at 22°C, 100°C, and 108°C.
b. Calculate K_c at each of those temperatures.

52. Given the following K_c values, calculate the corresponding values of K_p.

a. $CO_2(g) + H_2(g) \rightleftharpoons CO(g) + H_2O(g)$ $K_c = 0.0431$ at 327°C
b. $2SO_3(g) \rightleftharpoons 2SO_2(g) + O_2(g)$ $K_c = 3.2 \times 10^{-4}$ at 627°C
c. $2H_2(g) + S_2(g) \rightleftharpoons 2H_2S(g)$ $K_c = 2.0 \times 10^4$ at 1132°C

d. $NO(g) + \frac{1}{2}O_2(g) \rightleftharpoons NO_2(g)$ $K_c = 3.0 \times 10^3$ at 184°C

53. Calcium carbonate decomposes to calcium oxide and carbon dioxide. When one mole of solid calcium carbonate decomposes at 700°C, the pressure of the equilibrium mixture is 0.105 atm. What is K_p for the reaction?

54. Ammonium carbamate, $NH_4CO_2NH_2$, decomposes according to the equation

$$NH_4CO_2NH_2(s) \rightleftharpoons 2NH_3(g) + CO_2(g)$$

When equilibrium is reached, the partial pressures of the gases at 40°C are 0.342 atm for ammonia and 0.0606 atm for carbon dioxide. Calculate K_p for the system at 40°C.

55. For the reaction

$$2NO_2(g) \rightleftharpoons 2NO(g) + O_2(g)$$

K_p is 1.74×10^{-4} at 500 K. Calculate P_{NO} if $P_{O_2} = 0.127$ atm and $P_{NO_2} = 0.384$ atm at equilibrium.

56. Consider the reaction

$$2NO(g) + Cl_2(g) \rightleftharpoons 2NOCl(g)$$

At a certain temperature, $K_p = 0.30$. Calculate P_{Cl_2} if $P_{NO} = 0.65$ atm and $P_{NOCl} = 0.15$ atm at equilibrium.

57. For the formation of nitrogen oxide:

$$N_2(g) + O_2(g) \rightleftharpoons 2NO(g)$$

K_p is 2.5×10^{-3} at 2400°C. In a certain experiment, the original partial pressures of N_2 and O_2 are 0.35 atm. What is the partial pressure of NO when equilibrium is established?

58. Sodium bicarbonate is often used in household fire extinguishers. It decomposes to sodium carbonate, carbon dioxide, and steam according to the equation

$$2NaHCO_3(s) \rightleftharpoons Na_2CO_3(s) + CO_2(g) + H_2O(g)$$

At 125°C, K_p for the reaction is 0.25. What are the partial pressures of CO_2 and H_2O when equilibrium is established after decomposition?

Unclassified

59. Consider the equilibrium

$$C(s) + CO_2(g) \rightleftharpoons 2CO(g)$$

When this system is at equilibrium at 700°C in a 2.0-L container, there are 0.10 mol CO, 0.20 mol CO_2, and 0.40 mol C present. When cooled to 600°C, an additional 0.040 mol C(s) forms. Calculate K_c at 700°C and again at 600°C.

60. An examination question asked students to calculate K_c for the reaction

$$CS_2(g) + 4H_2(g) \rightleftharpoons CH_4(g) + 2H_2S(g)$$

given that, at equilibrium in a 5.0-L vessel, there are 5.5 mol CH_4, 1.25 mol H_2S, 1.5 mol CS_2, and 1.5 mol H_2.

Four students gave the following answers, all wrong. Explain the error(s) in their reasoning.

a. $K_c = \dfrac{(5.5)(1.25)^2}{(1.5)^5}$ b. $K_c = \dfrac{(0.30)(0.30)^4}{(1.1)(0.25)^2}$

c. $K_c = \dfrac{(1.1) \times (0.50)^2}{(0.30)(1.2)^4}$ d. $K_c = \dfrac{(1.1) + (0.25)^2}{(0.30) + (0.30)^4}$

61. Consider again the system referred to in Problem 60. If the original concentrations are all 0.60 M, which of the following are possible equilibrium concentrations?

	[CS_2]	[H_2]	[CH_4]	[H_2S]
a.	0.40	0.40	0.80	1.00
b.	0.50	0.20	0.70	0.80
c.	0.70	1.00	0.50	0.40
d.	0.70	0.70	0.50	0.50
e.	0.50	0.50	0.50	0.50

62. Phosphorus pentachloride dissociates to phosphorus trichloride and chlorine:

$$PCl_5(g) \rightleftharpoons PCl_3(g) + Cl_2(g)$$

A sealed 4.0-L vessel initially contains 0.0250 mol PCl_5. What is the total pressure at 250°C in the flask when equilibrium is reached? K_c at 250°C is 0.050.

63. Consider the statement, "The equilibrium constant of a roasting mixture of graphite, carbon dioxide, and carbon monoxide is 5.28." What information is missing from this statement?

64. At 1800 K, oxygen dissociates into gaseous atoms:

$$O_2(g) \rightleftharpoons 2O(g)$$

K_c for the system is 1.2×10^{-10}. If one mole of O_2 is placed in a 5.0-L flask and heated to 1800 K, what percent by mass of the oxygen dissociates? How many O atoms are there in the flask?

Challenge Problems

65. Hydrogen iodide gas decomposes to hydrogen gas and iodine gas:

$$2HI(g) \rightleftharpoons H_2(g) + I_2(g)$$

To determine the equilibrium constant of the system, identical one-liter glass bulbs are filled with 3.20 g of HI and maintained at a certain temperature. Each bulb is periodically opened and analyzed for iodine formation by titration with sodium thiosulfate, $Na_2S_2O_3$.

$$I_2(aq) + 2S_2O_3{}^{2-}(aq) \rightarrow S_4O_6{}^{2-}(aq) + 2I^-(aq)$$

It is determined that when equilibrium is reached, 37.0 mL of 0.200 M $Na_2S_2O_3$ is required to titrate the iodine. What is the K_c at the temperature of the experiment?

66. For the system $SO_3(g) \rightleftharpoons SO_2(g) + \frac{1}{2}O_2(g)$ at 1000 K, $K_c = 0.050$. Sulfur trioxide, originally at 1.00 atm pressure, partially dissociates to SO_2 and O_2 at 1000 K. What is the original concentration of SO_3 in moles per liter? What is its final concentration?

67. At a certain temperature the reaction $Xe(g) + 2F_2(g) \rightleftharpoons XeF_4(g)$ gives a 50% yield of XeF_4, starting with 0.40 mol Xe and 0.80 mol F_2 in a 2.0-L vessel. Calculate K_c at this temperature. How many additional moles of F_2 must be added to increase the yield of XeF_4 from Xe to 75%?

68. A student studying the equilibrium $I_2(g) \rightleftharpoons 2I(g)$ at a high temperature puts 0.10 mol I_2 in a 1.0-L container. He finds that the total pressure at equilibrium is 40% greater than it was originally. What is K_c for this reaction?

69. Sufficient N_2O_4 is put into a 10.0-L flask to establish an initial pressure of 1.00 atm at 25°C. Part of the N_2O_4 dissociates into NO_2; at equilibrium at 25°C, the total pressure is 1.17 atm. Calculate K_c at 25°C for:

$$N_2O_4(g) \rightleftharpoons 2NO_2(g)$$

$$CO_3{}^{2-} \ (aq) \ + \ 2H^+(aq) \ \rightarrow \ CO_2(g) \ + \ H_2O$$

Solutions of hydrochloric acid and sodium carbonate react to form carbon dioxide gas and water.

CHAPTER 14

ACIDS AND BASES

I n discussing reactions between ions in solution in Chapter 12, we talked about acid-base reactions. In this chapter we will take a closer look at the properties of acids and bases themselves. We start with the following working definitions referred to in Chapter 12:

1. An acid is a substance that, when added to water, produces **hydrogen ions (protons)**, H^+.*
2. A base is a substance that, when added to water, produces **hydroxide ions**, OH^-.

Acidic water solutions have certain properties in common. They react with active metals, such as zinc, to evolve hydrogen gas:

$$Zn(s) + 2H^+(aq) \rightarrow Zn^{2+}(aq) + H_2(g) \tag{14.1}$$

and with metal carbonates, such as $CaCO_3$, to form carbon dioxide gas (Fig. 14.1a, p. 470):

$$CaCO_3(s) + 2H^+(aq) \rightarrow Ca^{2+}(aq) + CO_2(g) + H_2O \tag{14.2}$$

They also have a sour taste, but that's not the way to test them in the lab

Acidic solutions also affect the color of certain organic dyes. For example, litmus turns from blue to red in acidic solution.

Water solutions of bases also have identifying properties. They feel slippery and turn the color of indicators. Litmus turns from red to blue in

---

*In water solution, the H^+ ion is hydrated and is often written as the hydronium ion, H_3O^+, which has the Lewis structure

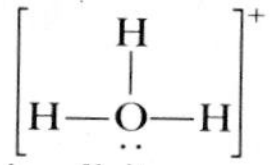

$$\left[H - \overset{\displaystyle H}{\underset{\displaystyle \cdot\cdot}{O}} - H \right]^+$$

For simplicity, we will ordinarily use H^+ as a shorthand representation for the H_3O^+ ion.

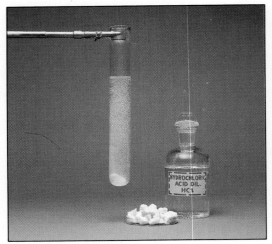 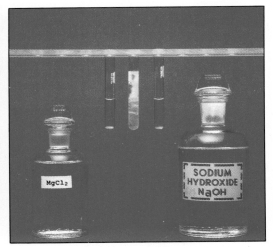

Figure 14.1 Chemical tests for acids and bases. Acids react with carbonates to evolve $CO_2(g)$; bases react with Mg^{2+} ions to precipitate $Mg(OH)_2$. (Marna G. Clarke)

basic solution. Basic solutions react with several cations, including Mg^{2+}, to form precipitates (Fig. 14.1b):

$$Mg^{2+}(aq) + 2OH^-(aq) \rightarrow Mg(OH)_2(s) \tag{14.3}$$

As you might suppose, acidic and basic solutions differ in concentration of H^+ or OH^- ions. In Section 14.1 we will look at the relationship between $[H^+]$ and $[OH^-]$ in all types of water solutions. We will also see how the acidity or basicity of a solution can be expressed in terms of pH, how pH can be measured, and how it can be calculated for solutions of strong acids or strong bases.

Section 14.2 focuses upon the properties of weak acids. In particular, we will look at the equilibrium between weak acids and their dissociation products in water solution. Weak bases are treated in Section 14.3; again, the emphasis is upon their equilibrium dissociation. In Section 14.4, we examine the acid-base properties of salt solutions and develop a method for predicting whether such solutions will be acidic, basic, or neutral. Finally, in Section 14.5, we will look at two more general models of acids and bases that can be applied to solvents other than water.

14.1
WATER DISSOCIATION; ACIDIC, NEUTRAL, AND BASIC SOLUTIONS

The acidic and basic properties of aqueous solutions are dependent upon an equilibrium that involves the solvent, water. Water, when pure or as a solvent, tends to dissociate to some extent into hydrogen ions and hydroxide ions:

$$H_2O \rightleftharpoons H^+(aq) + OH^-(aq) \tag{14.4}$$

The forward reaction proceeds only slightly before equilibrium is reached. Only a small fraction of the total number of water molecules is dissociated (about 1 in 500 million).

That's not many

Applying the general rules from Chapter 13 for equilibrium systems, we can write the equilibrium constant expression for Reaction 14.4 as

$$K_c = \frac{[H^+] \times [OH^-]}{[H_2O]}$$

In all dilute aqueous solutions the concentration of H_2O molecules is essentially the same, about 1000/18.0 or 55.5 mol/L. Hence, the term $[H_2O]$ can be combined with K_c to give a new constant, K_w, called the *dissociation constant* or *autoionization constant* of water:

$$K_c \times [H_2O] = K_w = [H^+] \times [OH^-] = 1.0 \times 10^{-14} \qquad (14.5)$$

This is an important equilibrium constant

At 25°C, K_w is 1.0×10^{-14}. This small value reflects the slight dissociation of water into H^+ and OH^- ions. In any aqueous solution or in pure water, the product of $[H^+]$ times $[OH^-]$ at 25°C is always 1.0×10^{-14}.

We can readily calculate $[H^+]$ and $[OH^-]$ in pure water at 25°C. From Equation 14.4, we see that equal amounts of these two ions form when water dissociates. Applying Equation 14.5 to pure water,

$$[H^+] = [OH^-]; \qquad [H^+] \times [OH^-] = [H^+]^2 = 1.0 \times 10^{-14}$$

$$[H^+] = 1.0 \times 10^{-7} M = [OH^-]$$

Any aqueous solution in which $[H^+]$ *equals* $[OH^-]$ is called a *neutral* solution. It has an $[H^+]$ equal to $1.0 \times 10^{-7} M$ at 25°C.

Ordinarily, the concentrations of H^+ and OH^- in a solution are not equal. Note that Equation 14.5 represents an inverse relation. As the concentration of one of the ions, for example, $[H^+]$, increases, that of the other, $[OH^-]$, must decrease. In this way the product $[H^+] \times [OH^-]$ remains constant at 1.0×10^{-14}. The inverse relationship between these two quantities is shown in Figure 14.2. Example 14.1 illustrates how we

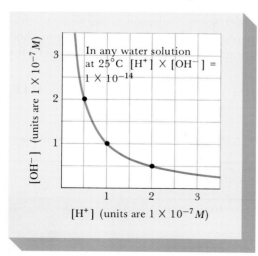

Figure 14.2 A graph of $[OH^-]$ vs. $[H^+]$ looks very much like a graph of gas volume vs. pressure. In both cases, the two variables are inversely proportional to one another. When $[H^+]$ gets larger, $[OH^-]$ gets smaller.

can calculate the concentration of one of these ions when we know that of the other one.

■ **EXAMPLE 14.1** _____

In a certain tap-water sample, $[H^+] = 3.0 \times 10^{-7} M$. What is the concentration of $[OH^-]$?

Solution

From Equation 14.5,

If we know $[H^+]$, we can always find $[OH^-]$

$$[OH^-] = \frac{K_w}{[H^+]} = \frac{1.0 \times 10^{-14}}{3.0 \times 10^{-7}} = \boxed{3.3 \times 10^{-8} M}$$

Exercise

What is $[H^+]$ in a seawater sample if $[OH^-] = 2.0 \times 10^{-6} M$? Answer: $5.0 \times 10^{-9} M$.

_____

An aqueous solution where $[H^+]$ is greater than $[OH^-]$ is termed acidic. An aqueous solution in which $[OH^-]$ is greater than $[H^+]$ is basic (alkaline). Therefore,

if $[H^+] > 1.0 \times 10^{-7} M$, $[OH^-] < 1.0 \times 10^{-7} M$, | solution is acidic

if $[OH^-] > 1.0 \times 10^{-7} M$, $[H^+] < 1.0 \times 10^{-7} M$, | solution is basic

Table 14.1 indicates some possible combinations of concentrations of these ions. The acidity (concentration of H^+) progressively decreases in solutions 1 through 9. Solutions 1 to 4 are decreasingly acidic; solution 5 is neutral, and solutions 6 to 9 are increasingly basic.

pH AND pOH

As we have seen, the acidity or basicity of a solution can be described in terms of its H^+ concentration. In 1909, Soren Sorensen, a biochemist working at the Carlsberg Brewery in Copenhagen, proposed an alternative method of specifying the acidity of a solution. He defined a term called pH (for "power of the hydrogen ion").

$$\textbf{pH} = -\textbf{log}_{10}\,[\textbf{H}^+] = \textbf{log}_{10}\,1/[\textbf{H}^+] \tag{14.6}$$

TABLE 14.1   Relations Between $[H^+]$, $[OH^-]$, and pH in Aqueous Solutions at 25°C

	SOLUTION								
	No. 1	*No. 2*	*No. 3*	*No. 4*	*No. 5*	*No. 6*	*No. 7*	*No. 8*	*No. 9*
$[H^+]$	10^0	10^{-2}	10^{-4}	10^{-6}	10^{-7}	10^{-8}	10^{-10}	10^{-12}	10^{-14}
$[OH^-]$	10^{-14}	10^{-12}	10^{-10}	10^{-8}	10^{-7}	10^{-6}	10^{-4}	10^{-2}	10^0
pH	0	2	4	6	7	8	10	12	14
		Acidic			Neutral		Basic		

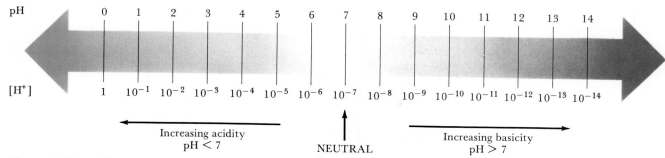

Figure 14.3 Acidity is inversely related to pH; the higher the H^+ ion concentration, the lower the pH. In neutral solution, $[H^+] = [OH^-] = 1.0 \times 10^{-7} M$; pH = 7.00.

Thus, we have

$[H^+] = 1 \times 10^{-4} M$;   pH $= -\log_{10} (1 \times 10^{-4}) = -(-4.0) = 4.0$

$[H^+] = 1 \times 10^{-7} M$;   pH $= -\log_{10} (1 \times 10^{-7}) = -(-7.0) = 7.0$

$[H^+] = 1 \times 10^{-10} M$;   pH $= -\log_{10} (1 \times 10^{-10}) = -(-10.0) = 10.0$

It's easier to say the pH is 4 than to say that $[H^+]$ is 1×10^{-4}

Most aqueous solutions have hydrogen ion concentrations between 1 and $10^{-14} M$. By Equation 14.6, such solutions have pH's lying between 0 and 14.

Looking at Table 14.1 or Figure 14.3, we see that $[H^+]$ and pH are inversely related. As $[H^+]$ decreases from 10^{-2} to $10^{-4} M$, pH increases from 2 to 4. In general, **the higher the pH, the less acidic (more basic) the solution**. A solution of pH 4 has a lower concentration of H^+ and a higher concentration of OH^- than does a solution of pH 2. Notice also that when pH increases by one unit, $[H^+]$ decreases by a factor of 10. A solution of pH 3 has a hydrogen ion concentration one-tenth that of a solution of pH 2 and ten times that of a solution of pH 4.

Previously, we used concentration of H^+ or OH^- to differentiate acidic, neutral, and basic solutions; pH can also be used for this purpose:

if pH < 7.0,   solution is acidic

if pH = 7.0,   solution is neutral

if pH > 7.0,   solution is basic

Table 14.2 shows the pH of some common solutions.

TABLE 14.2   pH of Some Common Liquids

Lemon juice	2.2–2.4	Urine, human	4.8–8.4
Wine	2.8–3.8	Cow's milk	6.3–6.6
Vinegar	3.0	Saliva, human	6.5–7.5
Tomato juice	4.0	Drinking water	5.5–8.0
Beer	4–5	Blood, human	7.3–7.5
Cheese	4.8–6.4	Seawater	8.3

Figure 14.4 Many common household items, including vinegar, lemon juice, and cola drinks, are acidic. Others, including ammonia and most detergents and cleaning agents, are basic. (Marna G. Clarke)

We use a similar approach for the hydroxide ion concentration. The pOH of a solution is defined as

$$pOH = -\log_{10} [OH^-] \tag{14.7}$$

Since $[H^+] \times [OH^-] = 1.0 \times 10^{-14}$ at 25°C, it follows that at this temperature:

$$pH + pOH = 14.00 \tag{14.8}$$

Thus a solution that has a pH of 6.20 must have a pOH of 7.80, and vice versa.

If you know the pH, the pOH is a piece of cake

■ **EXAMPLE 14.2** _____
Calculate

a. the pH and pOH of a lemon juice solution in which $[H^+]$ is 5.0×10^{-3} M.
b. the $[H^+]$ and $[OH^-]$ of human blood at pH 7.40.
c. the pH and pOH of a solution of baking soda, $NaHCO_3$, with an $[OH^-]$ of 2.5×10^{-6} M.

Solution
Problems of this type are most simply worked using a scientific calculator. The use of calculators and operations involving logarithms are discussed in Appendix 3. Note particularly the discussion of the use of significant figures in converting between logarithms and antilogarithms.

a. You should find on your calculator that

$$\log_{10} (5.0 \times 10^{-3}) = -2.30$$

Since pH $= -\log_{10} [H^+]$, then pH $= -(-2.30) =$ ⎡2.30⎤. We now calculate pOH using Equation 14.8.

$$14.00 = 2.30 + pOH$$

$$pOH = \boxed{11.70}$$

b. If the pH is 7.40,

$$\log_{10}[H^+] = -7.40; \quad [H^+] = 10^{-7.40}$$

To find $[H^+]$, enter -7.40 on your calculator. Then either

— punch the 10^x key, if you have one, or
— punch the INV and then the LOG key.

Either way, you should find that $[H^+] = \boxed{4.0 \times 10^{-8} \, M}$.

In the days before calculators, this problem wasn't so easy

Using the equilibrium expression for the dissociation of water (Equation 14.5) we calculate for $[OH^-]$:

$$K_w = 1.0 \times 10^{-14} = [H^+] \times [OH^-]$$

$$[OH^-] = \frac{1.0 \times 10^{-14}}{4.0 \times 10^{-8}} = \boxed{2.5 \times 10^{-7} \, M}$$

c. We first calculate pOH. You should find on your calculator that

$$\log_{10}(2.5 \times 10^{-6}) = -5.60$$

Since $pOH = -\log_{10}[OH^-]$, then $pOH = -(-5.60) = \boxed{5.60}$

Substituting into Equation 14.8,

$$14.00 = pH + 5.60$$

$$pH = \boxed{8.40}$$

Exercise
Calculate $[H^+]$ and $[OH^-]$ of a beer with pH 4.7. Answer: $[H^+] = 2 \times 10^{-5} \, M$, $[OH^-] = 5 \times 10^{-10} \, M$.

STRONG ACIDS AND STRONG BASES

You will recall from Chapter 12 that certain species (HCl, HBr, HI, HNO_3, $HClO_4$, H_2SO_4) are strong acids in the sense that they are completely dissociated to form H^+ ions in dilute water solution. It is ordinarily quite easy to calculate $[H^+]$ or pH of an aqueous solution of a strong acid.

Strong acids are strongly dissociated

■ **EXAMPLE 14.3** _____

Calculate the $[H^+]$, pH, and $[OH^-]$ of a 0.10 M solution of HCl.

Solution
Since we want to obtain $[H^+]$, we look for all species that furnish H^+ ions to the solution. HCl furnishes H^+ ions. In fact, all the HCl dissociates into H^+ and Cl^-. Thus

$$HCl(aq) \rightarrow H^+(aq) + Cl^-(aq)$$

$$0.10 \; M \; HCl \rightarrow 0.10 \; M \; H^+, \; 0.10 \; M \; Cl^-, \; 0.0 \; M \; HCl(aq)$$

H^+ is also furnished by the dissociation of water in the solution.

$$H_2O \rightleftharpoons H^+(aq) + OH^-(aq)$$

In pure water, $[H^+] = 1.0 \times 10^{-7} \; M$. Since $0.10 \gg 1.0 \times 10^{-7}$, we can ignore the H^+ ions coming from the water; $[H^+]$ in $0.10 \; M \; HCl$ is $1.0 \times 10^{-1} \; M$. The pH is:

$$-\log_{10}(0.10) = 1.00$$

The $[OH^-]$ can be obtained using Equation 14.5:

$$OH^- = \frac{1.0 \times 10^{-14}}{1.0 \times 10^{-1}} = 1.0 \times 10^{-13} \; M$$

Notice how far to the left the water equilibrium has shifted in the presence of the high concentration of H^+ ions produced by the strong acid HCl. The concentration of OH^- ions has dropped from $1.0 \times 10^{-7} \; M$ in pure water to $1.0 \times 10^{-13} \; M$ in $0.10 \; M \; HCl$.

Exercise
Calculate the pH of $1.0 \; M \; HClO_4$. Answer: 0.00.

The H^+ ions contributed by the water in aqueous solutions can most often be ignored, as in Example 14.3. With very dilute solutions, however, that may not be the case. Suppose, for example, you prepared a $1.0 \times 10^{-10} \; M$ solution of nitric acid, perhaps by repeated tenfold dilutions of $0.10 \; M \; HNO_3$. Here, the pH is not 10. In this case, the concentration of H^+ produced by the water is much larger than that contributed by HNO_3. The pH of $1.0 \times 10^{-10} \; M \; HNO_3$ is just a bit smaller than 7.

We can readily obtain the pH of a solution of a strong base (LiOH, NaOH, KOH, RbOH, CsOH; $Ca(OH)_2$, $Sr(OH)_2$, $Ba(OH)_2$) by first determining $[OH^-]$. Consider, for example, $0.10 \; M \; Ba(OH)_2$. Since there are two moles of OH^- ions per mole of $Ba(OH)_2$:

$$Ba(OH)_2(s) \rightarrow Ba^{2+}(aq) + 2OH^-(aq) \tag{14.9}$$

$$[OH^-] = 0.20 \; M; \; pOH = -\log_{10}(0.20) = 0.70; \; pH = 13.30$$

MEASURING pH

The pH of a solution can be measured accurately by an instrument called a pH-meter. A pH-meter translates the H^+ ion concentration of a solution into an electrical signal that is converted either into a digital display or into a deflection on a meter that reads pH directly (Fig. 14.5).

A less accurate but more colorful way to measure pH uses a "universal indicator," which is a mixture of dyes that shows changes in color at different pH values. A similar principle is used with "pH paper." Strips of

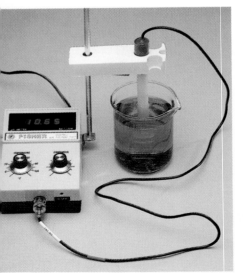

Figure 14.5 A pH meter with digital read-out. The pH of the solution in the beaker appears to be about 10.86. What do you suppose is in the solution? (Marna G. Clarke)

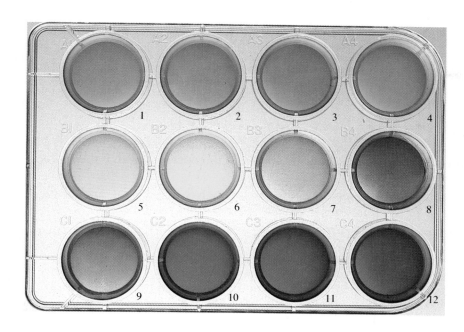

Universal indicator is deep red in strongly acidic solution (upper left). It changes to yellow and green at pH 6–8 and then to deep violet in strongly basic solution (lower right).

this paper are coated with a mixture of dyes; these strips are widely used to test the pH of biological fluids, soil, ground water, and foods. Depending upon the dyes used, a test strip can measure pH over a wide or a narrow range.

14.2
WEAK ACIDS AND THEIR DISSOCIATION IN WATER

A wide variety of solutes behave as weak acids; that is, they are only slightly dissociated in water. For convenience, weak acids can be divided into three categories: molecules, anions, and cations.

Weak acids are weakly dissociated

1. Molecules containing an ionizable hydrogen atom. We discussed this type of weak acid in Chapter 12. There are literally thousands of molecular weak acids, most of them organic in nature. When added to water, these weak acids form an equilibrium mixture consisting mostly of undissociated molecules, mixed with a low concentration of H^+ ions and anions. The equations for the dissociation of these weak acids are of the type:

$$HB(aq) \rightleftharpoons H^+(aq) + B^-(aq) \tag{14.10}$$

Examples include

$$HClO(aq) \rightleftharpoons H^+(aq) + ClO^-(aq) \qquad \text{(hypochlorous acid)}$$

$$HC_2H_3O_2(aq) \rightleftharpoons H^+(aq) + C_2H_3O_2^-(aq) \qquad \text{(acetic acid)}$$

2. Anions containing an ionizable hydrogen atom. Let us refer back for a moment to the dissociation of the strong acid, H_2SO_4. The first hydrogen atom is dissociated completely:

$$H_2SO_4(aq) \rightarrow H^+(aq) + HSO_4^-(aq)$$

Notice that the anion formed, HSO_4^-, contains a hydrogen atom. In water, the HSO_4^- ion undergoes further dissociation, producing an H^+ ion and a SO_4^{2-} ion. The dissociation takes place to a limited extent and is reversible, so we classify HSO_4^- as a weak acid.

$$HSO_4^-(aq) \rightleftharpoons H^+(aq) + SO_4^{2-}(aq) \qquad (14.11)$$

The HCO_3^- ion is actually a basic

In practice, very few anions give acidic solutions when added to water. The only other anion of this type that we need be concerned with is the $H_2PO_4^-$ ion.

3. Cations. The ammonium ion, NH_4^+, behaves as a weak acid in water because of the following reversible reaction:

$$NH_4^+(aq) \rightleftharpoons H^+(aq) + NH_3(aq) \qquad (14.12)$$

The products are an ammonia molecule, NH_3, and an H^+ ion that makes the solution acidic. Note that the behavior of the NH_4^+ ion in water is very similar to that of HB, the prototype for molecular weak acids (Equation 14.10) or the HSO_4^- ion (Equation 14.11). All three species contain hydrogen atoms that are converted to H^+ ion when the weak acid dissociates. The other product of dissociation is the anion or molecule left (ClO^-, $C_2H_3O_2^-$, SO_4^{2-}, NH_3) when a proton is removed from the acid.

The species formed when a proton is removed from an acid is referred to as the **conjugate base** of the acid. Thus ClO^- is the conjugate base of HClO; NH_3 is the conjugate base of NH_4^+, and so on. By the same token, the species formed when a proton is added to a base is called the **conjugate acid** of that base. The HClO molecule is the conjugate acid of the ClO^- ion; the NH_4^+ ion is the conjugate acid of the NH_3 molecule. In general:

$$\text{Conjugate acid} \rightleftharpoons H^+ + \text{Conjugate base}$$

You may be surprised to learn that most metal cations, except those of Groups 1 and 2, are weak acids. At first, it is not at all obvious how a cation such as Zn^{2+} can make a water solution acidic. To understand how this is possible, we must realize that this cation and others like it are *hydrated* in water solution. When $ZnCl_2$ or $Zn(NO_3)_2$ is added to water, the cation formed is $Zn(H_2O)_4^{2+}$. Here, a Zn^{2+} ion is bonded to four water molecules forming a complex ion. (Complex ions are discussed in detail in Chapter 16.) This complex cation is slightly dissociated in water, according to the following equation:

$$Zn(H_2O)_4^{2+}(aq) \rightleftharpoons H^+(aq) + Zn(H_2O)_3(OH)^+(aq) \qquad (14.13)$$

The H^+ ion, which makes the solution acidic, comes from the ionization of one of the H_2O molecules bonded to Zn^{2+}. The OH^- ion formed at the same time remains bonded to Zn^{2+} and so does not directly affect the pH of the solution. (It does, however, affect the charge of the complex cation, reducing it from $+2$ to $+1$.)

■ **EXAMPLE 14.4** _____

Write equations for the dissociation of the following weak acids in water:

a. HNO_2 b. $H_2PO_4^-$ c. $Fe(H_2O)_6^{3+}$

Solution

In each case, a proton (H^+) is formed to make the solution acidic. The other product is the residue from the weak acid after removal of the proton. All the reactions go to equilibrium, as indicated by a double arrow.

a. $HNO_2(aq) \rightleftharpoons H^+(aq) + NO_2^-(aq)$
b. $H_2PO_4^-(aq) \rightleftharpoons H^+(aq) + HPO_4^{2-}(aq)$
c. $Fe(H_2O)_6^{3+}(aq) \rightleftharpoons H^+(aq) + Fe(H_2O)_5(OH)^{2+}(aq)$

Exercise

HPO_4^{2-} and $Fe(H_2O)_5(OH)^{2+}$ can each dissociate to form an H^+ ion and another product. Give the formulas of the "other products," i.e., the conjugate bases of these weak acids. Answer: PO_4^{3-}, $Fe(H_2O)_4(OH)_2^+$.

THE EQUILIBRIUM CONSTANT FOR THE DISSOCIATION OF A WEAK ACID, K_a

The dissociation in water solution of a weak acid, HB, is expressed by the general equation

$$HB(aq) \rightleftharpoons H^+(aq) + B^-(aq)$$

Following the rules given in Chapter 13, we can write an expression for the equilibrium constant:

$$K_a = \frac{[H^+] \times [B^-]}{[HB]} \qquad (14.14)$$

The equilibrium constant, K_a, is called the *ionization constant* or *acid dissociation constant* of the weak acid HB.

> K_a is the symbol given to K_c for weak acid equilibria

The K_a values of some weak acids are given in Table 14.3, p. 480. These constants are a measure of the extent to which the acid dissociates in water solution. **The smaller the dissociation constant, the weaker the acid.** Notice that in Table 14.3 the acids are arranged in decreasing order of K_a. For example, acetic acid, $HC_2H_3O_2$ ($K_a = 1.8 \times 10^{-5}$) lies below hydrofluoric acid, HF ($K_a = 6.9 \times 10^{-4}$). This means that at a given concentration, let us say, $0.10\ M$,

— [H^+] is lower for $HC_2H_3O_2$ than for HF (1.3×10^{-3} vs. $8.0 \times 10^{-3}\ M$).
— pH is higher for $HC_2H_3O_2$ than for HF (2.87 vs. 2.10).

Both of these statements reflect the fact that acetic acid is weaker than hydrofluoric acid.

In comparing the extents of dissociation of weak acids, we sometimes use the term pK_a, which is analogous to pH:

$$pK_a = -\log_{10} K_a \qquad (14.15)$$

Thus, we have

$$HF: \quad pK_a = -\log_{10}(6.9 \times 10^{-4}) = -(-3.16) = 3.16$$

$$HC_2H_3O_2: \quad pK_a = -\log_{10}(1.8 \times 10^{-5}) = -(-4.74) = 4.74$$

In general, the larger the value of pK_a, the weaker the acid.

TABLE 14.3   Dissociation Constants of Weak Acids* at 25°C

	ACID	K_a
Sulfurous acid	H_2SO_3	1.7×10^{-2}
Hydrogen sulfate ion	HSO_4^-	1.0×10^{-2}
Chlorous acid	$HClO_2$	1.0×10^{-2}
Hydrofluoric acid	HF	6.9×10^{-4}
Nitrous acid	HNO_2	6.0×10^{-4}
Formic acid	$HCHO_2$	1.9×10^{-4}
Benzoic acid	$HC_7H_5O_2$	6.6×10^{-5}
Acetic acid	$HC_2H_3O_2$	1.8×10^{-5}
Propionic acid	$HC_3H_5O_2$	1.4×10^{-5}
Carbonic acid	H_2CO_3	4.4×10^{-7}
Hydrogen sulfide	H_2S	1.0×10^{-7}
Dihydrogen phosphate ion	$H_2PO_4^-$	6.2×10^{-8}
Hydrogen sulfite ion	HSO_3^-	6.0×10^{-8}
Hypochlorous acid	$HClO$	2.8×10^{-8}
Hypobromous acid	$HBrO$	2.6×10^{-9}
Hydrocyanic acid	HCN	5.8×10^{-10}
Ammonium ion	NH_4^+	5.6×10^{-10}
Hydrogen carbonate ion	HCO_3^-	4.7×10^{-11}
Hydrogen phosphate ion	HPO_4^{2-}	4.5×10^{-13}
Hydrogen sulfide ion	HS^-	1×10^{-13}

decreasing acid strength

*In each case the dissociation reaction is of the type:

$$HB(aq) \rightleftharpoons H^+(aq) + B^-(aq); \qquad K_a = \frac{[H^+] \times [B^-]}{[HB]}$$

The value of K_a for a weak acid must be found experimentally. There are several ways to do this. Perhaps the most common involves measuring $[H^+]$ or pH in a solution prepared by dissolving a known amount of the weak acid to form a given volume of solution. The calculation is illustrated in Example 14.5.

■ **EXAMPLE 14.5** _____

Acetylsalicyclic acid, more commonly known as aspirin, is a weak organic acid whose formula we will represent as HAsp. A water solution is prepared by dissolving 0.1000 mol HAsp per liter. The concentration of H^+ in this solution is found to be 0.0057 mol/L. Calculate K_a for aspirin.

Solution
The dissociation reaction for aspirin is

$$HAsp(aq) \rightleftharpoons H^+(aq) + Asp^-(aq); \qquad K_a = \frac{[H^+] \times [Asp^-]}{[HAsp]}$$

To calculate K_a, we need to know the *equilibrium* concentrations of H^+ ions, Asp^- ions, and HAsp molecules. The equilibrium concentration of H^+ is given as 0.0057 M. Virtually all of this comes from the dissociation of the weak acid; the amount coming from H_2O is negligible. Since one mole of Asp^- is produced with every mole of H^+:

$$[H^+] = [Asp^-] = 0.0057 \text{ mol/L}$$

For every mole of H^+ produced, a mole of HAsp must dissociate. This means that 0.0057 mol/L of HAsp must dissociate. Since the original concentration of HAsp was 0.1000 mol/L,

$$[HAsp] = 0.1000 \text{ mol/L} - 0.0057 \text{ mol/L} = 0.0943 \text{ mol/L}$$

Summarizing this reasoning in the form of a table, as in Chapter 13:

$$HAsp(aq) \rightleftharpoons H^+(aq) + Asp^-(aq)$$

	ORIG. CONC. (mol/L)	CHANGE IN CONC. (mol/L)	EQUIL. CONC. (mol/L)
HAsp	0.1000	−0.0057	0.0943
Asp^-	0.0000	+0.0057	0.0057
H^+	1.0×10^{-7}	+0.0057	0.0057

(Numbers in green are those given or implied in the statement of the problem; the other numbers are deduced using the dissociation equation.) We now have all the information we need to calculate K_a:

$$K_a = \frac{[H^+] \times [Asp^-]}{[HAsp]} = \frac{(0.0057)^2}{0.0943} = 3.4 \times 10^{-4}$$

Aspirin is more acidic than vinegar

Exercise

In a solution prepared by dissolving 0.100 mol of lactic acid per liter, $[H^+] = 3.7 \times 10^{-3}$ M. Calculate K_a for lactic acid. Answer: 1.4×10^{-4}.

Sometimes we use the term *percent dissociation* to describe the acidity of a solution prepared by dissolving a weak acid in water. Percent dissociation refers quite simply to the percentage of the weak acid molecules originally present that dissociate to form H^+ ions; that is,

$$\textbf{percent dissociation HB} = \frac{[H^+]}{\textbf{orig. conc. HB}} \times 100 \qquad (14.16)$$

For the solutions referred to in Example 14.5,

$$\text{percent dissociation 0.10 } M \text{ aspirin} = \frac{0.0057}{0.1000} \times 100 = 5.7$$

$$\text{percent dissociation 0.10 } M \text{ lactic acid} = \frac{3.7 \times 10^{-3}}{0.100} \times 100 = 3.7$$

As you might guess, percent dissociation at a given concentration is directly related to K_a. The larger the value of K_a, the greater the percent dissociation (Table 14.4). Percent dissociation also depends upon the concentration of weak acid, increasing as concentration decreases.

At very low molarities, weak acids are highly dissociated

TABLE 14.4   Percent Dissociation of Weak Acids

WEAK ACID	K_a	PERCENT DISSOCIATION ORIG. CONC. 1.0 M	ORIG. CONC. 0.10 M
Aspirin	3.4×10^{-4}	1.8	5.7
Lactic acid	1.4×10^{-4}	1.2	3.7
Acetic acid	1.8×10^{-5}	0.42	1.3
Carbonic acid	4.4×10^{-7}	0.066	0.21
Hypochlorous acid	2.8×10^{-8}	0.017	0.053
Ammonium ion	5.6×10^{-10}	0.0024	0.0075
Hydrocyanic acid	5.8×10^{-10}	0.0024	0.0076

decreasing acid strength ↓

CALCULATING [H⁺] IN A WATER SOLUTION OF A MONOPROTIC WEAK ACID

Given the dissociation constant of a weak acid and its original concentration, we can readily calculate the H^+ concentration in the weak acid solution. The approach used, illustrated in Example 14.6, is entirely analogous to that followed with gaseous equilibria in Chapter 13.

■ **EXAMPLE 14.6** _____

Nicotinic acid, $HC_6H_4O_2N$ ($K_a = 1.4 \times 10^{-5}$), is another name for niacin, an important member of the Vitamin B group. Determine $[H^+]$ in a solution prepared by dissolving 0.10 mol nicotinic acid, HNic, in water to form one liter of solution.

Solution

This is a classic problem, finding the pH of a solution of a weak acid

We start by writing the equilibrium expressions for the systems at equilibrium in the solution:

$$HNic(aq) \rightleftharpoons H^+(aq) + Nic^-(aq)$$

$$H_2O \rightleftharpoons H^+(aq) + OH^-(aq)$$

The equilibrium constant for each of these systems is

$$K_a \text{ HNic} = 1.4 \times 10^{-5}$$

$$K_w \text{ H}_2\text{O} = 1.0 \times 10^{-14}$$

Since K_a for nicotinic acid is so much larger than K_w, the dissociation of HNic is the governing reaction. The equilibrium expression for the system is:

$$K_a = 1.4 \times 10^{-5} = \frac{[H^+] \times [Nic^-]}{[HNic]}$$

We construct an equilibrium table in the usual way. The species included are those that appear in the equilibrium expression.

$$HNic(aq) \rightleftharpoons H^+(aq) + Nic^-(aq)$$

	ORIG. CONC. (M)	CHANGE IN CONC. (M)	EQUILIBRIUM CONC. (M)
HNic	0.10	$-x$	$0.10 - x$
H^+	0.00	$+x$	x
Nic^-	0.00	$+x$	x

We substitute the values in the last column into the equilibrium expression

$$1.4 \times 10^{-5} = \frac{(x)(x)}{0.10 - x}$$

This is a quadratic equation. It could be rearranged to the form $ax^2 + bx + c = 0$ and solved for x, using the quadratic formula (Example 14.7). Such a procedure is time-consuming, and, in this case, unnecessary. Nicotinic acid is a weak acid, only slightly dissociated in water. The equilibrium concentration of HNic, $0.10 - x$, is probably only very slightly less than its original concentration, $0.10\ M$. We therefore make the approximation $0.10 - x \approx 0.10$. This simplifies the equation:

$$1.4 \times 10^{-5} = \frac{x^2}{0.10}$$

$$x^2 = 1.4 \times 10^{-6}$$

You can't discard the x in the numerator, since that's what you're trying to find

Taking square roots we have

$$x = 1.2 \times 10^{-3}\ M = [H^+] = [Nic^-]$$

$$[HNic] = 0.10 - x = 0.10 - 0.0012 = 0.10\ M \text{ (2 sig. figs.)}$$

The fact that $[H^+]$, $0.0012\ M$, is so much less than the original concentration of nicotinic acid, $0.10\ M$, justifies the approximation that $0.10 - x \approx 0.10$.

Exercise
Calculate the percent dissociation of nicotinic acid in this example. Answer: 1.2%.

In general, the value of K_a is seldom known to better than $\pm 5\%$. Hence, in the expression

$$K_a = \frac{x^2}{a - x}$$

where $x = [H^+]$ and $a =$ *original* concentration of weak acid, you can neglect the x in the denominator if doing so does not introduce an error of more than 5%. In other words,

$$\text{if} \quad \frac{x}{a} = \frac{\text{percent dissociation}}{100} \leqslant 0.05$$

We refer to this as the 5% rule

$$\text{then we can take} \quad a - x = a$$

In most of the problems you will work, this condition holds. Looking back at Table 14.4, you can see that the dissociation is usually less than 5%. When this is the case, the approximation $a - x \approx a$ is valid and you can calculate $[H^+]$ quite simply.

Sometimes, however, you will find that the $[H^+]$ you calculate is greater than 5% of the original concentration of weak acid. If this happens, you can solve for x either by using the quadratic equation or by using the method of successive approximations. These approaches are illustrated in Example 14.7; a general approach is shown in Figure 14.6.

■ **EXAMPLE 14.7** _____

Calculate $[H^+]$ in a $0.100\ M$ solution of nitrous acid, HNO_2, for which $K_a = 6.0 \times 10^{-4}$.

Solution

Proceeding as in Example 14.6, we arrive at the equation

$$\frac{x^2}{0.100 - x} = 6.0 \times 10^{-4}$$

Making the same approximation as before, $0.100 - x \approx 0.100$,

$$x^2 = 0.100 \times 6.0 \times 10^{-4} = 6.0 \times 10^{-5}$$

$$x = 7.7 \times 10^{-3} \approx [H^+]$$

Again we check the validity of our assumption by calculating percent dissociation.

$$\% \text{ dissociation} = \frac{x}{HNO_2 \text{ conc.}} \times 100 = \frac{7.7 \times 10^{-3}}{0.100} \times 100 = 7.7$$

The assumption fails the 5% rule and cannot be used.

To obtain a better value for x, we have a choice of two approaches. One of these is called the method of successive approximations; the other uses the quadratic formula.

1. *The method of successive approximations.* We know that

$$[HNO_2] = 0.100 - x$$

Our first approximation was to take $x = 0$. We can do better by making a second approximation. Here, we use the value just calculated for x, 7.7×10^{-3}, or 0.0077. Now,

$$[HNO_2] = 0.100 - 0.0077 = 0.092\ M$$

Substituting in the expression for K_a,

$$\frac{x^2}{0.092} = 6.0 \times 10^{-4}; \qquad x^2 = 5.5 \times 10^{-5}$$

$$x = \boxed{7.4 \times 10^{-3}\ M \approx [H^+]}$$

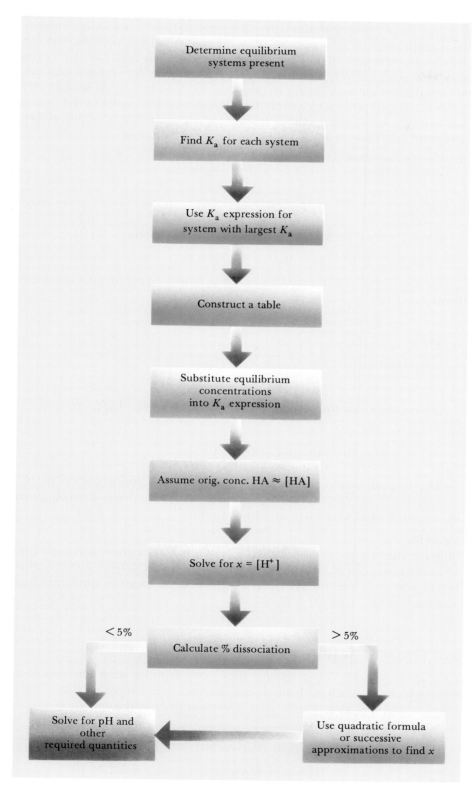

Figure 14.6 General approach for calculating the pH of a solution prepared by adding a weak acid to water. Usually, the $[H^+]$ from the water is negligibly small compared with that from the weak acid. Also, you usually find that the dissociation of the weak acid is less than 5%, which considerably simplifies the calculation.

This value is closer to the true $[H^+]$, since $0.092\ M$ is a better approximation for $[HNO_2]$ than was $0.100\ M$. If you're still not satisfied, you can go one step further. Using 7.4×10^{-3} for x instead of 7.7×10^{-3}, you can recalculate $[HNO_2]$ and solve again for x. If you do, you will find that your answer does not change. In other words, "you have gone about as far as you can go." This is generally true in calculations involving $[H^+]$ in a solution of a weak acid. Usually, the first approximation is sufficient; almost never do you have to go beyond a second approximation.

2. *The quadratic formula.* This gives an exact solution for x, but is more time-consuming. Here, we would rewrite the equation

$$\frac{x^2}{0.100 - x} = 6.0 \times 10^{-4}$$

in the form $ax^2 + bx + c = 0$. Doing this, we obtain

$$x^2 + (6.0 \times 10^{-4}\ x) - (6.0 \times 10^{-5}) = 0$$

Thus, $a = 1; b = 6.0 \times 10^{-4}; c = -6.0 \times 10^{-5}$. Applying the quadratic formula,

$$x = \frac{-b \pm \sqrt{b^2 - 4ac}}{2a}$$

$$= \frac{-6.0 \times 10^{-4} \pm \sqrt{(6.0 \times 10^{-4})^2 + (24.0 \times 10^{-5})}}{2}$$

If you carry out the arithmetic properly, you should get two answers for x:

$$x = \boxed{7.4 \times 10^{-3}\ M} \text{ and } -8.0 \times 10^{-3}$$

The second answer is physically ridiculous; the concentration of H^+ cannot be a negative quantity. The first answer is the same one we obtained by the method of successive approximations.

Exercise
Suppose the original concentration of HNO_2 in this example had been $1.00\ M$ instead of $0.100\ M$. Making the approximation $[HNO_2] = 1.00\ M$, what would you calculate for $[H^+]$? Would you need to go beyond this first approximation? Answer: $0.024\ M$; no.

POLYPROTIC ACIDS

Certain weak acids are *polyprotic;* they contain more than one ionizable hydrogen atom. Such acids dissociate in steps, with a separate dissociation constant for each step. Oxalic acid, a weak organic acid sometimes used to remove bloodstains, is *diprotic:*

$$H_2C_2O_4(aq) \rightleftharpoons H^+(aq) + HC_2O_4^-(aq) \qquad K_{a1} = 5.9 \times 10^{-2}$$

$$HC_2O_4^-(aq) \rightleftharpoons H^+(aq) + C_2O_4^{2-}(aq) \qquad K_{a2} = 5.2 \times 10^{-5}$$

Phosphoric acid, a common ingredient of cola drinks, is *triprotic:*

$$H_3PO_4(aq) \rightleftharpoons H^+(aq) + H_2PO_4^-(aq) \qquad K_{a1} = 7.1 \times 10^{-3}$$

$$H_2PO_4^-(aq) \rightleftharpoons H^+(aq) + HPO_4^{2-}(aq) \qquad K_{a2} = 6.2 \times 10^{-8}$$

$$HPO_4^{2-}(aq) \rightleftharpoons H^+(aq) + PO_4^{3-}(aq) \qquad K_{a3} = 4.5 \times 10^{-13}$$

The behavior of these acids is typical of that of all polyprotic acids in that:

— the conjugate base formed in one step (e.g., $HC_2O_4^-$, $H_2PO_4^-$) is the acid species that dissociates in the next step.

— the dissociation constant becomes smaller with each successive step.

$$K_{a1} > K_{a2} > K_{a3}$$

Putting it another way, **the acids formed in successive steps become progressively weaker.** This is reasonable; the dissociation of a proton, H^+, from a neutral molecule ($H_2C_2O_4$, H_3PO_4) leaves an anion ($HC_2O_4^-$, $H_2PO_4^-$). The energy requirement for the next dissociation is greater because a positive ion, H^+, must be removed from a negatively charged species.

Ordinarily, successive dissociation constants of weak acids decrease by a factor of at least 100 (Table 14.5). As a result, we ordinarily find that essentially all of the hydrogen ions produced by the acid come from the first dissociation. This makes it relatively easy to calculate the pH of a solution of a polyprotic acid.

TABLE 14.5   Dissociation Constants for Some Polyprotic Acids at 25°C

ACID	FORMULA	K_{a1}	K_{a2}	K_{a3}
Arsenic acid	H_3AsO_4	5.7×10^{-3}	1.8×10^{-7}	2.5×10^{-12}
Carbonic acid*	H_2CO_3	4.4×10^{-7}	4.7×10^{-11}	
Citric acid	$H_3C_5H_5O_7$	8.7×10^{-4}	1.8×10^{-5}	4.0×10^{-7}
Hydrogen sulfide	H_2S	1.0×10^{-7}	1×10^{-13}	
Oxalic acid	$H_2C_2O_4$	5.9×10^{-2}	5.2×10^{-5}	
Phosphoric acid	H_3PO_4	7.1×10^{-3}	6.2×10^{-8}	4.5×10^{-13}
Sulfurous acid	H_2SO_3	1.7×10^{-2}	6.0×10^{-8}	

*Carbonic acid is a water solution of carbon dioxide:

$$CO_2(g) + H_2O \rightleftharpoons H_2CO_3(aq)$$

The dissociation constants listed here are calculated assuming that all the carbon dioxide that dissolves is in the form of H_2CO_3.

■ **EXAMPLE 14.8** _____

Calculate the pH of a 4.0 M solution of arsenic acid, H_3AsO_4.

Solution

We follow the procedure outlined in Figure 14.6. In principle, there are four different sources of H^+:

$$H_3AsO_4(aq) \rightleftharpoons H^+(aq) + H_2AsO_4^-(aq) \qquad K_{a1} = 5.7 \times 10^{-3}$$

$$H_2AsO_4^-(aq) \rightleftharpoons H^+(aq) + HAsO_4^{2-}(aq) \qquad K_{a2} = 1.8 \times 10^{-7}$$

$$HAsO_4^{2-}(aq) \rightleftharpoons H^+(aq) + AsO_4^{3-}(aq) \qquad K_{a3} = 2.5 \times 10^{-12}$$

$$H_2O \rightleftharpoons H^+(aq) + OH^-(aq) \qquad K_w = 1.0 \times 10^{-14}$$

The pH of a solution of a polyprotic acid is fixed by the first dissociation reaction

In practice, the first equilibrium dominates, since K for this step is considerably larger than any of the other equilibrium constants. We assume then that $[H^+]$ can be calculated by treating arsenic acid as if it were monoprotic with a dissociation constant of 5.7×10^{-3}. In other words, we write:

$$\frac{[H^+] \times [H_2AsO_4^-]}{[H_3AsO_4]} \approx \frac{[H^+]^2}{4.0} = 5.7 \times 10^{-3}$$

Solving as in Example 14.6, we find $[H^+] = 0.15\ M$; pH = $\boxed{0.82}$. (The

percent dissociation is 3.8, so we are justified in taking $[H_3AsO_4] = 4.0\ M$.)

To check the assumption that virtually all the H^+ ions come from the first dissociation, let us estimate the concentration of H^+ from the second dissociation. Calling this concentration x, we have:

$$[H^+] \approx 0.15 + x; \quad [HAsO_4^{2-}] = x; \qquad [H_2AsO_4^-] = 0.15 - x$$

Setting up the expression for K_{a2}, we have:

$$\frac{(0.15 + x)(x)}{0.15 - x} = x \approx 1.8 \times 10^{-7}$$

Since $1.8 \times 10^{-7} \ll 0.15$, our approximation is indeed justified.

Exercise
Calculate $[H_3AsO_4]$, $[H_2AsO_4^-]$, $[HAsO_4^{2-}]$, and $[AsO_4^{3-}]$ in $4.0\ M$ H_3AsO_4.

Answer: $3.9\ M$, $0.15\ M$, $1.8 \times 10^{-7}\ M$, $3.0 \times 10^{-18}\ M$.

14.3
WEAK BASES AND THEIR DISSOCIATION IN WATER

Like the weak acids, there are a large number of solutes that act as weak bases. It is convenient to classify weak bases into two groups, molecules and anions.

1. Molecules. As pointed out in Chapter 12, there are many molecular weak bases, including the organic compounds known as amines. The simplest weak base is ammonia, whose reversible dissociation in water is represented by the equation:

$$NH_3(aq) + H_2O \rightleftharpoons NH_4^+(aq) + OH^-(aq) \tag{14.17}$$

2. Anions. Any anion that is the conjugate base of a weak acid behaves as a weak base in water. Typical examples of weak bases are the F^- and $C_2H_3O_2^-$ anions, derived from the weak acids HF and $HC_2H_3O_2$. The reactions of these anions with water are very similar to that of ammonia:

$$F^-(aq) + H_2O \rightleftharpoons HF(aq) + OH^-(aq) \qquad (14.18)$$

$$C_2H_3O_2^-(aq) + H_2O \rightleftharpoons HC_2H_3O_2(aq) + OH^-(aq) \qquad (14.19)$$

In each case, the products are an OH^- ion, which makes the solution basic, and the conjugate acid (NH_4^+, HF, $HC_2H_3O_2$) of the weak base. With weak bases the OH^- ion comes from the reaction with water, rather than directly from the base itself.

■ EXAMPLE 14.9

Write an equation to explain why each of the following species produces a basic water solution:

a. NO_2^- b. CO_3^{2-} c. HCO_3^-

Solution

In each case, the weak base reacts reversibly with a water molecule, picking up a proton from it. Two species are formed. One is an OH^- ion, which makes the solution basic. The other product is a molecule or ion formed by adding H^+ to the weak base.

a. $NO_2^-(aq) + H_2O \rightleftharpoons HNO_2(aq) + OH^-(aq)$
b. $CO_3^{2-}(aq) + H_2O \rightleftharpoons HCO_3^-(aq) + OH^-(aq)$
c. $HCO_3^-(aq) + H_2O \rightleftharpoons H_2CO_3(aq) + OH^-(aq)$

Exercise

Write an equation for the reaction with water of the weak base $C_2H_5NH_2$.
Answer:

$$C_2H_5NH_2(aq) + H_2O \rightleftharpoons C_2H_5NH_3^+(aq) + OH^-(aq)$$

EXPRESSION FOR K_b

Since most of the weak bases that we will deal with are anions, we might represent the general reaction with water as:

$$B^-(aq) + H_2O \rightleftharpoons HB(aq) + OH^-(aq) \qquad (14.20)$$

We can write an equilibrium constant expression for Reaction 14.20 in the usual way:

$$K_b = \frac{[HB] \times [OH^-]}{[B^-]} \qquad (14.21)$$

The concentration of water, which remains virtually constant at about 55.5 M, is incorporated into the equilibrium constant. The constant K_b is often called the *base dissociation constant*.

This general relation can be applied to ammonia, NH_3.

$$NH_3(aq) + H_2O \rightleftharpoons NH_4^+(aq) + OH^-(aq)$$

$$K_b = \frac{[NH_4^+] \times [OH^-]}{[NH_3]} \qquad (14.22)$$

This is not a very good name, but we are stuck with it

TABLE 14.6   Dissociation Constants of Some Weak Bases* at 25°C

BASE	FORMULA	K_b	
Hydrogen sulfite ion	HSO_3^-	5.9×10^{-13}	
Dihydrogen phosphate ion	$H_2PO_4^-$	1.4×10^{-11}	
Fluoride ion	F^-	1.4×10^{-11}	
Nitrite ion	NO_2^-	1.7×10^{-11}	
Formate ion	CHO_2^-	5.3×10^{-11}	
Benzoate ion	$C_7H_5O_2^-$	1.5×10^{-10}	
Aniline	$C_6H_5NH_2$	3.8×10^{-10}	
Acetate ion	$C_2H_3O_2^-$	5.6×10^{-10}	increasing basic strength
Propionate ion	$C_3H_5O_2^-$	7.1×10^{-10}	
Hydrogen carbonate ion	HCO_3^-	2.3×10^{-8}	
Hydrogen sulfide ion	HS^-	1.0×10^{-7}	
Hydrogen phosphate ion	HPO_4^{2-}	1.6×10^{-7}	
Sulfite ion	SO_3^{2-}	1.7×10^{-7}	
Hypochlorite ion	ClO^-	3.6×10^{-7}	
Hypobromite ion	BrO^-	3.8×10^{-6}	
Cyanide ion	CN^-	1.7×10^{-5}	
Ammonia	NH_3	1.8×10^{-5}	
Carbonate ion	CO_3^{2-}	2.1×10^{-4}	
Methyl amine	CH_3NH_2	4.2×10^{-4}	
Phosphate ion	PO_4^{3-}	2.2×10^{-2}	
Sulfide ion	S^{2-}	1×10^{-1}	

*In each case the dissociation reaction is of the type:

$$B^-(aq) + H_2O \rightleftharpoons OH^-(aq) + HB(aq); \quad K_b = \frac{[OH^-] \times [HB]}{[B^-]}$$

Values of K_b are listed in Table 14.6. In general, the larger the value of K_b, the stronger the base. Since K_b for NH_3 (1.8×10^{-5}) is larger than K_b for the acetate ion, $C_2H_3O_2^-$ (5.6×10^{-10}), ammonia is a stronger base than the acetate ion. The $[OH^-]$ and pH of 0.10 M NH_3 are higher than those of 0.10 M $NaC_2H_3O_2$.

DETERMINATION OF [OH⁻] IN A SOLUTION OF A WEAK BASE

We saw in Section 14.2 how the K_a of a weak acid is used to calculate $[H^+]$ in a solution of that acid. In a very similar way, we can use K_b to obtain $[OH^-]$ in a solution of a weak base. If desired, we can go a step further and calculate $[H^+]$ or pH (Example 14.10).

■ **EXAMPLE 14.10** _____

For the butyrate ion, But^-, K_b is 5.0×10^{-10}. For a 1.0 M solution of sodium butyrate, NaBut, calculate

a. $[OH^-]$ b. $[H^+]$ c. pH

Solution

a. We follow the scheme outlined in Section 14.2.

1. When sodium butyrate dissolves in water, it dissociates completely into Na^+ and But^- ions. The Na^+ ion does not react with water, but the butyrate ion does. The equilibrium systems are

$$But^-(aq) + H_2O \rightleftharpoons HBut(aq) + OH^-(aq)$$

$$H_2O \rightleftharpoons H^+(aq) + OH^-(aq)$$

2. The equilibrium constants for these reactions are

$$K_b \text{ for } But^- = 5.0 \times 10^{-10}$$

$$K_w = 1.0 \times 10^{-14}$$

The butyrate ion is a stronger base than water

The governing equilibrium system is

$$But^-(aq) + H_2O \rightleftharpoons HBut(aq) + OH^-(aq)$$

The K_b and equilibrium expression for the governing system are

$$K_b = 5.0 \times 10^{-10} = \frac{[HBut] \times [OH^-]}{[But^-]}$$

3. We construct a table as in Example 14.6, making x the change in concentration of OH^-.

$$But^-(aq) + H_2O \rightleftharpoons HBut(aq) + OH^-(aq)$$

	ORIG. CONC. (mol/L)	CHANGE IN CONC. (mol/L)	EQUIL. CONC. (mol/L)
But^-	1.0	$-x$	$1.0 - x$
HBut	0.00	$+x$	x
OH^-	0.00	$+x$	x

4. Substituting into the expression for K_b,

$$\frac{x^2}{1.0 - x} = 5.0 \times 10^{-10}$$

5. We assume $x \ll 1.0$, and $1.0 - x \approx 1.0$. Solving for x, we get

$$\frac{x^2}{1.0} = 5.0 \times 10^{-10}$$

$$x = [OH^-] = \boxed{2.2 \times 10^{-5} \, M}$$

6. We check the validity of our assumption by calculating percent dissociation.

$$\% \text{ dissociation} = \frac{2.2 \times 10^{-5}}{1.0} \times 100 = 2.2 \times 10^{-3}\%$$

The approximation is valid.

b. Recall from Equation 14.5 that $[H^+] \times [OH^-] = 1.0 \times 10^{-14}$; thus,

$$[H^+] = \frac{1.0 \times 10^{-14}}{2.2 \times 10^{-5}} = \boxed{4.5 \times 10^{-10}\ M}$$

c. $pH = -\log_{10}[H^+] = -\log_{10}(4.5 \times 10^{-10}) = \boxed{9.35}$

Exercise
What is $[OH^-]$ in a 0.10 M solution of a weak base that has a K_b of 1.0×10^{-9}? Answer: $1.0 \times 10^{-5}\ M$.

RELATION BETWEEN K_b AND K_a

It is possible to measure the dissociation constants of weak bases in the laboratory by procedures very much like those used for weak acids. In practice, this is seldom necessary. Instead, we take advantage of a simple mathematical relationship between K_b for a weak base and K_a for its conjugate acid. This relationship can be derived by adding together the equations for the dissociation of the weak acid HB, and the weak base B^-.

(1)	$HB(aq) \rightleftharpoons H^+(aq) + B^-(aq)$	$K_I = K_a$ of HB
(2)	$B^-(aq) + H_2O(aq) \rightleftharpoons HB(aq) + OH^-(aq)$	$K_{II} = K_b$ of B^-
(3)	$H_2O \rightleftharpoons H^+(aq) + OH^-(aq)$	$K_{III} = K_w$

Since Equation (1) + Equation (2) = Equation (3), we have, according to the Rule of Multiple Equilibria (Chap. 13),

$$K_I \times K_{II} = K_{III}$$

or

$$(K_a \text{ of HB})(K_b \text{ of } B^-) = K_w = 1.0 \times 10^{-14} \qquad (14.23)$$

From Equation 14.23 we see that the dissociation constants for a weak acid and its conjugate base are inversely proportional to one another. Qualitatively, this means that the stronger the acid, the weaker will be its conjugate base. Thus we have, comparing strengths of acids and bases,

Since HCN is a very weak weak acid, the CN^- ion is a fairly strong weak base

$$HF\ (K_a = 6.9 \times 10^{-4}) > HCN\ (K_a = 5.8 \times 10^{-10})$$

$$F^-\ (K_b = 1.4 \times 10^{-11}) < CN^-\ (K_b = 1.7 \times 10^{-5})$$

Equation 14.23 allows us to calculate K_b for a weak base if K_a is known, or vice versa (Example 14.11).

■ **EXAMPLE 14.11**
Butyric acid (HBut) is the substance responsible for the foul odor of rancid butter. It has a K_a of 2.0×10^{-5}. Calculate K_b for the butyrate ion, But^-.

Solution
Applying Equation 14.23

$$K_b\ But^- = \frac{1.0 \times 10^{-14}}{K_a\ HBut} = \frac{1.0 \times 10^{-14}}{2.0 \times 10^{-5}} = \boxed{5.0 \times 10^{-10}}$$

Exercise

Pyruvic acid, HPy, has a K_a of 3.3×10^{-3}. Which is the stronger acid, HBut or HPy? Which is the stronger base, But$^-$ or Py$^-$? Answer: HPy; But$^-$.

14.4
ACID-BASE PROPERTIES OF SALT SOLUTIONS

At this point, you should be able to predict correctly that an aqueous solution of HI or H_2SO_4 is acidic while a solution of NaOH or NH_3 is basic. Solutions of $NaNO_2$ or NH_4I might be more difficult for you to classify. These two compounds, and many others, such as NaCl, $Zn(NO_3)_2$, and $CuSO_4$, are **salts. A salt is an ionic compound containing a cation other than H$^+$ and an anion other than OH$^-$ or O^{2-}.**

In dilute water solution, a salt is completely dissociated into ions. A water solution labeled "$NaNO_2$" actually contains Na$^+$ and NO_2^- ions:

$$NaNO_2(s) \rightarrow Na^+(aq) + NO_2^-(aq) \tag{14.24}$$

It follows that the acid-base properties of a salt such as $NaNO_2$ are determined by the behavior of its ions. To decide whether a water solution of $NaNO_2$ is acidic, basic, or neutral, we must consider the effect of Na$^+$ and NO_2^- ions on the pH of water.

Some ions have no effect on the pH of water; in that sense they are neutral. Other ions are acidic because they dissociate to produce H$^+$ ions in water. Still other ions react with water to form OH$^-$ ions, making the solution basic. Table 14.7 summarizes the acid-base behavior of ions commonly present in water solution.

NEUTRAL IONS (NONBASIC AND NONACIDIC)

A neutral ion does not react with water to produce H$^+$ or OH$^-$ ions. Hence, it does not affect the pH. There are relatively few neutral ions. We see from Table 14.7 that

— **the neutral anions are those derived from strong acids.**

— **the neutral cations are those derived from strong bases.**

TABLE 14.7   Acid-Base Properties of Some Common Ions in Water Solution

	NEUTRAL		BASIC		ACIDIC	
Anion	Cl$^-$ Br$^-$ I$^-$	NO$_3^-$ ClO$_4^-$ SO$_4^{2-}$	C$_2$H$_3$O$_2^-$ F$^-$ CO$_3^{2-}$ S^{2-} PO$_4^{3-}$	CN$^-$ NO$_2^-$ HCO$_3^-$ HS$^-$ HPO$_4^{2-}$	HSO$_4^-$	H$_2$PO$_4^-$
Cation	Li$^+$ Na$^+$ K$^+$	Ca^{2+} Ba^{2+}	none		Mg^{2+} Al^{3+} NH$_4^+$ transition metal ions	

A typical neutral anion is the chloride ion, produced by the dissociation of hydrochloric acid:

$$HCl(aq) \rightarrow H^+(aq) + Cl^-(aq)$$

Since HCl is a strong acid, there is no tendency for the reverse of the above reaction to occur. Chloride ions do not combine with H^+ ions, whatever their source. In particular, Cl^- ions do not pick up H^+ ions from water. As a result, Cl^- ions and the other neutral anions listed in Table 14.7 do not change the $[H^+]$ or pH of water.

A similar argument applies to cations such as Na^+, produced by the dissociation of strong bases such as NaOH. Dissociation is complete:

$$NaOH(s) \rightarrow Na^+(aq) + OH^-(aq)$$

Hence, there is no tendency for the reverse reaction to occur; that is, Na^+ ions, regardless of their source, do not combine with OH^- ions in water. As a result, Na^+ and other cations derived from strong bases are neutral.

The pH of a solution of NaCl is 7. It is neutral

BASIC ANIONS

Recall from Section 14.3 that any anion derived from a weak acid acts as a weak base in water solution. There is a small army of such anions. Those listed in Table 14.7 are typical examples. In contrast, there are no common basic cations.

ACIDIC IONS

Acidic ions include

— all cations except those of the alkali metals and the heavier alkaline earths.
— the HSO_4^- and $H_2PO_4^-$ anions.

■ **EXAMPLE 14.12** _____

Consider the four salts

$$NH_4I, \; Zn(NO_3)_2, \; KClO_4, \; Na_3PO_4$$

For each salt,

a. indicate the ions present.
b. classify both anion and cation as acidic, basic, or neutral.
c. state whether the salt solution will be acidic, basic, or neutral.
d. write net ionic equations to explain the acidity or basicity of the salt.

Solution

a. NH_4^+, I^- ions; Zn^{2+}, NO_3^- ions; K^+, ClO_4^- ions; Na^+, PO_4^{3-} ions.

b. and c. It is convenient to prepare a table classifying both ions of the salt and then the salt itself as acidic, basic, or neutral.

SALT	CATION	ANION	SOLUTION OF SALT
NH_4I	NH_4^+ (acidic)	I^- (neutral)	acidic
$Zn(NO_3)_2$	Zn^{2+} (acidic)	NO_3^- (neutral)	acidic
$KClO_4$	K^+ (neutral)	ClO_4^- (neutral)	neutral
Na_3PO_4	Na^+ (neutral)	PO_4^{3-} (basic)	basic

It's easy to predict whether a salt solution is acidic, basic, or neutral

These predictions are confirmed by experiment (Fig. 14.7).

d. for NH_4I; $\quad NH_4^+(aq) \rightleftharpoons H^+(aq) + NH_3(aq)$
for $Zn(NO_3)_2$: $\quad Zn(H_2O)_4^{2+}(aq) \rightleftharpoons H^+(aq) + Zn(H_2O)_3(OH)^+(aq)$
for Na_3PO_4: $\quad PO_4^{3-}(aq) + H_2O \rightleftharpoons HPO_4^{2-}(aq) + OH^-(aq)$

Exercise

Write net ionic equations to explain why a solution of NaH_2PO_4 is acidic while a solution of Na_2HPO_4 is basic. Answer:

$$H_2PO_4^-(aq) \rightleftharpoons H^+(aq) + HPO_4^{2-}(aq)$$

$$HPO_4^{2-}(aq) + H_2O \rightleftharpoons OH^-(aq) + H_2PO_4^-(aq)$$

The procedure described in Example 14.12 is a general one. We did not, however, give an example of a salt made up of an acidic cation and a basic anion. This situation is somewhat more complicated than the ones we have considered. To decide whether the salt is acidic, basic, or neutral, we have to compare the K_a value of the cation with the K_b value for the anion. In general, for salts containing equal numbers of cations and anions:

If	Then	Example	K_a cation		K_b anion	Approx. pH
$K_a > K_b$	acidic	NH_4F	5.6×10^{-10}	>	1.4×10^{-11}	6.2
$K_a = K_b$	neutral	$NH_4C_2H_3O_2$	5.6×10^{-10}	=	5.6×10^{-10}	7.0
$K_a < K_b$	basic	$AlPO_4$	1×10^{-5}	<	2.2×10^{-2}	8.7

Figure 14.8, p. 496, is a diagram of the logical process you should go through when asked to predict the acid-base behavior of a salt solution.

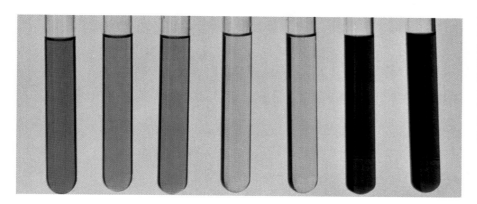

Figure 14.7 Colors of salt solutions with universal indicator, which is red in strong acid (*far left*) and purple in strong base (*far right*). NH_4I and $Zn(NO_3)_2$ are weakly acidic at about pH 5 (orange); $KClO_4$, like pure water, has a pH of 7 (yellow-green); Na_3PO_4 is basic at about pH 12 (purple).

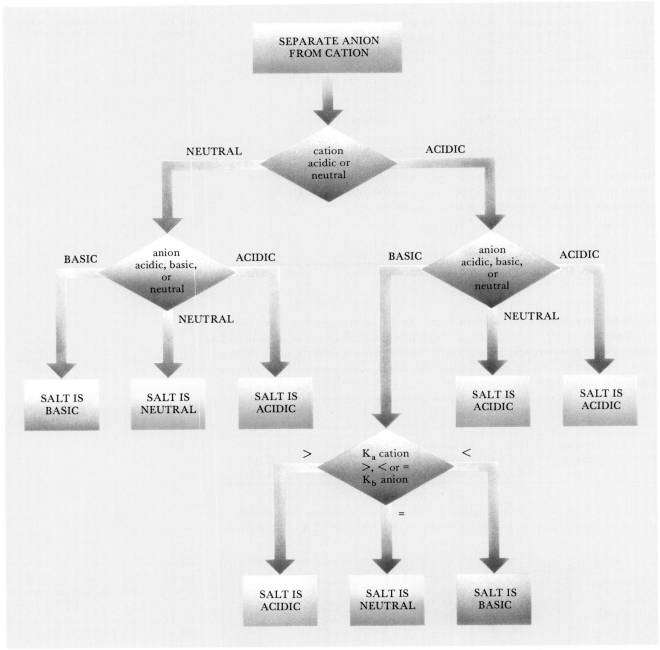

Figure 14.8 General approach to determine whether a given salt solution is acidic, basic, or neutral. If the cation is acidic and the anion basic, you need to know the values of K_a and K_b to make a decision.

14.5
GENERAL MODELS OF ACIDS AND BASES

Thus far in this chapter, we have considered an acid to be a substance that produces an excess of H^+ ions in water solution. A base was similarly defined to be a substance that, directly or indirectly, forms excess OH^- ions in water solution. This approach, first proposed by the Swedish chemist Svante Arrhenius in 1884, is a very practical one. In particular, it allows us to predict whether such substances as HCl, $Ca(OH)_2$, NH_4Cl, NaCl, and K_2CO_3 will form acidic, basic, or neutral solutions.

The Arrhenius model does, however, have one disadvantage. It greatly restricts the number of reactions that can be considered to be of the acid-base type. Over the years, many other, more general models of acids and bases have been proposed. In this section, we will discuss two such models. Curiously enough, they were both suggested in the same year, 1923. One of them was proposed independently by Brönsted in Denmark and Lowry in England. The other came from a man we have heard of before (Chap. 8), the American physical chemist G. N. Lewis.

Early chemists had a hard time understanding acids and bases. Many students still do

BRÖNSTED-LOWRY CONCEPT

According to this model, an acid-base reaction is one in which there is a *proton transfer* from one species to another. The species that gives up or **donates the proton** (H^+ ion) is referred to as an **acid**. The molecule or ion that **accepts the proton** (H^+ ion) is a **base**.

A simple example of a Brönsted-Lowry acid-base reaction is that between acetic acid and hydroxide ions:

$$HC_2H_3O_2(aq) + OH^-(aq) \rightarrow C_2H_3O_2^-(aq) + H_2O \qquad (14.25)$$

$\quad$ acid $\qquad\qquad$ base

Here, $HC_2H_3O_2$ gives up a proton to an OH^- ion. Hence, $HC_2H_3O_2$ is acting as an acid (proton donor), while OH^- is a base (proton acceptor). This conclusion is not particularly startling. We used almost the same words earlier in Chapter 12 in describing this reaction by the Arrhenius model.

Consider, however, the reverse of Reaction 14.25:

$$C_2H_3O_2^-(aq) + H_2O \rightarrow HC_2H_3O_2(aq) + OH^-(aq) \qquad (14.26)$$

$\quad$ base $\qquad$ acid

This is the equation written earlier to explain why a solution containing acetate ions is basic. At the time, we did not refer to it as an acid-base reaction. According to the Brönsted-Lowry model, it is. The $C_2H_3O_2^-$ anion accepts a proton from a water molecule and hence acts as a base. The H_2O molecule donates a proton to the acetate ion and so acts as a Brönsted-Lowry acid.

The Brönsted-Lowry model can be extended still further. Consider the equation written earlier for the dissociation of HCl in water:

$$HCl(aq) \rightarrow H^+(aq) + Cl^-(aq)$$

Figure 14.9 When HCl is added to water, there is a proton transfer from an HCl to an H_2O molecule, forming a Cl^- ion and an H_3O^+ ion. In this reaction, HCl acts as a Brönsted-Lowry acid, H_2O as a Brönsted-Lowry base.

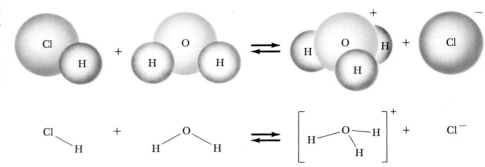

Here, although HCl appears to be the proton donor, there is not any obvious proton acceptor. According to Brönsted and Lowry, the proton acceptor is really a water molecule. They would rewrite this equation as

They would if they were alive

$$\underset{\text{acid}}{HCl(aq)} + \underset{\text{base}}{H_2O} \rightarrow Cl^-(aq) + H_3O^+(aq) \tag{14.27}$$

In this equation, HCl is donating a proton to H_2O (Fig. 14.9). Thus, HCl is acting as an acid and H_2O as a base. Equation 14.27 offers a plausible explanation as to why HCl dissociates in water. It does so by reacting with water to produce two more stable species, the Cl^- and H_3O^+ ions. In the Arrhenius model, the reason for the dissociation of HCl is not apparent.

You will note from this discussion that the H_2O molecule can act as either a Brönsted-Lowry acid (Equation 14.26) or base (Equation 14.27). When it acts as an acid, it donates a proton to another species and is converted to an OH^- ion. When water acts as a base, it accepts a proton, forming the hydronium ion, H_3O^+. The dissociation of water can be expressed in these terms. Here, water serves as both the acid and base:

We can consider the H_3O^+ ion to be a hydrated proton

$$\underset{\text{base}}{H_2O} + \underset{\text{acid}}{H_2O} \rightarrow H_3O^+(aq) + OH^-(aq) \tag{14.28}$$

The hydronium ion is formed by the acceptance of an H^+ ion by an H_2O molecule.

Several other molecules and ions can behave as both acids and bases in the Brönsted-Lowry sense. Among these is the HCO_3^- ion:

$$\underset{\text{base}}{HCO_3^-(aq)} + \underset{\text{acid}}{H_2O} \rightarrow H_2CO_3(aq) + OH^-(aq) \tag{14.29}$$

$$\underset{\text{acid}}{HCO_3^-(aq)} + \underset{\text{base}}{H_2O} \rightarrow CO_3^{2-}(aq) + H_3O^+(aq) \tag{14.30}$$

Species like H_2O or HCO_3^-, which can either donate or accept a proton, are referred to as being *amphiprotic*.

■ **EXAMPLE 14.13** _____

Consider the reaction

$$H_2PO_4^-(aq) + C_2H_3O_2^-(aq) \rightleftharpoons HPO_4^{2-}(aq) + HC_2H_3O_2(aq)$$

Classify each of the four species involved as a Brönsted-Lowry acid or base.

Solution

In the forward reaction, the $H_2PO_4^-$ ion donates a proton to the $C_2H_3O_2^-$ ion. In the reverse reaction, $HC_2H_3O_2$ donates a proton to HPO_4^{2-}. Hence,

$H_2PO_4^-$ and $HC_2H_3O_2$ are Brönsted-Lowry acids; $C_2H_3O_2^-$ and HPO_4^{2-} are Brönsted-Lowry bases.

Exercise

What is the conjugate base of $H_2PO_4^-$? the conjugate acid of PO_4^{3-}? Answer: HPO_4^{2-}.

By the Brönsted-Lowry theory, acid + base $\rightleftarrows$ conjugate base + conjugate acid

It is possible to compare the relative strengths of different Brönsted acids, i.e., their relative abilities to donate protons in an acid-base reaction. Such a comparison is given in Table 14.8. The Brönsted acids listed in the left column can be divided into three categories. In descending order, we have:

1. Those acids, like HNO_3, which are stronger than the H_3O^+ ion and are completely dissociated in water solution. The equilibrium

$$HNO_3(aq) + H_2O \rightleftarrows NO_3^-(aq) + H_3O^+(aq)$$

lies far to the right. To all intents and purposes the HNO_3 molecule does not exist as such in dilute water solution.

$K_a \gg 1$, essentially infinite

2. Acids such as NH_4^+, which are weaker than H_3O^+, but stronger than H_2O, are only slightly dissociated in water. The equilibrium

$$NH_4^+(aq) + H_2O \rightleftarrows NH_3(aq) + H_3O^+(aq)$$

lies far to the left. All the species referred to earlier as "weak acids" fall in this category.

TABLE 14.8   Relative Strengths of Brönsted-Lowry Acids and Bases

K_a	CONJUGATE ACID	CONJUGATE BASE	K_b
very large	$HClO_4$	ClO_4^-	very small
very large	HCl	Cl^-	very small
very large	HNO_3	NO_3^-	very small
	H_3O^+	**H_2O**	
6.9×10^{-4}	HF	F^-	1.4×10^{-11}
1.8×10^{-5}	$HC_2H_3O_2$	$C_2H_3O_2^-$	5.6×10^{-10}
1.4×10^{-5}	$Al(H_2O)_6^{3+}$	$Al(H_2O)_5(OH)^{2+}$	7.1×10^{-10}
4.4×10^{-7}	H_2CO_3	HCO_3^-	2.3×10^{-8}
2.8×10^{-8}	$HClO$	ClO^-	3.6×10^{-7}
5.6×10^{-10}	NH_4^+	NH_3	1.8×10^{-5}
4.7×10^{-11}	HCO_3^-	CO_3^{2-}	2.1×10^{-4}
	H_2O	**OH^-**	
very small	C_2H_5OH	$C_2H_5O^-$	very large
very small	OH^-	O^{2-}	very large
very small	H_2	H^-	very large

3. Species such as C_2H_5OH, which are weaker Brönsted acids than H_2O, show no tendency to ionize in water. From the Arrhenius viewpoint, ethyl alcohol, C_2H_5OH, is not an acid; its water solution is neutral.

Looking at the right column of Table 14.8, you can see that

The stronger the acid, the weaker its conjugate base.

Again, we can divide Brönsted bases into three categories so far as their behavior in water solution is concerned.

1. The species listed at the top of the right column, such as Cl^-, are weaker bases than H_2O. They show no tendency to react with water; their water solutions are neutral.

2. Bases which are stronger than H_2O but weaker than OH^- are partially dissociated in water. With species such as the fluoride ion

$$F^-(aq) + H_2O \rightleftarrows HF(aq) + OH^-(aq)$$

equilibrium lies far to the left. The "weak bases" referred to earlier in this chapter fall in this category.

3. The bases listed at the bottom of the right column are stronger than the OH^- ion. Species like the oxide ion react essentially completely with H_2O

$$O^{2-}(aq) + H_2O \rightleftarrows 2OH^-(aq)$$

As a result, the strong base O^{2-}, like the strong acid HNO_3, does not exist as such in water solution.

THE LEWIS CONCEPT

We have seen that the Brönsted-Lowry model extends the Arrhenius picture of acid-base reactions considerably. However, the Brönsted-Lowry model is restricted in one important respect. It can be applied only to reactions involving a proton transfer. For a species to act as a Brönsted-Lowry acid, it must contain an ionizable hydrogen atom.

The Lewis acid-base model removes this restriction. A **Lewis acid** is a species that in an acid-base reaction, **accepts** an **electron pair**. In this reaction, a **Lewis base donates** the **electron pair**.

From a structural point of view, the Lewis concept of a base does not differ in any essential way from the Brönsted-Lowry concept. In order for a species to accept a proton and thereby act as a Brönsted-Lowry base it must possess an unshared pair of electrons. Consider, for example, the NH_3 molecule, the H_2O molecule, and the F^- ion, all of which can act as Brönsted-Lowry bases:

$$H—\overset{..}{N}—H \qquad H—\overset{..}{\underset{..}{O}}—H \qquad (:\overset{..}{\underset{..}{F}}:)^-$$
$$\underset{H}{|}$$

Each of these species contains an unshared pair of electrons that is utilized in accepting a proton to form the NH_4^+ ion, the H_3O^+ ion, or the HF molecule:

Na₂O would react with water to produce Na^+ and OH^- ions

In the Lewis model, bases have unshared electron pairs

$$\left[\begin{array}{c} H \\ H:\overset{..}{N}:H \\ H \end{array}\right]^{+} \quad \left[\begin{array}{c} \overset{..}{H:\overset{..}{O}:H} \\ H \end{array}\right]^{+} \quad H:\overset{..}{\underset{..}{F}}:$$

Clearly, NH_3, H_2O, and F^- can also be Lewis bases, since they possess an unshared electron pair that can be donated to an acid. We see then that the Lewis concept does not significantly change the number of species that can behave as bases.

On the other hand, the Lewis concept greatly increases the number of species that can be considered to be acids. The substance that accepts an electron pair and therefore acts as a Lewis acid can be a proton:

$$\underset{\text{acid}}{H^+(aq)} + \underset{\text{base}}{H_2O} \rightarrow H_3O^+(aq) \tag{14.31}$$

$$\underset{\text{acid}}{H^+(aq)} + \underset{\text{base}}{NH_3(aq)} \rightarrow NH_4^+(aq) \tag{14.32}$$

It can equally well be a cation, such as Zn^{2+}, which can accept electron pairs from a Lewis base:

$$\underset{\text{acid}}{Zn^{2+}(aq)} + \underset{\text{base}}{4H_2O} \rightarrow Zn(H_2O)_4^{2+}(aq) \tag{14.33}$$

Here Zn^{2+} acts as a Lewis acid

$$\underset{\text{acid}}{Zn^{2+}(aq)} + \underset{\text{base}}{4NH_3} \rightarrow Zn(NH_3)_4^{2+}(aq) \tag{14.34}$$

We will discuss reactions of this type in greater detail in Chapter 16.

Another important class of Lewis acids comprises molecules containing an incomplete octet of electrons. A classic example is boric acid, an antiseptic sometimes found in eyewashes:

$$\begin{array}{c} \overset{..}{H-\overset{..}{O}-B-\overset{..}{O}-H} \\ | \\ :\overset{..}{O}: \\ | \\ H \end{array}$$

When boric acid is added to water, it acts as a Lewis acid, picking up an OH^- ion to complete the octet of boron and, at the same time, liberating a proton:

$$\underset{\substack{\text{Lewis} \\ \text{acid}}}{B(OH)_3(s)} + \underset{\substack{\text{Lewis} \\ \text{base}}}{H_2O} \rightarrow B(OH)_4^-(aq) + H^+(aq) \tag{14.35}$$

The Lewis model is commonly used in organic chemistry to consider the catalytic behavior of such Lewis acids as $ZnCl_2$ and BF_3. In general, when proton transfer reactions are involved, most chemists use the Arrhenius or Brönsted-Lowry concepts. Table 14.9, p. 502 summarizes the acid-base models we have discussed. As you might guess, others have been proposed since 1923. The three listed, however, will suffice for our purposes.

BF_3 is a classic Lewis acid

TABLE 14.9 **Alternative Definitions of Acids and Bases**

MODEL	ACID	BASE
Arrhenius	supplies H^+ to water	supplies OH^- to water
Brönsted-Lowry	H^+ donor	H^+ acceptor
Lewis	electron pair acceptor	electron pair donor

■ **SUMMARY PROBLEM**

Consider the salts ammonium chloride, NH_4Cl, and potassium cyanide, KCN.

1. Hydrochloric acid and ammonia can be used to prepare ammonium chloride. Potassium hydroxide and hydrogen cyanide can be used to prepare potassium cyanide. Classify each reactant as a strong or weak acid, strong or weak base.
2. Write equations for the dissociation of all acids and bases in (1).
3. Classify the salts as acidic or basic. Write net ionic equations to explain your answers.
4. Calculate the pH of 0.10 M HCl, KOH, NH_4Cl, and KCN (Tables 14.3, 14.4).
5. If the salt were ammonium cyanide, NH_4CN, would it be acidic, basic, or neutral?
6. What is the conjugate base for NH_4^+? The conjugate acid for CN^-?

Answers
1. HCl, strong acid; KOH, strong base; ammonia, weak base; HCN, weak acid.
2. $HCl(aq) \rightarrow H^+(aq) + Cl^-(aq)$
   $KOH(aq) \rightarrow K^+(aq) + OH^-(aq)$
   $NH_3(aq) + H_2O \rightleftharpoons NH_4^+(aq) + OH^-(aq)$
   $HCN(aq) \rightleftharpoons H^+(aq) + CN^-(aq)$
3. NH_4Cl is acidic; $NH_4^+(aq) \rightleftharpoons NH_3(aq) + H^+(aq)$
   KCN is basic; $CN^-(aq) + H_2O(aq) \rightleftharpoons HCN(aq) + OH^-(aq)$
4. HCl, pH = 1.0; KOH, pH = 13.0; NH_4Cl, pH = 5.13; KCN, pH = 11.11.
5. NH_4CN is basic.
6. NH_3; HCN.

As pointed out earlier in this chapter, most weak acids are organic in nature, i.e., they contain carbon and hydrogen atoms. Table 14.10 lists some of the organic acids found in foods. All of these compounds contain the *carboxyl* group:

$$-\overset{\displaystyle \parallel}{\underset{\displaystyle O}{C}}-O-H$$

The general equation for their reversible dissociation in water is:

$$RCOOH(aq) \rightleftharpoons RCOO^-(aq) + H^+(aq) \qquad (14.36)$$

Several different organic acids are formed in human metabolism, among them lactic acid. If you exercise strenuously, glucose may be converted to lactic acid in your muscles, producing tiredness and a painful sensation. Resting after vigorous exercise allows enough oxygen to reach the tissues and converts glucose to carbon dioxide and water.

Certain drugs, both prescription and over-the-counter, contain organic acids. Two of the most popular products of this type are the analgesics aspirin and ibuprofen (Advil, Nuprin, etc.).

TABLE 14.10 Some Naturally Occurring Organic Acids

NAME		SOURCE
Acetic acid	CH_3-COOH	Vinegar
Citric acid	$HOOC-CH_2-\overset{\displaystyle OH}{\underset{\displaystyle COOH}{C}}-CH_2-COOH$	Citrus fruits
Lactic acid	$CH_3-\underset{\displaystyle OH}{CH}-COOH$	Sour milk
Malic acid	$HOOC-CH_2-\underset{\displaystyle OH}{CH}-COOH$	Apples, watermelons, grape juice, wine
Oxalic acid	$HOOC-COOH$	Rhubarb, spinach, tomatoes
Quinic acid		Cranberries
Tartaric acid	$HOOC-\underset{\displaystyle OH}{CH}-\underset{\displaystyle OH}{CH}-COOH$	Grape juice, wine

Among the organic acids found in fruit juices are ascorbic acid (Vitamin C), quinic acid in cranberries, malic acid in apples, and tartaric acid in grapes. (Charles D. Winters)

$$\text{COOH}$$
$$\text{OCOCH}_3$$

aspirin

$$(CH_3)_2—CH—CH_2—\boxed{}—\underset{H}{\overset{CH_3}{C}}—COOH$$

ibuprofen

Because these compounds are acidic, they can cause stomach irritation unless taken with food or water. The other common analgesic, acetaminophen (Tylenol), is not an acid and does not have this effect.

To the general public, probably the best-known organic acid is ascorbic acid (Vitamin C):

ascorbic acid (Vitamin C)

Vitamin C plays a variety of roles in human nutrition, some of which have been discovered only recently. For one thing, it is involved in the formation of collagen, the principal protein in connective tissue. Beyond that, Vitamin C is required for the metabolism of several amino acids and for the absorption of iron. Moreover, since Vitamin C is easily oxidized, it serves to protect other vitamins (A and E) from oxidation. The recommended daily allowance of Vitamin C is 60 mg/day.

Many people, including Nobel laureate Linus Pauling, advocate taking megadoses of Vitamin C—two or more grams per day. In such large doses Vitamin C supposedly helps prevent colds and acts as an anticarcinogen. However, large doses of Vitamin C can be toxic to people with an enzyme deficiency that makes them susceptible to strong reducing agents. Perhaps 2 to 5% of the population falls into this category; for them megadoses of Vitamin C pose a very real health hazard.

QUESTIONS AND PROBLEMS

$[H^+]$, $[OH^-]$, pH, and pOH

1. Calculate $[OH^-]$ in solutions that have the following values for $[H^+]$:
 a. $4.0 \times 10^{-3}\,M$ b. $3.0\,M$
 c. $8.7 \times 10^{-8}\,M$ d. $3.9 \times 10^{-6}\,M$
2. Calculate $[H^+]$ in solutions in which $[OH^-]$ is
 a. $0.0010\,M$ b. $3.4 \times 10^{-4}\,M$
 c. $9.2 \times 10^{-10}\,M$ d. $4.5\,M$

3. Find the pH of solutions with the following $[H^+]$. Classify each as acidic or basic.
 a. $1 \times 10^{-2}\,M$
 b. $0.00010\,M$
 c. $4.0 \times 10^{-5}\,M$
 d. $6.2 \times 10^{-10}\,M$
4. Find the pH of the solutions with the following $[H^+]$. Classify each as acidic or basic.
 a. $0.1\,M$ b. $10\,M$
 c. $7.0 \times 10^{-3}\,M$ d. $8.2 \times 10^{-9}\,M$

5. Calculate $[H^+]$ and $[OH^-]$ in solutions with the following pH:
a. 4.0   b. 8.52   c. 0.00   d. 12.60

6. Find $[H^+]$ and $[OH^-]$ in solutions having the following pOH:
a. 9.0   b. 3.20   c. −1.05   d. 7.46

7. Solution A has $[OH^-] = 3.2 \times 10^{-4}\,M$. Solution B has $[H^+] = 6.9 \times 10^{-9}\,M$. Which solution is more basic? Which has the lower pH?

8. Solution 1 has pH 4.3. Solution 2 has $[OH^-] = 3.4 \times 10^{-7}\,M$. Which solution is more acidic? Which has the higher pOH?

9. One solution has a pH of 3.2. What must be the pH of another solution so that $[H^+]$ is four times as large? One fourth as large?

10. One solution has a pH of 3.5. Another has a pH of 5.5. What is the ratio of the H^+ ion concentrations in the two solutions? The ratio of OH^- ion concentrations?

11. Milk of magnesia has a pH of 10.5.
a. Calculate $[H^+]$.
b. Calculate the ratio of the $[H^+]$ concentration of gastric juice, pH 1.5, to that of milk of magnesia.

12. Unpolluted rain water has a pH of about 5.5. Acid rain has been shown to have a pH as low as 3.0. Calculate the $[H^+]$ ratio in acid rain to unpolluted rain.

13. Find $[H^+]$, and the pH of the following solutions:
a. 0.30 M HBr
b. A solution made by diluting 10.0 mL of 6.00 M HCl to 0.300 L with water.

14. Find $[H^+]$ and the pH of the following solutions:
a. 0.60 M HCl
b. A solution made by dissolving 75 g HNO_3 in water to make 2.0 L of solution.

15. What is the pH of a solution obtained by mixing 245 mL of 0.0235 M HI and 438 mL of 0.554 M HBr? Assume that volumes are additive.

16. What is the pH of a solution obtained by mixing 38.2 g $HClO_4$ and 28.9 g HCl in enough water to make 0.750 L of solution?

17. Find $[OH^-]$, $[H^+]$, and the pOH of the following solutions:
a. 0.50 M KOH
b. A solution made by dissolving 100.0 g NaOH in enough water to make 500.0 mL of solution.

18. Find $[OH^-]$, $[H^+]$, and the pOH of the following solutions:
a. 0.80 M NaOH
b. A solution made by diluting 8.0 mL of 6.0 M CsOH to a volume of 4.80×10^2 mL with water.

19. What is the pH of a solution obtained by adding 32.1 g of NaOH and 56.3 g of KOH to enough water to make 1.75 L of solution?

20. What is the pH of a solution obtained by adding 25.0 mL of 0.125 M rubidium hydroxide, RbOH, to 35.0 mL of 0.027 M barium hydroxide? Assume that volumes are additive.

Dissociation Expressions, Weak Acids

21. Write an equation for the dissociation in water of each of the following weak acids:
a. $Ni(H_2O)_5(OH)^+$ b. $Al(H_2O)_6^{3+}$ c. H_2S
d. $H_2PO_4^-$ e. $Cr(H_2O)_5(OH)^{2+}$ f. $HClO_2$

22. Write an equation for the dissociation in water of each of the following acids:
a. $Zn(H_2O)_3(OH)^+$ b. HSO_4^- c. HNO_2
d. $Fe(H_2O)_6^{2+}$ e. $Mn(H_2O)_6^{2+}$ f. $HC_2H_3O_2$

23. Write the dissociation equation and the K_a expression for each of the following acids:
a. HSO_3^- b. HPO_4^{2-} c. HNO_2

24. Write the dissociation equation and the K_a expression for each of the following acids:
a. PH_4^+ b. HS^- c. $HC_2O_4^-$

25. Referring to Table 14.3, give the pK_a value for the acids in Question 23.

26. Calculate the K_a for the weak acids that have the following pK_a values:
a. 3.0   b. 5.8   c. 9.67

27. Consider these acids:

Acid	A	B	C	D
K_a	3×10^{-4}	5×10^{-6}	2×10^{-3}	8×10^{-2}

a. Arrange the acids in order of increasing acid strength.
b. Which acid has the lowest pK_a value?

28. Consider these acids:

Acid	A	B	C	D
pK_a	8.3	2.7	12.9	5.6

a. List the acids in order of decreasing acid strength.
b. Which acid has the smallest K_a value?

Equilibrium Calculations, Weak Acids

29. Para-aminobenzoic acid (PABA), $HC_7H_6NO_2$, is used in some sunscreen agents. A solution is made by dissolving 0.030 mol PABA to form a liter of solution. The solution has $[H^+] = 8.1 \times 10^{-4}\,M$. Calculate K_a for para-aminobenzoic acid.

30. Caproic acid, $HC_6H_{11}O_2$, is found in coconut oil and is used in the making of artificial flavors. A solution prepared by dissolving 0.14 mol $HC_6H_{11}O_2$ to form 1.5 L of solution has $[H^+] = 1.1 \times 10^{-3}\,M$. Calculate K_a for caproic acid.

31. Benzoic acid, $HC_7H_5O_2$, is present in many berries. A benzoic acid solution prepared by dissolving 1.00 g of benzoic acid in 350.0 mL of solution has a pH of 2.91. What is the K_a for benzoic acid?

32. Phenol, HC_6H_5O, is a weak organic acid used in the manufacture of plastics. A solution prepared by dissolving 0.385 g of phenol in water to form a volume of 2.00 L has a pH of 6.29. What is the K_a of phenol?

33. Formic acid, $HCHO_2$, is the irritant in nettles and ants. Its K_a is 1.9×10^{-4}. Calculate the $[H^+]$ in solutions prepared by adding the following number of moles of formic acid to one liter of solution:

a. 2.0   b. 0.33

34. Follow the directions for Problem 33 for hypobromous acid, HBrO ($K_a = 2.6 \times 10^{-9}$).

35. Lactic acid ($K_a = 1.4 \times 10^{-4}$) is present in sore muscles after vigorous exercise. For a 1.3 M solution of lactic acid, calculate

a. $[H^+]$ b. $[OH^-]$ c. pH   d. % dissociation

36. Follow the directions of Problem 35 for a 1.25 M solution of ammonium chloride (K_a $NH_4^+ = 5.6 \times 10^{-10}$).

37. Chloroacetic acid, $ClCH_2COOH$, has a K_a of 1.4×10^{-3}. Calculate the pH of a 0.20 M solution of chloroacetic acid.

38. Chlorous acid, $HClO_2$, has a K_a of 1.0×10^{-2}. Calculate the pH of a 0.10 M solution of chlorous acid.

39. Using the K_a values listed in Table 14.5, calculate the pH of 0.10 M H_2CO_3.

40. Using the K_a values listed in Table 14.5, calculate the pH of a solution 0.10 M in H_2S.

41. For the solution referred to in Problem 39, estimate $[HCO_3^-]$ and $[CO_3^{2-}]$.

42. For the solution referred to in Problem 40, estimate $[HS^-]$ and $[S^{2-}]$.

Dissociation Expressions, Weak Bases

43. Write an equation for the dissociation in water of each of the following weak bases:

a. NH_3 b. NO_2^- c. $C_6H_5NH_2$
d. CO_3^{2-} e. F^- f. HCO_3^-

44. Write an equation for the dissociation in water of the weak base:

a. $(CH_3)_3N$ b. PO_4^{3-} c. HPO_4^{2-}
d. $H_2PO_4^-$ e. HS^- f. $C_2H_5NH_2$

45. Using the dissociation constants listed in Table 14.6, arrange the following solutions in order of decreasing pH:

a. 0.10 M NH_3 b. 0.10 M KOH
c. 0.10 M NaF   d. 0.10 M $C_6H_5NH_2$

46. Follow the directions of Question 45 for 0.10 M solutions of

a. HCl   b. NaOH   c. $NaNO_2$ d. NaClO

47. Which of the following are true regarding a 1 M solution of a strong base MOH?

a. The M^+ concentration is 1 M.
b. The concentration of MOH molecules is 1 M.
c. The sum of $[M^+]$ and $[OH^-]$ is 2 M.
d. The H^+ concentration is 1 M.
e. The pH is 14.0.

48. Which of the following are true regarding a 0.10 M solution of a weak base B^-?

a. The HB concentration is 0.10 M.
b. The $[OH^-] \approx [HB]$.
c. $[B^-] \gg [HB]$.
d. The pH is 13.
e. The H^+ concentration is 0.10 M.

Equilibrium Calculations, Weak Bases

49. Find the value of K_a for the conjugate acid of the following organic bases:

a. dimethyl amine, used to remove hair from hides; $K_b = 5.2 \times 10^{-4}$
b. aniline, an important dye intermediate; $K_b = 3.8 \times 10^{-10}$

50. Find the value of K_b for the conjugate base of the following acids:

a. gallic acid, present in tea; $K_a = 3.9 \times 10^{-5}$
b. mandelic acid, obtained from bitter almonds; $K_a = 4.4 \times 10^{-4}$

51. Write the net ionic equation for the reaction that makes solutions of sodium benzoate, $NaC_7H_5O_2$, basic. K_a for benzoic acid is 6.6×10^{-5}. Find

a. K_b for the reaction.
b. the pH of a 0.23 M solution of sodium benzoate.

52. Use Table 14.6 to determine the pH of a 0.50 M solution of sodium cyanide.

53. Using Table 14.6, calculate the pH of a household-cleaning ammonia solution prepared by dissolving enough ammonia gas in water to make a 0.30 M aqueous solution.

54. Cocaine is a weak base. Its ionization in water can be represented by the equation

$$Coc(aq) + H_2O \rightleftharpoons CocH^+(aq) + OH^-(aq)$$

A 0.0010 M solution of cocaine has a pH of 9.70. Calculate K_b for cocaine.

Salt Solutions

55. State whether 1 M solutions of the following salts in water would be acidic, basic, or neutral.

a. NH_4Cl b. NH_4CN (Tables 14.3, 14.6)
c. Na_3PO_4 d. KNO_3
e. $KHCO_3$ f. NaCN

56. State whether 1 M solutions of the following salts in water would be acidic, basic, or neutral.

a. $Al(NO_3)_3$ b. NH_4NO_2 (Tables 14.3, 14.6)
c. NaClO   d. NH_4NO_3
e. Na_2CO_3 f. $NaHSO_4$

57. Write net ionic equations to explain the acidity or basicity of the various salts listed in Question 55.

58. Write net ionic equations to explain the acidity or basicity of the various salts listed in Question 56.

59. Arrange the following 0.1 M aqueous solutions in order of increasing pH: KOH, $ZnCl_2$, NH_3, HBr, BaI_2.

60. Arrange the following 0.1 M aqueous solutions in order of decreasing pH: NH_4Br, $Ba(OH)_2$, $HClO_4$, K_2SO_4, LiCN.

61. Write formulas for four salts that
a. contain Fe^{3+} and are acidic.
b. contain NO_3^- and are neutral.
c. contain Ba^{2+} and are basic.
d. contain Cs^+ and are neutral.

62. Write formulas for four salts that
a. contain Li^+ and are basic.
b. contain Li^+ and are neutral.
c. contain SO_4^{2-} and are neutral.
d. contain SO_4^{2-} and are acidic.

Acid-Base Models

63. For each of the following reactions, indicate the Brönsted-Lowry acids and bases. What are the conjugate acid-base pairs?
a. $H_3O^+ + HSO_3^-(aq) \rightleftharpoons H_2SO_3(aq) + H_2O$
b. $HF(aq) + OH^-(aq) \rightleftharpoons F^-(aq) + H_2O$
c. $NH_4^+(aq) + H_2O \rightleftharpoons NH_3(aq) + H_3O^+$

64. Follow the directions for Question 63 for the following reactions:
a. $CN^-(aq) + H_2O \rightleftharpoons HCN(aq) + OH^-(aq)$
b. $HCO_3^-(aq) + H_3O^+(aq) \rightleftharpoons H_2CO_3(aq) + H_2O$
c. $HC_2H_3O_2(aq) + HS^-(aq) \rightleftharpoons C_2H_3O_2^-(aq) + H_2S(aq)$

65. According to the Brönsted-Lowry theory, which of the following would you expect to act as an acid? Which as a base?
a. HNO_2 b. OCl^- c. NH_2^-

66. According to the Brönsted-Lowry theory, which of the following would you expect to be an acid? Which a base?
a. NH_4^+ b. $CH_3NH_3^+$ c. $C_2H_3O_2^-$

67. Classify each of the following species as either a Lewis acid or a Lewis base. Draw Lewis structures where necessary.
a. CH_3OCH_3 b. PCl_3 c. H_2O

68. Classify each of the following species as either a Lewis acid or a Lewis base. Draw Lewis structures where necessary.
a. $H_3C-\underset{\underset{C_2H_5}{|}}{N}-CH_3$ b. BF_3 c. $BeCl_2$

69. Each of the following is a Lewis acid-base reaction. Which is the acid? Which is the base? Draw Lewis structures where necessary.
a. $SO_3(g) + H_2O \rightarrow H_2SO_4(aq)$
b. $CO_2(g) + LiOH(s) \rightarrow LiHCO_3(s)$

70. Each of the following is a Lewis acid-base reaction. Which is the acid? Which is the base? Draw Lewis structures where necessary.
a. $Al(OH)_3(s) + OH^-(aq) \rightarrow Al(OH)_4^-(aq)$
b. $SO_2(g) + CaO(s) \rightarrow CaSO_3(s)$

71. Determine whether each of the following substances is an Arrhenius acid/base, Brönsted-Lowry acid/base, or a Lewis acid/base. It is possible for a species to be in more than one category.

a. CO_3^{2-} b. H_3O^+ c. Fe^{3+} d. $H_2PO_4^-$

72. Follow the directions of Problem 71 for the following substances:
a. H_2O b. NO_3^- c. OCl^- d. HSO_4^-

Unclassified

73. Give two examples of
a. an acid stronger than HF.
b. a base weaker than NaOH
c. a salt that dissolves in water to yield an acidic solution.
d. a salt that dissolves in water to yield a basic solution.
e. a salt that dissolves in water to yield a solution with a pH of 7.

74. A solution is made by mixing 25.00 mL of 0.3500 M LiOH with 15.50 mL of 0.2225 M ammonia. Assuming that volumes are additive,
a. calculate the pH of the solution.
b. calculate the concentrations of all species in solution.

75. Consider the process

$$H_2O \rightleftharpoons H^+(aq) + OH^-(aq); \qquad \Delta H^0 = 55.8 \text{ kJ}$$

Will $[H^+]$ be greater or smaller than 1.0×10^{-7} at 40°C? Explain.

76. The K_w for water at 37°C, human body temperature, is 2.40×10^{-14}. Is a saline solution (0.1 M NaCl) acidic, basic, or neutral at 37°C?

77. There are 324 mg of acetylsalicylic acid (MM = 180.2 g/mol) per aspirin tablet. If two tablets are dissolved to give two ounces ($\frac{1}{16}$ quart) of solution, estimate the pH. K_a of acetylsalicylic acid is 3.4×10^{-4}.

78. Calculate the pH of a 0.025 M solution of aluminum nitrate. K_a for $Al(H_2O)_6^{3+}$ is 1.4×10^{-5}.

Challenge Problems

79. Using Table 14.5 and the Law of Multiple Equilibria, determine the equilibrium constant for the reaction

$$H_2S(aq) \rightleftharpoons 2H^+(aq) + S^{2-}(aq)$$

80. Silver hydroxide, AgOH, is insoluble in water. Describe a simple qualitative experiment that would enable you to determine whether AgOH is a strong or weak base.

81. Using the tables in Appendix 1, calculate ΔH for the reaction of
a. 1.00 L of 0.100 M NaOH with 1.00 L of 0.100 M HCl.
b. 1.00 L of 0.100 M NaOH with 1.00 L of 0.100 M HF, taking the heat of formation of HF(aq) to be -320.1 kJ/mol.

82. Show by calculation that when the concentration of a weak acid decreases by a factor of 10, its percent dissociation increases by a factor of $10^{1/2}$.

83. What is the freezing point of vinegar, which is an aqueous solution of 5.00% acetic acid, $HC_2H_3O_2$, by mass ($d = 1.006$ g/cm³)?

$$Bi^{3+}(aq) + 3OH^-(aq) \rightarrow Bi(OH)_3(s)$$
Precipitation of bismuth hydroxide

CHAPTER 15

ACID-BASE AND PRECIPITATION EQUILIBRIA

There is nothing in the Universe but alkali and acid,
From which Nature composes all things.

OTTO TACHENIUS (1671)

I n Chapter 14, we dealt with the equilibrium established when a single solute, either a weak acid or a weak base, is added to water. In this chapter, we will focus upon the equilibrium established when two different solutes are mixed in water solution. These solutes may be:

1. *A weak acid HB and its conjugate base B⁻.* Solutions produced in this way are called buffers; we will examine the equilibrium properties of buffer systems in Section 15.1.

2. *An acid and a base used in an acid-base titration.* This type of reaction was discussed in Chapter 12. Here, in Section 15.2, we will look at the equilibria involved and at the nature of the indicator used to determine the equivalence point of the titration.

3. *Two ionic solutes which react to form a precipitate.* Again, this type of reaction was discussed qualitatively in Chapter 12. In Section 15.3, we will examine the equilibrium between the precipitate and its ions in solution.

15.1
BUFFERS

There are many different ways in which we can establish an equilibrium in solution between a weak acid, HB, and its conjugate base, B⁻:

$$HB(aq) \rightleftharpoons H^+(aq) + B^-(aq)$$

We might, as discussed in Section 14.2, dissolve the weak acid in water. In that case, some of the HB molecules dissociate, producing H^+ and B^- ions.

Another common way of establishing this equilibrium is to add to water both the weak acid, HB, and its conjugate base, B^-. Thus, to establish the equilibrium

$$HC_2H_3O_2(aq) \rightleftharpoons H^+(aq) + C_2H_3O_2^-(aq)$$

we might dissolve both acetic acid, ($HC_2H_3O_2$ molecules) and sodium acetate (Na^+, $C_2H_3O_2^-$ ions) in water. This is the kind of system we will be talking about in this section.

A solution prepared by adding roughly equal amounts of a weak acid and its conjugate base to water acts as a **buffer**. Its pH is resistant to change (Fig. 15.1) upon the addition of small amounts of strong acid (H^+ ions) or strong base (OH^- ions). This is true because the weak acid, e.g., $HC_2H_3O_2$, reacts with OH^- ions:

$$HC_2H_3O_2(aq) + OH^-(aq) \rightarrow C_2H_3O_2^-(aq) + H_2O$$

while the conjugate base, the $C_2H_3O_2^-$ ion, reacts with H^+ ions:

$$C_2H_3O_2^-(aq) + H^+(aq) \rightarrow HC_2H_3O_2(aq)$$

That's why the buffer resists pH change

Both of these reactions go virtually to completion, so all the added OH^- or H^+ ions are consumed.

DETERMINATION OF [H⁺] IN A BUFFER SYSTEM

In principle, three different equilibria should be considered in an aqueous solution containing relatively large amounts of both a weak acid, HB, and its conjugate base B^-:

$$HB(aq) \rightleftharpoons H^+(aq) + B^-(aq)$$

$$B^-(aq) + H_2O \rightleftharpoons HB(aq) + OH^-(aq)$$

$$H_2O \rightleftharpoons H^+(aq) + OH^-(aq)$$

In practice, we can always calculate the $[H^+]$ or pH of a buffer solution by working with the equilibrium expression for the dissociation of the weak acid HB:

$$HB(aq) \rightleftharpoons H^+(aq) + B^-(aq); \qquad K_a = \frac{[H^+] \times [B^-]}{[HB]} \qquad (15.1)$$

Solving the K_a expression for $[H^+]$, we have:

$$[H^+] = K_a \times \frac{[HB]}{[B^-]} \qquad (15.2)$$

Figure 15.1 The three tubes at the left show the effect of adding a few drops of strong acid or strong base to water. The pH changes drastically, giving a pronounced color change with universal indicator. This experiment is repeated in the three tubes at the right, using a buffer of pH 7 instead of water. This time the pH changes only very slightly, and there is no change in the color of the indicator. (Marna G. Clarke)

Equation 15.2 is a general one, applicable to any buffer system. The calculation of $[H^+]$ can be further simplified if you keep in mind two general characteristics of such systems:

1. *We can always assume that equilibrium is established without appreciably changing the concentrations of either HB or B^-.* If we let $x = [H^+]$ produced by Reaction 15.1:

$$[HB] = (\text{orig. conc. HB} - x) \approx \text{orig. conc. HB}$$

$$[B^-] = (\text{orig. conc. } B^- + x) \approx \text{orig. conc. } B^-$$

The justification here is quite simple. We saw in Chapter 14 that when a weak acid is added to water, it is usually true that

$$\text{orig. conc. HB} \gg x$$

Here, since we have an appreciable concentration of B^- ions to start with, the weak acid HB will dissociate to an even smaller extent, making x very small indeed in comparison to the concentrations of HB or B^-.

2. Since the two species HB and B^- are present in the same solution, *the ratio of their concentrations is also their mole ratio.* That is:

$$\frac{[HB]}{[B^-]} = \frac{\text{no. moles HB}/V}{\text{no. moles } B^-/V} = \frac{\text{no. moles HB}}{\text{no. moles } B^-}$$

Hence we can rewrite Equation 15.2 as

$$[H^+] = K_a \times \frac{\text{no. moles HB}}{\text{no. moles } B^-} \tag{15.3}$$

Frequently we will find that Equation 15.3 is simpler to work with than 15.2.

The weak acid does *not* react appreciably with its conjugate base

■ **EXAMPLE 15.1**

Lactic acid, $C_3H_6O_3$, is a weak organic acid present in both sour milk and buttermilk. It is also a product of carbohydrate metabolism and is found in the blood after vigorous muscular activity. A buffer is prepared by dissolving 1.00 mol of lactic acid, HLac ($K_a = 1.4 \times 10^{-4}$), and 1.00 mol of sodium lactate, NaLac, in enough water to form one liter of solution. Calculate $[H^+]$ and the pH of this buffer.

Solution

We determine [HLac] and [Lac$^-$] and then use Equation 15.2 to find $[H^+]$:

$$[HLac] = \text{orig. conc. HLac} = 1.00 \text{ mol/L}$$

Sodium lactate, like all salts, is completely dissociated in water. Hence,

$$[Lac^-] = \text{orig. conc. NaLac} = 1.00 \text{ mol/L}$$

Thus,

$$[H^+] = K_a \times \frac{[HLac]}{[Lac^-]} = 1.4 \times 10^{-4} \times \frac{1.00}{1.00} = \boxed{1.4 \times 10^{-4} \, M}$$

$$pH = -\log_{10}(1.4 \times 10^{-4}) = \boxed{3.85}$$

Exercise
Suppose a buffer were prepared by dissolving 1.00 mol of acetic acid ($K_a = 1.8 \times 10^{-5}$) and 1.00 mol of sodium acetate in enough water to form a liter of solution. What would be the pH? Answer: 4.74.

Although, in Example 15.1, we specified the volume of solution to be "one liter," the pH is really independent of volume. As long as we have 1.00 mol of lactic acid and 1.00 mol of sodium lactate, the pH of the buffer will be 3.85, whether the volume of solution is 0.1 L, 1 L, 10 L, or any other quantity. This is generally true, as you can see from Equation 15.3; the pH of a buffer system is determined by the mole ratio of HB to B^-, which is independent of volume. From a slightly different point of view, we might say that *diluting a buffer system with water does not change its pH.* The mole ratio of HB to B^- stays the same on dilution, as does the concentration ratio.

Biologists and biochemists often calculate the pH of buffers by using a relation known as the Henderson-Hasselbalch equation. This is readily obtained from Equation 15.2. Taking the base 10 log of both sides gives

$$\log_{10} [H^+] = \log_{10} K_a + \log_{10} \frac{[HB]}{[B^-]}$$

Multiplying both sides by -1 and remembering that $-\log x = \log (1/x)$, we obtain the equation

$$-\log_{10} [H^+] = -\log_{10} K_a - \log_{10} \frac{[HB]}{[B^-]}$$

This equation is really Eqn. 15.2 in disguise

$$pH = pK_a + \log_{10} \frac{[B^-]}{[HB]} \qquad (15.4)$$

CHOOSING A BUFFER SYSTEM

In a buffer system, it is best if the concentrations of weak acid HB and its conjugate base B^- are roughly equal. Looking at Equation 15.2, we see that this means that the concentration of H^+ will be roughly equal to K_a of the weak acid. Putting it another way, the pH of a buffer will be rather close to the pK_a of the weak acid, ordinarily within ± 1 unit.

This principle must be taken into account in preparing buffer systems. To make up a buffer to a given pH, we choose a weak acid–weak base pair where pK_a of the acid is close to the desired pH (Table 15.1). If we want a pH of 7, a logical choice is the $H_2PO_4^- - HPO_4^{2-}$ system (pK_a $H_2PO_4^-$ = 7.21). Other reasonable choices would include the $H_2CO_3 - HCO_3^-$ system (pK_a H_2CO_3 = 6.36), and $HClO - ClO^-$ (pK_a HClO = 7.55). The $NH_4^+ - NH_3$ system would be a very poor choice, since the pK_a of the ammonium ion, 9.25, is far from the desired pH value of 7.

Once we have decided upon the appropriate buffer system to establish a particular pH, we must then determine the relative amounts of weak acid and weak base that should be used. Figure 15.2 shows a systematic procedure for this determination, outlined in Example 15.2.

TABLE 15.1 Buffer Systems at Different pH Values

DESIRED pH	BUFFER SYSTEM Weak Acid	Weak Base	K_a (WEAK ACID)	pK_a
4	Lactic acid (HLac)	Lactate ion (Lac$^-$)	1.4×10^{-4}	3.85
5	Acetic acid (HC$_2$H$_3$O$_2$)	Acetate ion (C$_2$H$_3$O$_2$$^-$)	1.8×10^{-5}	4.74
6	Carbonic acid (H$_2$CO$_3$)	Hydrogen carbonate ion (HCO$_3$$^-$)	4.4×10^{-7}	6.36
7	Dihydrogen phosphate ion (H$_2$PO$_4$$^-$)	Hydrogen phosphate ion (HPO$_4$$^{2-}$)	6.2×10^{-8}	7.21
8	Hypochlorous acid (HClO)	Hypochlorite ion (ClO$^-$)	2.8×10^{-8}	7.55
9	Ammonium ion (NH$_4$$^+$)	Ammonia (NH$_3$)	5.6×10^{-10}	9.25
10	Hydrogen carbonate ion (HCO$_3$$^-$)	Carbonate ion (CO$_3$$^{2-}$)	4.7×10^{-11}	10.32

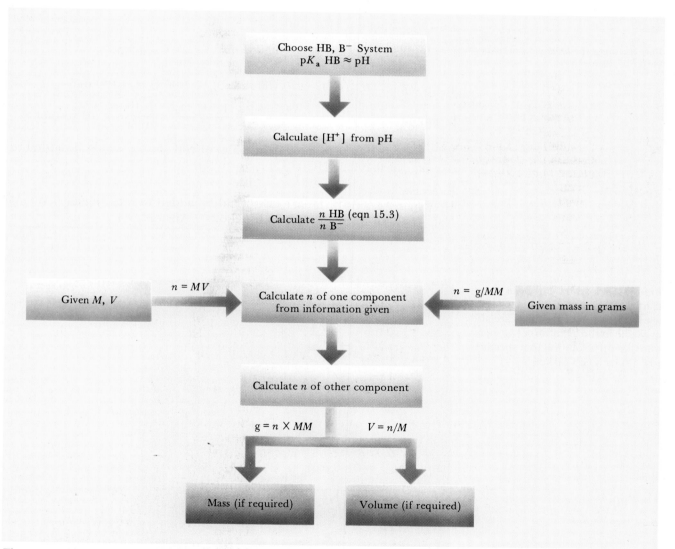

Figure 15.2 General procedure for making up a buffer system to a specified pH (n = no. moles; g = mass in grams; MM = molar mass; V = volume in liters; M = molarity)

■ EXAMPLE 15.2

Suppose you want to prepare an $HC_2H_3O_2$–$C_2H_3O_2^-$ buffer with a pH of 5.00. Taking K_a of $HC_2H_3O_2$ to be 1.8×10^{-5}, calculate the number of moles of sodium acetate, $NaC_2H_3O_2$, that should be added to one liter of $0.100\ M\ HC_2H_3O_2$ to prepare this buffer.

Solution

Following the steps indicated in Figure 15.2:

1. The buffer system is specified in the problem; it is an appropriate one, since $pK_a\ HC_2H_3O_2 = 4.74$, which is quite close to the desired pH of 5.00.

2. $[H^+] = 10^{-5.00} = 1.0 \times 10^{-5}\ M$

3. Using Equation 15.3:

This Example shows how to prepare a buffer of desired pH

$$\frac{\text{no. moles } HC_2H_3O_2}{\text{no. moles } C_2H_3O_2^-} = \frac{[H^+]}{K_a} = \frac{1.0 \times 10^{-5}}{1.8 \times 10^{-5}} = 0.56$$

4. no. moles $HC_2H_3O_2 = (0.100\ \text{mol/L}) \times 1.00\ \text{L} = 0.100\ \text{mol}$

5. no. moles $C_2H_3O_2^- = \dfrac{\text{no. moles } HC_2H_3O_2}{0.56} = \dfrac{0.100}{0.56} = \boxed{0.18\ \text{mol}}$

We conclude that 0.18 mol (about 15 g) of $NaC_2H_3O_2$ should be added to one liter of $0.100\ M\ HC_2H_3O_2$ to form this buffer.

Exercise

How many grams of $HC_2H_3O_2$ should be added to one liter of $0.100\ M$ $NaC_2H_3O_2$ to form a buffer with pH 5.00? Answer: 3.4 g.

Figure 15.3 Many products, including aspirin and blood plasma, are buffered. Buffer tablets are also available in the laboratory (*far right*) to make up a solution to a specified pH. (Marna G. Clarke)

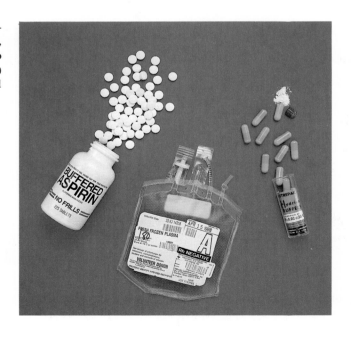

Many natural systems are buffered at appropriate pH's. In the applied perspective at the end of this chapter, we will consider some buffer systems of biological importance. Other applications of buffers are shown in Figure 15.3.

EFFECT OF ADDED H⁺ OR OH⁻ ON BUFFER SYSTEMS

As pointed out earlier, a buffer, consisting of a weak acid HB and its conjugate base, B⁻, can react with either strong base (OH⁻) or strong acid (H⁺).

$$HB(aq) + OH^-(aq) \rightarrow B^-(aq) + H_2O$$

$$B^-(aq) + H^+(aq) \rightarrow HB(aq)$$

Because of these reactions, added H⁺ or OH⁻ ions are consumed and do not directly affect the pH. However, the pH of a buffer does change slightly if a strong acid or strong base is added to it. By adding base, we convert a small amount of the weak acid HB to its conjugate base, B⁻. Addition of strong acid converts a small amount of B⁻ to HB. In both cases, the ratio [HB]/[B⁻] changes; this in turn changes the [H⁺] or pH of the buffer. The effect is ordinarily small, as indicated by Example 15.3.

■ **EXAMPLE 15.3** _____

Consider the buffer described in Example 15.1, where no. moles HLac = no. moles Lac⁻ = 1.00 (K_a HLac = 1.4×10^{-4}). You will recall that in this buffer the pH is 3.85. Calculate the pH after addition of

a. 0.10 mol HCl   b. 0.10 mol NaOH

Solution

a. When H⁺ ions are added, the following reaction occurs:

$$H^+(aq) + Lac^-(aq) \rightarrow HLac(aq)$$

The addition of 0.10 mol H⁺ produces 0.10 mol HLac and consumes 0.10 mol Lac⁻. Originally, we had 1.00 mol of both HLac and Lac⁻. Hence:

no. moles HLac = 1.00 + 0.10 = 1.10

no. moles Lac⁻ = 1.00 − 0.10 = 0.90

Using Equation 15.3:

$$[H^+] = K_a \times \frac{\text{no. moles HLac}}{\text{no. moles Lac}^-} = 1.4 \times 10^{-4} \times \frac{1.10}{0.90} = 1.7 \times 10^{-4}$$

pH = 3.77

b. Here the reaction is: $HLac(aq) + OH^-(aq) \rightarrow Lac^-(aq) + H_2O$. The reaction produces 0.10 mol of Lac⁻ and consumes 0.10 mol of HLac.

no. moles HLac $= 1.00 - 0.10 = 0.90$

no. moles Lac$^-$ $= 1.00 + 0.10 = 1.10$

Substituting in Equation 15.3:

$$[H^+] = 1.4 \times 10^{-4} \times \frac{0.90}{1.10} = 1.1 \times 10^{-4}; \quad pH = \boxed{3.94}$$

The pH again changes by less than 0.1 unit

Exercise
Suppose that in (a), we had added 0.50 mol H$^+$ instead of 0.10 mol. What would be the final pH? Answer: 3.38.

Example 15.3 and the accompanying exercise illustrate an important limitation of buffer systems: A buffer has a limited capacity to absorb H$^+$ or OH$^-$ ions. The amount of acid or base the buffer can react with before having a significant pH change is called the *buffer capacity*. Notice from Exercise 15.3 that if 0.50 mol H$^+$ were added to the buffer, it would produce a rather large change in pH, decreasing it from 3.85 to 3.38. Addition of 1.00 mol or more of H$^+$ would have an even more drastic effect. Indeed, it would consume all the Lac$^-$ ions, thereby destroying the buffer (Fig. 15.4).

We all have our limits, don't we?

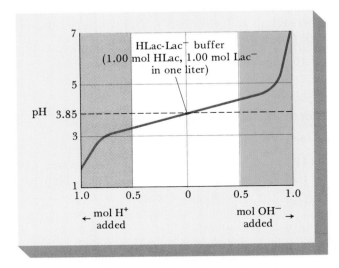

Figure 15.4   Effect of adding either acid or base to a lactic acid–sodium lactate buffer. The buffer itself has a pH of 3.85. Addition of OH$^-$ ions produces a small increase in pH until most of the lactic acid is gone. Then the pH rises sharply (*upper right of curve*). The buffer behaves similarly toward addition of H$^+$ ions (*lower left*). The pH decreases slowly until nearly all the lactate ions are gone, then drops sharply.

Calculations such as those in Example 15.3 are typical of "buffer" problems. Figure 15.5 is a schematic diagram of the thought process you should go through in solving such problems.

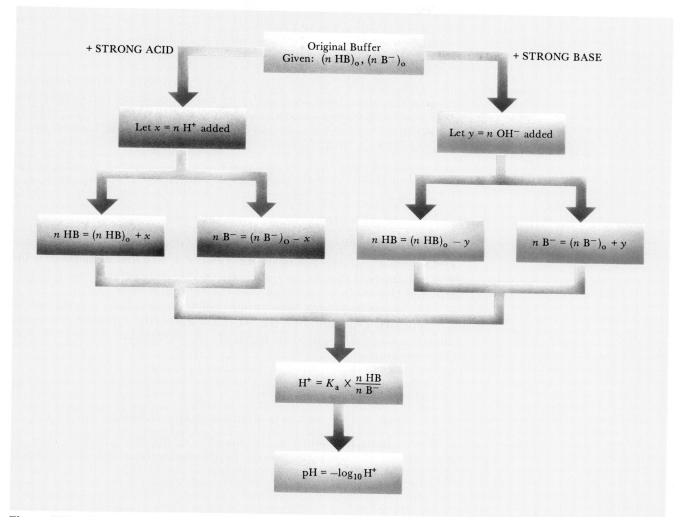

Figure 15.5 General procedure for calculating the pH of a buffer after addition of a strong acid or base. (n HB = no. moles HB after reaction; $(n$ HB$)_o$ = original no. moles HB; n B$^-$ = no. moles B$^-$ after reaction; $(n$ B$^-)_o$ = original no. moles B$^-$.)

15.2
ACID-BASE TITRATIONS

In this section we will apply the principles of solution stoichiometry (Chapter 12) and acid-base equilibrium (Section 15.1) to acid-base titrations. We will be particularly interested in the way that pH changes during the titration. Information of that sort allows us to choose the proper acid-base indicator.

INDICATORS

The end point of an acid-base titration is established by observing a change in color in an acid-base indicator, which is added in small amounts before carrying out the titration. The color of the indicator, which is ordinarily an organic dye, depends upon the H^+ ion concentration in the solution being titrated (Fig. 15.6). In this sense, it "indicates" the rather drastic change in pH that ordinarily occurs near the end point of an acid-base titration.

To illustrate how an indicator works, let us consider a typical example, bromthymol blue. Like most acid-base indicators, bromthymol blue is a weak acid; it has a dissociation constant of about 1×10^{-7}. Using HIn to represent the molecular acid and In^- the conjugate base formed when bromthymol blue dissociates in water, we have:

$$HIn(aq) \rightleftharpoons H^+(aq) + In^-(aq)$$

yellow   blue

$$K_a = \frac{[H^+] \times [In^-]}{[HIn]} = 1 \times 10^{-7}$$

The essential characteristic of bromthymol blue, like that of all other acid-base indicators, is that *the two species HIn and In^- have different colors.* In this case, the HIn molecule is yellow and the In^- ion is blue.

Let us consider what color bromthymol blue will have in solutions of different pH. Depending on the $[H^+]$, the ratio $[In^-]/[HIn]$ will take on whatever value is required to satisfy the K_a expression. Solving for that ratio, we get:

$$\frac{[In^-]}{[HIn]} = \frac{1 \times 10^{-7}}{[H^+]} \tag{15.5}$$

Using Equation 15.5, we can distinguish three possibilities:

1. If, in a particular solution, $[H^+]$ is 10^{-6} or greater (pH $\leq$ 6), most of the indicator will be in the form of the HIn molecule and the solution will appear yellow. Thus, at pH 6:

$$\frac{[In^-]}{[HIn]} = \frac{1 \times 10^{-7}}{1 \times 10^{-6}} = 0.1$$

and there will be only one In^- ion (blue) for every ten HIn molecules (yellow).

2. If $[H^+] = 10^{-7}$ (pH = 7), equal amounts of In^- and HIn will be present

$$\frac{[In^-]}{[HIn]} = \frac{1 \times 10^{-7}}{1 \times 10^{-7}} = 1$$

and the solution will appear green (equal amounts of blue and yellow).

3. If $[H^+]$ is 10^{-8} or less (pH $\geq$ 8), the In^- ion will predominate and the solution will appear blue.

$$\frac{[In^-]}{[HIn]} = \frac{1 \times 10^{-7}}{1 \times 10^{-8}} = 10$$

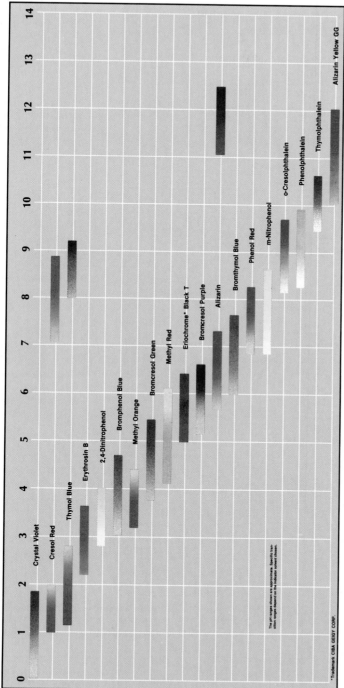

The pH ranges shown are approximate. Specific transition ranges depend on the indicator solvent chosen.

*Trademark CIBA GEIGY CORP.

Figure 15.6 Colors of indicators at various pH's (Hach Company)

Crystal Violet
Cresol Red
Thymol Blue
Erythrosin B
2,4-Dinitrophenol
Bromphenol Blue
Methyl Orange
Bromcresol Green
Methyl Red
Eriochrome* Black T
Bromcresol Purple
Alizarin
Bromthymol Blue
Phenol Red
m-Nitrophenol
o-Cresolphthalein
Phenolphthalein
Thymolphthalein
Alizarin Yellow GG

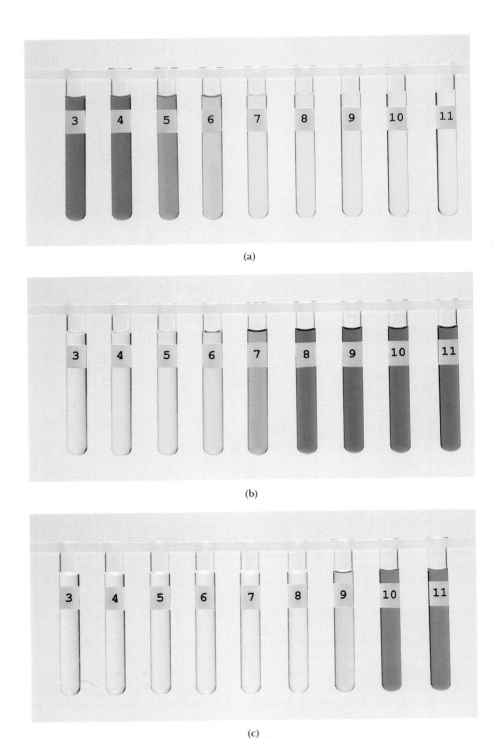

Figure 15.7   Methyl red (a) goes from red at low pH to orange at about pH 5 to yellow at high pH. Bromthymol blue (b) is yellow at low pH, blue at high pH, and green at about pH 7. Phenolphthalein (c) goes from colorless to pink at about pH 9. (Marna G. Clarke)

TABLE 15.2 Colors of Acid-Base Indicators

INDICATOR	K_a	pH = 3 4 5 6 7 8 9 10 11
Bromthymol blue	1×10^{-7}	⟵⟶
Methyl red	1×10^{-5}	⟵⟶
Phenolphthalein*	1×10^{-9}	⟵⟶

$$\frac{[In^-]}{[HIn]} = \frac{K_a}{[H^+]}$$

If $[H^+] \ll K_a$, color of In^- ion predominates.
If $[H^+] \gg K_a$, color of HIn predominates.

*Above pH 3, phenolphthalein changes from red to colorless.

In practice, bromthymol blue changes gradually from yellow to blue as the pH goes from 6 to 8. Like most indicators, it has a range of about 2 pH units over which it changes color. We say that bromthymol blue has an *end point* at pH 7, since at that pH the change in color is most easily detected.

The end point observed with any particular indicator will depend upon the magnitude of its acid dissociation constant, K_a. The pH at the end point will equal the pK_a of the indicator. Methyl red ($K_a = 1 \times 10^{-5}$) has an end point at pH 5; it changes color from red to yellow as the pH goes from 4 to 6. Phenolphthalein ($K_a = 1 \times 10^{-9}$) changes from colorless to pink in the pH range 8 to 10 (Fig. 15.7, p. 520, and Table 15.2).

In carrying out an acid-base titration, we try to select an indicator that changes color at or very near the equivalence point of the reaction, i.e., the point at which equal quantities of acid and base have been added. A successful titration, one with good quantitative results, is possible only if the end point as established by the indicator coincides with the equivalence point in the titration. Since the pH at the equivalence point in an acid-base reaction depends on the relative strengths of the acid and base involved, we cannot always use the same indicator for different kinds of acid-base titrations.

A good acid base indicator for a titration changes color at the equivalence point in the titration

STRONG ACID–STRONG BASE TITRATIONS

As we saw in Chapter 12, the net ionic equation for the reaction between a strong acid and a strong base is

$$H^+(aq) + OH^-(aq) \rightarrow H_2O \qquad (15.6)$$

Applying the Reciprocal Rule (Chap. 13), we can calculate K for Reaction 15.6:

$$K = 1/K_w = 1/(1.0 \times 10^{-14}) = 1.0 \times 10^{14}$$

The enormous value of K means that for all practical purposes this reaction goes to completion, consuming all of at least one reactant.

Consider now what happens when we titrate HCl, a typical strong acid, with NaOH. Figure 15.8 shows how the pH changes during the titration.

Figure 15.8 Titration of 50.00 mL of 1.000 M HCl with 1.000 M NaOH. The solution at the equivalence point is neutral (pH = 7). The pH rises so rapidly near the equivalence point that any of the three indicators methyl red (MR, end point pH = 5), bromthymol blue (BB, end point pH = 7), or phenolphthalein (PP, end point pH = 9) can be used.

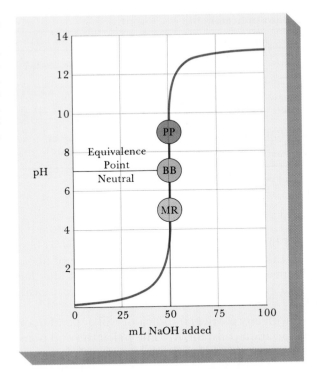

Two features of this curve are of particular importance:

1. At the equivalence point, when all the HCl has been neutralized by NaOH, we have a solution of NaCl, a neutral salt. The pH at the equivalence point is 7.

2. Near the equivalence point, the pH rises very rapidly. Indeed, the pH may increase by as much as six units (from 4 to 10) when a single drop of NaOH is added (Example 15.4).

■ **EXAMPLE 15.4** _____

If 50.00 mL of 1.000 M HCl is titrated with 1.000 M NaOH, find the pH of the solution after the following volumes of 1.000 M NaOH have been added:

a. 49.99 mL   b. 50.00 mL   c. 50.01 mL

Solution

a. At this point, we have

$$(50.00 - 49.99) \text{ mL} = 0.01 \text{ mL} = 1 \times 10^{-5} \text{ L}$$

of unneutralized 1.000 M HCl in a total volume of almost exactly 100 mL (0.100 L) of solution. Consequently,

$$[H^+] = \frac{1 \times 10^{-5} \text{ L} \times 1.000 \text{ mol/L}}{0.100 \text{ L}} = 1 \times 10^{-4} \, M; \quad pH = \boxed{4.0}$$

b. 50.00 mL of 1 M NaOH takes us to the equivalence point, giving a solution of NaCl with a $\boxed{\text{pH of 7}}$. Note that only 0.01 mL of base is required at the equivalence point to move the pH from 4 to 7. This volume is much less than that of 1 drop of reagent.

c. We are now past the equivalence point and OH^- is in excess. Specifically, we have

The pH changes fast at the equivalence point

$$(50.01 - 50.00) \text{ mL} = 0.01 \text{ mL} = 1 \times 10^{-5} \text{ L}$$

of unneutralized 1.000 M NaOH in a total volume just slightly greater than 100 mL (0.100 L) of solution:

$$[OH^-] = \frac{1 \times 10^{-5} \text{ L} \times 1.000 \text{ mol/L}}{0.100 \text{ L}} = 1 \times 10^{-4} M$$

$$[H^+] = \frac{1 \times 10^{-14}}{1 \times 10^{-4}} = 1 \times 10^{-10} M; \quad \text{pH} = \boxed{10.0}$$

Exercise

Calculate $[H^+]$ and the pH when 25.00 mL NaOH has been added. Answer:

$$[H^+] = \frac{25.00 \times 10^{-3} \text{ L} \times 1.000 \text{ mol/L}}{75.00 \times 10^{-3} \text{ L}} = 0.3333 \text{ mol/L};$$

$$\text{pH} = 0.48$$

From Figure 15.8 and Example 15.4, we conclude that any indicator that changes color between pH 4 and 10 should be satisfactory for a strong acid–strong base titration. Bromthymol blue (end point pH = 7) would work very well, but so would methyl red (end point pH = 5) or phenolphthalein (end point pH = 9).

Phenolphthalein is usually used in this titration; colorless to pink is easy to recognize

WEAK ACID–STRONG BASE TITRATION

A typical titration of this sort is that involving acetic acid and sodium hydroxide. We saw in Chapter 12 that the net ionic equation for this reaction is:

$$HC_2H_3O_2(aq) + OH^-(aq) \rightarrow C_2H_3O_2^-(aq) + H_2O \qquad (15.7)$$

Notice that this equation is the reverse of that for the dissociation of the weak base $C_2H_3O_2^-$ (the acetate ion) in water. It follows from the Reciprocal Rule that for Reaction 15.7:

$$K = 1/(K_b \ C_2H_3O_2^-) = 1/(5.6 \times 10^{-10}) = 1.8 \times 10^9$$

Here again, K is a very large number; the reaction of acetic acid with a strong base goes essentially to completion.

Let us consider how the pH changes when 50 mL of 1 M $HC_2H_3O_2$ is titrated with 1 M NaOH (Fig. 15.9, p. 524). Notice that:

1. The pH starts off at about 2.4, the pH of 1 M $HC_2H_3O_2$, a weak acid.

Figure 15.9 Titration of 50.00 mL of the weak acid $HC_2H_3O_2$ (1.000 M) with 1.000 M NaOH. The solution at the equivalence point is basic (pH = 9.22). Phenolphthalein is a suitable indicator. Methyl red would change color much too early, when only about 33 mL of NaOH had been added. Bromthymol blue would change color slightly too early.

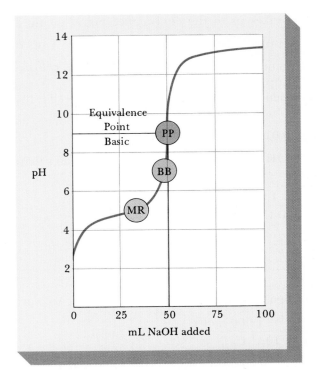

2. There is a region, centered around the halfway point of the titration (25 mL NaOH added), where the pH changes very slowly. In this region, there are appreciable amounts of two different species: unreacted $HC_2H_3O_2$ molecules and the $C_2H_3O_2^-$ ions produced by Reaction 15.7. Hence we have a buffer system that is resistant to change in pH when more strong base is added.

3. At the equivalence point (1 mol NaOH per mole $HC_2H_3O_2$), we have a solution of sodium acetate, $NaC_2H_3O_2$. As we saw in Chapter 14, this solution is basic because the $C_2H_3O_2^-$ ion is a weak base. Consequently, the pH at the equivalence point is greater than 7 (Example 15.5).

■ **EXAMPLE 15.5** _____

50.00 mL of 1.000 M acetic acid, $HC_2H_3O_2$, is titrated with 1.000 M NaOH. Find the pH of the solution after the following volumes of 1.000 M NaOH have been added:

a. 0.00 mL   b. 25.00 mL   c. 50.00 mL

Solution

a. Before any NaOH is added, we have a solution of acetic acid. The governing equilibrium system is:

$$HC_2H_3O_2(aq) \rightleftharpoons H^+(aq) + C_2H_3O_2^-(aq);$$

$$K_a = 1.8 \times 10^{-5} = \frac{[H^+] \times [C_2H_3O_2^-]}{[HC_2H_3O_2]}$$

If, as in Chapter 14, we let $x = [H^+] = [C_2H_3O_2^-]$ and $[HC_2H_3O_2] = 1.000 - x \approx 1.0000$, then

$$1.8 \times 10^{-5} = \frac{x^2}{1.000} ; \quad x = 4.2 \times 10^{-3} \quad \text{and} \quad pH = \boxed{2.37}$$

b. We first concentrate upon the stoichiometry of the titration. Originally we had:

$$0.05000 \text{ L} \times \frac{1.000 \text{ mol } HC_2H_3O_2}{1 \text{ L}} = 0.05000 \text{ mol } HC_2H_3O_2$$

We added:

$$0.02500 \text{ L} \times \frac{1.000 \text{ mol } OH^-}{1 \text{ L}} = 0.02500 \text{ mol } OH^-$$

All of the OH^- ions reacted, converting 0.02500 mol $HC_2H_3O_2$ to $C_2H_3O_2^-$ by Reaction 15.7. So, at this point we have

0.02500 mol $C_2H_3O_2^-$

$(0.05000 - 0.025000)$ mol $HC_2H_3O_2 = 0.02500$ mol $HC_2H_3O_2$

The solution here is a buffer

Since we are dealing with a buffer system, we can use Equation 15.3 to calculate $[H^+]$:

$$[H^+] = K_a \, HC_2H_3O_2 \times \frac{\text{no. moles } HC_2H_3O_2}{\text{no. moles } C_2H_3O_2^-}$$

$$= 1.8 \times 10^{-5} \times \frac{0.0250}{0.0250} = 1.8 \times 10^{-5}$$

$$pH = \boxed{4.74}$$

c. When 50.00 mL of 1.000 M NaOH is added, the equivalence point is reached, and all we have is 5.000×10^{-2} moles of acetate ion, $C_2H_3O_2^-$, in 100.00 mL of solution. The governing equilibrium is:

$$C_2H_3O_2^-(aq) + H_2O \rightleftharpoons HC_2H_3O_2(aq) + OH^-(aq)$$

We no longer have a buffer at the equivalence point

$$K_b = 5.6 \times 10^{-10} = \frac{[HC_2H_3O_2][OH^-]}{[C_2H_3O_2^-]}$$

Again, we let $x = [OH^-] = [HC_2H_3O_2]$ and $[C_2H_3O_2^-] = \dfrac{0.05000 - x}{0.1000} \approx 0.5000 \, M.$ Hence:

$$5.6 \times 10^{-10} = \frac{x^2}{0.5000} ; \quad x = 1.67 \times 10^{-5}; \quad pOH = 4.78;$$

$$pH = \boxed{9.22.}$$

Exercise

Calculate the pH when 10.00 mL of 1.000 M NaOH has been added.

Answer: $[H^+] = 1.8 \times 10^{-5} \times \dfrac{0.0400}{0.0100} = 7.2 \times 10^{-5}$; pH = 4.14.

From Figure 15.9 and Example 15.5, it should be clear that the indicator used in this titration must change color at about pH 9. Phenolphthalein (end point pH = 9) is satisfactory. Methyl red (end point pH = 5) is not suitable. If we used methyl red, we would stop the titration much too early, when reaction is only about 65% complete. This situation is typical of *weak acid–strong base titrations*. For such a titration we choose an indicator, such as phenolphthalein, that *changes color above pH 7*.

WEAK BASE–STRONG ACID TITRATION

A typical example of this type involves the titration of ammonia with hydrochloric acid. The essential features of this titration are developed in Example 15.6.

■ **EXAMPLE 15.6** ──────────────────────────────

50.00 mL of 1.000 M NH$_3$ is titrated with 1.000 M HCl

a. Write a net ionic equation for the reaction involved and calculate K for the reaction.
b. What is the pH before any HCl is added?
c. What is the pH after 25.00 mL of HCl has been added?
d. What is the pH at the equivalence point, when 50.00 mL of HCl has been added?

Solution

a. $NH_3(aq) + H^+(aq) \rightarrow NH_4^+(aq)$

Since this equation is the reverse of that for the dissociation of the weak acid NH_4^+:

For titration reactions, the K should be big

$$K = 1/(K_a\ NH_4^+) = 1/(5.6 \times 10^{-10}) = \boxed{1.8 \times 10^9}$$

This reaction, like all the acid-base reactions considered to this point, goes essentially to completion.

b. Before any HCl is added, we simply have a weak base with the following equilibrium reaction and expression:

$$NH_3(aq) + H_2O \rightleftharpoons NH_4^+(aq) + OH^-(aq)$$

$$K_b = \frac{[NH_4^+][OH^-]}{[NH_3]} = 1.8 \times 10^{-5}$$

Letting $x = [NH_4^+] = [OH^-]$ and $[NH_3] = 1.000 - x \approx 1.000$, we get

$$1.8 \times 10^{-5} = \frac{x^2}{1.000}; \quad x = [OH^-] = 4.2 \times 10^{-3} M$$

$$[H^+] = \frac{1.00 \times 10^{-14}}{4.2 \times 10^{-3}} = 2.4 \times 10^{-12} M; \quad pH = \boxed{11.62}$$

c. First let us consider the stoichiometry:

orig. no. moles NH_3 = 0.05000;   no. moles H^+ added = 0.025000

no. moles NH_4^+ formed = 0.02500

no. moles NH_3 left = 0.05000 − 0.02500 = 0.02500

Again, we have a buffer system. Applying Equation 15.3:

$$[H^+] = K_a\, NH_4^+ \times \frac{\text{no. moles } NH_4^+}{\text{no. moles } NH_3}$$

$$= 5.6 \times 10^{-10} \times \frac{0.02500}{0.02500} = 5.6 \times 10^{-10}$$

$$pH = \boxed{9.25}$$

d. At the equivalence point, all the ammonia is converted to NH_4^+. Thus we now have a solution of a weak acid, NH_4^+. There is 0.05000 mol NH_4^+ in a total volume of 100.00 mL; the concentration of NH_4^+ is 0.5000 M. The equilibrium reaction and expression are

The solution at the equivalence point is just 0.05 M NH_4Cl

$$NH_4^+(aq) \rightleftharpoons NH_3(aq) + H^+(aq)$$

$$K_a = 5.6 \times 10^{-10} = \frac{[H^+][NH_3]}{[NH_4^+]}$$

Again, we let $x = [H^+] = [NH_3]$ and $[NH_4^+] = 0.5000 - x \approx 0.5000$. Therefore

$$5.6 \times 10^{-10} = \frac{x^2}{0.5000}$$

$$x = [H^+] = 1.7 \times 10^{-5}; \quad pH = \boxed{4.77}$$

Exercise

Calculate the pH of the solution when 15.00 mL of 1.000 M HCl has been added. Answer: $[H^+] = 5.6 \times 10^{-10} \times \dfrac{0.01500}{0.03500} = 2.4 \times 10^{-10}$; pH = 9.62.

The calculations carried out in Example 15.6 are typical of those involved in weak base–strong acid titrations. We start (Fig. 15.10, p. 528) with a high pH. As strong acid is added, the pH drops, rapidly at first and then very slowly as we build up a buffer system (e.g., $NH_4^+–NH_3$). As more strong acid is added, the buffer capacity is exceeded and the pH again drops sharply through the equivalence point. At the equivalence

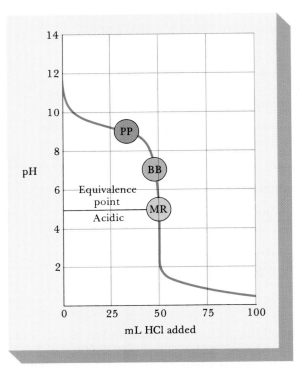

Figure 15.10 Titration of 50.00 mL of the weak base NH_3 (1.000 M) with 1.000 M HCl. The solution at the equivalence point is acidic, because of the NH_4^+ ion. Methyl red is a suitable indicator; phenolphthalein would change color much too early.

point, we have a solution of a weak acid (e.g., NH_4^+) with a pH less than 7. The appropriate indicator for the titration of a *weak base with a strong acid* is a species like methyl red (end point pH = 5), which *changes color on the acid side of pH 7.*

Table 15.3, on p. 530, summarizes important equilibrium considerations in acid-base titrations. Figure 15.11 is a flow chart indicating how one calculates the pH at different points in an acid-base titration. The relations listed for the beginning, intermediate, and equivalence points of the titration are those derived in Examples 15.4 to 15.6. Beyond the equivalence point, the pH is determined simply by the excess concentration of the strong base or strong acid used as a titrant.

Any solution made by mixing an acid with a base can be treated by the approach we have used with acid-base titrations. Frequently, we mix acidic and basic solutions, without doing a titration, and want to know the pH of the solution formed.

■ **EXAMPLE 15.7** _____

50.0 mL of 1.00 M NH_3 is mixed with 30.0 mL of 1.00 M HCl. Calculate the pH of the mixture.

Solution

Our approach is exactly the same as that followed in Example 15.6c. We first consider the stoichiometry:

$$NH_3(aq) + H^+(aq) \rightarrow NH_4^+(aq)$$

Orig. no. moles	0.0500	0.0000
Change	−0.0300	+0.0300
Equil. no. moles	0.0200	0.0300

Again , we have a buffer system.

$$[H^+] = K_a NH_4^+ \times \frac{\text{no. mol } NH_4^+}{\text{no. mol } NH_3} = 5.6 \times 10^{-10} \times \frac{0.0300}{0.0200}$$

$$= 8.4 \times 10^{-10}$$

$$pH = \underline{9.08}$$

Exercise

Suppose 40.0 mL of 1.00 M HCl were used instead of 30.0 mL. What would be the pH of the mixture? Answer: 8.65.

Titration Point	(1)	(2)	(3)
Original	$[H^+] = $ conc. HA	$[H^+] = (K_a \times$ conc. HB$)^{\frac{1}{2}}$	$[OH^-] = (K_b \times$ conc. B$^-)^{\frac{1}{2}}$
Intermediate	$[H^+] = \dfrac{n \text{ HA left}}{V}$	$[H^+] = K_a \times \dfrac{n \text{ HB}}{n \text{ B}^-}$	$[H^+] = K_a \times \dfrac{n \text{ HB}}{n \text{ B}^-}$
Equivalence	pH = 7.00	$[OH^-] = (K_b \times$ conc. B$^-)^{\frac{1}{2}}$	$[H^+] = (K_a \times$ conc. HB$)^{\frac{1}{2}}$
Excess	$[OH^-] = \dfrac{n \text{ OH}^- \text{ excess}}{V}$	$[OH^-] = \dfrac{n \text{ OH}^- \text{ excess}}{V}$	$[H^+] = \dfrac{n \text{ H}^+ \text{ excess}}{V}$

Figure 15.11 A given color indicates similar calculations. Calculation of pH during the titration of:

1. a strong base with a strong acid (e.g., HCl titrated with NaOH)
2. a weak acid with a strong base (e.g., $HC_2H_3O_2$ titrated with NaOH)
3. a weak base with a strong acid (e.g., NH_3 with HCl)

These relations, along with the general ones:

$$pH = -\log_{10}[H^+] \quad \text{and} \quad [H^+] \times [OH^-] = 1.0 \times 10^{-14}$$

will allow you to calculate pH at any point in an acid-base titration. The symbols have the following meanings: HA = strong acid, HB = weak acid of dissociation constant K_a, B$^-$ = weak base of dissociation constant K_b, n = no. moles, V = volume in liters.

TABLE 15.3   Characteristics of Acid-Base Titrations

Example	Equation	K	SPECIES EQUIV. PT.	pH EQUIV. PT.	INDICATOR*
		STRONG ACID–STRONG BASE			
NaOH–HCl	$H^+(aq) + OH^-(aq) \rightarrow H_2O$	$K = 1/K_w$ 1.0×10^{14}	Na^+, Cl^-	7.00	MR,BB,PP
Ba(OH)$_2$–HNO$_3$	$H^+(aq) + OH^-(aq) \rightarrow H_2O$	1.0×10^{14}	Ba^{2+}, NO_3^-	7.00	MR,BB,PP
		WEAK ACID–STRONG BASE			
HC$_2$H$_3$O$_2$–NaOH	$HC_2H_3O_2(aq) + OH^-(aq) \rightarrow$ $C_2H_3O_2^-(aq) + H_2O$	$K = 1/K_b$ 1.8×10^9	$Na^+, C_2H_3O_2^-$	9.22†	PP
HF–KOH	$HF(aq) + OH^-(aq) \rightarrow$ $F^-(aq) + H_2O$	6.9×10^{10}	K^+, F^-	9.42†	PP
		STRONG ACID–WEAK BASE			
NH$_3$–HCl	$NH_3(aq) + H^+(aq) \rightarrow NH_4^+(aq)$	$K = 1/K_a$ 1.8×10^9	NH_4^+, Cl^-	4.78†	MR
ClO$^-$–HCl	$ClO^-(aq) + H^+(aq) \rightarrow HClO(aq)$	3.6×10^7	$HClO, NO_3^-$	3.93†	MR

*MR = methyl red (end point pH = 5); BB = bromthymol blue (end point pH = 7); PP = phenolphthalein (end point pH = 9).

†When 1 M acid is titrated with 1 M base.

Example 15.7 illustrates the important point that a buffer can be prepared by adding a strong acid (e.g., HCl) to an excess of a weak base (e.g., NH$_3$). It would also be possible to make a buffer by adding a strong base (e.g., NaOH) to an excess of a weak acid (e.g., HF). This is indeed a common method of preparing buffers in the laboratory. It is often more convenient to make a buffer this way rather than by directly mixing a weak acid with its conjugate base, the approach discussed in Section 15.1.

15.3
EQUILIBRIA IN PRECIPITATION REACTIONS

In discussing precipitation reactions in Chapter 12, we assumed, in effect, that they went to completion. For the kind of calculations carried out in Section 12.4, that assumption is valid. We must keep in mind, however, that precipitation reactions, like all reactions, reach a position of equilibrium. Putting it another way, even the most "insoluble" electrolyte dissolves to at least a slight extent, thereby establishing equilibrium with its ions in solution.

There are two quite different ways in which we can establish equilibrium between a slightly soluble solid and its ions in solution. Consider, for example, the situation with strontium chromate:

$$SrCrO_4(s) \rightleftharpoons Sr^{2+}(aq) + CrO_4^{2-}(aq) \tag{15.8}$$

We might establish this equilibrium by adding pure strontium chromate to water and shaking or stirring to form a saturated solution (Fig. 15.12). In this case, Reaction 15.8 occurs in the forward direction, and we find that $[Sr^{2+}] = [CrO_4^{2-}]$.

Figure 15.12 A saturated solution of SrCrO$_4$ ($K_{sp} = 3.6 \times 10^{-5}$) can be prepared by stirring the yellow solid with water. (Charles D. Winters)

Alternatively, we can form strontium chromate by mixing solutions of $Sr(NO_3)_2$ and K_2CrO_4 (Fig. 15.13). In this case, Sr^{2+} and CrO_4^{2-} ions react by the reverse of Reaction 15.8 to precipitate $SrCrO_4$. Under these conditions, the concentrations of Sr^{2+} and CrO_4^{2-} ions remaining in solution above the precipitate are ordinarily not equal to each other. It turns out, however, that there is a simple mathematical relationship between these two concentrations. It involves a type of equilibrium constant given the symbol K_{sp}.

THE SOLUBILITY PRODUCT CONSTANT, K_{sp}

Consider once again the equilibrium between solid $SrCrO_4$ and its ions in solution. Ordinarily, we write the equation for such an equilibrium system with the solid on the left and aqueous ions on the right:

$$SrCrO_4(s) \rightleftharpoons Sr^{2+}(aq) + CrO_4^{2-}(aq)$$

The equilibrium constant expression for this system can be written by applying the rules cited in Chapter 13. It has the form

$$[Sr^{2+}] \times [CrO_4^{2-}] = K_{sp}$$

The symbol K_{sp} represents a particular type of equilibrium constant called a **solubility product constant**. Like all equilibrium constants, K_{sp} has a fixed value for a given system at a particular temperature. At 25°C, K_{sp} for $SrCrO_4$ is about 3.6×10^{-5}; that is,

$$[Sr^{2+}] \times [CrO_4^{2-}] = 3.6 \times 10^{-5}$$

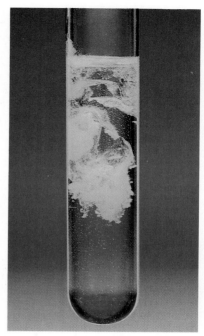

Figure 15.13 Another way to prepare a saturated solution of $SrCrO_4$ is to mix solutions of $Sr(NO_3)_2$ and K_2CrO_4. (Charles D. Winters)

This relation tells us that the product of the two ion concentrations at equilibrium must equal 3.6×10^{-5}, regardless of how equilibrium is established.

For other slightly soluble ionic solids, we can write K_{sp} expressions analogous to that for strontium chromate. For $PbCl_2$ and Ag_2CrO_4,

$$PbCl_2(s) \rightleftharpoons Pb^{2+}(aq) + 2Cl^-(aq)$$

$$[Pb^{2+}] \times [Cl^-]^2 = K_{sp} \ PbCl_2 = 1.7 \times 10^{-5} \qquad (15.9)$$

$$Ag_2CrO_4(s) \rightleftharpoons 2Ag^+(aq) + CrO_4^{2-}(aq)$$

$$[Ag^+]^2 \times [CrO_4^{2-}] = K_{sp} \ Ag_2CrO_4 = 1 \times 10^{-12} \qquad (15.10)$$

■ **EXAMPLE 15.8** _____

Write K_{sp} expressions for each of the following slightly soluble ionic solids:

a. CoS   b. CaF_2 c. As_2S_3

Solution
The dissociation equations are

$$CoS(s) \rightleftharpoons Co^{2+}(aq) + S^{2-}(aq)$$

$$CaF_2(s) \rightleftharpoons Ca^{2+}(aq) + 2F^-(aq)$$

$$As_2S_3(s) \rightleftharpoons 2As^{3+}(aq) + 3S^{2-}(aq)$$

The equilibrium constant expressions follow from the dissociation equations and the rule cited earlier:

a. K_{sp} CoS $= [Co^{2+}] \times [S^{2-}]$
b. K_{sp} CaF$_2$ $= [Ca^{2+}] \times [F^-]^2$
c. K_{sp} As$_2$S$_3$ $= [As^{3+}]^2 \times [S^{2-}]^3$

Exercise

Write K_{sp} expressions for $Fe(OH)_2$ and $Ca_3(PO_4)_2$. Answer:

$$K_{sp} \text{ Fe(OH)}_2 = [Fe^{2+}] \times [OH^-]^2$$

$$K_{sp} \text{ Ca}_3(PO_4)_2 = [Ca^{2+}]^3 \times [PO_4{}^{3-}]^2$$

Solubility product constants such as those listed in Table 15.4 can be used for a variety of purposes. In particular, we can use K_{sp} for a slightly soluble ionic solid to calculate

— the concentration of one ion in equilibrium with the solid when we know that of the other ion.
— whether or not the solid will precipitate when two solutions are mixed.
— the water solubility of the solid and the effect of a "common ion" upon that solubility.

K_{sp} AND THE EQUILIBRIUM CONCENTRATIONS OF IONS

We can use the relation

$$[Sr^{2+}] \times [CrO_4{}^{2-}] = K_{sp} \text{ SrCrO}_4 = 3.6 \times 10^{-5}$$

to calculate the equilibrium concentration of one ion when we know that of the other. Suppose, for example, we find that when SrCrO$_4$ is precipitated out of solution, the concentration of $CrO_4{}^{2-}$ in equilibrium with it is 2.0×10^{-3} M. It follows that

$$[Sr^{2+}] = \frac{K_{sp} \text{ SrCrO}_4}{[CrO_4{}^{2-}]} = \frac{3.6 \times 10^{-5}}{2.0 \times 10^{-3}} = 1.8 \times 10^{-2} \text{ } M$$

If in another case we know that $[Sr^{2+}] = 1.0 \times 10^{-4}$ M,

$$[CrO_4{}^{2-}] = \frac{K_{sp} \text{ SrCrO}_4}{[Sr^{2+}]} = \frac{3.6 \times 10^{-5}}{1.0 \times 10^{-4}} = 3.6 \times 10^{-1} \text{ } M$$

Example 15.9 illustrates the same kind of calculation for a different type of electrolyte.

■ EXAMPLE 15.9

A precipitate of PbCl$_2$ is formed by mixing solutions containing Pb^{2+} and Cl$^-$ ions. Use the K_{sp} value for PbCl$_2$ given in Table 15.4 to calculate

a. the concentration of Pb^{2+} in equilibrium with PbCl$_2$ if $[Cl^-] = 0.10$ M.
b. the concentration of Cl$^-$ in equilibrium with PbCl$_2$ if $[Pb^{2+}] = 0.10$ M.

Solution

The expression for K_{sp} is

$$[Pb^{2+}] \times [Cl^-]^2 = K_{sp}\ PbCl_2 = 1.7 \times 10^{-5}$$

a. Solving for Pb^{2+},

$$[Pb^{2+}] = \frac{K_{sp}\ PbCl_2}{[Cl^-]^2} = \frac{1.7 \times 10^{-5}}{(1.0 \times 10^{-1})^2} = \boxed{1.7 \times 10^{-3}\ M}$$

b. Solving for Cl^-,

$$[Cl^-]^2 = \frac{K_{sp}\ PbCl_2}{[Pb^{2+}]} = \frac{1.7 \times 10^{-5}}{1.0 \times 10^{-1}} = 1.7 \times 10^{-4}$$

$$[Cl^-] = (1.7 \times 10^{-4})^{1/2} = \boxed{1.3 \times 10^{-2}\ M}$$

Exercise

Using Table 15.4, calculate $[Mg^{2+}]$ in contact with $Mg(OH)_2$ if $[OH^-] = 1 \times 10^{-4}\ M$. Answer: $6 \times 10^{-4}\ M$.

TABLE 15.4 Solubility Product Constants at 25°C

		K_{sp}			K_{sp}
Acetates	$AgC_2H_3O_2$	1.9×10^{-3}	Hydroxides	$Al(OH)_3$	2×10^{-31}
				$Fe(OH)_2$	5×10^{-17}
Bromides	$AgBr$	5×10^{-13}		$Fe(OH)_3$	3×10^{-39}
	Hg_2Br_2	6×10^{-23}		$Mg(OH)_2$	6×10^{-12}
	$PbBr_2$	6.6×10^{-6}		$Tl(OH)_3$	2×10^{-44}
				$Zn(OH)_2$	4×10^{-17}
Carbonates	Ag_2CO_3	8×10^{-12}			
	$BaCO_3$	2.6×10^{-9}	Iodides	AgI	1×10^{-16}
	$CaCO_3$	4.9×10^{-9}		Hg_2I_2	5×10^{-29}
	$MgCO_3$	6.8×10^{-6}		PbI_2	8.4×10^{-9}
	$SrCO_3$	5.6×10^{-10}			
	$PbCO_3$	1×10^{-13}			
			Phosphates	Ag_3PO_4	1×10^{-16}
Chlorides	$AgCl$	1.8×10^{-10}		$AlPO_4$	1×10^{-20}
	Hg_2Cl_2	1×10^{-18}		$Ca_3(PO_4)_2$	1×10^{-33}
	$PbCl_2$	1.7×10^{-5}		$Mg_3(PO_4)_2$	1×10^{-24}
Chromates	Ag_2CrO_4	1×10^{-12}	Sulfates	$BaSO_4$	1.1×10^{-10}
	$BaCrO_4$	1.2×10^{-10}		$CaSO_4$	7.1×10^{-5}
	$PbCrO_4$	2×10^{-14}		$PbSO_4$	1.8×10^{-8}
	$SrCrO_4$	3.6×10^{-5}		$SrSO_4$	3.4×10^{-7}
Fluorides	BaF_2	1.8×10^{-7}	Sulfides	Ag_2S	1×10^{-49}
	CaF_2	1.5×10^{-10}		Bi_2S_3	1×10^{-99}
	MgF_2	7×10^{-11}		CdS	1×10^{-29}
	PbF_2	7.1×10^{-7}		CuS	1×10^{-36}
				Cu_2S	1×10^{-48}
				FeS	2×10^{-19}
				HgS	1×10^{-52}
				MnS	5×10^{-14}
				NiS	1×10^{-21}
				PbS	1×10^{-28}
				SnS	3×10^{-28}

K_{sp} AND PRECIPITATE FORMATION

In Chapter 12, we used the solubility rules (Table 12.1) to predict whether or not a precipitate will form when two solutions are mixed. That type of prediction is limited to the situation where the ions involved are at a concentration of 0.1 M or greater. If the ion concentrations are appreciably less than 0.1 M, a precipitate may not form even though the solid is listed as being "insoluble" in water.

We can use K_{sp} values such as those listed in Table 15.4 to make a more general prediction concerning precipitate formation, regardless of the concentrations of the ions involved. To do this, we compare a quantity called the *ion product*, Q, to the solubility product constant, K_{sp}. The form of the expression for Q is the same as that for K_{sp}. However, Q differs from K_{sp} in that it involves original rather than equilibrium concentrations. Thus, for $SrCrO_4$, we have

$$Q = (\text{orig. conc. } Sr^{2+}) \times (\text{orig. conc. } CrO_4^{2-});$$

$$K_{sp} = [Sr^{2+}] \times [CrO_4^{2-}]$$

The "original concentrations" are those in the solution before precipitation occurs. In contrast, the terms $[Sr^{2+}]$ and $[CrO_4^{2-}]$ refer to equilibrium concentrations, those that are established after precipitation. For $PbCl_2$, we would write

$$Q = (\text{orig. conc. } Pb^{2+}) \times (\text{orig. conc. } Cl^-)^2; \quad K_{sp} = [Pb^{2+}] \times [Cl^-]^2$$

We can distinguish three cases:

1. If $Q > K_{sp}$, the solution contains a higher concentration of ions than it can hold at equilibrium. In other words, the solution is supersaturated. **A precipitate forms**, decreasing the concentrations until the ion product becomes equal to K_{sp} and equilibrium is established.

There is no way that Q can get as big as K_{sp} under those conditions

2. If $Q < K_{sp}$, the solution contains a lower concentration of ions than is required for equilibrium with the solid. The solution is unsaturated. **No precipitate forms**; equilibrium is not established.

3. If $Q = K_{sp}$, the solution is just saturated with ions and is at the point of precipitation.

■ EXAMPLE 15.10

Chromate ions are added to a solution in which the original concentration of Sr^{2+} is $1.0 \times 10^{-3} M$. Assuming the concentration of Sr^{2+} stays constant,

a. will a precipitate of $SrCrO_4$ ($K_{sp} = 3.6 \times 10^{-5}$) form when conc. $CrO_4^{2-} = 3.0 \times 10^{-2} M$?

b. will a precipitate form when the concentration of CrO_4^{2-} is $5.0 \times 10^{-2} M$?

c. at what concentration of CrO_4^{2-} does a precipitate just start to form?

Solution

a. $\begin{aligned} Q &= (\text{orig. conc. } Sr^{2+}) \times (\text{orig. conc. } CrO_4^{2-}) \\ &= (1.0 \times 10^{-3}) \times (3.0 \times 10^{-2}) = 3.0 \times 10^{-5} \end{aligned}$

Since Q is less than K_{sp}, 3.6×10^{-5}, no precipitate forms.

You don't always get a precipitate when you mix the cation and the anion of an insoluble compound

b. $Q = (1.0 \times 10^{-3}) \times (5.0 \times 10^{-2}) = 5.0 \times 10^{-5}$; $Q > K_{sp}$

A precipitate forms, reducing the concentrations of Sr^{2+} and CrO_4^{2-} until the ion product becomes equal to 3.6×10^{-5}.

c. The precipitate starts to form when the ion product is equal to K_{sp}:

$$\text{conc. } CrO_4^{2-} = \frac{K_{sp} \, SrCrO_4}{\text{conc. } Sr^{2+}} = \frac{3.6 \times 10^{-5}}{1.0 \times 10^{-3}} = 3.6 \times 10^{-2} \, M$$

When the concentration of CrO_4^{2-} is less than $3.6 \times 10^{-2} \, M$, as in (a), no precipitate forms. When the concentration of CrO_4^{2-} exceeds $3.6 \times 10^{-2} \, M$, as in (b), a precipitate forms.

Exercise
Will PbI_2 ($K_{sp} = 8.4 \times 10^{-9}$) precipitate from a solution in which the concentrations of Pb^{2+} and I^- are both $1 \times 10^{-3} \, M$? Answer: No.

The calculations in Example 15.10 were simplified by assuming that the concentration of one ion (Sr^{2+}) stayed constant while the other ion (CrO_4^{2-}) was added. In reality, this assumption is unlikely to be valid. Mixing two solutions decreases the concentrations of the ions present because the volume increases. We have to take this effect into account before we can decide whether or not a precipitate forms. In particular, we have to calculate the concentrations of the ions after mixing but before precipitation. These are the "original concentrations" to be used in the expression for the ion product, Q. The calculations involved are shown in Example 15.11.

One solution dilutes the other when they are mixed

■ EXAMPLE 15.11

A student mixes 0.200 L of 0.0060 M $Sr(NO_3)_2$ solution with 0.100 L of 0.015 M K_2CrO_4 solution to give a final volume of 0.300 L. Determine

a. the concentration of Sr^{2+} after mixing.
b. the concentration of CrO_4^{2-} after mixing.
c. whether or not a precipitate of $SrCrO_4$ ($K_{sp} = 3.6 \times 10^{-5}$) will form under these conditions.

Solution

a. To find the concentration of Sr^{2+} after mixing, we need to know the number of moles of that ion present:

$$\text{no. moles } Sr^{2+} = 0.200 \text{ L} \times \frac{0.0060 \text{ mol } Sr(NO_3)_2}{1 \text{ L}} \times \frac{1 \text{ mol } Sr^{2+}}{1 \text{ mol } Sr(NO_3)_2}$$

$$= 1.2 \times 10^{-3} \text{ mol } Sr^{2+}$$

The volume after mixing is 0.300 L. Hence,

$$\text{conc. Sr}^{2+} = \frac{1.2 \times 10^{-3} \text{ mol}}{0.300 \text{ L}} = \boxed{4.0 \times 10^{-3} \text{ mol/L}}$$

b. Proceeding in the same way,

$$\text{no. moles CrO}_4^{2-} = 0.100 \text{ L} \times \frac{0.015 \text{ mol K}_2\text{CrO}_4}{1 \text{ L}} \times \frac{1 \text{ mol CrO}_4^{2-}}{1 \text{ mol K}_2\text{CrO}_4}$$

$$= 1.5 \times 10^{-3} \text{ mol CrO}_4^{2-}$$

$$\text{conc. CrO}_4^{2-} = \frac{1.5 \times 10^{-3} \text{ mol}}{0.300 \text{ L}} = \boxed{5.0 \times 10^{-3} \, M}$$

c. $Q = (\text{orig. conc. Sr}^{2+}) \times (\text{orig. conc. CrO}_4^{2-})$
$= (4.0 \times 10^{-3}) \times (5.0 \times 10^{-3}) = 2.0 \times 10^{-5}$

Since Q is less than K_{sp} (3.6×10^{-5}), we conclude that $\boxed{\text{no precipitate}}$

$\boxed{\text{should form.}}$

Exercise
Suppose the student ignored the effect of mixing the solutions upon ion concentrations and took orig. conc. $\text{Sr}^{2+} = 0.0060 \, M$, orig. conc. $\text{CrO}_4^{2-} = 0.015 \, M$. Would he conclude that a precipitate should form under these conditions? Answer: Yes.

K_{sp} AND WATER SOLUBILITY

To express the water solubility of a slightly soluble ionic solid, we most often cite the number of moles of the solid that dissolves in one liter of water. The solubility of an ionic compound and its solubility product constant are related but they are not equal. For example,

COMPOUND	WATER SOLUBILITY (mol/L)	SOLUBILITY PRODUCT CONSTANT, K_{sp}
$SrCrO_4$	6.0×10^{-3}	3.6×10^{-5}
$CaCO_3$	7.0×10^{-5}	4.9×10^{-9}
BaF_2	3.6×10^{-3}	1.8×10^{-7}

Knowing one of these two quantities, we can always, in principle at least,* calculate the other one, as shown in Examples 15.12 and 15.13.

---

*Experimentally, we usually find that the solubility is slightly greater than that predicted from K_{sp}. For example, the measured solubility of PbI_2 in water at 25°C is $1.7 \times 10^{-3} \, M$. This compares to a value of $1.4 \times 10^{-3} \, M$, calculated from the K_{sp} of PbI_2. The reason for this is that some of the lead in PbI_2 goes into solution in the form of species other than Pb^{2+}. For example, we can detect ions such as $Pb(OH)^+$ and PbI^+ in a water solution of lead iodide.

■ EXAMPLE 15.12 _____

a. When excess solid $SrCrO_4$ is shaken with water at 25°C, it is found that 6.0×10^{-3} mol dissolves per liter. Use this information to calculate K_{sp} for $SrCrO_4$.

b. The K_{sp} of $CaCO_3$ is 4.9×10^{-9}; what is its water solubility in moles per liter?

Solution

a. Strontium chromate dissolves in water according to the equation

$$SrCrO_4(s) \rightleftharpoons Sr^{2+}(aq) + CrO_4^{2-}(aq)$$

For every mole of $SrCrO_4$ that dissolves, one mole of Sr^{2+} and one mole of CrO_4^{2-} enter the solution. Hence, when 6.0×10^{-3} mol/L of solid dissolves, the equilibrium concentrations of Sr^{2+} and CrO_4^{2-} must both be 6.0×10^{-3} M:

$$K_{sp} = [Sr^{2+}] \times [CrO_4^{2-}] = (6.0 \times 10^{-3}) \times (6.0 \times 10^{-3})$$

$$= \boxed{3.6 \times 10^{-5}}$$

b. Calcium carbonate dissolves in water according to the equation

$$CaCO_3(s) \rightleftharpoons Ca^{2+}(aq) + CO_3^{2-}(aq)$$

For every mole of $CaCO_3$ that dissolves, one mole of Ca^{2+} and one mole of CO_3^{2-} are formed. Hence, if we let s be the solubility of $CaCO_3$ (that is, s = no. moles $CaCO_3$ dissolving per liter), then $[Ca^{2+}] = [CO_3^{2-}] = s$.

Hence,

$$K_{sp} = [Ca^{2+}] \times [CO_3^{2-}] = (s) \times (s) = s^2$$

Solving for s,

$$s = (K_{sp})^{1/2} = (4.9 \times 10^{-9})^{1/2} = \boxed{7.0 \times 10^{-5}\ M}$$

The solubility is not equal to the solubility product. The two terms have different meanings

Exercise

When $T\ell I$ is shaken with one liter of water, about 0.08 g dissolves. Calculate K_{sp} of $T\ell I$ ($mmT\ell I$ = 331.3 g/mol). Answer: 6×10^{-8}.

As Example 15.12 indicates, there is a simple relationship between K_{sp} and the water solubility, s, for an electrolyte like $SrCrO_4$ or $CaCO_3$, which dissolves to form equal numbers of cations and anions. In that case, $K_{sp} = s^2$. With other types of electrolytes, the relationship is quite different (Example 15.13).

■ EXAMPLE 15.13 _____

Taking K_{sp} of BaF_2 to be 1.8×10^{-7}, calculate its water solubility in moles per liter.

Solution

When barium fluoride dissolves in water, the following equilibrium is established:

$$BaF_2(s) \rightleftharpoons Ba^{2+}(aq) + 2F^-(aq)$$

When one mole of BaF_2 dissolves, one mole of Ba^{2+} ions and two moles of F^- ions are formed. It follows that if s moles of BaF_2 dissolves per liter, then s moles per liter of Ba^{2+} ions and $2s$ moles per liter of F^- ions are formed. In other words,

$$[Ba^{2+}] = s; \qquad [F^-] = 2s$$

where s is the water solubility in moles per liter of BaF_2.

To evaluate s, we substitute into the K_{sp} expression:

$$[Ba^{2+}] \times [F^-]^2 = K_{sp}\ BaF_2 = 1.8 \times 10^{-7}$$

$$(s) \times (2s)^2 = 4s^3 = K_{sp}\ BaF_2 = 1.8 \times 10^{-7}$$

$$s = \left(\frac{1.8 \times 10^{-7}}{4}\right)^{1/3} = \boxed{3.6 \times 10^{-3}\ M}$$

Exercise

Find the relationship between K_{sp} and s for Ag_2CrO_4; for Ag_3PO_4. Answer: $K_{sp} = 4s^3$; $K_{sp} = 27s^4$.

K_{sp} AND THE COMMON ION EFFECT

In our calculations of solubilities of ionic compounds, as in Examples 15.12 and 15.13, it was understood that the solid was dissolving in pure water. Sometimes, however, we dissolve an ionic solid in a solution that already contains ions. In particular, the solution may contain an ion in common with the solid. We might, for example, dissolve $CaCO_3$ in $0.1\ M\ Na_2CO_3$ solution or in $0.1\ M\ CaCl_2$. In the first case, the "common ion" is CO_3^{2-}; in the second case, it is Ca^{2+}.

No problem identifying the common ion

We can use K_{sp} for a slightly soluble electrolyte to estimate its solubility in a solution containing a common ion. Example 15.14 shows how this is done.

■ **EXAMPLE 15.14**

Taking K_{sp} of $CaCO_3$ to be 4.9×10^{-9}, estimate its solubility (moles per liter) in $0.10\ M\ Na_2CO_3$ solution.

Solution

Here, as in Example 15.12, we establish the equilibrium:

$$CaCO_3(s) \rightleftharpoons Ca^{2+}(aq) + CO_3^{2-}(aq)$$

For every mole of $CaCO_3$ that dissolves, one mole of Ca^{2+} and one mole of CO_3^{2-} are formed. If we let s be the solubility of $CaCO_3$ in moles per liter,

$$[Ca^{2+}] = s; \quad [CO_3^{2-}] = s + 0.10$$

since the original concentrations of Ca^{2+} and CO_3^{2-} were 0 and 0.10 mol/L, respectively.

Substituting into the K_{sp} expression, we have:

$$s(s + 0.10) = 4.9 \times 10^{-9}$$

This is a quadratic equation; as usual, we look for ways to avoid solving it by brute force. A suitable approximation here would appear to be:

$$s + 0.10 \approx 0.10$$

We expect the solubility to be much less than 0.10 mol/L, since $CaCO_3$ is a relatively insoluble solid. With that assumption, we have:

$$s(0.10) = 4.9 \times 10^{-9}$$

$$s = \frac{4.9 \times 10^{-9}}{1.0 \times 10^{-1}} = \boxed{4.9 \times 10^{-8} \text{ mol/L}}$$

We see that the solubility is indeed much, much less than 0.10 M, so the approximation is justified.

Exercise
Estimate the solubility of $CaCO_3$ in 0.10 M $CaCl_2$ solution. Answer: 4.9×10^{-8} M.

Comparing the results of Examples 15.12 and 15.14:

SOLVENT	SOLUBILITY OF $CaCO_3$ (mol/L)
Pure water	7.0×10^{-5}
0.1 M Na_2CO_3	4.9×10^{-8}
0.1 M $CaCl_2$	4.9×10^{-8}

we see that $CaCO_3$ is much less soluble in 0.1 M Na_2CO_3 or $CaCl_2$ solution than it is in pure water. The effect is a general one. *An ionic solid is always less soluble in a solution containing a common ion than it is in pure water.*

s is down by a factor of about 1000

Qualitatively, we can explain this effect in terms of Le Châtelier's Principle. Suppose we establish the equilibrium

$$CaCO_3(s) \rightleftharpoons Ca^{2+}(aq) + CO_3^{2-}(aq)$$

and then add Ca^{2+} ions (0.10 M $CaCl_2$) or CO_3^{2-} ions (0.10 M Na_2CO_3) to the solution. We expect the equilibrium to shift to the left, consuming part of the added ions and precipitating $CaCO_3$. This is just another way of saying that the solubility is reduced by adding a common ion. The more soluble the compound is, the greater the amount precipitated by adding a common ion (Fig. 15.14).

(a)

(b)

(c)

Figure 15.14 Sodium chloride can be precipitated from its saturated solution (6 M) by adding a solution containing either Na^+ or Cl^- ions at a concentration greater than 6 M. Both 12 M NaOH (a) and 12 M HCl (b) will bring about this reaction, illustrating the common ion effect. In contrast, nitric acid, HNO_3, has no effect (c), because it contains no ions in common with NaCl. (Marna G. Clarke)

■ **SUMMARY PROBLEM**

Consider the weak acid hydrogen fluoride, $K_a = 6.9 \times 10^{-4}$, and its conjugate base F^- ($K_b = 1.4 \times 10^{-11}$).

1. A buffer is made by dissolving 25.0 g of sodium fluoride, NaF, in one liter of a solution of 0.500 M HF.
   a. Calculate the pH of the buffer.
   b. Calculate the pH of the buffer after 0.100 mol HCl is added.
   c. Calculate the pH of the buffer after 0.100 mol NaOH is added.
   d. Calculate the pH of the buffer after 0.500 mol of NaOH is added.
2. Write the net ionic equations and calculate K for the reactions between
   a. HF and KOH
   b. F^- and HNO_3
   c. HF and NH_3 (K_b $NH_3 = 1.8 \times 10^{-5}$)
3. 50.00 mL of 1.000 M HF is titrated with 1.000 M KOH. Calculate the pH at the beginning of the titration and after 25.00 mL of KOH has been added.
4. Calcium fluoride ($K_{sp} = 1.5 \times 10^{-10}$) is formed when solutions of calcium nitrate and sodium fluoride are mixed.
   a. Will a precipitate form if 10.00 mL of 0.200 M calcium nitrate is added to 25.00 mL of 0.100 M sodium fluoride?
   b. Calculate the solubility of calcium fluoride (grams per liter) in water and in 0.100 M sodium fluoride.

Answers
1. **a.** 3.24 **b.** 3.08 **c.** 3.40 **d.** 8.59 (pH of 1.095 M NaF)
2. **a.** $HF(aq) + OH^-(aq) \rightleftharpoons F^-(aq) + H_2O$ $K = 7.1 \times 10^{10}$
   b. $F^-(aq) + H^+(aq)$ $\rightleftharpoons HF(aq)$ $K = 1.4 \times 10^3$
   c. $HF(aq) + NH_3(aq) \rightleftharpoons NH_4^+(aq) + F^-(aq)$ $K = 1.2 \times 10^6$
3. 1.58, 3.16   **4. a.** Yes   **b.** 0.026 g/L; 1.17×10^{-6} g/L

540

Blood, like many natural fluids, is buffered. Indeed, there are three different buffer systems in blood plasma that hold the pH at about 7.40. One of these comprises protein molecules (Chapter 27); another involves the $H_2PO_4^- - HPO_4^{2-}$ system (pK_a $H_2PO_4^- = 7.21$). By far the most important of the three systems is that involving H_2CO_3 (an aqueous solution of CO_2), and its conjugate base, the HCO_3^- ion. This system consumes H^+ or OH^- ions by the reactions:

$$HCO_3^-(aq) + H^+(aq) \rightarrow H_2CO_3(aq) \tag{15.11}$$

$$H_2CO_3(aq) + OH^-(aq) \rightarrow HCO_3^-(aq) + H_2O \tag{15.12}$$

Since the pH of blood is 7.40, we have:

$$[H^+] = 10^{-7.40} = 4.0 \times 10^{-8}\ M$$

The dissociation constant of carbonic acid is 4.4×10^{-7}; it follows that the concentration of HCO_3^- in the blood is more than ten times that of H_2CO_3:

$$\frac{[HCO_3^-]}{[H_2CO_3]} = \frac{K_a\ H_2CO_3}{[H^+]} = \frac{4.4 \times 10^{-7}}{4.0 \times 10^{-8}} = 11$$

Consequently, blood has a much greater capacity for absorbing H^+ ions (Reaction 15.11) than for absorbing OH^- ions (Reaction 15.12).

It needs this capacity, because exercise produces acids, such as lactic acid

If the pH of the blood drops significantly below 7.4, a condition called *acidosis* is created. The nervous system is depressed; fainting and even coma can result. Most often, acidosis is caused by a buildup of carbon dioxide concentration in the blood. You can produce a very mild case of acidosis by holding your breath. A more serious case can be the result of lung diseases such as asthma or emphysema. Diabetics are also susceptible to acidosis because their metabolic processes form organic acids such as

$$CH_3 - \overset{\displaystyle \|}{\underset{\displaystyle O}{C}} - CH_2 - COOH \qquad \text{acetoacetic acid}$$

If acetoacetic acid is produced in amounts greater than the HCO_3^- ions in the blood can react with, a diabetic coma can result.

Alkalosis, which results when the pH of the blood rises significantly above its normal value of 7.4, is much less common than acidosis (fortunately, since blood has a relatively small capacity to absorb OH^- ions). The symptoms of alkalosis are the inverse of those for acidosis; the nervous system is overstimulated, leading to muscle cramps and ultimately convulsions. Most often, alkalosis is caused by rapid or heavy breathing (hyperventilation). This can result from fever, infection, or the action of certain drugs. When a person breathes too deeply, carbon dioxide is expelled in large quantities from the blood stream and the pH rises.

When you think about it, the human body is remarkably sensitive to the concentration of H^+ ions in body fluids. The normal concentration of H^+ in the blood is 4.0×10^{-8} M; when this changes to 4.5×10^{-8} M, acidosis results. In the first place, this is a relatively small change, of the order of 10%. Moreover, the concentration itself is extremely low, considerably less than one part per million. The concentration of H^+ ion is much more critical than that of just about any other ion, because nearly all of the reactions that take place in the body involve either H^+ or OH^- ions. Since the concentrations of these species affect the rates of such reactions (Chap. 18), changes in the pH of body fluids can have a profound effect on bodily processes.

QUESTIONS AND PROBLEMS

Equilibrium constants required to solve these problems are listed in Table 14.3 (K_a of weak acids), 14.5 (polyprotic acids), 14.6 (K_b of weak bases), or 15.4 (solubility product constants). They are also found in Appendix 1.

Symbols, Formulas and Equations

1. Write a net ionic equation for the reaction between solutions of
  a. nitric acid and lithium hydroxide
  b. ammonia and hydrogen iodide
  c. hydrogen fluoride and potassium cyanide
  d. calcium hydroxide and nitrous acid, HNO_2

2. Follow the directions of Question 1 for:
  a. HCN and sodium hydroxide
  b. $NaClO_2$ and hydrochloric acid
  c. ammonium chloride and potassium hydroxide
  d. methylamine, CH_3NH_2, and hypochlorous acid, HClO

3. Write a balanced net ionic equation for the reaction of each of the following aqueous solutions with a strong acid.
  a. sodium formate ($NaCHO_2$)   b. calcium hydroxide
  c. ammonia

4. Write a balanced net ionic equation for the reaction of each of the following aqueous solutions with a strong acid.
  a. cesium hydroxide   b. potassium cyanide
  c. aniline, $C_6H_5NH_2$

Buffers

5. Calculate $[H^+]$ and pH in a solution in which $[NH_4^+]$ is 0.20 M and $[NH_3]$ is
  a. 0.50 M b. 0.20 M c. 0.10 M d. 0.010 M

6. Calculate $[OH^-]$ and pH in a solution in which [HClO] is 0.25 M and $[ClO^-]$ is
  a. 0.50 M b. 0.25 M c. 0.050 M d. 0.010 M

7. A solution is prepared by dissolving 0.020 mol of sodium nitrite, $NaNO_2$, in 250.0 mL of 0.040 M nitrous acid. Calculate the pH of this buffer.

8. A solution is prepared by dissolving 0.050 mol of potassium fluoride in 150.0 mL of 0.0275 M hydrogen fluoride. Calculate the pH of this buffer.

9. Consider the weak acids and their conjugate bases listed in Table 15.1. Which acid-base pair would be best for a buffer at a pH of
  a. 3.5?   b. 7.3?   c. 9.3?

10. Follow the instructions for Problem 9 for a pH of
  a. 2.0   b. 4.0   c. 8.5

11. To make a buffer using $HCHO_2$ and CHO_2^- in which the desired pH is 3.00:
  a. What must be the ratio $[HCHO_2]/[CHO_2^-]$?
  b. How many moles of $HCHO_2$ must be added to a liter of 0.200 M $NaCHO_2$ to give this pH?
  c. How many grams of $NaCHO_2$ must be added to a liter of 0.100 M $HCHO_2$ to give this pH?

12. A $NaHCO_3$–Na_2CO_3 buffer is to be prepared with a pH of 10.00.
  a. What must be the ratio $[HCO_3^-]/[CO_3^{2-}]$?
  b. What volume of 1.00 M $NaHCO_3$ should be added to a liter of 1.00 M Na_2CO_3 to form this buffer?

13. If a buffer solution is made of 15.00 g sodium acetate ($NaC_2H_3O_2$) and 12.50 g of acetic acid ($HC_2H_3O_2$) in 5.00×10^2 mL, what is the pH of the buffer? If the buffer is diluted to 1.50 L, what is the pH of the diluted buffer?

14. A buffer solution is prepared by adding 5.50 g of ammonium chloride to 150.0 mL of 0.125 M ammonia. What is the pH of the solution? If the buffer is diluted with water to a final volume of 2.00 L, what is the pH of the diluted buffer?

15. How many grams of ammonium nitrate should be added to 250.0 mL of 0.150 M ammonia to produce a buffer with pH 9.00?

16. It is desired to convert 250.0 mL of white vinegar ($d = 1.006$ g/mL), which is 5.00% acetic acid, $HC_2H_3O_2$, by mass, to a buffer. How many grams of sodium acetate, $NaC_2H_3O_2$, are required to produce a buffer of pH 4.00?

$HClO \rightarrow H^+ + ClO^-$

17. A solution prepared from 0.050 mol/L of a weak acid, HX, has a pH of 2.75. What is the pH of the solution after 0.035 mol of solid LiX has been dissolved in it?

18. A 0.045 M solution of a weak acid HX has a pH of 3.65. What is the pH of the solution after 0.015 mol/L of solid NaX has been dissolved in it?

19. A buffer is made up of one liter each of 0.20 M NH_3 and 0.20 M NH_4Cl. Calculate
   a. the pH of the buffer.
   b. the pH of the buffer after the addition of 0.0100 mol HCl.
   c. the pH of the buffer after the addition of 0.0200 mol KOH.

20. A buffer is made up of one liter each of 0.100 M $HC_2H_3O_2$ and 0.150 M $NaC_2H_3O_2$. Calculate
   a. the pH of the buffer.
   b. the pH of the buffer after the addition of 0.0100 mol HCl.
   c. the pH of the buffer after the addition of 0.0200 mol NaOH.

21. A buffer is prepared in which the ratio $[HCO_3^-]/[CO_3^{2-}]$ is 4.0.
   a. What is the pH of this buffer (K_a HCO_3^- = 4.7 × 10^{-11})?
   b. Enough strong acid is added to make the pH of the buffer 9.40. What is the ratio $[HCO_3^-]/[CO_3^{2-}]$ at this point?

22. Blood is buffered mainly by the HCO_3^-–H_2CO_3 system (K_a H_2CO_3 = 4.4 × 10^{-7}). The normal pH of blood is 7.40.
   a. What is the ratio $[H_2CO_3]/[HCO_3^-]$?
   b. What does the pH become if 10% of the HCO_3^- ions are converted to H_2CO_3?
   c. What does the pH become if 10% of the H_2CO_3 molecules are converted to HCO_3^-?

23. Which of the following would form a buffer if added to one liter of 0.20 M NaOH?
   a. 0.10 mol $HC_2H_3O_2$ b. 0.30 mol $HC_2H_3O_2$
   c. 0.10 mol $NaC_2H_3O_2$ d. 0.20 mol HCl
Explain your reasoning in each case.

24. Which of the following would form a buffer if added to one liter of 0.20 M $HC_2H_3O_2$?
   a. 0.10 mol $NaC_2H_3O_2$ b. 0.10 mol NaOH
   c. 0.30 mol NaOH   d. 0.10 mol HCl
Explain your answers.

25. Calculate the pH of a solution that is prepared by mixing 2.00 g of propionic acid, $HC_3H_5O_2$, and 0.45 g of NaOH in water.

26. Calculate the pH of a solution that is prepared by mixing 1.35 g of sodium nitrite, $NaNO_2$, with 100.0 mL of 0.095 M HCl.

Acid-Base Reaction and Titrations

27. Calculate K for the reactions given in Question 1.

28. Calculate K for the reactions given in Question 2.

29. Calculate K for the reactions given in Question 3.

30. Calculate K for the reactions in Question 4.

31. Given three acid-base indicators: methyl orange (end point at pH 4), bromthymol blue (end point at pH 7), and phenolphthalein (end point at pH 9), which would you select for the following acid-base titrations?
   a. formic acid with sodium hydroxide
   b. ethylamine, $C_2H_5NH_2$, with hydrochloric acid
   c. sodium acetate with hydrochloric acid
   d. perchloric acid with lithium hydroxide

32. Given the acid-base indicators in Question 31, select a suitable indicator for the following titrations:
   a. sulfuric acid with potassium hydroxide
   b. ammonia with hydrobromic acid
   c. hydrogen cyanide with barium hydroxide
   d. sodium nitrite with hydriodic acid

33. 35.00 mL of 0.2500 M sodium hydroxide is titrated with 0.4375 M HCl. Calculate the pH of the resulting solution (assuming volumes are additive) when the following amounts of acid are added:
   a. 10.00 mL   b. 20.00 mL   c. 30.00 mL

34. 75.00 mL of 0.1350 M perchloric acid is titrated with 0.3375 M KOH. Calculate the pH of the resulting solution (assuming volumes are additive) when the following amounts of base are added:
   a. 10.00 mL   b. 20.00 mL   c. 30.00 mL

35. Calculate the pH at the equivalence point in the titration of 50.0 mL of 0.100 M HNO_2 with 0.100 M KOH.

36. Calculate the pH at the equivalence point in the titration of 50.0 mL of 0.150 M ammonia with 0.150 M HCl.

37. 25.00 mL of 0.100 M formic acid, $HCHO_2$, is titrated with 0.150 M KOH. Calculate the pH
   a. before any KOH is added
   b. halfway to the equivalence point
   c. at the equivalence point
Of the three indicators listed in Table 15.2, which one would you use in this titration? Explain your reasoning.

38. 30.00 mL of 0.100 M propionic acid, $HC_2H_5O_2$, is titrated with 0.200 M KOH. Calculate the pH
   a. before any KOH is added
   b. halfway to the equivalence point
   c. at the equivalence point
Of the three indicators listed in Table 15.2, which one would you use in this titration? Explain your reasoning.

39. 30.00 mL of 0.200 M NaCN is titrated with 0.200 M HCl. Calculate the pH when the following volumes of acid have been added:
   a. 0.00 mL   b. 15.00 mL   c. 30.00 mL

40. 40.00 mL of 0.100 M methylamine, CH_3NH_2, is titrated with 0.100 M HCl. Calculate the pH when the following volumes of acid have been added:
   a. 0.00 mL   b. 20.00 mL   c. 40.00 mL

41. Consider the titration of 20.00 mL of 0.100 M hypochlorous acid, HClO, with 0.100 M NaOH.
   a. Write a net ionic equation for the reaction involved and calculate K for the reaction.
   b. Calculate the pH after 15.00 mL of NaOH is added.
   c. Calculate the pH after 20.00 mL of NaOH is added.
   d. Calculate the pH after 25.00 mL of NaOH is added.

42. Consider the titration of 20.00 mL of 0.100 M NaClO with 0.100 M HCl.
   a. Write a net ionic equation for the reaction involved and calculate K.
   b. Calculate the pH after 15.00 mL of HCl has been added.
   c. Calculate the pH after 20.00 mL of HCl has been added.
   d. Calculate the pH after 25.00 mL of HCl has been added.

Expression for K_{sp}

43. Write the dissociation equation and the K_{sp} expression for each of the following electrolytes:
   a. $Zr(OH)_4$ b. $PbBr_2$
   c. K_2SiF_6 d. Bi_2S_3

44. Write the dissociation equation and the K_{sp} expression for each of the following electrolytes:
   a. $Cu_2P_2O_7$ b. $Ni_3(AsO_4)_2$
   c. $Fe(OH)_3$ d. $Mg(NbO_3)_2$

45. Write the dissociation equations on which the following K_{sp} expressions are based.
   a. $[Tl^+] \times [I^-]$ b. $[Eu^{3+}] \times [OH^-]^3$
   c. $[Pb^{2+}]^3 \times [PO_4^{3-}]^2$ d. $[Zn^{2+}] \times [OH^-]^2$

46. Write the dissociation equations on which the following K_{sp} expressions are based.
   a. $[Ca^{2+}] \times [IO_3^-]^2$ b. $[Ag^+]^2 \times [SO_3^{2-}]$
   c. $[Mn^{2+}]^3 \times [AsO_4^{3-}]^2$ d. $[Pb^{2+}] \times [C_2O_4^{2-}]$

K_{sp} and Precipitation

47. Complete the following table (K_{sp} $Mg(OH)_2$ = 6×10^{-12}):

$[Mg^{2+}]$	$[OH^-]$
	$1 \times 10^{-4}\ M$
$6 \times 10^{-3}\ M$	
	$8 \times 10^{-5}\ M$
$2 \times 10^{-5}\ M$	

48. Complete the following table (K_{sp} Ag_2CO_3 = 8×10^{-12}):

$[Ag^+]$	$[CO_3^{2-}]$
$1 \times 10^{-3}\ M$	
	$5 \times 10^{-3}\ M$
$5 \times 10^{-4}\ M$	
	$1 \times 10^{-5}\ M$

49. Using Table 15.4, calculate the concentration of each of the following ions in equilibrium with $1.0 \times 10^{-4}\ M$ PO_4^{3-}.
   a. Ag^+ b. Ca^{2+} c. Al^{3+}

50. Using Table 15.4, calculate the concentration of each of the following ions in equilibrium with $1.0 \times 10^{-5}\ M$ S^{2-}.
   a. Bi^{3+} b. Cu^{2+} c. Cu^+

51. Silver nitrate is added to a solution of $1.00 \times 10^{-2}\ M$ sodium acetate (K_{sp} $AgC_2H_3O_2 = 1.9 \times 10^{-3}$).
   a. At what concentration of Ag^+ does a precipitate start to form?
   b. Enough silver nitrate is added to make $[Ag^+]$ = $0.500\ M$. What is $[C_2H_3O_2^-]$? What percentage of the acetate ion originally present remains in solution?

52. A solution contains $0.0100\ M$ Pb^{2+}. Potassium iodide is added to precipitate lead iodide ($K_{sp} = 8.4 \times 10^{-9}$).
   a. At what concentration of I^- does a precipitate start to form?
   b. When $[I^-] = 1.0 \times 10^{-3}\ M$, what is $[Pb^{2+}]$? What percentage of the Pb^{2+} originally present remains in solution?

53. Lead chromate, $PbCrO_4$, is the yellow paint pigment known as "chrome yellow." A solution is prepared by mixing a solution that is $1 \times 10^{-4}\ M$ in lead ion with a solution that is $5 \times 10^{-3}\ M$ in chromate ion. Ignoring dilution effects, would you expect a precipitate to form?

54. A ground-water source contains 2.5 mg of iodide ion per liter. Will lead iodide ($K_{sp} = 8.4 \times 10^{-9}$) precipitate when enough Pb^{2+} is added to make its concentration $4.0 \times 10^{-3}\ M$?

55. A solution is prepared by mixing 25.0 mL of $0.100\ M$ iron(II) nitrate with 50.0 mL of an NaOH solution of pH 9.00. Will a precipitate form?

56. 15.0 mg of magnesium nitrate is added to 50.0 mL of $0.025\ M$ sodium phosphate. Will a precipitate form?

Solubility

57. Calculate K_{sp} for $MnCO_3$ if 1.07 mg/L are required to make a saturated solution.

58. Calculate K_{sp} for strontium fluoride if 0.073 g/L are required to make a saturated solution.

59. Calculate the solubility (grams per liter) of magnesium phosphate ($K_{sp} = 1 \times 10^{-24}$) in
   a. pure water
   b. $0.010\ M$ $Mg(NO_3)_2$
   c. $0.020\ M$ Na_3PO_4

60. Calculate the solubility (grams per liter) of copper(II) sulfide in
   a. pure water
   b. $0.010\ M$ Na_2S
   c. $0.020\ M$ $CuSO_4$

Unclassified

61. Explain why

a. the pH increases when sodium benzoate is added to a benzoic acid solution.

b. the pH of 0.10 M HNO_2 is greater than 1.0.

c. a buffer resists changes in pH caused by addition of H^+ or OH^-.

62. Indicate whether each of the following statements is true or false; if it is false, correct it.

a. The acetate concentration in 0.10 M $HC_2H_3O_2$ is the same as in 0.10 M $NaC_2H_3O_2$.

b. A buffer can be destroyed by adding too much strong acid.

c. If K_a = 3.1 × 10^{-5}, K_b = 3.2 × 10^{-9}.

63. Predict what effect each of the following has on the position of the equilibrium

$$PbCl_2(s) \rightleftharpoons Pb^{2+}(aq) + 2Cl^-(aq); \Delta H = +23.4 \text{ kJ}$$

a. addition of $Pb(NO_3)_2$ solution.

b. increase in temperature.

c. addition of Ag^+, forming AgCl.

d. addition of hydrochloric acid.

64. A town adds 1.0 ppm of F^- ion to fluoridate its water supply (fluoridation of water reduces the incidence of dental caries). If the concentration of Ca^{2+} in the water is 2.0 × 10^{-4} M, will a precipitate of CaF_2 form when the water is fluoridated?

65. Using K_{sp} data in Table 15.4 to calculate K for the reaction, decide whether this reaction is likely to go to completion:

$$Ag_2CrO_4(s) + 2Cl^-(aq) \rightarrow 2AgCl(s) + CrO_4^{2-}(aq)$$

66. A solution is made up of 30.0 mL of 0.100 M HCl and 20.0 mL of 0.250 M $HC_2H_3O_2$. Calculate the pH after 50.0 mL of 0.100 M NaOH is added.

Challenge Problems

67. If 50.00 cm^3 of 1.000 M $HC_2H_3O_2$ (K_a = 1.8 × 10^{-5}) is titrated with 1.000 M NaOH, what is the pH of the solution after the following volumes of NaOH have been added:

a. 0.00 cm^3 b. 25.00 cm^3 c. 49.90 cm^3

d. 50.00 cm^3 e. 50.10 cm^3 f. 100.00 cm^3

Use your data to construct a plot similar to that shown in Figure 15.9 (pH vs. volume NaOH added).

68. Ammonium chloride solutions are slightly acidic, so they are better solvents than water for insoluble hydroxides such as $Mg(OH)_2$. Find the solubility of $Mg(OH)_2$ in moles/liter in 0.2 M NH_4Cl and compare with the solubility in water. *Hint:* Find K for the reaction

$$Mg(OH)_2(s) + 2NH_4^+(aq) \rightarrow Mg^{2+}(aq) + 2NH_3(aq) + 2H_2O$$

69. In a titration of 50.0 mL 1.00 M $HC_2H_3O_2$ with 1.00 M NaOH, a student used bromcresol green as an indicator (K_a = 1 × 10^{-5}). About how many mL of NaOH would it take to reach the end point with this indicator? What would be a better indicator for this titration?

70. What is the solubility of CaF_2 in a buffer solution containing 0.30 M $HCHO_2$ and 0.20 M $NaCHO_2$? (*Hint:* Consider the equation

$$CaF_2(s) + 2H^+(aq) \rightarrow Ca^{2+}(aq) + 2HF(aq)$$

and solve the equilibrium problem.)

71. What is the I^- concentration just as AgCl begins to precipitate when 1.0 M $AgNO_3$ is slowly added to a solution containing 0.020 M Cl^- and 0.020 M I^-?

72. The concentrations of various cations in seawater, in moles per liter, are

Ion	Na^+	Mg^{2+}	Ca^{2+}	Al^{3+}	Fe^{3+}
Molarity (M)	0.46	0.056	0.01	4 × 10^{-7}	2 × 10^{-7}

a. At what conc. of OH^- does $Mg(OH)_2$ start to precipitate?

b. At this concentration, will any of the other ions precipitate?

c. If enough OH^- is added to precipitate 50% of the Mg^{2+}, what percentage of each of the other ions will be precipitated?

d. Under the conditions in (c), what mass of precipitate will be obtained from one liter of seawater?

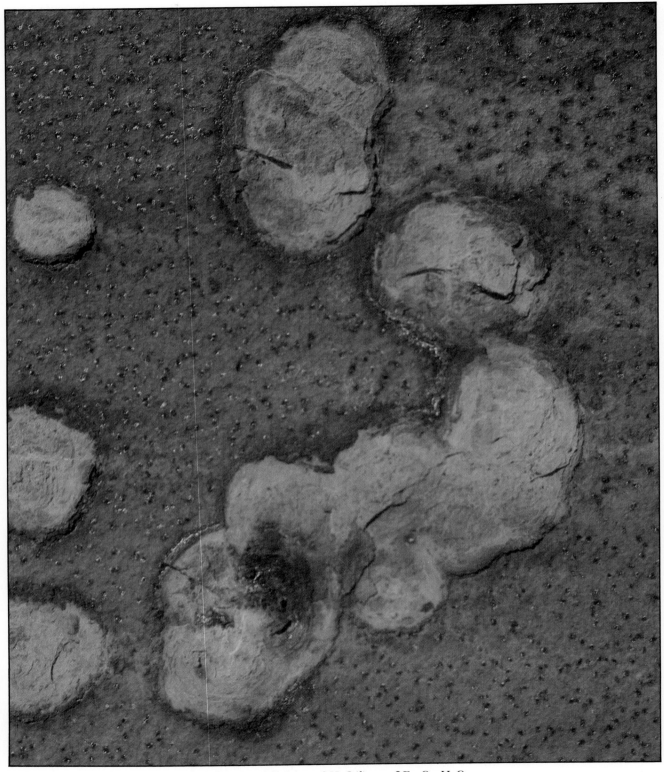

$$4Fe(s) \; + \; 3O_2(g) \; + \; 2H_2O(l) \; \rightarrow \; 2Fe_2O_3 \cdot H_2O$$

Formation of rust on an iron object exposed to water and oxygen.

COMPLEX IONS;
COORDINATION
COMPOUNDS

*Chromium ions are sensitive to their
Chemical environment.
They make the ruby red.
The emerald green.*

*The ruby and emerald are similar.
You say red, I say green.*

ANN RAE JONAS
THE CAUSES OF COLOR

In previous chapters we have referred from time to time to compounds of the transition metals. Many of these have relatively simple formulas such as $CuSO_4$, $CrCl_3$, and $Fe(NO_3)_3$. These compounds are ionic: the transition metal is present as a simple cation (Cu^{2+}, Cr^{3+}, Fe^{3+}). In that sense, they resemble the ionic compounds formed by the main-group metals, such as $CaSO_4$ and $Al(NO_3)_3$.

It has been known for more than a century, however, that transition metals also form a variety of ionic compounds with more complex formulas such as

$$[Cu(NH_3)_4]SO_4 \qquad [Cr(NH_3)_6]Cl_3 \qquad K_3[Fe(CN)_6]$$

In these so-called *coordination compounds,* the transition metal is present as a *complex ion,* enclosed within the square brackets.

This chapter is devoted to complex ions and the important role they play in inorganic chemistry. We will consider in turn

— the compositions and names of complex ions and the coordination compounds they form (Sections 16.1 and 16.2)

— the geometry of complex ions (Section 16.3)

— the electronic structure of the central metal atom in a complex (Section 16.4)

— the equilibrium constant for complex ion formation (Section 16.5)

16.1
COMPOSITIONS OF COMPLEX IONS AND
COORDINATION COMPOUNDS

A water solution containing the Cu^{2+} ion has a pale blue color. When aqueous ammonia, NH_3, is added, the color changes to a deep, almost

Figure 16.1 The $Cu(NH_3)_4^{2+}$ has an intense blue, almost violet, color, in contrast to the light blue $Cu(H_2O)_4^{2+}$ ion. (Charles D. Winters)

Many cations exist in water solution as hydrated complex ions

opaque, blue (Fig. 16.1). The color change is due to a chemical reaction in which four NH_3 molecules combine with a Cu^{2+} ion:

$$Cu^{2+}(aq) + 4NH_3(aq) \rightarrow Cu(NH_3)_4^{2+}(aq) \qquad (16.1)$$

light blue   deep blue

In electron-dot notation, we can represent this reaction as

The nitrogen atom of each NH_3 molecule contributes a pair of unshared electrons to form a covalent bond with the Cu^{2+} ion. This bond and others like it, where both electrons are contributed by the same atom, is referred to as a **coordinate** covalent bond. There are four such bonds in the $Cu(NH_3)_4^{2+}$ ion.

The $Cu(NH_3)_4^{2+}$ ion is commonly referred to as a **complex ion**. We use the term complex ion to indicate a charged species in which *a metal atom is joined by coordinate covalent bonds to neutral molecules and/or negative ions*. Species such as $Al(H_2O)_6^{3+}$ and $Zn(H_2O)_3(OH)^+$, found in previous chapters, are further examples of complex ions. The metals that show the greatest tendency to form complex ions are those that form small cations with a charge of $+2$ or greater. Typically, these are the metals toward the right of the transition series (in the first transition series, $_{24}Cr$ through $_{30}Zn$). Nontransition metals, including Al, Sn, and Pb, form a more limited number of stable complex ions.

The metal atom in a complex is called the *central atom*. The molecules or anions bonded directly to it are called **ligands**. The number of bonds formed by the central atom is called its **coordination number**. In the $Cu(NH_3)_4^{2+}$ ion, the central atom is copper(II). The ligands are NH_3 molecules. Since Cu^{2+} forms a total of four bonds, it has a coordination number of 4.

An ion such as $Cu(NH_3)_4^{2+}$ cannot exist by itself in the solid state. The $+2$ charge of this ion must be balanced by anions with a total charge of -2. A typical compound containing the $Cu(NH_3)_4^{2+}$ ion is

$[Cu(NH_3)_4]Cl_2$: $1 Cu(NH_3)_4^{2+}$ ion, $2 Cl^-$ ions

Compounds such as this, which contain a complex ion, are referred to as **coordination compounds**. The formula of the complex ion is set off by brackets, [ ], to make the structure of the compound clear.

Table 16.1 gives the formulas of a series of coordination compounds formed by platinum(II), which has a coordination number of 4. Notice that:

1. Compounds 1 and 2 contain complex cations with charges of $+2$ and $+1$, respectively. These coordination compounds are analogous to the sim-

TABLE 16.1 Coordination Compounds Containing Complexes of Pt^{2+}

COORDINATION COMPOUND	COMPLEX	CHARGE OF COMPLEX	ANALOGOUS SIMPLE IONIC COMPOUND
1. $[Pt(NH_3)_4]Cl_2$	$Pt(NH_3)_4^{2+}$	$+2$	$CaCl_2$
2. $[Pt(NH_3)_3Cl]Cl$	$Pt(NH_3)_3Cl^+$	$+1$	KCl
3. $[Pt(NH_3)_2Cl_2]$	$Pt(NH_3)_2Cl_2$	0	
4. $K[Pt(NH_3)Cl_3]$	$Pt(NH_3)Cl_3^-$	-1	KNO_3
5. $K_2[PtCl_4]$	$PtCl_4^{2-}$	-2	K_2SO_4

ple ionic compounds $CaCl_2$ and KCl. In both $[Pt(NH_3)_4]Cl_2$ and $CaCl_2$, two Cl^- ions are balanced by a $+2$ ion, $Pt(NH_3)_4^{2+}$ in one case, Ca^{2+} in the other.

2. Compounds 4 and 5 contain complex anions with charges of -1 and -2, respectively. In the solid, these are balanced by K^+ ions ($1K^+$ per $Pt(NH_3)Cl_3^-$ ion, $2K^+$ per $PtCl_4^{2-}$ ion).

3. Compound 3 is a neutral complex, with zero charge. There are no ions present.

The charge of a complex ion is readily determined. It is the algebraic sum of the oxidation number of the central atom and the charges of the anionic ligands. Applying this principle to the compounds in Table 16.1, where the species within the complexes are

Pt^{2+}(oxid. no. $= +2$); NH_3 molecules (charge $= 0$); Cl^- ions (charge $= -1$)

we obtain, for the charges of the complex ions,

Complex 1: $+2 + 4(0) = +2$

Complex 2: $+2 + 3(0) + 1(-1) = +1$

Complex 3: $+2 + 2(0) + 2(-1) = 0$

Complex 4: $+2 + 1(0) + 3(-1) = -1$

Complex 5: $+2 + 4(-1) = -2$

■ **EXAMPLE 16.1** _____

Determine the oxidation number of the central metal atom in each of the following:

a. $Zn(H_2O)_3(OH)^+$ b. $Pt(NH_3)_3Cl_3^-$ c. $Cr(CN)_6^{3-}$

Solution

The first step in each case is to identify the ligands and their charges. Then we use the fact that the charge of the complex is the sum of the oxidation number of the central atom and the charges of the ligands.

a. There are three H_2O molecules (0 charge) and an OH^- ion (-1 charge). The overall charge of the complex is $+1$. If we let x be the oxidation number of zinc:

$+1 = x + 3(0) + 1(-1)$

Solving for x, $x = +2$; (Zn^{2+})

b. The ligands are one NH_3 molecule (0 charge) and three Cl^- ions (-1 charge). The charge of the complex ion itself is -1:

$$-1 = x + 1(0) + 3(-1); \qquad x = \boxed{+2; (Pt^{2+})}$$

c. Here, the complex ion is an anion with a -3 charge. Each cyanide ion, CN^-, has a -1 charge:

$$-3 = x + 6(-1); \qquad x = \boxed{+3; (Cr^{3+})}$$

Exercise
In a certain complex ion, chromium(III) is bonded to two ammonia molecules, three water molecules, and a hydroxide ion. Give the formula and charge of the complex ion. Answer: $Cr(NH_3)_2(H_2O)_3 (OH)^{2+}$.

LIGANDS; CHELATING AGENTS

In principle, any molecule or anion with an unshared pair of electrons can donate them to a metal to form a coordinate covalent bond. In this sense, ligands are Lewis bases. The positive metal ion that accepts the electron pair for bonding is a Lewis acid. We expect species such as the ammonia molecule, the water molecule, the hydroxide ion, and the chloride ion to act as ligands:

In practice, a ligand usually contains an atom of one of the more electronegative elements (C, N, O, S, F, Cl, Br, I). Several hundred different ligands are known. Those most commonly encountered in general chemistry are NH_3 and H_2O molecules and Cl^- and OH^- ions.

Some ligands have more than one atom with an unshared pair of electrons. They therefore have several possible Lewis base sites. Ligands

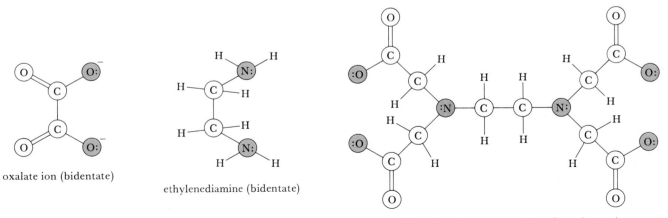

oxalate ion (bidentate)

ethylenediamine (bidentate)

ethylenediaminetetraacetate ion (hexadentate)

Figure 16.2 Structures of three chelating agents sites are shown in color.

are classified according to the number of electron pairs donated. *Monodentate* ligands provide one electron pair per ligand molecule or ion. The species NH_3, H_2O, Cl^-, and OH^- are of this type. *Bidentate* (two-toothed) ligands furnish two electron pairs per molecule or ion, a tridentate ligand furnishes three pairs, and so on. The general term *polydentate* is used for any ligand that supplies more than one pair of electrons. The structures of three important polydentate ligands are shown in Figure 16.2. These are the oxalate ion (ox), the ethylenediamine molecule (en), and the ethylenediaminetetraacetate anion (EDTA). Note that ethylenediamine and the oxalate ion are both bidentate; EDTA is hexadentate, forming six bonds per ligand.

EDTA usually forms 1:1 complexes with cations

The complexes formed by polydentate ligands are often called **chelates** (from the Greek *chela*, crab's claw). The structure of the chelates formed by copper(II) with ethylenediamine and with the oxalate ion are shown in Figure 16.3. Notice that in these complexes, as in $Cu(NH_3)_4^{2+}$, copper(II) has a coordination number of 4. When we write these complexes as $Cu(en)_2^{2+}$ and $Cu(ox)_2^{2-}$, we must remember that each ligand is forming two bonds.

■ **EXAMPLE 16.2** —————————————————————————————————
Give the coordination number of the central metal atom in

a. $Cu(en)_2(NH_3)_2^{2+}$ b. $Fe(en)(ox)Cl_2^-$

Solution
The coordination number is the number of bonds formed by the central metal atom.

a. Ammonia is a monodentate ligand; each NH_3 forms one bond. Each bidentate ethylenediamine (en) forms two bonds. Thus, there are six bonds to Cu^{2+}; its coordination number is 6.

b. Ethylenediamine (en) and oxalate (ox) are both bidentate (Fig. 16.2); each chloride is monodentate. Hence, in this case Fe^{3+} has a coordination number of 6.

Exercise
What is the coordination number of Pt in $Pt(en)_2^{2+}$? Answer: 4.

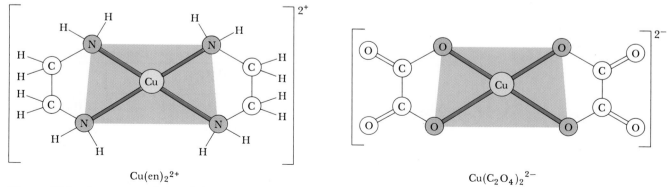

$Cu(en)_2^{2+}$ $Cu(C_2O_4)_2^{2-}$

Figure 16.3 Structures of the chelates formed by Cu^{2+} with the ethylenediamine molecule (en) and the oxalate ion $(C_2O_4^{2-})$.

For a ligand to act as a chelating agent, it must have at least two pairs of unshared electrons. Moreover, these electron pairs must be far enough removed from one another to give a chelate ring with a stable geometry. In chelates formed by ethylenediamine or the oxalate ion, such as those shown in Figure 16.3, the ring contains five atoms (four from the ligand plus the metal atom). Four-membered rings are less stable; three-membered rings do not occur. If we compare the three molecules,

$$H-\overset{..}{N}-CH_2-CH_2-\overset{..}{N}-H \qquad H-\overset{..}{N}-CH_2-\overset{..}{N}-H \qquad H-\overset{..}{N}-\overset{..}{N}-H$$

$$\underset{ethylenediamine}{\underset{|}{H} \qquad \underset{|}{H}} \qquad \underset{methylenediamine}{\underset{|}{H} \qquad \underset{|}{H}} \qquad \underset{hydrazine}{\underset{|}{H} \quad \underset{|}{H}}$$

we find that

— methylenediamine is a less effective chelating agent than ethylenediamine, since it would have to form four-membered as opposed to five-membered rings.
— hydrazine does not act as a chelating agent. To do so, it would have to form a three-membered ring with a metal atom. This would require a highly unstable 60° bond angle.

COORDINATION NUMBER

As shown in Table 16.2, the most common coordination number is 6. A coordination number of 4 is less common. A value of 2 is restricted largely to Cu^+, Ag^+, and Au^+. Odd coordination numbers (3, 5, 7, etc.) are relatively rare. One well-known example is $Fe(CO)_5$, a neutral, highly toxic complex.

A few cations show only one coordination number in their complexes. Thus, Co^{3+} always shows a coordination number of 6, as in

$$Co(NH_3)_6{}^{3+}, \qquad Co(NH_3)_4Cl_2{}^+, \qquad Co(en)_3{}^{3+}$$

Other cations, such as Al^{3+} and Ni^{2+}, have variable coordination numbers, depending upon the nature of the ligand. With molecules or very small anions as ligands, we usually observe the higher coordination number, as in

$$Al(H_2O)_6{}^{3+}, \qquad AlF_6{}^{3-}, \qquad Ni(H_2O)_6{}^{2+}, \qquad Ni(NH_3)_6{}^{2+}$$

It is not easy to predict the coordination numbers for cations, except that they are likely to be 2, 4, or 6

TABLE 16.2   Coordination Number and Geometry of Complex Ions*

METAL ION	COORDINATION NUMBER	GEOMETRY	EXAMPLE
Ag⁺, Au⁺, Cu⁺	2	linear	$Ag(NH_3)_2{}^+$
Cu²⁺, Ni²⁺, **Pd²⁺**, **Pt²⁺**	4	square planar	$Pt(NH_3)_4{}^{2+}$
Al³⁺, Au⁺, **Cd²⁺**, Co²⁺, Cu⁺, Ni²⁺, **Zn²⁺**	4	tetrahedral	$Zn(NH_3)_4{}^{2+}$
Al³⁺, Co²⁺, **Co³⁺**, **Cr³⁺**, Cu²⁺, **Fe²⁺**, **Fe³⁺**, Ni²⁺, **Pt⁴⁺**	6	octahedral	$Co(NH_3)_6{}^{3+}$

*Most common coordination number indicated by bold type.

With larger anions, the lower coordination number is more common:

$$AlCl_4^-, \qquad Al(OH)_4^-, \qquad Ni(CN)_4^{2-}$$

16.2 NAMING COMPLEX IONS AND COORDINATION COMPOUNDS

The nomenclature of compounds containing complex ions is more involved than that of the simple inorganic compounds considered in earlier chapters. We will see first how complex ions are named and then look at the nomenclature of coordination compounds.

COMPLEX CATIONS AND NEUTRAL COMPLEXES

To name a species like $Cu(NH_3)_4^{2+}$ or $Zn(H_2O)_2(OH)_2$, we need to specify

— the number and identity of each ligand (4 NH_3 molecules; 2 H_2O molecules, 2 OH^- ions).

— the oxidation number of the central metal atom: copper(II), zinc(II).

To accomplish this, we follow a set of rules.

1. The number of ligands of a particular type is ordinarily indicated by the Greek prefixes *di, tri, tetra, penta, hexa.* Thus we have

$Cu(NH_3)_4^{2+}$ tetraamminecopper(II)

$Zn(H_2O)_2(OH)_2$ diaquodihydroxozinc(II)

If the name of the ligand is itself complex (e.g., ethylenediamine), the number of such ligands is indicated by the prefixes *bis, tris, tetrakis, pentakis, hexakis.* The name of the ligand is enclosed in parentheses.

$Cu(en)_2^{2+}$ bis(ethylenediamine)copper(II)

$Cr(en)_3^{3+}$ tris(ethylenediamine)chromium(III)

2. The names of anions that act as ligands are obtained by substituting *-o* for the normal ending. Thus we have

Cl^-	chloro	SO_4^{2-}	sulfato
OH^-	hydroxo	NO_3^-	nitrato
CN^-	cyano	CO_3^{2-}	carbonato

Ordinarily, the names of molecules are not changed when they become ligands (e.g., ethylenediamine). There are three important exceptions:

     H_2O aquo NH_3 ammine   CO   carbonyl

3. When more than one type of ligand is present, they are named in alphabetical order (without regard to prefixes). Thus we have

$Zn(H_2O)_3(OH)^+$ triaquohydroxozinc(II)

$Co(NH_3)_4Cl_2^+$ tetraamminedichlorocobalt(III)

4. The oxidation number of the central metal atom is indicated by Roman numerals after the English name of the metal.

$Fe(en)_3^{3+}$ tris(ethylenediamine)iron(III)

$Pt(en)I_4$ ethylenediaminetetraiodoplatinum(IV)

COMPLEX ANIONS

If the complex is an anion, such as $Zn(OH)_4^{2-}$, this is indicated by inserting the suffix *ate* after the name of the metal.

$Zn(OH)_4^{2-}$ tetrahydroxozincate(II)

In a few cases, the Latin name of the metal is used in *anionic complexes*. Thus we have:

$Fe(CN)_6^{3-}$ hexacyanoferrate(III)

$CuCl_4^{2-}$ tetrachlorocuprate(II)

COORDINATION COMPOUNDS

Coordination compounds, like simple ionic compounds, are named in a straightforward way. The cation is named first, followed by the anion.

$[Cu(NH_3)_4]Cl_2$ tetraamminecopper(II) chloride

$[Fe(en)_3]PO_4$ tris(ethylenediamine)iron(III) phosphate

$K_3[Fe(CN)_6]$ potassium hexacyanoferrate(III)

Most complexes do not have common names, but when they do, we usually use them

The common name of $K_3[Fe(CN)_6]$ is potassium ferricyanide.

■ **EXAMPLE 16.3** _____

Name the following compounds:

a. $[Cr(NH_3)_4Cl_2]Cl$ b. $K_2[PtCl_4]$ c. $[Co(en)_3](NO_3)_3$

Solution

a. The cation is the complex ion. Its ligands are ammonia, NH_3, called ammine, and chloride, Cl, called chloro. Alphabetically, ammine comes first. The metal is chromium with an oxidation number of $+3$. The cation is therefore tetraamminedichlorochromium(III). The anion is chloride.

The name of the compound is   tetraamminedichlorochromium(III)

chloride.

b. Here the cation is simply K^+, potassium. The anion is the complex ion $PtCl_4^{2-}$. Platinum has an oxidation number of $+2$. The name of the compound is   potassium tetrachloroplatinate(II).

c. The cation is the complex $[Co(en)_3]^{3+}$. It is called tris(ethylene-diamine)cobalt(III). The anion is nitrate. The name of the compound is

tris(ethylenediamine)cobalt(III) nitrate.

Exercise
Give the formula for diaquobis(ethylenediamine)chromium(III) chloride.   WOW!
Answer: $[Cr(H_2O)_2(en)_2]Cl_3$.

16.3
GEOMETRY OF COMPLEX IONS

COORDINATION NUMBER = 2

Complex ions in which the central metal atom forms only two bonds to ligands are *linear;* that is, the two bonds are directed at 180° angles. The structures of $CuCl_2^-$, $Ag(NH_3)_2^+$, and $Au(CN)_2^-$ may be represented as

$$(Cl—Cu—Cl)^- \qquad \begin{bmatrix} H & & H \\ \diagdown & & \diagup \\ H—N—Ag—N—H \\ \diagup & & \diagdown \\ H & & H \end{bmatrix}^+ \qquad (N≡C—Au—C≡N)^-$$

COORDINATION NUMBER = 4

Four-coordinate metal complexes may have either of two different geometries (Fig. 16.4). The four bonds from the central metal atom may be directed toward the corners of a regular tetrahedron. Two common *tetra-hedral* complexes are $Zn(NH_3)_4^{2+}$ and $CoCl_4^{2-}$. In the other geometry,

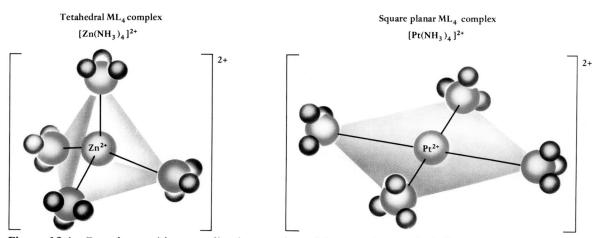

Tetrahedral ML$_4$ complex
$[Zn(NH_3)_4]^{2+}$

Square planar ML$_4$ complex
$[Pt(NH_3)_4]^{2+}$

Figure 16.4 Complexes with a coordination number of four can be tetrahedral, as is $Zn(NH_3)_4^{2+}$, or square planar, like $Pt(NH_3)_4^{2+}$.

$Cu(NH_3)_4^{2+}$ ion is also square planar

known as *square planar,* the four bonds are directed toward the corners of a square. The $Pt(NH_3)_4^{2+}$ and $Ni(CN)_4^{2-}$ ions are of this type.

Certain square planar complexes occur in two different forms with quite different properties. The complex $[Pt(NH_3)_2Cl_2]$ has two forms differing in absorption spectrum, water solubility, melting point, and chemical reactivity. In one form, made by reacting NH_3 with the $PtCl_4^{2-}$ ion, the two NH_3 ligands are at adjacent corners of the square. In the other form, prepared by reacting $Pt(NH_3)_4^{2+}$ with HCl, the NH_3 molecules are at opposite corners of the square. These two forms are referred to as **geometric isomers**. Their structures differ only in the spatial arrangement of ligands about the central metal atom. The form in which like ligands are as close as possible is called the *cis* isomer. The *trans* isomer has these groups as far apart as possible.

$$
\begin{array}{cc}
\text{Cl} \diagdown \quad \diagup \text{NH}_3 & \text{Cl} \diagdown \quad \diagup \text{NH}_3 \\
\text{Pt} & \text{Pt} \\
\text{Cl} \diagup \quad \diagdown \text{NH}_3 & \text{NH}_3 \diagup \quad \diagdown \text{Cl} \\
\textit{cis} & \textit{trans}
\end{array}
$$

The compounds *cis-* and *trans-*diamminedichloroplatinum(II) can be distinguished from one another by either physical or chemical means. The *cis* complex is polar because the two Cl atoms are located on the same side of the platinum; the *trans* complex is nonpolar. Experimentally, we find that the *cis* complex has a dipole moment, while the *trans* complex does not. The two compounds also differ in their biological activity, as we will see in Section 16.6.

Geometric isomerism can occur with any square planar complex of the type Ma_2b_2 or Ma_2bc, where M refers to the central metal and a, b, c are different ligands. Because all positions in a tetrahedral complex are equivalent, geometric isomerism cannot occur with such complexes.

COORDINATION NUMBER = 6

Octahedral geometry is characteristic of this coordination number. The six ligands surrounding the central metal atom in complexes such as $Fe(CN)_6^{3-}$ and $Co(NH_3)_6^{3+}$ are located at the corners of a regular octahedron. An octahedron is a geometric figure with six corners and eight faces, each of

Figure 16.5 The drawing at the left shows six ligands at the corners of an octahedron with a metal atom at the center. A simpler way to represent an octahedral complex is shown at the right.

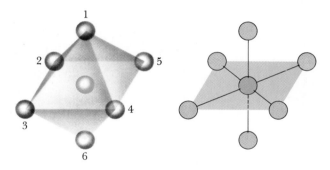

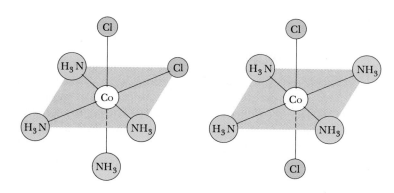

Figure 16.6 *Cis* and *trans* isomers of the $Co(NH_3)_4Cl_2^+$ ion. Note that the two Cl^- ions are closer to one another in the *cis* isomer (*left*) than in the *trans* isomer (*right*).

which is an equilateral triangle. The metal atom is located at the center of the octahedron (Fig. 16.5, p. 556). The six ligands are *equidistant from the metal atom at the center.* The skeleton structure at the right of Figure 16.5 is easier to draw and more commonly used. From this skeleton, we see that an octahedral complex can be regarded as a derivative of a square planar complex. The two extra ligands are located above and below the square, on a line perpendicular to the square at its center.

Geometric isomerism can occur in octahedral complexes. To see how this is possible, consider Figure 16.5. For any given position of a ligand, four equivalent positions are equidistant while a fifth is at a greater distance. Suppose, for example, we choose position 1 as a point of reference. Groups at positions 2, 3, 4, and 5 will be equidistant from 1; 6 is farther away. We may refer to positions 1 and 2, 1 and 3, 1 and 4, or 1 and 5 as being *cis* to each other. On the other hand, positions 1 and 6 are *trans*. Hence, a complex ion like $Co(NH_3)_4Cl_2^+$ can exist in two isomeric forms. In the *cis* isomer, the two Cl^- ions are close together. In the *trans* form, they are far apart (Fig. 16.6, above). The two isomers differ in their physical and chemical properties. The most striking difference is color. Compounds of Co^{3+} containing the *cis* complex ion tend to be violet, while those containing the *trans* isomer are often green (Fig. 16.7).

Figure 16.7 The purple compound shown at the left contains the *cis*-$Co(en)_2Cl_2^+$ cation; the green compound at the right is the *trans* isomer. (Charles D. Winters)

■ EXAMPLE 16.4

How many isomers are possible for the neutral complex $[Co(NH_3)_3Cl_3]$?

Solution

It is best to approach this problem systematically. We might start by putting two NH_3 molecules in *trans* positions, perhaps at the "top" and "bottom" of the octahedron (Fig. 16.8a). We then ask ourselves: In how many spatially different positions can we place the third NH_3 molecule? A moment's reflection should convince you that there is really only one choice. All four of the remaining positions are equivalent in that they are *cis* to the two groups that we have already located. Choosing one of these positions arbitrarily, we get our first isomer (Fig. 16.8b).

This kind of problem can be tricky

To see if there are other isomers we start again, this time locating two NH_3 molecules *cis* to each other (Fig. 16.8c). If we were to place the third NH_3 molecule at one of the other corners of the square, we would simply reproduce the first isomer. (Remember that the symmetry of a regular octahedron requires that the distance across the diagonal of the square be the same as that from "top" to "bottom.") We are left with two equivalent positions. Placing the third NH_3 molecule arbitrarily at the "top," we arrive at a second isomer (Fig. 16.8d), distinctly different from the first because all three NH_3 molecules are *cis* to one another.

Figure 16.8 Isomers of $Co(NH_3)_3Cl_3$ (Example 16.4). You may be tempted to write down additional structures, but you will find that they are equivalent to one of the two isomers shown here.

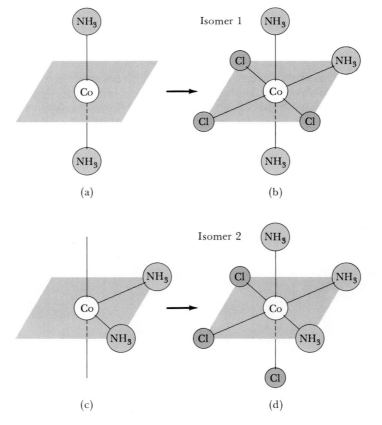

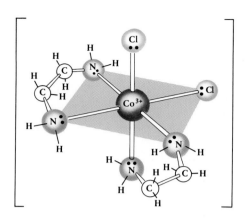

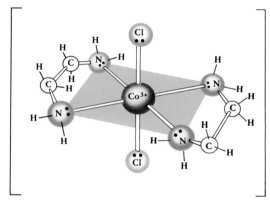

Figure 16.9 There are two isomers of the complex ion $Co(en)_2Cl_2^+$, which differ from one another in color (recall Fig. 16.7) and in certain chemical properties. The *cis* isomer reacts more readily with $C_2O_4^{2-}$ ions than does the *trans* isomer. (Why?) (photo, Charles D. Winters)

We have now exhausted in a logical manner all the possibilities for geometric isomerism, finding two isomers. There are no others.

Exercise
How many geometric isomers are there for $Cr(NH_3)_2Cl_4^-$? for the square planar complex $[Pt(NH_3)_2ClBr]$? Answer: Two; two.

Geometric isomerism can occur in chelated octahedral complexes (Fig. 16.9). Notice that an *ethylenediamine molecule*, here and indeed in all complexes, *can only bridge cis positions*. It is not long enough to connect to *trans* positions.

16.4
ELECTRONIC STRUCTURES OF COMPLEX IONS

To develop the electron distribution in a complex ion, a logical starting point is the electronic structure of the central metal. You may recall from Chapter 5 that in a transition metal cation, the inner d sublevel (for example, 3d) is lower in energy than the outer s sublevel (4s). From a slightly

TABLE 16.3   Electron Configurations of Transition Metal Atoms and Ions

ATOMIC NUMBER	CONFIGURATION OF ATOM		CONFIGURATION OF ION	
24	Cr	$[Ar]3d^54s^1$	Cr^{3+}	$[Ar]3d^3$
25	Mn	$[Ar]3d^54s^2$	Mn^{2+}	$[Ar]3d^5$
26	Fe	$[Ar]3d^64s^2$	Fe^{3+}	$[Ar]3d^5$
			Fe^{2+}	$[Ar]3d^6$
27	Co	$[Ar]3d^74s^2$	Co^{3+}	$[Ar]3d^6$
			Co^{2+}	$[Ar]3d^7$
28	Ni	$[Ar]3d^84s^2$	Ni^{2+}	$[Ar]3d^8$
29	Cu	$[Ar]3d^{10}4s^1$	Cu^{2+}	$[Ar]3d^9$
			Cu^+	$[Ar]3d^{10}$
30	Zn	$[Ar]3d^{10}4s^2$	Zn^{2+}	$[Ar]3d^{10}$

different point of view, we might say that, in forming a transition metal cation such as Cr^{3+} or Cu^{2+}, the outer s electrons are lost first. Any way you look at it, the following statement is valid:

There are no outer s electrons in a transition metal cation. Electrons beyond the preceding noble gas are located in an inner d sublevel.

This principle is illustrated in Table 16.3. If these electron configurations do not seem familiar to you, you may want to go back to Chapter 5 and review the structures of these ions.

With this background, we will look at two of the simpler models of the electronic structure of complex ions. One of these is the valence bond (atomic orbital) model, discussed in Chapter 9. The other is referred to as the crystal-field model.

VALENCE BOND (ATOMIC ORBITAL) MODEL

This model, for complex ions, starts with a simple assumption. It considers that electron pairs donated by ligands enter hybrid orbitals of the central metal. The particular orbitals that these electron pairs enter depend upon the coordination number and geometry of the complex. The total number of hybrid orbitals occupied by ligand electrons is equal to the coordination number.

The hybrid orbitals occupied by ligand electrons are listed in Table 16.4. Of these, the **sp**, **sp³**, and **sp³d²** sets were discussed in Chapter 9. You will recall that sp hybridization leads to linear geometry, while sp³ orbitals are directed tetrahedrally. In square planar complexes, the ligand electrons are accommodated by hybridizing an inner d orbital, an s orbital, and two p orbitals; hence the notation "**dsp²**." In the first transition series, these are a 3d, a 4s, and two 4p orbitals.

Two somewhat different hybridizations are possible for octahedral complexes. If inner d orbitals are involved (e.g., two 3d, one 4s, and three 4p orbitals), we get **d²sp³** hybridization. Octahedral complexes of this type, such as $Fe(CN)_6^{4-}$, are often referred to as *inner complexes*. In the $Fe(H_2O)_6^{2+}$ ion, outer d orbitals are hybridized (one 4s, three 4p, and two 4d orbitals).

The geometries of the hybrids are the same as in regular molecules

TABLE 16.4 Valence Bond Model Applied to Complex Ions

COORDINATION NUMBER	GEOMETRY	HYBRID ORBITALS OCCUPIED BY LIGAND ELECTRONS	EXAMPLE
2	linear	sp	$Cu(NH_3)_2{}^+$
4	tetrahedral	sp^3	$Zn(NH_3)_4{}^{2-}$
4	square planar	dsp^2	$Ni(CN)_4{}^{2-}$
6	octahedral	d^2sp^3	$Fe(CN)_6{}^{4-}$
6	octahedral	sp^3d^2	$Fe(H_2O)_6{}^{2+}$

Complexes of this type, where the hybridization is **sp^3d^2**, are called *outer complexes*.

Figure 16.10 shows the orbital diagrams of the complex ions listed in Table 16.4. Here, the hybrid orbitals are enclosed by horizontal lines. The bonding electrons within these orbitals, all of which are contributed by the ligands, are shown as colored arrows. The electrons shown in black are those contributed by the metal itself; none of these are involved in bonding. To save space, only those electrons beyond the argon core are shown; the inner 18 electrons of each metal are omitted.

The two complexes shown at the bottom of Figure 16.10 are both derived from the Fe^{2+} ion, which has the outer electron configuration $3d^6$. Notice, however, that the $Fe(CN)_6{}^{4-}$ and $Fe(H_2O)_6{}^{2+}$ ions differ in the number of unpaired electrons. In $Fe(H_2O)_6{}^{2+}$, where the inner d orbitals are not involved in bonding, the six electrons associated with Fe^{2+} are spread out according to Hund's rule, giving four unpaired electrons. The situation with $Fe(CN)_6{}^{4-}$ is quite different. Here, two inner d orbitals are used in bonding, so the six electrons of Fe^{2+} are crowded into the remaining three 3d orbitals. As a result, the $Fe(CN)_6{}^{4-}$ ion has no unpaired electrons.

In general, if a species such as Fe^{2+} forms both inner and outer complexes, the two types will differ in the number of unpaired electrons. The outer complex (sp^3d^2) will have a larger number of unpaired electrons than

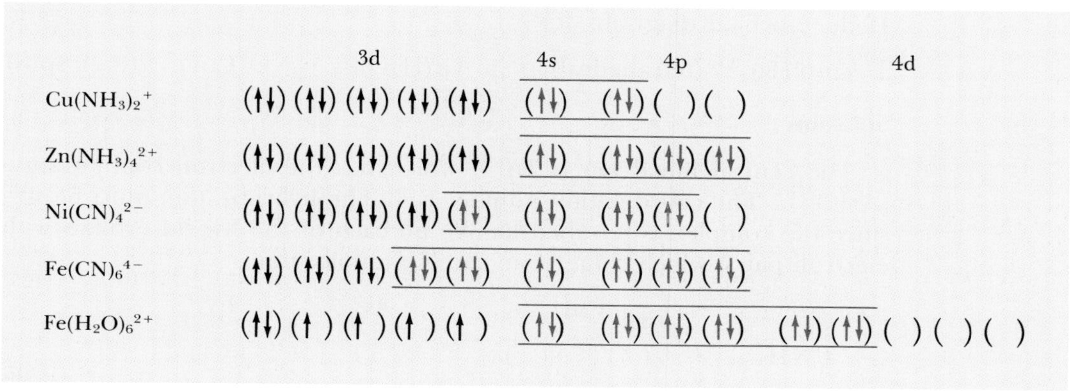

The bonding electrons in the hybrids are all furnished by the ligands

Figure 16.10 Orbital diagrams derived from the valence bond model for complexes of Cu^+ (linear), Zn^{2+} (tetrahedral), Ni^{2+} (square planar), and Fe^{2+} (octahedral). Electron pairs furnished by the ligands are shown in red.

the inner complex (d^2sp^3). The phrases "**high spin**" and "**low spin**" are sometimes used to distinguish between these two types of complexes: $Fe(H_2O)_6^{2+}$ is a high-spin complex (4 unpaired electrons), while $Fe(CN)_6^{4-}$ is a low-spin complex (0 unpaired electrons). Whenever these terms are used, "high-spin" refers to the complex with the greater number of unpaired electrons.

Orbital diagrams of complex ions can be derived by following a series of simple steps:

1. *Determine the electron configuration of the central metal* (Table 16.3). To do this, you may have to first deduce its oxidation number, as in Example 16.1.

2. *Determine the coordination number* (2, 4, or 6).

3. *Decide upon the hybridization.* For coordination number 2, the hybridization is sp. If the coordination number is 4, you will need further information to decide between dsp^2 and sp^3 hybridization. Most likely, you will be told the geometry: square planar (dsp^2) or tetrahedral (sp^3).

If the coordination number is 6, you can decide upon the hybridization using the following system:

  a. If the electron configuration of the central metal ion is d^0, d^1, d^2, or d^3, the hybridization is d^2sp^3. That is, inner d orbitals are involved in bonding.
  b. If the electron configuration of the central metal ion is d^4, d^5, d^6, or d^7, you will need further information to decide upon the hybridization. Most likely, you will be told whether the complex is low spin (d^2sp^3) or high spin (sp^3d^2).
  c. If the electron configuration of the central metal ion is d^8, d^9, or d^{10}, the hybridization is sp^3d^2.

4. *Place a pair of electrons* ($\uparrow\downarrow$) *in each of the hybrid orbitals found in Step 3.*

5. *Distribute the electrons of the central metal ion, as found in Step 1, among the available orbitals of lowest energy, following Hund's rule.*

■ **EXAMPLE 16.5** ───────────────────────────────────────

Deduce orbital diagrams for

a. $Zn(NH_3)_4^{2+}$ (tetrahedral)   b. $Cr(H_2O)_6^{3+}$ c. CoF_6^{3-} (high spin)

Solution

a. The central metal ion is Zn^{2+}, which has the electron configuration [Ar] $3d^{10}$. The coordination number is 4. The hybridization must be sp^3, since the complex is tetrahedral. We fill each of the hybrid orbitals with electron pairs, which come from the NH_3 molecules.

	3d	4s	4p
4 NH₃	( )( )( )( )( )	(↑↓)	(↑↓)(↑↓)(↑↓)

We distribute the ten outer electrons of Zn^{2+} (i.e., those beyond the argon configuration) among the 3d orbitals.

		3d	4s	4p

$Zn(NH_3)_4^{2+}$ [Ar] ($\uparrow\downarrow$)($\uparrow\downarrow$)($\uparrow\downarrow$)($\uparrow\downarrow$)($\uparrow\downarrow$) $\underline{(\uparrow\downarrow)}$ ($\uparrow\downarrow$)($\uparrow\downarrow$)($\uparrow\downarrow$)

b. The central metal ion is Cr^{3+}, electron configuration [Ar] $3d^3$. The coordination number is 6. Since we have only three d electrons, the hybridization must be d^2sp^3. The six pairs of bonding electrons are placed in these orbitals.

3d 4s 4p

6 H_2O ( )( )( )$\underline{(\uparrow\downarrow)(\uparrow\downarrow)}$ $\underline{(\uparrow\downarrow)}$ $\underline{(\uparrow\downarrow)(\uparrow\downarrow)(\uparrow\downarrow)}$

Putting the three outer electrons of Cr^{3+} in the available 3d orbitals, following Hund's rule, we obtain the orbital diagram of the complex.

3d 4s 4p

$Cr(H_2O)_6^{3+}$ [Ar] ($\uparrow$)($\uparrow$)($\uparrow$)$\underline{(\uparrow\downarrow)(\uparrow\downarrow)}$ $\underline{(\uparrow\downarrow)}$ $\underline{(\uparrow\downarrow)(\uparrow\downarrow)(\uparrow\downarrow)}$

c. The oxidation number of cobalt is +3; the central metal ion is Co^{3+} with the electron configuration $3d^6$. The coordination number is 6. Since this is a high-spin complex, the hybridization must be sp^3d^2. The orbital diagram is:

3d 4s 4p 4d

CoF_6^{3-} [Ar] ($\uparrow\downarrow$)($\uparrow$)($\uparrow$)($\uparrow$)($\uparrow$) $\underline{(\uparrow\downarrow)}$ $\underline{(\uparrow\downarrow)(\uparrow\downarrow)(\uparrow\downarrow)}$ $\underline{(\uparrow\downarrow)(\uparrow\downarrow)( )( )( )}$

Exercise
Write the orbital diagram for $Co(NH_3)_6^{3+}$ (low spin). Answer: Same as $Fe(CN)_6^{4-}$ (Fig. 16.10).

The valence bond model has been reasonably successful in explaining the magnetic behavior of complex ions. The structures shown in Figure 16.10 imply that $Cu(NH_3)_2^+$, $Zn(NH_3)_4^{2+}$, and $Ni(CN)_4^{2-}$ should all be *diamagnetic;* they have no unpaired electrons. In contrast, the $Cr(H_2O)_6^{3+}$ ion (Example 16.5), with three unpaired electrons, should be paramagnetic. These predictions are confirmed by experiment. The first three complexes are weakly repelled by a magnetic field. On the other hand, $Cr(H_2O)_6^{3+}$ is attracted into the field with a force corresponding to three unpaired electrons.

Paramagnetism is fairly common in complex ions

By introducing the idea of outer vs. inner complexes, the valence bond model can rationalize the existence of two different types of complexes with different numbers of unpaired electrons for an ion such as Fe^{2+}. However, it is by no means obvious why the CN^- ion should form one type of complex with Fe^{2+}, while the H_2O molecule forms a quite different type. More generally, the valence bond model does not offer a simple explanation for the fact that high-spin and low-spin complexes are restricted to ions with four to seven d electrons (d^4, d^5, d^6, d^7).

Figure 16.11 Most coordination compounds are brilliantly colored, a property that can be explained readily by the crystal-field model. (Marna G. Clarke)

In another area, valence bond theory has been notably deficient. It does not explain the most striking property of coordination compounds, their brilliant colors (Fig. 16.11). A quite different approach, known as the crystal-field model, has been much more successful here.

CRYSTAL-FIELD MODEL (OCTAHEDRAL COMPLEXES)

The crystal field model uses electro-static forces to hold the complex together

The crystal-field model assumes that the bonding in complexes such as $Fe(CN)_6^{4-}$ or $Fe(H_2O)_6^{2+}$ is primarily ionic rather than covalent. It is assumed that the only effect of the ligands is to create an electrostatic field around the central metal ion, which changes the relative energies of different d orbitals. In the isolated metal ion, before the ligands approach, all the d orbitals in a sublevel such as 3d have the same energy. After interaction, the d orbitals are split into two groups with different energies.

Although the crystal-field model can be applied to any type of complex, our discussion will be limited to octahedral complexes. To be specific, let us consider the formation of the octahedral $Fe(CN)_6^{4-}$ ion:

$$Fe^{2+} + 6CN^- \rightarrow Fe(CN)_6^{4-}$$

In the free ($3d^6$) Fe^{2+} ion, all the 3d orbitals are equal in energy. Hence, the six 3d electrons are distributed in accordance with Hund's rule:

$$3d$$
$$Fe^{2+} \quad (\uparrow\downarrow)(\uparrow)(\uparrow)(\uparrow)(\uparrow)$$

To form the octahedral complex, the six CN^- ions must approach along the x, y, and z axes (Fig. 16.12a). This causes a splitting of the five 3d orbitals into two sets:

1. A higher energy pair, referred to as $d_{x^2-y^2}$ and d_{z^2} orbitals.
2. A lower energy trio, called the d_{xy}, d_{yz}, and d_{xz} orbitals (Fig. 16.12b).

To understand why the splitting occurs, we look at the orientation of the electron density clouds associated with the five 3d orbitals (Fig. 16.13).

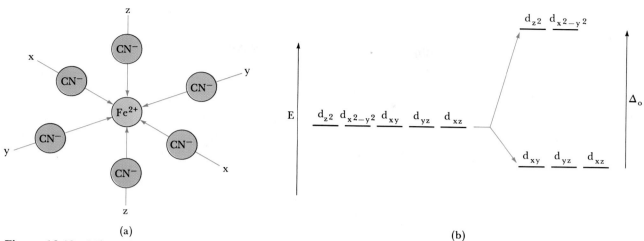

Figure 16.12 When six ligands (CN^- ions) approach a central metal ion (Fe^{2+}) along the x, y, and z axes, the five orbitals are split into two groups. Two of these are raised in energy, while the other three are lowered. The difference in energy between the two groups is the crystal-field splitting energy, Δ_o.

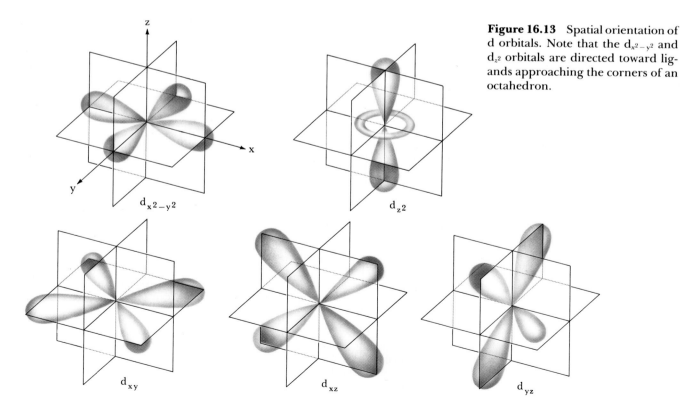

Figure 16.13 Spatial orientation of d orbitals. Note that the $d_{x^2-y^2}$ and d_{z^2} orbitals are directed toward ligands approaching the corners of an octahedron.

Notice that the two orbitals called $d_{x^2-y^2}$ and d_{z^2} have their maximum electron density directly along the x, y, and z axes, respectively. In contrast, the other d orbitals have their electron densities concentrated between the axes rather than along them. Hence, as CN^- ligands approach along the x, y, and z axes, their electrostatic field is felt more strongly by electrons in $d_{x^2-y^2}$ and d_{z^2} orbitals than by those in the d_{xy}, d_{yz}, and d_{xz} orbitals. The effect is to split the 3d orbitals into two sets of differing energy. The energy difference is known as the *crystal-field splitting energy*, Δ_0.

With $Fe(CN)_6^{4-}$, Δ_0 is *large enough* to cause the six 3d electrons of Fe^{2+} to *pair up* in the three lower energy orbitals (Fig. 16.14a). In the $Fe(H_2O)_6^{2+}$ ion, the situation is quite different. The splitting energy, Δ_0, is *smaller*. Indeed, it is *not* large enough to overcome the tendency for electrons to remain unpaired. In the $Fe(H_2O)_6^{2+}$ ion, as in the Fe^{2+} ion itself, the six 3d electrons are spread out over all orbitals (Fig. 16.14b).

Looking at Figure 16.14, we see that the crystal-field model explains why $Fe(CN)_6^{4-}$ is diamagnetic, while $Fe(H_2O)_6^{2+}$ is paramagnetic. In the first complex there are no unpaired electrons; in the second complex there are four. More generally, the crystal-field model explains in a simple way why certain transition metal ions can form both high-spin and low-spin complexes. Depending upon the nature of the ligand:

1. Δ_0 may be relatively large, in which case electrons pair up in the lower energy orbitals, giving a low-spin complex. This occurs with so-called "strong-field" ligands, e.g., CN^-.

2. Δ_0 may be relatively small, in which case electrons spread out over all the d orbitals in accordance with Hund's rule, giving a high-spin complex. Such complexes are formed with "weak-field" ligands, e.g., H_2O.

When Δ_0 is small, the electrons are located as in the free metal ion

Δ_0

When Δ_0 is large, the electrons don't have enough energy to reach the upper orbitals, so have to pair up in the lower ones

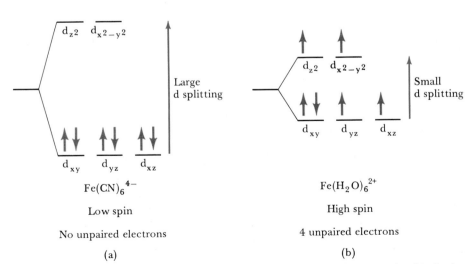

Figure 16.14 In the $Fe(CN)_6^{4-}$ ion, the energy difference between d orbitals, Δ_0, is large. Hence, the six 3d electrons of the Fe^{2+} ion pair up in the three orbitals of lower energy. In the $Fe(H_2O)_6^{2+}$ ion, Δ_0 is small, and the electrons spread out among the five orbitals with a maximum number (four) of unpaired electrons.

■ **EXAMPLE 16.6** _____

Using the crystal-field model, derive the structure of the Co^{2+} ion in low-spin and high-spin octahedral complexes.

Solution

We first determine the number of d electrons available. Since the atomic number of cobalt is 27, there are $27 - 2 = 25$ electrons in Co^{2+}. Of these, 18 are accounted for by the argon core, leaving 7 in the 3d orbitals. In a low-spin complex, formed with strongly interacting ligands, 6 of these electrons are crowded into the lower three energy levels, leaving only 1 unpaired electron. In a high-spin complex, where the splitting is small, Hund's rule is followed and there are 3 unpaired electrons:

$$(\uparrow)(\ ) \qquad\qquad (\uparrow)(\uparrow)$$
$$(\uparrow\downarrow)(\uparrow\downarrow)(\uparrow\downarrow) \qquad (\uparrow\downarrow)(\uparrow\downarrow)(\uparrow)$$
$$\text{low-spin} \qquad\qquad \text{high-spin}$$

Both types of complexes are known. The $Co(CN)_6^{4-}$ ion is of the low-spin type; $Co(H_2O)_6^{2+}$ is a high-spin complex.

Exercise

Consider a transition metal ion with four 3d electrons, such as Cr^{2+}. How many unpaired electrons are there in a high-spin complex of this ion? In a low-spin complex? Answer: Four; two.

_____

The crystal-field model offers a simple explanation of the fact that high-spin and low-spin complexes occur only with ions that have 4 to 7 d electrons (d^4, d^5, d^6, d^7). With three or fewer electrons, only one distribution is possible; the same is true with eight or more electrons.

$$(\ )(\ ) \qquad\qquad (\uparrow)(\uparrow)$$
$$(\uparrow)(\uparrow)(\uparrow) \qquad (\uparrow\downarrow)(\uparrow\downarrow)(\uparrow\downarrow)$$
$$\text{3 electrons} \qquad\qquad \text{8 electrons}$$

Crystal-field theory explains the color of complex ions in terms of electron transitions between energy levels. The energy difference between two sets of d orbitals in a complex is, in many cases, equivalent to a wavelength in the visible region. Hence, by absorbing visible light, an electron may be able to move from the lower energy set of d orbitals to the higher one. This removes some of the component wavelengths of white light, so that the light reflected or transmitted by the complex is colored.

The color of an object is that of the light that is transmitted or reflected

The situation is particularly simple in $_{22}Ti^{3+}$, where there is only one 3d electron. Consider, for example, the $Ti(H_2O)_6^{3+}$ ion, which has an intense purple color. This ion absorbs at 510 nm, in the green region. The purple (red-blue) color that we see when we look at a solution of $Ti(H_2O)_6^{3+}$

TABLE 16.5   Colors of Complex Ions of Co^{3+}

COMPLEX	COLOR OBSERVED	COLOR ABSORBED	APPROXIMATE WAVELENGTH (nm) ABSORBED
$Co(NH_3)_6^{3+}$	yellow	violet	430
$Co(NH_3)_5NCS^{2+}$	orange	blue	470
$Co(NH_3)_5H_2O^{3+}$	red	blue-green	500
$Co(NH_3)_5Cl^{2+}$	purple	yellow-green	530
trans-$Co(NH_3)_4Cl_2^+$	green	red	680

is what is left when the green component is subtracted from the visible spectrum. Using Equation 5.2 (Chapter 5), we can calculate the energy difference, ΔE, corresponding to a wavelength of 510 nm:

$$\Delta E = \frac{1.196 \times 10^5 \text{ kJ} \cdot \text{nm}}{\lambda} \frac{}{\text{mol}} = \frac{1.196 \times 10^5}{5.10 \times 10^2} \frac{\text{kJ}}{\text{mol}} = 234 \frac{\text{kJ}}{\text{mol}}$$

We conclude that this is the energy absorbed to raise the 3d electron from a lower to a higher orbital. In other words, in the $Ti(H_2O)_6^{3+}$ ion, the two sets of d orbitals are separated by this amount of energy. The splitting energy, Δ_0, is 234 kJ/mol.

The magnitude of the splitting energy, Δ_0, determines the wavelength of the light absorbed by a complex and hence its color. This effect is shown in Table 16.5 and Figure 16.15. Notice that when we substitute for NH_3 such ligands as NCS^-, H_2O, or Cl^-, which produce smaller values of Δ_0, the light absorbed shifts to longer wavelengths (lower energies). On the basis of these and other observations, we can arrange various ligands in order of decreasing tendency to split the d orbitals. A short version of such a *spectrochemical* series is

$$CN^- > NO_2^- > en > NH_3 > NCS^- > H_2O > F^- > Cl^-$$

$$\xrightarrow{\hspace{3cm}}$$

strong field   decreasing Δ_0 weak field

Figure 16.15   Color of complex ions of Co^{3+}. The five compounds form a spectrochemical series (Table 16.5)

As we move from left to right in the series, Δ_0 decreases, the wavelength of light absorbed increases, and there is an increasing tendency to form high-spin complexes.

Lest you suppose that crystal-field theory can explain all the properties of complex ions, we should point out one of its weaknesses. If the bonding in complex ions is primarily electrostatic, it is hard to see why certain molecules that are only slightly polar, such as CO, can act as ligands. To explain this, it is necessary to modify crystal-field theory to take into account covalent as well as ionic bonding. This is done in a more sophisticated approach known as *ligand-field theory*, which you may study in later chemistry courses.

16.5
FORMATION CONSTANTS OF COMPLEX IONS

For reactions involving complex ions in water solution (as indeed with all reactions), we must be concerned with the position of the equilibrium involved. We will consider the equilibrium constant for the formation of the complex ion, called the **formation constant** (or the stability constant) and given the symbol K_f. A typical example is

$$Cu^{2+}(aq) + 4NH_3(aq) \rightleftharpoons Cu(NH_3)_4{}^{2+}(aq); \; \mathbf{K_f} = \frac{\mathbf{[Cu(NH_3)_4{}^{2+}]}}{\mathbf{[Cu^{2+}] \times [NH_3]^4}} \quad (16.2)$$

Table 16.6, p. 570, lists formation constants of complex ions. In each case, K_f applies to the formation of the complex by a reaction similar to Reaction 16.2. Note that for each complex ion listed, K_f is a large number, 10^5 or greater. This means that equilibrium considerations strongly favor complex formation. Consider, for example, the system

$$Ag^+(aq) + 2NH_3(aq) \rightleftharpoons Ag(NH_3)_2{}^+(aq);$$

$$K_f = \frac{[Ag(NH_3)_2{}^+]}{[Ag^+] \times [NH_3]^2} = 1.7 \times 10^7$$

We interpret the large K_f value to mean that the forward reaction goes virtually to completion. Addition of ammonia to a solution of $AgNO_3$ will convert nearly all of the Ag^+ ions to the $Ag(NH_3)_2{}^+$ complex. From a slightly different point of view, we might say that since K_f is large, there will be very little tendency for the reverse reaction to occur. The $Ag(NH_3)_2{}^+$ ion is very stable toward decomposition to Ag^+ ions and NH_3 molecules.

The stabilities of different complexes of the same central metal ion are directly related to their formation constants: the larger the K_f value, the more stable the complex. For the reaction

$$Ag^+(aq) + 2S_2O_3{}^{2-}(aq) \rightleftharpoons Ag(S_2O_3)_2{}^{3-}(aq);$$

$$K_f = \frac{[Ag(S_2O_3)_2{}^{3-}]}{[Ag^+] \times [S_2O_3{}^{2-}]^2} = 1 \times 10^{13}$$

TABLE 16.6   Formation Constants of Complex Ions

COMPLEX ION	K_f	COMPLEX ION	K_f
$AgCl_2^-$	1.8×10^5	$CuCl_2^-$	1×10^5
$Ag(CN)_2^-$	2×10^{20}	$Cu(NH_3)_4^{2+}$	2×10^{12}
$Ag(NH_3)_2^+$	1.7×10^7	$FeSCN^{2+}$	9.2×10^2
$Ag(S_2O_3)_2^{3-}$	1×10^{13}	$Fe(CN)_6^{3-}$	4×10^{52}
$Al(OH)_4^-$	1×10^{33}	$Fe(CN)_6^{4-}$	4×10^{45}
$Cd(CN)_4^{2-}$	2×10^{18}	$Hg(CN)_4^{2-}$	2×10^{41}
$Cd(NH_3)_4^{2+}$	2.8×10^7	$Ni(NH_3)_6^{2+}$	9×10^8
$Cd(OH)_4^{2-}$	1.2×10^9	$PtCl_4^{2-}$	1×10^{16}
$Co(NH_3)_6^{2+}$	1×10^5	$t\text{-}Pt(NH_3)_2Cl_2$	3×10^{28}
$Co(NH_3)_6^{3+}$	1×10^{23}	$c\text{-}Pt(NH_3)_2Cl_2$	3×10^{29}
$Co(NH_3)_5Cl^{2+}$	2×10^{28}	$Zn(CN)_4^{2-}$	6×10^{16}
$Co(NH_3)_5NO_2^{2+}$	1×10^{24}	$Zn(NH_3)_4^{2+}$	3.6×10^8
		$Zn(OH)_4^{2-}$	3×10^{14}

This K_f is larger than that for $Ag(NH_3)_2^+$ ($K_f = 1.7 \times 10^7$). Hence, the $Ag(S_2O_3)_2^{3-}$ complex ion is more stable than $Ag(NH_3)_2^+$.

■ **EXAMPLE 16.7** _____

Determine the ratios

a. $[Ag(NH_3)_2^+]/[Ag^+]$ in 0.10 M NH_3
b. $[Ag(S_2O_3)_2^{3-}]/[Ag^+]$ in 0.10 M $S_2O_3^{2-}$

Solution
In each case, we write down the expression for K_f and solve for the desired ratio,

a. $\quad K_f = 1.7 \times 10^7 = \dfrac{[Ag(NH_3)_2^+]}{[Ag^+] \times [NH_3]^2}$

$\dfrac{[Ag(NH_3)_2^+]}{[Ag^+]} = K_f \times [NH_3]^2 = 1.7 \times 10^7(0.10)^2 = 1.7 \times 10^5$

Essentially all the silver is complexed   This means that in 0.10 M NH_3 there are 170,000 $Ag(NH_3)_2^+$ complex ions for every Ag^+ ion.

b. Proceeding as in (a),

$\dfrac{[Ag(S_2O_3)_2^{3-}]}{[Ag^+]} = K_f \times [S_2O_3^{2-}]^2 = 1 \times 10^{13}(0.1)^2 = 1 \times 10^{11}$

Note that this ratio is much higher than the comparable ratio for the $Ag(NH_3)_2^+$ complex. This is really what we mean when we say that the $Ag(S_2O_3)_2^{3-}$ complex is "more stable" than $Ag(NH_3)_2^+$.

Exercise
At what concentration of NH_3 is $[Ag(NH_3)_2^+] = [Ag^+]$? Answer: 2.4×10^{-4} M.

16.6
USES OF COORDINATION COMPOUNDS

Coordination compounds are used for a variety of purposes. In the laboratory, you frequently bring water-insoluble compounds into solution through complex ion formation. Examples include silver chloride and aluminum hydroxide.

$$\text{AgCl}(s) + 2\text{NH}_3(aq) \rightarrow \text{Ag(NH}_3)_2{}^+(aq) + \text{Cl}^-(aq) \qquad (16.3)$$

$$\text{Al(OH)}_3(s) + \text{OH}^-(aq) \rightarrow \text{Al(OH)}_4{}^-(aq) \qquad (16.4)$$

The stabilities of the complex ions $\text{Ag(NH}_3)_2{}^+$ and $\text{Al(OH)}_4{}^-$ are great enough to make these reactions spontaneous, even though AgCl and Al(OH)_3 are very insoluble in water. Indeed, the $\text{Al(OH)}_4{}^-$ ion is so stable ($K_f = 1 \times 10^{33}$) that aluminum metal dissolves in a solution of a strong base. The reaction is

Eqn. 16.3 affords us an easy way to dissolve AgCl

$$2\text{Al}(s) + 2\text{OH}^-(aq) + 6\text{H}_2\text{O} \rightarrow 2\text{Al(OH)}_4{}^-(aq) + 3\text{H}_2(g) \qquad (16.5)$$

This reaction has recently found practical application in the addition of finely divided aluminum to drain cleaners containing lye, NaOH. The bubbles of hydrogen formed are supposed to loosen deposits of grease or dirt.

Don't cook up some lye solution in your mother's favorite aluminum pot

In many natural products, a transition metal ion is tied up in a coordination complex. Compounds of this type include hemoglobin (Fe^{2+}), chlorophyll (Mg^{2+}), and vitamin B_{12} (Co^{2+}). Other coordination compounds, made in the laboratory, are used in medicine. In the remainder of this section, we will look at a few of the biological uses of coordination compounds.

HEMOGLOBIN AND OXYGEN TRANSPORT

The structure of heme, the pigment responsible for the color of blood, is shown in Figure 16.16. Notice that there is an Fe^{2+} ion at the center bonded to four nitrogen atoms at the corners of a square. Actually, the Fe^{2+} ion is part of an octahedral complex. A fifth ligand, not shown in the figure, is a large organic molecule called globin. In combination with heme, this gives the protein we refer to as hemoglobin. The sixth ligand, filling out the octahedron, is an H_2O molecule.

The water molecule in hemoglobin can be replaced reversibly by an O_2 molecule. The product formed is called oxyhemoglobin; it has a bright red color. In contrast, hemoglobin itself is blue. This explains why arterial blood is red (high concentration of O_2), while that in the veins is bluish (low concentration of O_2).

Nature makes good use of complexes

$$\text{hemoglobin·H}_2\text{O}(aq) + \text{O}_2(aq) \rightleftharpoons \text{hemoglobin·O}_2(aq) + \text{H}_2\text{O} \qquad (16.6)$$

blue

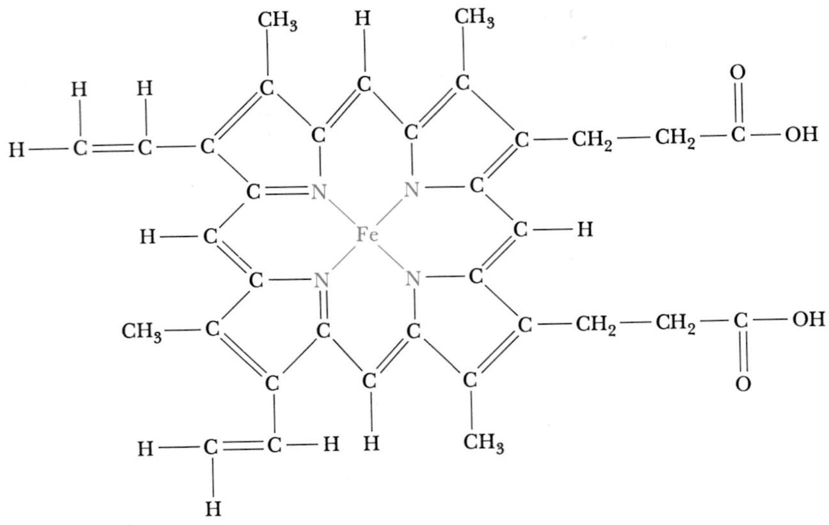

Figure 16.16 Structure of heme. The Fe^{2+} ion is at the center of an octahedron, surrounded by four nitrogen atoms, a globin molecule, and a water molecule.

The equilibrium in Reaction 16.6 is sensitive to the concentration of oxygen. In the lungs, where there is a large amount of O_2 available, the equilibrium shifts to the right. Oxygen in the form of the hemoglobin complex is taken up by red blood cells and carried to the tissues. There, where the concentration of dissolved O_2 is low, the reverse reaction occurs. This supplies the oxygen required for metabolism.

EDTA AND LEAD POISONING

EDTA, whose structure was shown in Figure 16.2, is a very effective chelating agent. It forms complexes with a large number of cations, including those of some of the main-group metals. The complex formed by calcium with EDTA is used to treat lead poisoning. When a solution containing the Ca–EDTA complex is given by injection, the calcium is displaced by lead. The more stable Pb–EDTA complex is eliminated in the urine. EDTA has also been used to remove radioactive isotopes of metals, notably plutonium, from body tissues.

So far, no one has claimed that beer is a cure for lead poisoning

You may have noticed Ca–EDTA on the list of ingredients of many prepared foods, ranging from beer to mayonnaise. Here, EDTA acts as a scavenger to pick up traces of metal ions that catalyze the chemical reactions responsible for flavor deterioration, loss of color, or rancidity. Typically, Ca–EDTA or Na–EDTA is added at a level of 30 to 800 ppm.

COORDINATION COMPOUNDS IN THE TREATMENT OF CANCER

The neutral complex *cis*-[Pt(NH$_3$)$_2$Cl$_2$] is an effective antitumor agent in cancer treatment (chemotherapy). The *trans* isomer is ineffective. The activity of the *cis* isomer is believed to reflect the ability of the two Cl atoms to interact with DNA, a molecule responsible for cell reproduction. A reaction occurs in which the Cl atoms are replaced by N atoms of the DNA molecule. The product is a Pt^{2+} complex that inhibits further cell growth (Fig. 16.17). This reaction cannot occur with the *trans* isomer, since the Cl atoms are too far apart.

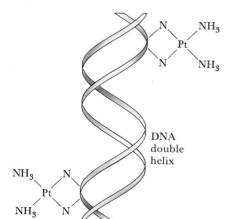

Figure 16.17 *cis*-$[Pt(NH_3)_2Cl_2]$ is effective in chemotherapy because of its ability to react with nitrogen atoms on DNA molecules, apparently causing the double helix to unwind. The *trans* isomer is ineffective.

■ SUMMARY PROBLEM

Consider the complex ion $Co(en)_2NH_3Cl^{2+}$

a. Identify the ligands and their charges.
b. What is the oxidation number of cobalt?
c. What is the name of this ion? The name and formula of its sulfate salt?
d. Suppose the ethylenediamine molecules were replaced by OH^- ions. Write the formula of the resulting complex, and the formula and name of the potassium salt of that complex.
e. What is the coordination number of cobalt in the complex?
f. Describe the geometry of the complex.
g. How many possible geometric isomers are there?
h. Give the electron configuration of cobalt(III).
i. Using the valence bond model, draw the orbital diagram of the complex (low spin).
j. Using the crystal-field model, show the electron distribution in the high-spin and low-spin complexes of cobalt(III).
k. If all the ligands in the complex were NH_3 molecules, the complex ion would be $Co(NH_3)_6{}^{3+}$. Write the expression for K for the formation of this complex. Taking $K_f = 1 \times 10^{23}$, calculate $[Co^{3+}]$ in a solution in which $[Co(NH_3)_6{}^{3+}] = [NH_3] = 0.10\ M$.
l. Use the Reciprocal Rule and the Rule of Multiple Equilibria to calculate K for the reaction

$$Co(OH)_3(s) + 6NH_3(aq) \rightleftharpoons Co(NH_3)_6{}^{3+}(aq) + 3OH^-(aq)$$

$K_{sp}\ Co(OH)_3 = 2 \times 10^{-43}$. Is this reaction likely to go to completion?

Answers

a. en = 0; NH_3 = 0; Cl = -1 **b.** $+3$

c. amminechlorobis(ethylenediamine)cobalt(III);
amminechlorobis(ethylenediamine)cobalt(III) sulfate; $[Co(en)_2NH_3Cl]SO_4$

d. $Co(NH_3)Cl(OH)_4{}^{2-}$; $K_2[Co(NH_3)Cl(OH)_4]$ potassium ammine-
chlorotetrahydroxocobaltate(III)   **e.** 6   **f.** octahedral   **g.** two

h. $[Ar]3d^6$

3d	4s	4p

i.   (↑↓)(↑↓)(↑↓) (↑↓)(↑↓)   (↑↓)   (↑↓)(↑↓)(↑↓)

j.   low spin: ( )( )   high spin: (↓)(↓)

(↑↓)(↑↓)(↑↓)   (↑↓)(↓)(↓)

k. $K_f = \dfrac{[Co(NH_3)_6{}^{3+}]}{[Co^{3+}] \times [NH_3]^6}$; $[Co(NH_3)_6{}^{3+}] = 1 \times 10^{-18}\,M$

l. $K = 2 \times 10^{-20}$; No

The basic ideas concerning the structure and geometry of complex ions presented in this chapter were developed by one of the most gifted individuals in the history of inorganic chemistry, Alfred Werner. His theory of coordination chemistry was published in 1893 when Werner was 26 years old, holding the equivalent of an associate professorship at the University of Zurich. In his paper Werner made the revolutionary suggestion that metal ions such as Co^{3+} could show two different kinds of valences. For the compound $Co(NH_3)_6Cl_3$, Werner postulated a central Co^{3+} ion joined by "primary valences" (ionic bonds) to three Cl^- ions and by "secondary valences" (coordinate covalent bonds) to six NH_3 molecules. Moreover, he made the inspired guess that the six secondary valences were directed toward the corners of a regular octahedron. In Table 16.7 we list the familiar Werner structures for the series of compounds $Co(NH_3)_xCl_3$, where $x = 6, 5, 4$, or 3. All these compounds were known at the time, and many of their properties had been established. In particular, it was known from conductivity studies and by precipitation with $AgNO_3$ that the first three members of the series yielded 3, 2, and 1 mol Cl^-, respectively, when dissolved in water. This evidence was, of course, in complete agreement with Werner's theory.

Werner's structures, which seem so obvious to us today, aroused little enthusiasm among his contemporaries. The opposition was led by Sophus Mads Jorgensen, a 56-year-old professor of chemistry at the University of Copenhagen. Jorgensen was convinced that Co^{3+} could form no more than three bonds. To rationalize the existence of "addition compounds" of $CoCl_3$ containing six, five, four, or three NH_3 molecules, he invoked the chain structures shown in Table 16.7, in which NH_3 molecules are

HISTORICAL
PERSPECTIVE

ALFRED WERNER AND
SOPHUS MADS
JORGENSEN

The chloride ions in the complex will not precipitate with $AgNO_3$

TABLE 16.7   Structure and Properties of the Compounds $Co(NH_3)_xCl_3$

x	STRUCTURE		MOLES Cl⁻ PER MOLE COMPOUND		
	WERNER	JORGENSEN	WERNER	JORGENSEN	OBSERVED
6	$[Co(NH_3)_6]^{3+}$, 3 Cl⁻	Co with NH₃—Cl, NH₃—NH₃—NH₃—NH₃—Cl, NH₃—Cl	3	3	3
5	$[Co(NH_3)_5Cl]^{2+}$, 2 Cl⁻	Co with Cl, NH₃—NH₃—NH₃—NH₃—Cl, NH₃—Cl	2	2	2
4	$[Co(NH_3)_4Cl_2]^{+}$, Cl⁻	Co with Cl, NH₃—NH₃—NH₃—NH₃—Cl, Cl	1	1	1
3	$[Co(NH_3)_3Cl_3]^{0}$	Co with Cl, NH₃—NH₃—NH₃—Cl, Cl	0	1	?

linked together much like CH_2 groups in hydrocarbons. The differing extents of ionization of these compounds in water were explained by assuming that only those chlorine atoms bonded to NH_3 groups could ionize. Chlorines attached directly to cobalt were supposed to be held so tightly that they could not ionize in water.

From the data in Table 16.7 it would appear that the controversy between Werner and Jorgensen could have been settled quite simply by studying the behavior in water solution of the compound $Co(NH_3)_3Cl_3$. Werner's structure required that this species be a nonelectrolyte with no ionizable chlorine. In contrast, the chain structure of Jorgensen implied one ionizable chlorine, i.e., a 1:1 electrolyte similar to NaCl. Unfortunately, the evidence was ambiguous: at 25°C, a water solution of $Co(NH_3)_3Cl_3$ has a conductivity intermediate between that of a nonelectrolyte and a 1:1 salt.

Werner's assumption of octahedral coordination around the Co^{3+} ion offered further opportunities for testing his ideas against those of Jorgensen. If Werner were correct, there should be two isomeric forms (*cis* and *trans*) of the compound $Co(NH_3)_4Cl_3$. At the time, only one compound of this formula was known. All of Werner's early attempts to prepare a second isomer failed, thereby weakening his position.

As the years passed, the weight of evidence began to shift toward Werner's structures. Studies at 0°C gave a very low value for the conductivity of $Co(NH_3)_3Cl_3$, which tended to support Werner's contention that the anomalous conductivity at 25°C was due to the reaction

$$Co(NH_3)_3Cl_3(s) + H_2O \rightarrow [Co(NH_3)_3(H_2O)Cl_2]^+(aq) + Cl^-(aq)$$

Moreover, Werner showed that the compound $Co(NH_3)_3(NO_2)_3$, which is entirely analogous to $Co(NH_3)_3Cl_3$, behaves as a true nonelectrolyte even at 25°C.

In 1907 Werner, after years of effort, finally succeeded in preparing a second isomer of the compound $Co(NH_3)_4Cl_3$. Jorgensen graciously accepted this new evidence as conclusive proof of Werner's structures, and the chain theory of coordination chemistry faded away. Six years later, in 1913, Alfred Werner received the Nobel Prize in chemistry.

QUESTIONS AND PROBLEMS

Composition of Complex Ions and Coordination Compounds

1. Consider the complex ion $[Co(en)_2(SCN)Cl]^+$.
 a. Identify the ligands and their charges.
 b. What is the oxidation number of cobalt?
 c. What is the formula of the sulfide salt of this ion?

2. Consider the complex ion $[Co(NH_3)_4Cl_2]^+$.
 a. Identify the ligands and their charges.
 b. What is the oxidation number of cobalt?
 c. What would be the formula and charge of the complex ion if the ammonia molecules were replaced by $C_2O_4^{2-}$ ions?

3. Pt^{2+} forms many complexes, among them those with the following ligands. Give the formula and charge of each complex.
 a. two ammonia molecules and one oxalate ion, $C_2O_4^{2-}$
 b. two ammonia molecules, one thiocyanate ion, SCN^- and one bromide ion

c. one ethylenediamine molecule and two nitrite ions, NO_2^-

4. Chromium(III) forms many complexes, among them those with the following ligands. Give the formula and charge of each chromium complex ion described below.
   a. two oxalate ions ($C_2O_4^{2-}$), and two water molecules
   b. five ammonia molecules and one sulfate ion
   c. one ethylenediamine molecule (en), two ammonia molecules, and two iodide ions

5. What is the coordination number of the central metal atom in the following complexes?
   a. $[Cu(en)_2(NH_3)_2]^{2+}$ b. $[Ni(CN)_4]^{2-}$
   c. $Ag(SCN)_2^-$ d. VCl_6^{4-}

6. What is the coordination number of the central metal atom in the following complexes?
   a. $[Fe(H_2O)_6]^{3+}$ b. $[Pt(NH_3)Br_3]^-$
   c. $[Ag(NH_3)_2]^+$ d. $[Co(en)_2(SCN)Cl]^+$

7. Refer to Table 16.2 to predict the formula of the complex formed by
   a. Co^{3+} with NH_3 b. Zn^{2+} with OH^-
   c. Ag^+ with CN^- d. Fe^{3+} with $C_2O_4^{2-}$

8. Refer to Table 16.2 to predict the formula of the complex formed by
   a. Ag^+ with en   b. Fe^{2+} with H_2O
   c. Zn^{2+} with CN^- d. Pt^{4+} with en

9. Which of the following would you expect to be effective chelating agents?
   a. CH_3CH_2OH
   b. $H_2N-(CH_2)_3-NH_2$
   c. HO—C(=O)—CH_2—C(=O)—OH
   d. PH_3

10. Classify the following ligands as monodentate, bidentate, etc.:
    a. $(CH_3)_3P$

    b. ⁻O—C(=O)—CH_2—N(—CH_2—C(=O)—O⁻)(—CH_2—C(=O)—O⁻)

    c. $H_2N-(CH_2)_2-NH-(CH_2)_2-NH_2$
    d. H_2O

Nomenclature

11. Write formulas for the following ions or compounds:
    a. Zinc hexachloroplatinate(IV)
    b. Dichlorobis(ethylenediamine)nickel(II)
    c. Diamminetriaquohydroxochromium(III) nitrate
    d. Ammonium pentachlorohydroxoferrate(III)

12. Write formulas for the following ions or compounds:
    a. Tetraammineaquochlorocobalt(III) chloride
    b. Pentaamminesulfatocobalt(III) bromide
    c. Potassium tetracyanonickelate(II)
    d. Pentaamminenitratocobalt(III)

13. Name the following compounds or ions:
    a. $Ru(NH_3)_5Cl^{2+}$ b. $Mn(NH_2CH_2CH_2NH_2)_3^{2+}$
    c. $K_2[PtCl_4]$ d. $[Cr(NH_3)_5I]I_2$

14. Name the following compounds or ions:
    a. $Na[Al(OH)_4]$ b. $[Co(C_2O_4)_2(H_2O)_2]^-$
    c. $[Ir(NH_3)_3Cl_3]$ d. $[Cr(en)(NH_3)_2Br_2]_2SO_4$

15. Name the compounds or ions in Question 5.

16. Name the compounds or ions in Question 6.

Geometry of Complex Ions

17. Sketch the geometry of
    a. *cis*-$Cu(H_2O)_2Br_4^{2-}$ b. $Zn(en)Cl_2$ (tetrahedral)
    c. *trans*-$Ni(NH_3)_2(en)_2^{2+}$ d. $Cu(C_2O_4)_3^{4-}$
    e. *trans*-$Ni(H_2O)_2Cl_2$

18. Sketch the geometry of
    a. $Ag(CN)_2^-$ b. $Co(H_2O)_2Cl_2$ (tetrahedral)
    c. *cis*-$Ni(H_2O)_2Cl_2$ d. *cis*-$Pt(en)_2Br_2^{2+}$
    e. *trans*-$Cr(H_2O)_4Cl_2^+$

19. The acetylacetonate ion ($acac^-$)

$$\left(\underset{CH_3-\overset{\overset{\displaystyle :O:}{\|}}{C}-CH=\overset{\overset{\displaystyle :\ddot{O}:}{|}}{C}-CH_3}{} \right)^-$$

forms complexes with many metal ions. Sketch the geometry of $Fe(acac)_3$.

20. The compound 1,2-diaminocyclohexane

(abbreviated "dech") is a ligand in the promising anticancer complex *cis*-$Pd(H_2O)_2(dech)^{2+}$. Sketch the geometry of this complex.

21. Which of the following octahedral complexes show geometric isomerism? If geometric isomers are possible, draw their structures.
    a. $[Cr(NH_3)_2(SCN)_4]^-$ b. $[Co(NH_3)_3(NO_2)_3]$
    c. $[Co(en)(NH_3)_2Cl_2]^+$

22. Follow the directions of Question 21 for
    a. $[Co(en)Cl_4]^-$ b. $[Co(en)_2ClBr]^+$
    c. $[Co(NH_3)_2(C_2O_4)_2]^-$

23. How many different octahedral complexes of Fe^{3+} can you write using only $C_2O_4^{2-}$ and/or Cl^- as ligands?

24. Draw as many structural formulas as possible for octahedral complexes in compounds of the formula $Fe(NH_3)_4Cl_2Br$.

Electronic Structure of Metal Ions

25. Give the electron configuration for
a. Ti^{3+} b. Cr^{2+} c. Ni^{3+}
d. Co^{2+} e. Ru^{4+}

26. Give the electron configuration for
a. Fe^{2+} b. V^{2+} c. Zn^{2+}
d. Cu^+ e. Nb^{2+}

27. Write an orbital diagram and determine the number of unpaired electrons in each species in Question 25.

28. Write an orbital diagram and determine the number of unpaired electrons in each species in Question 26.

Valence Bond Model

29. Using the valence bond model, draw orbital diagrams to indicate the electronic structure around the central metal atom in
a. $Cr(H_2O)_3Cl_3$ b. $Zn(H_2O)_3(OH)^+$ (tetrahedral)
c. *cis*-$Ni(H_2O)_2Cl_2$ d. $Co(en)_3{}^{2+}$ (high spin)

30. Using the valence bond model, draw orbital diagrams to indicate the electronic structure around the central metal atom in
a. $Cr(NH_3)_5Br^{2+}$
b. $Cu(H_2O)_3Cl^+$ (square planar)
c. $Zn(en)Cl_2$ (tetrahedral)
d. $Ag(SCN)_2{}^-$

31. Use the valence bond model and orbital diagrams to account for the fact that square planar Ni^{2+} complexes are diamagnetic, while the tetrahedral complexes are paramagnetic.

32. Use the valence bond model to show why the $CuCl_2{}^-$ ion is diamagnetic, while the $CuCl_4{}^{2-}$ ion is paramagnetic.

33. State the number of unpaired electrons you would expect to find in the tetrahedral complexes of the following ions, using the valence bond model:
a. Al^{3+} b. Ni^{2+} c. Cd^{2+}

34. State the number of unpaired electrons you would expect to find in the octahedral complexes of the following ions, using the valence bond model:
a. Al^{3+} b. Ni^{2+} c. Pt^{4+}
Which of these can form both high-spin and low-spin complexes?

35. Using valence bond theory, draw orbital diagrams showing the electron distribution around the central metal atom in each octahedral complex listed in Table 16.2, assuming low spin where there is a choice.

36. Follow the directions of Question 35, working with octahedral complexes, but assume high spin where there is a choice.

Crystal-Field Model

37. Using the crystal field model, give the electron distribution in low-spin and high-spin complexes of
a. Co^{3+} b. Mn^{2+}

38. Follow the direction of Question 37 for
a. Fe^{2+} b. Mn^{3+}

39. Using the crystal-field model, explain why Mn^{3+} forms high-spin and low-spin octahedral complexes but Mn^{4+} does not.

40. V^{3+} does not form high- and low-spin octahedral complexes. Use crystal-field theory to explain why this is so.

41. $Cr(CN)_6{}^{4-}$ is less paramagnetic than $Cr(H_2O)_6{}^{2+}$. Account for this using electron distribution and crystal-field theory.

42. Using the crystal-field model, account for the fact that $Co(NH_3)_6{}^{3+}$ is diamagnetic, while $CoF_6{}^{3-}$ is paramagnetic.

43. Give the number of unpaired electrons in octahedral complexes with strong-field ligands for
a. Mn^{3+} b. Co^{3+} c. Rh^{3+}
d. Ti^{2+} e. Mo^{2+}

44. For the species in Question 43 indicate the number of unpaired electrons with weak-field ligands.

45. The 460-nm absorption in $MnF_6{}^{2-}$ corresponds to its crystal-field splitting energy, Δ_0. Find Δ_0 in kJ/mol.

46. $Ti(NH_3)_6{}^{3+}$ has a d orbital electron transition where ΔE is 300 kJ/mol. Calculate the wavelength (nm) for this transition.

Formation Constants of Complex Ions

47. Using the data in Table 16.6, calculate the ratio $[Zn^{2+}]/[Zn(OH)_4{}^{2-}]$ at pH 1.0, 7.0, and 10.0.

48. At what concentration of $S_2O_3{}^{2-}$ is 99% of the Ag^+ in a solution converted to $Ag(S_2O_3)_2{}^{3-}$?

49. At what concentration of ammonia is
a. $[Ag^+] = [Ag(NH_3)_2{}^+]$?
b. $[Ni^{2+}] = [Ni(NH_3)_6{}^{2+}]$?

50. At what concentration of cyanide ion is
a. $[Zn^{2+}] = 10^{-8} \times [Zn(CN)_4{}^{2-}]$?
b. $[Fe^{3+}] = 10^{-20} \times [Fe(CN)_6{}^{3-}]$?

51. Calculate K for the reaction

$$AgCl(s) + Cl^-(aq) \rightarrow AgCl_2{}^-(aq)$$

How would you expect the solubility of AgCl in 1 M HCl to compare to that in 1 M NH_3?

52. Calculate K for the reaction

$$CuS(s) + 4NH_3(aq) \rightarrow Cu(NH_3)_4{}^{2+}(aq) + S^{2-}(aq)$$

Comment on the effectiveness of aqueous ammonia as a solvent for copper(II) sulfide.

53. Use Table 16.6, the reciprocal rule, and the rule of multiple equilibria to calculate K for the reaction

$$Co(NH_3)_6{}^{3+}(aq) + Cl^-(aq) \rightleftharpoons$$
$$Co(NH_3)_5Cl^{2+}(aq) + NH_3(aq)$$

54. Cyanide ions, CN^-, will displace thiosulfate ions, $S_2O_3{}^{2-}$, from the $Ag(S_2O_3)_2{}^{3-}$ complex ion. Show that this is true by calculating K for the reaction

$$Ag(S_2O_3)_2{}^{3-}(aq) + 2CN^-(aq) \rightleftharpoons$$
$$Ag(CN)_2{}^-(aq) + 2S_2O_3{}^{2-}(aq)$$

Unclassified

55. In your own words, explain why
a. $H_2N(CH_2)_3NH_2$ is a bidentate ligand.
b. AgCl dissolves in NH_3.
c. there are no geometric isomers of tetrahedral complexes.

56. Determine whether each of the following is true or false. If the statement is false, correct it.
a. The coordination number of Fe^{3+} in $Fe(H_2O)_4(C_2O_4)^+$ is 5.
b. Cu^+ has two unpaired electrons.
c. $Co(CN)_6{}^{3-}$ is expected to absorb at a longer wavelength than $Co(NH_3)_6{}^{3+}$.

57. Indicate each of the following statements as true or false. If false, correct the statement to make it true.
a. In $Pt(NH_3)_4Cl_4$, platinum has a $+4$ charge and a coordination number of 6.
b. Complexes of Cu^{2+} are brightly colored, while those of Zn^{2+} are colorless.
c. The K_f value for the octahedral complexes A and B are 1×10^{31} and 1×10^{24}, respectively. This means that complex B is more stable than complex A.

58. Explain why
a. oxalic acid solution removes rust stains.
b. EDTA is used to treat lead poisoning.
c. a pale green solution of nickel turns blue when NH_3 is added.

59. Analysis of a coordination compound gives the following results: 22.0% Co, 31.4% N, 6.78% H, and 39.8% Cl. One mole of the compound dissociates in water to form four moles of ions.
a. What is the formula of the compound?
b. Write an equation for its dissociation in water.

60. A chemist synthesizes two coordination compounds. One compound decomposes at 280°C, the other at 240°C.

Analysis of the compounds gives the same mass percent data: 52.6% Pt, 7.6% N, 1.63% H, and 38.2% Cl. Both compounds contain a $+4$ central ion.
a. What is the simplest formula of the compounds?
b. Draw structural formulas for the complexes present.

Challenge Problems

61. A certain coordination compound has the simplest formula $PtN_2H_6Cl_2$. It has a molar mass of about 600 g/mol and contains both a complex cation and a complex anion. What is its structure?

62. A child eats 10.0 g of paint containing 5.0% Pb. How many grams of the sodium salt of EDTA should he receive to bring the lead into solution?

63. Two coordination compounds decompose at different temperatures but have the same mass percent analysis data: 20.25% Cu, 15.29% C, 7.07% H, 26.86% N, 10.23% S, and 20.39% O. Each contains Cu^{2+}.
a. Determine the simplest formula of the compounds.
b. Draw the structural formulas of the complex ion in each case.

64. When solid aluminum hydroxide is stirred with a concentrated solution of NaOH, it dissolves.
a. Write an equation for the reaction involved.
b. Calculate the equilibrium constant for the reaction.

65. Consider the equilibrium

$$Zn(NH_3)_4{}^{2+}(aq) + 4OH^-(aq)$$
$$\rightarrow Zn(OH)_4{}^{2-}(aq) + 4NH_3(aq)$$

a. Calculate K for this reaction.
b. What is the ratio $[Zn(NH_3)_4{}^{2+}]/[Zn(OH)_4{}^{2-}]$ in a solution 1.0 M in NH_3 ($K_b = 1.8 \times 10^{-5}$)?

66. In the $Ti(H_2O)_6{}^{3+}$ ion, the splitting between the d levels, Δ_0, is 55 kcal/mol. What is the color of this ion, assuming the color results from a transition between upper and lower d levels?

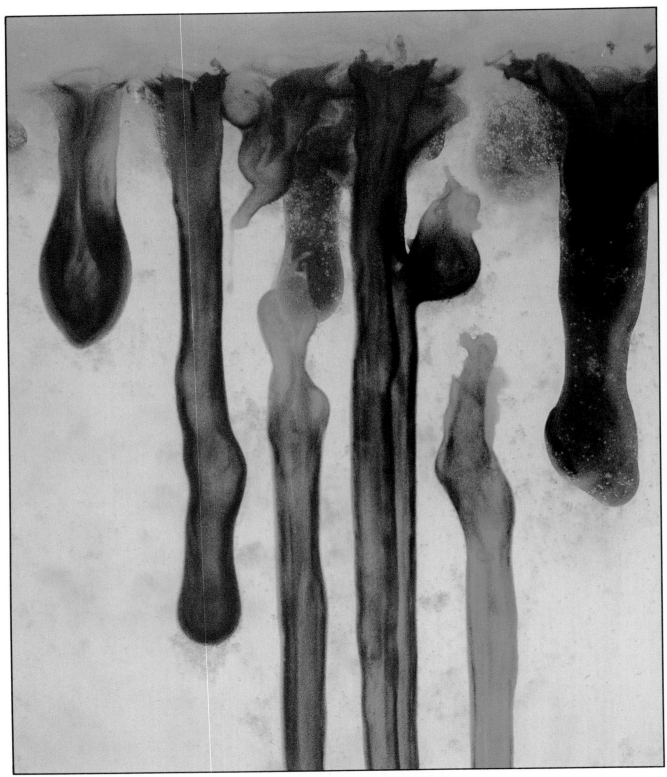

$$Cu^{2+}(aq) + 2OH^-(aq) \rightarrow Cu(OH)_2(s)$$

Formation of a blue precipitate of copper(II) hydroxide.

CHAPTER
17

QUALITATIVE ANALYSIS

Surrounded by beakers, by strange coils,
By ovens and flasks with twisted necks,
The chemist, fathoming the whims of attractions,
Artfully imposes on them their precise meetings.

SULLY-PRUDHOMME
The Naked World
(*translated by William Dock*)

I n earlier chapters, particularly in Chapter 12, we have seen how chemical reactions of different types can serve as a basis for quantitative analysis. Previously, we were interested in determining quantitatively how much of a species is present in a solution or solid mixture. Titrations involving precipitation reactions and acid-base reactions are particularly useful in quantitative analysis.

The subject of this chapter is *qualitative analysis*. Here our main goal is to identify each species in a mixture without too much concern for the relative amounts of different species. In particular, we will look at the qualitative analysis of cations and anions in water solution.

In qualitative analysis we just want to know what's there

Many physical methods can be used to detect ions in solution. Several such methods have been mentioned in previous chapters, including chromatography (Chapter 1) and emission spectroscopy (Chapter 5). The approach of this chapter is based upon chemical reactions. These reactions are used in a standard scheme of analysis designed to test for cations in water solution. They can also be used in spot testing to identify several different anions in aqueous solution. Many of the reactions used in qualitative analysis illustrate and reinforce the principles of solution chemistry discussed in the last several chapters. Indeed, a major purpose of this chapter is to review and extend concepts introduced in Chapter 12 and Chapters 14 through 16.

In Section 17.1, we will see how cations can be separated chemically into four different groups, each containing between three and seven ions. Then (Section 17.2) we will consider in some detail the analysis of one group (Ag^+, Pb^{2+}, and Hg_2^{2+} ions). In Section 17.3, we will focus on eight anions. We will see how they can be identified by adding simple reagents. In Section 17.4, we will look at the different kinds of reactions used in qualitative analysis and the equations written to represent them. This material should be familiar, since the reactions are the ones discussed in Chapter 12 and Chapters 14 through 16. In Section 17.5, we will examine solution equilibria in qualitative analysis. Here again, the equilibrium constants used are ones introduced in previous chapters.

17.1
ANALYSIS OF CATIONS: AN OVERVIEW

In the general chemistry laboratory, you will very likely carry out one or more experiments in the qualitative analysis of cations. Ordinarily, you will work with a water solution that may contain several different cations. Your task will be to separate and identify these ions. The scheme we will describe here is a general one that can deal with as many as 22 different cations, including just about all the common ones.

Figure 17.1 shows the metals whose cations are ordinarily included in the qualitative analysis scheme. Notice that they include the more common

— main-group metals (Na, K; Mg, Ca, Ba; Al).

— transition metals (Cr, Mn, Fe, Co, Ni, Cu, Zn; Ag, Cd; Hg).

— post-transition metals (Sn, Pb; Sb, Bi).

Other metals are ordinarily omitted either because they are too toxic (As, Tl), very expensive (Au, Pt), or relatively rare (Li, Rb, Sr, Ga, and a host of others).

The analysis for cations in a mixture requires a systematic approach. Generally, the procedure is to remove successive groups of cations by precipitation from solution. The standard scheme of qualitative analysis separates cations into four groups (Table 17.1). Concentrations of reagents and solution pH are adjusted so that only one group is affected by a given precipitating agent. The group precipitate is removed from the solution

These groups are removed from solution one after the other

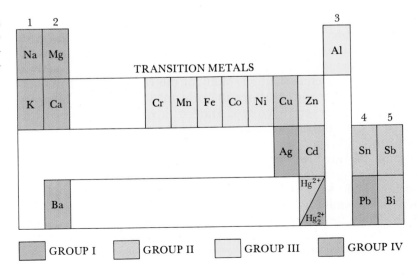

Figure 17.1 Metals whose cations are in the standard qualitative analysis scheme; the two cations of mercury, Hg_2^{2+} and Hg^{2+}, are in different groups.

TABLE 17.1 Cation Groups of Qualitative Analysis

GROUP	CATIONS	PRECIPITATING REAGENT/CONDITIONS
I	Ag^+, Pb^{2+}, Hg_2^{2+}	6 M HCl
II	Cu^{2+}, Bi^{3+}, Hg^{2+}, Cd^{2+}, Sn^{4+}, Sb^{3+}	0.1 M H_2S at a pH of 0.5
III	Al^{3+}, Cr^{3+}, Co^{2+}, Fe^{2+}, Mn^{2+}, Ni^{2+}, Zn^{2+}	0.1 M H_2S at a pH of 9
IV	Ba^{2+}, Ca^{2+}, Mg^{2+}; Na^+, K^+, NH_4^+	0.2 M $(NH_4)_2CO_3$ at a pH of 9.5. No precipitates with Na^+, K^+, NH_4^+; separate tests for identification

by centrifuging the mixture and pouring the remaining solution from the solid. Within a given group, cations are separated and identified by selective chemical reactions.

Many of the metals listed in Figure 17.1 form cations other than those listed in Table 17.1. For example, tin forms the Sn^{2+} ion as well as Sn^{4+}; iron forms Fe^{3+} as well as Fe^{2+}. The ion listed in Table 17.1 is the one present when the group is separated. Ordinarily, in qualitative analysis we do not attempt to distinguish between different ions of a given metal. For example, we do not try to distinguish between Fe^{2+} and Fe^{3+}.

GROUP I: Ag^+, Pb^{2+}, Hg_2^{2+}

These three ions are colorless. The silver (Ag^+) and lead (Pb^{2+}) ions are the only common cations of silver and lead. The Hg_2^{2+} ion is one of two cations formed by mercury. It has an unusual structure: there is a covalent bond between the two mercury atoms

$$(Hg : Hg)^{2+}$$

Salts containing this cation have formulas such as $Hg_2(NO_3)_2$ and Hg_2Cl_2.

The three ions in Group I are unique among the common cations in one respect: they form insoluble chlorides. Hence, they can be separated from other cations by adding 6 M HCl. The Group I cations precipitate as white chlorides (Fig. 17.2): AgCl, $PbCl_2$, and Hg_2Cl_2. These solids can be centrifuged off, leaving a solution that may contain cations in Groups II, III, and IV, and possibly some Pb^{2+} ions carried over from Group I ($PbCl_2$ has a relatively large K_{sp} value, 1.7×10^{-5}). The presence of H^+ ions in the precipitating solution is essential to prevent basic salts such as Sn(OH)Cl and BiOCl from coming down in Group I.

Figure 17.2 The cations of Group I are colorless in solution; their chlorides, AgCl, $PbCl_2$, and Hg_2Cl_2, are white precipitates. (Charles D. Winters)

GROUP II: Cu^{2+}, Bi^{3+}, Hg^{2+}, Cd^{2+}, Sn^{4+}, Sb^{3+}

Of these cations, only Cu^{2+} is colored (Fig. 17.3a, p. 584). Copper(II) salts are most often a pale blue, the color of the $Cu(H_2O)_4^{2+}$ complex ion. Of the other cations in the group, none have partially filled d orbitals. Thus, it is not too surprising that Bi^{3+}, Hg^{2+}, Cd^{2+}, Sn^{4+}, and Sb^{3+} often form white salts that dissolve to give colorless solutions.

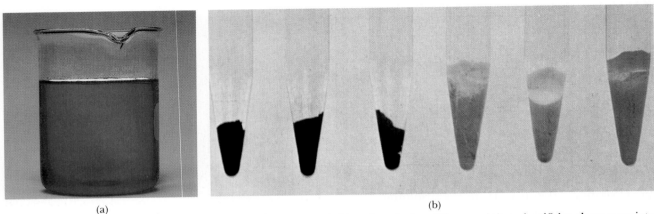

(a)   (b)

Figure 17.3   Of the Group II cations, only Cu^{2+} is colored (blue) in solution. The precipitated sulfides show a variety of colors. CuS, Bi_2S_3, and HgS are black, CdS is orange-yellow, SnS_2 is a pale yellow, and Sb_2S_3 is a brilliant red-orange. (a, Marna G. Clarke)

To separate the ions in Group II from those in Groups III and IV, the H^+ ion concentration is first adjusted to about 0.3 M (pH = 0.5). The solution is then saturated with hydrogen sulfide, a toxic, foul-smelling gas. At the rather high H^+ ion concentration, the equilibrium

$$H_2S(aq) \rightleftharpoons 2H^+(aq) + S^{2-}(aq)$$

lies far to the left. The concentration of S^{2-} is extremely low, about 10^{-20} M. However, this is sufficient to precipitate the very insoluble sulfides of the Group II cations, such as CuS ($K_{sp} = 1 \times 10^{-36}$). Several of these sulfides have characteristic colors (Fig. 17.3b). By observing the color of the Group II precipitate and that of the original solution, you can often deduce some of the ions likely to be present or absent.

GROUP III: Al^{3+}, Cr^{3+}, Co^{2+}, Fe^{2+}, Mn^{2+}, Ni^{2+}, Zn^{2+}

Of the cations in this group, only Al^{3+} and Zn^{2+} are colorless. The other five are transition metal ions with incomplete 3d sublevels. Solutions containing Ni^{2+}, Cr^{3+}, and Co^{2+} are most often brightly colored. The $Ni(H_2O)_6^{2+}$ ion imparts a characteristic green color to solutions of nickel(II) salts. Solutions containing chromium(III) and cobalt(II) can have different colors depending upon the nature of the complex ion present (Fig. 17.4a).

$$Cr(H_2O)_6^{3+} \quad Cr(H_2O)_5Cl^{2+} \quad Co(H_2O)_6^{2+} \quad CoCl_4^{2-}$$
purple   green   red   blue

The Group III cations form sulfides that are more soluble than those of Group II. Hence, they do not precipitate at the very low S^{2-} concentration (10^{-20} M) present at $[H^+]$ = 0.3 M. To bring the Group III cations out of solution, $[H^+]$ is reduced to about 10^{-9} M and the solution is saturated with H_2S. Under these conditions, the equilibrium

$$H_2S(aq) \rightleftharpoons 2H^+(aq) + S^{2-}(aq)$$

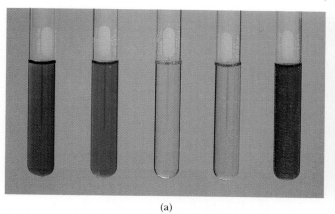
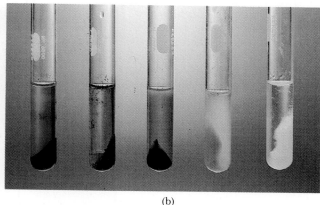

(a)
(b)

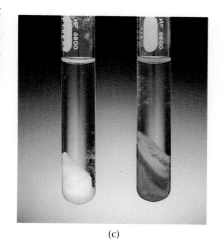

Figure 17.4 Several of the Group III cations are colored in solution (a). Among these are Co^{2+} (red), Fe^{2+} (pale green), Mn^{2+} (pale pink), Ni^{2+} (green), and Cr^{3+} (purple). Five Group III cations precipitate as sulfides (b); NiS, CoS, and FeS are all black, MnS is light pink, and ZnS is white. Two cations precipitate as hydroxides (c); $Al(OH)_3$ is white and $Cr(OH)_3$ is a grayish-green. (Charles D. Winters)

(c)

shifts to the right. The concentration of S^{2-} becomes high enough to precipitate CoS, FeS, MnS, NiS, and ZnS. Of these, CoS, FeS, and NiS are black; MnS, like the Mn^{2+} ion itself, is a pale pink. Zinc sulfide, ZnS, is the only white sulfide in either Group II or III (Fig. 17.4b).

Two ions in Group III, Al^{3+} and Cr^{3+}, precipitate as hydroxides rather than sulfides. Aluminum hydroxide, $Al(OH)_3$, and chromium hydroxide, $Cr(OH)_3$, both come down as gelatinous solids; $Al(OH)_3$ is white, while $Cr(OH)_3$ is green (Fig. 17.4c).

■ **EXAMPLE 17.1**

How would you analyze a solution that might contain Ag^+, Sn^{4+}, and Zn^{2+} but no other cations?

Solution

Ag^+ is in Group I, Sn^{4+} in Group II, and Zn^{2+} in Group III.

First add 6 M HCl; if a precipitate forms, it must be AgCl (white). Then adjust $[H^+]$ to 0.3 M and saturate with H_2S. If Sn^{4+} is present, it should precipitate as SnS_2 (yellow). Finally, adjust the pH to 9 and saturate again with H_2S; Zn^{2+} will precipitate as ZnS (white).

Exercise

What is the color of the original solution? Answer: Colorless.

GROUP IV: Ba^{2+}, Ca^{2+}, Mg^{2+}, Na^+, K^+, NH_4^+

All of these cations are colorless in solution. Their salts are typically white and generally more soluble than those of the transition metal ions. The Group IV cations do not form precipitates with Cl^- or H_2S. The alkaline

Figure 17.5 Flame tests are used for Na$^+$ (yellow) and K$^+$ (violet). A drop of solution is picked up on a platinum loop and immersed in the flame. The test for K$^+$ is best done with a filter that hides the strong Na$^+$ color.

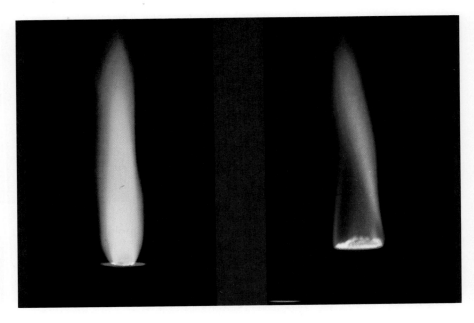

earth cations—Ba^{2+}, Ca^{2+}, and, to a lesser extent, Mg^{2+}—are precipitated as carbonates in the qualitative analysis scheme. The precipitating solution is buffered at pH 9.5 with NH$_4$$^+$ ions and NH$_3$ molecules.

The three remaining cations in Group IV—Na$^+$, K$^+$, and NH$_4$$^+$—stay in solution, since their carbonates are soluble. These ions are tested for individually. *The tests are carried out on the original solution*, rather than one from which Groups I, II, and III have been removed. One reason for this is that reagents added in the separation scheme commonly contain Na$^+$ and NH$_4$$^+$ ions.

The two alkali metal ions, Na$^+$ and K$^+$, are determined by flame tests (Fig. 17.5). The Na$^+$ ion gives an intense yellow color. The test for K$^+$ (violet) is much weaker. It is often observed through a cobalt-blue glass filter to minimize interference by Na$^+$.

To test for NH$_4$$^+$, a small sample of the original solution is heated with 6 M NaOH. An acid-base reaction occurs between the OH$^-$ ion and the weak acid NH$_4$$^+$:

$$NH_4^+(aq) + OH^-(aq) \rightarrow NH_3(aq) + H_2O \qquad (17.1)$$

Ammonia gas can be detected by its odor or its basic properties. It is the only common gas that turns moist red litmus blue.

It is not easy to precipitate any salts of Na$^+$, K$^+$, or NH$_4$$^+$

■ **EXAMPLE 17.2** _____

A solution may contain any of the ions in Groups I through IV. Addition of 6 M HCl gives no precipitate. The H$^+$ concentration is adjusted to 0.3 M and the solution is saturated with H$_2$S; no precipitate forms. However, when [H$^+$] is reduced to 10^{-9} M, and the solution is again saturated with H$_2$S, a precipitate forms. This precipitate is later shown to be a Group III hydroxide. In the Group IV analysis, a precipitate is formed with

$(NH_4)_2CO_3$. In this solution, what ions *may* be present from
a. Group I?   b. Group II?   c. Group III?   d. Group IV?

Solution

a.   No Group I ions   (would precipitate as chlorides).

If nothing happens, the ion or ions in question aren't there

b.   No Group II ions   (would precipitate as sulfides).

c. Al^{3+} or Cr^{3+}.   These are the only ions in Group III that precipitate as hydroxides in the group separation.

d.   Any Group IV ion may be present.   The carbonate precipitate could be $BaCO_3$, $CaCO_3$, or $MgCO_3$.

Exercise
How would you test the original solution for Na^+? NH_4^+? Answer: Flame test for Na^+ (yellow); heat with strong base and test for NH_3 to detect NH_4^+.

17.2
ANALYSIS FOR GROUP I

We will not attempt to describe the procedures used to separate and identify all the individual cations listed in Table 17.1. It is, however, useful to illustrate the general approach followed by considering in some detail the analysis for Group I. Here, only three ions are involved: Ag^+, Pb^{2+}, and Hg_2^{2+}.

A flow chart for the analysis of the Group I cations is shown in Fig. 17.6, p. 588. After precipitation with HCl, the next step involved is heating the solids with water. Lead chloride, $PbCl_2$, is considerably more soluble than either AgCl or Hg_2Cl_2. Its solubility increases with temperature to the extent that it dissolves fairly readily at 100°C:

$$PbCl_2(s) \rightarrow Pb^{2+}(aq) + 2Cl^-(aq) \qquad (17.2)$$

The hot solution is quickly centrifuged to separate it from AgCl or Hg_2Cl_2. The presence of Pb^{2+} in the solution is detected by adding a solution of potassium chromate, K_2CrO_4. A bright yellow precipitate of lead chromate forms if lead is present:

$$Pb^{2+}(aq) + CrO_4^{2-}(aq) \rightarrow PbCrO_4(s) \qquad (17.3)$$
$$\text{yellow}$$

This is the test for lead, and it's a good one

Any precipitate remaining after the removal of Pb^{2+} is treated with aqueous ammonia. If Hg_2Cl_2 is present, it undergoes the following disproportionation reaction with NH_3:

$$Hg_2Cl_2(s) + 2NH_3(aq) \rightarrow HgNH_2Cl(s) + Hg(l) + NH_4^+(aq) + Cl^-(aq) \quad (17.4)$$
$$\qquad\qquad\qquad\text{white}\qquad\qquad\text{black}$$

Figure 17.6 Flow chart for the analysis of the Group I cations.

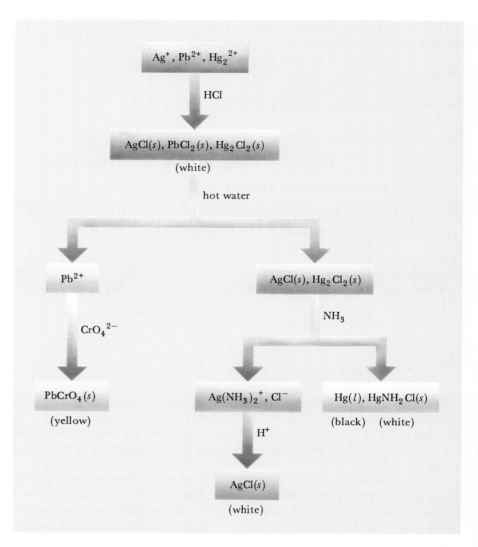

The gray precipitate formed is a mixture of finely divided mercury, which appears black, and $HgNH_2Cl$, which is white.

The addition of aqueous ammonia brings AgCl into solution. It does this by forming the stable $Ag(NH_3)_2^+$ complex ion. We might consider that this reaction occurs in two steps:

$$AgCl(s) \rightleftharpoons Ag^+(aq) + Cl^-(aq) \qquad (17.5a)$$

$$\underline{Ag^+(aq) + 2NH_3(aq) \rightleftharpoons Ag(NH_3)_2^+(aq) \qquad (17.5b)}$$

$$AgCl(s) + 2NH_3(aq) \rightleftharpoons Ag(NH_3)_2^+(aq) + Cl^-(aq) \qquad (17.5)$$

Since K_f for the formation of the complex ion $Ag(NH_3)_2^+$ is 2×10^7, Reaction 17.5b goes essentially to completion. Thus, in a sense, Ag^+ ions are "siphoned off" AgCl, shifting the equilibrium in Reaction 17.5a to the right. The net result is that AgCl dissolves via Reaction 17.5.

The solution formed by Reaction 17.5 is separated from any precipitate containing mercury by centrifuging. The solution must then be tested to establish the presence of silver. To do that, we add a strong acid, 6 M HNO_3. The H^+ ions of the acid destroy the $Ag(NH_3)_2^+$ complex ion by converting the NH_3 molecules to NH_4^+ ions. The Ag^+ ions formed combine with Cl^- ions in the solution to precipitate white AgCl. The overall reaction is:

$$Ag(NH_3)_2^+(aq) + 2H^+(aq) + Cl^-(aq) \rightarrow AgCl(s) + 2NH_4^+(aq) \quad (17.6)$$

The Group I procedure is similar to, but somewhat simpler than, those for Groups II and III

■ EXAMPLE 17.3

A Group I unknown gives a white precipitate with HCl. Treatment with hot water gives a solution to which K_2CrO_4 is added; a yellow precipitate forms. The white precipitate remaining from the hot water treatment dissolves completely in NH_3. What ions are present? absent?

Solution

Pb^{2+} is present (PbCl$_2$ dissolves in hot water and then precipitates $PbCrO_4$).

Hg_2^{2+} is absent (Hg_2Cl_2 would not dissolve in NH_3). Ag^+ is present

(AgCl dissolves in NH_3).

Exercise

If HNO_3 is added to the solution formed with NH_3, a white precipitate forms. Write the equation for the reaction involved. Answer: Equation 17.6.

Example 17.3 is typical of "paper unknowns" used to test your understanding of procedures in qualitative analysis. You should realize that in framing questions of this sort, confirmatory tests are ordinarily not included. Consider, for example, the unknown referred to in Example 17.3. In the laboratory you would ordinarily confirm the presence of silver by adding nitric acid to obtain a white precipitate of AgCl. This information is not included in the body of the example. In principle at least, the unknown must contain Ag^+; otherwise, the white precipitate would not dissolve in NH_3.

17.3
SPOT TESTS FOR ANIONS

It is possible to develop a systematic scheme of anion analysis analogous to that for cations. However, we will consider a much simpler analytical problem. Let us suppose that you are given an "unknown" solution and told that it contains *one and only one* of the following anions:

$$CO_3^{2-}, S^{2-}, CrO_4^{2-}, SO_4^{2-}, SCN^-, Br^-, I^-, NO_3^-$$

Your task is to identify the anion. This can be done by rather simple "spot tests" carried out on separate portions of the original solution.

Spot tests are often used to "spot" one ion in the presence of others

Carbonate, CO_3^{2-}, and Sulfide, S^{2-}

Of the eight anions just cited, only CO_3^{2-} and S^{2-} evolve a gas when treated with a dilute strong acid.

$$CO_3^{2-}(aq) + 2H^+(aq) \rightarrow CO_2(g) + H_2O \qquad (17.7)$$

$$S^{2-}(aq) + 2H^+(aq) \rightarrow H_2S(g) \qquad (17.8)$$

You can't miss the sulfide test

Carbon dioxide is odorless; hydrogen sulfide has the foul odor of rotten eggs. So, addition of dilute HCl to a sample of your unknown gives a definitive test for carbonate and sulfide ions. If you're not entirely sure that the odorless gas formed is CO_2, you can bring it into contact with a solution of barium hydroxide. A white precipitate of barium carbonate forms.

$$CO_2(g) + Ba^{2+}(aq) + 2OH^-(aq) \rightarrow BaCO_3(s) + H_2O \qquad (17.9)$$
$$\text{white}$$

Chromate, CrO_4^{2-}

The chromate ion is one of the few colored anions. If your unknown solution has a yellow color, the anion present is almost certainly CrO_4^{2-}. The color changes from yellow to red-orange upon addition of strong acid (Fig. 17.7).

$$2CrO_4^{2-}(aq) + 2H^+(aq) \rightarrow Cr_2O_7^{2-}(aq) + H_2O \qquad (17.10)$$
$$\text{yellow} \qquad\qquad\qquad \text{red}$$

If you're still not sure about the presence of chromate ion, you can add Ba^{2+} to a sample of your unknown: $BaCrO_4$ is yellow and insoluble.

$$Ba^{2+}(aq) + CrO_4^{2-}(aq) \rightarrow BaCrO_4(s) \qquad (17.11)$$
$$\text{yellow}$$

Sulfate, SO_4^{2-}

To test for sulfate, you first acidify the unknown with HCl and then add Ba^{2+}. Formation of a white precipitate of $BaSO_4$ indicates the presence of SO_4^{2-}.

 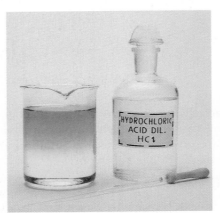 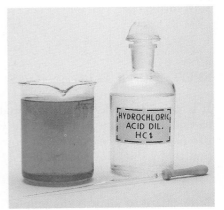

Figure 17.7 When acid is added to a solution containing the CrO_4^{2-} ion (yellow), the $Cr_2O_7^{2-}$ ion (red-orange) is produced. (Marna G. Clarke)

$$Ba^{2+}(aq) + SO_4^{2-}(aq) \rightarrow BaSO_4(s) \tag{17.12}$$
white

No other anion shows this property; $BaCrO_4$ and $BaCO_3$ are both soluble in strong acid.

Thiocyanate, SCN^-

The thiocyanate ion is readily identified by treating the unknown with Fe^{3+} ions. The formation of a blood-red complex $Fe(SCN)^{2+}$ is a conclusive test for SCN^- (Fig. 17.8).

$$SCN^-(aq) + Fe^{3+}(aq) \rightarrow Fe(SCN)^{2+}(aq) \tag{17.13}$$

Bromide, Br^-, and Iodide, I^-

To detect these ions, it is convenient to oxidize them to the free halogens, Br_2 and I_2, which can be identified by their colors. An oxidizing agent that will accomplish this is potassium permanganate in acid:

$$2MnO_4^-(aq) + 10Br^-(aq) + 16H^+(aq) \rightarrow 2Mn^{2+}(aq) + 5Br_2(aq) + 8H_2O \tag{17.14}$$

$$2MnO_4^-(aq) + 10I^-(aq) + 16H^+(aq) \rightarrow 2Mn^{2+}(aq) + 5I_2(aq) + 8H_2O \tag{17.15}$$

The free halogens are detected by extracting with hexane, C_6H_{14}, in which they are soluble (Fig. 17.9). Iodine, I_2, imparts a violet color to the hexane layer; bromine, Br_2, gives a reddish or yellow layer.

Nitrate, NO_3^-

The nitrate anion is one of the most difficult to detect. It does not react with H^+ ions, Ba^{2+} ions, or indeed any precipitating or oxidizing agent. It can, however, be reduced; the so-called "brown ring" test detects the

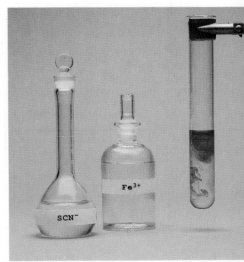

Figure 17.8 The SCN^- ion (or the Fe^{3+} ion) can be detected by the blood-red color of the $Fe(SCN)^{2+}$ complex. (Marna G. Clarke)

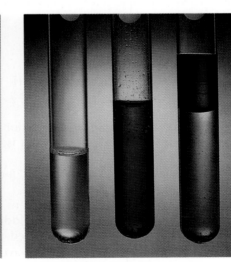

(a)   (b)

Figure 17.9 The tests for Br^- and I^- are shown in (a) and (b), respectively. In water solution (test tubes at left), both ions are colorless. Addition of an oxidizing agent gives aqueous solutions of Br_2 (light orange) and I_2 (red-brown). Extraction with an organic solvent (upper layer in the tubes at right) gives reddish-orange Br_2 and violet I_2. (Charles D. Winters)

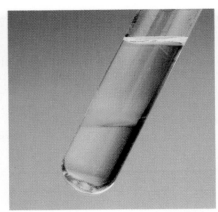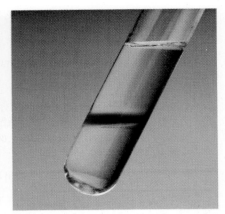

Figure 17.10   The nitrate ion reacts with Fe^{2+} to form a complex, $Fe(NO)^{2+}$, which appears as a brown ring at the junction between two layers. (Marna G. Clarke)

formation of a particular reduction product, nitrogen oxide, NO. The reducing agent used is iron(II) sulfate, which is added to the unknown in a test tube. Concentrated sulfuric acid, H_2SO_4, is carefully poured down the side of the test tube so that it forms a dense layer at the bottom. A brown ring of the $Fe(NO)^{2+}$ complex forms at the junction of the two layers (Fig. 17.10). The reaction is:

$$Fe^{2+}(aq) + NO(g) \rightarrow Fe(NO)^{2+}(aq) \qquad (17.16)$$
$$\text{brown}$$

■ **EXAMPLE 17.4** ——————————————
Write balanced net ionic equations for the reactions that occur when:

a. a solution of sodium carbonate is acidified.
b. sulfuric acid is added to a solution of barium chloride.
c. I^- ions are oxidized by Fe^{3+} ions, which are reduced to Fe^{2+}.

Solution

a. Reaction 17.7: $CO_3{}^{2-}(aq) + 2H^+(aq) \rightarrow CO_2(g) + H_2O$
b. Reaction 17.12: $Ba^{2+}(aq) + SO_4{}^{2-}(aq) \rightarrow BaSO_4(s)$

c. This is a rather simple redox reaction that can be balanced by the method described in Chapter 12. The result is:

$$2Fe^{3+}(aq) + 2I^-(aq) \rightarrow 2Fe^{2+}(aq) + I_2(s)$$

Exercise
Write an equation to explain why addition of OH^- ions to a solution of $K_2Cr_2O_7$ changes the color from red to yellow. Answer: $Cr_2O_7{}^{2-}(aq) + 2OH^-(aq) \rightarrow 2CrO_4{}^{2-}(aq) + H_2O$.

The spot tests we have described are nearly infallible given the starting assumptions, i.e., that only one anion is present and that it is one of the eight listed (CO_3^{2-}, S^{2-}, CrO_4^{2-}, SO_4^{2-}, SCN^-, Br^-, I^-, NO_3^-). If we allow more than one anion to be present or if we include other common anions such as Cl^- and PO_4^{3-}, the analytical problem becomes considerably more difficult. The spot tests can still be used, but interferences by other anions make the tests less easy to interpret. The brown ring test for nitrate ion is particularly sensitive to interferences.

17.4
REACTIONS IN QUALITATIVE ANALYSIS

A wide variety of different reactions are carried out in qualitative analysis. An acid or base may be added to adjust the pH of a solution before separating one group of cations from another. A cation can be oxidized or reduced in order to separate it from another. Most often, however, reactions are carried out in which

Qual analysis makes full use of the range of inorganic reactions

— a complex ion is formed or decomposed.
— a precipitate is formed or dissolved.

FORMATION OF COMPLEX IONS

In one type of reaction in qualitative analysis, a cation in water solution is converted to a complex ion. Usually, the complexing agent (ligand) is an NH_3 molecule or OH^- ion. Typical reactions include

$$Cu^{2+}(aq) + 4NH_3(aq) \rightarrow Cu(NH_3)_4^{2+}(aq) \tag{17.17}$$

$$Al^{3+}(aq) + 4OH^-(aq) \rightarrow Al(OH)_4^-(aq) \tag{17.18}$$

Cations in the qualitative analysis scheme that commonly form complexes with NH_3 or OH^- are listed in Table 17.2.

TABLE 17.2 Complexes of Cations with NH_3 and OH^-

CATION	NH_3 COMPLEX	OH^- COMPLEX
Ag^+	$Ag(NH_3)_2^+$	
Pb^{2+}		*$Pb(OH)_3^-$
Cu^{2+}	$Cu(NH_3)_4^{2+}$ (blue)	
Cd^{2+}	$Cd(NH_3)_4^{2+}$	
Sn^{4+}		$Sn(OH)_6^{2-}$
Sb^{3+}		$Sb(OH)_4^-$
Al^{3+}		$Al(OH)_4^-$
Ni^{2+}	$Ni(NH_3)_6^{2+}$ (blue)	
Zn^{2+}	$Zn(NH_3)_4^{2+}$	$Zn(OH)_4^{2-}$

*Most likely composition: $Pb(H_2O)_3(OH)_3^-$

■ **EXAMPLE 17.5**

Write a net ionic equation for the formation of a complex ion when a solution containing Zn^{2+} is treated with a solution of

a. NH_3 b. $NaOH$

Solution

a. Referring to Table 17.2, we note that the complex ion has the formula $Zn(NH_3)_4^{2+}$. The reactants are a Zn^{2+} ion and four NH_3 molecules, all in water solution:

$$Zn^{2+}(aq) + 4NH_3(aq) \rightarrow Zn(NH_3)_4^{2+}(aq)$$

b. $Zn^{2+}(aq) + 4OH^-(aq) \rightarrow Zn(OH)_4^{2-}(aq)$

Note that the Na^+ ions take no part in the reaction and therefore are not included in the equation.

Exercise

Write a net ionic equation for the reaction that occurs when a solution of silver nitrate is treated with ammonia. Answer: $Ag^+(aq) + 2NH_3(aq) \rightarrow Ag(NH_3)_2^+(aq)$.

If a complex ion is brightly colored, its formation may serve to identify a particular cation. In Group III, Fe^{3+} can be identified by adding SCN^- ions to form the blood-red complex $Fe(SCN)^{2+}$ (Reaction 17.13 and Fig. 17.8). The presence of Ni^{2+} is shown by adding the chelating agent dimethylglyoxime, which forms a complex (Fig. 17.11) that comes out of solution as a rose-red precipitate.

Some of the species used in qual tests have beautiful colors

Often in qualitative analysis, we want to destroy a complex ion formed in an earlier step. For example, in one step of the analysis for Group III, Al^{3+} is converted to $Al(OH)_4^-$. This separates aluminum from other ions in this group. To test for aluminum later on, the $Al(OH)_4^-$ complex ion must be converted back to Al^{3+}.

Complexes containing OH^- ligands can be decomposed back to the free cation by adding a strong acid. The H^+ ions of the strong acid react with the OH^- ions, converting them to water molecules. The behavior of the $Al(OH)_4^-$ ion is typical:

$$Al(OH)_4^-(aq) + 4H^+(aq) \rightarrow Al^{3+}(aq) + 4H_2O \qquad (17.19)$$

Complexes containing NH_3 ligands are also unstable in strong acid. Thus, we have

This is one of the reasons we use OH^- or NH_3 complexes

$$Ag(NH_3)_2^+(aq) + 2H^+(aq) \rightarrow Ag^+(aq) + 2NH_4^+(aq) \qquad (17.20)$$

In general, **complexes containing NH_3 molecules or OH^- ions as ligands are stable in basic solution but decompose in strong acid.**

CH₃—C≡N̈—O—H

$CH_3-C=\ddot{N}-O-H$

$CH_3-C=\underset{\cdot\cdot}{N}-O-H$

dimethylglyoxime, HDMG

nickel dimethylglyoxime, Ni(DMG)₂

Figure 17.11 Ni^{2+} reacts with dimethylglyoxime to form a chelate whose structure is shown at the top of the figure; it comes out of solution as a red precipitate. (Marna G. Clarke)

■ **EXAMPLE 17.6** _____

Write balanced equations for the reactions of H^+ ions with the complexes of Zn^{2+} given in Table 17.2.

Solution

With the OH^- complex, reaction with H^+ forms H_2O and the free Zn^{2+} ion:

$$Zn(OH)_4^{2-}(aq) + 4H^+(aq) \rightarrow Zn^{2+}(aq) + 4H_2O$$

The reaction with the ammonia complex is similar, except that NH_4^+ ions are formed rather than H_2O molecules:

$$Zn(NH_3)_4^{2+}(aq) + 4H^+(aq) \rightarrow Zn^{2+}(aq) + 4NH_4^+(aq)$$

Exercise

Write a balanced equation for the reaction of the hydroxide complex of Sn^{4+} with H^+. Answer: $Sn(OH)_6^{2-}(aq) + 6H^+(aq) \rightarrow Sn^{4+}(aq) + 6H_2O$.

FORMATION OF PRECIPITATES

In qualitative analysis, precipitation reactions are more common than any other type. As we have seen, all the group separations of cations involve some kind of precipitation. For example, the Group I cations, such as Pb^{2+}, are precipitated by adding Cl^- ions:

$$Pb^{2+}(aq) + 2Cl^-(aq) \rightarrow PbCl_2(s) \qquad (17.21)$$

The precipitating anion is usually in excess

The alkaline earth cations in Group IV are precipitated by adding CO_3^{2-} ions in basic solution. The reaction with Ba^{2+} is typical:

$$Ba^{2+}(aq) + CO_3^{2-}(aq) \rightarrow BaCO_3(s) \qquad (17.22)$$

Many of the spot tests used to detect anions also involve precipitation reactions. Recall, for example, that SO_4^{2-} and CrO_4^{2-} ions can be precipitated as $BaSO_4$ and $BaCrO_4$ (Reactions 17.12 and 17.11).

■ **EXAMPLE 17.7** _____

Write net ionic equations for the precipitation reactions that occur when

a. a solution of Na_2S is added to a qualitative analysis unknown containing Bi^{3+} ions.
b. a solution of NaOH is added to a qualitative analysis unknown containing Fe^{3+} ions.
c. a solution of $Sr(NO_3)_2$ is added to a qualitative analysis unknown containing carbonate (CO_3^{2-}) ions.

Solution
The products formed are bismuth (III) sulfide (Bi^{3+}, S^{2-} ions), iron (III) hydroxide (Fe^{3+}, OH^- ions), and strontium carbonate (Sr^{2+}, CO_3^{2-} ions). The principle of electrical neutrality requires that the formulas of these compounds be Bi_2S_3, $Fe(OH)_3$, and $SrCO_3$. The net ionic equations are

a. $2Bi^{3+}(aq) + 3S^{2-}(aq) \rightarrow Bi_2S_3(s)$
b. $Fe^{3+}(aq) + 3OH^-(aq) \rightarrow Fe(OH)_3(s)$
c. $Sr^{2+}(aq) + CO_3^{2-}(aq) \rightarrow SrCO_3(s)$

Note that the Na^+ ions in the sodium sulfide (Na_2S) and NaOH solutions and the nitrate ion (NO_3^-) in the $Sr(NO_3)_2$ solution are simply spectator ions and therefore are not included in the equations.

Exercise
Write balanced equations for the precipitation of Cd^{2+} by OH^-; by S^{2-}. Answer:

$$Cd^{2+}(aq) + 2OH^-(aq) \rightarrow Cd(OH)_2(s);$$
$$Cd^{2+}(aq) + S^{2-}(aq) \rightarrow CdS(s)$$

In qualitative analysis you often form precipitates in a somewhat more subtle way than that shown in Example 17.7. Consider, for example, the analysis for Cd^{2+} in Group II. Here you precipitate the sulfide, CdS, by adding hydrogen sulfide, H_2S, rather than S^{2-} ions. In this case, two different processes are involved:

dissociation of H_2S: $\quad H_2S(aq) \rightleftharpoons 2H^+(aq) + S^{2-}(aq) \quad$ (17.23a)

precipitation of CdS: $\quad Cd^{2+}(aq) + S^{2-}(aq) \rightarrow CdS(s) \quad$ (17.23b)

To obtain the net ionic equation for the overall reaction, we add Equations 17.23a and 17.23b to obtain

$$Cd^{2+}(aq) + H_2S(aq) \rightarrow CdS(s) + 2H^+(aq) \tag{17.23}$$

A similar situation applies when a water solution of ammonia, NH_3, is used as a source of OH^- ions. Recall (Chap. 12) that ammonia is a weak base. It undergoes the following reaction with water:

$$NH_3(aq) + H_2O \rightleftharpoons NH_4^+(aq) + OH^-(aq) \tag{17.24a}$$

The OH^- ions formed can then precipitate a cation such as Pb^{2+}:

$$Pb^{2+}(aq) + 2OH^-(aq) \rightarrow Pb(OH)_2(s) \tag{17.24b}$$

To obtain the net ionic equation for the reaction of Pb^{2+} with a water solution of ammonia, we multiply Equation 17.24a by 2 and add it to Equation 17.24b:

$$2NH_3(aq) + 2H_2O \rightleftharpoons 2NH_4^+(aq) + 2OH^-(aq)$$
$$\underline{Pb^{2+}(aq) + 2OH^-(aq) \rightarrow Pb(OH)_2(s)}$$
$$Pb^{2+}(aq) + 2NH_3(aq) + 2H_2O \rightarrow Pb(OH)_2(s) + 2NH_4^+(aq) \tag{17.24}$$

In practice, it is not necessary to add two different equations in precipitation reactions involving H_2S or NH_3 (Example 17.8).

■ **EXAMPLE 17.8** _____

Write net ionic equations for the precipitation reactions that occur when

a. H_2S is added to a solution containing Bi^{3+}.
b. NH_3 is added to a solution containing Fe^{3+}.

Solution

The precipitates are the same as those obtained in Example 17.7a and b, Bi_2S_3 and $Fe(OH)_3$. The equations, however, are somewhat more complex.

a. To form one mole of Bi_2S_3, we need two moles of Bi^{3+} and three moles of H_2S. The balanced equation is

$$2Bi^{3+}(aq) + 3H_2S(aq) \rightarrow Bi_2S_3(s) + 6H^+(aq)$$

Note that when H_2S is used to precipitate a cation, H^+ ions appear as a by-product; compare Equation 17.23.

b. To form one mole of $Fe(OH)_3$, we need one mole of Fe^{3+}. Three moles of NH_3 are required to furnish the three moles of OH^- ions needed. The equation is

$$Fe^{3+}(aq) + 3NH_3(aq) + 3H_2O \rightarrow Fe(OH)_3(s) + 3NH_4{}^+(aq)$$

When NH_3 is used to precipitate cations, $NH_4{}^+$ forms as by-product.

Exercise
Write balanced equations for the precipitation of Tl^+ by H_2S; of La^{3+} by NH_3.

$$2Tl^+(aq) + H_2S(aq) \rightarrow Tl_2S(s) + 2H^+(aq);$$

$$La^{3+}(aq) + 3NH_3(aq) + 3H_2O \rightarrow La(OH)_3(s) + 3NH_4{}^+(aq)$$

It is of interest to compare equations in Example 17.7a and b to those in Example 17.8. The cations are the same (Bi^{3+}, Fe^{3+}) in the two cases, as are the precipitates (Bi_2S_3, $Fe(OH)_3$). The sources of the anions are different. In Example 17.7, we worked with electrolyte solutions where S^{2-} and OH^- ions were principal species. Hence, those ions appeared directly in the equations we wrote for the precipitation reactions. In Example 17.8, we used the nonelectrolytes H_2S and NH_3 as a source of the precipitating anions. The principal species in these solutions are molecules, H_2S and NH_3, which appear in the equations as such. *In writing any net ionic equation, we always use formulas of the "principal species" in solution, i.e., ions or molecules present at high concentrations.*

Because

$$[H_2S] \gg [S^{2-}]$$
$$[NH_3] \gg [OH^-]$$

DISSOLVING PRECIPITATES

Very often in qualitative analysis, you will need to dissolve a precipitate. You may want to bring an ion into solution so you can test for it more readily. Many different methods can be used to do this. In the simplest case, it may be possible to bring a solid into solution by heating with water. This works with $PbCl_2$, which has a solubility of about 0.03 M at 100°C.

Two methods are used more often than any others to dissolve precipitates in qualitative analysis. One of these involves treating the solid with a strong acid, most often with 6 M HCl. The other uses a complexing agent, commonly 6 M NH_3 or 6 M NaOH. We will now consider the reactions involved in these methods.

1. Precipitates containing a basic anion can often be brought into solution with 6 M HCl. This works with all metal hydroxides in qualitative analysis. The H^+ ions of the HCl combine with the OH^- ions of the solid to form H_2O molecules. The cation of the hydroxide passes into solution. The reaction with zinc hydroxide is typical:

$$Zn(OH)_2(s) + 2H^+(aq) \rightarrow Zn^{2+}(aq) + 2H_2O \tag{17.25}$$

Notice that this equation is very similar to that for the neutralization of a strong base by a strong acid: $H^+(aq) + OH^-(aq) \rightarrow H_2O$. The difference is that the OH^- ions required for the reaction are tied up in the insoluble zinc hydroxide. The principal species is $Zn(OH)_2(s)$, so that is what we put in the equation.

Hydrochloric acid can also be used to dissolve many water-insoluble salts in which the anion is a weak base. In particular, we can use 6 M HCl to dissolve

— *all the carbonates* (CO_3^{2-}) in the qualitative analysis scheme. Here, the product is the weak acid H_2CO_3, which then decomposes into CO_2 and H_2O. The equation for the reaction of H^+ ions with $ZnCO_3$ may be written

$$ZnCO_3(s) + 2H^+(aq) \rightarrow Zn^{2+}(aq) + H_2CO_3(aq) \qquad (17.26)$$

— *many sulfides* (S^{2-}). The "driving force" behind this reaction is the formation of the weak acid H_2S, much of which evolves as a gas. The reaction in the case of zinc sulfide is

$$ZnS(s) + 2H^+(aq) \rightarrow Zn^{2+}(aq) + H_2S(aq) \qquad (17.27)$$

The Group III sulfides, including ZnS, dissolve in 6 M HCl, at least when freshly precipitated. In contrast, the Group II sulfides are much less soluble in strong acid (Fig. 17.12).

Reactions 17.26 and 17.27 are strong acid–weak base reactions quite similar to those discussed in Chapters 12 and 15. Addition of strong acid to solutions containing the weak bases carbonate (CO_3^{2-}) or sulfide (S^{2-}) gives the reactions

$$CO_3^{2-}(aq) + 2H^+(aq) \rightarrow H_2CO_3(aq)$$

$$S^{2-}(aq) + 2H^+(aq) \rightarrow H_2S(aq)$$

Equations 17.26 and 17.27 resemble those just written except that the source of the basic anion is an insoluble solid, $ZnCO_3$ or ZnS, rather than a water solution.

Figure 17.12 ZnS (white) is soluble in HCl; CuS (black) is not. (Charles D. Winters)

■ EXAMPLE 17.9

Write net ionic equations to explain why each of the following precipitates dissolves in 6 M HCl:

a. $Al(OH)_3$ b. $BaCO_3$ c. MnS

Solution

a. The OH^- ions in $Al(OH)_3$ are converted to H_2O molecules by reacting with H^+ ions. Three moles of H^+ ions are required to react with one mole of $Al(OH)_3$:

$$Al(OH)_3(s) + 3H^+(aq) \rightarrow Al^{3+}(aq) + 3H_2O$$

b. The CO_3^{2-} ion in $BaCO_3$ is converted to an H_2CO_3 molecule; two H^+ ions are required:

$$BaCO_3(s) + 2H^+(aq) \rightarrow Ba^{2+}(aq) + H_2CO_3(aq)$$

c. $$MnS(s) + 2H^+(aq) \rightarrow Mn^{2+}(aq) + H_2S(aq)$$

Exercise

Would you expect AgCl to dissolve in strong acid? Would PbI_2 react with H^+ ions? Answer: No; neither Cl^- nor I^- is a basic anion.

The complex ion is more stable than the solid

2. Precipitates containing a transition metal cation (or Al^{3+}, Sn^{4+}, or Pb^{2+}) can often be brought into solution by adding a complexing agent. The reagent, most often NH_3 or OH^-, forms a stable complex ion with the cation. This type of reaction is often used in qualitative analysis. For example, as we saw earlier, silver chloride can be brought into solution by treatment with ammonia, forming the $Ag(NH_3)_2^+$ complex (Equation 17.5). In another case, $Zn(OH)_2$ is dissolved by adding 6 M NaOH. The stable complex ion $Zn(OH)_4^{2-}$ is formed.

$$Zn(OH)_2(s) + 2OH^-(aq) \rightarrow Zn(OH)_4^{2-}(aq) \tag{17.28}$$

■ **EXAMPLE 17.10** _____

When a precipitate of $Cd(OH)_2$ is treated with excess 6 M NH_3, it dissolves. Write a net ionic equation for the reaction involved.

Solution

The reactants are $Cd(OH)_2(s)$ and $NH_3(aq)$. The products are the $Cd(NH_3)_4^{2+}$ ion (Table 17.2) and free OH^- ions, both in water solution. The net ionic equation for the reaction is

$$Cd(OH)_2(s) + 4NH_3(aq) \rightarrow Cd(NH_3)_4^{2+}(aq) + 2OH^-(aq)$$

Exercise

Write a net ionic equation to explain why $Pb(OH)_2$ dissolves in excess NaOH, using Table 17.2. Answer: $Pb(OH)_2(s) + OH^-(aq) \rightarrow Pb(OH)_3^-(aq)$.

17.5
SOLUBILITY EQUILIBRIA IN QUALITATIVE ANALYSIS

Every reaction that we deal with in qualitative analysis can be associated with an equilibrium constant. Sometimes, the constant is one that we considered in Chapters 14 to 16: K_{sp}, K_w, K_a, K_b, or K_f. Since precipitation reactions are among the most common in qualitative analysis, the solubility product constant is used most often.

■ **EXAMPLE 17.11** _____

At 100°C, K_{sp} of $PbCl_2$ is 1.3×10^{-4}.

a. What is the solubility, in moles per liter, of $PbCl_2$ in hot water at 100°C?
b. Will a precipitate of $PbCrO_4$ ($K_{sp} = 2 \times 10^{-14}$) form in Group I analysis

if enough K_2CrO_4 is added to a solution saturated with $PbCl_2$ at 100°C to make conc. $CrO_4^{2-} = 0.010\ M$?

Solution

a. The solubility equation is

$$PbCl_2(s) \rightleftharpoons Pb^{2+}(aq) + 2Cl^-(aq)$$

If we let s be the solubility of $PbCl_2$, then

$$K_{sp}\ PbCl_2 = 1.3 \times 10^{-4} = [Pb^{2+}] \times [Cl^-]^2 = (s) \times (2s)^2 = 4s^3$$

Solving for s,

$$s = \left(\frac{1.3 \times 10^{-4}}{4}\right)^{1/3} = \boxed{0.032\ mol/L}$$

This is roughly twice the calculated value at 25°C, which explains why $PbCl_2$ is heated with water to bring it into solution in Group I analysis.

b. $(conc.\ Pb^{2+}) \times (conc.\ CrO_4^{2-}) = (0.032) \times (0.010) = 3.2 \times 10^{-4}$
Since this product is much larger than K_{sp} of $PbCrO_4$ (2×10^{-14}), we conclude that $\boxed{PbCrO_4\ \text{should precipitate}}$ from the hot saturated solution of $PbCl_2$, as indeed it does.

Exercise
Calculate the solubility of $PbCl_2$ at 25°C, where $K_{sp} = 1.7 \times 10^{-5}$. Answer: 0.016 M.

PbCl$_2$ is reasonably soluble, even at 25°C

Often, the equilibrium constant for a reaction in qualitative analysis is not, by itself, one of those we have discussed. Instead, we have to derive the equilibrium constant for the reaction using one or more of the general relations introduced in Chapter 13:

1. The Coefficient Rule: $K' = K^n$ (17.29)
where K' refers to the equilibrium constant obtained when the coefficients of the balanced equation are multiplied through by a factor $n = \frac{1}{2}, 2$, etc.

2. The Reciprocal Rule: $K_f = 1/K_r$ (17.30)
where K_f and K_r refer to the equilibrium constants for forward and reverse reactions.

3. The Rule of Multiple Equilibria: $K_3 = K_1 \times K_2$ (17.31)
where Reaction 3 = Reaction 1 + Reaction 2

In the remainder of this section we will see how these relations can be applied to two important procedures in qualitative analysis:

— the selective precipitation of an ion or group of ions in the presence of another ion or group of ions.

— dissolving precipitates through reaction with acid or by complex ion formation.

SELECTIVE PRECIPITATION OF GROUP II VS. GROUP III SULFIDES

As we pointed out earlier, certain cations (Cu^{2+}, Cd^{2+}) are precipitated as sulfides in Group II, using H_2S in acidic solution. Other cations (e.g., Zn^{2+} and Ni^{2+}) stay in solution under these conditions but come down as sulfides when the pH is raised in Group III. To understand what is going on here, it is important to consider the equilibrium system:

$$H_2S(aq) \rightleftharpoons 2H^+(aq) + S^{2-}(aq) \tag{17.32}$$

This reaction is the sum of two consecutive weak acid dissociations:

$$
\begin{aligned}
H_2S(aq) &\rightleftharpoons H^+(aq) + HS^-(aq); & K_a\ H_2S &= 1 \times 10^{-7} \\
HS^-(aq) &\rightleftharpoons H^+(aq) + S^{2-}(aq) & K_a\ HS^- &= 1 \times 10^{-13} \\
\hline
H_2S(aq) &\rightleftharpoons 2H^+(aq) + S^{2-}(aq)
\end{aligned}
$$

Applying the Rule of Multiple Equilibria:

K for Reaction 17.32 = $(1 \times 10^{-7})(1 \times 10^{-13}) = 1 \times 10^{-20}$

Using this equilibrium constant, we can

— calculate the concentration of S^{2-} ions in Group II analysis (Example 17.12a).

— understand why certain cations precipitate in Group II as sulfides, while others carry over into Group III (Example 17.12b and c).

■ **EXAMPLE 17.12**

Taking K for the system $H_2S(aq) \rightleftharpoons 2H^+(aq) + S^{2-}(aq)$ to be 1×10^{-20}, determine

a. $[S^{2-}]$ in Group II analysis, where $[H_2S] = 0.1\ M$ and $[H^+] = 0.3\ M$.

b. whether CuS ($K_{sp} = 1 \times 10^{-36}$) will precipitate in Group II, taking conc. $Cu^{2+} = 0.02\ M$.

c. whether NiS ($K_{sp} = 1 \times 10^{-21}$) will precipitate in Group II, taking conc. $Ni^{2+} = 0.02\ M$.

Solution

a. $K = \dfrac{[H^+]^2 \times [S^{2-}]}{[H_2S]} = 1 \times 10^{-20}$

Solving for $[S^{2-}]$,

$$[S^{2-}] = 1 \times 10^{-20} \times \frac{[H_2S]}{[H^+]^2} = 1 \times 10^{-20} \times \frac{0.1}{(0.3)^2}$$

$$= \boxed{1 \times 10^{-20}\ M}$$

b. To determine whether a precipitate forms, we compare the concentration product, Q, to K_{sp}:

$$Q = (\text{conc. } Cu^{2+}) \times (\text{conc. } S^{2-}) = (2 \times 10^{-2}) \times (1 \times 10^{-20})$$
$$= 2 \times 10^{-22}$$

Since Q is greater than K_{sp} of CuS (1×10^{-36}), we conclude that CuS should precipitate. It does; CuS comes down in Group II.

c. $Q = (\text{conc. } Ni^{2+}) \times (\text{conc. } S^{2-}) = (2 \times 10^{-2}) \times (1 \times 10^{-20})$
$\quad = 2 \times 10^{-22}$

In this case, Q is less than K_{sp} of NiS (1×10^{-21}). Hence, NiS should not precipitate. Fortunately, it doesn't, since Ni^{2+} is in Group III.

Exercise
The K_{sp} value of a certain sulfide MS is 1×10^{-26}. Would M^{2+} fall in Group II or III? Answer: Group II.

The argument we have just gone through can be applied generally to the cations in Groups II and III. The sulfides of all the Group II cations are very insoluble (low K_{sp} values). This means that they are precipitated by H_2S in Group II, even though the concentration of S^{2-} is very low, about 10^{-20} M. In contrast, the Group III sulfides are more soluble (larger K_{sp} values). As a result, the Group III ions stay in solution when Group II is precipitated.

DISSOLVING PRECIPITATES

Relations 17.29 to 17.31 are perhaps most often used to find the equilibrium constant for the dissolving of a precipitate. Table 17.3, p. 604, shows this use in connection with the reactions

$$ZnCO_3(s) + 2H^+(aq) \rightleftharpoons Zn^{2+}(aq) + H_2CO_3(aq)$$

$$Zn(OH)_2(s) + 2H^+(aq) \rightleftharpoons Zn^{2+}(aq) + 2H_2O$$

$$Zn(OH)_2(s) + 2OH^-(aq) \rightleftharpoons Zn(OH)_4^{2-}(aq)$$

Notice that the calculated K's for dissolving the solids in a strong acid are very large $(1 \times 10^7, 4 \times 10^{11})$. Small wonder that both zinc carbonate (ZnCO$_3$) and zinc hydroxide (Zn(OH)$_2$) dissolve readily in strong acid! Indeed, we find that this situation applies to virtually all metal hydroxides and carbonates. Almost without exception, the equilibrium constant for their reaction with strong acid is very large. This explains the observation made in Section 17.4 that hydroxides and carbonates can be depended upon to dissolve in 6 M HCl.

The calculated K for dissolving Zn(OH)$_2$ in strong base is smaller than that for dissolving it in strong acid. However, given a high enough pH

Physically, we can say that adding H^+ reduces $[CO_3^{2-}]$ and $[OH^-]$ to the point where the solid dissolves

(large concentration of OH^- ions), $Zn(OH)_2$ will dissolve in strong base, too.

■ **EXAMPLE 17.13** _____

Given that $K = 1 \times 10^{-2}$ for the reaction

$$Zn(OH)_2(s) + 2OH^-(aq) \rightleftharpoons Zn(OH)_4^{2-}(aq)$$

a. Set up the expression for K.
b. Calculate the solubility (moles/liter) of $Zn(OH)_2$ in 2.0 M NaOH

Solution

a. $K = 1 \times 10^{-2} = \dfrac{[Zn(OH)_4^{2-}]}{[OH^-]^2}$

b. For every mole of $Zn(OH)_2$ that dissolves, one mole of $Zn(OH)_4^{2-}$ forms. If we let s = solubility of $Zn(OH)_2$, then:

$$[Zn(OH)_4^{2-}] = s$$

and we have: $s/[OH^-]^2 = 1 \times 10^{-2}$; $s = 4(1 \times 10^{-2}) = $ **0.04 mol/L**

We see that an appreciable amount of $Zn(OH)_2$ does indeed dissolve in strongly basic solution.

Exercise
At what $[OH^-]$ is the solubility of $Zn(OH)_2$ 0.01 mol/L? Answer: 1 M.

TABLE 17.3   Applications of the Rule of Multiple Equilibria (and the Reciprocal Rule)

1. *Dissolving zinc carbonate in strong acid* (H^+)

$ZnCO_3(s) \rightleftharpoons Zn^{2+}(aq) + CO_3^{2-}(aq)$	$K_1 = K_{sp} \, ZnCO_3$
$CO_3^{2-}(aq) + H^+(aq) \rightleftharpoons HCO_3^-(aq)$	$K_2 = 1/K_a \, HCO_3^-$
$HCO_3^-(aq) + H^+(aq) \rightleftharpoons H_2CO_3(aq)$	$K_3 = 1/K_a \, H_2CO_3$
$ZnCO_3(s) + 2H^+(aq) \rightleftharpoons Zn^{2+}(aq) + H_2CO_3(aq)$	K

$$K = K_1 \times K_2 \times K_3 = \frac{K_{sp}ZnCO_3}{K_a \, HCO_3^- \times K_a \, H_2CO_3} = \frac{2 \times 10^{-10}}{2.1 \times 10^{-17}} = 1 \times 10^7$$

2. *Dissolving zinc hydroxide in strong acid* (H^+)

$Zn(OH)_2(s) \rightleftharpoons Zn^{2+}(aq) + 2OH^-(aq)$	$K_1 = K_{sp} \, Zn(OH)_2$
$2H^+(aq) + 2OH^-(aq) \rightleftharpoons 2H_2O$	$K_2 = 1/(K_w)^2$
$Zn(OH)_2(s) + 2H^+(aq) \rightleftharpoons Zn^{2+}(aq) + 2H_2O$	K

$$K = K_1 \times K_2 = \frac{K_{sp}}{(K_w)^2} = \frac{4 \times 10^{-17}}{1 \times 10^{-28}} = 4 \times 10^{11}$$

3. *Dissolving zinc hydroxide in strong base* (OH^-)

$Zn(OH)_2(s) \rightleftharpoons Zn^{2+}(aq) + 2OH^-(aq)$	$K_1 = K_{sp} \, Zn(OH)_2$
$Zn^{2+}(aq) + 4OH^-(aq) \rightleftharpoons Zn(OH_4)^{2-}(aq)$	$K_2 = K_f \, Zn(OH)_4^{2-}$
$Zn(OH)_2(s) + 2OH^-(aq) \rightleftharpoons Zn(OH)_4^{2-}(aq)$	K

$$K = K_1 \times K_2 = K_{sp} \times K_f = (4 \times 10^{-17})(3 \times 10^{14}) = 1 \times 10^{-2}$$

The approach we followed with $Zn(OH)_2$ can be used for a reaction in which a precipitate dissolves in ammonia (Example 17.14).

■ **EXAMPLE 17.14** _____

Consider the reaction by which silver chloride dissolves in ammonia:
$$AgCl(s) + 2NH_3(aq) \rightleftharpoons Ag(NH_3)_2{}^+(aq) + Cl^-(aq)$$

a. Taking K_{sp} AgCl $= 1.8 \times 10^{-10}$ and K_f $Ag(NH_3)_2{}^+ = 1.7 \times 10^7$, calculate K for the above reaction.
b. Calculate the solubility (moles/liter) of AgCl in 6.0 M NH_3.

Solution

a. First we break the reaction down into two steps:

$$
\begin{array}{ll}
AgCl(s) \rightleftharpoons Ag^+(aq) + Cl^-(aq) & K_1 \\
Ag^+(aq) + 2NH_3(aq) \rightleftharpoons Ag(NH_3)_2{}^+(aq) & K_2 \\
\hline
AgCl(s) + 2NH_3(aq) \rightleftharpoons Ag(NH_3)_2{}^+(aq) + Cl^-(aq) & K = K_1 \times K_2
\end{array}
$$

Note that:

$$K_1 = K_{sp}\,AgCl = 1.8 \times 10^{-10}$$

$$K_2 = K_f\,Ag(NH_3)_2{}^+ = 1.7 \times 10^7$$

Hence,

$$K = (1.8 \times 10^{-10}) \times (1.7 \times 10^7) = \boxed{3.1 \times 10^{-3}}$$

b. The expression for K is

$$K = 3.1 \times 10^{-3} = \frac{[Ag(NH_3)_2{}^+] \times [Cl^-]}{[NH_3]^2}$$

If we let s = solubility AgCl, then:

$$[Ag(NH_3)_2{}^+] = [Cl^-] = s$$

Taking $[NH_3] = 6.0\ M$, we have:

$$s^2/36 = 3.1 \times 10^{-3}; \quad s^2 = 0.11; \quad s = \boxed{0.33\ mol/L}$$

Since K is not very big, $[NH_3]$ needs to be large to dissolve AgCl

Exercise

In solving this problem, we took the *equilibrium* concentration of NH_3 to be 6.0 M. Suppose we took the *original* concentration of NH_3, before any AgCl dissolved, to be 6.0 M. What would be $[NH_3]$? The calculated solubility of AgCl? Answer: $[NH_3] \approx 5.4\ M$; $s = 0.32\ mol/L$.

■ **SUMMARY PROBLEM**

20 mL of a yellow solution is known to contain four anions and 1.00 mg each of the nitrate salts of the following cations: silver, mercury(I), cadmium, aluminum, and sodium.

a. Name the reagent and write the equation for the precipitation of silver. How many drops (0.05 mL/drop) of a 0.0100 M solution of the reagent are required to precipitate all the silver in solution?

b. What reagent separates silver from mercury(I)? Write net ionic equations for the reactions of silver and mercury ions with this reagent.

c. Write the overall net ionic equation for the separation of cadmium as a Group II sulfide. What is K for this reaction? (K_{sp} CdS $= 1 \times 10^{-29}$).

d. In what form does Al^{3+} precipitate in Group III? Write the equation for dissolving this precipitate in strong acid. Calculate K for this reaction, taking K_{sp} Al(OH)$_3$ $= 2 \times 10^{-31}$. What pH is required to dissolve 1 mg/mL of precipitate by this reaction?

e. How would you test for the presence of sodium ion?

f. Separate spot tests for the anions are performed. Assume all interfering ions have been removed. The following results are obtained when 1 mL samples of unknown are tested. What ion is confirmed by each test? Write net ionic equations for all reactions that take place.

1. Treatment with iron(III) nitrate gives a blood-red solution.
2. Treatment with nitric acid turns the yellow solution red-orange.
3. Iron(II) sulfate is added and the solution is acidified. The test tube is tilted and concentrated sulfuric acid is added. A brown ring is formed.
4. An acidic solution of potassium permanganate is added. On addition of hexane, a red layer is obtained.

Answers

a. Reagent is HCl; $Ag^+(aq) + Cl^-(aq) \rightarrow AgCl(s)$; 12 drops

b. $NH_3(aq)$; $AgCl(s) + NH_3(aq) \rightarrow Ag(NH_3)_2^+(aq) + Cl^-(aq)$;
$Hg_2Cl_2(s) + 2NH_3(aq) \rightarrow Hg(l) + Hg(NH_2)Cl(s) + NH_4^+(aq) + Cl^-(aq)$

c. $Cd^{2+}(aq) + H_2S(aq) \rightarrow CdS(s) + 2H^+(aq)$; $K = 1 \times 10^9$

d. Al(OH)$_3$; $Al(OH)_3(s) + 3H^+(aq) \rightarrow Al^{3+}(aq) + 3H_2O$;
$K = 2 \times 10^{11}$; pH = 4.40

e. Flame test—An intense yellow flame indicates sodium.

f. 1. SCN^- present; $Fe^{3+}(aq) + SCN^-(aq) \rightarrow Fe(SCN)^{2+}(aq)$
2. CrO_4^{2-} present; $2CrO_4^{2-}(aq) + 2H^+(aq) \rightarrow Cr_2O_7^{2-}(aq) + H_2O$
3. NO_3^- present;
$Fe^{2+}(aq) + NO(g) \rightarrow Fe(NO)^{2+}(aq)$
4. Br^- present;
$10Br^-(aq) + 2MnO_4^-(aq) + 16H^+(aq) \rightarrow 5Br_2(l) + 2Mn^{2+}(aq) + 8H_2O$

Four of the cations in the qualitative analysis scheme are derived from the post-transition metals tin, lead, antimony, and bismuth. These metals follow the transition metals in the 5th and 6th periods of the Periodic Table.

Properties of the Metals

Some of the properties of these elements are listed in Table 17.4. Notice that their electrical conductivities are considerably smaller, by at least an order of magnitude, than that of silver, a "good" metallic conductor. All of these metals except lead have a nonmetallic allotrope with a conductivity much lower than the one listed. In the case of tin, this allotrope is referred to as grey tin; it has the diamond structure. When white tin (the metallic allotrope) is kept below the transition temperature, 13°C, for long periods of time, grey tin forms as a powder. Hence articles made of tin, notably organ pipes, sometimes crumble in very cold weather. The formation of grey tin seems to spread, like an infection, from a single point. This problem was common in the cold cathedrals of northern Europe in the nineteenth century. The organ pipes were said to suffer from "tin disease," for which there was no known "cure."

These metals are most familiar as their alloys, which include solder (Sn, Pb), bronze (Cu with 5–10% Sn), and pewter (>90% Sn plus Sb, Cu; Fig. 17.13). Wood's metal (50% Bi, 25% Pb, 13% Sn, 12% Cd) melts at about 70°C; it is used to make fusible plugs for automatic sprinkler systems. "Lead" storage batteries are made from a lead alloy containing about 3% antimony and a trace of arsenic.

Figure 17.13 Early American pewter handled church cup, made in Boston in the mid-18th century. (Pocumtuck Valley Memorial Association, Memorial Hall Museum, Deerfield, Massachusetts)

TABLE 17.4   Properties of Tin, Lead, Antimony, and Bismuth

	Sn*	Pb	Sb*	Bi*
Outer Electron Conf.	$5s^2 5p^2$	$6s^2 6p^2$	$5s^2 5p^3$	$6s^2 6p^3$
Abundance (%)	0.00021	0.0013	0.00002	0.0000008
Principal ores	SnO_2	PbS	Sb_2S_3	Bi_2S_3, Bi_2O_3
Melting point (°C)	232	328	631	271
Density (g/cm³)	7.27	11.34	6.70	9.81
Electrical Conductivity (Relative to Ag = 1)	0.14	0.072	0.038	0.013

*Properties listed are for the metallic allotrope.

Oxidation States

All of these metals show two different oxidation states. The lower state (+3 for Sb and Bi; +2 for Sn and Pb) results from the loss of two outer p electrons. If all the outer electrons, s and p alike, are lost, we obtain the higher oxidation state ($+5$ for Sb and Bi; $+4$ for Sn and Pb). The behavior of tin and antimony is representative:

$$_{50}Sn(0) \quad 4d^{10}5s^25p^2 \qquad _{51}Sb(0) \quad 4d^{10}5s^25p^3$$

$$_{50}Sn(II) \quad 4d^{10}5s^2 \qquad _{51}Sb(III) \quad 4d^{10}5s^2$$

$$_{50}Sn(IV) \quad 4d^{10} \qquad _{51}Sb(V) \quad 4d^{10}$$

In the higher oxidation state, the metal is covalently bonded to a more electronegative element, usually oxygen or a halogen. For example, $SnCl_4$ is a volatile liquid (bp = $114°C$); it has a molecular structure analogous to that of CCl_4.

As we move down in the Periodic Table (Sn $\rightarrow$ Pb; Sb $\rightarrow$ Bi), the higher oxidation state becomes less stable. Tin(IV) oxide, SnO_2, is a very stable compound found in nature as the mineral cassiterite. In contrast, PbO_2 (red lead) is relatively unstable; it is a powerful oxidizing agent, readily reduced to PbO or Pb. Among the Group 5 elements, the $+5$ state is quite common with antimony, as it is with nitrogen, phosphorus, and arsenic. With bismuth, the situation is quite different; the $+5$ state is effectively limited to the BiF_5 molecule and the BiO_3^- (bismuthate) ion.

These metals form cations only in the lower oxidation state (Sn^{2+}, Pb^{2+}, Sb^{3+}, Bi^{3+}). Some of the more important compounds and complex ions formed by these cations are listed in Table 17.5 and shown in Figure 17.14. The water-soluble compounds listed in the left column of the table are those you are likely to find in the chemistry storeroom.

Of the four cations listed in Table 17.5, only Pb^{2+} has an extensive and relatively straightforward aqueous solution chemistry. In water, lead(II) is most often present as a simple, hydrated cation, e.g., $Pb(H_2O)_6^{2+}$. Species analogous to this do not exist for the other three cations, except

SnCl₄ is not ionic, that's for sure

TABLE 17.5   Species Containing Sn^{2+}, Pb^{2+}, Sb^{3+}, and Bi^{3+}

CATION	WATER-SOLUBLE COMPOUNDS*	PRECIPITATES	STABLE COMPLEXES
Sn^{2+}	$SnCl_2$, $SnBr_2$, $SnSO_4$	SnI_2(red), SnO(black), SnS(brown)	$Sn(OH)_3^-$
Pb^{2+}	$Pb(NO_3)_2$, $Pb(C_2H_3O_2)_2$	$PbCl_2$, $PbBr_2$, PbI_2(yellow), PbS(black), $PbSO_4$, $PbCO_3$, PbO(orange), $PbCrO_4$(yellow)	$Pb(OH)_3^-$
Sb^{3+}	$SbCl_3$, $SbBr_3$, $Sb_2(SO_4)_3$	SbI_3(red), Sb_2S_3(orange), Sb_2O_3	$SbCl_4^-$, $Sb(OH)_4^-$
Bi^{3+}	$BiCl_3$, $Bi(NO_3)_3$, $BiBr_3$(yellow)	BiI_3(black), Bi_2S_3(black), $Bi(OH)_3$	$BiCl_4^-$

*The salts listed for Sn^{2+}, Sb^{3+}, and Bi^{3+} are soluble only in strong acid; upon dilution, basic salts such as Sn(OH)Cl, SbOCl, and BiOCl precipitate.

Figure 17.14 Among the insoluble compounds of the post-transition metals are $PbCrO_4$ (yellow), $Bi(OH)_3$ (white), and Sb_2S_3 (orange). (Charles D. Winters)

perhaps in very strong acid. For example, in a water solution of an Sb(III) salt, we find such species as $Sb(OH)^{2+}$, $Sb(OH)_2{}^+$, SbO^+, and many others too numerous (and too exotic) to mention. Salts of tin, antimony, and bismuth are typically insoluble in water. They usually dissolve in concentrated HCl or NaOH through complex ion formation (Table 17.5).

Toxicity

Of all the post-transition metal cations, Pb^{2+} has by far the worst reputation. Ingestion of as much as 3 mg (0.003 g) of Pb^{2+} ions per week can cause anemia. If intake of lead compounds occurs over a long period of time, damage to the central nervous system results. The principal source of lead in the environment is the organic compound $Pb(C_2H_5)_4$, used as an antiknock additive in gasoline. Since the use of leaded gasoline is being phased out, we can reasonably expect cases of chronic lead poisoning to decrease.

Antimony compounds are about as poisonous as those of lead, but fortunately much less common. One antimony compound, $KSbC_4H_4O_7$, is used in medicine; its common name is tartar emetic. Compounds of tin and bismuth are generally nontoxic. Organic tin derivatives are used in agriculture to destroy fungi, bacteria, and insects; the assumption is that they are not harmful to humans. Bismuth compounds are available in a few over-the-counter preparations for treating digestive upsets. Perhaps the best known of these is Pepto-Bismol, which contains "bismuth subsalicylate," a species with the simplest formula $BiC_7H_5O_4$.

They say you can't keep a good emetic down

QUESTIONS AND PROBLEMS

Analysis of Cations

1. Complete the following table for cations.

Cation	Analytical Group	Precipitating Agent	Precipitate Formed
Ag^+	_____	_____	_____
Bi^{3+}	_____	_____	_____
Co^{2+}	_____	_____	_____
Mg^{2+}	_____	_____	_____

2. Complete the following table for cations.

Species	Test/Reagent	Response to Test/Reagent
_____	flame test	yellow
Ag^+	_____	precipitate dissolves
Mg^{2+}	_____	_____
_____	H_2S	pale pink precipitate

3. Give the symbol for the cation that
   a. is in Group II and colored blue.
   b. is in Group III and colored green.
   c. is in Group I and forms a complex with ammonia.

4. Give the symbol for the cation that
   a. is in Group IV and gives a violet flame.
   b. is in Group II and forms a bright orange sulfide precipitate.
   c. forms the only white sulfide in Groups II and III.

5. Identify the general characteristic of
   a. Group I cations that allows for their separation from cations of Groups II through IV.
   b. Group IV cations that separates them from those of Groups I through III.

6. All Group II and most Group III cations precipitate as sulfides. Why not bring all these ions down as one group?

7. Explain why
   a. Mg^{2+} does not precipitate in Groups I, II, or III.
   b. heating a solution of NH_4Cl with $NaOH$ gives a gas with a pungent odor.
   c. gas bubbles form when $BaCO_3$ is treated with a strong acid.

8. What would happen if
   a. an unknown that had been tested for Groups I to III were tested for Na^+?
   b. a solution containing Cu^{2+} and Ni^{2+} were treated with H_2S at pH 9?
   c. acid were added to a solution containing the $Sn(OH)_6^{2-}$ ion?

9. In Group II analysis, tin is separated from antimony by the formation of an octahedral complex with the oxalate ion, $C_2O_4^{2-}$. Draw a sketch of the geometry of this complex.

10. Consider the $Al(OH)_4^-$ ion listed in Table 17.2.
   a. What is the coordination number of Al^{3+} in this complex?
   b. Write the electron configuration for Al^{3+}.
   c. Based on your answer to (b), what is the geometry of $Al(OH)_4^-$?

Group I Analysis

11. Write net ionic equations to explain the following observations:
   a. A precipitate forms when solutions of lead nitrate and sodium chloride are mixed.
   b. When ammonia is added to a mixture of silver chloride and mercury(I) chloride, the mixture turns black.
   c. If lead chloride is washed with hot water, a yellow precipitate forms when potassium chromate is added to the hot solution.

12. Write balanced net ionic equations to describe the following changes:
   a. Silver chloride can be dissolved in 6 M ammonia.
   b. Lead chloride is moderately soluble in hot water.
   c. When 6 M nitric acid is added in excess to a solution of silver chloride in 6 M ammonia, a white precipitate forms.

13. Suppose you are working with a Group I unknown that contains only mercury(I) and silver ions. State precisely what will be observed in each step of the analysis.

14. Suppose you have a Group I unknown that contains only lead and silver ions. Describe your observations in each step of the analysis.

15. Select a reagent used in the analysis of Group I that will in one step separate each of the following pairs:
   a. Hg_2Cl_2 and $PbCl_2$
   b. $AgCl$ and $CuCl_2$
   c. $PbCl_2$ and $AgCl$

16. Select a reagent in the analysis of Group I that will in one step separate the following pairs:
   a. Hg_2Cl_2 and $AgCl$ b. Hg_2^{2+} and Hg^{2+}
   c. $AgCl$ and $BiOCl$

17. A student, in analyzing a Group I unknown makes the following errors in procedure. What effect, if any, will they have on the results?
   a. To precipitate the ions, he adds HNO_3 instead of HCl.
   b. To the solution obtained by dissolving $AgCl$ in NH_3, he adds HCl instead of HNO_3.
   c. To a mixed precipitate of $AgCl$ and Hg_2Cl_2, he adds $NaOH$ instead of NH_3.

18. What effect, if any, will each of the following errors have on the analysis of a Group I unknown?

a. The solution obtained by heating $PbCl_2$ with water is allowed to cool to room temperature before being centrifuged.

b. The solution referred to in (a) is treated with HCl instead of K_2CrO_4.

c. The solution obtained by dissolving AgCl in NH_3 is treated with NaOH instead of HNO_3.

19. A solid unknown may contain one or more of the following:

$$AgCl, PbSO_4, Hg_2Cl_2, AgNO_3$$

The solid is stirred with water. A precipitate forms when the solution is treated with HCl. The solid remaining from the first step is treated with 6 M NH_3 and turns black; there is no precipitate when the NH_3 solution is acidified with HNO_3. State which solids are definitely present, which are absent, and which are in doubt.

20. A solid unknown may contain any of the following:

$$PbCl_2, Hg_2(NO_3)_2, AgI, PbCO_3$$

The unknown is treated with cold water; none of it dissolves. The mixture is heated to 100°C, and the liquid is drawn off. Treatment of that liquid with K_2CrO_4 gives a yellow solution but no precipitate. When the original solid is treated with nitric acid, a gas is evolved. State which solids are present, absent, and in doubt.

Tests for Anions

21. Write net ionic equations for the reaction between aqueous solutions of

a. iron(III) nitrate and potassium thiocyanate.

b. iron(II) sulfate and nitrogen oxide.

c. potassium permanganate and hydrogen bromide.

22. Write net ionic equations for the reaction between aqueous solutions of

a. carbonate ion and a strong acid.

b. barium nitrate and sulfuric acid.

c. potassium chromate and hydrochloric acid.

23. Describe the result obtained when

a. iron(II) sulfate is added to a nitrate solution acidified with dilute sulfuric acid. Concentrated sulfuric acid is then added to the solution with the tube in a tilted position.

b. H^+ is added to chromate ion.

c. An aqueous solution containing bromine is extracted with hexane.

24. Describe in words what happens when

a. Solutions of $BaCl_2$ and K_2CrO_4 are mixed.

b. A solution of sodium sulfide is treated with hydrochloric acid.

c. A strong acid is added to $BaSO_4(s)$.

25. You are given a pure salt. How would you determine by a simple test whether it is a sulfate? (Assume it is soluble in water.)

26. You are given a pure salt. How would you determine by a simple test whether it is a nitrate?

27. A solution in a test tube contains no anions other than CrO_4^{2-}, SO_4^{2-}, and CO_3^{2-}. When the solution is acidified with HCl, there is no color change but a gas is formed. Addition of $BaCl_2$ to the acidified solution fails to give a precipitate. However, when $BaCl_2$ is added to the original solution, a white precipitate forms. Which anions are present? Absent? In doubt?

28. Three test tubes contain only one anion each. The anions can be either bromide, carbonate, or thiocyanate. Test tube 1 turns red with iron(III) ion. Test tube 2 when treated with potassium permanganate and then hexane shows a colorless organic layer. Test tube 3 shows no change when treated with acid. What can you say about test tubes 1, 2, and 3?

29. What volume of a solution 0.010 M in Fe^{3+} is needed to oxidize 1.0 mg of NaI? The reaction is: $2Fe^{3+}(aq) + 2I^-(aq) \rightarrow 2Fe^{2+}(aq) + I_2(aq)$.

30. What volume of 0.10 M $KMnO_4$ is required to oxidize 5.0 mL of a solution 0.020 M in Br^- by Reaction 17.14?

Reactions and Equations

31. Write a net ionic equation for the formation of a complex ion when

a. a solution of aluminum chloride is made strongly basic with NaOH.

b. ammonia reacts with silver ion.

c. Sodium hydroxide is added to antimony(III) nitrate.

32. Write a net ionic equation for the formation of a complex ion when

a. a solution of zinc nitrate is treated with ammonia.

b. a solution of tin(IV) chloride is treated with sodium hydroxide.

c. ammonia is added to a solution of nickel(II) nitrate.

33. Write a net ionic equation for the reaction, if any, of each of the following complex ions with H^+:

a. $Ag(CN)_2^-$ b. $AgCl_2^-$

c. $Ni(NH_3)_6^{2+}$ d. $Zn(OH)_4^{2-}$

34. Follow the instructions for Question 33 for

a. $Pb(OH)_3^-$ b. $Ag(NH_3)_2^+$

c. $Al(OH)_4^-$ d. $Zn(NH_3)_4^{2+}$

35. Write net ionic equations to explain why a precipitate of $Cr(OH)_3$ forms when the following reagents are added to a solution containing Cr^{3+} ions.

a. NaOH   b. NH_3

c. Na_2S ($S^{2-} + H_2O \rightarrow OH^- + HS^-$)

36. Write net ionic equations to explain why a precipitate forms when solutions of the following are mixed.

a. Mn^{2+} and S^{2-} b. Mn^{2+} and OH^-

c. Mn^{2+} and NH_3 d. Mn^{2+} and H_2S

37. Write net ionic equations for the reactions of each of the following with strong acid.

a. FeS   b. $Bi(OH)_3$ c. $Cu(NH_3)_4^{2+}$ d. $MgCO_3$

38. Write net ionic equations for the reaction of H^+ with
  a. $SrCO_3$ b. Hg_2Cl_2
  c. $Co(OH)_3$ d. $Sb(OH)_4^-$

39. Write a net ionic equation for the reaction with ammonia by which
  a. silver chloride dissolves.
  b. aluminum ion forms a precipitate.
  c. copper(II) forms a complex ion.

40. Write a net ionic equation for the reaction with ammonia by which
  a. $Cu(OH)_2$ dissolves.
  b. Cd^{2+} forms a complex ion.
  c. Pb^{2+} forms a precipitate.

41. Write a net ionic equation for the reaction with OH^- by which
  a. lead ion forms a precipitate.
  b. lead(II) hydroxide dissolves when more OH^- is added.
  c. lead ion forms a complex ion.

42. Write a net ionic equation for the reaction with OH^- by which
  a. Ni^{2+} forms a precipitate.
  b. Pb^{2+} forms a complex ion.
  c. $Al(OH)_3$ dissolves.

43. Zinc(II) forms the complex ions $Zn(NH_3)_4^{2+}$, $Zn(OH)_4^{2-}$, and $Zn(CN)_4^{2-}$. Write net ionic equations to explain why $ZnCO_3$ dissolves in
  a. NH_3 b. $NaOH$ c. $NaCN$ d. HCl

44. Write net ionic equations to explain why the following precipitates dissolve in 6 M HCl:
  a. $CaCO_3$ b. $Mg(OH)_2$ c. ZnS

Solution Equilibria

45. A solution contains 0.010 M Pb^{2+} and 0.010 M Ag^+. If Cl^- is added to this solution, what is the concentration of Ag^+ when $PbCl_2$ just begins to precipitate? (K_{sp} $PbCl_2$ = 1.7 × 10^{-5}; K_{sp} AgCl = 1.8 × 10^{-10}.)

46. A solution of 10 mL of 0.10 M HCl is mixed with 20 mL of a solution 0.010 M in Pb^{2+}, giving a total volume of 30 mL.
  a. What are the concentrations of Cl^- and Pb^{2+} after mixing?
  b. Will a precipitate of $PbCl_2$ form (K_{sp} $PbCl_2$ = 1.7 × 10^{-5})?

47. Calculate the concentration of S^{2-} in an acid solution that is saturated with H_2S (0.1 M) and has a pH of 0.52.

48. At what pH is $[S^{2-}]$ = 1 × 10^{-10} M if $[H_2S]$ = 0.01 M?

49. In Group II precipitation, $[H^+]$ = 0.3 M, $[H_2S]$ = 0.10 M. Under these conditions, show by calculation whether the following ions will precipitate, assuming their original concentrations are 0.10 M.
  a. Hg^{2+} (K_{sp} HgS = 1 × 10^{-52})
  b. Fe^{2+} (K_{sp} FeS = 2 × 10^{-19})

50. A solution is 0.010 M in H^+ and 0.10 M in H_2S. Taking $[H^+]^2$ × $[S^{2-}]/[H_2S]$ = 1 × 10^{-20}, K_{sp} CdS = 1 × 10^{-29}, and K_{sp} ZnS = 1 × 10^{-20}, determine
  a. the concentration of S^{2-}.
  b. whether CdS will precipitate, taking conc. Cd^{2+} = 0.10 M.
  c. whether ZnS will precipitate, taking conc. Zn^{2+} = 0.10 M.

51. Calculate K for the reaction

$$Al(OH)_3(s) + OH^-(aq) \rightleftharpoons Al(OH)_4^-(aq)$$

$$K_{sp}\ Al(OH)_3 = 2 \times 10^{-31} \qquad K_f\ Al(OH)_4^- = 1 \times 10^{33}$$

52. Calculate K for the reaction

$$Cu(OH)_2(s) + 4NH_3(aq)$$
$$\rightleftharpoons Cu(NH_3)_4^{2+}(aq) + 2OH^-(aq)$$

$$K_{sp}\ Cu(OH)_2 = 2 \times 10^{-19} \qquad K_f\ Cu(NH_3)_4^{2+} = 2 \times 10^{12}$$

53. Using the value of K calculated in Problem 51, determine the solubility (moles/liter) of $Al(OH)_3$ at pH 11.0.

54. Using the value of K calculated in Problem 52, determine the solubility (moles/liter) of $Cu(OH)_2$ in 6.0 M NH_3.

55. Calculate K for the dissolving of magnesium hydroxide in strong acid (K_{sp} $Mg(OH)_2$ = 6 × 10^{-12}).

56. Calculate K for the dissolving of magnesium carbonate in strong acid (K_{sp} $MgCO_3$ = 6.8 × 10^{-6}, K_a H_2CO_3 = 4.4 × 10^{-7}, K_a HCO_3^- = 4.7 × 10^{-11}).

Unclassified

57. Explain why tests for Na^+, K^+, and NH_4^+ are carried out individually and on the original solution rather than that remaining after testing for Group I to IV cations.

58. Calculate the solubility of silver chloride in 6.0 M HCl (K_f $AgCl_2^-$ = 1.8 × 10^5, K_{sp} AgCl = 1.8 × 10^{-10}).

59. How many drops of 1.0 M aqueous ammonia will be required to react with the HCl in 3.0 mL of a solution to be analyzed for Group II that is 0.50 M in HCl? (20 drops = 1.0 mL)

60. What volume of ammonia gas measured at 22°C and 756 mm Hg will be evolved when a solution containing 25 mg of ammonium chloride is treated with NaOH and heated?

61. Calculate K for the reaction in which MnS is dissolved in a strong acid. Calculate $[Mn^{2+}]$ when the pH is 4.50 (K_{sp} MnS = 5 × 10^{-14}).

62. Calculate the solubility, in grams per 100 mL, of AgBr in 6.0 M NH_3.

Challenge Problems

63. How many moles of ammonia are required to dissolve 1.00 g $Cu(OH)_2$ (K_{sp} = 2 × 10^{-19}) suspended in 2.00 L of pure water? (K_f $Cu(NH_3)_4^{2+}$ = 2 × 10^{12}.)

64. Will 2.00 g of zinc hydroxide ($K_{sp} = 4 \times 10^{-17}$) dissolve in 1.00 L of 6.0 M NaOH?

65. A Group III unknown contains only Ni^{2+} and Al^{3+}. It is treated with aqueous ammonia to give a colored precipitate. As more NH_3 is added, part of the precipitate dissolves to give a deep blue solution. The precipitate remaining goes into solution when treated with excess NaOH. If acid is slowly added to this solution, a white precipitate forms that dissolves as more acid is added. Write a balanced net ionic equation for each reaction that takes place.

66. If a metal hydroxide is to dissolve in strong acid (H^+), K for the reaction must be at least 1×10^{-2}. On this basis, calculate the minimum value of K_{sp} for the hydroxide to dissolve if its general formula is

a. MOH   b. $M(OH)_2$ c. $M(OH)_3$

67. $Al(OH)_3$ is precipitated from an NH_3–NH_4^+ buffer in which $[NH_3] = [NH_4^+]$ and orig. conc. $Al^{3+} = 0.1\ M$.

What fraction of the Al^{3+} originally present remains in solution after precipitation? (K_b $NH_3 = 1.8 \times 10^{-5}$; K_{sp} $Al(OH)_3 = 2 \times 10^{-31}$).

68. Taking K_{sp} $Mn(OH)_2 = 2 \times 10^{-13}$, K_a $HC_2H_3O_2 = 2 \times 10^{-5}$, and $K_w = 1.0 \times 10^{-14}$, calculate K for the reaction

$$Mn(OH)_2(s) + 2HC_2H_3O_2(aq)$$
$$\rightleftharpoons Mn^{2+}(aq) + 2H_2O + 2C_2H_3O_2^-(aq)$$

Would you expect $Mn(OH)_2$ to be soluble in acetic acid?

69. Using equilibrium constants tabulated in previous chapters, calculate K for the reaction:

$$Ag(NH_3)_2^+(aq) + 2H^+(aq) + Cl^-(aq)$$
$$\rightleftharpoons AgCl(s) + 2NH_4^+(aq)$$

Explain how this calculation relates to the qualitative analysis of Group I cations.

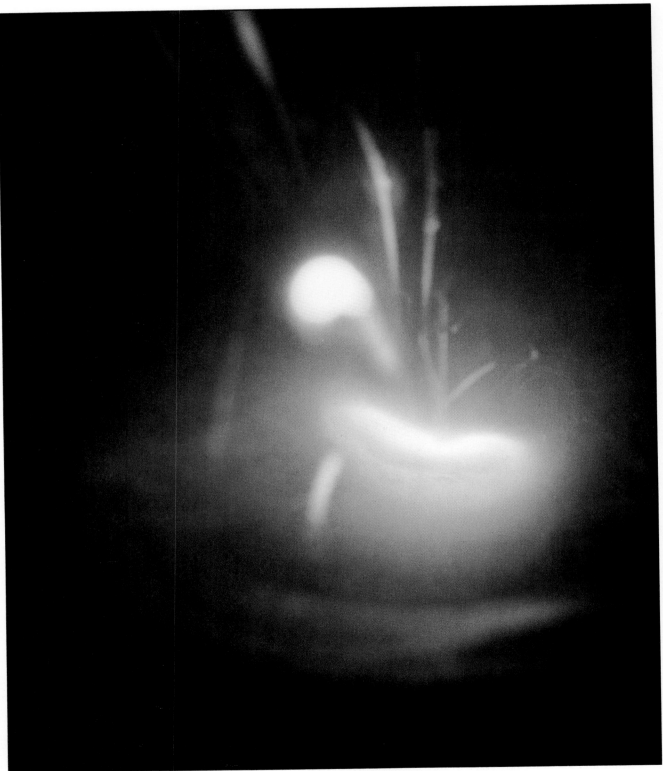

$$P_4(s) \ + \ 5O_2(g) \ \rightarrow \ P_4O_{10}(s)$$

In the presence of oxygen gas, white phosphorus glows and then bursts into flames.

CHAPTER 18

RATE OF REACTION

*Not every collision,
not every punctilious trajectory
by which billiard-ball complexes
arrive at their calculable meeting
places leads to reaction*

. . . .

*Men (and women) are not
as different from molecules
as they think.*

ROALD HOFFMAN
MEN AND MOLECULES

We saw in previous chapters that reactions with large equilibrium constants tend to go virtually to completion. Consider for example

$$CO(g) + NO(g) \rightarrow CO_2(g) + \tfrac{1}{2}N_2(g) \qquad (18.1)$$

$$Fe(s) + \tfrac{3}{2}O_2(g) + \tfrac{3}{2}H_2O(l) \rightarrow Fe(OH)_3(s) \qquad (18.2)$$

$$H_2(g) + \tfrac{1}{2}O_2(g) \rightarrow H_2O(l) \qquad (18.3)$$

At 25°C, the equilibrium constants (K_p) for Reactions 18.1 to 18.3 are 10^{60}, 10^{60}, and 10^{42}, respectively. Reaction 18.1 offers a way to remove the toxic gases CO and NO from automobile exhausts. Reaction 18.2 represents the rusting of iron, while Reaction 18.3 is the synthesis of water from the elements. At room temperature, all of these reactions occur very slowly. We have to wait essentially forever for CO and NO to be converted to CO_2 and N_2 and for hydrogen to react with oxygen (in the absence of a catalyst). It also takes a long time for iron to rust if it is kept clean and dry.

The situations just described are quite common. Many reactions that should in principle go to completion occur very, very slowly. Often, this is to our benefit. A case in point involves the fossil fuels—coal, petroleum, and natural gas. From equilibrium considerations, these fuels should be converted to CO_2 and H_2O upon exposure to air.

It's somewhat sobering to realize that people are thermodynamically unstable

We conclude that there is no correlation between the rate of a reaction and its equilibrium constant. To predict how rapidly a reaction will occur, we must become familiar with the principles of *chemical kinetics*. These principles, which focus on the rate of a reaction and the factors that affect it, are the subject of this chapter. All the reactions we will consider have very large equilibrium constants. Hence, we need only be concerned with the rate of the forward reaction and can ignore the reverse reaction.

The main emphasis in this chapter will be upon the factors that influence reaction rate. These include

— the concentrations of reactants (Sections 18.2 and 18.3)
— the nature of the reaction, as reflected in a quantity called the *activation energy* (Section 18.4)
— the reaction temperature (Section 18.5)
— the presence of a catalyst (Section 18.6)

Before discussing the effect of these factors, it is important to define reaction rate (Section 18.1).

18.1 MEANING OF REACTION RATE

To discuss reaction rate in a meaningful way, we must first understand precisely what this term means. *The rate of reaction is a positive quantity that tells us how the concentration of a reactant or product changes with time.* Most often, reaction rate is expressed in terms of reactant concentrations. Consider, for example, the decomposition of dinitrogen pentoxide, N_2O_5:

$$2N_2O_5(g) \rightarrow 4NO_2(g) + O_2(g) \tag{18.4}$$

for which we can write:

$$\text{rate of decomposition } N_2O_5 = \frac{-\Delta \text{ conc. } N_2O_5}{\Delta t} \tag{18.5}$$

where

$$\Delta \text{ conc. } N_2O_5 = \text{final conc. } N_2O_5 - \text{orig. conc. } N_2O_5$$

$$\Delta t = \text{final time} - \text{orig. time}$$

Rate of reaction is always positive

The minus sign in Equation 18.5 is required to make the rate a positive quantity; the concentration of N_2O_5 decreases with time.

Reaction rate has the units of concentration divided by time. We will always express concentration in moles per liter. On the other hand, time may be given in seconds, minutes, hours, days, or years. Thus, the units of reaction rate may be

$$\frac{\text{mol}}{\text{L·s}}, \quad \frac{\text{mol}}{\text{L·min}}, \quad \frac{\text{mol}}{\text{L·h}}, \dots$$

The magnitude of the reaction rate will depend upon the time unit used. The rate with time in hours will be 60 times the rate with time in minutes. Thus, a rate of 0.05 mol/L·min would be 3 mol/L·h:

Same rate, different units

$$0.05 \frac{\text{mol}}{\text{L·min}} \times \frac{60 \text{ min}}{1 \text{ h}} = 3 \frac{\text{mol}}{\text{L·h}}$$

MEASUREMENT OF RATE

In order to measure the rate of a reaction, we must find how concentration changes with time. To be specific, let us consider how the rate of Reaction 18.4 might be measured. We start by putting 0.160 mol N_2O_5 in a one-

TABLE 18.1   Rate of Decomposition of N_2O_5 at 67°C

Time (min)	0	1	2	3	4
Conc. N_2O_5 (mol/L)	0.160	0.113	0.080	0.056	0.040
Inst. rate (mol/L·min)	0.056	0.039	0.028	0.020	0.014

liter container at 67°C. At one-minute intervals, we withdraw small samples of the reaction mixture and analyze them for N_2O_5. In this way, we obtain the concentration-time data listed in Table 18.1 and plotted in Figure 18.1.

Using the data in Table 18.1, we can readily find the *average rate* of Reaction 18.4 over any desired time interval. Example 18.1 illustrates how this is done.

■ **EXAMPLE 18.1** _____

Calculate the average rate of the N_2O_5 decomposition (Equation 18.5) between

a. $t = 0$ and $t = 1$ min   b. $t = 1$ min and $t = 2$ min

Solution

a. $\text{rate} = \dfrac{-\Delta \text{ conc. } N_2O_5}{\Delta t} = \dfrac{-(0.113 \text{ mol/L} - 0.160 \text{ mol/L})}{1 \text{ min} - 0}$

$\text{rate} = \boxed{0.047 \dfrac{\text{mol}}{\text{L·min}}}$

b. $\text{rate} = \dfrac{-(0.080 \text{ mol/L} - 0.113 \text{ mol/L})}{2 \text{ min} - 1 \text{ min}} = \boxed{0.033 \dfrac{\text{mol}}{\text{L·min}}}$

The rate is not constant and decreases with time

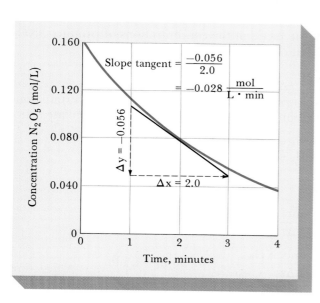

Figure 18.1   The rate of reaction can be determined by measuring concentration of reactant as a function of time and graphing the data. To determine the rate at a particular point, we draw a tangent to the curve and find its slope. This gives us the ratio Δ conc./Δt, which is numerically equal to the reaction rate.

Note that the rate decreases from an average value of 0.047 mol/L·min over the first minute to 0.033 mol/L·min during the second minute.

Exercise
What is the average rate between 0 and 2 min? Answer: 0.040 mol/L·min.

Ordinarily we are interested, not in the average rate, but in the **instantaneous rate** at a given time or concentration. To find the instantaneous rate of Reaction 18.4, we use Figure 18.1. If we draw a tangent to the curve at any point, its slope will equal Δ conc. $N_2O_5/\Delta t$ at that point. From Equation 18.5, we see that the rate of reaction is equal to $-\Delta$ conc. $N_2O_5/\Delta t$. Putting these relations together,

instantaneous rate $= -$(slope of tangent to conc. vs. time curve)

From Figure 18.1 we see that the slope of the tangent at $t = 2$ min is -0.028 mol/L·min. Hence, the rate at that point is

$$-\left(-0.028\ \frac{mol}{L\cdot min}\right) = 0.028\ \frac{mol}{L\cdot min}$$

Instantaneous rates at other points are obtained in a similar way.

So far, in discussing Reaction 18.4, we have focused upon the rate of decomposition of N_2O_5. We could equally well consider the rates of formation of the products, NO_2 or O_2. Each of these rates is related to the others through the coefficients of the balanced equation. Every time a mole of N_2O_5 decomposes, two moles of NO_2 and one-half mole of O_2 are formed. It follows that:

To speak meaningfully about the rate of a reaction, we need to say which conc. is being considered

$$\frac{\Delta\ conc.\ NO_2}{\Delta t} = -2\frac{\Delta\ conc.\ N_2O_5}{\Delta t}\ ;\ \frac{\Delta\ conc.\ O_2}{\Delta t} = -\frac{1}{2}\frac{\Delta\ conc.\ N_2O_5}{\Delta t}$$

rate of formation $NO_2 = -2 \times$ rate of decomposition N_2O_5

rate of formation $O_2 = -\frac{1}{2} \times$ rate of decomposition N_2O_5

(The minus sign reflects the fact that the concentrations of products, NO_2 and O_2, increase when that of the reactant, N_2O_5, decreases.)

■ **EXAMPLE 18.2**
Consider the reaction: $N_2(g) + 3H_2(g) \rightarrow 2NH_3(g)$. Under certain conditions, the rate of formation of NH_3 is 0.020 mol/L·min. Calculate:

a. the rate of change of N_2 concentration
b. the rate of change of H_2 concentration

Solution

a. $\quad\dfrac{\Delta\ conc.\ N_2}{\Delta t} = -\dfrac{1}{2}\dfrac{\Delta\ conc.\ NH_3}{\Delta t} = \boxed{-0.010\ mol/L\cdot min}$

b. $\quad\dfrac{\Delta\ conc.\ H_2}{\Delta t} = -\dfrac{3}{2}\dfrac{\Delta\ conc.\ NH_3}{\Delta t} = \boxed{-0.030\ mol/L\cdot min}$

Exercise
Suppose that Δ conc. $H_2/\Delta t = -0.090$ mol/L·h. What is the rate of disappearance of N_2? Answer: -0.030 mol/L·h.

18.2
REACTION RATE AND CONCENTRATION

In discussing the decomposition of N_2O_5, we pointed out that the rate of reaction decreases with time. From a slightly different point of view, the rate decreases as the concentration of N_2O_5 decreases. This behavior is typical of most reactions. We ordinarily find that reactions proceed more slowly as the concentration of reactant decreases. To increase the rate, we start with a higher concentration of reactant.

It turns out that there is a simple relation between concentration and rate for the decomposition of N_2O_5. You may be able to deduce this relation from the data in Table 18.1. Notice what happens when the concentration decreases by a factor of two (from 0.160 to 0.080 or from 0.080 to 0.040 mol/L). The instantaneous rate is cut in half (from 0.056 to 0.028 or from 0.028 to 0.014 mol/L·min). This suggests that the rate of this reaction is directly proportional to the concentration of N_2O_5. Indeed, this is true, as you can see from Figure 18.2. A plot of rate vs. concentration is a straight line. If extrapolated, it would pass through the origin (rate = 0 when conc. $N_2O_5 = 0$).

RATE EXPRESSION AND RATE CONSTANT

Since the rate of decomposition of N_2O_5 is directly proportional to its concentration, it follows that

$$\text{rate} = k(\text{conc. } N_2O_5) \tag{18.6}$$

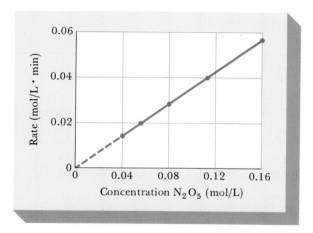

Figure 18.2 For the decomposition of N_2O_5, a plot of rate vs. conc. N_2O_5 is a straight line. The line, if extrapolated, passes through the origin. This means that rate is directly proportional to concentration; that is, rate = $k(\text{conc. } N_2O_5)$.

Equation 18.6 is referred to as the **rate expression** for the decomposition of N_2O_5. It tells us how the rate of the reaction

$$2N_2O_5(g) \rightarrow 4NO_2(g) + O_2(g)$$

depends upon the concentration of reactant. The proportionality constant k in Equation 18.6 is called a **rate constant**. It is independent of the other quantities in the equation.

The rate depends on conc., but the rate constant does not

We can calculate the value of k in Equation 18.6 when we know the rate of reaction at any given concentration of N_2O_5. Solving this equation for k, we have

$$k = \frac{\text{rate}}{\text{conc. } N_2O_5}$$

Note from Table 18.1 that rate = 0.056 mol/(L·min) when conc. N_2O_5 = 0.160 mol/L. Hence,

$$k = \frac{0.056 \text{ mol/L·min}}{0.160 \text{ mol/L}} = 0.35/\text{min}$$

ORDER OF A REACTION FOR A PROCESS INVOLVING A SINGLE REACTANT

Rate expressions have been established by experiment for a large number of reactions. For the process

$$a\, A(g) \rightarrow \text{products}$$

the rate expression has the general form

$$\text{rate} = k(\text{conc. A})^m \tag{18.7}$$

The power to which the concentration of reactant A is raised in the rate expression describes the **order** of the reaction. If m in Equation 18.7 is 0, we say that the reaction is "zero order." If $m = 1$, the reaction is "first order"; if $m = 2$, it is "second order"; and so on.

The most common orders for reactions are 1, 2, and 0

One way to determine the order of a reaction is to obtain the initial rate (i.e., the rate at $t = 0$) as a function of concentration of reactant. The reasoning involved is indicated in Example 18.3.

■ **EXAMPLE 18.3** _____

The initial rate of decomposition of acetaldehyde, CH_3CHO,

$$CH_3CHO(g) \rightarrow CH_4(g) + CO(g)$$

was measured at a series of different concentrations with the following results:

Conc. CH_3CHO (mol/L)	0.10	0.20	0.30	0.40
Rate (mol/L·s)	0.085	0.34	0.76	1.4

Using these data, determine the order of the reaction; that is, determine the value of m in the equation

$$\text{rate} = k(\text{conc. } CH_3CHO)^m$$

Solution

Let us write down the rate expression at two different concentrations:

$$\text{rate}_2 = k(\text{conc.}_2)^m$$
$$\text{rate}_1 = k(\text{conc.}_1)^m$$

Dividing the first equation by the second,

$$\frac{\text{rate}_2}{\text{rate}_1} = \left(\frac{\text{conc.}_2}{\text{conc.}_1}\right)^m$$

Now let us substitute data, taking $\text{conc.}_2 = 0.20\ M$, $\text{conc.}_1 = 0.10\ M$:

$$\frac{0.34}{0.085} = \left(\frac{0.20}{0.10}\right)^m$$

Simplifying,

$$4 = 2^m$$

Clearly, $m = 2$; that is, the reaction is second order.

This is the easiest way to find reaction order

Exercise

Repeat the calculation, using the data at 0.40 and 0.30 M. Answer: $1.4/0.76 = (0.40/0.30)^m$. Solving, $m = 2$ (approximately).

In case the value of m is not obvious (as it is in Example 18.3), it can be found by taking logarithms. That is:

$$\log(\text{rate } 2/\text{rate } 1) = m \log(\text{conc. } 2/\text{conc. } 1)$$

In this case:

$$\log 4 = m \log 2; \quad m = \frac{\log 4}{\log 2} = \frac{0.602}{0.301} = 2$$

Once the order of a reaction has been determined, the same rate data can be used to find the rate constant and then the rate at a new concentration (Example 18.4).

■ EXAMPLE 18.4 _____

Consider the rate data for the decomposition of CH_3CHO given in Example 18.3. Knowing that the reaction is second order, determine

a. the rate constant, k.

b. the rate of reaction when conc. $CH_3CHO = 0.50$ mol/L.

Solution

a. Solving for k and substituting data at the first concentration listed,

$$k = \frac{\text{rate}}{(\text{conc. } CH_3CHO)^2} = \frac{0.085 \text{ mol/L·s}}{(0.10 \text{ mol/L})^2} = 8.5 \frac{L}{\text{mol·s}}$$

b. rate $= 8.5 \dfrac{L}{mol \cdot s} \times \left(0.50 \dfrac{mol}{L}\right)^2 = 2.1 \dfrac{mol}{L \cdot s}$

Exercise
At what concentration of CH_3CHO is the rate of reaction 0.20 mol/L·s?
Answer: 0.15 mol/L.

___

Our discussion to this point illustrates the fact that *the order of a reaction must be determined experimentally. It cannot be deduced from the coefficients of the balanced equation.* The decomposition of acetaldehyde is second order even though the coefficient of CH_3CHO in the balanced equation is 1:

$$CH_3CHO(g) \rightarrow CH_4(g) + CO(g); \quad \text{rate} = k(\text{conc. } CH_3CHO)^2$$

Again, the decomposition of N_2O_5 is first order, even though the coefficient of N_2O_5 in the balanced equation is 2:

$$2N_2O_5(g) \rightarrow 4NO_2(g) + O_2(g); \quad \text{rate} = k(\text{conc. } N_2O_5)^1$$

ORDER OF REACTION FOR A PROCESS INVOLVING MORE THAN ONE REACTANT

Many reactions (indeed, most reactions) involve more than one reactant. For a reaction between A and B,

$$a\,A(g) + b\,B(g) \rightarrow \text{products}$$

the general form of the rate expression is

$$\text{rate} = k(\text{conc. A})^m \times (\text{conc. B})^n \tag{18.8}$$

Here we refer to m as "the order of the reaction with respect to A." Similarly, n is the order of the reaction with respect to B. The **overall order** of the reaction is the sum of the exponents $m + n$. Thus, for the reaction

$$CO(g) + NO_2(g) \rightarrow CO_2(g) + NO(g) \tag{18.9}$$

the experimentally determined rate expression above 600 K is

$$\text{rate} = k(\text{conc. CO}) \times (\text{conc. } NO_2)$$

Hence, we say that this reaction is

— first order with respect to CO ($m = 1$).
— first order with respect to NO_2 ($n = 1$).
— second order overall ($m + n = 2$).

When more than one reactant is involved in a reaction, the order is somewhat more difficult to determine experimentally. One rather straightforward approach involves holding the initial concentration of one reactant constant while varying that of the other reactant. From the rates measured under these conditions, we can find the order of the reaction with respect to the reactant whose initial concentration is changing.

TABLE 18.2 Initial Rates of Reaction (mol/L·s) at 55°C for
$(CH_3)_3CBr(aq) + OH^-(aq) \rightarrow (CH_3)_3COH(aq) + Br^-(aq)$

	SERIES 1			SERIES 2		
	$(CH_3)_3CBr$ (mol/L)	OH^- (mol/L)	RATE	$(CH_3)_3CBr$ (mol/L)	OH^- (mol/L)	RATE
Expt. 1	0.50	0.050	0.0050	1.0	0.050	0.010
Expt. 2	1.0	0.050	0.010	1.0	0.10	0.010
Expt. 3	1.5	0.050	0.015	1.0	0.15	0.010
Expt. 4	2.0	0.050	0.020	1.0	0.20	0.010

To illustrate this approach, let us consider the data in Table 18.2 for the reaction between tertiary butyl bromide and sodium hydroxide solution.

$$(CH_3)_3CBr(aq) + OH^-(aq) \rightarrow (CH_3)_3COH(aq) + Br^-(aq)$$

In the first series of experiments, we hold the initial concentration of hydroxide ion constant at 0.050 M and vary the initial concentration of $(CH_3)_3CBr$. If you look at the data in Series 1, it should be clear that the rate is directly proportional to the concentration of $(CH_3)_3CBr$. For example, when the concentration of $(CH_3)_3CBr$ is doubled (from 1.0 M to 2.0 M), the rate doubles (from 0.010 to 0.020 mol/L·s). This means that in the general rate expression

$$\text{rate} = k(\text{conc. } (CH_3)_3CBr)^m(\text{conc. } OH^-)^n$$

$m = 1$. To find the value of n, we examine the data in the Series 2 experiments. Here $(CH_3)_3CBr$ is held constant at 1.0 M while the concentration of OH^- varies. It should be apparent that the rate is not dependent on the concentration of hydroxide ion. We therefore conclude that in the rate expression above, $n = 0$. (Anything raised to the 0 power equals 1.) Putting the two series of experiments together, we find that

$$\text{rate} = k(\text{conc. } (CH_3)_3CBr)(\text{conc. } OH^-)^0 = k(\text{conc. } (CH_3)_3CBr)$$

■ **EXAMPLE 18.5**

For the reaction: $H_2(g) + 2NO(g) \rightarrow N_2O(g) + H_2O(g)$ at 800°C, the following data are obtained:

	EXPT. 1	EXPT. 2	EXPT. 3
Initial conc. H_2 (mol/L)	0.10	0.20	0.20
Initial conc. NO (mol/L)	0.10	0.10	0.20
Rate (mol/L·s)	0.12	0.24	0.96

Determine the order of reaction with respect to both H_2 and NO.

Solution

a. The concentration of NO is constant at 0.10 M for the first two experiments, so we use that data to determine the order with respect to H_2.

The conc. of NO cancels out

$$\frac{rate_2}{rate_1} = \left(\frac{conc. \ H_2 \ Expt. \ 2}{conc. \ H_2 \ Expt. \ 1}\right)^m; \frac{0.24}{0.12} = \left(\frac{0.20}{0.10}\right)^m; 2 = 2^m; \quad m = 1$$

The reaction is first order in hydrogen.

b. Here we use the second and third experiments, where conc. H_2 is constant at 0.20 M.

$$\frac{0.96}{0.24} = \left(\frac{0.20}{0.10}\right)^n; 4 = 2^n; \quad n = 2$$

The reaction is second order in NO.

Exercise
Calculate the rate constant k for the reaction. Answer: 120 $L^2/mol^2 \cdot s$.

18.3
REACTANT CONCENTRATION AND TIME

A rate expression such as Equation 18.6 tells us how the rate of a reaction changes with concentration. However, from a practical standpoint, we are usually more interested in the relation between concentration and time. Suppose, for example, you are studying the decomposition of N_2O_5. Most likely, you would want to know how much N_2O_5 is left after 5 min, 1 h, or several days. Equation 18.6 is not of much help for that purpose.

It is possible to develop algebraic equations relating reactant concentration to time. The form of the equation may be very simple or quite complex, depending upon the order of the reaction and the number of reactants.

FIRST-ORDER REACTIONS

For a first-order reaction:

$$a A \longrightarrow products; \ rate = k(conc. \ A)$$

it can be shown by using calculus that the relationship between concentration and time is:

$$\ln \frac{X_0}{X} = kt \tag{18.10}$$

where X_0 is the original concentration of A, X is its concentration at time t, k is the first-order rate constant, and the abbreviation "ln" refers to the natural logarithm.

Since $\ln a/b = \ln a - \ln b$, we can rewrite Equation 18.10 in the form:

$$\ln X_0 - \ln X = kt$$

Solving for $\ln X$, we have:

$$\ln X = \ln X_0 - kt \qquad (18.11)$$

Comparing Equation 18.11 to the general equation of a straight line:

$$y = b + mx \qquad (b = y \text{ intercept, } m = \text{slope}) \qquad (18.12)$$

we see that a plot of $\ln X$ ("y") vs. t ("x") should be a straight line with a y-intercept of $\ln X_0$ and a slope of $-k$. This is indeed the case, as you can see from Figure 18.3. Here we have plotted the data from Table 18.1 for the first-order decomposition of N_2O_5. The slope of the straight line is $-1.4/4.0$ min $= -0.35$/min. It follows that the rate constant must be 0.35/min, which is the same value we obtained by a quite different method in Section 18.2.

Equation 18.10 is a very useful one. For first-order reactions, we can use it to calculate

— the concentration of reactant remaining after a given time (Example 18.6a).

— the time required for reactant concentration to drop to a certain level (Example 18.6b and c).

■ **EXAMPLE 18.6** _____

Taking the first-order rate constant for the decomposition of N_2O_5 to be 0.35/min, calculate

a. the concentration of N_2O_5 after 4.0 min, starting with a concentration of 0.160 mol/L.
b. the time required for the concentration to drop from 0.160 to 0.100 mol/L.
c. the time required for half of a sample of N_2O_5 to decompose.

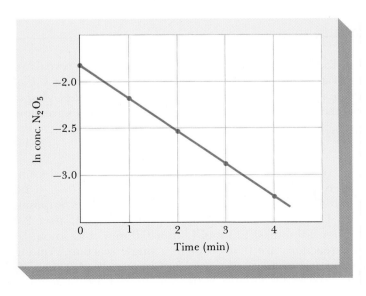

Figure 18.3 For a first-order reaction, a plot of ln conc. vs. time is a straight line.

Solution

a. Using the first-order rate law, Equation 18.10,

$$\ln \frac{0.160}{X} = kt = 0.35 \times 4.0 = 1.4$$

Hence

$$\frac{0.160}{X} = e^{1.4} = 4.0; \qquad X = \frac{0.160}{4.0} = \boxed{0.040 \text{ mol/L}}$$

If $\ln A = B$
$\qquad A = e^B$

Referring back to Table 18.1, we see that, starting with a concentration of N_2O_5 of 0.160 mol/L, the concentration after 4.0 minutes is indeed 0.040 M.

b. Solving Equation 18.10 for t, we have

$$t = \frac{1}{k} \ln \frac{X_0}{X} = \frac{1}{0.35} \ln \frac{0.160}{0.100} = \frac{\ln 1.60}{0.35} = \frac{0.47}{0.35} = \boxed{1.3 \text{ min}}$$

c. When half of the sample has decomposed,

$$X = X_0/2; \qquad X_0 = 2X; \qquad X_0/X = 2$$

Using the equation in (b),

$$t = \frac{1}{k} \ln 2 = \frac{0.693}{k}$$

With $k = 0.35/\text{min}$, we have

$$t = \frac{0.693}{0.35} = \boxed{2.0 \text{ min}}$$

Exercise
Looking back at the data in Table 18.1, how long does it take for the concentration of N_2O_5 to drop from 0.160 to 0.080 M? from 0.080 to 0.040 M? Answer: 2 min.

The analysis of Example 18.6c and the exercise that follows reveals an important feature of a first-order reaction: *The time required for one-half of a reactant to decompose via a first-order reaction has a fixed value, independent of concentration.* This quantity, called the half-life, is given by the expression

$$t_{1/2} = \frac{\ln 2}{k} = \frac{0.693}{k}, \quad \text{first-order reaction} \tag{18.13}$$

where k is the rate constant for the first-order reaction. For the decomposition of dinitrogen pentoxide (N_2O_5), where $k = 0.35/\text{min}$, $t_{1/2} = 2.0$ min. Thus, every two minutes, one-half of a sample of N_2O_5 decomposes. If we start out with 1.00 mol/L of N_2O_5, we have 0.500 mol/L after 2.0 min, 0.250 mol/L after 4.0 min, 0.125 mol/L after 6.0 min, and so on.

Many common reactions are first order in a single reactant. Among these is the process of radioactivity discussed in Chapter 25. Another reaction which follows first-order kinetics is the aquation of a complex ion,

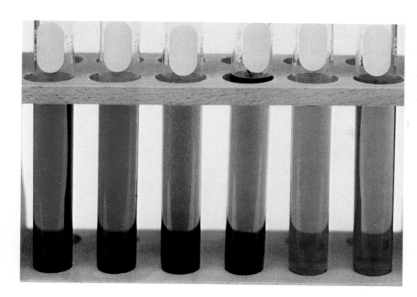

These solutions were prepared by dissolving t-$[Co(en)_2Cl_2]Cl$ in water at 10-minute intervals. The tube at the far left shows the characteristic green color of the t-$Co(en)_2Cl_2^+$ cation. As time passes, this is replaced by the red color of aquo complexes such as $Co(en)_2H_2OCl^+$.

in which one or more ligands are replaced by water molecules. A typical equation is:

$$t\text{-}Co(en)_2Cl_2^+(aq) + H_2O \rightarrow t\text{-}Co(en)_2(H_2O)Cl^{2+}(aq) + Cl^-(aq)$$

green   red

The rate of this reaction is readily followed by observing the color change. Experimentally, we find that the half-life is constant, about 30 min at 25°C, independent of the original concentration of complex. This shows that the reaction is first order in t-$Co(en)_2Cl_2^+$.

REACTIONS OF OTHER INTEGRAL ORDERS (ZERO ORDER, SECOND ORDER)

For a zero-order reaction, we can write

$$a\text{A} \rightarrow \text{products; rate} = k(\text{conc. A})^0 = k$$

since any quantity raised to the zero power, including conc. A, is equal to one. In other words, the rate of a zero-order reaction is constant, independent of concentration. Among the relatively few examples of zero-order reactions is the thermal decomposition of hydrogen iodide on a gold surface:

$$2HI(g) \xrightarrow{\text{Au}} H_2(g) + I_2(g) \tag{18.14}$$

This reaction occurs at a constant rate until all of the hydrogen iodide is gone.

We can readily obtain the concentration-time relation for a zero-order reaction without resorting to calculus. In general, we can say that:

$$-\frac{\Delta \text{ conc. A}}{\Delta t} = k \qquad \text{(zero-order)}$$

But $\Delta t = t - 0 = t$; $-\Delta$ conc. A $= -(X - X_0) = X_0 - X$. So we have:

$$\frac{X_0 - X}{t} = k$$

Solving for X, we obtain: $\qquad X_0 - X = kt \qquad\qquad$ (18.15)

where X is the concentration of reactant at time t, X_0 is the original concentration, and k is the zero-order rate constant.

Comparing Equation 18.15 to the general equation of a straight line

$$y = b + mx$$

we see that, for a zero-order reaction, a plot of concentration (X) vs. time (t) is a straight line with an intercept of X_0 and a slope of $-k$. To obtain an expression for the half-life, note that:

$$X = \tfrac{1}{2} X_0 \text{ when } t = t_{1/2}$$

Substituting in Equation 18.15:

$$\tfrac{1}{2} X_0 = X_0 - kt_{1/2}$$

$$t_{1/2} = \frac{X_0}{2k}, \qquad \text{zero-order reaction} \qquad\qquad (18.16)$$

And it's all over after two half-lives

For a zero-order reaction, the half-life is *directly proportional* to the original concentration.

For a second-order reaction involving a single reactant:

$$a\text{A}(g) \rightarrow \text{ products; rate } = k(\text{conc. A})^2$$

we again have to resort to calculus* to obtain the concentration–time relationship, which is:

$$\frac{1}{X} - \frac{1}{X_0} = kt \qquad\qquad (18.17)$$

where the symbols X, X_0, t, and k have their usual meanings. Comparing Equation 18.17 to Equation 18.12, we see that a plot of $1/X$ vs. t should be a straight line; the slope of that line is the rate constant k. To find the expression for the half-life, we substitute $X = \tfrac{1}{2} X_0$:

The more reactant you have initially, the shorter the half-life

$$\frac{1}{\tfrac{1}{2}X_0} - \frac{1}{X_0} = kt_{1/2}; \qquad t_{1/2} = \frac{1}{kX_0}, \text{ second-order reaction} \qquad (18.18)$$

---

*If you are taking a course in calculus, you may be surprised to learn how useful it can be in the real world (i.e., chemistry). The general rate expressions for zero, first, and second-order reactions are:

$$\frac{-dX}{dt} = k \qquad \frac{-dX}{dt} = kX \qquad \frac{-dX}{dt} = kX^2$$

Integrating these equations, from 0 to t and from X_0 to X, you should be able to derive Equations 18.10, 18.15, and 18.17.

TABLE 18.3 **Characteristics of Zero-, First-, and Second-Order Reactions of the Form $a\ A(g) \rightarrow$ products; X, X_0 = conc. A at t and $t = 0$, respectively**

ORDER	RATE EXPRESSION	CONC.–TIME RELATION	HALF-LIFE	LINEAR PLOT
0	rate $= k$	$X_0 - X = kt$	$X_0/2k$	X vs. t
1	rate $= kX$	$\ln \dfrac{X_0}{X} = kt$	$0.693/k$	$\ln X$ vs. t
2	rate $= kX^2$	$\dfrac{1}{X} - \dfrac{1}{X_0} = kt$	$1/kX_0$	$\dfrac{1}{X}$ vs. t

For a second-order reaction, the half-life is *inversely proportional* to the original concentration.

The characteristics of zero-, first-, and second-order reactions are summarized in Table 18.3. Example 18.7 illustrates how these characteristics can be used to determine the order of a reaction for which concentration–time data are available.

■ **EXAMPLE 18.7** ⎯⎯⎯⎯⎯⎯⎯⎯⎯⎯⎯⎯⎯⎯⎯⎯⎯⎯⎯⎯⎯⎯⎯⎯⎯

The following data were obtained for the gas-phase decomposition of hydrogen iodide:

time (h)	0	2	4	6
Conc. HI (M)	1.00	0.50	0.33	0.25

Is this reaction zero, first, or second order in HI?

Solution

It is useful to prepare a table in which we list X, $\ln X$, and $1/X$ as a function of time, letting $X =$ conc. HI.

t(h)	X	$\ln X$	$1/X$
0	1.00	0	1.0
2	0.50	−0.69	2.0
4	0.33	−1.10	3.0
6	0.25	−1.39	4.0

Pick the function that varies linearly with time

If it is not obvious from the table above that the only linear plot will be that of $1/X$ vs. time, that point should be clear from Fig. 18.4, p. 630. We conclude that we are dealing with a   second-order reaction.

Exercise

The half-life of a certain reaction is directly proportional to the original concentration of reactant. Using Table 18.3, determine the order of the reaction. Answer: Zero order.

Figure 18.4 Decomposition of HI (Example 18.7). Here, X is the concentration of HI at time t. If the reaction were zero-order, a plot of X vs. t would be linear; clearly it isn't. If the reaction were first-order, a plot of $\ln X$ vs. t would be linear. For a second-order reaction such as this one, a plot of $1/X$ vs. t is a straight line.

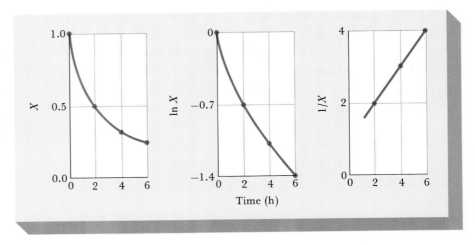

Time (h)

18.4 ACTIVATION ENERGY

This is a fundamental postulate of rate theory

Chemical reactions ordinarily occur as a result of collisions between reacting particles. This simple statement explains many of the characteristics of reaction rates. To illustrate the usefulness of this collision model, let us apply it to the reaction between carbon monoxide and nitrogen dioxide:

$$CO(g) + NO_2(g) \rightarrow CO_2(g) + NO(g)$$

We believe that this reaction takes place as the result of collisions between CO and NO_2 molecules. This is consistent with the rate expression

$$\text{rate} = k(\text{conc. CO}) \times (\text{conc. } NO_2)$$

If we double the concentration of CO, holding that of NO_2 constant, the number of these collisions in a given time doubles (Fig. 18.5). Doubling the concentration of NO_2 (holding that of CO constant) has the same effect. In general, the number of collisions per unit time is directly proportional to the concentration of carbon monoxide (CO) and to the concentration of nitrogen dioxide (NO_2). The fact that the rate is also directly proportional

Figure 18.5 Consider the reaction: $A + B \rightarrow$ products, where reaction occurs as the result of collision. If we double the number of A molecules (square at right), there will be twice as many collisions per unit time, so the rate will be doubled. In general, the rate will be directly proportional to the concentration of A. The same reasoning applies to B, so the rate law is rate = k(conc. A) $\times$ (conc. B).

A molecule   B molecule

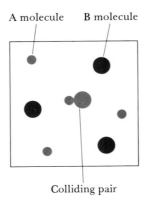
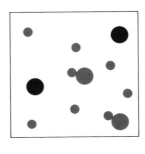

Colliding pair

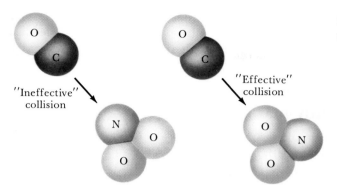

"Ineffective"
collision

"Effective"
collision

Figure 18.6 For a collision to result in reaction, the molecules must be properly oriented.

to these concentrations implies that reaction occurs as a direct result of collisions between CO and NO_2 molecules.

There is a restriction on this simple model for the $CO-NO_2$ reaction. We can easily show that *not every $CO-NO_2$ collision leads to reaction.* From the kinetic theory of gases, it is possible to calculate the rate at which molecules collide with each other. For a mixture of carbon monoxide and nitrogen dioxide at 700 K and concentrations of 0.10 mol/L, it turns out that every molecule should collide with about a billion molecules of the other reactant in one second. If every collision were effective, the reaction between CO and NO_2 should be over in a fraction of a second. By experiment, we find this is not the case. Under the conditions specified, the half-life of the reaction is about 10 s.

There are a couple of reasons why not every collision between CO and NO_2 molecules leads to reaction. For one thing, the molecules have to be properly oriented with respect to one another when they collide. Suppose, for example, that the carbon atom of a CO molecule strikes the nitrogen atom of an NO_2 molecule (Fig. 18.6). This is very unlikely to result in the transfer of an oxygen atom from NO_2 to CO, which is required for reaction to occur.

A more important factor that reduces the number of effective collisions has to do with the kinetic energy of reactant molecules. These molecules are held together by strong chemical bonds. When the molecules collide, part of their kinetic energy is converted to vibrational energy. If the total kinetic energy is large, the molecules will vibrate strongly enough so that the bonds within the molecules are weakened and eventually broken. Molecules with small kinetic energies bounce off one another and retain their identity. They do not react when they collide.

For every reaction, there is a certain minimum energy required to bring it about. This is referred to as the **activation energy**. It has the symbol E_a and is expressed in kilojoules. For the reaction between 1 mol CO and 1 mol NO_2, E_a is 134 kJ. The colliding molecules (CO and NO_2) must have a total kinetic energy of at least 134 kJ/mol if they are to react.

We find that the activation energy for a reaction

— *is a positive quantity* ($E_a > 0$).

— *depends upon the nature of the reaction.* Other factors being equal, we expect "fast" reactions to have a small activation energy. A reaction with a large

There would probably be an explosion

The high energy types get the work done

activation energy takes place slowly under ordinary conditions. The larger the value of E_a, the smaller will be the fraction of molecules having enough kinetic energy to react when they collide.

— *is independent of temperature or concentration.*

ACTIVATION ENERGY DIAGRAMS

Figure 18.7 is an energy diagram for the CO–NO_2 reaction. Reactants, CO and NO_2, are shown at the left. Products, CO_2 and NO, are at the right. They have an energy 226 kJ less than that of the reactants; ΔH for the reaction is -226 kJ. In the center of the figure is an intermediate called an *activated complex.* This is an unstable, high-energy species that must be formed before the reaction can occur. It has an energy 134 kJ greater than that of the reactants and 360 kJ greater than that of the products. The activation energy, 134 kJ, is absorbed in converting the reactants to the activated complex. Here, as in all reactions, the activation energy is a positive quantity. This is true even if, as in this case, the reaction itself is exothermic, i.e., the products have a lower energy than the reactants.

The activated complex has high energy

The exact nature of the activated complex is difficult to determine. For this reaction, the activated complex might be a "pseudomolecule" made up of CO and NO_2 molecules in close contact. The path of the reaction might be more or less as follows:

$$O{\equiv}C + O{-}N{\diagdown}_O \quad \rightarrow O{\equiv}C \cdots O \cdots N{\diagdown}_O \quad \rightarrow O{=}C{=}O + N{=}O$$

reactants   activated complex   products

The dotted lines stand for "partial bonds" in the activated complex. The $N{-}O$ bond in the NO_2 molecule has been partially broken. A new bond between carbon and oxygen has started to form.

Recent studies using molecular beams to mix reactant molecules have made it possible to establish, among other things, the required orientation

Figure 18.7 During the reaction initiation, 134 kJ—the activation energy E_a—must be furnished to the reactants for every mole of CO that reacts. This energy activates each CO–NO_2 complex to the point where reaction can proceed.

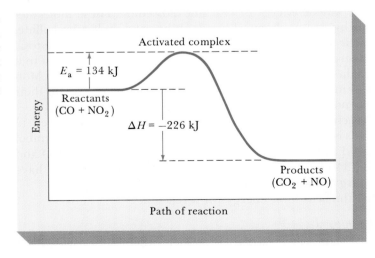

for reaction to occur between two molecules. The 1986 Nobel Prize in chemistry was awarded to John Polanyi, Douglas Herschbach, and Yuan Lee for their research in this area.

18.5
REACTION RATE AND TEMPERATURE

The rates of most reactions increase as the temperature rises. A person in a hurry to prepare dinner applies this principle in using a pressure cooker to raise the temperature for cooking potatoes, apples, or a pot roast (not all at the same time, we trust). By storing the leftovers in a refrigerator, we slow down the chemical reactions responsible for food spoilage. As a general and very approximate rule, it is often stated that an increase in temperature of 10°C doubles the reaction rate. If this rule holds, foods should cook twice as fast in a pressure cooker at 110°C as in an open saucepan, and deteriorate four times as rapidly at room temperature (25°C) as they do in a refrigerator at 5°C.

A freezer really slows things down

    The effect of temperature on reaction rate can be explained in terms of the kinetic theory of gases. Recall from Chapter 4 that raising the temperature greatly increases the fraction of molecules having very high kinetic energies. These are the molecules that are most likely to react when they collide. The higher the temperature, the larger the fraction of molecules that can provide the activation energy required for reaction. This effect is illustrated in Figure 18.8, where the distribution of kinetic energies among gas molecules is shown at two different temperatures. The shaded areas include those molecules having a kinetic energy equal to or greater than E_a. Note that this area is considerably larger at the higher temperature. This means that at the higher temperature, a larger fraction of the molecules will have sufficient energy to react when they collide. Hence, the reaction will go faster. Putting it another way, the rate constant, k, becomes larger as the temperature increases. Table 18.4 shows this effect for the

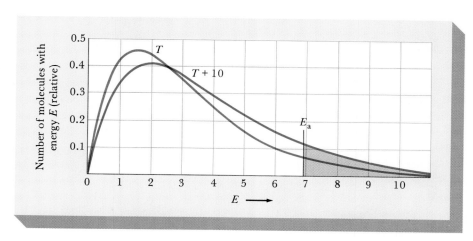

Figure 18.8 When the T is increased to $T + 10$, the fraction of molecules with very high energies increases sharply. Hence, many more molecules possess the activation energy, E_a, and reaction occurs more rapidly. If E_a is of the order of 50 kJ/mol, an increase in temperature of 10°C approximately doubles the number of molecules having energy E_a or greater, and thus doubles the reaction rate.

TABLE 18.4   Temperature Dependence of the Rate Constant for the Reaction
$$CO(g) + NO_2(g) \rightarrow CO_2(g) + NO(g)$$

T (K)	600	650	700	750	800
k (L/mol·s)	0.028	0.22	1.3	6.0	23

So at 800K, 1000 times as many molecules have enough energy to react as do at 600K

CO–NO_2 reaction. Notice that the rate constant increases by a factor of nearly a thousand when the temperature rises from 600 to 800 K.

RELATION BETWEEN k AND T

The argument we have just gone through can be made quantitative. Kinetic theory tells us that the fraction, f, of molecules having an energy equal to or greater than E_a is

$$f = e^{-E_a/RT} \tag{18.19}$$

where e is the base of natural logarithms, R is the gas constant, E_a is the activation energy, and T is the absolute temperature in K. If we assume that the rate constant, k, is directly proportional to f (which is approximately true),

$$k = cf = ce^{-E_a/RT} \tag{18.20}$$

where c is a constant. Taking the natural logarithm of both sides of Equation 18.20, we obtain:

$$\ln k = \ln c - E_a/RT$$

$$\ln k = A - E_a/RT \tag{18.21}$$

where A is a constant which is independent of the other quantities in the equation. This equation was first shown to be valid by the Swedish physical

Antimony powder reacts more rapidly with bromine at high temperatures (right photo). The reaction is: $2Sb(s) + 3Br_2(l) \rightarrow 2SbBr_3(s)$. (James Morgenthaler)

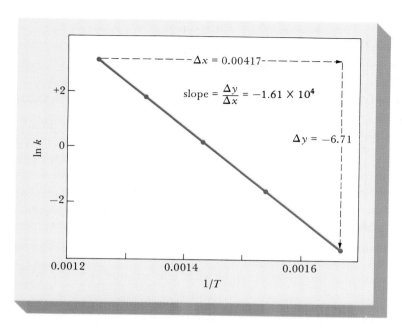

Figure 18.9 A plot of ln k vs. $1/T$ (k = rate constant, T = Kelvin temperature) is a straight line. From the slope of this line, the activation energy can be determined: $E_a = -R \times$ slope. For the CO–NO$_2$ reaction shown here, $E_a = -(8.31)(-1.61 \times 10^4)$ J/mol $= 1.34 \times 10^5$ J/mol = 134 kJ/mol.

chemist Svante Arrhenius in 1889 and is referred to as the Arrhenius equation.

Comparing Equation 18.21 to the general equation for a straight line (18.12), we see that a plot of ln k ("y") vs. $1/T$ ("x") should be a straight line with a slope of $-E_a/R$. In Figure 18.9, such a plot is shown for the CO–NO$_2$ reaction, using the data in Table 18.4; it is indeed a straight line. The activation energy can be determined from the slope:

$$E_a = -R \text{ (slope)} \tag{18.22}$$

Taking $R = 8.31$ J/mol·K, and the slope to be -1.61×10^4 K, we find:

$$E_a = \left(-8.31 \frac{J}{mol \cdot K}\right)(-1.61 \times 10^4 \text{ K})$$
$$= 1.34 \times 10^5 \text{ J/mol} = 134 \text{ kJ/mol}$$

"TWO-POINT" EQUATION RELATING k AND T

We can use Equation 18.21 to obtain a relation between rate constants, k_2 and k_1, at two different temperatures, T_2 and T_1. We follow the same procedure used with the Clausius-Clapeyron equation (Chapter 10). At the two temperatures,

$$\ln k_2 = A - \frac{E_a}{RT_2}$$

$$\ln k_1 = A - \frac{E_a}{RT_1}$$

Subtracting the second equation from the first,

$$\ln k_2 - \ln k_1 = \ln \frac{k_2}{k_1} = \frac{E_a}{R}\left(\frac{1}{T_1} - \frac{1}{T_2}\right)$$

or

$$\ln \frac{k_2}{k_1} = \frac{E_a}{R}\left(\frac{T_2 - T_1}{T_1 T_2}\right) \qquad (18.23)$$

In using Equation 18.23, we ordinarily take T_2 to be the higher temperature to avoid the use of negative logarithms. Remember that if we take $R = 8.31$ J/mol·K, the activation energy will be expressed in joules per mole.

■ EXAMPLE 18.8 _____

a. The activation energy of a certain reaction is 9.32×10^4 J/mol. At 27°C, $k = 1.25 \times 10^{-2}$ L/mol·s. Calculate k at 127°C.
b. For the reaction referred to in (a), at what temperature is $k = 2.50 \times 10^{-2}$ L/mol·s?
c. What must be the value of E_a for a reaction if the rate constant is to double when the temperature increases from 15°C to 25°C?

Solution
a. Using the "two-point" form of the Arrhenius equation, 18.23, we have

$$T_2 = 127 + 273 = 400 \text{ K}; \qquad T_1 = 27 + 273 = 300 \text{ K}$$

$$\ln \frac{k_2}{k_1} = \frac{9.32 \times 10^4}{8.31}\left(\frac{400 - 300}{400 \times 300}\right) = 9.35$$

This means that

INV LN of 9.35 = 1.15 × 10⁴

$$\frac{k_2}{k_1} = e^{9.35} = 1.15 \times 10^4$$

Hence,

$$k_2 = 1.15 \times 10^4 \, k_1 = (1.15 \times 10^4)(1.25 \times 10^{-2} \text{ L/mol·s})$$

$$= \quad 1.44 \times 10^2 \text{ L/mol·s}$$

b. Here we know k_2, k_1, E_a, and T_1; we need to calculate T_2:

$$k_2 = 2.50 \times 10^{-2} \text{ L/mol·s}; \qquad k_1 = 1.25 \times 10^{-2} \text{ L/mol·s}$$

$$T_1 = 300 \text{ K}; \qquad E_a = 9.32 \times 10^4 \text{ J/mol}$$

$$\ln \frac{2.50 \times 10^{-2}}{1.25 \times 10^{-2}} = 0.693 = \frac{9.32 \times 10^4 \, (T_2 - 300)}{(8.31)(300)T_2}$$

Simplifying,

$$\frac{T_2 - 300}{T_2} = \frac{0.693 \times 8.31 \times 300}{9.32 \times 10^4} = 0.0185$$

Solving,

$$0.9815\ T_2 = 300\ K; \qquad T_2 = \boxed{306\ K}$$

c. If $k_2/k_1 = 2$, then $\ln k_2/k_1 = \ln 2 = 0.693$. Substituting in Equation 18.23,

$$0.693 = \frac{E_a\ (298 - 288)}{(8.31)(298)(288)}$$

Working with this equation on your calculator is easy, once you have done it a few times

Solving,

$$E_a = 4.94 \times 10^4\ J/mol = \boxed{49.4\ kJ/mol}$$

Note that if E_a were appreciably greater than 50 kJ, k would more than double for a 10°C rise in temperature; if E_a were smaller than 50 kJ, k would increase by less than a factor of two. Clearly, the empirical rule that a temperature increase of 10°C doubles the reaction rate is at best a crude approximation.

Exercise

For the reaction referred to in parts (a) and (b), how large a temperature increase is required to double the rate? Answer: About 6°C.

18.6 CATALYSIS

The rate of a reaction can be increased by raising the temperature. However, this is not always feasible or practical. Some reactants and products decompose at high temperatures. From an economic standpoint, raising the temperature means increased energy costs. Fortunately, there are certain substances called *catalysts* that offer an alternative approach to speeding up a reaction.

A catalyst increases the rate of a reaction without being consumed by it. A catalyst operates by lowering the activation energy. Consider, for example, the decomposition of N_2O:

$$N_2O(g) \rightarrow N_2(g) + \tfrac{1}{2}O_2(g) \tag{18.24}$$

For the direct reaction, taking place in the gas phase, the activation energy is 250 kJ/mol. On a gold surface, which acts as a catalyst for this reaction, the activation energy is only 120 kJ/mol. Hence, the reaction occurs more rapidly on a gold surface.

The lowering of E_a speeds the reaction up fantastically

We see from Figure 18.10, p. 638, that a catalyst does not affect the relative energies of reactants and products. Neither does it change the equilibrium constant for the reaction. Adding a catalyst leaves the equilibrium composition unchanged, but it allows us to reach equilibrium more rapidly.

Figure 18.10   By changing the path by which a reaction occurs, a catalyst can lower the activation energy that is required and so speed up the reaction.

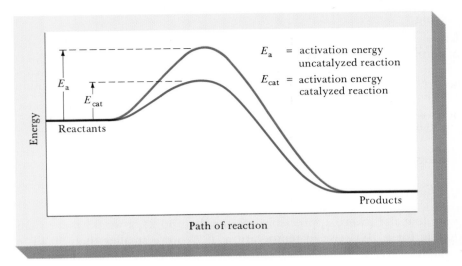

HETEROGENEOUS CATALYSIS

A heterogeneous catalyst is one that is in a different phase from the reaction mixture. Most commonly, the catalyst is a solid that increases the rate of a gas phase or liquid phase reaction. An example is the reaction previously cited:

$$N_2O(g) \xrightarrow{\text{Au}} N_2(g) + \tfrac{1}{2}O_2(g)$$

In the catalyzed decomposition, N_2O is chemically adsorbed on the surface of the solid. A chemical bond is formed between the oxygen atom of an N_2O molecule and a gold atom at the surface. This weakens the bond joining nitrogen to oxygen, making it easier for the N_2O molecule to break apart. We might show this process symbolically as:

$$N{\equiv}N{-}O(g) + Au(s) \rightarrow N{\equiv}N{\text{---}}O{\text{---}}Au(s) \rightarrow N{\equiv}N(g) + O(g) + Au(s)$$

where the broken lines represent weak covalent bonds.

Heterogeneous catalysis is widely used in industry to prepare such compounds as ammonia (Chap. 19), sulfuric acid (Chap. 24), and methyl alcohol (Chap. 26). It is also used in the catalytic converter of your automobile to reduce emissions of air pollutants. We will have more to say on that topic in Chapter 19.

Most industrial processes involve catalysis

HOMOGENEOUS CATALYSIS

A *homogeneous catalyst* is one that is present in the same phase as the reactants. It speeds up the reaction by forming a reactive intermediate that decomposes to give products. In this way, the catalyst provides an alternative path of lower activation energy for the reaction.

An example of a reaction that is subject to homogeneous catalysis is the decomposition of hydrogen peroxide:

$$2H_2O_2(aq) \rightarrow 2H_2O + O_2(g) \tag{18.25}$$

Under ordinary conditions, this reaction occurs very slowly. Dilute solutions of hydrogen peroxide sold in drugstores are stable for considerable periods of time. However, if a solution of sodium iodide, NaI, is added, reaction occurs almost immediately; you can see the bubbles of oxygen forming.

The homogeneous catalysis of Reaction 18.25 is believed to take place by a two-step process:

Step 1: $H_2O_2(aq) + I^-(aq) \rightarrow H_2O + IO^-(aq)$
Step 2: $\underline{H_2O_2(aq) + IO^-(aq) \rightarrow H_2O + O_2(g) + I^-(aq)}$
$\qquad\qquad\quad 2H_2O_2(aq) \rightarrow 2H_2O + O_2(g)$

> The catalyst participates in the reaction but is not used up

Notice that the end result is the same as in the direct reaction (Equation 18.25). The I^- ions are not consumed in the reaction. For every I^- ion used up in the first step, one is produced in the second step. The activation energy for this two-step process is much smaller than for the uncatalyzed reaction.

18.7
REACTION MECHANISMS

A reaction mechanism is a description of a path, or a sequence of steps, by which a reaction might occur at the molecular level. In the simplest case, only a single step is involved. This is a collision between two reactant molecules. This is the "mechanism" for the reaction of CO with NO_2 at high temperatures, above about 600 K:

$$CO(g) + NO_2(g) \rightarrow NO(g) + CO_2(g)$$

In practice, most reactions occur in more than one step. Consider, once again, the reaction between carbon monoxide and nitrogen dioxide. Research indicates that at low temperatures, below about 500 K, the reaction occurs in two steps:

Step 1: $NO_2(g) + NO_2(g) \rightarrow NO_3(g) + NO(g)$
Step 2: $\underline{NO_3(g) + CO(g) \rightarrow NO_2(g) + CO_2(g)}$
$\qquad\quad CO(g) + NO_2(g) \rightarrow NO(g) + CO_2(g)$

We have used color to distinguish the individual steps from the equation for the reaction.

ELEMENTARY STEPS

The individual steps that constitute a reaction mechanism are called *elementary steps*. An elementary step may involve the decomposition of a single molecule:

$A \rightarrow B + C$ (unimolecular step)

or a collision between two molecules, which may be the same or different:

Some reactions involve many steps

$$A + A \rightarrow B + C \quad \text{(bimolecular step)}$$

$$A + B \rightarrow C + D \quad \text{(bimolecular step)}$$

Simultaneous collisions of three molecules are also possible, although relatively rare.

$$A + B + C \rightarrow D + E \quad \text{(termolecular step)}$$

We can readily write the rate expression for an elementary step. **The rate of such a step is equal to a rate constant k multiplied by the concentration of each reactant molecule.** Thus we have:

$$A \rightarrow B + C; \text{ rate} = k(\text{conc. A}) \tag{18.26}$$

$$A + A \rightarrow B + C; \text{ rate} = k(\text{conc. A})(\text{conc. A})$$
$$= k(\text{conc. A})^2 \tag{18.27}$$

$$A + B \rightarrow C + D; \text{ rate} = k(\text{conc. A})(\text{conc. B}) \tag{18.28}$$

$$A + B + C \rightarrow D + E; \text{ rate} = k(\text{conc. A})(\text{conc. B})(\text{conc. C}) \tag{18.29}$$

The justification for this rule is straightforward. We pointed out earlier that for a process of the type

$$A + B \rightarrow C + D$$

the number of collisions per unit time is directly proportional to the concentrations of A and B (see Fig. 18.5). Since reaction occurs via collision, the rate must also be proportional to these concentrations, i.e.,

$$\text{rate} = \text{constant(conc. A)(conc. B)}$$

The validity of the other relations can be demonstrated by a similar but somewhat more involved line of reasoning (see Problem 73).

SLOW STEPS

Often, one step in a mechanism is much slower than any other. If this is the case, **the slow step is rate-determining**. That is, the rate of the overall reaction can be taken to be that of the slow step. Consider, for example, a three-step reaction:

Step 1:	A → B	fast
Step 2:	B → C	slow
Step 3:	C → D	fast

$$\overline{\quad A \rightarrow D \quad}$$

The rate at which A is converted to D (the overall reaction) is approximately equal to the rate of conversion of B to C (the slow step).

To understand the rationale behind this rule, and its limitations, consider an analogous situation. Suppose three people (A, B, and C) are assigned to grade general chemistry examinations that contain three questions. On the average, A spends 10 s grading Question 1 and B spends 15 s grading Question 2. In contrast, C, the ultimate procrastinator, takes 5 min to grade Question 3. The rate at which exams are graded is

Maybe he's a kind grader

$$\frac{1 \text{ exam}}{10 \text{ s} + 15 \text{ s} + 300 \text{ s}} = \frac{1 \text{ exam}}{325 \text{ s}} = 0.00308 \text{ exam/s}$$

This is approximately equal to the rate of the slow grader:

$$\frac{1 \text{ exam}}{300 \text{ s}} = 0.00333 \text{ exam/s}$$

Extrapolating from general chemistry exams to chemical reactions, we can say that

— the overall rate of the reaction cannot exceed that of the slowest step.

— if that step is by far the slowest, its rate will be approximately equal to that of the overall reaction.

The "slow step" rule explains why we cannot deduce the order of a reaction from the coefficients of the balanced equation. If the overall rate is that of the slowest step, it is the coefficients in that step that should appear in the rate expression. The chances are that these coefficients will *not* be the same as those in the balanced equation for the reaction.

The reacting species probably aren't the same either

DEDUCING A RATE EXPRESSION FROM A PROPOSED MECHANISM

As we have seen, rate expressions for reactions must be determined experimentally. Once this has been done, it is possible to imagine a plausible mechanism compatible with the observed rate expression. This, however, is a rather complex process and we will not attempt it here. Instead, we will consider the reverse process, which is much more straightforward. *Given a mechanism for a several-step reaction, how do we deduce the rate expression that corresponds to that mechanism?*

In principle at least we can do this quite readily using the rules just cited.

1. Find the slowest step and equate the rate of the overall reaction to the rate of that step.

2. Find the rate expression for the slowest step (Equations 18.26 to 18.29).

To illustrate this process, consider the low-temperature reaction between CO and NO_2, which occurs by the following mechanism:

Step 1: $NO_2 + NO_2 \rightarrow NO_3 + NO$ slow
Step 2: $NO_3 + CO \rightarrow NO_2 + CO_2$ fast
$$NO_2(g) + CO(g) \rightarrow NO(g) + CO_2(g)$$

Applying these rules in succession, we have:

rate of overall reaction = rate of Step 1 = $k(\text{conc. } NO_2)^2$

Experimentally, we find that below 500 K this reaction is indeed second order in NO_2 and zero order in CO, as the analysis predicts.

Sometimes when you follow this procedure you find that the rate expression involves a reactive intermediate, i.e., a species produced in one step of the mechanism and consumed in a later step. Ordinarily, concen-

Concentrations of intermediates are very low

trations of such species cannot be determined experimentally. Hence, they must be eliminated from the rate expression if it is to be compared with experiment. **The final rate expression must include only those species that appear in the balanced equation for the overall reaction.**

A reaction in which this situation arises is that between nitric oxide and chlorine. This reaction is believed to proceed by a two-step mechanism:

Step 1: $NO(g) + Cl_2(g) \rightleftharpoons NOCl_2(g)$ (fast)
Step 2: $NOCl_2(g) + NO(g) \rightarrow 2NOCl(g)$ (slow)

$$2NO(g) + Cl_2(g) \quad \rightarrow 2NOCl(g) \tag{18.30}$$

Frequently, intermediates exist at equilibrium concentrations

In the first step, the unstable intermediate $NOCl_2$ is formed, in equilibrium with NO and Cl_2. An $NOCl_2$ molecule then reacts with another molecule of NO in a slow, rate-determining second step.

Applying Rules 1 and 2, we find that the rate expression for Reaction 18.30 should be:

$$\text{rate} = k_2(\text{conc. } NOCl_2) \times (\text{conc. } NO) \tag{18.31}$$

The term (conc. $NOCl_2$) must be eliminated from the rate expression because the $NOCl_2$ molecule is an intermediate whose concentration cannot be determined experimentally. To eliminate (conc. $NOCl_2$), we take advantage of the fact that the first step in this reaction involves an equilibrium. Writing the equilibrium constant expression for that step, we have:

$$K_c = \frac{(\text{conc. } NOCl_2)}{(\text{conc. } NO) \times (\text{conc. } Cl_2)}$$

Solving for (conc. $NOCl_2$)

$$\text{conc. } NOCl_2 = K_c \times (\text{conc. } NO) \times (\text{conc. } Cl_2)$$

and substituting in Equation 18.31

$$\text{rate} = k_2 \times K_c \times (\text{conc. } NO)^2 \times (\text{conc. } Cl_2)$$

The product of the two constants, $k_2 \times K_c$, is simply the observed rate constant for Reaction 18.30. Calling that k, we predict the rate expression:

$$\text{rate} = k(\text{conc. } NO)^2 \times (\text{conc. } Cl_2) \tag{18.32}$$

Experimentally, the reaction between NO and Cl_2 is found to be second order in NO and first order in Cl_2, as Equation 18.32 predicts.

■ **EXAMPLE 18.9** ─────────────────────────────────

The decomposition of ozone, O_3, to diatomic oxygen, O_2, is believed to occur by a two-step mechanism:

Step 1: $O_3(g) \rightleftharpoons O_2(g) + O(g)$ (fast)
Step 2: $O_3(g) + O(g) \rightarrow 2O_2(g)$ (slow)

$$2O_3(g) \rightarrow 3O_2(g)$$

Obtain the rate expression corresponding to this mechanism.

Solution
According to Rule 1,

rate of reaction = rate of Step 2

According to Rule 2:

$$\text{rate} = k_2(\text{conc. O}_3) \times (\text{conc. O})$$

where k_2 is the rate constant for the second step. To obtain a valid rate expression, we must eliminate (conc. O), since atomic oxygen is an unstable intermediate. To eliminate (conc. O), we start by writing the equilibrium constant expression for Step 1:

$$K_c = \frac{(\text{conc. O}_2) \times (\text{conc. O})}{(\text{conc. O}_3)}$$

We now solve for (conc. O) and substitute:

$$\text{conc. O} = K_c \times \frac{(\text{conc. O}_3)}{(\text{conc. O}_2)}$$

$$\text{rate} = k_2 \times K_c \times \frac{(\text{conc. O}_3)^2}{(\text{conc. O}_2)} = k\frac{(\text{conc. O}_3)^2}{(\text{conc. O}_2)}$$

where k is the observed rate constant for the overall reaction. Notice that the concentration of O_2, a product in this reaction, appears in the denominator of the rate expression. This means that the rate is inversely proportional to the oxygen concentration. The more O_2 we have, the slower the reaction.

O_2 would be called an inhibitor of the reaction

Exercise

What happens to the rate of this reaction if the concentration of O_3 is doubled? If the concentration of O_2 is doubled? Answer: Rate increases by a factor of four; rate is cut in half.

We should take note of one of the limitations of mechanism studies. Usually more than one mechanism is compatible with the same experimentally obtained rate expression. To make a choice between alternative mechanisms, other evidence must be considered. A classic example of this situation is the reaction between hydrogen and iodine

$$H_2(g) + I_2(g) \rightarrow 2HI(g)$$

for which the observed rate expression is

$$\text{rate} = k (\text{conc. H}_2) \times (\text{conc. I}_2) \tag{18.33}$$

For many years, it was assumed that the H_2–I_2 reaction occurs in a single step, a collision between an H_2 molecule and an I_2 molecule. That would, of course, be compatible with the rate expression given by Equation 18.33. However, there is now evidence to indicate that a quite different and more complex mechanism is involved (see Problem 61).

■ SUMMARY PROBLEM

Hydrogen peroxide decomposes to water and oxygen according to the following reaction

$$2H_2O_2(aq) \rightarrow 2H_2O + O_2(g)$$

Its rate of decomposition is measured by titrating samples of the solution with potassium permanganate ($KMnO_4$) at certain time intervals.

a. If 0.100 mol/L of H_2O_2 is consumed in 72.0 min, what is the average rate of consumption? At what rates are water and oxygen gas produced?

b. Initial rate determinations at 40°C for the decomposition give the following data:

H_2O_2 (M)	INITIAL RATE (M min^{-1})
0.1000	1.93×10^{-4}
0.2000	3.86×10^{-4}
0.3000	5.79×10^{-4}

What is the order of the reaction? Write the rate equation for the decomposition. Calculate the rate constant and the half-life for the reaction at 40°C.

c. Hydrogen peroxide is sold commercially as a 30.0% solution. If the solution is kept at 40°C, how long will it take for the solution to become 10.0% H_2O_2?

d. It has been determined that at 50°C, the rate constant for the reaction is 4.32×10^{-3}/min. Calculate the activation energy for the decomposition of H_2O_2.

e. Manufacturers recommend that solutions of hydrogen peroxide be kept in a refrigerator at 4°C. How long would it take for a 30.0% solution to decompose to 10.0% if the solution is indeed kept in a refrigerator at 4°C?

f. The rate constant for the uncatalyzed reaction at 25°C is 5.21×10^{-4}/min. The rate constant for the catalyzed reaction at 25°C is 2.95×10^8/min. What is the half-life of the uncatalyzed reaction at 25°C? What is the half-life of the catalyzed reaction?

g. Hydrogen peroxide in basic solution oxidizes iodide ions to iodine. The proposed mechanism for this reaction is

$$H_2O_2(aq) + I^-(aq) \rightarrow HOI(aq) + OH^-(aq) \quad \text{slow}$$

$$HOI(aq) + I^-(aq) \rightleftharpoons I_2(aq) + OH^-(aq) \quad \text{fast}$$

Write the overall redox reaction. Write a rate law consistent with this proposed mechanism.

Answers

a. H_2O_2 consumption = 0.00139 M/min; H_2O production = 0.00139 M/min; O_2 production = 6.94×10^{-4} M/min

b. first order; rate = k(conc. H_2O_2); $k = 1.93 \times 10^{-3}$/min; $t_{1/2}$ = 5.98 h

c. 9.49 h   **d.** 68 kJ/mol   **e.** 2.8×10^2 h

f. $t_{1/2}$ uncatalyzed = 22.2 h; $t_{1/2}$ catalyzed = 2.35×10^{-9} min

g. $H_2O_2(aq) + 2I^-(aq) \rightarrow I_2(aq) + 2OH^-(aq)$

rate = k (conc. H_2O_2) (conc. I^-)

Many reactions that take place slowly under ordinary conditions occur readily in living organisms in the presence of catalysts called *enzymes*. Enzymes are protein molecules of high molar mass. An example of an enzyme-catalyzed reaction is that of sugar (sucrose) with oxygen:

$$C_{12}H_{22}O_{11}(s) + 12O_2(g) \rightarrow 12CO_2(g) + 11H_2O(l) \qquad (18.34)$$

This reaction is difficult to bring about directly, such as by heating a sample of sugar in a test tube. In the body, sugar is metabolized at 37°C (98.6°F) in a series of biochemical reactions. The end products are carbon dioxide and water. Each step in the sequence is catalyzed by a particular enzyme adapted for that purpose.

Most enzymes contain a nonprotein part called a *coenzyme* that must be present if the enzyme is to fulfill its function. In some cases the coenzyme is a metal cation such as Zn^{2+}, Cu^{2+}, or Co^{2+}. In others, it is an organic molecule, most often a vitamin. The B vitamins constitute an important group of coenzymes. For example, niacin is part of an enzyme that prevents pellagra. Your body requires only about 0.02 g of niacin per day, but it performs an essential function.

Many enzymes are extremely specific. For example, the enzyme maltase catalyzes the hydrolysis of maltose:

$$\underset{\text{maltose}}{C_{12}H_{22}O_{11}(aq)} + H_2O \xrightarrow{\text{maltase}} \underset{\text{glucose}}{2C_6H_{12}O_6(aq)} \qquad (18.35)$$

It's nice to be needed

This is the only function of maltase, but it is one that no other enzyme can perform. There are many such digestive enzymes required for the metabolism of carbohydrates, proteins, and fats. It has been estimated that without enzymes, it would take upwards of 50 years to digest a meal.

The existence and importance of enzymes were recognized by early chemists, who tried in a general way to describe how they worked. In 1894, Emil Fischer suggested the "lock and key" analogy shown in Figure 18.11. The reactant ("substrate") fits into a specific site on an enzyme surface, where it is held in position by intermolecular forces. The substrate-enzyme complex can then react with another species such as a water molecule. A somewhat more sophisticated explanation of enzyme

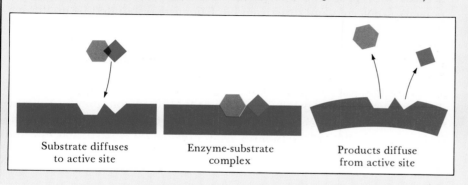

Substrate diffuses to active site   Enzyme-substrate complex   Products diffuse from active site

Figure 18.11 In enzyme catalysis, the substrate appears to fit on the enzyme in a "lock and key" arrangement. After adsorption, the enzyme configuration often changes, which assists in cleaving the crucial substrate bond, thereby increasing the reaction rate.

action involves the "induced fit" model. Here it is assumed that the active site of the enzyme can change its shape and size to accommodate a substrate, in somewhat the same way that a glove changes when you put it on. In recent years, x-ray diffraction studies of enzyme-substrate complexes have furnished truly remarkable information about the way enzymes selectively catalyze reactions.

Enzyme activity is diminished in the presence of substances known as *inhibitors*. Some inhibitors actually destroy enzymes; an example is the Pb^{2+} ion, which reacts irreversibly with —SH groups in certain enzymes. Symbolically, we can represent the process as:

$$R—SH(aq) + Pb^{2+}(aq) + R—SH(aq) \rightarrow R—S—Pb—S—R(aq) + 2H^+(aq) \quad (18.36)$$

where R is a complex organic group. This reaction is responsible for the toxicity of Pb^{2+} and indeed of many other heavy metal cations, including Hg^{2+}.

More commonly, an inhibitor operates by occupying sites on an enzyme molecule that are supposed to be reserved for the substrate. Frequently, inhibitors have geometries closely resembling those of the substrates they replace. For example, the metabolism of citric acid is inhibited by its close relative, fluorocitric acid, which can presumably fit into the same slot in an enzyme:

citric acid   fluorocitric acid

Antibiotics like penicillin operate by inhibiting enzymes that are required for bacteria to live and multiply. Penicillin, discovered accidentally by Alexander Fleming in 1928, was the wonder drug of the 1950s and 1960s. Then new bacterial strains appeared that were resistant to penicillin because they produced an enzyme (penicillinase) that attached to the penicillin molecule and deactivated it. Chemists responded by modifying the structure of the penicillin molecule to make it more resistant to attack by this enzyme. Indeed, they went a step further, using a drug called clavulinic acid, which inhibits the enzyme penicillinase. This explains why penicillin is still on the market today, along with many of its derivatives.

Enzymes, like all other catalysts, lower the activation energy for reaction. They can be enormously effective; it is not uncommon for the rate constant to increase by a factor of 10^{12} or more. However, from a commercial standpoint, enzymes have some drawbacks. A particular enzyme operates best over a narrow range of temperature and pH (the range

Bacteria are clever, but so are chemists

where the enzyme performs its function in the body). An increase in temperature frequently deactivates an enzyme by causing the molecule to "unfold," changing its characteristic shape. Recently, chemists have discovered that this effect can be prevented if the enzyme is immobilized by bonding to a solid support. Among the solids that have been used are synthetic polymers, porous glass, and even stainless steel. The development of immobilized enzymes has led to a host of new products. The sweetener aspartame, used in many diet soft drinks, is made using the enzyme aspartase in immobilized form.

QUESTIONS AND PROBLEMS

Meaning of Reaction Rates

1. Express the rate of the reaction

$$C_2H_6(g) \rightarrow C_2H_4(g) + H_2(g)$$

a. in terms of Δ conc. C_2H_6.
b. in terms of Δ conc. C_2H_4.

2. Express the rate of the reaction

$$2HI(g) \rightarrow H_2(g) + I_2(g).$$

a. in terms of Δ conc. H_2.
b. in terms of Δ conc. HI, if you want the same rate as in (a).

3. Consider the combustion of ethane:

$$2C_2H_6(g) + 7O_2(g) \rightarrow 4CO_2(g) + 6H_2O(g)$$

If the ethane is burning at a rate of 0.20 mol/L·s, at what rates are CO_2 and H_2O being produced?

4. For the reaction

$$5Br^-(aq) + BrO_3^-(aq) + 6H^+(aq)$$
$$\rightarrow 3Br_2(aq) + 3H_2O$$

it was found that at a particular instant bromine was being formed at the rate of 0.039 mol/L per second. At that instant,

a. at what rate was water being formed?
b. at what rate was bromide ion being oxidized?
c. at what rate was H^+ being consumed?

5. Nitrosyl chloride (NOCl) decomposes according to the reaction

$$2NOCl(g) \rightarrow 2NO(g) + Cl_2(g)$$

The initial concentration of nitrosyl chloride was 0.580 M. After 8.00 min, this concentration decreased to 0.238 M. Obtain the rate of reaction during this time interval in mol/L·s.

6. Dinitrogen pentoxide decomposes according to the following equation:

$$2N_2O_5(g) \rightarrow 4NO_2(g) + O_2(g)$$

The initial concentration of dinitrogen pentoxide is 0.384 M. It is determined that after one minute, 20.0% of the dinitrogen pentoxide has decomposed. What is the rate of the reaction in this time interval in mol/L·s?

7. Experimental data is listed for the hypothetical reaction

$$A + B \rightarrow C + D$$

Time (min)	0	1	2	3	4	5
Conc. A (M)	0.200	0.166	0.139	0.117	0.100	0.089

a. Calculate the average rate between 2 and 4 minutes.
b. Plot this data as in Figure 18.1. Draw a tangent to the curve to find the rate at 3 minutes.

8. Using the data in Problem 7, find
a. the average rate between 3 and 5 minutes.
b. the rate at 4 minutes using the plot referred to in Problem 7b.

Rate Expressions

9. What is the order with respect to each reactant and the overall order of the reactions described by the following rate expressions?

a. rate = k_1(conc. A)(conc. B)
b. rate = k_2(conc. A)
c. rate = k_3(conc. A)(conc. B)^2
d. rate = k_4

10. What is the order with respect to each reactant and the overall order of the reactions described by the following rate expressions?

a. rate = k_1(conc. A)(conc. B)
b. rate = k_2(conc. A)^2
c. rate = k_3(conc. A)(conc. B)^2
d. rate = k_4(conc. A)

11. What would be the units of the rate constants in Question 9 if rate is expressed in mol/L·s?

12. What would be the units of each of the rate constants in Question 10 if the rate is expressed in mol/L·s?

13. Complete the following table for the first order reaction

$A(g) \rightarrow$ products

CONC. A (M)	k (min^{-1})	RATE (mol/L·min)
0.750	2.0×10^{-3}	_____
0.035	_____	6.82
_____	0.0372	7.98×10^{-3}

14. Complete the following table for the reaction

$2A(g) + B(g) \rightarrow$ products

which is first order in both reactants.

CONC. A (M)	CONC. B (M)	k(L/mol·s)	RATE (mol/L·s)
0.200	0.300	1.5	_____
_____	0.029	0.78	0.025
0.450	0.520	_____	0.033

15. At 550 K, the decomposition of nitrogen dioxide is second order, with a rate constant of 0.31 L/mol·s.
   a. Write the rate expression.
   b. Calculate the rate when the concentration of nitrogen dioxide is 0.050 M.
   c. At what concentration of nitrogen dioxide will the rate be 1.5×10^{-3} mol/L·s?

16. The decomposition of ammonia on tungsten at 1100°C is zero order, with a rate constant of 2.5×10^{-4} mol/L·min.
   a. Write the rate expression.
   b. Calculate the rate when the concentration of ammonia is 0.080 M.
   c. At what concentration of ammonia is the rate equal to the rate constant?

17. The reaction

$2NO(g) + Br_2(g) \rightarrow 2NOBr(g)$

is second order in nitrogen oxide and first order in bromine. The rate of the reaction is 1.6×10^{-8} mol/L·min when the nitrogen oxide concentration is 0.020 M and the bromine concentration is 0.030 M.
   a. What is the value of k?
   b. At what concentration of bromine is the rate 2.0×10^{-6} mol/L·min when the nitrogen oxide concentration is 0.064 M?
   c. At what concentration of nitrogen oxide is the rate 4.5×10^{-7} mol/L·min if the bromine concentration is one-third of the nitrogen oxide concentration?

18. The reaction

$2ICl(g) + H_2(g) \rightarrow I_2(g) + 2HCl(g)$

is first order in both reactants. The rate of the reaction is 4.89×10^{-5} mol/L·s when the ICl concentration is 0.100 M and that of hydrogen gas is 0.0300 M.
   a. What is the value of k?
   b. At what concentration of hydrogen gas is the rate 3.00×10^{-3} mol/L·s when the ICl concentration is 0.150 M?
   c. At what concentration of iodine chloride is the rate 7.85 mol/L·s if the hydrogen gas concentration is twice that of ICl?

19. The greatest decrease in reaction rate for the reaction between A and D where rate $= k \times$ (conc. A)^2 (conc. D) is caused by
   a. halving conc. D.   b. halving conc. A.
   c. doubling conc. D.
   d. halving conc. A and conc. D.

20. The greatest increase in rate for the reaction between X and Z where rate $= k$(conc. X)(conc. Z)^2 will be caused by
   a. doubling conc. Z.   b. doubling conc. X.
   c. tripling conc. X.   d. lowering the temperature.

Determination of Reaction Order

21. For a reaction involving a single reactant, the following rate data are obtained:

Rate (mol/L·min)	0.020	0.020	0.019	0.021
Conc. reactant (M)	0.100	0.090	0.080	0.070

Determine the order of the reaction.

22. For a reaction involving a single reactant A, the following data are obtained:

Rate (mol/L·min)	0.020	0.016	0.013	0.010
Conc. A (M)	0.100	0.090	0.080	0.070

Determine the order of the reaction.

23. The rate of the reaction

$2HgCl_2(aq) + C_2O_4{}^{2-}(aq)$
$\rightarrow 2Cl^-(aq) + 2CO_2(g) + Hg_2Cl_2(s)$

is followed by measuring the number of moles of Hg_2Cl_2 that precipitate per liter per second. The following data are obtained:

CONC. HgCl$_2$ (M)	CONC. C$_2$O$_4{}^{2-}$ (M)	INITIAL RATE (mol/L·s)
0.10	0.10	1.3×10^{-7}
0.10	0.20	5.2×10^{-7}
0.20	0.20	1.0×10^{-6}
0.20	0.10	2.6×10^{-7}

a. What is the order of the reaction with respect to $HgCl_2$, with respect to $C_2O_4^{2-}$, and overall?

b. Write the rate expression for the reaction.

c. Calculate k for the reaction.

d. When the concentrations of both mercury(II) chloride and oxalate ion are 0.30 M, what is the rate of the reaction?

24. In solution at constant H^+ concentration, I^- reacts with H_2O_2 to produce I_2:

$$2H^+(aq) + 2I^-(aq) + H_2O_2(aq) \rightarrow I_2(aq) + 2H_2O$$

The reaction rate can be followed by monitoring iodine production. The following data apply:

CONC. I^- (M)	CONC. H_2O_2 (M)	INITIAL RATE (mol/L·s)
0.020	0.020	3.3×10^{-5}
0.040	0.020	6.6×10^{-5}
0.060	0.020	9.0×10^{-5}
0.040	0.040	1.3×10^{-4}

a. What is the order with respect to I^-?

b. What is the order with respect to H_2O_2?

c. What is the rate when conc. $I^- = 0.010$ M, conc. $H_2O_2 = 0.030$ M?

25. The following data refer to the reaction between A and B.

CONC. A (M)	CONC. B (M)	INITIAL RATE (mol/L·s)
0.10	0.10	3.0×10^{-3}
0.10	0.30	3.0×10^{-3}
0.30	0.30	2.7×10^{-2}
0.30	0.60	2.7×10^{-2}

a. What is the order of the reaction with respect to A, with respect to B, and overall?

b. Calculate k for the reaction.

c. At what concentration of A will the rate be 5.0×10^{-3} mol/L·s?

26. The following data refer to the equation

$$A + B + C \rightarrow \text{products}$$

CONC. A (M)	CONC. B (M)	CONC. C (M)	INITIAL RATE (mol/L·s)
0.100	0.100	0.100	2.50×10^{-4}
0.300	0.100	0.100	2.25×10^{-3}
0.300	0.200	0.100	2.25×10^{-3}
0.300	0.200	0.400	9.00×10^{-3}

a. What is the order of the reaction with respect to each of the three reactants, and overall?

b. Write the rate expression for the reaction.

c. Calculate k.

d. Calculate the rate of the reaction when (conc. A) = 2(conc. B) = $\frac{1}{3}$(conc. C) = 0.600 M

27. The following initial rate data were obtained for the reaction

$$2ClO_2(aq) + 2OH^-(aq) \rightarrow ClO_3^-(aq) + ClO_2^-(aq) + H_2O$$

CONC. ClO_2 (M)	CONC. OH^- (M)	INITIAL RATE (mol/L·s)
0.0575	0.0216	8.21×10^{-3}
0.0713	0.0216	1.26×10^{-2}
0.0575	0.0333	1.26×10^{-2}
0.0713	0.0333	1.95×10^{-2}

a. Write the rate expression.

b. Calculate k.

c. What is the pH when the concentration of ClO_2 is 0.100 M and the rate is 3.56×10^{-2} mol/L·s?

28. Consider the reaction

$$Cr(H_2O)_6^{3+}(aq) + SCN^-(aq) \rightarrow Cr(H_2O)_5SCN^{2+}(aq) + H_2O$$

The following data were obtained:

CONC. $Cr(H_2O)_6^{3+}$ (M)	SCN^- (M)	RATE (mol/L·h)
0.028	0.040	8.1×10^{-6}
0.028	0.055	1.1×10^{-5}
0.037	0.055	1.5×10^{-5}
0.037	0.040	1.1×10^{-5}

a. Write the rate expression for the reaction.

b. Calculate k.

c. What is the initial rate if 12 mg of KSCN is added to two liters of a solution 0.087 M in $Cr(H_2O)_6^{3+}$?

29. The following data are obtained for the decomposition of CH_3NO_2 at 500 K. Following the procedure used in Example 18.7, find the reaction order.

t (s)	0	300	600	900	1200
Conc. CH_3NO_2	0.200	0.145	0.105	0.076	0.055

30. The following data are obtained for the decomposition of acetaldehyde, CH_3CHO, at 700 K. Following the procedure used in Example 18.7, find the reaction order.

t (s)	0	50	100	150	200
Conc. CH_3CHO	0.400	0.333	0.286	0.250	0.222

31. For the reaction: A $\rightarrow$ products, it is found that the half-life is 27 s when the initial concentration is 0.50 M and 50 s when the initial concentration is 0.27 M. Is the reaction zero order, first order, or second order?

32. Consider the following data for the decomposition of a species B:

Conc. B (M)	0.160	0.080	0.040	0.020
t (s)	0	10.0	15.0	17.5

What is the half-life, starting at 0.160 M? 0.080 M? 0.040 M? What is the order of the reaction?

First-Order Reactions
33. The following data apply to the first-order decomposition of ethane, C_2H_6:

t (s)	CONC. (M)
0	0.01000
200	0.00916
400	0.00839
600	0.00768
800	0.00703

a. From a plot of conc. vs. time, estimate the rate at $t =$ 800 s.
b. From a plot of ln conc. vs. time, determine k, the rate constant. Use k to obtain the rate at $t = 800$ s.
c. Which of the rates just calculated in (a) or (b) do you think is more accurate? Explain.

34. The following data are for the gas-phase decomposition of ethyl chloride, C_2H_5Cl, at 740 K:

t (min)	CONC. (M)	t (min)	CONC. (M)
0	0.200	4	0.187
1	0.197	8	0.175
2	0.193	16	0.153
3	0.190		

a. By plotting the data, show that the reaction is first order.
b. From the graph, determine k.
c. Using k, find the time for the concentration to drop to one-fourth of the original concentration.

35. In the first-order decomposition of cyclobutane at 750 K, it is found that 25% of a sample has decomposed in eighty seconds.
a. What is the rate constant?
b. What is the half-life of the reaction?

36. In the first-order decomposition of acetone at 500°C, it was found that the concentration was 0.0300 M after 200 min and 0.0200 M after 400 min.
a. What is the rate constant?
b. What is the half-life?
c. What was the initial concentration of acetone?

37. The complex ion $[Cr(H_2O)_2(C_2O_4)_2]^-$ isomerizes (goes from the *trans* form to the *cis* form) by a first-order reaction. Its half-life is 64.2 min.

a. What is the rate constant for the isomerization?
b. How many hours are required for the concentration of the *trans* form to drop from 0.500 M to 0.125 M?
c. How long will it take to reduce the *trans* form to 5.00% of its original concentration?

38. The decomposition of dinitrogen pentoxide

$$2N_2O_5(g) \rightarrow 4NO_2(g) + O_2(g)$$

is first order. Its half-life is 0.363 h.
a. What is the rate constant for the decomposition?
b. What fraction of N_2O_5 will have decomposed after one hour?
c. How long will it take to decompose 15.0% of the dinitrogen pentoxide?

Zero- and Second-Order Reactions
39. The decomposition of nitrogen dioxide

$$2NO_2(g) \rightarrow 2NO(g) + O_2(g)$$

is a second-order reaction. It takes 125 s for the concentration of NO_2 to go from 0.800 M to 0.0104 M.
a. What is k for the reaction?
b. What is the half-life of the reaction when its initial concentration is 0.500 M?

40. The decomposition of hydrogen iodide

$$2HI(g) \rightarrow H_2(g) + I_2(g)$$

is second order. Its half-life is 85 s when the initial concentration is 0.15 M.
a. What is k for the reaction?
b. How long will it take to go from 0.300 M to 0.100 M?

41. The rate constant for the second-order reaction

$$2NOBr(g) \rightarrow 2NO(g) + Br_2(g)$$

is 48 L/mol·min at a certain temperature. How long will it take to decompose 90.0% of a 0.0200 M solution of nitrosyl bromide?

42. The decomposition of nitrosyl chloride

$$2NOCl(g) \rightarrow 2NO(g) + Cl_2(g)$$

is a second-order reaction. If it takes 0.20 min to decompose 15% of a 0.300 M solution of nitrosyl chloride, what is k for the reaction?

43. The rate constant for the zero-order decomposition of HI on a gold surface is 0.050 mol/L·s. How long will it take for the concentration of HI to drop from 1.00 M to 0.20 M?

44. For the zero-order decomposition of NH_3 on tungsten, it takes 32 min for the concentration of ammonia to drop from 0.500 M to 0.100 M. How long will it take for the rest of the ammonia to decompose? If the original concentration were 0.800 M, how long would it take for half of the NH_3 to decompose?

Activation Energy, Catalysis, and ΔH

45. Three reactions have the following activation energies:

Reaction	A	B	C
E_a (kJ)	145	210	48

Which reaction would you expect to be the fastest? The slowest? Explain.

46. Consider the data for several systems for the conversion of reactants to products at the same temperature.

SYSTEM	E_a (kJ)	ΔH (kJ)
1	50	+20
2	85	−20
3	12	−30

a. Which system has the highest forward rate?
b. In each case, determine the difference between the energy of the activated complex and that of the products.

47. For a certain reaction, E_a is 120 kJ and ΔH is 15 kJ. In the presence of a catalyst, the activation energy is lowered to 75 kJ. Draw a diagram similar to those in Figures 18.7 and 18.10 for this reaction.

48. The activation energy of a certain reaction is 85 kJ. In the presence of a catalyst, it is reduced to 40 kJ. For the reaction, ΔH is −50 kJ. Draw a diagram similar to those in Figures 18.7 and 18.10 for this reaction.

Reaction Rate and Temperature

49. The following data were obtained for the reaction

$$SiH_4(g) \rightarrow Si(s) + 2H_2(g)$$

k (s^{-1})	0.048	2.3	49	590
t (°C)	500	600	700	800

Plot these data (ln k vs. $1/T$) and find the activation energy for the reaction.

50. The following data are for the gas-phase decomposition of acetaldehyde:

$k \left(\dfrac{L}{mol \cdot s}\right)$	0.0105	0.101	0.60	2.92
T (K)	700	750	800	850

Plot these data and determine the activation energy for the reaction.

51. The activation energy of a certain enzyme-catalyzed reaction is 50.2 kJ/mol. By what percent is the rate of this reaction increased if you have a fever of 104.0°F, assuming that the enzyme works at a normal temperature of 98.6°F?

52. Raw milk will sour in about four hours at 28°C (82°F) but will last up to 48 hours in a refrigerator at 5°C. Assuming the rate to be inversely related to souring time, what is the activation energy for the reaction involving the souring of milk?

53. The chirping rate of a cricket, X, in chirps per minute, near room temperature is given by

$$X = 7.2\,t - 32$$

where t is the temperature in °C. Calculate the chirping rates at 25 and 35°C, and use them to estimate the activation energy for this reaction.

54. Cold-blooded animals decrease their body temperature in cold weather to match that of their environment. The activation energy of a certain enzyme-catalyzed reaction in a cold-blooded animal is 65 kJ/mol. By what percentage is the rate of this reaction decreased if the body temperature of the animal drops from 35 to 25°C?

55. For the reaction

$$HI(g) + CH_3I(g) \rightarrow CH_4(g) + I_2(g)$$

the rate constant is 0.28 L/mol·s at 300°C and 3.9×10^{-3} L/mol·s at 227°C.

a. What is the activation energy of the reaction?
b. What is k at 400°C?

56. For the reaction

$$C_2H_4(g) + H_2(g) \rightarrow C_2H_6(g)$$

the activation energy is 181 kJ/mol. The rate constant at 500°C is 2.5×10^{-2} L/mol·s.

a. At what temperature is the rate constant twice its value at 500°C?
b. What is the rate constant at 1000°C?

57. The decomposition of ethyl chloride is a first-order reaction. Its activation energy is 248 kJ/mol. At 470°C, it has a half-life of 40.8 min. What is its half-life at 400°C?

58. The decomposition of ethyl bromide is a first-order reaction. Its activation energy is 226 kJ/mol. At 720 K, its half-life is 78 s. What is its half-life at 800 K?

Reaction Mechanisms

59. Write the rate expression for each of the following elementary steps:

a. $N_2O_2 + H_2 \rightarrow N_2O + H_2O$
b. $ClCO + Cl_2 \rightarrow Cl_2CO + Cl$
c. $2NO_2 \rightarrow NO_3 + NO$

60. Write the rate expression for each of the following elementary steps:

a. $K + HCl \rightarrow KCl + H$
b. $NO_3 + CO \rightarrow NO_2 + CO_2$
c. $2NO_2 \rightarrow 2NO + O_2$

61. For the reaction between hydrogen and iodine:

$$H_2(g) + I_2(g) \rightarrow 2HI(g)$$

The rate expression is: rate = k(conc. H_2)(conc. I_2). Show that this expression is consistent with the mechanism:

$I_2(g) \rightleftharpoons 2I(g)$	(fast)
$H_2(g) + I(g) + I(g) \rightarrow 2HI(g)$	(slow)

62. For the reaction

$$2H_2(g) + 2NO(g) \rightarrow N_2(g) + 2H_2O(g)$$

the experimental rate expression is rate $= k(\text{conc. NO})^2$ (conc. H_2). The following mechanism is proposed:

$$2NO \rightleftharpoons N_2O_2 \qquad\qquad (\text{fast})$$
$$N_2O_2 + H_2 \rightarrow H_2O + N_2O \quad (\text{slow})$$
$$N_2O + H_2 \rightarrow N_2 + H_2O \quad (\text{fast})$$

Show that the mechanism is consistent with the rate expression.

63. Two mechanisms are proposed for the reaction

$$2NO(g) + O_2(g) \rightarrow 2NO_2(g)$$

Mechanism 1: $NO + O_2 \rightleftharpoons NO_3$ (fast)
$\qquad\qquad\quad NO_3 + NO \rightarrow 2NO_2$ (slow)
Mechanism 2: $NO + NO \rightleftharpoons N_2O_2$ (fast)
$\qquad\qquad\quad N_2O_2 + O_2 \rightarrow 2NO_2$ (slow)

Show that each of these mechanisms is consistent with the observed rate law, rate $= k(\text{conc. NO})^2 \times (\text{conc. O}_2)$.

64. At low temperatures, the rate law for the reaction $CO(g) + NO_2(g) \rightarrow CO_2(g) + NO(g)$ is as follows: rate $=$ constant $\times (\text{conc. NO}_2)^2$. Which of the following mechanisms is consistent with this rate law?

a. $CO + NO_2 \rightarrow CO_2 + NO$
b. $2NO_2 \rightleftharpoons N_2O_4$ (fast)
$\quad N_2O_4 + 2CO \rightarrow 2CO_2 + 2NO$ (slow)
c. $2NO_2 \rightarrow NO_3 + NO$ (slow)
$\quad NO_3 + CO \rightarrow NO_2 + CO_2$ (fast)
d. $2NO_2 \rightarrow 2NO + O_2$ (slow)
$\quad 2CO + O_2 \rightarrow 2CO_2$ (fast)

Unclassified

65. How does each of the following affect reaction rate?
a. The passage of time.
b. Decreasing container size for a gas phase reaction.
c. Adding a catalyst.

66. In your own words explain why
a. a decrease in temperature slows the rate of a reaction.
b. doubling the concentration of a reactant does not always double the rate of the reaction.
c. a flame lights a cigarette but the cigarette continues to burn after the flame is removed.
d. a catalyst does not change the equilibrium constant, K_c.

67. The decomposition of nitrosyl bromide

$$2NOBr(g) \rightarrow 2NO(g) + Br_2(g)$$

is a second-order reaction. Its rate constant at 10°C is 0.80 L/mol·s; ΔH_f^0 for NOBr is 82.2 kJ/mol; ΔH_f^0 for NO is

90.2 kJ/mol, for $Br_2(g)$, 30.9 kJ/mol. If the initial concentration of NOBr is 0.100 M, how much heat is evolved or absorbed per liter of solution in one minute?

68. Consider the decomposition of dinitrogen pentoxide in CCl_4 at 45°C:

$$2N_2O_5(g) \rightarrow 2N_2O_4(g) + O_2(g)$$

The rate constant for this first-order reaction is 0.037/min. A sample weighing 68.0 g is dissolved in CCl_4 and allowed to decompose.
a. How long will it take to reduce the amount of N_2O_5 to 10.0 g?
b. What total volume of O_2 at 45°C and 1.00 atm is produced if the reaction is stopped after 25 minutes?

69. How much faster would a reaction proceed at 25°C than at 10°C if the activation energy of the reaction is 100 kJ/mol?

Challenge Problems

70. Using calculus, derive the equation for
a. the concentration–time relation for a second-order reaction (see Table 18.3).
b. the concentration–time relation for a third-order reaction, 3 A $\rightarrow$ products.

71. The following data apply to the reaction

$$A(g) + 3B(g) + 2C(g) \rightarrow \text{products}$$

(concentrations are given in mol/L):

CONC. A	CONC. B	CONC. C	RATE
0.20	0.40	0.10	X
0.40	0.40	0.20	8X
0.20	0.20	0.20	X
0.40	0.40	0.10	4X

Determine the rate law for the reaction.

72. In a first-order reaction, let us suppose that a quantity, X, of reactant is added at regular intervals of time, Δt. At first the amount of reactant in the system builds up; eventually, however, it levels off at a "saturation value" given by the expression

$$\text{saturation value} = \frac{X}{1 - 10^{-a}}, \text{ where a} = 0.30\frac{\Delta t}{t_{1/2}}$$

This analysis applies to prescription drugs, where you take a certain amount each day. Suppose you take 0.100 g of a drug three times a day and that the half-life for elimination is 2.0 days. Using this equation, calculate the mass of the drug in the body at saturation. Suppose further that side effects show up when 0.500 g of the drug accumulates in the body. As a pharmacist, what is the maximum dosage you could assign to a patient for an 8-hour period without his suffering side effects?

73. Consider an elementary step in which two identical molecules collide with each other: A + A → B + C. If there are two such molecules in a container, there is only one way they can collide. If we call the molecules A_1 and A_2, we might represent the collision as A_1A_2. Suppose now that we have three molecules (A_1, A_2, A_3); three different collisions are possible (A_1A_2, A_1A_3, A_2A_3). How many different collisions are possible with four molecules? What is the general expression for the number of collisions with n molecules? Show that if n is large, the number of collisions is proportional to n^2, which means that the rate of reaction is proportional to the square of the concentration of A.

74. For a certain enzyme-catalyzed reaction the rate is measured as a function of substrate concentration with the following results (rate in mol/L·min):

Conc. S (M)	0.050	0.100	0.200	0.500	1.00	1.50	2.00
Rate	0.026	0.046	0.075	0.120	0.150	0.164	0.171

From a plot similar to that in Figure 18.2, determine the constants k and b in this general rate equation:

$$\text{rate} = \frac{kb(\text{conc. S})}{1 + b(\text{conc. S})}$$

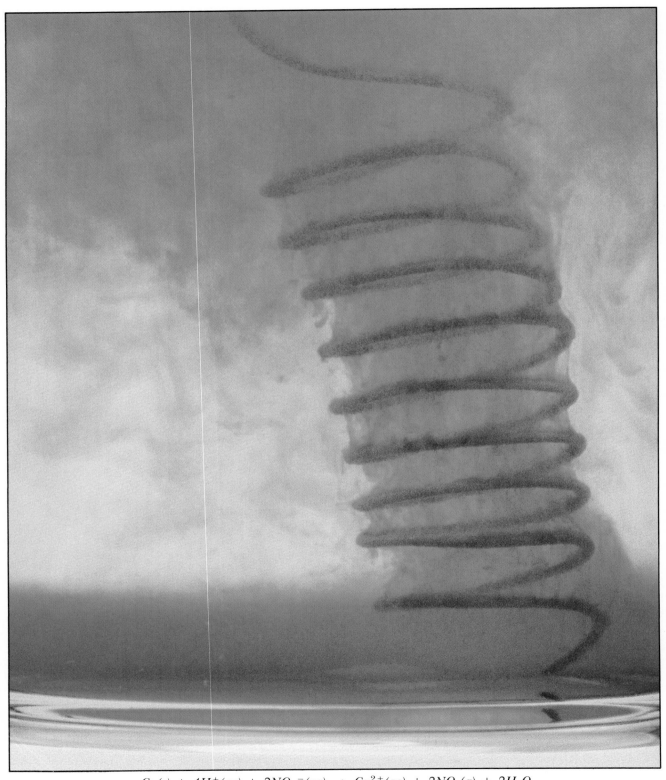

$$Cu(s) + 4H^+(aq) + 2NO_3^-(aq) \rightarrow Cu^{2+}(aq) + 2NO_2(g) + 2H_2O$$

Copper metal is oxidized to Cu^{2+} ions by concentrated nitric acid, which is reduced to $NO_2(g)$.

CHAPTER 19

THE ATMOSPHERE

We have met the enemy, and it is us.
WALT KELLY
POGO

Life on this planet depends upon the relatively thin layer of air that surrounds it. The atmosphere accounts for only about 0.0001% of the total mass of the earth, yet it is the reservoir from which we draw oxygen for metabolism, carbon dioxide for photosynthesis, and nitrogen, whose compounds are essential to plant growth. Our climate is governed by the movement of water vapor from the earth's surface into the atmosphere and back again.

Even trace components of the atmosphere can affect the delicate balance of life. Small amounts of ozone at a height of about 30 km absorb most of the harmful ultraviolet radiation of the sun. On the other hand, as little as 0.2 part per million of ozone near the earth's surface promotes smog formation.

Table 19.1 gives the mole fractions of gases in the atmosphere. Two species are omitted. One is water vapor, whose mole fraction may vary from 0.06 in the tropics to 0.0005 in polar regions. The other comprises suspended particles (e.g., dust, smoke), which vary in both concentration

TABLE 19.1   Composition of Clean, Dry Air at Sea Level

COMPONENT	MOLE FRACTION	COMPONENT	MOLE FRACTION	COMPONENT	MOLE FRACTION
N_2	0.7808	Ne	1.82×10^{-5}	SO_2	$<1 \times 10^{-6}$
O_2	0.2095	He	5.24×10^{-6}	O_3	$<1 \times 10^{-7}$
Ar	0.00934	CH_4	$2 \quad \times 10^{-6}$	NO_2	$<2 \times 10^{-8}$
CO_2	0.00034	Kr	1.14×10^{-6}	I_2	$<1 \times 10^{-8}$
		H_2	$5 \quad \times 10^{-7}$	NH_3	$<1 \times 10^{-8}$
		N_2O	$5 \quad \times 10^{-7}$	CO	$<1 \times 10^{-8}$
		Xe	8.7×10^{-8}	NO	$<1 \times 10^{-8}$

99% of the molecules in air are either N₂ or O₂

and chemical composition. There are five major components of air (N_2, O_2, Ar, H_2O, and CO_2). Together, their mole fractions total about 0.99998.

In this chapter, we will consider what might be called "selected topics" in atmospheric chemistry, including

— the liquefaction of air and separation of nitrogen and oxygen from liquid air (Section 19.1).
— the preparation of ammonia, NH_3, and nitric acid, HNO_3, by reactions involving the major component of the atmosphere, nitrogen (Section 19.2).
— the effect of water vapor and carbon dioxide on our weather and climate (Section 19.3).
— the chemistry of the upper atmosphere (Section 19.4).
— air pollution (Section 19.5).

Throughout the chapter, we will review and expand upon the principles introduced in earlier chapters. In the text discussion and examples, we will apply the concepts of chemical equilibrium and, in particular, chemical kinetics (Chap. 18).

19.1
SEPARATION OF ELEMENTS FROM LIQUID AIR

A total of six different elements are extracted commercially from the atmosphere. These include nitrogen and oxygen, the most abundant components, and four noble gases: neon, argon, krypton, and xenon. The other

Figure 19.1 Liquefaction of air. Carbon dioxide and water are removed by condensation at low temperature or passing the air through a drying chamber (A) packed with a solid adsorbent. The air is then compressed (B) and cooled by passing through a heat exchanger (C). Further cooling takes place upon expansion at a throttle valve (D). After going through the cycle several times, the air begins to condense at stage D.

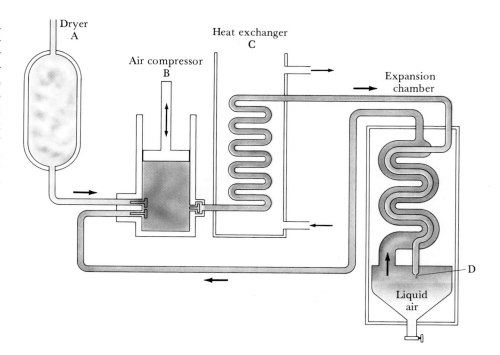

familiar noble gas, helium, is extracted from natural gas wells in the south-western United States, where it occurs at concentrations as high as 7 mole percent.

To separate the components of air, it is first cooled and passed through a drying chamber where CO_2 and H_2O are absorbed on the surface of a solid such as silica gel. The air, free of moisture and carbon dioxide, is then compressed to about 200 atm and cooled by passing through a heat exchanger (Fig. 19.1). The cooled compressed air is allowed to expand suddenly (stage D) to about 10 atm, causing its temperature to drop sharply. The cycles of compression, cooling, and expansion are continued until liquefaction occurs at about $-200°C$.

When liquid air at $-200°C$ is allowed to warm, the first substance that boils off is nitrogen, the lowest boiling species present (bp $N_2 = -196°C$). Careful fractionation gives nearly pure N_2 in the distillate. The residue is mostly oxygen, with about 5% argon and traces of the heavier noble gases krypton and xenon. Fractional distillation is used to separate these components. Small amounts of oxygen can be removed from the noble gases by sparking with hydrogen. The O_2 reacts with H_2 under these conditions to form water, which readily condenses out from the gas mixture.

About 2×10^7 metric tons of air are liquefied each year in the U.S.

19.2
NITROGEN FIXATION; AMMONIA AND NITRIC ACID

Combined ("fixed") nitrogen in the form of protein is essential to all forms of life. There is more than enough elementary nitrogen in the air, about 4×10^{18} kg, to meet all our needs. The problem is to convert the element to compounds that can be used by plants to make proteins. At room temperature and atmospheric pressure, N_2 does not react with any other element. Its inertness is due to the strength of the triple bond holding the N_2 molecule together:

$$:N \equiv N:(g) \rightarrow 2 \cdot \ddot{N} \cdot (g); \qquad \Delta H = 941 \text{ kJ}$$

The high stability of the bond implies that the activation energy for any reaction of N_2 is likely to be high. Consequently, we expect the rate of reaction to be slow.

The $N \equiv N$ bond is one of the strongest known

For thousands of years, nitrogen compounds have been added to the soil to increase the yield of food crops. Until about 100 years ago, the only way to do this was to add "organic nitrogen" (i.e., manure). Late in the nineteenth century, it became common practice in the United States and Western Europe to use sodium nitrate, $NaNO_3$, imported from Chile. Then, in 1908, Fritz Haber in Germany showed that atmospheric nitrogen could be fixed by reacting it with hydrogen to form ammonia.

THE HABER PROCESS FOR MAKING AMMONIA

The reaction that Haber used was

$$N_2(g) + 3H_2(g) \rightleftharpoons 2NH_3(g); \qquad \Delta H = -92.4 \text{ kJ} \qquad (19.1)$$

TABLE 19.2   Effect of Temperature and Pressure Upon the Yield of Ammonia in the Haber Process ($[H_2] = 3[N_2]$)

°C	K_c	MOLE PERCENT NH₃ IN EQUILIBRIUM MIXTURE				
		10 atm	*50 atm*	*100 atm*	*300 atm*	*1000 atm*
200	650	51	74	82	90	98
300	9.5	15	39	52	71	93
400	0.5	4	15	25	47	80
500	0.08	1	6	11	26	57
600	0.014	0.5	2	5	14	31

His research was supported by German industrialists, who wanted to convert the ammonia to nitric acid, a starting material for making explosives. In 1913, the first large-scale ammonia plant went into production. During World War I, the Haber process produced enough ammonia and nitric acid to make Germany independent of foreign supplies of sodium nitrate, which were cut off by the British blockade.

They used the nitrates to make their munitions

The Haber process is now the main synthetic source of fixed nitrogen in the world. Its feasibility depends upon choosing conditions under which nitrogen and hydrogen will react rapidly to give a high yield of ammonia. At room temperature and atmospheric pressure, the position of the equilibrium favors the formation of ammonia ($K_c = 5 \times 10^8$). However, the rate of reaction is virtually zero. Equilibrium can be reached more rapidly by increasing the temperature. However, since Reaction 19.1 is exothermic, high temperatures reduce K_c and hence the yield of ammonia. High pressures, on the other hand, have a favorable effect on both the rate of reaction and the position of the equilibrium (Table 19.2). An increase in pressure brings the gas molecules closer together. They collide more frequently, so equilibrium is reached more rapidly. High pressure also increases the relative amount of ammonia at equilibrium, since Reaction 19.1 results in a decrease in the number of moles of gas (4 mol → 2 mol).

Much of Haber's research involved finding a catalyst to make Reaction 19.1 take place at a reasonable rate without going to very high temperatures. Nowadays, the catalyst used is a special mixture of iron, potassium oxide (K_2O), and aluminum oxide (Al_2O_3). The hydrogen and nitrogen gases must be carefully purified to remove traces of sulfur compounds, which "poison" the catalyst. Reaction takes place at 450°C and at a pressure of 200 to 600 atm. The ammonia formed is passed through a cooling chamber. Since ammonia boils at −33°C, it is condensed out as a liquid, separating it from unreacted nitrogen and hydrogen. The yield is typically less than 50%, so the reactants are recycled to produce more ammonia.

Most of the N_2 obtained from fractionating air is used to make NH_3

■ **EXAMPLE 19.1** _____

Five liters of liquid nitrogen ($d = 0.808$ g/mL) and 5.00 L of liquid hydrogen ($d = 0.0710$ g/mL) are mixed in a 70.0-L catalytic reactor, which is heated to 500°C.

a. What is the pressure in the reactor when all the liquid N_2 and H_2 have vaporized?

b. How many moles of ammonia are produced if the yield at equilibrium is 40.0%?

c. What volume of liquid ammonia is obtained upon cooling (d $NH_3(l)$ = 1.45 g/mL)?

Solution

a. We first calculate the numbers of moles of H_2 and N_2 and then use the Ideal Gas Law to obtain the pressure:

$$n\ H_2 = 5.00\ L \times \frac{1000\ mL}{1\ L} \times \frac{0.0710\ g}{1\ mL} \times \frac{1\ mol}{2.016\ g} = 176\ mol$$

$$n\ N_2 = 5.00\ L \times \frac{1000\ mL}{1\ L} \times \frac{0.808\ g}{1\ mL} \times \frac{1\ mol}{28.01\ g} = 144\ mol$$

$$P_{tot} = \frac{n_{tot}RT}{V} = \frac{(320\ mol)(0.0821\ L{\cdot}atm/mol{\cdot}K)(773\ K)}{70.0\ L} = \boxed{290\ atm}$$

b. Let us first calculate the amount of ammonia that would be produced from each reactant if the yield were 100%.

$$H_2:\quad n\ NH_3 = 176\ mol\ H_2 \times \frac{2\ mol\ NH_3}{3\ mol\ H_2} = 117\ mol\ NH_3$$

$$N_2:\quad n\ NH_3 = 144\ mol\ N_2 \times \frac{2\ mol\ NH_3}{1\ mol\ N_2} = 288\ mol\ NH_3$$

Clearly hydrogen is the limiting reactant; 117 mol of NH_3 would be produced if the yield were 100%. Since the yield is 40.0%, we have

$$n\ NH_3 = 117\ mol \times 0.400 = \boxed{46.8\ mol}$$

$$c.\ 46.8\ mol\ NH_3 \times \frac{17.0\ g\ NH_3}{1\ mol\ NH_3} \times \frac{1\ mL}{1.45\ g} = \boxed{549\ mL}$$

Exercise

Assuming a 40.0% yield of NH_3, calculate $[NH_3]$, $[N_2]$, and $[H_2]$. What is K_c for Reaction 19.1 under these conditions? Answer: 0.669 mol/L, 1.72 mol/L, 1.51 mol/L; 0.0756.

As indicated in Example 19.1, the ammonia produced by the Haber process is separated out as a liquid. Upon warming to $-33°C$, it boils to form $NH_3(g)$ at one atmosphere pressure. You are most familiar with ammonia in the form of its water solution, where NH_3 molecules are in equilibrium with NH_4^+ and OH^- ions:

$$NH_3(aq) + H_2O \rightleftharpoons NH_4^+(aq) + OH^-(aq)$$

This equilibrium is established in a bottle of household ammonia, used as a cleaning agent, and in the laboratory reagent called "ammonium hydroxide" (Fig. 19.2). This reagent is often labeled NH_4OH, although there is no evidence that such a molecule actually exists.

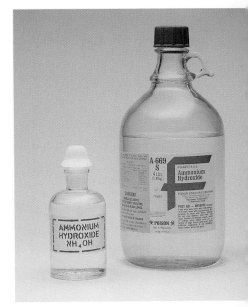

Figure 19.2   A bottle labeled "NH_4OH" or "ammonium hydroxide" contains a water solution of NH_3. (Marna G. Clarke)

Most of the ammonia produced today is used to make fertilizers. The pure liquid under pressure can be used directly. More commonly, it is converted to compounds containing the NH_4^+ ion. This is done by adding the ammonia to an acidic water solution:

$$NH_3(aq) + H^+(aq) \rightarrow NH_4^+(aq) \tag{19.2}$$

If the acid used is nitric acid, the final product is ammonium nitrate, NH_4NO_3. When sulfuric acid is used, ammonium sulfate, $(NH_4)_2SO_4$, is formed.

Ammonia produced by the Haber process is relatively expensive because of the cost of the hydrogen gas used. This has led chemists and chemical engineers to search for alternative methods of fixing nitrogen. One approach involves mimicking an important natural process. Certain bacteria found in the soil or on the roots of legumes (peas, beans, clover, alfalfa) "fix" nitrogen on a large scale. It is estimated that these bacteria produce 10^8 metric tons of NH_3 annually; this is five times the amount formed by the Haber process. The bacterial conversion is catalyzed by enzymes described collectively as "nitrogenase." The mechanism of the process is not completely understood. Two different enzymes seem to be involved. Both enzymes contain iron and sulfur atoms; one of them contains molybdenum as well. Molecular nitrogen forms a weak complex with the metal atoms of the enzymes. This is then converted to a more stable ammonia complex. This process has been simulated in the laboratory, using model compounds of molybdenum.

Genetic engineering may furnish us with corn and wheat that make their own fertilizer, like alfalfa does

THE OSTWALD PROCESS FOR MAKING NITRIC ACID

About 15% of the ammonia made by the Haber process is converted to nitric acid. The process by which this is done was developed by a German chemist, Wilhelm Ostwald. The reaction takes place in three steps. All of these are carried out at a relatively low pressure, 1 to 10 atm.

1. Ammonia is burned with ten times its volume of air. The desired reaction is

$$4NH_3(g) + 5O_2(g) \rightarrow 4NO(g) + 6H_2O(g) \tag{19.3}$$

However, a competing reaction also takes place

$$4NH_3(g) + 6NO(g) \rightarrow 5N_2(g) + 6H_2O(g)$$

With no catalyst you get $N_2 + H_2O$

This competing reaction changes available nitrogen atoms to very unreactive nitrogen molecules (N_2). To get Reaction 19.3 to occur, a gauze made of an alloy of platinum and rhodium metals (90% Pt, 10% Rh) is used as a catalyst. The reaction is carried out at 900°C. Under these conditions more than 95% of the ammonia is converted to nitrogen oxide, NO.

2. The gaseous mixture produced is mixed with more air. This lowers the temperature and brings about the reaction

$$2NO(g) + O_2(g) \rightleftharpoons 2NO_2(g) \tag{19.4}$$

3. The nitrogen dioxide produced in the second step is passed through water. In this way a solution of nitric acid is formed:

$$3NO_2(g) + H_2O(l) \rightarrow NO(g) + 2HNO_3(aq) \qquad (19.5)$$

The NO formed as a by-product is recycled in Reaction 19.4. The aqueous solution formed by Reaction 19.5 contains about 60 mass percent HNO_3. To obtain the anhydrous acid, H_2SO_4 is added and the mixture is distilled. Nearly pure nitric acid (bp = 83°C) boils off and is condensed.

■ **EXAMPLE 19.2**

Consider Reaction 19.4: $2NO(g) + O_2(g) \rightleftharpoons 2NO_2(g)$; $\Delta H = -114.0$ kJ. Applying the principles discussed in Chapters 13 and 18, what would you expect to happen to the rate and the yield of NO_2 at equilibrium if

a. the pressure were increased by compressing the system?
b. the temperature were increased?
c. a catalyst were used?

Solution

a. An increase in pressure should increase the rate (higher concentration of reactants). It should also increase the yield of NO_2 (3 mol gas → 2 mol gas).
b. One would expect an increase in temperature to increase the rate; k usually increases with T (see, however, Problems 15 and 16). An increase in temperature would decrease the yield of NO_2 since the forward reaction is exothermic.
c. A suitable catalyst would speed up the reaction but would have no effect upon the yield of NO_2.

Exercise

Suppose pure O_2 were used in this reaction instead of air (same total pressure). What effect would you expect this to have on the rate and yield of NO_2? Answer: Should increase both.

The major use of nitric acid today is in the manufacture of ammonium nitrate, NH_4NO_3. This compound is an important component of fertilizers and many explosives. Upon heating to 200°C or detonation, ammonium nitrate decomposes rapidly to gaseous products:

We didn't know NH_4NO_3 was an explosive until a boatload of it blew up in Texas City in 1947

$$NH_4NO_3(s) \rightarrow N_2O(g) + 2H_2O(g); \qquad \Delta H = -36.0 \text{ kJ} \qquad (19.6)$$

Nitric acid is also used to make the explosives nitroglycerine (Reaction 19.7) and trinitrotoluene (Reaction 19.8).

$$(19.7)$$

glycerine   nitroglycerine

$$\text{toluene} \qquad \text{trinitrotoluene (TNT)} \tag{19.8}$$

In 1867, Alfred Nobel found that nitroglycerine in porous silica was a powerful explosive that was relatively safe to work with; he called this explosive *dynamite*. Nobel became wealthy from his discovery; his estate is the source of the Nobel Prizes given out each year.

Pure nitroglycerine must be handled very carefully

19.3
WATER VAPOR AND CARBON DIOXIDE; WEATHER AND CLIMATE

The properties and composition of the atmosphere determine our weather over the short term and our climate over the years. Many factors affect both weather and climate. Among these are the presence in air of two gases: water vapor and carbon dioxide.

RELATIVE HUMIDITY

The concentration of water vapor in the air is often expressed in terms of *relative humidity:*

At 84% R.H., the air contains 84% of the water it could hold at that temperature

$$\text{R.H.} = \frac{P_{H_2O}}{P^0_{H_2O}} \times 100\% \tag{19.9}$$

where P_{H_2O} is the partial pressure of water vapor in the air and $P^0_{H_2O}$ is the equilibrium vapor pressure of water at the same temperature. On a day when the temperature is 25°C ($P^0_{H_2O}$ = 23.8 mm Hg) and the partial pressure of water vapor in the air is 20.0 mm Hg,

$$\text{R.H.} = \frac{20.0}{23.8} \times 100\% = 84.0\%$$

Our comfort depends upon relative humidity as well as temperature. At relative humidities below about 30%, evaporation of water from body surfaces is extensive and rapid. Mouth and nasal membranes dry out, allowing viruses to enter the lungs more readily. This may be why colds are so common in the winter months, when the relative humidity indoors is often quite low. Most of us are familiar with the uncomfortable effects of high relative humidities, above 80%. Perspiration fails to evaporate, leaving us feeling clammy and hot.

Changes in temperature can bring about marked variations in relative humidity. From Equation 19.9, we note that relative humidity is inversely related to the equilibrium vapor pressure of water, $P^0_{H_2O}$. Since $P^0_{H_2O}$ increases with temperature, this means that relative humidity tends to drop when the temperature rises (Example 19.3).

On a cold winter day, when the outside temperature is 0°C (32°F), the partial pressure of water in the air is 3.0 mm Hg. Calculate the relative humidity of the outside air and the relative humidity of the same air when it is brought into a house and warmed to 20°C (68°F).

The equilibrium vapor pressure of water is 4.6 mm Hg at 0°C and 17.5 mm Hg at 20°C (Appendix 1). Consequently,

$$\text{R.H. outside} = \frac{3.0}{4.6} \times 100\% = 65\%$$

$$\text{R.H. inside} = \frac{3.0}{17.5} \times 100\% = 17\%$$

Exercise

At what temperature would the relative humidity become 10%? Answer: 29°C (84°F), where vp H₂O is 30.0 mm Hg.

CLOUD FORMATION AND WEATHER MODIFICATION

If a warm air mass is suddenly cooled, $P^0_{H_2O}$ in Equation 19.9 drops and the relative humidity rises. When it reaches 100%, liquid water condenses. This is precisely the way in which clouds are formed in the atmosphere. Clouds consist of many billions of tiny droplets of liquid water, on the average perhaps 0.01 mm in diameter. Droplets of this size are too small to fall to the earth's surface as rain. The growth of small water droplets at temperatures above 0°C is ordinarily a very slow process. The rate of growth is increased in the presence of dust particles, which act as nuclei upon which small droplets can condense. This explains why volcanic eruptions are often followed by rainstorms.

More frequently, ice crystals formed in the colder upper regions of clouds act as nuclei for precipitation. In principle, ice crystals should form at 0°C; in practice, they seldom develop unless the temperature drops to at least −15°C. To stimulate the formation of ice crystals, clouds are sometimes seeded with dry ice. The sublimation of solid carbon dioxide absorbs enough heat from the cloud to reduce the temperature below that required for ice crystal formation. Another substance that is frequently used is finely divided silver iodide (Fig. 19.3), which has a crystal structure similar to that of ice. In general, cloud-seeding experiments have met with mixed success. Their major application nowadays is in fog dispersal at airports, where finely divided dry ice is quite effective.

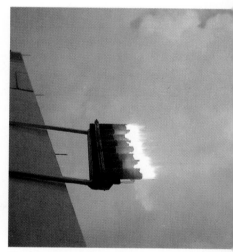

Figure 19.3 Finely dispersed silver iodide for cloud seeding experiments is produced by the reaction:

$$2Al(s) + AgIO_3(s) \rightarrow$$
$$AgI(s) + Al_2O_3(s)$$

This is carried out by igniting a mixture of powdered aluminum and silver iodate from burners in the wings of an aircraft. (Atmospherics, Inc.)

THE GREENHOUSE EFFECT

The mean global temperature at the earth's surface is about 15°C (59°F). This temperature is determined by a delicate balance between:

— the energy that is absorbed from the sun. This energy covers a broad spectrum of wavelengths from the ultraviolet (<400 nm) through the visible (400 to 800 nm) into the infrared (>800 nm).

— the energy emitted back into space by the earth. This consists of infrared radiation, mostly at wavelengths between 3000 and 30,000 nm.

Any change in the amount of energy absorbed or emitted by the earth could upset this balance, affecting our climate. In this connection, consider what happens to the infrared radiation given off by the earth. Part of it is absorbed by the atmosphere rather than being radiated to outer space. Two gases in the air, H_2O and CO_2, absorb infrared radiation (Fig. 19.4). In this way, they act as an insulating blanket to prevent heat from escaping; this is often referred to as a "greenhouse effect." Were it not for this effect, the earth's temperature would be much lower, of the order of $-20°C$. On Mars, which has a very thin atmosphere, the surface temperature is $-50°C$. In contrast, Venus, which has a dense atmosphere consisting mostly of CO_2, has a temperature of $400°C$. This difference in temperature is much greater than could be explained by the fact that Venus is closer to the sun; in large part it is due to the greenhouse effect.

Of the two gases, water vapor absorbs more infrared radiation than carbon dioxide because its concentration is higher. This property of water vapor accounts for the fact that the temperature drops less on nights when there is a heavy cloud cover. In desert regions, where there is very little water vapor, large variations between day and night temperatures are common.

The concentration of CO_2 in the earth's atmosphere is low, about 340 ppm,* but appears to be increasing exponentially. A century ago the CO_2 concentration was less than 300 ppm. The increase in CO_2 content has been caused by human activities. Increased consumption of fossil fuels is mainly responsible. Every gram of fossil fuel burned releases about three grams of carbon dioxide into the atmosphere. Part of this CO_2 is used by plants in photosynthesis or is absorbed by the oceans, but at least half of it remains. Extensive land clearing, which reduces the amount of carbon dioxide consumed by photosynthesis, is also a factor in raising the CO_2 content of the atmosphere. This is one of the adverse effects of the destruction of tropical rain forests for agricultural purposes.

It seems unlikely that we will ever colonize Mars or Venus

■ **EXAMPLE 19.4** _____

Assume that the composition of gasoline, a mixture of hydrocarbons, is approximated by that of octane, C_8H_{18} ($d = 0.70$ g/mL). Calculate the mass of carbon dioxide produced by the complete combustion, to CO_2 and H_2O, of one tankful (50.0 L) of gasoline.

_____

*For a gaseous species.

  parts per million (ppm) $= 10^6\ X$

where X is the mole fraction. Recall from Table 19.1 that the mole fraction of CO_2 in air is 0.00034. Hence, the concentration of CO_2 in parts per million is $0.00034 \times 10^6 = 340$ ppm.

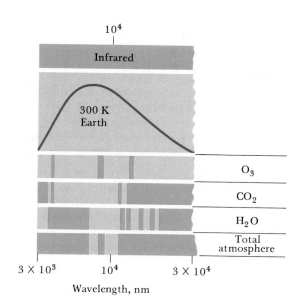

Figure 19.4 The earth at 300 K gives off radiation in the wavelength range 3000 to 30,000 nm, with a maximum emission close to 10,000 nm. A small amount of this is absorbed by ozone. Larger amounts are absorbed by CO$_2$ and H$_2$O, as indicated by the wide absorption bands (colored) for these two species. In the region around 10^4 nm, trace species such as CH$_4$ and CFCl$_3$ absorb strongly, contributing to the greenhouse effect.

Solution

The balanced equation for the combustion is:

$$C_8H_{18}(l) + \tfrac{25}{2}O_2(g) \rightarrow 8CO_2(g) + 9H_2O(l)$$

We determine the mass of one tankful of gasoline.

$$50.0 \text{ L} \times \frac{1000 \text{ mL}}{1 \text{ L}} \times \frac{0.70 \text{ g}}{1 \text{ mL}} = 3.5 \times 10^4 \text{ g}$$

To find out how much carbon dioxide is produced, we convert mass of gasoline to moles of C$_8$H$_{18}$, then to moles of CO$_2$ using the coefficients of the balanced equation, and finally to mass of CO$_2$.

$$\text{mass CO}_2 = 3.5 \times 10^4 \text{ g C}_8\text{H}_{18} \times \frac{1 \text{ mol C}_8\text{H}_{18}}{114 \text{ g C}_8\text{H}_{18}} \times \frac{8 \text{ mol CO}_2}{1 \text{ mol C}_8\text{H}_{18}} \times \frac{44.0 \text{ g CO}_2}{1 \text{ mol CO}_2}$$

$$= \boxed{1.1 \times 10^5 \text{ g CO}_2}$$

We see that the mass of CO$_2$ produced is roughly three times the mass of gasoline burned.

Exercise

Calculate the mass ratio of natural gas (CH$_4$) burned to CO$_2$ produced. Answer: 1:2.75.

It has been estimated that at the current rate of fossil fuel consumption, the CO$_2$ content of the air will double by the middle of the next century (~2050). This could increase the mean global temperature by 2°C–3°C; the increase is likely to be larger in the northern temperate regions, which include the United States, Canada, and most of Europe. Such an increase would have drastic effects on the world's climate. It is quite possible that the polar ice caps would melt, raising the level of the oceans by perhaps 5 m. That would flood many coastal areas, including half the state of Florida. On a more optimistic note, an increase in CO_2 concentration is known to promote photosynthesis. Doubling the amount of carbon dioxide in the air could increase the world's food supply by raising yields of such crops as rice, wheat, and soybeans.

Increased production of carbon dioxide is not the only factor that can change global temperatures. It is well known that suspended particles of dust or soot in the atmosphere have a cooling effect because they reflect sunlight back into space. This was discovered when the 1883 eruption of Mount Krakatoa in the South Pacific caused a July snowstorm in the northeastern United States. Some geologists have suggested that modern industrial processes that increase the number of suspended particles in the air could moderate or even cancel the climatic effects of an increase in carbon dioxide concentration.

Recently, there has been a great deal of speculation about the effect of a nuclear war on the earth's climate. Explosions and fires over a large fraction of the earth's surface could produce enough suspended particles to give a drastic cooling effect. The phrase "nuclear winter" has been used to describe a temperature drop variously estimated at from 5°C to 15°C. What effect this would have is difficult to estimate, given the fact that there are likely to be very few survivors of a nuclear war in the first place.

19.4
THE UPPER ATMOSPHERE

The species in the troposphere (near the surface of the earth) are mostly stable atoms and molecules such as N_2, O_2, Ar, CO_2, and H_2O. In the stratosphere (10–50 km above the earth) the situation is quite different. Exploration of the upper atmosphere by rockets and satellites reveals additional species, such as those listed in Table 19.3. Some of these are neutral particles (oxygen, nitrogen, and hydrogen atoms, and OH radicals). Others are positive ions formed from diatomic molecules (NO^+, O_2^+, N_2^+) or from atoms (O^+, H^+).

All the reactions listed in Table 19.3 are endothermic. They require either the breaking of a chemical bond or the removal of an electron. In

TABLE 19.3   Formation of High-Energy Species in the Upper Atmosphere

MOLECULAR FRAGMENTS			CATIONS		
REACTION	ΔE (kJ/mol)	λ_{max}* (nm)	REACTION	ΔE (kJ/mol)	λ_{max}* (nm)
$NO_2 \rightarrow NO + O$	+305	392	$NO \rightarrow NO^+ + e^-$	+892	134
$O_2 \rightarrow O + O$	+494	242	$O_2 \rightarrow O_2^+ + e^-$	+1165	103
$H_2O \rightarrow H + OH$	+502	238	$O \rightarrow O^+ + e^-$	+1314	91
$NO \rightarrow N + O$	+632	189	$H \rightarrow H^+ + e^-$	+1312	91
$N_2 \rightarrow N + N$	+941	127	$N_2 \rightarrow N_2^+ + e^-$	+1510	79

*Longest wavelength of light that can supply the energy indicated.

the upper atmosphere, sunlight is the source of energy for such reactions. The wavelengths of light given in Table 19.3 are calculated from the equation (Chap. 5)

$$\lambda_{max} = \frac{1.196 \times 10^5}{\Delta E} \frac{kJ \cdot nm}{mol} \qquad (19.10)$$

■ EXAMPLE 19.5 _____

Using Equation 19.10, calculate the longest wavelength of radiation that can bring about the reaction:

$$OH(g) \rightarrow O(g) + H(g)$$

given that the O—H bond energy is 464 kJ/mol.

Solution

$$\Delta E = \text{B.E. O—H} = +464 \text{ kJ/mol}$$

$$\lambda_{max} = \frac{1.196 \times 10^5}{464} \text{ nm} = \boxed{258 \text{ nm}}$$

In practice, radiation of shorter wavelength (higher energy) is absorbed to bring about this reaction. The excess energy is carried off by the products (O, H atoms) as kinetic energy.

Exercise

What is the bond energy of the strongest bond that can be broken by radiation with a wavelength of 205 nm? Answer: 583 kJ/mol.

_____

Notice that all of the wavelengths listed are in the ultraviolet region, below 400 nm. Very little of this high-energy radiation reaches the surface of the earth, so the reactions in Table 19.3 are not very likely to occur in the air around us. Above 30 km, significant amounts of radiation in the near ultraviolet (200 to 400 nm) are available, and species such as O, H, and OH begin to appear. At still higher altitudes, sunlight contains radiation in the far ultraviolet (<200 nm). This can bring about highly endothermic processes, forming such species as N, O_2^+, and N_2^+.

The UV is absorbed when these reactions occur

The concentrations of species such as those listed in Table 19.3 are much higher in the upper atmosphere than we would expect on the basis of equilibrium considerations. Consider, for example, the system

$$O_2(g) \rightleftharpoons 2O(g)$$

The equilibrium constant, K_c, for this reaction is a very small number, about 4×10^{-83}. On that basis, it would seem that the forward reaction should occur to only a tiny extent. This is indeed the case at the earth's surface, where virtually all elemental oxygen is in the form of O_2 rather than oxygen atoms. However, at 120 km, there are about as many O atoms as O_2 molecules. The concentration of atomic oxygen is far higher than we would expect from the value of K_c.

One way to explain this situation is to consider the distance between particles in the atmosphere. At the earth's surface, atoms and molecules in the air are close together. The *mean free path* (average distance traveled between collisions) is only of the order of 10^{-7} m. Collisions are frequent and equilibrium is rapidly established. At high altitudes, the situation is quite different. Particles are much farther apart; the mean free path at 120 km is 3 to 4 m. Once an oxygen atom is formed at this altitude, it may be a long time before it collides with another one to form an O_2 molecule. Moreover, at this altitude, oxygen atoms have very high kinetic energies. Even if they collide, two atoms may have too much energy to stick together. As a result, the concentration of O atoms relative to O_2 molecules is much higher than it should be from equilibrium considerations.

To get equilibrium you need lots of collisions

OZONE

In recent years, one component of the upper atmosphere has received more attention than any other. This is ozone, an allotropic form of oxygen with the molecular formula O_3. Ozone can be considered to be a resonance hybrid of the two structures

with a bond angle in good agreement with that predicted by VSEPR theory (117° vs. 120°). It is a pale blue gas with a characteristic odor that can be detected after lightning activity, in the vicinity of electric motors, or near a subway train.

From 95 to 99% of sunlight in the wavelength range of 200 to 300 nm is absorbed by ozone in the upper atmosphere. If this ultraviolet radiation were to reach the surface of the earth, it could have several adverse effects. A decrease in ozone concentration of only 5% could increase the incidence of skin cancer by 25%. It is also known that ultraviolet radiation is a factor in diseases of the eye, particularly cataract formation.

O_3 is a good UV absorber, much better than O_2 at 250 nm

The concentration of ozone in the earth's atmosphere passes through a maximum of about 10 ppm at an altitude of 30 km. The ozone in this region is formed by a two-step process. The first step involves the dissociation of an O_2 molecule:

$$O_2(g) \rightarrow 2O(g) \tag{19.11}$$

This is followed by a collision between an oxygen atom and an O_2 molecule:

$$O_2(g) + O(g) \rightarrow O_3(g) \tag{19.12}$$

Ozone molecules formed by Reaction 19.12 decompose by several mechanisms. One of the most important is

$$O_3(g) + O(g) \rightarrow 2O_2(g) \tag{19.13}$$

This reaction takes place at a rather slow rate by direct collision between an O_3 molecule and an O atom. It can occur more rapidly by a two-step process in which a trace component of the upper atmosphere acts as a catalyst. Two such catalysts that have received a great deal of attention are listed in Table 19.4.

TABLE 19.4   Mechanism for the Catalytic Decomposition of Ozone

	CATALYST	
	NO MOLECULE	Cl ATOM
Mechanism	$NO + O_3 \rightarrow NO_2 + O_2$	$Cl + O_3 \rightarrow ClO + O_2$
	$NO_2 + O \rightarrow NO + O_2$	$ClO + O \rightarrow Cl + O_2$
Overall reaction	$O_3 + O \rightarrow 2O_2$	$O_3 + O \rightarrow 2O_2$

Since NO acts as a catalyst for ozone decomposition, an increase in nitric oxide concentration in the upper atmosphere could cause the depletion of ozone. This was one factor that influenced the United States to abandon the development of the supersonic transport (SST). These planes burn jet fuel with air at high temperatures. It was suggested that combustion might produce significant amounts of NO by the reaction

$$N_2(g) + O_2(g) \rightarrow 2NO(g) \tag{19.14}$$

However, studies carried out in 1982 showed that ozone depletion by SST flights was not nearly as severe as had been feared. We now know that most of the NO in the upper atmosphere is formed from N_2O by the reaction:

$$N_2O(g) + O(g) \rightarrow 2NO(g) \tag{19.15}$$

The N_2O is formed by the decomposition of nitrogen-containing fertilizers.

In recent years, concern has focused on the Cl-catalyzed decomposition of ozone. At the time this mechanism was discovered, in 1973, it was believed to be unimportant. There was no known source of Cl atoms in the upper atmosphere. Less than a year later it was suggested that two organic compounds, $CFCl_3$ and CF_2Cl_2, might be a source of Cl. These compounds are widely used as refrigerants, in air conditioners, and as aerosol propellants. They decompose to form chlorine atoms when exposed to ultraviolet radiation at 200 nm:

$$CFCl_3(g) \rightarrow CFCl_2(g) + Cl(g) \tag{19.16}$$

$$CF_2Cl_2(g) \rightarrow CF_2Cl(g) + Cl(g) \tag{19.17}$$

A recent report by the National Aeronautics and Space Administration reveals that the Cl-catalyzed reaction has decreased the concentration of atmospheric ozone by 2.5% over the past decade. This report prompted DuPont, the leading manufacturer of chlorinated fluorocarbons, to call for a total phaseout of the production of these materials. As you can imagine, the development of replacements for $CFCl_3$ and CF_2Cl_2 is an active area of research. A leading possibility is:

$$
\begin{array}{ccc}
 & F & H \\
 & | & | \\
F- & C- & C-F \\
 & | & | \\
 & F & H
\end{array}
$$

which contains no chlorine atoms.

Another reason was that the SST is not energy efficient

The compounds are called Freons©. Both boil below 0°C, which makes them good refrigerants

Before Freons© were developed, we used SO_2 and NH_3 as refrigerants; any leaks in the system caused big problems, since these gases are poisonous

■ **EXAMPLE 19.6**

Consider the reaction mechanisms

$$O_3(g) + O(g) \rightarrow 2O_2(g); \qquad\qquad k_1 = 5 \times 10^6 \text{ L/mol·s}$$

$$O_3(g) + NO(g) \rightarrow NO_2(g) + O_2(g); \qquad k_2 = 1 \times 10^7 \text{ L/mol·s}$$

a. Write the rate expressions for the two mechanisms.
b. Calculate the ratio of the two rates (rate$_2$/rate$_1$) at an altitude of 40 km. Take the concentrations of O and NO to be 2×10^{-12} and 3×10^{-12} mol/L, respectively.

Solution

a. rate$_1$ = k_1(conc. O_3) × (conc. O)
    rate$_2$ = k_2(conc. O_3) × (conc. NO)

b. Dividing,

$$\frac{\text{rate}_2}{\text{rate}_1} = \frac{k_2 \times \text{conc. NO}}{k_1 \times \text{conc. O}} = \frac{1 \times 10^7 \times (3 \times 10^{-12})}{5 \times 10^6 \times (2 \times 10^{-12})} = \boxed{3}$$

This calculation suggests that, under these conditions, O_3 is reacting with NO three times as rapidly as with O atoms.

Exercise

Suppose the concentration of O were twice that of NO. How would the two rates compare in that case? Answer: Equal.

Figure 19.5 Depletion of atmospheric ozone in the Antartic, 1986. Dobson units, indicated in the color code at the right, are proportional to ozone concentration. The violet region at the center of the picture, representing the South Pole, has a very low concentration of ozone, about half the normal values, shown in blue at the edges of the map. (NASA)

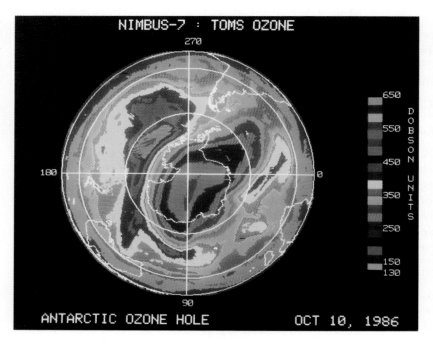

Recently, attention has focused on the "ozone hole" (Fig. 19.5) that has recently appeared in Antarctica each year in September and October, at the end of winter in the southern hemisphere. In 1985, the ozone concentration dropped to less than half of its normal value. Based on extensive research in 1986 and 1987, we are now quite sure why this occurs. All of the evidence points to the Cl-atom catalyzed decomposition of ozone as the culprit. It appears that this takes place as the result of a heterogeneous reaction on ice crystals in the clouds that surround the South Pole continuously in winter. Apparently this happens, to a lesser extent, in Arctic regions as well.

19.5
AIR POLLUTION

Since the discovery of fire, mankind has polluted the atmosphere with noxious gases and soot. When coal began to be used as a fuel in the fourteenth century, the problem became one of public concern. Increased fuel consumption by industry, concentration of population in urban areas, and the advent of motor vehicles have, over the years, made the problem worse. Today a major cause of pollution in our atmosphere is the gasoline engine.

Any substance whose addition to the atmosphere produces a measurable adverse effect on human beings or the environment can be called a pollutant. A host of materials fit this broad definition. Suspended particles (soot, dust, smoke) qualify, as do radioactive species produced by fallout from nuclear testing. In this chapter, we will limit our attention to a few major pollutants:

— the oxides of sulfur (SO_2, SO_3), formed when sulfur or sulfur compounds burn in air, and sulfuric acid (H_2SO_4), formed when sulfur trioxide reacts with water.

— carbon monoxide (CO), formed by the incomplete combustion of hydrocarbon fuels.

— the oxides of nitrogen (NO, NO_2), formed in high-temperature combustion processes.

SULFUR COMPOUNDS (SO_2, SO_3, H_2SO_4); ACID RAIN

Most of the coal burned in heating and power plants in the United States contains from 1% to 4% sulfur, much of which is in the form of minerals such as pyrite, FeS_2. Combustion converts the sulfur to sulfur dioxide:

$$4FeS_2(s) + 11O_2(g) \rightarrow 2Fe_2O_3(s) + 8SO_2(g) \qquad (19.18)$$

About two-thirds of the SO_2 that enters the atmosphere comes from the combustion of coal. Smaller amounts are formed from heating oil and from metallurgical processes including the "roasting" of sulfide ores.

Sulfur dioxide at concentrations as low as 0.3 ppm can cause acute injury to plants. Most healthy adults can tolerate considerably higher SO_2 levels without apparent adverse effects. However, individuals who suffer from chronic respiratory diseases such as bronchitis or asthma are much more sensitive. An increase in the concentration of SO_2 from 0.1 to 0.2 ppm can cause them to start coughing and experience severe difficulties in breathing.

Much of the sulfur dioxide in polluted air is converted to sulfur trioxide by reactions such as

$$SO_2(g) + \tfrac{1}{2}O_2(g) \rightarrow SO_3(g) \tag{19.19}$$

$$SO_2(g) + 2\ OH(g) \rightarrow SO_3(g) + H_2O(g) \tag{19.20}$$

Reaction 19.19 is catalyzed by suspended solids in the atmosphere such as those found in coal smoke. Reaction 19.20 involves the OH radical

$$H\!-\!\overset{\cdot\cdot}{\underset{\cdot\cdot}{O}}\cdot$$

produced by photochemical reactions of the type discussed in Section 19.4.

■ EXAMPLE 19.7 _____

The major source of the OH radicals involved in Reaction 19.20 is the reaction between oxygen atoms and water molecules:

$$O(g) + H_2O(g) \rightarrow 2\ OH(g)$$

The rate constant for this second-order reaction is 2.8×10^{-3} L/mol·s at 27°C. If the activation energy is 77 kJ/mol, calculate k for this reaction at an altitude of 3 km, where the temperature is -4°C.

Solution

We use the two-point form of the Arrhenius equation (Chapter 18):

$$\ln \frac{k_2}{k_1} = \frac{E_a(T_2 - T_1)}{RT_2T_1}$$

Here,

$$k_2 = 2.8 \times 10^{-3}\ \text{L/mol·s}; \qquad E_a = 7.7 \times 10^4\ \text{J/mol};\ R = 8.31\ \text{J/mol·K}$$

$$T_2 = 27 + 273 = 300\ \text{K}; \qquad T_1 = -4 + 273 = 269\ \text{K}$$

We substitute these values and solve for k_1:

$$\ln \frac{k_2}{k_1} = \frac{7.7 \times 10^4}{(8.31)}\left(\frac{300 - 269}{300 \times 269}\right) = 3.56$$

$$\frac{k_2}{k_1} = 35; \qquad k_1 = \frac{k_2}{35} = \frac{2.8 \times 10^{-3}}{35} = \boxed{8.0 \times 10^{-5}\ \text{L/mol·s}}$$

Exercise

At the lower edge of the stratosphere (altitude 11 km), the temperature is -55°C. Calculate k for this reaction at that temperature. Answer: 2.5×10^{-8} L/mol·s.

Most of the adverse effects of sulfur oxides in the atmosphere are caused by the sulfuric acid produced when sulfur trioxide (SO_3) reacts with water:

$$SO_3(g) + H_2O \rightarrow H_2SO_4(aq) \qquad (19.21)$$

The sulfuric acid forms as tiny droplets high in the atmosphere. These may be carried by prevailing winds as far as 1000 miles before falling to earth as *acid rain.*

Natural rainfall has a pH of about 5.5. It is slightly acidic because of dissolved carbon dioxide:

$$CO_2(g) + H_2O \rightleftharpoons H^+(aq) + HCO_3^-(aq) \qquad (19.22)$$

The presence of sulfuric acid and, to a lesser extent, nitric acid in the atmosphere has led to an average pH of about 4.4 in rainfall in the northeastern United States and southeastern Canada. In an extreme case, rain with a pH of 1.5 was recorded in Wheeling, West Virginia (distilled vinegar has a pH of 2.4). Coal-burning utilities in the midwestern United States appear to be largely responsible for acid rain (Fig. 19.6).

The effects of acid rain are particularly severe in areas where the bedrock is granite or other materials incapable of neutralizing H^+ ions. This situation prevails in the Adirondacks and the mountains of northern New England. Lakes in these regions are exposed annually to as much as 4 metric tons of sulfuric acid per square kilometer. As the concentration of acid builds up, marine life, from algae to brook trout, dies. The end

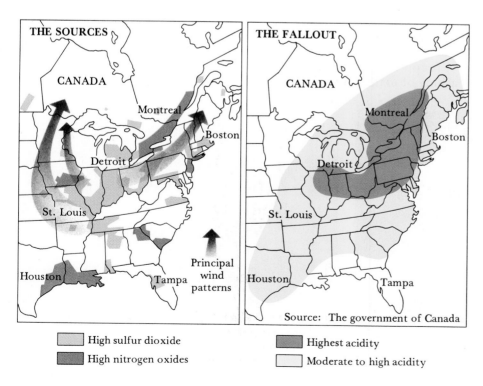

Figure 19.6 Acid rain is caused by dissolved H_2SO_4 and, to a lesser extent, HNO_3. Most of the sulfur oxides that are largely responsible for acid rain come from midwestern states such as Illinois, Indiana, and Ohio. Prevailing winds carry the acid droplets to the northeast, as far as the Maritime Provinces of Canada.

Figure 19.7 One way to offset the effect of acid rain on the pH of natural waters is to add calcium carbonate to react with excess H^+ ions. Here the $CaCO_3$ is being sprayed from a helicopter. (Ohio Edison)

product is a crystal clear, totally sterile lake. There are over 200 such dead lakes in the Adirondacks alone, all of them attributable to acid rain.

Sweden has pioneered a "solution" to the problem of acid lakes, pumping pulverized limestone into the lakes (Fig. 19.7). The calcium carbonate reacts with excess H^+ ions, reducing the pH of the lake water.

$$CaCO_3(s) + 2H^+(aq) \rightarrow Ca^{2+}(aq) + CO_2(g) + H_2O \qquad (19.23)$$

This has also been done on a small scale in the United States and Canada. The difficulty with this approach is that it is expensive and has to be repeated annually.

There is increasing evidence that acid rain has an adverse effect on trees as well as marine life (Fig. 19.8). Concerned by this effect, 21 European countries and Canada (but not the United States) agreed in 1985 to lower SO_2 emissions by 30% or more over a 10-year period. More than half of these countries have already achieved this goal.

It appears that much of the damage caused by acid rain is due to the leaching of toxic metal cations from the soil. For example, H^+ ions in acid

Figure 19.8 Effect of acid rain on evergreens at Camel's Hump, Vermont. The two photographs were taken 15 years apart. (EPA)

rain can react with insoluble aluminum compounds in the soil, bringing Al^{3+} ions into solution. The following reaction is typical:

$$Al(OH)_3(s) + 3H^+(aq) \rightarrow Al^{3+}(aq) + 3H_2O \qquad (19.24)$$

Al^{3+} ions in water solution are taken up by the marine life and the roots of trees, where they can have a lethal effect.

Sulfuric acid also attacks building materials such as limestone or marble (calcium carbonate):

$$CaCO_3(s) + H_2SO_4(aq) \rightarrow CaSO_4(s) + CO_2(g) + H_2O \qquad (19.25)$$

The calcium sulfate formed is soluble enough to be gradually washed away (Fig. 19.9). This process is responsible for the deterioration of the Greek ruins on the Acropolis in Athens. These structures have suffered more damage in the twentieth century than in the preceding 2000 years. Another effect that is due to sulfuric acid is the deterioration of the paper in books and documents. Manuscripts printed before 1750 are almost immune to sulfur oxides. At about that time, modern methods of papermaking were introduced. These leave traces of metal oxides, which catalyze the conversion of SO_2 to SO_3 and hence to sulfuric acid.

To reduce acid rain, it is most important to lower sulfur dioxide emissions from coal-burning power plants. One way to solve the problem is to remove sulfur compounds from the coal before it is burned. Pyrite, FeS_2, can be separated from coal either magnetically or by gravity (d FeS_2 = 4.9 g/cm³, d coal = 1.2 g/cm³).

A different approach involves adding a chemical to react with sulfur dioxide after it is formed. One possibility is to add limestone, $CaCO_3$, to the furnace where coal is burning. The limestone decomposes:

$$CaCO_3(s) \rightarrow CaO(s) + CO_2(g)$$

The calcium oxide formed reacts with sulfur dioxide and oxygen:

$$CaO(s) + SO_2(g) + \tfrac{1}{2}O_2(g) \rightarrow CaSO_4(s) \qquad (19.26)$$

Thousands of maple trees in Ontario have been killed by acid rain, presumably from too much Al^{3+} in the soil

Figure 19.9 Effect of pollutants on statuary. The photograph at the left was taken in about 1910 at Lincoln Cathedral in England; that at the right was taken in 1984. (Dean and Chapter of Lincoln)

Alternatively, a "scrubber" charged with a water solution of calcium hydroxide can be inserted directly into the smoke stack. A reaction very similar to Reaction 19.26 occurs, this time in aqueous solution:

$$Ca^{2+}(aq) + 2OH^-(aq) + SO_2(g) + \tfrac{1}{2}O_2(g) \rightarrow CaSO_4(s) + H_2O \quad (19.27)$$

Both of these approaches produce large amounts of calcium sulfate that must be disposed of. Partly for this reason, they add to the cost of the electrical energy produced by the combustion of coal.

CARBON MONOXIDE (CO)

The incomplete combustion of hydrocarbon fuels produces significant amounts of carbon monoxide. The principal culprit here is the automobile. If enough oxygen is present, all the carbon atoms in the fuel are converted to CO_2:

$$C_8H_{18}(l) + \tfrac{25}{2}O_2(g) \rightarrow 8CO_2(g) + 9H_2O(l) \quad (19.28)$$

However, if the fuel mixture is too "rich," that is, contains too much fuel and too little air, considerable amounts of CO may be formed:

$$C_8H_{18}(l) + \tfrac{23}{2}O_2(g) \rightarrow 6CO_2(g) + 2CO(g) + 9H_2O(l) \quad (19.29)$$

In principle, the carbon monoxide should be converted to CO_2 in the atmosphere:

$$CO(g) + \tfrac{1}{2}O_2(g) \rightarrow CO_2(g); \qquad K_c = 5 \times 10^{45} \text{ at } 25°C \quad (19.30)$$

but this reaction, like the conversion of SO_2 to SO_3, is ordinarily quite slow.

Carbon monoxide taken into the lungs reduces the ability of the blood to transport oxygen through the body. It does this by forming a complex with the hemoglobin of the blood that is more stable than that formed by oxygen (Chap. 16):

$$CO(g) + Hem \cdot O_2(aq) \rightleftharpoons O_2(g) + Hem \cdot CO(aq) \quad (19.31)$$

Blood absorbs CO about 200 times as readily as it does O_2

$$K_c = \frac{[Hem \cdot CO] \times [O_2]}{[Hem \cdot O_2] \times [CO]} = 210$$

The equilibrium constant for Reaction 19.31 is large enough so that low levels of carbon monoxide can convert significant amounts of hemoglobin to the CO complex. Symptoms of carbon monoxide poisoning show up when 10% of the hemoglobin is tied up by CO. When the fraction rises to 20%, death can result unless the victim is removed from the poisonous atmosphere.

■ EXAMPLE 19.8

Carbon monoxide is present in cigarette smoke. It is estimated that a heavy smoker is exposed to a CO concentration of 2.1×10^{-6} mol/L (50 ppm). Taking the concentration of O_2 to be 8.8×10^{-3} mol/L, calculate

a. the ratio $[CO]/[O_2]$.
b. the ratio $[Hem \cdot CO]/[Hem \cdot O_2]$.

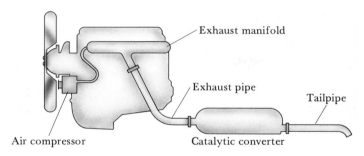

Air compressor   Catalytic converter

Figure 19.10 Catalytic converters contain a "three-way" catalyst designed to convert CO to CO_2, unburned hydrocarbons to CO_2 and H_2O, and NO to N_2. The active components of the catalysts are the precious metals platinum and rhodium; palladium is sometimes used as well.

Solution

a. $\dfrac{[CO]}{[O_2]} = \dfrac{2.1 \times 10^{-6} \text{ mol/L}}{8.8 \times 10^{-3} \text{ mol/L}} = \boxed{2.4 \times 10^{-4}}$

b. $K_c = \dfrac{[Hem \cdot CO] \times [O_2]}{[Hem \cdot O_2] \times [CO]} = 210$

Solving for the ratio $[Hem \cdot CO]/[Hem \cdot O_2]$,

$$\frac{[Hem \cdot CO]}{[Hem \cdot O_2]} = 210 \times \frac{[CO]}{[O_2]} = 210 \times 2.4 \times 10^{-4} = \boxed{0.050}$$

This calculation suggests that in the blood stream of a heavy smoker about 5% of the hemoglobin is tied up as the CO complex. As a result, smokers are more susceptible to carbon monoxide poisoning than nonsmokers.

Exercise

The equilibrium constant of Reaction 19.31 differs from one mammal to another. In a rabbit, K_c is about 100. What is the ratio $[Hem \cdot CO]/[Hem \cdot O_2]$ in a laboratory rabbit who is a heavy smoker (two packs a day)? Answer: 0.024.

At least three-fourths of the carbon monoxide in polluted air comes from automobiles. Since 1975, the principal method of reducing emissions of CO (and unburned hydrocarbons) has been to install catalytic converters in cars (Fig. 19.10). Typically, a catalytic converter contains from 1 to 3 g of platinum mixed with other heavy metals (Rh, Pd) embedded in a base of aluminum oxide, Al_2O_3. In the presence of platinum, CO and hydrocarbons in the exhaust are converted to carbon dioxide and water.

NITROGEN OXIDES (NO, NO₂)

The high-temperature combustion of fuels is the principal source of nitrogen oxide air pollutants. Detectable amounts of nitric oxide are produced by the reaction

$$N_2(g) + O_2(g) \rightarrow 2NO(g); \qquad \Delta H = +180.4 \text{ kJ}$$

The brown haze covering this city is pollution caused by NO₂. (National Center for Atmospheric Research)

In urban air, NO is converted to NO_2 by a mechanism that is poorly understood. Direct reaction with O_2 (Reaction 19.4) is too slow to account for the rapid buildup of NO_2 (see Example 19.9).

The oxides of nitrogen play a key role in the formation of *photochemical smog*. This type of air pollution was first noted in Los Angeles, but it is now common in cities from Honolulu to Washington, D.C. Typically, it develops on bright, sunny mornings when the concentration of NO_2 is relatively high (Fig. 19.11). It is called photochemical smog because light is needed to trigger the key step in its formation, the dissociation of nitrogen dioxide:

$$NO_2(g) \rightarrow NO(g) + O(g) \tag{19.32}$$

This occurs on exposure to light at the edge of the visible region (392 nm; Table 19.3). The oxygen atoms produced react with O_2 molecules:

$$O_2(g) + O(g) \rightarrow O_3(g)$$

O₃ is fine in the upper atmosphere, but not down here

The product, ozone, is a major component of photochemical smog. Ozone molecules, oxygen atoms, and nitric oxide molecules attack organic compounds in the air. Unsaturated hydrocarbons, containing multiple carbon-carbon bonds, such as ethylene and propylene, are particularly reactive:

$$
\begin{array}{cc}
\underset{\text{ethylene}}{\overset{\displaystyle H \quad H}{\underset{\displaystyle |\quad |}{H-C=C-H}}} &
\underset{\text{propylene}}{\overset{\displaystyle H \quad H}{\underset{\displaystyle |\quad |}{H-C=C-CH_3}}}
\end{array}
$$

The eye irritation associated with smoggy days is due to the formation of acrolein and peroxyacetyl nitrate (PAN) from the reaction of ozone with hydrocarbons:

$$
\underset{\text{acrolein}}{\overset{\displaystyle H-C=C-C=O}{\underset{\displaystyle |\;\;\;|\;\;\;|}{\quad H \;\; H \;\; H}}}
\qquad
\underset{\text{PAN}}{CH_3-\overset{\displaystyle O}{\underset{\displaystyle \|}{C}}-O-O-\overset{}{N}-O}
$$

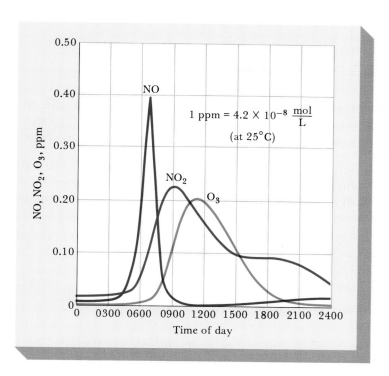

Figure 19.11 Average concentrations of the pollutants NO, NO_2, and O_3 on a smoggy day in Los Angeles. The concentration of NO from automobile exhaust builds up in the early morning rush hour. Later, NO_2 and O_3 are produced. The peak NO concentration of 0.40 ppm corresponds to about 1.7×10^{-8} mol/L at 25°C.

■ **EXAMPLE 19.9** _____

Consider the reaction $2NO(g) + O_2(g) \rightarrow 2NO_2(g)$, for which the rate expression is

rate = k(conc. NO)^2 × (conc. O_2), where $k = 2.5 \times 10^7$ L^2/mol$^2\cdot$h

a. What is the rate of this reaction at the peak NO concentration in smog formation (Fig. 19.11), 1.7×10^{-8} mol/L? The concentration of O_2 in the air is 8.8×10^{-3} mol/L.

b. At this rate, how long would it take for the concentration of NO_2 to build up to its peak value, 1.0×10^{-8} mol/L?

Solution

a. rate = $2.5 \times 10^7 \dfrac{L^2}{mol^2\cdot h} \times (1.7 \times 10^{-8}$ mol/L$)^2 \times (8.8 \times 10^{-3}$ mol/L$)$

= $\boxed{6.4 \times 10^{-11} \dfrac{mol}{L\cdot h}}$

b. time = $\dfrac{\text{conc. } NO_2}{\text{rate}} = \dfrac{1.0 \times 10^{-8} \text{ mol/L}}{6.4 \times 10^{-11} \text{ mol/L}\cdot h} = \boxed{160 \text{ h}}$

Clearly, this cannot be the reaction by which NO_2 is produced from NO in smog formation. It is much too slow. From Figure 19.11, about 3 h is required for the concentration of NO_2 to build up to its peak value.

Exercise
Using the calculated rate, determine the concentration of NO_2 produced by this reaction in 3.0 h. Answer: 1.9×10^{-10} mol/L.

Oxides of nitrogen also contribute to the formation of acid rain. Nitrogen dioxide reacts with water vapor in the atmosphere (recall Equation 19.5) to form nitric acid, HNO_3. Acid rainfall in the western United States appears to be due mainly to HNO_3 rather than H_2SO_4.

Leaded gasoline poisons the catalyst

Starting with 1981 model cars, a catalytic method has been used to lower NO emissions. This involves modifying the catalytic converter by introducing a small amount of rhodium, a rare and expensive transition metal. In the presence of a Pt–Rh catalyst, the reactions

$$CO(g) + NO(g) \rightarrow CO_2(g) + \tfrac{1}{2}N_2(g) \tag{19.33}$$

and, to a lesser extent

$$H_2(g) + NO(g) \rightarrow H_2O(g) + \tfrac{1}{2}N_2(g) \tag{19.34}$$

occur rapidly, drastically reducing the NO concentration.

■ EXAMPLE 19.10

Consider the reaction $2H_2(g) + 2NO(g) \rightarrow 2H_2O(g) + N_2(g)$ (Reaction 19.34). The initial rate data for this reaction is

EXPERIMENT	INITIAL PRESSURES (mm Hg) NO	H_2	INITIAL RATE (mm Hg/s)
1	750	750	16.1
2	750	375	8.04
3	375	375	2.01
4	150	375	0.743

Write the overall rate law for the reaction.

Solution
Since we are interested in concentration ratios, we can work with pressures; at constant temperature, the pressure of a gas is directly proportional to its concentration. To find the order with respect to H_2, we work with Experiments 1 and 2, where the pressure of NO is constant.

$$\frac{\text{rate}_1}{\text{rate}_2} = \left(\frac{P_1H_2}{P_2H_2}\right)^m; \; 16.1/8.04 = (750/375)^m; \; 2 = 2^m$$

Clearly $m = 1$; the reaction is first order with respect to H_2.
To find the order with respect to NO, we use Experiments 2 and 3 where the pressure of H_2 is constant at 375 mm Hg.

$$\frac{\text{rate}_2}{\text{rate}_3} = \left(\frac{P_2NO}{P_1NO}\right)^n; \; 8.04/2.01 = (750/375)^n; \; 4 = 2^n$$

Here, $n = 2$; the reaction is second order in NO. The rate law is:

$$\text{rate} = k(\text{conc. H}_2)(\text{conc. NO})^2$$

Exercise

Calculate the rate constant for the reaction. Answer: 3.81×10^{-8} $(\text{mm Hg})^{-2}\text{s}^{-1}$.

■ SUMMARY PROBLEM

The most abundant element in the atmosphere is nitrogen. The following problem involves nitrogen and some of its compounds which were discussed in this chapter.

a. What volume of liquid N_2 ($d = 0.808$ g/mL) can be obtained from 1.00 m³ of air at 750 mm Hg and 22°C, taking the mole fraction of N_2 to be 0.781?

b. Write a balanced equation for the reaction of nitrogen with hydrogen to form ammonia by the Haber process. Describe the effect, if any, of the following changes on both the rate of reaction and the yield of NH_3 at equilibrium:

—increasing the temperature from 300 to 400°C

—increasing the pressure from 400 to 500 atm

—poisoning the catalyst with sulfur compounds

c. At 300°C and a pressure of 100 atm, the mole fractions of NH_3, H_2, and N_2 in an equilibrium mixture are 0.52, 0.36, and 0.12, respectively. Calculate K_p for the reaction: $N_2(g) + 3H_2(g) \rightleftharpoons 2NH_3(g)$. Calculate K_c, using the relation: $K_p = K_c(RT)^{\Delta n_g}$

d. Write a series of three balanced equations to show how nitric acid is made from ammonia by the Ostwald process.

e. Nitrogen oxide, NO, is an air pollutant formed by the reaction: $N_2(g) + O_2(g) \rightarrow 2NO(g)$, taking place in automobile engines at high temperatures. Suppose the mechanism of this reaction were:

$$O_2(g) \rightleftharpoons 2O(g) \qquad \text{fast}$$

$$N_2(g) + O(g) \rightarrow NO(g) + N(g) \qquad \text{slow}$$

$$N(g) + O(g) \rightarrow NO(g) \qquad \text{fast}$$

What would be the order of the reaction with respect to N_2? O_2?

f. In the formation of photochemical smog, one reaction is:

$$N{=}O(g) \rightarrow N(g) + O(g)$$

Using the table of bond energies in Chapter 8, estimate ΔE for this reaction and calculate the maximum wavelength in nanometers at which the reaction can take place.

g. Consider the reaction: $NO_2(g) + O(g) \rightarrow NO(g) + O_2(g)$, which is first order in both reactants with a rate constant of 5.4×10^9 L/mol·s at 25°C. Determine the rate of reaction at 25°C if the concentrations of both NO_2 and O are 1×10^{-8} mol/L. Repeat this calculation at 50°C, taking the energy of activation to be 55 kJ/mol.

h. Give the formula of a compound of nitrogen that makes rainfall acidic. What mass of this compound would have to be present in 1.00 m³ of rainfall to lower the pH from 7.0 to 4.5?

Answers
a. 1.10 L   b. $N_2(g) + 3H_2(g) \rightarrow 2NH_3(g)$; incr. in T increases rate, decreases yield; increase in P increases both rate and yield; poisoning catalyst decreases rate, doesn't change yield.
c. $K_p = 4.8 \times 10^{-3}$; $K_c = 11$ d. Equations 19.3–19.5
e. 1st order in N_2, $\frac{1}{2}$ order in O_2 f. 607 kJ/mol; 197 nm
g. 5.4×10^{-7} mol/L·s; 3.0×10^{-6} mol/L·s   g. HNO_3; 2 g

Fritz Haber (1868–1934)

The scientific career of Fritz Haber was characterized by theoretical studies in areas of emerging practical importance. Training as an organic chemist and self-development as a physical chemist permitted Haber to relate chemistry to engineering. His most important achievement was the synthesis of ammonia from nitrogen and hydrogen.

Haber's approach to the synthesis of ammonia was not original. The French chemist Le Châtelier (Chap. 13) first studied this reaction, but an explosion led him to abandon the project. The German chemist Nernst, a contemporary of Haber, was the first to accomplish the high-pressure synthesis. Disagreement with Nernst's data led Haber to seek the optimum temperature, pressure, and catalyst. After obtaining about 8% ammonia at 600°C and 200 atm, Haber turned the process over to engineers. In 1918 he received the Nobel Prize for his work.

During World War I, Haber directed chemical warfare in Germany, introducing the use of chlorine and mustard gas into warfare. After the war, he attempted unsuccessfully to recover gold from the oceans. In this way, he hoped to repay Germany's war debts (which were never paid).

In 1912, Haber became director of the new Kaiser Wilhelm Institute for Chemistry. Under his influence, the Institute became world famous, attracting outstanding students and professors. In 1933 Haber was ordered by the Nazis to dismiss all Jewish workers at the Institute. He resigned instead. In his resignation letter he wrote, "For more than 40 years I have selected my collaborators on the basis of their intelligence and their character, not on the basis of their grandmothers. I am unwilling for the rest of my life to change this method which I have found so good." After a brief stay in England, Haber died en route to a research directorship in Israel.

Fritz Haber. (Professor John Stock, University of Connecticut, Storrs, Connecticut)

Friedrich Wilhelm Ostwald (1853–1932)

The conversion of ammonia to nitric acid was one of many contributions from this multifaceted individual. His early interest in chemistry, physics, painting, literature, music, and philosophy became lifelong pursuits. With Van't Hoff, Arrhenius, and Gibbs, Ostwald helped to establish the fledgling discipline of physical chemistry. He speculated that he became a physical chemist because his chemical education was carried out in Russia rather than Germany, where the emphasis was on organic chemistry. Early work on catalysis and reaction rates (1887) earned him the Nobel Prize in chemistry in 1909.

At 53, Ostwald retired as director of the University of Leipzig Physical Chemistry Institute and spent his remaining years in research and writing. From 1890 on, his work and personal philosophy were organized about his science of "energetics" (he even named his house *Energie*). This point of view made him a vigorous opponent of the concept of atoms and molecules! Matter, he believed, was a combination of energies occurring simultaneously at one place: "In fact, energy is the unique, real entity in

Friedrich Wilhelm Ostwald. (E.F. Smith Memorial Collection, CHOC, University of Pennsylvania)

the world and matter is not a bearer but a manifestation of energy." Eventually (1909), in a preface to his general chemistry text, Ostwald grudgingly accepted the atomic theory, in part because of the work of Perrin in kinetic molecular theory and discoveries by Thomson and Rutherford.

A pacifist and an advocate of the conservation of natural energy sources, Ostwald was a visionary. He believed in international harmony and composed Ido, an international language. His love of painting led him to spend the last 20 years of his life developing the science of color. He created color standards, dyes, and a quantitative theory of color. An inspiring teacher, his textbooks for general and analytical chemistry revolutionized the teaching of these subjects. When Ostwald died, a former student wrote, "He was loved and followed by more people than any chemist of our time."

Svante August Arrhenius (1859–1927)

Of the many contributions that Svante Arrhenius made to chemistry, the most important came in his Ph.D. thesis, written at the University of Uppsala in Sweden in 1884. Here Arrhenius proposed that salts, strong acids, and strong bases are completely dissociated in dilute water solution. Today, it seems quite reasonable that solutions of NaCl, HCl, and NaOH contain, respectively, Na^+ and Cl^- ions, H^+ and Cl^- ions, and Na^+ and OH^- ions. It did not seem nearly so obvious to the chemistry faculty at Uppsala in 1884. Arrhenius's dissertation received the lowest passing grade "approved not without praise." Ostwald, who visited Uppsala in 1884, said later:

> I can still plainly recall the scene in the chemistry laboratory at Uppsala, where the head (Cleve), himself an eminent chemist, heatedly asked me, pointing to a beaker containing an aqueous solution, "And you too believe that sodium atoms are swimming around there in this fashion?" When I agreed, he quickly looked at me in such a manner as if he had considerable doubt about my chemical rationality.

A few years later, when the colligative properties of electrolyte solutions were shown to be in agreement with the Arrhenius model, the situation changed. But the old ideas died hard in Uppsala. In 1901, Arrhenius was narrowly elected to the Swedish Academy of Science, over strong opposition. Two years later, he received the Nobel Prize in chemistry.

Among other contributions of Arrhenius, the most important were probably in the area of chemical kinetics (Chap. 18). In 1889 he derived the relation for the temperature dependence of the rate constant that we call the Arrhenius equation. The concepts of activation energy and the activated complex were first introduced by this versatile Swedish chemist. In quite a different area, in 1896 Arrhenius published an article entitled "On the Influence of Carbon Dioxide in the Air on the Temperature of the Ground." Here he presented the basic ideas of the greenhouse effect discussed in this chapter.

Svante August Arrhenius. (E.F. Smith Memorial Collection, CHOC, University of Pennsylvania)

In his later years, Arrhenius turned his attention to popularizing chemistry. He wrote several different textbooks that were well received. In 1925, under pressure from his publisher to submit a manuscript, Arrhenius started getting up at 4 A.M. to write. As might be expected, rising at such an early hour had an adverse effect on his health. Arrhenius suffered a physical breakdown in 1925, from which he never really recovered, dying two years later.

Publishers are like that

QUESTIONS AND PROBLEMS

Rate of Reaction

1. The average rate of oxygen production in the reaction: $2O_3(g) \rightarrow 3O_2(g)$ is found to be 0.0934 mol/L·s over a measured interval of time. What is the rate of ozone decomposition over that time interval?

2. In a catalytic converter, nitrogen oxide concentration is reduced by reaction with carbon monoxide: $2CO(g) + 2NO(g) \rightarrow 2CO_2(g) + N_2(g)$. If the rate of disappearance of NO is 5.60×10^{-2} mol/L·min, what is the rate of N_2 production?

3. The reaction

$$2NO_2(g) \rightarrow 2NO(g) + O_2(g)$$

is a second-order reaction. The rate constant is 0.498 L/mol·s at 600 K.

  a. What is the reaction rate at initial conc. $NO_2 = 0.010$ mol/L?

  b. What concentration of NO_2 is present when the reaction rate is reduced to half of that determined in (a)?

4. The reaction

$$NO(g) + N_2O(g) \rightarrow NO_2(g) + N_2(g)$$

is first order in each reactant. The rate is 1.92×10^{-15} mol/L·s when the NO and N_2O concentrations are each 5.0×10^{-8} mol/L.

  a. Calculate k for the forward reaction.

  b. At conc. $NO = 2.0 \times 10^{-8}$ mol/L, the rate is measured as 7.4×10^{-20} mol/L·s. What is conc. N_2O?

5. Formaldehyde, CH_2O, is a major eye irritant in smog. It may be formed by the reaction $O_3(g) + C_2H_4(g) \rightarrow 2CH_2O(g) + O(g)$; Δ conc. $CH_2O/\Delta t$ is first order in both O_3 and C_2H_4, with $k = 2 \times 10^3$ L/mol·s. The concentrations of O_3 and C_2H_4 in heavily polluted air are estimated to be about 5×10^{-8} and 1×10^{-8} mol/L, respectively. What is the rate of production of formaldehyde in mol/L·s? How long will it take to build up a formaldehyde concentration of 1×10^{-8} mol/L? This is the threshold above which eye irritation becomes noticeable. (Assume constant concentrations of O_3 and C_2H_4.)

6. The following reaction is involved in smog formation:

$$O_3(g) + NO(g) \rightarrow O_2(g) + NO_2(g)$$

This reaction has been shown to be first order in both ozone and NO with a rate constant of 1.2×10^7 L/mol·s. Calculate the concentration of NO_2 formed per second in polluted air, where O_3 and NO concentrations are both 2×10^{-8} mol/L. From the magnitude of your answer, would you expect the conversion of NO to be rapid or slow?

7. The rate constant for the decomposition of ozone by the mechanism

$$O_3(g) + O(g) \rightarrow 2O_2(g)$$

is 5.0×10^6 L/mol·s. Calculate the rate of ozone decomposition when the concentrations of O_3 and O are 3.0×10^{-8} and 1.2×10^{-14} mol/L, respectively.

8. If ozone is being produced by the mechanism $O_2(g) + O(g) \rightarrow O_3(g)$ at the same rate it is decomposing by the reaction in Problem 7, and if the concentrations of O_2 and O are 3.9×10^{-4} and 1.2×10^{-14} mol/L, respectively, what is the rate constant for the above reaction?

9. The rate constant for the NO-catalyzed decomposition of O_3 (Table 19.4) is 5.4×10^9 L/mol·s, as compared to 5.0×10^6 L/mol·s for the direct decomposition (Problem 7). The second step in the catalyzed decomposition is rate determining. If the concentration of NO_2 in the upper atmosphere is 1/2000 that of O_3, what is the ratio of the rate of the catalyzed reaction to that of the direct decomposition?

10. It is estimated that the amount of O_3 in the upper atmosphere decomposing by the NO-catalyzed reaction is four times that decomposing by the direct reaction. Using the rate constants given in Problem 9, estimate the ratio of the concentrations of NO_2 and O_3.

11. For the first-order thermal decomposition of ozone

$$O_3(g) \rightarrow O_2(g) + O(g)$$

$k = 3 \times 10^{-26}$ s^{-1} at 25°C. What is the half-life for this reaction in years? Comment on the likelihood that this reaction contributes to the depletion of the ozone layer.

12. For the reaction in Problem 11, how long would it take to lower the ozone concentration by 10.0%?

13. One of the reactions that destroys NO in the catalytic converter of an automobile is

$$CO(g) + NO(g) \rightarrow CO_2(g) + \tfrac{1}{2}N_2(g)$$

From the following data, determine the order of reaction with respect to both CO and NO.

RATE (mol/L·min)	CONC. CO (M)	CONC. NO (M)
3.2×10^{-9}	4.0×10^{-6}	4.0×10^{-8}
1.6×10^{-9}	2.0×10^{-6}	4.0×10^{-8}
0.90×10^{-9}	2.0×10^{-6}	3.0×10^{-8}
0.20×10^{-9}	1.0×10^{-6}	2.0×10^{-8}

Calculate the rate constant for the reaction.

14. Another reaction that lowers NO emissions with a catalytic converter is

$$H_2(g) + NO(g) \rightarrow H_2O(g) + \tfrac{1}{2}N_2(g)$$

Use the following data to find the order of reaction with respect to both H_2 and NO.

RATE (mol/L·min)	CONC. H_2 (M)	CONC. NO (M)
2.4×10^{-9}	4.0×10^{-6}	4.0×10^{-8}
1.8×10^{-9}	3.0×10^{-6}	4.0×10^{-8}
1.0×10^{-9}	3.0×10^{-6}	3.0×10^{-8}
0.30×10^{-9}	2.0×10^{-6}	2.0×10^{-8}

Calculate the rate constant for the reaction.

15. A possible mechanism for the reaction

$$2NO(g) + O_2(g) \rightarrow 2NO_2(g)$$

is

$$NO(g) + NO(g) \rightleftharpoons N_2O_2(g) \qquad (\text{fast})$$

$$N_2O_2(g) + O_2(g) \rightarrow NO_2(g) + NO_2(g) \qquad (\text{slow})$$

Derive the rate expression for this reaction, in terms of the concentrations of NO and O_2.

16. Experimentally, it is found that the rate constant for the formation of NO_2 from NO *decreases* as temperature increases. Using the rate expression derived in Problem 15, explain how this can happen. (ΔH for the formation of N_2O_2 from NO is a negative quantity.)

17. The following mechanism has been proposed for the uncatalyzed decomposition of O_3:

$$O_3(g) \rightleftharpoons O_2(g) + O(g) \qquad (\text{fast})$$

$$O(g) + O_3(g) \rightarrow 2O_2(g) \qquad (\text{slow})$$

Find the rate law associated with this mechanism. Express it in terms of the concentrations of O_3 and O_2.

18. Consider the following suggested mechanism:

$$NO_2(g) + NO_2(g) \rightleftharpoons N_2O_4(g) \qquad (\text{fast})$$

$$N_2O_4(g) + F_2(g) \rightarrow 2NO_2F(g) \qquad (\text{slow})$$

a. Write the equation for the overall reaction.
b. Write the rate law for the overall reaction in terms of the concentration of NO_2 and F_2.

Chemical Equilibrium

19. Discuss the effect of an increase in temperature and an increase in applied pressure on the position of the following equilibria:
   a. $NO(g) + O_3(g) \rightarrow NO_2(g) + O_2(g)$; $\Delta H = -199.8$ kJ
   b. $CH_4(g) + H_2O(g) \rightleftharpoons CO(g) + 3H_2(g)$; $\Delta H = +206.1$ kJ
   c. $N_2(g) + O_2(g) \rightleftharpoons 2NO(g)$; $\Delta H = +180.4$ kJ
   d. $2SO_2(g) + O_2(g) \rightleftharpoons 2SO_3(g)$; $\Delta H = -197.8$ kJ

20. How would you change temperature and applied pressure to obtain the maximum yield of products in each of the following reactions?
   a. $4NH_3(g) + 5O_2(g) \rightleftharpoons 4NO(g) + 6H_2O(g)$; $\Delta H = -905.6$ kJ
   b. $2NO_2(g) \rightleftharpoons 2NO(g) + O_2(g)$; $\Delta H = +114.0$ kJ
   c. $CO(g) + NO(g) \rightleftharpoons CO_2(g) + \tfrac{1}{2}N_2(g)$; $\Delta H = -373.2$ kJ
   d. $CO(g) + \tfrac{1}{2}O_2(g) \rightleftharpoons CO_2(g)$; $\Delta H = -283.0$ kJ

21. Consider the data in Table 19.2.
   a. Determine the number of moles of ammonia, nitrogen, and hydrogen in an equilibrium mixture containing one mole of gas at 300 atm and 400°C. Note that $[H_2] = 3[N_2]$.
   b. Use the Ideal Gas Law to determine the volume of one mole of gas under these conditions.
   c. Combine your answers from (a) and (b) to calculate the equilibrium concentrations (M) of ammonia, nitrogen, and hydrogen.
   d. From your answers in (c), calculate K_c at 400°C. Compare your value with that given in Table 19.2.

22. Consider the data in Table 19.2.
   a. Calculate the mole percent of all three gases at 500°C and 100 atm.
   b. Use the Ideal Gas Law to calculate the number of moles of gas in a 10.0-L container at 100 atm and 500°C.
   c. Use the results from (a) and (b) to calculate the equilibrium concentrations of all three gases in a 10.0-L vessel at 100 atm and 500°C.
   d. Use your answers in (c) to calculate K_c at 500°C. Compare your value with that in Table 19.2.

23. Much of the sulfur dioxide in polluted air is converted to sulfur trioxide. Sulfur trioxide dissolves in water in the atmosphere to give acid rain. The formation of SO_3 from SO_2 follows the equation:

$$2SO_2(g) + O_2(g) \rightleftharpoons 2SO_3(g)$$

A 1.00-L flask was filled with 0.0300 mol SO_2 and 0.0236 mol O_2. At equilibrium at 900 K, the flask contained 0.0256 mol SO_3. Calculate K_c for the reaction at 900 K.

24. The equilibrium constant K_c for the formation of ammonia by the Haber process

$$N_2(g) + 3H_2(g) \rightleftharpoons 2NH_3(g)$$

is 0.159 at 450°C. What is the equilibrium concentration of ammonia if at equilibrium there are 0.200 mol of nitrogen and 0.300 mol of hydrogen in a 10.0-L flask?

25. The reaction

$$2SO_2(g) + O_2(g) \rightleftharpoons 2SO_3(g)$$

has $K_p = 42.9$ at 627°C. What is K_c at this temperature?

26. Using Table 19.2, calculate K_p at each temperature for the reaction:

$$N_2(g) + 3H_2(g) \rightleftharpoons 2NH_3(g)$$

Relative Humidity

27. The relative humidity is 69% on a hot summer day when the temperature is 30°C.
  a. What is the pressure of water vapor in the air at this temperature? (See Appendix 1 for vapor pressures of water at different temperatures.)
  b. If the air is cooled to 25°C, what will the relative humidity become?

28. On a winter day, when the temperature is 0°C, the vapor pressure of water is 2.0 mm Hg.
  a. What is the relative humidity of the outside air?
  b. What does the relative humidity become if the air is warmed to 15°C?

29. The equilibrium vapor pressure of water at 20°C is 17.5 mm Hg. If the relative humidity at this time is 52%, what is the partial pressure of water in the air?

30. The equilibrium pressure of water at 10°C is 9.21 mm Hg. If the relative humidity at this temperature is 33%, what is the partial pressure of water in the air?

31. A 50.0-L sample of air contains 0.540 g $H_2O(g)$ at 25°C.
  a. Using the Ideal Gas Law, calculate the partial pressure of water vapor in the air.
  b. What is the relative humidity at 25°C?

32. The dew point is the temperature at which water vapor in the air condenses upon cooling at constant pressure. Estimate the dew point on a day when the temperature is 20°C and the relative humidity is 40%.

ΔE and λ

33. The photochemical decomposition of CCl_2F_2 produces Cl atoms that catalyze the decomposition of ozone. The reaction is:

$$CCl_2F_2(g) \rightarrow CClF_2(g) + Cl(g)$$

Taking the bond energy for the C—Cl bond to be 331 kJ/mol, calculate the maximum wavelength of radiation that can bring about the above reaction.

34. The dissociation energy of ozone is 106.5 kJ/mol. Calculate the maximum wavelength of a photon that could cause ozone to dissociate.

35. Radiation below 217 nm can dissociate SO_2. Use Equation 19.10 to calculate the dissociation energy of SO_2.

36. In the atmosphere, O_2 molecules are converted to oxygen atoms by ultraviolet radiation below 242 nm. Use Equation 19.10 to calculate the bond energy of oxygen molecules.

Stoichiometry and the Gas Laws

37. Assuming a yield of 100% in each of the three steps of the Ostwald process, what volume of NH_3 at 25°C and 720 mm Hg is required to produce one kilogram of nitric acid?

38. Suppose that in the first step of the Ostwald process, 95% of the ammonia is converted to NO. Suppose further that the yield of NO_2 in the second step is 75%, while that of HNO_3 from NO_2 in the third step is 60%. How many grams of nitric acid would be obtained under these conditions, starting with one hundred liters of ammonia gas at 10.0 atm and 200°C?

39. Ammonium nitrate is prepared by reacting ammonia with nitric acid:

$$NH_3(g) + HNO_3(l) \rightarrow NH_4NO_3(s)$$

If 10.0 L of ammonia at 10°C and 2.00 atm is added to 5.00 L of 0.100 M nitric acid, what is the theoretical yield in grams of ammonium nitrate?

40. Ammonia can be converted to urea, $(NH_2)_2CO$, a fertilizer, by the reaction:

$$2NH_3(g) + CO_2(g) \rightarrow (NH_2)_2CO(s) + H_2O(l)$$

Suppose 5.00 L of NH_3 and an equal volume of CO_2, both at 1.00 atm and 27°C, react with one another. Assuming a 100% yield, calculate the mass of urea formed and the volume of liquid water ($d = 1.00$ g/cm³).

41. One way to reduce NO emissions from smokestacks is to heat with methane:

$$CH_4(g) + 4NO(g) \rightarrow 2N_2(g) + CO_2(g) + 2H_2O(g)$$

What volume of pure CH_4 at STP is required to react with ten cubic meters of polluted air, also at STP, in which the mole fraction of NO is 2.5×10^{-6}?

42. In addition to lowering NO emissions, catalytic converters also catalyze undesirable reactions such as:

$$2NO(g) + 5H_2(g) \rightarrow 2H_2O(g) + 2NH_3(g)$$

How many grams of steam and ammonia can be produced by reacting 1.00 g of $H_2(g)$ with a 10.0-m³ sample of air in which the concentration of NO is 2.0×10^{-5} mol/L?

43. Sulfur dioxide can be removed from the effluents of smokestacks by reacting it with lime:

$$2CaO(s) + 2SO_2(g) + O_2(g) \rightarrow 2CaSO_4(s)$$

a. What mass of lime is required to remove one metric ton of SO_2?
b. Using Table 7.3 in Chapter 7, calculate the amount of heat evolved or absorbed in removing one metric ton of SO_2.

44. Repeat the calculations called for in Problem 43, this time substituting MgO for CaO.

Unclassified

45. Criticize the following statements:
a. In the Haber process, a high pressure is used primarily to increase reaction rate.
b. Ice crystals form in clouds when the temperature drops below 0°C.
c. Ozone is found only in the upper atmosphere.
d. Sulfur dioxide is a more serious air pollutant than sulfur trioxide.

46. Classify the following as true or false:
a. The nitrogen atoms in a sample of nitroglycerine were once part of N_2 molecules in the atmosphere.
b. If the temperature of an air mass drops, relative humidity increases.
c. Carbon dioxide in the atmosphere has a warming effect because it absorbs infrared radiation from the sun.

47. Describe the effects that may result from
a. an increase in atmospheric CO_2 concentration.
b. an increase in NO concentration in the morning rush hour.
c. a decrease in the concentration of ozone in the upper atmosphere.

48. Describe the possible effects of
a. an increase in Cl concentration in the upper atmosphere.
b. an increase in SO_3 concentration in the atmosphere.
c. an increase in suspended particles in air.

49. Explain why
a. chemical reactions that occur in the upper atmosphere (above 30 km) are not common near the surface of the earth.
b. a catalytic converter is used in an automobile exhaust system.
c. even a low-level concentration of CO may be dangerous.

50. Explain
a. the greenhouse effect.
b. why catalytic converters are expensive.
c. the relation between SO_3 production in air and acid rain.

51. Argon makes up 0.934 mol% of dry air. Calculate its concentration in
a. volume percent.
b. mass percent, taking the molar mass of dry air to be 29.0 g/mol.
c. moles per liter at 25°C and a total pressure of 750 mm Hg.
d. parts per million (mole basis).

52. The concentration of CO in a sample of polluted air is 11.2 ppm (mole basis). Calculate its concentration in
a. mole fraction.
b. partial pressure ($P_{tot} = 1.00$ atm).
c. moles per liter at 25°C ($P_{tot} = 1.00$ atm).
d. grams per cubic meter.

53. Discuss the effect of changes in pressure and temperature upon the rate and yield of ammonia in the Haber process.

54. What effect would increases in pressure and temperature have upon the rate of formation and yield of NO_2 by Reaction 19.4?

55. Write Lewis structures for
a. O_2^{2-} b. O_2^- c. O^{2-} d. CO

56. Write Lewis structures for
a. NH_3 b. NO_3^- c. O_3 d. SO_2

57. Give the geometry, bond angle, and hybridization around the central atom of all the species in Question 55.

58. Give the geometry, bond angle, and hybridization around the central atom of all the species in Question 56.

59. If the total volume of air in the Los Angeles basin is 6000 km^3, and the maximum permissible concentration of CO is 4×10^{-7} mol/L, calculate the maximum total mass of carbon monoxide in the basin.

60. Repeat the calculations in Problem 59 for hydrocarbons, taking the maximum permissible concentration to be 1×10^{-8} mol/L and the average molar mass of hydrocarbons to be 100 g/mol.

61. Carbon monoxide has a partial pressure of about 0.40 mm Hg in inhaled cigarette smoke mixed with air. If a smoker inhales 5.0×10^2 L of this gaseous mixture at 37°C and 755 mm Hg, how many grams of CO does he take in?

62. Breathing air that contains 350 ppm (mole basis) of CO can cause unconsciousness and eventually death. How many grams of carbon monoxide are there in a garage 16 ft × 15 ft × 8.0 ft that contains this concentration of carbon monoxide at 25°C and 750 mm Hg?

63. How much limestone, $CaCO_3$, is required to react with the H^+ ions in one liter of lake water, increasing the pH from 4.32 to 7.00?

64. A sample of coal contains 1.2% by mass of pyrite, FeS_2. What mass of calcium oxide is required to react with the SO_2 formed when one metric ton (10^3 kg) of this coal is burned?

Challenge Problems

65. Dinitrogen tetroxide, N_2O_4, is a colorless gas that dissociates to give nitrogen dioxide, NO_2, a reddish-brown gas responsible for the characteristic color of photochemical smog.

$$N_2O_4(g) \rightleftharpoons 2NO_2(g)$$

An equilibrium mixture of N_2O_4 and NO_2 is 52.0% NO_2 by mass at 2.23 atm and 25°C. Calculate K_c for this dissociation. (*Hint:* Assume 100.0 g of mixture and calculate the number of moles of each gas. Then find the total volume and finally the molar concentration of each gas.)

66. A 0.664-g sample of SO_3 is placed in a 1.00-L flask and heated to 1100 K. The SO_3 undergoes decomposition to SO_2 and O_2:

$$2SO_3(g) \rightleftharpoons 2SO_2(g) + O_2(g)$$

At equilibrium, the total pressure in the container is 1.034 atm. Find K_c for the reaction. (*Hint:* Calculate the total moles of gas first.)

67. Most of the ozone in the upper atmosphere exists in a layer at an altitude between 20 and 40 km.

a. Calculate the total volume in liters of this layer, taking the radius of the earth to be 6.4×10^3 km and using the following relation for the difference in volume between two concentric spheres of radii r_1 and r_2: $\Delta V \approx 4 \pi r_1^2(r_2 - r_1)$.

b. Taking the average pressure of air in this layer to be 0.013 atm and the temperature to be $-30°C$, calculate the total number of moles of gas in the layer.

c. Assuming the average concentration of ozone in the layer to be 7 ppm, what is the total mass in grams of the ozone in the upper atmosphere?

68. The ozone present in polluted air is formed by the two-step mechanism

$$NO_2(g) \xrightarrow{k_1} NO(g) + O(g) \qquad \text{(first order)}$$

$$O(g) + O_2(g) \xrightarrow{k_2} O_3(g) \qquad \text{(second order)}$$

Take $k_1 = 6.0 \times 10^{-3}$ s^{-1} and $k_2 = 1.0 \times 10^6$ L/mol·s. The concentrations of NO_2 and O_2 in polluted air are about 3.0×10^{-9} and 1.0×10^{-2} mol/L, respectively. One can assume that the concentration of atomic oxygen reaches a "steady state," a low constant concentration at which point it is being consumed in the second reaction at the same rate as it is being produced in the first.

a. Calculate the steady concentration of $O(g)$ in polluted air.

b. Calculate the rate of formation of O_3 in polluted air.

c. If the rate of formation of O_3 remains constant, how long would it take for its concentration to build up to one part per million in air at 25°C and 1 atm? Under such conditions air contains about 0.041 mol/L.

69. Suppose that a molecule of O_2 absorbs a photon at a wavelength of 220 nm and subsequently decomposes to 2 oxygen atoms. Taking ΔH for this reaction to be +498 kJ, how much "extra energy" in kilojoules will the oxygen atoms have? If all this energy is absorbed in raising the kinetic energy of $O(g)$ and, hence, its temperature, what will the final "apparent" temperature be? (The heat capacity of $O(g)$ is 22 J/mol·K.)

70. A certain type of coal contains 2.0% by mass of sulfur.

a. What mass of sulfur dioxide will be produced by burning one hundred grams of this coal?

b. What volume of SO_2 at 25°C and one atm will be produced?

c. If one thousand liters of air are used to burn the coal, what will be the concentration of SO_2 in the stack gas in ppm?

71. A storm deposits 0.50 in of rainfall over an area of 1.0×10^4 square miles. The pH of the rain is 4.35; that of normal rainfall is 5.60. Assuming that this reduction in pH was caused by sulfuric acid, what mass of H_2SO_4 was involved? (1 mol $H_2SO_4 \rightarrow$ 2 mol H^+)

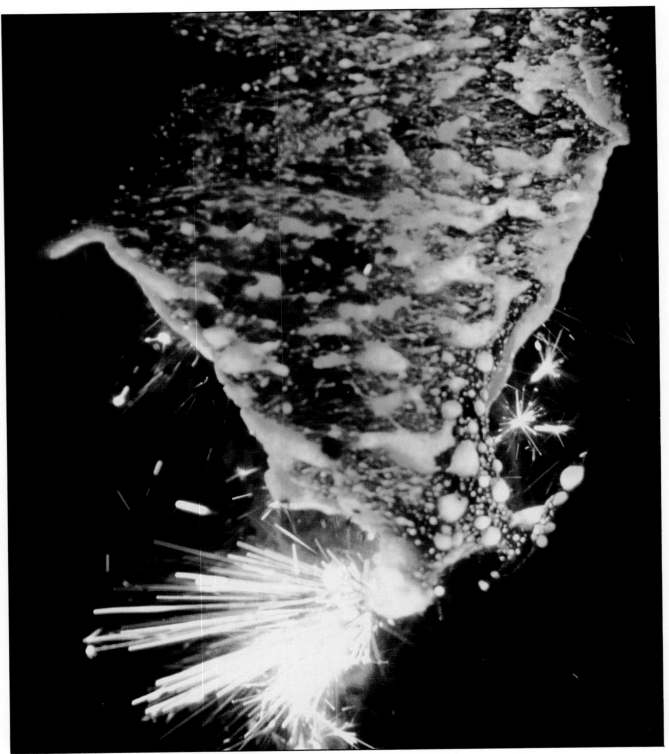

$$4Fe(s) \ + \ 3O_2(g) \ \rightarrow \ 2Fe_2O_3(s)$$

Hot steel wool bursts into flame when plunged into pure oxygen.

CHAPTER 20

SPONTANEITY OF REACTION: ΔS AND ΔG

I am a sleepless
Slowfaring eater,
Maker of rust and rot
In your bastioned fastenings,

I am the crumbler: tomorrow.

CARL SANDBURG
UNDER

In discussing thermochemistry in Chapter 7, we concentrated upon reactions taking place at constant temperature and pressure. Recall that under these conditions, the heat flow associated with the reaction is exactly equal to the difference in enthalpy, ΔH, between products and reactants.

In this chapter we return to a discussion of the energy changes accompanying reactions at constant temperature and pressure. This time we will be interested in one crucial question: how can we predict in advance whether a reaction will occur under these conditions, given sufficient time? Putting it another way, will the reaction at a certain temperature and pressure occur "by itself," without the exertion of any outside force? In that sense, is the reaction *spontaneous*?

To start with, we will consider some of the characteristics of spontaneous processes (Section 20.1). We continue by introducing a quantity called *entropy* and given the symbol S. We will see how to calculate the entropy change, ΔS, for a chemical reaction (Section 20.2) and how a basic law of nature, the Second Law of Thermodynamics, can be expressed in terms of entropy changes (Section 20.3).

By combining ΔH (Chap. 7) and ΔS in the proper way, it is possible to determine whether a reaction is spontaneous or not. To do this, in Section 20.4 we introduce a new function called the *free energy* and given the symbol G. As we will see, a process that is spontaneous at constant temperature and pressure is associated with a negative free energy change ($\Delta G < 0$). In the last several sections of this chapter, we will see how values of ΔG can be used to

— determine the effect of changes in temperature upon reaction spontaneity (Section 20.5).

A dropped brick will fall spontaneously to the ground

691

— determine whether or not "coupled reactions" will be spontaneous (Section 20.6).

— calculate the equilibrium constant for a reaction (Section 20.7).

— calculate the maximum amount of work that can be obtained from a reaction (Section 20.8).

20.1
SPONTANEOUS PROCESSES

As stated in the chapter introduction, a reaction that occurs "by itself," without exertion of any outside force, is said to be spontaneous. All of us are familiar with certain spontaneous processes. For example,

— an ice cube melts when added to a glass of water at room temperature.

— a mixture of hydrogen and oxygen burns when we set a match to it.

— an iron (steel) tool exposed to moist air rusts.

In other words, the following three reactions are spontaneous at 25°C and 1 atm:

$$H_2O(s) \rightarrow H_2O(l) \tag{20.1}$$

$$2H_2(g) + O_2(g) \rightarrow 2H_2O(l) \tag{20.2}$$

$$2Fe(s) + \tfrac{3}{2}O_2(g) + 3H_2O(l) \rightarrow 2Fe(OH)_3(s) \tag{20.3}$$

The word "spontaneous" does not imply anything about how rapidly a reaction occurs. Some spontaneous reactions, notably the rusting of iron by Reaction 20.3, are quite slow. Often a reaction that is potentially spontaneous does not occur without some sort of stimulus to provide the required activation energy. A mixture of hydrogen and oxygen shows no sign of reaction in the absence of a spark or match. Once started, though, a spontaneous reaction continues by itself without further input of energy from the outside.

We do allow using a spark, to start things off

If a reaction is spontaneous under a given set of conditions, the reverse reaction is nonspontaneous under the same conditions. Water in a beaker at room temperature does not spontaneously freeze by the reverse of Reaction 20.1; neither does it decompose to the elements by the reverse of Reaction 20.2. However, it is often possible to bring about a nonspontaneous reaction by supplying energy in the form of work. Electrolysis can be used to bring about the reaction

$$2H_2O(l) \rightarrow 2H_2(g) + O_2(g) \tag{20.4}$$

Here a spark won't do it

To do this, electrical energy must be furnished, perhaps from a storage battery. As with all other nonspontaneous processes, Reaction 20.4 stops immediately if the source of energy is cut off.

What are the requirements for spontaneity? What principle explains why these processes will go in one direction but not in the other? A hundred years ago many chemists felt that they had a general criterion for predicting reaction spontaneity. The prevailing idea, put forth by P. M. Berthelot in

Paris and Julius Thomsen in Copenhagen, was that all spontaneous reactions are exothermic. If this were true, all we would have to do to predict reaction spontaneity would be to calculate the enthalpy change, ΔH, and look at its sign. If ΔH turned out to be negative, we could assume that the reaction must be spontaneous; if ΔH were positive, the reaction could not occur by itself.

It turns out that almost all exothermic chemical reactions are spontaneous at 25°C and 1 atm. Consider, for example, the formation of water from the elements and the rusting of iron:

$$2H_2(g) + O_2(g) \rightarrow 2H_2O(l); \qquad \Delta H = -571.6 \text{ kJ}$$

$$2Fe(s) + \tfrac{3}{2}O_2(g) + 3H_2O(l) \rightarrow 2Fe(OH)_3(s); \qquad \Delta H = -780.6 \text{ kJ}$$

For both of these spontaneous reactions, ΔH is a negative quantity.

On the other hand, this simple rule fails for many familiar phase changes. An example is the melting of ice. This takes place spontaneously at 1 atm above 0°C, even though it is endothermic:

$$H_2O(s) \rightarrow H_2O(l); \qquad \Delta H = +6.0 \text{ kJ} \tag{20.5}$$

In another case we find that, above 100°C, liquid water vaporizes to steam at 1 atm. This process, like Reaction 20.5, absorbs heat:

$$H_2O(l) \rightarrow H_2O(g); \qquad \Delta H = +40.7 \text{ kJ} \tag{20.6}$$

There is still another basic objection to using the sign of ΔH as a general criterion for spontaneity. Endothermic reactions that are nonspontaneous at room temperature often become spontaneous when the temperature is raised. Consider, for example, the decomposition of limestone:

$$CaCO_3(s) \rightarrow CaO(s) + CO_2(g); \qquad \Delta H = +178.3 \text{ kJ} \tag{20.7}$$

At 25°C and 1 atm, this reaction is nonspontaneous. Witness the existence of the white cliffs of Dover and other limestone deposits over eons of time. However, if the temperature is raised to about 1100 K, the limestone decomposes to give off carbon dioxide gas at 1 atm. In other words, this endothermic reaction becomes spontaneous at high temperatures. This is true despite the fact that ΔH remains at +178.3 kJ, nearly independent of temperature.

We see then that the direction of spontaneous change is not always determined by the tendency for a system to go to a state of lower energy. There is another natural tendency that must be taken into account if we are to predict the direction of spontaneity. *Nature tends to move spontaneously from a state of lower probability to one of higher probability.* Or, as G. N. Lewis put it,

> Each system which is left to its own will, over time, change toward a condition of maximum probability.

To see what these statements mean, we should first explain what is meant by states of "low" and "high" probability. To do this, let us consider a pastime far removed from chemistry: tossing dice. If you've ever shot craps (and maybe even if you haven't), you know that when a pair of dice

Some states are much more probable than others. If you shake red and black marbles with each other, the random distribution at the left is much more probable than the highly ordered distribution at the right. (Charles Steele)

Figure 20.1 Certain "states" are more probable than others. For example, when you toss a pair of dice, a 7 is much more likely to come up than a 12.

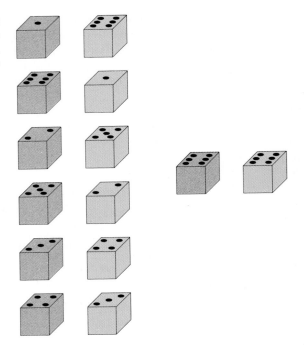

A 7 is six times as probable as a 12

are thrown, a 7 is much more likely to come up than a 12. Figure 20.1 shows why this is the case. There are six different ways to throw a 7 and only one way to throw a 12. Over time, a dice will come up 7 six times as often as 12. A 7 is a state of "high" probability; a 12 is a state of "low" probability.

The conclusion we just came to is hardly surprising; all of us have a feeling for relative probabilities. But, what do we mean when we say that "natural processes move toward a state of higher probability"? To understand this statement, consider the setup shown in Figure 20.2. Here we have two glass bulbs of equal volume separated by a stopcock, completely

Figure 20.2 Possible configurations of one or two molecules distributed between two flasks. Each configuration shown is equally likely.

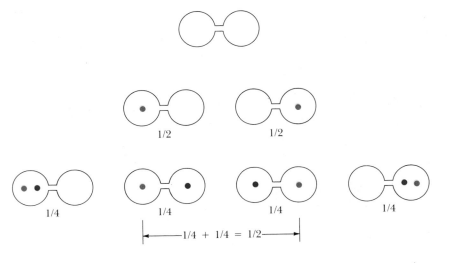

evacuated to begin with. If we add a gas molecule to this system, it is equally likely to be found in either bulb. The probability of finding it in the left bulb is $\frac{1}{2}$, equal to that of finding it in the right bulb. Suppose now that we add a second molecule. The probability of finding both molecules on the same side, either left or right, is $\frac{1}{2} \times \frac{1}{2} = \frac{1}{4}$. We are twice as likely to find one molecule in each bulb; there are two ways that can happen and only one way that they can both end up in a given bulb.

If we increase the number of molecules, the probability of finding them evenly distributed between the two bulbs as opposed to all in one bulb goes up rapidly. In Figure 20.3 we show the situation with four molecules. Here there are 16 possible ways of distributing the molecules (*configurations*). In six of these ways there are two molecules in each bulb; in contrast there is only one configuration with all the molecules in the left bulb. Since each of the 16 configurations is equally probable, we are six times as likely to find the molecules equally divided between the two bulbs as we are to find them all in the left bulb.

In the laboratory we would have far more than four molecules to work with; the number would be more like 10^{23}. In that case the probability of finding all the molecules in one bulb is

$$\left(\tfrac{1}{2}\right)^{10^{23}}$$

which is one small number. In contrast, the probability is essentially 100% that the molecules will distribute themselves so evenly that the pressures in the two bulbs will be equal.

Now let's consider a somewhat different experiment in which we again have two bulbs connected by a stopcock. This time, two different gases, perhaps H_2 and N_2, are involved (Fig. 20.4). To start with all the H_2 molecules are in the left bulb; an equal number of N_2 molecules are in the right bulb. We open the stopcock. Intuition tells us that if we wait long enough, the two different kinds of molecules will distribute themselves evenly between the two bulbs. Through diffusion, half of the H_2 molecules will end up in the left bulb and half in the right; the same holds for the N_2 molecules. Each gas achieves its own most probable distribution, independent of the presence of the other gas.

We could explain the results of this experiment the way we did before; the final distribution is clearly much more probable than the initial distri-

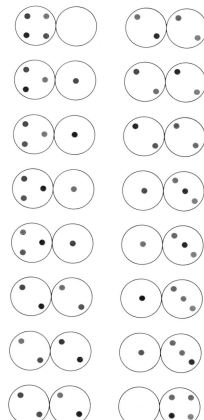

Figure 20.3 With four molecules distributed between two flasks, there is only 1 chance in 16 that all the molecules will be in the flask at the left. In contrast, there are 6 chances in 16 that the molecules will be evenly distributed, with two molecules in each flask.

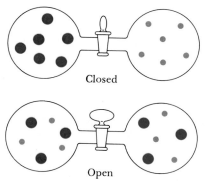

Figure 20.4 Different kinds of gas molecules mix by diffusion; the molecules go spontaneously from a more ordered to a more random state.

bution. There is however another way of looking at this process which we will find very useful. We have gone from a highly ordered state (all the H_2 molecules on the left, all the N_2 molecules on the right) to a more disordered or random state where the molecules are distributed evenly between the two bulbs. We could have interpreted the experiment shown in Figure 20.3 in the same way, albeit somewhat less obviously. In general, we find that **nature tends to move spontaneously from more ordered to more random states**.

This statement is quite easy for parents or students to understand. Your room tends to get messy because an ordered room has few options for objects to be moved around (socks on the floor is not an option for an orderly room). The comedian Bill Cosby insists that with an army of 80 two-year olds he could take over any country in the world, because they have a remarkable ability for disorganization.

■ **EXAMPLE 20.1** _____
Choose the state that is more random and thus more probable.

a. A chess game before the first move or a chess game in progress.
b. A solved jigsaw puzzle or the unassembled pieces in a box.
c. Humpty Dumpty after the fall, or Humpty Dumpty together again.

Solution

a. The chess game in progress is more random. There are many more ways the pieces on the board could be arranged when the game is going on.
b. A solved jigsaw puzzle is quite orderly. A box has all the pieces at random.
c. Humpty Dumpty after the fall is more probable. Putting him together again is nonspontaneous; all the king's horses and all the king's men couldn't do it.

Exercise
The New York Yankees and Boston Red Sox play a doubleheader (two games). Which is more likely: two victories by the Yankees, a sweep by the Red Sox, or an even split? Answer: Statistically at least, an even split.

Other factors are involved

20.2
ENTROPY

If we want to put spontaneity on a quantitative basis, we need a function whose value is directly related to the probability of a state. This quantity is called entropy and given the symbol S. In general, the more probable the state or the more random the distribution of molecules, the greater the entropy.

Entropy, like enthalpy (Chap. 7) is a state function. That is, the entropy depends only upon the properties of a system, not upon its history. The entropy change is determined by the entropies of the final and initial states, not upon the path by which we move from one state to another.

$$\Delta S = S_{final} - S_{initial} \tag{20.8}$$

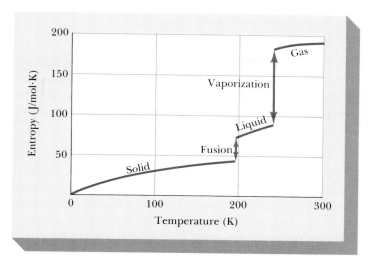

Figure 20.5 Molar entropy of NH_3 as a function of temperature. Note the large increases upon fusion and vaporization.

Several factors influence the amount of entropy that a system has in a particular state. In general:

— *a liquid has a higher entropy than the solid from which it is formed.* In a solid, the atoms, molecules, or ions are fixed in position; in the liquid these particles are free to move past one another. In that sense, the liquid structure is more random, the solid more ordered.

— *a gas has a higher entropy than the liquid from which it is formed.* Upon vaporization, the particles acquire greater freedom to move about. They are now distributed throughout the entire container instead of being restricted to a small volume.

— *increasing the temperature of a substance increases its entropy.* By raising the temperature, we increase the kinetic energy of the molecules (or atoms or ions) and hence their freedom of motion. In the solid, the molecules vibrate with a greater amplitude at a higher temperature. In a liquid or a gas, they move about more rapidly.

These effects are shown in Figure 20.5, where we plot the entropy of ammonia, NH_3, vs. temperature. Note that the entropy of the solid at 0 K is zero. This reflects the fact that molecules are completely ordered in the solid state at this temperature; there is only one way in which this system can be attained. More generally, the Third Law of Thermodynamics tells us that *a completely ordered pure crystalline solid has an entropy of zero at 0 K.*

Notice from the figure that an increase in temperature alone has relatively little effect on entropy. The slope of the curve is small in regions where only one phase is present. In contrast, there is a large jump in entropy when the solid melts and an even larger one when the liquid vaporizes. This behavior is typical of all substances; fusion and vaporization are accompanied by relatively large increases in entropy. This explains why these processes can be spontaneous even though they are endothermic. Ice melts above 0°C:

$$H_2O(s) \rightarrow H_2O(l); \qquad \Delta H > 0, \Delta S > 0$$

No disorder → zero entropy

and water vaporizes above 100°C at 1 atm:

$$H_2O(l) \rightarrow H_2O(g); \qquad \Delta H > 0, \Delta S > 0$$

Above 0°C, the entropy increase, which favors the melting of ice, outweighs the enthalpy increase, which tends to make the reaction nonspontaneous. Similarly, water at 1 atm vaporizes above 100°C because the driving force produced by the entropy increase is greater than that caused by the increase in enthalpy.

Many other spontaneous processes take place with an increase in entropy. Consider, for example, what happens when a tank of compressed helium is used to fill a balloon. Helium rushes out of the high-pressure tank into the balloon, which is at atmospheric pressure. We might explain this behavior by reasoning that the drop in pressure causes the helium atoms to spread out over a larger volume, thereby increasing their entropy. From a slightly different point of view, we could say that the gas molecules move spontaneously from a region of high concentration (high-pressure tank) to a region of low concentration (the balloon). Such a process is always accompanied by an increase in entropy. This is true whether we deal with a compressed gas escaping from a cylinder, or the natural process of diffusion illustrated in Fig. 20.4, p. 695. There, different gas molecules mix with one another, going from a region where their concentration is high to one where it is initially zero.

■ EXAMPLE 20.2

Predict whether ΔS is positive or negative for each of the following processes:

a. Taking dry ice from a freezer where its temperature is −60°C and allowing it to warm to room temperature.
b. Dissolving bromine in hexane.
c. Condensing gaseous bromine to liquid bromine.

Solution

a. This process involves an increase in temperature and a phase change from solid to gas; $\Delta S > 0$.

b. The solution is more random than the original liquids; $\Delta S > 0$.

c. This is a phase change from gas to liquid, from disorder to order; $\Delta S < 0$.

Exercise

Recall the process of osmosis, described in Chapter 11. Would you expect ΔS for the water undergoing osmosis to be positive or negative? Answer: positive.

As you might expect from our discussion to this point, we must consider both entropy and enthalpy effects when trying to decide whether a process will occur spontaneously. To do this, we need to know the magnitude of

TABLE 20.1   Standard Entropies (J/mol·K) of Elements and Compounds at 25°C, 1 atm

ELEMENTS							
$Ag(s)$	42.6	$Cl_2(g)$	223.0	$I_2(s)$	116.1	$O_2(g)$	205.0
$Al(s)$	28.3	$Cr(s)$	23.8	$K(s)$	64.2	$Pb(s)$	64.8
$Ba(s)$	62.8	$Cu(s)$	33.2	$Mg(s)$	32.7	$P_4(s)$	164.4
$Br_2(l)$	152.2	$F_2(g)$	202.7	$Mn(s)$	32.0	$S(s)$	31.8
$C(s)$	5.7	$Fe(s)$	27.3	$N_2(g)$	191.5	$Si(s)$	18.8
$Ca(s)$	41.4	$H_2(g)$	130.6	$Na(s)$	51.2	$Sn(s)$	51.6
$Cd(s)$	51.8	$Hg(l)$	76.0	$Ni(s)$	29.9	$Zn(s)$	41.6

COMPOUNDS							
$AgBr(s)$	107.1	$CaCl_2(s)$	104.6	$H_2O(g)$	188.7	$NH_4NO_3(s)$	151.1
$AgCl(s)$	96.2	$CaCO_3(s)$	92.9	$H_2O(l)$	69.9	$NO(g)$	210.7
$AgI(s)$	115.5	$CaO(s)$	39.8	$H_2O_2(l)$	109.6	$NO_2(g)$	240.0
$AgNO_3(s)$	140.9	$Ca(OH)_2(s)$	83.4	$H_2S(g)$	205.7	$N_2O_4(g)$	304.2
$Ag_2O(s)$	121.3	$CaSO_4(s)$	106.7	$H_2SO_4(l)$	156.9	$NaCl(s)$	72.1
$Al_2O_3(s)$	50.9	$CdCl_2(s)$	115.3	$HgO(s)$	70.3	$NaF(s)$	51.5
$BaCl_2(s)$	123.7	$CdO(s)$	54.8	$KBr(s)$	95.9	$NaOH(s)$	64.5
$BaCO_3(s)$	112.1	$Cr_2O_3(s)$	81.2	$KCl(s)$	82.6	$NiO(s)$	38.0
$BaO(s)$	70.4	$CuO(s)$	42.6	$KClO_3(s)$	143.1	$PbBr_2(s)$	161.5
$BaSO_4(s)$	132.2	$Cu_2O(s)$	93.1	$KClO_4(s)$	151.0	$PbCl_2(s)$	136.0
$CCl_4(l)$	216.4	$CuS(s)$	66.5	$KNO_3(s)$	133.0	$PbO(s)$	66.5
$CHCl_3(l)$	201.7	$Cu_2S(s)$	120.9	$MgCl_2(s)$	89.6	$PbO_2(s)$	68.6
$CH_4(g)$	186.2	$CuSO_4(s)$	107.6	$MgCO_3(s)$	65.7	$PCl_3(g)$	311.7
$C_2H_2(g)$	200.8	$Fe(OH)_3(s)$	106.7	$MgO(s)$	26.9	$PCl_5(g)$	364.5
$C_2H_4(g)$	219.5	$Fe_2O_3(s)$	87.4	$Mg(OH)_2(s)$	63.2	$SiO_2(s)$	41.8
$C_2H_6(g)$	229.5	$Fe_3O_4(s)$	146.4	$MgSO_4(s)$	91.6	$SnO_2(s)$	52.3
$C_3H_8(g)$	269.9	$HBr(g)$	198.6	$MnO(s)$	59.7	$SO_2(g)$	248.1
$CH_3OH(l)$	126.8	$HCl(g)$	186.8	$MnO_2(s)$	53.0	$SO_3(g)$	256.7
$C_2H_5OH(l)$	160.7	$HF(g)$	173.7	$NH_3(g)$	192.3	$ZnI_2(s)$	161.1
$CO(g)$	197.6	$HI(g)$	206.5	$N_2H_4(l)$	121.2	$ZnO(s)$	43.6
$CO_2(g)$	213.6	$HNO_3(l)$	155.6	$NH_4Cl(s)$	94.6	$ZnS(s)$	57.7

the entropy change as well as its sign. In the rest of this section, we will see how ΔS is calculated.

STANDARD MOLAR ENTROPIES OF ELEMENTS AND COMPOUNDS

The absolute entropy of a pure substance, unlike its energy or enthalpy, can be measured accurately. The details of how this is done are beyond the level of this text. However, the results for one particular substance, NH_3, are shown graphically in Figure 20.5. From such a plot, we can determine the **standard molar entropy** of a substance at 1 atm pressure and any given temperature, most often 25°C. This quantity is given the symbol S^0 and has the units of **joules per mole per kelvin** (J/mol·K). From Figure 20.5 we see that

$$S^0 \ NH_3(g) \text{ at } 25°C = 192 \text{ J/mol·K}$$

Standard molar entropies for a variety of substances are listed in Table 20.1. Notice that:

— standard molar entropies of substances are always positive quantities ($S^0 > 0$).

— *elements as well as compounds have nonzero standard entropies.* This is in contrast to the situation with heats of formation, where $\Delta H_f^0 = 0$ for elements in their stable states.

You will also notice that gases, as a group, have higher entropies than liquids or solids. Moreover, among substances of similar structure and physical state, entropy usually increases with molar mass. Compare, for example, the hydrocarbons:

$$CH_4(g) \qquad S^0 = 186.2 \text{ J/mol·K}$$

$$C_2H_6(g) \qquad S^0 = 229.5 \text{ J/mol·K}$$

$$C_3H_8(g) \qquad S^0 = 269.9 \text{ J/mol·K}$$

STANDARD MOLAR ENTROPIES OF IONS IN SOLUTION

It is possible to assign standard molar entropies to ions in water solution. To do this, we adopt a convention similar to that followed with heats of formation. The standard molar entropy of the H^+ ion in water solution is taken to be zero:

An arbitrary choice

$$S^0 H^+(aq) = 0$$

Standard entropies of ions based upon this convention are listed in Table 20.2. They are given in joules per mole per kelvin at 25°C and a concentration of 1 M.*

ΔS^0 FOR REACTIONS

Using Tables 20.1 and 20.2, it is possible to calculate the standard entropy change, ΔS^0, for a variety of reactions. Recall (Equation 20.8) that

$$\Delta S = S_{final} - S_{initial}$$

For a reaction system, the products constitute the final state, the reactants the initial state. Hence:

$$\Delta S^0_{reaction} = \Sigma S^0_{products} - \Sigma S^0_{reactants} \qquad (20.9)$$

As an example, consider the reaction

$$CaCO_3(s) \rightarrow CaO(s) + CO_2(g)$$

$$\Delta S^0 = S^0 \, CaO(s) + S^0 \, CO_2(g) - S^0 \, CaCO_3(s)$$
$$= 39.8 \text{ J/K} + 213.6 \text{ J/K} - 92.9 \text{ J/K} = +160.5 \text{ J/K}$$

---

*Notice that several ions have negative S^0 values, e.g., $S^0 \, F^- = -13.8$ J/mol·K. This is a consequence of the arbitrary way in which ionic entropies are defined, taking $S^0 \, H^+ = 0$. The fluoride ion has an entropy 13.8 units *less* than that of H^+.

TABLE 20.2 Standard Entropies (J/mol·K) of Aqueous Ions at 25°C, 1 M

CATIONS				ANIONS			
$Ag^+(aq)$	72.7	$Hg^{2+}(aq)$	-32.2	$Br^-(aq)$	82.4	$HPO_4^{2-}(aq)$	-33.5
$Al^{3+}(aq)$	-321.7	$K^+(aq)$	102.5	$CO_3^{2-}(aq)$	-56.9	$HSO_4^-(aq)$	131.8
$Ba^{2+}(aq)$	9.6	$Mg^{2+}(aq)$	-138.1	$Cl^-(aq)$	56.5	$I^-(aq)$	111.3
$Ca^{2+}(aq)$	-53.1	$Mn^{2+}(aq)$	-73.6	$ClO_3^-(aq)$	162.3	$MnO_4^-(aq)$	191.2
$Cd^{2+}(aq)$	-73.2	$Na^+(aq)$	59.0	$ClO_4^-(aq)$	182.0	$NO_2^-(aq)$	123.0
$Cu^+(aq)$	40.6	$NH_4^+(aq)$	113.4	$CrO_4^{2-}(aq)$	50.2	$NO_3^-(aq)$	146.4
$Cu^{2+}(aq)$	-99.6	$Ni^{2+}(aq)$	-128.9	$Cr_2O_7^{2-}(aq)$	261.9	$OH^-(aq)$	-10.8
$Fe^{2+}(aq)$	-137.7	$Pb^{2+}(aq)$	10.5	$F^-(aq)$	-13.8	$PO_4^{3-}(aq)$	-222
$Fe^{3+}(aq)$	-315.9	$Sn^{2+}(aq)$	-17.4	$HCO_3^-(aq)$	91.2	$S^{2-}(aq)$	-14.6
$H^+(aq)$	0.0	$Zn^{2+}(aq)$	-112.1	$H_2PO_4^-(aq)$	90.4	$SO_4^{2-}(aq)$	20.1

Values of ΔS^0 calculated in this way will, of course, have the units joules per kelvin (J/K), since these are the units listed for S^0. If we need to convert ΔS^0 to kilojoules per kelvin (kJ/K), all we need do is divide by 1000. For the decomposition of calcium carbonate,

$$\Delta S^0 = +160.5 \frac{J}{K} \times \frac{1 \text{ kJ}}{1000 \text{ J}} = +0.1605 \text{ kJ/K}$$

You will notice that ΔS^0 for the decomposition of calcium carbonate is a positive quantity. This is reasonable since the gas formed, CO_2, has a much higher molar entropy than either of the solids, CaO or $CaCO_3$. As a matter of fact, we almost always find that *a reaction that results in an increase in the number of moles of gas is accompanied by an increase in entropy. Conversely, if the number of moles of gas decreases, we expect ΔS^0 to be a negative quantity.* Consider, for example, the reaction

Gases in general have much higher entropies than liquids or solids

$$2H_2(g) + O_2(g) \rightarrow 2H_2O(l)$$

$$\Delta S^0 = 2 \, S^0 \, H_2O(l) - 2 \, S^0 \, H_2(g) - S^0 \, O_2(g)$$
$$= 139.8 \text{ J/K} - 261.2 \text{ J/K} - 205.0 \text{ J/K} = -326.4 \text{ J/K}$$

■ **EXAMPLE 20.3**

Calculate ΔS^0 for the reaction

$$HCl(g) \rightarrow H^+(aq) + Cl^-(aq)$$

Solution
Using values of S^0 from Tables 20.1 and 20.2,

$$\Delta S^0 = S^0 \, H^+(aq) + S^0 \, Cl^-(aq) - S^0 \, HCl(g)$$
$$= 0 + 56.5 \text{ J/K} - 186.8 \text{ J/K} = \boxed{-130.3 \text{ J/K}}$$

Exercise
Calculate ΔS^0 for the process $NaCl(s) \rightarrow Na^+(aq) + Cl^-(aq)$. Answer: $+43.4$ J/K.

Strictly speaking, calculations such as those just made are valid only at 25°C, the temperature at which standard entropies are recorded in Tables 20.1 and 20.2. In practice, though, ΔS^0 for a reaction is ordinarily nearly independent of temperature (provided we stay away from very low temperatures, near 0 K). Consider, for example, the reaction

$$CaCO_3(s) \rightarrow CaO(s) + CO_2(g)$$

where ΔS^0 is $+160.5$ J/K at 25°C. One can calculate that at 100°C, ΔS^0 for this reaction is about $+160.0$ J/K, only slightly less than the value at 25°C. The entropies of all three substances involved in the reaction (CaO, CO_2, and $CaCO_3$) increase with temperature, but the difference,

$$S^0\ CO_2(g) + S^0\ CaO(s) - S^0\ CaCO_3(s)$$

$$\Delta S^{\circ}_T \cong \Delta S^{\circ}_{298K}$$

remains nearly constant. In all calculations, we will *take ΔS^0 to be independent of temperature;* that is, we will assume that the value of ΔS^0 calculated from Tables 20.1 and 20.2 remains valid at temperatures other than 25°C.

On the other hand, the entropy change for a reaction often changes appreciably with the pressure of a gas or the concentration of an ion in solution. Consider, for example, the reaction

$$HCl(g) \rightarrow H^+(aq) + Cl^-(aq)$$

We saw in Example 20.3 that the standard entropy change for this reaction, ΔS^0, is -130.3 J/K. This is the entropy change when the HCl is at a pressure of 1 atm and the H^+ and Cl^- ions are at a concentration of 1 M. To emphasize this, we might write

$$HCl(g,\ 1\ atm) \rightarrow H^+(aq,\ 1\ M) + Cl^-(aq,\ 1\ M); \quad \Delta S^0 = -130.3\ \text{J/K}$$

The value of ΔS changes considerably if we change the pressure of the HCl gas:

$$HCl(g,\ 10\ atm) \rightarrow H^+(aq,\ 1\ M) + Cl^-(aq,\ 1\ M); \quad \Delta S = -111.2\ \text{J/K}$$

or the concentrations of the ions in solution:

$$HCl(g,\ 1\ atm) \rightarrow H^+(aq,\ 0.1\ M) + Cl^-(aq,\ 0.1\ M); \quad \Delta S = -92.1\ \text{J/K}$$

In all our calculations in this chapter, we will deal with the standard entropy change (1 atm, 1 M). This is given the special symbol ΔS^0 to distinguish it from the entropy change under other conditions, ΔS, which may be quite different.

20.3
THE SECOND LAW OF THERMODYNAMICS

Now that we have some idea of what entropy and entropy changes are, we can relate entropy change to spontaneity. This can be done through the Second Law of Thermodynamics, a basic principle of nature discovered more than a century ago. One way to state this law is to say that, **in a spontaneous process, there is a net increase in entropy**, taking into account both system and surroundings. That is:

$$\Delta S_{universe} = (\Delta S_{system} + \Delta S_{surroundings}) > 0 \qquad (20.10)$$

(Recall from Chapter 7 that the system is that portion of the universe upon which we focus attention; the surroundings include everything else.)

Notice that the Second Law refers to the total entropy change, involving both system and surroundings. To this point, we have been dealing only with ΔS_{system}. The system might be an ideal gas expanding into a vacuum, a substance undergoing a phase change, or a reaction mixture. For many spontaneous processes, the entropy change for the system is a negative quantity. Consider, for example, the rusting of iron, a spontaneous process:

$$2Fe(s) + \tfrac{3}{2}O_2(g) + 3H_2O(l) \rightarrow 2Fe(OH)_3(s) \qquad (20.11)$$

We can determine ΔS for this reaction system at 25°C and 1 atm from a table of standard entropies:

$$\Delta S_{system} = \Sigma S^0{}_{products} - \Sigma S^0{}_{reactants}$$

$$= 2\ S^0\ Fe(OH)_3 - 2\ S^0\ Fe - \tfrac{3}{2}\ S^0\ O_2 - 3\ S^0\ H_2O(l)$$

$$= 213.4\ J/K - 54.6\ J/K - 307.5\ J/K - 209.7\ J/K$$

$$= -358.4\ J/K$$

This result is entirely consistent with the Second Law; all it requires is that the entropy change of the surroundings be greater than 358.4 J/K, so that $\Delta S_{universe} > 0$. Clearly, we must be able to calculate $\Delta S_{surroundings}$ if Equation 20.10 is to be useful in predicting reaction spontaneity.

ΔS_{surr}

It turns out that the expression for the entropy change of the surroundings is a rather simple one. We can deduce the form of that expression by noting first that the *sign* of ΔS_{surr} depends upon the direction of heat flow. If heat flows out of the system into the surroundings, increasing thermal motion in the surroundings, we expect ΔS_{surr} to be a positive quantity. Conversely, if the surroundings give up heat to the system, ΔS_{surr} should be a negative quantity.

If the reaction gives off heat, $\Delta S_{surr} > 0$

The magnitude of ΔS_{surr} depends upon two factors. One of these is the amount of heat transferred between system and surroundings. The larger the heat flow, the greater the increase or decrease in thermal motion and hence the larger the value of ΔS_{surr}. Indeed, we expect ΔS_{surr} to be directly proportional to the amount of heat transferred, and it is.

The other factor that determines the magnitude of ΔS_{surr} is equally important but much less obvious. It is the absolute temperature, T. Figure 20.6 shows that for a particular spontaneous process, ΔS is inversely related to temperature. This is generally true for all types of processes. The entropy change produced when a given amount of heat is transferred is greater at low temperatures than at high temperatures.

If 100 J flow into the surroundings, ΔS_{surr} will be twice as great at 100K as it is at 200K

In general, the entropy change for the surroundings is given by the expression:

$$\Delta S_{surr} = q_{surr}/T$$

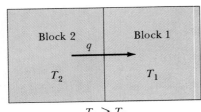

$T_2 > T_1$

Figure 20.6 *Relation between ΔS, q, and temperature.* Consider two Al blocks equal in mass. They are identical in every respect except that the block on the left is at a higher temperature (T_2) than the one at the right (T_1). Common sense tells us that heat will flow spontaneously from Block 2 to Block 1. According to the Second Law:

$$\Delta S(\text{Block 2}) + \Delta S(\text{Block 1}) > 0$$

Hence the magnitudes of the two entropy changes are different. But since the two blocks differ only in temperature, *the magnitude of ΔS must depend upon temperature.* Going a step further, since heat flows from Block 2 to Block 1:

$$\Delta S(\text{Block 2}) < 0; \ \Delta S(\text{Block 1}) > 0$$

For the total ΔS to be positive as required by the Second Law:

$$|\Delta S(\text{Block 1})| > |\Delta S(\text{Block 2})|$$

where $|\ |$ specifies the magnitude of ΔS, regardless of sign. But Block 1 is at a lower temperature than Block 2 ($T_1 < T_2$), so *the magnitude of ΔS must be inversely related to temperature.*

where q_{surr} is the heat flow into the surroundings, taking place at absolute temperature, T. You will recall from Chapter 7 that at constant temperature and pressure:

$$\Delta H = q_{\text{system}} = -q_{\text{surr}}$$

Hence we can write

$$\Delta S_{\text{surr}} = \frac{-\Delta H}{T} \qquad \text{(constant } T, P\text{)} \qquad\qquad (20.12)$$

This equation allows us to relate ΔS_{surr} to ΔH_{system}

In this expression, ΔH is the enthalpy change for the system and T is the Kelvin temperature. If the process is exothermic ($\Delta H < 0$), heat is transferred to the surroundings, and both Equation 20.12 and our previous discussion tell us that ΔS_{surr} is a positive quantity. Conversely, for an endothermic process taking place at constant T and P, the entropy of the surroundings decreases.

Equation 20.12 can be applied to any process taking place at a constant pressure (usually 1 atm) and constant temperature. Most often, we will apply it to chemical reactions. Consider, once again, the rusting of iron:

$$2\text{Fe}(s) + \tfrac{3}{2}\text{O}_2(g) + 3\text{H}_2\text{O}(l) \rightarrow 2\text{Fe(OH)}_3(s)$$

for which we showed earlier that ΔS_{system} at 1 atm and 25°C is -358.4 J/K or -0.3584 kJ/K. To calculate ΔS_{surr}, we first find ΔH for the reaction system. Using a table of heats of formation (Chap. 7, p. 225), we have:

$$\begin{aligned}
\Delta H^0 &= \Sigma \Delta H^0_{\text{f products}} - \Sigma \Delta H^0_{\text{f reactants}} \\
&= 2\ \Delta H^0_{\text{f}}\ \text{Fe(OH)}_3(s) - 3\ \Delta H^0_{\text{f}}\ \text{H}_2\text{O}(l) \\
&= -1646.0\ \text{kJ} + 857.4\ \text{kJ} = -788.6\ \text{kJ}
\end{aligned}$$

Now we can calculate ΔS_{surr} at 25°C and 1 atm:

$$\Delta S_{\text{surr}} = \frac{-\Delta H}{T} = \frac{788.6\ \text{kJ}}{298.15\ \text{K}} = 2.645\ \text{kJ/K}$$

It follows that the overall entropy change is a positive quantity:

$$\Delta S_{universe} = \Delta S_{system} + \Delta S_{surr} = -0.3584 \text{ kJ/K} + 2.645 \text{ kJ/K}$$
$$= +2.287 \text{ kJ/K}$$

The Second Law tells us what we already know from bitter experience: The rusting of iron is a spontaneous process.

■ **EXAMPLE 20.4** _____

Show by calculation that the vaporization of water at 120°C and 1 atm is spontaneous.

Solution

The equation for the phase change is:

$$H_2O(l) \rightarrow H_2O(g)$$

We first calculate ΔS_{system}, which is ΔS^0 for the reaction.

$$\Delta S_{system} = \Delta S^0 = S^0 \text{ } H_2O(g) - S^0 \text{ } H_2O(l)$$
$$= 188.7 \text{ J/K} - 69.9 \text{ J/K} = 118.8 \text{ J/K} = 0.1188 \text{ kJ/K}$$

As expected, ΔS_{system} is a positive quantity; we are going from a liquid to a gas. To calculate ΔS_{surr}, we first find ΔH. To do that, we use the table of heats of formation in Chapter 7.

$$\Delta H = \Delta H_f^0 \text{ } H_2O(g) - \Delta H_f^0 \text{ } H_2O(l)$$
$$= -241.8 \text{ kJ} + 285.8 \text{ kJ} = 44.0 \text{ kJ}$$

We now calculate ΔS_{surr}, using Equation 20.12.

$$\Delta S_{surr} = \frac{-\Delta H}{T} = \frac{-44.0 \text{ kJ}}{393 \text{ K}} = -0.112 \text{ kJ/K}$$

Substituting into Equation 20.10, we get

$$\Delta S_{universe} = \Delta S_{system} + \Delta S_{surr}$$

$$= 0.1188 \text{ kJ/K} - 0.112 \text{ kJ/K} = \boxed{0.007 \text{ kJ/K} > 0}$$

We conclude that the reaction is indeed spontaneous.

If you release the pressure of the water in a pressure cooker at 120°C, stand back as the steam comes pouring out

Exercise

Show by calculation that the same phase change is not spontaneous at 50°C and 1 atm. Answer: $\Delta S_{universe} = 0.1188 \text{ kJ/K} - 0.136 \text{ kJ/K} = -0.017 \text{ kJ/K}$.

ΔS and ΔH

Our discussion of these two processes (rusting of iron, vaporization of water) focused on two thermodynamic functions, ΔS and ΔH. Combining Equations 20.10 and 20.12, we see that for a spontaneous process at constant temperature and pressure:

$$\Delta S - \frac{\Delta H}{T} > 0$$

or:

$$(T\Delta S - \Delta H) > 0 \qquad \text{(spontaneous process, constant } T, P) \quad (20.13)$$

where ΔS and ΔH refer to the entropy and enthalpy changes of the *system*. Equation 20.13 tells us that if $(T\Delta S - \Delta H)$ is a positive quantity, we can expect the process under consideration to be spontaneous. For the two processes discussed:

	ΔS (kJ/K)	$T\Delta S$ (kJ)	ΔH (kJ)	$(T\Delta S - \Delta H)$ (kJ)
Rusting of iron at 25°C, 1 atm	−0.3584	−106.9	−788.6	+681.7
Vaporization of H_2O at 120°C, 1 atm	+0.1188	+46.7	+44.0	+2.7

Since the quantity $(T\Delta S - \Delta H)$ is positive in both cases, we conclude correctly that both processes are spontaneous.

At this point, we appear to have accomplished our original objective, stated at the beginning of this chapter. Equation 20.13 does indeed give us the criterion of spontaneity we have been looking for. It turns out though that we can express this relation in a somewhat simpler way, using a new function called free energy.

20.4
FREE ENERGY

The argument we just went through is not original with us; it was first developed by an authentic American genius, J. Willard Gibbs. Gibbs was a professor of mathematical physics at Yale at the turn of the century; his life and accomplishments are described in the historical perspective at the end of this chapter. It was Gibbs who first pointed out that

$$\text{if } (T\Delta S - \Delta H) > 0$$

or, alternatively,

$$\text{if } (\Delta H - T\Delta S) < 0 \qquad\qquad (20.14)$$

then a process carried out at constant temperature and pressure must be spontaneous. Gibbs went a step further and introduced a new thermodynamic function, called free energy and now given the symbol G in honor of Gibbs. The defining equation for free energy is:

The free energy depends on both H and S

$$G = H - TS \qquad\qquad (20.15)$$

The free energy of a system, like its enthalpy and entropy, is a state function. Its value is determined only by the state of a system, not by how it reached that state. In general:

$$\Delta G = G_{\text{final}} - G_{\text{initial}} \qquad\qquad (20.16)$$

For chemical reactions:

$$\Delta G = G_{\text{products}} - G_{\text{reactants}} \qquad (20.17)$$

Gibbs had a very good reason for defining free energy the way he did. Notice what happens to Equation 20.15 if we consider a process, perhaps a chemical reaction, taking place at constant temperature:

$$G_{\text{final}} = H_{\text{final}} - TS_{\text{final}}$$

$$G_{\text{initial}} = H_{\text{initial}} - TS_{\text{initial}}$$

Subtracting, we obtain:

$$\Delta G = \Delta H - T\Delta S \qquad (20.18)$$

Comparing Equations 20.18 and 20.14, we see that the free energy function, all by itself, is the criterion of spontaneity that we have been looking for. If ΔG for a process is negative, then $(\Delta H - T\Delta S)$ must be less than zero and the process carried out at constant temperature and pressure must be spontaneous.

Applying this reasoning to a chemical reaction carried out at constant temperature and pressure:

1. If ΔG is negative, the reaction is spontaneous.
2. If ΔG is positive, the reaction will not occur spontaneously. Instead, the reverse reaction will occur.
3. If ΔG is 0, the reaction system is at equilibrium; there is no tendency for it to occur in either direction.

Putting it another way, we can think of ΔG as a measure of the driving force of a reaction. **Reactions, at constant pressure and temperature, go in such a direction as to decrease the free energy of the system.** This means that the direction in which a reaction takes place is determined by the relative free energies of products and reactants. If the products have a lower free energy than the reactants ($G_{\text{products}} < G_{\text{reactants}}$), the forward reaction will occur (Fig. 20.7). If the reverse is true ($G_{\text{reactants}} < G_{\text{products}}$), the reverse reaction is spontaneous. Finally, if $G_{\text{products}} = G_{\text{reactants}}$, there is no driving force to make the reaction go in either direction.

A system at equilibrium has a minimum free energy at that T

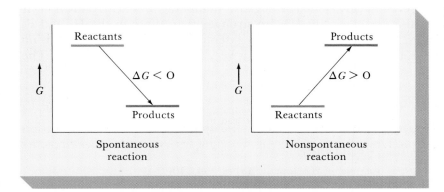

Spontaneous reaction

Nonspontaneous reaction

Figure 20.7 For a spontaneous reaction, the free energy of the products is less than that of the reactants: $\Delta G < 0$. For a nonspontaneous reaction, the reverse is true: $\Delta G > 0$.

RELATION AMONG ΔG, ΔH, AND ΔS

The relation we derived earlier

$$\Delta G = \Delta H - T\Delta S \qquad (20.18)$$

is called the Gibbs-Helmholtz equation. It is one of the most important equations in chemical thermodynamics. We will spend the rest of this section showing what it means and how it is used, especially to predict the direction of spontaneity of a chemical reaction.

The Gibbs-Helmholtz equation tells us that the driving force ΔG for a reaction depends upon two quantities. One of these is the enthalpy change due to the making and breaking of bonds, ΔH. The other is the product of the change in randomness, ΔS, times the absolute temperature, T.

The two factors that tend to make ΔG negative and hence lead to a spontaneous reaction are

1. A negative value of ΔH. Exothermic reactions ($\Delta H < 0$) tend to be spontaneous inasmuch as they contribute to a negative value of ΔG. On the molecular level, this means that there will be a tendency to form "strong" bonds at the expense of "weak" ones.

2. A positive value of ΔS. If the entropy change is positive ($\Delta S > 0$), the term $-T\Delta S$ will make a negative contribution to ΔG. Hence, there will be a tendency for a reaction to be spontaneous if the products are less ordered than the reactants.

What is the driving force for a falling brick?

In many physical processes, the entropy increase is the major driving force. This is the case with osmosis, where ΔH is zero and ΔS is a positive quantity. A similar situation applies when two liquids with similar inter-molecular forces, such as benzene and toluene, are mixed. There is no change in enthalpy, but the entropy increases because the pure substances become diluted when they form a solution.* In this and many other solution processes, it is the entropy change rather than the enthalpy change that accounts for spontaneity.

In certain reactions, ΔS is nearly zero and ΔH is the only important component of the driving force for spontaneity. An example is the synthesis of hydrogen fluoride from the elements:

$$\tfrac{1}{2}H_2(g) + \tfrac{1}{2}F_2(g) \rightarrow HF(g)$$

For this reaction, ΔH is a large negative number, -271.1 kJ. This reflects the fact that the bonds in HF are stronger than those in the H_2 and F_2 molecules. As we might expect for a gaseous reaction in which there is no change in the number of moles, ΔS is very small, about 0.0070 kJ/K. The

---

*Formation of a *water* solution is often accompanied by a *decrease* in entropy because of hydrogen bonding or hydration effects that lead to a highly ordered solution structure. Examples include:

$$C_2H_5OH(l) \rightarrow C_2H_5OH(aq); \qquad \Delta S^0 = -12.2 \text{ J/K}$$

$$CaCl_2(s) \rightarrow Ca^{2+}(aq) + 2Cl^-(aq); \qquad \Delta S^0 = -44.7 \text{ J/K}$$

free energy change, ΔG, at 1 atm is -273.2 kJ at 25°C, almost identical to ΔH. Even at very high temperatures, the difference between ΔG and ΔH is small, amounting to only about 14 kJ at 2000 K.

You can't break HF down to the elements by heating it up

The more common case is one in which both ΔH and ΔS make significant contributions to ΔG. To determine the sign of ΔG, we must consider the values of both ΔH and ΔS as well as the temperature. The approach we will follow is described below.

STANDARD FREE ENERGY CHANGE

The Gibbs-Helmholtz equation is valid under all conditions of temperature, pressure, and concentration. However, all our calculations using the equation will be restricted to *standard conditions*. As pointed out earlier, these are:

1 atm pressure for gases, pure liquids, and solids

1 M concentration for ions or molecules in water solution

In other words, we will use the equation in the form:

$$\Delta G^0 = \Delta H^0 - T\Delta S^0 \qquad (20.19)$$

where:

ΔG^0 is the standard free energy change (1 atm, 1 M)

ΔH^0 is the standard enthalpy change, which can be calculated from heats of formation, ΔH_f^0 (listed in Tables 7.3 and 7.4, Chapter 7)

ΔS^0 is the standard entropy change (Tables 20.1 and 20.2)

$\Delta G°$ will establish spontaneity only under standard conditions

CALCULATION OF ΔG^0 AT 25°C

To illustrate the use of Equation 20.19, let us apply it first at 25°C ($T = 298$ K). Consider the reaction

$$2H_2(g) + O_2(g) \rightarrow 2H_2O(l)$$

Using Tables 7.3 and 20.1, we find that for this reaction ΔH^0 is -571.6 kJ and ΔS^0 is -326.4 J/K. Hence, at 25°C,

$$\Delta G^0 = -571.6 \text{ kJ} - 298 \text{ K} \times (-0.3264 \text{ kJ/K})$$

$$= -571.6 \text{ kJ} + 97.3 \text{ kJ} = -474.3 \text{ kJ}$$

Notice that

— in making a calculation of this type, the units must be consistent. If, as is ordinarily the case, we want ΔG^0 in kilojoules, ΔS^0 must be expressed in **kilojoules per kelvin** (1 J/K $= 10^{-3}$ kJ/K).

— the quantity ΔG^0 for this reaction is negative. This is consistent with the fact that $H_2(g)$ and $O_2(g)$, both at 1 atm, react spontaneously at 25°C to form liquid water.

■ **EXAMPLE 20.5**
For the reaction $CaSO_4(s) \rightarrow Ca^{2+}(aq) + SO_4{}^{2-}(aq)$, calculate, using Tables 7.3, 7.4, 20.1, and 20.2,

a. ΔH^0 b. ΔS^0 c. ΔG^0 at 25°C

Solution

a. $\Delta H^0 = \Delta H_f^0 \, Ca^{2+}(aq) + \Delta H_f^0 \, SO_4{}^{2-}(aq) - \Delta H_f^0 \, CaSO_4(s)$

   $= -542.8 \text{ kJ} - 909.3 \text{ kJ} - (-1434.1 \text{ kJ}) = \boxed{-18.0 \text{ kJ}}$

b. $\Delta S^0 = S^0 \, Ca^{2+}(aq) + S^0 \, SO_4{}^{2-}(aq) - S^0 \, CaSO_4(s)$

   $= -53.1 \text{ J/K} + 20.1 \text{ J/K} - 106.7 \text{ J/K} = \boxed{-139.7 \text{ J/K}}$

   $= -0.1397 \text{ kJ/K}$

c. $\Delta G^0 = -18.0 \text{ kJ} - 298 \text{ K} \times (-0.1397 \text{ kJ/K}) = \boxed{+23.6 \text{ kJ}}$

We conclude that this process is not spontaneous at standard conditions at 25°C. In other words, calcium sulfate does not dissolve in water to produce a 1 M solution. This is indeed the case. The solubility of $CaSO_4$ at 25°C is considerably less than 1 mol/L, only about 0.02 mol/L.

Exercise
For the process $NaCl(s) \rightarrow Na^+(aq) + Cl^-(aq)$, $\Delta H^0 = +3.9 \text{ kJ}$, $\Delta S^0 = +43.4 \text{ J/K}$. Calculate ΔG^0 at 25°C. Answer: -9.0 kJ (NaCl dissolves spontaneously at 25°C to give a 1 M solution; its solubility at 25°C is about 5.4 mol/L).

The Gibbs-Helmholtz equation can be used to calculate the *standard free energy of formation* of a compound. This quantity, ΔG_f^0, is analogous to the heat of formation, ΔH_f^0. It is defined as the free energy change per mole when a compound is formed from the elements in their stable states.

■ **EXAMPLE 20.6**
Calculate the standard free energy of formation, ΔG_f^0, at 25°C, for $CH_4(g)$, using data in Tables 7.3 and 20.1.

Solution
By definition, ΔG_f^0 for CH_4 is ΔG^0 for the reaction

$$C(s) + 2H_2(g) \rightarrow CH_4(g)$$

For this reaction,

Most free energies of formation of compounds are found from $\Delta H°$ and $\Delta S°$

$\Delta H^0 = \Delta H_f^0 \, CH_4(g) = -74.8 \text{ kJ}$

$\Delta S^0 = S^0 \, CH_4(g) - S^0 \, C(s) - 2 \, S^0 \, H_2(g)$

$= 186.2 \text{ J/K} - 5.7 \text{ J/K} - 2(130.6 \text{ J/K}) = -80.7 \text{ J/K}$

$= -0.0807 \text{ kJ/K}$

At 25°C, that is, 298 K,

$$\Delta G^0 = \Delta H^0 - 298\ \Delta S^0 = -74.8\ \text{kJ} - 298\ \text{K} \times (-0.0807\ \text{kJ/K})$$
$$= -50.8\ \text{kJ}$$

We conclude that ΔG_f^0 $CH_4(g)$ at 25°C is -50.8 kJ/mol.

Exercise
What is the standard free energy of formation of liquid water at 25°C?
Answer: -237.2 kJ/mol.

Tables of standard free energies of formation at 25°C of compounds and ions in solution are given in Appendix 1 (along with standard heats of formation and standard entropies). Values of ΔG_f^0 found in these tables can be used to calculate free energy changes for reactions. The relationship here is entirely analogous to that given for enthalpies in Chapter 7.

$$\Delta G_{reaction}^0 = \Sigma \Delta G_{f\ products}^0 - \Sigma \Delta G_{f\ reactants}^0 \tag{20.20}$$

If you calculate ΔG^0 in this way, you should keep in mind an important limitation of Equation 20.20. **It is valid only at the temperature at which ΔG_f^0 data are tabulated, in this case 25°C.** Since ΔG^0 varies considerably with temperature, this approach is not even approximately valid at other temperatures.

ΔG° at 25°C is useful only at 25°C

■ **EXAMPLE 20.7**

Calculate ΔG^0 at 25°C for the reaction:

$$CH_4(g) + 2O_2(g) \rightarrow CO_2(g) + 2H_2O(l)$$

using:

a. standard free energies of formation (Appendix 1)
b. standard enthalpies of formation and standard entropies (Appendix 1)

Solution

a. ΔG^0 at 25°C $= \Delta G_f^0\ CO_2(g) + 2\ \Delta G_f^0\ H_2O(l) - \Delta G_f^0\ CH_4(g)$

$= -394.4$ kJ $- 474.4$ kJ $+ 50.7$ kJ $= -817.9$ kJ

(Note that the standard free energy of formation, like the standard enthalpy of formation, is zero for an element in its stable state.)

b. $\Delta H^0 = \Delta H_f^0 CO_2(g) + 2\Delta H_f^0 H_2O(l) - \Delta H_f^0 CH_4(g)$

$= -393.5$ kJ $- 571.6$ kJ $+ 74.8$ kJ $= -890.3$ kJ

$\Delta S^0 = S^0 CO_2(g) + 2S^0 H_2O(l) - S^0 CH_4(g) - 2S^0 O_2(g)$
$= 0.2136$ kJ/K $+ 0.1398$ kJ/K $- 0.1862$ kJ/K $- 0.4100$ kJ/K
$= -0.2428$ kJ/K

$\Delta G^0 = \Delta H^0 - T\Delta S^0 = -890.3$ kJ $+ 298(0.2428)$kJ $= -817.9$ kJ

The two answers for ΔG^0 are the same.

Exercise
Calculate ΔG^0 at 25°C for the ionization of one mole of liquid water. Answer: +80.0 kJ.

CALCULATION OF ΔG^0 AT OTHER TEMPERATURES

We can also use Equation 20.19 to calculate ΔG^0 at temperatures other than 25°C. To do this, we neglect the variations of ΔH^0 and ΔS^0 with temperature, which are ordinarily small.* In other words, we take the values of ΔH^0 and ΔS^0 from tables and simply insert them in Equation 20.19, using the appropriate value of T. Suppose, for example, we want to know ΔG^0 at 1000°C for the reaction

$$CaCO_3(s) \rightarrow CaO(s) + CO_2(g)$$

For this reaction, we showed earlier that $\Delta H^0 = +178.3$ kJ, $\Delta S^0 = +160.5$ J/K. Substituting $T = 1000 + 273 = 1273$ K, we have

$$\Delta G^0 = +178.3 \text{ kJ} - 1273 \text{ K} \times (0.1605 \text{ kJ/K})$$
$$= 178.3 \text{ kJ} - 204.3 \text{ kJ} = -26.0 \text{ kJ}$$

We conclude that this reaction is spontaneous at 1000°C and 1 atm, since ΔG^0 has a negative sign.

■ **EXAMPLE 20.8**
Calculate ΔG^0 for the reaction $Cu(s) + H_2O(g) \rightarrow CuO(s) + H_2(g)$ at 500 K.

Solution
From Table 7.3
$$\Delta H^0 = \Delta H_f^0 \; CuO(s) - \Delta H_f^0 \; H_2O(g) = -157.3 \text{ kJ} + 241.8 \text{ kJ}$$
$$= +84.5 \text{ kJ}$$

From Table 20.1
$$\Delta S^0 = S^0 \; CuO(s) + S^0 \; H_2(g) - S^0 \; Cu(s) - S^0 \; H_2O(g)$$
$$= 42.6 \text{ J/K} + 130.6 \text{ J/K} - 33.2 \text{ J/K} - 188.7 \text{ J/K}$$
$$= -48.7 \text{ J/K} = -0.0487 \text{ kJ/K}$$

Hence, $\Delta G^0 = +84.5 \text{ kJ} - 500 \text{ K} \times (-0.0487 \text{ kJ/K}) = \boxed{+108.9 \text{ kJ}}$

We conclude that this reaction is not spontaneous, since ΔG^0 is positive. Instead, the reverse reaction occurs at 1 atm and 500 K (227°C). Under

---

*As far as Equation 20.19 is concerned, there is another reason for ignoring the temperature dependence of ΔH and ΔS. These two quantities always change in the same direction as the temperature changes (that is, if ΔH becomes more positive, so does ΔS). Hence, the two effects tend to cancel each other. The true value of ΔG at any temperature is about the same as the one we calculate taking ΔH and ΔS to be constant.

these conditions, CuO is reduced to copper by being heated in a stream of H_2 gas.

Exercise
Calculate ΔG^0 for this reaction at 1000 K. Answer: $+133.2$ kJ (still nonspontaneous).

From Example 20.8 and the preceding discussion, it should be clear that ΔG^0, unlike ΔH^0 and ΔS^0, is strongly dependent upon temperature. This comes about, of course, because of the T in the Gibbs-Helmholtz equation:

$$\Delta G^0 = \Delta H^0 - T\Delta S^0$$

Comparing this equation to that of a straight line:

$$y = b + mx$$

we see that a plot of ΔG^0 vs. T should be linear, with a slope of $-\Delta S^0$ and a y intercept (at 0 K) of ΔH^0. Indeed it is (see Fig. 20.8, p. 714).

20.5
TEMPERATURE AND REACTION SPONTANEITY

When the temperature of a reaction system is increased, the direction in which the reaction proceeds spontaneously may or may not change. Whether it does or not depends upon the relative signs of ΔH^0 and ΔS^0. The four possible situations, deduced from the Gibbs-Helmholtz equation, are summarized in Table 20.3.

If ΔH^0 and ΔS^0 have opposite signs (Table 20.3, I and II), it is impossible to reverse the direction of spontaneity by a change in temperature alone. The two terms ΔH^0 and $-T\Delta S^0$ reinforce one another. Hence, ΔG^0

TABLE 20.3   Effect of Temperature on Reaction Spontaneity

	ΔH^0	ΔS^0	$\Delta G^0 = \Delta H^0 - T\Delta S^0$	REMARKS
I	−	+	always −	spontaneous at all T; reverse reaction always nonspontaneous
II	+	−	always +	nonspontaneous at all T; reverse reaction occurs
III	+	+	+ at low T − at high T	nonspontaneous at low T; becomes spontaneous as T is raised
IV	−	−	− at low T + at high T	spontaneous at low T; at high T, reverse reaction becomes spontaneous

Figure 20.8 $\Delta H°$ and $\Delta G°$ as a function of temperature for the reaction: $CaCO_3(s) \rightarrow CaO(s) + CO_2(g)$. Below about 1100 K, $\Delta G°$ is positive and the reaction is nonspontaneous. Above 1100 K, $\Delta G°$ is negative and $CaCO_3$ decomposes spontaneously to CaO and CO_2 at 1 atm.

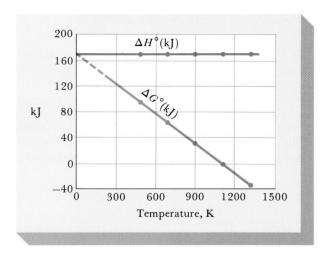

has the same sign at all temperatures. Reactions of this type are rather uncommon. One such reaction is that discussed in Example 20.8:

$$Cu(s) + H_2O(g) \rightarrow CuO(s) + H_2(g) \tag{20.21}$$

Here, ΔH^0 is $+84.5$ kJ and ΔS^0 is -0.0487 kJ/K. Hence,

$$\Delta G^0 = \Delta H^0 - T\Delta S^0$$
$$= +84.5 \text{ kJ} + T(0.0487 \text{ kJ/K})$$

Copper just won't react with water to make CuO and H$_2$

Clearly, ΔG^0 is positive at all temperatures. The reaction cannot take place spontaneously at 1 atm regardless of temperature.

It is more common to find that ΔH^0 and ΔS^0 have the same sign (Table 20.3, III and IV). When this happens, the enthalpy and entropy factors oppose each other. ΔG^0 changes sign as temperature increases, and the direction of spontaneity reverses. At low temperatures, ΔH^0 predominates and the exothermic reaction occurs. As the temperature rises, the quantity $T\Delta S^0$ increases in magnitude and eventually exceeds ΔH^0. At high temperatures, the reaction that leads to an increase in entropy occurs. In most cases, 25°C is a "low" temperature, at least at a pressure of 1 atm. This explains why exothermic reactions are usually spontaneous at room temperature and atmospheric pressure.

An example of a reaction for which ΔH^0 and ΔS^0 have the same sign is the decomposition of calcium carbonate:

$$CaCO_3(s) \rightarrow CaO(s) + CO_2(g)$$

Here, as pointed out previously, $\Delta H^0 = +178.3$ kJ and $\Delta S^0 = +160.5$ J/K. Hence,

$$\Delta G^0 = +178.3 \text{ kJ} - T(0.1605 \text{ kJ/K})$$

Figure 20.8, shows a plot of ΔG^0 vs. T. Notice that

— below about 1100 K (830°C), ΔH^0 predominates, ΔG^0 is positive, and the reaction is nonspontaneous at 1 atm.

— above about 1100 K, $T\Delta S^0$ predominates, and the reaction is spontaneous at 1 atm.

■ **EXAMPLE 20.9** _____

At what temperature does ΔG^0 become zero for the reaction

$$CaCO_3(s) \rightarrow CaO(s) + CO_2(g)?$$

Solution

In general, $\Delta G^0 = \Delta H^0 - T\Delta S^0$. When ΔG^0 is zero,

$$\Delta H^0 = T\Delta S^0 \qquad T = \frac{\Delta H^0}{\Delta S^0}$$

Here, $\Delta H^0 = +178.3$ kJ, $\Delta S^0 = 0.1605$ kJ/K, so that

$$T = \frac{178.3 \text{ kJ}}{0.1605 \text{ kJ/K}} = \boxed{1110 \text{ K}}$$

At this temperature, the reaction is at equilibrium at 1 atm pressure. If we put some $CaCO_3$ in a container and heat it to 1110 K, the pressure of CO_2 developed will be 1 atm.

In an open container at 1110 K, $CaCO_3$ will decompose to CaO and CO_2. That's how we make CaO

Exercise

Calculate the temperature at which ΔG^0 becomes zero for Reaction 20.21. Answer: The calculated value is -1740 K, which is nonsensical. There is no temperature at which ΔG^0 is zero for this reaction, because ΔH^0 and ΔS^0 have opposite signs.

_____

The development we went through in Example 20.9 is a general one. The temperature, T, at which a system is at equilibrium at *one atmosphere*, is given by the expression

$$T = \Delta H^0/\Delta S^0$$

This relationship is particularly useful for phase changes. We can relate the normal boiling point, T_b, to the heat and entropy of vaporization, ΔH^0_{vap} and ΔS^0_{vap}. At the normal boiling point, liquid and vapor are at equilibrium at 1 atm, so that

$$T_b = \frac{\Delta H^0_{vap}}{\Delta S^0_{vap}} \tag{20.22}$$

To illustrate the use of Equation 20.22, let us apply it to water. Taking data from Tables 7.3 and 20.1, we calculate

$$\Delta H^0_{vap} = \Delta H^0_f \ H_2O(g) - \Delta H^0_f \ H_2O(l) = +44.0 \text{ kJ}$$

$$\Delta S^0_{vap} = S^0 \ H_2O(g) - S^0 \ H_2O(l) = +118.8 \text{ J/K}$$

From these data, we can calculate the normal boiling point of water:

$$T_b = \frac{+44.0 \text{ kJ}}{0.1188 \text{ kJ/K}} = 370 \text{ K}$$

The error is due to variation of $\Delta S°$ and $\Delta H°$ with T

This calculated boiling point is very close to the observed value, 373 K (100°C).

PRESSURE, CONCENTRATION, AND SPONTANEITY

There is an important restriction upon all the calculations we have made in this section concerning reaction spontaneity. This restriction is implied when we say that if ΔG^0 is negative, the reaction is spontaneous at *standard pressure* (1 atm for gases) and *standard concentration* (1 M for ions in solution). The free energy change for a reaction varies with the pressure of a gas or the concentration of an ion. This means that the conclusions about spontaneity reached on the basis of the sign of ΔG^0 may not apply at other than standard conditions. A case in point involves the reaction

$$CaCO_3(s) \rightarrow CaO(s) + CO_2(g, 1 \text{ atm}); \quad \Delta G^0 = -26.0 \text{ kJ at } 1000°C$$

Since ΔG^0 is negative at 1000°C, we conclude that the reaction at this temperature is spontaneous at 1 atm. However, one can calculate that when the pressure of CO_2 is 20 atm, ΔG^0 is $+5.7$ kJ, and the reaction is not spontaneous. As another example, you will recall that we showed in Example 20.5 that

$$CaSO_4(s) \rightarrow Ca^{2+}(aq, 1 \text{ } M) + SO_4^{2-}(aq, 1 \text{ } M);$$

$$\Delta G^0 \text{ at } 25°C = +23.6 \text{ kJ}$$

The positive sign of ΔG^0 indicates that $CaSO_4$ will not dissolve to form a 1 M solution at 25°C. On the other hand, one can calculate that ΔG for the formation of a 0.001 M solution of $CaSO_4$ is -10.6 kJ. Calcium sulfate should, and does, form a 0.001 M solution; its solubility at 25°C is more than ten times that value.

In Example 20.9 we showed that for the decomposition of $CaCO_3$, $\Delta G^0 = 0$ at 1110 K. We interpreted this to mean that the reaction is at equilibrium at 1110 K when the pressure of CO_2 is 1 atm. You should realize that if the pressure changes, ΔG will no longer be zero, and the equilibrium temperature will change. Indeed, as we saw in Chapter 13, a reaction system such as this one will reach a position of equilibrium at any given temperature.

In the same way, a liquid-vapor system such as

$$H_2O(l) \rightleftharpoons H_2O(g)$$

can reach equilibrium over a wide range of temperatures. Depending on the pressure, a liquid can boil at any temperature between the triple point and the critical point. The normal boiling point is unique in only one respect; it is the temperature at which ΔG^0 is zero and the system is at equilibrium at *one atmosphere pressure*.

20.6
ADDITIVITY OF FREE ENERGY CHANGES: COUPLED REACTIONS

Free energy changes for reactions, like enthalpy or entropy changes, are additive. That is:

if Reaction 3 = Reaction 1 + Reaction 2

then $\Delta G_3 = \Delta G_1 + \Delta G_2$ (20.23)

Equation 20.23 can be regarded as the free-energy equivalent of Hess' Law. To illustrate its application, consider the synthesis of $CuCl_2$ from the elements:

$$Cu(s) + \tfrac{1}{2}Cl_2(g) \rightarrow CuCl(s); \quad \Delta G^0 \text{ at } 25°C = -119.9 \text{ kJ}$$
$$\underline{CuCl(s) + \tfrac{1}{2}Cl_2(g) \rightarrow CuCl_2(s); \quad \Delta G^0 \text{ at } 25°C = -55.8 \text{ kJ}}$$
$$Cu(s) + Cl_2(g) \rightarrow CuCl_2(s) \qquad\qquad\qquad (20.24)$$

For Reaction 20.24, ΔG^0 at $25°C = -119.9 \text{ kJ} - 55.8 \text{ kJ} = -175.7 \text{ kJ}$

Since free energy changes are additive, it is often possible to bring about a nonspontaneous reaction by coupling it with a reaction for which ΔG^0 is a large negative number. As an example, consider the preparation of iron metal from hematite ore. The reaction

$$Fe_2O_3(s) \rightarrow 2Fe(s) + \tfrac{3}{2}O_2(g); \quad \Delta G^0 \text{ at } 25°C = +742.2 \text{ kJ}$$

is clearly nonspontaneous; even at temperatures as high as 2000°C, ΔG^0 is a positive quantity. Suppose, though, that we "couple" this reaction with the spontaneous oxidation of carbon monoxide:

$$3CO(g) + \tfrac{3}{2}O_2(g) \rightarrow 3CO_2(g); \quad \Delta G^0 \text{ at } 25°C = -257.1 \text{ kJ}$$

The overall reaction is spontaneous:

$$Fe_2O_3(s) \rightarrow 2Fe(s) + \tfrac{3}{2}O_2(g)$$
$$\underline{3CO(g) + \tfrac{3}{2}O_2(g) \rightarrow 3CO;(g)}$$
$$Fe_2O_3(s) + 3CO(g) \rightarrow 2Fe(s) + 3CO_2(g) \qquad (20.25)$$

For Reaction 20.25 at 25°C:

$$\Delta G° = +742.2 \text{ kJ} + 3(-257.1 \text{ kJ}) = -29.1 \text{ kJ}$$

We see that although Fe_2O_3 does not spontaneously decompose, it can be converted to iron by reaction with carbon monoxide. This is in fact the reaction used in a blast furnace when iron ore consisting mainly of Fe_2O_3 is reduced to iron (Chap. 22).

Many industrial processes are based on coupled reactions

Coupled reactions are common in human metabolism. Spontaneous processes, such as the oxidation of glucose:

$$C_6H_{12}O_6(aq) + 6O_2(g) \rightarrow 6CO_2(g) + 6H_2O; \quad \Delta G^0 = -2870 \text{ kJ at } 25°C$$
$$(20.26)$$

ordinarily do not serve directly as a source of energy. Instead these reactions are used to bring about a nonspontaneous reaction:

$$ADP(aq) + HPO_4{}^{2-}(aq) + 2H^+(aq) \rightarrow ATP(aq) + H_2O$$

$$\Delta G^0 = +31 \text{ kJ at } 25°C \tag{20.27}$$

ADP (adenosine diphosphate) and ATP (adenosine triphosphate) are complex organic molecules (Fig. 20.9) that, in essence, differ only by the presence of an extra phosphate group in ATP. In the coupled reaction with glucose, about 38 mol of ATP are synthesized for every mole of glucose consumed. This gives an overall free energy change for the coupled reaction of

$$-2870 \text{ kJ} + 38(+31 \text{ kJ}) \approx -1700 \text{ kJ}$$

That's how your body gets energy in a hurry

In a very real sense, your body "stores" energy available from the metabolism of foods in the form of ATP. This molecule in turn supplies the energy required for all sorts of biochemical reactions taking place in the body. It does this by reverting to ADP, i.e., by reversing Equation 20.27. The amount of ATP consumed is amazingly large; a competitive sprinter may hydrolyze as much as 500 g of ATP per minute.

■ **EXAMPLE 20.10** _____

The lactic acid ($C_3H_6O_3(aq)$, $\Delta G_f^0 = -559$ kJ) produced in muscle cells after vigorous exercise eventually is absorbed into the blood stream, where

Figure 20.9 Structures of ADP and ATP.

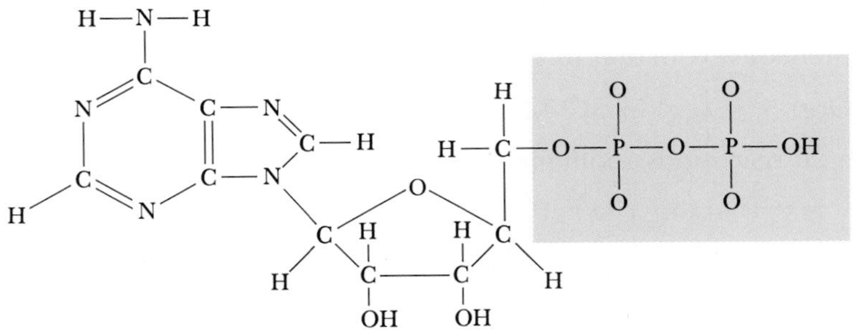

Adenosine diphosphate (ADP)

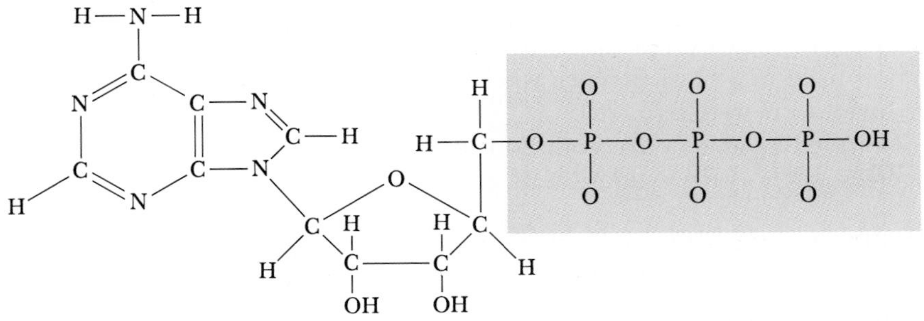

Adenosine triphosphate (ATP)

it is metabolized back to glucose ($\Delta G_f^0 = -919$ kJ) in the liver. The reaction is:

$$2C_3H_6O_3(aq) \rightarrow C_6H_{12}O_6(aq)$$

a. Calculate ΔG^0 for this reaction, using free energies of formation.
b. If the hydrolysis of ATP to ADP is coupled with this reaction, how many moles of ATP must react to make the process spontaneous?

Solution

a. $\Delta G^0 = \Delta G_f^0 \, C_6H_{12}O_6(aq) - 2 \, \Delta G_f^0 \, C_3H_6O_3(aq)$

$\qquad = -919$ kJ $+ 2(559$ kJ$) = \boxed{+199 \text{ kJ}}$

b. Since we get 31 kJ per mole of ATP:

$$199 \text{ kJ} \times \frac{1 \text{ mol ATP}}{31 \text{ kJ}} \approx \boxed{7 \text{ mol ATP}}$$

Exercise
The fermentation of glucose in grapes to form ethanol and carbon dioxide is a spontaneous reaction:

$$C_6H_{12}O_6(aq) \rightarrow 2C_2H_5OH(aq) + 2CO_2(g); \qquad \Delta G^0 = -219.2 \text{ kJ}$$

If the energy produced were used to synthesize ATP by Reaction 20.27, how many moles of ATP would be produced? Answer: 7.

20.7
FREE ENERGY CHANGE AND THE
EQUILIBRIUM CONSTANT

As we pointed out in Section 20.4, the free energy change ΔG is the basic criterion of spontaneity. A reaction occurs spontaneously if ΔG is a negative quantity. Earlier, in our discussion of equilibrium (Chaps. 13–15), we found that the magnitude of the equilibrium constant tells us the extent to which a reaction takes place. A large value of K is associated with a reaction that goes nearly to completion and is, in that sense, "spontaneous."

As you might expect, these two quantities, free energy change and equilibrium constant, are intimately related. By a thermodynamic argument that we will not attempt to go through here, it can be shown that

$$\Delta G^0 = -RT \ln K \qquad (20.28)$$

In this equation:

ΔG^0 is the standard free energy change when all gases are at a pressure of 1 atm and all species in aqueous solution are at a concentration of 1 M.

R is the gas constant, 8.31 J/K.

T is the kelvin temperature.

K applies to any reaction where concentrations of solutes are molarities and concentrations of gases are in atm

K is the equilibrium constant for the reaction. For reactions in aqueous solution, K can be any of the equilibrium constants referred to in Chapters 14 and 15: K_w, K_a, K_b, K_f. For reactions in the gas phase (Chap. 13), the equilibrium constant that appears in Equation 20.28 is K_p, not K_c. This is a consequence of the fact that ΔG^0 is the free energy change when gases are at 1 atm pressure; to be consistent, we must use an equilibrium constant involving pressures in atmospheres.

Looking at Equation 20.28, we can distinguish three possible situations:

1. If ΔG^0 is negative, ln K must be positive. This means that K is greater than one. The reaction proceeds spontaneously in the forward direction when all species are at unit concentrations.

2. If ΔG^0 is positive, ln K must be negative. Hence K is less than one and the reverse reaction is spontaneous when all species are at unit concentrations.

3. If perchance $\Delta G^0 = 0$, then ln $K = 0$, so $K = 1$. The reaction is at equilibrium when all species are at unit concentrations.

Equation 20.28 (and Table 20.4) also allow us to give meaning to the magnitude of the free energy change. If ΔG^0 is a large negative number, K will be much greater than 1. This means that the forward reaction will go virtually to completion. Conversely, if ΔG^0 is a large positive number, K will be much less than 1, and the reverse reaction will go nearly to completion. Only if ΔG^0 is relatively small, perhaps between $+50$ and -50 kJ, will the reaction yield an equilibrium mixture containing significant amounts of both products and reactants.

TABLE 20.4   Values of K Corresponding to ΔG^0 Values at 25°C

ΔG^0 (kJ)	K
-200	1×10^{35}
-100	3×10^{17}
-50	6×10^8
-25	2×10^4
0	1
$+25$	4×10^{-5}
$+50$	2×10^{-9}
$+100$	3×10^{-18}
$+200$	1×10^{-35}

■ **EXAMPLE 20.11** —————————————

Using ΔG_f^0 tables in Appendix 1, calculate the equilibrium constant for dissolving silver bromide (AgBr) in water at 25°C. Recall that the equation for this process is

$$AgBr(s) \rightleftharpoons Ag^+(aq) + Br^-(aq)$$

The equilibrium constant for this reaction is also known as K_{sp}.

Solution

To calculate the equilibrium constant, we first calculate ΔG^0:

$$\Delta G^0 = \Delta G_f^0 \, Ag^+(aq) + \Delta G_f^0 \, Br^-(aq) - \Delta G_f^0 \, AgBr(s)$$
$$= 77.1 \text{ kJ} - 104.0 + 96.9 = 70.0 \text{ kJ}$$

Now we use Equation 20.28 to find K_{sp}:

$$\ln K_{sp} = -\frac{\Delta G^0}{RT} = \frac{-70{,}000 \text{ J}}{8.31 \text{ J/K} \times 298 \text{ K}} = -28.3$$

$$K_{sp} = \boxed{5 \times 10^{-13}}$$

This is the value listed in Chapter 15 for K_{sp} of AgBr.

Exercise

Given that ΔG_f^0 of $NH_3(g) = -16.5$ kJ/mol at 25°C, calculate K_p for the system $N_2(g) + 3H_2(g) \rightleftharpoons 2NH_3(g)$ at 25°C. Then calculate K_c using the relation $K_p = K_c(RT)^{\Delta n_g}$. Answer: 6.0×10^5, 3.6×10^8.

■ **EXAMPLE 20.12**

Calculate ΔG^0 for the dissociation of hydrogen fluoride, HF, in water, taking $K_a = 6.9 \times 10^{-4}$ at 25°C.

Solution

Again, we apply Equation 20.28:

$$\Delta G^0 = -RT \ln K_a = 8.31 \text{ J/K} \times 298 \text{ K} \times \ln(6.9 \times 10^{-4})$$

$$= 1.8 \times 10^4 \text{ J} = \boxed{+18 \text{ kJ}}$$

Note that ΔG^0 is a positive number, which means that the reaction does not occur spontaneously to give H^+ and F^- ions at a concentration of 1 mol/L. This is not surprising; HF is a weak acid.

Exercise

Calculate ΔG^0 for the formation of the complex ion $Ag(NH_3)_2^+$, $K_f = 1.7 \times 10^7$ at 25°C. Answer: -41.2 kJ.

20.8
FREE ENERGY CHANGE AND WORK

To complete our discussion of the free energy change, we refer to a property of ΔG first recognized by Gibbs. *For a process taking place at constant temperature and pressure, ΔG is a measure of the maximum amount of useful work* *that can be obtained.* To illustrate what this statement means, consider the combustion of methane. Earlier in this chapter (Example 20.7) we showed that ΔG^0 for this reaction at 25°C is -818 kJ per mole of methane burned; that is:

$$CH_4(g) + 2O_2(g) \rightarrow CO_2(g) + 2H_2O(l); \qquad \Delta G^0 \text{ at } 25°C = -818 \text{ kJ}$$

The maximum amount of work that can be obtained from this reaction at 25°C and 1 atm is 818 kJ (per mole of methane burned). We may get much less work than that. When methane simply burns in an open flame, we get no work at all. If the combustion of methane is used to drive a gas turbine, we may get 200 to 300 kJ of electrical energy. In an electrochemical cell (Chap. 21), we can do much better, perhaps producing as much as 700 to 800 kJ of electrical work. No matter what we do, however, there is no way we can get more than 818 kJ of work from the combustion of a mole of methane. The value of ΔG sets an upper limit to the amount of work that can be obtained from the reaction.

This relation applies equally well to nonspontaneous reactions. Consider the decomposition of water:

$$H_2O(l) \rightarrow H_2(g) + \tfrac{1}{2}O_2(g); \qquad \Delta G^0 \text{ at } 25°C = +237 \text{ kJ}$$

Since ΔG is positive, work must be supplied to make this reaction go. Ordinarily, this takes the form of electrical energy, absorbed during the electrolysis of water. We must take at least 237 kJ of electrical energy from a storage battery or other source to electrolyze one mole of water.

If you think about it for a while, the argument we have just gone through leads to a very simple definition of spontaneity:

> **A process taking place at constant temperature and pressure is spontaneous if it is capable of doing useful work (ΔG negative). If work must be done on the system to make the process occur (ΔG positive), it is nonspontaneous.**

This analysis makes sense. When we say that a process is "nonspontaneous," we really mean that work must be done to make it go. If, on the other hand, a process is spontaneous, we should be able to harness it in such a way as to obtain useful work.

ΔG tells us not how much work a reaction will produce, but rather how much it could produce under optimum conditions

---

■ SUMMARY PROBLEM

Consider acetic acid, CH_3COOH, the main component of vinegar. Its name comes from the Latin name for vinegar, *acetum*. When wine is exposed to air, bacterial oxidation converts alcohol to acetic acid:

$$C_2H_5OH(aq) + O_2(g) \rightarrow CH_3COOH(aq) + H_2O(l)$$

a. Calculate ΔH^0 and ΔS^0 for this process. Use the following data in combination with tables of entropies and heats of formation: ΔH_f^0 $CH_3COOH(aq) = -485.8$ kJ/mol, ΔH_f^0 $C_2H_5OH(aq) = -288.3$ kJ/mol, S^0 $CH_3COOH(aq) = 178.7$ J/mol·K, S^0 $C_2H_5OH(aq) = +148.5$ J/mol·K.

b. Is the reaction spontaneous at 25°C? At 100°C?

c. The heat of fusion of acetic acid is 11.7 kJ/mol; its freezing point is 16.6°C. Calculate ΔS^0 for the reaction: $CH_3COOH(l) \rightarrow CH_3COOH(s)$.

d. What is the standard molar entropy of $CH_3COOH(s)$, taking S^0 $CH_3COOH(l)$ = 159.8 J/mol·K?

e. Calculate ΔG^0 for the ionization of acetic acid at 25°C (K_a = 1.8×10^{-5}).

Answers

a. -483.3 kJ, -104.9 J/K   b. yes, yes   c. -40.4 J/mol·K

d. 119.4 J/mol·K   e. 27 kJ

HISTORICAL
PERSPECTIVE

J. WILLARD GIBBS
(1839–1903)

A century ago chemistry was primarily an empirical science. The outstanding chemists of that era were experimentalists who isolated and characterized new substances. The principles of chemistry were descriptive or correlative in nature, as illustrated by the atomic theory of Dalton and the Periodic Table of Mendeleev. Two theoreticians working in the latter half of the nineteenth century changed the very nature of chemistry by deriving the mathematical laws that govern the behavior of matter undergoing physical or chemical change. One of these was James Clerk Maxwell, whose contributions to kinetic theory were discussed in Chapter 4. The other was J. Willard Gibbs, Professor of Mathematical Physics at Yale from 1871 until his death in 1903.

In 1876 Gibbs published the first portion of a remarkable paper in the *Transactions of the Connecticut Academy of Sciences* entitled "On the Equilibrium of Heterogeneous Substances." When the paper was completed in 1878 (it was 323 pages long), the foundation was laid for the science of chemical thermodynamics. Here, for the first time, the concept of free energy appeared. Included as well were the basic principles of chemical equilibrium (Chap. 13), phase equilibrium (Chap. 10), and the relations governing energy changes in electrical cells (Chap. 21).

If Gibbs had never published another paper, this single contribution would have placed him among the greatest theoreticians in the history of science. Generations of experimental scientists have established their reputations by demonstrating in the laboratory the validity of the relationships that Gibbs derived at his desk. Many of these relationships were rediscovered by others; an example is the Gibbs-Helmholtz equation developed in 1882 by Helmholtz, a prestigious German physiologist and physicist who was completely unaware of Gibbs's work.

In the 25 years that remained to him, Gibbs made substantial contributions in chemistry, astronomy, and mathematics. Among these were two papers published in 1881 and 1884 that established the discipline known today as vector analysis. His last work, published in 1901, was a book entitled *Elementary Principles in Statistical Mechanics*. Here Gibbs used the statistical principles that govern the behavior of systems to develop thermodynamic equations that he had derived from an entirely different point of view at the beginning of his career. Here, too, we find the "randomness" interpretation of entropy that has received so much attention in the social as well as the natural sciences.

J. Willard Gibbs is often cited as an example of the "prophet without honor in his own country." His colleagues in New Haven and elsewhere in the United States seem not to have realized the significance of his work until late in his life. During his first 10 years as a professor at Yale he received no salary. In 1920, when he was first proposed for the Hall of Fame of Distinguished Americans at New York University, he received 9 votes out of a possible 100. Not until 1950 was he elected to that body. Even today the name of J. Willard Gibbs is generally unknown among educated Americans outside of those interested in the natural sciences.

Admittedly, Gibbs himself was largely responsible for the fact that for many years his work did not attract the attention it deserved. He made little effort to publicize it; the *Transactions of the Connecticut Academy of Sciences* was hardly the leading scientific journal of its day. Gibbs was one of those rare individuals who seem to have no inner need for recognition by contemporaries. His satisfaction came from solving a problem in his mind; having done so, he was ready to proceed to other problems. His papers are not easy to read; he seldom cites examples to illustrate his abstract reasoning. Frequently, the implications of the laws that he derives are left for the readers to grasp on their own. One of his colleagues at Yale confessed many years later that none of the members of the Connecticut Academy of Sciences understood his paper on thermodynamics; as he put it, "We knew Gibbs and took his contributions on faith."

Gibbs achieved recognition in Europe long before his work was generally appreciated in this country. Maxwell read Gibbs's paper on thermodynamics, saw its significance, and referred to it repeatedly in his own publications. Wilhelm Ostwald, who said of Gibbs, "To physical chemistry, he gave form and content for a hundred years," translated the paper into German in 1892. Seven years later, Le Châtelier translated it into French.

Maxwell was perhaps the first to recognize Gibbs' genius

QUESTIONS AND PROBLEMS

Spontaneity and Probability

1. Which of the following processes are spontaneous?
 a. Glass shattering when it is dropped.
 b. Outlining your chemistry notes.
 c. Perfume aroma from an open bottle filling the air.

2. Which of the following processes are spontaneous?
 a. Building a house of cards.
 b. Sugar dissolving in water.
 c. Raking leaves into a pile.

3. Based on your experience, predict whether the following reactions are spontaneous:
 a. $C_2H_2(g) + \frac{5}{2}O_2(g) \rightarrow 2CO_2(g) + H_2O(l)$
 b. $NaCl(s) + H_2O(l) \rightarrow NaOH(s) + HCl(g)$
 c. $C(s) + O_2(g) \rightarrow CO_2(g)$
 d. $C_2H_5OH(l) \rightarrow C_2H_5OH(s)$ at 25°C

4. Follow the directions for Question 3 for
 a. $Na^+(aq) + OH^-(aq) + \frac{1}{2}H_2(g) \rightarrow Na(s) + H_2O$
 b. $H_2O(s) \rightarrow H_2O(l)$ at $-10°C$
 c. $Mg(s) + \frac{1}{2}O_2(g) \rightarrow MgO(s)$
 d. $Zn(s) + 2H^+(aq) \rightarrow Zn^{2+}(aq) + H_2(g)$

5. State the relative probability of getting a 6 vs. a 7 in throwing a pair of dice.

6. Suppose you toss a coin five times. In how many ways can you get 3 heads and 2 tails? 4 heads and 1 tail? 5 heads and 0 tails? Which of these distributions is the most likely?

7. Suppose the bulb in Figure 20.2 contained five molecules to start with. What is (are) the most probable final state(s)?

8. Show that the probability of finding 8 molecules in the left bulb of Figure 20.2 and 0 molecules in the right bulb, with the stopcock open, is less than 0.01.

Entropy; ΔS^0

9. Predict the sign of ΔS for
 a. the freezing of water.
 b. evaporation of a seawater sample to dryness.
 c. ammonia vapor condensing.
 d. weeding a garden.

10. Predict the sign of ΔS for
 a. a candle burning.
 b. butter melting.
 c. separating air into its components.
 d. tea dissolving in water.

11. Predict the sign of ΔS^0 for each of the following reactions:
 a. $N_2(g) + 3H_2(g) \rightarrow 2NH_3(g)$
 b. $C(s) + H_2O(g) \rightarrow CO(g) + H_2(g)$
 c. $2H_2(g) + O_2(g) \rightarrow 2H_2O(l)$
 d. $S(s) + O_2(g) \rightarrow SO_2(l)$

12. Predict the sign of ΔS^0 for each of the following four reactions:
 a. $H_2(g) + O_2(g) \rightarrow H_2O_2(l)$
 b. $CO(g) + 3H_2(g) \rightarrow CH_4(g) + H_2O(g)$

c. $NH_3(g) + HCl(g) \rightarrow NH_4Cl(s)$
d. $K(s) + O_2(g) \rightarrow KO_2(s)$

13. Predict the sign of ΔS^0 for each of the following reactions:
  a. $H_2(g) + Cu^{2+}(aq) \rightarrow 2H^+(aq) + Cu(s)$
  b. $2Cl(g) \rightarrow Cl_2(g)$
  c. $CaCl_2(s) + 6H_2O(g) \rightarrow CaCl_2 \cdot 6H_2O(s)$

14. Predict the sign of ΔS^0 for each of the following reactions:
  a. $CuSO_4 \cdot 5H_2O(s) \rightarrow CuSO_4(s) + 5H_2O(g)$
  b. $6H^+(aq) + 2Al(s) \rightarrow 2Al^{3+}(aq) + 3H_2(g)$
  c. $2SO_3(g) \rightarrow 2SO_2(g) + O_2(g)$

15. Use Table 20.1 to calculate ΔS^0 for each of the following reactions:
  a. $4NH_3(g) + 7O_2(g) \rightarrow 4NO_2(g) + 6H_2O(g)$
  b. $2H_2O_2(l) + N_2H_4(l) \rightarrow N_2(g) + 4H_2O(g)$
  c. $C(s) + O_2(g) \rightarrow CO_2(g)$
  d. $CH_4(g) + 3Cl_2(g) \rightarrow CHCl_3(l) + 3HCl(g)$

16. Use Table 20.1 to calculate ΔS^0 for each of the following reactions:
  a. $CO(g) + 2H_2(g) \rightarrow CH_3OH(l)$
  b. $N_2(g) + O_2(g) \rightarrow 2NO(g)$
  c. $BaCO_3(s) \rightarrow BaO(s) + CO_2(g)$
  d. $2NaCl(s) + F_2(g) \rightarrow 2NaF(s) + Cl_2(g)$

17. Use Tables 20.1 and 20.2 to calculate ΔS^0 for each of the following reactions:
  a. $Zn(s) + 2H^+(aq) \rightarrow Zn^{2+}(aq) + H_2(g)$
  b. $H^+(aq) + OH^-(aq) \rightarrow H_2O(l)$
  c. $NH_3(g) + H_2O(l) \rightarrow NH_4^+(aq) + OH^-(aq)$

18. Use Tables 20.1 and 20.2 to calculate ΔS^0 for each of the following reactions:
  a. $2Na(s) + 2H_2O(l) \rightarrow 2Na^+(aq) + 2OH^-(aq) + H_2(g)$
  b. $Zn(s) + 2Ag^+(aq) \rightarrow 2Ag(s) + Zn^{2+}(aq)$
  c. $2NO_3^-(aq) + 8H^+(aq) + 3Cu(s) \rightarrow 3Cu^{2+}(aq) + 2NO(g) + 4H_2O(l)$

19. Use Tables 20.1 and 20.2 to calculate ΔS^0 for each of the following reactions:
  a. $2H_2S(g) + 3O_2(g) \rightarrow 2H_2O(g) + 2SO_2(g)$
  b. $Ag(s) + 2H^+(g) + NO_3^-(aq) \rightarrow Ag^+(aq) + H_2O(l) + NO_2(g)$

20. Use Tables 20.1 and 20.2 to calculate ΔS^0 for each of the following reactions:
  a. $2HNO_3(l) + 3H_2S(g) \rightarrow 4H_2O(l) + 2NO(g) + 3S(s)$
  b. $PCl_5(g) + 4H_2O(l) \rightarrow 6H^+(aq) + 5Cl^-(aq) + H_2PO_4^-(aq)$
  c. $MnO_4^-(aq) + 3Fe^{2+}(aq) + 4H^+(aq) \rightarrow 3Fe^{3+}(aq) + MnO_2(s) + 2H_2O(l)$

ΔG^0 and the Gibbs-Helmholtz Equation

21. Calculate ΔG^0 at 25°C for reactions for which
  a. $\Delta H^0 = +210$ kJ; $\Delta S^0 = +32.5$ J/K
  b. $\Delta H^0 = +638$ kJ; $\Delta S^0 = -215.2$ J/K
  c. $\Delta H^0 = +7.34$ kJ; $\Delta S^0 = +0.337$ kJ/K

22. Calculate ΔG^0 at 25°C for reactions for which

a. $\Delta H^0 = +79.6$ kJ; $\Delta S^0 = +433.1$ J/K
b. $\Delta H^0 = -837.4$ kJ; $\Delta S^0 = +173.8$ J/K
c. $\Delta H^0 = -34.9$ kJ; $\Delta S^0 = +0.039$ kJ/K

23. Calculate ΔG^0 at 400°C for each of the reactions in Problem 15. State whether the reactions are spontaneous or not.

24. Calculate ΔG^0 at 400 K for each of the reactions in Problem 16. State whether the reactions are spontaneous or not.

25. Using values of ΔG_f^0 given in Appendix 1, calculate ΔG^0 at 25°C for each of the reactions in Problem 15.

26. Follow the directions of Problem 25 for each of the reactions in Problem 16.

27. Calculate ΔG_f^0 at 25°C and 1 atm, using standard entropies and heats of formation, for
  a. calcium carbonate
  b. manganese(II) oxide
  c. phosphorus pentachloride
Which of these compounds are stable with respect to the elements?

28. Follow the directions of Problem 27 for the following compounds:
  a. tin(IV) oxide (s)
  b. silver nitrate (s)
  c. hydrogen peroxide (l)

29. A student warned his friends not to swim in a river close to an electric plant. He claimed that the ozone produced by the plant turned the river water to hydrogen peroxide, which would bleach hair. The reaction is

$$O_3(g) + H_2O(l) \rightarrow H_2O_2(aq) + O_2(g)$$

Show by calculation whether his claim is plausible, assuming the river water is at 25°C, and all species are at standard concentrations. Take ΔG_f^0 $O_3(g)$ at 25°C to be $+163.2$ kJ/mol and ΔG_f^0 $H_2O_2(aq) = -134.0$ kJ/mol.

30. It has been proposed that wood alcohol, CH_3OH, a relatively inexpensive fuel to produce, be decomposed to produce methane. Methane is a natural gas commonly used for heating homes. Is the decomposition of wood alcohol to methane and oxygen thermodynamically feasible at 25°C and 1 atm?

31. Sodium carbonate, also called "washing soda," can be made by heating sodium hydrogen carbonate:

$$2NaHCO_3(s) \rightarrow Na_2CO_3(s) + CO_2(g) + H_2O(g)$$

$$\Delta H^0 = +135.6 \text{ kJ}; \Delta G^0 = +34.6 \text{ kJ at } 25°C$$

  a. Calculate ΔS^0 for this reaction. Is the sign reasonable?
  b. Calculate ΔG^0 at 0 K; at 1000 K.

32. Oxygen can be made in the laboratory by reacting sodium peroxide and water:

$$2Na_2O_2(s) + 2H_2O(l) \rightarrow 4NaOH(s) + O_2(g)$$

$$\Delta H^0 = -109.0 \text{ kJ}; \Delta G^0 = -148.4 \text{ kJ at } 25°C$$

a. Calculate ΔS^0 for this reaction. Is this sign reasonable?
b. Calculate S^0 for Na_2O_2.
c. Calculate ΔH_f^0 for Na_2O_2.

33. Consider the reaction

$$2CuCl(s) + 2OH^-(aq) \rightarrow Cu_2O(s) + 2Cl^-(aq) + H_2O(l)$$

$$\Delta H^0 = -54.3 \text{ kJ}; \Delta S^0 = +125.1 \text{ J/K}$$

a. Calculate ΔG^0 for this reaction at 25°C.
b. Determine ΔH_f^0 for $CuCl(s)$.
c. Calculate S^0 for $CuCl(s)$ (see Tables 20.1 and 20.2).

34. Phosgene, a poisonous gas, can be formed by the reaction of chloroform, $CHCl_3(l)$, with oxygen:

$$2CHCl_3(l) + O_2(g) \rightarrow 2COCl_2(g) + 2HCl(g)$$

$$\Delta H^0 = -353.2 \text{ kJ}; \Delta G^0 = -452.4 \text{ kJ at } 25°C$$

a. Calculate ΔS^0 for the reaction. Is the sign reasonable?
b. Calculate S^0 for phosgene.
c. Calculate ΔH_f^0 for phosgene.

Temperature Dependence of Spontaneity

35. Discuss the effect of temperature change upon the spontaneity of the following reactions at 1 atm.
a. $Al_2O_3(s) + 2Fe(s) \rightarrow 2Al(s) + Fe_2O_3(s)$
   $\Delta H^0 = 851.5 \text{ kJ}; \Delta S^0 = 38.5 \text{ J/K}$
b. $N_2H_4(l) \rightarrow N_2(g) + 2H_2(g)$
   $\Delta H^0 = -50.6 \text{ kJ}; \Delta S^0 = 0.3315 \text{ kJ/K}$
c. $SO_3(g) \rightarrow SO_2(g) + \frac{1}{2}O_2(g)$
   $\Delta H^0 = 98.9 \text{ kJ}; \Delta S^0 = 0.0939 \text{ kJ/K}$

36. Discuss the effect of temperature change upon the spontaneity of the following reactions at 1 atm.
a. $2PbO(s) + 2SO_2(g) \rightarrow 2PbS(s) + 3O_2(g)$
   $\Delta H^0 = +830.8 \text{ kJ}; \Delta S^0 = 168 \text{ J/K}$
b. $2As(s) + 3F_2(g) \rightarrow 2AsF_3(l)$
   $\Delta H^0 = -1643 \text{ kJ}; \Delta S^0 = -0.316 \text{ kJ/K}$
c. $CO(g) \rightarrow C(s) + \frac{1}{2}O_2(g)$
   $\Delta H^0 = 110.5 \text{ kJ}; \Delta S^0 = -89.4 \text{ J/K}$

37. At what temperature does ΔG^0 become zero for each of the reactions in Problem 35? Explain the significance of your answers.

38. At what temperature does ΔG^0 become zero for each of the reactions in Problem 36? Explain the significance of your answers.

39. For the reaction

$$NH_4Cl(s) \rightarrow NH_3(g) + HCl(g)$$

Calculate the temperature at which $\Delta G^0 = 0$.

40. For the reaction

$$SO_2(g) + 2H_2S(g) \rightarrow 3S(s) + 2H_2O(g)$$

Calculate the temperature at which $\Delta G^0 = 0$.

41. For the decomposition of Ag_2O,

$$2Ag_2O(s) \rightarrow 4Ag(s) + O_2(g)$$

a. Using Tables 7.3 and 20.1, obtain an expression for ΔG^0 as a function of temperature. Use it to prepare a table of ΔG^0 values at 100 K intervals between 100 K and 500 K.
b. Calculate the temperature at which ΔG^0 becomes zero.

42. Earlier civilizations smelted iron from ore by heating it with charcoal from a wood fire:

$$2Fe_2O_3(s) + 3C(s) \rightarrow 4Fe(s) + 3CO_2(g)$$

a. Obtain an expression for ΔG^0 as a function of temperature. Prepare a table of ΔG^0 values at 200 K intervals between 200 and 1000 K.
b. Calculate the lowest temperature at which the smelting could be carried out.

43. Two possible ways of producing iron from iron ore are
a. $Fe_2O_3(s) + \frac{3}{2}C(s) \rightarrow 2Fe(s) + \frac{3}{2}CO_2(g)$
b. $Fe_2O_3(s) + 3H_2(g) \rightarrow 2Fe(s) + 3H_2O(g)$
Which of these reactions would proceed spontaneously at the lower temperature?

44. It is desired to produce tin from its ore cassiterite, SnO_2, at as low a temperature as possible. The ore could be

a. decomposed by heating, producing tin and oxygen.
b. heated with hydrogen gas, producing tin and water vapor.
c. heated with carbon, producing tin and carbon dioxide.

Based solely on thermodynamic principles, which method would you recommend? Show calculations.

45. Given the following data for mercury:

$$Hg(l): S^0 = 76.0 \text{ J/mol·K}$$

$$Hg(g): S^0 = 175.0 \text{ J/mol·K}; \quad \Delta H_f^0 = 61.32 \text{ kJ/mol}$$

Estimate the normal boiling point of mercury.

46. Diethyl ether, $(C_2H_5)_2O$, boils at 35.0°C and 1 atm. Its heat of vaporization is 26.0 kJ/mol. Calculate ΔS^0 for the reaction

$$(C_2H_5)_2O(l) \rightarrow (C_2H_5)_2O(g)$$

47. Given the following data for boron trichloride, BCl_3, at 25°C:

$$BCl_3(l): \Delta H_f^0 = -427.2 \text{ kJ/mol}$$

$$BCl_3(g): \Delta H_f^0 = -403.8 \text{ kJ/mol}; \quad S^0 = 290 \text{ J/mol·K}$$

Normal boiling point = 13°C

Calculate S^0 for $BCl_3(l)$.

48. Given the following data for ethanol, C_2H_5OH:

Normal bp = 78.4°C; ΔS^0_{vap} = 122.0 J/mol·K

ΔH^0_f $C_2H_5OH(l)$ = −277.7 kJ/mol

Calculate ΔH^0_f for $C_2H_5OH(g)$.

49. Red phosphorus is formed by heating white phosphorus. Calculate the temperature at which the two forms are at equilibrium given

white P: ΔH^0_f = 0.00 kJ/mol; S^0 = 41.09 J/mol·K

red P: ΔH^0_f = −17.6 kJ/mol; S^0 = 22.80 J/mol·K

50. Tin organ pipes in unheated churches develop tin "disease," in which white tin is converted to gray tin. Given

white Sn: ΔH^0_f = 0.00 kJ/mol; S^0 = 51.55 J/mol·K

gray Sn: ΔH^0_f = −2.09 kJ/mol; S^0 = 44.14 J/mol·K

calculate the temperature for the transition

$$Sn_{white}(s) \rightarrow Sn_{gray}(s)$$

Additivity of ΔG; Coupled Reactions

51. Given that, at 25°C:

$$Fe(s) + Cl_2(g) \rightarrow FeCl_2(s); \quad \Delta G^0 = -302.3 \text{ kJ}$$

$$Fe(s) + \tfrac{3}{2}Cl_2(g) \rightarrow FeCl_3(s); \quad \Delta G^0 = -334.0 \text{ kJ}$$

Calculate ΔG^0 at 25°C for the reaction:

$$2FeCl_2(s) + Cl_2(g) \rightarrow 2FeCl_3(s)$$

52. Given that, at 25°C:

$$2Cu(s) + \tfrac{1}{2}O_2(g) \rightarrow Cu_2O(s); \quad \Delta G^0 = -146.0 \text{ kJ}$$

$$Cu_2O(s) + \tfrac{1}{2}O_2(g) \rightarrow 2CuO(s); \quad \Delta G^0 = -113.4 \text{ kJ}$$

Calculate ΔG^0_f $CuO(s)$ at 25°C.

53. How many moles of ATP must be converted to ADP by the reverse of Reaction 20.27 to bring about a nonspontaneous biochemical reaction in which $\Delta G^0 = +372$ kJ?

54. Consider Reactions 20.26 and 20.27. Write a balanced equation for a coupled reaction between glucose and ADP in which $\Delta G^0 = -390$ kJ.

Free Energy and Equilibrium

55. Consider the reaction

$$H_2O(l) \rightleftharpoons H^+(aq) + OH^-(aq)$$

Using the appropriate tables, calculate
a. ΔG^0 at 25°C.   b. K_w at 25°C.

56. Consider the reaction

$$CaCO_3(s) \rightleftharpoons CaO(s) + CO_2(g)$$

Using the appropriate tables, calculate
a. ΔG^0 at 500°C.   b. K_p at 500°C.

57. Calculate ΔG^0 at 25°C for the following reactions:
a. $NH_4Cl(s) \rightleftharpoons NH_3(g) + HCl(g); \quad K_p = 1.1 \times 10^{-16}$

b. $2NO_2(g) \rightleftharpoons 2NO(g) + O_2(g); \quad K_p = 4.5 \times 10^{-13}$
c. $H_2(g) + I_2(g) \rightleftharpoons 2HI(g); \quad K_p = 6.2 \times 10^2$

58. For the reaction

$$Cl_2(g) \rightleftharpoons 2Cl(g)$$

K_p is 1.0×10^{-37} at 25°C and 4.2×10^{-5} at 1000°C. Calculate ΔG^0 at each of these temperatures.

59. Given that ΔH^0_f for HF(aq) is −320.1 kJ/mol and S^0 for HF(aq) is 88.7 J/mol·K, find K_a for HF at 25°C.

60. Using the appropriate tables in Appendix 1, calculate K_{sp} for barium sulfate at 25°C. Compare with the value given in Chapter 15, Table 15.4.

Unclassified

61. Which of the following quantities can be taken to be independent of temperature? Independent of pressure?
a. ΔH for a reaction   b. ΔS for a reaction
c. ΔG for a reaction   d. S for a substance

62. Criticize each of the following statements:
a. An exothermic reaction is spontaneous.
b. When ΔG^0 is positive, the reaction cannot occur.
c. ΔS^0 is positive for a reaction in which there is an increase in the number of moles.
d. If ΔH^0 and ΔS^0 are both positive, ΔG^0 will be positive.

63. In your own words, explain why
a. ΔS^0 is negative for a reaction in which the number of moles of gas decreases.
b. We take ΔS^0 to be independent of T, even though entropy increases with T.
c. A solid has lower entropy than its corresponding liquid.

64. Fill in the blanks:
a. ΔH^0 and ΔG^0 become equal at _____ K.
b. At equilibrium, ΔG is _____ .
c. In a spontaneous reaction, ΔG is _____ .
d. S^0 for ice is _____ than S^0 for liquid water.

65. For the equilibrium system at 1000 K:

$$CO_2(g) + H_2(g) \rightleftharpoons CO(g) + H_2O(g)$$

the reaction mixture in a 10.0-L flask contains 0.3231 mol CO, 0.1562 mol H_2O, 0.2242 mol CO_2, and 0.3417 mol H_2. Calculate ΔG^0 at 1000 K.

66. Some bacteria use light energy to convert carbon dioxide and water to glucose and oxygen:

$$6CO_2(g) + 6H_2O(l) \rightarrow C_6H_{12}O_6(aq) + 6O_2(g);$$
$$\Delta G^0 = +2870 \text{ kJ at } 25°C$$

Other bacteria, those that do not have light available to them, couple the reaction

$$H_2S(g) + \tfrac{1}{2}O_2(g) \rightarrow H_2O(l) + S(s)$$

to the glucose synthesis above. Coupling the two reactions, the overall reaction is

$$24H_2S(g) + 6CO_2(g) + 6O_2(g)$$
$$\rightarrow C_6H_{12}O_6(s) + 18H_2O(l) + 24S(s)$$

Show that the overall reaction is spontaneous at 25°C.

67. How many grams of lead chloride will dissolve in 200.0 mL of water at 100°C? (*Hint:* Use the appropriate tables to calculate the K_{sp} of $PbCl_2$ at 100°C.)

68. For acetic acid(*aq*), $\Delta H_f^0 = -485.8$ kJ/mol and $S^0 = 178.7$ J/mol·K. For the acetate ion, $\Delta H_f^0 = -486.0$ kJ/mol and $S^0 = 86.6$ J/mol·K. Calculate K_a for acetic acid at 100°C. Compare with the K_a at 25°C.

Challenge Problems

69. The normal boiling point of benzene is 80.1°C. Thermodynamic data for benzene are

$$S^0\ C_6H_6(l) = 172.80\ \text{J/K};\ S^0\ C_6H_6(g) = 268.2\ \text{J/K}$$

At what temperature is the vapor pressure of benzene 1.50×10^2 mm Hg?

70. ΔH_f^0 for iodine gas is 62.4 kJ/mol, while S^0 is 260.7 J/mol·K. Calculate the equilibrium concentrations of $I_2(g)$, $H_2(g)$, and $HI(g)$ for the equilibrium system

$$2HI(g) \rightleftharpoons H_2(g) + I_2(g)$$

at 520°C, if initially there are 0.100 mol HI, I_2, and H_2 in a 5.00-L flask.

71. The heat of fusion of ice is 333 J/g. For the process $H_2O(s) \rightarrow H_2O(l)$, determine

  a. ΔH^0 b. ΔG^0 at 0°C   c. ΔS^0
  d. ΔG^0 at -10°C   e. ΔG^0 at 10°C

72. The overall reaction that occurs when sugar is metabolized is

$$C_{12}H_{22}O_{11}(s) + 12O_2(g) \rightarrow 12CO_2(g) + 11H_2O(l)$$

For this reaction, ΔH^0 is -5650 kJ and ΔG^0 is -5790 kJ at 25°C.

  a. If 30% of the free energy change is actually converted to useful work, how many kilojoules of work could be obtained when one gram of sugar is metabolized at body temperature, 37°C?

  b. How many grams of sugar would you have to eat to get the energy to climb a mountain 1610 meters high? ($w = 9.79 \times 10^{-3}\ mh$, where w = work in kilojoules, m is body mass in kilograms, and h is height in meters.)

73. Hydrogen has been suggested as the fuel of the future. One way to store it is to convert it to a compound that can then be heated to release the hydrogen. One such compound is calcium hydride, CaH_2. This compound has a heat of formation of -186.2 kJ/mol and a standard entropy of 42.0 J/mol·K. What is the minimum temperature to which calcium hydride would have to be heated to produce hydrogen at one atmosphere pressure?

74. When a copper wire is exposed to air at room temperature, it becomes coated with a black oxide, CuO. If the wire is heated above a certain temperature, the black oxide is converted to a red oxide, Cu_2O. At a still higher temperature, the oxide coating disappears. Explain these observations in terms of the thermodynamics of the reactions

$$2CuO(s) \rightarrow Cu_2O(s) + \tfrac{1}{2}O_2(g)$$

$$Cu_2O(s) \rightarrow 2Cu(s) + \tfrac{1}{2}O_2(g)$$

and estimate the temperatures at which the changes occur.

$$Ni(s) + Cu^{2+}(aq) \rightarrow Ni^{2+}(aq) + Cu(s)$$

Nickel metal reacts spontaneously with Cu^{2+} ions to form copper metal and Ni^{2+} ions (green).

CHAPTER 21

ELECTROCHEMISTRY

If by fire
Of sooty coal th' empiric Alchymist
Can turn, or holds it possible to turn
Metals of drossiest ore to perfect
gold.

JOHN MILTON

Electrochemistry is the study of the interconversion of electrical and chemical energy. This conversion takes place in an electrochemical cell. There are two different types of cells:

1. *Voltaic cells* use a spontaneous chemical reaction to generate electrical energy. We will see how this is accomplished in Section 21.2. From the standpoint of chemistry, the most important characteristic of such a cell is its voltage, which is a measure of the spontaneity of the reaction. The voltage depends upon the nature of the cell reaction (Section 21.3) and upon the concentrations of reactants and products (Section 21.5). From the cell voltage, it is possible to calculate the free energy change and equilibrium constant for the cell reaction (Section 21.4).

Beyond their application to chemistry, voltaic cells serve as a practical source of electrical energy. We will look at some commercial voltaic cells, with emphasis on the lead storage battery, in Section 21.6.

2. *Electrolytic cells* use electrical energy to make a nonspontaneous reaction take place. We considered such cells briefly in Chapter 6, where we described the production of sodium and aluminum by electrolysis of molten ionic compounds (NaCl, Al_2O_3). In Section 21.7, we will look at the electrolysis of water solutions.

Before discussing electrochemical cells, it will be helpful to consider some of the terms and definitions used in electrochemistry (Section 21.1).

21.1
ELECTROCHEMICAL TERMS

The chemical reactions carried out in electrochemical cells are of the oxidation-reduction type (Chap. 12). At one electrode, called the **cathode**, a **reduction** half-reaction occurs. **Oxidation** occurs simultaneously at the other electrode, called the **anode**, (to remember these relations, note that both reduction and cathode start with a consonant; both oxidation and anode begin with a vowel.) Within an electrochemical cell positive ions (*cations*) move toward the cathode; negative ions (*anions*) move to the anode.

The electrical energy produced or consumed in an electrochemical cell is ordinarily measured in **joules** (J). A joule is the amount of energy absorbed or evolved when one **coulomb** (C) of electrical charge moves through a potential difference of one **volt** (V). To calculate the amount of electrical energy involved, we use the relation:

$$\text{no. of joules} = (\text{no. of coulombs}) \times (\text{no. of volts}) \tag{21.1}$$

The practical unit of electrical energy, which appears on your electric bill is the kilowatt-hour (kWh). By definition, one watt·second (W·s) is equal to one joule. So:

$$1 \text{ kWh} = 1000 \text{ W} \times 3600 \text{ s} \times \frac{1 \text{ J}}{1 \text{ W·s}} = 3.6 \times 10^6 \text{ J} \tag{21.2}$$

As described in Chapter 2, an electron has a charge of 1.6022×10^{-19} C. This means that the charge of one mole of electrons is:

$$(6.0220 \times 10^{23} \ e^-) \times (1.6022 \times 10^{-19} \ C/e^-) = 9.6485 \times 10^4 \ C$$

In other words, one mole of electrons carries a total charge of 96,485 coulombs. In that sense:

$$1 \text{ mol } e^- = 96,485 \text{ C} \tag{21.3}$$

The constant that appears in this equation, 96,485 C/mol, is called the Faraday constant, in honor of Michael Faraday, who did extensive research on electrochemical cells.

The rate at which electrical charge moves through a circuit is commonly measured in **amperes** (A). An ampere is a current such that one coulomb passes a given point in one second. Alternatively, we can say that

$$\text{no. of amperes} = (\text{no. of coulombs})/(\text{no. of seconds}) \tag{21.4}$$

■ **EXAMPLE 21.1** _____

A lead storage battery when fully charged has a voltage of 12.0 V. Suppose the battery furnishes a current of 10.0 A for 15.0 min. Calculate

a. the number of coulombs flowing through the circuit.
b. the number of moles of electrons.
c. the electrical energy in joules and kilojoules.

Solution

a. no. of coulombs = (no. of amperes) × (no. of seconds)

$$= 10.0 \ \frac{C}{s} \times 15.0 \text{ min} \times 60 \ \frac{s}{\text{min}} = \boxed{9.00 \times 10^3 \text{ C}}$$

b. no. mol electrons $= 9.00 \times 10^3 \text{ C} \times \dfrac{1 \text{ mol } e^-}{96,485 \text{ C}} = \boxed{0.0933 \text{ mol } e^-}$

c. energy $= 12.0 \text{ V} \times 9.00 \times 10^3 \text{ C} = \boxed{1.08 \times 10^5 \text{ J} = 108 \text{ kJ}}$

Margin notes:

1 J = 1 volt-coulomb
1 watt = 1 J/s

1 ampere = 1 coulomb/s

Exercise

If the current were supplied for one hour at the same voltage, what would be the energy in kilowatt hours? Answer: 0.120 kWh.

21.2 VOLTAIC CELLS

In principle, at least, any spontaneous redox reaction can serve as a source of electrical energy in a voltaic cell. The problem is to design the cell in such a way that oxidation occurs at one electrode (anode) with reduction at the other electrode (cathode). Electrons must be transferred from anode to cathode, doing electrical work as they pass through an external circuit. To understand how a voltaic cell operates, we start with some simple cells that are easily made in the general chemistry laboratory.

Figure 21.1 When a strip of zinc is placed in a solution containing Cu^{2+} ions (*left*), a spontaneous redox reaction occurs. The final result is shown at the right. Copper metal plates out and the blue color due to Cu^{2+} fades. (Marna G. Clarke)

THE Zn–Cu²⁺ CELL (Zn/Zn²⁺ ‖ Cu²⁺/Cu)

When a piece of zinc is added to a water solution containing Cu^{2+} ions, the following redox reaction takes place:

$$Zn(s) + Cu^{2+}(aq) \rightarrow Zn^{2+}(aq) + Cu(s) \qquad (21.5)$$

In this reaction, copper metal plates out on the surface of the zinc. The blue color of the aqueous Cu^{2+} ion fades as it is replaced by the colorless aqueous Zn^{2+} ion (Fig. 21.1). From Equation 21.5, we see that the reaction amounts to electron transfer from a zinc atom to a Cu^{2+} ion.

To design an electrical cell using Reaction 21.5 as a source of electrical energy, the electron transfer must occur indirectly; that is, the electrons given off by zinc atoms must be made to pass through an external electric circuit before they reduce Cu^{2+} ions to copper atoms. One way to do this is shown in Figure 21.2. Let us trace the flow of electric current through this cell.

Otherwise the cell is "shorted out"

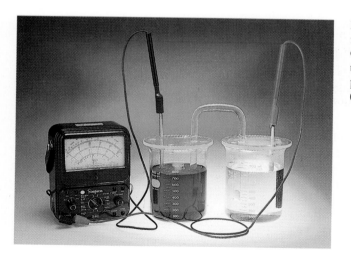

Figure 21.2 In this voltaic cell, the following spontaneous redox reaction takes place: $Zn(s) + Cu^{2+}(aq) \rightarrow Zn^{2+}(aq) + Cu(s)$. The salt bridge allows ions to pass from one solution to the other to complete the circuit. At the same time, it prevents direct contact between Zn atoms and Cu^{2+} ions. (Charles D. Winters)

1. At the zinc *anode,* electrons are produced by the *oxidation* half-reaction

$$Zn(s) \rightarrow Zn^{2+}(aq) + 2e^- \quad (21.5a)$$

This electrode, which "pumps" electrons into the external circuit, is ordinarily marked as the negative pole of the cell.

The anode has excess electrons, so it is −

2. Electrons generated by Reaction 21.5a move through the external circuit (right to left in Fig. 21.2). This part of the circuit may be a simple resistance wire, a light bulb, an electric motor, an electrolytic cell, or some other device that consumes electrical energy.

3. Electrons pass from the external circuit to the copper *cathode,* where they are consumed in the *reduction* of Cu^{2+} ions from the surrounding solution at the electrode surface:

$$Cu^{2+}(aq) + 2e^- \rightarrow Cu(s) \quad (21.5b)$$

The copper electrode, which "pulls" electrons from the external circuit, is considered to be the positive pole of the cell.

The Cu^{2+} ions soak up electrons from the electrode, making it +

4. To complete the circuit, ions must move through the aqueous solutions in the cell. As Reactions 21.5a and 21.5b proceed, a surplus of positive ions (Zn^{2+}) tends to build up around the zinc electrode. The region around the copper electrode tends to become deficient in positive ions as Cu^{2+} ions are consumed. To maintain electrical neutrality, cations must move toward the copper cathode or, alternatively, anions must move toward the zinc anode. In practice, both migrations occur.

In the cell shown in Figure 21.2, movement of ions occurs through a *salt bridge.* In its simplest form a salt bridge may consist of an inverted U-tube, plugged with glass wool at each end. The tube is filled with a solution of a salt that takes no part in the electrode reactions; potassium nitrate, KNO_3, is frequently used. As current is drawn from the cell, K^+ ions move from the salt bridge into the copper half-cell to compensate for the Cu^{2+} ions consumed at the cathode. At the same time, NO_3^- ions move into the zinc half-cell to compensate for the charge on the Zn^{2+} ions formed at the anode.

The cell shown in Figure 21.2 is often abbreviated as

$$Zn/Zn^{2+} \parallel Cu^{2+}/Cu$$

In this notation,

— the **anode** reaction (**oxidation**) is shown at the left. Zn atoms are oxidized to Zn^{2+} ions.

Anode ‖ Cathode

— the salt bridge (or other means of separating the half-cells) is indicated by the symbol ‖.

— the **cathode** reaction (**reduction**) is shown at the right. Cu^{2+} ions are reduced to Cu atoms.

Notice that the anode half-reaction comes first in the cell notation, just as the letter *a* comes before *c*.

OTHER SALT BRIDGE CELLS

Cells similar to that shown in Figure 21.2 can be set up for many different spontaneous redox reactions. Consider, for example, the reaction

$$Ni(s) + Cu^{2+}(aq) \rightarrow Ni^{2+}(aq) + Cu(s) \qquad (21.6)$$

For this reaction the voltaic cell would closely resemble that in Figure 21.2. Indeed, the Cu^{2+}/Cu half-cell and the salt bridge would be identical to the one shown. The only difference would be in the oxidation half-cell. Here, we would use a nickel electrode, surrounded by a solution containing Ni^{2+} ions, such as $NiSO_4$ or $Ni(NO_3)_2$. The half-cell reactions would be

anode: $\qquad Ni(s) \rightarrow Ni^{2+}(aq) + 2e^- \qquad$ (oxidation)

cathode: $\qquad Cu^{2+}(aq) + 2e^- \rightarrow Cu(s) \qquad$ (reduction)

The cell notation would be $Ni/Ni^{2+} \parallel Cu^{2+}/Cu$.

Another spontaneous redox reaction that can serve as a source of electrical energy is that between zinc metal and H^+ ions:

$$Zn(s) + 2H^+(aq) \rightarrow Zn^{2+}(aq) + H_2(g) \qquad (21.7)$$

A voltaic cell using this reaction is shown in Figure 21.3. The Zn/Zn^{2+} half-cell and the salt bridge are the same as those in Figure 21.2. Since no metal is involved in the cathode half-reaction, we use an *inert* electrode; that is, the cathode is made of an unreactive material that conducts an electric current. In this case, it is convenient to use a special electrode made of platinum (graphite rods and Nichrome wires are often used as inert elec-

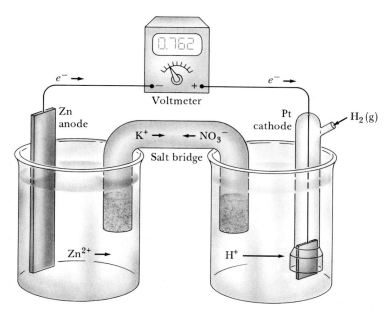

Figure 21.3 A voltaic cell in which the reaction: $Zn(s) + 2H^+(aq) \rightarrow Zn^{2+}(aq) + H_2(g)$ occurs. Hydrogen gas is bubbled over a specially prepared platinum electrode, which is surrounded by a solution containing H^+ ions.

The Pt serves as an electron carrier

trodes instead of platinum). Hydrogen gas is bubbled over the Pt electrode, which is surrounded by a solution containing H^+ ions (e.g., a solution of HCl).

The half-reactions occurring in the cell shown in Figure 21.3 are

anode: $Zn(s) \rightarrow Zn^{2+}(aq) + 2e^-$ (oxidation)

cathode: $2H^+(aq) + 2e^- \rightarrow H_2(g)$ (reduction)

The cell notation is $Zn/Zn^{2+} \parallel (Pt)H^+/H_2$. The symbol (Pt) is used to indicate the presence of an inert platinum electrode.

■ **EXAMPLE 21.2** _____

When chlorine gas is bubbled through a water solution of NaBr, a spontaneous redox reaction occurs:

$$Cl_2(g) + 2Br^-(aq) \rightarrow 2Cl^-(aq) + Br_2(l)$$

This reaction can serve as a source of electrical energy in the voltaic cell shown in Figure 21.4. In this cell,

a. what is the cathode reaction? the anode reaction?
b. which way do electrons move in the external circuit?
c. which way do anions move within the cell? cations?

Solution

a. cathode: $Cl_2(g) + 2e^- \rightarrow 2Cl^-(aq)$ (reduction)

   anode: $2Br^-(aq) \rightarrow Br_2(l) + 2e^-$ (oxidation)

Figure 21.4 In this voltaic cell, the spontaneous redox reaction is $Cl_2(g) + 2Br^-(aq) \rightarrow 2Cl^-(aq) + Br_2(l)$. Both electrodes are made of platinum.

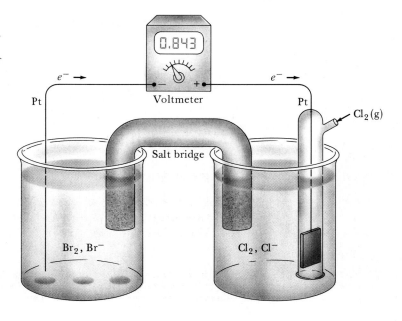

b. From anode to cathode (left to right in Fig. 21.4).

c. Anions move to the anode (right to left); cations move to the cathode (left to right).

Exercise

What is the notation for the cell shown in Figure 21.4? Answer: $(Pt)Br^-/Br_2 \parallel (Pt)Cl_2/Cl^-$.

Given the cell notation or the nature of the redox reaction occurring in a voltaic cell, you should be able to draw a diagram of the cell, similar to those in Figures 21.3 and 21.4. The key points to remember are:

1. There are always two separate half-cells connected by a wire and a salt bridge.

2. One of the compartments is the anode and the other is the cathode. All species (reactants and products) except water in the reduction half-reaction are shown in the cathode compartment. Similarly, all species in the oxidation half-reaction are shown in the anode compartment.

3. If a metal participates in a cell reaction (either as product or reactant) it is ordinarily chosen as an electrode. If no metal is involved in the half-reaction, an electrically conducting solid like platinum (Pt) or graphite (C) should be used.

4. Electron flow, from anode to cathode, is shown by an arrow in the external circuit.

5. Flow of ions through the salt bridge is shown by an arrow; cations move to the cathode, anions to the anode.

21.3
CELL VOLTAGES; STANDARD POTENTIALS

The driving force behind the spontaneous reaction taking place within a voltaic cell is determined by the cell voltage. The larger the voltage of the cell, the more spontaneous the cell reaction.

Cell voltage depends upon two factors. One of these is the concentrations of species taking part in the cell reaction; the other is the nature of the cell reaction itself. In this section, we will concentrate upon the second factor. We will see how to calculate the standard cell voltage, i.e., the voltage when reactants and products are at standard concentrations.

STANDARD VOLTAGES; DEFINITIONS

The **standard voltage** for a given cell is that measured when *all ions and molecules in solution are at a concentration of 1 M and all gases are at a pressure of 1 atm.** * To illustrate, consider the Zn–Cu cell shown in Figure 21.2. Let

---

*Strictly speaking, the standard voltage is observed when all ions, molecules, and gases are at unit *activity*. We can take activity to be the effective concentration of a species; to a good degree of approximation, a gas at 1 atm or a species in solution at 1 M has an activity of 1.

us suppose that the half-cells are set up in such a way that the concentrations of Zn^{2+} and Cu^{2+} are both 1 M. Under these conditions we find that the cell voltage at very low current flow is +1.101 V. This quantity is referred to as the standard voltage and is given the symbol E^0_{tot}.

$$Zn(s) + Cu^{2+}(aq, 1\ M) \rightarrow Zn^{2+}(aq, 1\ M) + Cu(s); E^0_{tot} = +1.101\ V$$

As another example, consider the $Zn/Zn^{2+} \parallel (Pt)H^+/H_2$ cell shown in Figure 21.3. Here we find that the standard voltage is +0.762 V. This is the voltage measured when

— the concentration of Zn^{2+} in the Zn/Zn^{2+} half-cell is 1 M.
— the concentration of H^+ in the H^+/H_2 half-cell is 1 M and the pressure of H_2 gas is 1 atm.

$$Zn(s) + 2H^+(aq, 1\ M) \rightarrow Zn^{2+}(aq, 1\ M) + H_2(g, 1\ atm); \qquad (21.7)$$
$$E^0_{tot} = +0.762\ V$$

E^0_{ox} AND E^0_{red}

You will recall that any redox reaction can be split into two half-reactions. One of these is an oxidation, and the other a reduction. For Reaction 21.7,

oxidation: $\qquad Zn(s) \rightarrow Zn^{2+}(aq, 1\ M) + 2e^- \qquad$ (21.7a)

reduction: $\qquad 2H^+(aq, 1\ M) + 2e^- \rightarrow H_2(g, 1\ atm) \qquad$ (21.7b)

It is possible to associate standard voltages with half-reactions such as these. Such a voltage is a measure of the driving force behind the half-reaction. The standard voltage for the oxidation half-reaction is given the symbol E^0_{ox}. That for the reduction half-reaction is written as E^0_{red}. The standard voltage for the cell reaction is the sum of these two quantities. Thus, we have

$$
\begin{array}{ll}
Zn(s) \rightarrow Zn^{2+}(1\ M) + 2e^- & E^0_{ox}\ (Zn \rightarrow Zn^{2+}) \\
2H^+(1\ M) + 2e^- \rightarrow H_2(1\ atm) & E^0_{red}\ (H^+ \rightarrow H_2) \\
\hline
Zn(s) + 2H^+(1\ M) \rightarrow Zn^{2+}(1\ M) + H_2(1\ atm) & E^0_{tot} = E^0_{ox} + E^0_{red} \\
& = +0.762\ V
\end{array}
$$

We would like to be able to calculate standard voltages for half-reactions such as those listed above. There is, however, a problem. As you can see from the equation just written, we have one known, E^0_{tot} (+0.762 V), and two unknowns, E^0_{ox} and E^0_{red}. No matter what we do, we cannot determine an individual half-reaction voltage experimentally. To resolve this dilemma, we make an arbitrary decision. We assign the value 0 to the standard voltage for reduction of H^+ ions to H_2 gas:

$$2H^+(aq, 1\ M) + 2e^- \rightarrow H_2(g, 1\ atm); \quad E^0_{red}\ (H^+ \rightarrow H_2) = 0.000\ V$$

Using this convention, and knowing that E^0_{tot} for the $Zn/Zn^{2+} \parallel (Pt)H^+/H_2$ cell is +0.762 V, it follows that the standard voltage for the oxidation of zinc must be +0.762 V, that is,

An equation in two unknowns cannot be solved

$$Zn(s) \rightarrow Zn^{2+}(aq, 1\ M) + 2e^-; \quad E_{ox}^0\ (Zn \rightarrow Zn^{2+}) = +0.762\ V$$

As soon as one voltage is established, others can be determined from measurements on appropriate cells. Suppose, for example, we want to determine the standard voltage for the reduction of Cu^{2+} to Cu. One way to do this is to set up a $Zn/Zn^{2+} \parallel Cu^{2+}/Cu$ cell, using 1 M solutions of Zn^{2+} and Cu^{2+}, and measure the voltage. We find the standard voltage of this cell to be $+1.101$ V.

$$Zn(s) + Cu^{2+}(aq, 1\ M) \rightarrow Zn^{2+}(aq, 1\ M) + Cu(s); \quad E_{tot}^0 = +1.101\ V$$

In this cell, zinc is being oxidized and Cu^{2+} ions reduced. Hence,

$$E_{ox}^0\ (Zn \rightarrow Zn^{2+}) + E_{red}^0\ (Cu^{2+} \rightarrow Cu) = +1.101\ V$$

Since the standard voltage for the oxidation of zinc must be the same here as in the Zn–H$^+$ cell, $+0.762$ V,

$$+0.762\ V + E_{red}^0\ (Cu^{2+} \rightarrow Cu) = +1.101\ V$$

$$E_{red}^0\ (Cu^{2+} \rightarrow Cu) = +1.101\ V - 0.762\ V = +0.339\ V$$

Standard half-cell voltages are ordinarily obtained from a list of *standard potentials* such as those in Table 21.1, p. 740. **The potentials listed give us directly the standard voltages for reduction half-reactions.** For example, since the standard potentials listed in the table for $Zn^{2+} \rightarrow Zn$ and $Cu^{2+} \rightarrow Cu$ are -0.762 V and $+0.339$ V, respectively, we see immediately that

$$Zn^{2+}(aq) + 2e^- \rightarrow Zn(s); \quad E_{red}^0 = -0.762\ V$$

$$Cu^{2+}(aq) + 2e^- \rightarrow Cu(s); \quad E_{red}^0 = +0.339\ V$$

Standard voltages for oxidation half-reactions are obtained by changing the sign of the standard potential listed in Table 21.1. Thus, we have

$$Zn(s) \rightarrow Zn^{2+}(aq) + 2e^-; \quad E_{ox}^0 = +0.762\ V$$

$$Cu(s) \rightarrow Cu^{2+}(aq) + 2e^-; \quad E_{ox}^0 = -0.339\ V$$

In general, standard voltages for forward and reverse half-reactions are equal in magnitude but opposite in sign.

    In the remainder of this section, we will consider some of the applications of standard voltages. We start by showing how they can be used to compare the strengths of different oxidizing and reducing agents. Later, we will consider their use in determining whether or not a given redox reaction will occur spontaneously in the laboratory.

$E_{ox}^0 = -E_{red}^0$ for a half-reaction

STRENGTH OF OXIDIZING AND REDUCING AGENTS

As pointed out in Chapter 12, an oxidizing agent is a species that can gain electrons; it is a reactant in a reduction half-reaction. Since Table 21.1 lists reduction half-reactions from left to right, it follows that oxidizing agents are located in the left column of the table. All of the species listed in that column (Li$^+$, . . . , F$_2$) are, in principle, oxidizing agents. A "strong" *oxidizing agent* is one that has a strong attraction for electrons and hence *can readily*

Good oxidizing agents want to get reduced

TABLE 21.1   Standard Potentials in Water Solution at 25°C

OXIDIZING AGENT	REDUCING AGENT	E^0_{red} (V)
$Li^+(aq) + e^-$	$\rightarrow Li(s)$	-3.040
$K^+(aq) + e^-$	$\rightarrow K(s)$	-2.936
$Ba^{2+}(aq) + 2e^-$	$\rightarrow Ba(s)$	-2.906
$Ca^{2+}(aq) + 2e^-$	$\rightarrow Ca(s)$	-2.869
$Na^+(aq) + e^-$	$\rightarrow Na(s)$	-2.714
$Mg^{2+}(aq) + 2e^-$	$\rightarrow Mg(s)$	-2.357
$Al^{3+}(aq) + 3e^-$	$\rightarrow Al(s)$	-1.68
$Mn^{2+}(aq) + 2e^-$	$\rightarrow Mn(s)$	-1.182
$Zn^{2+}(aq) + 2e^-$	$\rightarrow Zn(s)$	-0.762
$Cr^{3+}(aq) + 3e^-$	$\rightarrow Cr(s)$	-0.744
$Fe^{2+}(aq) + 2e^-$	$\rightarrow Fe(s)$	-0.409
$Cr^{3+}(aq) + e^-$	$\rightarrow Cr^{2+}(aq)$	-0.408
$Cd^{2+}(aq) + 2e^-$	$\rightarrow Cd(s)$	-0.402
$PbSO_4(s) + 2e^-$	$\rightarrow Pb(s) + SO_4^{2-}(aq)$	-0.356
$Tl^+(aq) + e^-$	$\rightarrow Tl(s)$	-0.336
$Co^{2+}(aq) + 2e^-$	$\rightarrow Co(s)$	-0.282
$Ni^{2+}(aq) + 2e^-$	$\rightarrow Ni(s)$	-0.236
$AgI(s) + e^-$	$\rightarrow Ag(s) + I^-(aq)$	-0.152
$Sn^{2+}(aq) + 2e^-$	$\rightarrow Sn(s)$	-0.141
$Pb^{2+}(aq) + 2e^-$	$\rightarrow Pb(s)$	-0.127
$2H^+(aq) + 2e^-$	$\rightarrow H_2(g)$	0.000
$AgBr(s) + e^-$	$\rightarrow Ag(s) + Br^-(aq)$	0.073
$S(s) + 2H^+(aq) + 2e^-$	$\rightarrow H_2S(aq)$	0.144
$Sn^{4+}(aq) + 2e^-$	$\rightarrow Sn^{2+}(aq)$	0.154
$SO_4^{2-}(aq) + 4H^+(aq) + 2e^-$	$\rightarrow SO_2(g) + 2H_2O$	0.155
$Cu^{2+}(aq) + e^-$	$\rightarrow Cu^+(aq)$	0.161
$Cu^{2+}(aq) + 2e^-$	$\rightarrow Cu(s)$	0.339
$Cu^+(aq) + e^-$	$\rightarrow Cu(s)$	0.518
$I_2(s) + 2e^-$	$\rightarrow 2I^-(aq)$	0.534
$Fe^{3+}(aq) + e^-$	$\rightarrow Fe^{2+}(aq)$	0.769
$Hg_2^{2+}(aq) + 2e^-$	$\rightarrow 2Hg(l)$	0.796
$Ag^+(aq) + e^-$	$\rightarrow Ag(s)$	0.799
$2Hg^{2+}(aq) + 2e^-$	$\rightarrow Hg_2^{2+}(aq)$	0.908
$NO_3^-(aq) + 4H^+(aq) + 3e^-$	$\rightarrow NO(g) + 2H_2O$	0.964
$AuCl_4^-(aq) + 3e^-$	$\rightarrow Au(s) + 4Cl^-(aq)$	1.001
$Br_2(l) + 2e^-$	$\rightarrow 2Br^-(aq)$	1.077
$O_2(g) + 4H^+(aq) + 4e^-$	$\rightarrow 2H_2O$	1.229
$MnO_2(s) + 4H^+(aq) + 2e^-$	$\rightarrow Mn^{2+}(aq) + 2H_2O$	1.229
$Cr_2O_7^{2-}(aq) + 14H^+(aq) + 6e^-$	$\rightarrow 2Cr^{3+}(aq) + 7H_2O$	1.33
$Cl_2(g) + 2e^-$	$\rightarrow 2Cl^-(aq)$	1.360
$ClO_3^-(aq) + 6H^+(aq) + 5e^-$	$\rightarrow \frac{1}{2}Cl_2(g) + 3H_2O$	1.458
$Au^{3+}(aq) + 3e^-$	$\rightarrow Au(s)$	1.498
$MnO_4^-(aq) + 8H^+(aq) + 5e^-$	$\rightarrow Mn^{2+}(aq) + 4H_2O$	1.512
$PbO_2(s) + SO_4^{2-}(aq) + 4H^+(aq) + 2e^-$	$\rightarrow PbSO_4(s) + 2H_2O$	1.687
$H_2O_2(aq) + 2H^+(aq) + 2e^-$	$\rightarrow 2H_2O$	1.763
$Co^{3+}(aq) + e^-$	$\rightarrow Co^{2+}(aq)$	1.953
$F_2(g) + 2e^-$	$\rightarrow 2F^-(aq)$	2.889

OXIDIZING AGENT	REDUCING AGENT	E^0_{red} (V)
Basic Solution		
$Fe(OH)_2(s) + 2e^-$	$\rightarrow Fe(s) + 2OH^-(aq)$	-0.891
$2H_2O + 2e^-$	$\rightarrow H_2(g) + 2OH^-(aq)$	-0.828
$Fe(OH)_3(s) + e^-$	$\rightarrow Fe(OH)_2(s) + OH^-(aq)$	-0.547
$S(s) + 2e^-$	$\rightarrow S^{2-}(aq)$	-0.445
$NO_3^-(aq) + 2H_2O + 3e^-$	$\rightarrow NO(g) + 4OH^-(aq)$	-0.140
$NO_3^-(aq) + H_2O + 2e^-$	$\rightarrow NO_2^-(aq) + 2OH^-(aq)$	0.004
$ClO_4^-(aq) + H_2O + 2e^-$	$\rightarrow ClO_3^-(aq) + 2OH^-(aq)$	0.398
$O_2(g) + 2H_2O + 4e^-$	$\rightarrow 4OH^-(aq)$	0.401
$ClO_3^-(aq) + 3H_2O + 6e^-$	$\rightarrow Cl^-(aq) + 6OH^-(aq)$	0.614
$ClO^-(aq) + H_2O + 2e^-$	$\rightarrow Cl^-(aq) + 2OH^-(aq)$	0.890

oxidize other species. In contrast, a "weak" oxidizing agent does not gain electrons readily. It is capable of reacting only with those species that are very easily oxidized.

The strength of an oxidizing agent is directly related to its E^0_{red} value. **The more positive E^0_{red} is, the stronger the oxidizing agent.** Looking at Table 21.1, we see that oxidizing strength increases as we move down the table. The Li^+ ion, at the top of the left column, is a very weak oxidizing agent. E^0_{red} for Li^+ is a large *negative* number, which implies that it has little tendency to gain electrons:

Strong oxidizing agents have large positive E^0_{red} values

$$Li^+(aq) + e^- \rightarrow Li(s); \quad E^0_{red} = -3.040 \text{ V}$$

In practice, cations of the Group 1 metals (Li^+, Na^+, K^+, . . .) and the Group 2 metals (Mg^{2+}, Ca^{2+}, . . .) never act as oxidizing agents in water solution. Further down the list, the H^+ ion has a greater tendency to gain electrons:

$$2H^+(aq) + 2e^- \rightarrow H_2(g); \quad E^0_{red} = 0.000 \text{ V}$$

It is capable of oxidizing metals such as Mg or Zn (to Mg^{2+} or Zn^{2+}). The strongest oxidizing agents are those at the bottom of the left column. Species such as $Cr_2O_7^{2-}$ ($E^0_{red} = +1.33$ V), Cl_2 ($E^0_{red} = +1.360$ V), and MnO_4^- ($E^0_{red} = +1.512$ V) are commonly used as oxidizing agents in redox reactions. The fluorine molecule, F_2, is in principle the strongest of all oxidizing agents:

$$F_2(g) + 2e^- \rightarrow 2F^-(aq); \quad E^0_{red} = +2.889 \text{ V}$$

In practice, fluorine is seldom used as an oxidizing agent because it is too dangerous to work with. The F_2 molecule takes electrons away from just about anything, including water, often with explosive violence.

The argument we have just gone through can be applied, in reverse, to reducing agents. These species are listed in the column to the right of Table 21.1 (Li, . . . , F^-). In principle, at least, all of them can supply electrons to another species in a redox reaction. Their strength as reducing agents is directly related to their E^0_{ox} values. **The more positive E^0_{ox} is, the stronger the reducing agent.** Looking at the values, remembering that $E^0_{ox} = -E^0_{red}$,

$$Li(s) \rightarrow Li^+(aq) + e^-; \qquad E_{ox}^0 = +3.040 \text{ V}$$

$$H_2(g) \rightarrow 2H^+(aq) + 2e^-; \qquad E_{ox}^0 = 0.000 \text{ V}$$

$$2F^-(aq) \rightarrow F_2(g) + 2e^-; \qquad E_{ox}^0 = -2.889 \text{ V}$$

we conclude that reducing strength decreases as we move down the table. The strongest reducing agents are located at the upper right (Li, . . .), and the weakest at the lower right (. . . , F^-).

These principles are illustrated in Example 21.3. Remember that in comparing oxidizing agents, you look at E_{red}^0 values; the strength of a reducing agent is directly related to its E_{ox}^0 value.

■ **EXAMPLE 21.3** _____
Consider the following species: $Cr_2O_7{}^{2-}$, $NO_3{}^-$, Br^-, Mg, and Sn^{2+}. Using Table 21.1,

a. classify each of these as an oxidizing and/or reducing agent.
b. arrange the oxidizing agents in order of increasing strength.
c. do the same with the reducing agents.

Solution

a. Remember that oxidizing agents are located in the left column; reducing agents are in the right column. Scanning the left column, we find Sn^{2+}, $NO_3{}^-$, and $Cr_2O_7{}^{2-}$; these are oxidizing agents. In the right column, we find the reducing agents Br^-, Sn^{2+}, and Mg. Note that Sn^{2+} can act either as an oxidizing agent or as a reducing agent because it is in an intermediate oxidation state, between Sn (0) and Sn^{4+}.
b. We look at E_{red}^0 values to rank oxidizing agents according to strength:

$$Sn^{2+}: \qquad E_{red}^0 (Sn^{2+} \rightarrow Sn) = -0.141 \text{ V}$$

$$NO_3{}^-: \qquad E_{red}^0 (NO_3{}^- \rightarrow NO) = +0.964 \text{ V}$$

$$Cr_2O_7{}^{2-}: \qquad E_{red}^0 (Cr_2O_7{}^{2-} \rightarrow Cr^{3+}) = +1.33 \text{ V}$$

Recalling that the criterion for strength is the voltage, we get the following ranking:

$$\boxed{Sn^{2+} < NO_3{}^- < Cr_2O_7{}^{2-}}$$

c. Looking at E_{ox}^0 values for the reducing agents:

$$Mg: \qquad E_{ox}^0 (Mg \rightarrow Mg^{2+}) = +2.357 \text{ V}$$

$$Sn^{2+}: \qquad E_{ox}^0 (Sn^{2+} \rightarrow Sn^{4+}) = -0.154 \text{ V}$$

$$Br^-: \qquad E_{ox}^0 (Br^- \rightarrow Br_2) = -1.077 \text{ V}$$

The ranking is $\boxed{Br^- < Sn^{2+} < Mg.}$

Exercise
Give the formula of a cation that is a stronger reducing agent than Sn^{2+}; a stronger oxidizing agent than Sn^{2+}. Answer: Cr^{2+}; Pb^{2+}, H^+, Sn^{4+}, ...

CALCULATION OF E_{tot}^0 FROM E_{red}^0 AND E_{ox}^0

As pointed out earlier, the standard voltage for a redox reaction is the sum of the standard voltages of the two half-reactions, reduction and oxidation; that is,

$$E_{tot}^0 = E_{red}^0 + E_{ox}^0 \qquad (21.8)$$

This simple relation makes it possible, using Table 21.1, to calculate standard voltages for more than 3000 different redox reactions. Example 21.4 illustrates how this is done.

■ **EXAMPLE 21.4** _____

Using Table 21.1, calculate the standard voltage for the reaction

$$2Fe^{3+}(aq) + 2I^-(aq) \rightarrow 2Fe^{2+}(aq) + I_2(s)$$

Solution
Splitting the reaction into two half-reactions, finding the appropriate values of E_{red}^0 and E_{ox}^0 from Table 21.1, and adding, we have

reduction:	$2Fe^{3+}(aq) + 2e^- \rightarrow 2Fe^{2+}(aq)$	$E_{red}^0 = 0.769$ V
oxidation:	$2I^-(aq) \rightarrow I_2(s) + 2e^-$	$E_{ox}^0 = -0.534$ V
		$E_{tot}^0 = +0.235$ V

Notice that E_{red}^0 for Fe^{3+} is taken directly from Table 21.1, where we find

$$Fe^{3+}(aq) + e^- \rightarrow Fe^{2+}(aq); \quad E_{red}^0 = +0.769 \text{ V}$$

We do *not* multiply the voltage by two just because two Fe^{3+} ions appear in the balanced equation for the reaction. E_{tot}^0, E_{red}^0, or E_{ox}^0 for a given reaction is independent of the number of electrons transferred.

We can't change a cell voltage by writing a number on a piece of paper

Exercise
Calculate E_{tot}^0 for the cell $Cu/Cu^{2+} \parallel Ag^+/Ag$. Answer: $+0.460$ V.

_____

You will notice that in all the examples we have worked, E_{tot}^0 is a positive quantity. This is generally true for reactions taking place in a voltaic cell. A spontaneous reaction taking place within the cell generates a positive voltage. If the calculated voltage is negative, the reaction as written cannot serve as a source of energy in a voltaic cell. It may, however, be possible to carry out the reaction in an electrolytic cell (Section 21.7).

SPONTANEITY OF REDOX REACTIONS

As we have seen, the voltage of a cell in which a spontaneous redox reaction is taking place is always positive. Such a reaction will take place directly and spontaneously in the laboratory, perhaps in a test tube, beaker, or other container. This means that:

If the calculated voltage for a redox reaction is a positive quantity, the reaction will occur spontaneously in the laboratory. If the calculated

voltage is negative, the reaction will not occur; instead, the reverse reaction will be spontaneous.

Ordinarily, we apply this principle at standard concentrations, where the calculated voltage is E_{tot}^0. Under these conditions, we can say that

— **if $E_{tot}^0 > 0$, the reaction is spontaneous.**

— **if $E_{tot}^0 < 0$, the reaction is nonspontaneous;** the reverse reaction will tend to occur.

— **if, perchance, $E_{tot}^0 = 0$, the reaction is at equilibrium at standard concentrations;** there is no tendency for reaction to occur in either direction.

To illustrate this principle, consider the problem of reducing Ni^{2+} ions to nickel ($E_{red}^0 = -0.236$ V). It should be possible to do this with zinc ($E_{ox}^0 = 0.762$ V) but not with copper ($E_{ox}^0 = -0.339$ V):

$$Zn(s) + Ni^{2+}(aq) \rightarrow Zn^{2+}(aq) + Ni(s)$$

$$E_{tot}^0 = E_{ox}^0 + E_{red}^0 = +0.762 \text{ V} - 0.236 \text{ V} = +0.526 \text{ V} \quad \text{(spontaneous)}$$

$$Cu(s) + Ni^{2+}(aq) \rightarrow Cu^{2+}(aq) + Ni(s)$$

$$E_{tot}^0 = E_{ox}^0 + E_{red}^0 = -0.339 \text{ V} - 0.236 \text{ V} = -0.575 \text{ V}$$

$$\text{(nonspontaneous)}$$

Sure enough, if we add zinc metal to a solution of nickel chloride ($NiCl_2$) at 1 M concentration, a reaction occurs. Nickel metal plates out on zinc, and the green color of the Ni^{2+} ion fades. If, on the other hand, we add copper to a solution of $NiCl_2$, there is no evidence of reaction. In contrast, the reverse reaction

$$Ni(s) + Cu^{2+}(aq) \rightarrow Ni^{2+}(aq) + Cu(s); \quad E_{tot}^0 = +0.575 \text{ V}$$

takes place spontaneously. Nickel metal reacts with a solution of $CuCl_2$ to plate out copper and form Ni^{2+} ions in solution (Fig. 21.5).

Figure 21.5 Nickel metal reacts spontaneously with Cu^{2+} ions, producing Cu metal and Ni^{2+} ions. Copper plates out on the surface of the nickel, and the blue color of Cu^{2+} is replaced by the green color of Ni^{2+}. (Marna G. Clarke)

■ **EXAMPLE 21.5** _____

Using the standard potentials listed in Table 21.1, decide whether

a. Fe(s) will be oxidized to Fe^{2+} by treatment with 1 M hydrochloric acid (HCl).

b. Cu(s) will be oxidized to Cu^{2+} by treatment with 1 M hydrochloric acid.

c. Cu(s) will be oxidized to Cu^{2+} by treatment with 1 M nitric acid (HNO_3).

Solution

a. In order for iron to be oxidized, some species must be reduced. In hydrochloric acid, the only reducible species is the H^+ ion. Looking up the appropriate potentials,

$$Fe(s) \rightarrow Fe^{2+}(aq) + 2e^- \qquad E_{ox}^0 = +0.409 \text{ V}$$

$$\underline{2H^+(aq) + 2e^- \rightarrow H_2(g) \qquad E_{red}^0 = 0.000 \text{ V}}$$

$$Fe(s) + 2H^+(aq) \rightarrow Fe^{2+}(aq) + H_2(g) \qquad E_{tot}^0 = +0.409 \text{ V}$$

Since the calculated voltage is positive, we deduce that the reaction should occur. In the laboratory, we find that it does. Steel wool dropped into hydrochloric acid dissolve, with the evolution of $H_2(g)$ (Fig. 21.6).

b. Proceeding in the same way,

$$
\begin{array}{ll}
Cu(s) \rightarrow Cu^{2+}(aq) + 2e^- & E^0_{ox} = -0.339 \text{ V} \\
\underline{2H^+(aq) + 2e^- \rightarrow H_2(g)} & \underline{E^0_{red} = 0.000 \text{ V}} \\
Cu(s) + 2H^+(aq) \rightarrow Cu^{2+}(aq) + H_2(g) & E^0_{tot} = -0.339 \text{ V}
\end{array}
$$

We find, as predicted, that no reaction occurs when copper is added to 1 M hydrochloric acid.

c. In HNO_3, there is another possible oxidizing agent, the NO_3^- ion. Combining the proper half-equations,

$$
\begin{array}{l}
3[Cu(s) \rightarrow Cu^{2+}(aq) + 2e^-] \\
\underline{2[NO_3^-(aq) + 4H^+(aq) + 3e^- \rightarrow NO(g) + 2H_2O]} \\
3Cu(s) + 2NO_3^- + 8H^+(aq) \rightarrow 3Cu^{2+}(aq) + 2NO(g) + 4H_2O
\end{array}
$$

$$E^0_{tot} = E^0_{ox} + E^0_{red} = -0.339 \text{ V} + 0.964 \text{ V} = +0.625 \text{ V}$$

As predicted, nitric acid does indeed oxidize copper metal to Cu^{2+}; the reduction product is one of the lower oxidation states of nitrogen (such as NO or NO_2) rather than $H_2(g)$. The reaction is shown in Figure 21.7.

Exercise

Which of the metals Zn, Cd, and Hg will react with HCl, based on E^0 values? Answer: Zn($E^0_{ox} = +0.762$ V) and Cd ($E^0_{ox} = +0.402$ V).

Strictly speaking, the conclusions reached in Example 21.5 apply only at standard concentrations (1 M for species in aqueous solution, 1 atm for gases). Only under these conditions is the cell voltage equal to that calculated from the standard potentials listed in Table 21.1.

Figure 21.6 Finely divided iron in the form of steel wool reacts with hydrochloric acid to evolve hydrogen: $Fe(s) + 2H^+(aq) \rightarrow Fe^{2+}(aq) + H_2(g)$. (Charles D. Winters)

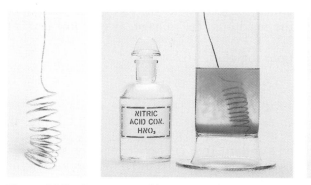 

Figure 21.7 Copper metal is comparatively inactive, but it reacts with concentrated nitric acid. The brown fumes are $NO_2(g)$, a reduction product of HNO_3. The copper is oxidized to Cu^{2+} ions, which impart their color to the solution. (Marna G. Clarke)

21.4
RELATIONS BETWEEN E_{tot}^0, ΔG^0, AND K

In Section 21.3, we pointed out that E_{tot}^0 is a criterion of spontaneity of redox reactions. If E_{tot}^0 is positive, the reaction is spontaneous at standard concentrations (1 M for species in solution, 1 atm for gases). If E_{tot}^0 is negative, the reaction is nonspontaneous; energy must be supplied to make it take place.

In earlier chapters, we discussed two other indicators of spontaneity:

1. *The standard free energy change, ΔG^0.* If ΔG^0 is negative, the reaction is spontaneous at standard concentrations. If ΔG^0 is positive, the reaction is nonspontaneous.

2. *The equilibrium constant, K.* If K is greater than 1, reaction occurs spontaneously from left to right at standard concentrations (1 M, 1 atm). If K is less than 1, the reverse reaction is spontaneous at standard concentrations.

As you might expect, E_{tot}^0 is related to both the standard free energy change, ΔG^0, and the equilibrium constant, K. In this section, we will discuss the nature of these relationships.

E_{tot}^0 AND ΔG^0

We pointed out in Chapter 20 that the free energy change, ΔG, is a measure of the maximum amount of useful work that can be obtained from a reaction. Specifically, at constant T and P:

$$\Delta G = -w_{max} \tag{21.9}$$

where w_{max} refers to the work *produced* by the reaction system. When a reaction is carried out in a voltaic cell, the "useful work" is the electrical energy that is generated. As noted in Section 21.1, electrical energy in joules is the product of coulombs times volts. This means that

$$w_{max} = nFE \tag{21.10}$$

where n is the number of moles of electrons that pass through the cell and F is the Faraday constant, which converts moles of electrons to coulombs. The symbol E stands for the maximum cell voltage, at very low current flow.

Putting these two relations together, we obtain

$$\Delta G = -nFE \tag{21.11}$$

At standard conditions, this equation becomes

$$\Delta G^0 = -nFE_{tot}^0 \tag{21.12}$$

In this equation, ΔG^0 is the standard free energy change for the redox reaction carried out in the voltaic cell and E_{tot}^0 is the standard voltage for that reaction. Moreover, as indicated above:

— n is the number of moles of electrons transferred in the chemical equation written for the redox reaction.

— F is the Faraday constant, 96,485 C/mol = 96,485 J/V·mol. If we use Equation 21.12 with F = 96,485 J/V·mol, we obtain ΔG^0 for the redox reaction in joules. Most often, we want ΔG^0 in kilojoules. Since 1 kJ = 10^3 J,

$$\Delta G^0 \text{ (kJ)} = -96.5 \, nE^0_{tot} \qquad\qquad (21.13)$$

We can find ΔG^0 experimentally this way for some reactions

From Equation 21.13, we see that ΔG^0 and E^0_{tot} have opposite signs. This is reasonable; remember that a spontaneous reaction has a *negative* free energy change and a *positive* voltage. Equation 21.13 can readily be used to calculate ΔG^0 from E^0_{tot} or vice versa (Example 21.6).

■ EXAMPLE 21.6

For the reaction $Cl_2(g) + 2Br^-(aq) \rightarrow 2Cl^-(aq) + Br_2(l)$, calculate

a. E^0_{tot}, using Table 21.1.   b. ΔG^0, using equation 21.13.

Solution

a. $E^0_{tot} = E^0_{red} \, Cl_2(g) + E^0_{ox} \, Br^-(aq) = +1.360 \text{ V} - 1.077 \text{ V} = \boxed{+0.283 \text{ V}}$

Since E^0_{tot} is positive, we conclude that the reaction is spontaneous. Indeed it is; one way to make bromine in the laboratory is to shake a solution of NaBr with a saturated solution of chlorine gas.

b. In this equation, n = 2; that is,

$$Cl_2(g) + 2e^- \rightarrow 2Cl^-(aq); \quad 2Br^-(aq) \rightarrow Br_2(l) + 2e^-$$

Hence, we have

$$\Delta G^0 \text{ (in kJ)} = -2(96.5)(0.283) = \boxed{-54.6 \text{ kJ}}$$

As we would expect, ΔG^0 is negative for this spontaneous reaction.

Exercise

Calculate E^0_{tot} for a reaction for which ΔG^0 is +10.0 kJ and n = 2. Answer: -0.0518 V.

E^0_{tot} AND K

Recall from Chapter 20 that for any reaction:

$$\Delta G^0 = -RT \ln K \qquad\qquad (21.14)$$

where K is the equilibrium constant and ΔG^0 is the standard free energy change. Combining Equations 21.12 and 21.14, we see that

$$nFE^0_{tot} = RT \ln K$$

or: $\ln K = \dfrac{nFE^0_{tot}}{RT}$

If we can find E^0_{tot} for a reaction, we can find K

Substituting R = 8.31 J/mol·K, F = 96,485 J/V·mol, and taking T = 298 K, we obtain:

$$\ln K = \frac{nE^0_{tot}}{0.0257} \quad \text{(at 25°C)} \tag{21.15}$$

Equation 21.15, which relates E^0_{tot} to K, tells us that

if E^0_{tot} is a positive quantity, ln K is positive, and K is greater than 1.

if E^0_{tot} is a negative quantity, ln K is negative, and K is less than 1.

■ EXAMPLE 21.7

For the reaction:

$3Ag(s) + NO_3^-(aq) + 4H^+(aq) \rightarrow 3Ag^+(aq) + NO(g) + 2H_2O$, calculate

a. E^0_{tot} using Table 21.1.   b. K, using Equation 21.15.

Solution

a. $E^0_{tot} = E^0_{red} \, NO_3^- + E^0_{ox} \, Ag = +0.964 \text{ V} - 0.799 \text{ V} = \boxed{+0.165 \text{ V}}$

b. To find n, it is helpful to break the equation into half-equations:

$$3Ag(s) \rightarrow 3Ag^+(aq) + 3e^-$$

$$NO_3^-(aq) + 4H^+(aq) + 3e^- \rightarrow NO(g) + 2H_2O$$

Clearly, $n = 3$. Substituting in Equation 21.15:

$$\ln K = \frac{3(+0.165)}{0.0257} = 19.3 \quad \boxed{K = 2 \times 10^8}$$

Exercise

Suppose $n = 2$ and $E^0_{tot} = 3.00$ V; calculate K. Answer: $K = 10^{101}$.

Table 21.2 is a list of values of K calculated from Equation 21.15, corresponding to various values of E^0_{tot} with $n = 2$. It turns out that if E^0_{tot} is greater than about 0.2 V, K is very large, of the order of 10^7 or greater. In such a case, the redox reaction goes virtually to completion under ordinary conditions. At the opposite extreme, note what happens when E^0_{tot} is less than about -0.2 V. Here, K is very small, 10^{-7} or less, and in essence the reaction does not occur at all. Only if E^0_{tot} falls within a quite narrow range, perhaps -0.2 to $+0.2$ V, will we get an equilibrium mixture containing appreciable amounts of both products and reactants.

Most redox reactions go either to completion or not at all

TABLE 21.2   Relation between E^0_{tot} and K ($n = 2$)

E^0_{tot}	ln K	K	E^0_{tot}	ln K	K
1.00	77.8	6×10^{33}	-0.05	-3.9	0.02
0.80	62	1×10^{27}	-0.10	-7.8	0.0004
0.60	47	2×10^{20}	-0.20	-16	2×10^{-7}
0.40	31	3×10^{13}	-0.40	-31	3×10^{-14}
0.20	16	6×10^6	-0.60	-47	5×10^{-21}
0.10	7.8	2×10^3	-0.80	-62	9×10^{-28}
0.05	3.9	50	-1.00	-77.8	2×10^{-34}
0	0	1			

21.5
EFFECT OF CONCENTRATION UPON VOLTAGE

To this point, we have dealt only with standard voltages: E_{tot}^0, E_{red}^0, E_{ox}^0. These, you will recall, apply at standard concentrations (1 M for species in solution, 1 atm for gases). As we have seen, standard voltages are useful for many purposes. However, we often carry out redox reactions where the concentrations of one or more species are far removed from 1 M. Under these conditions, we need to consider the effect of concentration upon voltage.

Qualitatively, we can readily predict the direction in which voltage will shift when concentrations are changed. Recalling that cell voltage is directly related to reaction spontaneity, we predict that

1. Voltage will *increase* if the concentration of a reactant is increased or that of a product is decreased. Either of these changes makes the reaction more spontaneous (Table 21.3).

These predictions are similar to those we get with Le Chatelier's Principle

2. Voltage will *decrease* if the concentration of a reactant is decreased or that of a product is increased. Either of these changes makes the reaction less spontaneous; they tend to favor the reverse reaction.

When a voltaic cell operates, supplying electrical energy, the concentration of reactants decreases and that of the products increases. As time passes, the voltage drops steadily. Eventually it becomes zero, and we say that the cell is "dead." At that point, the redox reaction taking place within the cell is at equilibrium and there is no driving force to produce a voltage.

That's what happens if you leave your car lights on

TABLE 21.3 Effect of Concentration upon Voltage for
$Zn(s) + Cu^{2+}(aq) \rightarrow Zn^{2+}(aq) + Cu(s)$;
$E_{tot}^0 = +1.101$ V

CONC. Cu^{2+} (M)	CONC. Zn^{2+} (M)	E
10	0.0010	1.219 V
1.0	0.010	1.160 V
0.10	0.10	1.101 V
0.010	1.0	1.042 V
0.0010	10	0.983 V

NERNST EQUATION

Quantitatively, we use a relation called the *Nernst equation* to determine the effect of concentration upon voltage. To find this relation, we work through the free energy change for the cell reaction. Recall from Section 21.4 that

$$\Delta G = -nFE$$

where ΔG and E refer to the free energy change and voltage at any given concentrations. At standard concentrations:

$$\Delta G^0 = -nFE^0_{tot}$$

Subtracting the first equation from the second:

$$nF(E - E^0_{tot}) = \Delta G^0 - \Delta G$$

$$E = E^0_{tot} + \frac{\Delta G^0 - \Delta G}{nF} \tag{21.16}$$

Consider now a general redox reaction represented by the chemical equation

$$aA + bB \rightarrow cC + dD$$

Here, the small letters (a, b, c, d) represent coefficients in the balanced equation. The letters A, B, C, D represent species involved in the reaction (e.g., Zn^{2+}, Cu^{2+}). It can be shown by a thermodynamic argument that we will not attempt to go through, that

$$\Delta G = \Delta G^0 + RT \ln \frac{(\text{conc. C})^c \times (\text{conc. D})^d}{(\text{conc. A})^a \times (\text{conc. B})^b} \tag{21.17}$$

Substituting for $\Delta G^0 - \Delta G$ in Equation 21.1 and simplifying, we obtain the Nernst equation:

$$E = E^0_{tot} - \frac{RT}{nF} \ln \frac{(\text{conc. C})^c \times (\text{conc. D})^d}{(\text{conc. A})^a \times (\text{conc. B})^b} \tag{21.18}$$

In this equation,

E is the voltage at a given concentration.

E^0_{tot} is the standard voltage.

n is the number of moles of electrons transferred for the equation as written.

R is the gas law constant, 8.31 J/mol·K.

T is the absolute temperature in K.

F is the Faraday constant, 96,485 J/V·mol.

Substituting values for the constants R and F, and taking $T = 298$ K, the Nernst equation has the following form at 25°C:

$$E = E^0_{tot} - \frac{(8.31)(298)}{n(96,485)} \ln \frac{(\text{conc. C})^c \times (\text{conc. D})^d}{(\text{conc. A})^a \times (\text{conc. B})^b}$$

$$E = E^0_{tot} - \frac{0.0257}{n} \ln \frac{(\text{conc. C})^c \times (\text{conc. D})^d}{(\text{conc. A})^a \times (\text{conc. B})^b} \tag{21.19}$$

The Nernst equation is often expressed in terms of base 10 logarithms:

$$E = E^0_{tot} - \frac{0.0591}{n} \log_{10} \frac{(\text{conc. C})^c \times (\text{conc. D})^d}{(\text{conc. A})^a \times (\text{conc. B})^b} \tag{21.20}$$

We should emphasize the following points about the Nernst equation:

Concentrations of species in aqueous solution are expressed as molarity, M.

Concentrations of gases are expressed as partial pressures (P), in atmospheres. Terms for solids (or pure liquids) do not appear.

To understand the use of the Nernst equation, consider the reaction

$$Zn(s) + Cu^{2+}(aq) \rightarrow Zn^{2+}(aq) + Cu(s); \quad E^0_{tot} = +1.101 \text{ V}$$

Here, $n = 2$ (two moles of electrons are produced by the oxidation of one mole of Zn and are consumed by the reduction of one mole of Cu^{2+}). Hence,

$$E = 1.101 \text{ V} - \frac{0.0257}{2} \ln \frac{(\text{conc. } Zn^{2+})}{(\text{conc. } Cu^{2+})}$$

You can readily check that this expression leads to the cell voltages listed in Table 21.3. Consider, for example, the calculation when conc. $Zn^{2+} = 10 \, M$, conc. $Cu^{2+} = 0.0010 \, M$:

$$E = 1.101 \text{ V} - \frac{0.0257}{2} \ln \frac{10}{0.0010} = 1.101 \text{ V} - 0.118 \text{ V} = 0.983 \text{ V}$$

■ **EXAMPLE 21.8** _____

Consider a voltaic cell in which the following reaction occurs:

$$O_2(g) + 4H^+(aq) + 4Br^-(aq) \rightarrow 2H_2O + 2Br_2(l)$$

a. Set up the Nernst equation for the cell, relating E to E^0_{tot}.
b. Determine E^0_{tot}.
c. Calculate E when $P \, O_2 = 1.0$ atm, conc. $H^+ =$ conc. $Br^- = 0.10 \, M$.

Solution

a. To find n, it is helpful to break the reaction down into two half-reactions:

$$O_2(g) + 4H^+(aq) + 4e^- \rightarrow 2H_2O$$

$$4Br^-(aq) \rightarrow 2Br_2(l) + 4e^-$$

Clearly, $n = 4$; four electrons are transferred from Br^- ions to O_2 molecules. The Nernst equation must then be

$$E = E^0_{tot} - \frac{0.0257}{4} \ln \frac{1}{(P \, O_2) \times (\text{conc. } H^+)^4 \times (\text{conc. } Br^-)^4}$$

b. $E^0_{tot} = E^0_{red} \, O_2 + E^0_{ox} \, Br^- = +1.229 \text{ V} - 1.077 \text{ V} = \boxed{+0.152 \text{ V}}$

c. $E = +0.152 \text{ V} - \dfrac{0.0257}{4} \ln \dfrac{1}{1.0(0.10)^4(0.10)^4}$

$\quad = +0.152 \text{ V} - \dfrac{0.0257}{4} \ln (1.0 \times 10^8)$

$\quad = +0.152 \text{ V} - \dfrac{0.0257}{4} (18.4) = \quad + 0.034 \text{ V}$

Exercise
Calculate E for this cell under the same conditions except that $P \, O_2 = 10$ atm. Answer: $+0.048$ V.

The Nernst equation can also be used to determine the effect of changes in concentration upon the voltage of an individual half-cell, E_{red}^0 or E_{ox}^0. Consider, for example, the half-reaction

$$MnO_4^-(aq) + 8H^+(aq) + 5e^- \rightarrow Mn^{2+}(aq) + 4H_2O; \quad E_{red}^0 = +1.512 \text{ V}$$

Here the Nernst equation takes the form

$$E_{red} = +1.512 \text{ V} - \frac{0.0257}{5} \ln \frac{(\text{conc. Mn}^{2+})}{(\text{conc. MnO}_4^-)(\text{conc. H}^+)^8}$$

where E_{red} is the observed reduction voltage corresponding to any given concentrations of Mn^{2+}, MnO_4^-, and H^+.

USE OF THE NERNST EQUATION TO DETERMINE ION CONCENTRATIONS

In chemistry, the most important use of the Nernst equation lies in the experimental determination of the concentration of ions in solution. Suppose we measure the cell voltage E and know the concentrations of all but one species in the two half-cells. It should then be possible to calculate the concentration of that species by using the Nernst equation (Example 21.9).

■ EXAMPLE 21.9 ────────────────────────────
Consider the cell shown in Figure 21.8, where the reaction is

$$Zn(s) + 2Ag^+(aq) \rightarrow Zn^{2+}(aq) + 2Ag(s)$$

a. Set up the Nernst equation for this cell reaction.
b. Determine E_{tot}^0.

Figure 21.8 The reaction in this voltaic cell is: $Zn(s) + 2Ag^+(aq) \rightarrow Zn^{2+}(aq) + 2Ag(s)$. From the measured cell voltage, it is possible to calculate the concentration of Ag^+ in the half-cell at the right, as shown in Example 21.9. If excess Cl^- ions are added to that half-cell, precipitating AgCl, the concentration of Ag^+ drops sharply, as does the voltage. The K_{sp} value of AgCl can be calculated from the measured voltage (Example 21.10).

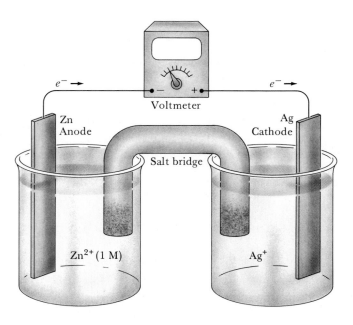

c. Suppose the concentration of Zn^{2+} in the Zn/Zn^{2+} half-cell is maintained at 1.0 M. If the measured cell voltage is 1.210 V, what must be the concentration of Ag^+ in the Ag/Ag^+ half-cell?

Solution

a. Here, $n = 2$; two electrons are transferred from a Zn atom to two Ag^+ ions:

$$E = E_{tot}^0 - \frac{0.0257}{2} \ln \frac{(\text{conc. } Zn^{2+})}{(\text{conc. } Ag^+)^2}$$

b. $E_{tot}^0 = E_{red}^0 Ag^+ + E_{ox}^0 Zn = +0.799 \text{ V} + 0.762 \text{ V} = \boxed{+1.561 \text{ V}}$

c. The equation is

$$1.210 \text{ V} = 1.561 \text{ V} - \frac{0.0257}{2} \ln \frac{1}{(\text{conc. } Ag^+)^2}$$

Since $\ln 1/x^2 = -\ln x^2 = -2 \ln x$, we can write

$$1.210 = 1.561 + \frac{0.0257}{2} (2 \ln \text{conc. } Ag^+)$$

$$= 1.561 + 0.0257 \ln (\text{conc. } Ag^+)$$

Solving,

$$\ln (\text{conc. } Ag^+) = \frac{1.210 - 1.561}{0.0257} = -13.7$$

$$\text{conc. } Ag^+ = \boxed{1.2 \times 10^{-6} \text{ } M}$$

Voltaic cells are particularly useful when you want to determine small conc.

Exercise

Suppose conc. $Zn^{2+} = 0.10$ M. If $E = +1.210$ V, what is conc. Ag^+?
Answer: 3.7×10^{-7} M.

The approach suggested by Example 21.9 is widely used to determine the concentration of ions in solution. It is very useful in cases where the concentration is low, perhaps 10^{-3} M or less. At such concentrations other approaches, such as titrations, are difficult to carry out accurately. To determine the concentration of an ion in solution using the Nernst equation, we need the following equipment:

1. A vacuum-tube voltmeter or potentiometer capable of measuring E to at least 0.01 V.

No current flows, so E has its maximum value

2. A reference half-cell of known voltage. In Example 21.9 and Figure 21.8, this is the Zn/Zn^{2+} half-cell, $E_{ox}^0 = +0.762$ V.

3. A half-cell whose voltage depends upon the concentration of the ion involved. In the simplest case (Fig. 21.8), this may consist of a metal wire or strip (Ag metal) dipping into a solution containing the cation (Ag^+) whose concentration is to be measured. More sophisticated half-cells are often used. Figure 21.9 shows a *glass* electrode used in the pH meter referred to in Chapter 14. This electrode is immersed in the solution whose $[H^+]$ or pH is to be measured. The voltage across the thin, fragile glass membrane is a linear function of the pH of the solution outside the membrane.

USE OF THE NERNST EQUATION TO DETERMINE EQUILIBRIUM CONSTANTS

As we have just seen, the Nernst equation can be used to find the concentration of an ion in solution. With a properly designed cell, this concentration can in turn be used to calculate the value of an equilibrium constant involving that ion. For example, we might determine $[Ag^+]$ in a solution in which the concentration of Cl^- in equilibrium with $AgCl(s)$ is known. From that information, we could calculate the solubility product constant of silver chloride (Example 21.10).

■ **EXAMPLE 21.10** _____

Consider again the voltaic cell referred to in Example 21.9 and shown in Figure 21.8, where conc. $Zn^{2+} = 1.0\ M$. Suppose that, to the Ag/Ag^+ half-cell, we add an excess of hydrochloric acid, precipitating AgCl and making the concentration of Cl^- ion over the precipitate $0.10\ M$. Under these conditions, the cell voltage is found to be 1.040 V. Calculate

a. the concentration of Ag^+ in the Ag/Ag^+ half-cell.
b. K_{sp} of AgCl.

Solution

a. As before, we have

$$E = E^0_{tot} - \frac{0.0257}{2} \ln \frac{1.0}{(\text{conc. } Ag^+)^2}$$

Substituting $E = 1.040$ V, $E^0_{tot} = 1.561$ V, and solving as in Example 21.9, we obtain

$$\ln (\text{conc. } Ag^+) = \frac{1.040 - 1.561}{0.0257} = -20.3$$

$$\text{conc. } Ag^+ = \boxed{1.5 \times 10^{-9}\ M}$$

b. $K_{sp} = [Ag^+] \times [Cl^-]$. The concentration of Cl^- was stated to be $0.1\ M$; we have just calculated $[Ag^+] = 1.5 \times 10^{-9}\ M$. Hence,

$$K_{sp}\ AgCl = (1.5 \times 10^{-9}) \times (1.0 \times 10^{-1}) = \boxed{1.5 \times 10^{-10}}$$

Exercise

Suppose enough HI were added to the cell to precipitate AgI and make conc. $I^- = 0.10\ M$. Would the cell voltage be greater or less than 1.040 V ($K_{sp}\ AgI = 1 \times 10^{-16}$)? Answer: Less (about 0.67 V).

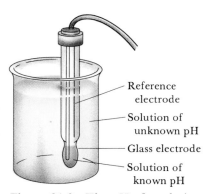

Figure 21.9 The pH of a solution can be determined with the aid of a "glass electrode." The voltage between the glass electrode and the reference electrode is directly related to pH. The leads from the electrode are connected to a pH meter of the type discussed in Chapter 14.

Reference electrode
Solution of unknown pH
Glass electrode
Solution of known pH

Most of the solubility product constants listed in Table 15.4, Chapter 15, were determined with voltaic cells using the approach illustrated in Example 21.10.

21.6
COMMERCIAL VOLTAIC CELLS

As we saw earlier in the chapter, salt bridge cells yield valuable information about the spontaneity of redox reactions. However, they have too high an internal resistance to be used commercially. When a significant amount of current is drawn from a salt bridge cell, its voltage drops sharply. Commercial voltaic cells are designed to supply large amounts of current, at steady voltages, at least for a short time.

These cells are called batteries and can be classified as primary (or nonrechargeable) cells and secondary (storage, or rechargeable) cells. We will discuss the dry cell and mercury cell (primary batteries) and the lead storage and nickel-cadmium (Nicad) batteries. We will also look briefly at fuel cells.

Figure 21.10 Section of an ordinary Zn–MnO$_2$ dry cell. This cell produces 1.5 V and will deliver a current of about half an ampere for 6 hours.

PRIMARY CELLS

The construction of the ordinary dry cell (Leclanché cell) used in flashlights is shown in Figure 21.10. The zinc wall of the cell is the anode. The graphite rod through the center of the cell is the cathode. The space between the electrodes is filled with a moist paste. This contains MnO_2, $ZnCl_2$, and NH_4Cl. When the cell operates, the half-reaction at the anode is

$$Zn(s) \rightarrow Zn^{2+}(aq) + 2e^- \tag{21.21a}$$

At the cathode, manganese dioxide is reduced to species in which Mn is in the +3 oxidation state, such as Mn_2O_3:

$$2MnO_2(s) + 2NH_4^+(aq) + 2e^- \rightarrow Mn_2O_3(s) + 2NH_3(aq) + H_2O \tag{21.21b}$$

The overall reaction occurring in this voltaic cell is

$$Zn(s) + 2MnO_2(s) + 2NH_4^+(aq) \rightarrow Zn^{2+}(aq) + Mn_2O_3(s) + 2NH_3(aq) + H_2O \tag{21.21}$$

If too large a current is drawn from a Leclanché cell, the ammonia forms a gaseous insulating layer around the carbon cathode. When this happens the voltage drops sharply, and then returns slowly to its normal value of 1.5 V. This problem can be avoided by using an "alkaline" dry cell, in which the paste between the electrodes contains KOH rather than NH_4Cl. In this case the overall cell reaction is simply

A flashlight draws about 1 amp and can run for about an hour before the battery runs down

$$Zn(s) + 2MnO_2(s) \rightarrow ZnO(s) + Mn_2O_3(s) \tag{21.22}$$

No gas is produced. The alkaline dry cell, although more expensive than the Leclanché cell, has a longer shelf-life and provides more current.

Another important primary battery is the mercury cell. It usually comes in very small sizes and is used in hearing aids, watches, cameras, and some

Figure 21.11 A mercury battery, used in hearing aids and watches.

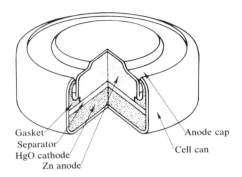

Gasket
Separator
HgO cathode
Zn anode

Anode cap

Cell can

Modern calculators use very little energy, so small batteries are practical

calculators. The anode of this cell is a zinc-mercury amalgam; the reacting species is zinc. The cathode is a plate made up of mercury(II) oxide, HgO. The electrolyte is a paste containing HgO and sodium or potassium hydroxide. Figure 21.11 shows a schematic diagram of this cell. The electrode reactions are:

$$
\begin{array}{lll}
\text{anode:} & Zn(s) + 2\,OH^-(aq) & \rightarrow Zn(OH)_2(s) + 2e^- \\
\text{cathode:} & HgO(s) + H_2O + 2e^- & \rightarrow Hg(l) + 2\,OH^-(aq) \\
\hline
& Zn(s) + HgO(s) + H_2O & \rightarrow Zn(OH)_2(s) + Hg(l) \qquad (21.23)
\end{array}
$$

Notice that the overall reaction does not involve any ions in solution, so there are no concentration changes when current is drawn. As a result, the battery maintains a constant voltage of about 1.3 V throughout its life.

The Leclanché cell, alkaline battery, and mercury battery are not rechargeable. Indeed it is dangerous to attempt to charge them, using a battery charger. A buildup of gas during the recharging process could result in an explosion.

STORAGE OR SECONDARY CELLS

A storage cell, unlike an ordinary dry cell, can be recharged repeatedly. This can be accomplished because the products of the reaction are deposited directly on the electrodes. By passing a current through a storage cell, it is possible to reverse the electrode reactions and restore the cell to its original condition.

The best-known voltaic cell of this type is the lead storage battery. The 12-V battery used in automobiles consists of six voltaic cells of the type shown in Figure 21.12. A group of lead plates, the grills of which are filled with spongy gray lead, forms the anode of the cell. The multiple cathode consists of another group of plates of similar design filled with lead(IV) oxide, PbO_2. These two sets of plates alternate through the cell. They are immersed in a water solution of sulfuric acid, H_2SO_4, which acts as the electrolyte.

A 12-V lead storage battery can deliver about 300 amperes for a minute or so, about 5 horsepower

When a lead storage battery is supplying current, the lead in the anode grids is oxidized to Pb^{2+} ions. These immediately react with SO_4^{2-} ions in the electrolyte, precipitating $PbSO_4$ (lead sulfate) on the plates. At the

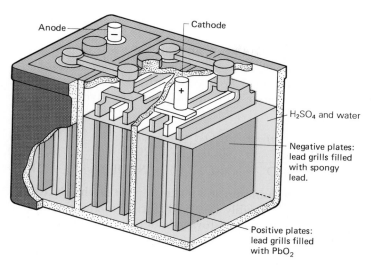

Anode · Cathode

H₂SO₄ and water

Negative plates: lead grills filled with spongy lead.

Positive plates: lead grills filled with PbO_2

Figure 21.12 One cell of a lead storage battery. Three advantages of the lead storage battery are its ability to deliver large amounts of energy for a short time, the ease of recharging, and a nearly constant voltage from full charge to discharge. A disadvantage is its high mass/energy ratio.

cathode, lead dioxide is reduced to Pb^{2+} ions, which also precipitate as $PbSO_4$:

$$Pb(s) + SO_4^{2-}(aq) \rightarrow PbSO_4(s) + 2e^- \qquad (21.24a)$$
$$PbO_2(s) + 4H^+(aq) + SO_4^{2-}(aq) + 2e^- \rightarrow PbSO_4(s) + 2H_2O \qquad (21.24b)$$
$$\overline{Pb(s) + PbO_2(s) + 4H^+(aq) + 2SO_4^{2-}(aq) \rightarrow 2PbSO_4(s) + 2H_2O} \qquad (21.24)$$

Deposits of lead sulfate slowly build up on the plates, partially covering and replacing the lead and lead dioxide. As the cell discharges, the concentration of sulfuric acid decreases. For every mole of lead reacting, two moles of H_2SO_4 (4 H^+, 2 SO_4^{2-}) are replaced by two moles of water. The state of charge of a storage battery can be checked by measuring the density of the electrolyte. When fully charged, the density is in the range of 1.25 to 1.30 g/cm³. A density below 1.20 g/cm³ indicates a low sulfuric acid concentration and hence a partially discharged cell.

A lead storage battery can be recharged and thus restored to its original condition. To do this, a direct current is passed through the cell in the reverse direction. While a storage battery is being charged, it acts as an electrolytic cell. The half-reactions 21.24a and 21.24b are reversed:

$$2PbSO_4(s) + 2H_2O \rightarrow Pb(s) + PbO_2(s) + 4H^+(aq) + 2SO_4^{2-}(aq)$$

The electrical energy required to bring about this nonspontaneous reaction in an automobile is furnished by an alternator equipped with a rectifier to convert alternating to direct current.

As you may have found from experience, lead storage batteries do not endure forever, particularly if they are allowed to stand for some time when discharged. Repeated quick-charging can cause Pb, PbO_2, and $PbSO_4$ to flake off the electrodes. This collects as a sludge at the bottom of the battery, often short-circuiting one or more cells. Discharged batteries are also susceptible to freezing, since the sulfuric acid concentration is low. If

If you don't let the battery run down, it can last over 6 years

freezing occurs, the electrodes may warp and come in contact with one another.

Another type of rechargeable voltaic cell is the "Nicad" storage battery, used for small appliances, tools, and calculators. The anode in this cell is made of cadmium metal and the cathode contains nickel(IV) oxide, NiO_2. The electrolyte is a concentrated solution of potassium hydroxide. The discharge reactions are

anode: $Cd(s) + 2OH^-(aq) \rightarrow Cd(OH)_2(s) + 2e^-$ (21.25a)

cathode: $NiO_2(s) + 2H_2O + 2e^- \rightarrow Ni(OH)_2(s) + 2OH^-(aq)$ (21.25b)

$$Cd(s) + NiO_2(s) + 2H_2O \rightarrow Cd(OH)_2(s) + Ni(OH)_2(s)$$ (21.25)

In this battery, the Ni in NiO_2 is in the rare +4 state

The insoluble hydroxides of cadmium and nickel deposit on the electrodes. Hence, the half-reactions are readily reversed during recharging. Nicad batteries are more expensive than lead storage batteries for a given amount of electrical energy delivered but also have a longer life.

FUEL CELLS

Most of our electrical energy today is produced by generators driven by steam turbines operating on the heat produced by combustion of coal, oil, or natural gas. Here, the conversion of chemical into electrical energy is indirect, in that the chemical energy is first converted to heat, which is then used to make steam. The indirect process is considerably less efficient than the direct conversion that occurs in a voltaic cell. The best power plants convert only about 30 to 40% of the heat of combustion of a fuel into electrical energy. The remainder is dissipated to the air or to bodies of water, where it contributes to thermal pollution.

Since the combustion of a fuel is an oxidation-reduction reaction, there is no reason in principle why it could not be carried out in a voltaic cell. Commercial cells designed to accomplish this are referred to as *fuel cells*. Here, the reactants, which are ordinarily gases, are fed continuously to the electrodes, where they undergo half-reactions of oxidation and reduction.

Figure 21.13 shows a fuel cell that uses the reaction between hydrogen and oxygen to produce electrical energy. It has been used in the space program as a source of energy and also as a secondary source of water. The electrodes are porous tubes made of carbon and impregnated with a catalyst such as platinum or palladium. The electrolyte is potassium hydroxide. The electrode reactions are:

anode: $H_2(g) + 2OH^-(aq) \rightarrow 2H_2O + 2e^-$

cathode: $\frac{1}{2}O_2(g) + H_2O + 2e^- \rightarrow 2OH^-(aq)$

$$H_2(g) + \frac{1}{2}O_2(g) \rightarrow H_2O$$ (21.26)

This cell produces about 200 kJ of electrical energy per mole of hydrogen consumed. If the hydrogen were burned and the heat used to generate electrical energy in a conventional power plant, only about half as much energy would be obtained; the rest would be wasted as heat. The major disadvantage of the H_2–O_2 and other fuel cells is their high cost. In part, this is due to the high operating temperatures required, 200°C–300°C.

They also require high pressure, so leaks are a problem

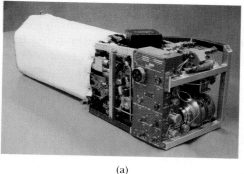

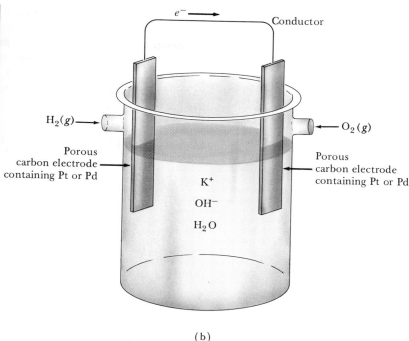

(a)

Figure 21.13 A hydrogen/oxygen fuel cell is used in spacecraft (*left*) (United Technologies); at the right is a schematic diagram of the cell.

(b)

Current research is concentrated upon developing electrode catalysts that will allow the half-reactions to proceed more rapidly at lower temperatures.

New York City now has a power plant with stacks of hydrogen-oxygen fuel cells that can respond to fluctuating demands for power. Tokyo also has a power plant operating with fuel cells.

21.7
ELECTROLYTIC CELLS

In an electrolytic cell, a nonspontaneous redox reaction is made to occur by pumping electrical energy into the system. A generalized diagram for such a cell is shown in Figure 21.14. The storage battery at the left provides a source of direct electric current. From the terminals of the battery, two wires lead to the electrolytic cell. This consists of two electrodes, A and C, dipping into a liquid containing ions M^+ and X^-.

The battery acts as an electron pump, pushing electrons into the *cathode*, C, and removing them from the *anode*, A. To maintain electrical neutrality, some process within the cell must consume electrons at C and liberate them at A. This process is an oxidation-reduction reaction; when carried out in an electrolytic cell, it is called electrolysis. At the cathode, an ion or molecule undergoes reduction by accepting electrons. At the anode, electrons are produced by the oxidation of an ion or molecule.

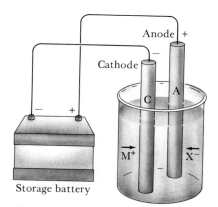

Figure 21.14 Schematic diagram of an electrolytic cell.

In Chapter 6, we considered several examples of electrolysis in the molten state at high temperatures. In this chapter, we will consider electrolytic processes taking place in water solution at room temperature. It is less expensive and more convenient to use a water solution as an electrolyte rather than a molten salt. The products of electrolysis, however, may be quite different in the two cases.

CATHODE HALF-REACTIONS

There are several possible reduction half-reactions that can occur when a water solution is electrolyzed. One of these is the reduction of a metal cation in solution:

$$K^+(aq) + e^- \rightarrow K(s); \qquad E^0_{red} = -2.936 \text{ V}$$

$$Cu^{2+}(aq) + 2e^- \rightarrow Cu(s); \qquad E^0_{red} = +0.339 \text{ V} \qquad (21.27)$$

Another possibility is the reduction of a water molecule:

$$2H_2O + 2e^- \rightarrow H_2(g) + 2OH^-(aq); \qquad E^0_{red} = -0.828 \text{ V} \qquad (21.28)$$

Ordinarily, the species reduced is the one that has the most favorable reduction potential.

Looking at the E^0_{red} values just cited, you can deduce correctly that Cu^{2+} ions will be reduced in preference to H_2O molecules. When a water solution of $CuSO_4$ is electrolyzed (Fig. 21.15), copper metal plates out at the cathode. On the other hand, water molecules are reduced in preference to K^+ ions. Hence, when a water solution of KI is electrolyzed, the cathode half-reaction is Reaction 21.28 (Fig. 21.16). Hydrogen gas forms and the solution around the cathode becomes strongly basic.

You can't make K(s) by electrolysis of an aqueous solution of KI

Figure 21.15 When a water solution of $CuSO_4$ is electrolyzed, copper plates out at the cathode. Water molecules are oxidized at the anode, forming $O_2(g)$ and H^+ ions. (Charles D. Winters)

ANODE HALF-REACTIONS

The anode half-reaction in an electrolytic cell can be the oxidation of an anion:

$$2F^-(aq) \rightarrow F_2(g) + 2e^-; \qquad E^0_{ox} = -2.889 \text{ V}$$

$$2I^-(aq) \rightarrow I_2(s) + 2e^-; \qquad E^0_{ox} = -0.534 \text{ V} \qquad (21.29)$$

Alternatively, a water molecule may be oxidized to give oxygen gas:

$$H_2O \rightarrow \tfrac{1}{2}O_2(g) + 2H^+(aq) + 2e^-; \qquad E^0_{ox} = -1.229 \text{ V} \qquad (21.30)$$

Again, we expect the half-reaction that is most spontaneous to occur. When a water solution of KI is electrolyzed, iodine forms at the anode because the I^- ion is more readily oxidized than the H_2O molecule. In contrast, electrolysis of a water solution of KF produces $O_2(g)$ and H^+ ions by Reaction 21.30; the F^- ion is so difficult to oxidize that $F_2(g)$ is never formed in water solution.

Two other anions that are essentially impossible to oxidize are the NO_3^- and SO_4^{2-} ions. In both cases, the central nonmetal atom is in its highest oxidation state (+5 for nitrogen in NO_3^-, +6 for sulfur in SO_4^{2-}). Consequently, in the electrolysis of metal nitrates or sulfates, the anode half-reaction is Reaction 21.30. Oxygen gas is evolved and the solution around the anode becomes acidic.

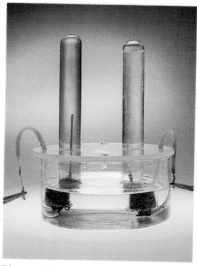

Figure 21.16 When a water solution of KI is electrolyzed, the characteristic reddish color of I_2 appears at the anode (*left*). At the cathode (*right*), H_2O molecules are reduced to $H_2(g)$ and OH^- ions, which turn phenolphthalein pink. (Charles D. Winters)

OVERALL CELL REACTION

To obtain the net cell reaction in electrolysis, we simply add the appropriate half-reactions. Thus we have

1. Electrolysis of KI in water:

$2H_2O + 2e^- \rightarrow H_2(g) + 2OH^-(aq)$	$E^0_{red} = -0.828$ V	
$2I^-(aq) \rightarrow I_2(s) + 2e^-$	$E^0_{ox} = -0.534$ V	
$2H_2O + 2I^-(aq) \rightarrow H_2(g) + 2OH^-(aq) + I_2(s);$	$E^0_{tot} = -1.362$ V	(21.31)

2. Electrolysis of $CuSO_4$ in water:

$Cu^{2+}(aq) + 2e^- \rightarrow Cu(s)$	$E^0_{red} = +0.339$ V	
$H_2O \rightarrow \tfrac{1}{2}O_2(g) + 2H^+(aq) + 2e^-$	$E^0_{ox} = -1.229$ V	
$Cu^{2+}(aq) + H_2O \rightarrow Cu(s) + \tfrac{1}{2}O_2(g) + 2H^+(aq);$	$E^0_{tot} = -0.890$ V	(21.32)

Ordinarily, the calculated voltage, E^0_{tot}, for the cell reaction is negative, as it is in these cases. This makes sense; we have to supply energy to make a nonspontaneous reaction like Reaction 21.31 or 21.32 occur in an electrolytic cell. A voltage at least equal to the calculated E^0_{tot} must be supplied, perhaps from a storage battery, to make these reactions occur. We would have to apply a voltage of at least 1.362 V to electrolyze a KI solution and 0.890 V for the electrolysis of a $CuSO_4$ solution.

■ **EXAMPLE 21.11**

Consider the electrolysis of a water solution of NaBr.

a. Write a balanced equation for the cathode half-reaction.
b. Write a balanced equation for the anode half-reaction.
c. What is the overall cell reaction and the minimum voltage that must be applied to bring it about?

Solution

a. The possible reductions are:

$$Na^+(aq) + e^- \rightarrow Na(s); \qquad\qquad E^0_{red} = -2.714 \text{ V}$$

$$2H_2O + 2e^- \rightarrow H_2(g) + 2OH^-(aq); \; E^0_{red} = -0.828 \text{ V}$$

Water is reduced rather than Na^+ ions; the second half-reaction takes place.

b. The possible oxidations are

$$2Br^-(aq) \rightarrow Br_2(l) + 2e^-; \qquad\qquad E^0_{ox} = -1.077 \text{ V}$$

$$H_2O \rightarrow \tfrac{1}{2}O_2(g) + 2H^+(aq) + 2e^-; \; E^0_{ox} = -1.229 \text{ V}$$

Br^- ions are oxidized rather than H_2O molecules; the first half-reaction occurs.

c.

$$
\begin{array}{ll}
2H_2O + 2e^- \rightarrow H_2(g) + 2OH^-(aq) & E^0_{red} = -0.828 \text{ V} \\
2Br^-(aq) \rightarrow Br_2(l) + 2e^- & E^0_{ox} = -1.077 \text{ V} \\
\hline
2H_2O + 2Br^-(aq) \rightarrow H_2(g) + Br_2(l) + 2OH^-(aq) & E^0_{tot} = -1.905 \text{ V}
\end{array}
$$

The minimum voltage required for electrolysis is 1.905 V.

Exercise

What is the overall cell reaction in the electrolysis of a water solution of NaF, assuming the solution is stirred so the products at anode and cathode come in contact with each other? Answer: $H_2O \rightarrow H_2(g) + \tfrac{1}{2}O_2(g)$.

In practice, the voltage required to carry out an electrolysis reaction is usually somewhat higher than that calculated from standard voltages. The difference between required and calculated voltages is referred to as *overvoltage*. It arises because electrode processes, particularly those involving gases, typically occur very slowly at the calculated voltage. For reactions involving H_2 or O_2, the overvoltage may be as high as 0.6 V, depending upon the nature of the electrode and other factors. This means that to electrolyze a water solution of $CuSO_4$ (Reaction 21.32), the required voltage may be in the vicinity of 1.5 V rather than the 0.89 V calculated from standard voltages.

We can sometimes reduce the overvoltage by using various additives

The phenomenon of overvoltage sometimes changes the nature of electrode reactions. The classic case involves the electrolysis of solutions containing the Cl^- ion, e.g., solutions of hydrochloric acid or sodium chloride. From standard voltages:

$$2Cl^-(aq) \rightarrow Cl_2(g) + 2e^-; \qquad E^0_{ox} = -1.360 \text{ V}$$

$$H_2O \rightarrow \tfrac{1}{2}O_2(g) + 2H^+(aq) + 2e^-; \qquad E^0_{ox} = -1.229 \text{ V}$$

we would expect oxygen gas to be formed. In practice, unless the solution is very dilute, the product at the anode is $Cl_2(g)$ rather than O_2. This happens because of the high overvoltage for oxygen generation, which makes it more difficult to oxidize water molecules.

ELECTROLYSIS OF AQUEOUS NaCl

From a commercial standpoint, the most important electrolysis carried out in water solution is that of sodium chloride, where the reactions are:

cathode: $\quad 2H_2O + 2e^- \rightarrow H_2(g) + 2OH^-(aq)$

anode: $\quad\quad\quad 2Cl^-(aq) \rightarrow Cl_2(g) + 2e^-$

$$2H_2O + 2Cl^-(aq) \rightarrow H_2(g) + Cl_2(g) + 2OH^-(aq) \quad (21.33)$$

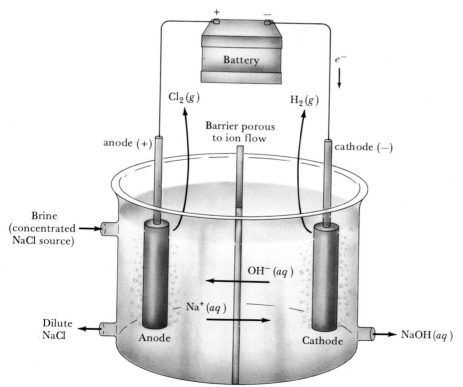

Figure 21.17 Schematic diagram for the electrolysis of aqueous NaCl (brine). Migration of ions through the membrane maintains charge balance.

The products of electrolysis have a variety of uses. Chlorine is used to purify drinking water; large quantities of it are consumed in making plastics such as polyvinyl chloride (PVC). Hydrogen, prepared in this and many other industrial processes, is used chiefly in the synthesis of ammonia (Chap. 19). Sodium hydroxide (lye), obtained on evaporation of the electrolyte, is used in processing pulp and paper, in the purification of aluminum ore, in the manufacture of glass and textiles, and for many other purposes. Over 10 million metric tons of NaOH are produced annually in the United States, almost all of it by Reaction 21.33.

The electrolysis cell used is shown schematically in Figure 21.17. The anode product, Cl_2 gas, is separated from the cathode products (H_2 gas, OH^- ions) by a porous asbestos diaphragm through which ions but not gas molecules can move.

QUANTITATIVE ASPECTS OF ELECTROLYSIS

There is a simple relationship between the amount of electricity passed through an electrolytic cell and the amounts of substances produced at the electrodes. The nature of that relationship should be clear from the half-equation for the electrode process:

HALF-EQUATION	QUANTITY OF CHARGE	AMOUNT OF PRODUCT
$Ag^+(aq) + e^- \rightarrow Ag(s)$	1 mol e^-	1 mol Ag = 107.9 g Ag
$Cu^{2+}(aq) + 2e^- \rightarrow Cu(s)$	2 mol e^-	1 mol Cu = 63.6 g Cu
$Au^{3+}(aq) + 3e^- \rightarrow Au(s)$	3 mol e^-	1 mol Au = 197.0 g Au

In other words, in these half-equations,

1 mol $e^- \simeq$ 1 mol Ag (107.9 g Ag)

2 mol $e^- \simeq$ 1 mol Cu (63.6 g Cu)

3 mol $e^- \simeq$ 1 mol Au (197.0 g Au)

Recall from Section 21.1 that

1 mol e^- = 96,485 C   1 A = 1 C/s

These relations, in combination with the half-equations for electrode processes, can be used in many practical calculations involving electrolytic cells.

■ **EXAMPLE 21.12** ————————————————————
Chromium metal can be plated from an acidic solution containing chromium(VI) oxide, CrO_3 (Figure 21.18).

a. Write a balanced half-equation for the reduction of CrO_3 to chromium.
b. How many grams of chromium will be plated by 1.00×10^4 C?
c. How long will it take to plate one gram of chromium using a current of 6.00 A?

Solution

a. The unbalanced half-equation is

$$CrO_3(aq) \rightarrow Cr(s)$$

Proceeding as indicated in Chapter 12, we first add three H_2O molecules to the right to balance oxygen. This requires that we add 6 H^+ to the left side of the equation to balance hydrogen. Finally, to balance the charge, we add $6e^-$ to the left:

$$CrO_3(aq) + 6H^+(aq) + 6e^- \rightarrow Cr(s) + 3H_2O$$

b. From the coefficients of the balanced half-equation, we see that 6 mol of electrons are required for 1 mol of chromium (52.0 g Cr):

$$6 \text{ mol } e^- \simeq 52.0 \text{ g Cr}$$

To find the number of grams of chromium plated, we need only convert the 1.00×10^4 C given to moles of electrons (1 mol $e^- = 9.6485 \times 10^4$ C) and then to grams of chromium (52.0 g Cr $\simeq$ 6 mol e^-):

$$\text{mass Cr} = 1.00 \times 10^4 \text{ C} \times \frac{1 \text{ mol } e^-}{9.6485 \times 10^4 \text{ C}} \times \frac{52.0 \text{ g Cr}}{6 \text{ mol } e^-}$$

$$= \boxed{0.898 \text{ g Cr}}$$

c. The indicated procedure here is first to convert grams of chromium to moles of electrons and then to coulombs. From the number of coulombs and the number of amperes (6.00), we can readily calculate the time in seconds:

$$\text{no. coulombs} = 1.00 \text{ g Cr} \times \frac{6 \text{ mol } e^-}{52.0 \text{ g Cr}} \times \frac{9.6485 \times 10^4 \text{ C}}{1 \text{ mol } e^-}$$

$$= 1.11 \times 10^4 \text{ C}$$

$$\text{time} = \frac{\text{coulombs}}{\text{amperes}} = \frac{1.11 \times 10^4}{6.00}$$

$$= \boxed{1.85 \times 10^3 \text{ s or about 30.8 min}}$$

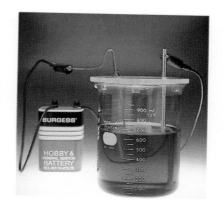

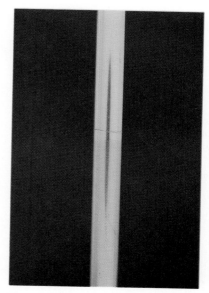

Figure 21.18 Chromium metal can be plated on copper by electrolysis of a deep red water solution of CrO_3. (Charles D. Winters)

Exercise

How long does it take to prepare 1.00 g Al by the electrolysis of Al_2O_3, using a current of 1.00 A? Answer: 1.07×10^4 s.

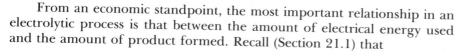

Commercial electrolysis occurs at very high currents

From an economic standpoint, the most important relationship in an electrolytic process is that between the amount of electrical energy used and the amount of product formed. Recall (Section 21.1) that

$$\text{no. of joules} = (\text{no. of volts}) \times (\text{no. of coulombs})$$

$$1 \text{ kWh} = 3.6 \times 10^6 \text{ J}$$

■ **EXAMPLE 21.13**

Consider the electroplating of chromium, referred to in Example 21.12. If the applied voltage is 4.5 V, calculate the amount of electrical energy absorbed in plating 1.00 g Cr, first in joules and then in kilowatt-hours.

Solution

In Example 21.12, we found that 1.11×10^4 C was required to plate one gram of chromium. If the voltage is 4.5 V,

$$\text{no. joules} = (4.5) \times (1.1 \times 10^4) = \boxed{5.0 \times 10^4 \text{ J}}$$

$$\text{energy in kWh} = 5.0 \times 10^4 \text{ J} \times \frac{1 \text{ kWh}}{3.6 \times 10^6 \text{ J}} = \boxed{1.4 \times 10^{-2} \text{ kWh}}$$

Exercise

Suppose the cost of electrical energy is 6.0¢ per kilowatt-hour. How many grams of chromium can be plated for one dollar? Answer: 1.2×10^3 g.

In working Examples 21.12 and 21.13, we have in effect assumed that the electrolyses were 100% efficient in converting electrical energy into chemical energy. In practice, this is almost never the case. Some electrical energy is wasted in side reactions at the electrodes and in the form of heat. This means that the actual yield of products is less than the "theoretical yield" (Example 21.12), or that the amount of electrical energy consumed is greater than the theoretical amount (Example 21.13).

■ **SUMMARY PROBLEM**

An electrolytic cell contains a solution of $Cr(NO_3)_3$. Assume that chromium metal plates out at one electrode and oxygen gas is evolved at the other electrode.

a. Write the anode half-reaction, the cathode half-reaction, and the overall reaction for the cell.

b. How long will it take to deposit 15.0 g of chromium metal, using a current of 4.50 A?

c. A current of 4.50 A is passed through the cell for 30.0 min. If we start out with 250 mL of 1.00 M $Cr(NO_3)_3$, what is the concentration of Cr^{3+} after electrolysis? The pH of the solution? Assume 100% current efficiency and no change in volume during electrolysis.

A voltaic cell consists of two half-cells, one of which contains a Pt electrode surrounded by Cr^{3+} and $Cr_2O_7^{2-}$ ions. The other half-cell contains a Pt electrode surrounded by Mn^{2+} ions and $MnO_2(s)$. Assume the cell reaction, which produces a positive voltage, involves both Cr^{3+} and Mn^{2+} ions.

d. Write the anode half-reaction, the cathode half-reaction, and the overall equation for the cell.

e. Write the cell description in abbreviated notation.

f. Calculate E^0_{tot} for the cell.

g. For the redox reaction in (d), calculate K and ΔG^0.

h. Calculate the voltage of the cell when all ionic species except H^+ have a concentration of $0.300\ M$ and the solution has a pH of 3.5.

Answers

a. cathode: $Cr^{3+}(aq) + 3e^- \rightarrow Cr(s)$
anode: $2H_2O \rightarrow 4H^+(aq) + O_2(g) + 4e^-$
overall: $4Cr^{3+}(aq) + 6H_2O \rightarrow 4Cr(s) + 12H^+(aq) + 3O_2(g)$

b. 5.15 h   **c.** $0.888\ M$ $Cr(NO_3)_3$; pH $= 0.47$

d. anode: $Mn^{2+}(aq) + 2H_2O \rightarrow MnO_2(s) + 4H^+(aq) + 2e^-$
cathode: $Cr_2O_7^{2-}(aq) + 14H^+(aq) + 6e^- \rightarrow 2Cr^{3+}(aq) + 7H_2O$
overall: $3Mn^{2+}(aq) + Cr_2O_7^{2-}(aq) + 2H^+(aq) \rightarrow 2Cr^{3+}(aq) + 3MnO_2(s) + H_2O$

e. $(Pt)Mn^{2+}/MnO_2 \parallel (Pt)Cr_2O_7^{2-}/Cr^{3+}$

f. 0.10 V   **g.** $K = 1.4 \times 10^{10}$; $\Delta G^0 = -58$ kJ   **h.** 0.02 V

MODERN PERSPECTIVE

THE CORROSION OF IRON

It is estimated that in the United States the annual cost of corrosion of ferrous metals exceeds 80 billion dollars. We see the results of corrosion all around us in junk piles and automobile graveyards. Perhaps as much as 20% of all the iron produced each year in this country goes to replace products whose usefulness has been destroyed by rust.

To understand how iron corrodes, consider what happens when a sheet of iron is exposed to a water solution containing dissolved oxygen. The iron tends to oxidize according to the half-reaction

$$Fe(s) \rightarrow Fe^{2+}(aq) + 2e^- \qquad (21.34a)$$

At the same time, oxygen molecules in the solution are reduced:

$$\tfrac{1}{2}O_2(g) + H_2O + 2e^- \rightarrow 2OH^-(aq) \qquad (21.34b)$$

Adding these two half-equations, and noting that iron(II) hydroxide is insoluble, we obtain for the primary corrosion reaction

$$Fe(s) + \tfrac{1}{2}O_2(g) + H_2O \rightarrow Fe(OH)_2(s) \qquad (21.34)$$

Reaction 21.34 appears to be the reaction involved in the first step of the corrosion of iron or steel. Ordinarily, iron(II) hydroxide is further oxidized in a second step:

This reaction occurs easily in moist air

$$2Fe(OH)_2(s) + \tfrac{1}{2}O_2(g) + H_2O \rightarrow 2Fe(OH)_3(s) \qquad (21.35)$$

The final product is the loose, flaky deposit that we call rust. It has the reddish-brown color of iron(III) hydroxide, $Fe(OH)_3$.

Experimentally, we find that the two half-reactions 21.34a and 21.34b do not occur at the same location. The rust on a nail extracted from an old building is concentrated near the head, which has been in contact with moist air (Fig. 21.19). The most serious pitting, caused by oxidation, is found along the shank of the nail, which is embedded in the wood. These observations suggest that oxidation is occurring along a surface some distance away from the point where oxygen is being reduced.

The fact that oxidation and reduction half-reactions take place at different locations suggests that corrosion occurs by an electrochemical

Figure 21.19 Corrosion of an iron nail driven into wood. Rust collects near the head of the nail, but pitting occurs along its length.

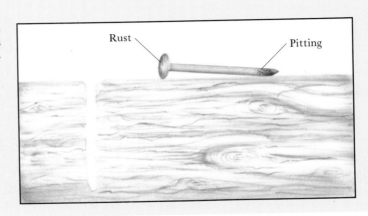

mechanism. The surface of a piece of corroding iron may be visualized as consisting of a series of tiny voltaic cells. At *anodic areas*, iron is oxidized to Fe^{2+} ions; at *cathodic areas*, elementary oxygen is reduced to OH^- ions. Electrons are transferred through the iron, which acts like the external conductor of an ordinary voltaic cell. The electrical circuit is completed by the flow of ions through the water solution or film covering the iron.

Many characteristics of corrosion are most readily explained in terms of an electrochemical mechanism. A perfectly dry metal surface is not attacked by oxygen; iron exposed to dry air does not corrode. This seems plausible if corrosion occurs through a voltaic cell, which requires a water solution through which ions can move to complete the circuit. The fact that corrosion occurs more readily in seawater than in fresh water has a similar explanation. The dissolved salts in seawater supply the ions necessary for the conduction of current.

The existence of discrete cathodic and anodic areas on a piece of corroding iron requires that adjacent surface areas differ from each other chemically. There are several ways in which one small area on a piece of iron or steel can become anodic or cathodic with respect to an adjacent area. Two of the most important are the following:

1. *The presence of impurities at scattered locations along the metal surface.* A tiny crystal of a less active metal such as copper or tin embedded in the surface of the iron acts as a cathode at which oxygen molecules are reduced. The iron atoms in the vicinity of these impurities are anodic and undergo oxidation to Fe^{2+} ions. This process took place within the Statue of Liberty, which has a copper skin supported by an iron framework. Over the course of a century, many of the iron rods were reduced to half of their original diameter by oxidation.

2. *Differences in oxygen concentration along the metal surface.* To illustrate this effect, consider what happens when a drop of water adheres to the surface of a piece of iron exposed to the air (Fig. 21.20). The metal around the edges of the drop is in contact with water containing a high concentration of dissolved oxygen. The water touching the metal beneath the center of the drop is depleted in oxygen, since it is cut off from contact with air. As a result, a small oxygen concentration cell is set up. The area around the edge of the drop, where the oxygen concentration is high, becomes

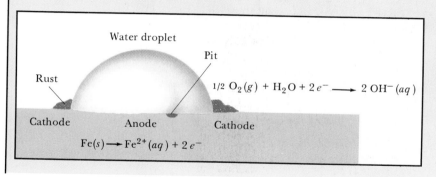

Figure 21.20 Corrosion of iron under a drop of water. The Fe^{2+} ions migrate toward the edge of the drop, where they precipitate as $Fe(OH)_2$, which later forms $Fe(OH)_3$.

cathodic; oxygen molecules are reduced there via Reaction 21.34b. Directly beneath the drop is an anodic area where the iron is oxidized. A particle of dirt on the surface of an iron object can act in much the same way as a drop of water to cut off the supply of oxygen to the area beneath it and thereby establish anodic and cathodic areas. This explains why garden tools left covered with soil are particularly susceptible to rusting.

Iron or steel objects can be protected from corrosion in several different ways.

1. *Covering the surface with a protective coating.* This may be a layer of paint that cuts off access to moisture and oxygen. Under more severe conditions, it may be desirable to cover the surface of the iron or steel with a layer of another metal. Metallic plates, applied electrically (Cr, Ni, Cu, Ag, Zn, Sn) or by immersion at high temperatures (Zn, Sn), are ordinarily more resistant to heat and chemical attack than the organic coating left when paint dries. If the plating metal is more active than iron (for example, Zn), it, rather than iron, will be oxidized if the surface is broken. However, if the plating metal is less active than iron, there is a danger that cracks

Figure 21.21 Cathodic protection of the Alaska oil pipeline. (Alyeska Oil Pipeline Service Company)

in its surface may enhance the corrosion of the iron or steel. This problem can arise with "tin cans," which are made by applying a layer of tin over a steel base. If the food in the can contains citric acid, some of the tin plate may dissolve, exposing the steel beneath.* When the can is opened, exposing the interior to the air, rust forms spontaneously on the iron surrounding the breaks in the tin surface. A thin coating of lacquer is ordinarily applied over the tin to prevent corrosive effects of this type.

2. *Bringing the object into electrical contact with a more active metal* such as magnesium or zinc. Under these conditions, the iron becomes cathodic and hence, is protected against rusting; the more active metal serves as a sacrificial anode in a large-scale corrosion cell. This method of combating corrosion, known as **cathodic protection**, is particularly useful for steel objects such as cables or pipelines that are buried under soil or water (Fig. 21.21).

Sometimes an external voltage is applied to make the pipe negative

---

*Tin forms an extremely stable complex with citrate ion and, hence, is attacked more readily by citric acid than by many stronger inorganic acids.

QUESTIONS AND PROBLEMS

Voltaic Cells

1. Write a balanced chemical equation for the overall cell reaction represented as
a. (Pt) H_2/H^+ ‖ (Pt) Fe^{3+}/Fe^{2+}
b. Cd/Cd^{2+} ‖ Ni^{2+}/Ni
c. (Pt) Cl^-/Cl_2 ‖ (Pt) MnO_4^-/Mn^{2+}

2. Write a balanced chemical equation for the overall cell reaction represented as
a. Zn/Zn^{2+} ‖ Cr^{3+}/Cr
b. Sn/Sn^{2+} ‖ (Pt) O_2/H_2O
c. Al/Al^{3+} ‖ (Pt) I_2/I^-

3. Draw a diagram for a salt bridge cell for each of the following reactions. Label the anode and cathode, and indicate the direction of current flow throughout the circuit.
a. $Sn(s) + 2Ag^+(aq) \rightarrow Sn^{2+}(aq) + 2Ag(s)$
b. $H_2(g) + Hg_2Cl_2(s) \rightarrow 2H^+(aq) + 2Cl^-(aq) + 2Hg(l)$
c. $Pb(s) + PbO_2(s) + 4H^+(aq) +$
   $2SO_4^{2-}(aq) \rightarrow 2PbSO_4(s) + 2H_2O$

4. Follow the directions for Question 3 for
a. $Zn(s) + Cd^{2+}(aq) \rightarrow Zn^{2+}(aq) + Cd(s)$
b. $2AuCl_4^-(aq) + 3Cu(s) \rightarrow 2Au(s) + 8Cl^-(aq) + 3Cu^{2+}(aq)$
c. $Fe(s) + Cu(OH)_2(s) \rightarrow Cu(s) + Fe(OH)_2(s)$

5. Consider a salt bridge cell in which the anode is a manganese rod immersed in an aqueous manganese(II) sulfate solution. The cathode is a chromium strip immersed in an aqueous chromium(III) sulfate solution. Sketch a diagram of the cell, indicating the flow of current throughout. Write the half-equations for the electrode reactions, the overall equation, and the abbreviated notation for the cell.

6. Follow the directions for Question 5 for a salt bridge cell in which the anode is a platinum rod immersed in an aqueous solution of sodium iodide containing solid iodine crystals. The cathode is another platinum rod immersed in a solution of sodium bromide with bromine liquid.

Strength of Oxidizing and Reducing Species

7. Which species in each pair is the better oxidizing agent?
a. Br_2 or H_2O_2 b. SO_4^{2-} or MnO_4^-
c. $AgBr(s)$ or Ag^+
d. O_2 in acidic solution or O_2 in basic solution

8. Which species in each pair is the better reducing agent?
a. Au or Ag   b. Cl^- or I^-
c. Mn^{2+} or Fe^{2+}
d. H_2 in acidic solution or H_2 in basic solution

9. Using Table 21.1, arrange the following reducing agents in order of increasing strength.

Br^-, Zn, Co, $PbSO_4$, H_2S

10. Using Table 21.1, arrange the following oxidizing agents in order of increasing strength:

Al^{3+}, AgBr, F_2, ClO_3^- (acidic), Ni^{2+}

11. Consider the following species:

Cu^+, Zn, Ni^{2+}, Fe^{2+}, H^+

Classify each species as an oxidizing agent, reducing agent, or both. Arrange the oxidizing agents in order of increasing strength. Do the same for the reducing agents.

12. Follow the directions for Question 11 for the following species:

Cr^{3+}, Hg(l), H_2, Sn^{2+}, Br_2

13. Use Table 21.1 to select
   a. a reducing agent that will convert Pb^{2+} to Pb but not Tl^+ to Tl.
   b. an oxidizing agent to convert Fe to Fe^{2+} but not Co to Co^{2+}.
   c. a reducing agent that converts Au^{3+} to Au but not $AuCl_4^-$ to Au.

14. Use Table 21.1 to select
   a. an oxidizing agent that converts I^- to I_2 but not Cl^- to Cl_2.
   b. a reducing agent capable of converting Co^{2+} to Co but not Zn^{2+} to Zn.
   c. an oxidizing agent capable of converting Cl^- to Cl_2 but not F^- to F_2.

Calculation of E_{tot}^0

15. Calculate E_{tot}^0 for the following voltaic cells:
   a. $Pb(s) + 2Ag^+(aq) \rightarrow Pb^{2+}(aq) + 2Ag(s)$
   b. $O_2(g) + 4Fe^{2+}(aq) + 4H^+(aq) \rightarrow 2H_2O + 4Fe^{3+}(aq)$
   c. a $Cd–Cd^{2+}$ half-cell and a $Zn–Zn^{2+}$ half-cell

16. Calculate E_{tot}^0 for the following voltaic cells:
   a. $MnO_2(s) + 4H^+(aq) + 2I^-(aq) \rightarrow Mn^{2+}(aq) + 2H_2O + I_2(s)$
   b. $H_2(g) + 2OH^-(aq) + S(s) \rightarrow 2H_2O + S^{2-}(aq)$
   c. an $Ag–Ag^+$ half-cell and an $Au–AuCl_4^-$ half-cell.

17. Using Table 21.1, calculate E_{tot}^0 for
   a. the reaction of chromium(II) ions with tin(IV) ions to produce chromium(III) ions and tin(II) ions.
   b. the reaction between manganese(II) ions and hydrogen peroxide to produce solid manganese dioxide (MnO_2) and hydrogen ions.
   c. the reaction between iron and oxygen in base.

18. Using Table 21.1, calculate E_{tot}^0 for the reaction between
   a. iron and water to produce iron(II) hydroxide and hydrogen gas.
   b. iron and iron(III) ions to give iron(II) ions.
   c. iron(II) hydroxide with oxygen in basic solution.

19. Calculate E_{tot}^0 for the cells
   a. $Al/Al^{3+} \parallel (Pt) NO_3^-/NO$
   b. (Pt) $Cr^{2+}/Cr^{3+} \parallel$ (Pt) $O_2(acidic)/H_2O$
   c. $Cu/Cu^{2+} \parallel I_2/I^-$

20. Calculate E_{tot}^0 for the cells
   a. $Pb/Pb^{2+} \parallel$ (Pt) Sn^{4+}/Sn^{2+}
   b. $Cu/Cu^{2+}(s) \parallel$ (Pt) NO_3^-/NO
   c. $Mn/Mn^{2+} \parallel Ni^{2+}/Ni$

21. Suppose E_{red}^0 for $H^+ \rightarrow H_2$ were taken to be 0.500 V instead of 0.000 V. What would be
   a. E_{ox}^0 for H_2?
   b. E_{red}^0 for $Cu^{2+} \rightarrow Cu$?
   c. E_{tot}^0 for the cell $Zn/Zn^{2+} \parallel Cu^{2+}/Cu$?

22. Suppose E_{red}^0 of $Ag^+ \rightarrow Ag$, instead of that of $H^+ \rightarrow H_2$, were set equal to zero. What would be
   a. E_{red}^0 for $H^+ \rightarrow H_2$?
   b. E_{ox}^0 of $Ca \rightarrow Ca^{2+}$?
   c. E_{tot}^0 for the cell $Zn/Zn^{2+} \parallel Cu^{2+}/Cu$?

Spontaneity and E_{tot}^0

23. Which of the following reactions are spontaneous at standard conditions?
   a. $AuCl_4^-(aq) + 3Fe^{2+}(aq) \rightarrow Au(s) + 4Cl^-(aq) + 3Fe^{3+}(aq)$
   b. $Zn(s) + 2Fe^{3+}(aq) \rightarrow Zn^{2+}(aq) + 2Fe^{2+}(aq)$
   c. $Cu(s) + 2H^+(aq) \rightarrow Cu^{2+}(aq) + H_2(g)$

24. Which of the following reactions are spontaneous at standard conditions?
   a. $O_2(g) + 4H^+(aq) + 4Cl^-(aq) \rightarrow 2H_2O + 2Cl_2(g)$
   b. $2NO_3^-(aq) + 8H^+(aq) + 6Cl^-(aq) \rightarrow 2NO(g) + 4H_2O + 3Cl_2(g)$
   c. $I_2(s) + 2Br^-(aq) \rightarrow Br_2(l) + 2I^-(aq)$

25. Using Table 21.1, calculate E_{tot}^0 and decide whether the following ions will oxidize chloride ions to chlorine gas in acidic solution at standard concentrations.
   a. permanganate ion   b. dichromate ion
   c. nitrate ion

26. Using Table 21.1, calculate E_{tot}^0 and decide whether the following ions will reduce chlorate ion (ClO_3^-) to chlorine gas in acidic solution at standard concentrations.
   a. iodide ion   b. fluoride ion   c. copper(I) ion

27. Under standard conditions, write the equation for the reaction that occurs, if any, when each of the following experiments are performed:
   a. Crystals of iodine are added to a solution of sodium bromide.
   b. Bromine is added to a solution of sodium chloride.
   c. A chromium wire is placed into a solution of nickel(II) chloride.

28. Under standard conditions, write the equation for the reaction that occurs, if any, when each of the following experiments are performed:
   a. Sulfur is added to a solution of iron(II) nitrate.
   b. Manganese dioxide in acidic solution is added to liquid mercury.
   c. Aluminum metal is added to a solution of potassium ions.

29. Which of the following metals will react with 1 M HCl?
   a. Au   b. Fe   c. Cu   d. Pb

30. Which of the following will be oxidized by 1 M HNO_3?
   a. I^- b. Mg   c. Ag   d. F^-

31. Using Table 21.1, decide what reaction, if any, will occur when the following are mixed (standard concentrations):
a. Co^{2+}, Co, Fe^{2+}
b. Fe^{2+}, Fe^{3+}, Ag^+
c. ClO_3^-, H^+, $Hg(l)$

32. Predict what reaction, if any, will occur when liquid bromine is added to an acidic aqueous solution of each of the following (standard concentrations):
a. $Ca(NO_3)_2$ b. FeI_2 c. AgF

E_{tot}^0, ΔG^0, and K

33. Calculate ΔG^0 and K at 25°C for cell reactions in which $n = 3$ and E_{tot}^0 is:
a. 0.00 V   b. +0.500 V   c. −0.500 V

34. Calculate ΔG^0 and K at 25°C for cell reactions in which $n = 6$ and E_{tot}^0 is
a. −1.00 V   b. 1.00 V   c. −2.00 V

35. For a certain cell, ΔG^0 is −38.7 kJ. Calculate E_{tot}^0 if n is
a. 1   b. 2   c. 4

36. For a certain cell, $n = 3$. Calculate E_{tot}^0 if ΔG^0 is
a. −42.3 kJ   b. +42.3 kJ

37. Calculate E_{tot}^0, ΔG^0, and K at 25°C for the reaction

$$NO_3^-(aq) + H_2O + 2Fe(OH)_2(s) \rightarrow 2Fe(OH)_3(s) + NO_2^-(aq)$$

38. Calculate E_{tot}^0, ΔG^0, and K at 25°C for the reaction

$$4ClO_3^-(aq) \rightarrow Cl^-(aq) + 3ClO_4^-(aq)$$

in basic solution.

39. Calculate ΔG^0 for each of the reactions referred to in Question 15 (assume smallest whole-number coefficients in part c).

40. Calculate ΔG^0 for each of the reactions referred to in Question 16 (assume smallest whole-number coefficients in part c).

41. Calculate K for each of the reactions referred to in Question 17 (assume smallest whole-number coefficients).

42. Calculate K for each of the reactions referred to in Question 18 (assume smallest whole-number coefficients).

Nernst Equation

43. Consider a voltaic cell in which the following reaction takes place:

$$2Cr(s) + 6H^+(aq) \rightarrow 2Cr^{3+}(aq) + 3H_2(g)$$

a. Write the Nernst equation for this cell.
b. Calculate E_{tot}^0.
c. Calculate E under the following conditions: $H^+ = 1.00 \times 10^{-3}$ M, $Cr^{3+} = 1.00$ M, $PH_2 = 1.00$ atm.

44. Consider a voltaic cell in which the following reaction occurs:

$$Br_2(l) + 2I^-(aq) \rightarrow 2Br^-(aq) + I_2(s)$$

a. Write the Nernst equation for this cell.
b. Calculate E_{tot}^0.
c. Calculate E when the concentration of iodide ion is twice the concentration of bromide ion.

45. Consider the half-reaction for the reduction of dichromate ion to chromium(III) ion. Using the Nernst equation, calculate E_{red} when all the ionic species, except H^+, are at 0.100 M and $H^+ = 1.00 \times 10^{-4}$ M.

46. Consider the half-reaction for the oxidation of lead sulfate to lead dioxide, sulfate ion, and hydrogen ion. Using the Nernst equation, calculate E_{ox} when the sulfate ion concentration is 0.0100 M and the hydrogen ion concentration is 2.00×10^{-3} M.

47. Calculate the voltages of cells under the following conditions:
a. Zn/Zn^{2+} (0.50 M) ‖ Cd^{2+} (0.020 M)/Cd
b. Cu/Cu^{2+} (0.0010 M) ‖ H^+ (0.010 M)/H_2 (1 atm)
Are the cell reactions spontaneous?

48. Calculate the voltages of cells under the following conditions:
a. Fe/Fe^{2+} (0.010 M) ‖ Cu^{2+} (0.10 M)/Cu
b. Sn^{2+} (0.10 M)/Sn^{4+} (0.010 M) ‖ Co^{2+} (0.10 M)/Co
Are the cell reactions spontaneous?

49. Consider the reaction:

$$O_2(g) + 4H^+(aq) + 4Cl^-(aq) \rightarrow 2Cl_2(g) + 2H_2O$$

a. Calculate E_{tot}^0.
b. At what pressure of Cl_2 is the voltage zero, if all other species are at standard concentrations?

50. Consider the reaction:

$$2Fe^{3+}(aq) + 2I^-(aq) \rightarrow 2Fe^{2+}(aq) + I_2(s)$$

a. Calculate E_{tot}^0.
b. At what concentration of I^- is the voltage zero, if all other species are at standard concentrations?

51. Consider the cell:

$$Zn(s) + 2H^+(aq) \rightarrow Zn^{2+}(aq) + H_2(g)$$

At what pH is the voltage 0.500 V, taking conc. $Zn^{2+} = 1.0$ M, P $H_2 = 1.0$ atm?

52. Consider the cell:

$$Sn^{2+}(aq) + \tfrac{1}{2}O_2(g) + 2H^+(aq) \rightarrow Sn^{4+}(aq) + H_2O$$

At what pH is the voltage 0.820 V, assuming all species other than H^+ are at standard concentrations?

53. Consider a cell in which the reaction is:

$$Pb(s) + 2H^+(aq) \rightarrow Pb^{2+}(aq) + H_2(g)$$

a. Calculate E_{tot}^0.
b. Chloride ions are added to the Pb/Pb^{2+} half-cell to precipitate $PbCl_2$. The voltage is measured to be

+0.210 V. Taking conc. $H^+ = 1.0\ M$ and $P\ H_2 = 1.0$ atm, calculate conc. Pb^{2+}.

c. Taking $[Cl^-]$ in (b) to be 0.10 M, calculate K_{sp} for $PbCl_2$.

54. Consider a cell in which the reaction is:

$$2Ag(s) + Cu^{2+}(aq) \rightarrow 2Ag^+(aq) + Cu(s)$$

a. Calculate E^0_{tot} for this cell.
b. Chloride ions are added to the Ag/Ag^+ half-cell to precipitate AgCl. The measured voltage is $+0.060$ V. Taking conc. $Cu^{2+} = 1.0\ M$, calculate conc. Ag^+.
c. Taking $[Cl^-]$ in (b) to be 0.10 M, calculate K_{sp} of AgCl.

Electrolytic Cells

55. Consider the electrolysis of a water solution of $Cu(NO_3)_2$. Using standard voltages:
a. Write an equation for the cathode half-reaction; the anode half-reaction.
b. Write the equation for the overall cell reaction and calculate the minimum applied voltage required for electrolysis.

56. Follow the directions of Question 55 for a solution of $MgBr_2$.

57. In the electrolysis of an aqueous solution of NaCl, 2.61×10^{22} electrons pass through a cell.
a. How many coulombs does this represent?
b. What masses of $H_2(g)$, $Cl_2(g)$, and OH^- are produced, assuming 100% yield?

58. An electrolytic cell is producing aluminum from Al_2O_3 at the rate of one kilogram per day. Assuming a yield of 100%,
a. how many electrons must pass through the cell in one day?
b. what is the current passing through the cell?
c. how much oxygen is being produced simultaneously?

59. A spoon with an area of 4.00 cm^2 is plated with silver from a $Ag(CN)_2^-$ solution, using a current of 0.500 amperes for two hours.
a. If the current efficiency is 80.0%, how many grams of silver are plated?
b. What is the thickness of the silver plate formed (d Ag $= 10.5$ g/cm^3)?

60. It is desired to plate a coin, which has a diameter of 1.05 in and a thickness of 0.0600 in, with a layer of gold 0.0010 in thick.
a. How many grams of gold ($d = 19.3$ g/cm^3) are required?
b. How long will it take to plate the coin from AuCN, using a current of 0.200 ampere and assuming 100% yield?

61. A lead storage battery delivers a current of 2.00 A for one hour at a voltage of 12.0 V.
a. How many grams of Pb are converted to $PbSO_4$?

b. How much electrical energy is produced (kilowatt-hours)?

62. Aluminum is produced by the electrolysis of Al_2O_3, using a voltage of 6.0 V.
a. How many joules of electrical energy are required to form 1.00 kg Al?
b. What is the cost of the electrical energy in (a) at the rate of 6.0¢ per kilowatt-hour?

Unclassified

63. Consider the electrolysis of $CuCl_2$ to form $Cu(s)$ and $Cl_2(g)$. Calculate the minimum voltage required to carry out this reaction. If a voltage of 1.50 V is actually used, how many kilojoules of electrical energy is consumed in producing 1.00 g of Cu?

64. When a 1.50 V dry cell is used to "pump" $2.01 \times 10^{23}\ e^-$ through an electrolytic cell, determine
a. the number of coulombs passed through the cell.
b. the current in amperes if the time required is 10.0 min.
c. the energy in kilojoules and kilowatt-hours.

65. Explain why
a. sulfuric acid is used in a lead storage battery.
b. a salt bridge is used in a voltaic cell.
c. Cl_2 rather than O_2 is formed in the electrolysis of aqueous NaCl.

66. Identify each of the following statements as true or false. If false, correct it to make it true.
a. Electrolysis of aqueous $NiCl_2$ produces nickel and chlorine.
b. K^+ ions from a salt bridge move to the anode during operation of a voltaic cell.
c. When a lead storage battery is recharged, lead sulfate is deposited.

67. A current of 2.00 A is drawn from a dry cell for a period of ten minutes. Assuming that the only reaction taking place is Equation 21.21,
a. how many grams of zinc are consumed?
b. what volume of ammonia gas at 25°C and 1.00 atm would be produced if the solution were heated to drive off all the ammonia?

68. A cell used to electrolyse sodium chloride solution is filled with 1.00 L of 2.00 M NaCl. A current of two amperes is passed through this solution for 8.00 h. Assuming that the only reaction that occurs is Reaction 21.33,
a. What volume of chlorine gas is produced at 25°C and 1.00 atm?
b. What is the pH of the solution after eight hours?

69. A voltaic cell consists of a copper electrode immersed in 1.0 M $CuSO_4$ connected via a salt bridge to a solution of 1.0 M silver nitrate, in which is immersed a silver electrode. By what amount will the voltage change if
a. the concentration of Cu^{2+} is decreased to 0.00100 M?

b. enough Cl^- is added to precipitate AgCl ($K_{sp} = 1.8 \times 10^{-10}$) and leave $[Cl^-] = 1.0\ M$?

c. the area of the copper electrode is doubled?

70. Atomic masses can be determined by electrolysis. In one hour, a current of 0.600 A deposits 2.42 g of a certain metal, M, which is present in solution as M^+ ions. What is the atomic mass of the metal?

71. The standard potential for the reduction of AgSCN is 0.0895 V.

$$AgSCN(s) + e^- \rightarrow Ag(s) + SCN^-(aq)$$

Find another electrode potential to use together with the above value and calculate K_{sp} for AgSCN.

72. Consider the following reaction at 25°C.

$$O_2(g) + 4H^+(aq) + 4Br^-(aq) \rightarrow 2H_2O + 2Br_2(l)$$

If $[H^+]$ is adjusted by adding a buffer that is 0.100 M in sodium acetate and 0.100 M in acetic acid, the pressure of O_2 is 1.00 atm, and the bromide concentration is 0.100 M, what is the calculated cell voltage? (K_a acetic acid = 1.8×10^{-5}.)

Challenge Problems

73. In a fully charged lead storage battery, the electrolyte consists of 38% sulfuric acid by mass. The solution has a density of 1.286 g/cm³. Calculate E for the cell. (Assume all the H^+ ions come from the first dissociation of H_2SO_4, which is complete; K_a $HSO_4^- = 1.0 \times 10^{-2}$.)

74. Consider a voltaic cell in which the following reaction occurs:

$$Zn(s) + Sn^{2+}(aq) \rightarrow Zn^{2+}(aq) + Sn(s)$$

a. Calculate E^0_{tot} for the cell.

b. When the cell operates, what happens to the concentration of Zn^{2+}? the concentration of Sn^{2+}?

c. When the cell voltage drops to zero, what is the ratio of the concentration of Zn^{2+} to that of Sn^{2+}?

d. If the concentration of both cations is 1.0 M originally, what are their concentrations when the voltage drops to zero?

75. In biological systems, acetate ion is converted to ethyl alcohol in a two-step process:

$$CH_3COO^-(aq) + 3H^+(aq) + 2e^- \rightarrow CH_3CHO(aq) + H_2O;$$
$$E^{0'} = -0.581\ V$$

$$CH_3CHO(aq) + 2H^+(aq) + 2e^- \rightarrow C_2H_5OH(aq);$$
$$E^{0'} = -0.197\ V$$

($E^{0'}$ is the standard reduction voltage at 25°C and a pH of 7.00).

a. Calculate $\Delta G^{0'}$ for each step and for the overall conversion.

b. Calculate $E^{0'}$ for the overall conversion.

76. When 5.0 mL of 0.10 M Ce^{4+} is added to 5.0 mL of 0.20 M Fe^{2+}, the following reaction occurs:

$$Fe^{2+}(aq) + Ce^{4+}(aq) \rightarrow Fe^{3+}(aq) + Ce^{3+}(aq)$$

Taking the standard reduction voltage for Ce^{4+} to be +1.610 V, calculate:

a. K for the system.

b. the equilibrium concentration of each species.

77. Consider the cell: (Pt) H_2/H^+ ∥ (Pt) H^+/H_2. In the anode half-cell, hydrogen gas at 1.0 atm is bubbled over a platinum electrode dipping into a solution that has a pH of 7.0. The other half-cell is identical to the first except that the solution around the platinum electrode has a pH of 0.0. What is the cell voltage?

$$2Ag(s) + S(s) \rightarrow Ag_2S(s)$$

Silver becomes tarnished when exposed to sulfur compounds.

CHAPTER 22

CHEMISTRY OF THE TRANSITION METALS

*The fields of Nature long prepared and fallow
    the silent, cyclic chemistry
The slow and steady ages plodding,
the unoccupied surface ripening,
the rich ores forming beneath.*

WALT WHITMAN
SONG OF THE REDWOOD TREE

In Chapter 6, we discussed the general properties of metals and went on to consider the chemistry of the main-group metals in Groups 1, 2, and 3 of the Periodic Table. This chapter is devoted to the chemistry of some of the more important transition metals, the elements in the center of the Periodic Table (Fig. 22.1).

We will not attempt to discuss in detail the chemistry of these metals; to do so would require much more than one textbook chapter. Instead, we will look at some of the more important types of reactions involving the transition metals and the cations and oxyanions derived from them. Our goal will be to correlate their chemical behavior, attempting to draw general principles from what may seem to be a bewildering array of chemical facts.

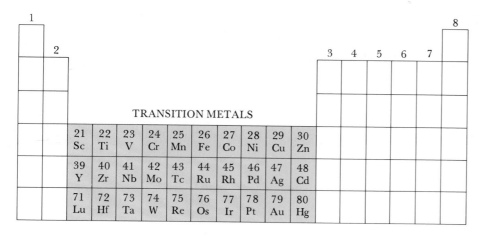

TRANSITION METALS

		21 Sc	22 Ti	23 V	24 Cr	25 Mn	26 Fe	27 Co	28 Ni	29 Cu	30 Zn
39 Y	40 Zr	41 Nb	42 Mo	43 Tc	44 Ru	45 Rh	46 Pd	47 Ag	48 Cd
71 Lu	72 Hf	73 Ta	74 W	75 Re	76 Os	77 Ir	78 Pt	79 Au	80 Hg

Figure 22.1 The transition elements are those in the center of the 4th, 5th, and 6th periods. The more familiar transition metals, whose chemistry we will refer to in this chapter, are shown in red.

Our discussion will be organized around the chemistry of

— the metals themselves (Sections 22.1, 22.2)
— the cations, such as Cr^{3+} and Mn^{2+}, derived from these metals (Section 22.3)
— oxyanions of the transition metals, such as CrO_4^{2-} and MnO_4^{-} (Section 22.4)

The chapter ends with a brief discussion of the uses of the transition metals and their alloys (Section 22.5).

22.1 METALLURGY OF THE TRANSITION METALS

Table 22.1 summarizes the methods used to extract the more common transition metals from their ores. Three metals (cobalt, silver, and cadmium) are not listed since they are formed mainly as by-products in the extraction of other metals (Co with Ni, Cd with Zn, Ag with Cu). The processes used to extract a metal depend in large part upon the nature of its ore (Fig. 22.2). Transition metal ores typically fall into one of three categories:

> An ore is a deposit containing appreciable amounts of the desired metal

— *native metals* such as gold, which is found in nature in elemental form, mixed with large amounts of rocky material.
— *oxides*, notably Fe_2O_3, the principal ore of iron.
— *sulfides*, including Cu_2S, from which copper metal is extracted.

NATIVE METALS

For countless centuries, gold has been extracted by taking advantage of its high density (19.3 g/cm^3). In ancient times, gold-bearing sands were washed over sheepskins which retained the gold; this is believed to be the source

TABLE 22.1   Metallurgy of the Transition Metals

METAL	PRINCIPAL ORE	EXTRACTION PROCESS
Cr	$Cr_2O_3 \cdot FeO$	heat with C to give Fe–Cr alloy; Cr_2O_3 reduced with Al to give pure Cr
Mn	MnO_2	heat with C and Fe_2O_3 to give Mn–Fe alloy; $MnSO_4$ electrolyzed to give pure Mn
Fe	Fe_2O_3	heat with C; CO reduces ore to Fe
Ni	NiS	flotation, roasting to NiO, reduction with CO
Cu	Cu_2S, $CuFeS_2$	flotation, roasting in air to give Cu
Zn	ZnS	flotation, roasting to ZnO, reduction with CO
Au	Au	treatment with CN^{-} (see text discussion)
Hg	HgS	roasting in air to form Hg

(a) Malachite, $Cu(OH)_2 \cdot CuCO_3$

(b) Cinnabar, HgS

(c) Hematite, $Fe_2O_3 \cdot xH_2O$

(d) Pyrite, FeS_2

Figure 22.2 Ores of transition metals. (Marna G. Clarke)

of the "Golden Fleece" of Greek mythology. The forty-niners in California obtained gold by swirling gold-bearing sands with water in a pan. Less dense impurities were washed away, leaving gold nuggets or flakes at the bottom of the pan.

Nowadays, the gold content of ores is much too low for these simple mechanical separation methods to be effective. Instead, the ore is treated with very dilute (0.01 M) sodium cyanide solution, through which air is blown. The following redox reaction takes place:

Gold at a concentration of 10g per ton can be recovered profitably

$$4Au(s) + 8CN^-(aq) + O_2(g) + 2H_2O \rightarrow 4Au(CN)_2{}^-(aq) + 4OH^-(aq) \quad (22.1)$$

The oxidizing agent is O_2, which takes gold to the +1 state. The cyanide ion acts as a complexing ligand, forming the stable $Au(CN)_2{}^-$ ion. Metallic gold is recovered from solution by adding zinc; the gold in the complex ion is reduced to the metal.

$$Zn(s) + 2Au(CN)_2{}^-(aq) \rightarrow Zn(CN)_4{}^{2-}(aq) + 2Au(s) \quad (22.2)$$

OXIDE ORES

A very pure metal can be produced from an oxide ore by using aluminum as a reducing agent. The reaction with chromium(III) oxide is typical:

$$Cr_2O_3(s) + 2Al(s) \rightarrow 2Cr(s) + Al_2O_3(s); \qquad \Delta H = -536 \text{ kJ} \qquad (22.3)$$

This type of reaction, known as the thermite process, is highly exothermic (Fig. 22.3). In principle, it will work with almost any oxide ore (except Al_2O_3!). In practice, it is seldom used on a large scale because of the relatively high cost of aluminum.

Carbon, in the form of coke, or, more exactly, carbon monoxide formed from the coke, is the most commonly used reducing agent in metallurgical processes. Its most important application is in the reduction of hematite ore, which consists largely of iron(III) oxide, Fe_2O_3, mixed with silicon dioxide, SiO_2 (sand). Reduction occurs in a blast furnace (Fig. 22.4), typically 30 m high and 10 m in diameter. The solid charge admitted at the top of the furnace consists of iron ore, coke, and limestone ($CaCO_3$). To get the process started, a blast of compressed air or pure O_2 at 500°C is blown into the furnace through nozzles near the bottom. Several different reactions occur, of which three are most important:

1. *Conversion of carbon to carbon monoxide.* In the lower part of the furnace coke burns to form carbon dioxide, CO_2. As the CO_2 rises through the solid mixture, it reacts further with the coke to form carbon monoxide, CO. The overall reaction is

$$2C(s) + O_2(g) \rightarrow 2CO(g); \qquad \Delta H = -221 \text{ kJ} \qquad (22.4)$$

Figure 22.3 The thermite reaction, between Al and Fe_2O_3 to produce Al_2O_3 and Fe, is highly exothermic; molten iron is produced, melting the crucible in which the reaction is carried out. (Charles D. Winters)

The heat given off by this reaction maintains a high temperature within the furnace.

2. *Reduction of Fe^{3+} ions to Fe.* The CO produced by Reaction 22.4 reacts with the iron(III) oxide in the ore:

$$Fe_2O_3(s) + 3CO(g) \rightarrow 2Fe(l) + 3CO_2(g) \qquad (22.5)$$

Molten iron, formed at a temperature of 1600°C, collects at the bottom of the furnace. Four or five times a day, it is drawn off. The daily production of iron from a single blast furnace is about 1000 metric tons. This is enough to make 500 Cadillacs or 1000 Volkswagens.

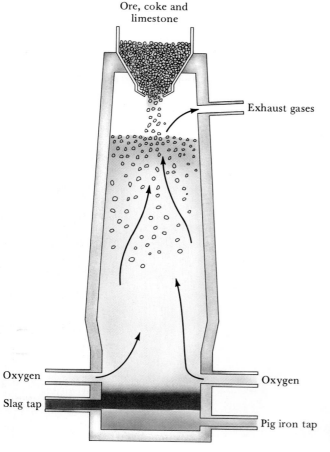 Once they get it started, they don't want to turn the furnace off

3. *Formation of slag.* The limestone added to the furnace decomposes at about 800°C:

$$CaCO_3(s) \rightarrow CaO(s) + CO_2(g) \qquad (22.6)$$

The calcium oxide formed reacts with impurities in the iron ore to form a glassy material called slag. The main reaction is with SiO_2 to form calcium silicate, $CaSiO_3$:

$$CaO(s) + SiO_2(s) \rightarrow CaSiO_3(l) \qquad (22.7)$$

The slag, which is less dense than molten iron, forms a layer on the surface of the metal. This makes it possible to draw off the slag through an opening

Ore, coke and limestone

Exhaust gases

Oxygen

Oxygen

Slag tap

Pig iron tap

Figure 22.4 Blast furnace for production of iron. Oxygen reacting with coke furnishes the high temperatures required for reduction of Fe_2O_3. The reducing agent is carbon monoxide, CO.

in the furnace above that used to remove the iron. The slag is used to make cement and as a base in road construction.

■ EXAMPLE 22.1

Write balanced chemical equations for the reduction of each of the following oxide ores by carbon monoxide:

a. ZnO   b. MnO_2 c. Fe_3O_4

Solution

In each case, the products are the metallic element and $CO_2(g)$. The equations are most simply balanced by the "trial and error" process described in Chapter 3.

Zinc is also made commercially by electrolysis

a. $$ZnO(s) + CO(g) \rightarrow Zn(s) + CO_2(g)$$

b. $$MnO_2(s) + 2CO(g) \rightarrow Mn(s) + 2CO_2(g)$$

c. $$Fe_3O_4(s) + 4CO(g) \rightarrow 3Fe(s) + 4CO_2(g)$$

Exercise

What mass of carbon is required to form 1.00 g of iron from Fe_2O_3 (hematite)? from magnetite, Fe_3O_4? Answer: 0.323 g, 0.287 g.

The product that comes out of the blast furnace, called "pig iron," is highly impure. On the average, it contains about 4% carbon along with

Figure 22.5   Pouring molten steel from a basic oxygen furnace.

lesser amounts of silicon, manganese, and phosphorus. To make steel from pig iron, it is necessary to lower the carbon content below 2%. Most of the carbon is burned to carbon dioxide:

$$C(s) + O_2(g) \rightarrow CO_2(g) \qquad (22.8)$$

By controlling the amount of oxygen used, it is possible to adjust the carbon content of the steel within very narrow limits.

Most of the steel produced in the world today is made by the "basic oxygen" process (Fig. 22.5). This is the only process used in Japan and Western Europe and accounts for more than 60% of steel production in the United States. The "converter" is filled with a mixture of about 70% molten iron from the blast furnace, 25% scrap iron or steel, and 5% limestone. Pure oxygen under a pressure of about 10 atm is blown through the molten metal. Reaction 22.8 occurs rapidly; impurities such as silicon are converted to oxides which react with limestone to form a slag. When the carbon content drops to the desired level, the supply of oxygen is cut off. (Too much oxygen would oxidize the iron, a reaction we don't want). At this stage, the steel is ready to be poured. The whole process takes from 30 min to 1 h and yields about 200 metric tons of steel in a single "blow."

The carbon content is checked to make sure it is right

SULFIDE ORES

Sulfide ores, after preliminary treatment, are most often "roasted," i.e., heated with air or pure oxygen. Typically, the product formed is a metal oxide, as with zinc:

$$2ZnS(s) + 3O_2(g) \rightarrow 2ZnO(s) + 2SO_2(g) \qquad (22.9)$$

The oxide produced by roasting is then reduced to the metal as described earlier (recall Example 22.1). In some cases, the metal oxide is unstable at the roasting temperature, decomposing to the metal. In that case, roasting of the sulfide gives the metal and sulfur dioxide as products. This happens when cinnabar, HgS, the sulfide ore of mercury, is heated in air:

$$HgS(s) + O_2(g) \rightarrow Hg(g) + SO_2(g) \qquad (22.10)$$

■ EXAMPLE 22.2 _____

A major ore of bismuth is the sulfide, Bi_2S_3.

a. Assuming Bi_2S_3 is converted to bismuth(III) oxide, write a balanced equation for the roasting reaction.
b. Given that

$$Bi_2O_3(s) \rightarrow 2Bi((s) + \tfrac{3}{2}O_2(g); \quad \Delta H^0 = +574 \text{ kJ}; \quad \Delta S^0 = +270 \text{ J/K}$$

show by calculation whether Bi_2O_3 will decompose to the elements at the roasting temperature, 1000°C, and 1 atm.

Solution

a. The by-product is $SO_2(g)$. The balanced equation is:

$$2Bi_2S_3(s) + 9O_2(g) \rightarrow 2Bi_2O_3(s) + 6SO_2(g)$$

b. Let us calculate ΔG^0 at 1000°C (1273 K):

$$\Delta G^0 = \Delta H^0 - T\Delta S^0 = +574 \text{ kJ} - 1273 \text{ K}(0.270 \text{ kJ/K}) = +230 \text{ kJ}$$

Since ΔG^0 is a large positive number, we conclude that Bi_2O_3 will be stable at the roasting temperature, or indeed at any temperature below about 1800°C. This means that roasting of Bi_2S_3 will give the oxide rather than the metal.

Exercise

For the reaction: $HgO(s) \rightarrow Hg(g) + \frac{1}{2}O_2(g)$, $\Delta H^0 = +175$ kJ, $\Delta S^0 = +207$ J/K. Above what temperature will the roasting of HgS give mercury vapor rather than HgO? Answer: 845 K (572°C).

Among the several ores of copper, one of the most important is chalcocite, which contains copper(I) sulfide, Cu_2S, in highly impure form. Rocky material typically lowers the fraction of copper in the ore to 1% or less. The Cu_2S is concentrated by a process called *flotation* (Fig. 22.6), which raises the fraction of copper to 20–40%. The concentrated ore is then converted to the metal by blowing air through it at a high temperature, typically above 1000°C. (Pure O_2 is often used instead of air.) The overall reaction that occurs is a simple one:

$$Cu_2S(s) + O_2(g) \rightarrow 2Cu(s) + SO_2(g) \tag{22.11}$$

The solid produced is called "blister copper." It has an irregular appearance due to air bubbles that enter the copper while it is still molten. Blister copper is impure, containing small amounts of several other metals.

Copper is purified by electrolysis. The anode, which may weigh as much as 300 kg, is made of blister copper. The electrolyte is 0.5 to 1.0 M

Because roasting can produce Cu, it was one of the first metals known

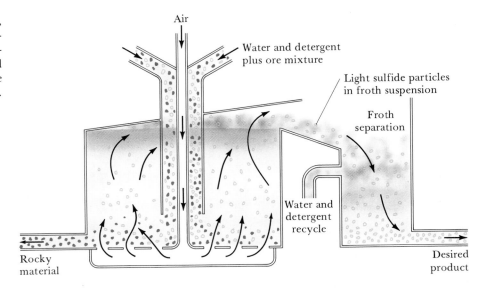

Figure 22.6 Low-grade sulfide ores, including Cu_2S, are often concentrated by flotation. The finely divided sulfide particles are trapped in soap bubbles, while the rocky waste sinks to the bottom and is discarded.

Air

Water and detergent plus ore mixture

Light sulfide particles in froth suspension

Froth separation

Water and detergent recycle

Rocky material

Desired product

$CuSO_4$, adjusted to a pH of about 0 with sulfuric acid. The cathode is a piece of pure copper, weighing perhaps 150 kg. The half-reactions are:

oxidation: $Cu(s, \text{impure}) \rightarrow Cu^{2+}(aq) + 2e^-$

reduction: $Cu^{2+}(aq) + 2e^- \rightarrow Cu(s, \text{pure})$

The overall reaction, obtained by adding these two half-reactions, is

$$Cu(s, \text{impure}) \rightarrow Cu(s, \text{pure}) \tag{22.12}$$

Thus, the net effect of electrolysis is to transfer copper metal from the impure blister copper used as one electrode to the pure copper sheet used as the other electrode. Electrolytic copper is 99.95% pure.

More active metals such as zinc and iron, which are present as impurities in blister copper, are oxidized to cations (Zn^{2+}, Fe^{2+}), which remain in solution. Less active metals (Ag, Au, Pt) are not oxidized; instead they drop off and collect at the bottom of the cell. These metals are recovered, separated from each other, and purified. Their value may exceed the cost of the energy used in the electrolysis.

■ **EXAMPLE 22.3** _____

Suppose the current in a cell used to refine copper is 120 A. If the cell operates 24 hours a day at 100% efficiency, how many kilograms of pure copper are produced in a year?

Solution

The number of moles of electrons is:

$$\frac{120 \, \dfrac{C}{s} \times \dfrac{60 \, s}{1 \, min} \times \dfrac{60 \, min}{1 \, h} \times \dfrac{24 \, h}{1 \, d} \times \dfrac{365 \, d}{1 \, yr}}{96485 \, C/mol \, e^-} = 3.92 \times 10^4 \, mol \, e^-/yr$$

The mass of copper produced in a year is:

$$3.92 \times 10^4 \, mol \, e^- \times \frac{1 \, mol \, Cu}{2 \, mol \, e^-} \times \frac{63.5 \, g \, Cu}{1 \, mol \, Cu} \times \frac{1 \, kg}{10^3 \, g} = \boxed{1.25 \times 10^3 \, kg}$$

Exercise

How many metric tons of ore, containing 0.78% by mass of Cu_2S, must be processed to obtain this mass of copper? Answer: 2.0×10^2 t.

22.2 REACTIONS OF THE TRANSITION METALS

The transition metals, unlike those in Groups 1 and 2 of the Periodic Table, typically show several oxidation numbers in their compounds (Fig. 22.7, p. 786). This tends to make their chemistry more complex (and more interesting) than that of the main-group metals. We should emphasize that only in the lower oxidation states ($+1$, $+2$, $+3$) are the transition metals present as simple cations (e.g., Ag^+, Zn^{2+}, Fe^{3+}). In higher oxidation states,

Figure 22.7 Oxidation states of the transition metals. The more common or stable states are shown in red.

21 Sc +3	22 Ti +4 +3 +2	23 V +5 +4 +3 +2	24 Cr +6 +3 +2	25 Mn +7 +6 +4 +3 +2	26 Fe +3 +2	27 Co +3 +2	28 Ni +2	29 Cu +2 +1	30 Zn +2
39 Y +3	40 Zr +4	41 Nb +5 +4	42 Mo +6 +5 +4 +3	43 Tc +7 +6 +4	44 Ru +8 +6 +4 +3	45 Rh +3 +2	46 Pd +4 +2	47 Ag +1	48 Cd +2
71 Lu +3	72 Hf +4	73 Ta +5	74 W +6 +5 +4	75 Re +7 +6 +5 +4	76 Os +8 +6 +4	77 Ir +4 +3	78 Pt +4 +2	79 Au +3 +1	80 Hg +2 +1

As the oxidation state goes up, the bonding to the metal becomes more covalent

the transition metal atom is covalently bonded to a nonmetal atom, most often oxygen or fluorine. Compounds such as MnO_2, CrO_3, and Mn_2O_7 are either network-covalent solids or, in a few cases, volatile molecular species. Manganese(VII) oxide, Mn_2O_7, is a green, oily liquid (mp = 6°C) which detonates upon heating to 95°C. It has a molecular structure entirely similar to that of the dichromate ion, $Cr_2O_7^{2-}$, shown in Figure 22.13 (Section 22.4).

REACTIONS WITH NONMETALS

Table 22.2 lists the formulas of the compounds formed when the more common transition metals react with the nonmetals in Groups 6 and 7 of the Periodic Table. Notice that, in several cases, the transition metal shows

TABLE 22.2 Products of Reactions of the Transition Metals with Nonmetals

	$O_2(g)$	$S(s)$	$Cl_2(g)$	$Br_2(l)$	$I_2(s)$
Cr	Cr_2O_3	Cr_2S_3	$CrCl_3$	$CrBr_3$	CrI_2
Mn	Mn_3O_4	MnS	$MnCl_2$	$MnBr_2$	MnI_2
Fe	Fe_2O_3, Fe_3O_4	FeS	$FeCl_3$	$FeBr_3$	FeI_2
Co	Co_3O_4	CoS	$CoCl_2$	$CoBr_2$	CoI_2
Ni	NiO	NiS	$NiCl_2$	$NiBr_2$	NiI_2
Cu	Cu_2O, CuO	Cu_2S	$CuCl_2$	$CuBr_2$	CuI
Zn	ZnO	ZnS	$ZnCl_2$	$ZnBr_2$	ZnI_2
Ag	—	Ag_2S	$AgCl$	$AgBr$	AgI
Cd	CdO	CdS	$CdCl_2$	$CdBr_2$	CdI_2
Au	—	—	$AuCl_3$	$AuBr_3$	—
Hg	HgO	HgS	$HgCl_2$	$HgBr_2$	HgI_2

a lower oxidation state in the sulfide and iodide than in the oxide, chloride, and bromide. Thus iron reacts with sulfur to form iron(II) sulfide, FeS, and with iodine to form iron(II) iodide, FeI_2; in contrast, Fe is present in the +3 state in Fe_2O_3, $FeCl_3$, and $FeBr_3$. This reflects the fact that S and I_2 are considerably weaker oxidizing agents than O_2, Cl_2, and Br_2 (recall Table 21.1, Chapter 21).

Note that Table 22.2 lists only those compounds formed by the direct reaction of the elements with each other. Many other compounds are known and can be prepared by indirect means. For example, although silver does not react directly with oxygen, the compound Ag_2O can be prepared by treating a solution of a silver salt with a strong base:

$$2Ag^+(aq) + 2OH^-(aq) \rightarrow Ag_2O(s) + H_2O \qquad (22.13)$$

In another case, cobalt(II) oxide can be prepared by heating the carbonate in the absence of air:

$$CoCO_3(s) \rightarrow CoO(s) + CO_2(g) \qquad (22.14)$$

The olive green CoO, upon exposure to air, is converted to black Co_3O_4, slowly at room temperature, more rapidly upon heating. In Co_3O_4 and in other oxides with the general formula M_3O_4, there are two kinds of cations: +2 and +3. To be specific, there are twice as many +3 as +2 cations; we might show the composition of Co_3O_4 as:

$$1\ Co^{2+} : 2\ Co^{3+} : 4\ O^{2-}$$

Most oxides and sulfides of the transition metals are *nonstoichiometric;* the atom ratio differs slightly from that required by the formula. An example is cobalt(II) oxide, which is deficient in cobalt. The atom ratio of Co to O is closer to 0.99 than to the 1.00 value implied by CoO. About one in every hundred Co^{2+} ions is missing, leaving a "hole" in the lattice (Fig. 22.8). To maintain electrical neutrality, two Co^{2+} ions are converted to Co^{3+}, giving a composition of:

$$2\ Co^{3+} : 97\ Co^{2+} : 100\ O^{2-}$$

Co^{2+}	O^{2-}	Co^{2+}	O^{2-}	Co^{2+}	O^{2-}
O^{2-}	Co^{2+}	O^{2-}	Co^{2+}	O^{2-}	Co^{2+}
Co^{2+}	O^{2-}		O^{2-}	Co^{3+}	O^{2-}
O^{2-}	Co^{3+}	O^{2-}	Co^{2+}	O^{2-}	Co^{2+}
Co^{2+}	O^{2-}	Co^{2+}	O^{2-}	Co^{2+}	O^{2-}
O^{2-}	Co^{2+}	O^{2-}	Co^{2+}	O^{2-}	Co^{2+}

Figure 22.8 Structure (schematic) of nonstoichiometric cobalt(II) oxide.

Cobalt(II) oxide is a semiconductor, with properties like those of semiconductors derived from silicon (Chapter 10).

Superconducting materials (Chapter 3) are nonstoichiometric oxides containing at least three different metals, e.g., $YBa_2Cu_3O_7$. In this compound, the number of oxygen atoms varies from 6.5 (the stoichiometric value assuming Y, Ba, and Cu have oxidation numbers of $+3$, $+2$, and $+2$, respectively) to 7.2. It appears that some of the copper is present as $+3$ rather than $+2$ ions.

REACTIONS WITH ACIDS

Any metal with a positive standard oxidation voltage, E^0_{ox}, can be oxidized by the H^+ ions present in a 1 M solution of a strong acid. All of the transition metals in the left column of Table 22.3 react spontaneously, albeit slowly, with dilute solutions of such strong acids as HCl, HBr, and H_2SO_4. The products are hydrogen gas and a cation of the transition metal. A typical reaction is that of nickel:

$$Ni(s) + 2H^+(aq) \rightarrow Ni^{2+}(aq) + H_2(g) \tag{22.15}$$

$$E^0_{tot} = E^0_{ox} Ni = +0.236 \text{ V}; \qquad \Delta G^0 = -45.5 \text{ kJ}$$

With metals that can form more than one cation, such as iron, the product upon reaction with H^+ in the absence of air is ordinarily the cation of lower charge, e.g., Fe^{2+}:

$$Fe(s) + 2H^+(aq) \rightarrow Fe^{2+}(aq) + H_2(g) \tag{22.16}$$

Reactions such as 22.15 and 22.16 offer a convenient way to prepare simple salts of the transition metals. Thus, to prepare nickel(II) chloride, we might first dissolve the metal in dilute hydrochloric acid; evaporation of the resulting solution would then give the solid salt:

$$Ni(s) + 2H^+(aq) + 2Cl^-(aq) \rightarrow Ni^{2+}(aq) + 2Cl^-(aq) + H_2(g)$$
$$\underline{Ni^{2+}(aq) + 2Cl^-(aq) \rightarrow NiCl_2(s)} \tag{22.17}$$
$$Ni(s) + 2H^+(aq) + 2Cl^-(aq) \rightarrow NiCl_2(s) + H_2(g)$$

TABLE 22.3   Ease of Oxidation of Transition Metals

METAL		CATION	E^0_{ox} (V)	METAL		CATION	E^0_{ox} (V)
Mn	$\rightarrow$	Mn^{2+}	$+1.182$	Cu	$\rightarrow$	Cu^{2+}	-0.339
Cr	$\rightarrow$	Cr^{2+}	$+0.912$	Ag	$\rightarrow$	Ag^+	-0.799
Zn	$\rightarrow$	Zn^{2+}	$+0.762$	Hg	$\rightarrow$	Hg^{2+}	-0.852
Fe	$\rightarrow$	Fe^{2+}	$+0.409$	Au	$\rightarrow$	Au^{3+}	-1.498
Cd	$\rightarrow$	Cd^{2+}	$+0.402$				
Co	$\rightarrow$	Co^{2+}	$+0.282$				
Ni	$\rightarrow$	Ni^{2+}	$+0.236$				

Figure 22.9 Nickel reacts slowly with hydrochloric acid to form $H_2(g)$ and Ni^{2+} ions in solution. Evaporation of the solution formed gives green crystals of $NiCl_2 \cdot 6H_2O$. (Charles D. Winters)

(The product actually obtained from water solution is the deep green hexahydrate, $NiCl_2 \cdot 6H_2O$, shown in Figure 22.9. It can be purified by recrystallization).

Metals with negative values of E^0_{ox}, listed at the right of Table 22.3, are too inactive to react with hydrochloric acid. The H^+ ion is not a strong enough oxidizing agent to convert a metal such as copper ($E^0_{ox} = -0.339$ V) to a cation. However, copper can be oxidized by nitric acid. Here, the oxidizing agent is the nitrate ion, NO_3^-, which may be reduced to NO_2 or NO:

Essentially all metals can be dissolved in one acid or another

$$3Cu(s) + 8H^+(aq) + 2NO_3^-(aq) \rightarrow 3Cu^{2+}(aq) + 2NO(g) + 4H_2O \quad (22.18)$$

$$E^0_{tot} = E^0_{ox} \, Cu + E^0_{red} \, NO_3^- = -0.339 \text{ V} + 0.964 \text{ V} = +0.625 \text{ V}$$

Silver can also be oxidized by heating with nitric acid (Example 22.4).

■ **EXAMPLE 22.4** _____

Consider the reaction of silver with nitric acid.

a. Assuming the products are $Ag^+(aq)$ and $NO(g)$, write a balanced redox equation for the reaction.
b. Describe in words how solid silver nitrate can be prepared from silver.

Solution

a. We proceed as described in Chapter 12. The half-equations are:

oxidation: $Ag(s) \rightarrow Ag^+(aq)$

reduction: $NO_3^-(aq) \rightarrow NO(g)$

Balancing the half-equations, we obtain

oxidation: $Ag(s) \rightarrow Ag^+(aq) + e^-$

reduction: $NO_3^-(aq) + 4H^+(aq) + 3e^- \rightarrow NO(g) + 2H_2O$

Multiplying the oxidation half-equation by three and adding to the reduction half-equation gives

$$3Ag(s) + NO_3^-(aq) + 4H^+(aq) \rightarrow 3Ag^+(aq) + NO(g) + 2H_2O$$

b. The anion present in the solution formed by dissolving silver in nitric acid is, of course, the NO_3^- ion. Evaporation of the solution gives silver nitrate, which is purified by recrystallization to yield the solid $AgNO_3$ that you obtain from the general chemistry storeroom.

Exercise
Write a balanced equation for the reaction of silver with nitric acid if the reduction product is NO_2 rather than NO. Answer: $Ag(s) + NO_3^-(aq) + 2H^+(aq) \rightarrow Ag^+(aq) + NO_2(g) + H_2O$.

Gold ($E_{ox}^0 = -1.498$ V) is too inactive to be oxidized by nitric acid alone ($E_{red}^0 = +0.964$ V). It does, however, go into solution in *aqua regia*, a 3:1 mixture by volume of 12 M HCl and 16 M HNO$_3$:

$$Au(s) + 4H^+(aq) + 4Cl^-(aq) + NO_3^-(aq) \rightarrow$$
$$AuCl_4^-(aq) + NO(g) + 2H_2O \quad (22.19)$$

This reaction is promoted by the formation of the very stable complex ion $AuCl_4^-$ ($K_f = 1 \times 10^{25}$). The function of the hydrochloric acid is to furnish the Cl^- ions needed to form this complex.

22.3
REACTIONS OF TRANSITION METAL CATIONS

When we write equations for the reactions of transition metal salts in water solution, we ordinarily show the cation as a simple species, e.g., Ni^{2+}, Cu^{2+}, Co^{2+}. It is important to realize, however, that transition metal cations in solution are always in the form of complex ions of the type discussed in Chapter 16. The colors associated with solutions of nickel(II), copper(II), and cobalt(II) salts are those of the hydrated ions:

$Ni(H_2O)_6^{2+}$	$Cu(H_2O)_4^{2+}$	$Co(H_2O)_6^{2+}$
green	blue	pink

Often, solutions of transition metal salts contain complex ions other than hydrated ions (Fig. 22.10). The green color of a solution of $CrCl_3$ is associated with such species as $Cr(H_2O)_5Cl^{2+}$ and $Cr(H_2O)_4Cl_2^+$; the $Cr(H_2O)_6^{3+}$ ion, found in solutions of $Cr(NO_3)_3$, is violet. In another case, the characteristic yellow color of solutions of iron(III) salts is due to the presence of the $Fe(H_2O)_5(OH)^{2+}$ ion, formed by the acid dissociation of the hexahydrated ion:

If you want just $Fe(H_2O)_6^{3+}$, make the solution acidic

$$Fe(H_2O)_6^{3+}(aq) \rightleftharpoons H^+(aq) + Fe(H_2O)_5(OH)^{2+}(aq); \quad K_a = 2 \times 10^{-3}$$
light purple   yellow

(a)

(b)

Figure 22.10 Addition of HCl to aqueous solutions of transition metal compounds often results in a color change because it converts an aquo to a chloro complex. This happens with Co^{2+} (pink to blue) shown in (a) and Cu^{2+} (b), where the color change is from blue to green. (Marna G. Clarke)

PRECIPITATION REACTIONS: SULFIDES AND HYDROXIDES

With very few exceptions, the nitrates, chlorides, and sulfates of transition metal cations are water-soluble. Indeed, it's a safe bet that if you are working in the general chemistry laboratory with a water solution of a transition metal salt, the anion is either NO_3^-, Cl^-, or SO_4^{2-}. For example, the salts $AgNO_3$, $CoCl_2 \cdot 6H_2O$ and $CuSO_4 \cdot 5H_2O$ are the common sources of the cations Ag^+, Co^{2+}, and Cu^{2+}.

On the other hand, many compounds of transition metal cations are water-insoluble. The situation with cobalt(II) salts is representative; the sulfide, hydroxide, carbonate, and phosphate, are all insoluble in water. Addition of these anions to a solution containing Co^{2+} gives a precipitate, e.g.:

$$Co^{2+}(aq) + CO_3^{2-}(aq) \rightarrow CoCO_3(s) \tag{22.20}$$

$$3Co^{2+}(aq) + 2PO_4^{3-}(aq) \rightarrow Co_3(PO_4)_2(s) \tag{22.21}$$

In general chemistry, particularly in qualitative analysis, the precipitates you work with most often are sulfides and hydroxides. Typically, sulfides are precipitated by saturating a solution of the appropriate cation with hydrogen sulfide. When H_2S is bubbled through a solution of a Cu^{2+} salt, the following reaction occurs:

$$Cu^{2+}(aq) + H_2S(aq) \rightarrow CuS(s) + 2H^+(aq) \tag{22.22}$$

Sometimes when you add sulfide ions to a solution of a transition metal ion, you get a surprise. Consider, for example, what happens when the cation is Cr^{3+}. The precipitate formed is $Cr(OH)_3$ rather than Cr_2S_3. The OH^- ions in the precipitate come from the reaction of S^{2-} ions with water:

$$3S^{2-}(aq) + 3H_2O \rightarrow 3HS^-(aq) + 3OH^-(aq)$$
$$\underline{Cr^{3+}(aq) + 3OH^-(aq) \rightarrow Cr(OH)_3(s)}$$
$$Cr^{3+}(aq) + 3S^{2-}(aq) + 3H_2O \rightarrow Cr(OH)_3(s) + 3HS^-(aq) \tag{22.23}$$

Transition metal hydroxides are ordinarily precipitated by adding either sodium hydroxide or ammonia. With Fe^{3+} the reactions are

NaOH: $Fe^{3+}(aq) + 3OH^-(aq) \rightarrow Fe(OH)_3(s)$ (22.24)

NH_3: $Fe^{3+}(aq) + 3NH_3(aq) + 3H_2O \rightarrow Fe(OH)_3(s) + 3NH_4^+(aq)$ (22.25)

All metal hydroxides dissolve in strong acid (for example, 6 M HCl). The reaction with iron(III) hydroxide is

$$Fe(OH)_3(s) + 3H^+(aq) \rightarrow Fe^{3+}(aq) + 3H_2O \qquad (22.26)$$

Water-insoluble hydroxides can be dissolved in 6 M NH_3 if the cation forms a stable complex with ammonia. The behavior of copper(II) hydroxide is typical:

$$Cu(OH)_2(s) + 4NH_3(aq) \rightarrow Cu(NH_3)_4{}^{2+}(aq) + 2OH^-(aq) \qquad (22.27)$$

If the cation forms a stable complex with OH^- ions, 6 M NaOH can bring a metal hydroxide into solution:

$$Zn(OH)_2(s) + 2OH^-(aq) \rightarrow Zn(OH)_4{}^{2-}(aq) \qquad (22.28)$$

Hydroxides, such as $Zn(OH)_2$, that dissolve in both strong acid (6 M HCl) and strong base (6 M NaOH) are referred to as being *amphoteric*.

■ **EXAMPLE 22.5** _____

Write net ionic equations for the reactions by which

a. Ni^{2+} is precipitated by H_2S.
b. Ni^{2+} is precipitated by adding a solution of NaOH.
c. the precipitate in (b) dissolves in NH_3.

Solution

a. The precipitate is NiS. The equation is

$$Ni^{2+}(aq) + H_2S(aq) \rightarrow NiS(s) + 2H^+(aq)$$

b. $$Ni^{2+}(aq) + 2OH^-(aq) \rightarrow Ni(OH)_2(s)$$

c. The product is the complex ion $Ni(NH_3)_6{}^{2+}$. The equation is

$$Ni(OH)_2(s) + 6NH_3(aq) \rightarrow Ni(NH_3)_6{}^{2+}(aq) + 2OH^-(aq)$$

Figure 22.11 Addition of NH_3 to a solution containing the green Ni^{2+} ion first gives a precipitate of green $Ni(OH)_2$. Upon addition of excess NH_3, the precipitate dissolves to form the blue complex $Ni(NH_3)_6{}^{2+}$. (Marna G. Clarke)

Exercise

When ammonia is added to a solution of $Ni(NO_3)_2$, a precipitate first forms and then dissolves (Fig. 22.11). Write net ionic equations for the two reactions involved. Answer:

$$Ni^{2+}(aq) + 2NH_3(aq) + 2H_2O \rightarrow Ni(OH)_2(s) + 2NH_4^+(aq)$$

$$Ni(OH)_2(s) + 6NH_3(aq) \rightarrow Ni(NH_3)_6^{2+}(aq) + 2OH^-(aq)$$

OXIDATION-REDUCTION REACTIONS

Several transition metals form more than one cation, e.g., Fe^{2+}, Fe^{3+}. Table 22.4 lists standard reduction voltages, E_{red}^0, for several such systems. In general, we expect cations for which E_{red}^0 is a large positive number to be easily reduced. These cations tend to be unstable in water solution. A case in point is the Mn^{3+} ion:

$$Mn^{3+}(aq) + e^- \rightarrow Mn^{2+}(aq); \qquad E_{red}^0 = +1.559 \text{ V}$$

We would not expect Mn^{3+} ions to stay around very long in water solution, and indeed they don't. They react with H_2O molecules and are reduced to Mn^{2+} ions:

$$2Mn^{3+}(aq) + H_2O \rightarrow 2Mn^{2+}(aq) + \tfrac{1}{2}O_2(g) + 2H^+(aq) \qquad (22.29)$$

$$E_{tot}^0 = E_{red}^0 \text{ Mn}^{3+} + E_{ox}^0 \text{ H}_2\text{O} = +1.559 \text{ V} - 1.229 \text{ V} = +0.330 \text{ V}$$

Manganese(III) is found only in insoluble oxides and hydroxides such as Mn_2O_3 and $MnO(OH)$.

There are not many Mn(III) compounds

TABLE 22.4   Ease of Reduction of Transition Metal Cations

Chromium	Cr^{3+}	$\xrightarrow{-0.408 \text{ V}}$	Cr^{2+}	$\xrightarrow{-0.912 \text{ V}}$	Cr
Manganese	Mn^{3+}	$\xrightarrow{+1.559 \text{ V}}$	Mn^{2+}	$\xrightarrow{-1.182 \text{ V}}$	Mn
Iron	Fe^{3+}	$\xrightarrow{+0.769 \text{ V}}$	Fe^{2+}	$\xrightarrow{-0.409 \text{ V}}$	Fe
Cobalt	Co^{3+}	$\xrightarrow{+1.953 \text{ V}}$	Co^{2+}	$\xrightarrow{-0.282 \text{ V}}$	Co
Copper	Cu^{2+}	$\xrightarrow{+0.161 \text{ V}}$	Cu^+	$\xrightarrow{+0.518 \text{ V}}$	Cu
Gold	Au^{3+}	$\xrightarrow{+1.400 \text{ V}}$	Au^+	$\xrightarrow{+1.695 \text{ V}}$	Au
Mercury	Hg^{2+}	$\xrightarrow{+0.908 \text{ V}}$	Hg_2^{2+}	$\xrightarrow{+0.796 \text{ V}}$	Hg

From Table 22.4, you might think that the Co^{3+} ion, like Mn^{3+}, would be unstable toward reduction:

$$Co^{3+}(aq) + e^- \rightarrow Co^{2+}(aq); \qquad E^0_{red} = +1.953 \text{ V}$$

Experimentally, we find that simple cobalt salts (such as $CoCl_2$, $CoSO_4$, . . .) always contain the Co^{2+} ion, never Co^{3+}. However, as you may recall from Chapter 16, complexes of cobalt(III), such as $Co(NH_3)_6^{3+}$, are quite common. The voltage for reduction of cobalt(III) changes drastically in the presence of a complexing agent such as ammonia:

$$Co(NH_3)_6^{3+}(aq) + e^- \rightarrow Co(NH_3)_6^{2+}(aq); \qquad E^0_{red} = +0.1 \text{ V}$$

The difference in E^0_{red} values reflects the much greater stability of cobalt(III) complexes as compared to cobalt(II). Cobalt(III) complexes are *not* reduced by H_2O. Indeed, in the presence of a complexing agent like ammonia or ethylenediamine, Co^{2+} is readily oxidized to Co^{3+} by dissolved oxygen.

The Fe^{3+} ion ($E^0_{red} = +0.769$ V) is less readily reduced than Mn^{3+} or Co^{3+}. In the presence of a suitable reducing agent, however, it can be converted to Fe^{2+} (Example 22.6).

■ **EXAMPLE 22.6** _____

When solutions of $FeCl_3$ and NaI are mixed, a redox reaction occurs and the solution takes on a reddish color. Suggest what this reaction might be and, using Table 21.1, calculate E^0_{tot}.

Solution
The Fe^{3+} ion is reduced to Fe^{2+}; the reddish color suggests that I_2 has been produced by the oxidation of I^- ions. The reaction is

$2Fe^{3+} + 2e^- \rightarrow 2Fe^{2+}(aq)$	$E^0_{red} = +0.769 \text{ V}$
$2I^-(aq) \rightarrow I_2(s) + 2e^-$	$E^0_{ox} = -0.534 \text{ V}$
$2Fe^{3+}(aq) + 2I^-(aq) \rightarrow 2Fe^{2+}(aq) + I_2(s)$	$E^0_{tot} = +0.235 \text{ V}$

Since the overall voltage is positive, we expect the reaction to be spontaneous, as it is.

Exercise
Would you expect Fe^{3+} ions to be reduced by Br^- ions? Answer: No, $E^0_{tot} = -0.308$ V.

Usually Fe in its common compounds is in the +3 state

The cations in the center column of Table 22.4 (Cr^{2+}, Mn^{2+}, . . .) are in an intermediate oxidation state. They can either be oxidized to a cation of higher charge ($Cr^{2+} \rightarrow Cr^{3+}$) or reduced to the metal ($Cr^{2+} \rightarrow Cr$). With certain cations of this type, these two half-reactions occur simultaneously. Consider, for example, the Cu^+ ion. In water solution, copper(I) salts *disproportionate*, undergoing reduction to copper metal and oxidation to Cu^{2+}:

$$2Cu^+(aq) \rightarrow Cu(s) + Cu^{2+}(aq) \qquad (22.30)$$

As a result, the only stable copper(I) species are insoluble compounds such as CuCN ($K_{sp} = 1 \times 10^{-19}$) or complex ions such as $Cu(CN)_2^-$ ($K_f = 1 \times 10^{24}$).

■ EXAMPLE 22.7

For the disproportionation given by Equation 22.30, calculate

a. E_{tot}^0, using Table 22.4.
b. the equilibrium constant for the reaction.
c. the concentration of Cu^+ in equilibrium with 0.1 M Cu^{2+}.

Solution

a. $E_{tot}^0 = E_{red}^0 \; Cu^+ \rightarrow Cu + E_{ox}^0 \; Cu^+ \rightarrow Cu^{2+}$

$\qquad = +0.518\ V - 0.161\ V = \boxed{+0.357\ V}$

b. $\ln K = \dfrac{nE_{tot}^0}{0.0257}$

Since $Cu^+ \rightarrow Cu$ and $Cu^+ \rightarrow Cu^{2+}$ are one-electron changes, $n = 1$:

$\qquad \ln K = \dfrac{(1)(+0.357)}{0.0257} = +13.9; \qquad K = \boxed{1 \times 10^6}$

c. $K = \dfrac{[Cu^{2+}]}{[Cu^+]^2} = 1 \times 10^6; \; [Cu^+]^2 = \dfrac{[Cu^{2+}]}{1 \times 10^6} = \dfrac{1 \times 10^{-1}}{1 \times 10^6} = 1 \times 10^{-7}$

$\qquad [Cu^+] = (1 \times 10^{-7})^{1/2} = \boxed{3 \times 10^{-4}\ M}$

Exercise

What is ΔG^0 for this reaction? Answer: -34.4 kJ.

Figure 22.12 *Oxidation States of Chromium:* Compounds of Cr^{3+} are most often violet due to the presence of such complexes as $Cr(H_2O)_6^{3+}$. Solutions of $CrCl_3$ are an exception; species such as $Cr(H_2O)_5Cl^{2+}$ and $Cr(H_2O)_4Cl_2^+$ shift the color to green. In the +6 state, shown at the right of the figure, chromium forms two anions, CrO_4^{2-} (yellow) and $Cr_2O_7^{2-}$ (red). (Charles D. Winters)

The Cu^+ ion is one of the few species in the center column of Table 22.4 that disproportionates in water (there is one other such species in the table; can you find it?). However, ions of this type may be unstable for a quite different reason. Water ordinarily contains dissolved air; the O_2 in air may oxidize the cation. When a blue solution of a chromium(II) salt is exposed to air, the color quickly changes to violet or green (Fig. 22.12) as the Cr^{3+} ion is formed by the reaction

$$2Cr^{2+}(aq) \rightarrow 2Cr^{3+}(aq) + 2e^- \qquad E_{ox}^0 = +0.408\ V$$
$$\underline{\tfrac{1}{2}O_2(g) + 2H^+(aq) + 2e^- \rightarrow H_2O \qquad E_{red}^0 = +1.229\ V}$$
$$2Cr^{2+}(aq) + \tfrac{1}{2}O_2(g) + 2H^+(aq) \rightarrow 2Cr^{3+}(aq) + H_2O \qquad E_{tot}^0 = +1.637\ V \qquad (22.31)$$

As a result of this reaction, chromium(II) salts are difficult to prepare and even more difficult to store.

The Fe^{2+} ion ($E_{ox}^0 = -0.769\ V$) is much more stable toward oxidation than Cr^{2+}. However, iron(II) salts in water solution are slowly converted to iron(III) by dissolved oxygen. In acidic solution, the reaction is

$$2Fe^{2+}(aq) + \tfrac{1}{2}O_2(g) + 2H^+(aq) \rightarrow 2Fe^{3+}(aq) + H_2O \qquad (22.32)$$

$$E^0_{tot} = E^0_{ox}\ Fe^{2+} + E^0_{red}\ O_2 = -0.769\ V + 1.229\ V = +0.460\ V$$

A similar reaction takes place in basic solution. Iron(II) hydroxide is pure white when first precipitated but, in the presence of air, it turns first green and then brown as it is oxidized by O_2:

$$2Fe(OH)_2(s) + \tfrac{1}{2}O_2(g) + H_2O \rightarrow 2Fe(OH)_3(s) \qquad (22.33)$$

22.4
OXYANIONS OF THE TRANSITION METALS

In general chemistry, you are most likely to encounter the higher oxidation states of the transition metals in the form of *oxyanions*. These are polyatomic negative ions in which a transition metal is bonded to oxygen. In this section, we will focus attention upon the oxyanions of chromium and manganese.

CHROMIUM (CrO_4^{2-}, $Cr_2O_7^{2-}$)

In the margin: *In solutions containing these species there are no Cr^{+6} ions*

Chromium in the +6 state forms two different oxyanions. These are the yellow chromate ion, CrO_4^{2-}, and the red dichromate ion, $Cr_2O_7^{2-}$. The geometries of these ions are shown in Figure 22.13. Notice that, in both cases, chromium is bonded tetrahedrally to oxygen atoms. The CrO_4^{2-} ion is stable in basic or neutral solution; in acid it is converted to the $Cr_2O_7^{2-}$ ion:

$$\underset{\text{yellow}}{2CrO_4^{2-}(aq)} + 2H^+(aq) \rightleftharpoons \underset{\text{red}}{Cr_2O_7^{2-}(aq)} + H_2O; \quad K = 3 \times 10^{14} \quad (22.34)$$

The dichromate ion in acidic solution is a powerful oxidizing agent:

$$Cr_2O_7^{2-}(aq) + 14H^+(aq) + 6e^- \rightarrow 2Cr^{3+}(aq) + 7H_2O; \quad E^0_{red} = +1.33\ V$$

In strong acid, $Cr_2O_7^{2-}$ oxidizes I^-, Br^-, or Fe^{2+} ions:

$$Cr_2O_7^{2-}(aq) + 14H^+(aq) + 6Fe^{2+}(aq) \rightarrow 2Cr^{3+}(aq) + 7H_2O + 6Fe^{3+}(aq) \quad (22.35)$$

Figure 22.13 Structures and geometries of the chromate and dichromate ions. In both ions the Cr atom is at the center of a tetrahedron, on the corners of which are oxygen atoms.

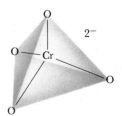

Chromate ion

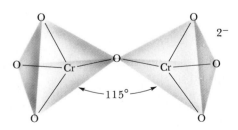

Dichromate ion

The oxidizing strength of $Cr_2O_7^{2-}$ decreases as pH increases (Example 22.8).

■ EXAMPLE 22.8

Using the Nernst equation, calculate E_{red} for $Cr_2O_7^{2-}$ at pH 4.00, taking conc. Cr^{3+} = conc. $Cr_2O_7^{2-}$ = 1.0 M.

Solution

Applying the Nernst equation to the half-reaction for the reduction of $Cr_2O_7^{2-}$,

$$E_{red} = E_{red}^0 - \frac{0.0257}{6} \ln \frac{(\text{conc. } Cr^{3+})^2}{(\text{conc. } Cr_2O_7^{2-})(\text{conc. } H^+)^{14}}$$

Substituting numbers (conc. H^+ = 1.0×10^{-4} M, since pH = 4.00):

$$E_{red} = +1.33 \text{ V} - \frac{0.0257}{6} \ln \frac{1}{(1 \times 10^{-4})^{14}}$$

$$= +1.33 \text{ V} - \frac{3.31}{6} = \boxed{+0.78 \text{ V}}$$

Note that the value of E_{red} decreases by 0.55 V (from 1.33 to 0.78 V) when the pH increases from 0 (conc. H^+ = 1.0 M) to 4. The oxidizing strength of the $Cr_2O_7^{2-}$ ion decreases accordingly.

Exercise

At what pH does E_{red} = +1.00 V? Answer: 2.4.

The $Cr_2O_7^{2-}$ ion can act as an oxidizing agent in the solid state as well as in water solution. In particular, it can oxidize the NH_4^+ ion to molecular nitrogen. When a pile of ammonium dichromate is ignited, a spectacular reaction occurs (Fig. 22.14).

$$\underset{\text{red}}{(NH_4)_2Cr_2O_7(s)} \rightarrow N_2(g) + 4H_2O(g) + \underset{\text{green}}{Cr_2O_3(s)} \qquad (22.36)$$

Figure 22.14 Ammonium dichromate, $(NH_4)_2Cr_2O_7$, has an orange color due to the presence of the $Cr_2O_7^{2-}$ ion. When ignited, it decomposes to give finely divided Cr_2O_3, which is green, nitrogen gas, and water vapor. (Charles D. Winters)

The ammonium dichromate resembles a tiny volcano as it burns, emitting hot gases, sparks, and a voluminous green dust of chromium(III) oxide.

Chromates and dichromates are gradually disappearing from chemistry teaching laboratories because of concern about their toxicity. Long-term exposure of industrial workers to dust containing chromates has, in a few cases, been implicated in lung cancer. More commonly, repeated contact with chromate salts leads to skin disorders; a few people are extremely allergic to CrO_4^{2-} and $Cr_2O_7^{2-}$ ions, breaking into a rash upon first exposure.

Also poison ivy and ring worm

MANGANESE (MnO_4^-, MnO_4^{2-})

The permanganate ion, MnO_4^-, is tetrahedral. It has an intense purple color (Fig. 22.15), easily visible even in very dilute solution.* Crystals of solid potassium permanganate, $KMnO_4$, have a deep purple, almost black color. This compound is used to treat such diverse ailments as "athlete's foot" and rattlesnake bites. These applications depend upon the fact that the MnO_4^- ion is a very powerful oxidizing agent. This is especially true in acidic solution, where MnO_4^- is reduced to Mn^{2+}:

$$MnO_4^-(aq) + 8H^+(aq) + 5e^- \rightarrow Mn^{2+}(aq) + 4H_2O; \qquad E^0_{red} = +1.512 \text{ V}$$

In basic solution, MnO_4^- is reduced to MnO_2, with a considerably smaller value of E^0_{red}:

$$MnO_4^-(aq) + 2H_2O + 3e^- \rightarrow MnO_2(s) + 4OH^-(aq); \qquad E^0_{red} = +0.596 \text{ V}$$

---

*The deep purple color of glass that has been exposed to the sun for a long time is believed to be due to MnO_4^- ions. These are formed when ultraviolet light brings about the oxidation of manganese oxides in the glass.

Figure 22.15 *Oxidation States of Manganese:* The Mn^{2+} ion has a pink color in concentrated solution; MnO_2 is a black, insoluble solid. In the +6 state, manganese exists as the MnO_4^{2-} ion, which is green. The familiar purple color of the MnO_4^- ion (oxid. state Mn = +7) is shown at the far right. (Charles D. Winters)

However, even in basic solution, MnO_4^- can oxidize water:

$$4MnO_4^-(aq) + 2H_2O \rightarrow 4MnO_2(s) + 3O_2(g) + 4OH^-(aq) \quad (22.37)$$

$$E_{tot}^0 = E_{red}^0\, MnO_4^- + E_{ox}^0\, H_2O = +0.596\ V - 0.401\ V = +0.195\ V$$

This reaction accounts for the fact that laboratory solutions of $KMnO_4$ slowly decompose, producing a brownish solid (MnO_2) and gas bubbles (O_2).

> Mn has some of the most interesting chemistry of all the transition metals

The manganate ion, MnO_4^{2-}, in which manganese has an oxidation number of $+6$, has a deep green color. It can be prepared either by reducing MnO_4^- (oxid. no. Mn $= +7$) or oxidizing MnO_2 (oxid. no. Mn $= +4$). In either case, the reaction must be carried out in highly basic solution. The MnO_4^{2-} ion is unstable in acid, undergoing disproportionation:

$$3MnO_4^{2-}(aq) + 4H^+(aq) \rightarrow 2MnO_4^-(aq) + MnO_2(s) + 2H_2O \quad (22.38)$$

■ **EXAMPLE 22.9**

One way to prepare the MnO_4^{2-} ion is to oxidize MnO_2 in basic solution with molecular oxygen. Write a balanced equation for the redox reaction involved.

Solution

The half-reactions are

oxidation: $MnO_2(s) \rightarrow MnO_4^{2-}(aq)$

reduction: $O_2(g) \rightarrow H_2O$

The balanced half-equations in *acidic* solution are

oxidation: $MnO_2(s) + 2H_2O \rightarrow MnO_4^{2-}(aq) + 4H^+(aq) + 2e^-$

reduction: $O_2(g) + 4H^+(aq) + 4e^- \rightarrow 2H_2O$

To obtain the balanced equation in acidic solution, we multiply the oxidation half-equation by two and add to the reduction half-equation. After simplification, we obtain

$$2MnO_2(s) + O_2(g) + 2H_2O \rightarrow 2MnO_4^{2-}(aq) + 4H^+(aq)$$

To find the equation in basic solution, we add 4 OH^- ions to each side, simplify, and obtain

$$2MnO_2(s) + O_2(g) + 4OH^-(aq) \rightarrow 2MnO_4^{2-}(aq) + 2H_2O$$

In practice, this reaction is ordinarily carried out by strongly heating a mixture of MnO_2 and KOH in the presence of air.

Exercise

What volume of $O_2(g)$ at 27°C and 1.00 atm is required to react with 1.00 mol MnO_2? Answer: 12.3 L.

Iron shows ferromagnetism (Charles Steele)

22.5
USES OF THE TRANSITION METALS AND THEIR COMPOUNDS

Some of the applications of the transition metals and their compounds are listed in Table 22.5. A few of the more prosaic uses have been omitted. Notice that many of the metals are most useful in the form of alloys with other metals. We will have more to say on that topic later in this section.

Some of the uses of cobalt and nickel depend upon a property that these metals share with iron: *ferromagnetism.* These three metals are strongly attracted into a magnetic field and retain their magnetism when removed from the field. Atoms of Fe, Co, and Ni, like those of most transition metals, have unpaired 3d electrons. This alone would make them paramagnetic, i.e., weakly attracted into a magnetic field. The much stronger ferromagnetism comes about because the magnetic moments of neighboring atoms line up so as to reinforce one another. A large number of atoms with parallel moments may line up to form a "domain" (Fig. 22.16). In an unmagnetized sample, individual domains are oriented randomly (Fig. 22.16a). When the sample is brought into a magnetic field, the domains tend to line up parallel to one another (Fig. 22.16b). This creates the strong interaction that we call ferromagnetism. When the sample is removed from the field, the domains remain aligned, so the metal retains its magnetism.

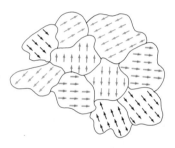

Unmagnetized (a)

Magnetized (b)

Figure 22.16 In a ferromagnetic substance, the magnetic moments of individual atoms (shown by arrows) are lined up parallel to one another through a large region (domain) of the solid. When the solid is magnetized, the individual domains line up parallel to one another as shown at the bottom.

TABLE 22.5   Uses of the Transition Metals and Their Compounds

	METAL	COMPOUNDS
Cr	Chrome plate, alloys such as stainless steel, nichrome	Cr^{3+} in tanning leather, pigments; CrO_2 in magnetic tape
Mn	Alloy steels (rails, safes)	MnO_2 in dry cells, decolorizing glass; MnO_4^- as germicide, water treatment
Fe	Structural metal, magnets	Fe^{2+} in treating anemia; Fe_2O_3 as red pigment (rouge)
Co	Alloy steels, alnico magnets, industrial catalyst	Co^{2+} as humidity indicator (red-blue); blue glass
Ni	Alloys such as alnico, monel, coinage	NiO_2 in Nicad battery
Cu	Electrical wiring, plumbing, alloys (Table 22.6)	Cu^{2+} as algicide, fungicide, wood preservative (Cuprinol)
Zn	Galvanized iron, dry cells, brass	$ZnCl_2$ as soldering flux, ZnS in fluorescent screens, ZnO in rubber manufacture
Ag	Tableware, jewelry	AgBr in photography, $Ag(NH_3)_2^+$ for mirrors, AgCl in photochromic glass
Cd	Alloys such as Wood's metal, Nicad battery	CdS in yellow pigments, CdTe in semiconductors
Au	Jewelry, gold leaf, electronic contacts	Treatment of arthritis
Hg	Thermometers, barometers, fluorescent lights, amalgams	Hg_2Cl_2 in medicine, HgO in dry cells, $Hg(ONC)_2$ as detonator for explosives

The three metals in the copper subgroup (Cu, Ag, Au) have long been referred to as the "coinage" metals. Today this phrase is somewhat misleading, at least as far as United States coinage is concerned. Gold coins were taken out of circulation in 1934. The rising price of silver and the increasing demand for this metal in photography led to the elimination of silver coins in 1971. The present 10¢, 25¢, and 50¢ coins consist of an alloy of 75% Cu and 25% Ni, sandwiched around a copper core. The same Cu–Ni alloy is used to make the "nickel" coin. In 1982 the composition of the penny, which had been 95% copper, was changed. Pennies now being minted are 98% zinc, covered with a thin copper plate. Perhaps we should refer to Ni, Cu, and Zn as the "coinage" metals.

The largest use by far of silver is in photography. The thin, light-sensitive coating on photographic film contains a silver halide, most often AgBr, dispersed in gelatin. Upon exposure to light, a few of the Ag^+ ions in each grain of silver bromide are reduced to Ag atoms. Fewer than 100 Ag^+ ions per grain are reduced, but enough silver is formed to make up what is known as the "latent image." After exposure, the film is treated in the dark with an organic reducing agent. The silver bromide grains that have been sensitized by exposure are reduced to silver metal. This process, known as development, can be represented by the half-equation

$$AgBr(s) + e^- \rightarrow Ag(s) + Br^-$$

where the electron is furnished by the reducing agent. The small number of silver atoms making up the latent image act as a catalyst for this reaction.

After development, a film shows dark areas of silver metal where it was exposed and light areas of unchanged AgBr in the unexposed regions. This silver bromide must be removed so that the finished negative will not be light-sensitive. This is accomplished by dipping the negative into a water solution of sodium thiosulfate, $Na_2S_2O_3$. The reaction that occurs is referred to as fixing; it involves complex ion formation to bring silver bromide into solution:

$$AgBr(s) + 2S_2O_3^{2-}(aq) \rightarrow Ag(S_2O_3)_2^{3-}(aq) + Br^-(aq) \qquad (22.39)$$

The negative is thoroughly washed, dried, and used to prepare a positive (Fig. 22.17, p. 802).

ALLOYS

An important characteristic of metallic elements is their ability to form alloys. An alloy is a material with metallic properties that contains two or more elements, at least one of which is a metal. Solid alloys are ordinarily prepared by melting the elements together, stirring the molten mixture until it is homogeneous, and allowing it to cool. Many alloys, notably bronze, brass, and pewter, have been made for centuries by this method.

The properties of alloys are often quite different from those of their component metals. By alloying a metal with another element, we usually

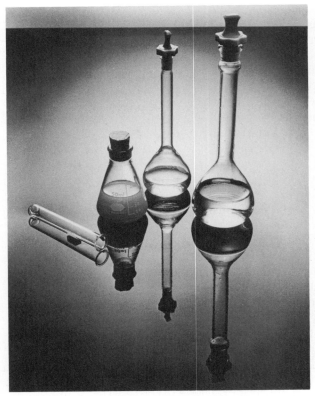
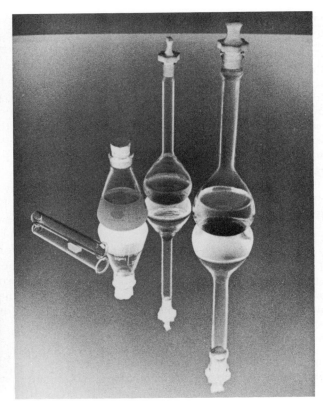

Figure 22.17 To prepare a positive print, the negative is placed over a piece of printing paper coated with AgBr. Light is passed through the negative to the printing paper. The amount of light reaching the print is inversely related to the thickness of the silver deposit on the negative. In this way, the light and dark areas of the negative are reversed.

— *lower the melting point.* Wood's metal, an alloy of Bi, Pb, Sn, and Cd, has a melting point, 70°C, which is much lower than that of any of these metals. It is used in fusible plugs that melt to set off automatic sprinkler systems.

— *increase the hardness.* A small amount of copper is present in sterling silver, which is much harder than pure silver. Gold is alloyed with silver and copper to form a metal hard enough to be used in jewelry. The lead plates used in storage batteries contain small amounts of antimony to prevent them from bending under stress.

— *lower the electrical and thermal conductivity.* Copper used in electrical wiring must be extremely pure; as little as 0.03% of arsenic can lower its conductivity by 15%. Sometimes we take advantage of this effect. High-resistance nichrome wire (Ni, Cr) is used in the heating elements of hair dryers and electric toasters.

Table 22.6 lists typical compositions and uses of some familiar alloys, most of which contain at least one transition metal. Several of these are alloy steels, such as stainless steel (Cr, Ni, Fe), which is very resistant to corrosion and tarnishing. Steel containing 5–20% manganese, called spiegeleisen, is a very tough, hard metal used in steel armor plate and safes.

TABLE 22.6   Some Commercially Important Alloys

COMMON NAME	COMPOSITION, ELEMENT (MASS PERCENT)	USES
Alnico	Fe(50), Al(20), Ni(20), Co(10)	magnets
Aluminum bronze	Cu(90), Al(10)	crankcases, hinges
Brass	Cu(67–90), Zn(10–33)	plumbing, hardware
Bronze	Cu(70–95), Zn(1–25), Sn(1–18)	bearings, bells, medals
Cast iron	Fe(96–97), C(3–4)	castings
Coinage, US	Cu(75), Ni(25)	5¢, 10¢, 25¢, 50¢ coins
Dental amalgam	Hg(50), Ag(35), Sn(15)	dental fillings
Duriron	Fe(84), Si(14), C(1), Mn(1)	pipes, kettles, condensers
German silver	Cu(60), Zn(25), Ni(15)	tea pots, jugs, faucets
Gold, 18 carat	Au(75), Ag(10–20), Cu(5–15)	jewelry
Gold, 10 carat	Au(42), Ag(12–20), Cu(38–46)	jewelry
Gunmetal	Cu(88), Sn(10), Zn(2)	gun barrels, machine parts
Monel	Ni(60–70), Cu(25–35), Fe, Mn	instruments, machine parts
Nichrome	Ni(60), Fe(25), Cr(15)	electrical resistance wire
Pewter	Sn(70–95), Sb(5–15), Pb(0–15)	tableware
Silver solder	Ag(63), Cu(30), Zn(7)	high-melting solder
Solder	Pb(67), Sn(33)	joining metals
Spiegeleisen	Fe(80–95), Mn(5–20), C(0–1)	safes, armor plate, rails
Stainless steel	Fe(73–79), Cr(14–18), Ni(7–9)	instruments, sinks
Steel	Fe(98–99.5), C(0.5–2)	structural metal
Sterling silver	Ag(92.5), Cu(7.5)	tableware, jewelry
White gold	Au(90), Pd(10)	jewelry
Wood's metal	Bi(50), Pb(25), Sn(13), Cd(12)	automatic sprinkler systems

■ SUMMARY EXAMPLE

Consider the transition element manganese

a. Give the formula of at least four manganese compounds in each of which Mn is in a different oxidation state.

b. What is the color of the $Mn(H_2O)_6^{2+}$ ion? the MnO_4^{2-} ion? the MnO_4^- ion?

c. Write a balanced equation for the reaction of manganese metal with sulfur; with chlorine; with a solution of hydrochloric acid.

d. Write a balanced equation for the reaction of the Mn^{2+} ion with OH^- ions; with NH_3 to form a precipitate; with H_2S.

e. Which of the following compounds, $MnCl_2$, $Mn(NO_3)_2$, $MnCO_3$, $Mn(OH)_2$, would you expect to be water-soluble?

f. Write a balanced equation for the reduction of the principal ore of manganese, pyrolusite, MnO_2, with CO; with Al. What mass of aluminum is required to form one kilogram of Mn metal?

g. For the reaction: $MnO(s) + H_2(g) \rightarrow Mn(s) + H_2O(g)$, $\Delta H^0 = +143$ kJ, $\Delta S^0 = +0.0297$ kJ/K. Would it be feasible to reduce MnO to the metal by heating with hydrogen?

h. Given that $E^0_{red}\ Mn^{3+} \rightarrow Mn^{2+} = +1.559$ V, $E^0_{red}\ Mn^{2+} \rightarrow Mn = -1.182$ V, and $E^0_{red}\ O_2(g) \rightarrow H_2O = +1.229$ V, show by calculation whether Mn^{2+} in water solution will disproportionate; whether it will be oxidized to Mn^{3+} by dissolved oxygen; whether it will be reduced by hydrogen gas to the metal.

i. Given: $MnO_4^-(aq) + 2H_2O + 3e^- \rightarrow MnO_2(s) + 4OH^-(aq)$; $E_{red}^0 = +0.59$ V, write a balanced equation for the reaction of MnO_4^- in basic solution with Fe^{2+} to form $Fe(OH)_3$. Calculate E_{red} for the MnO_4^- ion at pH 9.0, taking conc. $MnO_4^- = 1M$.

j. Given that K_{sp} MnS $= 5 \times 10^{-14}$ and K for the process: $H_2S(aq) \rightleftharpoons 2H^+(aq) + S^{2-}(aq)$ is 1×10^{-20}, determine the $[H^+]$ at which the solubility of MnS is 0.1 mol/L; the reaction is: $MnS(s) + 2H^+(aq) \rightleftharpoons Mn^{2+}(aq) + H_2S(aq)$.

Answers

a. $MnCl_2$, MnO_2, K_2MnO_4, $KMnO_4$ $----$ **b.** pink, green, purple

c. $Mn(s) + S(s) \rightarrow MnS(s)$; $Mn(s) + Cl_2(g) \rightarrow MnCl_2(s)$

$Mn(s) + 2H^+(aq) \rightarrow Mn^{2+}(aq) + H_2(g)$

d. $Mn^{2+}(aq) + 2OH^-(aq) \rightarrow Mn(OH)_2(s)$;

$Mn^{2+}(aq) + 2NH_3(aq) + 2H_2O \rightarrow Mn(OH)_2(s) + 2NH_4^+(aq)$;

$Mn^{2+}(aq) + H_2S(aq) \rightarrow MnS(s) + 2H^+(aq)$

e. $MnCl_2$, $Mn(NO_3)_2$

f. $MnO_2(s) + 2CO(g) \rightarrow Mn(s) + 2CO_2(g)$

$3MnO_2(s) + 4Al(s) \rightarrow 3Mn(s) + 2Al_2O_3(s)$; 655 g Al

g. no; $T_{calc} \approx 4800$ K   **h.** no; in all cases, E_{tot}^0 is negative

i. $MnO_4^-(aq) + 3Fe^{2+}(aq) + 5OH^-(aq) + 2H_2O \rightarrow MnO_2(s) + 3Fe(OH)_3(s)$

$E_{red} = +0.98$ V

j. 7×10^{-5} M

able 22.7 lists the ten trace elements known to be essential in human nutrition; seven of these are transition metals. Of five other trace elements essential to at least some life form (As, B, Si, Ni, and V), two are transition metals. For the most part, transition metals in biochemical compounds are present as complex ions, chelated by organic ligands. You will recall (Chapter 16) that hemoglobin has such a structure with Fe^{2+} as the central ion of the complex. The Co^{3+} ion occupies a similar position in Vitamin B-12, octahedrally coordinated to organic molecules somewhat similar to those in hemoglobin.

As you can judge from Table 22.7, transition metal cations are frequently found in enzymes. The Zn^{2+} ion alone is known to be a component of at least 70 different enzymes. One of these, referred to as "alcohol dehydrogenase" is concentrated in the liver, where it acts to break down alcohols. Another zinc-containing enzyme is involved in the normal functioning of oil glands in the skin, which accounts for the use of Zn^{2+} compounds in the treatment of acne.

Although Zn^{2+} is essential to human nutrition, compounds of the two elements below zinc in the Periodic Table, Cd and Hg, are extremely toxic. This reflects the fact that Cd^{2+} and Hg^{2+}, in contrast to Zn^{2+}, form very stable complexes with ligands containing sulfur atoms. As a result, these two cations react with and thereby deactivate enzymes containing —SH groups.

Over the years, compounds of the transition metals have been used for a variety of medicinal purposes. Calomel, Hg_2Cl_2, was the wonder drug of the early nineteenth century, prescribed for everything from constipation to pneumonia. It probably killed more patients than it cured. Quite recently, gold salts have been found effective in the treatment of rheumatoid arthritis. On a more prosaic level, iron(II) compounds such as $FeSO_4$ are used in treating anemia.

TABLE 22.7 Essential Trace Elements in Human Nutrition

ELEMENT	AMOUNT IN BODY (mg)	DAILY REQUIREMENT (mg)	RICH SOURCES*	FUNCTION
Fe^{2+}, Fe^{3+}	5000	15	Liver, meat, clams, spinach	Component of hemoglobin, myoglobin
Zn^{2+}	3000	15	Oysters, crab, meat, nuts	Component of many enzymes, hormones
F^-	3000	3	Fluoridated water, tea	Component of bones, teeth
Cu^{2+}, Cu^+	100	3	Liver, lobster, cherries	Iron metabolism, component of enzymes
I^-	30	0.15	Iodized salt, seafood	Synthesis of thyroxine in thyroid gland
Mn^{2+}	15	3	Beet greens, nuts, blueberries	Metabolism of carbohydrates, lipids
Mo(IV,V,VI)	10	0.3	Legumes, green vegetables	Fe, N metabolism; component of enzymes
Se(2−)	10	0.1	Coconut, onions, herring, cereals	Associated with Vitamin E; possible anticarcinogen
Cr^{3+}	5	0.1	Corn, clams, whole grains	Glucose metabolism; affects action of insulin
Co^{2+}, Co^{3+}	1	0.005	Liver, shellfish	Component of Vitamin B-12

*Other, more mundane, sources can be found in any nutrition textbook.

QUESTIONS AND PROBLEMS

Reactions and Equations

1. Write a balanced equation to represent
a. the roasting of nickel(II) sulfide to form nickel(II) oxide.
b. the reduction of nickel(II) oxide by carbon monoxide.
c. the reaction of nickel with hydrochloric acid.

2. Write a balanced equation to represent
a. the roasting of copper(I) sulfide to form "blister copper."
b. the reaction of copper with nitric acid with NO as one of the products.
c. the reaction of copper with nitric acid with NO_2 as one of the products.

3. Write a balanced equation for the reaction of iron with
a. sulfur   b. chlorine   c. bromine   d. oxygen

4. Write a balanced equation for the reaction of chromium with
a. sulfur   b. chlorine   c. bromine   d. oxygen

5. Write a balanced equation for the formation of a complex ion when
a. $Fe(H_2O)_6^{3+}$ ion acts as a weak acid.
b. zinc hydroxide dissolves in 6 M NaOH.
c. gold reacts with cyanide ion, oxygen, and water.

6. Write a balanced equation for the formation of a complex ion when
a. $Fe(H_2O)_5(OH)^{2+}$ is made strongly acidic.
b. cadmium hydroxide is dissolved in ammonia.
c. the green nickel(II) complex, $Ni(H_2O)_6^{2+}$, is treated with ammonia.

7. Write a balanced equation for the formation of a precipitate when
a. H_2S is added to a solution of zinc nitrate.
b. ammonia is added to a solution of iron(III) chloride.
c. sodium hydroxide is added to a solution of copper(II) nitrate.

8. Write a balanced equation for the formation of a precipitate when
a. H_2S is added to a solution of cadmium sulfate.
b. S^{2-} ions are added to a solution of chromium(III) nitrate.
c. ammonia is added to a solution of cobalt chloride.

9. Write a balanced equation to show
a. the reaction of chromate ion with a strong acid.
b. the acid behavior of $Co(H_2O)_6^{2+}$
c. the dissolving of $Fe(OH)_3$ in strong acid.

10. Write a balanced equation to show
a. the oxidation of water to oxygen gas by permanganate ion in basic solution.
b. the precipitation of silver oxide from silver ion in base.
c. the reduction half-reaction of chromate ion to chromium(III) hydroxide in basic solution.

11. Addition of concentrated NaOH solution to $Zn(NO_3)_2$ solution forms a precipitate that later dissolves as more NaOH is added. Write a net ionic equation for each of the two reactions involved.

12. When ammonia is added to aqueous $CuSO_4$, a precipitate first forms and then dissolves. Write a net ionic equation for
a. precipitate formation.
b. the dissolving of the precipitate.

13. Write an equation (net ionic, if necessary) to account for
a. the formation of slag in a blast furnace.
b. the formation of gas bubbles when cobalt reacts with hydrochloric acid.
c. the disappearance of the pink color of permanganate when a solution of $KMnO_4$ is added to $FeSO_4$.

14. Write an equation (net ionic where appropriate) to account for
a. the reduction of iron ore in a blast furnace.
b. the blue color that appears when excess ammonia is added to Ni^{2+}.
c. the fact that iron filings dissolve in hydrochloric acid.

15. Write a balanced redox equation for the reaction of mercury with aqua regia, assuming the products include $HgCl_4^{2-}$ and $NO_2(g)$.

16. Write a balanced redox equation for the reaction of cadmium with aqua regia, assuming the products include $CdCl_4^{2-}$ and $NO(g)$.

17. Balance the following redox equations:
a. $Cu(s) + NO_3^-(aq) \rightarrow Cu^{2+}(aq) + NO_2(g)$ (acidic)
b. $Cr(OH)_3(s) + ClO^-(aq) \rightarrow CrO_4^{2-}(aq) + Cl^-(aq)$ (basic)

18. Balance the following redox equations:
a. $Fe(s) + NO_3^-(aq) \rightarrow Fe^{3+}(aq) + NO_2(g)$ (acidic)
b. $Cr(OH)_3(s) + O_2(g) \rightarrow CrO_4^{2-}(aq)$ (basic)

19. Balance the following redox equations:
a. $Cd(s) + NO_3^-(aq) \rightarrow Cd^{2+}(aq) + NO(g)$ (acidic)
b. $MnO_2(s) + ClO^-(aq) \rightarrow MnO_4^-(aq) + Cl^-(aq)$ (basic)

20. Balance the following redox equations:
a. $Cr^{2+}(aq) + O_2(g) \rightarrow Cr^{3+}(aq)$ (acidic)
b. $Mn^{2+}(aq) + BiO_3^-(aq) \rightarrow MnO_4^-(aq) + Bi^{3+}(aq)$ (acidic)

Equilibrium

21. Taking K_{sp} CdS = 1 × 10^{-29} and K for: $H_2S(aq) \rightleftharpoons 2H^+(aq) + S^{2-}(aq)$ to be 1 × 10^{-20}, calculate the solubility (mol/L) of CdS at pH 1.0.

22. Calculate the solubility of CuS (K_{sp} = 1 × 10^{-36}) in a solution in which the pH is 2.0.

23. The equilibrium constant for the reaction

$$2CrO_4^{2-}(aq) + 2H^+(aq) \rightleftharpoons Cr_2O_7^{2-}(aq) + H_2O$$

is 3×10^{14}. What must the pH be so that the concentrations of chromate and dichromate ion are both 0.10 M?

24. A solution is prepared by dissolving 0.200 mol Na_2CrO_4 in 1.00 L of a buffer solution at pH 6.00. What will be the $[CrO_4{}^{2-}]$ and $[Cr_2O_7{}^{2-}]$ in the solution? K for the conversion of $CrO_4{}^{2-}$ to $Cr_2O_7{}^{2-}$ is 3×10^{14}.

Thermodynamics

25. Using data in Appendix 1, estimate the temperature at which Fe_2O_3 can be reduced to iron, using hydrogen gas as a reducing agent (assume $H_2O(g)$ is the other product).

26. Repeat the calculation of Problem 25, substituting Fe_3O_4 for Fe_2O_3.

27. Using data in Appendix 1, calculate ΔG^0 (smallest whole-number coefficients) at 25°C for the redox reaction:

$MnO_4{}^-(aq) + S(s) \rightarrow Mn^{2+}(aq) + SO_4{}^{2-}(aq)$ (acidic solution)

28. Using data in Appendix 1, calculate ΔG^0 at 25°C for the reaction:

$Cu(s) + NO_3{}^-(aq) \rightarrow Cu^{2+}(aq) + NO_2(g)$ (acidic solution)

Electrochemistry

29. When 1.00 g of chromium metal is plated from a solution of CrO_3, assuming 100% efficiency,
   a. how many coulombs are required?
   b. what current must be used if the electrolysis is to be carried out in 15.0 min?

30. When gold is plated, the electrode reactions are:

$AuCl_4{}^-(aq) + 3e^- \rightarrow Au(s) + 4Cl^-(aq)$

$2H_2O \rightarrow O_2(g) + 4H^+(aq) + 4e^-$

   a. Calculate the minimum voltage required for the electrolysis at standard concentrations (Table 21.1, Ch. 21).
   b. How many coulombs must pass through the cell to plate one gram of gold?
   c. What is the minimum amount of electrical energy in kilojoules that must be used to plate one gram of gold?

31. Using Table 21.1, Chapter 21, calculate E_{tot}^0 for
   a. $2Co^{3+}(aq) + H_2O \rightarrow 2Co^{2+}(aq) + \frac{1}{2}O_2(g) + 2H^+(aq)$
   b. $2Cr^{2+}(s) + I_2(s) \rightarrow 2Cr^{3+}(aq) + 2I^-(aq)$

Which reactions are spontaneous at standard conditions?

32. Of the cations listed in Table 22.4, which ones would you expect to oxidize I^- ions to I_2 ($E_{ox}^0 = -0.534$ V)?

33. Consider the reaction

$2Ag^+(aq) \rightarrow Ag(s) + Ag^{2+}(aq)$

for which $E_{tot}^0 = -1.18$ V. Use the Nernst equation to calculate

   a. E when conc. $Ag^+ = 1.0 \times 10^{-4} M = 5 \times$ conc. Ag^{2+}.
   b. Conc. Ag^{2+} when conc. $Ag^+ = 1.0 M$ and $E = 0.00$ V.

34. Consider

$2Co(NH_3)_6{}^{2+}(aq) + \frac{1}{2}O_2(g) + H_2O(l) \rightarrow$
$2Co(NH_3)_6{}^{3+}(aq) + 2OH^-(aq)$

$E_{red}^0 \; Co(NH_3)_6{}^{3+} = +0.1$ V.

Use the Nernst equation to calculate E when the complex ions are 1 M, $P \; O_2 = 1$ atm, and the pH is
   a. 4.00   b. 7.00

35. Calculate the voltage for the reduction of $MnO_4{}^-$ to Mn^{2+} ($E_{red}^0 = +1.512$ V) at a pH of 7.00 when conc. $MnO_4{}^- = 0.20 M$ and conc. $Mn^{2+} = 0.10 M$.

36. Consider the reaction

$2Cu^+(aq) \rightarrow Cu^{2+}(aq) + Cu(s)$

for which $E_{tot}^0 = +0.357$ V. Using the Nernst equation, calculate
   a. E when conc. $Cu^{2+} =$ conc. $Cu^+ = 1 \times 10^{-4} M$.
   b. conc. Cu^+ when conc. $Cu^{2+} = 1 M$ and $E = 0.00$ V.

37. Using Table 22.4, calculate, for the disproportionation of Fe^{2+},
   a. the equilibrium constant, K.
   b. the concentration of Fe^{3+} in equilibrium with 0.10 M Fe^{2+}.

38. Using Table 22.4, calculate, for the disproportionation of Au^+,
   a. K
   b. the concentration of Au^+ in equilibrium with 0.10 M Au^{3+}.

39. Using Table 22.4, calculate ΔG^0 for
   a. $Fe(s) + 2Fe^{3+}(aq) \rightarrow 3Fe^{2+}(aq)$
   b. $Hg(l) + Hg^{2+}(aq) \rightarrow Hg_2{}^{2+}(aq)$

40. For the reaction

$2Au(s) + AuCl_4{}^-(aq) + 2Cl^-(aq) \rightarrow 3AuCl_2{}^-(aq)$

ΔG^0 at 25°C is $+44.2$ kJ. Calculate E_{tot}^0 and K.

Unclassified

41. Of the cations listed in Table 22.4, show by calculation which one (besides Cu^+) will disproportionate at standard conditions.

42. Consider the elements Cr, Mn, Fe, Co, Ni, Cu, and Zn.
   a. Which of these elements form ions that give colorless solutions in water?
   b. Which oxidation state is common to all the elements?
   c. Which element has a purple oxyanion?
   d. Which elements are used in the manufacture of steel?
   e. Which elements are used to plate steel parts to prevent rusting?

43. A 0.500-g sample of a zinc–copper alloy was treated with dilute hydrochloric acid. The hydrogen gas evolved was collected by water displacement at 27°C and a total pressure of 755 mm Hg. The volume of the water displaced by the gas is 105.7 mL. What is the percent composition, by mass, of the alloy? (Vapor pressure of H_2O at 27°C is 26.74 mm Hg.) Assume only the zinc reacts.

44. One type of stainless steel contains 22% by mass nickel. How much nickel sulfide ore, NiS, is required to produce one metric ton of stainless steel?

45. Vitamin B-12 is a coordination compound of cobalt with the molecular formula $C_{63}H_{90}O_{14}N_{14}PCo$. If the U.S. recommended daily allowance (USRDA) for Vitamin B-12 is 6.0 micrograms (1 microgram = 1×10^{-6} g), how much cobalt does this involve?

46. The K_{sp} for silver phosphate is 1×10^{-16}. How many micrograms of silver phosphate will dissolve in one liter of pure water, and how many will dissolve in one liter of water that contains 1.00 g of silver nitrate? [1 microgram $(\mu g) = 1 \times 10^{-6}$ g.]

47. How many cubic feet of air (assume 21% by volume of oxygen in air) at 25°C and 1.00 atm are required to react with coke to form the CO needed to convert one metric ton of hematite ore (92% Fe_2O_3) to iron?

48. Zinc is produced by electrolytic refining. The electrolytic process, which is similar to that for copper, can be represented by the two half-reactions

$$Zn\ (impure,\ s) \rightarrow Zn^{2+} + 2e^-$$

$$Zn^{2+} + 2e^- \rightarrow Zn\ (pure,\ s)$$

For this process, a voltage of 3.0 V is used. How many kilowatt hours are needed to produce one thousand kg of pure zinc?

49. When 2.876 g of a certain metal sulfide is roasted in air, 2.368 g of the metal oxide are formed. If the metal has an oxidation number of +2, what is its molar mass?

50. Chalcopyrite, $CuFeS_2$, is an important source of copper. A typical chalcopyrite ore contains about 0.75% Cu. What volume of sulfur dioxide at 25°C and 1.00 atm pressure is produced when one boxcar load (4.00×10^3 ft^3) of chalcopyrite ore ($d = 2.6$ g/cm^3) is roasted? Assume all the sulfur in the ore is converted to SO_2 and no other source of sulfur is present.

51. If silver is obtained in the same manner as gold, using NaCN solution and O_2, describe with appropriate equations the extraction of silver from argentite ore, Ag_2S. (The products are SO_2 and $Ag(CN)_2^-$, which is reduced with zinc.)

52. Write the equation for the reaction of zinc metal with water and OH^- ions to form hydrogen gas and $Zn(OH)_4^{2-}$. What volume of hydrogen gas at STP would be liberated and what would be the final pH if 1.00 g of Zn were added to 100.0 mL of 6.0 M NaOH?

53. A 2.50-g sample of iron combined with 0.833 L of oxygen measured at 755 mm Hg and 27°C. Determine the simplest formula of the oxide formed.

54. Iron(II) can be oxidized to iron(III) by permanganate ion in acidic solution. The permanganate ion is reduced to manganese(II) ion.
  a. Write the oxidation half-reaction, the reduction half-reaction, and the overall redox equation.
  b. Calculate E_{tot}^0 for the reaction.
  c. Calculate the percent Fe in an ore if a 0.3500-g sample is dissolved and the Fe^{2+} formed requires for titration 55.63 mL of a 0.0200 M solution of $KMnO_4$.

55. The surface area of a car bumper is 1.8×10^2 in^2. If the bumper is to be chrome-plated with a thickness of 0.00050 in, using a Cr(VI) solution for an hour, what must the current be? Assume 25% efficiency. The density of chromium is 7.20 g/cm^3.

56. A sample of magnetite ore (impure Fe_3O_4) weighing 0.5000 g is treated so that the iron is precipitated as iron(III) hydroxide. The precipitate is heated and converted to Fe_2O_3. The mass of Fe_2O_3 obtained is 0.4980 g. What is the percent iron in the ore? What is the % Fe_3O_4?

57. Pyrolusite ore contains manganese dioxide (MnO_2). A 0.6500-g sample is treated with 1.000 g of oxalic acid ($H_2C_2O_4$) in an acidic medium. The reaction is

$$H_2C_2O_4(aq) + MnO_2(s) + 2H^+(aq) \rightarrow$$
$$Mn^{2+}(aq) + 2H_2O + 2CO_2(g)$$

The excess oxalic acid is titrated with 12.39 mL of 0.1500 M $KMnO_4$. The redox reaction for this titration is

$$5H_2C_2O_4(aq) + 2MnO_4^-(aq) + 6H^+(aq) \rightarrow$$
$$2Mn^{2+}(aq) + 8H_2O + 10CO_2(g)$$

What is the percent MnO_2 in the sample?

58. Of the cations listed on the center column of Table 22.4, which one is the
  a. strongest reducing agent?
  b. strongest oxidizing agent?
  c. weakest reducing agent?
  d. weakest oxidizing agent?

Challenge Problems

59. For the reaction

$$2Cu^+(aq) \rightleftharpoons Cu(s) + Cu^{2+}(aq); \qquad K = 1 \times 10^6$$

The formation constants of $Cu(NH_3)_2^+$ and $Cu(NH_3)_4^{2+}$ are 5×10^{10} and 2×10^{12}, respectively. Calculate K for the reaction

$$2Cu(NH_3)_2^+(aq) \rightleftharpoons Cu(s) + Cu(NH_3)_4^{2+}(aq)$$

Would you expect Cu^+ to disproportionate in a solution 1 M in NH_3?

60. Taking K_{sp} $Fe(OH)_3$ = 3×10^{-39}, K_a $HC_2H_3O_2$ = 2×10^{-5}, and K_w = 1.0×10^{-14}, calculate K for the reaction

$$Fe(OH)_3(s) + 3HC_2H_3O_2(aq) \rightleftharpoons$$
$$Fe^{3+}(aq) + 3H_2O + 3C_2H_3O_2^-(aq)$$

Would you expect $Fe(OH)_3$ to be soluble in acetic acid?

61. Rust, which you can take to be $Fe(OH)_3$, can be dissolved by treating it with oxalic acid. An acid-base reaction occurs, and a complex ion is formed.

a. Write a balanced equation for the reaction.

b. What volume of 0.10 M $H_2C_2O_4$ would be required to remove a rust stain weighing 1.0 g?

62. Consider the complex $Fe(NO)^{2+}$.

a. Assuming it consists of an Fe^{2+} ion and an NO molecule, calculate the number of unpaired electrons if it is a low-spin complex; a high-spin complex. (Remember that there is one unpaired electron in the NO molecule.)

b. Experimentally, it is found that the complex contains three unpaired electrons. Can you suggest an explanation for this number? (*Hint:* Consider the possibility of electron exchange between Fe^{2+} and NO.)

63. A 0.500-g sample of steel is analyzed for manganese. The sample is dissolved in acid and the manganese is oxidized to permanganate ion. A measured excess of Fe^{2+} is added to reduce MnO_4^- to Mn^{2+}. The excess Fe^{2+} is determined by titration with $K_2Cr_2O_7$. If 75.00 mL of 0.125 M $FeSO_4$ is added and the excess requires 13.50 mL of 0.100 M $K_2Cr_2O_7$ to oxidize Fe^{2+}, calculate the percent of Mn in the sample.

64. Calculate the temperature in °C at which the equilibrium constant (K_p) for the following reaction is 1.00:

$$MnO_2(s) \rightarrow Mn(s) + O_2(g)$$

$Cu(s) + Cl_2(g) \rightarrow CuCl_2(s)$

Hot copper foil in the presence of chlorine gas reacts vigorously, giving off sparks

CHAPTER
23

NONMETALS:
THE HALOGENS

He takes up the waters of the sea in his hand, leaving the salt;
He disperses it in mist through the skies;
He recollects and sprinkles it like grain in six-rayed snowy stars over the earth,
There to lie till he dissolves its bonds again.

HENRY DAVID THOREAU
JOURNAL JANUARY 5, 1856

In this chapter we will look at the chemistry of the elements in Group 7 of the Periodic Table, the nonmetals called the halogens. The reactions we will be discussing occur, for the most part, in water solution. You will need to be familiar with the principles governing these reactions, as covered in Chapters 14–16 and, in particular, Chapter 21, dealing with redox reactions.

We will not attempt an encyclopedic coverage of halogen chemistry. Instead, we will concentrate upon a few important topics, all of which were introduced earlier. This chapter is organized around the chemistry of:

— the halogens themselves: F_2, Cl_2, Br_2, and I_2 (Section 23.1)
— the hydrogen halides: HF, HCl, HBr, HI (Section 23.2)
— the halide ions in aqueous solution: F^-, Cl^-, Br^-, I^- (Section 23.3)
— the oxyanions and oxyacids of the halogens (Section 23.4)

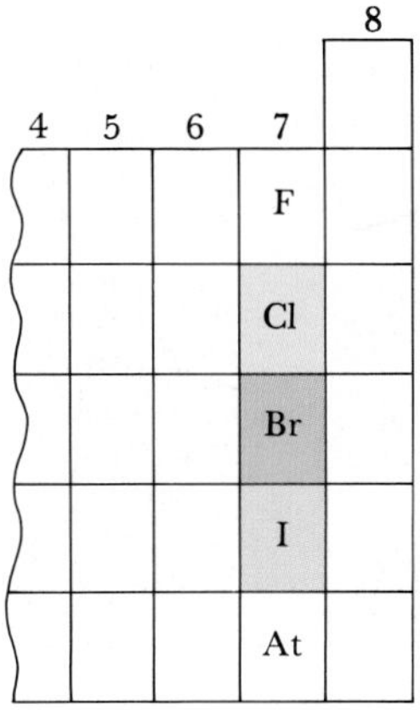

	4	5	6	7	8
				F	
				Cl	
				Br	
				I	
				At	

Throughout our discussion, we will emphasize the compounds of chlorine, since these are the ones you are most likely to encounter in the general chemistry laboratory.

23.1
THE HALOGEN MOLECULES (F_2, Cl_2, Br_2, I_2)

Table 23.1 lists some of the properties of the Group 7 elements. The colors given are those of the vapors (Fig. 23.1); solid iodine consists of shiny, black crystals. Notice the steady increase in melting and boiling point with molar mass, referred to earlier in Chapter 10. This trend reflects the fact that dispersion forces between these nonpolar molecules increase as the molecules get larger ($F_2 < Cl_2 < Br_2 < I_2$).

The heaviest halogen, astatine, is not listed in Table 23.1 for a very good reason; we know next to nothing about its chemistry. All of the isotopes of astatine are radioactive. There is perhaps 25 mg of the element on the surface of the earth at any given moment, making it the rarest of all naturally occurring elements.

CHEMICAL REACTIVITY

Fluorine is the most reactive of all elements, combining with every element except He, Ne, and Ar. With a few metals, it forms a surface film of metal fluoride which adheres tightly enough to prevent further reaction. This is the case with nickel, where the product is NiF_2. Fluorine gas is ordinarily stored in containers made of a nickel alloy, such as stainless steel (Fe, Cr, Ni) or monel (Ni,Cu). Fluorine also reacts with a great many compounds. Among these is water, which is oxidized by fluorine to give a mixture of O_2, O_3, H_2O_2, and OF_2. This means that reactions of fluorine with other species cannot be carried out in water solution.

The high reactivity of fluorine is explained in terms of two of its properties. Thermodynamically, it is a very powerful oxidizing agent; its standard reduction potential, $+2.889$ V, is higher than that of any other substance. From a kinetic standpoint, reactions of fluorine occur rapidly,

Pure F_2 reacts with just about everything, including all organic substances

TABLE 23.1   Properties of the Halogens

	F_2	Cl_2	Br_2	I_2
Molar mass (g/mol)	38.00	70.91	159.81	253.81
State at 25°C, 1 atm	gas	gas	liquid	solid
Melting point (°C)	-220	-101	-7	114
Boiling point (°C)	-188	-34	59	184
Color (gas)	pale yellow	yellow-green	red	violet
Abundance (%)	0.027	0.19	0.00016	0.00003
Bond energy (kJ/mol)	153	243	193	151
Electronegativity	4.0	3.0	2.8	2.5
E^0_{red} (V)	$+2.889$	$+1.360$	$+1.077$	$+0.534$

Figure 23.1 Flasks containing Cl_2, Br_2, and I_2 show a gradation in color from greenish-yellow through deep red to violet. The colors shown for bromine and iodine are those of the vapors in equilibrium with $Br_2(l)$ and $I_2(s)$. (Marna G. Clarke)

often violently, because the F—F bond is among the weakest of covalent bonds. The first step in a reaction involving F_2 is usually the breaking of this bond. Its weakness means that the activation energy for reactions involving fluorine is usually low, making these reactions occur rapidly, as well as spontaneously.

Chlorine is somewhat less reactive than fluorine. Although it reacts with nearly all metals (Fig. 23.2), heating is often required. This reflects the relatively strong bond in the Cl_2 molecule (B.E. Cl—Cl = 243 kJ/mol). In principle, chlorine should oxidize water:

$$2Cl_2(g) + 2H_2O \rightarrow 2Cl^-(aq) + 2H^+(aq) + O_2(g)$$

$$E^0_{tot} = E^0_{red}\ Cl_2 + E^0_{ox}\ H_2O = 1.360\ V - 1.229\ V = +0.131\ V$$

Figure 23.2 When a heated piece of copper foil is plunged into a cylinder containing chlorine gas, it reacts vigorously, giving off sparks. The equation for the reaction is: $Cu(s) + Cl_2(g) \rightarrow CuCl_2(s)$. (Charles D. Winters)

In practice, this reaction is seldom observed. Instead, chlorine *disproportionates* in water, forming Cl^- ions (oxid. no. $Cl = -1$) and HClO molecules (oxid. no. $Cl = +1$).

$$Cl_2(g) + H_2O \rightleftharpoons Cl^-(aq) + H^+(aq) + HClO(aq) \qquad (23.1)$$

Reaction 23.1 accounts for the high water solubility of Cl_2 and also for the use of "chlorine water" as a bleach and disinfectant. Hypochlorous acid, HClO, is a powerful oxidizing agent ($E^0_{red} = +1.630$ V); it kills bacteria, apparently by destroying certain enzymes essential to their metabolism. The taste and odor that we associate with "chlorinated water" is actually due to compounds such as CH_3NHCl, produced by the action of hypochlorous acid on bacteria.

■ EXAMPLE 23.1

For Reaction 23.1:

$$K = \frac{[H^+] \times [Cl^-] \times [HClO]}{(P\ Cl_2)} = 2.7 \times 10^{-5}$$

Calculate the concentration of HClO in equilibrium with $Cl_2(g)$ at 1.0 atm.

Solution

If we let $x = [HClO]$, we see from the balanced equation that:

$$[H^+] = [Cl^-] = [HClO] = x$$

Hence we have: $\dfrac{x^3}{1.0} = 2.7 \times 10^{-5}$; $x = (2.7 \times 10^{-5})^{1/3} = \boxed{0.030}$ We conclude that the concentration of HClO in a solution formed by bubbling chlorine gas through water should be about 0.03 mol/L.

Exercise

What is the pH of the equilibrium mixture in Reaction 23.1, assuming essentially all the H^+ ions come from the strong acid HCl? Answer: 1.52.

As you would expect on the basis of standard potentials, chlorine readily oxidizes Br^- or I^- ions in solution:

$$Cl_2(g) + 2Br^-(aq) \rightarrow 2Cl^-(aq) + Br_2(l) \qquad (23.2)$$

$$E^0_{tot} = E^0_{red}\ Cl_2 + E^0_{ox}\ Br^- = 1.360\ V - 1.077\ V = +0.283\ V$$

$$Cl_2(g) + 2I^-(aq) \rightarrow 2Cl^-(aq) + I_2(s) \qquad (23.3)$$

$$E^0_{tot} = E^0_{red}\ Cl_2 + E^0_{ox}\ I^- = 1.360\ V - 0.534\ V = +0.826\ V$$

Bromine, in turn, can oxidize iodide ions:

$$Br_2(l) + 2I^-(aq) \rightarrow 2Br^-(aq) + I_2(s);\ E^0_{tot} = +0.543\ V \qquad (23.4)$$

Reactions 23.2 and 23.3 are often used to test for the presence of Br^- or I^- ions in solution (Fig. 23.3). Addition of chlorine to a solution containing either of these ions gives the free halogens, Br_2 or I_2. After addition of chlorine, the solution is shaken with an organic solvent. The free halogen

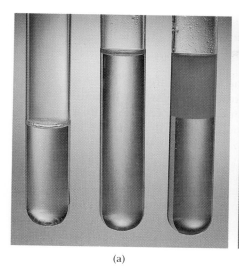

(a)

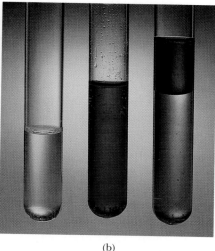

(b)

Figure 23.3 The reaction of chlorine with Br^- and I^- ions is shown in (a) and (b), respectively. The anions themselves (test tubes at *left*) are colorless. Oxidation by chlorine yields the free halogens, Br_2 and I_2, which are colored in water solution (*center* tubes) and more intensely colored in an organic solvent (upper layer in the tubes at the *right*). (Charles D. Winters)

enters the organic layer, in which it is more soluble. It gives that layer its characteristic color: reddish-yellow (bromine) or violet (iodine).

The oxidizing power of the halogens makes them hazardous to work with. Fluorine is the most dangerous, but it is very unlikely that you will ever come across it in a teaching laboratory. Chlorine you are most likely to encounter as its saturated water solution, called "chlorine water." Remember that the pressure of chlorine gas over this solution (if it is freshly prepared) is 1 atm and that chlorine was used as a poison gas in World War I. Use small quantities of chlorine water and don't breathe the vapors. Bromine, although not as strong an oxidizing agent as chlorine, can cause severe burns if it comes in contact with your skin, particularly if it gets under your fingernails.

Of the four halogens, iodine is the weakest oxidizing agent. When the authors of this textbook were young, "tincture of iodine," a 10% solution of I_2 in alcohol, was widely used as an antiseptic. Today, hospitals use a product called "povidone-iodine," a quite powerful iodine-containing antiseptic and disinfectant, which can be diluted with water to the desired strength. These applications of molecular iodine should not delude you into thinking that the solid is harmless. On the contrary, if $I_2(s)$ is allowed to remain in contact with your skin, it can cause painful burns that are slow to heal.

The colors can be used to identify the halogen

PREPARATION

Fluorine is made commercially* by electrolytic oxidation. The cell (Fig. 23.4) operates at about 100°C; it contains a liquid mixture of HF and KF

---

*Until very recently, fluorine had never been prepared by chemical oxidation. In 1986 the element was made chemically by heating K_2MnF_6 with SbF_5 at 150°C. (See *Chemical and Engineering News*, Sept. 15, 1986, p. 23.)

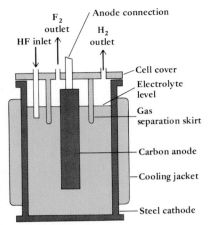

Figure 23.4 Fluorine gas is produced by the electrolysis of hydrogen fluoride in molten potassium fluoride.

in a mole ratio of approximately 2HF:1KF. Fluorine is generated at the carbon anode by the decomposition of hydrogen fluoride:

$$2HF(l) \rightarrow H_2(g) + F_2(g) \qquad \Delta G^0 \approx +540 \text{ kJ} \qquad (23.5)$$

The potassium fluoride furnishes the ions required to carry the electric current. The cell is designed to avoid contact between H_2 and F_2, which would react spontaneously (and violently) by the reverse of Reaction 23.5.

As noted in Chapter 21, chlorine is ordinarily prepared by the electrolysis of aqueous sodium chloride:

$$2Cl^-(aq) + 2H_2O \rightarrow Cl_2(g) + H_2(g) + 2OH^-(aq) \qquad (23.6)$$

Smaller amounts of Cl_2 are obtained as by-products in the electrolytic preparation of sodium and magnesium metal from their molten chlorides.

$$2NaCl(l) \rightarrow 2Na(l) + Cl_2(g) \qquad (23.7)$$

$$MgCl_2(l) \rightarrow Mg(l) + Cl_2(g) \qquad (23.8)$$

■ **EXAMPLE 23.2**

Suppose $Cl_2(g)$ is prepared by Reaction 23.6 at the rate of 1.00 kg per hour.

a. What voltage must be used (recall Table 21.1, Ch. 21)?
b. How many coulombs of electricity must pass through the cell?
c. What current must be used, in amperes?

Solution

a. $E^0_{tot} = E^0_{ox} \, Cl^- + E^0_{red} \, H_2O = -1.360 \text{ V} - 0.828 \text{ V} = -2.188 \text{ V}$

The negative sign means that a voltage of at least 2.188 V must be applied to make this nonspontaneous reaction occur in an electrolytic cell.

b.   no. moles $e^- = 1000$ g $Cl_2 \times \dfrac{1 \text{ mol } Cl_2}{70.91 \text{ g } Cl_2} \times \dfrac{2 \text{ mol } e^-}{1 \text{ mol } Cl_2} = 28.2$ mol e^-

   no. coulombs $= 28.2$ mol $e^- \times \dfrac{9.65 \times 10^4 C}{1 \text{ mol } e^-} = 2.72 \times 10^6$ C

c.   no. amperes $= \dfrac{\text{no. coulombs}}{\text{no. seconds}} = \dfrac{2.72 \times 10^6 \text{ C}}{3600 \text{ s}} = 756$ A

Exercise

How many kilojoules of electrical energy must be supplied to produce one kilogram of $Cl_2(g)$ by Reaction 23.6? Answer: 5.95×10^3 kJ.

Seawater was used as a source of Br_2 for many years

Bromine and iodine are produced commercially from Br^- and I^- ions by oxidation with chlorine (Reactions 23.2 and 23.3). The source of bromide ions in the United States is a series of deep brine wells in Arkansas, where the concentration of Br^- is about 0.05 mol/L. Iodide ions are obtained from salt water in certain wells in Michigan; the concentration of I^- is of the order of 0.001 mol/L. About 1×10^3 metric tons of I_2 and 2×10^5 metric tons of Br_2 are produced annually in the United States, as compared to 1×10^7 metric tons of Cl_2. These numbers give you an idea of the relative economic importance of these three halogens.

23.2
THE HYDROGEN HALIDES (HF, HCl, HBr, HI)

Table 23.2 lists some of the properties of the hydrogen halides. One number that stands out is the high boiling point of HF, 20°C. This is at least a hundred degrees higher than the value obtained by extrapolation from the boiling points of the other hydrogen halides. This is one of the effects of hydrogen bonding (Chapter 10), which is found in HF but not in HCl, HBr, or HI. Hydrogen fluoride is unusual in that hydrogen bonding persists in the gas phase. The vapor density at the normal boiling point, 20°C and 1 atm, is 3.11 g/L; the calculated molar mass is:

$$\text{MM} = \frac{dRT}{P} = \frac{3.11 \times 0.0821 \times 293}{1.00} \text{ g/mol} = 74.8 \text{ g/mol}$$

This is nearly four times the molar mass of unassociated HF, 20.0 g/mol.

That means that there are a lot of $(HF)_n$ molecules in the gas, where n_{ave} is about 4

Water solutions of the hydrogen halides are referred to as *hydrohalic* acids (e.g., hydrofluoric acid, hydrochloric acid). You may be curious as to why HF is the only weak acid of the group; if so you are not alone. There is no simple explanation for the anomalous behavior of HF. In part, it can be attributed to the unusually strong bond in the HF molecule (Table 23.2) which tends to inhibit its dissociation. A more important factor may be hydrogen bonding between the F^- ion and H_2O molecules in solution. This leads to a highly ordered solution structure with an unusually low entropy, which in turn makes the dissociation of HF nonspontaneous.

CHEMICAL REACTIVITY

The hydrohalic acids react readily with bases; the reactions with OH^- ions are typical:

$$H^+(aq) + OH^-(aq) \rightarrow H_2O \quad \text{(with HCl, HBr, HI)} \qquad (23.9)$$

$$HF(aq) + OH^-(aq) \rightarrow H_2O + F^-(aq) \qquad (23.10)$$

In these equations, as in all equations, we include only the *principal species* taking part in the reaction. With the strong acids HCl, HBr, and HI, this

TABLE 23.2   Properties of the Hydrogen Halides

	HF	HCl	HBr	HI
Molar mass (g/mol)	20.01	36.46	80.91	127.91
State at 25°C, 1 atm	gas	gas	gas	gas
Melting point (°C)	-83	-115	-89	-51
Boiling point (°C)	$+20$	-84	-67	-35
Bond energy (kJ/mol)	565	431	368	297
K_a	6.9×10^{-4}	$\sim 10^7$	$\sim 10^9$	$\sim 10^{10}$

is the H^+ ion; the halide ions (Cl^-, Br^-, I^-) do not enter into the reaction and so do not appear in the equation. The principal species in the weak hydrofluoric acid is the HF molecule; there are relatively few H^+ ions present, so we don't put them in the equation.

■ EXAMPLE 23.3

Write net ionic equations for the acid-base reactions between solutions of:

a. hydrochloric acid and ammonia
b. hydrochloric acid and sodium cyanide, NaCN
c. hydrofluoric acid and sodium acetate, $NaC_2H_3O_2$

Solution

In each case, a weak base (NH_3, CN^-, $C_2H_3O_2^-$) is converted to the conjugate acid (NH_4^+, HCN, $HC_2H_3O_2$). The Na^+ ions do not take part in the reactions; neither do the halide ions from the strong acids HCl and HI.

a. $H^+(aq) + NH_3(aq) \rightarrow NH_4^+(aq)$

b. $H^+(aq) + CN^-(aq) \rightarrow HCN(aq)$

c. $HF(aq) + C_2H_3O_2^-(aq) \rightarrow HC_2H_3O_2(aq) + F^-(aq)$

Exercise

Suppose hydrobromic acid is added to a solution of sodium fluoride. Write a net ionic equation for the reaction that takes place. Answer: $H^+(aq) + F^-(aq) \rightarrow HF(aq)$.

Another base which reacts readily with the hydrohalic acids is the carbonate ion, CO_3^{2-}. With excess acid, carbon dioxide gas is formed. The equations for the reactions are:

$$2H^+(aq) + CO_3^{2-}(aq) \rightarrow H_2O + CO_2(g) \quad \text{(with HCl, HBr, HI)} \quad (23.11)$$
$$2HF(aq) + CO_3^{2-}(aq) \rightarrow H_2O + CO_2(g) + 2F^-(aq) \quad (23.12)$$

The strong hydrohalic acids react with water-insoluble hydroxides and carbonates. The acid-base reactions are essentially identical to Equations 23.9 and 23.11, but the equations look different because we start with an insoluble solid instead of an ion in solution. The reactions of hydrochloric acid with magnesium hydroxide and magnesium carbonate are represented by the equations:

$$Mg(OH)_2(s) + 2H^+(aq) \rightarrow 2H_2O + Mg^{2+}(aq) \quad (23.13)$$

$$MgCO_3(s) + 2H^+(aq) \rightarrow H_2O + CO_2(g) + Mg^{2+}(aq) \quad (23.14)$$

Reactions such as these can be used in the laboratory to convert a metal hydroxide or carbonate to the corresponding halide. Suppose, for example, you want to make iron(III) chloride, $FeCl_3$, from the hydroxide, $Fe(OH)_3$. To start with, you should add enough hydrochloric acid (Fig. 23.5) to

Figure 23.5 Addition of hydrochloric acid to red, insoluble $Fe(OH)_3$ produces a solution containing Fe^{3+} and Cl^- ions. Evaporation of this solution gives crystals of hydrated $FeCl_3$. (Marna G. Clarke)

dissolve the hydroxide. Including the Cl^- ions, because they will be needed later, the equation for this reaction becomes:

$$Fe(OH)_3(s) + 3H^+(aq) + 3Cl^-(aq) \rightarrow 3H_2O + Fe^{3+}(aq) + 3Cl^-(aq) \quad (23.15a)$$

At this point, you have a water solution of iron(III) chloride. Evaporation of the solution gives the solid (most likely in the form of a hydrate):

$$Fe^{3+}(aq) + 3Cl^-(aq) \rightarrow FeCl_3(s) \qquad (23.15b)$$

Adding the equations for the two steps gives the overall equation for the conversion of iron(III) hydroxide to the chloride:

$$Fe(OH)_3(s) + 3H^+(aq) + 3Cl^-(aq) \rightarrow FeCl_3(s) + 3H_2O \qquad (23.15)$$

■ **EXAMPLE 23.4** _____

Describe how cobalt(II) bromide can be prepared from cobalt(II) carbonate, a water-insoluble solid. Write a balanced equation for the preparation.

Solution

Add hydrobromic acid with stirring until all the $CoCO_3$ goes into solution. Heat to drive off the water and any excess HBr.

$$CoCO_3(s) + 2H^+(aq) + 2Br^-(aq) \rightarrow H_2O + CO_2(g) + Co^{2+}(aq) + 2Br^-(aq)$$
$$Co^{2+}(aq) + 2Br^-(aq) \rightarrow CoBr_2(s)$$

$$\overline{CoCO_3(s) + 2H^+(aq) + 2Br^-(aq) \rightarrow CoBr_2(s) + H_2O + CO_2(g)}$$

Exercise

Write the equation for the preparation of magnesium iodide from magnesium hydroxide. Answer:
$$Mg(OH)_2(s) + 2H^+(aq) + 2I^-(aq) \rightarrow MgI_2(s) + 2H_2O.$$

Concentrated hydrofluoric acid reacts with glass, which we can consider to be a mixture of SiO_2 and ionic silicates such as calcium silicate, $CaSiO_3$:

$$SiO_2(s) + 4HF(aq) \rightarrow SiF_4(g) + 2H_2O \qquad (23.16)$$

$$CaSiO_3(s) + 6HF(aq) \rightarrow SiF_4(g) + CaF_2(s) + 3H_2O \qquad (23.17)$$

As you might guess, HF solutions are never stored in glass bottles; plastic is used instead. Reactions 23.16 and 23.17 are sometimes used to etch glass. The glass object is first covered with a thin protective coating of wax or plastic. Then the coating is removed from the area to be etched and the glass is exposed to the HF solution. Thermometer stems and burets can be etched or light bulbs frosted in this way.

Hydrogen fluoride is a very unpleasant chemical to work with. If spilled on the skin, it removes Ca^{2+} ions from the tissues, forming insoluble CaF_2. A white patch forms which is agonizingly painful to the touch. To make matters worse, HF is a local anaesthetic, so a person may be unaware of what's happening until it's too late.

We used it in the old days, in the glass etching experiment

Rest assured that you will never have occasion to use hydrofluoric acid in the general chemistry laboratory. Neither are you likely to come in contact with HBr or HI, both of which are relatively expensive. Hydrochloric acid, on the other hand, is a "workhorse" chemical of the teaching laboratory. It is commonly available as "dilute HCl" (6 M) or "concentrated HCl" (12 M). You will use it as a source of H^+ ions for such purposes as:

— dissolving insoluble hydroxides and carbonates (Reactions 23.13, 23.14)

— converting a weak base such as NH_3 to its conjugate acid (Example 23.3)

— generating a gas, which might be CO_2 (Reaction 23.11) or hydrogen:

$$Zn(s) + 2H^+(aq) \rightarrow Zn^{2+}(aq) + H_2(g) \tag{23.18}$$

or even hydrogen sulfide:

$$FeS(s) + 2H^+(aq) \rightarrow Fe^{2+}(aq) + H_2S(g) \tag{23.19}$$

Occasionally, you may use hydrochloric acid to furnish Cl^- ions, as in the precipitation of the Group I cations (Chapter 17).

PREPARATION

There are two general methods of preparing the hydrogen halides.

1. *Direct reaction between the elements*

$$H_2(g) + X_2(g) \rightarrow 2HX(g) \tag{23.20}$$

This reaction is favored thermodynamically; except with HI, the equilibrium constant is so large that the product is virtually pure. The rate of reaction decreases in the order HF > HCl > HBr > HI. The reaction between H_2 and F_2 is virtually impossible to control; it occurs instantly and violently when the elements come in contact with each other. In contrast, a mixture of H_2 and Cl_2 can be kept in the dark indefinitely; upon exposure to light, HCl is formed rapidly, but at a controllable rate. With Br_2 and I_2, hydrogen reacts quite slowly; the commercial preparation of HBr and HI by Reaction 23.20 is carried out at 200–400°C using a platinum catalyst.

■ **EXAMPLE 23.5** —————————————————————
Use the data in Table 23.3 to calculate

a. ΔH^0 and ΔS^0 for the reaction of H_2 with I_2.
b. K at 25°C and 200°C for the reaction of H_2 with I_2.

TABLE 23.3   Thermodynamic Data for the Hydrogen Halides

REACTION	ΔG^0 at 25°C	ΔG^0 at 200°C
$H_2(g) + F_2(g) \rightarrow 2HF(g)$	−546 kJ	−549 kJ
$H_2(g) + Cl_2(g) \rightarrow 2HCl(g)$	−191 kJ	−194 kJ
$H_2(g) + Br_2(g) \rightarrow 2HBr(g)$	−110 kJ	−114 kJ
$H_2(g) + I_2(g) \rightarrow 2HI(g)$	−16 kJ	−20 kJ

Solution

a. Applying the Gibbs-Helmholtz equation, $\Delta G^0 = \Delta H^0 - T\Delta S^0$, at the two different temperatures and subtracting:

$$-16 \text{ kJ} = \Delta H^0 - 298 \ K(\Delta S^0)$$
$$-20 \text{ kJ} = \Delta H^0 - 473 \ K(\Delta S^0)$$

$$4 \text{ kJ} = 175 \ K(\Delta S^0); \quad \boxed{\Delta S^0 = +0.02 \text{ kJ/K}}$$

Substituting this value of ΔS^0 into either of the equations and solving for ΔH^0, we obtain: $\boxed{\Delta H^0 = -10 \text{ kJ.}}$

b. The appropriate relation is: $\Delta G^0 = -0.00831 \ T \ln K$

At 25°C: $\ln K = \dfrac{16}{0.00831 \times 298} = 6.5$; $\boxed{K = 6 \times 10^2}$

At 200°C: $\ln K = \dfrac{20}{0.00831 \times 473} = 5.1$; $\boxed{K = 2 \times 10^2}$

Exercise
Can you suggest why ΔG^0 is nearly independent of temperature? Answer: ΔS^0 is very small because there is no change in the number of moles of gas.

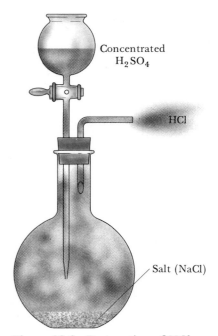

Figure 23.6 Preparation of HCl, a colorless gas, from NaCl, a white solid.

2. *Heating a halide salt with a nonvolatile acid.* This reaction offers the only feasible way of making hydrogen fluoride. Sulfuric acid and the mineral fluorspar, CaF_2, are heated together at 200°C:

$$CaF_2(s) + H_2SO_4(l) \rightarrow 2HF(g) + CaSO_4(s) \qquad (23.21)$$

Hydrogen fluoride, which boils at 20°C, is driven off as a gas.

Hydrogen chloride was prepared by the alchemists by heating sodium chloride with sulfuric acid (Fig. 23.6). At 200°C, the products are hydrogen chloride and sodium hydrogen sulfate:

$$NaCl(s) + H_2SO_4(l) \rightarrow HCl(g) + NaHSO_4(s) \qquad (23.22)$$

The old name was muriatic acid

Many years ago, this method of preparing HCl gave way to the direct reaction of the elements with each other (Reaction 23.20). Today, more than 90% of the hydrogen chloride made in the United States is a by-product of the chlorination of organic compounds, e.g:

$$C_2H_4(g) + Cl_2(g) \rightarrow C_2H_3Cl(g) + HCl(g)$$

Figure 23.7 When sulfuric acid is added to solid sodium iodide (white), a redox reaction occurs, producing violet iodine vapor and hydrogen sulfide gas. (Charles D. Winters)

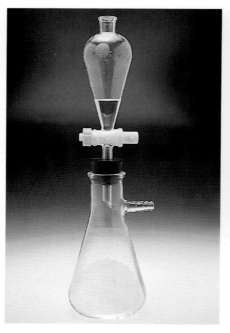 

Hydrogen bromide and hydrogen iodide cannot be prepared by heating NaBr or NaI with sulfuric acid; the Br^- and I^- ions are oxidized by SO_4^{2-} (Fig. 23.7, Example 23.6). HBr and HI are made using phosphoric acid, H_3PO_4, which is a much weaker oxidizing agent than H_2SO_4.

$$NaBr(s) + H_3PO_4(l) \rightarrow HBr(g) + NaH_2PO_4(s) \tag{23.23}$$

$$NaI(s) + H_3PO_4(l) \rightarrow HI(g) + NaH_2PO_4(s) \tag{23.24}$$

■ EXAMPLE 23.6

When sodium iodide is heated with sulfuric acid in water solution, the products include iodine and hydrogen sulfide. Write a balanced equation for the redox reaction.

Solution

The balanced half-equations are:

oxidation: $2I^-(aq) \rightarrow I_2(s) + 2e^-$

reduction: $SO_4^{2-}(aq) + 10H^+(aq) + 8e^- \rightarrow H_2S(g) + 4H_2O$

Combining these equations in the usual way, we obtain:

$$8I^-(aq) + SO_4^{2-}(aq) + 10H^+(aq) \rightarrow 4I_2(s) + H_2S(g) + 4H_2O$$

Exercise

What volume of $H_2S(g)$, at 25°C and 1.00 atm, is produced from 1.00 g of NaI, using excess H_2SO_4? Answer: 20.4 mL.

23.3
THE HALIDE IONS (F^-, Cl^-, Br^-, I^-)

In this section, we will consider the chemistry of the halide ions in water solution. They can take part in three quite different types of reactions.

1. *Lewis acid-base reactions.* A halide ion can act as a Lewis base, supplying an electron pair to form a coordinate covalent bond with a Lewis acid. The electron-pair acceptor may be a molecule, like BF_3:

$$\ddot{\text{F}} - \text{B} \;+\; (\ddot{\text{F}})^- \;\longrightarrow\; (\ddot{\text{F}} - \text{B} - \ddot{\text{F}})^- \tag{23.25}$$

or I_2:

$$\ddot{\text{I}} - \ddot{\text{I}} \;+\; (\ddot{\text{I}})^- \;\longrightarrow\; (\ddot{\text{I}} - \ddot{\text{I}} - \ddot{\text{I}})^- \tag{23.26}$$

This reaction, leading to the formation of the triiodide ion, I_3^-, accounts for the fact that iodine is much more soluble in KI solution than in pure water.

More frequently, the Lewis acid is a cation like Al^{3+} or Hg^{2+}, in which case the product is a complex ion of the type discussed in Chapter 16.

$$Al^{3+}(aq) + 6F^-(aq) \longrightarrow AlF_6^{3-}(aq) \tag{23.27}$$

$$Hg^{2+}(aq) + 4I^-(aq) \longrightarrow HgI_4^{2-}(aq) \tag{23.28}$$

The AlF_6^{3-} ion is found in cryolite, Na_3AlF_6. The formation of the HgI_4^{2-} ion explains why mercury(II) iodide is more soluble in KI solution than in pure water.

The fluoride ion can supply an electron pair to a proton, forming the weak acid molecule HF:

$$H^+(aq) + F^-(aq) \longrightarrow HF(aq) \tag{23.29}$$

This reaction does not occur with Cl^-, Br^-, or I^- ions, because HCl, HBr, and HI are all strong acids.

2. *Precipitation reactions.* You will recall from Chapter 17 that the Cl^- ion is used to precipitate the Group I cations:

$$Ag^+(aq) + Cl^-(aq) \longrightarrow AgCl(s) \tag{23.30}$$

$$Pb^{2+}(aq) + 2Cl^-(aq) \longrightarrow PbCl_2(s) \tag{23.31}$$

$$Hg_2^{2+}(aq) + 2Cl^-(aq) \longrightarrow Hg_2Cl_2(s) \tag{23.32}$$

Similar reactions occur with Br^- and I^- (Table 23.4, p. 824). The F^- ion behaves quite differently. Silver fluoride, AgF, is soluble in water; on the other hand, the fluorides of the alkaline earth metals are insoluble. If solutions of sodium fluoride and calcium chloride are mixed, a white precipitate forms:

$$Ca^{2+}(aq) + 2F^-(aq) \longrightarrow CaF_2(s) \tag{23.33}$$

TABLE 23.4   Solubility Products of Insoluble Halides

FLUORIDES	K_{sp}	CHLORIDES	K_{sp}	BROMIDES	K_{sp}	IODIDES	K_{sp}
MgF_2	7×10^{-11}	$AgCl$	1.8×10^{-10}	$AgBr$	5×10^{-13}	AgI	1×10^{-16}
CaF_2	1.5×10^{-10}	$PbCl_2$	1.7×10^{-5}	$PbBr_2$	6.6×10^{-6}	PbI_2	8.4×10^{-9}
SrF_2	4.4×10^{-9}	Hg_2Cl_2	1×10^{-18}	Hg_2Br_2	6×10^{-23}	Hg_2I_2	5×10^{-29}
BaF_2	1.8×10^{-7}						
PbF_2	7.1×10^{-7}						
Hg_2F_2	3.2×10^{-6}						

■ **EXAMPLE 23.7**

Calcium fluoride occurs as the mineral fluorspar. Ground water which comes in contact with fluorspar becomes saturated with CaF_2. What is the concentration of F^- ions (mol/L) in this water?

Solution

Letting s = solubility of CaF_2, we have:

$$[Ca^{2+}] = s; \qquad [F^-] = 2s; \qquad (s)(2s)^2 = K_{sp}\ CaF_2 = 1.5 \times 10^{-10}$$

$$s = \left(\frac{1.5 \times 10^{-10}}{4}\right)^{1/3} = 3.3 \times 10^{-4}\ \text{mol/L}$$

$$[F^-] = 2s = \boxed{6.6 \times 10^{-4}\ \text{mol/L}}$$

This corresponds to about 13 mg F^-/L; the concentration of F^- added to water to reduce tooth decay is only about 1 mg/L.

Exercise

At low pH, the solubility of CaF_2 is considerably higher than the value just calculated. Explain with a chemical equation.
Answer: $CaF_2(s) + 2H^+(aq) \rightarrow Ca^{2+}(aq) + 2HF(aq)$.

3. *Oxidation-reduction reactions.* The lowest oxidation number for a halogen is -1. It follows that **in any redox reaction in which a halide ion, X^-, is involved, it must be oxidized and hence act as a reducing agent.** Ordinarily, oxidation produces the free halogens (oxid. no. = 0); reducing strength increases in the order:

$$\begin{array}{ccccccc} & F^- & < & Cl^- & < & Br^- & < & I^- \\ E^0_{ox}\ (V) & -2.889 & & -1.360 & & -1.077 & & -0.534 \end{array}$$

The fluoride ion never participates in redox reactions in water solution; it is too difficult to oxidize. The other halide ions can be oxidized by MnO_4^- ions in water solution. The reaction with Cl^- is typical:

$$2MnO_4^-(aq) + 16H^+(aq) + 10Cl^-(aq) \rightarrow 2Mn^{2+}(aq) + 8H_2O + 5Cl_2(g) \quad (23.34)$$

Manganese dioxide can also be used as an oxidizing agent (Fig. 23.8), even though the standard voltage with Cl^- is a negative quantity (Example 23.8):

$$MnO_2(s) + 4H^+(aq) + 2Cl^-(aq) \rightarrow Mn^{2+}(aq) + Cl_2(g) + 2H_2O \quad (23.35)$$

$$E^0_{tot} = E^0_{red}\ MnO_2 + E^0_{ox}\ Cl^- = 1.229\ \text{V} - 1.360\ \text{V} = -0.131\ \text{V}$$

Figure 23.8 Addition of HBr (dropping funnel) to MnO_2 produces bromine (red fumes). A similar reaction (Equation 23.35) occurs with HCl to produce chlorine gas. (Charles D. Winters)

■ EXAMPLE 23.8

Calculate E for Reaction 23.35 taking place in concentrated HCl (12 M), assuming $P\ Cl_2(g) = 1$ atm and conc. $Mn^{2+} = 0.10\ M$.

Solution

We use the Nernst equation, with $E^0_{tot} = -0.131$ V and $n = 2$:

$$E = -0.131\ \text{V} - \frac{0.0257}{2} \ln \frac{(\text{conc. } Mn^{2+}) \times P\ Cl_2}{(\text{conc. } H^+)^4 \times (\text{conc. } Cl^-)^2}$$

$$= -0.131\ \text{V} - \frac{0.0257}{2} \ln \frac{(0.10)(1.0)}{12^4 \times 12^2}$$

$$= -0.131\ \text{V} + 0.221\ \text{V} = \boxed{+0.090\ \text{V}}$$

It takes conc. HCl to make this reaction go

This voltage is a positive quantity, as it must be if the reaction is to occur.

Exercise

Calculate the concentration of HCl at which $E = 0$. Answer: 3.7 M.

Bromide and iodide ions are much more readily oxidized than chloride ions. As pointed out earlier, Cl_2 oxidizes Br^- and I^- to Br_2 and I_2. In principle, Br^- and I^- ions should be oxidized by molecular oxygen, which is nearly as strong an oxidizing agent as chlorine ($E^0_{red}\ O_2 = +1.229$ V).

Solutions of NaBr appear to be stable in the presence of dissolved air, but solutions containing I^- ions slowly take on a yellow color as the following reaction occurs:

$$O_2(g) + 4H^+(aq) + 4I^-(aq) \rightarrow 2H_2O + 2I_2(s) \tag{23.36}$$

$$E^0_{tot} = E^0_{red} \, O_2 + E^0_{ox} \, I^- = +0.695 \text{ V}$$

23.4
OXYACIDS AND OXYANIONS

Table 23.5 lists the oxyacids and oxyanions of the heavier halogens. Only one such compound, HFO, is known for fluorine. Hypofluorous acid can be prepared as a pale yellow liquid by passing fluorine over ice at $-40°C$. At room temperature, it decomposes rapidly to HF and O_2.

The structures of the oxyacids of chlorine are shown in Figure 23.9. Notice that in each case the hydrogen atom is bonded to oxygen rather than chlorine. This is generally true for oxyacids; *only those hydrogen atoms bonded to oxygen dissociate to form H^+ ions in water.*

Of the many oxyacids of the halogens, only four ($HClO_4$, HIO_4, H_5IO_6, HIO_3) can be isolated as pure compounds. Of these, by far the most important is perchloric acid, $HClO_4$. This substance can be prepared by heating (very carefully!) a metal perchlorate with sulfuric acid:

$$KClO_4(s) + H_2SO_4(l) \rightarrow KHSO_4(s) + HClO_4(l) \tag{23.37}$$

The pure acid and its concentrated water solution (above 70% $HClO_4$) are explosive and unsafe to work with. In cold, dilute solution, perchloric acid is a stable, very strong acid:

$$HClO_4(aq) \rightarrow H^+(aq) + ClO_4^-(aq) \tag{23.38}$$

Putting it another way, the ClO_4^- ion is an extremely weak Lewis base. Not only does it not combine with a H^+ ion; it shows little or no tendency to form complexes with metal cations.

$HClO_4(l)$ is a very dangerous chemical, one to stay away from

NOMENCLATURE

Table 23.6, p. 227, lists names of oxyacids and oxyanions of chlorine. The system used illustrates the general procedure for naming oxyacids and oxyanions of the nonmetals. The rules are as follows:

1. When a nonmetal forms two oxyanions, the suffix *-ate* is used for the anion in which the nonmetal is in the higher oxidation state (larger number

TABLE 23.5 Oxyacids and Oxyanions of Cl, Br, and I

OXID. STATE	ACID	ANION
+7	$HClO_4$, $HBrO_4$, HIO_4, H_5IO_6	ClO_4^-, BrO_4^-, IO_4^-, IO_6^{5-}
+5	$HClO_3$, $HBrO_3$, HIO_3	ClO_3^-, BrO_3^-, IO_3^-
+3	$HClO_2$	ClO_2^-, BrO_2^-
+1	$HClO$, $HBrO$, HIO	ClO^-, BrO^-, IO^-

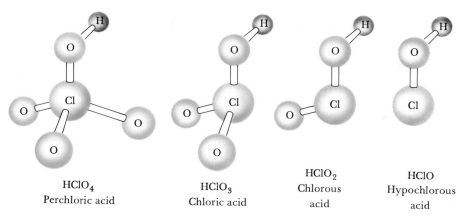

HClO₄
Perchloric acid

HClO₃
Chloric acid

HClO₂
Chlorous
acid

HClO
Hypochlorous
acid

Figure 23.9 In perchloric acid, $HClO_4$, the Cl atom is at the center of a tetrahedron, bonded to four oxygen atoms; there is a hydrogen atom bonded to one of the oxygens. In $HClO_3$, $HClO_2$, and $HClO$, successive oxygen atoms are removed from corners of the tetrahedron. In all of these acids, hydrogen is bonded to oxygen, which explains why you often see formulas like $HOCl$ (for hypochlorous acid).

of oxygen atoms). The suffix *-ite* is used for the anion containing the nonmetal in the lower oxidation state (fewer oxygen atoms). Thus we have

SO_4^{2-} sulfate NO_3^- nitrate

SO_3^{2-} sulfite NO_2^- nitrite

2. When a nonmetal forms more than two oxyanions, the prefixes *per-* (highest oxidation state) and *hypo-* (lowest oxidation state) are used as well. This is necessary with the oxyanions of chlorine, bromine, and iodine.

BrO_4^- perbromate
BrO_3^- bromate
BrO_2^- bromite
BrO^- hypobromite

The *-ate* anions are derived from *-ic* acids; sulfate from sulfuric acid

3. The name of an oxyacid is directly related to that of the corresponding anion. The suffix *-ate* is replaced by *-ic*; *-ite* is replaced by *-ous*.

H_2SO_4 sulfuric acid HNO_3 nitric acid

H_2SO_3 sulfurous acid HNO_2 nitrous acid

The sulf*ite* ion comes from sulfur*ous* acid

ACID STRENGTH

The dissociation constants of the oxyacids of the halogens are listed in Table 23.7, p. 828. Notice that the values of K_a increase with

— *increasing oxidation number of the central atom* ($HClO < HClO_2 < HClO_3$)
— *increasing electronegativity of the central atom* ($HIO < HBrO < HClO$)

TABLE 23.6 Names of the Oxyacids and Oxyanions of Chlorine

OXID. STATE	ACID		ANION	
+7	$HClO_4$	perchloric acid	ClO_4^-	perchlorate
+5	$HClO_3$	chloric acid	ClO_3^-	chlorate
+3	$HClO_2$	chlorous acid	ClO_2^-	chlorite
+1	$HClO$	hypochlorous acid	ClO^-	hypochlorite

TABLE 23.7   Dissociation Constants of Oxyacids

OXID. STATE		K_a		K_a		K_a
+7	$HClO_4$	$\sim 10^7$	$HBrO_4$	$\sim 10^6$	*HIO_4	1.4×10^1
+5	$HClO_3$	$\sim 10^3$	$HBrO_3$	3.0	HIO_3	1.6×10^{-1}
+3	$HClO_2$	1.0×10^{-2}				
+1	$HClO$	2.8×10^{-8}	$HBrO$	2.6×10^{-9}	HIO	2.4×10^{-11}

*Estimated; in water solution the stable species is H_5IO_6, whose first dissociation constant is 5×10^{-4}.

These trends are general ones, observed with other oxyacids of the nonmetals. Recall, for example, that nitric acid, HNO_3 (oxid. no. $N = +5$), is a strong acid, completely dissociated in water. In contrast, nitrous acid, HNO_2 (oxid. no. $N = +3$) is a weak acid ($K_a = 6.0 \times 10^{-4}$). The electronegativity effect shows up when we compare the strengths of the oxyacids of sulfur and selenium:

$$K_a \; H_2SO_3 = 1.7 \times 10^{-2} \qquad K_a \; H_2SeO_3 = 2.7 \times 10^{-3}$$

These trends in acid strength can be related to molecular structure. In an oxyacid molecule, the hydrogen atom that dissociates is bonded to oxygen, which in turn is bonded to a nonmetal atom, X. We might represent the structure of an oxyacid as H—O—X and its dissociation in water as

$$H—O—X(aq) \rightleftharpoons H^+(aq) + XO^-(aq)$$

For a proton, with its +1 charge, to separate from the molecule, the electron density around the oxygen should be as low as possible. This will weaken the O—H bond and favor dissociation. The electron density around the oxygen atom is decreased when:

1. X is a highly electronegative atom such as Cl. This draws electrons away from the oxygen atom and makes hypochlorous acid stronger than hypoiodous acid.

2. Additional, strongly electronegative oxygen atoms are bonded to X. These tend to draw electrons away from the oxygen atom bonded to H. Thus, we would predict that the ease of dissociation of a proton, and hence K_a, should increase in the following order, from left to right:

The higher the oxid. no. of X, the stronger the acid

$$X—O—H \quad < \quad O—X—O—H \quad < \quad O—X—O—H \quad < \quad O—X—O—H$$

oxid. no. X = +1   +3   +5   +7

PRINCIPAL SPECIES IN WATER SOLUTION

When we say that perchloric acid is a strong acid, we mean in effect that the $HClO_4$ molecule does not exist in dilute water solution. Putting it another way, the ClO_4^- ion is the only species present in solution in the +7 oxidation state of chlorine. This is true regardless of pH; when we make the solution strongly acidic (e.g., pH = 0) or strongly basic (e.g., pH = 14), we find only ClO_4^- ions, never $HClO_4$ molecules.

The situation is quite different with the $+3$ state of chlorine. Since chlorous acid is weak, the $HClO_2$ molecule is a stable species in water. The amount of this molecule relative to the ClO_2^- ion depends upon pH. Since K_a of chlorous acid is 1.0×10^{-2}, we have:

$$\frac{[H^+] \times [ClO_2^-]}{[HClO_2]} = 1.0 \times 10^{-2}$$

Solving for the ratio $[HClO_2]/[ClO_2^-]$:

$$\frac{[HClO_2]}{[ClO_2^-]} = \frac{[H^+]}{1.0 \times 10^{-2}} \tag{23.39}$$

We can distinguish three possible conditions so far as the relative concentrations of $HClO_2$ and ClO_2^- are concerned:

1. At pH = 2.00 (conc. $H^+ = 1.0 \times 10^{-2}$), the ratio $[HClO_2]/[ClO_2^-] = 1.0$, and there are equal amounts of the two species. This condition would apply, for example, in a solution made by mixing equal amounts of 0.10 M $HClO_2$ and 0.10 M ClO_2^-.

2. At pH < 2.00 (conc. $H^+ > 1.0 \times 10^{-2}$), the ratio $[HClO_2]/[ClO_2^-] > 1.0$, and the molecule $HClO_2$ is the *principal species*. This situation would apply if we added strong acid to the $HClO_2$—ClO_2^- system.

3. At pH > 2.00 (conc. $H^+ < 1.0 \times 10^{-2}$), the ratio $[HClO_2]/[ClO_2^-] < 1.0$, and the anion ClO_2^- is the *principal species*. This condition would apply in a solution formed by adding strong base to the $HClO_2$—ClO_2^- system.

These relationships are shown in Figure 23.10 and Table 23.8. Similar relations hold for any weak acid–weak base system. At "low pH," the dominant species is the weak acid, HX. At "high pH," the weak base X^- will be the principal species. At a certain intermediate pH, whose value is fixed

$HClO_2$ is one of the strongest of the weak acids

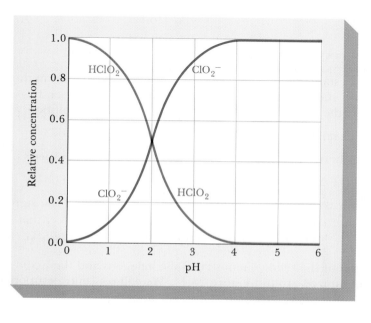

Figure 23.10 Relative concentrations of $HClO_2$ ($K_a = 1.0 \times 10^{-2}$) and ClO_2^- as a function of pH.

TABLE 23.8   Principal Species in the +3 State of Chlorine

	K_a $HClO_2 = 1.0 \times 10^{-2}$				
$[H^+]$	1.0	0.10	0.010	0.0010	0.00010
pH	0.00	1.00	2.00	3.00	4.00
$[HClO_2]/[ClO_2^-]$	100	10	1.0	0.10	0.010
Principal species	$HClO_2$	$HClO_2$	—	ClO_2^-	ClO_2^-

by the dissociation constant of the weak acid, the two species are present in equal amounts.

$$[HX] = [X^-], \text{ when } [H^+] = K_a \text{ of HX, or pH} = pK_a \qquad (23.40)$$

■ **EXAMPLE 23.9** ——————————————————————
Consider the +1 state of chlorine, where K_a $HClO = 2.8 \times 10^{-8}$. Calculate the ratio $[HClO]/[ClO^-]$ at pH 7.00. What is the principal species present?

Solution
The appropriate relation is:

$$\frac{[HClO]}{[ClO^-]} = \frac{[H^+]}{2.8 \times 10^{-8}}$$

At pH 7.00: $[HClO]/[ClO^-] = (1.0 \times 10^{-7})/(2.8 \times 10^{-8}) =$ ⎡3.6.⎤ The HClO molecule is the principal species; there are nearly four times as many HClO molecules as ClO^- ions.

Exercise
At what pH is $[HClO] = [ClO^-]$? Answer: 7.55. (This is the approximate pH maintained in swimming pools in which HClO and ClO^- are present as disinfectants.)

REDOX REACTIONS

Figure 23.11 shows standard reduction potentials for various species in the different oxidation states of chlorine, bromine, and iodine. We see from the figure that for the reduction of +3 chlorine to +1 in acid solution:

$$HClO_2(aq) + 2H^+(aq) + 2e^- \rightarrow HClO(aq) + H_2O; E^0_{red} = +1.673 \text{ V}$$

In basic solution, where the ClO_2^- and ClO^- ions are the principal species:

$$ClO_2^-(aq) + H_2O + 2e^- \rightarrow ClO^-(aq) + 2OH^-(aq); E^0_{red} = +0.681 \text{ V}$$

Standard oxidation voltages are obtained, as usual, by reversing the sign of E^0_{red}:

$$ClO^-(aq) + 2OH^-(aq) \rightarrow ClO_2^-(aq) + H_2O + 2e^-; E^0_{ox} = -0.681 \text{ V}$$

Our discussion of the redox chemistry of the halogens and their compounds will be organized around Figure 23.11. We will concentrate upon three features of the figure.

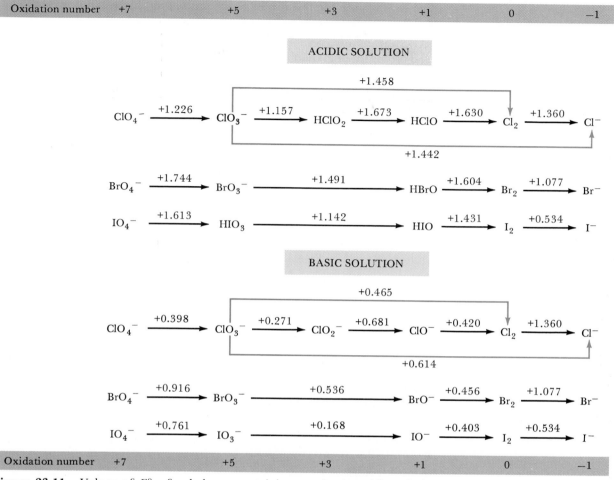

Figure 23.11 Values of E^0_{red} for halogen-containing species in acidic solution (*upper diagram*) and basic solution (*lower diagram*).

1. Species in which a halogen is in its highest oxidation state, $+7$ (ClO_4^-, BrO_4^-, IO_4^-) **can only be reduced and hence can only act as oxidizing agents.** All other oxyanions and oxyacids can, at least in principle, act as either oxidizing or reducing agents. Consider, for example, the ClO_3^- ion.

a. It can act as an oxidizing agent, usually being reduced to the -1 state, as in the oxidation of Br^- ions:

$$ClO_3^-(aq) + 6H^+(aq) + 6e^- \rightarrow Cl^-(aq) + 3H_2O$$
$$\underline{6Br^-(aq) \rightarrow 3Br_2(l) + 6e^-}$$
$$ClO_3^-(aq) + 6H^+(aq) + 6Br^-(aq) \rightarrow Cl^-(aq) + 3H_2O + 3Br_2(l) \quad (23.41)$$

b. It can be oxidized to ClO_4^- ions:

$$ClO_3^-(aq) + H_2O \rightarrow ClO_4^-(aq) + 2H^+(aq) + e^-$$

This half-reaction can be carried out in an electrolytic cell. Indeed, this is the method used commercially to prepare perchlorates such as $KClO_4$, the solid responsible for the bright flash and ear-splitting boom in fireworks displays.

In general these species are strong oxidizing agents

2. Oxyanions and oxyacids of the halogens (e.g., ClO_3^-, $HClO_2$) **are much stronger oxidizing than reducing agents.** All of the standard reduction potentials listed in Figure 23.11 are positive; it follows that all the standard oxidation voltages are negative. The chlorate ion is typical:

Any oxyanion is a better oxidizing agent in acidic solution

$$ClO_3^-(aq) + 6H^+(aq) + 6e^- \rightarrow Cl^-(aq) + 3H_2O; \quad E^0_{red} = +1.442 \text{ V}$$

$$ClO_3^-(aq) + H_2O \rightarrow ClO_4^-(aq) + 2H^+(aq) + e^-; \quad E^0_{ox} = -1.226 \text{ V}$$

Notice also from Figure 23.11 that **oxyanions are stronger oxidizing agents in acidic than in basic solution.** Again, using ClO_3^- as an example:

$$ClO_3^-(aq) + 6H^+(aq) + 6e^- \rightarrow Cl^-(aq) + 3H_2O; \quad E^0_{red} = +1.442 \text{ V}$$

$$ClO_3^-(aq) + 3H_2O + 6e^- \rightarrow Cl^-(aq) + 6OH^-(aq); \quad E^0_{red} = +0.614 \text{ V}$$

The effect is substantial. Although Br^- is readily oxidized by ClO_3^- in acid solution (Reaction 23.41), the redox reaction cannot take place in base:

$$\text{acid:} \quad E^0_{tot} = E^0_{red} \, ClO_3^- + E^0_{ox} \, Br^-$$
$$= +1.442 \text{ V} - 1.077 \text{ V} = +0.355 \text{ V}$$

$$\text{base:} \quad E^0_{tot} = E^0_{red} \, ClO_3^- + E^0_{ox} \, Br^-$$
$$= +0.614 \text{ V} - 1.077 \text{ V} = -0.463 \text{ V}$$

The effect of acidity on oxidizing strength is readily explained. When an oxyanion is reduced, H^+ ion is a reactant, e.g.,

$$ClO_3^-(aq) + 6H^+(aq) + 6e^- \rightarrow Cl^-(aq) + 3H_2O$$

It follows that the higher the concentration of H^+ the more spontaneous the reaction. Conversely, reducing the concentration lowers the voltage (Example 23.10 and Fig. 23.12).

■ **EXAMPLE 23.10** _____

For the reduction of ClO_3^-:

$$ClO_3^-(aq) + 6H^+(aq) + 6e^- \rightarrow Cl^-(aq) + 3H_2O; \quad E^0_{red} = +1.442 \text{ V}$$

Holding the concentrations of Cl^- and ClO_3^- constant at 1.0 M, calculate E_{red} at

a. pH = 7.00   b. pH = 14.00

Solution
Here, since we are dealing with pH $= -\log_{10}[H^+]$, it is simplest to use the Nernst equation in the base 10 form:

$$E_{red} = +1.442 \text{ V} - \frac{0.0591}{6} \log_{10} \frac{(\text{conc. } Cl^-)}{(\text{conc. } ClO_3^-) \times (\text{conc. } H^+)^6}$$

Taking conc. $Cl^- =$ conc. $ClO_3^- = 1.0$ M and using the relation: log $1/x = -\log x$:

$$E_{red} = +1.442 \text{ V} + \frac{0.0591}{6} \log_{10}(\text{conc. } H^+)^6$$

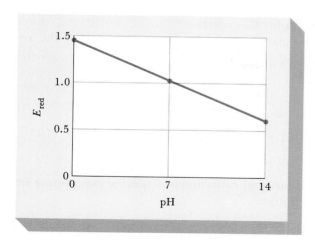

Figure 23.12 Dependence of E_{red} for the ClO_3^- ion on pH. Like most oxyanions, the ClO_3^- ion is a much stronger oxidizing agent in acidic than in basic solution.

Since $\log x^n = n \log x$, and $pH = -\log(\text{conc. } H^+)$:

$$E_{red} = +1.442 \text{ V} + \frac{0.0591}{6} (6) \log \text{conc. } H^+$$
$$= +1.442 \text{ V} - 0.0591 \text{ pH}$$

Clearly, E_{red} is a linear function of pH, decreasing as pH increases (Fig. 23.12).

a. $E_{red} = +1.442 \text{ V} - 0.0591(7.00) = \boxed{+1.028 \text{ V}}$

This is the reduction voltage in pure water (pH = 7).

b. $E_{red} = +1.442 \text{ V} - 0.0591(14.00) = \boxed{+0.614 \text{ V}}$

This is the standard reduction voltage in basic solution, where conc. $OH^- = 1.0 \, M$. Indeed, this is how the value $+0.614$ V appearing in Figure 23.11 is calculated.

Exercise
Why is the standard reduction voltage for $Cl_2(g)$ the same in basic as in acidic solution? Answer: Because there are no H^+ or OH^- ions in the reduction half-equation: $Cl_2(g) + 2e^- \rightarrow 2Cl^-(aq)$.

3. *Many of the species in Figure 23.11 are unstable with respect to* **disproportionation**, i.e., simultaneous oxidation and reduction.

   The +3 state of chlorine, whether we deal with the $HClO_2$ molecule or the ClO_2^- ion, spontaneously disproportionates (Example 23.11). Ordinarily these reactions occur rather rapidly. This explains why you don't find compounds containing +3 chlorine in the general chemistry laboratory. Compounds containing +3 bromine and iodine are even more unstable.

■ **EXAMPLE 23.11** ————————————————————————

Consider the disproportionation of chlorous acid, $HClO_2$, to hypochlorous acid, $HClO$, and chlorate ion, ClO_3^-.

KClO₃ in water would have this E_{red}

a. Write a balanced equation for this reaction.
b. Using Figure 23.11, calculate the standard voltage, E^0_{tot}, for the reaction.

Solution

a. The balanced half-equations, obtained in the usual way, are

reduction: $HClO_2(aq) + 2H^+(aq) + 2e^- \rightarrow HClO(aq) + H_2O$

oxidation: $HClO_2(aq) + H_2O \rightarrow ClO_3^-(aq) + 3H^+(aq) + 2e^-$

Adding the two half-equations and canceling out species that appear on both sides ($2e^-$, $2H^+$, H_2O) gives

$$2HClO_2(aq) \rightarrow HClO(aq) + ClO_3^-(aq) + H^+(aq)$$

b. $E^0_{tot} = E^0_{red}(HClO_2 \rightarrow HClO) + E^0_{ox}(HClO_2 \rightarrow ClO_3^-)$

From Figure 23.11, we read $E^0_{red} = +1.673$ V directly. To obtain E^0_{ox}, we note that E^0_{red} for the reverse reaction is $+1.157$ V. Changing the sign, we get $E^0_{ox} = -1.157$ V:

$$E^0_{tot} = +1.673 \text{ V} - 1.157 \text{ V} = +0.516 \text{ V}$$

The fact that E^0_{tot} is positive means that the $HClO_2$ molecule (oxid. no. Cl = +3) is unstable in water solution. $HClO_2$ disproportionates to HClO (oxid. no. Cl = +1) and ClO_3^- (oxid. no. Cl = +5).

Exercise
Determine E^0_{tot} for the disproportionation of the ClO_2^- ion (in basic solution) to give ClO^- and ClO_3^-. Answer: $+0.410$ V.

From Figure 23.11 it appears that the chlorine molecule should disproportionate in basic solution:

base: E^0_{red} $Cl_2 \rightarrow Cl^- + E^0_{ox}$ $Cl_2 \rightarrow ClO^- = +1.360$ V $- 0.420$ V $= +0.940$ V

but not in acid solution, at least at standard concentrations:

acid: E^0_{red} $Cl_2 \rightarrow Cl^- + E^0_{ox}$ $Cl_2 \rightarrow HClO = +1.360$ V $- 1.630$ V $= -0.270$ V

The explanation is quite simple. Hydroxide ions are involved as a reactant in the oxidation half-reaction, so it occurs more readily in basic solution:

oxidation: $\frac{1}{2}Cl_2(g) + 2OH^-(aq) \rightarrow ClO^-(aq) + H_2O + e^-$
reduction: $\frac{1}{2}Cl_2(g) + e^- \rightarrow Cl^-(aq)$

$$Cl_2(g) + 2OH^-(aq) \rightarrow ClO^-(aq) + Cl^-(aq) + H_2O \quad (23.42)$$

Reaction 23.42 is widely used to prepare the household bleach and cleaner known under such trade names as "Clorox" or "liquid bleach." Commercially, the preparation is carried out by electrolyzing a *cold*, stirred water

This is the most common strong oxidizing agent used in the home

solution of sodium chloride. Recall that the electrolysis of an NaCl solution gives Cl_2 molecules and OH^- ions; stirring ensures that these species react with each other. The active ingredient of the resulting solution is the hypochlorite ion, a potent oxidizing agent:

$$ClO^-(aq) + H_2O + 2e^- \rightarrow Cl^-(aq) + 2OH^-(aq); \qquad E^0_{red} = +0.890 \text{ V}$$

Many solid cleaning powders and disinfectants contain the ClO^- ion as their active ingredient. One of them, called "bleaching powder," is made by reacting chlorine gas with slaked lime, calcium hydroxide:

$$3Ca(OH)_2(s) + 2Cl_2(g) \rightarrow Ca(OCl)_2 \cdot CaCl_2 \cdot Ca(OH)_2 \cdot 2H_2O(s) \qquad (23.43)$$
bleaching powder

This compound is used in some home swimming pool chlorinating agents

Referring back to Figure 23.11, it is evident that another, quite different disproportionation is possible for chlorine, again in basic solution:

$$\frac{1}{2}Cl_2(g) + 6\,OH^-(aq) \rightarrow ClO_3^-(aq) + 3H_2O + 5e^-$$
$$\underline{\frac{5}{2}Cl_2(g) + 5e^- \rightarrow 5Cl^-(aq)}$$
$$3Cl_2(g) + 6\,OH^-(aq) \rightarrow ClO_3^-(aq) + 5Cl^-(aq) + 3H_2O \qquad (23.44)$$
$$E^0_{tot} = -0.465 \text{ V} + 1.360 \text{ V} = +0.895 \text{ V}$$

This reaction, in competition with 23.42, is favored by high temperatures. When a hot, stirred solution of a metal chloride is electrolyzed, Reaction 23.44 occurs, producing a mixture of ClO_3^- and Cl^- ions. Typically, potassium chloride is used as the electrolyte. Potassium chlorate, $KClO_3$, is much less soluble than KCl (0.27 mol/L vs. 3.3 mol/L at 0°C), so it crystallizes out of solution upon cooling. Safety matches contain sulfur and potassium chlorate; the special surface against which they are struck contains red phosphorus and powdered glass. Friction sets off a redox reaction between red phosphorus and potassium chlorate.

This is how we make $KClO_3$ commercially

■ SUMMARY EXAMPLE

Throughout this chapter, we have emphasized the chemistry of chlorine, the most common halogen. The following questions involve bromine, whose chemistry is quite similar to that of chlorine.

a. Write equations for the reaction of bromine with I^- ions; for the reaction of Br_2 with water (disproportionation).

b. Write equations for the preparation of bromine by the electrolysis of aqueous NaBr; by the reaction of manganese(IV) oxide with hydrobromic acid.

c. Write equations for the preparation of HBr from the elements; from NaBr using phosphoric acid.

d. Write equations for the reaction of hydrobromic acid with OH^- ions; with CO_3^{2-} ions; with ammonia.

e. Write equations for the reaction of Br^- ions with Ag^+ to form a precipitate; with Hg^{2+} to form a complex; with O_2 in a redox reaction.

f. Consider the reaction: $Br_2(l) + H_2O \rightleftharpoons HBrO(aq) + H^+(aq) + Br^-(aq)$. Calculate E^0_{tot} and K for this reaction. What is the equilibrium concentration of HBrO produced by the reaction?

g. Using Table 23.3, estimate the value of K for the formation of HBr from the gaseous elements at 100°C. ($H_2(g) + Br_2(g) \rightleftharpoons 2HBr(g)$)

h. When sodium bromide is heated with a concentrated solution of sulfuric acid, the products include liquid bromine and sulfur dioxide. Write a balanced equation for the redox reaction involved.

i. What is the equilibrium concentration of Br^- in a solution prepared by shaking $PbBr_2$ with water? ($K_{sp} = 6.6 \times 10^{-6}$)

j. Calculate the ratio $[HBrO]/[BrO^-]$ at pH 10.00. (K_a HBrO $= 2.6 \times 10^{-9}$)

k. Which of the bromine-containing species in Figure 23.11 undergo disproportionation at standard concentrations?

l. For the reduction of HBrO to Br^- ions, what is the change in voltage when the pH increases by one unit?

Answers

a. $Br_2(l) + 2I^-(aq) \rightarrow 2Br^-(aq) + I_2(s)$
   $Br_2(l) + H_2O \rightarrow HBrO(aq) + H^+(aq) + Br^-(aq)$

b. $2Br^-(aq) + 2H_2O \rightarrow Br_2(l) + H_2(g) + 2OH^-(aq)$
   $MnO_2(s) + 4H^+(aq) + 2Br^-(aq) \rightarrow Mn^{2+}(aq) + Br_2(l) + 2H_2O$

c. $H_2(g) + Br_2(g) \rightarrow 2HBr(g)$
   $NaBr(s) + H_3PO_4(l) \rightarrow HBr(g) + NaH_2PO_4(s)$

d. $H^+(aq) + OH^-(aq) \rightarrow H_2O$; $2H^+(aq) + CO_3^{2-}(aq) \rightarrow CO_2(g) + H_2O$
   $H^+(aq) + NH_3(aq) \rightarrow NH_4^+(aq)$

e. $Ag^+(aq) + Br^-(aq) \rightarrow AgBr(s)$; $Hg^{2+}(aq) + 4Br^-(aq) \rightarrow HgBr_4^{2-}(aq)$
   $O_2(g) + 4H^+(aq) + 4Br^-(aq) \rightarrow 2Br_2(l) + 2H_2O$

f. $E_{tot}^0 = -0.527$ V; 1.2×10^{-9}; 0.0011 M g. 5×10^{15}

h. $2Br^-(aq) + SO_4^{2-}(aq) + 4H^+(aq) \rightarrow Br_2(l) + SO_2(g) + 2H_2O$

i. 0.024 M j. 0.038 k. HBrO; Br_2 in base   l. -0.0296 V

Of the four halide ions, three (F^-, Cl^-, I^-) are essential to human nutrition. Fluoride and iodide ions are trace species (recall Table 22.7, Ch. 22). In contrast, our body fluids contain 100 g or more of Cl^- ions. Much of this is in the gastrointestinal tract; hydrochloric acid is secreted in the stomach during digestion. We need to take in 2–5 g of chloride ions daily; most of us get far more than that in the form of sodium chloride.

Iodide ions concentrate in the thyroid gland, where they are essential for the synthesis of iodine-containing growth hormones such as thyroxine:

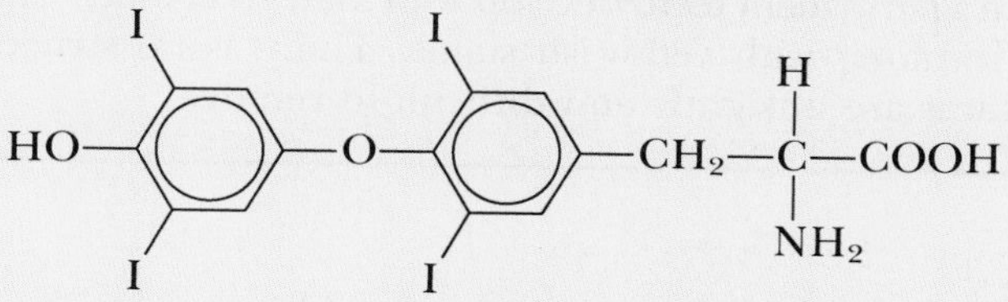

Iodine deficiency leads to abnormal enlargement of the thyroid, a condition known as goiter. This disease has been virtually wiped out in the United States and Canada by the use of iodized salt (0.01% KI), but it is estimated that goiter still affects 200 million people worldwide.

Curiously enough, excessive intakes of iodine can also cause an enlargement of the thyroid gland which resembles goiter. This occurs when daily consumption exceeds 2–3 mg per day. In the 1970s, average consumption in the United States rose rapidly, approaching 1 mg/day. This came about because the baking industry used sodium iodate, $NaIO_3$, as a dough conditioner and dairy farmers used iodine compounds as medicinals and disinfectants. When these practices were curtailed, average iodide intake dropped, but it is still at least four times the recommended daily allowance.

The principal mineral component of bones and teeth is hydroxyapatite, whose formula may be written as $Ca(OH)_2 \cdot 3Ca_3(PO_4)_2$. An OH^- ion in this compound is readily replaced by a F^- ion, which has the same charge as OH^- and nearly the same size. The product is fluorapatite, $CaF_2 \cdot 3Ca_3(PO_4)_2$. On the average, about 2% of the OH^- ions in hydroxyapatite in bones and teeth are replaced by F^- ions. The fluoride ions come mostly from drinking water, where they occur naturally (recall Example 23.7) or are added at the 1 ppm level in the form of sodium fluoride. About half of the people in the United States and Canada drink fluoridated water. This has been shown to reduce the incidence of tooth decay in children by 50–70%. In areas where fluoridated water is not available, other approaches are possible. Most toothpastes contain a small amount of a metal fluoride like tin(II) fluoride, SnF_2, or sodium fluoride, NaF.

As you might expect, fluorapatite is much more resistant to attack by acid than is hydroxyapatite. The F^- ion is a much weaker base than the OH^- ion:

$$F^-(aq) + H^+(aq) \rightarrow HF(aq); \; K = 1.4 \times 10^3$$

$$OH^-(aq) + H^+(aq) \rightarrow H_2O; \quad K = 1.0 \times 10^{14}$$

This may explain why F^- ions are effective in preventing tooth decay. Acids produced by plaque bacteria are a major culprit in destroying tooth enamel. Interestingly enough, drinking fluoridated water has also been found to reduce the incidence of osteoporosis, a degenerative bone disease that affects many older people.

If the concentration of F^- ions in drinking water is too high, perhaps 5–10 ppm, children's permanent teeth develop with mottled enamel. Chalky white patches form along with yellowish stains. There is no structural damage but the stains are unsightly and difficult to remove.

QUESTIONS AND PROBLEMS

Formulas, Equations, and Reactions

1. Name the following compounds.
 a. HIO_4 b. BrO_2^- c. HIO d. $NaClO_3$

2. Name the following compounds.
 a. $HBrO_3$ b. KIO c. $NaClO_2$ d. $NaBrO_4$

3. Write the formula for each of the following compounds.
 a. chloric acid   b. periodic acid
 c. hypobromous acid   d. hydriodic acid

4. Write the formula for each of the following compounds.
 a. potassium bromite   b. calcium bromide
 c. sodium periodate   d. magnesium hypochlorite

5. Name the acid for which each of the following is the conjugate base.
 a. IO_4^- b. ClO_2^- c. BrO_3^- d. F^-

6. Name the base for which each of the following is the conjugate acid.
 a. $HBrO$ b. HIO_3 c. $HClO_4$ d. $HBrO_2$

7. Write a balanced net ionic equation for
 a. the electrolytic decomposition of hydrogen fluoride.
 b. the oxidation of iodide ion to iodine by hydrogen peroxide in acidic solution. Hydrogen peroxide is reduced to water.
 c. the formation of barium bromide from barium hydroxide.

8. Write a balanced net ionic equation for
 a. the formation of zinc iodide from zinc carbonate.
 b. the oxidation of iodide to iodine by sulfate ion in acidic solution. Sulfur dioxide gas is also produced.
 c. the preparation of hydrogen iodide from an iodide salt and phosphoric acid.

9. Write a balanced net ionic equation for the disproportionation reaction
 a. of iodine to give iodate and iodide ions in basic solution.
 b. of chlorine gas to chloride and perchlorate ions in basic solution.

10. Write a balanced net ionic equation for the disproportionation reaction
 a. of hypochlorous acid to chlorine gas and chlorous acid in acidic solution.
 b. chlorate ion to perchlorate and chlorite ions.

11. Complete and balance the following equations. If no reaction occurs, write NR.
 a. $Cl_2(g) + I^-(aq) \rightarrow$ b. $F_2(g) + Br^-(aq) \rightarrow$
 c. $I_2(s) + Cl^-(aq) \rightarrow$ d. $Br_2(l) + I^-(aq) \rightarrow$

12. Complete and balance the following equations. If no reaction occurs, write NR.
 a. $Cl_2(g) + Br^-(aq) \rightarrow$ b. $I_2(s) + Cl^-(aq) \rightarrow$
 c. $I_2(s) + Br^-(aq) \rightarrow$ d. $Br_2(l) + Cl^-(aq) \rightarrow$

13. Write a balanced net ionic equation for the reaction of hydrofluoric acid with
 a. a water solution of sodium carbonate
 b. a water solution of ammonia
 c. $SiO_2(s)$ d. $Pb^{2+}(aq)$

14. Write a balanced net ionic equation for the reaction of hydrochloric acid with
 a. a water solution of Na_2S
 b. a water solution of potassium acetate
 c. $Fe(OH)_3(s)$ d. $Pb^{2+}(aq)$

15. Describe in words how you would prepare
 a. Cl_2 from $NaCl$
 b. HBr from $NaBr$
 c. $NiCl_2$ from $Ni(OH)_2$

16. Describe in words how you would prepare
a. Br_2 from NaBr   b. HCl from NaCl
c. NaI from NaOH

Stoichiometry

17. The average concentration of bromine (as bromide) in seawater is 65 ppm. Calculate
a. the volume of seawater (d = 64.0 lb/ft^3) in cubic feet required to produce one kilogram of liquid bromine.
b. the volume of chlorine gas in liters, measured at 20°C and 762 mm Hg, required to react with this volume of seawater.

18. A 425-gallon tank is filled with water containing 175 g of sodium iodide. How many liters of chlorine gas at 758 mm Hg and 25°C will be required to oxidize all the iodide to iodine?

19. When antimony is added to a bottle containing chlorine gas, a white powder containing 47% by mass chlorine is formed. What is the empirical formula of the antimony chloride?

20. A sample of nickel chloride with mass 2.835 g gave 1.284 g of nickel. What is the formula of nickel chloride?

21. Iodine can be prepared by allowing an aqueous solution of hydrogen iodide to react with manganese dioxide, MnO_2. The reaction is

$$2I^-(aq) + 4H^+(aq) + MnO_2(s)$$
$$\rightarrow Mn^{2+}(aq) + 2H_2O + I_2(s)$$

If an excess of hydrogen iodide is added to 0.200 g of MnO_2, how many grams of iodine are obtained, assuming 100% yield?

22. How many grams of sodium iodide must react with excess phosphoric acid to produce enough HI to form 3.00 L of 3.33 M solution?

23. What volume of hydrogen iodide gas at 27°C and 0.984 atm would be required to prepare 500.0 mL of 0.1250 M HI solution?

24. When a solution of hydrogen bromide is prepared, 1.283 L of HBr gas at 25°C and 0.974 atm is bubbled into 250.0 mL of water. Assuming all the HBr dissolves with no volume change, what is the molarity of the hydrobromic acid solution produced?

25. Titanium metal reacts with HF(g) at 250°C to produce a solid compound that is 45.67% titanium. Hydrogen gas is also produced. Write a balanced equation for the reaction.

26. Iodine reacts with liquid chlorine at −40°C to give an orange solid containing 54.5% iodine. Write a balanced equation for the reaction.

27. How many grams of sodium chloride must react completely with sulfuric acid to produce 225.0 mL of an HCl solution that has a density of 1.170 g/cm^3 and has 33.50% HCl by mass?

28. What volume of a hydrochloric acid solution which has a density of 1.153 g/cm^3 and contains 30.00% HCl by mass is required to react with 125 g of magnesium?

Equilibria

29. The equilibrium constant at 25°C for the reaction

$$Br_2(l) + H_2O \rightleftharpoons H^+(aq) + Br^-(aq) + HBrO(aq)$$

is 1.2×10^{-9}. This is the system present in a bottle of "bromine water." Assuming that HBrO does not ionize appreciably, what is the pH of the bromine water?

30. Calculate the pH and the equilibrium concentration of HClO in a 0.10 M solution of hypochlorous acid. K_a HClO = 2.8×10^{-8}.

31. Calculate the pH of an aqueous solution of hydrofluoric acid that is 50.0% HF by mass and has a density of 1.155 g/cm^3. K_a HF = 6.9×10^{-4}.

32. Calculate the pH of an aqueous solution of hydrobromic acid which has a density of 1.48 g/cm^3 and that is 47.5% HBr by mass.

33. At equilibrium, a gas mixture has a partial pressure of 0.7324 atm for HBr and 2.80×10^{-3} atm for both hydrogen and bromine gases. What is K_c for the formation of 2 moles of HBr from H_2 and Br_2?

34. Given

$$HF(aq) \rightleftharpoons H^+(aq) + F^-(aq) \qquad K_a = 6.9 \times 10^{-4}$$
$$HF(aq) + F^-(aq) \rightleftharpoons HF_2^-(aq) \qquad K = 2.7$$

Calculate K for the reaction

$$2HF(aq) \rightleftharpoons H^+(aq) + HF_2^-(aq)$$

35. What is the concentration of fluoride ion in a water solution saturated with BaF_2, $K_{sp} = 1.8 \times 10^{-7}$?

36. Calculate the solubility in grams per 100 mL of BaF_2 in 0.10 M $BaCl_2$ solution.

37. For HBrO, $K_a = 2.6 \times 10^{-9}$. Calculate
a. the ratio $[HBrO]/[BrO^-]$ at pH 8.00, 10.00, and 12.00.
b. the pH at which $[HBrO] = 2.00 \times [BrO^-]$.

38. Repeat the calculations of Problem 37 for HIO, where $K_a = 2.4 \times 10^{-11}$.

Thermodynamics

39. Determine whether the following redox reaction is spontaneous at 25°C:

$$2KIO_3(s) + Cl_2(g) \rightarrow 2KClO_3(s) + I_2(s)$$

Use data in Appendix 1 and the following information: ΔH_f^0 $KIO_3(s)$ = −501.4 kJ/mol, S^0 $KIO_3(s)$ = 151.5 J/mol·K. What is the lowest temperature at which the reaction is spontaneous?

40. Follow the directions for Problem 39 for the reaction

$$2KBrO_3(s) + Cl_2(g) \rightarrow 2KClO_3(s) + Br_2(l)$$

The following thermodynamic data may be useful:

$\Delta H_f^0 \, KBrO_3 = -360.2 \, kJ/mol; \, S^0 \, KBrO_3 = 149.2 \, J/mol \cdot K$

41. Consider the equilibrium system:

$$HF(aq) \rightleftharpoons H^+(aq) + F^-(aq)$$

Given: $\Delta H_f^0 \, HF(aq) = -320.1 \, kJ/mol; \, \Delta H_f^0 \, F^-(aq) = -332.6 \, kJ/mol$

$S^0 \, F^-(aq) = -13.8 \, J/mol \cdot K; \, K_a \, HF = 6.9 \times 10^{-4}$ at 25°C

Calculate S^0 for HF(aq).

42. Using the Tables in Appendix 1 for the reaction

$$4HCl(g) + O_2(g) \rightarrow 2Cl_2(g) + 2H_2O(l)$$

determine
a. whether the reaction is spontaneous at 25°C.
b. K_p for the reaction at 25°C.

Electrochemistry

43. In acidic solution, $Cr_2O_7^{2-}$ oxidizes iodide ion to iodine and is itself reduced to Cr^{3+} ion.
a. Write a balanced equation for the redox reaction and calculate E_{tot}^0.
b. Would E be larger at pH 2 or pH 4, assuming all other concentrations remain constant?

44. In acidic solution, MnO_4^- oxidizes bromide ion to bromine; the other product is Mn^{2+}.
a. Write the balanced equation for the redox reaction and calculate E_{tot}^0.
b. By how much does the voltage change when the pH increases by one unit?

45. In the electrolysis of a KI solution, using 5.00 V, how much electrical energy in kilojoules is consumed when one mole of I_2 is formed?

46. If an electrolytic cell producing fluorine uses a current of 7.00×10^3 A (at 10.0 V), how many grams of fluorine gas can be produced in two days (assuming that the cell operates continuously at 95% efficiency)?

47. Sodium hypochlorite is produced by the electrolysis of cold sodium chloride solution. How long must a cell operate to produce 1.500×10^3 L of 5.00% NaClO by mass if the cell current is 2.00×10^3 A? Assume that the density of the solution is $1.00 \, g/cm^3$.

48. Sodium perchlorate is produced by the electrolysis of sodium chlorate. If a current of 1.50×10^3 A passes through an electrolytic cell, how many kilograms of sodium perchlorate are produced in an eight-hour run?

49. For the reaction

$$2HClO_2(aq) \rightarrow HClO(aq) + ClO_3^-(aq) + H^+(aq)$$

$E_{tot}^0 = +0.516$ V. Calculate E when conc. $HClO_2 = 0.01 \, M$, conc. $HClO = 0.10 \, M$, conc. $ClO_3^- = 0.10 \, M$, pH = 5.

50. For the reaction

$$2ClO_2^-(aq) \rightarrow ClO^-(aq) + ClO_3^-(aq)$$

$E_{tot}^0 = +0.410$ V. Calculate E when conc. $ClO_2^- = $ conc. $ClO^- = $ conc. $ClO_3^- = 0.10 \, M$.

51. For the reaction

$$3BrO_3^-(aq) \rightarrow 2BrO_4^-(aq) + BrO^-(aq)$$

$\Delta G^0 = +147.0 \, kJ$. Calculate E_{tot}^0 and compare to the value obtained from Figure 23.11.

52. For the reaction

$$3BrO_3^-(aq) + H^+(aq) \rightarrow 2BrO_4^-(aq) + HBrO(aq)$$

$\Delta G^0 = +98.0 \, kJ$. Calculate E_{tot}^0 and compare to the value obtained from Figure 23.11.

Unclassified

53. As one goes down Group VII in the Periodic Table, describe the trend in
a. atomic radius   b. electronegativity
c. ionization energy   d. molar mass
e. melting point   f. boiling point
g. E_{red}^0 for $X_2 \rightarrow X^-$ h. acid strength of HXO

54. Draw the Lewis structure for I_3^-.

55. Choose the strongest acid from each group.
a. HClO, HBrO, HIO   b. HIO, HIO_3, HIO_4
c. HIO, HBrO_2, HBrO_4

56. What intermolecular forces are present in the following?
a. Cl_2 b. HBr   c. HF   d. $HClO_4$ e. MgI_2

57. Explain why
a. fluorine gas is stored in containers made of nickel alloys.
b. hydrogen fluoride is never stored in a glass bottle.
c. fluoride ions are effective in preventing tooth decay.
d. iodine is more soluble in a solution of KI than in pure water.
e. E_{red}^0 for I_2 is the same in both basic and acidic solutions.

58. Consider the equilibrium present in a bottle labelled "bromine water."

$$Br_2(l) + H_2O \rightleftharpoons HBrO(aq) + H^+(aq) + Br^-(aq)$$

Without adding more bromine or changing the amount of water in the solution,
a. how can the concentration of HBrO be increased?
b. how can the concentration of HBrO be decreased?

59. Calculate the molarity of a brine (saturated sodium chloride solution), used to produce chlorine, that is 26.0% NaCl by mass and has a density of $1.201 \, g/cm^3$.

60. The vapor pressure of bromine at 25°C is 221 mm Hg. How many grams of bromine are there in a liter of air saturated with bromine at 25°C?

61. Arrange the halogens in order of increasing
a. atomic radii   b. ionic radii

c. electronegativity   d. boiling points
e. oxidizing activity   f. depth of color

62. Bromine has an atomic mass of 79.904 amu. It is made up of two isotopes. Bromine-79 has an atomic mass of 78.9183 amu. Calculate the percent abundance of bromine-81 which has an atomic mass of 80.9163 amu.

63. Show by calculation which hydrogen halide (HX) diffuses 0.395 times as fast as HF.

64. Based on the potentials given in Figure 23.11, which iodine species is

a. the strongest oxidizing agent in acidic solution?
b. the strongest oxidizing agent in basic solution?
c. the strongest reducing agent in acidic solution?
d. the strongest reducing agent in basic solution?

Challenge Problems

65. Perbromates have only recently been prepared. For the half-reaction

$$2BrO_4^-(aq) + 16H^+(aq) + 14e^- \rightarrow Br_2(l) + 8H_2O(l)$$
$$E_{red}^0 = 1.579 \text{ V}$$

Using Appendix 1 and E_{red}^0 given in Figure 23.11 for

$$BrO_4^-(aq) + 2H^+(aq) + 2e^- \rightarrow BrO_3^-(aq) + H_2O(l)$$

calculate ΔG_f^0 for $BrO_3^-(aq)$.

66. Oxidation of iodate ion by chlorine gas in basic solution yields a paraperiodate ion, $H_3IO_6^{2-}(aq)$, and chloride ion. The paraperiodate ion is then made to react with silver ion to get the precipitate Ag_3IO_5 according to the equation

$$H_3IO_6^{2-}(aq) + Ag^+(aq) \rightarrow Ag_3IO_5(s) + H_2O + H^+(aq)$$

A solution of paraperiodic acid is formed when an aqueous suspension of Ag_3IO_5 is treated with chlorine gas.

$$Ag_3IO_5(s) + Cl_2(g) + H_2O$$
$$\rightarrow H_5IO_6(aq) + AgCl(s) + O_2(g)$$

The paraperiodic acid can be crystallized from this solution.

a. Write out and balance all the equations given and described.
b. What mass of H_5IO_6 can be prepared from one hundred grams of sodium iodate?

67. The amount of sodium hypochlorite in a bleach solution can be determined by using a given volume of bleach to oxidize excess iodide ion to iodine; ClO^- is reduced to Cl^-. The amount of iodine produced by the redox reaction is determined by titration with sodium thiosulfate, $Na_2S_2O_3$; I_2 is reduced to I^-. The sodium thiosulfate is oxidized to sodium tetrathionate, $Na_2S_4O_6$. In this analysis, potassium iodide was added in excess to 5.00 mL of bleach ($d = 1.00$ g/cm^3). If 25.00 mL of 0.0700 M $Na_2S_2O_3$ was required to reduce all the iodine produced by the bleach back to iodide, what is the mass percent of NaClO in the bleach?

68. Prepare a graph similar to Figure 23.10 for HClO ($K_a = 2.8 \times 10^{-8}$).

69. The reaction

$$4HF(aq) + SiO_2(s) \rightarrow SiF_4(aq) + 2H_2O$$

can be used to release gold that is distributed in certain quartz (SiO_2) veins of hydrothermal origin. If the quartz contains $1.0 \times 10^{-3}\%$ Au by weight and the gold has a market value of $425 per troy ounce, would the process be economically feasible if commercial HF (50% by weight, $d = 1.17$ g/cm^3) costs 75¢ a liter? (1 troy ounce = 31.1 g.)

$S(s) \; + \; O_2(g) \; \rightarrow \; SO_2(g)$
Combustion of sulfur in oxygen

CHAPTER 24

THE NONMETALS: NITROGEN, PHOSPHORUS, OXYGEN, AND SULFUR

This chapter, the second dealing with the nonmetals, focuses on the four most important elements in Groups 5 and 6. We will start (Section 24.1) by discussing the molecular structures of these elements and their extraction from natural sources. Then we will consider some of the compounds they form with hydrogen (Section 24.2) and oxygen (Section 24.3). Finally, we will look at the oxyacids of nitrogen, sulfur, and phosphorus (Section 24.4).

The approach here will be similar to that followed in other descriptive chapters; a major objective is to review principles covered earlier. If you find parts of the discussion familiar, that is encouraging; it means that you have mastered these principles and can apply them to new situations.

3	4	5	6	7	8
		N	O		
		P	S		
		As	Se		
		Sb	Te		
		Bi	Po		

The elements shown in green are covered in the body of the text in this chapter. Those shown in red are discussed in the Applied Perspective (pp. 872–874).

24.1
THE ELEMENTS (N, P, O, S)

Table 24.1 lists some of the properties of these four elements. The first to be discovered was sulfur; biblical references to "brimstone" describe sulfur deposits associated with volcanos. Phosphorus was isolated from urine by the alchemist Henning Brandt, who was searching for the philosopher's stone in what would seem a most unlikely source. Nitrogen and oxygen were among the first gaseous elements isolated, at the dawn of modern chemistry, late in the eighteenth century.

ALLOTROPY

The allotropic form of oxygen known as ozone, O_3, was discussed in Chapter 19 in connection with air pollution. You will recall that it can be considered to be a resonance hybrid of the two structures:

Commercially, ozone is prepared by passing O_2 gas through a high-voltage (10^4 V) electric discharge. At atmospheric pressure, the reaction:

$$2O_3(g) \rightarrow 3O_2(g); \Delta H^0 = -285.4 \text{ kJ}, \Delta S^0 = +137.5 \text{ J/K} \qquad (24.1)$$

is thermodynamically spontaneous at all temperatures. Kinetically, however, ozone stays around long enough to find some application as a substitute for chlorine in disinfecting municipal water supplies. The end prod-

TABLE 24.1 Properties of Nitrogen, Phosphorus, Oxygen, and Sulfur

	NITROGEN	PHOSPHORUS	OXYGEN	SULFUR
Discovery	Rutherford (1772)	Brandt (1669)	Priestley (1774)	ancient
Major source	Air	Phosphate rock $Ca_3(PO_4)_2$	Air	Elemental form, H_2S, metal sulfides
Molecular formula[a]	N_2	P_4	O_2	S_8
Physical state 25°C, 1 atm	gas	solid	gas	solid
Melting point (°C)	−210	44	−218	119
Boiling point (°C)	−196	280	−183	444
Allotropy (gas)		P_4, P_2, P	O_2, O_3[c]	S_8, S_2[d]
(liquid)		P_4		S_8, S_x
(solid)		white (P_4)		rhombic (S_8)
		red (P_x)[b]		monoclinic (S_8)
		black (P_x)		

[a]For the species melting and boiling at the listed temperatures.

[b]Red and black phosphorus are polymeric, as is liquid sulfur above 160°C.

[c]Ozone, O_3, is highly unstable as a liquid or solid, decomposing violently to O_2.

[d]Many other allotropes of sulfur are known in both the gaseous and solid states.

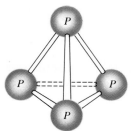

Figure 24.1 At the top of the figure are shown the two most common allotropes of phosphorus, the white and red forms. Below are their structures. White phosphorus is molecular, formula P_4. Red phosphorus has a network structure, shown here in simplified form. (photo, Charles D. Winters)

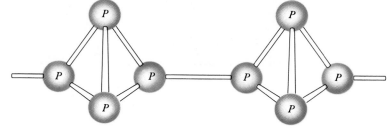

White phosphorus   Red phosphorus

ucts of ozone oxidation are less hazardous: chlorine can react with organic compounds in water to form suspected carcinogens such as $CHCl_3$. On the other hand, ozone is more expensive than chlorine. It also decomposes more rapidly, so that it offers little or no protection against bacteria that enter the water supply after treatment.

One good feature is that there is no chlorine taste

Phosphorus forms several allotropes in the solid state, of which the two most common are:

1. *White phosphorus,* which consists of P_4 molecules with the structure shown in Figure 24.1. It is a soft waxy substance with a low melting point (44°C) and boiling point (280°C). Like most molecular substances, white phosphorus is readily soluble in such nonpolar solvents as CCl_4. The chemical reactivity of white phosphorus is so great that it is stored under water to protect it from O_2. A piece of P_4 exposed to air in a dark room glows because of the light given off upon oxidation (Fig. 24.2, p. 846). White phosphorus is extremely toxic. As little as 0.1 g taken internally can be fatal. Direct contact with the skin produces painful burns.

White phosphorus is used in napalm, one of the nastiest chemical weapons

2. *Red phosphorus,* which is the form in which the element is usually found in the laboratory. This allotrope has properties quite different from those of white phosphorus. It is much higher melting (mp = 590°C at 43 atm) and is insoluble in common solvents. The low volatility of red phosphorus makes it much less toxic than the white form. It is also less reactive and must be heated to 250°C to burn in air. These properties are consistent with the structure of red phosphorus, which is known to be network covalent (Fig. 24.1). This allotrope can be made by heating white phosphorus in the absence of air to about 300°C. Actually, red phosphorus is the favored thermodynamic form at 25°C, but the transition time at that temperature is too long (essentially forever!).

845

Figure 24.2 When white phosphorus is exposed to oxygen gas, it first glows (phosphorescence) and then bursts into flame. The reaction is: $P_4(s) + 5O_2(g) \rightarrow P_4O_{10}(s)$. (Charles D. Winters)

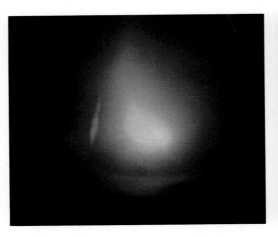

■ **EXAMPLE 24.1**

For the allotropic conversion: P(white) $\rightleftharpoons$ P(red)
ΔH^0 is -17.6 kJ and ΔS^0 is -18.3 J/K. Calculate:

a. ΔG^0 at 25°C and 300°C.
b. the temperature at which the two allotropes are in equilibrium at 1 atm.

Solution

a. At 25°C: $\Delta G^0 = -17.6$ kJ + 298 K (0.0183 kJ/K) = -12.1 kJ
    At 300°C: $\Delta G^0 = -17.6$ kJ + 573 K (0.0183 kJ/K) = -7.1 kJ

b. $T = \dfrac{\Delta H^0}{\Delta S^0} = \dfrac{-17.6 \text{ kJ}}{-0.0183 \text{ kJ/K}} = 962$ K $\approx$ 690°C

   At any temperature below about 690°C, red phosphorus is the stable allotrope.

Exercise

Referring back to the data for Reaction 24.1, at what temperature are O_3 and O_2 in equilibrium at 1 atm pressure? Answer: at no temperature.

In the solid state, sulfur can have more than 20 different allotropic forms. You will no doubt be relieved to learn that we will only talk about two of these, rhombic and monoclinic sulfur (Fig. 24.3). Both consist of ring molecules of formula S_8. They differ only in the way that molecules are packed in the solid, reflected in different crystal structures.

At room temperature, rhombic sulfur is the stable allotrope. However, since the process

$$S_8(\text{rhombic}) \rightarrow S_8(\text{monoclinic}); \qquad \Delta H^0 = +2.6 \text{ kJ}$$

is endothermic, the monoclinic form is stable at high temperatures. The two forms are in equilibrium at 96°C; if rhombic sulfur is heated to that temperature, it slowly converts to monoclinic sulfur. More commonly, the monoclinic allotrope is prepared by freezing liquid sulfur at the melting

Figure 24.3 Naturally occurring crystals of monoclinic sulfur (*left*) and rhombic sulfur (*right*). (left, Fletcher, W.K. Photoreachers, Inc.; right, Allen B. Smith, Tom Stack and Assoc.)

point (119°C) and then cooling quickly to room temperature. Typically, at 25°C, monoclinic crystals stay around for a day or more before converting to the rhombic form.

The free-flowing, pale yellow liquid formed when sulfur melts contains S_8 molecules. However, upon heating to 160°C, a striking change occurs. The liquid becomes so viscous that it cannot be poured readily. At the same time its color changes to a deep reddish-brown. These effects reflect a change in molecular structure. The S_8 rings break apart and then link to one another to form long chains such as

$$\cdot S \underset{S}{\overset{S}{\diagup}} \underset{S}{\overset{S}{\diagup}} \underset{S}{\overset{S}{\diagup}} \underset{S}{\overset{S}{\diagup}} \underset{S}{\overset{S}{\diagup}} \underset{S}{\overset{S}{\diagup}} \underset{S}{\overset{S}{\diagup}} \underset{S}{\overset{S}{\diagup}} S\cdot$$

Liquid sulfur between 160 and 250°C contains a high proportion of such chains. They vary in length from eight to perhaps a million atoms. The chains become tangled, producing a highly viscous liquid. The deep color is due to the absorption of light by the unpaired electrons at the ends of the chains.

Under these conditions, sulfur is a polymer

If liquid sulfur at 200°C is quickly poured into water, a rubbery mass results (Fig. 24.4). This is referred to as "plastic sulfur." It consists of long-

Figure 24.4 When liquid sulfur at 200°C is poured into cold water, a rubbery material called "plastic sulfur" forms. Here S. Ruven Smith shows some of its (and his) properties.

chain molecules that did not have time to rearrange to the S_8 molecules stable at room temperature. Within a few hours, the plastic sulfur loses its elasticity as it converts to rhombic crystals.

PREPARATION

As pointed out in Chapter 19, nitrogen and oxygen are obtained commercially by the fractional distillation of liquid air. Small quantities of nitrogen are sometimes prepared in the laboratory by heating a water solution containing ammonium and nitrite ions:

$$NH_4^+(aq) + NO_2^-(aq) \rightarrow N_2(g) + 2H_2O \qquad (24.2)$$

The laboratory preparation of oxygen involves heating potassium chlorate to about 270°C, using MnO_2 as a catalyst:

$$2KClO_3(s) \rightarrow 2KCl(s) + 3O_2(g) \qquad (24.3)$$

The element phosphorus is prepared by heating a mineral known as "phosphate rock," which is mostly calcium phosphate, with sand and coke in an electric furnace at 1500°C. The equation for the reaction can be written as:

$$2Ca_3(PO_4)_2(s) + 6SiO_2(s) + 10C(s) \rightarrow P_4(g) + 10CO(g) + 6CaSiO_3(l) \quad (24.4)$$

Most of the phosphorus is burned to make P_4O_{10}, from which we make phosphoric acid

Phosphorus distils out of the furnace; white phosphorus is obtained when the gas condenses.

■ **EXAMPLE 24.2**

A typical charge to the furnace used to make phosphorus is 5.0 metric tons of $Ca_3(PO_4)_2$, 3.5 tons of SiO_2, and 1.5 tons of C. What is the theoretical yield of $P_4(g)$?

Solution
We follow the approach described originally in Chapter 3, calculating the yield to be expected if each reactant in turn is limiting:

(1) $Ca_3(PO_4)_2$ limiting:

$$5.0 \times 10^6 \text{ g } Ca_3(PO_4)_2 \times \frac{124 \text{ g } P_4}{620 \text{ g } Ca_3(PO_4)_2} = 1.0 \times 10^6 \text{ g } P_4$$

(2) SiO_2 limiting:

$$3.5 \times 10^6 \text{ g } SiO_2 \times \frac{124 \text{ g } P_4}{361 \text{ g } SiO_2} = 1.2 \times 10^6 \text{ g } P_4$$

(3) C limiting:

$$1.5 \times 10^6 \text{ g C} \times \frac{124 \text{ g } P_4}{120 \text{ g C}} = 1.6 \times 10^6 \text{ g } P_4$$

The theoretical yield is 1.0×10^6 g, i.e., one metric ton, of phosphorus.

Exercise

What is the mass of calcium silicate slag formed in the process? Answer: 5.6 metric tons.

At the close of the Civil War in 1865, oil prospectors in Louisiana discovered (to their disgust) elemental sulfur in the caprock of vast salt domes, as much as 20 km² in area. The sulfur lies 60 to 600 m below the surface of the earth. It is believed to have been formed from the SO_4^{2-} ions of $CaSO_4$ by bacterial action.

The process used to mine sulfur is named after its inventor, Herman Frasch, an American chemical engineer. A diagram of the Frasch process is shown in Figure 24.5. The sulfur is heated to its melting point (119°C) by pumping superheated water at 165°C down one of three concentric pipes. Compressed air is used to bring the sulfur to the surface. The air and sulfur form a frothy mixture that rises through the middle pipe. Upon cooling, the sulfur solidifies, filling huge vats that may be 0.5 km long. The sulfur obtained in this way has a purity approaching 99.9%.

Considerable amounts of sulfur are now being recovered from natural gas, which contains some hydrogen sulfide, H_2S. The hydrogen sulfide is separated and then burned in a limited amount of air. The overall reaction is

$$2H_2S(g) + O_2(g) \rightarrow 2S(s) + 2H_2O(l) \qquad (24.5)$$

It's tricky to keep SO_2 from forming

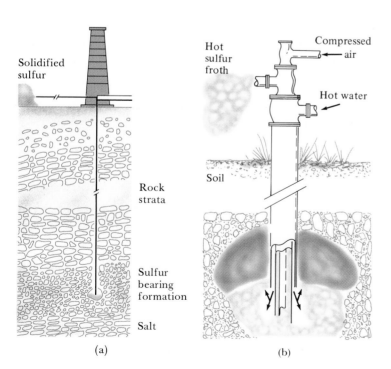

Solidified sulfur

Hot sulfur froth

Compressed air

Hot water

Rock strata

Soil

Sulfur bearing formation

Salt

(a)

(b)

Figure 24.5 Frasch process for mining sulfur. Superheated water at 165°C is sent down through the outer pipe to form a pool of molten sulfur (mp = 119°C) at the base. Compressed air, pumped down the inner pipe, brings the sulfur to the surface. Sulfur deposits are often 100 m or more beneath the earth's surface, covered with quicksand and rock.

24.2
HYDROGEN COMPOUNDS OF N, P, O, AND S

These elements form a variety of compounds with hydrogen (Table 24.2). Notice that the boiling points of the hydrides of nitrogen and oxygen are higher than those of the corresponding compounds of phosphorus and sulfur. This can be attributed to hydrogen bonding in compounds such as N_2H_4 and H_2O.

As you can see from Table 24.2, all the hydrides of the Group 5 elements, except NH_3, have positive free energies of formation. This means that the compounds PH_3, N_2H_4, P_2H_4, and HN_3 are all potentially unstable with respect to decomposition to the elements. Hydrogen azide, HN_3, is a dangerous explosive, even in water solution. Hydrazine, N_2H_4, has been used as a rocket propellant because of the large amount of energy given off when it decomposes or burns.

TABLE 24.2   Properties of Nonmetal Hydrides

	bp (°C)	PHYS. STATE 25°C, 1 atm	ΔG_f^0 (kJ) 25°C, 1 atm
NH_3	−33	gas	−16.5
PH_3	−88	gas	+13.4
N_2H_4	114	liquid	+149.2
P_2H_4	52	liquid	>0
HN_3	37	liquid	+327.3
H_2O	100	liquid	−237.2
H_2S	−61	gas	−33.6
H_2O_2	151	liquid	−120.4

■ **EXAMPLE 24.3**
Consider hydrogen peroxide, H_2O_2.

a. Write a reasonable Lewis structure for H_2O_2.
b. Predict the bond angle.

Solution

a. Hydrogen atoms can form only one bond. The most reasonable skeleton would be a symmetrical one in which a hydrogen is bonded to each oxygen atom:

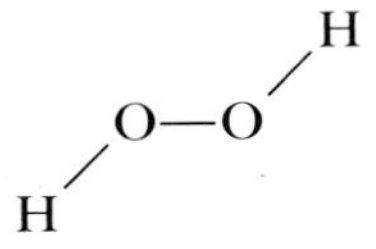

We start with a total of $2(1) + 2(6) = 14$ valence electrons. With three single bonds in the skeleton (six valence electrons), that leaves eight valence electrons to be distributed. Putting two unshared pairs on each oxygen, we arrive at the correct Lewis structure:

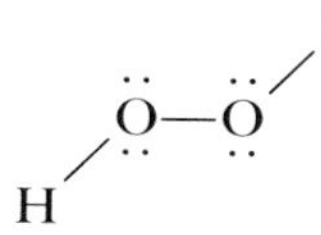

The molecule is not planar

b. The four electron pairs around each oxygen should be directed toward the corners of a regular tetrahedron. The predicted bond angle is 109.5°. Experimentally, it is found to be 105°.

Exercise

The skeleton of hydrazine, N_2H_4, is similar to that of hydrogen peroxide. Write a reasonable Lewis structure for N_2H_4. Answer:

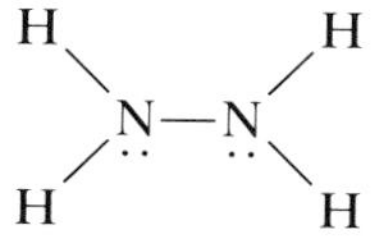

PROPERTIES OF AMMONIA, NH₃

The NH_3 molecule acts as a *Brönsted base* in water, accepting a proton from a strong acid such as HCl:

$$NH_3(aq) + H^+(aq) \rightarrow NH_4^+(aq) \qquad (24.6)$$

Ammonia can also act as a *Lewis base* when it reacts with a metal cation to form a complex ion:

$$2NH_3(aq) + Ag^+(aq) \rightarrow Ag(NH_3)_2^+(aq) \qquad (24.7)$$

In qualitative analysis (Chapter 17), ammonia is often used as a *precipitating agent* to form an insoluble hydroxide. The reaction with Al^{3+} is typical:

$$Al^{3+}(aq) + 3NH_3(aq) + 3H_2O \rightarrow Al(OH)_3(s) + 3NH_4^+(aq) \qquad (24.8)$$

Since Al^{3+} forms a stable hydroxo complex, $Al(OH)_4^-$, with NaOH, ammonia is a better choice for the precipitation of aluminum hydroxide.

Nitrogen cannot have an oxidation number lower than -3, which means that when NH_3 takes part in a redox reaction, it always acts as a *reducing agent*. Ammonia may be oxidized to elementary nitrogen or to a compound of nitrogen. An important redox reaction of ammonia is that with hypochlorite ion:

$$2NH_3(aq) + ClO^-(aq) \rightarrow N_2H_4(aq) + Cl^-(aq) + H_2O \qquad (24.9)$$

Hydrazine, N_2H_4, is made commercially by this process. Certain byproducts of this reaction, notably NH_2Cl and $NHCl_2$, are both toxic and explosive, so solutions of household bleach and ammonia should never be mixed with one another.

They usually wake you up in a hurry

As you know from your experience in the laboratory, ammonia gas has a sharp, irritating odor. Ammonia is a heart stimulant, which makes it dangerous at high concentrations. On a more positive note, "smelling salts," used to revive people who have fainted, generate small quantities of ammonia by the decomposition of an ammonium salt. A typical reaction is:

$$NH_4HCO_3(s) \rightarrow NH_3(g) + CO_2(g) + H_2O \tag{24.10}$$

PROPERTIES OF HYDROGEN SULFIDE, H_2S

In water solution, hydrogen sulfide acts as a *Brönsted acid;* it can donate a proton to a water molecule:

$$H_2S(aq) + H_2O \rightleftharpoons HS^-(aq) + H_3O^+(aq) \tag{24.11}$$

You will recall that in qualitative analysis H_2S is used as a *precipitating agent* for the cations of Groups II and III (Fig. 24.6):

$$2Sb^{3+}(aq) + 3H_2S(aq) \rightarrow Sb_2S_3(s) + 6H^+(aq) \tag{24.12}$$

Like ammonia, hydrogen sulfide (oxid. no. S = -2) can act only as a *reducing agent* when it takes part in redox reactions. Most often, the H_2S is oxidized to elementary sulfur, as in the reaction with Fe^{3+} ions:

$H_2S(aq)$ is slowly oxidized if exposed to air

$$2Fe^{3+}(aq) + H_2S(aq) \rightarrow 2Fe^{2+}(aq) + S(s) + 2H^+(aq) \tag{24.13}$$

$$E^0_{tot} = E^0_{red} \, Fe^{3+} + E^0_{ox} \, H_2S = +0.625 \text{ V}$$

Figure 24.6 Sulfides of Group II and III cations.

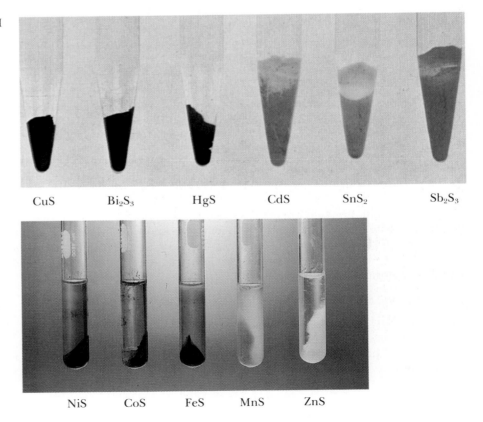

CuS   Bi$_2$S$_3$ HgS   CdS   SnS$_2$ Sb$_2$S$_3$

NiS   CoS   FeS   MnS   ZnS

The spontaneity of this reaction explains why, when H_2S is added to a Group III unknown, any Fe^{3+} present is precipitated as FeS rather than Fe_2S_3.

If you've worked with H_2S in the laboratory, you won't soon forget its rotten-egg odor. In a sense, it's fortunate that hydrogen sulfide has such a distinctive odor. The gas is highly toxic, as poisonous as HCN. At a concentration of 10 parts per million, H_2S can cause headaches and nausea; at 100 ppm it can be fatal.

PROPERTIES OF HYDROGEN PEROXIDE, H_2O_2

In hydrogen peroxide, oxygen has an oxidation number of -1, intermediate between the extremes for the element, 0 and -2. This means that H_2O_2 can act as either an oxidizing agent, in which case it is reduced to H_2O, or as a reducing agent, where it is oxidized to O_2. In practice, hydrogen peroxide is an extremely strong oxidizing agent:

$$H_2O_2(aq) + 2H^+(aq) + 2e^- \rightarrow 2H_2O; \quad E^0_{red} = +1.763 \text{ V}$$

but a very weak reducing agent:

$$H_2O_2(aq) \rightarrow O_2(g) + 2H^+(aq) + 2e^-; \quad E^0_{ox} = -0.695 \text{ V}$$

Hydrogen peroxide tends to decompose in water, which explains why its solutions soon lose their oxidizing power. The reaction involved is *disproportionation*, combining the two half-reactions referred to above:

$$
\begin{array}{ll}
H_2O_2(aq) + 2H^+(aq) + 2e^- \rightarrow 2H_2O & E^0_{red} = +1.763 \text{ V} \\
\underline{H_2O_2(aq) \rightarrow O_2(g) + 2H^+(aq) + 2e^-} & \underline{E^0_{ox} = -0.695 \text{ V}} \\
2H_2O_2(aq) \rightarrow O_2(g) + 2H_2O & E^0_{tot} = +1.068 \text{ V} \quad (24.14)
\end{array}
$$

This reaction is catalyzed by a wide variety of materials, including I^- ions, MnO_2, metal surfaces (Pt, Ag), and even by traces of OH^- ions dissolved from glass (Fig. 24.7).

Figure 24.7 Hydrogen peroxide in water solution is stable in the absence of a catalyst. When I^- ions are added, some of them are oxidized to I_2 molecules (yellow color). However, the principal function of I^- ions is to catalyze the decomposition of hydrogen peroxide: $H_2O_2(aq) \rightarrow H_2O + \frac{1}{2}O_2(g)$. (Charles D. Winters)

■ **EXAMPLE 24.4**

Taking account of the fact that H_2O_2 can act as either an oxidizing agent ($E^0_{red} = +1.763$ V) or a reducing agent ($E^0_{ox} = -0.695$ V), determine whether the following reactions will occur (standard concentrations):

a. $H_2O_2(aq) + 2Fe^{2+}(aq) + 2H^+(aq) \rightarrow 2H_2O + 2Fe^{3+}(aq)$
b. $H_2O_2(aq) + 2Fe^{3+}(aq) \rightarrow O_2(g) + 2H^+(aq) + 2Fe^{2+}(aq)$
c. $H_2O_2(aq) + I_2(s) \rightarrow O_2(g) + 2H^+(aq) + 2I^-(aq)$
d. $H_2O_2(aq) + 2I^-(aq) + 2H^+(aq) \rightarrow 2H_2O + I_2(s)$

Use Table 21.1, Chapter 21, to find necessary values of E^0_{red} or E^0_{ox}.

Solution
In each case, we calculate E^0_{tot} and note its sign.

a. H_2O_2 is reduced; Fe^{2+} ions are oxidized:

$$E^0_{tot} = E^0_{red}\ H_2O_2 + E^0_{ox}\ Fe^{2+}$$

$$= +1.763\ V - 0.769\ V = \boxed{+0.994\ V} \qquad \text{spontaneous}$$

b. H_2O_2 is oxidized; Fe^{3+} ions are reduced:

$$E^0_{tot} = E^0_{ox}\ H_2O_2 + E^0_{red}\ Fe^{3+}$$

$$= -0.695\ V + 0.769\ V = \boxed{+0.074\ V} \qquad \text{spontaneous}$$

c. $$E^0_{tot} = E^0_{ox}\ H_2O_2 + E^0_{red}\ I_2$$

$$= -0.695\ V + 0.534\ V = \boxed{-0.161\ V} \qquad \text{nonspontaneous}$$

d. $$E^0_{tot} = E^0_{red}\ H_2O_2 + E^0_{ox}\ I^-$$

$$= +1.763\ V - 0.534\ V = \boxed{+1.229\ V} \qquad \text{spontaneous}$$

Exercise
Will hydrogen peroxide oxidize the Mn^{2+} ion? Will it reduce Mn^{2+}? Answer: Yes; no.

The 3% solution is a good gargle and disinfectant

You are most likely to come across hydrogen peroxide as its water solution. Two concentrations are available; one of these, containing 3 mass percent H_2O_2, is sold in drugstores. The other solution contains 30 mass percent H_2O_2. Both solutions contain stabilizers to prevent Reaction 24.14 from taking place during storage. Hydrogen peroxide is used as a disinfectant (cuts, sore throats) or as a bleach (cloth, paper, hair, etc.).

24.3
OXIDES OF NITROGEN, PHOSPHORUS, AND SULFUR

Table 24.3 lists some of the properties of the nonmetal oxides of the Group 5 and Group 6 elements. Notice that all of the oxides of nitrogen have positive free energies of formation. All of these compounds are thermo-

TABLE 24.3   Properties of Oxides of N, P, and S

	bp (°C)	PHYS. STATE 25°C, 1 atm	ΔG_f^0 (kJ) 25°C, 1 atm
N_2O_5	subl. 32	solid	+113.8
N_2O_4	21	gas	+97.9
NO_2	21	gas	+51.3
N_2O_3	2	gas	+139.5
NO	−152	gas	+86.6
N_2O	−88	gas	+104.2
P_4O_{10}	subl. 359	solid	−2697.8
P_4O_6	175	solid	~−1470
SO_3	45	liquid	−371.1
SO_2	−10	gas	−300.2

Sulfur burns in oxygen with a blue flame to form $SO_2(g)$. (Charles D. Winters)

dynamically unstable with respect to decomposition into the elements. For example:

$$2N_2O(g) \rightarrow 2N_2(g) + O_2(g); \Delta G^0 \text{ at } 25°C = -2\Delta G_f^0 N_2O(g) = -208.4 \text{ kJ}$$

In practice, N_2O stays around for a long time at room temperature; the rate of this reaction is extremely slow under ordinary conditions.

■ **EXAMPLE 24.5** _____

Calculate ΔG^0 at 25°C for the reactions:

a. $N_2O_4(g) \rightarrow 2NO_2(g)$
b. $N_2O_3(g) \rightarrow NO(g) + NO_2(g)$
c. $2SO_2(g) + O_2(g) \rightarrow 2SO_3(g)$

Solution
We use the general relation:

$$\Delta G_{reaction}^0 = \Sigma \Delta G_f^0 \text{ products} - \Sigma \Delta G_f^0 \text{ reactants}$$

a. $\Delta G^0 = 2(+51.3 \text{ kJ}) - 97.9 \text{ kJ} = \boxed{+4.7 \text{ kJ}}$

b. $\Delta G^0 = 86.6 \text{ kJ} + 51.3 \text{ kJ} - 139.5 \text{ kJ} = \boxed{-1.6 \text{ kJ}}$. At 25°C and 1 atm, N_2O_3 tends to decompose to NO and NO_2.

c. $\Delta G^0 = 2(-371.1 \text{ kJ}) - 2(-300.2 \text{ kJ}) = \boxed{-141.8 \text{ kJ}}$. At 25°C and 1 atm, SO_3 is the thermodynamically stable oxide of sulfur. However, as pointed out in Chapter 19, the rate of reaction is slow at room temperature in the absence of a catalyst.

Exercise
Using the relation: $\Delta G^0 = -RT \ln K_p$, with $R = 8.31$ J/mol·K, calculate K_p for each of these reactions at 25°C. Answer: 0.15, 1.9, 7×10^{24}.

Figure 24.8 Lewis structures of the oxides of nitrogen and sulfur. Many other resonance forms are possible.

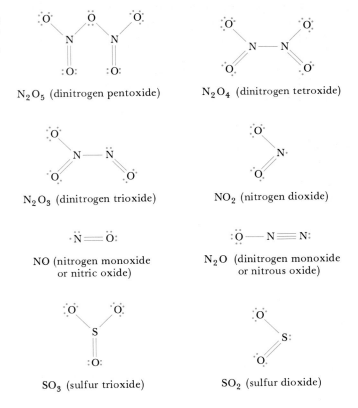

N_2O_5 (dinitrogen pentoxide)

N_2O_4 (dinitrogen tetroxide)

N_2O_3 (dinitrogen trioxide)

NO_2 (nitrogen dioxide)

NO (nitrogen monoxide or nitric oxide)

N_2O (dinitrogen monoxide or nitrous oxide)

SO_3 (sulfur trioxide)

SO_2 (sulfur dioxide)

MOLECULAR STRUCTURES

The Lewis structures of the oxides of nitrogen and sulfur are shown in Figure 24.8. Two of these species, NO and NO_2, are paramagnetic, with one unpaired electron. When nitrogen dioxide is cooled, it dimerizes; the unpaired electrons combine to form a single bond between the two nitrogen atoms:

NO₂ is brown; N₂O₄ is colorless

$$2NO_2(g) \rightleftharpoons N_2O_4(g)$$

A similar reaction occurs when an equimolar mixture of NO and NO_2 is cooled. Two odd electrons, one from each molecule, pair off to form an N—N bond:

$$NO_2(g) + NO(g) \rightleftharpoons N_2O_3(g)$$

At $-20°C$, dinitrogen trioxide separates from the mixture as a blue liquid.

Perhaps the best known oxide of nitrogen is N_2O, commonly called nitrous oxide or "laughing gas." Nitrous oxide is frequently used as an anaesthetic, particularly in dentistry. It is also the propellant gas used in whipped cream containers; N_2O is nontoxic, virtually tasteless, and quite soluble in vegetable oils. The N_2O molecule, like all those in Figure 24.8, can be represented as a resonance hybrid.

In the early days of N_2O, they gave laughing gas parties

■ **EXAMPLE 24.6** _____

Consider the N_2O molecule shown in Figure 24.8.

a. Draw another resonance form of N_2O.
b. What is the bond angle in N_2O?
c. Is the N_2O molecule polar or nonpolar?

Solution

a. $\ddot{O}{=}N{=}\ddot{N}$ or $:O{\equiv}N{-}\ddot{N}:$

b. In any of the resonance forms, the central nitrogen atom, insofar as geometry is concerned, would behave as if it were surrounded by two electron pairs. The bond angle is   180° ; the molecule is linear, like BeF_2.

c.   Polar (unsymmetrical).

Exercise

What is the hybridization about the central nitrogen atom in N_2O? Answer: sp.

_____

Of the two oxides of phosphorus, P_4O_6 and P_4O_{10}, the latter is the more stable; it is formed when white phosphorus burns in air:

$$P_4(s) + 5O_2(g) \rightarrow P_4O_{10}(s) \qquad (24.15)$$

Note from Figure 24.9 that in both P_4O_6 and P_4O_{10}, as in the P_4 molecule, the four phosphorus atoms form a tetrahedron. We can visualize P_4O_6 as being derived from P_4 by inserting an oxygen atom between each pair of phosphorus atoms. In P_4O_{10}, an extra oxygen atom is bonded to each phosphorus.

REACTION WITH WATER

Many nonmetal oxides react with water to form acids. Compounds which behave in this way are referred to as *acid anhydrides*. Looking at the reaction:

$$SO_3(g) + H_2O(l) \rightarrow H_2SO_4(l) \qquad (24.16)$$

Figure 24.9 Structures of the oxides of phosphorus, simplest formulas P_2O_3 and P_2O_5.

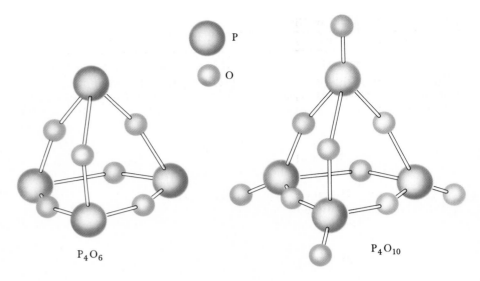

P_4O_6 P_4O_{10}

we see that sulfur trioxide is the acid anhydride of sulfuric acid. Notice that, in this reaction, the nonmetal does not change oxidation number; sulfur is in the $+6$ state in both SO_3 and H_2SO_4. Other acid anhydrides include N_2O_5 and N_2O_3:

$$+5 \text{ nitrogen:}\quad N_2O_5(s) + H_2O(l) \rightarrow 2HNO_3(l) \tag{24.17}$$

$$+3 \text{ nitrogen:}\quad N_2O_3(g) + H_2O(l) \rightarrow 2HNO_2(aq) \tag{24.18}$$

The products here are nitric acid, HNO_3, and an aqueous solution of nitrous acid, HNO_2.

One of the most important reactions of this type involves the $+5$ oxide of phosphorus, P_4O_{10}. Here the product is phosphoric acid, H_3PO_4:

$$P_4O_{10}(s) + 6H_2O(l) \rightarrow 4H_3PO_4(s) \tag{24.19}$$

This reaction is used to prepare high-purity phosphoric acid and salts of that acid for use in food products. Phosphoric acid, H_3PO_4, is added to cola and root beer to give them a tart taste.

24.4
OXYACIDS OF NITROGEN, PHOSPHORUS, AND SULFUR

Table 24.4 lists the formulas and molecular structures of some of the more important oxyacids of these elements. Notice that, in all but one case, the hydrogen atoms are bonded to oxygen, a characteristic structural feature of oxyacids. The exception is phosphorous acid, H_3PO_3, where one of the hydrogen atoms is bonded to phosphorus. Only the hydrogens bonded to oxygen form H^+ ions in water, so phosphorous acid is a *diprotic acid:*

$$H_3PO_3(aq) \rightleftharpoons H^+(aq) + H_2PO_3^-(aq); \quad K_a = 5 \times 10^{-2} \tag{24.20}$$

$$H_2PO_3^-(aq) \rightleftharpoons H^+(aq) + HPO_3^{2-}(aq); \quad K_a = 2 \times 10^{-7} \tag{24.21}$$

TABLE 24.4 Oxyacids of Nitrogen, Phosphorus, and Sulfur

OXID. STATE	FORMULA	STRUCTURE	FORMULA	STRUCTURE
+6	H_2SO_4 sulfuric acid	$$\begin{array}{c} O \\ \| \\ HO-S-OH \\ \| \\ O \end{array}$$		
+5	HNO_3 nitric acid	$$\begin{array}{c} HO-N=O \\ \| \\ O \end{array}$$	H_3PO_4 phosphoric acid	$$\begin{array}{c} O \\ \| \\ HO-P-OH \\ \| \\ O \end{array}$$
+4	H_2SO_3 sulfurous acid	$$\begin{array}{c} HO-S-OH \\ \| \\ O \end{array}$$		
+3	HNO_2 nitrous acid	$HO-N=O$	H_3PO_3 phosphorous acid	$$\begin{array}{c} O \\ \| \\ HO-P-OH \\ \| \\ H \end{array}$$

The molecules shown in Table 24.4 do not by any means represent all of the oxyacids of these elements. In particular, there is an entire series of condensed phosphoric acids, formed by eliminating water from successive molecules of H_3PO_4. Two of the simpler species of this type are the molecules $H_4P_2O_7$ and $H_5P_3O_{10}$:

$$\begin{array}{cc} O \quad\quad O \\ \| \quad\quad \| \\ HO-P-O-P-OH \quad\quad HO-P-O-P-O-P-OH \\ \| \quad\quad \| \\ OH \quad\quad OH \end{array}$$

You have probably used salts of these acids, perhaps without realizing it. The compound $Na_4P_2O_7$ is an ingredient of instant pudding mixes which require no cooking; the $P_2O_7^{4-}$ ion forms a gel with soluble Ca^{2+} salts. Sodium tripolyphosphate, $Na_5P_3O_{10}$, is a major component of many detergents, added to complex Mg^{2+} and Ca^{2+} ions, preventing them from forming precipitates. This compound also promotes the growth of algae in lakes and streams; for that reason $Na_5P_3O_{10}$, and indeed phosphates in general, have been banned or restricted in many areas.

NITRIC ACID, HNO₃

Nitric acid is a strong acid, completely dissociated to H^+ and NO_3^- ions in dilute water solution:

$$HNO_3(aq) \rightarrow H^+(aq) + NO_3^-(aq) \tag{24.22}$$

Nitric acid is used to prepare metal nitrates from the corresponding oxides, hydroxides, or carbonates. To make zinc nitrate from zinc oxide, we add nitric acid and evaporate the resulting solution:

$$ZnO(s) + 2H^+(aq) + 2NO_3^-(aq) \rightarrow Zn^{2+}(aq) + 2NO_3^-(aq) + H_2O$$
$$Zn^{2+}(aq) + 2NO_3^-(aq) \rightarrow Zn(NO_3)_2(s)$$
$$\overline{ZnO(s) + 2H^+(aq) + 2NO_3^-(aq) \rightarrow Zn(NO_3)_2(s) + H_2O} \quad (24.23)$$

In another case, we can prepare calcium nitrate by adding nitric acid to calcium carbonate. The overall reaction is:

$$CaCO_3(s) + 2H^+(aq) + 2NO_3^-(aq) \rightarrow Ca(NO_3)_2(s) + H_2O + CO_2(g) \quad (24.24)$$

Concentrated nitric acid (16 M) is colorless when pure. In sunlight, it turns yellow (Fig. 24.10) because it decomposes to $NO_2(g)$:

$$4HNO_3(aq) \rightarrow 4NO_2(g) + 2H_2O + O_2(g) \quad (24.25)$$

The yellow color that appears on your skin if it comes in contact with nitric acid has quite a different explanation. Nitric acid reacts with proteins to give a yellow material called xanthoprotein.

The concentrated acid is a strong oxidizing agent. It can be reduced to a variety of different species; NO_2 is most common. This is the product when 16 M acid reacts with copper (Fig. 24.11, p. 861):

$$Cu(s) + 4H^+(aq) + 2NO_3^-(aq) \rightarrow Cu^{2+}(aq) + 2NO_2(g) + 2H_2O \quad (24.26)$$

Dilute nitric acid (6 M) is a weaker oxidizing agent than 16 M HNO_3. It also gives a wider variety of reduction products, depending upon the nature of the reducing agent. With inactive metals such as copper ($E_{ox}^0 = -0.339$ V), the major product is usually NO (oxid. no. N = +2):

$$3Cu(s) + 2NO_3^-(aq) + 8H^+(aq) \rightarrow 3Cu^{2+}(aq) + 2NO(g) + 4H_2O \quad (24.27)$$

With very dilute acid and a strong reducing agent such as zinc ($E_{ox}^0 = +0.762$ V) reduction may go all the way to the NH_4^+ ion (oxid. no. N = -3):

$$4Zn(s) + NO_3^-(aq) + 10H^+(aq) \rightarrow 4Zn^{2+}(aq) + NH_4^+(aq) + 3H_2O \quad (24.28)$$

Above 16 M HNO_3 you can always smell NO_2

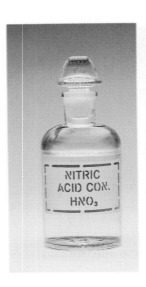
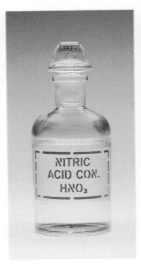

Figure 24.10 A water solution of nitric acid slowly turns yellow because of the $NO_2(g)$ formed by decomposition. Nitric acid also reacts with proteins (casein in milk (test tube) and albumen in eggs) to give a characteristic yellow color. (a,b, Marna G. Clarke; c, Charles D. Winters)

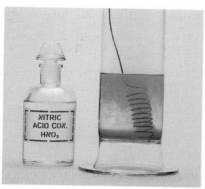

Figure 24.11 Copper metal is comparatively inactive, but it reacts with concentrated nitric acid. The brown fumes are $NO_2(g)$, a reduction product of HNO_3. The copper is oxidized to Cu^{2+} ions, which impart their color to the solution. (Marna G. Clarke)

As you can guess from Equations 24.26 to 24.28, hydrogen gas is seldom produced when a metal reacts with nitric acid; the NO_3^- ion is reduced rather than the H^+ ion. A few metals, notably Al, Cr, Fe, Co, and Ni, are "passive" to concentrated nitric acid, even though they react readily with the dilute acid. It is generally supposed that failure to react with 16 M acid is due to the formation of a tightly adherent oxide film (e.g., Al_2O_3, Fe_3O_4).

■ **EXAMPLE 24.7** _____

Write balanced equations for

a. the reaction of dilute nitric acid with a solution of barium hydroxide.
b. the preparation of $Mg(NO_3)_2(s)$ from $Mg(OH)_2(s)$.
c. the reaction of concentrated nitric acid with $CuS(s)$ to form Cu^{2+}, $S(s)$, and NO_2.

Solution

a. Since nitric acid is a strong acid, the reaction is simply:

$$H^+(aq) + OH^-(aq) \rightarrow H_2O$$

Neither the NO_3^- ion nor the Ba^{2+} ion take part in the reaction.
b. Nitric acid is used to effect the conversion; the overall equation is:

$$Mg(OH)_2(s) + 2H^+(aq) + 2NO_3^-(aq) \rightarrow Mg(NO_3)_2(s) + 2H_2O$$

c. The balanced half-equations are:

$$CuS(s) \rightarrow Cu^{2+}(aq) + S(s) + 2e^-$$

$$NO_3^-(aq) + 2H^+(aq) + e^- \rightarrow NO_2(g) + H_2O$$

Combining these half-equations in the usual way, we obtain:

$$CuS(s) + 2NO_3^-(aq) + 4H^+(aq) \rightarrow Cu^{2+}(aq) + S(s) + 2NO_2(g) + 2H_2O$$

Exercise
Write an equation for the conversion of $Al_2O_3(s)$ to $Al(NO_3)_3(s)$. Answer:
$$Al_2O_3(s) + 6H^+(aq) + 6NO_3^-(aq) \rightarrow 2Al(NO_3)_3(s) + 3H_2O$$

You will recall from Chapter 19 that nitric acid is made commercially from ammonia by the Ostwald process. The ammonia is oxidized, first to NO and then to NO_2. When nitrogen dioxide is bubbled through water, nitric acid is formed. Before ammonia became available through the Haber process after World War I, nitric acid was made by the reaction between "saltpeter" and sulfuric acid:

$$KNO_3(s) + H_2SO_4(l) \rightarrow HNO_3(g) + KHSO_4(s) \tag{24.29}$$

This preparation depends upon the fact that nitric acid has a lower boiling point than sulfuric acid (86°C vs. 338°C).

SULFURIC ACID, H_2SO_4

You will recall from Chapter 12 that H_2SO_4 is a strong acid, completely dissociated to H^+ and HSO_4^- ions in dilute water solution. The HSO_4^- ion dissociates further to give H^+ and SO_4^{2-} ions:

$$H_2SO_4(aq) \rightarrow H^+(aq) + HSO_4^-(aq)$$
$$HSO_4^-(aq) \rightleftharpoons H^+(aq) + SO_4^{2-}(aq); \qquad K_a = 1.0 \times 10^{-2}$$

The dissociation constant of the HSO_4^- ion is relatively large. This explains why in writing equations for the reactions of sulfuric acid, we often consider it to consist of $2H^+(aq) + SO_4^{2-}(aq)$. For example, to describe the reaction of barium hydroxide with dilute sulfuric acid, we ordinarily write:

$$Ba(OH)_2(s) + 2H^+(aq) + SO_4^{2-}(aq) \rightarrow BaSO_4(s) + 2H_2O \tag{24.30}$$

Sulfuric acid is a relatively weak oxidizing agent; most often it is reduced to sulfur dioxide:

$$SO_4^{2-}(aq) + 4H^+(aq) + 2e^- \rightarrow SO_2(g) + 2H_2O; E_{red}^0 = +0.155 \text{ V}$$

This is the case when copper metal is oxidized by hot concentrated sulfuric acid:

$$Cu(s) + 4H^+(aq) + SO_4^{2-}(aq) \rightarrow Cu^{2+}(aq) + 2H_2O + SO_2(g) \tag{24.31}$$

When dilute sulfuric acid (3 M) reacts with metals, it is ordinarily the H^+ ion rather than the SO_4^{2-} ion which is reduced. For example, zinc reacts with dilute sulfuric acid to form hydrogen gas:

$$Zn(s) + 2H^+(aq) \rightarrow Zn^{2+}(aq) + H_2(g) \tag{24.32}$$

The sulfate ion is not involved in this reaction.

Figure 24.12 Crystals of hydrated metal sulfates have a characteristic glassy appearance. They are often colored; $CuSO_4 \cdot 5H_2O$ ("blue vitriol") is blue, $FeSO_4 \cdot 7H_2O$ ("green vitriol") and $NiSO_4 \cdot 6H_2O$ are green, while $CoSO_4 \cdot 7H_2O$ is red. (Charles D. Winters)

Concentrated sulfuric acid (98% H_2SO_4, 18 M) was called "oil of vitriol" by the alchemists. "Oil" refers to the high viscosity of the concentrated acid, a result of hydrogen bonding between H_2SO_4 molecules. "Vitriol" refers to the shiny, glassy appearance of many metal sulfates (Fig. 24.12).

Concentrated sulfuric acid, in addition to being an acid and an oxidizing agent, is also a dehydrating agent. Small amounts of water can be removed from organic liquids such as gasoline by extraction with sulfuric acid. Sometimes it is even possible to remove the elements of water from a compound by treating it with 18 M H_2SO_4. This happens with table sugar, $C_{12}H_{22}O_{11}$; the product is a black char that is mostly carbon (Fig. 24.13).

$$C_{12}H_{22}O_{11}(s) \rightarrow 12C(s) + 11H_2O(l) \tag{24.33}$$

Figure 24.13 When sulfuric acid is added to sugar, sucrose, an exothermic reaction occurs. The elements of water are removed from the sugar, $C_{12}H_{22}O_{11}$, leaving a mass of black carbon. (Charles D. Winters)

When concentrated sulfuric acid dissolves in water, a great deal of heat is given off, nearly 100 kJ per mole of H_2SO_4. Sometimes enough heat is evolved to bring the solution to the boiling point. To prevent this and to avoid spattering, the acid should be added slowly to water, with constant stirring. If it comes in contact with the skin, concentrated sulfuric acid can cause painful chemical burns.

Be very careful when working with conc. H_2SO_4

Commercially, sulfuric acid is made by the three-step "contact" process.

1. Elemental sulfur is burned in air to form sulfur dioxide:

$$S(s) + O_2(g) \rightarrow SO_2(g) \tag{24.34}$$

2. Sulfur dioxide is converted to sulfur trioxide by bringing it into "contact" with oxygen on the surface of a solid catalyst:

$$SO_2(g) + \tfrac{1}{2}O_2(g) \rightleftharpoons SO_3(g); \qquad \Delta H = -98.9 \text{ kJ} \tag{24.35}$$

This is the key step in the process and the most difficult to carry out. The catalyst used today is an oxide of vanadium, V_2O_5. Platinum is equally effective but is more expensive and more easily "poisoned" (made ineffective) by impurities in the gases.

Since Reaction 24.35 is exothermic, we expect the equilibrium constant to decrease as the temperature rises. This is indeed the case (Table 24.5). Hence, according to equilibrium principles, a low temperature is preferred. Too low a temperature, however, reduces the rate to the point where Reaction 24.35 becomes impractical. The temperature actually used in the contact process represents a compromise between equilibrium and rate considerations. A mixture of SO_2 and O_2 is first passed over the catalyst at 600°C; about 80% of the SO_2 is quickly converted to SO_3. The sulfur trioxide is removed from the gas mixture and the remaining SO_2 and O_2 are recycled over a second catalyst bed. This time the temperature is lower, about 450°C. It takes a while under these conditions, but eventually the yield of SO_3 is raised to above 99%.

3. The sulfur trioxide formed by Reaction 24.35 is converted to sulfuric acid by reaction with water:

$$SO_3(g) + H_2O \rightarrow H_2SO_4(aq) \tag{24.36}$$

This reaction cannot be carried out directly. If SO_3 is bubbled through water, H_2SO_4 is formed as a fog of tiny particles that are difficult to condense. Instead, SO_3 is absorbed in concentrated sulfuric acid to form an intermediate product, $H_2S_2O_7$, called pyrosulfuric acid. Subsequent addition of water to $H_2S_2O_7$ forms H_2SO_4:

$$\begin{aligned} SO_3(g) + H_2SO_4(l) &\rightarrow H_2S_2O_7(l) \\ \underline{H_2S_2O_7(l) + H_2O} &\underline{\rightarrow 2H_2SO_4(aq)} \\ SO_3(g) + H_2O &\rightarrow H_2SO_4(aq) \end{aligned} \tag{24.37}$$

TABLE 24.5   Equilibrium Constant for the Reaction
$SO_2(g) + \tfrac{1}{2}O_2(g) \rightleftharpoons SO_3(g); \qquad \Delta H = -98.9 \text{ kJ}$

t (°C)	25	200	400	500	600	700	800
K_c	9.2×10^{12}	5.0×10^6	2300	400	70	20	7

■ EXAMPLE 24.8

At 600°C, K_c for Reaction 24.35 is 70 (Table 24.5). Suppose that, at this temperature, $[SO_2] = 0.020$ mol/L and $[O_2] = 0.010$ mol/L. Calculate the equilibrium concentration of SO_3 under these conditions.

Solution

The expression for K_c is

$$K_c = \frac{[SO_3]}{[SO_2] \times [O_2]^{1/2}} = 70$$

Solving for $[SO_3]$,

$$[SO_3] = 70 \times [SO_2] \times [O_2]^{1/2}$$

Substituting for $[SO_2]$ and $[O_2]$,

$$[SO_3] = 70 \times 0.020 \times (0.010)^{1/2} = \boxed{0.14 \text{ mol/L}}$$

Exercise

Assume the original mixture contained only SO_2 and O_2. What must the original concentration of SO_2 have been? What percentage of the SO_2 was converted to SO_3? Answer: 0.16 M; 88%.

NITROUS ACID, HNO₂

A water solution of nitrous acid, HNO_2, is made by adding a strong acid such as HCl to sodium nitrite, $NaNO_2$. Since HNO_2 is a weak acid ($K_a = 6.0 \times 10^{-4}$), the following acid-base reaction goes nearly to completion:

$$H^+(aq) + NO_2^-(aq) \rightarrow HNO_2(aq) \tag{24.38}$$

If enough acid is added to bring the pH to 0, nearly all the NO_2^- ions are converted to HNO_2 molecules:

$$\frac{[HNO_2]}{[NO_2^-]} = \frac{[H^+]}{K_a} = \frac{1.0}{6.0 \times 10^{-4}} = 1.7 \times 10^3$$

In HNO_2, nitrogen is in an intermediate oxidation state, $+3$. Hence, at least in principle, nitrous acid can act as either an oxidizing or reducing agent. Nitrous acid is a strong oxidizing agent

$$HNO_2(aq) + H^+(aq) + e^- \rightarrow NO(g) + H_2O; \quad E^0_{red} = +1.036 \text{ V}$$

but a weak reducing agent

$$HNO_2(aq) + H_2O \rightarrow NO_3^-(aq) + 3H^+(aq) + 2e^-; \quad E^0_{ox} = -0.928 \text{ V}$$

Looking at the equations just written, you might guess that HNO_2 would disproportionate in water solution. Indeed it does; the overall reaction, obtained by combining these two half-equations, is

$$3HNO_2(aq) \rightarrow 2NO(g) + NO_3^-(aq) + H_2O + H^+(aq) \quad E^0_{tot} = +0.108 \text{ V} \tag{24.39}$$

Ordinarily, this reaction occurs rather slowly, so that the HNO_2 stays around long enough for its properties to be studied. The pure acid cannot be isolated, however.

Sodium nitrite, $NaNO_2$, and sodium nitrate, $NaNO_3$, are commonly added in small quantities (~100 parts per million) to many processed meats, including ham, bacon, and frankfurters. The NO_3^- and NO_2^- ions are reduced to NO, which retards the oxidation of hemoglobin. This allows these meats to retain their pink color, rather than darkening. More importantly, NO_2^- and NO_3^- ions prevent the growth of bacteria that cause botulism, a potentially fatal type of food poisoning.

There is, however, a potential hazard associated with these food additives. It is known that, under certain conditions, NO_2^- ions can be converted to organic compounds called nitrosamines, which have the general formula:

$$\begin{array}{c} R \\ \diagdown \\ \diagup N\!-\!N\!=\!O \\ R' \end{array}$$

where R and R' are organic groups such as CH_3, C_2H_5, ———. Nitrosamines are carcinogenic; the possibility that they might be formed from NO_2^- and NO_3^- ions has led to restrictions on the amounts of these ions that can be used in foods.

So far we have found no good substitutes for nitrites as preservatives

SULFUROUS ACID, H_2SO_3

Sulfur dioxide, SO_2, is very soluble in water. The concentration of its saturated solution at 25°C is about 1.3 mol/L. The high solubility is explained in part by a reversible reaction with H_2O to form sulfurous acid, H_2SO_3:

SO_2 is an acid anhydride

$$SO_2(g) + H_2O \rightleftharpoons H_2SO_3(aq) \tag{24.40}$$

The equilibrium constant for this reaction is not known, but it appears that much of the SO_2 remains unreacted.

Sulfurous acid is a weak acid that ionizes in two steps:

$$H_2SO_3(aq) \rightleftharpoons H^+(aq) + HSO_3^-(aq); \qquad K_a = 1.7 \times 10^{-2}$$

$$HSO_3^-(aq) \rightleftharpoons H^+(aq) + SO_3^{2-}(aq); \qquad K_a = 6.0 \times 10^{-8}$$

The species present in the +4 state in water solution depends upon the pH of the solution. Figure 24.14 shows the relative concentrations of H_2SO_3 (or SO_2), HSO_3^-, and SO_3^{2-} at different pH's (the calculations are indicated in Example 24.9). Note that as pH increases, the species present in highest concentration, the *principal species*, shifts from H_2SO_3 to HSO_3^- to SO_3^{2-}.

Compounds containing HSO_3^- or SO_3^{2-} ions are prepared by bubbling SO_2 through a strongly basic solution (for example, 1 M NaOH). As sulfur dioxide reacts with OH^- ions, the pH decreases. If the reaction is stopped when the pH has dropped to about 10 (Fig. 24.14), the product is the SO_3^{2-} ion:

$$SO_2(g) + 2OH^-(aq) \rightarrow SO_3^{2-}(aq) + H_2O \tag{24.41}$$

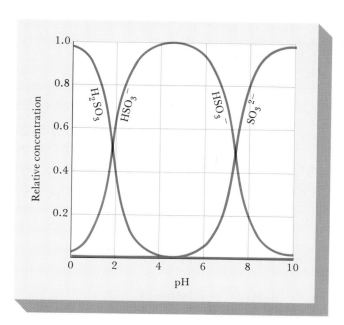

Figure 24.14 In the +4 state, sulfur can exist as H_2SO_3, HSO_3^-, or SO_3^{2-}, depending upon pH. In strongly acidic solution, H_2SO_3 dominates. Between pH 3 and pH 7, HSO_3^- is the major species present. In basic solution, above pH 7, the SO_3^{2-} ion dominates.

If more SO_2 is added, until the pH decreases to about 5, further reaction occurs. The principal species under these conditions is the hydrogen sulfite ion, HSO_3^-:

$$SO_2(g) + SO_3^{2-}(aq) + H_2O \rightarrow 2HSO_3^-(aq) \qquad (24.42)$$

Reactions 24.41 and 24.42 are used to prepare Na_2SO_3 and $NaHSO_3$. The latter compound is used in the "sulfite process" for making paper. Hot aqueous solutions containing the HSO_3^- ion (and some unreacted SO_2) dissolve lignin in pulp and allow separation and bleaching of wood fibers. Bleaching in this case occurs through the reduction of colored compounds (Fig. 24.15) by the HSO_3^- ion, which is itself oxidized to the +6 state:

$$HSO_3^-(aq) + H_2O \rightarrow SO_4^{2-}(aq) + 3H^+(aq) + 2e^-; \quad E_{ox}^0 = -0.105 \text{ V}$$

When the bleached material is once more exposed to air and light, it may be re-oxidized to the original colored compound. This happens with newsprint, which turns yellow upon aging.

Sulfites are also used as food preservatives. At one time, they were sprayed on salad bars to prevent discoloration of lettuce, spinach, and other vegetables. Their use for this purpose was banned by the FDA in 1986 because some people are highly allergic to sulfites. Such people suffer outbreaks of hives, asthmatic attacks, even sharp drops in blood pressure after eating foods treated with sulfites.

■ **EXAMPLE 24.9**

Consider a solution containing +4 sulfur in which the pH has been adjusted to 3.00. Taking K_a $H_2SO_3 = 1.7 \times 10^{-2}$ and K_a $HSO_3^- = 6.0 \times 10^{-8}$, calculate:

a. $[H_2SO_3]/[HSO_3^-]$ b. $[SO_3^{2-}]/[HSO_3^-]$
c. the percentages of H_2SO_3, HSO_3^-, and SO_3^{2-}

Figure 24.15 Sulfur dioxide, SO_2, and the HSO_3^- and SO_3^{2-} ions in solution, frequently act as bleaching agents, in this case with a rose. (Charles D. Winters)

Solution

a. $K_a\ H_2SO_3 = \dfrac{[H^+] \times [HSO_3^-]}{[H_2SO_3]};$

$\dfrac{[H_2SO_3]}{[HSO_3^-]} = \dfrac{[H^+]}{K_a\ H_2SO_3} = \dfrac{1.0 \times 10^{-3}}{1.7 \times 10^{-2}} = \boxed{0.059}$

Clearly there are relatively few H_2SO_3 molecules as compared to HSO_3^- ions

b. $K_a\ HSO_3^- = \dfrac{[H^+] \times [SO_3^{2-}]}{[HSO_3^-]};$

$\dfrac{[SO_3^{2-}]}{[HSO_3^-]} = \dfrac{K_a\ HSO_3^-}{[H^+]} = \dfrac{6.0 \times 10^{-8}}{1.0 \times 10^{-3}} = \boxed{6.0 \times 10^{-5}}$

There are very, very few SO_3^{2-} ions as compared to HSO_3^- ions.

c. If we let x = % HSO_3^- ions in the mixture:

$\quad$ % $H_2SO_3 = 0.059\ x;\qquad$ % $SO_3^{2-} = 0.000060\ x$

Hence: $x + 0.059x + 0.000060\ x = 100$

$\quad$ Solving, x = % $HSO_3^- = \boxed{94.43}$

$\quad$ % $H_2SO_3 = 0.059x = \boxed{5.6}$;

$\quad$ % $SO_3^{2-} = 6.0 \times 10^{-5}\ x = \boxed{0.0057}$

At pH 3, more than 94% of the sulfur is in the form of HSO_3^-; virtually all of the remainder is in the form of H_2SO_3, with essentially no SO_3^{2-} ions.

Exercise
Calculate the ratios in (a) and (b) at pH = 8.00. Answer: 5.9×10^{-7}; 6.0 (the principal species at pH 8 is the SO_3^{2-} ion).

PHOSPHORIC ACID, H₃PO₄

Phosphoric acid is a weak *triprotic* acid:

$H_3PO_4(aq) \rightleftharpoons H^+(aq) + H_2PO_4^-(aq) \qquad K_1 = 7.1 \times 10^{-3}$

$H_2PO_4^-(aq) \rightleftharpoons H^+(aq) + HPO_4^{2-}(aq) \qquad K_2 = 6.2 \times 10^{-8}$

$HPO_4^{2-}(aq) \rightleftharpoons H^+(aq) + PO_4^{3-}(aq) \qquad K_3 = 4.5 \times 10^{-13}$

In a water solution containing +5 phosphorus, four different phosphorus-containing species can be present, depending upon pH: H_3PO_4, $H_2PO_4^-$, HPO_4^{2-}, and PO_4^{3-}. Figure 24.16 shows how the relative amounts of these species change with pH. The figure can be used for a variety of purposes, a couple of which we have not discussed before (Example 24.10).

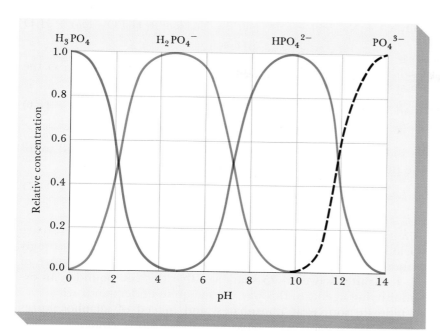

Figure 24.16 In water solution, phosphorus in the $+5$ state can exist as H_3PO_4, $H_2PO_4^-$, HPO_4^{2-}, or PO_4^{3-}, depending upon pH. The H_3PO_4 molecule dominates in strongly acidic solution; the PO_4^{3-} ion is the principal species in strongly basic solution. The $H_2PO_4^-$ ion has its maximum concentration at pH 5, the HPO_4^{2-} ion at pH 10.

■ **EXAMPLE 24.10**

Referring to Figure 24.16:

a. Estimate the pH of a solution of NaH_2PO_4; Na_2HPO_4.
b. How would you prepare a "phosphate buffer" with a pH of 7?

Solution

a. The $H_2PO_4^-$ ion has its maximum concentration at about pH 5. This is the approximate pH of a solution of NaH_2PO_4 (regardless of its concentration); this salt is weakly acidic. In contrast, Na_2HPO_4 has a pH of a little more than 9.

b. At pH 7, it appears that about 60% of the $+5$ phosphorus is in the form of the $H_2PO_4^-$ ion; 40% is present as HPO_4^{2-}. (An exact calculation indicates that the ratio is about 62% $H_2PO_4^-$, 38% HPO_4^{2-}). You might make up such a buffer by, for example, adding 60 mL of 1.0 M NaH_2PO_4 to 40 mL of 1.0 M Na_2HPO_4 .

This is a common buffer

Exercise

What is the approximate pH when one mole of NaOH is added to one mole of H_3PO_4? two moles of NaOH per mole H_3PO_4? Answer: 5, 9.

From Figure 24.16 (and Example 24.10) we conclude that to make NaH_2PO_4, we should add enough base to bring the pH to about 5. At that point, the $H_2PO_4^-$ ion is the principal species:

$$H_3PO_4(aq) + Na^+(aq) + OH^-(aq) \rightarrow Na^+(aq) + H_2PO_4^-(aq) + H_2O$$
$$pH \approx 5$$

Further addition of NaOH converts $H_2PO_4^-$ to HPO_4^{2-}, which is the dominant species around pH 9:

$$Na^+(aq) + H_2PO_4^-(aq) + Na^+(aq) + OH^-(aq) \rightarrow 2Na^+(aq) + HPO_4^{2-}(aq) + H_2O$$
$$pH \approx 9$$

To make sodium phosphate, Na_3PO_4, enough base must be added to increase the pH to above 12:

$$2Na^+(aq) + HPO_4^{2-}(aq) + Na^+(aq) + OH^-(aq) \rightarrow 3Na^+(aq) + PO_4^{3-}(aq) + H_2O$$
$$pH \approx 13$$

Of the three compounds NaH_2PO_4, Na_2HPO_4, and Na_3PO_4, two are used as cleaning agents: NaH_2PO_4 in acid-type cleaners, Na_3PO_4 in strongly basic cleaners. The principal use of Na_2HPO_4 is in the manufacture of cheese; J. L. Kraft discovered 75 years ago that this compound is an excellent emulsifying agent; to this day no one quite understands why.

The ultimate source of phosphates, and indeed of all phosphorus compounds, is "phosphate rock." This mineral is mostly calcium phosphate, $Ca_3(PO_4)_2$. Treatment of phosphate rock with sulfuric acid brings about the reaction:

$$Ca_3(PO_4)_2(s) + 2H_2SO_4(l) + 4H_2O(l) \rightarrow Ca(H_2PO_4)_2(s) + 2[CaSO_4 \cdot 2H_2O(s)] \quad (24.43)$$

The product, a 1:2 mol mixture of $Ca(H_2PO_4)_2$ and gypsum, $CaSO_4 \cdot 2H_2O$, is a fertilizer known as "superphosphate of lime." The active ingredient is calcium dihydrogen phosphate, $Ca(H_2PO_4)_2$. This compound is much more soluble in water than $Ca_3(PO_4)_2$ and so is a better source of phosphorus for growing plants. Nearly pure $Ca(H_2PO_4)_2$ can be made by treating phosphate rock with phosphoric acid:

$Ca_3(PO_4)_2$ would be useless as a fertilizer

$$Ca_3(PO_4)_2(s) + 4H_3PO_4(l) \rightarrow 3Ca(H_2PO_4)_2(s) \quad (24.44)$$

The product is often referred to as "triple superphosphate of lime." It contains a much higher percentage of phosphorus than the mixture produced by Reaction 24.43. In this respect, it is a better fertilizer; it is also more expensive.

■ SUMMARY EXAMPLE

Consider Figure 24.17 and the compounds listed there.

a. Write a balanced equation for the disproportionation of HNO_2 and calculate E^0_{tot}.

b. Write a balanced half-equation for the reduction of N_2O to N_2 in acidic solution. Calculate E_{red} at pH 7.00, taking all species except H^+ to be at unit concentration.

c. List the species in Figure 24.17 which are weak acids; weak bases; contain an unpaired electron.

d. Which of the species listed can act only as oxidizing agents in redox reactions? only as reducing agents?

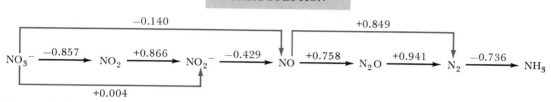

Figure 24.17 Values of E_{red}° for nitrogen-containing species.

e. Consider HNO_2 ($K_a = 6.0 \times 10^{-4}$). What is the principal species, HNO_2 or NO_2^-, at pH 3.00? pH 5.00?

f. How would you prepare a HNO_2–NO_2^- buffer with a pH of 4.00 (i.e., what should be the ratio of HNO_2 to NO_2^-)?

g. Consider the NO_3^- ion. Draw its Lewis structure, give the bond angles, state the hybridization of nitrogen, and indicate whether or not the ion is a dipole.

h. Draw two different resonance structures for the NO_2^- ion.

i. Referring back to (a), calculate the volume of $NO(g)$ formed at 25°C and 1.00 atm from 1.00 g of HNO_2.

j. Write balanced equations for the reaction of NO_2^- ions with hydrochloric acid; with I^- ions, assuming the products are NO and I_2.

k. Give the formula of the product formed when aqueous ammonia reacts with HF; with Cu^{2+}; with Fe^{3+}.

l. At 25°C, ΔG_f^0 NO_2 = +51.8 kJ, ΔG_f^0 NO = +86.7 kJ. Calculate K_p at 25°C for the reaction: $2NO(g) + O_2(g) \rightarrow 2NO_2(g)$.

Answers

a. $3HNO_2(aq) \rightarrow NO_3^-(aq) + 2NO(g) + H_2O + H^+(aq)$; +0.108 V

b. $N_2O(g) + 2H^+(aq) + 2e^- \rightarrow N_2(g) + H_2O$; +1.355 V

c. HNO_2, NH_4^+; NO_2^-, NH_3; NO_2, NO   **d.** NO_3^-; NH_4^+, NH_3

e. HNO_2, NO_2^- **f.** $[HNO_2]/[NO_2^-] = 0.17$

g. $:\overset{..}{O}\!-\!\overset{..}{N}\!-\!\overset{..}{O}:$, 120°, sp^2, no   **h.**
$$\overset{..}{N} \quad\leftrightarrow\quad \overset{..}{N}$$
$$:\overset{..}{O}:$$

i. 0.347 L

j. $NO_2^-(aq) + H^+(aq) \rightarrow HNO_2(aq)$
$2NO_2^-(aq) + 2I^-(aq) + 4H^+(aq) \rightarrow 2NO(g) + I_2(s) + 2H_2O$

k. NH_4^+, F^-; $Cu(NH_3)_4^{2+}$, $Fe(OH)_3$ **l.** 2×10^{12}

7	8
N	O
14.00674	15.9994
15	16
P	S
30.973762	32.066
33	34
As	Se
74.92159	78.96

Figure 24.18 Arsenic and selenium fall directly below phosphorus and sulfur in the Periodic Table.

Selenium imparts a brilliant red color to glass. (Brian Parker, Tom Stack and Assoc.)

The two elements which lie directly below phosphorus and sulfur in the Periodic Table (Fig. 24.18) are of interest for a variety of reasons. Arsenic is a true metalloid. A metallic allotrope, called gray arsenic, has an electrical conductivity approaching that of lead. Another allotrope, yellow arsenic, is distinctly nonmetallic; it has the molecular formula As_4, analogous to white phosphorus, P_4. Selenium is properly classified as a nonmetal, although one of its allotropes has a somewhat metallic appearance and is a semiconductor. Another allotrope of selenium has the molecular formula Se_8, analogous to sulfur. Both of these elements occur in a wide variety of relatively uncommon minerals (Table 24.6); they are ordinarily obtained as by-products in the metallurgy of copper or lead.

The principal use of elemental arsenic is in its alloys with lead. The "lead" storage battery contains a trace of arsenic along with 3% antimony. Lead shot, which are formed by allowing drops of molten metal to fall through air, contain from 0.5 to 2.0% arsenic. It was found many years ago that adding arsenic makes the shot more nearly spherical. Nowadays, increasing amounts of arsenic are being used to make gallium arsenide semiconductors: GaAs is becoming competitive with silicon.

Selenium has long been used as an additive in glassmaking. Particles of colloidally dispersed selenium give the ruby red color seen in stained glass windows and traffic lights. The principal use of selenium today takes advantage of its *photoconductivity;* the electrical conductivity of selenium increases by a factor of 1000 when it is exposed to light. Modern photocopiers use a plate in which a thin film of a photoconductive material, usually selenium, is deposited on an aluminum base. A high potential is placed on the plate, which is then exposed to a light and dark image pattern, obtained by illuminating the article to be copied. In the light areas, the electrostatic charge on the plate is reduced due to a photoconductive discharge; the dark areas retain their original charge. An image is formed on the plate when toner particles (black or colored) are attracted to the high-charge areas. The image is then transferred to a piece of plain paper by charging it sufficiently to pull the toner particles off the plate.

Compounds of As and Se

The chemistry of arsenic strongly resembles that of phosphorus; selenium resembles sulfur. Not only do the compounds of As and P or Se and S have similar formulas; oftentimes they have similar properties as well. For example, hydrogen selenide, H_2Se, is a poisonous gas with an odor even more offensive than that of H_2S, if that's possible. Arsine, AsH_3, like phosphine, PH_3, is an unstable compound which readily decomposes to the elements upon heating:

$$2AsH_3(g) \rightarrow 2As(s) + 3H_2(g) \tag{24.45}$$

This reaction served as the basis for the famous Marsh test for arsenic poisoning (Fig. 24.19), which brought a host of murderers (and a few

TABLE 24.6   Properties of Arsenic and Selenium

	As*	Se*
Outer Electr. Conf.	$4s^24p^3$	$4s^24p^4$
Abundance (%)	2×10^{-4}	5×10^{-6}
Principal Ores	$FeAsS$, As_2S_3, As_4S_4	Cu_2Se, Ag_2Se, $PbSe$
Melting Point (°C)	816	217
Boiling Point (°C)	sublimes	685
Density (g/cm³)	5.778	4.189

*Data for the more metallic allotrope.

analytical chemists) to an untimely end. The arsenic deposited as a metallic ring on a glass tube.

The major difference in chemistry between As and P, or Se and S, relates directly to the positions of these elements in the Periodic Table. For the heavier elements (As and Se), the highest oxidation state (+5 or +6) is considerably less stable than it is for the lighter elements. For example, the SeO_4^{2-} ion is much less stable than the SO_4^{2-} ion. In particular, it is a much stronger oxidizing agent in acidic solution:

$$SeO_4^{2-}(aq) + 4H^+(aq) + 2e^- \rightarrow H_2SeO_3(aq) + H_2O; \quad E^0_{red} = +1.150 \text{ V}$$

$$SO_4^{2-}(aq) + 4H^+(aq) + 2e^- \rightarrow SO_2(g) + 2H_2O; \quad E^0_{red} = +0.155 \text{ V}$$

Physiological Properties

The "arsenic poison" referred to in true crime dramas is actually the oxide As_4O_6; the element itself is nontoxic. Most arsenic compounds in the +3 oxidation state are poisonous, except for As_2S_3, which has a very low water

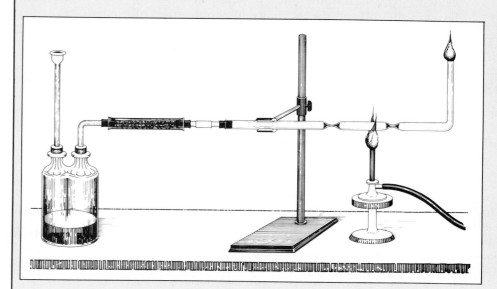

Figure 24.19 Until about 40 years ago, the Marsh test was the best way to test samples for arsenic; chemistry texts spent several pages discussing the nuances of this test.

solubility. The classic symptoms of acute arsenic poisoning include various unpleasant gastrointestinal disturbances, severe abdominal pain, and burning of the mouth and throat. It appears that the toxic action of As^{3+}, like that of Pb^{2+} and Hg^{2+}, involves reaction with the —SH groups of enzymes that are vital for body processes.

When arsenic is taken into the body, it is not evenly distributed. In particular, it tends to concentrate in the hair, which may show ten times as high an As^{3+} level as body fluids. In the modern forensic laboratory, arsenic poisoning is detected by analysis of hair samples, using atomic absorption spectroscopy (Chapter 5) or neutron activation analysis (Chapter 25). If the concentration is higher than about 3 ppm, poisoning is indicated; normal arsenic levels are much lower than that. By analyzing different sections of a piece of hair, it's even possible to estimate the time intervals over which the arsenic was administered. Application of this technique to a lock of hair taken from Napoleon on St. Helena suggests that he may have been a victim of arsenic poisoning.

It has long been known that high concentrations of selenium are toxic. The cattle disease known picturesquely as "blind staggers" arises from grazing on grass from soil with a high selenium content. Recently, though, selenium was shown to be an essential element. One of the enzymes in the body that destroys harmful peroxides contains selenium. Recommended selenium intake is 0.05–0.2 mg/day. This is just about the amount of selenium you'll get from foods, provided you eat a balanced diet.

There is considerable evidence to suggest that selenium compounds are anticarcinogens. For one thing, tests with laboratory animals show that the incidence and size of malignant tumors is reduced when a solution containing Na_2SeO_3 is injected at the part per million level. Beyond that evidence, statistical studies show an inverse correlation between selenium levels in the soil and the incidence of certain types of cancer.

QUESTIONS AND PROBLEMS

Reactions, Equations, and Formulas

1. Name the following oxyacids and anions:
   a. PO_4^{3-} b. NO_2^- c. H_3PO_2

2. Name the following oxyacids and oxyanions:
   a. SO_3^{2-} b. HSO_4^- c. HPO_4^{2-}

3. Write the formulas of the following compounds:
   a. ammonia   b. laughing gas
   c. hydrogen peroxide   d. sulfur trioxide

4. Write the formula for the following compounds:
   a. hydrogen azide   b. sulfurous acid
   c. hydrazine   d. sodium dihydrogen phosphate

5. Write balanced equations for the disproportionation in acidic solution of
   a. nitrogen to nitrogen oxide, NO, and ammonium ion.

   b. sulfur dioxide to hydrogen sulfide and sulfate ion.
   c. nitrous acid to nitrate ion and nitrogen oxide.

6. Write balanced equations for the disproportionation in basic solution of
   a. sulfur to sulfide and sulfite ions.
   b. nitrite ion to nitrate ion and nitrogen oxide, NO.
   c. thiosulfate ion ($S_2O_3^{2-}$) to sulfate ion and sulfur.

7. Write balanced equations for the oxidation of the following species by $Cr_2O_7^{2-}$, which is reduced to Cr^{3+} in acidic solution.
   a. HNO_2 to NO_3^- b. H_2S to S   c. H_2O to H_2O_2

8. Write balanced equations for the reduction of the following species by iron(II) hydroxide, which is oxidized to iron(III) hydroxide in basic solution.
   a. S to S^{2-} b. NO_3^- to NO_2^- c. SO_3^{2-} to S

9. Write the formula of a compound of hydrogen with
a. nitrogen which is a gas at 25°C and 1 atm.
b. phosphorus which is a liquid at 25°C and 1 atm.
c. oxygen which contains an O—O bond.

10. Write the formula of a compound of hydrogen with
a. sulfur.
b. nitrogen which is a liquid at 25°C and 1 atm.
c. phosphorus which is a poisonous gas at 25°C and 1 atm.

11. Write an equation for
a. the formation of nitrogen from ammonium and nitrite ions.
b. the laboratory preparation of oxygen from potassium chlorate.
c. the preparation of P_4 from phosphate rock.
d. the preparation of sulfur by burning the hydrogen sulfide present in natural gas.

12. Write an ionic equation for the formation of
a. calcium nitrate from calcium hydroxide.
b. barium nitrate from barium carbonate.
c. zinc nitrate from zinc oxide.

13. Give the formula of
a. an anion in which S has an oxidation number of -2.
b. two anions in which S has an oxidation number of $+4$.
c. three different acids of sulfur.

14. Give the formula of a compound of nitrogen that is
a. a weak base.   b. a strong acid.
c. a weak acid.   d. capable of oxidizing copper.

Stoichiometry

15. When ammonium nitrate explodes, nitrogen, steam, and oxygen gas are produced. If the explosion is carried out by heating one kilogram of ammonium nitrate sealed in a rigid bomb with a volume of one liter, what is the total pressure produced by the gases before the bomb ruptures? Assume the reaction goes to completion and the final temperature is 500°C.

16. Sulfur dioxide can be removed from the smoke stack emissions of power plants by reacting it with aqueous hydrogen sulfide, producing sulfur and water. What volume of hydrogen sulfide at 27°C and 755 mm Hg is required to remove the sulfur dioxide produced by a power plant that burns one metric ton of coal containing 5.0% sulfur by mass? How many grams of sulfur are produced by the reaction of H_2S with SO_2?

17. Striking a match involves the combustion of P_4S_3 to produce P_4O_{10} and SO_2 gases. What volume of SO_2 at 25°C and 1.07 atm is produced from the combustion of 0.200 g of P_4S_3?

18. Calculate the molarity and molality of a concentrated solution of nitric acid containing 52.81% HNO_3 with a density of 1.333 g/cm³.

19. A 1.500-g sample containing sodium nitrate was heated to form $NaNO_2$ and O_2. The oxygen evolved was collected over water at 23°C and 752 mm Hg; its volume was 125.0 mL. Calculate the percentage of $NaNO_3$ in the sample. The vapor pressure of water at 23°C is 21.07 mm Hg.

20. Chlorine can remove the foul smell of H_2S in water. The reaction is

$$H_2S(aq) + Cl_2(aq) \rightarrow 2H^+(aq) + 2Cl^-(aq) + S(s)$$

If the contaminated water has 5.0 ppm hydrogen sulfide by mass, what volume of chlorine gas at STP is required to remove all the H_2S from 1.00×10^3 gallons of water? What is the pH of the solution after treatment with chlorine?

21. Sodium sulfite can be used to remove dissolved oxygen from water used in steam boilers. The reaction is:

$$2SO_3{}^{2-}(aq) + O_2(g) \rightarrow 2SO_4{}^{2-}(aq)$$

If the solubility of $O_2(g)$ is 3.08 mL (at STP) per 100.0 mL of water, how many grams of sodium sulfite are required to remove the dissolved oxygen in 5.00×10^4 L of saturated water solution?

22. Mercury reacts with hot nitric acid to produce mercury(II) ions and nitrogen oxide gas, NO. How many grams of mercury are required to react with 60.0 mL of a nitric acid solution in which the pH is -0.80?

23. What volume of air (21% O_2 by volume) at 20°C and 1.00 atm is required to react with one kilogram of sulfur to produce sulfur dioxide?

24. What mass of sodium sulfite requires 27.35 mL of 0.1000 M $KMnO_4$ for oxidation to sulfate ion in acidic solution? $MnO_4{}^-$ is reduced to Mn^{2+}.

Equilibrium

25. At a certain temperature, a 1.00-L vessel originally contains 2.00 mol SO_2 and 2.50 mol O_2. Once equilibrium is reached, it is determined that 75.0% of the SO_2 is converted to SO_3. Calculate K_c for the formation of two moles of SO_3.

26. Using the data in Table 24.5, calculate the concentration of O_2 in an equilibrium mixture at 800°C in which $[SO_3] = 2.00 \times [SO_2]$.

27. A buffer is prepared by adding 10.0 g of potassium nitrite to 1.00 L of 0.250 M nitrous acid. Taking K_a of HNO_2 to be 6.0×10^{-4}, calculate the pH of the buffer.

28. How many grams of KH_2PO_4 should be added to one liter of 0.100 M K_2HPO_4 to form a buffer with a pH of 7.50? (K_a $H_2PO_4{}^- = 6.2 \times 10^{-8}$.)

29. Calculate the pH of a 1.00 M solution of K_3PO_4 (K_b $PO_4{}^{3-} = 2.2 \times 10^{-2}$). Use successive approximations.

30. Hydrogen peroxide is a weak acid in aqueous solution. Write a balanced net ionic equation showing how H_2O_2 acts as a Brönsted acid in water. If $K_a = 1.78 \times 10^{-12}$ for the dissociation of the first H^+, what is the pH of a 0.500 M H_2O_2 solution?

31. Using Figure 24.14
   a. estimate the pH of a solution of $NaHSO_3$.
   b. describe how you would prepare a $HSO_3^- -SO_3^{2-}$ buffer with a pH of 7.00.

32. Taking the successive ionization constants of H_3PO_4 to be 7.1×10^{-3}, 6.2×10^{-8}, and 4.5×10^{-13}, show by calculation what the principal species is (H_3PO_4, $H_2PO_4^-$, HPO_4^{2-}, or PO_4^{3-}) at pH
   a. 5.00.   b. 10.50.

Thermodynamics

33. Consider the reaction:

$$4NH_3(g) + 5O_2(g) \rightarrow 4NO(g) + 6H_2O(g)$$

   a. Calculate ΔH^0 for this reaction. Is it exothermic or endothermic?
   b. Would you expect ΔS^0 to be positive or negative? Calculate ΔS^0.
   c. Is the reaction spontaneous at 25°C and 1 atm?
   d. At what temperature, if any, is the reaction at equilibrium at 1 atm pressure?

34. Data is given in Appendix 1 for white phosphorus, $P_4(s)$. $P_4(g)$ has the following thermodynamic values: $\Delta H_f^0 = 58.9$ kJ/mol, $S^0 = 280.0$ J/K·mol. What is the temperature at which white phosphorus sublimes at 1 atm pressure?

35. Consider the formation of nitrogen oxide by the equation

$$N_2(g) + O_2(g) \rightarrow 2NO(g)$$

Calculate the standard free energy change, K_c, and K_p for the reaction at 25°C.

36. Dinitrogen trioxide is unstable at 25°C. It decomposes by the reaction

$$N_2O_3(g) \rightarrow NO(g) + NO_2(g)$$

The dissociation of one mole of N_2O_3 absorbs 39.7 kJ of heat and has $\Delta G^0 = -1.60$ kJ at 25°C. Calculate, using appropriate thermodynamic tables
   a. ΔH_f^0 of $N_2O_3(g)$.   b. ΔS^0 for the reaction.
   c. K_p for the reaction.

37. The production of sulfuric acid is an exothermic, three-step process:

$$S(s) + O_2(g) \rightarrow SO_2(g)$$
$$SO_2(g) + \tfrac{1}{2}O_2(g) \rightarrow SO_3(g)$$
$$SO_3(g) + H_2O(l) \rightarrow H_2SO_4(l)$$

Calculate, using appropriate tables, the amount of heat evolved per kilogram of sulfuric acid made by this process. What volume of water at 25°C can be converted to steam at 100°C by absorbing this amount of heat? Take the density of water to be 1.00 g/mL, its specific heat to be 4.18 J/g·°C, and its heat of vaporization to be 40.7 kJ/mol.

38. Hydrazine and hydrogen peroxide are used together as a rocket propellant. The reaction is

$$N_2H_4(l) + 2H_2O_2(l) \rightarrow N_2(g) + 4H_2O(g)$$

Calculate the amount of heat evolved if 1.00 g of N_2H_4 reacts with 1.00 g of H_2O_2. Use Appendix 1.

Electrochemistry

39. Based on the potentials given in Figure 24.17, which nitrogen species is
   a. the strongest oxidizing agent in acidic solution?
   b. the strongest oxidizing agent in basic solution?
   c. the strongest reducing agent in acidic solution?
   d. the strongest reducing agent in basic solution?

40. Use Figure 24.17 to decide whether the following processes will occur at standard conditions:
   a. $NO_3^-(aq) + HNO_2(aq) \rightarrow NO_2(g)$
   b. $NO_3^-(aq) + NO_2^-(aq) \rightarrow NO_2(g)$
   c. $NO_3^-(aq) + HNO_2(aq) \rightarrow NO(g)$
   d. $NO(g) \rightarrow HNO_2(aq) + N_2O(g)$

41. Taking E_{ox}^0 $H_2O_2 = -0.695$ V, determine which of the following species will be reduced by hydrogen peroxide (use Table 21.1 to find E_{red}^0 values).
   a. $Cr_2O_7^{2-}$ b. Fe^{2+} c. I_2 d. Br_2

42. Taking E_{red}^0 $H_2O_2 = +1.763$ V, determine which of the following species will be oxidized by hydrogen peroxide (use Table 21.1 to find E_{ox}^0 values).
   a. Co^{2+} b. Cl^- c. Fe^{2+} d. Sn^{2+}

43. Consider the reduction of nitrate ion in acidic solution to nitrogen oxide ($E_{red}^0 = 0.964$ V) by sulfur dioxide which is oxidized to sulfate ion ($E_{red}^0 = 0.155$ V). Calculate the voltage of a cell involving this reaction in which all the gases have pressures of 1.00 atm, all the ionic species (except H^+) are at 0.100 M, and the pH is 4.30.

44. For the reaction in Problem 43, if gas pressures are at 1.00 atm and ionic species are at 0.100 M (except H^+), at what pH will the voltage be 1.000 V?

45. The reduction half-reaction for hydrazine in acid solution ($N_2H_5^+$) is

$$N_2(g) + 5H^+(aq) + 4e^- \rightarrow N_2H_5^+(aq)$$
$$E_{red}^0 = -0.214 \text{ V}$$

Calculate E_{tot}^0 for the reaction between hydrazine and chlorate ion in acid solution to give nitrogen and chlorine gases.

46. $H_2PO_2^-$ salts are excellent reducing agents in basic solution. They are used to plate metals onto plastic surfaces. The reduction half-reaction is

$$HPO_3^{2-}(aq) + H_2O + 2e^-$$
$$\rightarrow H_2PO_2^-(aq) + 3OH^-(aq) \quad E_{red}^0 = -1.60 \text{ V}$$

Will $H_2PO_2^-$ reduce $Fe(OH)_2$ to Fe? What about $Fe(OH)_3$ to $Fe(OH)_2$? (Use Table 21.1.)

Unclassified

47. State the oxidation number of N in
   a. NO_2^- b. NO_2 c. HNO_3 d. NH_4^+

48. Write the Lewis structures and describe the geometry of
   a. $H_2PO_4^-$ b. HSO_4^- c. SO_3^{2-} d. $[PCl_4]^+$
   e. NO_2^-

49. Explain why
   a. concentrated sulfuric acid converts table sugar to a black char.
   b. in preparing dilute sulfuric acid, it is important to add the concentrated acid to water with constant stirring.
   c. nitric acid solutions often have a yellow or brown color.

50. Naturally occurring nitrogen has an average atomic mass of 14.00674 amu. Nitrogen-14 has a mass of 14.00307 amu, while N-15 has a mass of 15.00011 amu. Calculate the percent N-15 in a sample of naturally occurring nitrogen.

51. The density of sulfur vapor at one atmosphere pressure and 973 K is 0.8012 g/L. What is the molecular formula of the vapor?

52. Consider the compound dinitrogen oxide. Draw its Lewis structure. Determine its polarity, bond angle, hybridization, and number of sigma and pi bonds.

53. At 20°C, enough ammonia can be dissolved in water to get a solution of 33.1% NH_3 by mass. If the density of this solution is 0.890 g/mL, calculate the molarity and molality.

54. The solubility product constant of calcium phosphate is 1×10^{-33}. What is the concentration (m) of $Ca_3(PO_4)_2$ in a saturated solution of calcium phosphate?

55. Dichromate ion is reduced to chromium(III) ions by hydrogen peroxide. A 30.0-mL sample of hydrogen peroxide is titrated with 32.50 mL of 0.175 M $K_2Cr_2O_7$ in acidic solution. What is the molarity of H_2O_2? If the solution's density is 1.00 g/cm³, what is the mass percent of H_2O_2 in the solution?

56. Which compounds in Table 24.3 have resonance structures?

57. Consider the reaction

$$O_3(g) \rightarrow O_2(g) + O(g)$$

   a. Write the Lewis structure for ozone.
   b. Using bond energies, calculate the energy required for the decomposition of ozone (assume a double bond in O_2).
   c. Calculate the maximum wavelength of light that will cause this reaction.

Challenge Problems

58. The compound hyponitrous acid has the molecular formula $H_2N_2O_2$. Draw a reasonable Lewis structure for this molecule. Are there isomers of $H_2N_2O_2$ in which the atoms are bonded in a different pattern?

59. Taking the Lewis structure of the SCN^- ion to be

$$: \overset{..}{S} = C = \overset{..}{N} :$$

find the oxidation number of each atom (assign bonding electrons to the more electronegative atom).

60. Of the 45 numbered equations in this chapter, how many involve oxidation and reduction?

61. Using data in Table 14.5, Chapter 14, prepare a graph similar to Figure 24.16 for the +5 state of arsenic.

62. You may have noticed from Figure 24.17 that E_{red}^0 values are not additive. That is, if $A \rightarrow B \rightarrow C$, then $E_{red}^0 A \rightarrow C \neq E_{red}^0 A \rightarrow B + E_{red}^0 B \rightarrow C$. There is, however, a simple general relation that applies here, taking into account the number of electrons involved in each reduction. Can you discover this relation, using the data given in this figure? Check your relation with similar figures in Chapter 23. Can you show, working with ΔG^0 values, why this relation is valid?

$$Fe^{3+} (aq) + 3OH^- (aq) \rightarrow Fe(OH)_3(s)$$

Precipitation of red iron(III) hydroxide

CHAPTER
25

NUCLEAR
REACTIONS

The soul, perhaps, is a gust of gas
And wrong is a form of right—
But we know that Energy equals
Mass
By the Square of the Speed of
Light.

MORRIS BISHOP
$E = MC^2$

The "ordinary chemical reactions" discussed to this point involve changes in the outer electronic structure of atoms or molecules. In contrast, nuclear reactions result from changes taking place within atomic nuclei. You will recall (Chapter 2) that atomic nuclei are represented by symbols such as

$^{12}_{6}C$ $\qquad$ $^{14}_{6}C$

Here, the atomic number Z (number of protons in the nucleus), is shown as a left subscript. The mass number, A (number of protons + number of neutrons in the nucleus), appears as a left superscript. Nuclei with the same number of protons but different numbers of neutrons are called *isotopes*. The symbols written above represent two isotopes of the element carbon (at. no. = 6). One isotope has 6 neutrons in the nucleus and hence has a mass number of $6 + 6 = 12$. The heavier isotope has 8 neutrons and hence a mass number of 14.

We start this chapter by discussing briefly the factors that contribute to nuclear stability (Section 25.1). Unstable nuclei, whether formed in nature or in the laboratory, decompose by a type of nuclear reaction called radioactivity (Section 25.2). One of the most important features of this process is the rate at which radioactive nuclei decompose or "decay" (Section 25.3). Two other important types of nuclear reactions are fission (Section 25.5) and fusion (Section 25.6).

Nuclear reactions are accompanied by energy changes that greatly exceed those associated with ordinary chemical reactions. The energy evolved when one gram of radium undergoes radioactive decay is 500,000 times as great as that given off when the same amount of radium reacts with chlorine to form $RaCl_2$. Still larger amounts of energy are given off in nuclear fission and nuclear fusion. This energy is related to the change in

Nuclear reactions differ from chemical reactions in many ways. They are isotope-dependent and often involve transmutation of elements

879

mass that accompanies a nuclear reaction. We will examine the nature of this relationship in Section 25.4.

25.1
NUCLEAR STABILITY

They are not chemical bonds, that's for sure

As you know, atomic nuclei consist of protons, which are positively charged, and neutrons, which have zero charge. According to classical electrostatics, we would expect the protons to repel one another and the nucleus to fly apart. It turns out, however, that at the very short distances of separation characteristic of atomic nuclei, about 10^{-15} m, there are strong attractive forces between nuclear particles. The stability of a nucleus depends upon the balance between these forces and those of electrostatic repulsion.

Despite a great deal of research, we do not have a clear understanding of the nature of nuclear forces or the way in which those forces lead to nuclear stability. Our knowledge of nuclear structure today leaves a great deal to be desired; all we really have is a series of empirical rules for predicting which nuclei will be most stable. The more important of these rules are listed below.

1. The neutron-to-proton ratio required for stability varies with atomic number (Fig. 25.1). For light elements, this ratio is close to one. For example, the isotopes $^{12}_{6}C$, $^{14}_{7}N$, and $^{16}_{8}O$ are stable. As atomic number increases, the ratio increases; the "belt of stability" shifts to higher numbers of neutrons. With very heavy isotopes such as $^{206}_{82}Pb$, the stable neutron-to-proton ratio is about 1.5:

$$(206 - 82)/82 = 124/82 = 1.51$$

2. Nuclei containing more than 83 protons are unstable. Putting it another way, no element beyond bismuth (at. no. = 83) has a stable isotope; all the isotopes of such elements are radioactive.

3. Nuclei with an even number of *nucleons*, either protons or neutrons, tend to be more stable than those with an odd number of nuclear particles. Consider, for example, the stable isotopes of the elements of atomic number 50–53:

Nature likes evens more than odds, here and elsewhere

	AT. NO.	MASS NO. OF STABLE ISOTOPE
Sn	50	112, 114, 115, 116, 117, 118, 119, 120, 122, 124
Sb	51	121, 123
Te	52	120, 122, 123, 124, 125, 126, 128, 130
I	53	127

Of the 21 stable isotopes of these elements, 18 have an even number of protons, 16 an even number of neutrons. Thirteen of 21 have an even number of both protons and neutrons; statistically, we would expect only 5 to fall in this category.

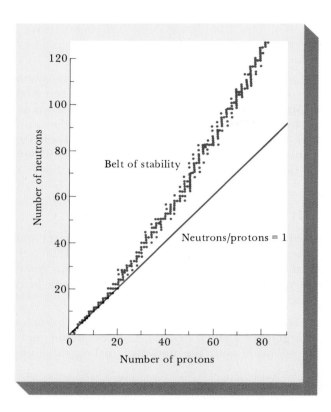

Figure 25.1 Stable isotopes (red dots) have neutron-to-proton ratios that fall within a narrow range, referred to as a "belt of stability." For light isotopes of small atomic number, the stable ratio is 1.0; with heavier isotopes it increases to about 1.5. There are no stable isotopes for elements of atomic number greater than 83 (Bi).

4. Certain numbers of protons and neutrons appear to be particularly stable: 2, 8, 20, 28, 50, 82, and 126. These *magic numbers* were first suggested by Maria Mayer, who developed a nuclear shell model similar to, but much less regular than, the electron shell model discussed in Chapter 5. Mayer won a Nobel prize for her work in physics in 1963.

■ **EXAMPLE 25.1** ————————————————————————————

For each pair of nuclei, list which one is more stable.

a. $^6_3$Li or $^9_3$Li   b. $^{204}_{82}$Pb or $^{209}_{85}$At

Solution

a. $^6_3$Li is stable; it has a neutron:proton ratio of 3:3 = 1. In contrast, $^9_3$Li, with a neutron:proton ratio of 2, is unstable; it lies well above the belt of stability.

b. $^{204}_{82}$Pb is more stable because its atomic number, Z, is less than 83. All elements with atomic number greater than 83 are known to be radioactive.

Exercise
Which is more stable, $^{138}_{55}$Cs or $^{136}_{54}$Xe? Answer: $^{136}_{54}$Xe.

25.2 RADIOACTIVITY

An unstable nucleus undergoes a reaction called radioactive decomposition or decay. A few such nuclei occur in nature; their decomposition is referred to as *natural radioactivity*. Many more unstable nuclei have been made in the laboratory; the process by which such nuclei decompose is called *induced radioactivity*. In this section, we will look at the characteristics of these reactions. We will be particularly interested in the nature of the radiation given off and the effect it has on human beings.

NATURAL RADIOACTIVITY

This process was discovered, almost accidentally, by Henri Becquerel and his colleagues Marie and Pierre Curie at the Sorbonne in Paris in 1896 (see the historical perspective at the end of this chapter). The radiation given off in natural radioactivity can be separated by an electric or magnetic field into three distinct parts (Fig. 25.2):

1. **Alpha radiation** consists of a stream of positively charged particles (alpha particles) with a charge of $+2$ and a mass of 4 on the atomic mass scale. These particles are identical with the nuclei of ordinary helium atoms, ^4_2He.

α's are He nuclei

When an alpha particle is ejected from the nucleus, the atomic number decreases by two units; the mass number decreases by four units. Consider, for example, the loss of an alpha particle by a uranium-238 atom:

$$^{238}_{92}\text{U} \rightarrow {}^4_2\text{He} + {}^{234}_{90}\text{Th} \tag{25.1}$$

Here, as in all nuclear equations, there is a balance of both atomic number $(90 + 2 = 92)$ and mass number $(4 + 234 = 238)$ on the two sides.

2. **Beta radiation** is made up of a stream of negatively charged particles (beta particles) identical in their properties to electrons. The ejection of a

β's are electrons

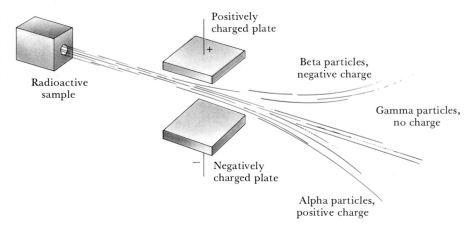

Figure 25.2 The direction in which beta particles are deflected shows that they are negatively charged. Alpha particles move so as to indicate that they carry a positive charge. Gamma rays are undeflected and so must be uncharged.

beta particle (mass $\approx$ 0, charge = -1) converts a neutron (mass = 1, charge = 0) in the nucleus into a proton (mass = 1, charge = $+1$). Hence, beta emission leaves the mass number unchanged but increases the atomic number by one unit. An example of beta emission is the radioactive decay of thorium-234:

$$^{234}_{90}\text{Th} \rightarrow \,^{0}_{-1}e + \,^{234}_{91}\text{Pa} \qquad\qquad\qquad (25.2)$$

The symbol $^{0}_{-1}e$ is written to stand for a beta particle (electron).

3. Gamma radiation consists of high-energy photons of very short wavelength (λ = 0.0005 to 0.1 nm). The emission of gamma radiation accompanies most nuclear reactions. Since gamma radiation changes neither the atomic number nor the mass number, it is ordinarily omitted in writing nuclear equations.

In radioactive decay, the unstable nucleus is frequently called the *parent* nucleus. It decays to form a *daughter* nucleus, which may or may not be stable (Example 25.2).

■ **EXAMPLE 25.2** _____

Thorium-232 undergoes radioactive decay in a three-step process:

a. $^{232}_{90}\text{Th} \rightarrow \text{Q} + \,^{4}_{2}\text{He}$ b. $\text{Q} \rightarrow \text{R} + \,^{0}_{-1}e$ c. $\text{R} \rightarrow \text{T} + \,^{228}_{90}\text{Th}$

Write a balanced nuclear equation for each step, identifying Q, R, and T by their nuclear symbols.

Many decay processes involve several steps. $^{238}_{92}\text{U} \rightarrow \,^{206}_{82}\text{Pb}$ in 14 steps

Solution

The mass numbers and atomic numbers must balance on the two sides of the equation.

a. Alpha emission by $^{232}_{90}\text{Th}$ decreases the mass number by four and the atomic number by two. The mass number of Q is $232 - 4 = 228$; its atomic number is $90 - 2 = 88$. Locating element 88 in the Periodic Table, we find it to be radium, symbol Ra. Hence, the balanced nuclear equation is

$$^{232}_{90}\text{Th} \rightarrow \,^{4}_{2}\text{He} + \,^{228}_{88}\text{Ra}$$

b. When $^{228}_{88}\text{Ra}$ emits an electron, the mass number is unchanged but the atomic number increases by 1 to 89. The daughter product is an isotope of actinium, Ac:

$$^{228}_{88}\text{Ra} \rightarrow \,^{0}_{-1}e + \,^{228}_{89}\text{Ac}$$

c. The unbalanced equation in this case is

$$^{228}_{89}\text{Ac} \rightarrow \text{T} + \,^{228}_{90}\text{Th}$$

For the equation to balance, T must have a mass number of 0 and an atomic number of -1; T must be an electron. The balanced equation is

$$^{228}_{89}\text{Ac} \rightarrow \,^{0}_{-1}e + \,^{228}_{90}\text{Th}$$

Exercise

The first two steps in the conversion of ^{234}Pa to ^{206}Pb are beta emission followed by alpha emission. Write a balanced nuclear equation for each step. Answers:

$$^{234}_{91}\text{Pa} \rightarrow \,_{-1}^{0}e + \,^{234}_{92}\text{U}; \qquad ^{234}_{92}\text{U} \rightarrow \,^{4}_{2}\text{He} + \,^{230}_{90}\text{Th}$$

INDUCED RADIOACTIVITY; BOMBARDMENT REACTIONS

During the past 50 years, more than 1500 radioactive isotopes have been prepared in the laboratory. The number of such isotopes per element ranges from one (hydrogen and boron) to 34 (indium). They are all prepared by bombardment reactions in which a stable nucleus is converted to one that is radioactive. A typical reaction is the one that occurs when the stable isotope of aluminum, $^{27}_{13}$Al, absorbs a neutron to form $^{28}_{13}$Al. The latter is unstable, decaying by electron emission to give a stable isotope of silicon, $^{28}_{14}$Si. The two steps involved in the process are

neutron bombardment: $^{27}_{13}\text{Al} + \,^{1}_{0}n \rightarrow \,^{28}_{13}\text{Al}$ (25.3)

radioactive decay: $^{28}_{13}\text{Al} \rightarrow \,^{28}_{14}\text{Si} + \,_{-1}^{0}e$ (25.4)

The first radioactive isotopes to be made in the laboratory were prepared in 1934 by Irene (daughter of Marie and Pierre) Curie and her husband, Frederic Joliot. They achieved this by bombarding certain stable isotopes with high-energy alpha particles. One reaction was

The α particles came from radium

$$^{27}_{13}\text{Al} + \,^{4}_{2}\text{He} \rightarrow \,^{30}_{15}\text{P} + \,^{1}_{0}n \qquad (25.5)$$

The product, phosphorus-30, is radioactive. It decays by emitting a **positron**:

$$^{30}_{15}\text{P} \rightarrow \,^{30}_{14}\text{Si} + \,^{0}_{1}e \qquad (25.6)$$

The positron is a particle that has the same mass as the electron, but a charge of $+1$ rather than -1. Positron emission converts a proton in the nucleus to a neutron (15 p, 15 n in P-30; 14 p, 16 n in Si-30). This same result can be achieved by **K-electron capture**. Here an electron in the innermost energy level (**n** = 1) "falls" into the nucleus:

$$^{30}_{15}\text{P} + \,_{-1}^{0}e \rightarrow \,^{30}_{14}\text{Si} \qquad (25.7)$$

BIOLOGICAL EFFECTS OF RADIATION

The harmful effect of radiation on human beings is caused by its ability to ionize and ultimately destroy the organic molecules of which body cells are composed. The extent of damage depends mainly upon two factors. These are the amount of radiation absorbed and the type of radiation. The former is commonly expressed in *rads* (radiation *a*bsorbed *d*ose). A rad corresponds to the absorption of 10^{-2} J of energy per kilogram of tissue

$$1 \text{ rad} = 10^{-2} \text{ J/kg} \qquad (25.8)$$

TABLE 25.1 Effect of Exposure to a Single Dose of Radiation

DOSE (REMS)	PROBABLE EFFECT
0 to 25	no observable effect
25 to 50	small decrease in white blood cell count
50 to 100	lesions, marked decrease in white blood cells
100 to 200	nausea, vomiting, loss of hair
200 to 500	hemorrhaging, ulcers, possible death
500+	fatal

The total biological effect of radiation is expressed in *rems* (radiation *equiv*-alent for *man*). The number of rems is found by multiplying the number of rads by an appropriate factor, n, for the particular type of radiation:

no. rems = n (no. rads)

$n = 1$ for β, γ, x-rays

$n = 10$ for α rays, high-energy neutrons   (25.9)

Table 25.1 lists some of the effects to be expected when a person is exposed to a single dose of radiation at various levels.

Different types of radiation differ not only in their ionizing effect but also in their penetrating ability (Table 25.2). Notice that alpha particles, which have very high ionizing power ($n = 10$ in Equation 25.9) have very little penetrating ability. In practical terms, this means that alpha emitters are relatively safe to work with unless they are taken into the body by ingestion or inhalation, in which case they can be very damaging.

Small doses of radiation repeated over long periods of time can have very serious consequences. Many of the early workers in the field of ra-dioactivity developed cancer in this way. Cases are known in which cancers developed as long as 40 years after initial exposure. Studies have shown an abnormally large number of cases of leukemia among the survivors of the atomic bombs dropped on Hiroshima and Nagasaki.

Radiation can also have a genetic effect; that is, it can produce mu-tations in plants and animals by bringing about changes in chromosomes. There is every reason to suppose that similar effects can arise in human beings. Surveys of the children of radiologists show an increased frequency of congenital defects. This is confirmed by studies of children born to the survivors of Nagasaki and Hiroshima. A disturbing aspect of this problem is that there seems to be no lower limit, or "threshold," below which the genetic effects of radiation become negligible. Even a small increase in

A nuclear war would have many long-term effects, all bad

TABLE 25.2 Penetrating Abilities of Radiation Emissions

EMISSION	APPROXIMATE DEPTH OF PENETRATION (DEPENDENT ON INITIAL ENERGY)		
	Dry Air	**Animal Tissue**	**Lead**
α-rays	4 cm	skin deep	does not penetrate
β-rays	5–300 cm	1 cm	5×10^{-4} to 3×10^{-2} cm
γ-rays	4×10^4 cm	50 cm	3 cm

background radiation can be expected to produce a proportional increase in undesirable mutations.

Table 25.3 lists average exposures to radiation of people living in the United States. Notice that about two thirds of the radiation exposure comes from natural sources. The level depends upon location. Cosmic radiation is much more intense at high elevations. A person living in Denver is exposed to about 100 millirems/year from this source. This is twice the national average.

Radiation listed in Table 25.3 as coming from the earth is mostly in the form of $^{222}_{86}$Rn, a radioactive isotope of radon, which is a decay product of uranium-238. Because it is gaseous and chemically inert, radon seeps through cracks in concrete and masonry from the ground into houses. There its concentration builds up, particularly if the house is tightly insulated. Inhalation of radon-222 can cause health problems because both it and its decay products (Po-218, Po-214) are alpha emitters. A concentration of less than a million Rn atoms per liter ($\sim 6 \times 10^{-19}$ mol/L) can increase the risk of lung cancer by at least 5%.

The largest exposure to man-made sources of radiation is due to x-rays. The numbers shown in Table 25.3 assume only occasional x-ray examinations. A single dental x-ray corresponds to 20 millirems, and a chest x-ray to 50 to 200 millirems. These amounts can be much higher if the operator of the x-ray machine is inexperienced or if the machine itself is defective or old. (Newer x-ray machines have lower radiation emissions.) Clearly, excessive diagnostic use of x-rays should be avoided.

It wasn't until after WWII that the dangers from radiation were clearly recognized

TABLE 25.3   Typical Radiation Exposures in the United States (1 Millirem = 10^{-3} rem)

SOURCES	MILLIREMS/YR
I. Natural	
A. External to the body	
1. From cosmic radiation	50
2. From the earth	47
3. From building materials	3
B. Inside the body	
1. Inhalation of air	5
2. In human tissues (mostly $^{40}_{19}$K)	21
Total from natural sources	**126**
II. Man-made	
A. Medical procedures	
1. Diagnostic x-rays	50
2. Radiotherapy x-rays, radioisotopes	10
3. Internal diagnosis, therapy	1
B. Nuclear power industry	0.2
C. Luminous watch dials, TV tubes, industrial wastes	2
D. Radioactive fallout (nuclear tests)	4
Total from manmade sources	**67**
Total	**193**

USES OF RADIOACTIVE ISOTOPES

A large number of radioactive isotopes have been used both in industry and in many areas of basic and applied research. A few of these are discussed below.

Medicine The high-energy radiation given off by radium was used for many years in the treatment of cancer. Nowadays, cobalt-60, which is cheaper than radium and gives off even more powerful radiation, is used for this purpose. Certain types of cancer can be treated internally with radioactive isotopes. If a patient suffering from cancer of the thyroid drinks a solution of NaI containing radioactive iodide ions (^{131}I or ^{123}I), the iodine moves preferentially to the thyroid gland. There, the radiation destroys malignant cells without affecting the rest of the body.

This approach is also used to treat hyperactive thyroid

Trace amounts of radioactive samples injected into the blood can be used to detect circulatory disorders. For example, a sodium chloride solution containing a small amount of radioactive sodium may be injected into the leg of a patient. By measuring the build-up of radiation in the foot, a physician can quickly find out whether the circulation in that area is abnormal.

Positron emission tomography (PET) is a technique used to study brain disorders. The patient is given a dose of glucose ($C_6H_{12}O_6$) containing a small amount of carbon-11, a positron emitter. The brain is then scanned to detect positron emission from the radioactive, "labeled" glucose. In this way, differences in glucose uptake and metabolism in the brains of normal and abnormal patients are established. For example, PET scans have determined that the brain of a schizophrenic metabolizes only about 20% as much glucose as that of a normal individual.

Technetium-99 is used to develop images of internal organs. Iodine-131 is preferentially absorbed by tumor cells and therefore can be used to locate tumors, particularly in the brain. Medicine seems to come up with many imaginative uses for these radioactive isotopes.

Chemistry Radioactive isotopes can be used to trace the path of an element as it passes through various steps from reactant to final product. Organic chemists have learned a great deal about the mechanism of complex reactions by using carbon-14 as a tracer. One such reaction is the natural process of photosynthesis. The overall reaction can be represented as

$$6CO_2(g) + 6H_2O(l) \rightarrow C_6H_{12}O_6(s) + 6O_2(g) \qquad (25.10)$$

This reaction proceeds through a series of steps in which successively more complex organic molecules are formed. To study the path of reaction, plants are exposed to CO_2 containing carbon-14. At various time intervals, the plants are analyzed to determine which organic compounds contain carbon-14 and hence are early products of photosynthesis. Research along these lines by Melvin Calvin at the University of California at Berkeley led to a Nobel prize in chemistry in 1961.

The CO_2 is said to be "tagged"

Radioactive isotopes are used extensively in chemical analysis. One technique of particular importance is *neutron activation analysis*. This procedure depends upon the phenomenon of induced radioactivity. A sample is bombarded by neutrons, bringing about such reactions as:

$$^{84}_{38}\text{Sr} + ^{1}_{0}n \rightarrow ^{85}_{38}\text{Sr} + \gamma \tag{25.11}$$

Ordinarily the element retains its chemical identity, but the isotope formed is radioactive, decaying by gamma emission. The magnitude of the energy change and hence the wavelength of the gamma ray varies from one element to another and so can serve for the qualitative analysis of the sample. The intensity of the radiation depends upon the amount of the element present in the sample; this permits quantitative analysis of the sample. Neutron activation analysis can be used to analyze for 50 different elements in amounts as small as one picogram (10^{-12} g).

> A very sensitive method indeed, 0.001 ppb

One application of this technique is in the field of archaeology. By measuring the amount of strontium in the bones of prehistoric humans, it is possible to get some idea of their diet. Plants contain considerably more strontium than animals, so a high strontium content suggests a largely vegetarian diet. Strontium analyses of bones taken from ancient farming communities consistently show a difference by sex; women have higher strontium levels than men. Apparently, in those days, women did most of the farming; men spent a lot of time away from home hunting and eating their kill.

Commercial Applications   Most smoke alarms (Fig. 25.3) use a radioactive isotope, typically americium-241. A tiny amount of this isotope is placed in a small ionization chamber; decay of Am-241 ionizes air molecules within the chamber. Under the influence of a potential applied by a battery, these ions move across the chamber, producing an electric current. If smoke particles get into the chamber, the flow of ions is impeded and the current drops. This is detected by electronic circuitry and an alarm sounds. The alarm also goes off if the battery voltage drops, indicating that it needs to be replaced.

Another potential application of radioactive isotopes is in food preservation (Fig. 25.4). It is well known that gamma rays can kill insects, larvae, and parasites like trichina that cause trichinosis in pork. Radiation can also

Figure 25.3 Most smoke detectors use a tiny amount of a radioactive isotope to produce a current flow which drops off sharply in the presence of smoke particles, emitting an alarm in the process. (Marna G. Clarke)

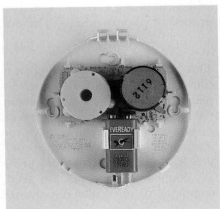

Figure 25.4 Strawberries irradiated with gamma rays from radioactive isotopes are still fresh after 15 days storage at 4°C (*right*) while those not irradiated are moldy (*left*) (International Atomic Energy Agency).

inhibit sprouting of onions and potatoes. Perhaps most important from a commercial standpoint, it can extend the shelf lives of many foods for weeks or even months. Since many chemicals used to preserve foods have later been shown to have adverse health effects, irradiation would seem to be an attractive alternative.

There are, however, some problems with food irradiation. For one thing, the taste of some fruits and all milk products is affected adversely. More generally, there is concern that irradiation produces a variety of reaction products, at least a few of which could be harmful.

25.3
KINETICS OF RADIOACTIVE DECAY

The rate at which a radioactive sample decays can be measured by counting the number of nuclei disintegrating per unit time. Instruments for measuring radioactivity, such as a Geiger counter and a scintillation counter (Fig. 25.5) do this automatically. In using such instruments, it is necessary to correct for the "background" radiation given off by natural sources.

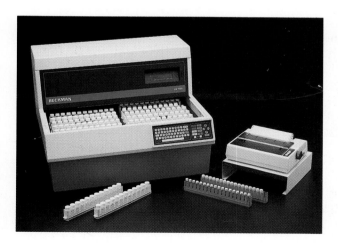

Figure 25.5 Liquid scintillation counter used to detect radiation and measure disintegrations per minute quickly and accurately. (Beckman)

Nuclear decay is a first-order reaction. That is, the rate of decay is directly proportional to the amount of radioactive isotope present. Mathematically:

$$\text{rate} = k\,N_t \tag{25.12}$$

In this equation:

N_t represents the amount of radioactive isotope present at time t. Most often, N_t is taken to be the number of atoms in the sample.

k is the first-order rate constant and has the units of reciprocal time (e.g., s^{-1}, d^{-1}, yr^{-1}).

The **rate** of decay, often referred to as the **activity** of the sample, is most often expressed in terms of the number of atoms decaying in unit time. Frequently, it is stated in curies (Ci); a **curie** corresponds to the decay of 3.700×10^{10} atoms per second.

$$\textbf{1 Ci} = \textbf{3.700} \times \textbf{10}^{10} \textbf{ atoms/s} \tag{25.13}$$

If a scintillation counter indicates that 750 atoms disintegrate over a ten-second interval, the rate of decay in curies is

$$\frac{7.50 \times 10^2 \text{ atoms}}{10.0 \text{ s}} \times \frac{1 \text{ Ci}}{3.700 \times 10^{10} \text{ atoms/s}} = 2.02 \times 10^{-9} \text{ Ci}$$

If we measure the rate of decay of a known mass of a radioactive isotope, we can readily calculate the decay constant, k.

■ **EXAMPLE 25.3** _____

Cobalt-60 is used in cancer therapy especially to reduce the size of inoperable brain tumors. A 1.00-g sample of Co-60 (molar mass = 59.92 g/mol) has an activity of 1.1×10^3 Ci. Find the rate constant k for the decay of Co-60.

Solution

We work with Equation 25.12, finding first the number of atoms in the one-gram sample, then the decay rate in atoms per second, and finally the value of k.

1. $N_t = 1.00 \text{ g Co-60} \times \dfrac{1 \text{ mol}}{59.92 \text{ g}} \times \dfrac{6.022 \times 10^{23} \text{ atoms}}{1 \text{ mol}}$

    $= 1.01 \times 10^{22}$ atoms

2. $\text{rate} = 1.1 \times 10^3 \text{ Ci} \times \dfrac{3.700 \times 10^{10} \text{ atoms/s}}{1 \text{ Ci}} = 4.1 \times 10^{13}$ atoms/s

3. $k = \dfrac{\text{rate}}{N_t} = \dfrac{4.1 \times 10^{13} \text{ atoms/s}}{1.01 \times 10^{22} \text{ atoms}} = \boxed{4.1 \times 10^{-9}/\text{s}}$

Exercise

The rate constant for the decay of radium-226 (MM = 225.98 g/mol) is 1.372×10^{-11}/s. Calculate the rate in curies for the decay of a 1.000-g sample of radium-226. Answer: 0.9882 Ci. (The curie was originally defined so that 1.00 g of radium would have a decay rate or activity of 1 Ci.)

Notice from Example 25.3 that the expression for the first-order rate constant, k, is:

$$k = \frac{\text{rate}}{N_t} = \frac{\text{no. of atoms decomposing per second}}{\text{total no. atoms}}$$

From a slightly different point of view, we can think of the rate constant as representing *the fraction of the sample that decomposes in unit time.* When we see that the rate constant for a first-order reaction is, for example, 0.010/s, we conclude that 1% of the atoms or molecules in the sample are disintegrating per second. In another case, if $k = 0.15/yr$, the rate is such that 15% of the sample decomposes per year.

Applying the first-order rate law (Ch. 18) to radioactive decay, we obtain the following relationship:

$$\ln \frac{X_0}{X} = kt \tag{25.14}$$

where X_0 is the original amount of radioactive material, X is the amount left after time t, and k is the first-order rate constant referred to above. The amount of a radioactive isotope can be expressed in terms of moles, grams, or number of atoms. Indeed, X_0 and X can even refer to counting rates (counts per minute, counts per second, etc.), since the rate of decay is directly proportional to amount.

Decay rates of radioactive isotopes are most often expressed in terms of their half-lives, $t_{1/2}$, rather than the first-order rate constant, k. As noted in Chapter 18, these two quantities are related by the equation

$$k = \frac{0.693}{t_{1/2}} \tag{25.15}$$

The application of these equations to radioactive decay processes is illustrated in Example 25.4.

■ **EXAMPLE 25.4** _____

Plutonium-240, produced in nuclear reactors, has a half-life of 6.58×10^3 years. Calculate

a. the first-order rate constant for the decay of plutonium-240.
b. the fraction of a sample that will remain after 100 (1.00×10^2) years.

Solution

a. $k = 0.693/(6.58 \times 10^3 \text{ yr}) = \boxed{1.05 \times 10^{-4}/yr}$

b. $\ln \dfrac{X_0}{X} = 1.05 \times 10^{-4} \text{ yr}^{-1} \times 100 \text{ yr} = 1.05 \times 10^{-2}$

Taking the inverse of $\ln \dfrac{X_0}{X}$ we get $e^{0.00105} = 1.01$. The fraction remaining is $X_0/X = 1/1.01 = 0.99$; that is, $\boxed{\text{99\% remains.}}$ This problem illustrates the difficulty inherent in storing or disposing of a long-lived radioactive species. Such isotopes cannot be released into the environment in the naive

It takes a long time for some nuclei to "cool off"

hope that they will decompose rapidly. In the case of plutonium-240, its radiation level would be virtually unchanged a century from now.

Exercise
How long does it take for three fourths of a ^{240}Pu sample to decay? Answer: 1.32×10^4 years.

Half-lives can be interpreted in terms of the level of radiation of the corresponding isotopes. Since uranium has a very long half-life (4.5×10^9 yr), it gives off radiation very slowly. At the opposite extreme is fermium-258, which decays with a half-life of 3.8×10^{-4} s. You would expect the rate of decay to be quite high. Within a second all the radiation from this isotope is gone. Species such as this produce a very high level of radiation during their brief existence.

AGE OF ROCKS

Certain radioactive isotopes act as "natural clocks"; that is, they help us to determine the time at which rock deposits solidified. The time elapsed since then is referred to as the "age" of the rock. To see how this information is obtained, consider a uranium-bearing rock formed billions of years ago by solidification from a molten mass. Once the rock became solid, the products of radioactive decay of uranium could no longer diffuse away. Hence, they were incorporated into the rock. Over time these products, all of which have short half-lives, were converted to lead-206. The overall equation for the decay process can be written

$$^{238}_{92}U \rightarrow {}^{206}_{82}Pb + 8 \, {}^4_2He + 6 \, {}^0_{-1}e; \qquad t_{1/2} = 4.5 \times 10^9 \text{ yr} \qquad (25.16)$$

Knowing the half-life for this process, we can calculate the time that has elapsed since the rock has solidified. We do so by measuring the ratio of lead-206 to uranium-238 in the rock today. If we should find, for example, that equal numbers of atoms of these two isotopes were present, we would infer that the rock must be about 4.5×10^9 (4.5 billion) years old.

Ages of rocks determined by this method range from 3 to 4.5×10^9 years. The larger number is often taken as an approximate value for the age of the earth. Analyses of rock samples from the moon indicate ages in the same range. This argues against the once prevalent idea that the moon was torn from the earth's surface by a violent event a long time after the earth solidified.

That's one of the best values of the earth's age we can get

AGE OF ORGANIC MATERIAL

During the 1950s, Professor W. F. Libby of the University of Chicago and others worked out a method for determining the age of organic material. It is based upon the decay rate of carbon-14. The method can be applied to objects from a few hundred up to 50,000 years old. It has been used to determine the authenticity of canvases of Renaissance painters and to check the ages of relics left by prehistoric cavemen.

Carbon-14 is produced in the atmosphere by the interaction of neutrons from cosmic radiation with ordinary nitrogen atoms:

$$^{14}_{7}N + ^{1}_{0}n \rightarrow ^{14}_{6}C + ^{1}_{1}H \tag{25.17}$$

The carbon-14 formed by this nuclear reaction is eventually incorporated into the carbon dioxide of the air. A steady-state concentration, amounting to about one atom of carbon-14 for every 10^{12} atoms of carbon-12, is established in atmospheric CO_2. A living plant, taking in carbon dioxide, has this same $^{14}C/^{12}C$ ratio, as do plant-eating animals or human beings.

Everything we eat has this ratio

When a plant or animal dies, the intake of radioactive carbon stops. Consequently, the radioactive decay of carbon-14

$$^{14}_{6}C \rightarrow ^{14}_{7}N + ^{0}_{-1}e \qquad (t_{1/2} = 5720 \text{ yr}) \tag{25.18}$$

takes over and the ratio $^{14}C/^{12}C$ drops. By measuring this ratio and comparing it to that in living plants, we can estimate the time at which the plant or animal died (Example 25.5).

■ **EXAMPLE 25.5** ——————————————————

A tiny piece of paper taken from the Dead Sea Scrolls, believed to date back to the first century A.D., was found to have a $^{14}C/^{12}C$ ratio of 0.795 times that in a living plant. Estimate the age of the scrolls.

Solution

Knowing the half-life of carbon-14 ($t_{1/2} = 5720$ years), we can calculate the first-order rate constant from Equation 25.15. Then, using Equation 25.14, we can obtain the elapsed time:

$$k = \frac{0.693}{5720 \text{ yr}} = 1.21 \times 10^{-4}/\text{yr}$$

$$\ln \frac{X_0}{X} = (1.21 \times 10^{-4}/\text{yr}) \times t$$

Since $X = 0.795 X_0$,

$$\ln \frac{X_0}{X} = \ln \frac{1.000}{0.795} = 0.229$$

Hence,

$$0.229 = (1.21 \times 10^{-4}/\text{yr}) \times t; \quad t = 1.89 \times 10^3 \text{ yr}$$

Exercise

What is the age of a piece of charcoal in which the $^{14}C/^{12}C$ ratio is 0.400 times that in a living plant? Answer: 7.57×10^3 years.

——————————————————

In practice, the $^{14}C/^{12}C$ ratio is determined by measuring the activity of the sample and calculating the number of atoms decaying per minute per gram of carbon. In a living plant, the $^{14}C/^{12}C$ ratio is such that, in a one-gram sample of carbon, about 15.3 atoms of C-14 decay per minute.

TABLE 25.4   Representative Radiocarbon Dates*

SOURCE	AGE (YR)
Charcoal from the Lascaux cave in France, containing early cave paintings	15,500 ± 900
Bison bones from Lubbock, Texas, associated with prehistoric man	9,900 ± 350
Charcoal from a tree burned during the upheaval that formed Crater Lake, Oregon	6,500 ± 250
Wheat and barley from ancient Egypt	6,100 ± 250
Charcoal associated with early period from Stonehenge, England	3,800 ± 275
Wood from coffin from Ptolemaic period in Egypt	2,200 ± 450
Linen wrappings used for the Book of Isaiah in the Dead Sea Scrolls	1,900 ± 200
Ancient Manchurian lotus seeds, still fertile	1,000 ± 210

*Taken from Charles Compton, *Inside Chemistry*, New York, McGraw-Hill, 1979.

If we were to find that, with a particular sample, the activity was 13.0 atoms/min·g C, we would conclude that the $^{14}C/^{12}C$ ratio must be

$$13.0/15.3 = 0.850$$

or about 85% of that in a living plant.

As you can imagine, it is not easy to determine accurately activities of the order of 15 atoms decaying per minute, about one "event" every four seconds. Elaborate precautions have to be taken to exclude background radiation. Moreover, relatively large samples must be used to increase the counting rate. Recently, a technique has been developed whereby C-14 atoms can be counted very accurately in a specially designed mass spectrometer. This method is now being used to date the Shroud of Turin, using six samples with a total mass of about 0.1 g. Using the conventional approach with a scintillation counter, a piece of the shroud the size of a handkerchief would have been required.

The C-14 method has been applied to date a wide variety of organic objects (Table 25.4). It depends upon an assumption that cannot be proved. For carbon-14 dating to be valid, the $^{14}C/^{12}C$ ratio in the atmosphere must have remained constant over the centuries. Ages estimated by this method agree within ± 10% with historical records, suggesting that the assumption is valid.

The Shroud appears to be from the 14th century (10/88)

Tree ring patterns are also useful in checking age up to about 1000 years

25.4
MASS-ENERGY RELATIONS

We pointed out at the beginning of this chapter that the energy change in nuclear reactions is much greater than that for ordinary chemical reactions. The energy change can be calculated from Einstein's equation:

$$\Delta E = \Delta mc^2 \qquad (25.19)$$

where Δm is the change in mass,* ΔE is the change in energy, and c is the speed of light. If we substitute for c its value in meters per second

$$c = 3.00 \times 10^8 \text{ m/s}$$

and use the basic definition of the joule (Appendix 1)

$$1 \text{ J} = \frac{1 \text{ kg} \cdot \text{m}^2}{\text{s}^2}$$

we obtain: $\Delta E = \left(3.00 \times 10^8 \frac{\text{m}}{\text{s}}\right)^2 \times \left(\frac{1 \text{ J} \cdot \text{s}^2}{1 \text{ kg} \cdot \text{m}^2}\right) \Delta m$

$$\Delta E = 9.00 \times 10^{16} \frac{\text{J}}{\text{kg}} (\Delta m) \tag{25.20}$$

Using this equation, we can readily calculate the energy change in joules given the mass change in kilograms. In dealing with nuclear reactions, we usually want ΔE in *kilojoules* corresponding to a mass change in *grams*. Using the relations

$$1 \text{ kJ} = 10^3 \text{ J}; \qquad 1 \text{ kg} = 10^3 \text{ g}$$

we obtain an alternate, more useful form of Equation 25.20:

$$\Delta E = 9.00 \times 10^{16} \frac{\text{J}}{\text{kg}} \times \frac{1 \text{ kg}}{10^3 \text{ g}} \times \frac{1 \text{ kJ}}{10^3 \text{ J}} \times \Delta m$$

$$\Delta E = 9.00 \times 10^{10} \frac{\text{kJ}}{\text{g}} \times \Delta m \tag{25.21}$$

Using Equation 25.21 along with the appropriate nuclear masses (Table 25.5, p. 896), we can calculate the energy change accompanying a nuclear reaction (Example 25.6).

■ EXAMPLE 25.6

For the radioactive decay of radium, $^{226}_{88}\text{Ra} \rightarrow {}^{222}_{86}\text{Rn} + {}^{4}_{2}\text{He}$, calculate ΔE in kilojoules when

a. one mole of radium decays.   b. one gram of radium decays.

Solution

a. We first calculate Δm for the reaction and then obtain ΔE from Equation 25.21. For the decay of one mole of Ra (using Table 25.5)

$$\Delta m = \text{mass 1 mol } {}^{4}_{2}\text{He} + \text{mass 1 mol } {}^{222}_{86}\text{Rn} - \text{mass 1 mol } {}^{226}_{88}\text{Ra}$$
$$= 4.0015 \text{ g} + 221.9703 \text{ g} - 225.9771 \text{ g}$$
$$= -0.0053 \text{ g}$$

---

*Specifically, Δm = mass of products − mass of reactants; ΔE = energy of products − energy of reactants. In spontaneous nuclear reactions, the products weigh less than the reactants (Δm negative). In this case, the energy of the products is less than that of the reactants (ΔE negative), and energy is evolved to the surroundings.

TABLE 25.5 Nuclear Masses on the ^{12}C Scale*

	AT. NO.	MASS NO.	MASS (amu)		AT. NO.	MASS NO.	MASS (amu)
e	0	0	0.00055	Br	35	79	78.8992
n	0	1	1.00867		35	81	80.8971
H	1	1	1.00728		35	87	86.9028
	1	2	2.01355	Rb	37	89	88.8913
	1	3	3.01550	Sr	38	90	89.8869
He	2	3	3.01493	Mo	42	99	98.8846
	2	4	4.00150	Ru	44	106	105.8832
Li	3	6	6.01348	Ag	47	109	108.8790
	3	7	7.01436	Cd	48	109	108.8786
Be	4	9	9.00999		48	115	114.8791
	4	10	10.01134	Sn	50	120	119.8748
B	5	10	10.01019	Ce	58	144	143.8817
	5	11	11.00656		58	146	145.8868
C	6	11	11.00814	Pr	59	144	143.8809
	6	12	11.99671	Sm	62	152	151.8857
	6	13	13.00006	Eu	63	157	156.8908
	6	14	13.99995	Er	68	168	167.8951
O	8	16	15.99052	Hf	72	179	178.9065
	8	17	16.99474	W	74	186	185.9138
	8	18	17.99477	Os	76	192	191.9197
F	9	18	17.99601	Au	79	196	195.9231
	9	19	18.99346	Hg	80	196	195.9219
Na	11	23	22.98373	Pb	82	206	205.9295
Mg	12	24	23.97845		82	207	206.9309
	12	25	24.97925		82	208	207.9316
	12	26	25.97600	Po	84	210	209.9368
Al	13	26	25.97977		84	218	217.9628
	13	27	26.97439	Rn	86	222	221.9703
	13	28	27.97477	Ra	88	226	225.9771
Si	14	28	27.96924	Th	90	230	229.9837
S	16	32	31.96329	Pa	91	234	233.9934
Cl	17	35	34.95952	U	92	233	232.9890
	17	37	36.95657		92	235	234.9934
Ar	18	40	39.95250		92	238	238.0003
K	19	39	38.95328		92	239	239.0038
	19	40	39.95358	Np	93	239	239.0019
Ca	20	40	39.95162	Pu	94	239	239.0006
Ti	22	48	47.93588		94	241	241.0051
Cr	24	52	51.92734	Am	95	241	241.0045
Fe	26	56	55.92066	Cm	96	242	242.0061
Co	27	59	58.91837	Bk	97	245	245.0129
Ni	28	59	58.91897	Cf	98	248	248.0186
Zn	30	64	63.91268	Es	99	251	251.0255
	30	72	71.91128	Fm	100	252	252.0278
Ge	32	76	75.90380		100	254	254.0331
As	33	79	78.90288				

*Note that these are *nuclear masses*. The masses of the corresponding atoms can be calculated by adding the masses of each extranuclear electron (0.000549). For example, for an *atom* of ^4_2He we have

$$4.00150 + 2(0.000549) = 4.00260$$

Similarly, for an atom of $^{12}_6C$,

$$11.99671 + 6(0.000549) = 12.00000$$

(Note that since Δm is extremely small, it is necessary to know the masses of products and reactants very accurately to obtain the mass difference to two significant figures.)

$$\Delta E = 9.00 \times 10^{10} \frac{\text{kJ}}{\text{g}} (-0.0053 \text{ g}) = \boxed{-4.8 \times 10^8 \text{ kJ}}$$

b. Since one mole of radium weighs 226 g, we have

$$\Delta E = 1.00 \text{ g Ra} \times \frac{(-4.8 \times 10^8 \text{ kJ})}{226 \text{ g Ra}} = \boxed{-2.1 \times 10^6 \text{ kJ}}$$

In studying a reaction like this one, we work with only a few atoms, so the total amount of energy released is small

Exercise

Calculate ΔE for the decay of one mole of $^{239}_{94}\text{Pu}$: $^{239}_{94}\text{Pu} \rightarrow {}^{4}_{2}\text{He} + {}^{235}_{92}\text{U}$. Answer: -5.1×10^8 kJ.

Energy changes in ordinary chemical reactions are of the order of 50 kJ/g or less. For example, in the combustion of petroleum, about 46 kJ of heat is evolved per gram of fuel burned. Looking at Example 25.6b, we see that ΔE for the radioactive decay of radium is about 50,000 times as great.

NUCLEAR BINDING ENERGY

Using the numbers listed in Table 25.5, we always find that *a nucleus weighs less than the individual protons and neutrons of which it is composed.* Consider, for example, the $^{6}_{3}\text{Li}$ nucleus, which contains three protons and three neutrons. From Table 25.5 we see that one mole of Li-6 nuclei weighs 6.01348 g. The mass of the corresponding number of protons and neutrons (3 mol protons, 3 mol neutrons) is:

$$\text{mass of protons} = 3 \times 1.00728 \text{ g} = 3.02184 \text{ g}$$
$$\text{mass of neutrons} = 3 \times 1.00867 \text{ g} = \underline{3.02601 \text{ g}}$$
$$\text{total mass nucleus} = 6.04785 \text{ g}$$

We see that one mole of Li-6 nuclei weighs 0.03437 g less than the protons and neutrons making up these nuclei. For the process: $^{6}_{3}\text{Li} \rightarrow 3\,{}^{1}_{0}n + 3\,{}^{1}_{1}\text{H}$:

$$\Delta m = 6.04785 \text{ g} - 6.01348 \text{ g} = 0.03437 \text{ g}$$

The quantity just calculated is referred to as the *mass defect;* we say that the mass defect of Li-6 is 0.03437 g/mol. The corresponding energy, calculated from Equation 25.21, is

$$\Delta E = 9.00 \times 10^{10} \frac{\text{kJ}}{\text{g}} \times 0.03437 \frac{\text{g}}{\text{mol}} = 3.09 \times 10^9 \text{ kJ/mol}$$

This energy is referred to as the **binding energy**; we say that the binding energy of Li-6 is 3.09×10^9 kJ/mol. This means that 3.09×10^9 kJ of energy would have to be absorbed ($\Delta m > 0$) to decompose one mole of Li-6 nuclei into protons and neutrons:

$$^{6}_{3}\text{Li} \rightarrow 3\,{}^{1}_{0}n + 3\,{}^{1}_{1}\text{H}; \qquad \Delta E = 3.09 \times 10^9 \text{ kJ/mol Li-6}$$

It takes a lot of energy to blow a nucleus apart

By the same token, 3.09×10^9 kJ of energy would be evolved when one mole of Li-6 is formed from protons and neutrons.

■ **EXAMPLE 25.7** ――――――――――――――――――――――――――
Consider the nucleus of $^9_4$Be. Calculate its

a. mass defect in grams per mole.
b. binding energy in kilojoules per mole.

Solution

a. mass of 4 mol protons $= 4 \times 1.00728$ g $= 4.02912$ g
    mass of 5 mol neutrons $= 5 \times 1.00867$ g $= \underline{5.04335 \text{ g}}$

    total mass of nucleus $= 9.07247$ g

    mass of 1 mol Be-9 $= 9.00999$ g

    $\Delta m = 9.07247$ g $- 9.00999$ g $= 0.06248$ g;

    mass defect $=$ $\boxed{0.06248 \text{ g/mol}}$

Note that mass defect is ordinarily taken to be a positive quantity, as is binding energy.

b. $\Delta E = 9.00 \times 10^{10} \, \dfrac{\text{kJ}}{\text{g}} \times 0.06248 \, \dfrac{\text{g}}{\text{mol}} = \boxed{5.62 \times 10^9 \text{ kJ/mol}}$

Exercise
Calculate the binding energy per mole of nuclear particles in Be-9. Answer: 6.24×10^8 kJ/mol nucleons.

――――――――――――――――――――――――――――――――――――

The binding energy of a nucleus is, in a sense, a measure of its stability. The greater the binding energy, the more difficult it would be to decompose the nucleus into protons and neutrons. A better measure of the relative stabilities of different nuclei is the *binding energy per mole of nuclear particles* (*nucleons*). This quantity is calculated by dividing the binding energy per mole of nuclei by the number of particles per nucleus. Thus we have

> A nucleon is a proton or a neutron

$$\text{Li-6: } 3.09 \times 10^9 \, \frac{\text{kJ}}{\text{mol Li-6}} \times \frac{1 \text{ mol Li-6}}{6 \text{ mol nucleons}} = 5.15 \times 10^8 \, \frac{\text{kJ}}{\text{mol}}$$

$$\text{Be-9: } 5.62 \times 10^9 \, \frac{\text{kJ}}{\text{mol Be-9}} \times \frac{1 \text{ mol Be-9}}{9 \text{ mol nucleons}} = 6.24 \times 10^8 \, \frac{\text{kJ}}{\text{mol}}$$

Figure 25.6 shows a plot of this quantity, binding energy per mole of nucleons, vs. mass number. Notice that the curve has a broad maximum in the vicinity of mass numbers 50 to 80. Consider what would happen if a heavy nucleus such as $^{235}_{92}$U were to split into smaller nuclei with mass numbers near the maximum. This process, referred to as *nuclear fission*, should result in an evolution of energy. The same effect would be obtained if very light nuclei such as $^2_1$H were to combine with one another. Indeed,

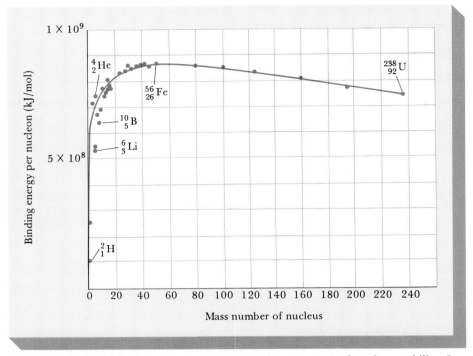

Figure 25.6 The binding energy per nucleon is a measure of nuclear stability. It has its maximum value for nuclei of intermediate mass, falling off for very heavy or very light nuclei. The form of this curve accounts for the fact that both fission and fusion give off large amounts of energy.

this process, called *nuclear fusion,* should evolve even more energy since the binding energy per nucleon increases very sharply at the beginning of the curve.

25.5
NUCLEAR FISSION

The process of nuclear fission was discovered more than half a century ago in 1938 in Germany. With the outbreak of World War II a year later, interest focused on the enormous amount of energy released in the process. At Los Alamos, in the mountains of New Mexico, a group of scientists led by J. Robert Oppenheimer worked feverishly to produce the fission, or "atomic," bomb. Many of the members of this group were exiles from Nazi Germany. They were spurred on by the fear that Hitler would obtain the bomb first. Their work led to the explosion of the first atomic bomb in the New Mexico desert at 5:30 A.M. on July 16, 1945. Less than a month later (August 6, 1945), the world learned of this new weapon when another bomb was exploded over Hiroshima. This bomb killed 70,000 people and completely devastated an area of ten square kilometers. Three days later

Some scientists tried to prevent the use of these bombs, to no avail

Nagasaki and its inhabitants met a similar fate. On August 14th, Japan surrendered and World War II was over.

THE FISSION PROCESS ($^{235}_{92}U$)

Several isotopes of the heavy elements undergo fission if bombarded by neutrons of high enough energy. In practice, attention has centered upon two particular isotopes, $^{235}_{92}U$ and $^{239}_{94}Pu$. Both of these can be split into fragments by low-energy neutrons.

Our discussion will concentrate upon the uranium-235 isotope. It makes up only about 0.7% of naturally occurring uranium. The more abundant isotope, uranium-238, does not undergo the fission reaction. During World War II, several different processes were studied for the separation of these two isotopes. The most successful technique was that of gaseous effusion (Chap. 4), using the volatile compound UF_6, which sublimes at 56°C.

Fission Products   When a uranium-235 atom undergoes fission, it splits into two unequal fragments and a number of neutrons and beta particles. The fission process is complicated by the fact that different uranium-235 atoms split up in many different ways. For example, while one atom of $^{235}_{92}U$ is splitting to give isotopes of rubidium (at. no. = 37) and cesium (at. no. = 55), another may break up to give isotopes of bromine (at. no. = 35) and lanthanum (at. no. = 57), while still another atom yields isotopes of zinc (at. no. = 30) and samarium (at. no. = 62):

$$^{90}_{37}Rb + {}^{144}_{55}Cs + 2\,{}^{1}_{0}n \tag{25.22}$$

Fission occurs when a ^{235}U nucleus absorbs a neutron

$$^{1}_{0}n + {}^{235}_{92}U \rightarrow {}^{87}_{35}Br + {}^{146}_{57}La + 3\,{}^{1}_{0}n \tag{25.23}$$

$$^{72}_{30}Zn + {}^{160}_{62}Sm + 4\,{}^{1}_{0}n \tag{25.24}$$

More than 200 isotopes of 35 different elements have been identified among the fission products of uranium-235.

The stable neutron-to-proton ratio near the middle of the Periodic Table, where the fission products are located, is considerably smaller ($\sim$1.2) than that of uranium-235 (1.5). Hence, the immediate products of the fission process contain too many neutrons for stability. In the case of rubidium-90, three steps are required to reach a stable nucleus:

$$^{90}_{37}Rb \rightarrow {}^{90}_{38}Sr + {}^{0}_{-1}e; \qquad t_{1/2} = 2.8 \text{ min}$$

$$^{90}_{38}Sr \rightarrow {}^{90}_{39}Y + {}^{0}_{-1}e; \qquad t_{1/2} = 29 \text{ yr}$$

$$^{90}_{39}Y \rightarrow {}^{90}_{40}Zr + {}^{0}_{-1}e; \qquad t_{1/2} = 64 \text{ h}$$

The radiation hazard associated with nuclear fallout arises from the formation of radioactive isotopes such as these. One of the most dangerous is strontium-90. In the form of strontium carbonate, $SrCO_3$, it is incorporated into the bones of animals and human beings.

You will notice from Equations 25.22 to 25.24 that two to four neutrons are produced by fission for every one consumed. Once a few atoms of uranium-235 split, the neutrons produced can bring about the fission of many more uranium-235 atoms. This creates the possibility of a chain reaction. This is precisely what happens in the atomic bomb. The energy evolved in successive fissions escalates to give a tremendous explosion within a few seconds.

For nuclear fission to result in a chain reaction, the sample must be large enough so that most of the neutrons are captured internally. If the sample is too small, most of the neutrons escape, breaking the chain. The *critical mass* of uranium-235 required to maintain a chain reaction in a bomb appears to be about 1 to 10 kg. In the bomb dropped on Hiroshima, the critical mass was achieved by using a conventional explosive to fire one piece of uranium-235 into another.

Fission Energy   The evolution of energy in nuclear fission is directly related to the decrease in mass that takes place. About 80,000,000 kJ of energy is given off for every gram of $^{235}_{92}U$ that reacts. This is about 40 times as great as the energy change for simple nuclear reactions such as radioactive decay. The heat of combustion of coal is only about 30 kJ/g; the energy given off when TNT explodes is still smaller, about 2.8 kJ/g. Putting it another way, the fission of one gram of $^{235}_{92}U$ produces as much energy as the combustion of 2700 kg of coal or the explosion of 30 metric tons (3×10^4 kg) of TNT.

The Hiroshima bomb was equivalent to 20,000 tons of TNT, so about 7 kg of $^{235}_{92}U$ reacted

NUCLEAR REACTORS

Even before the first atomic bomb exploded, scientists and political leaders began to speculate on the use of fission as a peacetime energy source. Nuclear reactors that convert the heat produced by the fission of uranium-235 into electrical energy are now a reality. More than 100 such reactors supply 12% of the electrical energy used in the United States.

The type of reactor that is most common in the United States today is shown in Figure 25.7. Fuel rods alternate with control rods in a con-

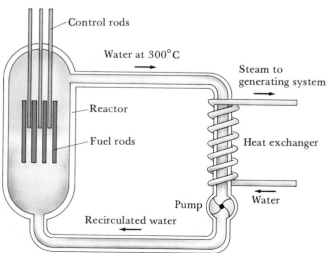

Figure 25.7   Nuclear reactor of the pressurized water type. The control rods are made of a material such as cadmium or boron, which absorb neutrons effectively. The fuel rods contain uranium oxide, usually enriched in U-235.

tainment chamber. The *fuel rods* are cylinders that contain fissionable material, uranium dioxide (UO_2) pellets, in a zirconium alloy tube. Natural uranium contains only about 0.7% U-235, the isotope that undergoes fission. The uranium in these reactors is "enriched" so that it contains about 3% U-235. The *control rods* are cylinders composed of substances, such as boron and cadmium, that absorb neutrons. Increased absorption of neutrons slows down the chain reaction. By varying the depth of the control rods within the fuel-rod assembly, the speed of the chain reaction can be controlled. If the control rods are dropped all the way down into the fuel-rod assembly, the chain reaction stops. Water at a pressure of 140 atm is passed through the reactor to absorb the heat given off by fission. The water, coming out of the reactor core at 320°C, circulates through a closed loop containing a heat exchanger. A second stream of water at lower pressure passes through the heat exchanger and is converted to steam at 270°C. This steam is used to drive a turbogenerator that produces electrical energy.

In this type of reactor, called a "light water reactor" (LWR), the circulating water serves another purpose in addition to heat transfer. It acts to slow down or *moderate* the neutrons given off by fission. This is necessary if the chain reaction is to continue; fast neutrons are not readily absorbed by U-235. Reactors used in Canada substitute "heavy water," D_2O, for ordinary water. Although heavy water is much more expensive, it has an important advantage over H_2O. Its moderating properties are such that naturally occurring uranium can be used as a fuel; enrichment in U-235 is not necessary. Still another substance used to moderate neutrons is graphite; this is widely used in the Soviet Union but is found in only two commercial reactors in the United States. Graphite-moderated reactors can operate with fuel enriched to about 1.8% U-235.

Nuclear Accidents: Three Mile Island and Chernobyl   To operate a nuclear reactor safely, it is necessary to prevent significant amounts of radioactive fission products from escaping into the environment. This can happen if the temperature of the nuclear fuel rises to the melting point ("meltdown"), leading to an explosion. Every reactor is designed with elaborate safety systems to prevent overheating. Beyond that, the reactor core is separated from the surroundings by being isolated within a reinforced containment building.

The most likely cause of meltdown is a loss of cooling water. If this happens in a light water reactor, fission stops because the unmoderated neutrons move too fast to continue the chain reaction. However, large amounts of heat are given off by the radioactive decay of fission products (recall Example 25.6). This happened at Three Mile Island in Pennsylvania in March of 1979. Through operator error, water was lost from the cooling system. For some time, the fuel rods were uncovered and there was a real danger of meltdown. Beyond that, at the high temperature reached (>2000°C), the zirconium tubes holding the nuclear fuel reacted with steam:

$$Zr(s) + 2H_2O(g) \rightarrow ZrO_2(s) + 2H_2(g) \tag{25.25}$$

The hydrogen formed raised the specter of an explosion that could have spewed out radioactive gases into the environment. Fortunately, there was not enough oxygen present for this to occur. The total amount of radiation released was quite small; there were no fatalities.

A much more severe nuclear accident took place at Chernobyl, near Kiev in the Soviet Union, in April of 1986. It released 100 million curies of radiation, much of it carried westward by the prevailing winds to other European countries. There were 25 fatalities from acute radiation exposure; estimates of the number of delayed cancers from lower level radiation range from 2000 to 75,000.

The accident at Chernobyl, like that at Three Mile Island, was due to operator error. Somehow, almost all of the control rods were withdrawn from the reactor core. Very quickly, the cooling water was converted to superheated steam. In the absence of liquid water, more and more neutrons reached the graphite moderator and were slowed down enough to keep the chain reaction going. A steam explosion blew off a 1000-ton concrete slab on top of the reactor core. At that point, the graphite caught fire. Ten days later the fire was finally extinguished, after 40 tons of boron carbide, 800 tons of dolomite ($CaCO_3 \cdot MgCO_3$), and 2400 tons of lead had been dumped on top of the reactor. The purpose of the boron carbide was to stop the fission process; dolomite decomposed to produce $CO_2(g)$, which eventually extinguished the burning graphite. The lead acted to absorb heat and later served as a radiation shield over the reactor core.

Many design changes have been proposed to prevent such accidents from happening again. Most of these changes are aimed at eliminating the "human factor" that was responsible for the Three Mile Island and Chernobyl accidents. One possibility is to incorporate within the reactor core such natural heat transfer processes as convection and thermal radiation. This would allow heat to dissipate rapidly in the event that normal cooling fails. This passive means of cooling would prevent overheating and core damage, eliminating the need for external forced-cooling systems and plant operators to activate them.

Nuclear Waste When a nuclear reactor operates, the fuel rods undergo physical and chemical changes due to the enormous amount of radiation to which they are exposed. Each year, on the average, one fourth of these intensely radioactive rods must be replaced. Some of the spent fuel rods are reprocessed to recover uranium or plutonium; others are stored without treatment. Either approach generates a large amount of dangerous nuclear waste, of the order of 2×10^3 metric ions per year. In 1982, a federal program was established for the disposal of such waste. It is supposed to be buried in an underground site.

Several factors enter into the selection and design of such a deep underground nuclear waste facility. To prevent radioactivity from reaching the surface, the site must be geologically stable, dry, and sufficiently deep to prevent inadvertent mining. Current proposals call for encasing the concentrated wastes in glass or ceramic containers for burial. The containers must withstand ground-water leaching and extensive radiation bombardment and extreme heat from the radioactive decay of the waste, especially if it is "new" (see Example 25.4). Calculations indicate that waste from reprocessed fuel requires 20,000 years to decay to a "safe" level.

The town of Chernobyl has now been razed; it would have been uninhabitable for decades if not centuries

Unprocessed waste takes 100 times longer. Data like these do not tend to encourage people to accept the idea of having nuclear wastes stored in their backyards.

25.6
NUCLEAR FUSION

Recall (Fig. 25.6) that very light isotopes, such as those of hydrogen, are unstable with respect to fusion into heavier isotopes. Indeed, the energy available from nuclear fusion is considerably greater than that given off in the fission of an equal mass of a heavy element (Example 25.8).

■ **EXAMPLE 25.8** _____

Calculate the amount of energy evolved, in kilojoules per gram of reactants, in

a. a fusion reaction, $^2_1H + ^2_1H \rightarrow ^4_2He$.
b. a fission reaction, $^{235}_{92}U \rightarrow ^{90}_{38}Sr + ^{144}_{58}Ce + ^1_0n + 4\ ^0_{-1}e$.

Solution

a. We first calculate the change in mass per mole of product, using Table 25.5:

$$\Delta m = 4.00150 \text{ g} - 2(2.01355 \text{ g}) = -0.02560 \text{ g}$$

Converting to kilojoules,

$$\Delta E = -2.56 \times 10^{-2} \text{ g} \times 9.00 \times 10^{10} \text{ kJ/g} = -2.30 \times 10^9 \text{ kJ}$$

Since 4.03 g of deuterium is involved, ΔE per gram of reactant is

$$\Delta E = \frac{-2.30 \times 10^9 \text{ kJ}}{4.03} = \boxed{-5.71 \times 10^8 \text{ kJ}}$$

b. Proceeding as in (a), we find that, per mole of uranium reacting,

$$\begin{aligned} \Delta m = &\ 89.8869 \text{ g} + 143.8817 \text{ g} + 1.0087 \text{ g} \\ &+ 4(0.00055 \text{ g}) - 234.9934 \text{ g} \\ = &\ -0.2139 \text{ g} \end{aligned}$$

Hence, for one mole of $^{235}_{92}U$,

$$\Delta E = -0.2139 \text{ g} \times 9.00 \times 10^{10} \text{ kJ/g} = -1.93 \times 10^{10} \text{ kJ}$$

For one gram of ^{235}U,

$$\Delta E = \frac{-1.93 \times 10^{10} \text{ kJ}}{235} = \boxed{-8.21 \times 10^7 \text{ kJ}}$$

Comparing the answers to (a) and (b), we conclude that the fusion reaction produces about seven times as much energy per gram of starting material (57.1×10^7 vs. 8.21×10^7 kJ) as does the fission reaction. This factor varies from about three to ten, depending upon the particular reactions chosen to represent the fusion and fission processes.

Exercise
Calculate ΔE per gram of reactant for $^2_1H + ^3_1H \rightarrow ^4_2He + ^1_0n$. Answer: -3.38×10^8 kJ.

As an energy source, nuclear fusion possesses several additional advantages over nuclear fission. For one thing, fusion is a "clean" process in the sense that final products are stable species such as ^4_2He rather than the radioactive isotopes formed by fission. Equally important, light isotopes suitable for fusion are far more abundant than the heavy isotopes required for fission. We can calculate, for example (Problem 64), that the fusion of only 2×10^{-9} percent of the deuterium (^2_1H) in seawater would meet the total annual energy requirements of the world.

Unfortunately, fusion processes, unlike neutron-induced fission, have very high activation energies. In order to overcome the electrostatic repulsion between two deuterium nuclei and cause them to react, they have to be accelerated to velocities of about 10^6 m/s, about 10,000 times greater than ordinary molecular velocities at room temperature. The corresponding temperature for fusion, as calculated from kinetic theory (Problem 63), is of the order of 10^9 °C. In the hydrogen bomb, temperatures of this magnitude were achieved by using a fission reaction to trigger nuclear fusion. If fusion reactions are to be used to generate electricity, it will be necessary to develop equipment in which very high temperatures can be maintained long enough to allow fusion to occur and give off energy. In any conventional container, the reactant nuclei would quickly lose their high kinetic energies by collisions with the walls.

That's how a hydrogen bomb works

One fusion reaction currently under study is a two-step process involving deuterium and lithium as the basic starting materials:

$$^2_1H + ^3_1H \rightarrow ^4_2He + ^1_0n$$
$$\underline{^6_3Li + ^1_0n \rightarrow ^4_2He + ^3_1H}$$
$$^2_1H + ^6_3Li \rightarrow 2\,^4_2He \qquad\qquad (25.26)$$

This process is attractive because it has a lower activation energy than other fusion reactions. On the other hand, the neutrons produced as an intermediate in the first step cause some serious problems. They react with the metal shell surrounding the fusion reaction, weakening the metal and producing hazardous radioactive isotopes at the same time. Beyond that, lithium is not a pleasant substance to work with. It burns violently when it comes in contact with either air or water.

Promising results have been obtained with Reaction 25.26 using "magnetic bottles" to confine the reactant nuclei. A prototype reactor, the Tokomak nuclear reactor at Princeton University (Fig. 25.8) uses this principle. It has been able to sustain the reaction for a fraction of a second. To achieve a net evolution of energy, this time must be extended to at least one second. At least another 25 years will be required to develop commer-

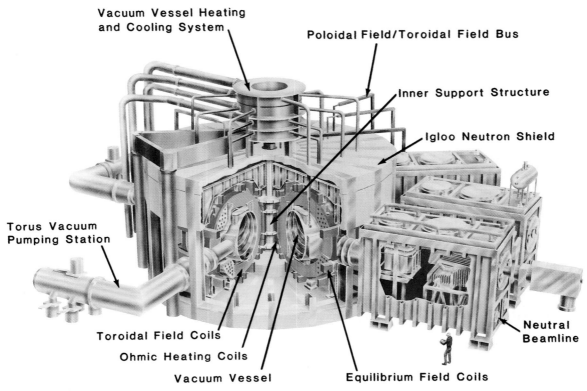

Figure 25.8 Tokomac fusion test reactor (magnetic confinement). (Department of Energy)

cial fusion reactors capable of making a significant contribution to our energy needs.

Another approach to nuclear fusion is shown in Figure 25.9. Tiny glass pellets (about 0.1 mm in diameter) filled with frozen deuterium and tritium are illuminated by a powerful laser beam. In principle at least, the neutrons produced should be able to react with lithium to complete Reaction 25.26 and give off energy. Unfortunately, the economic feasibility of this approach is extremely dubious. The glass pellets are expensive to produce and the laser beam has a very short life expectancy.

Perhaps the ultimate irony of our time is the fact that we have made so little use of the energy produced in a nuclear fusion process that has been going on since the universe was formed. The energy given off by the sun and other stars results from fusion reactions in which ordinary hydrogen is converted to helium. One mechanism that has been suggested for this process is

$$\begin{aligned}
{}^1_1H + {}^1_1H &\rightarrow {}^2_1H + {}^0_1e \\
{}^2_1H + {}^1_1H &\rightarrow {}^3_2He \\
{}^3_2He + {}^1_1H &\rightarrow {}^4_2He + {}^0_1e \\
\hline
4\,{}^1_1H &\rightarrow {}^4_2He + 2\,{}^0_1e; \quad \Delta E = -6.0 \times 10^8 \text{ kJ/g reactant} \quad (25.27)
\end{aligned}$$

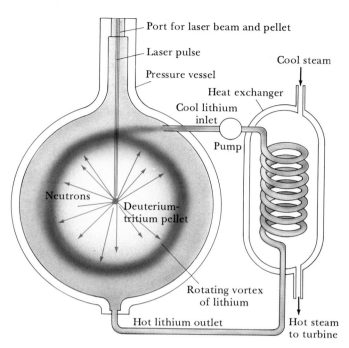

Port for laser beam and pellet

Laser pulse

Cool steam

Pressure vessel

Heat exchanger

Cool lithium inlet

Pump

Neutrons

Deuterium-tritium pellet

Rotating vortex of lithium

Hot lithium outlet

Hot steam to turbine

Figure 25.9 Schematic diagram of laser-induced fusion apparatus. A pellet of deuterium-tritium undergoes fusion on absorbing energy from a laser pulse. Neutrons from the fusion reaction interact with molten lithium (Reaction 25.26).

■ SUMMARY PROBLEM

Consider the isotopes of copper.

a. Write the nuclear symbol for Cu-64, which is used medically to scan for brain tumors. How many protons are there in the nucleus? How many neutrons?

b. Write the reaction for the β-decay of Cu-64.

c. When Cu-65 is bombarded with carbon-12, three neutrons and another particle are produced. Write the equation for this reaction.

d. Calculate ΔE in kilojoules when 1.00 g of Cu-63 (molar mass = 62.91367 g/mol) is formed, along with a proton, by neutron bombardment of Zn-63 (molar mass = 62.91674 g/mol).

e. What is the mass defect and binding energy of Cu-63?

f. A one-milligram sample of Cu-64 has an activity of 3.82×10^3 Ci. How many atoms are there in the sample (molar mass Cu-64 = 63.91 g/mol)? What is the decay rate in atoms per second? What is the rate constant (s^{-1})? the half-life in seconds?

g. How long will it take for 20.0% of a sample of Cu-64 to decay?

Solution

a. $^{64}_{29}\text{Cu}$, 29 protons, 35 neutrons   **b.** $^{64}_{29}\text{Cu} \rightarrow {}^{0}_{-1}e + {}^{64}_{30}\text{Zn}$

c. $^{65}_{29}\text{Cu} + {}^{12}_{6}\text{C} \rightarrow 3\,{}^{1}_{0}n + {}^{74}_{35}\text{Br}$ **d.** -6.38×10^6 kJ

e. mass defect = 0.5922 g; binding energy = 5.33×10^{10} kJ

f. 9.42×10^{18}; 1.41×10^{14}; 1.50×10^{-5}/s; 4.62×10^4 s

g. 1.49×10^4 s

Throughout this book, all the historical perspectives have involved men. It seems appropriate, in discussing nuclear chemistry, to focus upon three women who made major contributions in this area. These are Marie Curie (1867–1934), Irene Joliot-Curie (1897–1956), and Lise Meitner (1878–1968). We will consider their contributions chronologically.

1891–1902 Marie Curie: The Early Years

Marie Curie was born Maria Sklodowska in Warsaw, Poland, then a part of the Russian empire. In 1891 she emigrated to Paris to study at the Sorbonne, where she met and married a French physicist, Pierre Curie. The Curies were associates of Henri Becquerel, the man who discovered that uranium salts are radioactive. They showed that thorium, like uranium, is radioactive and that the amount of radiation emitted is directly proportional to the amount of uranium or thorium in the sample.

In 1898, Marie and Pierre Curie isolated two new radioactive elements, which they named radium and polonium. To obtain a few milligrams of these elements, they started with several tons of pitchblende ore and carried out a long series of tedious separations. Their work was done in a poorly equipped, unheated shed where the temperature reached 6°C (43°F) in winter. Four years later, in 1902, Marie determined the atomic mass of radium to within 0.5%, working with a tiny sample.

1903–1911 Marie Curie: Triumph and Tragedy

In 1903, the Curies received the Nobel Prize in physics (with Becquerel) for the discovery of radioactivity. Three years later, Pierre Curie died at the age of 46, the victim of a tragic accident. He stepped from behind a carriage in a busy Paris street and was run down by a truck. That same year, Marie became the first woman instructor at the Sorbonne. In 1911, she narrowly missed election to the French Academy of Science for quite irrelevant reasons. She did, however, receive the Nobel Prize in chemistry for the discovery of radium and polonium, thereby becoming the first person to win two Nobel Prizes.

1914–1918 The War Years

When Europe exploded into war in 1914, scientists largely abandoned their studies to go to the front. Marie Curie, with her daughter Irene, then 17 years old, organized medical units equipped with x-ray machinery. These were used to locate foreign metallic objects in wounded soldiers. Many of the wounds were to the head; French soldiers came out of the trenches without head protection because their government had decided that helmets looked too German. In November of 1918, the Curies celebrated the end of World War I; France was victorious and Marie's beloved Poland was free again.

By a curious coincidence, Lise Meitner was also a radiological nurse during the war. Born in Austria, she treated Austrian and German soldiers. In the war years, she also worked part-time with Otto Hahn at the Kaiser Wilhelm Institute in Berlin. Their research led to the discovery of protactinium. In 1918, Meitner became head of the physics department at the Institute. In the same year, in Paris, Marie Curie was appointed director of the newly created Radium Institute. Her daughter Irene became her assistant and held that position for many years.

Marie and Irene Curie. (E.F. Smith Memorial Collection, CHOC, University of Pennsylvania)

1926–1935 Irene Curie: Years of Discovery

In 1921, Irene Curie began research at the Radium Institute. Five years later she married Frederic Joliot, a brilliant young physicist who was also an assistant at the Institute. In 1931, they began a research program in nuclear chemistry that led to several important discoveries and at least one near miss. The Joliot-Curies were the first to demonstrate induced radioactivity. They also discovered the positron, a particle that scientists had been seeking for many years. They narrowly missed finding another, more fundamental particle, the neutron. That honor went to Chadwick in England. In 1935, Irene and Frederic Joliot-Curie received the Nobel Prize in physics. The award came too late for Irene's mother, who had died of leukemia a year earlier.

1935–1939 Irene Curie and Lise Meitner: Nuclear Fission

In the late 1930s several different research groups were working on the reactions undergone by uranium under neutron bombardment. One team was led by Irene Curie at the Radium Institute in Paris. She showed that one of the products was a radioactive isotope which resembled actinium (at. no. = 89) in its chemical properties. Years later, this species was shown to be lanthanum (at. no. = 57). Once again, Irene had narrowly missed an epochal discovery; this time it was nuclear fission.

Another group working in this area was led by Lise Meitner and Otto Hahn in Berlin. In 1937, Meitner, who was Jewish, was forced to leave Germany. She went first to Holland, then to Bohr's laboratory in Copenhagen, and finally, in 1938, to Stockholm. In that same year Hahn and Strassman in Germany isolated a compound of a Group 2 element, which they originally believed to be radium (at. no. = 88), from the neutron bombardment of uranium. Later they showed that this element was barium (at. no. = 56), indicating that a uranium atom had been split into fragments. Hahn's first reaction to this discovery was one of disbelief. He later stated, in January 1939, "As chemists, we should replace the symbol Ra . . . by Ba . . . [but], as nuclear chemists, closely associated with physics, we cannot decide to take this step in contradiction to all previous experience in nuclear physics."

If Hahn was reluctant to admit the possibility of an entirely new type of nuclear reaction, his former colleague, Lise Meitner, was not. In a letter published with O. R. Frisch in January 1939, she stated: "At first sight, this result seems very hard to understand. . . . On the basis, however, of present ideas about the behavior of heavy nuclei, an entirely different picture of these new disintegration processes suggests itself. . . . It seems possible that the uranium nucleus . . . may, after neutron capture, divide itself into nuclei of roughly equal size." This revolutionary suggestion was confirmed by experiments carried out all over the world.

QUESTIONS AND PROBLEMS

Nuclear Equations

1. Lead-210 is used to prepare eyes for corneal transplants. Its daughter product is bismuth-210. Identify the emission from lead-210.

2. The blood volume of a patient can be measured by using chromium-51, a positron emitter, administered as a solution of sodium chromate. Write the nuclear equation for the decay of chromium-51.

3. Write balanced nuclear equations for
  a. the loss of an alpha particle by Th-230.
  b. the loss of an electron by lead-210.
  c. the fusion of two C-12 nuclei to give another nucleus and a neutron.
  d. the fission of U-235 to give Ba-140, another nucleus and an excess of two neutrons.

4. Write balanced nuclear equations for
  a. the alpha emission resulting in the formation of Pa-233.
  b. the loss of a positron by Y-85.
  c. the fusion of two C-12 nuclei to give sodium-23 and another particle.
  d. the fission of Pu-239 to give tin-130, another nucleus, and an excess of three neutrons.

5. Rubidium-87, a beta emitter, is the product of positron emission. Identify
  a. the daughter product of ^{87}Rb decay.
  b. the parent nucleus of ^{87}Rb.

6. Thorium-231 is the product of alpha emission and is radioactive, emitting beta radiation. Determine
  a. the parent nucleus of ^{231}Th.
  b. the daughter product of ^{231}Th decay.

7. Write balanced nuclear equations for the bombardment of
  a. Fe-54 with an alpha particle to produce another nucleus and two protons.
  b. Mo-96 with deuterium ($^2_1$H) to produce a neutron and another nucleus.
  c. Ar-40 with an unknown particle to produce potassium-43 and a proton.

  d. a nucleus with a neutron to produce a proton and P-31.

8. Write balanced nuclear equations for the bombardment of
  a. U-238 to produce Fm-249 and five neutrons.
  b. Al-26 with an alpha particle to produce P-30.
  c. Cu-63 to produce Zn-63 and a neutron.
  d. Al-27 with deuterium ($^2_1$H) to produce an alpha particle and another nucleus.

9. Balance the following nuclear equations by filling in the blanks:
  a. $^{121}_{51}$Sb + $^4_2$He $\rightarrow$ _____ + $^1_1$H
  b. $^{238}_{92}$U + $^1_0$n $\rightarrow$ $^0_{-1}$e + _____
  c. $^{14}_7$N + _____ $\rightarrow$ $^{17}_8$O + $^1_1$H
  d. _____ + $^4_2$He $\rightarrow$ $^{27}_{14}$Si + $^1_0$n

10. Balance the following nuclear equations by filling in the blanks:
  a. $^{238}_{92}$U + $^1_1$H $\rightarrow$ $^{238}_{93}$Np + _____
  b. $^{241}_{95}$Am + $^4_2$He $\rightarrow$ _____ + 2 $^1_0$n
  c. _____ + $^4_2$He $\rightarrow$ $^1_0$n + $^{12}_6$C
  d. $^{27}_{13}$Al + _____ $\rightarrow$ $^{24}_{11}$Na + $^4_2$He

Nuclear Stability

11. Which isotope in each of the following pairs should be more stable?
  a. $^{12}_6$C or $^{13}_6$C   b. $^{19}_9$F or $^{20}_9$F   c. $^{16}_7$N or $^{14}_7$N

12. Which isotope in each of the following pairs is more stable?
  a. $^{28}_{14}$Si or $^{29}_{14}$Si   b. $^6_3$Li or $^8_3$Li   c. $^{23}_{11}$Na or $^{20}_{11}$Na

13. For each pair of elements listed, predict which one has more stable isotopes.
  a. Co or Fe   b. F or Ge   c. Ag or Pd

14. For each pair of elements listed, predict which one has more stable isotopes.
  a. Ni or Cu   b. Se or Sb   c. Cd or Au

Rate of Nuclear Decay

15. Thallium-206 decays to lead-206 and has a half-life of 4.20 minutes. How many milligrams of thallium-206 remain after four half-lives starting with 10.0 mg?

16. Strontium-90 has a half-life of 28.8 years. How much strontium-90 was present initially, if after 144 years 10.0 g remain?

17. Argon-41 is used to measure the rate of gas flow. It has a half-life of 110 minutes.
   a. Calculate k.
   b. How long will it take before only 15.0% of the original amount of Ar-41 remains?

18. Strontium-90 was one of the isotopes present after the Chernobyl nuclear disaster. Cow's milk in Amsterdam was found to have high levels of Sr-90 after the explosion. Suppose that a year-old toddler had drunk contaminated milk. Calculate the fraction of Sr-90 left in his body when he reaches 80 years of age, assuming no loss of Sr-90 except by radioactive decay. (k for Sr-90 = 0.024 yr^{-1}.)

19. A sample of sodium-24 chloride containing 0.075 mg of Na-24 is used to study sodium balance in an animal. After 8.0 h, it is determined that there are 0.052 mg of Na-24 left. What is the half-life of Na-24?

20. A sample of Br-82 was found to have an activity of 8.7×10^4 disintegrations per minute at 1:00 P.M., November 20, 1987. At 10:15 A.M. November 22, 1987, its activity was redetermined and found to be 3.6×10^4 disintegrations/minute. Calculate the half-life of Br-82.

21. I-131 is used to locate tumors in the thyroid glands. A 1.00-g sample of I-131 (at. mass = 130.9 amu) has an activity of 1.3×10^5 Ci. Calculate its decay constant and half-life in days.

22. Technetium-99 (at. mass = 98.9 amu) is used for bone scans. It has a half-life of 6.0 hours. How many disintegrations/s can you expect from 1.00 mg of technicium-99?

23. Np-237 (at. mass = 237.0 amu) has a half-life of 2.20×10^6 years. This isotope decays by alpha particle emission. Write a balanced nuclear equation for the decay of Np-237 and calculate the activity in millicuries of a 0.500-g sample.

24. Lead-210 has a half-life of 20.4 years. This isotope decays by beta particle emission. Write a balanced nuclear equation for the decay of Pb-210 and calculate the activity in millicuries of a 0.500-g sample.

25. A sample of a wooden artifact gives 5.0 disintegrations/min/g of carbon. The half-life of C-14 is 5720 years and the activity of wood just cut down from a tree is 15.3 disintegrations/minute/g of carbon. How old is the wooden artifact?

26. A sample of a beam from the tomb of an ancient Egyptian king was analyzed in 1987 and gave 7.0 counts per minute (cpm) in a scintillation counter. A sample of freshly cut wood containing the same amount of carbon gave 15.3 cpm. In what year, approximately, did the king die? ($t_{1/2}$ C-14 = 5720 yr)

27. The radioactive isotope tritium, ^3_1H, is produced in nature in much the same way as $^{14}_6C$. Its half-life is 12.3 yr. Estimate the age of a sample of Scotch whiskey that has a tritium content 0.59 times that of the water in the area where the whiskey was produced.

28. An oil painting supposed to be by Rembrandt (1606–1669 A.D.) is checked by ^{14}C dating. The ^{14}C content ($t_{1/2} = 5720$ yr) of the canvas is 0.961 times that in a living plant. Could the painting have been by Rembrandt?

29. What is the approximate age of a rock in which the mole ratio of U-238 to Pb-206 is 1.10? Take the half-life of U-238 to be 4.5×10^9 yr.

30. What is the approximate age of a rock in which the mass ratio (grams) of U-238 to Pb-206 is 1.10?

31. One way of dating rocks is to determine the relative amounts of ^{40}K and ^{40}Ar; the decay of ^{40}K has a half-life of 1.26×10^9 yr. Analysis of a certain lunar sample gives the following results in mole ratios:

$$^{40}Ar/^{40}K = 4.13 \qquad (t_{1/2} \; ^{40}K = 1.26 \times 10^9 \text{ yr})$$

$$^{206}Pb/^{238}U = 0.66 \qquad (t_{1/2} \; ^{238}U = 4.5 \times 10^9 \text{ yr})$$

$$^{87}Sr/^{87}Rb = 0.049 \qquad (t_{1/2} \; ^{87}Rb = 4.8 \times 10^{10} \text{ yr})$$

Using these data, obtain the best possible value for the age of the sample. Can you suggest why the K–Ar method gives a low result?

32. The ^{87}Rb to ^{87}Sr method of dating rocks was used to analyze lunar samples from the Apollo-15 mission. Estimate the age of the lunar sample in which
   a. the mole ratio of ^{87}Rb to ^{87}Sr is 25.0 (see Problem 31).
   b. the mole ratio of ^{87}Rb to ^{87}Sr is 20.0.

Mass-Energy Changes

33. For the reaction $^{230}_{90}Th \rightarrow ^{226}_{88}Ra + ^4_2He$,
   a. calculate Δm in grams when one mole of $^{230}_{90}Th$ decays.
   b. calculate ΔE in joules when one mole of $^{230}_{90}Th$ decays; one gram of $^{230}_{90}Th$ decays.

34. Bk-245 decays by alpha emission. For one gram of ^{245}Bk, calculate Δm (grams) and ΔE (kilojoules).

35. For carbon-14, calculate
   a. the mass defect.
   b. the binding energy.

36. Which has the larger binding energy, fluorine-19 or oxygen-17?

37. Some of the sun's energy comes from the reaction

$$4 \, ^1_1H \rightarrow ^4_2He + 2 \, ^0_1e$$

Calculate the energy change in this reaction per gram of hydrogen.

38. Compare the energies given off per gram of reactant in the two fusion processes considered to occur during the formation of a star:
   a. $^2_1H + ^1_1H \rightarrow ^3_2He$
   b. $2 \, ^3_2He \rightarrow ^4_2He + 2 \, ^1_1H$

39. The sun radiates energy into space at the rate of 3.9×10^{26} J/s. Calculate the rate of mass loss by the sun.

40. Calculate the change in mass when 1.00 kg of carbon burns to form carbon dioxide, taking the heat of formation of $CO_2(g)$ to be -393.5 kJ/mol.

41. For the fission reaction

$$_0^1n + {}_{92}^{235}U \rightarrow {}_{37}^{89}Rb + {}_{58}^{144}Ce + 3\ _{-1}^{0}e + 3\ _0^1n$$

a. how much energy (in kJ) is given off per gram of $_{92}^{235}U$?

b. how many kilograms of TNT must be detonated to produce the same amount of energy? ($\Delta E = -2.76$ kJ/g.)

42. Consider the fission reaction

$$_{94}^{239}Pu + {}_0^1n \rightarrow {}_{58}^{146}Ce + {}_{38}^{90}Sr + 2\ _{-1}^{0}e + 4\ _0^1n$$

a. How many grams of $_{94}^{239}Pu$ would have to react to produce 1.00 kJ of energy?

b. How many atoms of $_{94}^{239}Pu$ would have to react to produce 1.00 kJ of energy?

43. Show by calculation whether the following reaction is spontaneous.

$$_1^1H + {}_8^{18}O \rightarrow {}_9^{19}F$$

44. Will Al-26 decay spontaneously by positron emission? Show by calculation.

45. Consider the fission reaction where U-235 is bombarded with a neutron. Cerium-144, bromine-87, electrons, and neutrons are produced.

a. Write a balanced nuclear equation for the reaction.

b. Calculate ΔE when one gram of U-235 undergoes fission.

c. The decomposition of ammonium nitrate, an explosive, evolves 37.0 kJ/mol. How many kilograms of NH_4NO_3 are required to produce the same amount of energy as one milligram of U-235?

46. Using the equation in Problem 45, compare the energy produced by the fission of one gram of U-235 to the energy produced by one gram of He-3 in the fusion reaction

$$2\ _2^3He \rightarrow {}_2^4He + 2\ _1^1H$$

Unclassified

47. Consider the interaction of beta and slow neutron radiation with human cells.

a. Calculate the number of rems absorbed by a human being exposed to 15 rads of slow neutrons ($n = 3$) and 10 rads of beta radiation ($n = 1$). Assume the dosage is additive.

b. Using Table 25.1, describe the effect of the absorption of the radiation in part a.

48. Classify each of the following statements as true or false. If false, correct the statement to make it true.

a. The mass number increases in beta emission.

b. A radioactive species with a large rate constant, k, decays very slowly.

c. Fusion gives off less energy per gram of fuel than fission.

49. Explain how

a. alpha and beta radiation are separated by an electric field.

b. radioactive ^{14}C can be used as a tracer to study steps in photosynthesis.

c. a self-sustaining chain reaction occurs in nuclear fission.

50. Suppose the $^{14}C/^{12}C$ ratio in plants a thousand years ago was 10% higher than it is today. What effect, if any, would this have on the calculated age of an artifact found by the C-14 method to be a thousand years old?

51. The amount of oxygen dissolved in a sample of water can be determined by using thallium metal containing a small amount of the isotope Tl-204. When excess thallium is added to oxygen-containing water, the following reaction occurs:

$$2Tl(s) + \tfrac{1}{2}O_2(g) + H_2O \rightarrow 2Tl^+(aq) + 2OH^-(aq)$$

It is found that, after reaction, the activity of a 25.0 mL water sample is 745 counts per minute (cpm), caused by the presence of Tl^+-204 ions. The activity of Tl-204 is 5.53×10^5 cpm per gram of thallium metal. Assuming O_2 is the limiting reactant in the above equation, calculate its concentration in moles per liter.

52. A 35-mL sample of 0.050 M $AgNO_3$ is mixed with 35 mL of 0.050 M NaI labeled with I-131. The following reaction occurs:

$$Ag^+(aq) + I^-(aq) \rightarrow AgI(s)$$

The filtrate is found to have an activity of 2.50×10^3 counts per minute per milliliter. The 0.050 M NaI solution had an activity of 1.25×10^{10} counts per minute per milliliter. Calculate K_{sp} for AgI.

53. A 50.0-g sample of water containing tritium, $_1^3H$, emits 2.89×10^3 beta particles per second. Tritium has a half-life of 12.3 years. What percent of all the hydrogen atoms in the water sample is tritium?

54. Using the half-life of tritium in Problem 53, calculate the activity in curies of 1.00 mL of $_1^3H_2$ at STP.

55. In order to measure the volume of the blood in an animal's circulatory system, the following experiment was performed. A 5.0-mL sample of an aqueous solution containing 1.7×10^5 counts per minute (cpm) of tritium was injected into the bloodstream. After an adequate period of time to allow for the complete circulation of the tritium, a 5.0-mL sample of blood was withdrawn and found to have 1.3×10^3 cpm on the scintillation counter. What is the volume of the animal's circulatory system if we assume that only a negligible amount of tritium has decayed during the experiment?

56. One of the causes of the explosion at Chernobyl may have been the reaction between zirconium, which coated the fuel rods, and steam (Reaction 25.25). If half a metric ton of zirconium reacted, what pressure was exerted by the hydrogen gas produced at 55°C in the containment chamber of volume 2.0×10^4 L?

57. Consider the fission reaction

$$\,_0^1 n + \,_{92}^{235}U \rightarrow \,_{37}^{89}Rb + \,_{58}^{144}Ce + \,_{-1}^0 e + 3 \,_0^1 n$$

How many liters of methane at 25°C and 1.00 atm pressure must be burned to $CO_2(g)$ and $H_2O(l)$ to produce as much energy as the fission of one gram of U-235 fuel?

58. A chelate of Cr^{3+} and $C_2O_4^{2-}$ is made by a reaction that involves Na_2CrO_4, a reducing agent, and oxalic acid, $H_2C_2O_4$. The sodium chromate has an activity of 765 counts per minute per gram, from Cr-51. The oxalic acid has an activity of 512 counts per minute per gram; it is labeled with C-14. Since Cr-51 and C-14 emit different particles during decay, their activities can be counted independently. A sample of the chelate was found to have a Cr-51 count of 314 cpm and a C-14 count of 235 cpm. How many oxalate ions are bound to each Cr^{3+} ion?

59. Polonium-210 decays to Pb-206 by alpha emission. Its half-life is 138 days. What volume of helium at 25°C and 755 mm Hg would be obtained from a 10.00-g sample of Po-210 left to decay for 62 hours?

60. Radium-226 decays by alpha emission to radon-222. Suppose that 15.0% of the energy given off by one gram of radium is converted to electrical energy. What is the minimum amount of chromium that would be needed for the voltaic cell $Cr/Cr^{3+} \| Cu^{2+}/Cu$, at standard concentrations, to produce the same amount of energy?

Challenge Problems

61. An activity of 20 picocuries (20×10^{-12} Ci) of radon-222 per liter of air in a house constitutes a health hazard to anyone living in the house. Taking the half-life of radon-222 to be 3.82 d, calculate the concentration of radon in air (moles per liter) corresponding to this activity.

62. Plutonium-239 decays by the reaction $\,_{94}^{239}Pu \rightarrow \,_{92}^{235}U + \,_2^4 He$, with a rate constant of 5.5×10^{-11}/min. In a one-gram sample of ^{239}Pu,

a. how many grams decompose in 10 min?

b. how much energy in kilojoules is given off in 10 min?

c. what radiation dosage in rems is received by a 70-kg man exposed to a gram of ^{239}Pu for 10 min?

63. It is possible to estimate the activation energy for fusion by calculating the energy required to bring two deuterons close enough to one another to form an alpha particle. This energy can be obtained by using Coulomb's Law in the form: $E = 8.99 \times 10^9 \, q_1 q_2 / r$, where q_1 and q_2 are the charges of the deuterons (1.60×10^{-19} C), r is the radius of the He nucleus, about 2×10^{-15} m, and E is the energy in joules.

a. Estimate E in joules per alpha particle.

b. Using the equation $E = mv^2/2$, estimate the velocity (meters per second) each deuteron must have if a collision between two of them is to supply the activation energy for fusion (m is the mass of the deuteron in kilograms).

c. Using Equation 4.26, Chapter 4, estimate the temperature that would have to be reached to achieve fusion (recall Example 4.13).

64. Consider the reaction $2 \,_1^2 H \rightarrow \,_2^4 He$.

a. Calculate ΔE in kilojoules per gram of deuterium fused.

b. How much energy is potentially available from the fusion of all the deuterium atoms in seawater? The percentage of deuterium in water is about 0.0017%. The total mass of water in the oceans is 1.3×10^{24} g.

c. What fraction of the deuterium in the oceans would have to be consumed to supply the annual energy requirements of the world (2.3×10^{17} kJ)?

$$2Ni^{2+}(aq) + 6NH_3(aq) + 2OH^-(aq) \rightarrow Ni(OH)_2(s) + Ni(NH_3)_6^{2+}(aq)$$

Precipitation of green nickel hydroxide. Upon the addition of excess NH_3, the precipitate dissolves to form the blue complex, $Ni(NH_3)_6^{2+}$.

CHAPTER
26

ORGANIC CHEMISTRY

No single thing abides; but all things flow
Fragment to fragment clings—the things thus grow
Until we know and name them. By degrees
They melt, and are no more the things we know.

TITUS LUCRETIUS CARUS (92–52 B.C.)
NO SINGLE THING ABIDES
(Translated by W. H. Mallock)

O rganic chemistry deals with the compounds of carbon, of which there are literally millions. More than 90% of all known compounds contain carbon atoms. There is a simple explanation for this remarkable fact. Carbon atoms bond to one another to a far greater extent than do atoms of any other element. Carbon atoms may link together to form chains or rings

The bonds may be single (one electron pair), double (two electron pairs), or triple (three electron pairs).

In this chapter, we will consider a variety of different organic compounds. They will have quite different structures and properties. However, all these substances have certain features in common. In particular

1. *Organic compounds are molecular rather than ionic.* Most of the compounds we will discuss consist of small, discrete molecules. Many of them are gases or liquids at room temperature. A few, generally of high molar mass, are solids.

2. *Each carbon atom forms a total of four covalent bonds.* This is illustrated by the structures written above. A particular carbon atom may form four single bonds, two single bonds and a double bond, or one single bond and a triple bond. One way or another, though, the bonds add up to four.

It was once thought that organic molecules could only be made by living organisms

915

3. *Carbon atoms may be bonded to each other or to other nonmetal atoms, most often hydrogen, a halogen, oxygen, or nitrogen.* In all the organic compounds that we will consider

— a hydrogen or halogen atom (F, Cl, Br, I) forms one covalent bond, —H, —X

— an oxygen atom forms two covalent bonds, —O— or =O

— a nitrogen atom forms three covalent bonds, $-\overset{|}{N}-$, $=N-$, or $\equiv N$

Throughout this chapter, we will represent molecules by *structural formulas*, which show all the bonds present. Thus we have

$$
\begin{array}{ccc}
\text{H} \quad \text{H} & \text{H} \quad \text{H} & \text{H} \qquad \text{H} \\
| \quad | & | \quad | & | \qquad | \\
\text{H}-\text{C}-\text{C}-\text{H} & \text{H}-\text{C}-\text{C}-\text{O}-\text{H} & \text{H}-\text{C}-\text{O}-\text{C}-\text{H} \\
| \quad | & | \quad | & | \qquad | \\
\text{H} \quad \text{H} & \text{H} \quad \text{H} & \text{H} \qquad \text{H} \\
\text{ethane} & \text{ethanol} & \text{dimethyl ether}
\end{array}
$$

To save space we often write condensed structural formulas such as

$$CH_3CH_3 \qquad\qquad CH_3CH_2OH \qquad\qquad CH_3-O-CH_3$$
$$\text{or } C_2H_5OH \qquad\qquad \text{or } (CH_3)_2O$$

In the first half of this chapter, we will consider the simplest type of organic compound, called a **hydrocarbon**. These compounds contain only two kinds of atoms: hydrogen and carbon. They can be classified into several different categories, including alkanes (Section 26.1), alkenes and alkynes (Section 26.2), and aromatics (Section 26.3). In Section 26.4, we will look at the structures of compounds containing halogen, oxygen, or nitrogen atoms in addition to carbon and hydrogen.

In the last two sections of this chapter, we will look at some general aspects of organic chemistry. One of these is the phenomenon of **isomerism** (Section 26.5), the existence of different compounds with the same molecular formula. Ethanol and dimethyl ether, whose structures are shown above, are isomers; they both have the molecular formula C_2H_6O. In Section 26.6, we will survey different types of organic reactions.

There are several kinds of isomerism

26.1
SATURATED HYDROCARBONS: ALKANES

One large and structurally simple class of hydrocarbons includes those substances in which all the carbon—carbon bonds are single bonds. These are called *saturated* hydrocarbons or **alkanes**. In the alkanes the carbon atoms are bonded to each other in chains, which may be long or short, straight or branched.

The simplest alkanes are methane (CH_4), ethane (C_2H_6), and propane (C_3H_8):

$$
\begin{array}{ccc}
\quad\;\; \text{H} & \quad\;\; \text{H}\;\; \text{H} & \quad\;\; \text{H}\;\; \text{H}\;\; \text{H}\\
\quad\;\; | & \quad\;\; |\;\;\; | & \quad\;\; |\;\;\; |\;\;\; |\\
\text{H}-\text{C}-\text{H} & \text{H}-\text{C}-\text{C}-\text{H} & \text{H}-\text{C}-\text{C}-\text{C}-\text{H}\\
\quad\;\; | & \quad\;\; |\;\;\; | & \quad\;\; |\;\;\; |\;\;\; |\\
\quad\;\; \text{H} & \quad\;\; \text{H}\;\; \text{H} & \quad\;\; \text{H}\;\; \text{H}\;\; \text{H}\\
\text{methane} & \text{ethane} & \text{propane}
\end{array}
$$

Around the carbon atoms in these molecules and indeed in any saturated hydrocarbon, there are four single bonds involving sp³ hybrid orbitals. As we would expect from VSEPR theory, these bonds are directed toward the corners of a regular tetrahedron. The bond angles are 109.5°, the tetrahedral angle. This means that in propane (C_3H_8) and in the higher alkanes, the carbon atoms are arranged in a "zigzag" pattern. The three-dimensional structures of CH_4, C_2H_6, and C_3H_8 are indicated in Figure 26.1.

The outer surfaces of these molecules contain mainly H atoms

Two different alkanes are known with the molecular formula C_4H_{10}. In one of these, called butane, the four carbon atoms are linked in a "straight chain." In the other, called 2-methylpropane, there is a "branched chain." The longest continuous chain in the molecule contains three carbon atoms; there is a CH_3 branch from the central carbon atom. The geometries

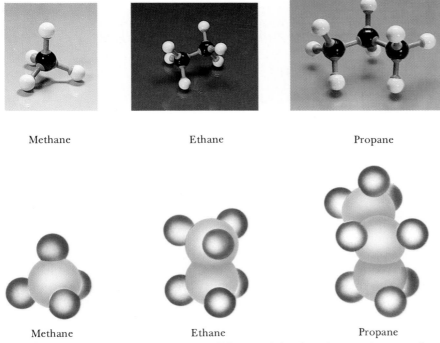

| Methane | Ethane | Propane |

Figure 26.1 Ball-and-stick and space-filling models of methane, ethane, and propane. The bond angles in all of these compounds are 109.5°, the tetrahedral angle. (Charles Steele)

Figure 26.2 Ball-and-stick and space-filling models of butane and 2-methylpropane, the isomers of C_4H_{10}. (Charles Steele)

Butane

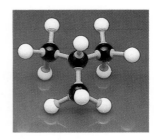

2-methylpropane

Butane

2-methylpropane

of these molecules are shown in Figure 26.2; the two-dimensional structural formulas are

```
       H                    H
       |                    |
  H — C — H           H — C — H
       |                    |
  H — C — H           H — C — CH₃
       |                    |
  H — C — H           H — C — H
       |                    |
  H — C — H                 H
       |
       H
    butane          2-methylpropane
```

Compounds having the same molecular formula but different molecular structures are called **structural isomers**. Butane and 2-methylpropane are referred to as structural isomers of C_4H_{10}. They are two distinct compounds with their own characteristic physical and chemical properties. Isomerism of this and other types is common among hydrocarbons and indeed among organic compounds in general. For small molecules, it is quite easy to identify the various isomers.

The left isomer has a 4-carbon chain. The one on the right has only 3-carbon chains

■ **EXAMPLE 26.1** _____

Draw structural formulas for the isomers of C_5H_{12}.

Solution

First sketch the straight-chain structure

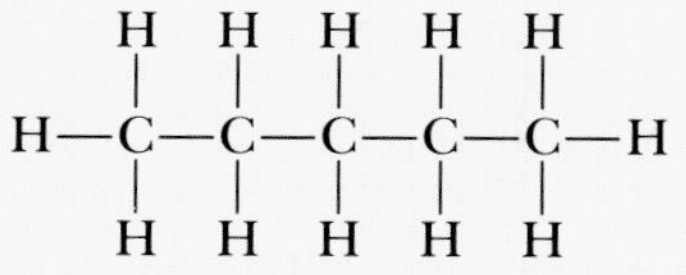

$$
\begin{array}{ccccc}
\text{H} & \text{H} & \text{H} & \text{H} & \text{H} \\
| & | & | & | & | \\
\text{H—C—C—C—C—C—H} \\
| & | & | & | & | \\
\text{H} & \text{H} & \text{H} & \text{H} & \text{H}
\end{array}
$$

I

Counting hydrogens, we find there are 12, as required. Since each carbon atom has four bonds and each hydrogen one, this is a correct Lewis structure.

Having found one correct structure, we need to determine all the *nonequivalent* alternate structures. Two such structures are shown below (only the carbon skeletons are shown):

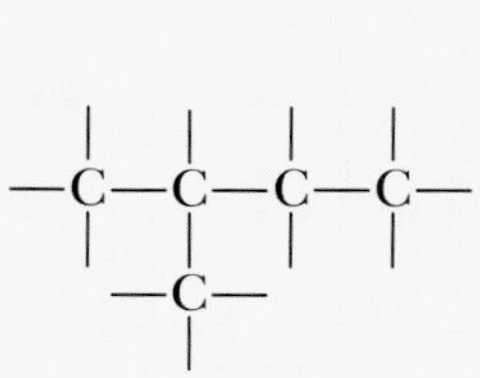

II 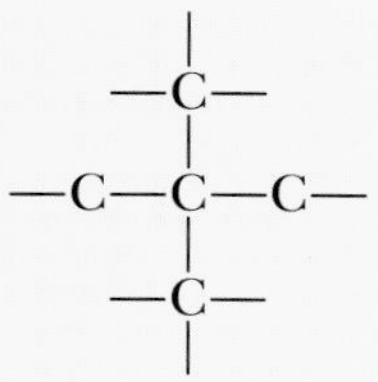   III

In structure II the longest continuous carbon chain consists of only four atoms, so that II and I are clearly different. Similarly, in structure III, there are only three carbon atoms in the longest carbon chain, so it differs from both I and II. Looking at structures I, II, and III you will notice another important difference that distinguishes one from another. In I, no carbon atom is attached to more than two other carbon atoms. In II there is one carbon atom that is bonded to three other carbons. Finally, in III, the carbon atom in the center is bonded to four other carbon atoms.

At this point, working only with pencil and paper, you might be tempted to draw other structures, such as

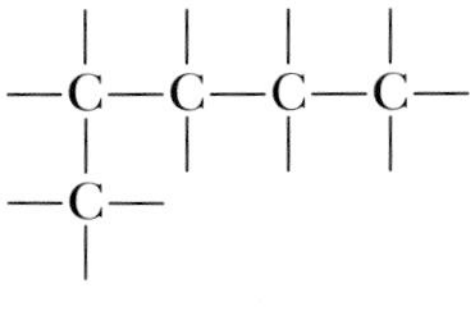

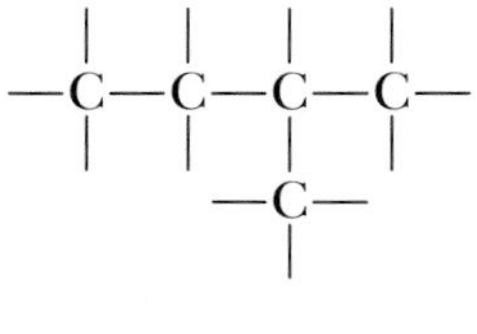

However, a few moments' reflection (or access to a molecular model kit) should convince you that these are in fact equivalent to structures written previously. In particular, the first one, like I, has a five-carbon chain in which no carbon atom is attached to more than two other carbons. The second structure, like II, has a four-carbon chain with one carbon atom bonded to three other carbons. Structures I, II, and III represent the three possible isomers of C_5H_{12}; there are no others.

There is free rotation around single bonds

Exercise
How many isomers are there of C_6H_{14}? Answer: Five (one has a six-carbon chain, two have five-carbon chains, and two have four-carbon chains).

These are all fuels

Figure 26.3 Petroleum refinery. (Standard Oil)

There are two major sources of alkanes: natural gas and petroleum. Natural gas consists largely of methane (80–90%) with smaller amounts of C_2H_6, C_3H_8, and C_4H_{10}. The higher alkanes are most often obtained from petroleum, a dark brown, viscous liquid dispersed through porous rock deposits. Distillation of petroleum (Fig. 26.3) gives a series of fractions of different boiling points. The most important of these is gasoline, discussed in the applied perspective at the end of this chapter.

NOMENCLATURE

As organic chemistry developed, it became apparent that some systematic way of naming compounds was needed. About 50 years ago, the International Union of Pure and Applied Chemistry (IUPAC) devised a system that could be used for all organic compounds. To illustrate this system, we will show how it works with alkanes.

For straight-chain alkanes such as

$$CH_3—CH_2—CH_3 \qquad CH_3—CH_2—CH_2—CH_3$$
propane   butane

the IUPAC name consists of a single word. These names, for up to eight carbon atoms, are listed in Table 26.1.

With alkanes containing a branched chain, such as

2-methylpropane

the name is more complex. A branched-chain alkane such as 2-methylpropane can be considered to be derived from a straight-chain alkane by

TABLE 26.1   Nomenclature of Alkanes

STRAIGHT-CHAIN ALKANES		ALKYL GROUPS	
Methane	CH_4	Methyl	$CH_3—$
Ethane	CH_3CH_3	Ethyl	$CH_3—CH_2—$
Propane	$CH_3CH_2CH_3$	Propyl	$CH_3—CH_2—CH_2—$
Butane	$CH_3(CH_2)_2CH_3$		
Pentane	$CH_3(CH_2)_3CH_3$	Isopropyl	$CH_3—\overset{H}{\underset{CH_3}{C}}—$
Hexane	$CH_3(CH_2)_4CH_3$		
Heptane	$CH_3(CH_2)_5CH_3$		
Octane	$CH_3(CH_2)_6CH_3$		
		Butyl	$CH_3—CH_2—CH_2—CH_2—$

replacing one or more hydrogen atoms by alkyl groups. The name consists of two parts:

— a **suffix** that identifies the parent straight-chain alkane. To find the suffix, count the number of carbon atoms in the longest continuous chain. For a three-carbon chain, the suffix is *propane;* for a four-carbon chain it is *butane,* and so on.

— a **prefix** that identifies the branching alkyl group (Table 26.1) and indicates by a number the carbon atom where branching occurs. In 2-methylpropane, referred to above, the methyl group is located at the second carbon from the end of the chain:

$$C_1-C_2-C_3$$
$$|$$

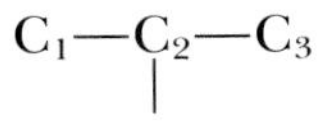

Following this system, the IUPAC names of the isomers of pentane are

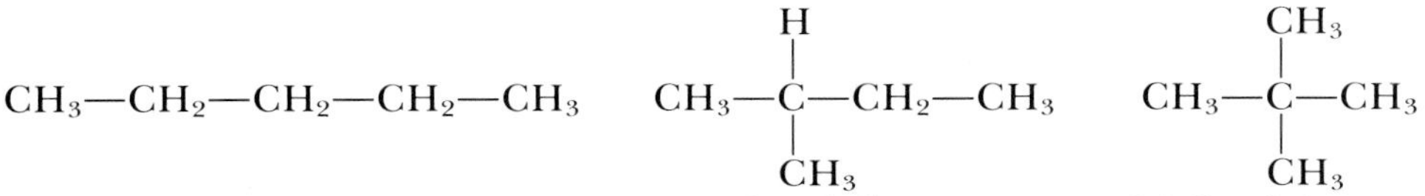

$$CH_3-CH_2-CH_2-CH_2-CH_3 \qquad CH_3-\overset{\displaystyle H}{\underset{\displaystyle CH_3}{C}}-CH_2-CH_3 \qquad CH_3-\overset{\displaystyle CH_3}{\underset{\displaystyle CH_3}{C}}-CH_3$$

$$\text{pentane} \qquad\qquad \text{2-methylbutane} \qquad\qquad \text{2,2-dimethylpropane}$$

Notice that

— if the same alkyl group is at two branches, the prefix *di-* is used (2,2-dimethylpropane). If there were three methyl branches, we would write trimethyl, and so on.

— the number in the name is made as small as possible. Thus, we write 2-methylbutane, numbering the chain from the left

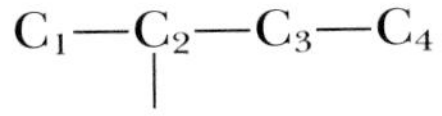

$$C_1-C_2-C_3-C_4$$
$$|$$

rather than from the right.

■ **EXAMPLE 26.2**

Assign IUPAC names to the following

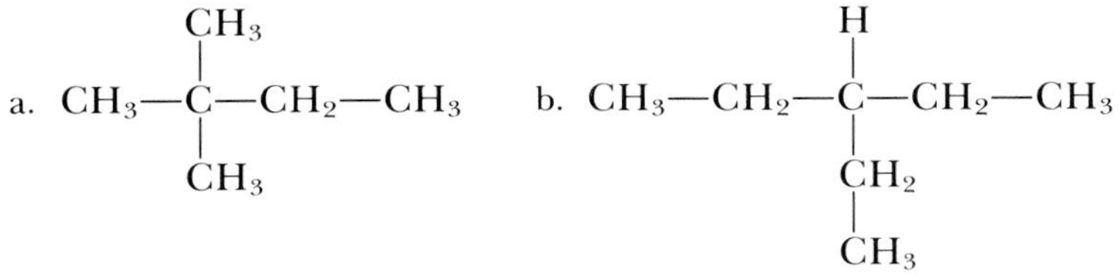

$$\text{a.} \quad CH_3-\overset{\displaystyle CH_3}{\underset{\displaystyle CH_3}{C}}-CH_2-CH_3 \qquad \text{b.} \quad CH_3-CH_2-\overset{\displaystyle H}{\underset{\displaystyle \underset{\displaystyle CH_3}{\overset{\displaystyle |}{CH_2}}}{C}}-CH_2-CH_3$$

Solution

a. The longest chain contains four carbon atoms (butane). There are two CH_3 (methyl) groups branching at the second carbon from the end of the chain (2). The correct name is

2,2-dimethylbutane

b. The longest chain, however you count it, contains five carbon atoms. There is a $CH_3—CH_2$ branch at the number three carbon, whichever end of the chain you start from. The IUPAC name is

3-ethylpentane

Exercise

How would you name the compound: $CH_3—CH_2—\overset{\displaystyle CH_3}{\underset{\displaystyle CH_3}{\overset{|}{\underset{|}{C}}}}—CH_3$? Answer: 2,2-dimethylbutane.

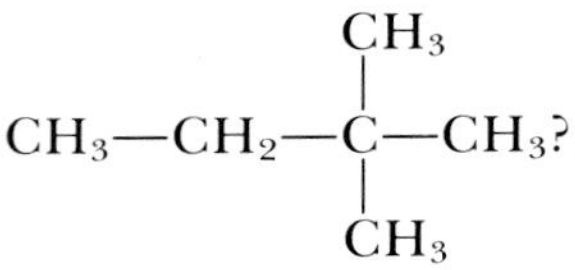

26.2
UNSATURATED HYDROCARBONS: ALKENES AND ALKYNES

In an unsaturated hydrocarbon, at least one of the carbon—carbon bonds in the molecule is a multiple bond. There are many types of unsaturated hydrocarbons, only two of which will be discussed here:

— *alkenes*, in which there is one carbon—carbon double bond in the molecule

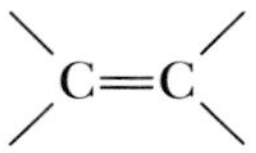

$$\overset{\diagdown}{\underset{\diagup}{C}}=\overset{\diagup}{\underset{\diagdown}{C}}$$

— *alkynes*, in which there is one carbon—carbon triple bond in the molecule

$$—C\equiv C—$$

ALKENES

The simplest alkene is ethene, C_2H_4 (common name, ethylene). Its structural formula is:

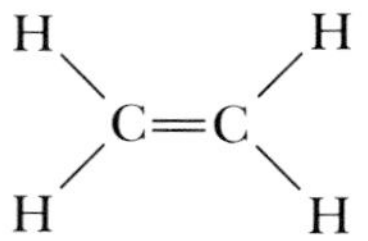

$$\underset{H}{\overset{H}{\diagdown}}C=C\underset{H}{\overset{H}{\diagup}}$$

ethene

You may recall that we discussed the bonding in ethene in Chapter 9. The double bond in ethene and other alkenes consists of a sigma bond and a pi bond. The ethene molecule is planar. There is no rotation about the double bond, since that would require "breaking" the pi bond. The bond angle in ethene is 120°, corresponding to sp² hybridization about each carbon atom. The geometries of ethene and the next member of the alkene series, C_3H_6, are shown in Figure 26.4.

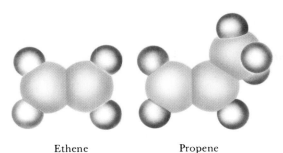

Figure 26.4 Space-filling models of ethene and propene. The ethene molecule is planar. Propene contains three carbon atoms, two of which are joined by a double bond.

Ethene   Propene

Ethene is produced in larger amounts than any other organic chemical, about 1.5×10^7 metric tons annually in the United States. It is made by heating ethane to about 700°C in the presence of a catalyst.

$$C_2H_6(g) \rightarrow C_2H_4(g) + H_2(g) \qquad (26.1)$$

Ethene is used to make a host of organic compounds; it is also the starting material for the preparation of polyethylene (Chapter 27). Since it is a plant hormone, ethene finds application in agriculture. It is used to ripen fruit which has been picked green to avoid spoilage in shipping. Exposure to ethene at very low concentrations produces the colors we associate with ripe bananas and oranges.

The names of alkenes are derived from those of the corresponding alkanes with the same number of carbon atoms per molecule. There are two modifications.

— the ending *-ane* is replaced by *-ene*

$$CH_3—CH_3 \qquad CH_2=CH_2$$
    ethane   ethene

— where necessary, a number is used to designate the double bonded carbon; the number is made as small as possible.

$$CH_2=CH—CH_2—CH_3 \qquad CH_3—CH=CH—CH_3$$
       1-butene   2-butene

$$CH_2=C—CH_2—CH_3 \qquad CH_3—C=C—CH_3$$
$$\quad\;\; | \qquad\qquad\qquad\qquad\qquad | \;\;\; |$$
$$\quad\;\; CH_3 \qquad\qquad\qquad\qquad\; H_3C \;\; H$$
   2-methyl-1-butene   2-methyl-2-butene

ALKYNES

The IUPAC names of alkynes are derived from those of the corresponding alkenes by replacing the suffix *-ene* with *-yne*. Thus we have

$$H—C\equiv C—H \qquad H—C\equiv C—CH_3$$
    ethyne   propyne

$$H—C\equiv C—CH_2—CH_3 \qquad CH_3—C\equiv C—CH_3$$
     1-butyne   2-butyne

The most common alkyne by far is the first member of the series, commonly called acetylene. Recall from Chapter 9 that the C_2H_2 molecule is linear, with 180° bond angles. The triple bond consists of a sigma bond and two pi bonds; each carbon atom is sp-hybridized. The geometries of acetylene and the next member of the series, C_3H_4, are shown in Figure 26.5.

Ethyne (Acetylene)   **Propyne (Methylacetylene)**

Figure 26.5   In acetylene and methylacetylene, two carbon atoms are linked by a triple bond. Both molecules contain four atoms on a straight line.

Acetylene can be made in the laboratory by allowing water to drop on calcium carbide.

$$CaC_2(s) + H_2O(l) \rightarrow C_2H_2(g) + CaO(s) \tag{26.2}$$

The gas produced in this way has a garlic-like odor due to traces of phosphine, PH_3. Commercially, acetylene is made by heating methane to about 1500°C in the absence of air:

$$2CH_4(g) \rightarrow C_2H_2(g) + 3H_2(g) \tag{26.3}$$

Thermodynamically, acetylene is unstable with respect to decomposition to the elements:

$$C_2H_2(g) \rightarrow 2C(s) + H_2(g); \qquad \Delta G^0 = -209.2 \text{ kJ at } 25°C \tag{26.4}$$

At high pressures, this reaction can occur explosively. For that reason, cylinders of acetylene do not contain the pure gas. Instead the cylinder is packed with an inert, porous material that holds a solution of acetylene gas in acetone.

You are probably most familiar with acetylene as a gaseous fuel used in welding and cutting metals. When mixed with pure oxygen in a torch, acetylene burns at temperatures above 2000°C. The heat comes from the reaction

$$C_2H_2(g) + \tfrac{5}{2}O_2(g) \rightarrow 2CO_2(g) + H_2O(l); \qquad \Delta H = -1300 \text{ kJ} \tag{26.5}$$

Figure 26.6 The headlamp used by miners and spelunkers uses the light given off by the combustion of acetylene. (Charles D. Winters, photographer; John Reynolds, spelunker)

The reaction gives off a brilliant white light, which served as a source of illumination in the headlights of early automobiles and is still used occasionally by spelunkers (Fig. 26.6).

26.3
AROMATIC HYDROCARBONS AND THEIR DERIVATIVES

Hydrocarbons of this type, sometimes referred to as *arenes,* can be considered to be derived from benzene, C_6H_6. Benzene is a clear, pleasant-smelling liquid (mp = 5°C, bp = 80°C) discovered by Michael Faraday in 1825. Its molecular structure was discussed in Chapter 9. In molecular orbital notation, we represent the molecule by the symbol

Here the sigma bond framework is outlined by the hexagon; the three delocalized pi bonds are represented by the circle in the center of the hexagon.

For many years, benzene and other aromatic hydrocarbons were obtained from coal tar, a product obtained by heating coal to 1000°C in the absence of air. Nowadays, benzene is obtained mostly from gasoline, where it is a relatively minor constituent. Annual production of benzene in the United States is about 5×10^3 metric tons. Nearly all of it is converted to a variety of different aromatic compounds.

Until quite recently benzene was widely used as a solvent, both commercially and in research and teaching laboratories. Its use for that purpose has been largely abandoned because of its toxicity. Chronic exposure to benzene vapor leads to various blood disorders and, in extreme cases, aplastic anemia and leukemia. It appears that the culprits here are oxidation products of the aromatic ring, formed in an attempt to solubilize benzene and thus eliminate it from the body.

Benzene is a component of gasoline

DERIVATIVES OF BENZENE

Monosubstituted benzenes are ordinarily named as derivatives of benzene.

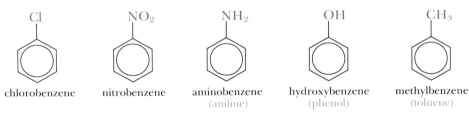

| chlorobenzene | nitrobenzene | aminobenzene (aniline) | hydroxybenzene (phenol) | methylbenzene (toluene) |

The last three compounds listed are always referred to by their common names, shown in red. Phenol was the first commercial antiseptic; its introduction into hospitals in the 1870s led to a dramatic decrease in deaths from postoperative infections. Its use for this purpose has long since been abandoned because phenol burns exposed tissue, but many modern antiseptics are phenol derivatives. Toluene has largely replaced benzene as a solvent because it is much less toxic. Oxidation of toluene in the body gives

The old name for phenol was carbolic acid

benzoic acid, C_6H_5COOH, which is readily eliminated and has none of the toxic properties of oxidation products of benzene.

When there are two groups attached to the benzene ring, three isomers are possible. These are designated by the prefixes *ortho-*, *meta-*, and *para-*, often abbreviated as *o-*, *m-*, and *p-*.

o-dichlorobenzene   m-dichlorobenzene   p-dichlorobenzene

Numbers can also be used; these three compounds may be referred to as 1,2-dichlorobenzene, 1,3-dichlorobenzene, and 1,4-dichlorobenzene, respectively. When three or more substituents are present, numbers become mandatory.

1,3,5-trinitrobenzene   2,4,6-trinitrotoluene (TNT)

Notice that the numbers used are as small as possible.

■ EXAMPLE 26.3

Name the following compounds as derivatives of phenol.

a.   b.

Solution

a.   *meta*-methylphenol.   The common name of this compound is *meta*-cresol. It is one of the constituents of creosote, a black oily liquid used as a wood preservative and fungicide.

b.   1,3,5-trinitrophenol,   common name picric acid, an explosive.

Exercise
Write the structure of 2,3,4-trichlorotoluene. Answer:

CH₃
Cl   Cl
Cl

CONDENSED RING STRUCTURES

In another type of aromatic hydrocarbon, two or more benzene rings are fused together. Naphthalene, the solid that gives mothballs their aromatic odor, is the simplest compound of this type. Fusion of three benzene rings gives two different isomers, anthracene and phenanthrene. Cholesterol and other steroids can be considered to be derivatives of phenanthrene.

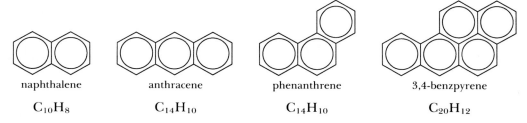

naphthalene   anthracene   phenanthrene   3,4-benzpyrene

$C_{10}H_8$ $C_{14}H_{10}$ $C_{14}H_{10}$ $C_{20}H_{12}$

   Certain compounds of this type are potent carcinogens. One of the most dangerous is 3,4-benzpyrene, which has been detected in cigarette smoke. It is believed to be a cause of lung cancer, to which heavy smokers are susceptible. Enzymatic oxidation of 3,4-benzpyrene gives a very reactive oxygen derivative which reacts with DNA (Chapter 27), causing undesirable mutations.

26.4
FUNCTIONAL GROUPS

Many organic molecules can be considered to be derived from hydrocarbons by substituting a functional group for a hydrogen atom. The functional group can be a nonmetal atom or small group of atoms that is bonded to carbon. Table 26.2, p. 928, lists the types of functional groups commonly found in organic compounds. Those shown in boldface will be discussed in this section.

ORGANIC HALOGEN COMPOUNDS

The simplest compounds of this type are derived from methane, CH_4, by replacing hydrogen atoms with halogens. Examples include

$$
\begin{array}{cccc}
& \text{Cl} & \text{Cl} & \text{Cl} & \text{F} \\
& | & | & | & | \\
\text{H}-\text{C}-\text{H} & \text{H}-\text{C}-\text{Cl} & \text{Cl}-\text{C}-\text{Cl} & \text{Cl}-\text{C}-\text{Cl} \\
& | & | & | & | \\
& \text{Cl} & \text{Cl} & \text{Cl} & \text{F}
\end{array}
$$

| dichloromethane | trichloromethane | tetrachloromethane | dichlorodifluoromethane |
| (methylene chloride) | (chloroform) | (carbon tetrachloride) | (Freon-12) |

The chloromethanes can be prepared by heating methane with chlorine. The compound CCl_2F_2, known commercially as Freon-12, is made by the fluorination of CCl_4.

$$CCl_4(l) \;+\; 2HF(g) \;\xrightarrow{\text{SbF}_5}\; CCl_2F_2(g) \;+\; 2HCl(g) \qquad (26.6)$$

These compounds have rather agreeable odors

Substances such as $CHCl_3$ and CCl_2F_2 belong to a class of organic compounds known as *alkyl halides*, in which one or more of the hydrogen atoms of an alkane are replaced by halogens. The IUPAC names of these compounds are obtained by considering them to be substituted alkanes; a number is used to designate the carbon atom to which the halogen is bonded. Thus we have

$$
\begin{array}{cc}
\text{H} & \text{H} \\
| & | \\
\text{CH}_3-\text{C}-\text{CH}_2-\text{CH}_3 & \text{CH}_3-\text{C}-\text{CH}_3 \\
| & | \\
\text{Cl} & \text{CH}_2\text{Cl}
\end{array}
$$

| 2-chlorobutane | 1-chloro-2-methylpropane |

TABLE 26.2   Common Functional Groups

GROUP	CLASS	EXAMPLE	NAME*
—F, —Cl, —Br, —I	**Halides**	C_2H_5Cl	Chloroethane (ethyl chloride)
—OH	**Alcohols**	C_2H_5OH	Ethanol (ethyl alcohol)
—O—	Ethers	CH_3-O-CH_3	Dimethyl ether
$\overset{\text{O}}{\overset{\|}{-\text{C}}}-\text{H}$	Aldehydes	$CH_3-\overset{\text{O}}{\overset{\|}{\text{C}}}-H$	Ethanal (acetaldehyde)
$\overset{\text{O}}{\overset{\|}{-\text{C}}}-$	Ketones	$CH_3-\overset{\text{O}}{\overset{\|}{\text{C}}}-CH_3$	Propanone (acetone)
$\overset{\text{O}}{\overset{\|}{-\text{C}}}-\text{OH}$	**Carboxylic acids**	$CH_3-\overset{\text{O}}{\overset{\|}{\text{C}}}-OH$	Ethanoic acid (acetic acid)
$\overset{\text{O}}{\overset{\|}{-\text{C}}}-\text{O}-$	**Esters**	$CH_3-\overset{\text{O}}{\overset{\|}{\text{C}}}-OCH_3$	Methyl acetate
—N—	**Amines**	CH_3NH_2	Aminomethane (methyl amine)
$\overset{\text{O H}}{\overset{\| \;\;\;\|}{-\text{C}-\text{N}}}-$	Amides	$CH_3-\overset{\text{O}}{\overset{\|}{\text{C}}}-NH_2$	Ethanamide (acetamide)

*Common names shown in red.

As usual, the numbers are made as small as possible. If more than one halogen is involved, they are named in alphabetical order. Thus CCl_2F_2 is named dichlorodifluoromethane rather than difluorodichloromethane.

■ **EXAMPLE 26.4** ⎯⎯⎯⎯⎯⎯⎯⎯⎯⎯⎯⎯⎯⎯⎯⎯⎯⎯⎯⎯
Draw the structures of

a. 1,2-diiodobutane.   b. 3-chloro-2-methylbutane.

Solution
a. First we draw the butane chain and number it.

$$-\overset{\displaystyle |}{\underset{\displaystyle |}{C_1}}-\overset{\displaystyle |}{\underset{\displaystyle |}{C_2}}-\overset{\displaystyle |}{\underset{\displaystyle |}{C_3}}-\overset{\displaystyle |}{\underset{\displaystyle |}{C_4}}-$$

Now we attach iodine atoms at the first and second carbon atoms.

$$I-\overset{\displaystyle |}{\underset{\displaystyle |}{C_1}}-\overset{\displaystyle |}{\underset{\displaystyle I}{C_2}}-\overset{\displaystyle |}{\underset{\displaystyle |}{C_3}}-\overset{\displaystyle |}{\underset{\displaystyle |}{C_4}}-$$

Finally we insert hydrogen atoms.

$$I-\overset{\displaystyle H}{\underset{\displaystyle H}{C}}-\overset{\displaystyle H}{\underset{\displaystyle I}{C}}-\overset{\displaystyle H}{\underset{\displaystyle H}{C}}-\overset{\displaystyle H}{\underset{\displaystyle H}{C}}-H$$

b. The structure of 2-methylbutane is:

$$CH_3-\overset{\displaystyle H}{\underset{\displaystyle CH_3}{C}}-CH_2-CH_3$$

We substitute a chlorine atom for hydrogen on the third carbon atom

$$CH_3-\overset{\displaystyle H}{\underset{\displaystyle H_3C}{C}}-\overset{\displaystyle H}{\underset{\displaystyle Cl}{C}}-CH_3$$

Exercise
Write the structure of 1-fluoro-3-iodopentane. Answer:

$$CH_3-CH_2-\overset{\displaystyle H}{\underset{\displaystyle I}{C}}-CH_2-CH_2F$$

ALCOHOLS

An alcohol can be considered to be derived from a hydrocarbon by replacing one or more H atoms by —OH groups. Alcohols are named by substituting the suffix -*ol* for the -*ane* suffix of the corresponding alkane. For the first four members of the series we have (common names shown in red):

$$CH_3OH \qquad CH_3\text{—}CH_2OH \qquad CH_3\text{—}CH_2\text{—}CH_2OH \qquad CH_3\text{—}\underset{\underset{OH}{|}}{CH}\text{—}CH_3$$

methanol   ethanol   1-propanol   2-propanol

(methyl alcohol)   (ethyl alcohol)   (propyl alcohol)   (isopropyl alcohol)

About 3×10^9 kg of methanol are produced annually in the United States from water gas, a mixture of carbon monoxide and hydrogen:

$$CO(g) + 2H_2(g) \xrightarrow[250 \text{ atm}, 350°C]{ZnO, Cr_2O_3} CH_3OH(g) \qquad (26.7)$$

It is also formed as a by-product when charcoal is made by heating wood in the absence of air. For this reason, methanol is sometimes called wood alcohol. Methanol is used in jet fuels and as a solvent, gasoline additive, and starting material for several industrial syntheses. It is poisonous, causing blindness or death. Although it is an intoxicant, it must not be used in alcoholic beverages.

Ethanol, the most common alcohol, can be prepared by the fermentation of grains or sugar (Fig. 26.7). A typical reaction is that of glucose, $C_6H_{12}O_6$:

$$C_6H_{12}O_6(aq) \rightarrow 2C_2H_5OH(aq) + 2CO_2(g) \qquad (26.8)$$

Figure 26.7 Wine (*far right*) is produced from the glucose in grape juice (*left*) by fermentation. The reaction is: $C_6H_{12}O_6(aq) \rightarrow 2C_2H_5OH(aq) + 2CO_2(g)$. The purpose of the bubble chamber in the fermentation jug (*center*) is to allow the carbon dioxide to escape but prevent oxygen from entering and oxidizing ethanol to acetic acid. (Charles D. Winters)

A mixture of 95% ethanol and 5% water can be separated from the fermentation products by distillation. *Absolute* (100%) *alcohol* can be made by adding quicklime, CaO, to 95% ethanol; the quicklime converts water to $Ca(OH)_2$, which can be filtered off. *Denatured alcohol* contains a toxic additive such as methanol or benzene which makes it unfit to drink. Laboratories often use denatured alcohol to avoid paying the high federal tax on 95% alcohol. Industrial ethanol is made from ethene by the reaction:

95% ethanol is a strong dehydrating agent. You should never drink it straight.

$$\underset{\substack{H \\ | \\ H}}{\overset{\substack{H \\ | \\ H}}{C}}=\underset{\substack{H \\ | \\ H}}{\overset{\substack{H \\ | \\ H}}{C} \quad + \quad H-O-H \quad \xrightarrow[\text{200 atm, 300°C}]{H_3PO_4} \quad H-\underset{\substack{| \\ H}}{\overset{\substack{H \\ |}}{C}}-\underset{\substack{| \\ H}}{\overset{\substack{H \\ |}}{C}}-OH \quad (26.9)}$$

Ethanol is the active ingredient of alcoholic beverages. There it is present in various concentrations (4% to 8% in beer, 12% to 15% in wine, and 40% or more in distilled spirits). The "proof" of an alcoholic beverage is twice the volume percent of ethanol. Thus, an 86 proof bourbon whiskey contains 43% ethanol. The taste of a distilled beverage depends upon the percentage of ethanol and, more generally, upon the nature and concentration of impurities. Vodka is virtually pure alcohol and nearly tasteless. Gin, made by treating alcohol with juniper berries, has very few impurities. The characteristic tastes of rum, scotch, or bourbon depend upon small amounts of organic compounds (impurities) that distil over with the alcohol or enter it during aging.

To keep people under the influence of alcohol from driving, police use breath analyzers to determine the percentage of alcohol in the blood. An alcohol content of more than 0.10% is often considered evidence of intoxication. A common method of breath analysis (Fig. 26.8) uses a redox reaction with potassium dichromate:

$$3C_2H_5OH(aq) + 2Cr_2O_7{}^{2-}(aq) + 16H^+(aq)$$
$$\text{orange}$$

$$\to 3CH_3COOH(aq) + 4Cr^{3+}(aq) + 11H_2O \quad (26.10)$$
$$\text{green}$$

Figure 26.8 A breathalyzer gives an estimate of the alcohol content of an individual's breath and, indirectly, of the percent alcohol in the blood. The ethanol reduces potassium dichromate (orange) to chromium(III) compounds, which are green. The extent of the color change is a measure of the alcohol content. (Charles D. Winters)

The extent of the color change, from orange to green, is an indication of the alcohol level in the breath and, presumably, in the bloodstream. The color change can be detected visually or spectrophotometrically.

Certain alcohols contain two or more —OH groups per molecule. Perhaps the most familiar compounds of this type are ethylene glycol and glycerol:

H   H   H   H   H
| | | | |
H—C—C—H   H—C—C—C—H
| | | | |
OH OH   OH OH OH
ethylene glycol   glycerol

Glycerol is used in the antifreeze for R.V. water systems

Ethylene glycol is the antifreeze for cars in the U.S.

Ethylene glycol is widely used as an antifreeze. Glycerol is formed as a by-product in making soaps and detergents. It is a viscous, sweet-tasting liquid, used in making drugs, antibiotics, plastics, and explosives (nitroglycerin).

CARBOXYLIC ACIDS

Carboxylic acids can be considered to be derived from hydrocarbons by replacing one or more H atoms by a carboxyl group, —C—OH, often

$$\overset{||}{\underset{O}{}}$$

abbreviated —COOH. The systematic names of these compounds are obtained by adding the suffix -oic to the stem of the name of the corresponding alkanes. In practice, these names are seldom used for the first two members of the series which are commonly referred to as formic acid and acetic acid.

H—C—OH   CH₃—C—OH
|| ||
O   O
methanoic acid   ethanoic acid
formic acid   acetic acid

Acetic acid is the active ingredient of vinegar, responsible for its sour taste. "White" vinegar is made by adding pure acetic acid to water, forming a 5% solution. "Brown" or "apple cider" vinegar is made from apple juice; the ethanol formed by fermentation is oxidized to acetic acid:

Homemade wine often contains some vinegar

$$C_2H_5OH(aq) + O_2(g) \rightarrow CH_3COOH(aq) + H_2O \qquad (26.11)$$

The most important chemical property of carboxylic acids is implied by their name; they act as weak acids in water solution.

$$RCOOH(aq) \rightleftharpoons H^+(aq) + RCOO^-(aq) \qquad (26.12)$$

Recalling our earlier discussion of the relative strengths of different oxy-acids (Chapter 23), you might expect that the more electronegative the R group, the stronger the acid should be. This is indeed the case, as you can see by comparing the dissociation constants of halogen-containing acids (Table 26.3). The value of K_a increases with the electronegativity of the halogen atom and with the number of these atoms in the molecule. Tri-

TABLE 26.3   Strengths of Carboxylic Acids

ACID	FORMULA	K_a	pK_a
Acetic acid	CH_3COOH	1.8×10^{-5}	4.74
Iodoacetic acid	CH_2ICOOH	6.6×10^{-4}	3.18
Bromoacetic acid	$CH_2BrCOOH$	1.2×10^{-3}	2.92
Chloroacetic acid	$CH_2ClCOOH$	1.5×10^{-3}	2.82
Fluoroacetic acid	CH_2FCOOH	2.6×10^{-3}	2.59
Dichloroacetic acid	$CHCl_2COOH$	5.0×10^{-2}	1.30
Trichloroacetic acid	CCl_3COOH	2.0×10^{-1}	0.70

chloroacetic acid is one of the strongest organic acids, used by dentists to cauterize gums and by doctors to treat warts.

Treatment of a carboxylic acid with the strong base NaOH forms the sodium salt of the acid. With acetic acid, the acid-base reaction is

$$CH_3-\overset{\underset{\|}{O}}{C}-OH(aq) + OH^-(aq) \rightarrow CH_3-\overset{\underset{\|}{O}}{C}-O^-(aq) + H_2O \qquad (26.13)$$

Evaporation gives the salt sodium acetate, which contains Na^+ and CH_3COO^- ions. Many of the salts of carboxylic acids have important uses. Sodium and calcium propionate (Na^+ or Ca^{2+} ions, $CH_3CH_2COO^-$ ions) are added to bread, cake, and cheese to inhibit the growth of mold. *Soaps* are sodium salts of long-chain carboxylic acids such as stearic acid:

$$CH_3(CH_2)_{16}-\overset{\underset{\|}{O}}{C}-OH \qquad\qquad Na^+, CH_3(CH_2)_{16}-\overset{\underset{\|}{O}}{C}-O^-$$

stearic acid   sodium stearate, a soap

The cleaning action of soap reflects the nature of the long-chain carboxylate anion (Fig. 26.9). The long hydrocarbon group is a good solvent for greases and oils, while the ionic COO^- group gives high water solubility.

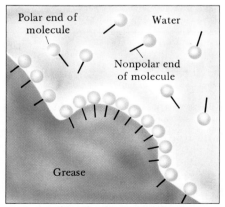

Soap molecules on surface of grease

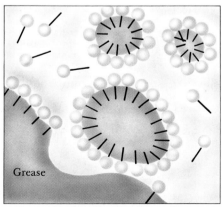

Grease droplets dispersed in wash water

Figure 26.9 The long hydrocarbon chain of a soap ion is a good solvent for grease. The polar end of the ion is soluble in water. In the washing process, the grease is dispersed into the water as droplets surrounded by soap ions.

ESTERS

The reaction between a carboxylic acid and an alcohol forms an ester, containing the functional group $-\overset{\displaystyle O}{\underset{\displaystyle ||}{C}}-O-$, often abbreviated —COO—.

The reaction between acetic acid and methyl alcohol is typical:

$$CH_3-\overset{||}{\underset{O}{C}}-OH(aq) + HO-CH_3(aq) \xrightarrow{H^+} CH_3-\overset{||}{\underset{O}{C}}-O-CH_3(aq) + H_2O \quad (26.14)$$

acetic acid   methyl alcohol   methyl acetate

Most esters have a pleasant odor; they are commonly found in natural and synthetic fragrances (Table 26.4). Esters are also used as industrial solvents and starting materials for plastics (Chapter 27).

TABLE 26.4 Properties of Esters

ESTER	STRUCTURE	ODOR, FLAVOR
Ethyl formate	$CH_3CH_2-O-\overset{||}{\underset{O}{C}}-H$	rum
Isobutyl formate	$(CH_3)_2-CHCH_2-O-\overset{||}{\underset{O}{C}}-H$	raspberry
Methyl butyrate	$CH_3-O-\overset{||}{\underset{O}{C}}-(CH_2)_2CH_3$	apple
Ethyl butyrate	$CH_3CH_2-O-\overset{||}{\underset{O}{C}}-(CH_2)_2CH_3$	pineapple
Isopentyl acetate	$(CH_3)_2-CH-(CH_2)_2-O-\overset{||}{\underset{O}{C}}-CH_3$	banana
Octyl acetate	$CH_3-(CH_2)_7-O-\overset{||}{\underset{O}{C}}-CH_3$	orange
Pentyl propionate	$CH_3-(CH_2)_4-O-\overset{||}{\underset{O}{C}}-CH_2CH_3$	apricot

■ **EXAMPLE 26.5** _____

Give the structural formula of

a. the three-carbon alcohol with an —OH group at the end of the chain.
b. the three-carbon carboxylic acid.
c. the ester formed when these two compounds react.

Solution

a. $CH_3CH_2CH_2-OH$ b. $CH_3CH_2-\overset{||}{\underset{O}{C}}-OH$

c. 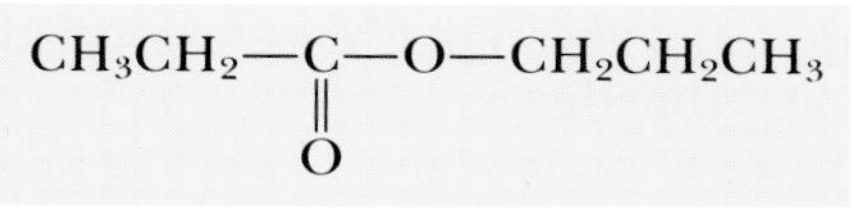 CH_3CH_2—$\overset{\displaystyle \underset{\displaystyle O}{\|}}{C}$—O—$CH_2CH_2CH_3$

Exercise

Which one of the following is *neither* an alcohol, a carboxylic acid, nor an ester? CH_3CH_2—OH; H—$\overset{\displaystyle \underset{\displaystyle O}{\|}}{C}$—OH; CH_3—O—CH_3; H—$\overset{\displaystyle \underset{\displaystyle O}{\|}}{C}$—O—$CH_3$.

Answer: CH_3—O—CH_3.

Animal fats and vegetable oils are esters of long-chain carboxylic acids with glycerol. A typical fat molecule might have the structure

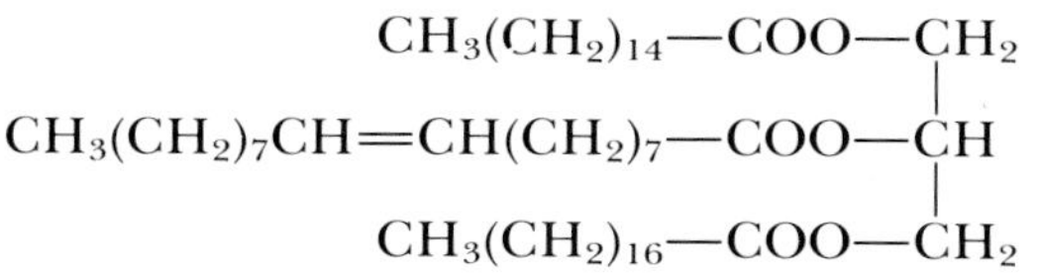

$$CH_3(CH_2)_{14}\text{—COO—CH}_2$$
$$CH_3(CH_2)_7CH\text{=}CH(CH_2)_7\text{—COO—CH}$$
$$CH_3(CH_2)_{16}\text{—COO—CH}_2$$

The three carboxylic acids from which this fat is derived are

— palmitic acid, $CH_3(CH_2)_{14}COOH$.
— oleic acid, $CH_3(CH_2)_7CH\text{=}CH(CH_2)_7COOH$.
— stearic acid, $CH_3(CH_2)_{16}COOH$.

These acids are typical of those found in fats. Some "fatty acids" are *saturated*, such as palmitic and stearic acid; the hydrocarbon chain contains no multiple bonds. Others, such as oleic acid, are *unsaturated;* there are one or more carbon–carbon multiple bonds in the molecule.

Fats are sometimes called triglycerides. They are also esters

AMINES

Amines are perhaps visualized most simply as derivatives of ammonia, NH_3,

$$H\text{—}\overset{\displaystyle \overset{\displaystyle H}{|}}{N}\text{—}H$$

in which one or more of the H atoms is replaced by hydrocarbon groups. We distinguish between

— *primary* amines, in which one H atom of NH_3 is replaced.
— *secondary* amines, in which two H atoms of NH_3 are replaced.
— *tertiary* amines, in which all three H atoms of NH_3 are replaced.

Examples include

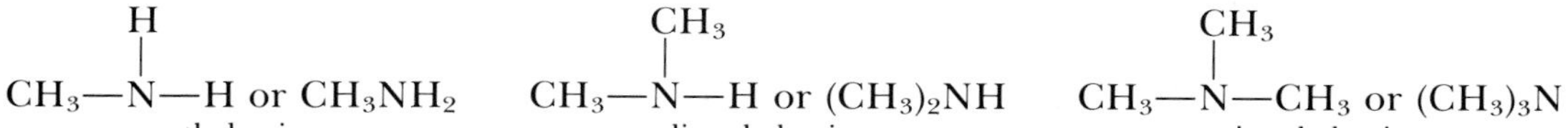

CH_3—$\overset{\displaystyle \overset{\displaystyle H}{|}}{N}$—H or CH_3NH_2 CH_3—$\overset{\displaystyle \overset{\displaystyle CH_3}{|}}{N}$—H or $(CH_3)_2NH$ CH_3—$\overset{\displaystyle \overset{\displaystyle CH_3}{|}}{N}$—$CH_3$ or $(CH_3)_3N$
methylamine (primary)   dimethylamine (secondary)   trimethylamine (tertiary)

These compounds and other amines of low molar mass are volatile, water-soluble, weak bases. For example, trimethylamine, $(CH_3)_3N$, is a gas at room temperature (bp = 3°C) with a solubility of about 40 g/100 g water. It is a somewhat stronger base than ammonia:

$$(CH_3)_3N(aq) + H_2O \rightleftharpoons (CH_3)_3NH^+(aq) + OH^-(aq) \qquad (26.15)$$

$$K_b = 5.9 \times 10^{-5}$$

Its odor is distinctly unpleasant, somewhere between that of ammonia and spoiled fish.

Alkaloids such as caffeine, nicotine, morphine, and coniine (Fig. 26.10) form an important class of naturally occurring amines. Caffeine occurs in tea leaves, coffee beans, and cola nuts used to make cola soft drinks. Nicotine is the most abundant alkaloid in tobacco. The painkillers morphine and codeine are obtained from unripe opium poppy seed pods. They are narcotics that, with repeated usage, cause addiction. Coniine, extracted from hemlock, is the alkaloid that killed Socrates. He was sentenced to drink a brew made from hemlock because of his unconventional teaching methods and religious practices.

Figure 26.10 Molecular structures of four alkaloids.

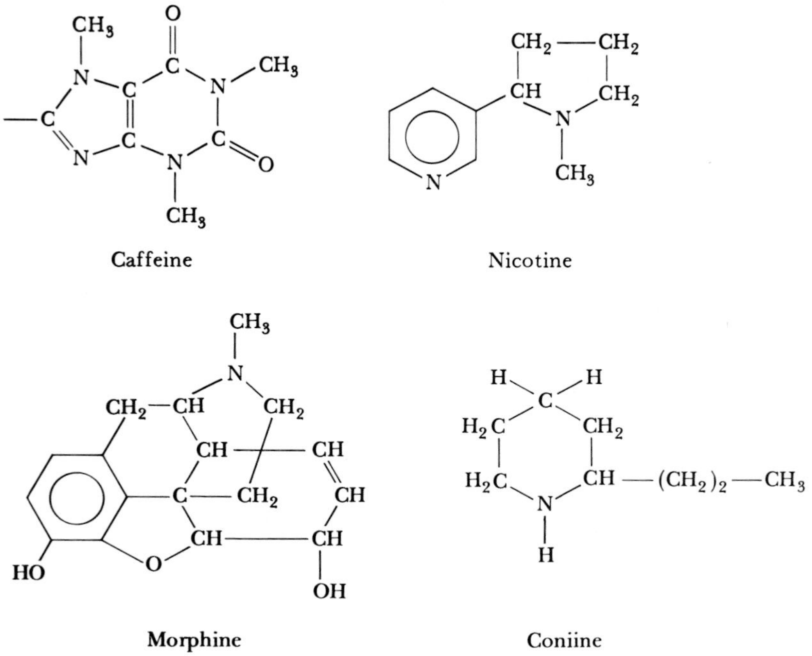

Caffeine

Nicotine

Morphine

Coniine

■ **EXAMPLE 26.6**

Classify each of the following compounds as an alcohol, carboxylic acid, ester, or amine. More than one functional group may be present.

a. pyridine

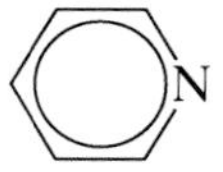

b. novocain   H$_2$N—⟨benzene ring⟩—C(=O)—O—(CH$_2$)$_2$—N(C$_2$H$_5$)$_2$

c. aspirin   ⟨benzene ring⟩ with O—C(=O)—CH$_3$ and COOH

Solution

a. Pyridine is a tertiary amine; there are no hydrogen atoms bonded to nitrogen.

b. There are two amine groups and an ester group in novocain.

c. Aspirin has a carboxyl group and an ester group.

Exercise
Of the amine groups shown in Figure 26.10, how many are primary? secondary? tertiary? Answer: 0, 1, 7.

26.5
ISOMERISM IN ORGANIC COMPOUNDS

Isomers are distinctly different compounds, with different properties, that have the same molecular formula. In Section 26.1, we considered structural isomers of alkanes. You will recall that butane and 2-methylpropane have the same molecular formula, C_4H_{10}, but different structural formulas. In these, as in all structural isomers, the order in which the atoms are bonded to each other differs.

Structural isomerism is common among all types of organic compounds. Consider the following examples:

1. The three structural isomers of the alkene C_4H_8:

$$CH_3, H \quad C=C \quad CH_3, H$$
2-butene   2-methylpropene   1-butene

2. The two structural isomers of the three-carbon alcohol C_3H_7OH:

$$CH_3CH_2—C(H)(H)—OH \qquad CH_3—C(H)(OH)—CH_3$$
1-propanol   2-propanol

3. The two structural isomers dimethyl ether and ethyl alcohol, both of which have the molecular formula C_2H_6O:

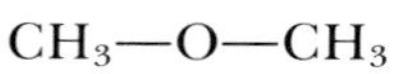

$$CH_3-O-CH_3 \qquad CH_3-\overset{\overset{\displaystyle H}{|}}{\underset{\underset{\displaystyle H}{|}}{C}}-OH$$

dimethyl ether 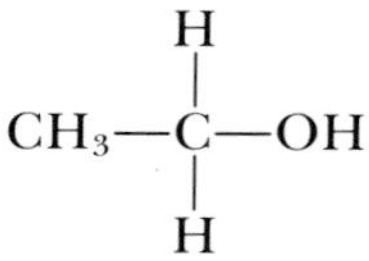   ethanol

■ **EXAMPLE 26.7** _____

Consider the molecule $C_3H_6Cl_2$, which is derived from propane by substituting two Cl atoms for H atoms. Draw the structural isomers of $C_3H_6Cl_2$.

Solution

There are four isomers. In one, both Cl atoms are bonded to a carbon atom at the end of the chain. In another, both Cl atoms are bonded to the central carbon atom:

It can be tricky finding all the isomers

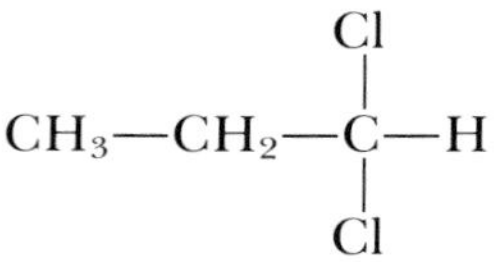

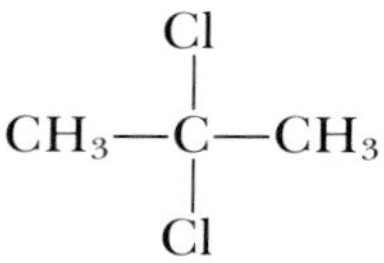

$$CH_3-CH_2-\overset{\overset{\displaystyle Cl}{|}}{\underset{\underset{\displaystyle Cl}{|}}{C}}-H \qquad CH_3-\overset{\overset{\displaystyle Cl}{|}}{\underset{\underset{\displaystyle Cl}{|}}{C}}-CH_3$$

In the other two isomers, the Cl atoms are bonded to different carbons:

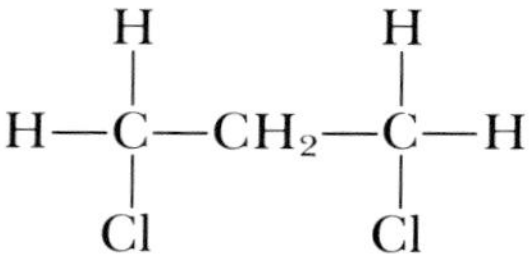

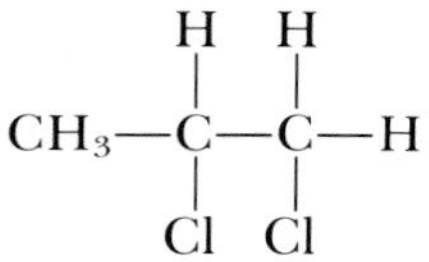

$$H-\overset{\overset{\displaystyle H}{|}}{\underset{\underset{\displaystyle Cl}{|}}{C}}-CH_2-\overset{\overset{\displaystyle H}{|}}{\underset{\underset{\displaystyle Cl}{|}}{C}}-H \qquad CH_3-\overset{\overset{\displaystyle H}{|}}{\underset{\underset{\displaystyle Cl}{|}}{C}}-\overset{\overset{\displaystyle H}{|}}{\underset{\underset{\displaystyle Cl}{|}}{C}}-H$$

Exercise

How many structural isomers are there of $C_3H_5Cl_3$? Answer: Five.

GEOMETRIC ISOMERS

As we have seen, there are three structural isomers of the alkene C_4H_8. You may be surprised to learn that there are actually *four* different alkenes with this molecular formula. The "extra" compound arises because of a phenomenon called **geometric isomerism**. There are two different 2-butenes:

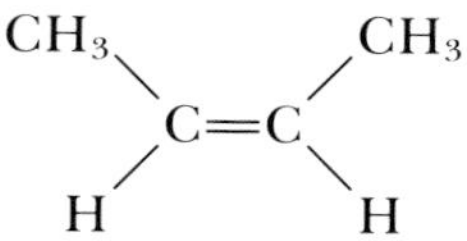

cis-2-butene   *trans*-2-butene
bp = 4°C   bp = 1°C

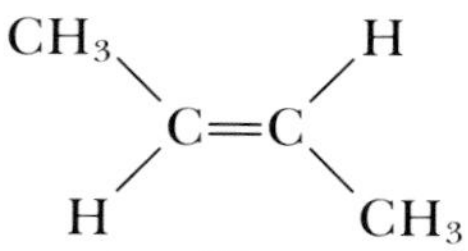

In the *cis* isomer, the two CH_3 groups (or the two H atoms) are as close to one another as possible. In the *trans* isomer, the two identical groups are farther apart. The two forms exist because there is no free rotation about the carbon—carbon double bond. The situation here is analogous to that with *cis-trans* isomers of square planar complexes (Ch. 16). In both cases, the difference in geometry is responsible for isomerism; the atoms are bonded to each other in the same way.

  Geometric, or *cis-trans*, isomerism is common among alkenes. Indeed, it occurs with all alkenes *except those in which two identical atoms or groups are attached to one of the double-bonded carbons.* Thus, although 2-butene has *cis* and *trans* isomers, 1-butene and 2-methylpropene (p. 937) do not.

If there were free rotation, these isomers would not exist

■ EXAMPLE 26.8

Draw all the isomers of the molecule $C_2H_2Cl_2$ in which two of the H atoms of ethylene are replaced by Cl atoms.

Name the compound:

H   Pa
   C=C
Ma   H

Ans.: transparent

Solution

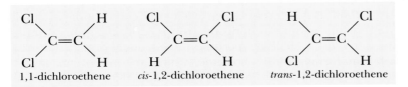

1,1-dichloroethene *cis*-1,2-dichloroethene *trans*-1,2-dichloroethene

Notice that 1,1-dichloroethene, in which the two Cl atoms are bonded to the same carbon, does not show geometric isomerism.

Exercise
Would you expect to find geometric isomers of

CH₃   Br?
  C=C
Cl   H

Answer: Yes. The other isomer is:

CH₃   H
  C=C
Cl   Br

  Cis- and *trans-* isomers differ from one another in their physical, and, to a lesser extent, their chemical properties. They may also differ in their physiological behavior. For example, the compound *cis*-9-tricosene

CH₃—(CH₂)₇   (CH₂)₁₂—CH₃
    C=C
   H   H

is a sex attractant secreted by the female housefly. The *trans* isomer is totally ineffective in this area. One way to control houseflies and other insects is to synthesize sex attractants and use them to trap male insects. A chemist working on such a project has to be sure the product is the proper isomer.

Figure 26.11 The two optical isomers of CHClBrI are mirror images of each other.

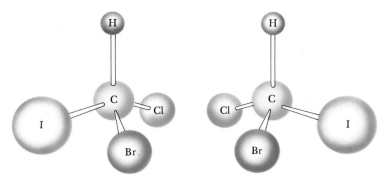

OPTICAL ISOMERISM

Optical isomerism arises because of the tetrahedral nature of the bonding around a carbon atom. It occurs when at least one carbon atom in a molecule is bonded to four different atoms or groups. Consider, for example, the methane derivative CHClBrI. As you can see from Figure 26.11, there are two different forms of this molecule, which are mirror images of one another. The mirror images are not superimposable; that is, you cannot place one molecule over the other so that identical groups are touching. In this sense, the two isomers resemble right- and left-hand gloves. Optical isomers differ from geometric isomers in that the latter are not mirror images of one another.

A carbon atom with four different atoms or groups attached to it is referred to as a **chiral center**. A molecule containing such a carbon atom shows optical isomerism. It exists in two different forms that are nonsuperimposable mirror images. These forms are referred to as optical isomers or *enantiomers*. Molecules may contain more than one chiral center, in which case there can be more than two enantiomers.

■ **EXAMPLE 26.9** _____

In the following structural formulas, locate each chiral carbon atom:

$$
\text{a. } CH_3 - \underset{\underset{Cl}{|}}{\overset{\overset{Cl}{|}}{C}} - \underset{\underset{OH}{|}}{\overset{\overset{H}{|}}{C}} - CH_3 \qquad
\text{b. } HO - \underset{\underset{O}{\|}}{C} - \underset{\underset{NH_2}{|}}{\overset{\overset{H}{|}}{C}} - \underset{\underset{H}{|}}{\overset{\overset{H}{|}}{C}} - OH
$$

$$
\text{c. } Cl - \underset{\underset{OH}{|}}{\overset{\overset{H}{|}}{C}} - \underset{\underset{OH}{|}}{\overset{\overset{CH_3}{|}}{C}} - Cl
$$

Solution

A chiral carbon has four different atoms or groups attached to it. In (a), the C atom bonded to OH is chiral. In (b), the C atom bonded to NH₂ is chiral. In (c), both C atoms are chiral.

Exercise
Consider the glycerol molecule, whose structural formula is shown on p. 932. Would you expect glycerol to show optical isomerism? Answer: No.

The term "optical isomerism" comes from the effect that enantiomers have upon plane-polarized light, such as that produced by a Polaroid lens (Fig. 26.12). When this light is passed through a solution containing a single enantiomer, the plane is rotated from its original position. One isomer rotates it to the right (clockwise) and the other to the left (counterclockwise). If both isomers are present in equal amounts, we obtain what is known as a racemic mixture. In this case, the two rotations offset each other and there is no effect on plane-polarized light.

Enantiomers ordinarily resemble each other closely in their physical and chemical properties. For example, the two forms of lactic acid have the same melting point (52°C), density (1.25 g/cm^3), and acid dissociation constant ($K_a = 1.4 \times 10^{-4}$):

$$(I) \quad HO-\underset{\underset{CH_3}{|}}{\overset{\overset{COOH}{|}}{C}}-H \qquad\qquad (II) \quad H-\underset{\underset{CH_3}{|}}{\overset{\overset{COOH}{|}}{C}}-OH$$

enantiomers of lactic acid

It seems reasonable that these molecules would have similar properties

On the other hand, enantiomers frequently differ in their physiological activity. This was discovered by Louis Pasteur, the father of modern biochemistry. Working with a mixture of the optical isomers of lactic acid, he found that mold growth occurred only with enantiomer II. Apparently, the mold was unable to metabolize enantiomer I. A more modern example

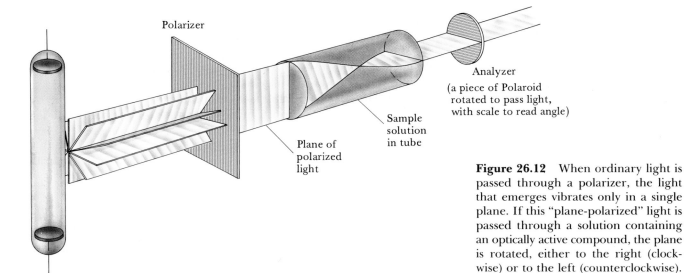

Polarizer

Analyzer
(a piece of Polaroid rotated to pass light, with scale to read angle)

Plane of polarized light

Sample solution in tube

Figure 26.12 When ordinary light is passed through a polarizer, the light that emerges vibrates only in a single plane. If this "plane-polarized" light is passed through a solution containing an optically active compound, the plane is rotated, either to the right (clockwise) or to the left (counterclockwise).

of this type involves amphetamine, often used illicitly as an "upper" or "pep pill." Amphetamine consists of two enantiomers:

$$
\text{(I)} \quad \bigcirc - CH_2 - \underset{\underset{CH_3}{|}}{\overset{\overset{H}{|}}{C}} - NH_2 \qquad\qquad \text{(II)} \quad NH_2 - \underset{\underset{CH_3}{|}}{\overset{\overset{H}{|}}{C}} - CH_2 - \bigcirc
$$

enantiomers of amphetamine

Enantiomer I, called Dexedrine, is by far the stronger stimulant. It is from two to four times as active as Benzedrine, the racemic mixture of the two isomers.

26.6
ORGANIC REACTIONS

The reactions that take place in organic chemistry differ in important ways from those discussed earlier in this text. In particular, the species taking part in organic reactions are most often molecules rather than ions. Again, most organic reactions take place between pure substances or in nonpolar solvents as opposed to water. Generally speaking, organic reactions occur more slowly than inorganic ones. In this section we will look at four general types of organic reactions known as addition, elimination, condensation, and substitution.

Ions tend to react fast

ADDITION REACTIONS

Double and triple bonds are much more reactive than single bonds

In an addition reaction, a small molecule (e.g., H_2, Cl_2, HCl, H_2O) adds across a double or triple bond. A simple example is the addition of hydrogen gas to ethene in the presence of a nickel catalyst.

$$
H - \underset{\underset{H}{|}}{\overset{}{C}} = \underset{\underset{H}{|}}{\overset{}{C}} - H \;+\; H - H \;\rightarrow\; H - \underset{\underset{H}{|}}{\overset{\overset{H}{|}}{C}} - \underset{\underset{H}{|}}{\overset{\overset{H}{|}}{C}} - H \tag{26.16}
$$

With ethyne (acetylene), one or two moles of H_2 may add, depending upon the catalyst and the conditions used.

As we saw earlier (Section 26.4), ethanol can be made from ethene by "adding" water. A similar reaction occurs directly with the hydrogen halides (HF, HCl, HBr, HI).

$$
H - \underset{\underset{H}{|}}{\overset{}{C}} = \underset{\underset{H}{|}}{\overset{}{C}} - H \;+\; H - Cl \;\rightarrow\; H - \underset{\underset{H}{|}}{\overset{\overset{H}{|}}{C}} - \underset{\underset{Cl}{|}}{\overset{\overset{H}{|}}{C}} - H
$$

ethene   chloroethane

With ethene only one product is possible; it does not matter to which carbon atom the chlorine is bonded. When propene, C_3H_6, reacts with HCl, two products are possible:

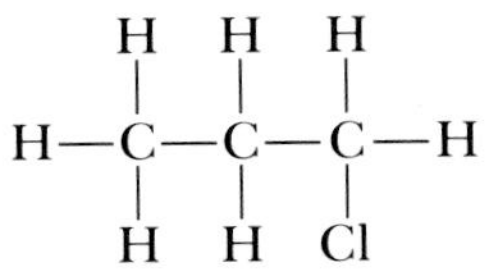

1-chloropropane 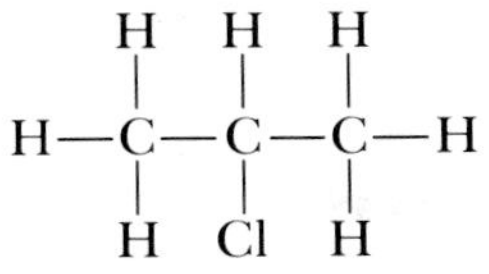   2-chloropropane

In practice, the product is almost exclusively the second isomer, 2-chloropropane. This is an example of the rule formulated by the Russian chemist Vladimir Markovnikov more than a century ago:

When a polar molecule (e.g., HX) adds to a multiple bond, the electropositive part of the molecule (e.g., H) bonds to the carbon atom that holds the largest number of hydrogen atoms. This rule is further illustrated by the following examples:

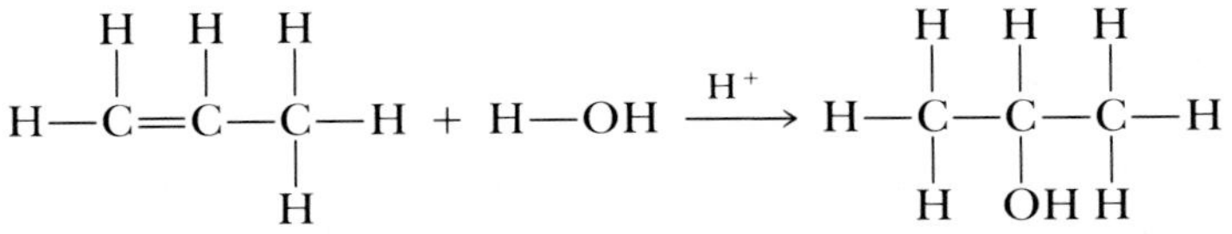

$$\text{propene} + \text{H—OH} \xrightarrow{\text{H}^+} \text{2-propanol} \qquad (26.17)$$

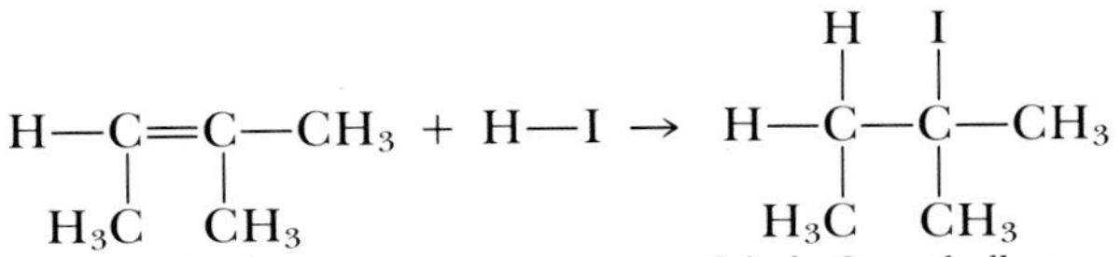

2-methyl-2-butene   + H—I → 2-iodo-2-methylbutane   (26.18)

■ **EXAMPLE 26.10**

Methyl methacrylate, the starting material for the preparation of the polymers Lucite and Plexiglas, has the structure

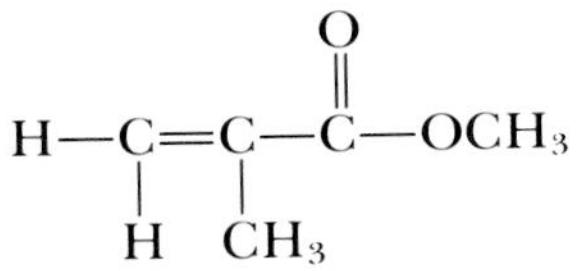

Show the structures of the products formed when H_2, H_2O, and HCl add to methyl methacrylate, assuming Markovnikov's rule is followed.

Solution

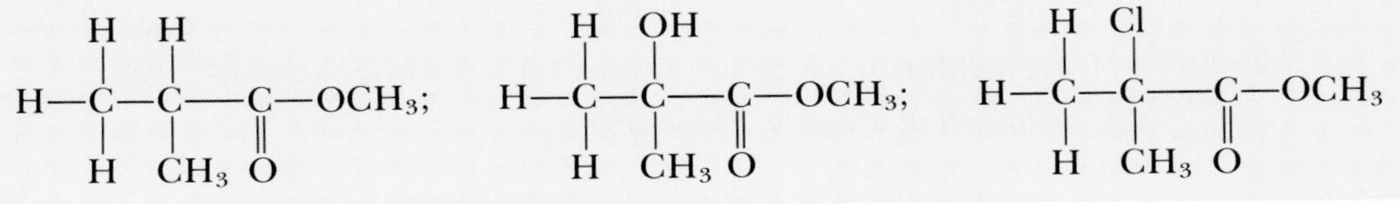

Exercise

Of these products, which ones can show structural isomerism? geometric isomerism? optical isomerism? Answer: all; none; all.

ELIMINATION AND CONDENSATION REACTIONS

An elimination reaction is, in a sense, the reverse of an addition reaction. It involves the elimination of two groups from adjacent carbon atoms, converting a saturated molecule into one that is unsaturated. An example is the dehydration of ethanol, which occurs when it is heated with sulfuric acid:

The H_2SO_4 absorbs the H_2O and drives the reaction to the right

$$H-\underset{\underset{\text{H}}{|}}{\overset{\overset{\text{H}}{|}}{C}}-\underset{\underset{\text{OH}}{|}}{\overset{\overset{\text{H}}{|}}{C}}-H \rightarrow H-\overset{\overset{\text{H}}{|}}{C}=\overset{\overset{\text{H}}{|}}{C}-H + H-OH \qquad (26.19)$$

$$\quad\text{ethanol} \qquad\qquad\qquad \text{ethene}$$

This is, of course, the reverse of the addition reaction (Equation 26.9) by which ethanol is formed.

A somewhat similar reaction, referred to as condensation, occurs when two molecules combine by splitting out a small molecule such as H_2O. The molecules that combine may be the same or different:

$$CH_3-OH + HO-CH_3 \rightarrow CH_3-O-CH_3 + H-OH \qquad (26.20)$$

$$\quad\text{methanol} \qquad \text{methanol} \qquad\quad \text{dimethyl ether}$$

$$CH_3-OH + HO-\underset{\underset{\text{O}}{\|}}{C}-CH_3 \rightarrow CH_3-O-\underset{\underset{\text{O}}{\|}}{C}-CH_3 + H-OH$$

$$\quad\text{methanol} \qquad \text{acetic acid} \qquad\qquad \text{methyl acetate} \qquad (26.21)$$

Reaction 26.20 offers a general way of preparing ethers from alcohols. The second reaction is that of esterification; compare Equation 26.14.

SUBSTITUTION REACTIONS

A substitution reaction is one in which an atom or group of atoms in a molecule is replaced by a different atom or group. Substitution reactions are very common in organic chemistry; we will consider only three examples.

1. *Chlorination of alkanes*

It's not easy to control the number of Cl atoms that substitute

$$CH_4(g) + Cl_2(g) \rightarrow CH_3Cl(g) + HCl(g) \qquad (26.22)$$

$$\text{methane} \qquad\qquad \text{chloromethane}$$

Here, a chlorine atom substitutes for a hydrogen atom.

2. *Nitration of aromatic hydrocarbons*

$$C_6H_6(l) + HNO_3(l) \xrightarrow{H_2SO_4} C_6H_5NO_2(l) + H_2O(l) \qquad (26.23)$$

$$\text{benzene} \qquad\qquad\qquad \text{nitrobenzene}$$

In this case, the nitro group, $-NO_2$, substitutes for a hydrogen atom on the benzene ring.

3. *Conversion of an alkyl halide to an alcohol*

$$C_2H_5Br(l) + OH^-(aq) \rightarrow C_2H_5OH(aq) + Br^-(aq) \qquad (26.24)$$
bromoethane   ethanol

A bromine atom is replaced by an —OH group.

■ **EXAMPLE 26.11** _____

Classify each of the following as an addition, elimination, condensation, or substitution reaction.

a. $C_2H_5OH(aq) + H^+(aq) + Br^-(aq) \rightarrow C_2H_5Br(l) + H_2O$
b. $C_6H_{12}(l) + Cl_2(g) \rightarrow C_6H_{12}Cl_2(l)$
c. $C_7H_8(l) + 2Cl_2(g) \rightarrow C_7H_6Cl_2(l) + 2HCl(g)$
d. $C_5H_{11}Cl(l) \rightarrow C_5H_{10}(l) + HCl(g)$

Solution

a. substitution (Br for OH)	b. addition (C_6H_{12} is an alkene)
c. substitution (Cl for H)	d. elimination (HCl lost)

Exercise
How would you classify the reaction: $Cl_2(g) + H_2O \rightarrow HOCl(aq) + HCl(g)$?
Answer: condensation.

■ **SUMMARY PROBLEM**

Consider the alkyl halide which has the structure:

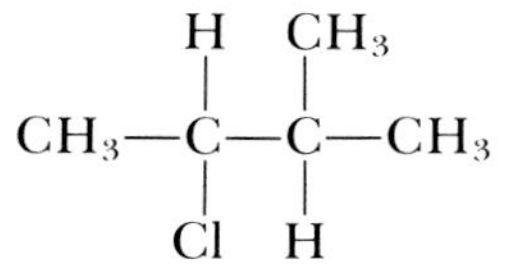

$$\begin{array}{ccc} & H & CH_3 \\ & | & | \\ CH_3\!-\!&C\!-\!C\!-\!&CH_3 \\ & | & | \\ & Cl & H \end{array}$$

a. Give the name of the alkyl halide and that of the alkane from which it is derived.
b. Write the structure of the alkene formed when HCl is eliminated from the alkyl halide; name the alkene.
c. Name all of the alkanes isomeric with the one referred to in (a).
d. Name all of the alkenes isomeric with the one referred to in (b).
e. Draw the structure of the species formed when this alkyl halide undergoes substitution with an OH^- ion.
f. Draw the structure of the ester formed when the species in (e) reacts with formic acid.
g. How many chiral carbon atoms are there in this alkyl halide?

Answers

a. 3-chloro-2-methylbutane; 2-methylbutane

b.

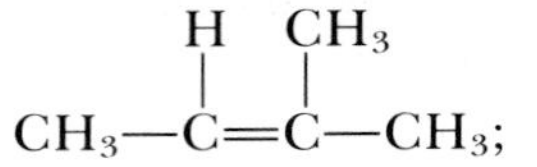

$$CH_3 - \overset{\overset{\displaystyle H}{|}}{C} = \overset{\overset{\displaystyle CH_3}{|}}{C} - CH_3; \qquad \text{2-methyl-2-butene}$$

c. pentane, 2,2-dimethylpropane

d. 1-pentene, *cis-* and *trans-*2-pentene, 2-methyl-1-butene, 3-methyl-1-butene

e.

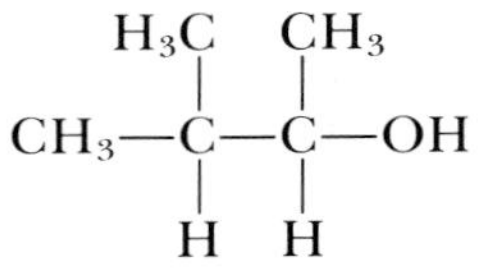

$$CH_3 - \overset{\overset{\displaystyle H_3C}{|}}{\underset{\underset{\displaystyle H}{|}}{C}} - \overset{\overset{\displaystyle CH_3}{|}}{\underset{\underset{\displaystyle H}{|}}{C}} - OH$$

f.

$$CH_3 - \overset{\overset{\displaystyle H_3C}{|}}{\underset{\underset{\displaystyle H}{|}}{C}} - \overset{\overset{\displaystyle CH_3}{|}}{\underset{\underset{\displaystyle H}{|}}{C}} - O - \overset{}{\underset{\underset{\displaystyle O}{\|}}{C}} - H$$

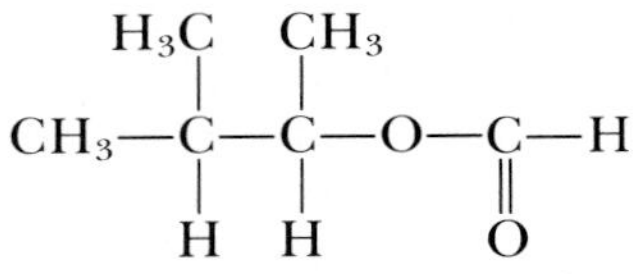

g. 1

Gasoline, a complex mixture of hydrocarbons most of which are alkanes, is produced by the fractional distillation of petroleum. Distillation of a liter of crude oil gives about 250 cm^3 of "straight-run" gasoline. It is possible to double the yield of gasoline by converting higher or lower boiling fractions to hydrocarbons in the gasoline range. Several processes are available for doing this.

One problem with the internal combustion engine is that gasoline-air mixtures tend to ignite prematurely, or "knock," rather than burn smoothly. The *octane number* of a gasoline is a measure of its resistance to knock. It is determined by comparing the knocking characteristics of a gasoline sample to those of "isooctane" and heptane:

$$CH_3-\underset{\underset{CH_3}{|}}{\overset{\overset{CH_3}{|}}{C}}-CH_2-\underset{\underset{CH_3}{|}}{\overset{\overset{H}{|}}{C}}-CH_3 \qquad CH_3-CH_2-CH_2-CH_2-CH_2-CH_2-CH_3$$

"isooctane"   heptane
(2,2,4-trimethylpentane)

Isooctane, which is highly branched, burns smoothly with little knocking and is assigned an octane number of 100. Heptane, being unbranched, knocks badly. It is given an octane number of zero. Gasoline with the same knocking properties as a mixture of 90% isooctane and 10% heptane is rated as "90 octane."

To obtain "premium" gasoline of octane number above 80, it is necessary to use additives of one type or another. Until the mid-1970s, the principal anti-knock agent was tetraethyllead, $(C_2H_5)_4Pb$. Its use was phased out because it poisons catalytic converters and contaminates the environment with lead compounds. Among the additives in use today are two oxygen compounds:

$$CH_3-\underset{\underset{CH_3}{|}}{\overset{\overset{CH_3}{|}}{C}}-OH \qquad and \qquad CH_3-\underset{\underset{CH_3}{|}}{\overset{\overset{CH_3}{|}}{C}}-O-CH_3$$

2-methyl-2-propanol   methyl-*t*-butyl ether
(*t*-butyl alcohol)   (MTBE)

Octane number can also be increased by blending in highly branched alkanes or aromatic hydrocarbons such as benzene and toluene.

You are probably familiar with "gasohol," a motor fuel introduced in the 1970s to stretch gasoline supplies. It contains up to 10% by volume of ethanol. Nowadays methanol is a more popular additive; when you see the word "alcohol" on a gas pump, the reference is almost certainly to CH_3OH. Methanol is cheaper than ethanol, increases octane number, and reduces emission of pollutants.

On the other hand, the high water solubility of methanol can cause problems. Small amounts of moisture bring about a separation into two phases, which in turn accelerates engine corrosion. To prevent this, *t*-butyl alcohol and corrosion inhibitors are added to the methanol–gasoline mixture. There is some controversy as to how effective these compounds are; some automobile manufacturers warn that use of gasoline containing methanol can void engine warranties.

The "energy crisis" of the 1970s prompted scientists and engineers to look beyond petroleum for other sources of gasoline. The most promising source is coal, of which the United States has vast reserves. One way to make gasoline from coal, the so-called Fischer-Tropsch process, was described in the Applied Perspective at the end of Chapter 13. You may recall that it involves the conversion of a carbon monoxide–hydrogen mixture to hydrocarbons in the gasoline range. The process can be represented by the general equations:

$$nCO(g) + (2n + 1)H_2(g) \rightarrow C_nH_{2n+2}(l) + nH_2O(l) \qquad (26.27)$$

$$nCO(g) + 3nH_2(g) \rightarrow C_nH_{2n}(l) + nH_2O(l) \qquad (26.28)$$

The product is a mixture of alkanes and alkenes; *n* ranges from 5 to 20.

A somewhat different method of making synthetic gasoline from coal or natural gas was developed by the Mobil Corporation in the mid-1970s. A carbon monoxide–hydrogen mixture is first converted to methanol:

$$CO(g) + 2H_2(g) \rightarrow CH_3OH(l)$$

Methanol is then passed over a synthetic zeolite catalyst to give a liquid which upon distillation gives a gasoline fraction with an octane number in the high 90s. A plant in New Zealand using this process is currently producing about 2 million liters of gasoline per day. New Zealand has abundant natural gas from offshore wells but no oil.

QUESTIONS AND PROBLEMS

Nomenclature

1. Name the following alkanes.

a. $CH_3-CH_2-CH-CH_3$
          |
          CH_3

b. $CH_3-CH_2-CH-CH_3$
          |
          CH_2
          |
          CH_3

c. $CH_3-CH-CH-CH_3$
        | |
       CH_3 CH_2
           |
           CH_3

d. $CH_3-CH-CH_2-CH-CH_3$
        | |
       CH_3 $H-C-CH_3$
             |
             CH_3

2. Name the following alkanes.

a. $CH_3-(CH_2)_5-CH-CH_3$
             |
             CH_3

b. $(CH_3)_4C$

c. $CH_3-CH-CH_2-C(CH_3)_3$
        |
       CH_3

d. H
         |
     $CH_3-C-(CH_2)_2-CH-CH_3$
        | |
       CH_2 CH_3
        |
       CH_3

3. Write structural formulas for the following alkanes.
 a. 3-ethylpentane   b. 2,2-dimethylbutane
 c. 2-methyl-3-ethylheptane   d. 2,3-dimethylpentane

4. Write structural formulas for the following alkanes.
 a. 2,2,4-trimethylpentane   b. 2,2-dimethylpropane
 c. 4-isopropyloctane   d. 2,3,4-trimethylheptane

5. The following names are incorrect; draw a reasonable structure for the alkane and give the proper IUPAC name.
a. 5-isopropyloctane b. 2-ethylpropane
c. 1,2-dimethylpropane

6. Follow the directions of Problem 5 for the following names.
a. 2-dimethylbutane b. 4-methylpentane
c. 2-ethylpropane

7. Name the following alkenes.
a. $CH_2{=}C{-}(CH_3)_2$
b. $(CH_3)_2{-}C{=}C{-}(CH_3)_2$
c. $CH_3{-}CH{=}CH{-}CH_2CH_3$
d. $CH_3{-}\underset{\underset{\displaystyle CH_3}{|}}{\underset{\displaystyle CH_2}{|}}{C}{=}CH_2$

8. Write structural formulas for the following alkynes.
a. 2-pentyne b. 4-methyl-2-pentyne
c. 2-methyl-3-hexyne d. 3,3-dimethyl-l-butyne

9. Name the following compounds as derivatives of nitrobenzene.

a. 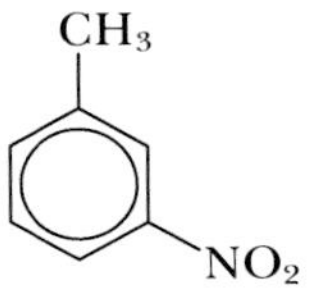

b. 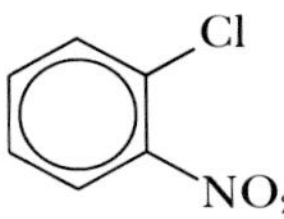

c. 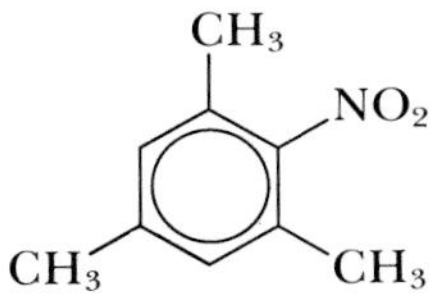

10. Name the following compounds as derivatives of toluene.

a. 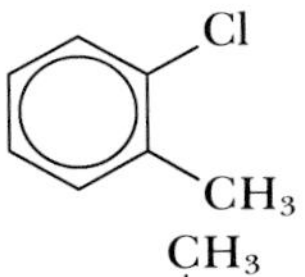

b. 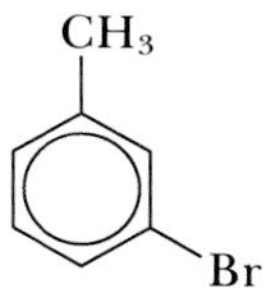

c. 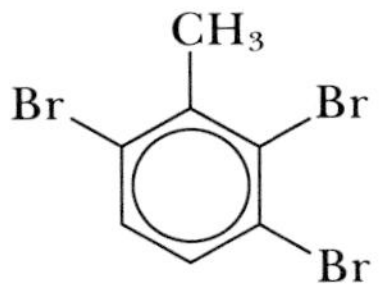

11. Give the IUPAC and common names of
a. CH_3OH b. CH_3COOH c. $CH_3CH_2CH_2OH$

12. Give the IUPAC and common names of
a. HCOOH b. CH_3CH_2OH
c. $CH_3{-}\underset{\displaystyle OH}{\underset{|}{C}H}{-}CH_3$

13. Name the following alkyl halides.
a. $CH_3{-}CH_2{-}CHCl_2$ b. $(CH_3)_3C{-}Cl$
c. $CH_3{-}CHI{-}CH_2Cl$

14. Write structural formulas for the following alkyl halides.
a. 2-chloro-2-iodopropane
b. 3-fluoro-2-methylbutane
c. 1-bromo-2,2-dimethylpropane

15. Give the molecular formula of
a. toluene b. naphthalene
c. *o*-dichlorobenzene d. TNT

16. Give the molecular formula of
a. propanoic acid b. 1-butanol
c. isopropyl alcohol d. anthracene

17. Draw the structural formula of
a. methyl formate b. ethyl propionate
c. propyl acetate d. isopropyl acetate

18. Name the following esters.
a. $CH_3{-}O{-}\underset{\displaystyle O}{\underset{\|}{C}}{-}CH_3$
b. $H{-}\underset{\displaystyle O}{\underset{\|}{C}}{-}O{-}CH_2CH_3$
c. $CH_3{-}O{-}\underset{\displaystyle O}{\underset{\|}{C}}{-}CH_2CH_3$

19. Write structural formulas for
a. ammonia b. dimethylamine
c. ethylamine d. ethyl chloride

20. Write structural formulas for
a. ethylenediamine b. chloroform
c. methylene chloride d. triethylamine

Structural Isomerism

21. Draw the structural isomers of the alkane C_6H_{14}.

22. Draw the structural isomers of the alkene C_4H_8.

23. Draw the structural isomers of C_4H_9Cl in which one hydrogen atom of a C_4H_{10} molecule has been replaced by chlorine.

24. Draw the structural isomers of $C_3H_6Cl_2$ in which two of the hydrogen atoms of C_3H_8 have been replaced by chlorine atoms.

25. There are three compounds with the formula C_6H_4ClBr in which two of the hydrogen atoms of the benzene molecule have been replaced by halogen atoms. Draw structural formulas for these compounds.

26. There are three compounds with the formula $C_6H_3Cl_3$ in which three of the hydrogen atoms of the benzene molecule have been replaced by chlorine atoms. Draw structural formulas for these compounds.

27. Draw structural formulas for all the isomers of $C_2H_3Cl_3$.

28. Draw structural formulas for all the isomers of $C_2H_4Br_2$.

29. Write structural formulas for all the structural isomers of double-bonded compounds with the molecular formula C_5H_{10}.

30. Write structural formulas for all the structural isomers of compounds with the molecular formula C_4H_6ClBr in which Cl and Br are bonded to a double-bonded carbon.

31. Draw structural formulas for all the alcohols with molecular formula $C_5H_{12}O$.

32. Draw structural formulas for all the saturated carboxylic acids with four carbon atoms per molecule.

33. How many isomers are there corresponding to the following molecular formulas?
a. C_2H_7N b. C_2H_6O c. $C_3H_6Cl_2$

34. How many isomers are there corresponding to the following molecular formulas?
a. C_3H_9N b. $C_4H_{10}O$ c. $C_3H_5Cl_3$

Geometric and Optical Isomerism

35. Of the compounds in Problem 29, which ones show geometric isomerism? Draw the *cis* and *trans* isomers.

36. Of the compounds in Question 30, which ones show geometric isomerism? Draw the *cis* and *trans* isomers.

37. Maleic acid and fumaric acid are the *cis* and *trans* isomers, respectively, of $C_2H_2(COOH)_2$, a dicarboxylic acid. Draw and label their structural formulas.

38. For which of the following is geometric isomerism possible?
a. $(CH_3)_2C{=}CCl_2$ b. $CH_3ClC{=}CCH_3Cl$
c. $CH_3BrC{=}CCH_3Cl$

39. Draw the structural formulas for all the isomers, structural and geometric, of all the alkenes of molecular formula C_5H_{10}.

40. Draw the structures of all the alcohols containing four carbon atoms per molecule and one double bond; consider both structural and geometric isomers.

41. Which of the following can show optical isomerism?
a. 2-bromo-2-chlorobutane
b. 2-methylpropane
c. 2,2-dimethyl-1-butanol
d. 2,2,4-trimethylpentane

42. Which of the following compounds can show optical isomerism?
a. dichloromethane
b. 1,2-dichloroethane
c. bromochlorofluoromethane
d. 1-bromoethanol

43. Locate the chiral carbon(s), if any, in the following molecules:

a.
$$HO{-}\overset{\displaystyle H}{\underset{\displaystyle O}{C}}{-}\overset{\displaystyle H}{\underset{\displaystyle OH}{C}}{-}\overset{\displaystyle H}{\underset{\displaystyle OH}{C}}{-}\overset{\displaystyle H}{\underset{\displaystyle O}{C}}{-}H$$

b.
$$CH_3{-}\overset{}{\underset{\displaystyle O}{C}}{-}\overset{}{\underset{\displaystyle O}{C}}{-}OH$$

c.
$$CH_3{-}CH_2{-}\overset{\displaystyle H}{\underset{\displaystyle NH_2}{C}}{-}COOH$$

44. Locate the chiral carbon(s), if any, in the following molecules:

a.
$$CH_3{-}\overset{\displaystyle H}{\underset{\displaystyle OH}{C}}{-}\overset{\displaystyle H}{\underset{\displaystyle OH}{C}}{-}H$$

b.
$$H{-}\overset{\displaystyle H}{\underset{\displaystyle H}{C}}{=}\overset{\displaystyle H}{\underset{\displaystyle H}{C}}{-}CH_2{-}OH$$

c.
$$CH_3{-}\overset{\displaystyle Cl}{\underset{\displaystyle Cl}{C}}{-}\overset{\displaystyle F}{\underset{\displaystyle H}{C}}{-}Cl$$

Functional Groups

45. Classify each of the following as a carboxylic acid, ester, and/or alcohol.
a. $HO{-}CH_2{-}CH_2{-}CH_2{-}OH$

b.
$$\text{(benzene ring)}{-}COOH,\ {-}O{-}\overset{}{\underset{\displaystyle O}{C}}{-}CH_3$$

c. $CH_3{-}\overset{\displaystyle H}{\underset{\displaystyle OH}{C}}{-}COOH$

46. Classify each of the following as a carboxylic acid, ester, and/or alcohol.
a. $CH_3{-}(CH_2)_3{-}OH$

b. $CH_3{-}CH_2{-}\overset{}{\underset{\displaystyle O}{C}}{-}O{-}CH_2{-}CH_3$

c. $CH_3{-}CH_2{-}O{-}\overset{}{\underset{\displaystyle O}{C}}{-}(CH_2)_6{-}COOH$

47. Give the structural formula for
a. a four-carbon alcohol with an —OH group not at the end of the chain.
b. a five-carbon carboxylic acid.
c. the ester formed when these two compounds react.

48. Give the structural formula of
a. a three-carbon alcohol with the —OH group on the center carbon.
b. a four-carbon carboxylic acid.
c. the ester formed when these two compounds react.

49. Write the structural formula of the ester formed by methyl alcohol with
a. formic acid   b. acetic acid
c. $CH_3{-}(CH_2)_6{-}COOH$

50. Write the structural formulas of all the esters that can be formed by ethylene glycol with formic and acetic acids. (One or both of the —OH groups of ethylene glycol can react.)

51. Using three carbon atoms and the required number of hydrogen atoms, write the structural formula for
a. a primary amine
b. a secondary amine

52. Classify each of the following as a primary, secondary, or tertiary amine:

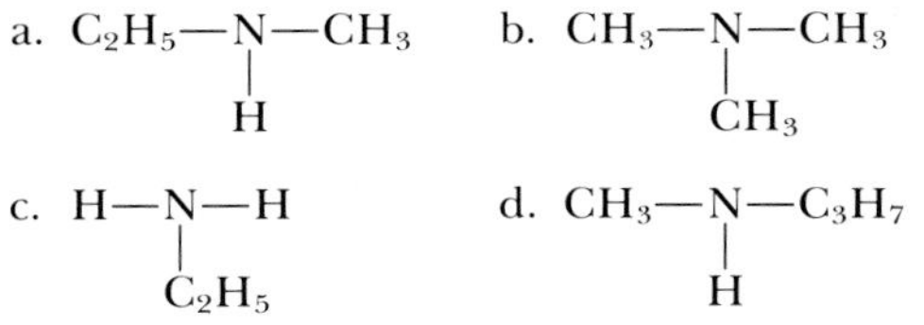

a. CH_3—N—CH_3
   |
   H
   (shown as C_2H_5—N—CH_3 with H below)

b. CH_3—N—CH_3
   |
   CH_3

c. H—N—H
   |
   C_2H_5

d. CH_3—N—C_3H_7
   |
   H

Types of Reactions

53. Classify the following reactions as addition, substitution, elimination, or condensation.

a. $C_2H_2(g) + HBr(g) \rightarrow C_2H_3Br(l)$

b. $C_6H_5OH(s) + HNO_3(l) \rightarrow$
   $C_6H_4(OH)(NO_2)(s) + H_2O(l)$

c. $C_2H_5OH(aq) + HCOOH(aq) \rightarrow$
   C_2H_5—O—C—H$(l) + H_2O(l)$
   $\quad\quad\quad\quad\; \| $
   $\quad\quad\quad\quad\; O$

54. Classify the following reactions according to the categories listed in Question 53.

a. $PCl_3(g) + Cl_2(g) \rightarrow PCl_5(g)$

b. $C_3H_7OH(l) \rightarrow C_3H_6(g) + H_2O(l)$

c. $CH_3OH(l) + C_2H_5OH(l) \rightarrow$
   CH_3—O—$C_2H_5(l) + H_2O(l)$

55. Name the products formed when the following reagents add to 1-butene.

a. H_2 b. HCl   c. H_2O

56. Name the products obtained when the following reagents add to 2-methyl-2-butene.

a. Cl_2 b. HI   c. H_2O

Unclassified

57. Give the formula of at least one starting material used to make

a. ethene   b. acetylene
c. ethanol   d. acetic acid

58. Which of the following compounds is *least* toxic?

a. benzene   b. methanol
c. chloroform   d. toluene

59. The general formula of an alkane is C_nH_{2n+2}. What is the general formula of an

a. alkene?   b. alkyne?
c. alcohol derived from an alkane?

60. Explain what each of the following terms means.

a. denatured alcohol   b. wood alcohol
c. soap   d. unsaturated fat

61. Arrange the following anions in order of increasing base strength.

acetate ion, chloroacetate ion, iodoacetate ion, dichloroacetate ion

62. Write a balanced net ionic equation for the reaction of

a. $(CH_3)_2NH$ with hydrochloric acid.
b. acetic acid with a solution of barium hydroxide.
c. 2-chloropropane with a solution of sodium hydroxide.

Challenge Problems

63. Heroin is a diester derivative of morphine (Fig. 26.10). It can be considered to be the reaction product of two moles of acetic acid per mole of morphine. What are the mass percents of the elements present in heroin?

64. Draw structures for all the alcohols with molecular formula $C_6H_{14}O$.

65. Write an equation for the reaction of chloroacetic acid ($K_a = 1.5 \times 10^{-3}$) with trimethylamine ($K_b = 5.9 \times 10^{-5}$). Calculate the equilibrium constant for the reaction. If 0.10 M solutions of these two species are mixed, what will be their concentrations at equilibrium?

66. Assume that the alcohol content of the air in a driver's lungs is 0.15 mole percent. If he expels 0.500 L of air at 37°C and 750 mm Hg into a breathalyzer, what mass of $K_2Cr_2O_7$ will be reduced? (Equation 26.10)

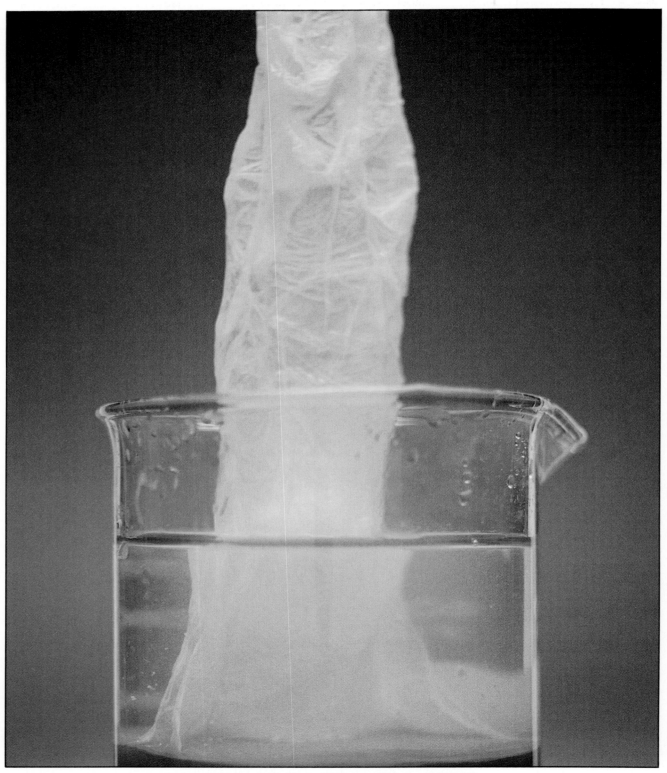

$$H_2N\text{—}(CH_2)_6\text{—}NH_2(l)\ +\ HOOC\text{—}(CH_2)_4\text{—}COOH(l)\ \rightarrow$$
$$\cdots\ \text{—}(CH_2)_4\text{—}CONH\text{—}(CH_2)_6\text{—}CONH\text{—}(CH_2)_4\text{—}\ \cdots$$

Hexamethylenediamine reacts with adipic acid to form nylon.

CHAPTER 27

NATURAL AND SYNTHETIC POLYMERS

I n previous chapters we have discussed the chemical and physical properties of many kinds of substances. For the most part, these materials were made up of either small molecules or simple ions. In this chapter, we will be concerned with an important class of compounds containing large molecules. We call these compounds polymers.

A polymer is made up of a large number of small molecular units, called *monomers,* combined together chemically. A typical polymer molecule contains a chain of monomers several thousand units long. The monomer units of which a given polymer is composed may be the same or different.

We start with synthetic polymers. Since about 1930, a wide variety of synthetic polymers have been made available by the chemical industry. The monomer units are joined together either by addition (Section 27.1) or by condensation (Section 27.2). They are used to make cups, plates, fabrics, automobile tires, even artificial hearts. They contain, in addition to carbon and hydrogen, such elements as oxygen, nitrogen, sulfur, and the halogens. Ordinarily, they contain at most two different types of monomer units.

These polymers are usually called plastics

The remainder of this chapter will deal with natural polymers. These are large molecules, produced by plants and animals, that carry out the many life-sustaining processes in a living cell. The cell membranes of plants and the woody structure of trees are composed in large part of cellulose, a polymeric carbohydrate. We will look at the structures of a variety of different carbohydrates in Section 27.3. Another class of natural polymers are the proteins. Section 27.4 deals with these polymeric materials that make up our tissues, bone, blood, and even hair. Section 27.5 describes nucleic acids. They carry in coded form the template for making proteins. They also control the inherited characteristics of the next generation.

The subject matter of this chapter is broad; chemists and biologists spend lifetimes studying one small aspect of it. In this chapter we can only present an overview of the material. You may learn more about polymeric molecules in later courses in chemistry and/or biology.

27.1
SYNTHETIC ADDITION POLYMERS

When monomer units add directly to one another, the result is an addition polymer. Table 27.1 lists some of the more familiar synthetic addition polymers. You will notice that each of these is derived from a monomer

TABLE 27.1 Some Common Addition Polymers

MONOMER	NAME	POLYMER	USES
$H_2C{=}CH_2$ (ethylene structure)	ethylene	polyethylene	bags, coatings, toys
$H_2C{=}CHCH_3$ (propylene structure)	propylene	polypropylene	beakers, milk cartons
$H_2C{=}CHCl$ (vinyl chloride structure)	vinyl chloride	polyvinyl chloride, PVC	floor tile, raincoats, pipe, phonograph records
$H_2C{=}CHCN$ (acrylonitrile structure)	acrylonitrile	polyacrylonitrile, PAN	rugs; Orlon and Acrilan are copolymers with other monomers.
$H_2C{=}CH{-}C_6H_5$ (styrene structure)	styrene	polystyrene	cast articles using a transparent plastic
$H_2C{=}C(CH_3){-}C({=}O){-}OCH_3$ (methyl methacrylate structure)	methyl methacrylate	Plexiglas, Lucite, acrylic resins	high-quality transparent objects, latex paints
$F_2C{=}CF_2$ (tetrafluoroethylene structure)	tetrafluoroethylene	Teflon	gaskets, insulation, bearings, pan coatings

containing a carbon—carbon double bond. Upon polymerization, the double bond is converted to a single bond:

$$\underset{/}{\overset{\backslash}{}}C=C\underset{\backslash}{\overset{/}{}} \rightarrow -\overset{|}{\underset{|}{C}}-\overset{|}{\underset{|}{C}}-$$

and successive monomer units add to one another.

POLYETHYLENE

Perhaps the most familiar addition polymer is polyethylene, a solid derived from the monomer ethene (common name ethylene). We might represent the polymerization process as:

$$n\left(\underset{H}{\overset{H}{}}C=C\underset{H}{\overset{H}{}}\right) \rightarrow \left(-\overset{H}{\underset{H}{C}}-\overset{H}{\underset{H}{C}}-\right)_n \qquad (27.1)$$

ethylene   polyethylene

In an addition polymer, all the reactants end up in the polymer

where n is a very large number, of the order of 2000.

There are several different ways to initiate Reaction 27.1. One method uses a small amount of a very reactive species called a *free radical,* which contains an unpaired electron. Such a species can be produced by adding to ethylene a small amount of an organic compound known as benzoyl peroxide. Upon heating, the benzoyl peroxide decomposes to form a free radical:

$$\bigcirc-\overset{}{\underset{O}{C}}-O-O-\overset{}{\underset{O}{C}}-\bigcirc \rightarrow 2 \bigcirc-\overset{}{\underset{O}{C}}-O\cdot \qquad (27.2)$$

benzoyl peroxide   a free radical, X·

This free radical, which we will represent simply as X·, reacts rapidly with ethylene to form a species which still contains an unpaired electron:

A free radical is not an anarchist on parole.

$$X\cdot + \underset{H}{\overset{H}{}}C=C\underset{H}{\overset{H}{}} \rightarrow X-\overset{H}{\underset{H}{C}}-\overset{H}{\underset{H}{C}}\cdot \qquad (27.3)$$

The product of this reaction is able to add another ethylene molecule, and then another, and then another, ... Eventually, very long chains are formed, each with an unpaired electron at one end. Two of these chains can combine to terminate the polymerization process. The product formed by free radical polymerization of ethylene is one in which most of the chains are branched. Its structure might be shown as

$$
\begin{array}{cccccccccccccc}
& \mathrm{H} & \mathrm{H} & \mathrm{H} & \mathrm{H} && \mathrm{H} & \mathrm{H} & \mathrm{H} & \mathrm{H} & \mathrm{H} & \mathrm{H} & \mathrm{H} && \mathrm{H} \\
& | & | & | & | && | & | & | & | & | & | & | && | \\
-&\mathrm{C}&-\mathrm{C}&-\mathrm{C}&-\mathrm{C}&-&\mathrm{C}&-\mathrm{C}&-\mathrm{C}&-\mathrm{C}&-\mathrm{C}&-\mathrm{C}&-\mathrm{C}&-&\mathrm{C}- \\
& | & | & | & | && | & | & | & | & | & | & | && | \\
& \mathrm{H} & \mathrm{H} & \mathrm{H} & \mathrm{CH_2} && \mathrm{H} & \mathrm{H} & \mathrm{H} & \mathrm{H} & \mathrm{H} & \mathrm{H} & \mathrm{CH_2} && \mathrm{H} \\
&&&& | &&&&&&&& | \\
&&&& \mathrm{CH_3} &&&&&&&& \mathrm{CH_3}
\end{array}
$$

In this polymer, known as *branched* polyethylene, neighboring chains are arranged in a somewhat random fashion, often overlapping each other. This produces a soft, flexible solid, used mostly in films and coatings where a pliable material is needed.

The plastic bags at the vegetable counter are made of this material

With a different type of catalyst, it is possible to produce *linear* polyethylene. This consists almost entirely of unbranched chains:

$$
\begin{array}{cccccccccccccccc}
& \mathrm{H} & \mathrm{H} & \mathrm{H} & \mathrm{H} & \mathrm{H} & \mathrm{H} & \mathrm{H} & \mathrm{H} & \mathrm{H} & \mathrm{H} & \mathrm{H} & \mathrm{H} & \mathrm{H} & \mathrm{H} & \mathrm{H} & \mathrm{H} \\
& | & | & | & | & | & | & | & | & | & | & | & | & | & | & | & | \\
-&\mathrm{C}&-\mathrm{C}&-\mathrm{C}&-\mathrm{C}&-\mathrm{C}&-\mathrm{C}&-\mathrm{C}&-\mathrm{C}&-\mathrm{C}&-\mathrm{C}&-\mathrm{C}&-\mathrm{C}&-\mathrm{C}&-\mathrm{C}&-\mathrm{C}&-\mathrm{C}- \\
& | & | & | & | & | & | & | & | & | & | & | & | & | & | & | & | \\
& \mathrm{H} & \mathrm{H} & \mathrm{H} & \mathrm{H} & \mathrm{H} & \mathrm{H} & \mathrm{H} & \mathrm{H} & \mathrm{H} & \mathrm{H} & \mathrm{H} & \mathrm{H} & \mathrm{H} & \mathrm{H} & \mathrm{H} & \mathrm{H}
\end{array}
$$

Neighboring chains in linear polyethylene line up nearly parallel to each other. This gives a polymer that approaches a crystalline material. It is used for bottles, toys, and other semirigid objects (Fig. 27.1).

POLYVINYL CHLORIDE AND OTHER ADDITION POLYMERS

Vinyl chloride, in contrast to ethylene, is an unsymmetrical molecule. We might refer to the CH_2 group in vinyl chloride as the "head" of the molecule and the $CHCl$ group as the "tail":

$$
\begin{array}{ccc}
\mathrm{H} & & \mathrm{H} \\
\diagdown & & \diagup \\
& \mathrm{C}{=}\mathrm{C} & \\
\diagup & & \diagdown \\
\mathrm{H} & & \mathrm{Cl} \\
\text{head} & & \text{tail}
\end{array}
\qquad \text{vinyl chloride}
$$

In principle at least, vinyl chloride molecules can add to one another in any of three ways to form:

1. A *head-to-tail* polymer, in which there is a Cl atom on every other C atom in the chain:

$$
\begin{array}{cccccccccc}
\mathrm{H} & \mathrm{H} & \mathrm{H} & \mathrm{H} & \mathrm{H} & \mathrm{H} & \mathrm{H} & \mathrm{H} & \mathrm{H} & \mathrm{H} \\
| & | & | & | & | & | & | & | & | & | \\
-\mathrm{C}-\mathrm{C}-\mathrm{C}-\mathrm{C}-\mathrm{C}-\mathrm{C}-\mathrm{C}-\mathrm{C}-\mathrm{C}-\mathrm{C}- \\
| & | & | & | & | & | & | & | & | & | \\
\mathrm{H} & \mathrm{Cl} & \mathrm{H} & \mathrm{Cl} & \mathrm{H} & \mathrm{Cl} & \mathrm{H} & \mathrm{Cl} & \mathrm{H} & \mathrm{Cl}
\end{array}
$$

2. A *head-to-head, tail-to-tail* polymer, in which Cl atoms occur in pairs on adjacent carbon atoms in the chain:

Figure 27.1 The polyethylene bottle at the left is made of pliable, branched polyethylene. The one at the right is made of semirigid, linear polyethylene. (Marna G. Clarke)

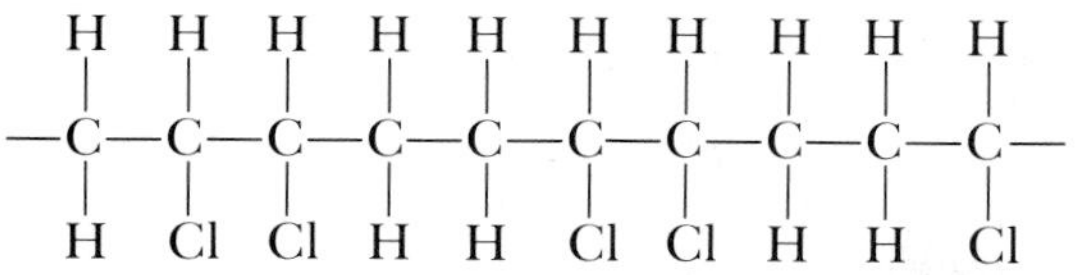

$$\begin{array}{cccccccccc}
H & H & H & H & H & H & H & H & H & H \\
| & | & | & | & | & | & | & | & | & | \\
-C-&C-&C-&C-&C-&C-&C-&C-&C-&C- \\
| & | & | & | & | & | & | & | & | & | \\
H & Cl & Cl & H & H & Cl & Cl & H & H & Cl
\end{array}$$

PVC is a stiff, rugged, cheap polymer

3. A *random* polymer:

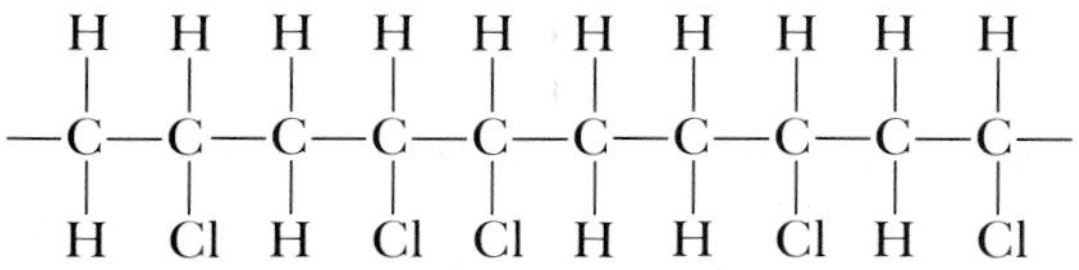

$$\begin{array}{cccccccccc}
H & H & H & H & H & H & H & H & H & H \\
| & | & | & | & | & | & | & | & | & | \\
-C-&C-&C-&C-&C-&C-&C-&C-&C-&C- \\
| & | & | & | & | & | & | & | & | & | \\
H & Cl & H & Cl & Cl & H & H & Cl & H & Cl
\end{array}$$

Similar arrangements are possible with polymers made from other unsymmetrical monomers (Example 27.1). In practice, the addition polymer is usually of the head-to-tail type. This is the case with polyvinyl chloride and polypropylene.

■ **EXAMPLE 27.1** _____

Sketch a polymer derived from propene:

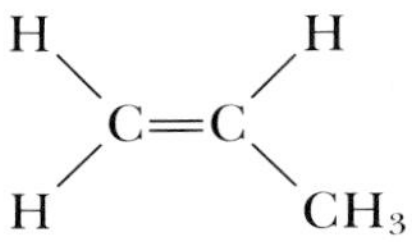

$$\begin{array}{cc}
H & H \\
\diagdown & \diagup \\
& C=C \\
\diagup & \diagdown \\
H & CH_3
\end{array}$$

assuming it to be a

a. head-to-tail polymer;   b. head-to-head, tail-to-tail polymer.

Solution

a.
$$\begin{array}{cccccccccc}
H & H & H & H & H & H & H & H & H & H \\
| & | & | & | & | & | & | & | & | & | \\
-C-&C-&C-&C-&C-&C-&C-&C-&C-&C- \\
| & | & | & | & | & | & | & | & | & | \\
H & CH_3 & H & CH_3 & H & CH_3 & H & CH_3 & H & CH_3
\end{array}$$

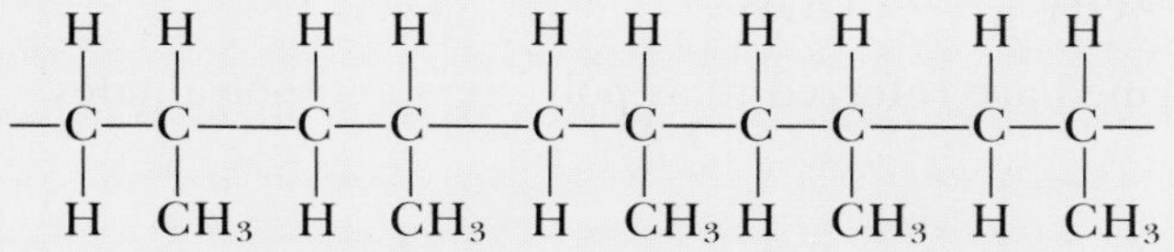

b.
$$\begin{array}{cccccccccc}
H & H & H & H & H & H & H & H & H & H \\
| & | & | & | & | & | & | & | & | & | \\
-C-&C-&C-&C-&C-&C-&C-&C-&C-&C- \\
| & | & | & | & | & | & | & | & | & | \\
H & CH_3 & CH_3 & H & H & CH_3 & CH_3 & H & H & CH_3
\end{array}$$

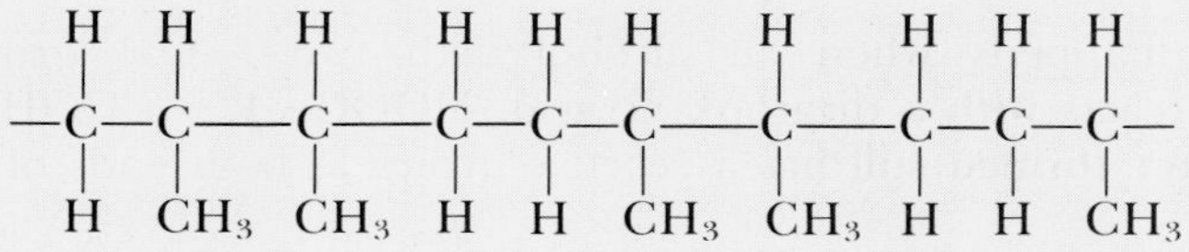

Exercise

Of the seven monomers listed in Table 27.1, how many can form three different polymers of the type discussed above? Answer: Five.

Most polymers decompose at high temperatures. For example, polyvinyl chloride starts to decompose at 100°C, liberating HCl. Teflon, a polymer of tetrafluoroethylene (Table 27.1) behaves quite differently. A hard waxy solid, Teflon is useful from about -70 to $250°C$. Teflon is also extremely unreactive. It resists chemical attack by all known reagents except molten alkali metals. Its best known use is as a nonstick coating for frying pans (Fig. 27.2). In recent years, waterproof Goretex products, which use Teflon-coated fabrics, have become popular with outdoor sports enthusiasts.

Figure 27.2 (Courtesy of E.I. duPont de Nemours and Company.)

27.2
SYNTHETIC CONDENSATION POLYMERS

As we pointed out in Chapter 26, a condensation reaction is one in which two molecules combine by splitting out a small molecule such as water. The reaction of an alcohol, ROH, or an amine, RNH_2, with a carboxylic acid $R'COOH$ is a condensation:

$$R—OH + HO—\underset{\underset{O}{\|}}{C}—R' \rightarrow R—O—\underset{\underset{O}{\|}}{C}—R' + H_2O \qquad (27.4)$$

alcohol   acid   ester

$$R—\underset{\underset{H}{|}}{N}—H + HO—\underset{\underset{O}{\|}}{C}—R' \rightarrow R—\underset{\underset{H}{|}}{N}—\underset{\underset{O}{\|}}{C}—R' + H_2O \qquad (27.5)$$

amine   acid   amide

Condensation polymers are formed in a similar way; the molecule split out is most often water. In order to produce a condensation polymer, *the monomers involved must have functional groups at both ends of the molecule.* Most often these groups are

$$—NH_2 \quad —OH \quad —COOH$$

The products formed are referred to as polyesters and polyamides.

POLYESTERS

Consider what happens when an alcohol with two —OH groups, HO—R—OH, reacts with a dicarboxylic acid, HOOC—R'—COOH. In this case the ester formed still has a reactive group at both ends of the molecule.

$$HO—R—OH + HO—\underset{\underset{O}{\|}}{C}—R'—\underset{\underset{O}{\|}}{C}—OH$$

dihydroxy   dicarboxylic
alcohol   acid

$$\rightarrow HO—R—O—\underset{\underset{O}{\|}}{C}—R'—\underset{\underset{O}{\|}}{C}—OH + H_2O \qquad (27.6)$$

ester with active
end groups

Figure 27.3 The *Gossamer Albatross*, the first human-powered craft to fly across the English Channel (1979). Light-weight, durable polyesters cover the wings and body. (Dupont Corporation)

The COOH group at one end of the ester molecule can react with another alcohol molecule. The OH group at the other end can react with an acid molecule. This process can continue, leading eventually to a long-chain polymer containing 500 or more ester groups. The general structure of the polyester can be represented as

$$-\overset{\displaystyle\|}{\underset{\displaystyle O}{C}}-R'-\overset{\displaystyle\|}{\underset{\displaystyle O}{C}}-O-R-O-\overset{\displaystyle\|}{\underset{\displaystyle O}{C}}-R'-\overset{\displaystyle\|}{\underset{\displaystyle O}{C}}-O-R-O-$$

section of a polyester molecule

A thin polyester film was used to cover the wings and pilot compartment of the *Gossamer Albatross*, the first human-powered aircraft to cross the English Channel (Fig. 27.3).

One of the most familiar polyesters is Dacron, in which the monomers are ethylene glycol and terephthalic acid:

HO—CH$_2$—CH$_2$—OH   HO—$\overset{\displaystyle\|}{\underset{\displaystyle O}{C}}$—⬡—$\overset{\displaystyle\|}{\underset{\displaystyle O}{C}}$—OH

ethylene glycol   terephthalic acid

In developing polymers, chemists try all kinds of monomers. Some work well, most don't

■ **EXAMPLE 27.2** ────────────────

a. Write the structural formula of the ester formed when one molecule of ethylene glycol reacts with one molecule of terephthalic acid.
b. Draw the structure of a section of the Dacron (Mylar) polymer.

Solution

a. $HO-CH_2-CH_2-O-\overset{\displaystyle C}{\underset{\displaystyle O}{\|}}-⟨\text{benzene ring}⟩-\overset{\displaystyle C}{\underset{\displaystyle O}{\|}}-OH$

b. $-\overset{\displaystyle C}{\underset{\displaystyle O}{\|}}-⟨\text{benzene ring}⟩-\overset{\displaystyle C}{\underset{\displaystyle O}{\|}}-O-CH_2-CH_2-O-\overset{\displaystyle C}{\underset{\displaystyle O}{\|}}-⟨\text{benzene ring}⟩-\overset{\displaystyle C}{\underset{\displaystyle O}{\|}}-O-CH_2-CH_2-O-$

Exercise

Kodel is a polyester made from terephthalic acid and the alcohol

$$HO-CH_2-CH⟨\begin{array}{c}CH_2-CH_2\\CH_2-CH_2\end{array}⟩CH-CH_2-OH$$

Draw the structure of a portion of the Kodel polymer. Answer:

$$-\overset{\displaystyle C}{\underset{\displaystyle O}{\|}}-⟨\text{benzene ring}⟩-\overset{\displaystyle C}{\underset{\displaystyle O}{\|}}-O-CH_2-CH⟨\begin{array}{c}CH_2-CH_2\\CH_2-CH_2\end{array}⟩CH-CH_2-O-$$

POLYAMIDES

When a diamine (molecule containing two NH_2 groups) reacts with a dicarboxylic acid (two COOH groups), a **polyamide** is formed. This condensation polymerization is entirely analogous to that used to make polyesters. In this case, the NH_2 group of the diamine reacts with the COOH group of the dicarboxylic acid:

$$NH_2-R-\overset{\displaystyle N}{\underset{\displaystyle H}{|}}-H + HO-\overset{\displaystyle C}{\underset{\displaystyle O}{\|}}-R'-\overset{\displaystyle C}{\underset{\displaystyle O}{\|}}-OH \rightarrow NH_2-R-\overset{\displaystyle N}{\underset{\displaystyle H}{|}}-\overset{\displaystyle C}{\underset{\displaystyle O}{\|}}-R'-\overset{\displaystyle C}{\underset{\displaystyle O}{\|}}-OH + H_2O \qquad (27.7)$$

Condensation can continue to form a long-chain polymer (Example 27.3).

■ **EXAMPLE 27.3**

Consider the diamine $H_2N-(CH_2)_6-NH_2$ and the dicarboxylic acid $HOOC-(CH_2)_4-COOH$. Give the

a. structural formula of the dimer formed when one molecule of diamine reacts with one molecule of dicarboxylic acid.

b. structure of a section of the polyamide formed by these two monomers.

Solution

a. H₂N—(CH₂)₆—N—C—(CH₂)₄—C—OH
                 | || ||
                 H O   O

b. —C—(CH₂)₄—C—N—(CH₂)₆—N—C—(CH₂)₄—C—N—(CH₂)₆—N—
   || || | | || || | |
   O   O H   H O   O H   H

Exercise
Consider the dimer

H₂N—(CH₂)₅—N—C—⟨benzene ring⟩—C—OH
               | || ||
               H O   O

Write structural formulas for the amine and carboxylic acid used to make this amide. Answer:

H₂N—(CH₂)₅—NH₂ and HOOC—⟨benzene ring⟩—COOH

Figure 27.4 Nylon can be made by the reaction between hexamethyl-enediamine, H₂N—(CH₂)₆—NH₂ and adipic acid, HOOC—(CH₂)₄—COOH. It forms at the interface between the two reagents. (Charles D. Winters)

The polyamide in Example 27.3 is Nylon 66 (the numbers indicate the number of carbon atoms in the amine and acid monomers). This polyamide was first made in 1935 by Wallace Carothers, working at DuPont (Fig. 27.4). Since then, other nylons, all polyamides, have been synthesized. The earliest use of nylon was as a textile, especially for women's hosiery. Careful control of chain length during manufacturing gives nylon fibers of the desired sheerness and luster. The elasticity of nylon fibers is due in part to hydrogen bonds between adjacent polymer chains. These hydrogen bonds join carboxyl oxygen atoms on one chain to NH groups on adjacent chains. Bulk nylon can be molded or cut into a variety of shapes. Its high wear resistance and slight slipperiness make bulk nylon ideal for door latches and gears.

Polyurethanes are condensation polymers related to polyesters and polyamides. A typical polyurethane has the structure

              O   O   O   O
              || || || ||
—O—CH₂—O—C—N—(CH₂)₄—N—C—O—CH₂—O—C—N—(CH₂)₄—N—C—
                   | | | |
                   H   H   H   H

Polyurethanes are used as foams in mattresses, portable ice chests, and many other products. The foaming agent is carbon dioxide, formed in the polymerization process. Polyurethanes are also added to varnishes to give a tough, shiny surface. A polyurethane fabric is used in the Jarvik-7 artificial heart (Fig. 27.5).

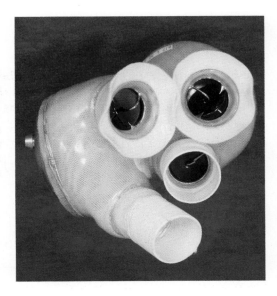

Figure 27.5 The Jarvik-7 artificial heart; the base is aluminum and the valves are pyrolitic graphite and titanium. The base, valves, and valve-holding rings are coated with the polyurethane Biomer. Layers of this polymer serve as the diaphragm in this device. (University of Utah Medical Center)

27.3
CARBOHYDRATES

Carbohydrates, which comprise one of the three basic classes of foodstuffs, contain carbon, hydrogen, and oxygen atoms. Their general formula, $C_n(H_2O)_m$, is the basis for their name. They can be classified as:

1. *Monosaccharides,* which cannot be broken down chemically to simpler carbohydrates. The most familiar monosaccharides, including glucose and fructose, contain six carbon atoms per molecule and have the molecular formula $C_6H_{12}O_6$.

2. *Disaccharides,* which are dimers formed when two monosaccharide units combine with the elimination of H_2O. The monosaccharides may be the same (two glucose units in maltose) or different (a glucose and fructose unit in sucrose).

3. *Polysaccharides,* which are condensation polymers containing from several hundred to several thousand monosaccharide units. Cellulose and starch are the most common polysaccharides.

GLUCOSE

By far the most important monosaccharide is glucose. In water solution, glucose exists primarily as six-membered ring molecules, shown below. The rings are not planar but have a three-dimensional "chair" form. The carbon atoms at each corner of the ring are not shown in the structure; they are represented by the numbers 1 to 5. The sixth atom in the ring is oxygen. There is a CH_2OH group bonded to carbon atom 5. There is an H atom bonded to each of the ring carbons; an OH group is bonded to carbon atoms 1 through 4. The heavy lines indicate the front of the ring, projected toward you.

Glucose has many isomers and several chiral centers

The bonds to the H and OH groups from the ring carbons are oriented either in the plane of the ring (*equatorial*, shown in blue) or perpendicular to it (*axial*). In β-glucose, all four OH groups are equatorial. This keeps these rather bulky groups out of each other's way. In α-glucose, the OH group on carbon atom 1 is axial; all the others are equatorial. In water solution, glucose exists as an equilibrium mixture, about 37% in the alpha form and 63% in the beta form.

How in the world could we prove that?
Ans.: Optical rotation

α-glucose   β-glucose

As you can see from their structures, α- and β-glucose have several chiral carbon atoms. Both isomers are optically active; they are *not* enantiomers (mirror images of one another) because they differ in configuration only at carbon atom 1. As it happens, both α- and β-glucose rotate the plane of polarized light to the right (clockwise).

Glucose is also called "blood sugar." It is absorbed readily into the bloodstream and is normally found there at concentrations ranging from 0.004 to 0.008 mol/L. If the concentration of glucose drops below 0.003 M, a condition called hypoglycemia is created, with symptoms ranging from nervousness to loss of consciousness. If the glucose level rises above 0.01 M, as can easily happen with diabetics, it is excreted in the urine.

Although glucose can exist as a simple sugar, it is most often found in nature in combined form, as a disaccharide or polysaccharide. Several glucose-containing disaccharides are known. We will consider two of these, maltose and sucrose.

MALTOSE AND SUCROSE

Maltose, a decomposition product of starch, is a dimer of two glucose molecules. These are combined head-to-tail; carbon atom 1 of one molecule is joined through an oxygen atom to carbon atom 4 of the second molecule. To form maltose, the two OH groups on these carbon atoms react, condensing out H_2O and leaving the O atom bridge.

α-glucose   α-glucose   α-maltose $+ H_2O$ (27.8)

Reaction 27.8 can be reversed either by using the enzyme maltase as a catalyst or by heating with acid. Two molecules of glucose are formed when this happens.

Sucrose, the compound we call "sugar," is the most common disaccharide. One of the monomer units in sucrose is α-glucose. The other is fructose, a monosaccharide found in honey and natural fruit juices.

In the body, Reaction 27.9 is reversed by the enzyme sucrase. This occurs in digestion, which makes glucose and fructose available for absorption into the blood. Honey bees also carry an enzyme which can hydrolyze sucrose. Honey consists mostly of a 1:1 mol mixture of glucose and fructose with a small amount of unreacted sucrose.

■ **EXAMPLE 27.4**
Another common disaccharide is lactose, the sugar found in milk. Lactose is a dimer of glucose and galactose, a monosaccharide which is identical to glucose except that the positions of the H and OH groups at carbon atom 4 are switched.

a. Draw the structure of β-galactose.
b. Draw the structure of lactose, which is broken down by the enzyme lactase to give equimolar amounts of β-galactose and β-glucose. Linkage occurs between carbon atom 1 of galactose and 4 of glucose.

Solution

ı. The structure of β-galactose is identical to that of β-glucose except at carbon atom 4. Showing equatorial groups in color:

b.

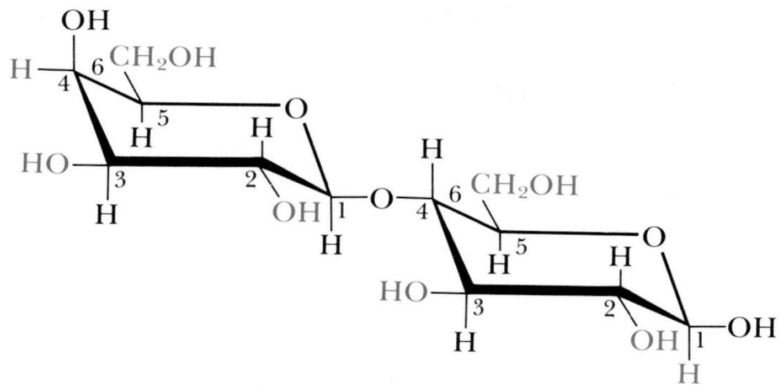

Exercise

How many chiral atoms are there in lactose? What is its molecular formula?
Answer: 10, $C_{12}H_{22}O_{11}$.

STARCH AND GLYCOGEN

Starch is a polysaccharide found in many plants, where it is stored in roots and seeds. It is particularly abundant in corn and potatoes, the major sources of commercial starch. Perhaps as much as 50% of our food energy comes from starch, mostly in the form of wheat products.

Starch is actually a mixture of two types of α-glucose polymers. One of these, called amylose, is insoluble in water and comprises about 20% of natural starch. It consists of long single chains of 1000 or more α-glucose units joined head-to-tail, as in maltose (Fig. 27.6). The other glucose polymer found in starch is amylopectin, which is soluble in water. The linkage between α-glucose units is the same as in amylose. However, amylopectin has a highly branched structure. Short chains of 20 to 25 glucose units are linked through oxygen atom bridges. The oxygen atom joins a number 1 carbon atom at the end of one chain to a number 6 carbon atom on an adjacent chain.

Starch and cellulose are both condensation polymers of glucose

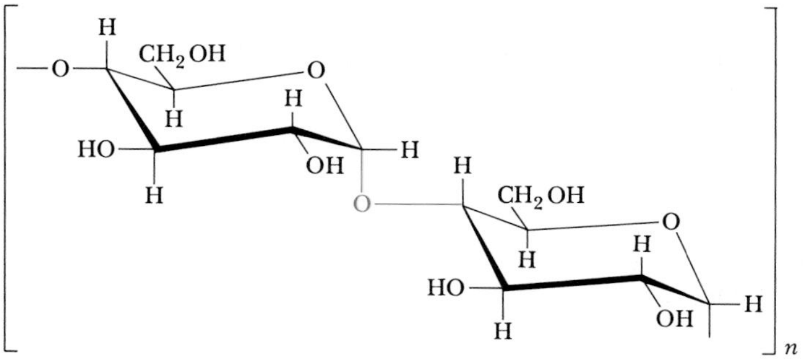

Amylose

Figure 27.6 In starch, glucose molecules are joined head-to-tail through oxygen atoms. A thousand or more glucose molecules may be linked in this way, either in long single chains (amylose) or branched chains (amylopectin).

When you eat "starchy" foods, they are broken down into glucose by enzymes. The process starts in your mouth with the enzyme amylase found in saliva. This explains why, if you chew a piece of bread long enough, it starts to taste sweet. The breakdown of starch molecules continues in other parts of the digestive system. Within 1 to 4 hours after eating, all the starch in food is converted into glucose.

■ **EXAMPLE 27.5**
What is the simplest formula of amylose?

Solution
Amylose is a condensation polymer of glucose, whose formula is $C_6H_{12}O_6$. Every time a glucose unit is added to the polysaccharide chain a molecule of H_2O is eliminated. Hence, the unit added is not $C_6H_{12}O_6$, but rather

$C_6H_{10}O_5$.   This is the simplest formula of amylose.

Exercise
Amylose has a molar mass of about 3.00×10^5 g/mol. How many glucose units does an amylose molecule contain? Answer: 1850.

In animals (including human beings), glucose not needed immediately for energy is polymerized to *glycogen* and stored in the liver. This polysaccharide has a structure similar to amylopectin except that it is more highly branched. The branching is thought to make the breakdown of glycogen to glucose easier. This commonly occurs several hours after eating, when blood glucose has dropped to a relatively low level.

Glycogen breakdown helps stabilize blood sugar levels

CELLULOSE

Cellulose contains long, unbranched chains of glucose units, about 10,000 per chain. It differs from starch in the way the glucose units are joined to each other. In starch, the oxygen bridge between the units is in the alpha position; in cellulose it is in the beta position (Fig. 27.7). Because humans lack the enzymes required to catalyze the hydrolysis of beta linkages, we cannot digest cellulose. The cellulose that we take in from fruits and vegetables remains undigested as "fiber." Ruminants, like cows and deer, and termites have bacteria in their intestines capable of breaking cellulose down into glucose.

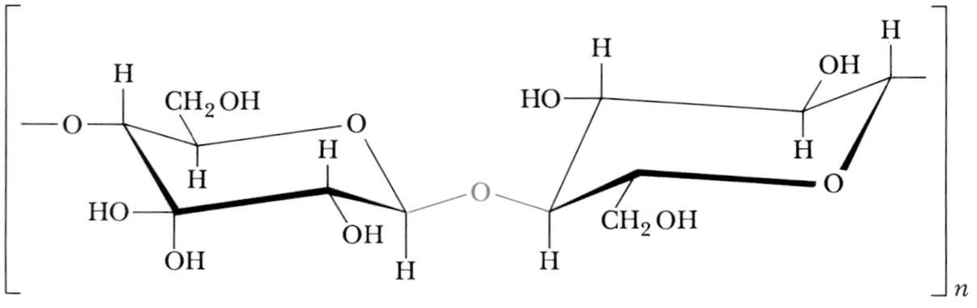

Figure 27.7 The bonding between glucose rings in cellulose is through oxygen bridges in the β position for each ring to the left of the bridge. This structure allows for ordered hydrogen bonding between chains and formation of long strong fibers. Cotton and wood fiber have structures like this.

The molecular structure of cellulose, unlike that of starch, allows for strong hydrogen bonding between polymer chains. This results in the formation of strong water-resistant fibers such as those found in cotton, which is 98% cellulose. Cotton actually has a tensile strength greater than that of steel. The major industrial source of cellulose is wood (~50% cellulose). Wood chips can be treated with hot NaOH solution, which dissolves some of the wood components and partially breaks down the polymer chains. The insoluble residue is impure cellulose in the form of wood pulp. It can be used directly to make paper or treated further to make various plastics. Although cellulose can be degraded completely to glucose, the process has not been practical economically. If it could be made so, the glucose could be used for food, or converted to ethanol by fermentation. Both such uses would be important, in view of world shortages of food and energy. Genetic engineers are trying to "tailor-make" bacteria that will decompose cellulose to glucose.

27.4 PROTEINS

The natural polymers known as proteins make up about 15% by mass of our bodies. They serve many functions. Fibrous proteins are the main components of hair, muscle, and skin. Other proteins found in body fluids transport oxygen, fats, and other substances needed for metabolism. Still others, such as insulin and vasopressin, are hormones. Enzymes, which catalyze reactions in the body, are chiefly protein.

Protein is an important component of most foods. Nearly everything we eat contains at least a small amount of protein. Lean meats and vegetables such as peas and beans are particularly rich in protein. In our digestive system, proteins are broken down into small molecules called *α-amino acids*. These molecules can then be reassembled in cells to form other proteins required by the body.

Proteins are condensation polymers of α-amino acids

α-AMINO ACIDS

The monomers from which proteins are derived have the general structure shown below. These compounds are called α-amino acids. They have an NH_2 group attached to the carbon atom (the α carbon) adjacent to a COOH group.

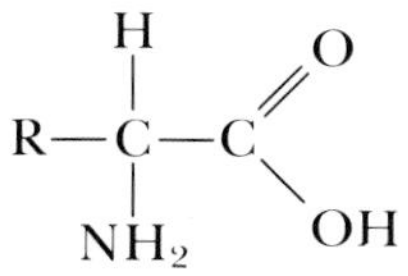

R—C—C
α-amino acid

Natural proteins can be broken down into about 20 different α-amino acids (Table 27.2). These molecules differ in the nature of the R group attached to the alpha carbon. As you can see from the table, R can be:

TABLE 27.2   The Common α-Amino Acids (names and abbreviations below structures)

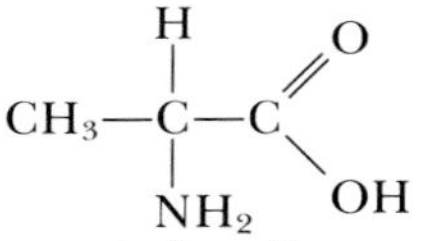

alanine Ala

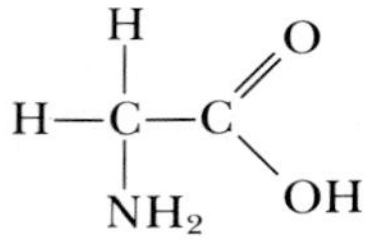

glycine Gly

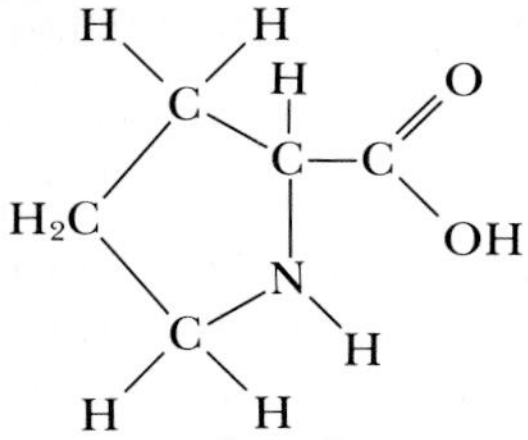

proline Pro

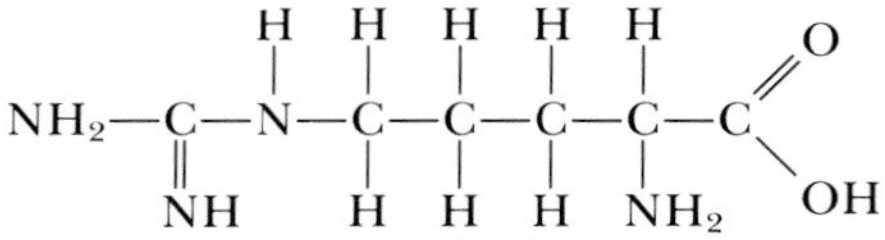

arginine* Arg

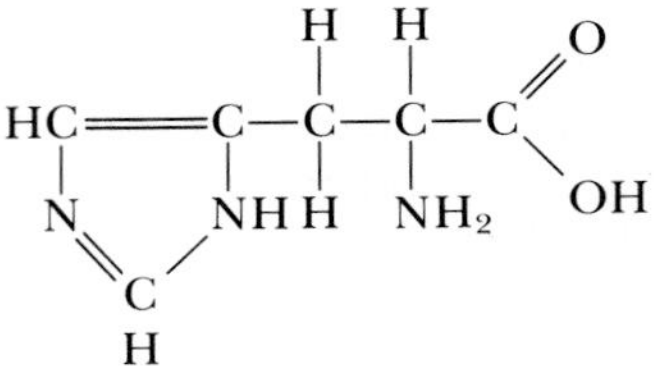

histidine His

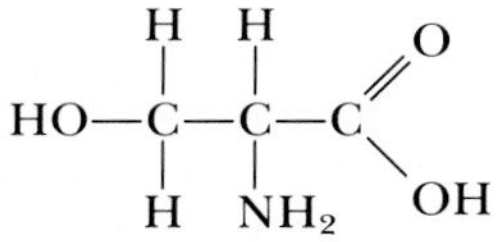

serine Ser

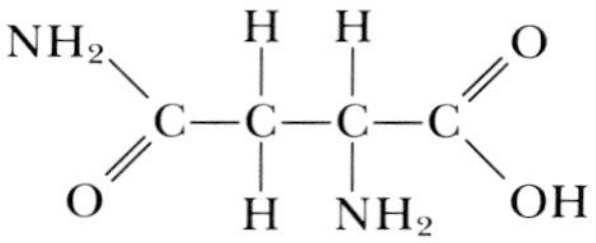

asparagine Asn

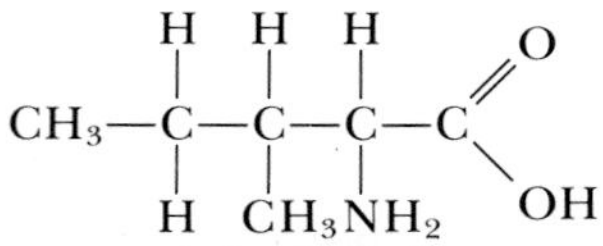

isoleucine* Ile

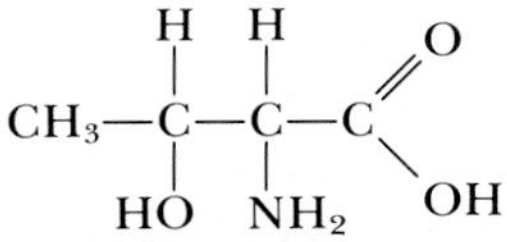

threonine* Thr

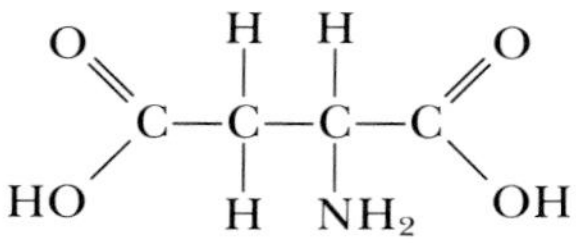

aspartic acid Asp

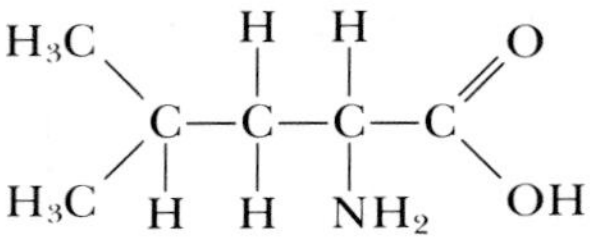

leucine* Leu

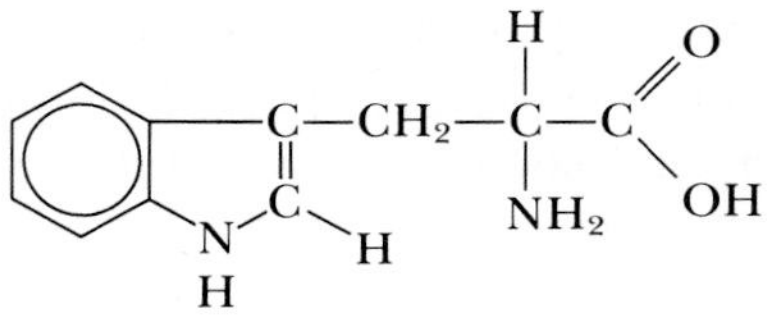

tryptophan* Trp

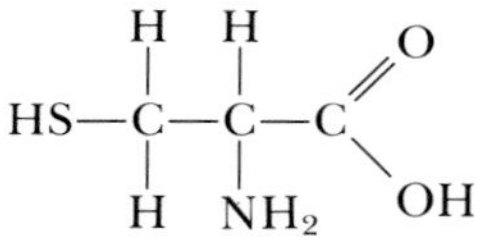

cysteine Cys

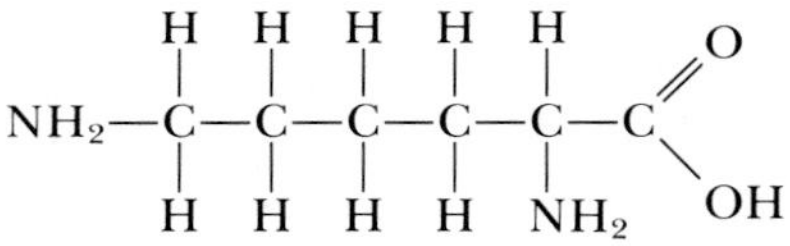

lysine* Lys

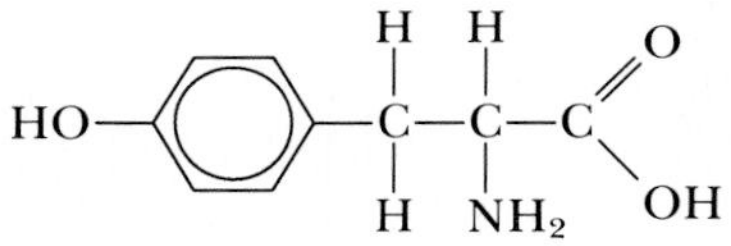

tyrosine Tyr

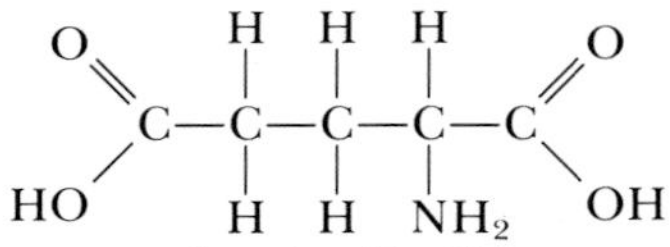

glutamic acid Glu

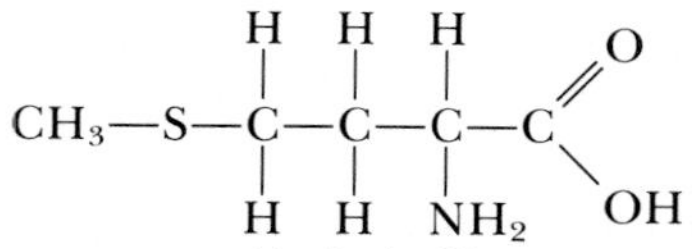

methionine* Met

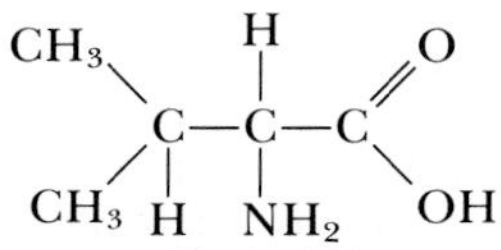

valine* Val

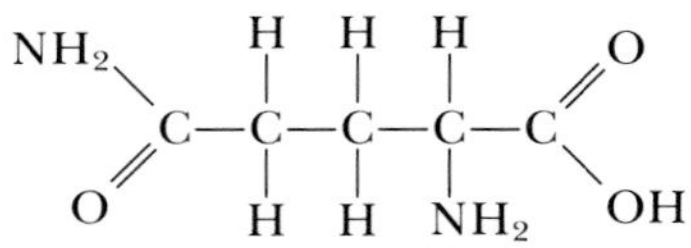

glutamine Gln

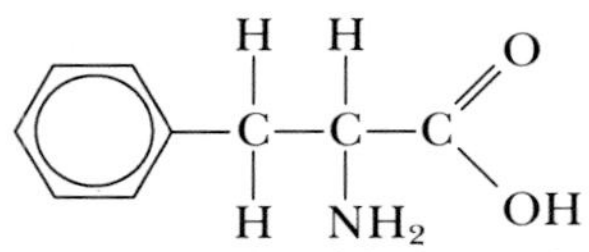

phenylalanine* Phe

*These are essential amino acids that cannot be synthesized by the body and therefore must be obtained from the diet.

— a H atom (glycine).

— a simple hydrocarbon group (alanine, valine, . . .).

— a more complex group containing one or more atoms of oxygen (serine, aspartic acid, . . .), nitrogen (lysine, . . .), or sulfur (methionine, . . .).

In all the amino acids shown in Table 27.2 except glycine, the α-carbon is chiral. This means that these compounds are optically active. With alanine, for example, there should be two optical isomers:

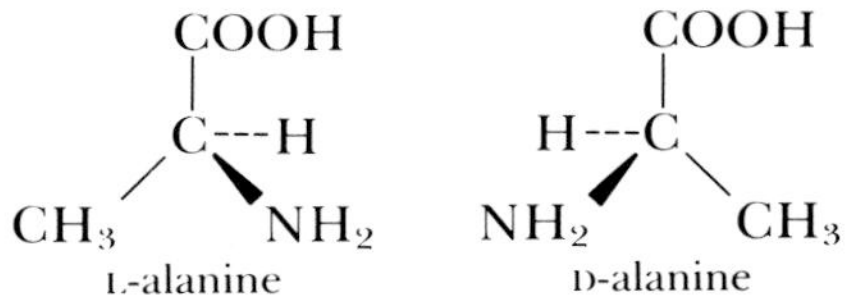

COOH   COOH

C---H   H---C

CH₃   NH₂   NH₂   CH₃

L-alanine   D-alanine

All of the amino acids found naturally in proteins have the L-configuration.

Of the 20 common amino acids, eight are "essential" in the sense that they are required by adult humans for protein construction and cannot be synthesized in the body. These are designated by an asterisk in Table 27.2; one other amino acid, histidine, is required by infants but not by adults. Clearly the proteins we eat should contain these essential amino acids. Beyond that, a "high-quality" protein contains the various amino acids in the approximate ratio required by the body for protein synthesis. Generally speaking, animal proteins are of higher quality than plant proteins, with eggs at the head of the class. This is something to keep in mind if you're a vegetarian.

ACID-BASE PROPERTIES OF AMINO ACIDS

The structures shown in Table 27.2 are those ordinarily written for α-amino acids. However, molecular structures such as these fail to explain the physical properties of this class of compounds. Glycine, for example, is a solid melting at 232°C; beyond that, it is very soluble in water but almost completely insoluble in nonpolar organic solvents such as benzene. This is hardly the behavior we would expect for a small molecule with a molar mass of 75 g/mol.

These and other observations lead us to believe that in the solid state, α-amino acids exist in the form of dipolar ions, often called **zwitterions**, formed by the transfer of a proton from a —COOH group to an —NH₂ group:

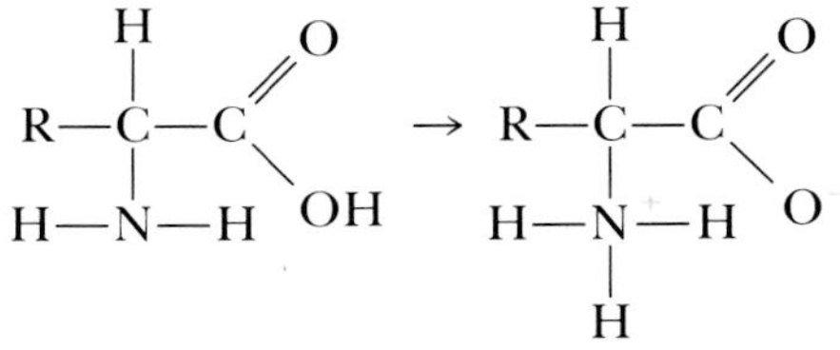

amino acid   zwitterion

A zwitterion behaves like a polar molecule. Within it, there is a +1 charge at the nitrogen atom and a −1 charge at one of the oxygen atoms of the COOH group. Overall, the zwitterion has no net charge.

In water zwitterions are stable only over a certain pH range. At low pH (high H^+ concentration), the COO^- group picks up a proton. The product, shown at the left below, has a +1 charge. At high pH (low H^+ concentration), the NH_3^+ group loses a proton. The species formed, shown at the right below, has a −1 charge.

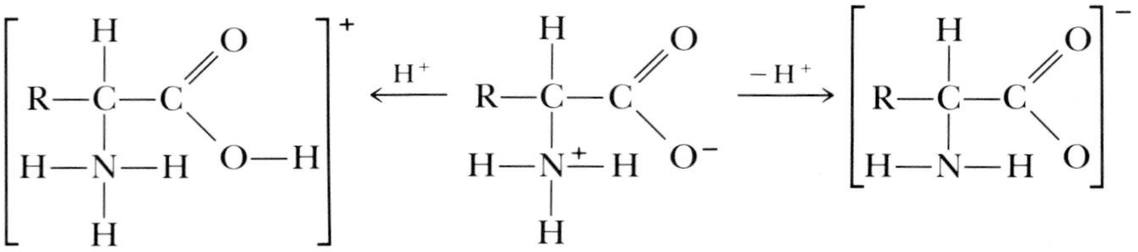

$$
\begin{array}{ccc}
\text{cation} & \text{zwitterion} & \text{anion}
\end{array}
$$

In a very real sense, a zwitterion is analogous to a species such as the HCO_3^- ion, which accepts a proton at low pH and donates a proton at high pH:

$$
H_2CO_3(aq) \xleftarrow{H^+} HCO_3^-(aq) \xrightarrow{-H^+} CO_3^{2-}(aq)
$$
$$
\quad\quad \text{low pH} \quad\quad\quad\quad\quad\quad\quad \text{high pH}
$$

Both the zwitterion and the HCO_3^- ion are amphoteric, capable of acting as either a Brönsted acid or a Brönsted base.

To treat acid-base equilibria involving zwitterions, it is convenient to consider the cation stable at low pH to be a diprotic acid (analogous to H_2CO_3), which ionizes in two steps. Using the symbols C^+, Z, and A^- to stand for the cation, zwitterion, and anion respectively, we have:

$$
C^+(aq) \rightleftharpoons H^+(aq) + Z(aq); \qquad K_{a1} = \frac{[H^+] \times [Z]}{[C^+]} \tag{27.10}
$$

$$
Z(aq) \rightleftharpoons H^+(aq) + A^-(aq); \qquad K_{a2} = \frac{[H^+] \times [A^-]}{[Z]} \tag{27.11}
$$

For glycine:

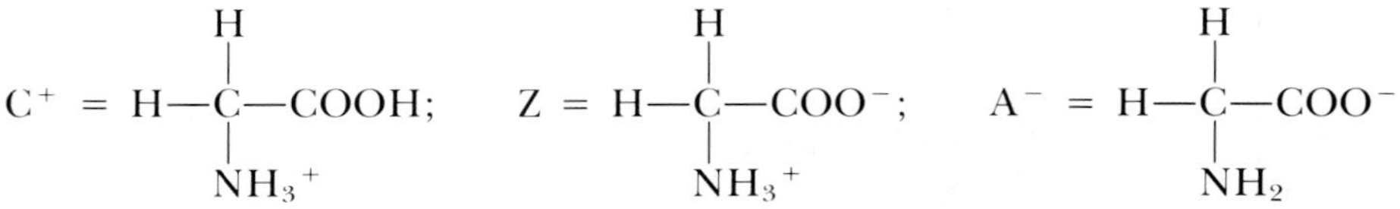

$$
K_{a1} = 4.6 \times 10^{-3}; \qquad K_{a2} = 2.5 \times 10^{-10}
$$

■ **EXAMPLE 27.6**

Using these acid dissociation constants for glycine, calculate the ratios $[Z]/[C^+]$ and $[Z]/[A^-]$ at pH

a. 2.00.   b. 6.00.   c. 10.00.

Solution

Using the general equations 27.10 and 27.11 with the appropriate equilibrium constants, we have:

$$\frac{[Z]}{[C^+]} = \frac{K_{a1}}{[H^+]} = \frac{4.6 \times 10^{-3}}{[H^+]} \; ; \qquad \frac{[Z]}{[A^-]} = \frac{[H^+]}{K_{a2}} = \frac{[H^+]}{2.5 \times 10^{-10}}$$

Using these relationships, we can readily calculate the required ratios at various pH's.

a. Here, the concentration of H^+ is $1.0 \times 10^{-2} \, M$, so we have

$$\frac{[Z]}{[C^+]} = \frac{4.6 \times 10^{-3}}{1.0 \times 10^{-2}} = 0.46;$$

$$\frac{[Z]}{[A^-]} = \frac{1.0 \times 10^{-2}}{2.5 \times 10^{-10}} = 4.0 \times 10^{7}$$

At this low pH, there is more cation than zwitterion, $[Z] < [C^+]$, and essentially no anion, $[Z] \gg [A^-]$.

b.
$$\frac{[Z]}{[C^+]} = \frac{4.6 \times 10^{-3}}{1.0 \times 10^{-6}} = 4.6 \times 10^{3} \; ;$$

$$\frac{[Z]}{[A^-]} = \frac{1.0 \times 10^{-6}}{2.5 \times 10^{-10}} = 4.0 \times 10^{3}$$

If you dissolve glycine in water, you get mostly zwitterion

At this intermediate pH, the zwitterion is the principal species present. Its concentration is much higher than that of either cation or anion.

c.
$$\frac{[Z]}{[C^+]} = \frac{4.6 \times 10^{-3}}{1.0 \times 10^{-10}} = 4.6 \times 10^{7} \; ;$$

$$\frac{[Z]}{[A^-]} = \frac{1.0 \times 10^{-10}}{2.5 \times 10^{-10}} = 0.40$$

At this high pH, the anion is the principal species:

$$[C^+] \ll [Z] < [A^-]$$

Exercise

At what pH is $[Z] = [C^+]$? $[Z] = [A^-]$? Answer: 2.34, 9.60.

From Example 27.6 or Figure 27.8, p. 972, it is clear that the zwitterion of glycine is the principal species over a wide pH range, certainly from pH 3 to pH 9. The maximum concentration of zwitterion occurs at about pH 6; this is referred to as the *isoelectric point* of glycine. At this pH, glycine does not migrate in an electric field, since the zwitterion is a neutral species. At low pH, the cation moves to the cathode; at high pH the anion migrates to the anode. In general, for an amino acid like glycine, where there is only one —COOH group and one —NH$_2$ group:

$$\text{pH at isoelectric point} = \frac{pK_{a1} + pK_{a2}}{2}$$

Figure 27.8 Relative concentrations of different glycine species as a function of pH. Between about pH 3 and 9, the zwitterion is the principal species; it has its maximum concentration at pH 6, the isoelectric point. Below pH 2, the cation dominates; above pH 10, the anion is the principal species.

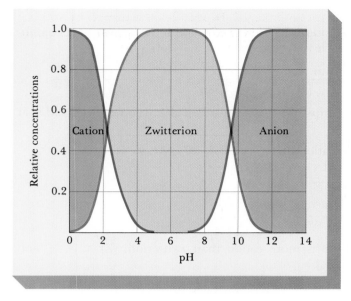

where K_{a1} and K_{a2} are the successive dissociation constants of the cation of the amino acid.

POLYPEPTIDES

As noted in Section 27.2, molecules containing NH_2 and $COOH$ groups can undergo condensation polymerization. Amino acids contain both groups in the same molecule. Hence, two amino acid molecules can combine by the reaction of the $COOH$ group in one molecule with the NH_2 group of the other molecule. If the acids involved are different, two different products are possible:

$$H_2N-CH_2-\underset{\underset{O}{\|}}{C}-OH + H-\underset{\underset{H}{|}}{N}-\underset{\underset{CH_3}{|}}{\overset{\overset{H}{|}}{C}}-COOH \rightarrow H_2N-CH_2-\underset{\underset{O}{\|}}{C}-\underset{\underset{H}{|}}{N}-\underset{\underset{CH_3}{|}}{\overset{\overset{H}{|}}{C}}-COOH + H_2O \qquad (27.12)$$

glycine   alanine   glycylalanine (Gly-Ala)

$$H_2N-\underset{\underset{CH_3}{|}}{\overset{\overset{H}{|}}{C}}-\underset{\underset{O}{\|}}{C}-OH + H-\underset{\underset{H}{|}}{N}-CH_2-COOH \rightarrow H_2N-\underset{\underset{CH_3}{|}}{\overset{\overset{H}{|}}{C}}-\underset{\underset{O}{\|}}{C}-\underset{\underset{H}{|}}{N}-CH_2-COOH + H_2O \qquad (27.13)$$

alanine   glycine   alanylglycine (Ala-Gly)

The products are referred to as *dipeptides;* they contain the group

This linkage is similar to that in nylon

$$-\underset{\underset{O}{\|}}{C}-\underset{\underset{H}{|}}{N}-$$

commonly called a "peptide linkage." **Dipeptides are named by writing first (far left) the amino acid which retains a free NH$_2$ group; the amino acid with a free COOH group is written last (far right).**

■ **EXAMPLE 27.7** _____

Draw the structural formula of glycylserine (Gly-Ser).

Solution

The COOH group of glycine condenses with the NH$_2$ group of serine:

$$H_2N—CH_2—\underset{\underset{O}{\parallel}}{C}—OH + H—\underset{\underset{H}{\mid}}{N}—\underset{\overset{CH_2OH}{\mid}\atop\underset{H}{\mid}}{C}—COOH \rightarrow$$

glycine   serine

$$H_2N—CH_2—\underset{\underset{O}{\parallel}}{C}—\underset{\underset{H}{\mid}}{N}—\underset{\overset{CH_2OH}{\mid}\atop\underset{H}{\mid}}{C}—COOH + H_2O$$

glycylserine (Gly-Ser)

Exercise

Draw the structural formula of the other dipeptide that can be formed from glycine and serine. Answer:

$$H_2N—\underset{\overset{CH_2OH}{\mid}\atop\underset{H}{\mid}}{C}—\underset{\underset{O}{\parallel}}{C}—\underset{\underset{H}{\mid}}{N}—CH_2—COOH$$

_____

The dipeptides formed by Reactions 27.12 and 27.13 have reactive groups at both ends of the molecule (NH$_2$ at one end, COOH at the other). Hence they can combine further to give tripeptides (three amino acid units), tetrapeptides, . . . , and eventually long-chain polymers called polypeptides. Proteins are polypeptides containing a large number of amino acid units:

$$—\underset{\underset{H}{\mid}}{N}—\underset{\underset{R_1}{\mid}}{C}—\underset{\underset{O}{\parallel}}{C}—\underset{\underset{H}{\mid}}{N}—\underset{\underset{R_2}{\mid}}{C}—\underset{\underset{O}{\parallel}}{C}—\underset{\underset{H}{\mid}}{N}—\underset{\underset{R_3}{\mid}}{C}—\underset{\underset{O}{\parallel}}{C}—\underset{\underset{H}{\mid}}{N}—\underset{\underset{R_4}{\mid}}{C}—\underset{\underset{O}{\parallel}}{C}—$$

(peptide linkages in color).

Proteins differ from the other polymers we have discussed in that they may contain up to 20 different monomer units. This means that there are a huge number of possible proteins. Using the amino acids listed in Table 27.2, we could make

$20 \times 20 = 400$ different dipeptides

$20 \times 20 \times 20 = 8000$ different tripeptides

20^n polypeptides containing n monomer units

For a relatively simple protein containing only 50 monomer units, we have

$20^{50} = 1 \times 10^{65}$ possibilities

PROTEINS; PRIMARY STRUCTURE

As you can imagine, it is not an easy task to identify all the amino acid units present in a protein chain. To go further and determine the **primary structure** of the protein, the sequence in which these units are arranged, might seem next to impossible. In recent years, however, this type of analysis has become possible. The primary structure of some very complex proteins has been determined. The first protein for which this was done was insulin. The structure for insulin is shown in Figure 27.9. This structure was established by Frederick Sanger, who received the 1958 Nobel prize in chemistry for his work.

The amino acid sequence in hemoglobin, with 574 units, is known

In general, the amino acid sequence in proteins is determined by two different methods.

1. *End-group analysis.* A reagent is added to the protein which reacts either with the free —NH$_2$ group at one end of the chain or the free —COOH group at the other end, but not with peptide linkages within the protein chain. The protein is then broken down into its constituent amino acids and the "tagged" amino acid at one end of the chain is identified. Sanger used the reagent 2,4-dinitrofluorobenzene, which reacts with an end —NH$_2$ group:

$$NO_2-\text{⬡}-F(aq) \;+\; R-NH_2(aq) \;+\; OH^-(aq) \;\rightarrow\; NO_2-\text{⬡}-\overset{H}{\underset{}{N}}-R(aq) \;+\; F^-(aq) \;+\; H_2O$$

(the first ring bearing NO$_2$ substituent; the product ring bearing NO$_2$ substituent)

Nowadays, it is possible to tag an end group with a reagent, split off and identify the end amino acid, tag the next group in the chain, and so on. By this means, it is possible to establish the identity and position of up to 20 amino acids at one end of the chain.

Column chromatography is very useful in this kind of research

2. *Hydrolysis of the protein chain into fragments containing 2-20 amino acids.* This is necessary with long-chain proteins containing hundreds or even thousands of amino acids. The compositions of the fragments can then be established by end group analysis. The protein chain can be split selectively by using a reagent that attacks it only at certain points. For example, the enzyme trypsin cleaves a chain only at the carboxyl end of a lysine or arginine group. It would split the polypeptide

Ala—Gly—Tyr—Trp—Ser—Lys—Gly—Leu—Arg—Met

into the three fragments:

    Ala—Gly—Tyr—Trp—Ser—Lys;   Gly—Leu—Arg;   Met

Remember that, by convention, the NH_2 group of each amino acid is at the *left;* the COOH group is at the *right.* Other reagents split the chain at different points. By correlating the results of different hydrolysis experiments, it is possible to deduce the primary structure of the protein.

■ **EXAMPLE 27.8** —————————————————————————

Consider the hydrolysis experiment referred to above, where the fragments obtained with trypsin are:

    Ala—Gly—Tyr—Trp—Ser—Lys;   Gly—Leu—Arg;   Met

a. Give two possible primary structures for the polypeptide, based on this information alone.

b. In another experiment with a different reagent the fragments obtained are

    Ala—Gly;   Tyr—Trp—Ser—Lys—Gly;   Leu—Arg—Met

What is the primary structure?

Solution

a. In principle the three fragments could be arranged in six different ways. In practice, only two arrangements are possible; the "Met" group must be at the far right of the polypeptide, since trypsin can split the chain only at the —COOH group of a "Lys" or "Arg" unit:

    (1) Ala—Gly—Tyr—Trp—Ser—Lys—Gly—Leu—Arg—Met

    (2) Gly—Leu—Arg—Ala—Gly—Tyr—Trp—Ser—Lys—Met

b. These fragments are consistent only with (1), which must be the correct structure.

Exercise

Assume that the reagent referred to in (b) can only split the chain at the carboxyl end of a glycine molecule. What fragments would be obtained if the second structure listed in (a) were correct? Answer: Gly; Leu—Arg—Ala—Gly; Tyr—Trp—Ser—Lys—Met.

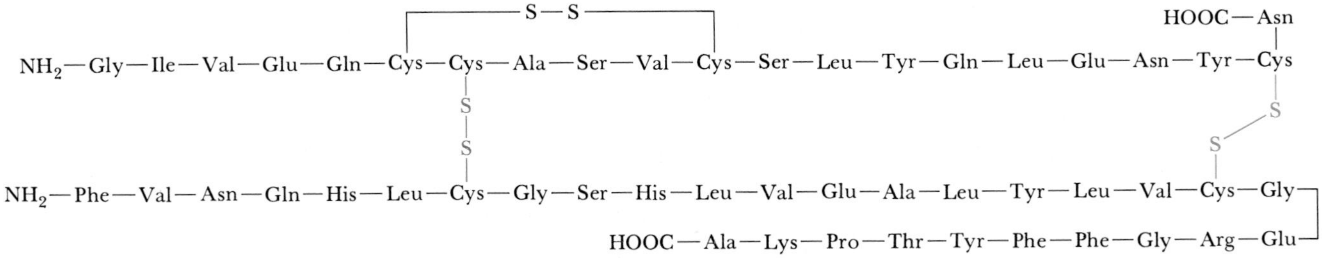

Figure 27.9 The primary structure of insulin. One chain contains 21 amino acids; the other has 30. The two chains are linked by disulfide (—S—S—) bridges.

In the body, proteins are built up by a series of reactions that in general produce a specific sequence of amino acids. Even tiny errors in this sequence may have serious effects. Among the genetic diseases known to be caused by improper sequencing are hemophilia, sickle cell anemia, and albinism. Sickle cell anemia is caused by the substitution of *one* valine unit for a glutamic acid unit in a chain containing 146 monomers.

Fortunately nature doesn't make too many mistakes

PROTEINS; SECONDARY AND TERTIARY STRUCTURES

Protein chains can align themselves so that certain patterns are repeated. These repeating patterns establish what we call the **secondary structure** of the protein. The nature of the pattern is determined in large part by hydrogen bonding. Oxygen atoms on C=O groups can interact with H atoms in nearby N—H groups to form these bonds. This can occur within a single protein chain or between neighboring chains.

This is a rather special kind of H bond

There are two ways in which a protein chain can be oriented to give maximum hydrogen bonding. When amino acid units with small R groups are present, as in glycine or alanine, hydrogen bonding leads to a pleated-sheet structure (Fig. 27.10). The sheet shown lies in the plane of the paper. It consists of many parallel chains held together by hydrogen bonds between peptide linkages in adjacent chains. By following the middle chain in Figure 27.10 you can see that hydrogen bonding occurs at each amino acid unit. This pleated-sheet structure is found in muscles and silk fibers.

Most proteins have amino acid units with R groups that are too bulky for a pleated-sheet structure. These proteins form a coil called an *α-helix* to maximize hydrogen bonding between peptide linkages. At the left of Figure 27.11 is the outline of the protein chain forming the helix. The more complete structure is given at the right. This shows where the hydrogen bonds form between amino acid units in the protein chain. Notice

Figure 27.10 β-keratin structure of silk fiber. When the R groups on the amino acid residues are small, the protein chains can hydrogen-bond in the roughly planar structure shown. Since the bonding around the N and noncarbonyl carbon atoms is tetrahedral (shown in blue), the sheet is pleated. Hydrogen bonds are shown by red dots in the drawing.

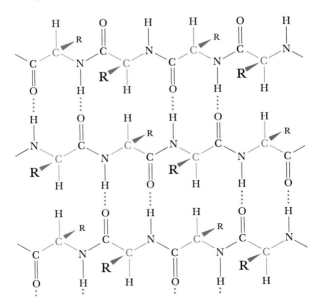

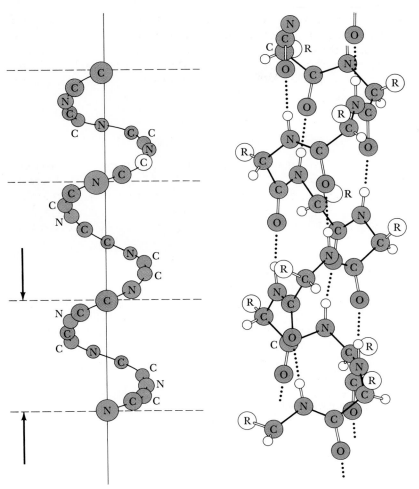

Figure 27.11 α-helix structure of proteins. The main atom chain in the helix is shown schematically on the left. The sketch on the right more nearly represents the actual positions of atoms and shows where intrachain hydrogen bonding occurs. Wool and many other fibrous proteins have the α-helix structure.

that the bulky R groups are all on the outside of the helix, where they have the most room. The actual structure of the helix is nearly independent of the nature of the R groups. The dimensions of the helix correspond closely to those observed in such fibrous proteins as wool, hair, skin, feathers, and fingernails.

The three-dimensional conformation of a protein is called its **tertiary structure**. An α-helix can either be twisted, folded, or folded and twisted into a definite geometric pattern. These structures are stabilized by dispersion forces, hydrogen bonding, and other intermolecular forces.

Collagen, the principal fibrous protein in mammalian tissue, has a tertiary structure made up of twisted α-helices. Three polypeptide chains, each of which is a left-handed helix, are twisted into a right-handed super helix to form an extremely strong tertiary structure. It has remarkable tensile strength which makes it important in the structure of bones, tendons, teeth, and cartilage.

Globular proteins are those proteins whose helical structure is both twisted and folded. They get this way because the nonpolar side chains

The conformation of a protein appears to be fixed mainly by the amino acid sequence in the chain

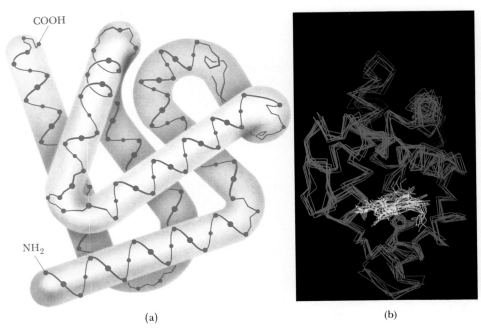

Figure 27.12 The structure of the protein myoglobin. The primary structure is shown in the diagram at the left as a sequence of dots representing the amino acids. The secondary structure is the clearly visible helical arrangement down the length of the chain. The tertiary structure is the folding and wrapping of the chain into its actual conformation. The photo (b) is a computerized drawing of myoglobin. (James Kilkelly/DOT)

(a)

(b)

tend to fold into the center of the chain to get away from water. Globular proteins like myoglobin (Fig. 27.12) tend to be proteins used for transport and metabolism rather than for structure. Myoglobin, for example, is used to transport oxygen to muscle. It is found in the thighs and wings of birds and is responsible for the color of "dark meat" in fowl.

Human hair consists largely of the protein called keratin, which contains a large number of cysteine molecules:

$$H_2N-\overset{\overset{\displaystyle H}{|}}{\underset{\underset{\displaystyle SH}{|}}{\underset{\underset{\displaystyle CH_2}{|}}{C}}}-COOH$$

cysteine

Cysteine molecules in adjacent keratin chains form S—S bridges:

978

The number and location of these bridges determines in large measure whether hair is straight or curly. When women (and an increasing number of men) get a permanent wave, a reducing agent called a "relaxer" is first applied. This breaks the —S—S— bridges, restoring the —SH groups and making the strands highly flexible. Curlers are then used to fix the hair in position, whereupon an oxidizing agent called a "setting agent" is added. This forms —S—S— bridges in the desired location.

$$\{-S-H + H-S-\} + \tfrac{1}{2}O_2(g) \rightarrow \{-S-S-\} + H_2O$$

27.5
NUCLEIC ACIDS

Living cells contain polymers called nucleic acids. These polymers were first isolated from cell nuclei, which explains their name, but are now known to be distributed throughout all parts of the cell. They play a vital role in the transmission of genetic information from one generation to another. Beyond that, nucleic acids are responsible for the synthesis of specific proteins, assuring that the requisite amino acids are assembled in the proper order. There are two general types of nucleic acids, known as DNA and RNA.

STRUCTURES OF DNA AND RNA

All nucleic acids can be hydrolyzed into three basic components.

1. *Phosphoric acid*

```
        H
        |
        O
        |
HO—P—O
        |
        O
        |
        H
```

In the polymer, the two hydrogen atoms shown in color are missing, allowing for bond formation through the oxygen atoms.

2. *A five-carbon sugar.* For DNA, the sugar is deoxyribose (thus the name deoxyribonucleic acid). In RNA, the sugar is ribose (hence ribonucleic acid). The structures of these sugars are shown in Table 27.3. Notice that they differ only at carbon atom 2, where the OH group of ribose is replaced by an H atom in deoxyribose.

3. *Organic, nitrogen-containing bases* (amines) containing five- and/or six-membered rings. The structures of the bases most commonly found in nucleic acids are shown in Table 27.3. These bases can be considered to be derived from the parent bases pyrimidine and purine:

pyrimidine 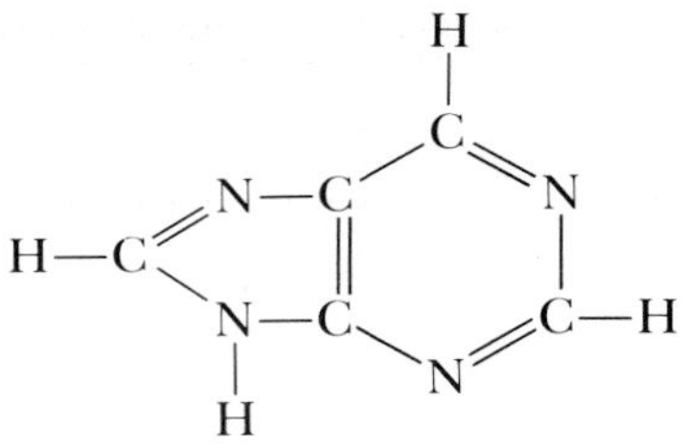   purine

Notice that three bases are common to both DNA and RNA; these are adenine (A), cytosine (C), and guanine (G). Thymine (T) is found only in DNA; uracil (U), which differs from thymine only in the absence of a methyl group, is found only in RNA.

The backbone of a nucleic acid polymer consists of a chain of alternating phosphate groups and sugar molecules. Attached to each sugar molecule is one of the bases shown in Table 27.3. Schematically, we can represent a section of a nucleic acid molecule as:

—phosphate—sugar—phosphate—sugar—phosphate—sugar—
              | | |
            base   base   base

This backbone is common to all molecules of DNA and to all molecules of RNA. One DNA molecule is distinguished from another by the sequence in which the bases (A, C, G, T) occur. The same is true for RNA molecules; there are many, many different ways in which the bases A, C, G, and U can be ordered.

The primary structures of DNA and RNA are illustrated in somewhat more detail in Figure 27.13, p. 982. Notice that:

1. A phosphate group joins carbon atom 3 of one sugar molecule to carbon atom 5 of another sugar molecule. Putting it in a slightly different way, a sugar molecule bonds to phosphate at both the 3 and 5 positions.

2. The base is attached to the sugar at carbon atom 1. So far as the base is concerned, it is always a secondary amine that bonds to the sugar, losing an H atom in the process.

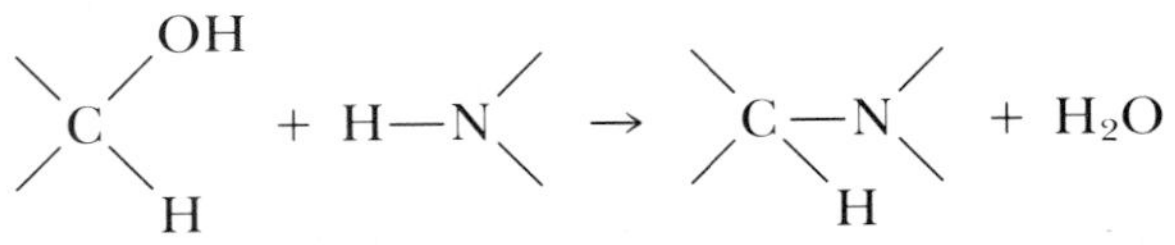

So far as their primary structures are concerned, DNA and RNA differ only in the nature of the sugar molecule present (deoxyribose vs. ribose) and in the identity of one of the bases present (thymine vs. uracil). However, these two types of nucleic acids have quite different secondary structures. In DNA, two molecules are entwined around each other in a double helix of the type shown in Figure 27.14, p. 983.

The double helix model for DNA was first proposed by Watson and Crick in 1953; its acceptance can be gauged from the fact that they won

TABLE 27.3   Components of Each Type of Nucleic Acid*

NUCLEIC ACID	COMPONENTS		
	Purine Base	*Pyrimidine Base*	*Sugar*
DNA only		thymine	deoxyribose
DNA and RNA	guanine	cytosine	adenine
RNA only		uracil	ribose

*The carbon atoms in the sugars are numbered; hydrogen atoms lost in bonding are shown in color.

the 1962 Nobel Prize for their work. In part, they were influenced by the α-helix model for proteins discovered earlier by Pauling. Beyond that, they were intrigued by the results of analyses which showed that in DNA the ratio of adenine to thymine molecules is almost exactly 1:1, as is the ratio of cytosine to guanine:

$$A/T = C/G = 1.00$$

This behavior is readily explained by the double helix; an A molecule in one strand is always hydrogen-bonded to a T molecule in the second strand.

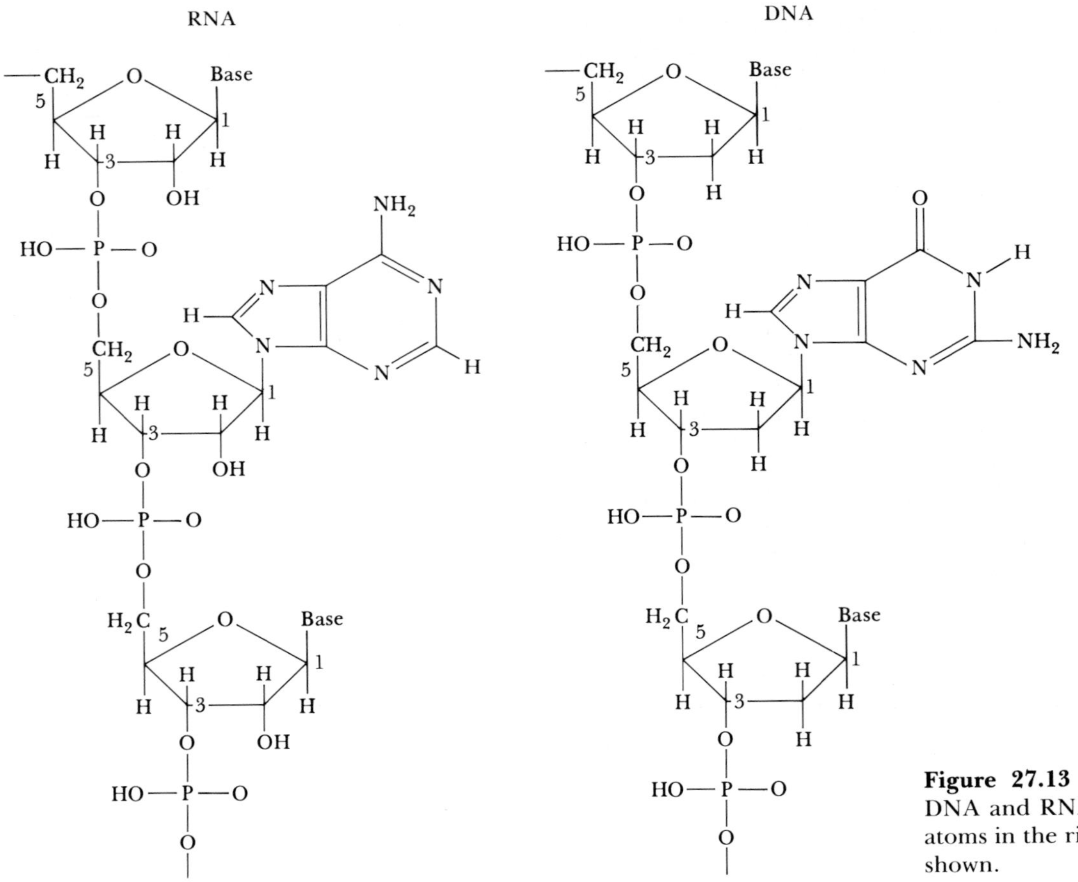

Figure 27.13 The structures of DNA and RNA molecules. Carbon atoms in the ring molecules are not shown.

Similarly, a C molecule in one strand is situated properly to form a hydrogen bond with a G molecule in the other strand (Fig. 27.14).

The double helix model provides a simple explanation for cell division and reproduction. In the reproduction process, the two DNA chains unwind from each other. As this happens, a new matching chain of DNA is synthesized on each of the original ones, creating two double helices. Since the base pairs in each new double helix must match in the same way as in the original, the two new double helices must be identical to the original. Exact replication of genetic data is thereby accomplished, however complex that data may be.

PROTEIN SYNTHESIS AND THE GENETIC CODE

The nucleic acids DNA and RNA are responsible for the synthesis of those proteins which distinguish one species from another and indeed one human being from another. Whether you have blue eyes or brown, curly hair or straight, is determined ultimately by the structure of the proteins synthe-

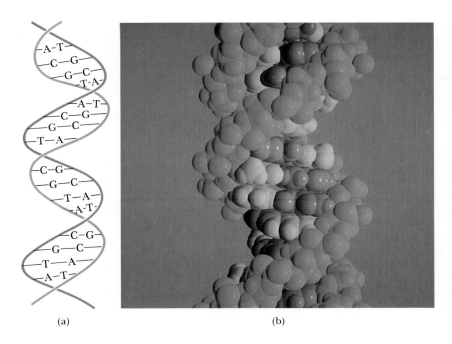

(a)   (b)

Figure 27.14 The double-helical strand of a DNA molecule. The diagram at the left shows the hydrogen bonding between base pairs adenine-thymine and cytosine-guanine that hold the strands together. (photo, NIH)

sized in your body. We will not attempt to describe protein synthesis in detail, but it may be of interest to look at some of the general features of the process.

Protein manufacture starts when the base sequence of a DNA molecule in the nucleus of a cell is transcribed onto a molecule of RNA, synthesized in much the same way as in the replication of DNA. This molecule, called messenger RNA (mRNA), carries the information from the cell nucleus to the cytoplasm where proteins are being synthesized. There the base sequence that distinguishes one nucleic acid from another is translated into the amino acid sequence that distinguishes one protein from another. This is accomplished with the aid of two other types of RNA, known as transfer RNA (tRNA) and ribosomal RNA (rRNA).

Translation from the language of nucleic acids to that of proteins occurs through what is known as the genetic code. The principle is a simple one:

A particular sequence of three bases attached to the backbone of a mRNA molecule calls for the insertion of a particular amino acid into the protein chain. For example, the sequence "GGA" (guanine, guanine, adenine) calls for the synthesis of the amino acid glycine (Gly). A base sequence of GCA (guanine, cytosine, adenine) translates to the amino acid alanine (Ala). Thus, if at one end of a mRNA molecule the base sequence is:

G—G—A—G—C—A—G—G—A

the protein synthesized will start with the sequence:

Gly—Ala—Gly—

TABLE 27.4   Messenger RNA Codes for Amino Acids

AMINO ACID	ABBREVIATION	BASE CODE ON mRNA
Alanine	Ala	GCA, GCG, GCC, GCU
Arginine	Arg	AGA, AGG, CGA, CGG, CGC, CGU
Asparagine	Asn	AAC, AAU
Aspartic acid	Asp	GAC, GAU
Cysteine	Cys	UGC, UGU
Glutamic acid	Glu	GAA, GAG
Glutamine	Gln	CAA, CAG
Glycine	Gly	GGA, GGC, GGG, GGU
Histidine	His	CAC, CAU
Isoleucine	Ile	AUA, AUC, AUU
Leucine	Leu	CUA, CUC, CUG, CUU, UUA, UUG
Lysine	Lys	AAA, AAG
Methionine	Met	AUG
Phenylalanine	Phe	UUU, UUC
Proline	Pro	CCA, CCC, CCG, CCU
Serine	Ser	AGC, AGU, UCA, UCC, UCG, UCU
Threonine	Thr	ACA, ACG, ACC, ACU
Tryptophane	Trp	UGG
Tyrosine	Tyr	UAC, UAU
Valine	Val	GUA, GUG, GUC, GUU
End or stop codes		UAA, UAG, UGA

The entire genetic code is listed in Table 27.4. As you can see, a given amino acid usually has more than one base code. This is reasonable. There are only 20 amino acids, while the number of different base codes is

$$4 \times 4 \times 4 = 64$$

since four different bases are available for each of three positions. Actually, only 61 of these 64 sequences call for inserting particular amino acids; three (UAA, UAG, UGA) signal the termination of the protein molecule.

■ **EXAMPLE 27.9** _____

Suppose the base sequence in a mRNA molecule is

a. C—C—A—A—A—A—C—A—A—C—C—C—

b. C—C—G—A—A—G—C—A—G—C—C—G—

What is the corresponding sequence in the protein molecule synthesized on this template?

Solution

You should find that in both cases the sequence is:

Pro—Lys—Gln—Pro—

This points out an interesting feature of the genetic code. An "error" in the base code, which is estimated to happen once in a million times, need not necessarily lead to a different protein.

Exercise

Suppose that in (a) the last base in the sequence were uracil instead of cytosine. What would be the protein sequence? Answer: Pro—Lys—Gln—Pro—.

Earlier we mentioned that sickle cell anemia results from the replacement of glutamic acid in adult hemoglobin by valine. The base codes for glutamic acid are GAA and GAG. Those for valine are GUA, GUG, GUC, and GUU. Substitution of the middle base A by U changes the meaning of the code words for glutamic acid to code words for valine. All the harm and misery that sickle cell anemia brings can be attributed to two mistakes in a code that has about 600 letters!

As you can imagine, it was not easy to unravel the genetic code. The key experiments were carried out by Marshall Nirenberg and co-workers. They started with a synthetic mRNA containing only the base uracil (U). This was added to a solution containing all the ingredients for protein synthesis except mRNA. Analysis showed that the new protein formed contained only one amino acid, phenylalanine. In another experiment, they worked with mRNA containing only cytosine (C) and obtained a protein consisting only of proline. The first two words had been decoded:

Genetic engineering became possible once we understood the genetic code

$$UUU \rightarrow Phe; \quad CCC \rightarrow Pro$$

By 1967, the genetic code had been completely deciphered.

■ SUMMARY PROBLEM

All of these questions pertain to six-carbon molecules.

a. Sketch the head-to-tail and tail-to-tail addition polymers derived from

b. If the molar mass of the polymer is 1.2×10^4 g/mol, how many monomer units are present? What is the mass percent of carbon in the polymer?

c. Write the structure of the condensation polymer formed when a straight-chain, saturated, six-carbon dialcohol reacts with a straight-chain, saturated, six-carbon dicarboxylic acid.

d. Write the structure of the disaccharide formed when α-glucose and β-glucose, both of formula $C_6H_{12}O_6$, combine to form an isomer of maltose.

e. Draw the structures of the two polypeptides derived from the six-carbon amino acids lysine and leucine.

f. Write at least one set of base codes which would lead to the synthesis of the polypeptides listed in (e).

g. Which, if any, of the component molecules of DNA and RNA contain six carbon atoms?

Answers

a.

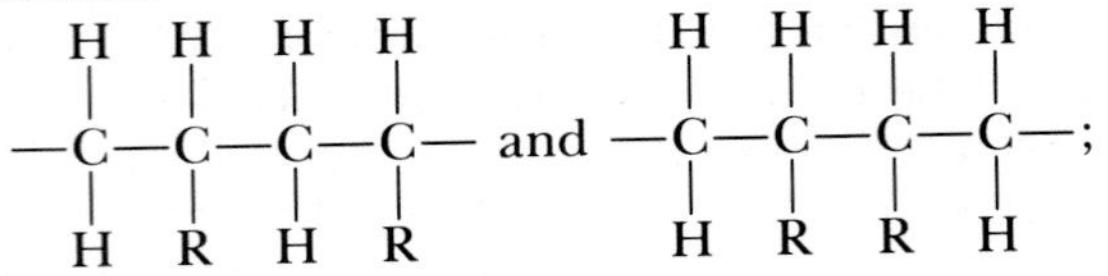

$$-\overset{\overset{\displaystyle H}{|}}{\underset{\underset{\displaystyle H}{|}}{C}}-\overset{\overset{\displaystyle H}{|}}{\underset{\underset{\displaystyle R}{|}}{C}}-\overset{\overset{\displaystyle H}{|}}{\underset{\underset{\displaystyle H}{|}}{C}}-\overset{\overset{\displaystyle H}{|}}{\underset{\underset{\displaystyle R}{|}}{C}}-$$
and
$$-\overset{\overset{\displaystyle H}{|}}{\underset{\underset{\displaystyle H}{|}}{C}}-\overset{\overset{\displaystyle H}{|}}{\underset{\underset{\displaystyle R}{|}}{C}}-\overset{\overset{\displaystyle H}{|}}{\underset{\underset{\displaystyle R}{|}}{C}}-\overset{\overset{\displaystyle H}{|}}{\underset{\underset{\displaystyle H}{|}}{C}}-;$$
$R = -(CH_2)_3CH_3$

b. 140; 85.6

c. $-O-(CH_2)_6-O-\overset{}{\underset{\|}{C}}-(CH_2)_4-\overset{}{\underset{\|}{C}}-$
(with O double-bonded below each C)

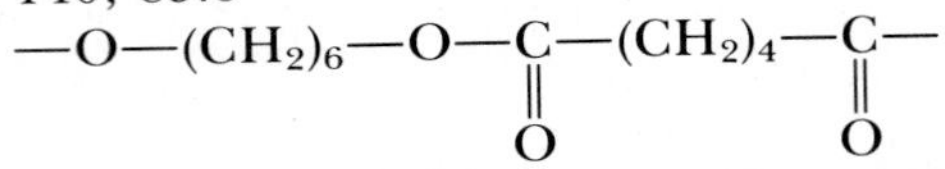

d. See maltose structure, (Equ. 27.8); invert OH and H on carbon atom at lower right.

e.

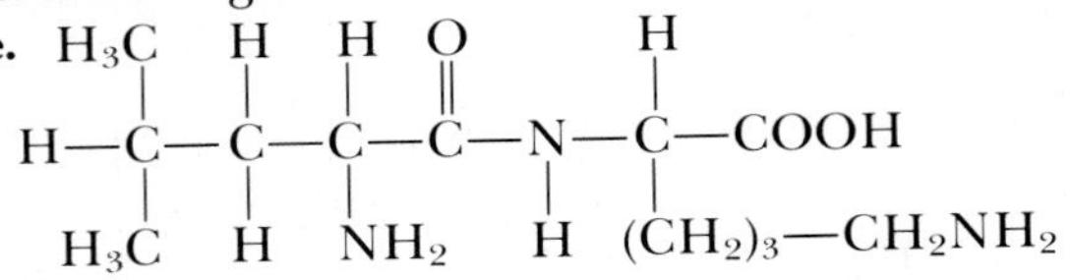

$$H-\overset{\overset{\displaystyle H_3C}{|}}{\underset{\underset{\displaystyle H_3C}{|}}{C}}-\overset{\overset{\displaystyle H}{|}}{\underset{\underset{\displaystyle H}{|}}{C}}-\overset{\overset{\displaystyle H}{|}}{\underset{\underset{\displaystyle NH_2}{|}}{C}}-\overset{O}{\underset{\|}{C}}-\overset{\overset{\displaystyle H}{|}}{\underset{\underset{\displaystyle H}{|}}{N}}-\overset{\overset{\displaystyle H}{|}}{\underset{\underset{\displaystyle (CH_2)_3-CH_2NH_2}{|}}{C}}-COOH$$
Leu-Lys

$$NH_2-(CH_2)_4-\overset{\overset{\displaystyle H}{|}}{\underset{\underset{\displaystyle NH_2}{|}}{C}}-\overset{O}{\underset{\|}{C}}-\overset{\overset{\displaystyle H}{|}}{\underset{\underset{\displaystyle H}{|}}{N}}-\overset{\overset{\displaystyle H}{|}}{\underset{\underset{\displaystyle CH_2-CH-(CH_3)_2}{|}}{C}}-COOH$$
Lys-Leu

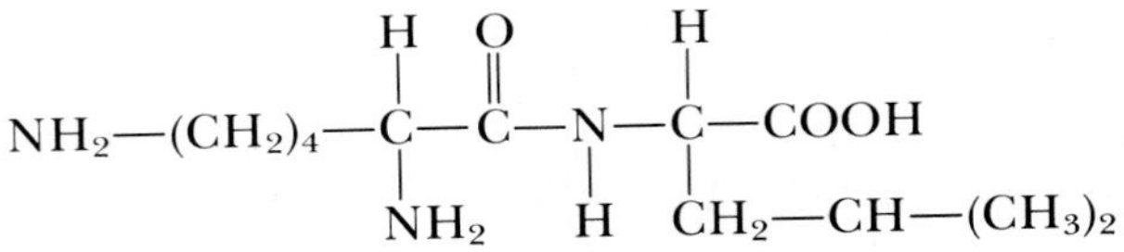

f. CUAAAA; AAGCUC   **g.** none

A great many different compounds have a sweet taste, among them lead acetate, $Pb(C_2H_3O_2)_2$ ("sugar of lead"), which is a deadly poison. It is possible to establish a sweetness scale (Table 27.5), using a taste panel. To most people, a 0.60 M solution of lactose is about as sweet as a 0.10 M sucrose solution, so we say that lactose is $0.10/0.60 = 0.16$ times as sweet as sucrose. Going in the other direction, 0.01 M aspartame and 1 M sucrose seem equally sweet, so aspartame is said to be $1/0.01 = 100$ times as sweet as sugar.

The principal sweetener in our diet is sucrose; per capita consumption in the United States is about 50 kg per year. About 70% of this comes from processed foods, some of which may surprise you: soups, salad dressings, and processed meats (read the label!). So far as we know, the only adverse effect of sugar is on teeth; there is a direct correlation between sugar consumption and tooth decay. Actually, the decay is caused by an acid by-product of bacterial action. Bacteria feed upon sugar and other simple carbohydrates, which explains the correlation. Brushing your teeth or drinking a glass of water after eating will get rid of most of the sugar.

Nutritionists often refer to sugar as a source of "empty calories." Sugar serves only as a source of energy; it contributes nothing so far as vitamins, minerals, or other nutrients are concerned. This can be a problem for dieters who have a limited calorie intake. With diabetics, blood glucose levels can soar when they take in food or beverages containing sugar. Artificial sweeteners are a must in their case.

The first artificial sweetener, saccharin, was discovered more than a century ago. It has the structure:

Saccharin

Saccharin has almost no energy value and need be consumed only in very small quantities, since it is so sweet (Table 27.5). It does have a somewhat

MODERN
PERSPECTIVE

SUGAR AND OTHER
SWEETENERS

TABLE 27.5 Relative Sweetness of Different Substances

SUBSTANCE	MOLECULAR FORMULA	RELATIVE SWEETNESS
Lactose	$C_{12}H_{22}O_{11}$	0.16
Maltose	$C_{12}H_{22}O_{11}$	0.33
Glucose	$C_6H_{12}O_6$	0.75
Sucrose	$C_{12}H_{22}O_{11}$	1.0
Fructose	$C_6H_{12}O_6$	1.7
Aspartame	$C_{14}H_{18}O_5N$	100
Saccharin	$C_7H_5O_3SN$	400

bitter aftertaste, at least for some people. Beyond that, some experiments have indicated that it causes cancer in laboratory animals when fed to them in massive amounts. Saccharin is still sold in the United States, albeit with a warning label; Canada has banned its use.

The most common artificial sweetener today is aspartame, a dipeptide formed from aspartic acid and the methyl ester of phenylalanine.

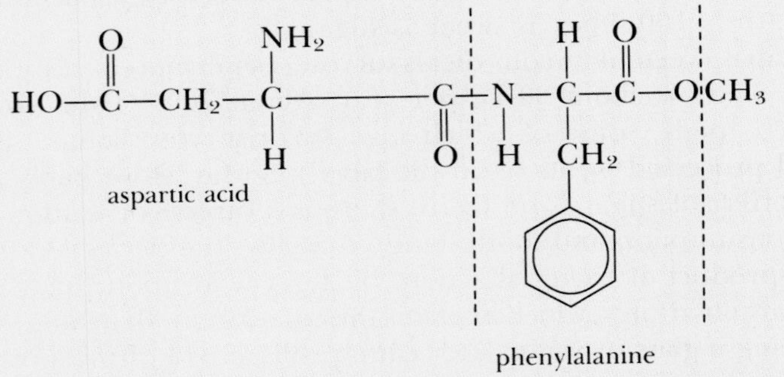

aspartic acid

phenylalanine

Discovered in 1969, it was approved for use in the United States in 1981. Since it acquires a sour taste when heated, it cannot be used for cooking or baking. However, it is widely used in diet soft drinks, cold cereals, and as a packaged sugar substitute. Aspartame does have an energy value (17 kJ or 4 kcal per gram), but since only small amounts need be consumed, this is of minor concern. Like saccharin, aspartame is not a carbohydrate, so it poses no risk to diabetics.

QUESTIONS AND PROBLEMS

Addition Polymers

1. Consider a polymer made from tetrachloroethylene.
   a. Draw a portion of the polymer chain.
   b. What is the molar mass of the polymer if it contains 3.2×10^3 tetrachloroethylene molecules?
   c. What are the mass percents of C and Cl in the polymer?

2. Consider Teflon, the polymer made from tetrafluoroethylene.
   a. Draw a portion of the Teflon molecule.
   b. Calculate the molar mass of a Teflon molecule that contains 5.0×10^4 CF_2 units.
   c. What are the mass percents of C and F in Teflon?

3. Sketch a portion of the acrylonitrile polymer (Table 27.1) assuming it is a
   a. head-to-tail polymer.
   b. head-to-head, tail-to-tail polymer.

4. Styrene, $H_2C=C$..., forms a head-to-tail addition polymer. Sketch a portion of a polystyrene molecule.

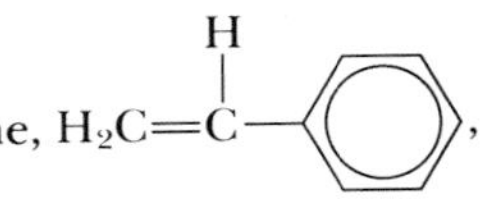

5. The polymer whose structure is shown below is made from two different monomers. Identify the monomers.

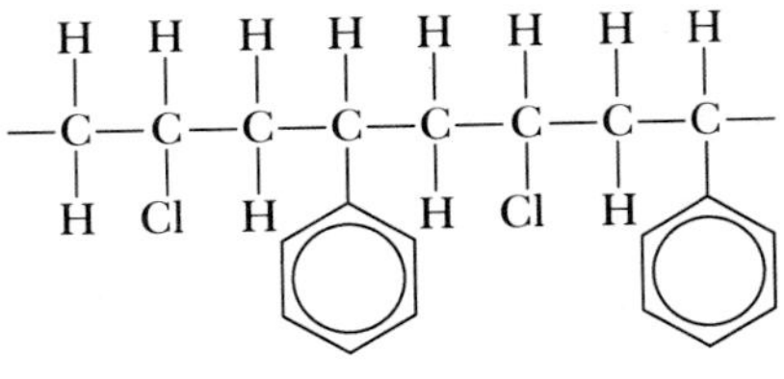

6. Show the structure of the monomer used to make the following addition polymers:

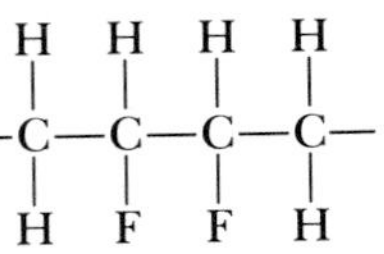

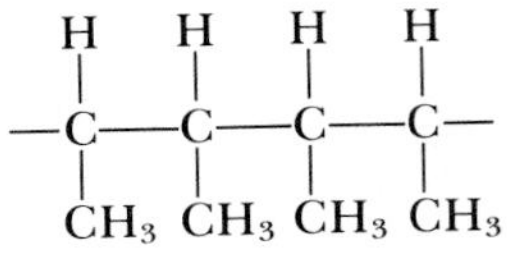

Consider the polymers referred to in Table 27.1. If each one contains the same number of monomer units, which one has the largest molar mass?

8. Of the polymers referred to in Table 27.1, which one contains the highest percentage by mass of carbon?

Condensation Polymers

9. A rather simple polymer can be made from ethylene glycol, $HO—CH_2—CH_2—OH$, and oxalic acid, $HO—\underset{\underset{O}{\|}}{C}—\underset{\underset{O}{\|}}{C}—OH$. Sketch a portion of the polymer chain obtained from these monomers.

10. Lexan is a very rugged polyester in which the monomers can be taken to be carbonic acid, $HO—\underset{\underset{O}{\|}}{C}—OH$,

and

Sketch a section of the Lexan chain.

11. *Para*-aminobenzoic acid is an "essential vitamin" for many bacteria:

Sketch a portion of a polyamide polymer made from this monomer.

12. Nylon 6 is made from a single monomer:

$$H_2N—(CH_2)_5—\underset{\underset{O}{\|}}{C}—OH$$

Sketch a section of the polymer chain in Nylon 6.

13. The following condensation polymer is made from a single monomer. Identify the monomer.

$$—\underset{\underset{H}{|}}{N}—CH_2—\underset{\underset{O}{\|}}{C}—\underset{\underset{H}{|}}{N}—CH_2—\underset{\underset{O}{\|}}{C}—$$

14. Identify the monomers from which the following condensation polymers are made:

a. $—\underset{\underset{H}{|}}{N}—CH_2—CH_2—\underset{\underset{H}{|}}{N}—\underset{\underset{O}{\|}}{C}—CH_2—\underset{\underset{O}{\|}}{C}—$

b.

Carbohydrates

15. Write a chemical equation, using molecular formulas, for the reaction of sucrose with water to form glucose and fructose.

16. Write a chemical equation, using molecular formulas, for the reaction of maltose with water to form glucose.

17. Cellulose consists of about 10,000 $C_6H_{10}O_5$ units linked together.

a. What are the mass percents of C, H, and O in cellulose?

b. What is the molar mass of cellulose?

18. Starch has the same empirical formula as cellulose and a molar mass of about 1.0×10^5 g/mol.

a. What are the mass percents of C, H, and O in starch?

b. How many $C_6H_{10}O_5$ units are linked together in a starch molecule?

19. Mannose has the same molecular formula as glucose and the same geometry except at carbon-2, where the H and OH groups are interchanged. Draw the structures of α- and β-mannose.

20. Draw the structure of the disaccharide formed by two moles of α-mannose (see Question 19).

21. How many chiral carbon atoms are there in α-glucose? in fructose?

22. How many chiral carbon atoms are there in sucrose? maltose?

Amino Acids and Proteins

23. Give the structural formulas of two different dimers formed between arginine and serine.

24. Give the structural formulas of two different dimers formed between leucine and lysine.

25. a. How many tripeptides can be made from glycine, alanine, and leucine, using each amino acid only once per tripeptide?

b. Write the structural formulas of these tripeptides and name them in the shorthand abbreviation used for showing amino acid sequences.

26. A tripeptide contains valine, lysine, and phenylalanine residues.

a. How many tripeptides are possible from these amino acids?

b. Draw a structural formula for a possible form of the tripeptide and name it, using the shorthand form.

27. Consider the cysteine molecule shown in Table 27.2. Write structural formulas for

a. the zwitterion of cysteine.

b. the cation formed in acid.

c. the anion formed in base.

28. Follow the directions of Question 27 for serine.

29. For alanine, $K_{a1} = 5.1 \times 10^{-3}$, $K_{a2} = 1.8 \times 10^{-10}$. Calculate the ratios $[Z]/[C^+]$ and $[Z]/[A^-]$ at pH

a. 2.00.   b. 6.00.   c. 10.50.

What is the principal species at each pH?

30. Using the information given in Problem 29, calculate the pH

a. when $[Z] = [C^+]$.

b. when $[Z] = [A^-]$.

c. at the isoelectric point.

31. On complete hydrolysis, a polypeptide gives two alanine, one leucine, one methionine, one phenylalanine, and one valine residue. Partial hydrolysis gives the following fragments: Ala-Phe, Leu-Met, Val-Ala, Phe-Leu. It is known that the first amino acid in the sequence is valine and the last one is methionine. What is the complete sequence of amino acids?

32. Suppose that, in the polypeptide referred to in Question 31, the first amino acid is alanine and the last one is also alanine. What is the complete sequence of amino acids?

33. How would you explain the fact that the β-keratin structure present in silk is much less common in proteins than the α-helix structure?

34. In β-keratin, how many hydrogen bonds are present per amino acid residue (Fig. 27.10)?

35. What amino acid sequence is specified by the base sequence (from left to right) in the following mRNA strand?

U C G A U C C U U A G C A A C

36. What base sequence of an mRNA strand would produce a polypeptide having the amino acid sequence

Ala—Pro—Asp—Tyr—Ile—Gly

37. What sequence on an mRNA strand would produce two polypeptides

Leu—Pro—Gly—Tyr—Trp; and Asp—Cys—His—Glu

38. What amino acid sequence is specified by the base sequence (from left to right) in the following mRNA strand?

A U G A A C C A A G U G U U U A G C

Unclassified

39. Which of the following monomers could form an addition polymer? a condensation polymer?
a. C_2H_6
b. C_2H_4
c. $HO—CH_2—CH_2—OH$
d. $HO—CH_2—CH_3$

40. How would you explain to a young science student how to decide whether a given compound might be useful as a monomer for addition polymerization? condensation polymerization?

41. Explain the difference between
a. a synthetic and natural polymer.
b. a polyester and polyamide.
c. α- and β-glucose.

42. Explain the difference between
a. linear and branched polyethylene.
b. glucose and fructose.
c. maltose and sucrose.

43. Draw the structures of the monomers that could be used to make the following polymers

a.
```
   Br  H  Br  H  Br  H
   |   |  |   |  |   |
 —C—C—C—C—C—C—
   |   |  |   |  |   |
   H   H  H   H  H   H
```

b.
```
       H           H
       |           |
 —N—C—C—N—C—C—
   |  |  ||  |  |  ||
   H  H  O  H  H  O
```

c.
```
   O              O   H              H
   ||             ||  |              |
 —C—(CH_2)_4—C—N—(CH_2)_6—N—
```

44. What monomers would be used to make the following polymers?

a.
```
   O                    O
   ||                   ||
 —C—⬡—C—O—(CH_2)_2—O—
```

b.
```
    H   H  H   H  H   H
    |   |  |   |  |   |
 —C—C—C—C—C—C—
    |   |  |   |  |   |
    Cl  Cl Cl  Cl Cl  Cl
```

c.
```
   O   H              O   H
   ||  |              ||  |
 —C—N—(CH_2)_5—C—N—(CH_2)_5—
```

45. Sketch the tetrapeptide obtained from four molecules of the α-amino acid glycine.

46. Sketch a portion of a head-to-head, tail-to-tail, and a head-to-tail polymer made from acrylonitrile, $H_2C{=}CHCN$.

47. Using bond energies, estimate ΔH for the hydrolysis of maltose to glucose.

48. Sketch the form in which leucine would exist in acid solution; in basic solution.

49. How many tripeptides could one make from glycine, valine, and lysine, using any number of each amino acid?

50. A 1.00-mg sample of a pure protein yielded on hydrolysis 0.0165 mg of leucine and 0.0248 mg of isoleucine. What is the minimum possible molar mass of the protein? (MM leucine = MM isoleucine = 131 g/mol)

51. Describe what is meant by
a. the primary structure of a protein.
b. the secondary structure of a protein.
c. the tertiary structure of a protein.

52. Discuss the similarities and differences between DNA and RNA.

53. Plants synthesize carbohydrates from CO_2 and H_2O by the process of photosynthesis. For example

$$6CO_2(g) + 6H_2O(l) \rightarrow C_6H_{12}O_6(aq) + 6O_2(g)$$

$\Delta G^0 = 2.87 \times 10^3$ kJ at pH 7.0 and 25°C. What is K for the reaction at 25°C?

54. Glycolysis is the process by which glucose is broken down to lactic acid according to the equation

$$C_6H_{12}O_6(aq) \rightarrow 2C_3H_6O_3(aq)$$
$$\Delta G^0 = -198 \text{ kJ at pH 7.0 and 25°C}$$

Glycolysis is the source of energy in human red blood cells. In these cells, the concentration of glucose is 5.0×10^{-3} M, while that of lactic acid is $2.9 \times 10^{-3} M$. Calculate ΔG for glycolysis in human blood cells under these conditions. Use the equation: $\Delta G = \Delta G^0 + RT \ln Q$, where Q is the concentration quotient, analogous to K.

Challenge Problems

55. Glycerol, $C_3H_5(OH)_3$, and orthophthalic acid,

—COOH, form a cross-linked polymer in which

COOH

adjacent polymer chains are linked together; this polymer is used in floor coverings and dentures.
  a. Write the structural formula for a portion of the polymer chain.
  b. Use your answer in (a) to show how cross-linking can occur between the polymer chains to form a water-insoluble, network covalent solid.

56. Determine the mass percents of the elements in Nylon whose structure is shown in Example 27.3.

57. Using bond energies, estimate ΔH for protein formation, per mole of amino acid added to the chain. Does this value seem reasonable?

58. One of the earliest, and still one of the most important, polymers is the material known as Bakelite. Bakelite is a condensation polymer of phenol, C_6H_5OH, and formaldehyde, $H_2C{=}O$. Formaldehyde will react with phenol to produce the following compounds when the ratio is 1:1:

and

These species, which can be taken to be monomers, on being heated condense with each other and themselves, eliminating water (formed from the OH groups on CH_2OH and ring hydrogen atoms) and linking benzene rings by CH_2 groups. If the phenol:formaldehyde ratio is 1:1, a linear polymer forms. If the ratio is 1:2, two $H_2C{=}O$ molecules react with each ring, and the chains cross-link at every benzene ring, forming the infusible, insoluble, hard, brittle solid we know as Bakelite. Sketch the linear chain polymer and the cross-linked solid.

59. Aspartic acid acts as a triprotic acid with successive dissociation constants of 8.0×10^{-3}, 1.4×10^{-4}, and 1.5×10^{-10}. Depending upon pH, aspartic acid can exist in four different forms in water solution. Draw these forms and calculate the pH range over which each form is the principal species.

APPENDIX 1

CONSTANTS, REFERENCE DATA, SI UNITS

CONSTANTS

Acceleration of gravity (standard)	9.8066 m/s^2
Atomic mass unit (amu)	$1.6606 \times 10^{-24} \text{ g}$
Avogadro's number	6.0220×10^{23}
Electronic charge	$1.6022 \times 10^{-19} \text{ C}$
Electronic mass	$9.1095 \times 10^{-28} \text{ g}$
Faraday constant	$9.6485 \times 10^4 \text{ J/V}$
Gas constant	$0.082057 \text{ L·atm/(mol·K)}$
	8.3144 J/(mol·K)
	$8.3144 \text{ kg·m}^2/(\text{s}^2\text{·mol·K})$
Planck's constant	$6.6262 \times 10^{-34} \text{ J·s}$
Velocity of light	$2.9979 \times 10^8 \text{ m/s}$
π	3.1416
e	2.7183
$\ln x$	$2.3026 \log_{10} x$

VAPOR PRESSURE OF WATER (mm Hg)

T(°C)	vp	T(°C)	vp	T(°C)	vp	T(°C)	vp
0	4.58	21	18.65	35	42.2	92	567.0
5	6.54	22	19.83	40	55.3	94	610.9
10	9.21	23	21.07	45	71.9	96	657.6
12	10.52	24	22.38	50	92.5	98	707.3
14	11.99	25	23.76	55	118.0	100	760.0
16	13.63	26	25.21	60	149.4	102	815.9
17	14.53	27	26.74	65	187.5	104	875.1
18	15.48	28	28.35	70	233.7	106	937.9
19	16.48	29	30.04	80	355.1	108	1004.4
20	17.54	30	31.82	90	525.8	110	1074.6

THERMODYNAMIC DATA

	ΔH_f^0 (kJ/mol)	S^0 (kJ/K·mol)	ΔG_f^0 (kJ/mol) at 25°C		ΔH_f^0 (kJ/mol)	S^0 (kJ/K·mol)	ΔG_f^0 (kJ/mol) at 25°C
$Ag(s)$	0.0	+0.0426	0.0	$CrO_4^{2-}(aq)$	−881.2	+0.0502	−727.8
$Ag^+(aq)$	+105.6	+0.0727	+77.1	$Cr_2O_3(s)$	−1139.7	+0.0812	−1058.1
$AgBr(s)$	−100.4	+0.1071	−96.9	$Cr_2O_7^{2-}(aq)$	−1490.3	+0.2619	−1301.1
$AgCl(s)$	−127.1	+0.0962	−109.8	$Cu(s)$	0.0	+0.0332	0.0
$AgI(s)$	−61.8	+0.1155	−66.2	$Cu^+(aq)$	+71.7	+0.0406	+50.0
$AgNO_3(s)$	−124.4	+0.1409	−33.4	$Cu^{2+}(aq)$	+64.8	−0.0996	+65.5
$Ag_2O(s)$	−31.0	+0.1213	−11.2	$CuO(s)$	−157.3	+0.0426	−129.7
$Al(s)$	0.0	+0.0283	0.0	$Cu_2O(s)$	−168.6	+0.0931	−146.0
$Al^{3+}(aq)$	−531.0	−0.3217	−485.0	$CuS(s)$	−53.1	+0.0665	−53.6
$Al_2O_3(s)$	−1675.7	+0.0509	−1582.3	$Cu_2S(s)$	−79.5	+0.1209	−86.2
$Ba(s)$	0.0	+0.0628	0.0	$CuSO_4(s)$	−771.4	+0.1076	−661.9
$Ba^{2+}(aq)$	−537.6	+0.0096	−560.8	$F_2(g)$	0.0	+0.2027	0.0
$BaCl_2(s)$	−858.6	+0.1237	−810.4	$F^-(aq)$	−332.6	−0.0138	−278.8
$BaCO_3(s)$	−1216.3	+0.1121	−1137.6	$Fe(s)$	0.0	+0.0273	0.0
$BaO(s)$	−553.5	+0.0704	−525.1	$Fe^{2+}(aq)$	−89.1	−0.1377	−78.9
$BaSO_4(s)$	−1473.2	+0.1322	−1362.3	$Fe^{3+}(aq)$	−48.5	−0.3159	−4.7
$Br_2(l)$	0.0	+0.1522	0.0	$Fe(OH)_3(s)$	−823.0	+0.1067	−696.6
$Br^-(aq)$	−121.6	+0.0824	−104.0	$Fe_2O_3(s)$	−824.2	+0.0874	−742.2
$C(s)$	0.0	+0.0057	0.0	$Fe_3O_4(s)$	−1118.4	+0.1464	−1015.5
$CCl_4(l)$	−135.4	+0.2164	−65.3	$H_2(g)$	0.0	+0.1306	0.0
$CHCl_3(l)$	−134.5	+0.2017	−73.7	$H^+(aq)$	0.0	0.0000	0.0
$CH_4(g)$	−74.8	+0.1862	−50.7	$HBr(g)$	−36.4	+0.1986	−53.4
$C_2H_2(g)$	+226.7	+0.2008	+217.1	$HCl(g)$	−92.3	+0.1868	−95.3
$C_2H_4(g)$	+52.3	+0.2195	+68.1	$HCO_3^-(aq)$	−692.0	+0.0912	−586.8
$C_2H_6(g)$	−84.7	+0.2295	−32.9	$HF(g)$	−271.1	+0.1737	−273.2
$C_3H_8(g)$	−103.8	+0.2699	−23.5	$HI(g)$	+26.5	+0.2065	+1.7
$CH_3OH(l)$	−238.7	+0.1268	−166.3	$HNO_3(l)$	−174.1	+0.1556	−80.8
$C_2H_5OH(l)$	−277.7	+0.1607	−174.9	$H_2O(g)$	−241.8	+0.1887	−228.6
$CO(g)$	−110.5	+0.1976	−137.2	$H_2O(l)$	−285.8	+0.0699	−237.2
$CO_2(g)$	−393.5	+0.2136	−394.4	$H_2O_2(l)$	−187.8	+0.1096	−120.4
$CO_3^{2-}(aq)$	−677.1	−0.0569	−527.8	$H_2PO_4^-(aq)$	−1296.3	+0.0904	−1130.3
$Ca(s)$	0.0	+0.0414	0.0	$HPO_4^{2-}(aq)$	−1292.1	−0.0335	−1089.2
$Ca^{2+}(aq)$	−542.8	−0.0531	−553.6	$H_2S(g)$	−20.6	+0.2057	−33.6
$CaCl_2(s)$	−795.8	+0.1046	−748.1	$H_2SO_4(l)$	−814.0	+0.1569	−690.1
$CaCO_3(s)$	−1206.9	+0.0929	−1128.8	$HSO_4^-(aq)$	−887.3	+0.1318	−755.9
$CaO(s)$	−635.1	+0.0398	−604.0	$Hg(l)$	0.0	+0.0760	0.0
$Ca(OH)_2(s)$	−986.1	+0.0834	−898.5	$Hg^{2+}(aq)$	+171.1	−0.0322	+164.4
$CaSO_4(s)$	−1434.1	+0.1067	−1321.8	$HgO(s)$	−90.8	+0.0703	−58.6
$Cd(s)$	0.0	+0.0518	0.0	$I_2(s)$	0.0	+0.1161	0.0
$Cd^{2+}(aq)$	−75.9	−0.0732	−77.6	$I^-(aq)$	−55.2	+0.1113	−51.6
$CdCl_2(s)$	−391.5	+0.1153	−344.0	$K(s)$	0.0	+0.0642	0.0
$CdO(s)$	−258.2	+0.0548	−228.4	$K^+(aq)$	−252.4	+0.1025	−283.3
$Cl_2(g)$	0.0	+0.2230	0.0	$KBr(s)$	−393.8	+0.0959	−380.7
$Cl^-(aq)$	−167.2	+0.0565	−131.2	$KCl(s)$	−436.7	+0.0826	−409.1
$ClO_3^-(aq)$	−104.0	+0.1623	−8.0	$KClO_3(s)$	−397.7	+0.1431	−296.3
$ClO_4^-(aq)$	−129.3	+0.1820	−8.5	$KClO_4(s)$	−432.8	+0.1510	−303.2
$Cr(s)$	0.0	+0.0238	0.0	$KNO_3(s)$	−369.8	+0.1330	−394.9

	ΔH_f^0 (kJ/mol)	S^0 (kJ/K·mol)	ΔG_f^0 (kJ/mol) at 25°C		ΔH_f^0 (kJ/mol)	S^0 (kJ/K·mol)	ΔG_f^0 (kJ/mol) at 25°C
Mg(s)	0.0	+0.0327	0.0	Ni^{2+}(aq)	−54.0	−0.1289	−45.6
Mg^{2+}(aq)	−466.8	−0.1381	−454.8	NiO(s)	−239.7	+0.0380	−211.7
$MgCl_2$(s)	−641.3	+0.0896	−591.8	O_2(g)	0.0	+0.2050	0.0
$MgCO_3$(s)	−1095.8	+0.0657	−1012.1	OH^-(aq)	−230.0	−0.0108	−157.2
MgO(s)	−601.7	+0.0269	−569.4	P_4(s)	0.0	+0.1644	0.0
$Mg(OH)_2$(s)	−924.5	+0.0632	−833.6	PCl_3(g)	−287.0	+0.3117	−267.8
$MgSO_4$(s)	−1284.9	+0.0916	−1170.7	PCl_5(g)	−374.9	+0.3645	−305.0
Mn(s)	0.0	+0.0320	0.0	PO_4^{3-}(aq)	−1277.4	−0.222	−1018.7
Mn^{2+}(aq)	−220.8	−0.0736	−228.1	Pb(s)	0.0	+0.0648	0.0
MnO(s)	−385.2	+0.0597	−362.9	Pb^{2+}(aq)	−1.7	+0.0105	−24.4
MnO_2(s)	−520.0	+0.0530	−465.2	$PbBr_2$(s)	−278.7	+0.1615	−261.9
MnO_4^-(aq)	−541.4	+0.1912	−447.2	$PbCl_2$(s)	−359.4	+0.1360	−314.1
N_2(g)	0.0	+0.1915	0.0	PbO(s)	−219.0	+0.0665	−188.9
NH_3(g)	−46.1	+0.1923	−16.5	PbO_2(s)	−277.4	+0.0686	−217.4
NH_4^+(aq)	−132.5	+0.1134	−79.3	S(s)	0.0	+0.0318	0.0
NH_4Cl(s)	−314.4	+0.0946	−203.0	S^{2-}(aq)	+33.1	−0.0146	+85.8
NH_4NO_3(s)	−365.6	+0.1511	−184.0	SO_2(g)	−296.8	+0.2481	−300.2
N_2H_4(l)	+50.6	+0.1212	+149.2	SO_3(g)	−395.7	+0.2567	−371.1
NO(g)	+90.2	+0.2107	+86.6	SO_4^{2-}(aq)	−909.3	+0.0201	−744.5
NO_2(g)	+33.2	+0.2400	+51.3	Si(s)	0.0	+0.0188	0.0
NO_2^-(aq)	−104.6	+0.1230	−32.2	SiO_2(s)	−910.9	+0.0418	−856.7
NO_3^-(aq)	−205.0	+0.1464	−108.7	Sn(s)	0.0	+0.0516	0.0
N_2O_4(g)	+9.2	+0.3042	+97.9	Sn^{2+}(aq)	−8.8	−0.0174	−27.2
Na(s)	0.0	+0.0512	0.0	SnO_2(s)	−580.7	+0.0523	−519.6
Na^+(aq)	−240.1	+0.0590	−261.9	Zn(s)	0.0	+0.0416	0.0
NaCl(s)	−411.2	+0.0721	−384.2	Zn^{2+}(aq)	−153.9	−0.1121	−147.1
NaF(s)	−573.6	+0.0515	−543.5	ZnI_2(s)	−208.0	+0.1611	−209.0
NaOH(s)	−425.6	+0.0645	−379.5	ZnO(s)	−348.3	+0.0436	−318.3
Ni(s)	0.0	+0.0299	0.0	ZnS(s)	−206.0	+0.0577	−201.3

BOND ENERGIES (kJ/mol)

Br—Br	193	C—H	414	Cl—N	201	H—S	339
Br—C	276	C—I	218	Cl—O	205	I—I	151
Br—Cl	218	C—N	293	Cl—S	255	I—O	201
Br—F	255	C=N	615	F—F	153	N—N	159
Br—H	368	C≡N	890	F—H	565	N=N	418
Br—I	180	C—O	351	F—I	277	N≡N	941
Br—N	243	C=O	715	F—N	272	N—O	222
Br—O	201	C≡O	1075	F—O	184	N=O	607
Br—S	213	C—S	259	F—S	285	O—O	138
C—C	347	C=S	477	H—H	436	O=O	498
C=C	612	Cl—Cl	243	H—I	297	O—S	347
C≡C	820	Cl—F	255	H—N	389	O=S	498
C—Cl	331	Cl—H	431	H—O	464	S—S	226
C—F	485	Cl—I	209				

EQUILIBRIUM CONSTANTS

SOLUBILITY PRODUCT CONSTANTS, K_{sp}

$AgBr$	5×10^{-13}	$Co_3(PO_4)_2$	1×10^{-35}	$Ni_3(PO_4)_2$	1×10^{-32}
$AgC_2H_3O_2$	1.9×10^{-3}	$CuCl$	1.7×10^{-7}	NiS	1×10^{-21}
Ag_2CO_3	8×10^{-12}	$CuBr$	6.3×10^{-9}	$PbBr_2$	6.6×10^{-6}
$AgCl$	1.8×10^{-10}	CuI	1.2×10^{-12}	$PbCO_3$	1×10^{-13}
Ag_2CrO_4	1×10^{-12}	$Cu_3(PO_4)_2$	1×10^{-37}	$PbCl_2$	1.7×10^{-5}
AgI	1×10^{-16}	CuS	1×10^{-36}	$PbCrO_4$	2×10^{-14}
Ag_3PO_4	1×10^{-16}	Cu_2S	1×10^{-48}	PbF_2	7.1×10^{-7}
Ag_2S	1×10^{-49}	$FeCO_3$	3.1×10^{-11}	PbI_2	8.4×10^{-9}
$AgSCN$	1.0×10^{-12}	$Fe(OH)_2$	5×10^{-17}	$Pb(OH)_2$	1×10^{-20}
AlF_3	1×10^{-18}	$Fe(OH)_3$	3×10^{-39}	PbS	1×10^{-28}
$Al(OH)_3$	2×10^{-31}	FeS	2×10^{-19}	$Pb(SCN)_2$	2.1×10^{-5}
$AlPO_4$	1×10^{-20}	GaF_3	2×10^{-16}	$PbSO_4$	1.8×10^{-8}
$BaCO_3$	2.6×10^{-9}	$Ga(OH)_3$	1×10^{-35}	$Sc(OH)_3$	2×10^{-31}
$BaCrO_4$	1.2×10^{-10}	$GaPO_4$	1×10^{-21}	SnS	3×10^{-28}
BaF_2	1.8×10^{-7}	Hg_2Br_2	6×10^{-23}	$SrCO_3$	5.6×10^{-10}
$BaSO_4$	1.1×10^{-10}	Hg_2Cl_2	1×10^{-18}	$SrCrO_4$	3.6×10^{-5}
Bi_2S_3	1×10^{-99}	Hg_2I_2	5×10^{-29}	SrF_2	4.3×10^{-9}
$CaCO_3$	4.9×10^{-9}	HgS	1×10^{-52}	$SrSO_4$	3.4×10^{-7}
CaF_2	1.5×10^{-10}	Li_2CO_3	8.2×10^{-4}	$TlCl$	1.9×10^{-4}
$Ca_3(PO_4)_2$	1×10^{-33}	$MgCO_3$	6.8×10^{-6}	$TlBr$	3.7×10^{-6}
$CaSO_4$	7.1×10^{-5}	MgF_2	7×10^{-11}	TlI	5.6×10^{-8}
$CdCO_3$	6×10^{-12}	$Mg(OH)_2$	6×10^{-12}	$Tl(OH)_3$	2×10^{-44}
$Cd(OH)_2$	5×10^{-15}	$Mg_3(PO_4)_2$	1×10^{-24}	Tl_2S	1×10^{-20}
$Cd_3(PO_4)_2$	1×10^{-33}	$Mn(OH)_2$	2×10^{-13}	$ZnCO_3$	1.1×10^{-10}
CdS	1×10^{-29}	MnS	5×10^{-14}	$Zn(OH)_2$	4×10^{-17}
$Co(OH)_2$	2×10^{-16}	$NiCO_3$	1.4×10^{-7}		

DISSOCIATION CONSTANTS, WEAK ACIDS, K_a

H_3AsO_4	5.7×10^{-3}	HNO_2	6.0×10^{-4}	$N_2H_5^+$	1.0×10^{-8}
$H_2AsO_4^-$	1.8×10^{-7}	H_3PO_4	7.1×10^{-3}	$Al(H_2O)_6^{3+}$	1.2×10^{-5}
$HAsO_4^{2-}$	2.5×10^{-12}	$H_2PO_4^-$	6.2×10^{-8}	$Ag(H_2O)_2^+$	1.2×10^{-12}
$HBrO$	2.6×10^{-9}	HPO_4^{2-}	4.5×10^{-13}	$Ca(H_2O)_6^{2+}$	2.2×10^{-13}
$HCHO_2$	1.9×10^{-4}	H_2S	1.0×10^{-7}	$Cd(H_2O)_4^{2+}$	4.0×10^{-10}
$HC_2H_3O_2$	1.8×10^{-5}	HS^-	1×10^{-13}	$Fe(H_2O)_6^{3+}$	6.7×10^{-3}
HCN	5.8×10^{-10}	H_2SO_3	1.7×10^{-2}	$Fe(H_2O)_6^{2+}$	1.7×10^{-7}
H_2CO_3	4.4×10^{-7}	HSO_3^-	6.0×10^{-8}	$Mg(H_2O)_6^{2+}$	3.7×10^{-12}
HCO_3^-	4.7×10^{-11}	HSO_4^-	1.0×10^{-2}	$Mn(H_2O)_6^{2+}$	2.8×10^{-11}
$HClO_2$	1.0×10^{-2}	H_2Se	1.5×10^{-4}	$Ni(H_2O)_6^{2+}$	2.2×10^{-10}
$HClO$	2.8×10^{-8}	H_2SeO_3	2.7×10^{-3}	$Pb(H_2O)_6^{2+}$	6.7×10^{-7}
HF	6.9×10^{-4}	$HSeO_3^-$	5.0×10^{-8}	$Sc(H_2O)_6^{3+}$	1.1×10^{-4}
HIO	2.4×10^{-11}	$CH_3NH_3^+$	2.4×10^{-11}	$Zn(H_2O)_4^{2+}$	3.3×10^{-10}
HN_3	2.4×10^{-5}	NH_4^+	5.6×10^{-10}		

$k_aK_b=K_w$

DISSOCIATION CONSTANTS, WEAK BASES, K_b					
AsO_4^{3-}	4.0×10^{-3}	N_3^-	4.2×10^{-10}	$HSeO_3^-$	3.7×10^{-12}
$HAsO_4^{2-}$	5.6×10^{-8}	NH_3	1.8×10^{-5}	$AlOH^{2+}$	8.3×10^{-10}
$H_2AsO_4^-$	1.8×10^{-12}	N_2H_4	1.0×10^{-6}	$AgOH$	8.3×10^{-3}
BrO^-	3.8×10^{-6}	NO_2^-	1.7×10^{-11}	$CaOH^+$	4.5×10^{-2}
CH_3NH_2	4.2×10^{-4}	PO_4^{3-}	2.2×10^{-2}	$CdOH^+$	2.5×10^{-5}
CHO_2^-	5.3×10^{-11}	HPO_4^{2-}	1.6×10^{-7}	$FeOH^{2+}$	1.5×10^{-12}
$C_2H_3O_2^-$	5.6×10^{-10}	$H_2PO_4^-$	1.4×10^{-12}	$FeOH^+$	5.9×10^{-8}
CN^-	1.7×10^{-5}	S^{2-}	1×10^{-1}	$MgOH^+$	2.7×10^{-3}
CO_3^{2-}	2.1×10^{-4}	HS^-	1.0×10^{-7}	$MnOH^+$	3.6×10^{-4}
HCO_3^-	2.3×10^{-8}	SO_3^{2-}	1.7×10^{-7}	$NiOH^+$	4.5×10^{-5}
ClO_2^-	1.0×10^{-12}	HSO_3^-	5.9×10^{-13}	$PbOH^+$	1.5×10^{-8}
ClO^-	3.6×10^{-7}	SO_4^{2-}	1.0×10^{-12}	$ScOH^{2+}$	9.1×10^{-11}
F^-	1.4×10^{-11}	HSe^-	6.7×10^{-11}	$ZnOH^+$	3.0×10^{-5}
IO^-	4.2×10^{-4}	SeO_3^{2-}	2.0×10^{-7}		

FORMATION CONSTANTS					
$AgBr_2^-$	2×10^7	$Cu(CN)_2^-$	1×10^{24}	$PdBr_4^{2-}$	6×10^{13}
$AgCl_2^-$	1.8×10^5	$Cu(NH_3)_4^{2+}$	2×10^{12}	$PtBr_4^{2-}$	6×10^{17}
$Ag(CN)_2^-$	2×10^{20}	$FeSCN^{2+}$	9.2×10^2	$PtCl_4^{2-}$	1×10^{16}
AgI_2^-	5×10^{10}	$Fe(CN)_6^{3-}$	4×10^{52}	$Pt(CN)_4^{2-}$	1×10^{41}
$Ag(NH_3)_2^+$	1.7×10^7	$Fe(CN)_6^{4-}$	4×10^{45}	$c\text{-}Pt(NH_3)_2Cl_2$	3×10^{29}
$Al(OH)_4^-$	1×10^{33}	$HgBr_4^{2-}$	1×10^{21}	$t\text{-}Pt(NH_3)_2Cl_2$	3×10^{28}
$Cd(CN)_4^{2-}$	2×10^{18}	$HgCl_4^{2-}$	1×10^{15}	$c\text{-}Pt(NH_3)_2(H_2O)_2^{2+}$	4×10^{23}
CdI_4^{2-}	4.0×10^5	$Hg(CN)_4^{2-}$	2×10^{41}	$c\text{-}Pt(NH_3)_2I_2$	2×10^{33}
$Cd(NH_3)_4^{2+}$	2.8×10^7	HgI_4^{2-}	1×10^{30}	$t\text{-}Pt(NH_3)_2I_2$	5×10^{32}
$Cd(OH)_4^{2-}$	1.2×10^9	$Hg(NH_3)_4^{2+}$	2×10^{19}	$c\text{-}Pt(NH_3)_2(OH)_2$	1×10^{39}
$Co(NH_3)_6^{2+}$	1×10^5	$Ni(CN)_4^{2-}$	1×10^{30}	$Pt(NH_3)_4^{2+}$	2×10^{35}
$Co(NH_3)_6^{3+}$	1×10^{23}	$Ni(NH_3)_6^{2+}$	9×10^8	$Zn(CN)_4^{2-}$	6×10^{16}
$Co(NH_3)_5Cl^{2+}$	2×10^{28}	PbI_4^{2-}	1.7×10^4	$Zn(NH_3)_4^{2+}$	3.6×10^8
$Co(NH_3)_5NO_2^{2+}$	1×10^{24}	$Pb(OH)_3^-$	8×10^{13}	$Zn(OH)_4^{2-}$	3×10^{14}

SI UNITS

BASE UNITS

The International System of Units or *Système International* (SI), which represents an extension of the metric system, was adopted by the 11th General Conference of Weights and Measures in 1960. It is constructed from seven base units, each of which represents a particular physical quantity (Table I).

TABLE I SI Base Units

PHYSICAL QUANTITY	NAME OF UNIT	SYMBOL
Length	metre	m
Mass	kilogram	kg
Time	second	s
Temperature	kelvin	K
Amount of substance	mole	mol
Electric current	ampere	A
Luminous intensity	candela	cd

Of the seven units listed in Table I, the first five are particularly useful in general chemistry. They are defined as follows:

1. The *metre* was redefined in 1983 to be equal to the distance light travels in a vacuum in 1/299 792 458 second.

2. The *kilogram* represents the mass of a platinum-iridium block kept at the International Bureau of Weights and Measures at Sevres, France.

3. The *second* was redefined in 1967 as the duration of 9 192 631 770 periods of a certain line in the microwave spectrum of cesium-133.

4. The *kelvin* is 1/273.16 of the temperature interval between the absolute zero and the triple point of water (0.01°C = 273.16 K).

5. The *mole* is the amount of substance that contains as many entities as there are atoms in exactly 0.012 kg of carbon-12.

PREFIXES USED WITH SI UNITS

Decimal fractions and multiples of SI units are designated by using the prefixes listed in Table II. Those most commonly used in general chemistry are underlined.

TABLE II   SI Prefixes

FACTOR	PREFIX	SYMBOL	FACTOR	PREFIX	SYMBOL
10^{12}	tera	T	10^{-1}	deci	d
10^{9}	giga	G	10^{-2}	centi	c
10^{6}	mega	M	10^{-3}	milli	m
10^{3}	kilo	k	10^{-6}	micro	μ
10^{2}	hecto	h	10^{-9}	nano	n
10^{1}	deca	da	10^{-12}	pico	p
			10^{-15}	femto	f
			10^{-18}	atto	a

DERIVED UNITS

In the International System of Units, all physical quantities are represented by appropriate combinations of the base units listed in Table I. To choose a particularly simple example, the SI unit for volume, the cubic metre, represents the volume of a cube one metre on an edge. Again, in SI, the density of a substance can be expressed by dividing its mass in kilograms by its volume in cubic metres. A list of the derived units most frequently used in general chemistry is given in Table III.

TABLE III   SI Derived Units

PHYSICAL QUANTITY	NAME OF UNIT	SYMBOL	DEFINITION
Area	square metre	m^2	
Volume	cubic metre	m^3	
Density	kilogram per cubic metre	kg/m^3	
Force	newton	N	$kg \cdot m/s^2$
Pressure	pascal	Pa	N/m^2
Energy	joule	J	$kg \cdot m^2/s^2$
Electric charge	coulomb	C	$A \cdot s$
Electric potential difference	volt	V	$J/(A \cdot s)$

Perhaps the least familiar of these units to the beginning chemistry student are the ones used to represent force, pressure, and energy.

The *newton* is defined as the force required to impart an acceleration of one metre per second squared to a mass of one kilogram (recall that Newton's second law can be stated as force = mass × acceleration).

The *pascal* is defined as the pressure exerted by a force of one newton acting upon an area of one square metre (recall that pressure = force/area). Commonly, pressures are expressed in kilopascals:

$$1 \text{ kPa} = 10^3 \text{ Pa}$$

Typically, atmospheric pressure near sea level is in the vicinity of 100 kPa.

The **joule** is defined as the work done when a force of one newton ($kg \cdot m/s^2$) acts through a distance of one metre (recall that work = force × distance). Commonly, energies are expressed in kilojoules:

$$1 \text{ kJ} = 10^3 \text{ J}$$

In terms of more familiar units, a kilowatt-hour is 3600 kJ; a kilocalorie is 4.184 kJ.

CONVERSIONS BETWEEN SI AND OTHER UNITS

Table IV lists conversion factors for translating units from other systems to the International System.

TABLE IV   Conversion Factors

QUANTITY	SI UNIT	OTHER UNIT	
Area	m^2	ft^2	$1\ ft^2 = 0.092\ 903\ 04\ m^2$
		acre	$1\ acre = 4.046\ 856 \times 10^3\ m^2$
		cm^2	$1\ cm^2 = 10^{-4}\ m^2$
		hectare	$1\ hectare = 10^4\ m^2$
Density	kg/m^3	g/cm^3	$1\ g/cm^3 = 10^3\ kg/m^3$
		lb/ft^3	$1\ lb/ft^3 = 16.018\ 46\ kg/m^3$
Electric charge	coulomb (C)	mole electrons	$1\ mol\ e^- = 9.6485 \times 10^4\ C$
Electric potential	volt (V)	joule/coulomb	$1\ V = 1\ J/C$
Energy	joule (J)	calorie	$1\ cal = 4.184\ J$
		L·atm	$1\ L\cdot atm = 101.3\ J$
		erg	$1\ erg = 10^{-7}\ J$
		kilowatt-hour	$1\ kWh = 3.6 \times 10^6\ J$
		BTU	$1\ BTU = 1.055 \times 10^3\ J$
Entropy	J/K	cal/K	$1\ cal/K = 4.184\ J/K$
Force	newton (N)	dyne	$1\ dyn = 10^{-5}\ N$
Frequency	hertz (Hz)	cycle/second	$1\ Hz = 1\ cycle/s$
Length	metre (m)	inch	$1\ in = 0.0254\ m$
		mile	$1\ mile = 1.609\ 344\ km$
		angstrom	$1\ Å = 10^{-10}\ m = 10^{-1}\ nm$
		micron	$1\ micron = 10^{-6}\ m$
Mass	kilogram (kg)	pound	$1\ lb = 0.453\ 592\ 37\ kg$
		metric ton (t)	$1\ t = 10^3\ kg$
		ton (short)	$1\ ton = 2000\ lb = 9.071\ 847\ 4 \times 10^2\ kg$
Power	watt (w)	joule/second	$1\ W = 1\ J/s$
Pressure	pascal (Pa)	atmosphere	$1\ atm = 101.325\ kPa$
		bar	$1\ bar = 10^5\ Pa$
		mm Hg	$1\ mm\ Hg = 133.322\ Pa$
		lb/in^2	$1\ lb/in^2 = 6.894\ 757\ kPa$
		torr	$1\ torr = 133.322\ P$
Temperature*	kelvin (K)	Celsius degree	$1°C = 1\ K$
		Fahrenheit degree	$1°F = 5/9\ K$
Surface tension	N/m	dynes/cm	$1\ dyn/cm = 10^{-3}\ N/m$
Volume	m^3	litre	$1\ L = 1\ dm^3 = 10^{-3}\ m^3$
		cm^3	$1\ cm^3 = 1\ mL = 10^{-6}\ m^3$
		ft^3	$1\ ft^3 = 28.316\ 85\ dm^3$
		gallon (US)	$1\ gal = 4\ qt = 3.785\ 412\ dm^3$

*Temperatures in degrees Celsius (t_c) or degrees Fahrenheit (t_f) can be converted to temperatures in Kelvin (T) by using the equations

$$t_c = T - 273.15; \qquad t_f = \frac{9}{5}T - 459.67$$

PROPERTIES OF THE ELEMENTS

ELEMENT	AT. NO.	mp(°C)	bp(°C)	E.N.	ION. ENER. (kJ/mol)	AT. RAD. (nm)	ION. RAD. (nm)
H	1	−259	−253	2.1	1312	0.037	(−1)0.208
He	2	−272	−269		2372	0.05	
Li	3	186	1326	1.0	520	0.152	(+1)0.060
Be	4	1283	2970	1.5	900	0.111	(+2)0.031
B	5	2300	2550	2.0	801	0.088	
C	6	3570	subl.	2.5	1086	0.077	
N	7	−210	−196	3.0	1402	0.070	
O	8	−218	−183	3.5	1314	0.066	(−2)0.140
F	9	−220	−188	4.0	1681	0.064	(−1)0.136
Ne	10	−249	−246		2081	0.070	
Na	11	98	889	0.9	496	0.186	(+1)0.095
Mg	12	650	1120	1.2	738	0.160	(+2)0.065
Al	13	660	2327	1.5	578	0.143	(+3)0.050
Si	14	1414	2355	1.8	786	0.117	
P	15	44	280	2.1	1012	0.110	
S	16	119	444	2.5	1000	0.104	(−2)0.184
Cl	17	−101	−34	3.0	1251	0.099	(−1)0.181
Ar	18	−189	−186		1520	0.094	
K	19	64	774	0.8	419	0.231	(+1)0.133
Ca	20	845	1420	1.0	590	0.197	(+2)0.099
Sc	21	1541	2831	1.3	631	0.160	(+3)0.081
Ti	22	1660	3287	1.5	658	0.146	
V	23	1890	3380	1.6	650	0.131	
Cr	24	1857	2672	1.6	653	0.125	(+3)0.064
Mn	25	1244	1962	1.5	717	0.129	(+2)0.080
Fe	26	1535	2750	1.8	759	0.126	(+2)0.075
Co	27	1495	2870	1.9	758	0.125	(+2)0.072
Ni	28	1453	2732	1.9	737	0.124	(+2)0.069

ELEMENT	AT. NO.	mp(°C)	bp(°C)	E.N.	ION. ENER. (kJ/mol)	AT. RAD. (nm)	ION. RAD. (nm)
Cu	29	1083	2567	1.9	746	0.128	(+1)0.096
Zn	30	420	907	1.6	906	0.133	(+2)0.074
Ga	31	30	2403	1.6	579	0.122	(+3)0.062
Ge	32	937	2830	1.8	762	0.122	
As	33	814	subl.	2.0	944	0.121	
Se	34	217	685	2.4	941	0.117	(−2)0.198
Br	35	−7	59	2.8	1140	0.114	(−1)0.195
Kr	36	−157	−152		1351	0.109	
Rb	37	39	688	0.8	403	0.244	(+1)0.148
Sr	38	770	1380	1.0	550	0.215	(+2)0.113
Y	39	1509	2930	1.2	616	0.180	(+3)0.093
Zr	40	1852	3580	1.4	660	0.157	
Nb	41	2468	5127	1.6	664	0.143	
Mo	42	2610	5560	1.8	685	0.136	
Tc	43	2200	4700	1.9	702	0.136	
Ru	44	2430	3700	2.2	711	0.133	
Rh	45	1966	3700	2.2	720	0.134	
Pd	46	1550	3170	2.2	805	0.138	
Ag	47	961	2210	1.9	731	0.144	(+1)0.126
Cd	48	321	767	1.7	868	0.149	(+2)0.097
In	49	157	2000	1.7	558	0.162	(+3)0.081
Sn	50	232	2270	1.8	709	0.140	
Sb	51	631	1380	1.9	832	0.141	
Te	52	450	990	2.1	869	0.137	(−2)0.221
I	53	114	184	2.5	1009	0.133	(−1)0.216
Xe	54	−112	−107		1170	0.130	
Cs	55	28	690	0.7	376	0.262	(+1)0.169
Ba	56	725	1640	0.9	503	0.217	(+2)0.135
La	57	920	3469	1.1	538	0.187	(+3)0.115
Ce	58	795	3468	1.1	528	0.182	(+3)0.101
Pr	59	935	3127	1.1	523	0.182	(+3)0.100
Nd	60	1024	3027	1.1	530	0.182	(+3)0.099
Pm	61	1027	2727	1.1	536	0.181	
Sm	62	1072	1900	1.1	543	0.180	
Eu	63	826	1439	1.1	547	0.204	(+2)0.097
Gd	64	1312	3000	1.1	592	0.179	(+3)0.096
Tb	65	1356	2800	1.1	564	0.177	(+3)0.095
Dy	66	1407	2600	1.1	572	0.177	(+3)0.094
Ho	67	1461	2600	1.1	581	0.176	(+3)0.093
Er	68	1497	2900	1.1	589	0.175	(+3)0.092
Tm	69	1356	2800	1.1	597	0.174	(+3)0.091
Yb	70	824	1427	1.1	603	0.193	(+3)0.089
Lu	71	1652	3327	1.1	524	0.174	(+3)0.089
Hf	72	2225	5200	1.3	654	0.157	
Ta	73	2980	5425	1.5	761	0.143	
W	74	3410	5930	1.7	770	0.137	
Re	75	3180	5885	1.9	760	0.137	
Os	76	2727	4100	2.2	840	0.134	
Ir	77	2448	4500	2.2	880	0.135	
Pt	78	1769	4530	2.2	870	0.138	
Au	79	1063	2966	2.4	890	0.144	(+1)0.137

ELEMENT	AT. NO.	mp(°C)	bp(°C)	E.N.	ION. ENER. (kJ/mol)	AT. RAD. (nm)	ION. RAD. (nm)
Hg	80	−39	357	1.9	1007	0.155	(+2)0.110
Tl	81	304	1457	1.8	589	0.171	(+3)0.095
Pb	82	328	1750	1.9	716	0.175	
Bi	83	271	1560	1.9	703	0.146	
Po	84	254	962	2.0	812	0.165	
At	85	302	334	2.2			
Rn	86	−71	−62		1037	0.14	
Fr	87	27	677	0.7			
Ra	88	700	1140	0.9	509	0.220	
Ac	89	1050	3200	1.1	490	0.20	
Th	90	1750	4790	1.3	590	0.180	
Pa	91	1600	4200	1.4	570		
U	92	1132	3818	1.4	590	0.14	

APPENDIX

3

REVIEW OF
MATHEMATICS

The mathematics you will use in general chemistry is relatively simple. You will, however, be expected to

— make calculations involving exponential numbers, such as 6.022×10^{23} or 1.6×10^{-10}.

— work with logarithms or antilogarithms, particularly in problems involving pH:

$$pH = -\log_{10} (\text{conc. } H^+)$$

In this appendix, we will review each of these topics briefly. To start with, it will be helpful to comment on electronic calculators, which are very useful for all kinds of calculations in general chemistry.

ELECTRONIC CALCULATORS

A simple "scientific calculator" selling for $20 or less is entirely adequate for general chemistry. Like any calculator, it will allow you to carry out such simple operations as addition, subtraction, multiplication, and division. Beyond that, make sure that the scientific calculator you buy can be used to

— enter and perform operations on numbers expressed in exponential (scientific) notation.

— find a base 10 or natural logarithm or antilogarithm (number corresponding to a given logarithm).

— raise a number to any power, n, or extract the nth root of a number.

The first thing you should do after buying a calculator is to learn how to use it. Read the instruction manual and work with the calculator until you become familiar with it. To get started, try carrying out the following operations:

a. $2.2 \times 6.1 = 13.42$
b. $8.1/2.7 = 3$
c. $(64)^{1/2} = 8$
d. $(27)^{1/3} = 3$
e. $3^4 = 81$

(In d and e, you will need to use the $\boxed{y^x}$ or $\boxed{x^y}$ key. Refer to your instruction manual for the sequence of operations, which differs depending upon the brand of calculator.)

f. $\dfrac{16 \times 9}{3 \times 8} = 6$

(This is carried out as a single operation; you do *not* solve for intermediate answers.)

In working with a calculator, you should be aware of one of its limitations. It does not indicate the number of significant figures in the answer. Consider, for example, the operations in a and b above. Assume that 2.2, 6.1, 8.1, and 2.7 represent experimentally measured quantities. Following the rules for significant figures (Chap. 1), the answers should be 13 and 3.0, in that order, *not* 13.42 and 3, the numbers appearing on the calculator. In another case, if you are asked to obtain the reciprocal of 3.68, your answer should be

$$1/3.68 = 0.272$$

not 0.2717391 . . . , or whatever other number appears on your calculator.

EXPONENTIAL NOTATION

In chemistry we frequently deal with very large or very small numbers. In one gram of the element carbon there are

50,140,000,000,000,000,000,000 atoms of carbon.

At the opposite extreme, the mass of a single atom is

0.00000000000000000000001994 g

Numbers such as these are very awkward to work with. For example, neither of the numbers just written could be entered directly on a calculator. To simplify operations involving very large or very small numbers, we use what is known as **exponential** or **scientific notation**. To express a number in exponential notation, we write it in the form

$$C \times 10^n$$

where C is a number between 1 and 10 (for example, 1, 2.62, 5.8) and n is a positive or negative integer such as 1, -1, -3. To find n, we count the number of places that the decimal point must be moved to give the coefficient, C. If the decimal point must be moved to the *left*, n is a *positive* integer; if it must be moved to the *right*, n is a *negative* integer. Thus, we have

$$26.23 = 2.623 \times 10^1 \qquad \text{(decimal point moved 1 place to left)}$$

$$5609 = 5.609 \times 10^3 \qquad \text{(decimal point moved 3 places to left)}$$

$$0.0918 = 9.18 \times 10^{-2} \qquad \text{(decimal point moved 2 places to right)}$$

Numbers written in exponential notation can be given a very simple interpretation. Recognizing that $10^1 = 10$, $10^3 = 1000$, and $10^{-2} = 1/100 = 0.01$, we could express the three exponentials written above as

$$2.623 \times 10^1 \;\; = 2.623 \times 10$$

$$5.609 \times 10^3 \;\; = 5.609 \times 1000$$

$$9.18 \times 10^{-2} = 9.18 \times 0.01$$

The magnitude of a number written in exponential notation depends upon the values of both the coefficient, C, and the exponent, n. Suppose we compare two numbers that have the same value of n, such as 2.6×10^2 and 3.8×10^2. Here, the larger number is the one with the larger coefficient:

$$3.8 \times 10^2 > 2.6 \times 10^2; \qquad 380 > 260$$

Suppose now that we compare two exponential numbers with different values of n. Here, *the larger number is the one that has the larger value of n*:

$$2.6 \times 10^2 > 4.8 \times 10^1 \qquad 260 > 48$$

$$3.2 \times 10^1 > 8.0 \times 10^{-1} \qquad 32 > 0.80$$

$$2 \times 10^{-2} > 4 \times 10^{-3} \qquad 0.02 > 0.004$$

MULTIPLICATION AND DIVISION

A major advantage of exponential notation is that it simplifies the processes of multiplication and division. To *multiply*, we *add exponents:*

$$10^1 \times 10^2 = 10^{1+2} = 10^3; \qquad 10^6 \times 10^{-4} = 10^{6+(-4)} = 10^2$$

To *divide*, we *subtract* exponents:

$$10^3/10^2 = 10^{3-2} = 10^1; \qquad 10^{-3}/10^6 = 10^{-3-6} = 10^{-9}$$

To multiply one exponential number by another, we can first multiply the coefficients in the usual manner and then add exponents. To divide one exponential number by another, we can find the quotient of the coefficients and then subtract exponents. For example,

$$(5.00 \times 10^4) \times (1.60 \times 10^2) = (5.00 \times 1.60) \times (10^4 \times 10^2)$$

$$= 8.00 \times 10^6$$

$$(6.01 \times 10^{-3})/(5.23 \times 10^6) = \frac{6.01}{5.23} \times \frac{10^{-3}}{10^6} = 1.15 \times 10^{-9}$$

It often happens that multiplication or division yields an answer that is not in standard exponential notation. Thus, we might have

$$(5.0 \times 10^4) \times (6.0 \times 10^3) = (5.0 \times 6.0) \times 10^4 \times 10^3 = 30 \times 10^7$$

The product is not in standard exponential notation since the coefficient, 30, does not lie between 1 and 10. To correct this situation, we could rewrite the coefficient as 3.0×10^1 and then add exponents:

$$30 \times 10^7 = (3.0 \times 10^1) \times 10^7 = 3.0 \times 10^8$$

In another case,

$$0.526 \times 10^3 = (5.26 \times 10^{-1}) \times 10^3 = 5.26 \times 10^2$$

EXPONENTIAL NOTATION ON THE CALCULATOR

On all scientific calculators, it is possible to enter numbers in exponential notation. The method used depends upon the brand of calculator. Most often, it involves using a key labeled $\boxed{\textbf{EXP}}$, $\boxed{\textbf{EE}}$, or $\boxed{\textbf{EEX}}$. Check your instruction manual for the procedure to be followed. To make sure you understand it, try entering the following numbers:

$$2.4 \times 10^6; \quad 3.16 \times 10^{-8}; \quad 6.2 \times 10^{-16}$$

Multiplication, division, and many other operations can be carried out directly on your calculator. Try the following exercises, using your calculator:

a. $(6.0 \times 10^2) \times (4.2 \times 10^{-4}) = ?$

b. $\dfrac{6.0 \times 10^2}{4.2 \times 10^{-4}} = ?$

c. $(2.50 \times 10^{-9})^{1/2} = ?$

d. $3.6 \times 10^{-4} + 4 \times 10^{-5} = ?$

The answers, expressed in exponential notation and following the rules of significant figures, are as follows: a. 2.5×10^{-1} b. 1.4×10^6 c. 5.00×10^{-5} d. 4.0×10^{-4}.

LOGARITHMS AND ANTILOGARITHMS

The logarithm of a number n to the base m is defined as the power to which m must be raised to give the number n. Thus:

if $m^x = n$, then $\log_m n = x$

In general chemistry, you will encounter two kinds of logarithms.

1. *Common logarithms*, where the base is 10, ordinarily denoted as $\log_{10}$. If $10^x = n$, then $\log_{10} n = x$. Examples include the following:

$$\log_{10} 100 = 2.000 \qquad (\text{since } 10^2 = 100)$$

$$\log_{10} 1 = 0.000 \qquad (\text{since } 10^0 = 1)$$

$$\log_{10} 0.001 = -3.000 \qquad (\text{since } 10^{-3} = 0.001)$$

2. *Natural logarithms,* where the base is the quantity $e = 2.718. \ldots$ Many of the equations used in general chemistry are expressed most simply in terms of natural logarithms, denoted as ln. If $e^x = n$, then $\ln n = x$.

$$\ln 100 = 4.606 \qquad (\text{i.e., } 100 = e^{4.606})$$

$$\ln 1 = 0 \qquad (\text{i.e., } 1 = e^0)$$

$$\ln 0.001 = -6.908 \qquad (\text{i.e., } 0.001 = e^{-6.908})$$

Notice that

a. $\log_{10} 1 = \ln 1 = 0.$ The logarithm of 1 to any base is zero, since any number raised to the zero power is 1. That is

$$10^0 = e^0 = 2^0 = \cdots = n^0 = 1$$

b. Numbers larger than 1 have a positive logarithm; numbers smaller than 1 have a negative logarithm. For example:

$$\log_{10} 100 = 2; \qquad \ln 100 = 4.606$$

$$\log_{10} 10 = 1; \qquad \ln 10 = 2.303$$

$$\log_{10} 0.1 = -1; \qquad \ln 0.1 = -2.303$$

$$\log_{10} 0.01 = -2; \qquad \ln 0.01 = -4.606$$

c. The common and natural logarithms of a number are related by the expression:

$\ln n = 2.303 \log_{10} n$

That is, the natural logarithm of a number is 2.303 . . . times its base 10 logarithm.

An *antilogarithm* is, quite simply, the number corresponding to a given logarithm. In general

if $m^x = n$, then $\text{antilog}_m x = n$

Thus we have:

$$10^2 = 100; \qquad \text{antilog}_{10} 2 = 100$$

$$10^0 = 1; \qquad \text{antilog}_{10} 0 = 1$$

$$10^{-3} = 0.001; \qquad \text{antilog}_{10} (-3) = 0.001$$

In other words, the numbers whose base 10 logarithms are 2, 0, and -3 are 100, 1, and 0.001, respectively. The same reasoning applies to natural logarithms.

$$e^{4.606} = 100; \qquad \text{antilog}_e 4.606 = 100$$

$$e^0 = 1; \qquad \text{antilog}_e 0 = 1$$

$$e^{-6.908} = 0.001; \qquad \text{antilog}_e -6.908 = 0.001$$

The numbers whose natural logarithms are 4.606, 0, and -6.908 are 100, 1, and 0.001, respectively. *Notice that a positive antilogarithm corresponds to a number larger than 1, while a negative antilogarithm indicates a number less than 1.*

FINDING LOGARITHMS ON A CALCULATOR

To obtain a base 10 logarithm using a calculator, all you need do is enter the number and press the $\boxed{\textbf{LOG}}$ key. This way you should find that

$$\log_{10} 2.00 = 0.301 \ldots$$
$$\log_{10} 0.526 = -0.279 \ldots$$

Similarly, to find a natural logarithm, you enter the number and press the $\boxed{\textbf{LN X}}$ key.

$$\ln 2.00 = 0.693 \ldots$$
$$\ln 0.526 = -0.642 \ldots$$

To find the logarithm of an exponential number, you simply enter the number in exponential form and take the logarithm in the usual way. This way you should find that:

$$\log_{10} 2.00 \times 10^3 = 3.301 \ldots \qquad \ln 2.00 \times 10^3 = 7.601 \ldots$$
$$\log_{10} 5.3 \times 10^{-12} = -11.28 \ldots \qquad \ln 5.3 \times 10^{-12} = -25.96 \ldots$$

The base 10 logarithm of an exponential number can be found in a somewhat different way by applying the relation:

$$\log_{10}(C \times 10^n) = n + \log_{10} C$$

Thus

$$\log_{10} 2.00 \times 10^3 = 3 + \log_{10} 2.00 = 3 + 0.301 = 3.301$$
$$\log_{10} 5.3 \times 10^{-12} = -12 + \log_{10} 5.3 = -12 + 0.72 = -11.28$$

FINDING ANTILOGARITHMS ON A CALCULATOR

The method used to find antilogarithms depends upon the type of calculator. On certain calculators, you enter the number and then press, in succession, the $\boxed{\textbf{INV}}$ and either $\boxed{\textbf{LOG}}$ or $\boxed{\textbf{LN X}}$ keys. With other calculators, you press the $\boxed{\textbf{10}^x}$ or $\boxed{\textbf{e}^x}$ key. Either way, you should find that

$$\text{antilog}_{10} 1.632 = 42.8 \ldots \qquad \text{antilog}_e 1.632 = 5.11 \ldots$$
$$\text{antilog}_{10} -8.82 = 1.5 \times 10^{-9} \qquad \text{antilog}_e -8.82 = 1.5 \times 10^{-4}$$

SIGNIFICANT FIGURES IN LOGARITHMS AND ANTILOGARITHMS

As with other operations, a calculator does not indicate the number of significant figures to be retained in a logarithm or antilogarithm. In the examples just worked, you probably found several more digits in the calculator display, beyond those listed. For base 10 logarithms, the rules governing significant figures are quite simple:

1. In taking the logarithm of a number, retain after the decimal point in the log as many digits as there are significant figures in the number. (This part of the logarithm is often referred to as the *mantissa;* digits that precede the decimal point comprise the *characteristic* of the logarithm.) To illustrate this rule, consider the following:

$$\log_{10} 2.00 = 0.301 \qquad \log_{10}(2.00 \times 10^3) = 3.301$$
$$\log_{10} 2.0 = 0.30 \qquad \log_{10}(2.0 \times 10^1) = 1.30$$
$$\log_{10} 2 = 0.3 \qquad \log_{10}(2 \times 10^{-3}) = 0.3 - 3 = -2.7$$

2. In taking the antilogarithm of a number, retain as many significant figures in the antilogarithm as there are after the decimal point in the number. Thus,

$$\text{antilog}_{10}\ 0.301 = 2.00 \qquad \text{antilog}_{10}\ 3.301 = 2.00 \times 10^3$$
$$\text{antilog}_{10}\ 0.30 = 2.0 \qquad \text{antilog}_{10}\ 1.30 = 2.0 \times 10^1$$
$$\text{antilog}_{10}\ 0.3 = 2 \qquad \text{antilog}_{10}\ -2.7 = 2 \times 10^{-3}$$

These rules take into account the fact that, as mentioned earlier:

$$\log_{10}(C \times 10^n) = n + \log_{10}C$$

The digits that appear before (to the left of) the decimal point specify the value of n, i.e., the power of 10 involved in the expression. In that sense, they are not experimentally significant. In contrast, the digits that appear after (to the right of) the decimal point specify the value of the logarithm of C; the number of such digits reflects the uncertainty in C. Thus:

$$\log_{10} 209 = 2.320 \qquad \text{(3 sig. fig.)}$$
$$\log_{10} 209.0 = 2.3201 \qquad \text{(4 sig. fig.)}$$
$$\log_{10} 209.00 = 2.32015 \qquad \text{(5 sig. fig.)}$$

The rules for significant figures involving natural logarithms and antilogarithms are somewhat more complex than those for base 10 logs. However, for simplicity, we will assume that the rules listed above apply here as well. Thus:

$$\ln 209 = 5.342 \qquad \text{(3 sig. fig.)}$$
$$\ln 209.0 = 5.3423 \qquad \text{(4 sig. fig.)}$$
$$\ln 209.00 = 5.34233 \qquad \text{(5 sig. fig.)}$$

OPERATIONS INVOLVING LOGARITHMS

Since logarithms are exponents, the rules governing the use of exponents apply here as well. The rules that follow are valid for all types of logarithms, regardless of the base. We illustrate the rules with natural logarithms, since that is where you are most likely to use them in working with this text.

Multiplication: $\ln(xy) = \ln x + \ln y$
  Example: $\ln(2.50 \times 1.25) = \ln 2.50 + \ln 1.25 = 0.916 + 0.223 = 1.139$

Division: $\ln(x/y) = \ln x - \ln y$
  Example: $\ln(2.50/1.25) = 0.916 - 0.223 = 0.693$

Raising to a Power: $\ln(x^n) = n \ln x$
  Example: $\ln(2.00)^4 = 4 \ln 2.00 = 4(0.693) = 2.772$

Extracting a Root: $\ln(x^{1/n}) = \dfrac{1}{n} \ln x$

  Example: $\ln(2.00)^{1/3} = \dfrac{\ln 2.00}{3} = \dfrac{0.693}{3} = 0.231$

Taking a Reciprocal: $\ln(1/x) = -\ln x$
  Example: $\ln(1/2.00) = -\ln 2.00 = -0.693$

APPENDIX 4

ANSWERS TO PROBLEMS

Chapter 1

2. **(a)** chlorine, Cl   **(b)** copper, Cu
4. **(a)** manganese   **(b)** sodium   **(c)** arsenic
   (d) tungsten   **(e)** phosphorus
6. **(a)** 2   **(b)** 8   **(c)** 8   **(d)** 18   **(e)** 18
8. **(a)** length   **(b)** mass   **(c)** energy   **(d)** volume
   (e) density   **(f)** pressure   **(g)** energy
10. **(a)** 27.12 g   **(b)** 35 cm^3 **(c)** 2.87 g/L   **(d)** 525 mm
12. **(a)** 40^{b}C   **(b)** 313 K
14. **(a)** $-320.4°F$ **(b)** 77.4 K
16. **(a)** 5   **(b)** 4   **(c)** 3   **(d)** 5   **(e)** 6   **(f)** 2 or 3
18. 0.0270 nm^3
20. **(a)** 4   **(b)** 3   **(c)** 3   **(d)** 4
22. **(a)** 1.390×10^{-6} mile^2 **(b)** 3.599×10^{-6} km^2
   (c) 3.599 m^2
24. **(a)** 1.000×10^3 J   **(b)** 239.0 cal   **(c)** 9.869 L·atm
26. 1.7×10^9
28. 2.3×10^2
30. **(a)** 1157   **(b)** 0.260 lb; 3.12 oz
32. 117 g
34. 8.769 g Al, 13.00 g O, 0.819 g H, 5.761 g Cl
36. **(a)** P   **(b)** P   **(c)** C   **(d)** P
38. 13.6 g/cm^3
40. 1.6 g/cm^3
42. 7.0 lb
44. 737 g
46. **(a)** 0.00724 g   **(b)** 3.45×10^3 g
48. **(a)** 25 g/100 g water   **(b)** 1.2×10^3 g   **(c)** 4.0 g
50. **(a)** 85 g   **(b)** 0 g   **(c)** 11
51. **(a)** chemical property observed during reaction
   (b) solute is one component of solution
   (c) normal bp at 1 atm
   (d) compound is pure substance
52. **(a)** compound contains two or more elements
   (b) element is pure substance
   (c) solution has uniform composition
   (d) distillation requires vaporization
53. b, c
54. a, d
55. 0.11 cm^3
56. 4.1×10^5 g
57. 2.48 cm
58. $\$7.15 \times 10^5$
59. 7.946 g/cm^3
60. $-24.6°C$
61. 2.0 km^2
62. 172 cm
63. 8.1×10^{-3} g

Chapter 2

2. **(a)** NaCl   **(b)** CaCl$_2$ **(c)** CaO   **(d)** Na$_2$O
4. **(a)** yellow solid   **(b)** red solid
6. See discussion, Section 2.1
8. **(a)** Conservation of Mass   **(b)** Constant Composition
10. **(a)** g H/g C in ethane = 0.252; g H/g C in
   ethene = 0.168; 0.252/0.168 = 3/2
   (b) could be CH$_3$ and CH$_2$; C$_2$H$_6$, C$_2$H$_4$
12. % Hg $= \dfrac{48.43}{52.30} \times 100 = 92.60$;

   % Hg $= \dfrac{15.68}{16.93} \times 100 = 92.62$
14. $^{222}_{86}$Rn
16. Differ in number of neutrons; $^{63}_{29}$Cu and $^{65}_{29}$Cu
18. **(a)** 53   **(b)** 78   **(c)** 53   **(d)** 78 n, 53 p
20. $^{79}_{35}$Br   0   35   44   35
   $^{14}_{7}$N^{3-} -3 7   7   10
   $^{75}_{33}$As^{5+} $+5$ 33   42   28
   $^{90}_{40}$Zr   0   40   50   40
22. **(a)** 3.08256   **(b)** 0.631131   **(c)** 0.37351
24. **(a)** 79.911 amu   **(b)** 16.0007 amu
   (c) 196.98310 amu

26. 151.960 amu
28. 0.36%
30. 4.1 % Cr-50, 84.0% Cr-52
32. (a) 2 (b) 36, 38   (c) peaks at masses 36, 38
34. (a) 60, 60 (b) 100, 100   (c) 15, 18   (d) 15, 10
36. 33, 28; 46, 46; 53, 54; 10, 10
38. (a) Li_2S, Li_3N (b) FeO, Fe_2O_3
40. (a) 6.7×10^{-7} g (b) 3.002×10^{12} atoms
42. (a) 3.441×10^{-10} g   (b) 8.240×10^{22}
44. 7.579×10^{24}
46. (a) 10   (b) 6.022×10^{24} (c) 1.206×10^{26}
    (d) 5.974×10^{23}
48. (a) 17.03056 g/mol   (b) 84.007 g/mol
    (c) 190.2 g/mol
50. (a) 0.0877 mol (b) 0.0105 mol   (c) 187 mol
52. (a) 427 g   (b) 1.07×10^3 g   (c) 314 g
54. 0.2500, 0.004304, 2.592×10^{21}, 1.555×10^{22}
    11.62, 0.2000, 1.204×10^{23}, 7.226×10^{23}
    2.4×10^{-12}, 4.2×10^{-14}, 2.5×10^{10}, 1.5×10^{11}
    1.3×10^{-7}, 2.2×10^{-9}, 1.3×10^{15}, 8.0×10^{15}
56. (a) Dissolve 35.1 g KOH to form 0.500 L solution
    (b) Dissolve 419 g $CuSO_4$ to form 0.750 L solution
58. (a) 0.256 g NaCl, 12.6 g $HC_2H_3O_2$
    (b) 0.683 L; 0.0142 L
60. 1.01 M, 1.58 g, 0.0173 L, 2.57 g
61. (a) 6.08 mol   (b) 132 g
62. (a) 285.34 g/mol   (b) 0.425   (c) N
63. 2.0×10^{14}
64. a < e < c < d < b
65. 1.7%
66. 801 cm³
67. 2.7×10^{20}
68. (a) 22.989219 g   (b) 126.90502 g   (c) 149.89424 g
69. 894 g/mol
70. 6.01×10^{23}
71. 2×10^{14}
72. (a) 2.5×10^{24} (b) 2.3×10^{-20} (c) appr. 290
73. 43.62 g = 43.62 g

Chapter 3

2. (a) Sb(s), gray   (b) $CuSO_4 \cdot 5H_2O(s)$, blue
   (c) $CoCl_2 \cdot 4H_2O(s)$, purple
4. (a) $Na_2Cr_2O_7$ (b) SnS   (c) $AlPO_4$ (d) $AuCl_3$
   (e) Cr_2O_3 (f) HNO_3 (g) $Ca_3(PO_4)_2$ (h) $Ca(ClO_4)_2$
6. cesium hydroxide, $NaMnO_4$, $Li_2Cr_2O_7$, ammonium
   chloride, $Al_2(SO_4)_3$, barium nitrate
8. hydrazine, H_2O_2, xenon tetrafluoride, tetrasulfur
   tetranitride, NF_3, CCl_4
10. 30.93% Al, 45.86% O, 2.889% H, 20.32% Cl
12. 193.9 kg

14. 39.9%
16. 38.40% C, 1.50% H, 52.28% Cl, 7.8% O
18. Cr_2O_3
20. (a) C_3H_8O (b) $C_5H_8O_4NNa$ (c) $C_2H_3Cl_3O_2$
22. $C_8H_8O_3$
24. $C_2H_6N_2O$
26. C_3H_8N, $C_6H_{16}N_2$
28. CuO
30. 52.68; 16
32. (a) $UO_2(s) + 4\,HF(l) \rightarrow UF_4(s) + 2H_2O(l)$
    (b) $4PH_3(g) + 8\,O_2(g) \rightarrow P_4O_{10}(s) + 6H_2O(g)$
    (c) $2C_2H_3Cl(l) + 5\,O_2(g) \rightarrow$
    $\qquad\qquad 4CO_2(g) + 2H_2O(g) + 2HCl(g)$
34. (a) $2Al(s) + 3Cl_2(g) \rightarrow 2AlCl_3(s)$
    (b) $Sr(s) + Cl_2(g) \rightarrow SrCl_2(s)$
    (c) $2Li(s) + Cl_2(g) \rightarrow 2LiCl(s)$
    (d) $2Cr(s) + 3Cl_2(g) \rightarrow 2CrCl_3(s)$
    (e) $2Ag(s) + Cl_2(g) \rightarrow 2AgCl(s)$
36. (a) $NH_4NO_3(s) \rightarrow N_2O(g) + 2H_2O(g)$
    (b) $4NH_3(g) + 3\,O_2(g) \rightarrow 2N_2(g) + 6H_2O(g)$
    (c) $2KClO_3(s) \rightarrow 2KCl(s) + 3\,O_2(g)$
    (d) $B_2O_3(s) + 3C(s) + 3Cl_2(g) \rightarrow$
    $\qquad\qquad\qquad 2BCl_3(g) + 3CO(g)$
    (e) $2C_6H_6(l) + 15\,O_2(g) \rightarrow 12CO_2(g) + 6H_2O(l)$
38. (a) 51.7   (b) 9.29   (c) 0.0541   (d) 14.95
40. (a) 333 g   (b) 0.02499 mol   (c) 4.01 g   (d) 0.8149 g
42. (a) $NCl_3(g) + 3H_2O(l) \rightarrow NH_3(g) + 3HClO(aq)$
    (b) 5.09 mol   (c) 105 g
44. (a) 294 cm³   (b) 584 g
46. 9.2×10^3 L
48. 125
50. (a) $H_2(g) + Cl_2(g) \rightarrow 2HCl(g)$
    (b) H_2 (c) 15.0 mL   (d) 1.50 mol
52. 193 g; 49.0%
54. 0.65 kg
56. 1.50×10^3 g; 277 cm³
57. (a) could be in exact ratio required for reaction
    (b) Co not equal to CO
    (c) true only for such compounds as Na_2SO_4
    (d) no molecules in $CaCl_2$
58. (a) not true   (b) true   (c) not always true
    (d) not true
59. (a) 2.10 mol   (b) 1.01 mol   (c) 0.1884 mol
60. (a) 5.34   (b) 2.50   (c) 0.02104
61. $2C_3H_6(g) + 9\,O_2(g) \rightarrow 6CO_2(g) + 6H_2O(l)$
62. 65.95% Ba, 34.05% Cl
63. NH_4NO_3
64. 127 g
65. 1.4 M
66. 3.657 g, 2.973 g
67. 34.8%
68. (a) V_2O_5, V_2O_3 (b) 2.270 g
69. 28%

Chapter 4

2. 11.35 L, 14.8 g, 300 K

4. 0.208 atm, 21.1 kPa; 604 mm Hg, 80.5 kPa; 9.60×10^2 mm Hg, 1.26 atm

6. (a) 414 mm Hg   (b) 2.69×10^3 mm Hg

8. 696°C

10. 140°F

12. 529 mL

14. 808 mm Hg

16. 0.10 m^3

18. 33 atm

20. 0.00516 mol

22.
735 mm Hg	5.01 L	15°C	0.205 mol	18.9 g
15.8 atm	489 mL	38°C	0.302 mol	27.8 g
1.49 atm	0.885 L	45°C	0.0505 mol	4.65 g
239 kPa	2.75 L	−111°C	0.489 mol	45.0 g

24. (a) 1.32 g/L   (b) 1.89 g/L   (c) 11.0 g/L

26. 54.8 g/L vs. 1.80 g/L

28. C_2H_3NO

30. (a) 28.6 g/mol   (b) 1.17 g/L vs. 1.18 g/L

32. Cl

34. (a) $2NF_3(g) + 3H_2O(g) \rightarrow 6HF(g) + NO(g) + NO_2(g)$
(b) 2.00 L

36. 10.6 mL

38. (a) 2.53 g   (b) 1.79 L

40. 0.225 atm; CH_4

42. (a) 736 mm Hg   (b) 0.160 g

44. 0.083 atm

46. 0.2821

48. 46.8 s

50. (a) 15°C   (b) 1300°C

52. 2.3 K

54. (a) more ideally   (b) less ideally

56. (a) 179 atm   (b) 1.5×10^2 atm

57. 13.1 m

58. (a)   (b)

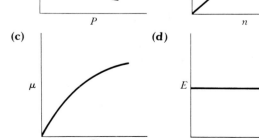

(c)   (d)

59. (a)   (b)

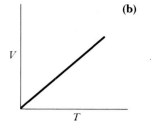

(c)   (d)

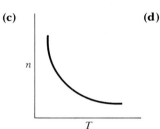

60. (a) same   (b) Cl_2 (c) Cl_2 (d) Ne

61. (a) 1.00   (b) 1.36   (c) 1.00   (d) 1.00

62. (a) 4.5×10^{-3} mol   (b) 0.038 g

63. 3.3×10^4 L

64. 46 g

65. 32.0% C, 6.7% H, 18.7% N, 42.6% O; $C_2H_5NO_2$

66. 0.572 g/L; 28.0 g/mol

67. (a) II   (b) I

68. 1.8 ft

69. 0.0456 L·atm/mol·K

70. 24.1% Al

71. 6.72 L

72. 0.821 atm

73. $V_a/V = n_a/n$; Vol % = mol %; gases have different molar masses

Chapter 5

2. higher energy in excited state

4. (a) 5.25×10^4 nm   (b) 3.78×10^{-21} J/particle
(c) 2.28 kJ/mol

6. (a) 1.498×10^4 nm   (b) IR   (c) 7.984 kJ/mol

8. (a) 4.87×10^{-8} m   (b) 48.7 nm   (c) 6.16×10^{15}/s

10. (a) 3.82×10^{14}/s   (b) 2.53×10^{-19} J

12. absorbed (1), (2); emitted (3), (4)

14.
n	1	2	3	4
E	−1312	−328	−146	−82

Lyman: n = 2,3,4 to n = 1;
Balmer: n = 3,4 to n = 2

16. 1005 nm

18. 7456 nm

20. 145.8 kJ/mol

22. 0.243 nm; 1.32×10^{-4} nm

24. 3.7×10^{-24} nm

26. (a) 0   (b) 3,2,1,0,$-1,-2,-3$
    (c) $\ell = 2$, $\mathbf{m}_\ell = 2,1,0,-1,-2$; $\ell = 1$,
       $\mathbf{m}_\ell = 1,0,-1$; $\ell = 0$, $\mathbf{m}_\ell = 0$
28. (a) 3s   (b) 4p   (c) 1s   (d) 4d
30. (a) p   (b) d   (c) s
32. (a) 32   (b) 10   (c) 2
34. (a) 3   (b) d   (c) 7   (d) 3
36. b < d < e < c = f < a
38. (a) $1s^22s^22p^63s^2$ (b) $1s^22s^22p^63s^23p^64s^23d^7$
    (c) $1s^22s^22p^63s^23p^64s^1$
    (d) $1s^22s^22p^63s^23p^64s^23d^{10}4p^65s^24d^{10}5p^1$
40. (a) [He]$2s^22p^4$ (b) [Xe]$6s^2$ (c) [Ar]$4s^23d^8$
    (d) [Kr]$5s^24d^{10}5p^4$
42. (a) Cu   (b) Nb   (c) Cl   (d) H
44. (a) 12/20   (b) 24/55   (c) 18/44
46. (a) ground   (b) excited   (c) ground   (d) impossible
    (e) impossible   (f) excited
48.   1s   2s   2p   3s   3p
    (a) (⇅)   (⇅)   (⇅)(⇅)(⇅)   (⇅)   (↑)(↑)( )
    b–d: 1s, 2s, 2p, 3s, 3p, 4s, 3d, 4p filled, plus:
        5s   4d   5p
    (b) (⇅)   (↑)(↑)(↑)( )( )
    (c) (⇅)
    (d) (⇅)   (⇅)(⇅)(⇅)(⇅)(⇅)   (↑)(↑)(↑)
50. (a) Mg   (b) P   (c) O
52. (a) Mo, Tc   (b) Rb, Y, Ag, In   (c) Kr   (d) Sb
54. (a) 2   (b) 5   (c) 1
56. (a) 1, 2, 3, 4, 5   (b) 6   (c) 7   (d) 8
58.

n	1	1	2	2	2	2	2	2	2
ℓ	0	0	0	0	1	1	1	1	1
$\mathbf{m}_\ell$	0	0	0	0	1	1	0	0	-1
$\mathbf{m}_s$	$\frac12$	$-\frac12$	$\frac12$	$-\frac12$	$\frac12$	$-\frac12$	$\frac12$	$-\frac12$	$\frac12$

60. (a) $\mathbf{n} = 4,4$; $\ell = 0,0$; $\mathbf{m}_\ell = 0,0$; $\mathbf{m}_s = \frac12, -\frac12$

(b)

n	3	3	3	3	3
ℓ	2	2	2	2	2
$\mathbf{m}_\ell$	2	1	0	-1	-2
$\mathbf{m}_s$	$\frac12$	$\frac12$	$\frac12$	$\frac12$	$\frac12$

(c)

n	3	3	3	3	3	3
ℓ	1	1	1	1	1	1
$\mathbf{m}_\ell$	1	1	0	0	-1	-1
$\mathbf{m}_s$	$\frac12$	$-\frac12$	$\frac12$	$-\frac12$	$\frac12$	$-\frac12$

62. (a) 1   (b) 10   (c) 2
64. (a) $1s^22s^22p^3$; $1s^22s^22p^6$
    (b) $1s^22s^22p^63s^23p^64s^1$; $1s^22s^22p^63s^23p^6$
    (c) $1s^22s^22p^63s^23p^64s^23d^1$; $1s^22s^22p^63s^23p^6$
    (d) $1s^22s^22p^63s^23p^63d^6$; $1s^22s^22p^63s^23p^63d^5$
66. (a) 2   (b) 0   (c) 0   (d) 0   (e) 0
67. 1.5×10^{19}
68. 5.34×10^{19}
69. (a) specification of position of electron
    (b) $\nu = c/\lambda$
    (c) paramagnetic species attracted into field, diamagnetic repelled
    (d) at 90° angles

70. (a) no two electrons can have same set of quantum numbers
    (b) maximum number of unpaired electrons
    (c) wavelength of emission
    (d) $\mathbf{n} = 1, 2, 3, --$
71. (a) false; emits energy   (b) false; longer λ (c) true
    (d) true
72. (a) directly proportional
    (b) inversely proportional to $\mathbf{n}^2$
    (c) 4s fills before 3d
73. $\mathbf{n} = 5$; $\mathbf{m}_\ell = 4,3,2,1,0,-1,-2,-3,-4$; 18
74. 2.952×10^3 kJ/mol
75. $E_\mathbf{n} = -1312/\mathbf{n}^2$; $E_2 = -1312/4$

$$E_{hi} - E_{lo} = 1312\left[\frac{1}{4} - \frac{1}{\mathbf{n}^2}\right] = \frac{1.196 \times 10^5}{\lambda}$$

$$\lambda = \frac{1.196 \times 10^5}{1312} \times \frac{4\mathbf{n}^2}{(\mathbf{n}^2 - 4)} = 364.6\frac{\mathbf{n}^2}{(\mathbf{n}^2 - 4)}$$

76.

n	1				
ℓ	0		1		
$\mathbf{m}_\ell$	0	1	0	1	2
$\mathbf{m}_s$	$\pm\frac12$	$\pm\frac12$	$\pm\frac12$	$\pm\frac12$	$\pm\frac12$

	2					
0		1		2		
0	1	0	1	2	0 1 2 3
$\pm\frac12$ $\pm\frac12$		$\pm\frac12$ $\pm\frac12$ $\pm\frac12$		$\pm\frac12$ $\pm\frac12$ $\pm\frac12$ $\pm\frac12$		

$1s^41p^4$

77. (a) 3, 9, 15   (b) 27   (c) $1s^32s^32p^2$; $1s^32s^32p^93s^2$
78. (a) 3.42×10^{-19} J   (b) 581 nm

Chapter 6

2. (a) KO_2 (b) $NaHCO_3$ (c) Al_2O_3 (d) $CaCO_3$
4. (a) $MgCl_2(l) \rightarrow Mg(s) + Cl_2(g)$
   (b) $CaSO_4 \cdot \frac12 H_2O(s) + \frac32 H_2O(l) \rightarrow CaSO_4 \cdot 2H_2O(s)$
   (c) $CaO(s) + H_2O(l) \rightarrow Ca(OH)_2(s)$
6. (a) horizontal row   (b) Group 7 element
   (c) at. no. 57–70   (d) at. no. 89–102
   (e) properties intermediate between metal, nonmetal
8. (a) 118   (b) 112
10. (a) francium, Fr   (b) thallium, Tl   (c) bismuth, Bi
    (d) actinium, Ac
12. (a) $6s^26p^6$ (b) $7s^1$ (c) $5s^25p^4$ (d) $6s^26p^3$
14. (a) Cl < S < Mg   (b) Mg < S < Cl
    (c) Cl < S < Mg
16. (a) Rb   (b) Br   (c) Br
18. (a) metalloid   (b) metal   (c) nonmetal   (d) metal
    (e) metal
20. (a) O, S, Se   (b) Li, Na, K, Rb   (c) He, Mg, Ca, Sr, Ba, Ra   (d) B, Al, Ga, In, Tl

22. 5.00 g/cm^3

24. (a) N^{3-} (b) Ba   (c) Cr   (d) Cu$^+$

26. (a) Co $>$ Co^{2+} $>$ Co^{3+} (b) Br$^-$ $>$ Cl$^-$ $>$ Cl

28. (a) potassium nitride, K$_3$N   (b) potassium iodide, KI
(c) potassium hydroxide, KOH
(d) potassium hydride, KH
(e) potassium sulfide, K$_2$S

30. (a) Na$_2$O$_2$(s) + 2H$_2$O $\rightarrow$ 2Na$^+$(aq) + 2OH$^-$(aq) +
H$_2$O$_2$(aq); hydrogen peroxide
(b) 2Ca(s) + O$_2$(g) $\rightarrow$ 2CaO(s); calcium oxide
(c) Rb(s) + O$_2$(g) $\rightarrow$ RbO$_2$(s); rubidium
superoxide
(d) SrH$_2$(s) + 2H$_2$O $\rightarrow$ Sr^{2+}(aq) + 2OH$^-$(aq) +
H$_2$(g); strontium hydroxide

32. (a) Al$_2$(SO$_4$)$_3$ (b) AlBr$_3$ (c) Al(OH)$_3$
(d) KAl(SO$_4$)$_2$·12H$_2$O

34. 2Ca(s) + O$_2$(g) $\rightarrow$ 2CaO(s)
CaO(s) + H$_2$O(l) $\rightarrow$ Ca(OH)$_2$(s)
Ca(s) + 2H$_2$O $\rightarrow$ Ca^{2+}(aq) + 2OH$^-$(aq) + H$_2$(g)

36. (a) Na   (b) Cs (Fr)   (c) Na   (d) K

38. (a) outer s electron lost   (b) inner p electron lost

40. 2.91 g Al

42. 780.2 nm; Rb

43. (a) added electrons repel
(b) has 3 more e^- than Ar
(c) Na^{2+} requires removal of inner electron
(d) Cl has much greater attraction for e^-

44. (a) closely spaced energy levels
(b) effective nuclear charge increases
(c) metals have 2e^- beyond noble gas
(d) elements become less metallic

45. (a) absorbed   (b) atomic number   (c) organic liquid

46. (a) false; increases   (b) false; ns^{2}np^5
(c) false (superoxide, peroxide)

47. See text discussion

48. See text discussion

49. (a) phosphorus, P   (b) astatine, At   (c) titanium, Ti
(d) francium, Fr   (e) argon, Ar

50. (a) X is a metal, Y a nonmetal   (b) Y   (c) X   (d) X
(e) X   (f) X^{2+}, Y^{2-}

51. 3.25 atm

52. 5.13 g

53. 208 g

54. 1.85×10^3 g Al; 189 L

55. 11%

56. (l); Br, Hg

57. 216 kJ/mol

58. 285.2 nm; UV

59. (a) [Rn] 7s^{2}5f^{14}6d^{10}7p^4 (b) Po

60. (a) 0.15 nm   (b) $\sim300°$C   (c) $\sim400°$C

61. 2% BaO$_2$

62. 1.26 g/cm^3; 2.08 g/cm^3

63. 0.32; 0.76 L

Chapter 7

2. (a) Hg(l) $\rightarrow$ Hg(s); $\Delta H = -2.33$ kJ
(b) C$_{10}$H$_8$(l) $\rightarrow$ C$_{10}$H$_8$(g); $\Delta H = 43.3$ kJ
(c) C$_6$H$_6$(s) $\rightarrow$ C$_6$H$_6$(l); $\Delta H = 9.84$ kJ

4. (a) K(s) + O$_2$(g) $\rightarrow$ KO$_2$(s)
(b) N$_2$(g) + 2H$_2$(g) $\rightarrow$ N$_2$H$_4$(l)
(c) N$_2$(g) + 4H$_2$(g) + C(s) + $\frac{3}{2}$O$_2$(g) $\rightarrow$
(NH$_4$)$_2$CO$_3$(s)
(d) N$_2$(g) + 2O$_2$(g) $\rightarrow$ N$_2$O$_4$(g)

6. $25.4°$C

8. 67.8 g

10. (a) KBr(s) $\rightarrow$ K$^+$(aq) + Br$^-$(aq)
(b) endothermic   (c) +223 J   (d) +19.8 kJ

12. -1.38×10^3 kJ

14. 4302 J/°C

16. $21.81°$C

18. (a) 3.87 kJ/°C   (b) -50.0 kJ   (c) -1.40×10^3 kJ

20. (a) Ni(CO)$_4$(g) $\rightarrow$ Ni(s) + 4CO(g)
(b) endothermic   (c) products above reactants
(d) 0.9413 kJ   (e) 10.62 g

22. (a) -84.4 kJ   (b) -0.449 kJ

24. (a) 2Mg(s) + CO$_2$(g) $\rightarrow$ C(s) + 2MgO(s);
$\Delta H = -812$ kJ
(b) 8.77 kJ

26. 1.548×10^4 kJ; 3.700×10^6 cal; 3.700×10^3 kcal

28. condensation C$_6$H$_6$(g)

30. -4.9 kJ

32. SiO$_2$(s) + 2Mg(s) + 2Cl$_2$(g) + 2C(s) $\rightarrow$
Si(s) + 2MgCl$_2$(s) + 2CO(g)
-592.7 kJ; exothermic

34. -226.7 kJ

36. (a) -1675.7 kJ/mol   (b) -205.4 kJ

38. 3.67 kJ

40. (a) $+354.8$ kJ   (b) -7.1 kJ   (c) $+264.4$ kJ

42. (a) -633.2 kJ   (b) -1530.4 kJ

44. (a) C$_2$H$_4$(g) + 2HCl(g) + $\frac{1}{2}$O$_2$(g) $\rightarrow$
C$_2$H$_4$Cl$_2$(l) + H$_2$O(l); $\Delta H = -318.7$ kJ
(b) -165.2 kJ/mol

46. -919.9 kJ

48. $\Delta H = -2.91$ kJ

50. (a) $+77$ J   (b) $+22$ J

52. $+37.6$ kJ

54. (a) C$_3$H$_8$(g) + 5O$_2$(g) $\rightarrow$ 3CO$_2$(g) + 4H$_2$O(l);
$\Delta H = -2219.9$ kJ   (b) 50.2 kJ

56. 2.74×10^5 g

58. 1.6×10^{16} kJ; 20%

60. (a) 7.5×10^3 kJ   (b) 12¢

61. 50.6 kJ/mol

62. 8.453×10^4 kJ

63. 34 miles

64. $23.1°$C

65. $83.9°$C

66. 56% C_2H_6
67. 5.7×10^3 kJ
68. (a) 171 kJ (b) 514 g
69. $\frac{1}{4}$
70. (a) -851.5 kJ (b) 6.6×10^3 °C (c) yes

Chapter 8

2. (a) H—P̈—H (b) :F̈—S̈i—F̈:
with H below P; F above and below Si with :F̈: groups

(c) H—Ö—N̈=Ö: (d) [H—N̈—H]$^+$ with H above and below N

4. (a) $(:Ö—I̋—Ö:)^-$ with :Ö: below (b) $(:F̈—C̈l—F̈:)^+$

(c) $(:Ö—S̈—Ö:)^{2-}$ with :Ö: below (d) $(:Ö—Äs—Ö:)^{3-}$ with :Ö: below

6. (a) :F̈—B̈r—F̈: with :F̈: below (b) :F̈—Xe—Ö: with :Ö: above and :F̈: below

(c) $(:I̋—I̋—I̋:)^-$ (d) central I with five :F̈: groups (:Ẍe) arranged

8. (a) $(H—Ö—C̈—Ö:)^-$ with :Ö: below (b) H—C—C—Ö—H with H above and below first C, :Ö: on second C

(c) H—Ö—S̈—Ö—H with :Ö: below

10. H₂C=Ö—Ö: (other structures are possible), with two H on C

12. H—C—C—Ö—Ö—N with H on first C, :Ö: below second C, O. groups on N

14. H—C—C—Ö—H, H—C—Ö—C—H (with H atoms around each C)

16. (a) Cl_2 (b) H_2SO_4 (c) CH_4 (d) CCl_4

18. (a) $(:C̈l—B—C̈l:)^-$ with :Cl̈: above and below (b) $(:Ö—S̈—S̈:)^{2-}$ with :Ö: above and below

(c) $(H—Ö—P—Ö—H)^-$ with :Ö: above and below (d) $(:Ö—C̈l—Ö:)^-$ with :Ö: above and below

20. (a) ·N̈ with O. and O: (b) H—Be—H

(c) $(:Ö—S̈—Ö:)^-$ (d) $(·C≡O:)^-$

22. (a) $(:S̈—C≡N:)^- \leftrightarrow (:S̈=C=N̈:)^- \leftrightarrow (:S≡C—N̈:)^-$

(b) $(H—C—Ö:)^- \longleftrightarrow (H—C=Ö)^-$ with :Ö: below

(c) $H—Ö—N̈—Ö: \longleftrightarrow H—Ö=N—Ö: \longleftrightarrow$ with :Ö: below

$H—Ö—N̈=O$ with :Ö: below

24. (a) $H—N≡N—N̈:$, $H—N̈—N≡N:$
(b) no; different skeletons

26. (two six-membered B₃N₃H₆ ring structures with H atoms)

28. N—F > N—Se > N—S > N—N
30. Cl—O
32. (a) N—H (b) N—S (c) P—O

34. (a) C (b) S (c) H (d) I
36. a > b > c
38. a; c
40. (a) −941 kJ (b) −1656 kJ (c) −1167 kJ
42. +98 kJ
44. −78 kJ
46. +101 kJ
47. (a) increases (b) closer (c) electron pair
  (d) decreases
48. (a) need electron pairs (b) see text discussion
  (c) polarity of H—F bond
49. (a) electrons in outermost energy level
  (b) not involved in bond formation
  (c) more than 8 valence e^- around atom
  (d) 2 or 3 electron pairs between bonded atoms
50. (a) 8 electrons
  (b) single Lewis structure insufficient
  (c) electron pair shared by atoms
  (d) odd number of valence e^-

51. (a) $(:\!\ddot{O}\!-\!\overset{\displaystyle :\ddot{O}:}{\underset{\displaystyle :\ddot{O}:}{S}}\!-\!\ddot{O}\!:)^{2-}$ (b) $(:\!\ddot{O}\!-\!\overset{\displaystyle :\ddot{O}:}{\underset{\displaystyle :\ddot{O}:}{P}}\!-\!\ddot{O}\!:)^{3-}$

  (c) $(:\!\ddot{O}\!-\!\overset{\displaystyle}{\underset{\displaystyle :\ddot{O}:}{Cl}}\!-\!\ddot{O}\!:)^{-}$

52. $(:\!\ddot{O}\!-\!\overset{\displaystyle :\ddot{O}:}{\underset{\displaystyle :\ddot{O}:}{Cr}}\!-\!\ddot{O}\!-\!\overset{\displaystyle :\ddot{O}:}{\underset{\displaystyle :\ddot{O}:}{Cr}}\!-\!\ddot{O}\!:)^{2-}$

53. (ring structure of S₈ with eight sulfur atoms, each with two lone pairs)

54. −644 kJ; −633.2 kJ
55. 0.114 nm, 0.176 nm
56. b, d
57. strength of bond, paramagnetism, no

58. (two benzene-like resonance structures of C₆H₄ with cumulated/triple bonds) ↔ etc.

59. $:\!\ddot{F}\!-\!N\!=\!N\!-\!\ddot{F}\!:$ ↔ $:\!\ddot{F}\!=\!N\!-\!\ddot{N}\!-\!\ddot{F}\!:$ ↔
  $:\!\ddot{F}\!-\!\ddot{N}\!-\!N\!=\!\ddot{F}\!:$

  $:\!\ddot{F}\!-\!\ddot{F}\!-\!N\!\equiv\!N\!:$ ↔ $:\!\ddot{F}\!-\!\ddot{F}\!=\!N\!-\!\ddot{N}\!:$ ↔
  $:\!\ddot{F}\!=\!\ddot{F}\!-\!\ddot{N}\!-\!\ddot{N}\!:$

  and many others
60. (a) 137 g/mol (b) P (c) $:\!\ddot{C}l\!-\!\overset{\displaystyle \ddot{\;\;}}{\underset{\displaystyle :\ddot{C}l:}{P}}\!-\!\ddot{C}l\!:$

Chapter 9

2. (a) linear (b) bent (c) bent (d) tetrahedral
4. (a) tetrahedral (b) bent (c) equilateral triangle
  (d) triangular pyramid
6. (a) see-saw (b) octahedral (c) linear
  (d) T-shaped
8. (a) 120° (b) 120° (c) 120°, 109.5°

10. (a) $H\!-\!C\!\equiv\!C\!-\!\ddot{C}l\!:$ linear

  (b) $(H\!-\!\overset{\displaystyle :\ddot{O}:}{C}\!=\!\ddot{O})^{-}$ equilateral triangle

  (c) $H\!-\!\overset{\displaystyle H}{\underset{\displaystyle H}{C}}\!-\!\ddot{C}l\!:$ tetrahedral

  (d) $(:\!\ddot{C}l\!-\!\overset{\displaystyle :\ddot{C}l:}{\underset{\displaystyle :\ddot{C}l:}{P}}\!-\!\ddot{C}l\!:)^{+}$ tetrahedral

12. (a) square planar (b) square pyramid
  (c) triangular bipyramid

14. (a) $\overset{\displaystyle :\ddot{C}l}{\underset{\displaystyle :\ddot{C}l}{}}\!C\!=\!\ddot{O}$ equilateral triangle

  (b) $:\!\ddot{I}\!-\!\overset{\displaystyle}{\underset{\displaystyle :\ddot{I}:}{N}}\!-\!\ddot{I}\!:$ triangular pyramid

  (c) $(:\!\ddot{O}\!-\!\overset{\displaystyle}{\underset{\displaystyle :\ddot{O}:}{P}}\!-\!\ddot{O}\!:)^{3-}$ triangular pyramid

  (d) $\overset{\displaystyle \ddot{O}}{\underset{\displaystyle :\ddot{O}\diagdown\;\diagup\ddot{O}:}{}}$ bent

16. (a) 0 in MnO_4^-, 2 in O_3^{2-}, 0 in CO_3^{2-}, 1 in ClO_3^-
(b) 4, 2, 3, 3
(c) 109.5°, 109.5°, 120°, 109.5°

18. (a)

$$H-\underset{\underset{H}{|}}{\overset{\overset{H}{|}}{C}}-\underset{}{\overset{\overset{:O:}{\|}}{C}}-\overset{..}{\underset{..}{O}}-\overset{..}{\underset{..}{O}}-\underset{}{\overset{\overset{:O:}{\|}}{C}}-\underset{\underset{H}{|}}{\overset{\overset{H}{|}}{C}}-H$$

(b) 109.5° in CH_3 groups, around central O atoms; 120° around C=O
20. (a) 109.5° **(b)** 109.5°, 180° **(c)** 109.5°, 120°
22. PH_3, H_2S; unshared pairs present
24. b, c, d
26. d
28. 1st molecule polar
30. 5.7%
32. (a) sp **(b)** sp^2 **(c)** sp^2 **(d)** sp^3
34. (a) sp^3 **(b)** sp^3 **(c)** sp^2 **(d)** sp^3
36. (a) 5 in $SeCl_4$, 6 in SF_5Cl, 5 in KrF_2, 5 in IF_3
(b) sp^3d, sp^3d^2, sp^3d, sp^3d
38. (a) 6, sp^3d^2 **(b)** 6, sp^3d^2 **(c)** 5, sp^3d
40. All carbons sp^2; nitrogen sp^2; one oxygen sp^2, the other sp^3
42. (a) sp^3 **(b)** sp^2 **(c)** sp **(d)** sp^2
44. (a) sp^3 **(b)** sp^2 **(c)** sp^3
46. (a) 4 sigma **(b)** 3 sigma, 1 pi **(c)** 2 sigma, 2 pi
(d) 3 sigma, 1 pi
48. (a) BCl_3 **(b)** O_2 **(c)** NH_4^+ **(c)** CO_3^{2-}
50.

	σ_{2s}	σ_{2s}^*	π_{2p}	π_{2p}	σ_{2p}	π_{2p}^*	π_{2p}^*	σ_{2p}^*
(a)	2	2	2	2	2	1	1	0
(b)	2	2	2	2	1	0	0	0
(c)	2	2	2	2	2	2	2	0

Bond Order	Unpaired e^-
2	2
$\frac{5}{2}$	1
1	0

52.

Species	Li_2^+	Be_2^+	B_2^+	C_2^+	N_2^+	O_2^+	F_2^+
Bond Order	$\frac{1}{2}$	$\frac{1}{2}$	$\frac{1}{2}$	$\frac{3}{2}$	$\frac{5}{2}$	$\frac{5}{2}$	$\frac{3}{2}$
Unpaired e^-	1	1	1	1	1	1	1

54. (a) 1 **(b)** $\frac{1}{2}$ **(c)** 2 **(d)** 1
55. (a) See Figure 9.6
(b) See Table 9.1, Figure 9.6
(c) See text discussion
56. (a) "Extra bonds" of multiple bond are pi bonds
(b) Determine whether molecule is symmetric
(c) Draw Lewis structure; put any extra electrons around central atom
57. (a) species with + and − poles
(b) product of charge of + pole × distance between poles
(c) orbital made by mixing s, p, or d orbitals
(d) electron density symmetric about bond axis

58. bond order = 0
59. (a)

$$\left(:\overset{..}{\underset{..}{O}}-\overset{\overset{:O:}{|}}{\underset{\underset{:O:}{|}}{S}}-\overset{..}{\underset{..}{S}}:\right)^{2-}$$

(b) tetrahedral, 109.5°
(c) yes **(d)** sp^3

60. (a)

$$:\overset{:\overset{..}{F}:\;:\overset{..}{F}:}{\underset{:\overset{..}{F}:\;:\overset{..}{F}:}{\overset{..}{F}-S-\overset{..}{F}:}}$$

(b) octahedral **(c)** no **(d)** sp^3d^2

61. (a) 5 pi, 21 sigma **(b)** 109.5°, 120°, 109.5°
(c) sp^2, sp^3, sp^2
62.

AX_2E_2	2	2	bent	sp^3	(p)
AX_3	3	0	equilateral triangle	sp^2	(np)
AX_4E_2	4	2	square planar	sp^3d^2	(np)
AX_5	5	0	triangular bipyramid	sp^3d	(np)

63. 3, T-shaped, polar, 90°, sp^3d, 3 sigma

64. $H-\overset{\overset{..}{}}{\underset{\underset{H}{|}}{N}}-\underset{\underset{H}{|}}{N}-H$; 109.5° bond angles; yes

65. ClF_3 could be equilateral triangle with Cl in center; XeF_2 could be bent. Structures found are those in which unshared pairs occupy more space than bonded pairs.
66. 6, sp^3d^2, octahedral

67. (a) $H-\underset{\underset{H}{|}}{\overset{\overset{H}{|}}{C}}-\overset{\overset{H}{|}}{C}=\underset{\underset{H}{|}}{C}-H$ and [cyclopropane-type structure]

(b) all sp^3 except sp^2 for double-bonded carbons
(c) 109.5°, 120° in first structure, 60° in second
68. 7 pairs of e^- around Xe; pentagonal bipyramid

Chapter 10

2. (a) 6.26 mg **(b)** 0.15 mm Hg **(c)** 0.466 mm Hg
4. (a) yes **(b)** 19.83 mm Hg
6. (a) 40.6 kJ **(b)** 17 mm Hg
8. 3.8×10^2 mm Hg
10. 120°C
12. 32 kJ
14. b, c
16. (a) liquid, vapor **(b)** vapor **(c)** liquid
18. (a) sublimes **(b)** melts **(c)** condenses
(d) condenses

20. (a, b)

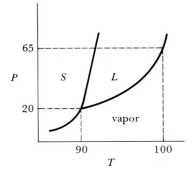

(c) liquid vaporizes

22. a > d > c > b

24. (a) dispersion   **(b)** dispersion, dipole
   (c) dispersion, dipole   **(d)** dispersion, dipole

26. b, d

28. (a) molecular vs. ionic   **(b)** hydrogen bonding
   (c) hydrogen bonding
   (d) dispersion forces increase with MM

30. a, c, d

32. (a) PH_3 (lower MM)   **(b)** C_6H_6 (lower MM)
   (c) PH_3 (no H bonds)
   (d) C_3H_8 (molecular vs. ionic)

34. (a) H bonds   **(b)** dispersion   **(c)** dispersion
   (d) dispersion

36. (a) network covalent or ionic   **(b)** molecular
   (c) ionic

38. (a) molecular, network covalent, metallic
   (b) network covalent, ionic   **(c)** metallic

40. (a) metallic   **(b)** molecular   **(c)** network covalent
   (d) ionic   **(e)** molecular

42. (a) $C_{12}H_{22}O_{11}$ **(b)** Na_2CO_3 **(c)** SiC   **(d)** steel

44. (a) C atoms   **(b)** Si and C atoms
   (c) Fe^{2+}, Cl^- ions   **(d)** C_2H_2 molecules

46. 0.219 nm

48. 0.0793 nm^3

50. (a) 0.151 nm   **(b)** no

52. $s(3)^{1/2} = 2(r_{cation} + r_{anion})$

54. 3.66 g/cm^3

56. (a) 4
   (b) 8 for Cl^- at corner; 2 for Cl^- in center of face

57. (a) independent of volume
   (b) logarithmic function
   (c) only for molecular substances
   (d) covalent bonds intact during melting

58. a

59. (a) atom at center of each face; atom at center of
   cube
   (b) $s \rightarrow l$; $s \rightarrow g$ **(c)** normal bp at 1 atm
   (d) VP curve is one line on phase diagram

60. (a) intramolecular forces are covalent bonds
   (b) See Figure 10.8

(c) H bond is special type of dipole force
(d) See Figure 10.8

61. BaO < MgO; BaO > CsF

62. 226, 225, 217, 208

63. (a) 14.8 cm^3 **(b)** 1.5×10^{10} cm^3 **(c)** 9.39 cm^3
   (d) 63.4%, 6.3×10^{-8}%

64. (a) H bonds in ethyl alcohol   **(b)** H bonds in HF
   (c) LiF ionic

65. network covalent

66. 0.0436 atm

67. ethane, carbon disulfide

68. (a) vapor and liquid   **(b)** 26.7 mm Hg

69. yes

70. 80 atm; 0.60°C; not likely; heat conduction

71. 0.414

72. 52.36; 74.04

Chapter 11

2. (a) $C_6H_{12}O_6$ **(b)** H_2SO_4 **(c)** K_2CrO_4

4. (a) $KClO_4(s) \rightarrow K^+(aq) + ClO_4^-(aq)$
   (b) $Sc_2(SO_4)_3(s) \rightarrow 2Sc^{3+}(aq) + 3SO_4^{2-}(aq)$
   (c) $CaBr_2(s) \rightarrow Ca^{2+}(aq) + 2Br^-(aq)$
   (d) $Ni(ClO_3)_2(s) \rightarrow Ni^{2+}(aq) + 2ClO_3^-(aq)$

6. (a) $Br_2(l) \rightarrow Br_2(aq)$
   (b) $KMnO_4(s) \rightarrow K^+(aq) + MnO_4^-(aq)$
   (c) $(NH_4)_3PO_4(s) \rightarrow 3NH_4^+(aq) + PO_4^{3-}(aq)$
   (d) $C_6H_{12}O_6(s) \rightarrow C_6H_{12}O_6(aq)$

8. (a) 0.99   **(b)** 0.99   **(c)** 0.99   **(d)** 1.65

10. (a) 11.2   **(b)** 87.7   **(c)** 0.0251

12. 0.707 m

14. 2.7×10^{-3} mol

16. (a) 0.139, 0.305   **(b)** 3.38 g, 0.272 L
   (c) 78.2 g, 0.868

18. (a) 0.8544, 1.200×10^5, 0.01516
   (b) 0.00367, 0.0586, 0.0000661
   (c) 5.85, 5.85×10^4, 0.00696
   (d) 3.13, 33.3, 3.33×10^5

20. (a) Dilute 1.67 L of 0.750 M acid to 5.00 L with
   water
   (b) Dilute 3.57 L of 0.350 M acid to 5.00 L with
   water

22. (a) 0.0383 M, 0.115 M, 0.0383 M
   (b) 0.690 M, 0.230 M

24. 14.6 M, 57.8 m, $X = 0.510$

26. (a) 0.940, 11.0   **(b)** 2.26, 2.66   **(c)** 2.71, 29.1

28. (a) NaF (ionic)   **(b)** NH_3 (H bonds)
   (c) CO_2 (molecular)   **(d)** CH_3OH (H bonds)

30. (a) greater   **(b)** less   **(c)** greater   **(d)** less

32. (a) 0.23 M, 1.1×10^{-5} M

34. (a) 5.26 mm Hg   **(b)** 52.6 mm Hg
   (c) 63.1 mm Hg

36. 345 g
38. 1.50×10^{-4} atm
40. **(a)** 14.5 g, 100.42°C   **(b)** 15.5 g, 100.42°C
42. -33.5°C; yes
44. 1.70°C/m
46. $C_{12}H_{26}O$
48. 6.12
50. 0.16 M
52. 6.50×10^4 g/mol
54. **(a)** -0.56°C   **(b)** -0.37°C   **(c)** -0.74°C
56. nonionized
57. To prepare supersaturated solution, bubble CO_2 through water and warm gently.
58. Add a small crystal of KNO_3. If nothing happens, solution is saturated. If crystal dissolves, it is unsaturated. If precipitate forms, it is supersaturated.
59. **(a)** 3 mol ions vs. 2   **(b)** ions free to move
   (c) solution process endothermic
   (d) osmosis is a spontaneous process; p must exceed π
60. **(a)** more ions   **(b)** solution process exothermic
   (b) ions formed in water
   (d) 1 L solution contains about 1 kg water
61. **(a)** must be isotonic with blood
   (b) higher conc. O_2
   (c) CO_2 comes out of solution as P drops
   (d) See defining equations
62. **(a)** dissolved solutes lower fp
   (b) depends on concentration rather than type of solute
   (c) crystallization of excess solute
   (d) air dissolves in blood under pressure
63. **(a)** may contain very little solute if solubility is low
   (b) in some cases, it increases
   (c) not in concentrated solution
   (d) approximately 3/2 as great
   (e) NaCl higher because more particles are present
64. **(a)** test conductivity
   (b) CO_2 comes out of solution
   (c) no. kg solvent much less than no. moles solution
   (d) vapor pressure lowered
65. **(a)** 0.8060   **(b)** 0.973   **(c)** 17.24 mm Hg
   (d) -1.81°C
66. **(a)** 2.2   **(b)** 2.4   **(c)** 2.2×10^2 atm   **(d)** 105.0°C
67. 28 atm
68. Add about 1030 g water
69. $m = \dfrac{\text{no. moles solute}}{\text{no. kg solvent}}$;

for 1 L soln., no. moles solute $= M$, and

no. kg solvent $= \dfrac{1000\, d - M(MM)}{1000} = \dfrac{d - M(MM)}{1000}$

In dilute solution, $m \rightarrow M/d$; for water, $d \rightarrow 1.00$ g/mL

70. 48
71. 0.0018 g/cm³
72. **(a)** 2.08   **(b)** 1.872   **(c)** 47.4 L
73. $V_{\text{gas}} = \dfrac{n_{\text{gas}} \times RT}{P_{\text{gas}}}$; $n_{\text{gas}} = k \times P_{\text{gas}}$; $V_{\text{gas}} = kRt$

Chapter 12

1. **(a)** K_2CO_3 (soluble)   **(b)** $Mg(OH)_2$ (insoluble)
   (c) $Ce(NO_3)_3$ (soluble)   **(d)** $NiSO_4$ (soluble)
4. **(a)** Add solution of NaOH, filter off $Cu(OH)_2$
   (b) Add solution of $SrCl_2$, filter off $SrSO_4$
   (c) Add HCl, filter off Hg_2Cl_2
6. **(a)** $Fe^{3+}(aq) + 3\,OH^-(aq) \rightarrow Fe(OH)_3(s)$
   (b) $Cd^{2+}(aq) + S^{2-}(aq) \rightarrow CdS(s)$
      $Sr^{2+}(aq) + SO_4{}^{2-}(aq) \rightarrow SrSO_4(s)$
8. **(a)** $Cu^{2+}(aq) + 2\,OH^-(aq) \rightarrow Cu(OH)_2(s)$
   (b) $Cu^{2+}(aq) + S^{2-}(aq) \rightarrow CuS(s)$
   (c) $Hg_2{}^{2+}(aq) + 2Cl^-(aq) \rightarrow Hg_2Cl_2(s)$
   (d) $Fe^{3+}(aq) + 3\,OH^-(aq) \rightarrow Fe(OH)_3(s)$
      $Ba^{2+}(aq) + SO_4{}^{2-}(aq) \rightarrow BaSO_4(s)$
   (e) $Ni^{2+}(aq) + CO_3{}^{2-}(aq) \rightarrow NiCO_3(s)$
10. **(a)** no reaction
   (b) $Ca^{2+}(aq) + CO_3{}^{2-}(aq) \rightarrow CaCO_3(s)$
   (c) $Pb^{2+}(aq) + S^{2-}(aq) \rightarrow PbS(s)$
   (d) $Fe^{3+}(aq) + 3\,OH^-(aq) \rightarrow Fe(OH)_3(s)$
12. **(a)** HNO_2 **(b)** H^+ **(c)** $HC_2H_3O_2$
   (d) H^+ **(e)** $HC_3H_5O_2$
14. **(a)** OH^- **(b)** $(CH_3)_2NH$ **(c)** OH^- **(d)** C_5H_5N
16. **(a)** weak acid   **(b)** strong base
   (c) weak base   **(d)** strong acid
18. **(a)** $H^+(aq) + OH^-(aq) \rightarrow H_2O$
   (b) $HNO_2(aq) + OH^-(aq) \rightarrow NO_2{}^-(aq) + H_2O$
   (c) $H^+(aq) + C_6H_5NH_2(aq) \rightarrow C_6H_5NH_3{}^+(aq)$
20. **(a)** $H^+(aq) + C_2H_5NH_2(aq) \rightarrow C_2H_5NH_3{}^+(aq)$
   (b) $H^+(aq) + OH^-(aq) \rightarrow H_2O$
   (c) $HC_2H_3O_2(aq) + OH^-(aq) \rightarrow$
      $\qquad\qquad\qquad\qquad\quad C_2H_3O_2{}^-(aq) + H_2O$
   (d) $H^+(aq) + OH^-(aq) \rightarrow H_2O$
   (e) $H^+(aq) + OH^-(aq) \rightarrow H_2O$
22. **(a)** $N = +3$, $O = -2$ **(b)** $Te = +6$, $F = -1$
   (c) $N = +3$, $O = -2$ **(d)** $S = +2$, $O = -2$
   (e) $Cl = +7$, $O = -2$
24. **(a)** $Ca = +2$, $C = +3$, $O = -2$
   (b) $H = +1$, $S = +6$, $O = -2$
   (c) $Na = +1$, $Fe = +3$, $O = -2$
   (d) $N = +3$, $O = -2$, $F = -1$
   (e) $N = -2$, $H = +1$
26. **(a)** R   **(b)** O   **(c)** R   **(d)** O
28. **(a)** FeS oxidized, $NO_3{}^-$ reduced; FeS is reducing agent, $NO_3{}^-$ oxidizing agent
   (b) C_2H_4 oxidized, O_2 reduced; C_2H_4 is reducing agent, O_2 oxidizing agent

30. (a) $3FeS(s) + 8NO_3^-(aq) + 8H^+(aq) \rightarrow$
$8NO(g) + 3SO_4^{2-}(aq) + 3Fe^{2+}(aq) + 4H_2O$
(b) $C_2H_4(g) + 3O_2(g) \rightarrow 2CO_2(g) + 2H_2O(l)$

32. (a) $P_4(s) + 6H_2O \rightarrow$
$2PH_3(g) + 2HPO_3^{2-}(aq) + 4H^+(aq)$
(b) $3H_3AsO_3(aq) + BrO_3^-(aq) \rightarrow$
$3H_3AsO_4(aq) + Br^-(aq)$
(c) $2MnO_4^-(aq) + 5HSO_3^-(aq) + H^+(aq) \rightarrow$
$2Mn^{2+}(aq) + 5SO_4^{2-}(aq) + 3H_2O$
(d) $2Sn^{2+}(aq) + O_2(g) + 4H^+(aq) \rightarrow$
$2Sn^{4+}(aq) + 2H_2O$
(e) $3Pt(s) + 4NO_3^-(aq) + 18Cl^-(aq) + 16H^+(aq) \rightarrow$
$3PtCl_6^{2-}(aq) + 4NO(g) + 8H_2O$

34. (a) $Ag(s) + NO_3^-(aq) + 2H^+(aq) \rightarrow$
$Ag^+(aq) + NO_2(g) + H_2O$
(b) $3CuS(s) + 2NO_3^-(aq) + 8H^+(aq) \rightarrow$
$3Cu^{2+}(aq) + 2NO(g) + 3S(s) + 4H_2O$
(c) $4Sn^{2+}(aq) + IO_4^-(aq) + 8H^+(aq) \rightarrow$
$4Sn^{4+}(aq) + I^-(aq) + 4H_2O$

36. (a) $S_2O_3^{2-}(aq) + 4I_2(s) + 10OH^-(aq) \rightarrow$
$2SO_4^{2-}(aq) + 8I^-(aq) + 5H_2O$
(b) $3CN^-(aq) + 2MnO_4^-(aq) + H_2O \rightarrow$
$3CNO^-(aq) + 2MnO_2(s) + 2OH^-(aq)$
(c) $2Cr(OH)_3(s) + ClO_3^-(aq) + 4OH^-(aq) \rightarrow$
$2CrO_4^{2-}(aq) + Cl^-(aq) + 5H_2O$

38. (a) $0.00831\ M$　**(b)** 0.0563 g
40. (a) 35.0 mL　**(b)** 107 mL　**(c)** 249 mL
42. 16.6 mL
44. (a) 9.78 mL　**(b)** 21.9 mL　**(c)** 512 mL
46. (a) $2MnO_4^-(aq) + 5C_2O_4^{2-}(aq) + 16H^+(aq) \rightarrow$
$2Mn^{2+}(aq) + 10CO_2(g) + 8H_2O$
(b) 0.571 M
48. (a) $4Fe(OH)_2(s) + O_2(g) + 2H_2O \rightarrow 4Fe(OH)_3(s)$
(b) 4.76 g　**(c)** 0.269 L
50. (a) 53.0 mL　**(b)** 0.988 g
52. 29.7%
54. 85.9%
56. 1
58. 41.8% Fe, 59.8% Fe_2O_3
60. $0.206\ M$
61. (a) precipitation　**(b)** acid-base　**(c)** redox
(d) redox
62. 20.4 mL
63. (a) $Au(s) + NO_3^-(aq) + 4Cl^-(aq) + 4H^+(aq) \rightarrow$
$AuCl_4^-(aq) + NO(g) + 2H_2O$
(b) 4 HCl:1 HNO₃　**(c)** 17 mL HCl, 3.2 mL HNO₃
64. 0.379%
65. 93.8%
66. 0.794 g; yes
67. 0.30 L
68. 6.5×10^7 g
69. $0.0980\ M$ Fe^{2+}; $0.0364\ M$ Fe^{3+}

Chapter 13

2. (a) $0.350\ M$　**(b)** faster at 45 s, equal at 90 s
4. (a) $3A(g) \rightleftharpoons B(g)$　**(b)** no; conc. still changing
6. (a) $K_c = \dfrac{[F_2]^{1/2} \times [I_2]^{1/2}}{[IF]}$　**(b)** $K_c = \dfrac{[C_{10}H_{12}]}{[C_5H_6]^2}$
(c) $K_c = \dfrac{[POCl_3]^{10}}{[P_4O_{10}] \times [PCl_5]^6}$
8. (a) $K_c = \dfrac{[Ni(CO)_4]}{[CO]^4}$　**(b)** $K_c = [O_2]$
(c) $K_c = \dfrac{[O_2]^3}{[H_2O]^2}$
10. (a) $N_2(g) + Na_2CO_3(s) + 4C(s) \rightarrow$
$3CO(g) + 2NaCN(s); K_c = [CO]^3/[N_2]$
(b) $Mg_3N_2(s) + 6H_2O(g) \rightarrow$
$3Mg(OH)_2(s) + 2NH_3(g); K_c = [NH_3]^2/[H_2O]^6$
(c) $BaCO_3(s) \rightarrow BaO(s) + CO_2(g); K_c = [CO_2]$
12. (a) $2HBr(g) \rightleftharpoons H_2(g) + Br_2(g)$
(b) $CO(g) + 2H_2(g) \rightleftharpoons CH_3OH(g)$
(c) $C_2H_6(g) \rightleftharpoons H_2(g) + C_2H_4(g)$
(d) $SO_2(g) + \frac{1}{2}O_2(g) \rightleftharpoons SO_3(g)$
14. (a) 0.018　**(b)** 3.1×10^3
16. 2.0×10^{-10}
18. 1.1×10^3
20. 3.92
22. (a) $CO(g) + 2H_2(g) \rightarrow CH_3OH(g)$　**(b)** 10.5
24. 55
26. 2.6×10^{-4}
28. (a) no; $Q > K_c$　**(b)** ←
30. (a) →　**(b)** ←
32. some of both
34. $0.16\ M$
36. $0.063\ M$
38. $[CO] = [H_2O] = 0.234\ M$;
$[CO_2] = [H_2] = 0.266\ M$
40. $[COBr_2] = 0.0167\ M$, $[CO] = 0.0673\ M$,
$[Br_2] = 0.0473\ M$
42. (a) 0.16　**(b)** $[CO] = [H_2] = 0.059\ M$
44. (a) increase　**(b)** increase　**(c)** no effect
(d) increase　**(e)** increase
46. (a) ←　**(b)** ←　**(c)** no effect
48. increase; $\Delta H > 0$
50. c or d
52. (a) 0.0431　**(b)** 0.024　**(c)** 1.7×10^2　**(d)** 4.9×10^2
54. 7.09×10^{-3}
56. 0.18 atm
58. 0.50 atm
59. 0.025; 8.3×10^{-4}
60. (a) used no. moles rather than concentration
(b) upside down　**(c)** multiplied by coefficients
(d) added instead of multiplied
61. b, c (at different T)

62. 0.51 atm

63. chemical equation, type of K, temperature

64. 0.0012; 1.5×10^{19}

65. 0.0442

66. 0.0122 M, 0.0063 M

67. 25; 0.50

68. 0.11

69. 0.0057

Chapter 14

2. **(a)** $1.0 \times 10^{-11}\ M$ **(b)** $2.9 \times 10^{-11}\ M$
(c) $1.1 \times 10^{-5}\ M$ **(d)** $2.2 \times 10^{-15}\ M$

4. **(a)** 1.0; acidic **(b)** -1.0; acidic **(c)** 2.15; acidic
(d) 8.09; basic

6. **(a)** $1 \times 10^{-9}\ M$; $1 \times 10^{-5}\ M$
(b) $6.3 \times 10^{-4}\ M$; $1.6 \times 10^{-11}\ M$
(c) $11\ M$; $9.1 \times 10^{-16}\ M$
(d) $3.5 \times 10^{-8}\ M$; $2.9 \times 10^{-7}\ M$

8. Soln. 1; Soln. 1

10. 10^2; 10^{-2}

12. 3×10^2

14. **(a)** 0.60 M; 0.22 **(b)** 0.60 M; 0.22

16. -0.194

18. **(a)** 0.80 M, $1.2 \times 10^{-14}\ M$, 0.10
(b) 0.10 M, $1.0 \times 10^{-13}\ M$, 1.00

20. 12.92

22. **(a)** $Zn(H_2O)_3(OH)^+(aq) \rightleftharpoons$
$$H^+(aq) + Zn(H_2O)_2(OH)_2(aq)$$
(b) $HSO_4^-(aq) \rightleftharpoons H^+(aq) + SO_4^{2-}(aq)$
(c) $HNO_2(aq) \rightleftharpoons H^+(aq) + NO_2^-(aq)$
(d) $Fe(H_2O)_6^{2+}(aq) \rightleftharpoons$
$$H^+(aq) + Fe(H_2O)_5(OH)^+(aq)$$
(e) $Mn(H_2O)_6^{2+}(aq) \rightleftharpoons$
$$H^+(aq) + Mn(H_2O)_5(OH)^+(aq)$$
(f) $HC_2H_3O_2(aq) \rightleftharpoons H^+(aq) + C_2H_3O_2^-(aq)$

24. **(a)** $PH_4^+(aq) \rightleftharpoons H^+(aq) + PH_3(aq)$;

$$K_a = \frac{[H^+] \times [PH_3]}{[PH_4^+]}$$

(b) $HS^-(aq) \rightleftharpoons H^+(aq) + S^{2-}(aq)$;

$$K_a = \frac{[H^+] \times [S^{2-}]}{[HS^-]}$$

(c) $HC_2O_4^-(aq) \rightleftharpoons H^+(aq) + C_2O_4^{2-}(aq)$;

$$K_a = \frac{[H^+] \times [C_2O_4^{2-}]}{[HC_2O_4^-]}$$

26. **(a)** 1×10^{-3} **(b)** 2×10^{-6} **(c)** 2.1×10^{-10}

28. **(a)** B > D > A > C **(b)** C

30. 1.3×10^{-5}

32. 1.3×10^{-10}

34. **(a)** $7.2 \times 10^{-5}\ M$ **(b)** $2.9 \times 10^{-5}\ M$

36. **(a)** $2.6 \times 10^{-5}\ M$ **(b)** $3.8 \times 10^{-10}\ M$ **(c)** 4.59
(d) $2.1 \times 10^{-3}\%$

38. 1.58

40. 4.00

42. $1.0 \times 10^{-4}\ M$; $1 \times 10^{-13}\ M$

44. **(a)** $(CH_3)_3N(aq) + H_2O \rightleftharpoons$
$$OH^-(aq) + (CH_3)_3NH^+(aq)$$
(b) $PO_4^{3-}(aq) + H_2O \rightleftharpoons OH^-(aq) + HPO_4^{2-}(aq)$
(c) $HPO_4^{2-}(aq) + H_2O \rightleftharpoons OH^-(aq) + H_2PO_4^-(aq)$
(d) $H_2PO_4^-(aq) + H_2O \rightleftharpoons OH^-(aq) + H_3PO_4(aq)$
(e) $HS^-(aq) + H_2O \rightleftharpoons OH^-(aq) + H_2S(aq)$
(f) $C_2H_5NH_2(aq) + H_2O \rightleftharpoons$
$$OH^-(aq) + C_2H_5NH_3^+(aq)$$

46. b > d > c > a

48. b, c

50. **(a)** 2.6×10^{-10} **(b)** 2.3×10^{-11}

52. 11.47

54. 2.6×10^{-6}

56. **(a)** A **(b)** A **(c)** B **(d)** A **(e)** B **(f)** A

58. **(a)** $Al(H_2O)_6^{3+}(aq) \rightleftharpoons$
$$H^+(aq) + Al(H_2O)_5(OH)^{2+}(aq)$$
(b) $NH_4^+(aq) \rightleftharpoons NH_3(aq) + H^+(aq)$
$NO_2^-(aq) + H_2O \rightleftharpoons HNO_2(aq) + OH^-(aq)$
(c) $ClO^-(aq) + H_2O \rightleftharpoons OH^-(aq) + HClO(aq)$
(d) $NH_4^+(aq) \rightleftharpoons H^+(aq) + NH_3(aq)$
(e) $CO_3^{2-}(aq) + H_2O \rightleftharpoons OH^-(aq) + HCO_3^-(aq)$
(f) $HSO_4^-(aq) \rightleftharpoons H^+(aq) + SO_4^{2-}(aq)$

60. $Ba(OH)_2 > LiCN > K_2SO_4 > NH_4Br > HClO_4$

62. **(a)** LiF, Li_2CO_3, Li_2S, $LiNO_2$
(b) $LiCl$, $LiBr$, LiI, $LiNO_3$
(c) Na_2SO_4, K_2SO_4, Li_2SO_4, $CaSO_4$
(d) $CuSO_4$, $Al_2(SO_4)_3$, $ZnSO_4$, $(NH_4)_2SO_4$

64. **(a)** acids: H_2O, HCN; bases: CN^-, OH^-
pairs: HCN, CN^- and H_2O, OH^-
(b) acids: H_3O^+, H_2CO_3; bases: HCO_3^-, H_2O
pairs: H_3O^+, H_2O and H_2CO_3, HCO_3^-
(c) acids: $HC_2H_3O_2$, H_2S; bases: HS^-, $C_2H_3O_2^-$
pairs: $HC_2H_3O_2$, $C_2H_3O_2^-$ and H_2S, HS^-

66. **(a)** A **(b)** A **(c)** B

68. **(a)** B **(b)** A **(c)** A

70. **(a)** $Al(OH)_3$ is the acid, OH^- is the base
(b) SO_2 is the acid, CaO is the base

72. **(a)** BA, BB, LB **(b)** BB, LB (in principle)
(c) AB, BB, LB **(d)** AA, BA, BB, LB

73. **(a)** HCl **(b)** NH_3 **(c)** NH_4Cl **(d)** Na_2CO_3
(e) NaCl

74. **(a)** 13.34
(b) $[Li^+] = [OH^-] = 0.2160\ M$; $[NH_3] = 0.0852\ M$;
$[NH_4^+] = 7.1 \times 10^{-6}\ M$

75. greater; $\Delta H > 0$

76. neutral

77. 2.36

78. 3.23

79. 1×10^{-20}

80. Measure pH of $AgNO_3$ solution

81. (a) -5.58 kJ (b) -6.83 kJ

82. % diss. $= \dfrac{[H^+]}{[HA]} \times 100 \approx \dfrac{K_a^{1/2} \times [HA]^{1/2}}{[HA]} \times 100$

$\qquad = \dfrac{K_a^{1/2}}{[HA]^{1/2}} \times 100$

% diss. directly proportional to $K_a^{1/2}$

83. $-1.64°C$

Chapter 15

2. (a) $HCN(aq) + OH^-(aq) \rightarrow H_2O + CN^-(aq)$
(b) $H^+(aq) + ClO_2^-(aq) \rightarrow HClO_2(aq)$
(c) $NH_4^+(aq) + OH^-(aq) \rightarrow NH_3(aq) + H_2O$
(d) $CH_3NH_2(aq) + HClO(aq) \rightarrow$
$\qquad\qquad\qquad CH_3NH_3^+(aq) + ClO^-(aq)$

4. (a) $H^+(aq) + OH^-(aq) \rightarrow H_2O$
(b) $H^+(aq) + CN^-(aq) \rightarrow HCN(aq)$
(c) $H^+(aq) + C_6H_5NH_2(aq) \rightarrow C_6H_5NH_3^+(aq)$

6. (a) $7.1 \times 10^{-7}\ M$, 7.85 (b) $3.6 \times 10^{-7}\ M$, 7.55
(c) $7.1 \times 10^{-8}\ M$, 6.85 (d) $1.4 \times 10^{-8}\ M$, 6.15

8. 4.24

10. (a) none in Table 15.1 (pK_a should be about 2)
(b) $HLac - Lac^-$
(c) $HClO - ClO^-$ or $NH_4^+ - NH_3$

12. (a) 2.1 (b) 2.1 L

14. 8.51

16. 3.1 g

18. 5.48

20. (a) 4.92 (b) 4.85 (c) 5.07

22. (a) 0.091 (b) 7.03 (c) 7.44

24. a, b

26. 3.25

28. (a) 5.9×10^4 (b) 1.0×10^2 (c) 5.6×10^4
(d) 1.2×10^3

30. (a) 1.0×10^{14} (b) 1.7×10^9 (c) 3.8×10^4

32. (a) any (b) MO (c) PP (d) MO

34. (a) 1.1001 (b) 1.4494 (c) 7.00

36. 5.19

38. (a) 2.92 (b) 4.85 (c) 8.84

40. (a) 11.81 (b) 10.62 (c) 5.96

42. (a) $H^+(aq) + ClO^-(aq) \rightarrow HClO(aq)$; 3.6×10^7
(b) 7.08 (c) 4.43 (d) 1.95

44. (a) $Cu_2P_2O_7(s) \rightarrow 2Cu^{2+}(aq) + P_2O_7^{4-}(aq)$
$\qquad K_{sp} = [Cu^{2+}]^2 \times [P_2O_7^{4-}]$
(b) $Ni_3(AsO_4)_2(s) \rightarrow 3Ni^{2+}(aq) + 2AsO_4^{3-}(aq)$
$\qquad K_{sp} = [Ni^{2+}]^3 \times [AsO_4^{3-}]^2$

(c) $Fe(OH)_3(s) \rightarrow Fe^{3+}(aq) + 3OH^-(aq)$
$\qquad K_{sp} = [Fe^{3+}] \times [OH^-]^3$
(d) $Mg(NbO_3)_2(s) \rightarrow Mg^{2+}(aq) + 2NbO_3^-(aq)$
$\qquad K_{sp} = [Mg^{2+}] \times [NbO_3^-]^2$

46. (a) $Ca(IO_3)_2(s) \rightarrow Ca^{2+}(aq) + 2IO_3^-(aq)$
(b) $Ag_2SO_3(s) \rightarrow 2Ag^+(aq) + SO_3^{2-}(aq)$
(c) $Mn_3(AsO_4)_2(s) \rightarrow 3Mn^{2+}(aq) + 2AsO_4^{3-}(aq)$
(d) $PbC_2O_4(s) \rightarrow Pb^{2+}(aq) + C_2O_4^{2-}(aq)$

48. $8 \times 10^{-6}\ M$, $4 \times 10^{-5}\ M$, $3 \times 10^{-5}\ M$,
$9 \times 10^{-4}\ M$

50. (a) $1 \times 10^{-42}\ M$ (b) $1 \times 10^{-31}\ M$
(c) $3 \times 10^{-22}\ M$

52. (a) $9.2 \times 10^{-4}\ M$ (b) $8.4 \times 10^{-3}\ M$; 84%

54. no

56. yes

58. 8.4×10^{-10}

60. (a) 1×10^{-16} g/L (b) 1×10^{-32} g/L
(c) 5×10^{-33} g/L

61. (a) benzoate ion is a weak base
(b) weak acid (c) reacts with H^+ or OH^-

62. (a) false; much less (b) true
(c) false; $K_b = 3.2 \times 10^{-10}$

63. (a) $\leftarrow$ (b) $\rightarrow$ (c) $\rightarrow$ (d) $\leftarrow$

64. no

65. 3×10^7; yes

66. 4.57

67. (a) 2.38 (b) 4.74 (c) 7.44 (d) 9.23
(e) 11.00 (f) 13.52

68. $K = 0.02$; $s = 0.06\ M$ vs. $1 \times 10^{-4}\ M$

69. about 30 mL; PP

70. $1.8 \times 10^{-4}\ M$

71. $1 \times 10^{-8}\ M$

72. (a) $1 \times 10^{-5}\ M$ (b) yes, Al^{3+} and Fe^{3+}
(c) virtually all (d) 1.6 g

Chapter 16

2. (a) 4 NH_3 molecules, 2 Cl^- ions (b) $+3$
(c) $[Co(C_2O_4)_2Cl_2]^{3-}$

4. (a) $Cr(C_2O_4)_2(H_2O)_2^-$ (b) $Cr(NH_3)_5SO_4^+$
(c) $Cr(en)(NH_3)_2I_2^+$

6. (a) 6 (b) 4 (c) 2 (d) 6

8. (a) $Ag(en)^+$ (b) $Fe(H_2O)_6^{2+}$ (c) $Zn(CN)_4^{2-}$
(d) $Pt(en)_3^{4+}$

10. (a) mono (b) tetra (c) tri (d) mono

12. (a) $[Co(NH_3)_4(H_2O)Cl]\ Cl_2$ (b) $[Co(NH_3)_5SO_4]\ Br_2$
(c) $K_2[Ni(CN)_4]$ (d) $Co(NH_3)_5NO_3^{2+}$

14. (a) sodium tetrahydroxoaluminate(III)
(b) diaquodioxalatocobaltate(III)
(c) triamminetrichloroiridium(III)
(d) diamminedibromoethylenediaminechromium(III)
sulfate

16. **(a)** hexaaquoiron(III)
   (b) amminetribromoplatinate(II)
   (c) diamminesilver(I)
   (d) chlorobis(ethylenediamine)thiocyanatocobalt(III)

18. **(a)** CN—Ag—CN   **(b)**

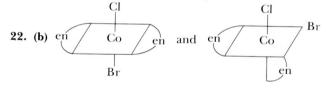

(c)   **(d)**

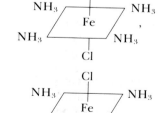

(e)

20.

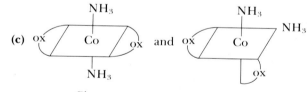

22. **(b)** en—Co—en and en—Co—Br
         Br   en

   (c) ox—Co—ox and ox—Co—NH₃
         NH₃   NH₃ ox

24.

26. **(a)** [Ar] 3d⁶   **(b)** [Ar] 3d³   **(c)** [Ar] 3d¹⁰
   (d) [Ar] 3d¹⁰   **(e)** [Kr] 4d³

28.
                    3d
   (a) (↑↓)(↑)(↑)(↑)(↑)   4 unp. e⁻
   (b) (↑)(↑)(↑)( )( )   3 unp. e⁻
   (c) (↑↓)(↑↓)(↑↓)(↑↓)(↑↓)   0 unp. e⁻
   (d) (↑↓)(↑↓)(↑↓)(↑↓)(↑↓)   0 unp. e⁻
                    4d
   (e) (↑)(↑)(↑)( )( )   3 unp. e⁻

30.
        3d   4s   4p
   (a) (↑)(↑)(↑)(↑↓)(↑↓) (↑↓) (↑↓)(↑↓)(↑↓)
   (b) (↑↓)(↑↓)(↑↓)(↑↓)(↑↓) (↑↓) (↑↓)(↑↓)(↑)
   (c) (↑↓)(↑↓)(↑↓)(↑↓)(↑↓) (↑↓) (↑↓)(↑↓)(↑↓)
        4d   5s   5p
   (d) (↑↓)(↑↓)(↑↓)(↑↓)(↑↓) (↑↓) (↑↓)( )( )

32.
                    3d   4s   4p
   CuCl₂⁻ (Cu⁺)   (↑↓)(↑↓)(↑↓)(↑↓)(↑↓) (↑↓) (↑↓)( )( )
   CuCl₄²⁻(Cu²⁺) (↑↓)(↑↓)(↑↓)(↑↓)(↑↓) (↑↓) (↑↓)(↑↓)(↑)

34. **(a)** 0   **(b)** 2   **(c)** 0 (low spin) or 4 (high spin)

36.
            3s   3p   3d
   Al³⁺   (↑↓) (↑↓)(↑↓)(↑↓) (↑↓)(↑↓)( )( )( )

            3d   4s   4p   4d
   Co²⁺ (↑↓)(↑↓)(↑)(↑)(↑) (↑↓) (↑↓)(↑↓)(↑↓) (↑↓)(↑↓)( )( )( )
   Co³⁺ (↑↓)(↑)(↑)(↑)(↑)   (↑↓) (↑↓)(↑↓)(↑↓) (↑↓)(↑↓)( )( )( )
   Cr³⁺ (↑)(↑)(↑)(↑↓)(↑↓) (↑↓) (↑↓)(↑↓)(↑↓) ( )( )( )( )( )

            3d   4s   4p   4d
   Cu²⁺ (↑↓)(↑↓)(↑↓)(↑↓)(↑) (↑↓) (↑↓)(↑↓)(↑↓) (↑↓)(↑↓)( )( )
   Fe²⁺ (↑↓)(↑)(↑)(↑)(↑)   (↑↓) (↑↓)(↑↓)(↑↓) (↑↓)(↑↓)( )( )
   Fe³⁺ (↑)(↑)(↑)(↑)(↑)   (↑↓) (↑↓)(↑↓)(↑↓) (↑↓)(↑↓)( )( )
   Ni²⁺ (↑↓)(↑↓)(↑↓)(↑)(↑) (↑↓) (↑↓)(↑↓)(↑↓) (↑↓)(↑↓)( )( )

            5d   6s   6p   6d
   Pt⁴⁺ (↑↓)(↑)(↑)(↑)(↑)   (↑↓) (↑↓)(↑↓)(↑↓) (↑↓)(↑↓)( )( )( )

38. **(a)** ( )( )   and (↑)(↑)
         (↑↓)(↑↓)(↑↓)   (↑↓)(↑)(↑)
   (b) ( )( )   and (↑)( )
         (↑↓)(↑)(↑)   (↑)(↑)(↑)

40. Two 3d electrons

42.   ( )( )   (↑)(↑)
     (↑↓)(↑↓)(↑↓)   (↑↓)(↑)(↑)
     Co(NH₃)₆³⁺   CoF₆³⁻

44. **(a)** 4 **(b)** 4 **(c)** 4 **(d)** 2 **(e)** 4
46. 399 nm
48. 3 × 10⁻⁶ M
50. **(a)** 6 × 10⁻³ M   **(b)** 4 × 10⁻⁶ M
52. 2 × 10⁻²⁴; ineffective
54. 2 × 10⁷
55. **(a)** 2 unshared pairs, separated by three atoms
   (b) forms Ag(NH₃)₂⁺   **(c)** all positions equivalent
56. **(a)** false; 6   **(b)** false; 0   **(c)** false; shorter
57. **(a)** true   **(b)** true   **(c)** false
58. **(a)** forms complex with Fe³⁺
   (b) forms stable complex with Pb²⁺
   (c) Ni(NH₃)₆²⁺ is blue

59. (a) $CoN_6H_{18}Cl_3$
(b) $[Co(NH_3)_6]Cl_3(s) \rightarrow Co(NH_3)_6^{3+}(aq) + 3Cl^-(aq)$

60. (a) $PtN_2H_6Cl_4$ **(b)**

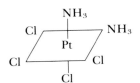

and

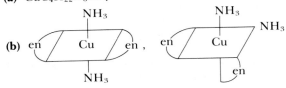

61. $[Pt(NH_3)_4][PtCl_4]$ or $[Pt(NH_3)_3Cl][Pt(NH_3)Cl_3]$
62. 0.92 g
63. (a) $CuC_4H_{22}N_6SO_4$

(b)

64. (a) $Al(OH)_3(s) + OH^-(aq) \rightarrow Al(OH)_4^-(aq)$
(b) $K = 2 \times 10^2$
65. (a) 8×10^5 **(b)** 4×10^3
66. purple

Chapter 17

2. Na^+; NH_3; $(NH_4)_2CO_3$, white ppt.; Mn^{2+}
4. (a) K^+ **(b)** Sb^{3+} **(c)** Zn^{2+}
6. Would make separation within group much more difficult
8. (a) positive Na^+ test
(b) both Cu^{2+} and Ni^{2+} precipitate
(c) forms Sn^{4+}
10. (a) 4 **(b)** $1s^2 2s^2 2p^6$ **(c)** tetrahedral; sp^3
12. (a) $AgCl(s) + 2NH_3(aq) \rightarrow$
$$Ag(NH_3)_2^+(aq) + Cl^-(aq)$$
(b) $PbCl_2(s) \rightarrow Pb^{2+}(aq) + 2Cl^-(aq)$
(c) $Ag(NH_3)_2^+(aq) + 2H^+(aq) + Cl^-(aq) \rightarrow$
$$AgCl(s) + 2NH_4^+(aq)$$
14. White ppt. with HCl. Partially soluble in hot water; addition of CrO_4^{2-} gives yellow ppt. Addition of NH_3 to remaining white ppt. gives clear solution; ppt. re-forms upon addition of acid.
16. (a) NH_3 **(b)** Cl^- **(c)** NH_3
18. (a) $PbCl_2$ will reprecipitate
(b) white rather than yellow ppt.
(c) will not destroy complex to form ppt.

20. $Hg_2(NO_3)_2$ absent, $PbCl_2$ absent, $PbCO_3$ present, AgI in doubt
22. (a) $2H^+(aq) + CO_3^{2-}(aq) \rightarrow CO_2(g) + H_2O$
(b) $Ba^{2+}(aq) + SO_4^{2-}(aq) \rightarrow BaSO_4(s)$
(c) $2CrO_4^{2-}(aq) + 2H^+(aq) \rightarrow Cr_2O_7^{2-}(aq) + H_2O$
24. (a) yellow ppt. forms
(b) foul-smelling gas evolved
(c) nothing happens
26. brown-ring test
28. 1 contains SCN^-; 2 does not contain Br^-; 3 does not contain CO_3^{2-}
30. 0.20 mL
32. (a) $Zn^{2+}(aq) + 4NH_3(aq) \rightarrow Zn(NH_3)_4^{2+}(aq)$
(b) $Sn^{4+}(aq) + 6OH^-(aq) \rightarrow Sn(OH)_6^{2-}(aq)$
(c) $Ni^{2+}(aq) + 6NH_3(aq) \rightarrow Ni(NH_3)_6^{2+}(aq)$
34. (a) $Pb(OH)_3^-(aq) + 3H^+(aq) \rightarrow Pb^{2+}(aq) + 3H_2O$
(b) $Ag(NH_3)_2^+(aq) + 2H^+(aq) \rightarrow$
$$Ag^+(aq) + 2NH_4^+(aq)$$
(c) $Al(OH)_4^-(aq) + 4H^+(aq) \rightarrow Al^{3+}(aq) + 4H_2O$
(d) $Zn(NH_3)_4^{2+}(aq) + 4H^+(aq) \rightarrow$
$$Zn^{2+}(aq) + 4NH_4^+(aq)$$
36. (a) $Mn^{2+}(aq) + S^{2-}(aq) \rightarrow MnS(s)$
(b) $Mn^{2+}(aq) + 2OH^-(aq) \rightarrow Mn(OH)_2(s)$
(c) $Mn^{2+}(aq) + 2NH_3(aq) + 2H_2O \rightarrow$
$$Mn(OH)_2(s) + 2NH_4^+(aq)$$
(d) $Mn^{2+}(aq) + H_2S(aq) \rightarrow MnS(s) + 2H^+(aq)$
38. (a) $SrCO_3(s) + 2H^+(aq) \rightarrow$
$$Sr^{2+}(aq) + CO_2(g) + H_2O$$
(b) no reaction
(c) $Co(OH)_3(s) + 3H^+(aq) \rightarrow Co^{3+}(aq) + 3H_2O$
(d) $Sb(OH)_4^-(aq) + 4H^+(aq) \rightarrow Sb^{3+}(aq) + 4H_2O$
40. (a) $Cu(OH)_2(s) + 4NH_3(aq) \rightarrow$
$$Cu(NH_3)_4^{2+}(aq) + 2OH^-(aq)$$
(b) $Cd^{2+}(aq) + 4NH_3(aq) \rightarrow Cd(NH_3)_4^{2+}(aq)$
(c) $Pb^{2+}(aq) + 2NH_3(aq) + 2H_2O \rightarrow$
$$Pb(OH)_2(s) + 2NH_4^+(aq)$$
42. (a) $Ni^{2+}(aq) + 2OH^-(aq) \rightarrow Ni(OH)_2(s)$
(b) $Pb^{2+}(aq) + 3OH^-(aq) \rightarrow Pb(OH)_3^-(aq)$
(c) $Al(OH)_3(s) + OH^-(aq) \rightarrow Al(OH)_4^-(aq)$
44. (a) $CaCO_3(s) + 2H^+(aq) \rightarrow$
$$Ca^{2+}(aq) + CO_2(g) + H_2O$$
(b) $Mg(OH)_2(s) + 2H^+(aq) \rightarrow Mg^{2+}(aq) + 2H_2O$
(c) $ZnS(s) + 2H^+(aq) \rightarrow Zn^{2+}(aq) + H_2S(aq)$
46. (a) 0.033 M, 0.0067 M **(b)** no
48. 6.0
50. (a) $1 \times 10^{-17}\ M$ **(b)** yes **(c)** yes
52. 4×10^{-7}
54. 0.05 M
56. 3.3×10^{11}
57. reagents often contain these ions
58. $1.9 \times 10^{-4}\ M$
59. 30

60. 11 mL
61. 5×10^6; 0.07 M
62. 0.4 g/100 mL
63. 2
64. yes
65. $Ni^{2+}(aq) + 2NH_3(aq) + 2H_2O \rightarrow$
$\qquad Ni(OH)_2(s) + 2NH_4^+(aq)$
$Al^{3+}(aq) + 3NH_3(aq) + 3H_2O \rightarrow$
$\qquad Al(OH)_3(s) + 3NH_4^+(aq)$
$Ni(OH)_2(s) + 6NH_3(aq) \rightarrow$
$\qquad Ni(NH_3)_6^{2+}(aq) + 2OH^-(aq)$
$Al(OH)_3(s) + OH^-(aq) \rightarrow Al(OH)_4^-(aq)$
$Al(OH)_4^-(aq) + H^+(aq) \rightarrow Al(OH)_3(s) + H_2O$
$Al(OH)_3(s) + 3H^+(aq) \rightarrow Al^{3+}(aq) + 3H_2O$
66. (a) 1×10^{-16} (b) 1×10^{-30} (c) 1×10^{-44}
67. 3×10^{-16}
68. 8×10^5; yes
69. 1.0×10^{21}

Chapter 18

2. (a) Δ conc. $H_2/\Delta t$ (b) $-\frac{1}{2}\Delta$ conc. $HI/\Delta t$
4. (a) 0.039 mol/L·s (b) 0.065 mol/L·s
(c) 0.078 mol/L·s
6. 0.00128 mol/L·s
8. (a) 0.014 mol/L·min (b) 0.014 mol/L·min
10. (a) 1, 1, 2 (b) 2 (c) 1, 2, 3 (d) 1
12. (a) L/mol·s (b) L/mol·s (c) L^2/mol^2·s (d) s^{-1}
14. 0.090 mol/L·s; 1.1 M; 0.14 L/mol·s
16. (a) rate = k (b) 2.5×10^{-4} mol/L·min
(c) all concentrations
18. (a) 0.0163 L/mol·s (b) 1.23 M c. 15.5 M
20. a
22. 2nd
24. (a) 1st (b) 1st (c) 2.5×10^{-5} mol/L·s
26. (a) 2nd in A, 0 in B, 1st in C
(b) rate = k(conc. A)^2(conc. C)
(c) 0.250 L^2/mol^2·s (d) 0.162 mol/L·s
28. (a) rate = k(conc. $Cr(H_2O)_6^{3+}$)(conc. SCN^-)
(b) 0.0072 L/mol·s (c) 3.9×10^{-8} mol/L·s
30. 2nd
32. 10.0 s, 5.0 s, 2.5 s; zero order
34. (a) plot of ln conc. vs. t is straight line
(b) 0.017/min (c) 82 min
36. (a) 2.03×10^{-3}/min (b) 341 min (c) 0.0450 M
38. (a) 1.91/h (b) 0.852 (c) 0.101 h
40. (a) 0.078 L/mol·s (b) 85 s
42. 3.0 L/mol·min
44. 8.0 min; 32 min
46. (a) 3 (b) 30 kJ, 105 kJ, 42 kJ

48.

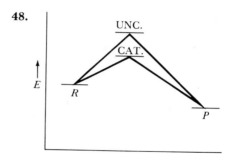

50. 1.9×10^2 kJ
52. 75 kJ
54. 57%
56. (a) 519°C (b) 1.6×10^3 L/mol·s
58. 1.8 s
60. (a) rate = k(conc. K)(conc. HCl)
(b) rate = k(conc. NO_3)(conc. CO)
(c) rate = k(conc. NO_2)^2
62. rate = k_2(conc. N_2O_2)(conc. H_2) =
k_2K(conc. NO)^2(conc. H_2)
64. c, d
65. (a) usually decreases (b) increases (c) increases
66. (a) fewer molecules have E_a
(b) not always 1st order (c) supplies E_a
(d) does not change relative energies of products,
reactants
67. 1.9 kJ absorbed
68. (a) 52 min (b) 5.0 L
69. 8
70. (a) $-\dfrac{dX}{dt} = kX^2; \; -\dfrac{dX}{X^2} = kdt; \; \dfrac{1}{X} - \dfrac{1}{X_0} = kt$

(b) $-\dfrac{dX}{dt} = kX^3; \; -\dfrac{dX}{X^3} = kdt; \; \dfrac{1}{2X^2} - \dfrac{1}{2X_0^2} = kt$

71. rate = k(conc. A)^2(conc. B)(conc. C)
72. 0.90 g; 0.055 g
73. 6; $n(n-1)/2 \rightarrow n^2/2$ if n is large
74. $k = 0.20, b = 3.0$

Chapter 19

2. 2.80×10^{-2} mol/L·min
4. (a) 0.77 L/mol·s
(b) 4.8×10^{-12} mol/L
6. 5×10^{-9} mol/L·s; all gone in 4 s!
8. 3.8×10^2 L/mol·s
10. 3.7×10^{-3}
12. 1×10^{17} years
14. rate = k(conc. H_2)(conc. NO)^2;
3.8×10^{11} L^2/mol^2·min

16. rate $= k_2K$(conc. O_2)(conc. NO)^2
    K decreases as T increases
18. (a) $2NO_2(g) + F_2(g) \rightarrow 2NO_2F(g)$
    (b) rate $= k_2K$(conc. F_2)(conc. NO_2)^2
20. (a) decrease T, decrease P
    (b) increase T, decrease P
    (c) decrease T, increase P
    (d) decrease T, increase P
22. (a) 11% NH_3, 67% H_2, 22% N_2
    (b) 15.8 mol
    (c) $[NH_3] = 0.17\ M$, $[H_2] = 1.1\ M$, $[N_2] = 0.35$ M
    (d) 0.062
24. $2.93 \times 10^{-4}\ M$
26. 0.43, 0.0043, 0.0002, 2×10^{-5}, 2.7×10^{-6}
28. (a) 44% (b) 16%
30. 3.0 mm Hg
32. 6°C
34. 1123 nm
36. 494 kJ/mol
38. 4.6×10^2 g
40. 6.10 g; 1.83 mL
42. 3.58 g H_2O, 3.38 g NH_3
44. (a) 6.291×10^5 g MgO
    (b) $\Delta H = -6.032 \times 10^6$ kJ
45. (a) also increases yield
    (b) usually supersaturated
    (c) involved in smog formation
    (d) $SO_3 \rightarrow H_2SO_4$
46. (a) T (b) T (c) F
47. (a) increase T (b) more smog
    (c) more skin cancers
48. (a) loss O_3 (b) lower pH rain
    (c) lower T
49. (a) K is unfavorable (b) removes HC, CO, NO
    (c) reacts with hemoglobin; K favorable
50. (a) absorption of IR radiation from earth
    (b) contain rare transition metals
    (c) $SO_3 \rightarrow H_2SO_4$
51. (a) 0.934% (b) 1.29%
    (c) $3.77 \times 10^{-4}\ M$ (d) 9.34×10^3 ppm
52. (a) 1.12×10^{-5} (b) 1.12×10^{-5} atm
    (c) $4.58 \times 10^{-7}\ M$ (d) 1.28×10^{-2} g/m^3
53. increase in P increases yield and rate; increase in T
    increases rate, decreases yield
54. increase in P increases yield and rate; increase in T
    increases rate, decreases yield
55. (a) $(:\overset{..}{O}\!-\!\overset{..}{O}:)^{2-}$ (b) $(:\overset{..}{O}\!-\!\overset{..}{O}:)^{-}$
    (c) $(:\overset{..}{\underset{..}{O}}:)^{2-}$ (d) $:C\!\equiv\!O:$
56. (a) 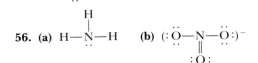 (b) $(:\overset{..}{O}\!-\!\overset{..}{N}\!-\!\overset{..}{O}:)^{-}$
                                              $\overset{\|}{\underset{:O:}{}}$

(c) 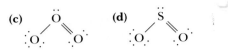 (d)

57. (a) linear, sp^3 (b) linear, sp^3 (c) sp^3
    (d) linear, sp
58. (a) triangular pyramid, 109°, sp^3
    (b) equilateral triangle, 120°, sp^2
    (c) bent, 120°, sp^2 (d) bent, 120°, sp^2
59. 7×10^{10} g
60. 6×10^9 g
61. 0.29 g
62. 21 g
63. 0.0024 g
64. 1.1×10^4 g
65. 0.135
66. 0.027
67. (a) 1.0×10^{22} L (b) 6.5×10^{18} mol
    (c) 2.2×10^{15} g
68. (a) $1.8 \times 10^{-15}\ M$ (b) 1.8×10^{-11} mol/L·s
    (c) 2.3×10^3 s
69. 46 kJ; $\Delta t \approx 1100$ K
70. (a) 4.0 g (b) 1.5 L (c) 1.5×10^3
71. 6.9×10^8 g

Chapter 20

2. b
4. (a) no (b) no (c) yes (d) yes
6. 10; 5; 1   3H, 2T
8. $(\frac{1}{2})^8 = 1/256 = 0.0039$
10. (a) + (b) + (c) − (d) +
12. (a) − (b) − (c) − (d) −
14. (a) + (b) + (c) +
16. (a) −332.0 J/K (b) +24.9 J/K (c) +171.9 J/K
    (d) −20.9 J/K
18. (a) −15.2 J/K (b) −213.9 J/K (c) +9.8 J/K
20. (a) −131.9 J/K (b) −271.2 J/K (c) −533.0 J/K
22. (a) −49.5 kJ (b) −889.2 kJ (c) −46 kJ
24. (a) +4.6 kJ (NS) (b) +170.4 kJ (NS)
    (c) +200.5 kJ (NS) (d) −316.4 kJ (S)
26. (a) −29.1 kJ (b) +173.2 kJ
    (c) +218.1 kJ (d) −318.6 kJ
28. (a) −519.8 kJ (b) −33.5 kJ (c) −120.5 kJ
30. no
32. (a) +132 J/K; yes, gas produced
    (b) +96 J/K (c) −510.9 kJ/mol
34. (a) +333 J/K; yes (b) +284 J/K
    (c) −218.8 kJ/mol
36. (a) more spont. at high T
    (b) less spont. at high T
    (c) nonspont. at all T

38. (a) 4950 K; nonspontaneous at any reasonable T
(b) 5200 K; spontaneous at any reasonable T
(c) no T
40. 780 K
42. (a) $\Delta G° = 467.9 - 0.5581\ T$

T	200	400	600	800	1000
$\Delta G°$	356.3	244.7	133.0	21.4	-90.2

(b) 838 K
44. (a) $T = 2840$ K　**(b)** 841 K (best)　**(c)** 903 K
46. 84.4 J/K
48. -234.8 kJ/mol
50. 282 K
52. -129.7 kJ/mol
54. $C_6H_{12}O_6(aq) + 6\,O_2(g) + 80ADP(aq) + 80HPO_4{}^{2-}(aq)$
$+ 160H^+(aq) \rightarrow 80ATP(aq) + 86H_2O + 6CO_2(g)$
56. (a) 54.2 kJ　**(b)** 2.2×10^{-4}
58. $+211$ kJ; $+107$ kJ
60. 1.0×10^{-10}
61. (a) T, P　**(b)** T　**(c)** neither　**(d)** neither
62. (a) only if $T\Delta S$ can be ignored
(b) at standard concentrations　**(c)** moles of gas
(d) only at low T
63. (a) randomness increases
(b) ΔS^0 is difference between products, reactants
(c) more ordered
64. (a) 0 K　**(b)** 0　**(c)** <0　**(d)** less
65. $+3.47$ kJ
66. $+2870$ kJ $+ 24(-203.6$ kJ$) = -2016$ kJ
67. 1.7 g
68. 1.6×10^{-5}
69. 309 K
70. $[H_2] = [I_2] = 0.006\ M$: $[HI] = 0.048\ M$
71. (a) $+6.00$ kJ　**(b)** 0　**(c)** $+22.0$ J/K　**(d)** $+0.21$ kJ
(e) -0.23 kJ
72. (a) 5.1 kJ　**(b)** $\sim$190 g (if m $\approx$ 60 kg)
73. $\sim$1160°C
74. (1) $\Delta G^0 = 146.0 - 0.1104\ T$;
becomes spont. at high T
(2) $\Delta G^0 = 168.6 - 0.0758\ T$;
becomes spont. at higher T

Chapter 21

2. (a) $3Zn(s) + 2Cr^{3+}(aq) \rightarrow 3Zn^{2+}(aq) + 2Cr(s)$
(b) $2Sn(s) + O_2(g) + 4H^+(aq) \rightarrow$
$2Sn^{2+}(aq) + 2H_2O$
(c) $2Al(s) + 3I_2(s) \rightarrow 2Al^{3+}(aq) + 6I^-(aq)$
4. (a) Zn anode, Cd cathode; e^- move from Zn to Cd.
Anions move to Zn, cations to Cd.
(b) Cu anode, Au cathode; e^- move from Cu to
Au. Anions move to Cu, cations to Au.
(c) Fe anode, Cu cathode; e^- move from Fe to Cu.
Anions move to Fe, cations to Cu.

6. cathode:　　$Br_2(l) + 2e^- \rightarrow 2Br^-(aq)$
anode:　　$\dfrac{2I^-(aq) \rightarrow I_2(s) + 2e^-}{Br_2(l) + 2I^-(aq) \rightarrow 2Br^-(aq) + I_2(s)}$
cell notation: (Pt) $I^-/I_2 \parallel$ (Pt) Br_2/Br^-
Electrons move from Pt anode to Pt cathode; anions
to anode, cations to cathode
8. (a) Ag　**(b)** I^-　**(c)** Fe^{2+}　**(d)** H_2 in basic solution
10. $Al^{3+} < Ni^{2+} < AgBr < ClO_3{}^- < F_2$
12. oxid. agents: $Cr^{3+} < Sn^{2+} < Br_2$
red. agents: $Cr^{3+} < Hg < Sn^{2+} < H_2$
14. (a) Fe^{3+}, $Hg_2{}^{2+}$, $-$ $-$　**(b)** Tl, Pb, $-$ $-$
(c) $ClO_3{}^-$, Au, $-$ $-$
16. (a) $+0.695$ V　**(b)** $+0.383$ V　**(c)** $+0.202$ V
18. (a) $+0.063$ V　**(b)** $+1.178$ V　**(c)** $+0.948$ V
20. (a) $+0.281$ V　**(b)** $+0.625$ V　**(c)** $+0.946$ V
22. (a) -0.799 V　**(b)** $+3.668$ V　**(c)** $+1.101$ V
24. none
26. (a) $+0.924$ V; yes　**(b)** -1.431 V; no
(c) $+1.297$ V; yes
28. (a) no reaction
(b) $MnO_2(s) + 4H^+(aq) + 2Hg(l) \rightarrow$
$Mn^{2+}(aq) + Hg_2{}^{2+}(aq) + 2H_2O$
(c) no reaction
30. a, b, c
32. (a) no reaction
(b) $2Fe^{2+}(aq) + Br_2(l) \rightarrow 2Fe^{3+}(aq) + 2Br^-(aq)$
$2I^-(aq) + Br_2(l) \rightarrow I_2(s) + 2Br^-(aq)$
(c) no reaction
34. (a) $+579$ kJ, 10^{-101}　**(b)** -579 kJ, 10^{101}
(c) 1160 kJ, 10^{-203}
36. (a) $+0.146$ V　**(b)** -0.146 V
38. $+0.216$ V, -125 kJ, 8×10^{21}
40. (a) -134 kJ　**(b)** -73.9 kJ　**(c)** -58.5 kJ
42. (a) 1.3×10^2　**(b)** 7×10^{39}　**(c)** 1×10^{64}
44. (a) $E = E_{tot}^0 - \dfrac{0.0257}{2} \ln \dfrac{(\text{conc. } Br^-)^2}{(\text{conc. } I^-)^2}$
(b) $+0.543$ V　**(c)** $+0.561$ V
46. -1.308 V
48. (a) $+0.778$ V, yes　**(b)** -0.436 V, no
50. (a) $+0.235$ V　**(b)** $1.1 \times 10^{-4}\ M$
52. 4.31
54. (a) -0.460 V　**(b)** $1.6 \times 10^{-9}\ M$　**(c)** 1.6×10^{-10}
56. (a) cathode:　$2H_2O + 2e^- \rightarrow H_2(g) + 2OH^-(aq)$
anode:　　$2Br^-(aq) \rightarrow Br_2(l) + 2e^-$
(b) $2H_2O + 2Br^-(aq) \rightarrow$
$H_2(g) + Br_2(l) + 2OH^-(aq)$; $E = 1.905$ V
58. (a) 6.696×10^{25}　**(b)** 124.2 A　**(c)** 889.5 g
60. (a) 0.61 g　**(b)** 1.5×10^3 s
62. (a) 6.4×10^7 J　**(b)** \$1.1
63. 1.021 V; 4.56 kJ
64. (a) 3.22×10^4　**(b)** 53.7 A
(c) 48.3 kJ; 0.0134 kWh
65. (a) serves as source of H^+, $SO_4{}^{2-}$ ions

(b) allows for movement of ions but prevents direct reaction

(c) high O_2 overvoltage

66. (a) T **(b)** F; to cathode **(c)** F; $PbSO_4$ consumed

67. (a) 0.407 g **(b)** 0.304 L

68. (a) 7.30 L **(b)** 13.78

69. (a) increases 0.089 V **(b)** decreases 0.577 V
(c) no change

70. 108 g

71. 1×10^{-12}

72. -0.188 V

73. 2.007 V

74. (a) $+0.621$ V **(b)** Zn^{2+} increases, Sn^{2+} decreases
(c) 1×10^{21}
(d) conc. $Zn^{2+} = 2.0\ M$,
   conc. $Sn^{2+} = 2 \times 10^{-21}\ M$

75. (a) $+112$ kJ; $+38.0$ kJ; $+150$ kJ **(b)** -0.389 V

76. (a) 2×10^{14}
(b) $[Fe^{3+}] = [Fe^{2+}] = [Ce^{3+}] = 0.050\ M$;
   $[Ce^{4+}] = 3 \times 10^{-16}\ M$

77. 0.414 V

Chapter 22

2. (a) $Cu_2S(s) + O_2(g) \rightarrow 2Cu(s) + SO_2(g)$
(b) $3Cu(s) + 2NO_3^-(aq) + 8H^+(aq) \rightarrow$
   $3Cu^{2+}(aq) + 2NO(g) + 4H_2O$
(c) $Cu(s) + 2NO_3^-(aq) + 4H^+(aq) \rightarrow$
   $Cu^{2+}(aq) + 2NO_2(g) + 2H_2O$

4. (a) $2Cr(s) + 3S(s) \rightarrow Cr_2S_3(s)$
(b) $2Cr(s) + 3Cl_2(g) \rightarrow 2CrCl_3(s)$
(c) $2Cr(s) + 3Br_2(l) \rightarrow 2CrBr_3(s)$
(d) $4Cr(s) + 3O_2(g) \rightarrow 2Cr_2O_3(s)$

6. (a) $Fe(H_2O)_5(OH)^{2+}(aq) + H^+(aq) \rightarrow$
   $Fe(H_2O)_6^{3+}(aq)$
(b) $Cd(OH)_2(s) + 4NH_3(aq) \rightarrow$
   $Cd(NH_3)_4^{2+}(aq) + 2OH^-(aq)$
(c) $Ni(H_2O)_6^{2+}(aq) + 6NH_3(aq) \rightarrow$
   $Ni(NH_3)_6^{2+}(aq) + 6H_2O$

8. (a) $Cd^{2+}(aq) + H_2S(aq) \rightarrow CdS(s) + 2H^+(aq)$
(b) $Cr^{3+}(aq) + 3S^{2-}(aq) + 3H_2O \rightarrow$
   $Cr(OH)_3(s) + 3HS^-(aq)$
(c) $Co^{2+}(aq) + 2NH_3(aq) + 2H_2O \rightarrow$
   $Co(OH)_2(s) + 2NH_4^+(aq)$

10. (a) $4MnO_4^-(aq) + 2H_2O \rightarrow$
   $4MnO_2(s) + 3O_2(g) + 4OH^-(aq)$
(b) $2Ag^+(aq) + 2OH^-(aq) \rightarrow Ag_2O(s) + H_2O$
(c) $CrO_4^{2-}(aq) + 3e^- + 4H_2O \rightarrow$
   $Cr(OH)_3(s) + 5OH^-(aq)$

12. (a) $Cu^{2+}(aq) + 2NH_3(aq) + 2H_2O \rightarrow$
   $Cu(OH)_2(s) + 2NH_4^+(aq)$
(b) $Cu(OH)_2(s) + 4NH_3(aq) \rightarrow$
   $Cu(NH_3)_4^{2+}(aq) + 2OH^-(aq)$

14. (a) $Fe_2O_3(s) + 3CO(g) \rightarrow 2Fe(l) + 3CO_2(g)$
(b) $Ni^{2+}(aq) + 6NH_3(aq) \rightarrow Ni(NH_3)_6^{2+}(aq)$
(c) $Fe(s) + 2H^+(aq) \rightarrow Fe^{2+}(aq) + H_2(g)$

16. $3Cd(s) + 12Cl^-(aq) + 2NO_3^-(aq) + 8H^+(aq) \rightarrow$
   $3CdCl_4^{2-}(aq) + 2NO(g) + 4H_2O$

18. (a) $Fe(s) + 3NO_3^-(aq) + 6H^+(aq) \rightarrow$
   $Fe^{3+}(aq) + 3NO_2(g) + 3H_2O$
(b) $4Cr(OH)_3(s) + 3O_2(g) + 8OH^-(aq) \rightarrow$
   $4CrO_4^{2-}(aq) + 10H_2O$

20. (a) $4Cr^{2+}(aq) + O_2(g) + 4H^+(aq) \rightarrow$
   $4Cr^{3+}(aq) + 2H_2O$
(b) $2Mn^{2+}(aq) + 5BiO_3^-(aq) + 14H^+(aq) \rightarrow$
   $2MnO_4^-(aq) + 5Bi^{3+}(aq) + 7H_2O$

22. $1 \times 10^{-10}\ M$

24. $[CrO_4^{2-}] = 0.02\ M$; $[Cr_2O_7^{2-}] = 0.09\ M$

26. 900 K

28. -88.9 kJ

30. (a) 0.228 V **(b)** 1.47×10^3 C **(c)** 0.335 kJ

32. Mn^{3+}, Fe^{3+}, Co^{3+}, Au^{3+}, Au^+, Hg^{2+}, Hg_2^{2+}

34. (a) 0.9 V **(b)** 0.7 V

36. (a) $+0.120$ V **(b)** $9.6 \times 10^{-4}\ M$

38. (a) 9×10^9 **(b)** $2 \times 10^{-4}\ M$

40. -0.229 V; 2×10^{-8}

41. Au^+; $E_{tot}^0 = +0.295$ V

42. (a) Zn **(b)** $+2$ **(c)** Mn
(d) Cr, Mn, Fe, Co, Ni **(e)** Cr, Ni, Cu, Zn

43. 53.8% Zn, 46.2% Cu

44. 0.34 t

45. 0.26 μg

46. $2 \times 10^4\ \mu$g; 0.2 μg

47. 3.6×10^4 ft^3

48. 2.5×10^3 kWh

49. 58.9 g/mol

50. 1.7×10^6 L

51. $2Ag_2S(s) + 8CN^-(aq) + 3O_2(g) + 2H_2O \rightarrow$
   $4Ag(CN)_2^-(aq) + 2SO_2(g) + 4OH^-(aq)$
   $2Ag(CN)_2^-(aq) + Zn(s) \rightarrow Zn(CN)_4^{2-}(aq) + 2Ag(s)$

52. $Zn(s) + 2H_2O + 2OH^-(aq) \rightarrow$
   $Zn(OH)_4^{2-}(aq) + H_2(g)$; 0.343 L; 14.74

53. Fe_2O_3

54. (a) $Fe^{2+}(aq) \rightarrow Fe^{3+}(aq) + e^-$
   $\underline{MnO_4^-(aq) + 8H^+(aq) + 5e^- \rightarrow Mn^{2+}(aq) + 4H_2O}$
   $5Fe^{2+}(aq) + MnO_4^-(aq) + 8H^+(aq) \rightarrow 5Fe^{3+}(aq)$
   $+ Mn^{2+}(aq) + 4H_2O$
(b) $+0.743$ V **(c)** 88.8%

55. 1.3×10^2 A

56. 69.66% Fe; 96.27% Fe_3O_4

57. 86.4%

58. (a) Cr^{2+} **(b)** Au^+ **(c)** Co^{2+} **(d)** Mn^{2+}

59. 8×10^{-4}; no

60. 2×10^{-11}; no

61. (a) $Fe(OH)_3(s) + 3H_2C_2O_4(aq) \rightarrow$
   $Fe(C_2O_4)_3^{3-}(aq) + 3H_2O + 3H^+(aq)$
(b) 0.28 L

62. (a) 1, 5 **(b)** Fe^+, NO^-
63. 2.80%
64. 2.83×10^3 K

Chapter 23

2. (a) bromic acid **(b)** potassium hypoiodite
(c) sodium chlorite **(d)** sodium perbromate
4. (a) $KBrO_2$ **(b)** $CaBr_2$ **(c)** $NaIO_4$ **(d)** $Mg(ClO)_2$
6. (a) hypobromite ion **(b)** iodate ion
(c) perchlorate ion **(d)** bromite ion
8. (a) $ZnCO_3(s) + 2H^+(aq) + 2I^-(aq) \rightarrow$
$$ZnI_2(s) + CO_2(g) + H_2O$$
(b) $2I^-(aq) + SO_4^{2-}(aq) + 4H^+(aq) \rightarrow$
$$I_2(s) + SO_2(g) + 2H_2O$$
(c) $NaI(s) + H_3PO_4(l) \rightarrow HI(g) + NaH_2PO_4(s)$
10. (a) $3HClO(aq) \rightarrow Cl_2(g) + HClO_2(aq) + H_2O$
(b) $2ClO_3^-(aq) \rightarrow ClO_4^-(aq) + ClO_2^-(aq)$
12. (a) $Cl_2(g) + 2Br^-(aq) \rightarrow 2Cl^-(aq) + Br_2(l)$
(b) NR **(c)** NR **(d)** NR
14. (a) $2H^+(aq) + S^{2-}(aq) \rightarrow H_2S(aq)$
(b) $H^+(aq) + C_2H_3O_2^-(aq) \rightarrow HC_2H_3O_2(aq)$
(c) $3H^+(aq) + Fe(OH)_3(s) \rightarrow Fe^{3+}(aq) + 3H_2O$
(d) $Pb^{2+}(aq) + 2Cl^-(aq) \rightarrow PbCl_2(s)$
16. (a) react with MnO_4^- **(b)** react with H_2SO_4
(c) neutralize with HI, evaporate
18. 14.3 L
20. $NiCl_2$
22. 1.50×10^3 g
24. 0.204 M
26. $I_2(s) + 3 Cl_2(g) \rightarrow 2ICl_3(s)$
28. 1.08 L
30. 4.28; 0.10 M
32. -0.939
34. 1.9×10^{-3}
36. 0.012 g/100 mL
38. (a) 4.2×10^2, 4.2, 0.042 **(b)** 10.32
40. 904 K
42. (a) yes **(b)** 2×10^{16}
44. (a) $2MnO_4^-(aq) + 16H^+(aq) + 10Br^-(aq) \rightarrow$
$$2Mn^{2+}(aq) + 5Br_2(l) + 8H_2O$$
(b) decreases by 0.0947 V
46. 2.26×10^5 g
48. 27.4 kg
50. $+0.410$ V
52. -0.254 V
53. (a) increases **(b)** decreases **(c)** decreases
(d) increases **(e)** increases **(f)** increases
(g) decreases **(h)** decreases
54. $(\ddot{\overset{..}{I}} — \overset{..}{I} — \ddot{\overset{..}{I}}:)^-$
55. (a) HClO **(b)** HIO_4 **(c)** $HBrO_4$
56. (a) dispersion **(b)** dispersion, dipole

(c) dispersion, H bonds **(d)** dispersion, dipole
(e) no molecules
57. (a) NiF_2 forms protective coating
(b) reacts with glass
(c) converts hydroxyapatite to fluorapatite
(d) forms I_3^-
(e) equation does not involve H^+ or OH^-
58. (a) Add Ag^+ to precipitate AgBr
(b) Add acid (or base!)
59. 5.34 M
60. 1.90 g
61. (a) $F < Cl < Br < I$ **(b)** $F^- < Cl^- < Br^- < I^-$
(c) $I < Br < Cl < F$ **(d)** $F_2 < Cl_2 < Br_2 < I_2$
(e) $I_2 < Br_2 < Cl_2 < F_2$ **(f)** $F_2 < Cl_2 < Br_2 < I_2$
62. 49.3%
63. HI
64. (a) IO_4^- **(b)** IO_4^- **(c)** I^- **(d)** IO^-
65. $+18$ kJ/mol
66. (a) $IO_3^-(aq) + Cl_2(g) + 3OH^-(aq) \rightarrow$
$$H_3IO_6^{2-}(aq) + 2Cl^-(aq)$$
$$H_3IO_6^{2-}(aq) + 3Ag^+(aq) \rightarrow$$
$$Ag_3IO_5(s) + H_2O + H^+(aq)$$
$$4Ag_3IO_5(s) + 6Cl_2(g) + 10H_2O \rightarrow$$
$$4H_5IO_6(aq) + 12AgCl(s) + 3O_2(g)$$
(b) 115.2 g
67. 1.30%
68. Typical data:

pH	5	6	7
$[HClO]/[ClO^-]$	360	36	3.6
f HClO	0.997	0.973	0.783

	8	9	10
	0.36	0.036	0.0036
	0.265	0.035	0.004

69. no

Chapter 24

2. (a) sulfite ion **(b)** hydrogen sulfate ion
(c) hydrogen phosphate ion
4. (a) HN_3 **(b)** H_2SO_3 **(c)** N_2H_4 **(d)** NaH_2PO_4
6. (a) $3S(s) + 6OH^-(aq) \rightarrow$
$$2S^{2-}(aq) + SO_3^{2-}(aq) + 3H_2O$$
(b) $3NO_2^-(aq) + H_2O \rightarrow$
$$NO_3^-(aq) + 2NO(g) + 2OH^-(aq)$$
(c) $3S_2O_3^{2-}(aq) + H_2O \rightarrow$
$$2SO_4^{2-}(aq) + 4S(s) + 2OH^-(aq)$$
8. (a) $2Fe(OH)_2(s) + S(s) + 2OH^-(aq) \rightarrow$
$$2Fe(OH)_3(s) + S^{2-}(aq)$$
(b) $2Fe(OH)_2(s) + NO_3^-(aq) + H_2O \rightarrow$
$$2Fe(OH)_3(s) + NO_2^-(aq)$$
(c) $4Fe(OH)_2(s) + SO_3^{2-}(aq) + 3H_2O \rightarrow$
$$4Fe(OH)_3(s) + S(s) + 2OH^-(aq)$$

10. (a) H_2S (b) N_2H_4, HN_3 (c) PH_3

12. (a) $Ca(OH)_2(s) + 2H^+(aq) + 2NO_3^-(aq) \rightarrow$
$$Ca(NO_3)_2(s) + 2H_2O$$

(b) $BaCO_3(s) + 2H^+(aq) + 2NO_3^-(aq) \rightarrow$
$$Ba(NO_3)_2(s) + CO_2(g) + H_2O$$

(c) $ZnO(s) + 2H^+(aq) + 2NO_3^-(aq) \rightarrow$
$$Zn(NO_3)_2(s) + H_2O$$

14. (a) NH_3, N_2H_4 (b) HNO_3 (c) HNO_2 (d) HNO_3

16. 7.7×10^4 L; 1.5×10^5 g

18. 11.17 M, 17.76 m

20. 12 L; 3.53

22. 29 g

24. 0.8615 g

26. 0.08 M

28. 7.1 g

30. $H_2O_2(aq) + H_2O \rightleftharpoons HO_2^-(aq) + H_3O^+(aq)$; 6.00

32. (a) $H_2PO_4^-$ (b) HPO_4^{2-}

34. 510 K

36. (a) $+83.7$ kJ/mol (b) 139 J/K (c) 1.91

38. 9.44 kJ

40. (a) no (b) no (c) impossible (d) yes

42. b, c, d

44. 4.6

46. yes, yes

47. (a) $+3$ (b) $+4$ (c) $+5$ (d) -3

48. (a) H—O—P—O—H (with :O: groups) tetrahedral around P
(b) H—O—S—O (with :O: groups) tetrahedral around S
(c) :O—S—O: (with :O:) triangular pyramid
(d) :Cl—P—Cl: (with :Cl: groups) tetrahedral
(e) N with :O: :O: bent

49. (a) removes elements of H_2O
(b) dissipates heat
(c) decomposes to NO_2

50. 0.368%

51. S_2

52. $:N{\equiv}N{-}\ddot{O}:$, polar, 180°, sp around N, 2 sigma, 2 pi

53. 17.3 M, 29.1 m

54. 1×10^{-7} M

55. 0.570 M; 1.94%

56. all except P_4O_{10}, P_4O_6

57. (a) O with :O: :O: structure (b) 138 kJ (c) 867 nm

58. $H{-}\ddot{O}{-}N{=}N{-}\ddot{O}{-}H$; yes

59. S = 0, C = +2, N = −3

60. 18

61. Typical data:

pH	0	2	4	6
f H_3AsO_4	0.99	0.64	0.02	0.00
f $H_2AsO_4^-$	0.01	0.36	0.98	0.85
f $HAsO_4^{2-}$	0.00	0.00	0.00	0.15
f AsO_4^{3-}	0.00	0.00	0.00	0.00

pH	8	10	12	14
f H_3AsO_4	0.00	0.00	0.00	0.00
f $H_2AsO_4^-$	0.05	0.00	0.00	0.00
f $HAsO_4^{2-}$	0.95	0.98	0.29	0.00
f AsO_4^{3-}	0.00	0.02	0.71	1.00

62. $E_{tot}^0 = \dfrac{n_1 E_1^0 + n_2 E_2^0}{n_1 + n_2}$

$\Delta G_1^0 = n_1(96.5)E_1^0$; $\Delta G_2^0 = n_2(96.5)E_2^0$

$\Delta G_{tot}^0 = 96.5(n_1 E_1^0 + n_2 E_2^0) = (n_1 + n_2)(96.5)E_{tot}^0$

Chapter 25

2. $^{51}_{24}Cr \rightarrow {}^{0}_{1}e + {}^{51}_{23}V$

4. (a) $^{237}_{93}Np \rightarrow {}^{4}_{2}He + {}^{233}_{91}Pa$ (b) $^{85}_{39}Y \rightarrow {}^{0}_{1}e + {}^{85}_{38}Sr$
(c) $^{12}_{6}C + {}^{12}_{6}C \rightarrow {}^{23}_{11}Na + {}^{1}_{1}H$
(d) $^{239}_{94}Pu + {}^{1}_{0}n \rightarrow {}^{130}_{50}Sn + 4\,{}^{1}_{0}n + {}^{106}_{44}Ru$

6. (a) $^{235}_{92}U$ (b) $^{231}_{91}Pa$

8. (a) $^{238}_{92}U + {}^{16}_{8}O \rightarrow {}^{249}_{100}Fm + 5\,{}^{1}_{0}n$
(b) $^{26}_{13}Al + {}^{4}_{2}He \rightarrow {}^{30}_{15}P$
(c) $^{63}_{29}Cu + {}^{1}_{1}H \rightarrow {}^{63}_{30}Zn + {}^{1}_{0}n$
(d) $^{27}_{13}Al + {}^{2}_{1}H \rightarrow {}^{4}_{2}He + {}^{25}_{12}Mg$

10. (a) $^{1}_{0}n$ (b) $^{243}_{97}Bk$ (c) $^{9}_{4}Be$ (d) $^{1}_{0}n$

12. (a) $^{28}_{14}Si$ (b) $^{6}_{3}Li$ (c) $^{23}_{11}Na$

14. (a) Ni (b) Se (c) Cd

16. 3.20×10^2 g

18. 0.15

20. 36 h

22. 2.0×10^{14}

24. $^{210}_{82}Pb \rightarrow {}^{210}_{83}Bi + {}^{0}_{-1}e$; 4.16×10^4 mCi

26. ~4500 B.C.

28. yes

30. 4.7×10^9 yr

32. (a) 2.7×10^9 yr (b) 3.4×10^9 yr

34. -2.8×10^{-5} g; -2.5×10^6 kJ

36. F-19

38. (a) -1.76×10^8 kJ (b) -2.06×10^8 kJ

40. -3.64×10^{-7} g

42. (a) 1.33×10^{-8} g (b) 3.34×10^{13} atoms

44. yes; $\Delta m = -0.00322$ g/mol

46. 6.65×10^7 kJ vs. 2.06×10^8 kJ

47. (a) 55 rems (b) lesions

48. (a) F; unchanged　**(b)** F; rapidly
　　(c) F; more energy
49. (a) opposite charges
　　(b) See discussion Section 25.2
　　(c) See discussion Section 25.5
50. Actual age would be greater than that calculated
51. 6.59×10^{-5} mol/L
52. 1.0×10^{-16}
53. $4.84 \times 10^{-11}\%$
54. 2.60 Ci
55. 6.5×10^2 mL
56. 15 atm
57. 2.13×10^6 L
58. 2
59. 0.015 L
60. 5.3×10^4 g
61. 5.8×10^{-19} mol/L
62. (a) 5.5×10^{-10} g　**(b)** 1.2×10^{-3} kJ　**(c)** 17 rems
63. (a) 1×10^{-13} J　**(b)** 6×10^6 m/s　**(c)** 3×10^9 K
64. (a) -5.72×10^8 kJ　**(b)** 1.3×10^{28} kJ
　　(c) 1.8×10^{-11}

Chapter 26

2. (a) 2-methyloctane　**(b)** 2,2-dimethylpropane
　　(c) 2,2,4-trimethylpentane
　　(d) 2,5-dimethylheptane

4. (a)
$$CH_3-\underset{\underset{CH_3}{|}}{\overset{\overset{CH_3}{|}}{C}}-CH_2-\underset{\underset{CH_3}{|}}{\overset{\overset{H}{|}}{C}}-CH_3$$
(b)
$$CH_3-\underset{\underset{CH_3}{|}}{\overset{\overset{CH_3}{|}}{C}}-CH_3$$

(c)
$$CH_3-(CH_2)_2-\underset{\underset{\underset{\underset{H}{|}}{CH_3-\overset{}{C}-CH_3}}{|}}{\overset{\overset{H}{|}}{C}}-(CH_2)_3-CH_3$$

(d)
$$CH_3-\underset{\underset{H}{|}}{\overset{\overset{CH_3}{|}}{C}}-\underset{\underset{CH_3}{|}}{\overset{\overset{H}{|}}{C}}-\underset{\underset{H}{|}}{\overset{\overset{CH_3}{|}}{C}}-(CH_2)_2-CH_3$$

6. (a)
$$CH_3-\underset{\underset{CH_3}{|}}{\overset{\overset{CH_3}{|}}{C}}-CH_2-CH_3$$
2,2-dimethylbutane

(b)
$$CH_3-(CH_2)_2-\underset{\underset{CH_3}{|}}{\overset{}{CH}}-CH_3$$
2-methylpentane

(c)
$$CH_3-\underset{\underset{CH_3}{|}}{\overset{}{CH}}-CH_2$$
$$\qquad\qquad |$$
$$\qquad\qquad CH_3$$
2-methylbutane

8. (a) $CH_3-C\equiv C-CH_2-CH_3$
(b) $CH_3-C\equiv C-\underset{\underset{CH_3}{|}}{\overset{}{CH}}-CH_3$
(c) $CH_3-\underset{\underset{}{}}{\overset{\overset{CH_3}{|}}{CH}}-C\equiv C-CH_2-CH_3$
(d) $H-C\equiv C-\underset{\underset{CH_3}{|}}{\overset{\overset{CH_3}{|}}{C}}-CH_3$

10. (a) *o*-chlorotoluene　**(b)** *m*-bromotoluene
　　(c) 1,2,5-tribromotoluene
12. (a) methanoic acid; formic acid
　　(b) ethanol; ethyl alcohol
　　(c) 2-propanol; isopropyl alcohol

14. (a) $CH_3-\underset{\underset{Cl}{|}}{\overset{\overset{I}{|}}{C}}-CH_3$　**(b)** $CH_3-\underset{\underset{CH_3}{|}}{\overset{\overset{H}{|}}{C}}-\underset{\underset{F}{|}}{\overset{\overset{H}{|}}{C}}-CH_3$
　　(c) $CH_3-\underset{\underset{CH_3}{|}}{\overset{\overset{CH_3}{|}}{C}}-CH_2-Br$

16. (a) $C_3H_6O_2$　**(b)** $C_4H_{10}O$
　　(c) C_3H_8O　**(d)** $C_{14}H_{10}$
18. (a) methyl acetate　**(b)** ethyl formate
　　(c) methyl propionate
20. (a) $NH_2-CH_2-CH_2-NH_2$
　　(b) $H-\underset{\underset{Cl}{|}}{\overset{\overset{Cl}{|}}{C}}-Cl$　**(c)** $Cl-\underset{\underset{H}{|}}{\overset{\overset{H}{|}}{C}}-Cl$
　　(d) $CH_3-CH_2-\underset{\underset{CH_2-CH_3}{|}}{\overset{\overset{H}{|}}{N}}-CH_2-CH_3$

22.
$$\underset{\diagup}{\overset{\diagup}{C}}=C-\overset{|}{\underset{|}{C}}-\overset{|}{\underset{|}{C}}-,\quad -\overset{|}{\underset{|}{C}}-C=C-\overset{|}{\underset{|}{C}}-,\quad -\overset{|}{\underset{|}{C}}-\overset{|}{\underset{\|}{C}}-\overset{|}{\underset{|}{C}}-$$
$$\qquad\qquad\qquad\qquad\qquad\qquad\qquad\qquad C$$

24.

$$-\overset{|}{\underset{|}{C}}-\overset{|}{\underset{Cl}{C}}-\overset{|}{\underset{|}{C}}-Cl, \quad -\overset{|}{\underset{|}{C}}-\overset{Cl}{\underset{Cl}{C}}-\overset{|}{\underset{|}{C}}-,$$

$$-\overset{|}{\underset{Cl}{C}}-\overset{|}{\underset{|}{C}}-\overset{|}{\underset{Cl}{C}}-, \quad -\overset{|}{\underset{|}{C}}-\overset{|}{\underset{Cl}{C}}-\overset{|}{\underset{Cl}{C}}-$$

26.

(1,2,3-trichlorobenzene) , (1,2,4-trichlorobenzene) , (1,3,5-trichlorobenzene)

28.

$$-\overset{|}{\underset{|}{C}}-\overset{Br}{\underset{|}{C}}-Br, \quad -\overset{Br}{\underset{|}{C}}-\overset{Br}{\underset{|}{C}}-$$

30.

$$Br-\overset{Cl}{\underset{}{C}}=\overset{}{C}-\overset{|}{\underset{|}{C}}-\overset{|}{\underset{|}{C}}-, \quad Br-\overset{}{C}=\overset{Cl}{\underset{}{C}}-\overset{|}{\underset{|}{C}}-\overset{|}{\underset{|}{C}}-,$$

$$Cl-\overset{}{C}=\overset{Br}{\underset{}{C}}-\overset{|}{\underset{|}{C}}-\overset{|}{\underset{|}{C}}-, \quad -\overset{}{C}-\overset{Br}{\underset{}{C}}=\overset{Cl}{\underset{}{C}}-\overset{|}{\underset{|}{C}}-,$$

$$-\overset{|}{\underset{|}{C}}-\overset{-C-}{\underset{}{C}}=\overset{}{C}-Br$$
(with Cl below)

32.

$$-\overset{|}{\underset{|}{C}}-\overset{|}{\underset{|}{C}}-\overset{|}{\underset{\overset{\|}{O}}{C}}-C-OH, \quad -\overset{|}{\underset{|}{C}}-\overset{|}{\underset{HO-C=O}{C}}-\overset{|}{\underset{|}{C}}-$$

34. (a) 4   **(b)** 7   **(c)** 5

36.

$$\overset{Cl}{\underset{Br}{>}}C=C\overset{H}{\underset{C_2H_5}{<}} \quad and \quad \overset{Br}{\underset{Cl}{>}}C=C\overset{H}{\underset{C_2H_5}{<}} ;$$

$$\overset{Br}{\underset{H}{>}}C=C\overset{Cl}{\underset{C_2H_5}{<}} \quad and \quad \overset{H}{\underset{Br}{>}}C=C\overset{Cl}{\underset{C_2H_5}{<}} ;$$

$$\overset{Cl}{\underset{H}{>}}C=C\overset{Br}{\underset{C_2H_5}{<}} \quad and \quad \overset{H}{\underset{Cl}{>}}C=C\overset{Br}{\underset{C_2H_5}{<}} ;$$

$$\overset{Br}{\underset{CH_3}{>}}C=C\overset{Cl}{\underset{CH_3}{<}} \quad and \quad \overset{CH_3}{\underset{Br}{>}}C=C\overset{Cl}{\underset{CH_3}{<}}$$

38. b, c

40.

$$\overset{H}{\underset{HO}{>}}C=C\overset{H}{\underset{C_2H_5}{<}} \quad and \quad \overset{HO}{\underset{H}{>}}C=C\overset{H}{\underset{C_2H_5}{<}} ;$$

$$\overset{H}{\underset{H}{>}}C=C\overset{OH}{\underset{C_2H_5}{<}} ; \quad \overset{H}{\underset{H}{>}}C=C\overset{H}{\underset{\overset{|}{C}-CH_3}{<}}$$ (with H and OH below)

$$\overset{H}{\underset{H}{>}}C=C\overset{H}{\underset{CH_2CH_2OH}{<}} ;$$

$$\overset{H}{\underset{CH_2OH}{>}}C=C\overset{H}{\underset{CH_3}{<}} \quad and \quad \overset{CH_2OH}{\underset{H}{>}}C=C\overset{H}{\underset{CH_3}{<}} ;$$

$$\overset{CH_3}{\underset{HO}{>}}C=C\overset{CH_3}{\underset{H}{<}} \quad and \quad \overset{HO}{\underset{CH_3}{>}}C=C\overset{CH_3}{\underset{H}{<}} ;$$

$$\overset{H}{\underset{HO}{>}}C=C\overset{CH_3}{\underset{CH_3}{<}} ; \quad \overset{H}{\underset{H}{>}}C=C\overset{CH_2OH}{\underset{CH_3}{<}}$$

42. c, d

44. (a) center carbon   **(c)** carbon atom at right

46. (a) alcohol   **(b)** ester   **(c)** ester, acid

48. (a)

$$CH_3-\overset{H}{\underset{OH}{C}}-CH_3$$

(b)

$$CH_3-CH_2-CH_2-\overset{}{\underset{\|}{C}}-OH$$ (with O below C)

or

$$CH_3-\overset{H}{\underset{CH_3}{C}}-\overset{O}{\underset{\|}{C}}-OH$$

(c)

$$CH_3-CH_2-CH_2-\overset{}{\underset{\overset{\|}{O}}{C}}-O-\overset{H}{\underset{CH_3}{C}}-CH_3$$

or

$$CH_3-\overset{CH_3}{\underset{H}{C}}-\overset{}{\underset{\overset{\|}{O}}{C}}-O-\overset{H}{\underset{CH_3}{C}}-CH_3$$

50.

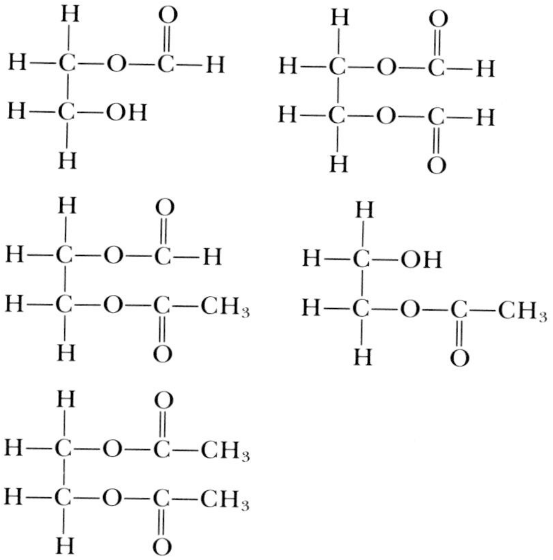

H—C—O—C—H (with H—C—OH and H below)

H—C—O—C—H (with H—C—O—C—H)

H—C—O—C—H (with H—C—O—C—CH₃ and O below)

H—C—OH (with H—C—O—C—CH₃ and O below)

H—C—O—C—CH₃ (with H—C—O—C—CH₃ and H, O below)

52. (a) secondary **(b)** tertiary **(c)** primary
  (d) secondary
54. (a) addition **(b)** elimination **(c)** condensation
56. (a) 2,3-dichloro-2-methylbutane
  (b) 2-iodo-2-methylbutane **(c)** 2-methyl-2-butanol
57. (a) C_2H_6 **(b)** CaC_2
  (c) C_2H_4, $C_6H_{12}O_6$ **(d)** C_2H_5OH
58. d
59. (a) C_nH_{2n} **(b)** C_nH_{2n-2} **(c)** $C_nH_{2n+2}O$
60. (a) ethanol with toxic additive **(b)** methanol
  (c) sodium salt long-chain acid
  (d) glyceryl ester of unsaturated, long-chain acid
61. $Cl_2CHCOO^- < ClCH_2COO^- < ICH_2COO^- <$
  CH_3COO^-
62. (a) $(CH_3)_2NH(aq) + H^+(aq) \rightarrow (CH_3)_2NH_2{}^+(aq)$
  (b) $CH_3COOH(aq) + OH^-(aq) \rightarrow$
  $CH_3COO^-(aq) + H_2O$

  (c) 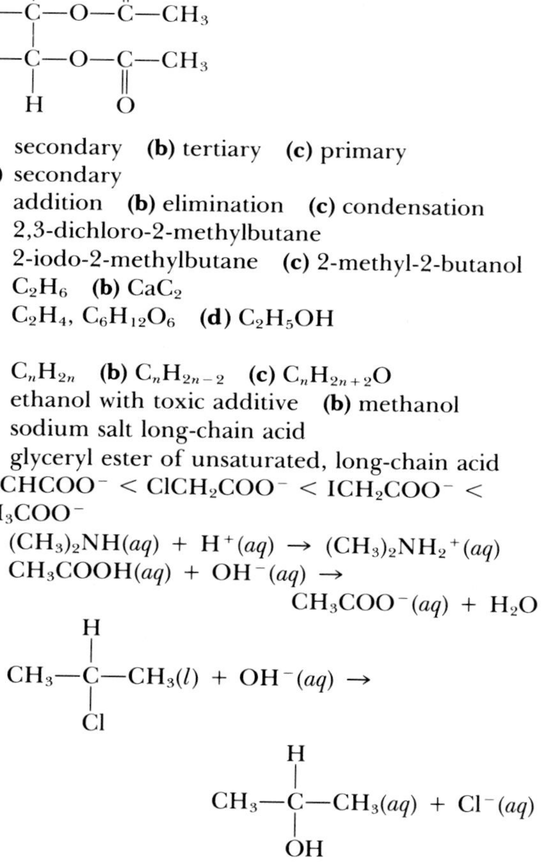 CH_3—C(—CH₃)(—Cl)(—H)$(l) + OH^-(aq) \rightarrow$

  CH_3—C(—CH₃)(—OH)(—H)$(aq) + Cl^-(aq)$

63. $C_{21}H_{23}O_5N$: 68.28% C, 6.28% H, 21.66% O,
  3.79% N
64. 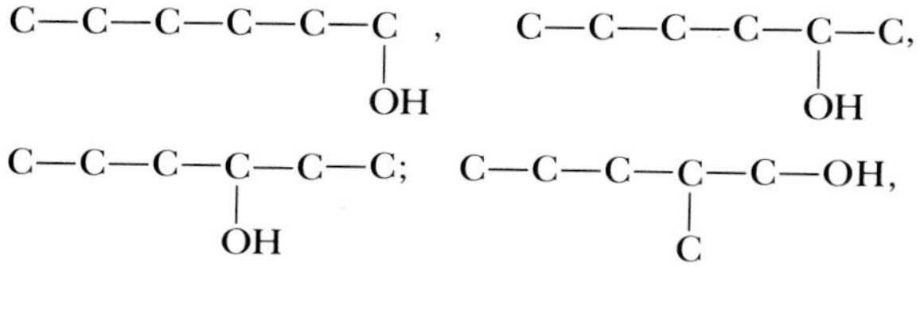 C—C—C—C—C—C(—OH) , C—C—C—C—C—C(—OH),
  C—C—C—C—C—C(—OH); C—C—C—C—C—C—OH(—C)

64 (right column):

C—C—C—C(—OH)(—C)—C, C—C—C(—OH)—C—C

C—C—C—C(—OH)—C (with C below), HO—C—C—C—C—C (with C below)

C—C—C(—OH)—C—C (with C below), C—C—C—C—C (with C—OH below)

C—C—C—C—C (with C, OH below), C—C—C—C—C—OH (with C below)

C—C—C(—OH)—C (with C, C below), C—C—C—C—OH (with C, C below)

C—C(—C)—C—C (with C—H below), C—C(—C)—C—C (with C, OH below)

C—C(—C)—C—C—OH (with C below)

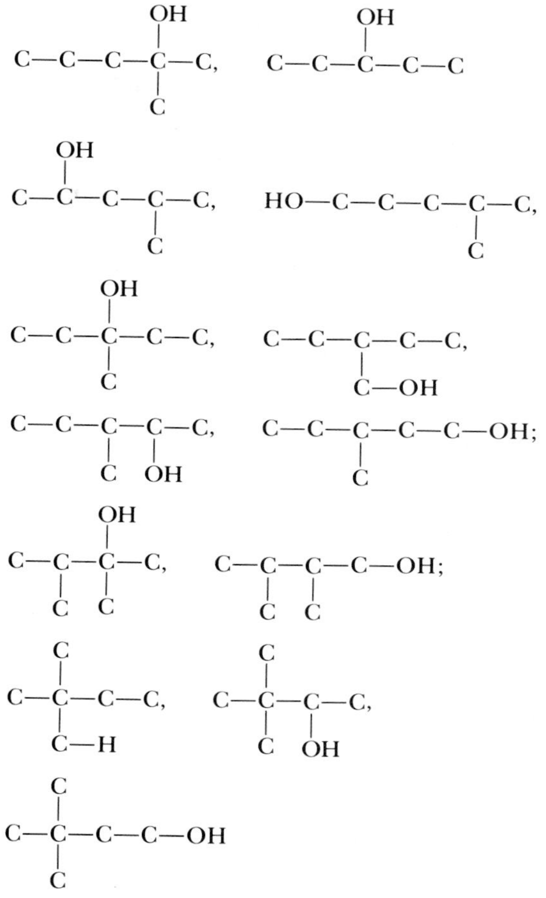

65. $ClCH_2COOH(aq) + (CH_3)_3N(aq) \rightarrow$
  $ClCH_2COO^-(aq) + (CH_3)_3NH^+(aq)$
  $K = 8.8 \times 10^6$; 3.4×10^{-5} M
66. 0.0057 g

Chapter 27

2. (a)

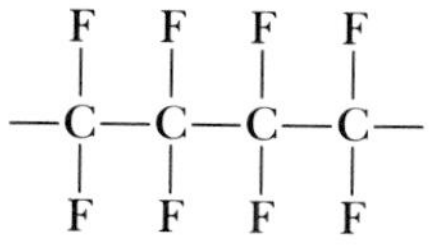

—C(F)(F)—C(F)(F)—C(F)(F)—C(F)(F)—

  (b) 2.5×10^6 g/mol
  (c) 24.02% C, 75.98% F

4.

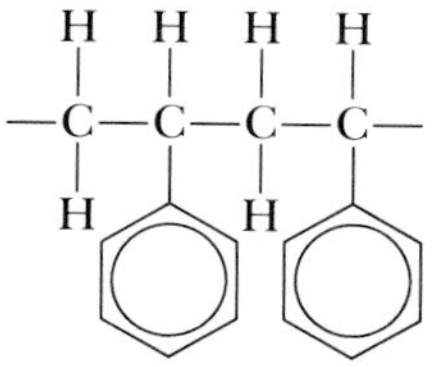

—C(H)(H)—C(H)(phenyl)—C(H)(H)—C(H)(phenyl)—

6. (a) $H_2C = CHF$

(b) $CH_3 - C = C - CH_3$ (with H, H below the two central carbons)

8. polystyrene

10. $-O-\overset{\text{O}}{\underset{\|}{C}}-O-$ 〈benzene ring〉$-\overset{CH_3}{\underset{CH_3}{C}}-$ 〈benzene ring〉$-O-$

12. $-\overset{H}{\underset{|}{N}}-(CH_2)_5-\overset{O}{\underset{\|}{C}}-\overset{H}{\underset{|}{N}}-(CH_2)_5-\overset{O}{\underset{\|}{C}}-$

14. (a) $H_2N-CH_2-CH_2-NH_2$ and $HOOC-CH_2-COOH$

(b) $HOOC-$〈benzene ring〉$-COOH$

and $HO-CH_2-\overset{H}{\underset{CH_3}{C}}-OH$

16. $C_{12}H_{22}O_{11}(aq) + H_2O \rightarrow 2C_6H_{12}O_6(aq)$

18. (a) 44.44% C, 6.22% H, 49.34% O **(b)** 6.2×10^2

20.

[structure of disaccharide with HO, CH_2OH groups]

22. 9; 10

24. Leu-Lys:

$(CH_3)_2-CH-CH_2-\overset{H}{\underset{NH_2}{C}}-\overset{O}{\underset{\|}{C}}-\overset{H}{\underset{H}{N}}-\overset{H}{\underset{(CH_2)_4-NH_2}{C}}-COOH$

Lys-Leu:

$NH_2-(CH_2)_4-\overset{H}{\underset{NH_2}{C}}-\overset{O}{\underset{\|}{C}}-\overset{H}{\underset{H}{N}}-\overset{H}{\underset{CH_2-CH(CH_3)_2}{C}}-COOH$

26. (a) 6

(b) $(CH_3)_2CH-\overset{H}{\underset{NH_2}{C}}-\overset{O}{\underset{\|}{C}}-\overset{H}{\underset{(CH_2)_4-NH_2}{N}}-\overset{O}{\underset{\|}{C}}-\overset{H}{\underset{CH_2-\text{〈ring〉}}{N}}-\overset{H}{C}-COOH$

Val-Lys-Phe

28. (a) $HO-CH_2-\overset{H}{\underset{NH_3^+}{C}}-COO^-$

(b) $HO-CH_2-\overset{H}{\underset{NH_3^+}{C}}-COOH$

(c) $HO-CH_2-\overset{H}{\underset{NH_2}{C}}-COO^-$

30. (a) 2.29 **(b)** 9.74 **(c)** 6.02

32. Ala-Phe-Leu-Met-Val-Ala

34. 1

36. GCA CCA GAC UAC AUA GGA (among many others)

38. Met-Asn-Gln-Val-Phe-Ser

39. (a) neither **(b)** addition **(c)** condensation **(d)** neither

40. Double bond for addition; two functional groups for condensation

41. (a) natural polymer has more complex structure

(b) polyester has ester linkage, polyamide has amide linkage

(c) differ in configuration at carbon-1

42. (a) linear has straight chain with no branches

(b) See Equation 27.9

(c) Compare Equations 27.8, 27.9

43. (a) $H_2C=CHBr$ **(b)** H_2N-CH_2-COOH

(c) $HOOC-(CH_2)_4-COOH$ and $H_2N-(CH_2)_6-NH_2$

44. (a) $HOOC-$〈benzene ring〉$-COOH$ and $HO-(CH_2)_2-OH$

(b) $ClHC=CHCl$ **(c)** $H_2N-(CH_2)_5-COOH$

45. $H_2N-CH_2-C\underset{O}{\overset{}{\|}}\left(N\underset{H}{-}CH_2-C\underset{O}{\overset{}{\|}}\right)_2 N-CH_2-COOH$
$\quad\quad\quad\quad\quad\quad\quad\quad\quad\quad\quad\quad\quad\quad\quad\quad H$

46. Head to head:
$$-\overset{H}{\underset{H}{C}}-\overset{H}{\underset{CN}{C}}-\overset{H}{\underset{CN}{C}}-\overset{H}{\underset{H}{C}}-$$

Head to tail:
$$-\overset{H}{\underset{CN}{C}}-\overset{H}{\underset{H}{C}}-\overset{H}{\underset{CN}{C}}-\overset{H}{\underset{H}{C}}-$$

47. 0

48. $(CH_3)_2CH-CH_2-\overset{H}{\underset{NH_3^+}{C}}-COOH$;

$(CH_3)_2CH-CH_2-\overset{H}{\underset{NH_2}{C}}-COO^-$

49. 27
50. 1.59×10^4 g/mol
51. See discussion, Section 27.4
52. See discussion, Section 27.5
53. 10^{-503}
54. -214 kJ

55. (a) $-\overset{}{\underset{O}{C}}\overset{}{\underset{O}{\underset{\|}{C}}}-O-CH_2-\overset{H}{\underset{OH}{C}}-CH_2-O-$

(b) orthophthalic acid condenses with OH groups in two adjacent chains

56. 63.68% C, 9.80% H, 14.14% O, 12.38% N
57. -17 kJ

58. linear:

cross-linked:

59. $HOOC-CH_2-\overset{H}{\underset{NH_3^+}{C}}-COOH$, I

$HOOC-CH_2-\overset{H}{\underset{NH_3^+}{C}}-COO^-$, II

$HOOC-CH_2-\overset{H}{\underset{NH_2}{C}}-COO^-$, III

$^-OOC-CH_2-\overset{H}{\underset{NH_2}{C}}-COO^-$ IV

I, pH < 2.10; II, pH 2.10–3.85; III, pH 3.85–9.82; IV pH > 9.82

APPENDIX
5

CHAPTER SUMMARIES, IMPORTANT EQUATIONS, KEY WORDS

CHAPTER 1

Chemistry is the study of matter and its composition, properties, structure, and reactions. Matter is composed of elements and compounds, which may occur as pure substances or as mixtures. A compound is a substance in which two or more elements are combined chemically. An element is unique; no two elements have the same symbol or exactly the same properties. The Periodic Table, shown inside the front cover, is a convenient tabulation of the elements. It is arranged by periods (horizontal rows) and groups (vertical columns). Elements in a group are similar chemically.

Substances are identified by their properties. These include density (Examples 1.7, 1.8), melting point, boiling point, and solubility (Example 1.9). Differences in solubility between solids serve as the basis for fractional crystallization, a process by which a pure solid is isolated from a mixture. Other separation techniques include distillation (Figure 1.7) and filtration (Figure 1.10).

Many different kinds of measurements have been discussed in this chapter. All measured quantities have an uncertainty whose magnitude depends upon the instrument used and the skill of the person using it. Significant figures indicate the degree of uncertainty in a measurement (Examples 1.2 and 1.3). Measured quantities may be expressed in various units (Tables 1.1 and 1.2). To convert a quantity from one set of units to another, we use conversion factors. Their use is illustrated in Examples 1.4 through 1.6. We will use the conversion factor approach over and over again to solve problems in future chapters, so you should become familiar with it now. Other problems will be solved by substituting into equations such as those relating temperature scales (Example 1.1).

IMPORTANT EQUATIONS

Temperature conversions $°F = 1.8(°C) + 32°$; $K = °C + 273.15$

KEY WORDS

alkali metals	family	milli
alkaline earth metals	filtrate	millimeter of mercury
atmosphere	filtration	mixture
boiling point	fractional crystallization	nano-
Celsius degree	group	noble gases
centi-	halogens	pascal
chemical property	heterogeneous	period
compound	homogeneous	Periodic Table
conversion factor	joule	physical property
cubic centimeter	kilo	SI unit
density	liter	significant figure
element	melting point	solubility
Fahrenheit degree	meter	solution

CHAPTER 2

The three basic components of atoms are protons, neutrons, and electrons. Protons and neutrons are in the small, positively charged nucleus of an atom. Electrons, which carry a negative charge, surround the nucleus. The proton carries a positive charge equal in magnitude to that of the electron. A neutron has zero charge. In a neutral atom, there are the same number of electrons as protons. The number of protons in the nucleus (atomic number) is characteristic of a particular element. In the Periodic Table, elements are arranged in order of increasing atomic number as we move from left to right in a horizontal row (period).

The mass number of an atom is found by adding the number of neutrons and protons. Atoms of the same element (same number of protons) that differ in the number of neutrons are called isotopes (Example 2.1). The atomic mass is a number that tells us how heavy an atom is relative to a $^{12}_{6}C$ atom, which is assigned an atomic mass of exactly 12 amu. The atomic mass of an element reflects the masses and abundances of its component isotopes (Example 2.3). This information is obtained experimentally, using a mass spectrometer (Figure 2.4).

When an atom loses electrons, a positive ion (cation) is formed (Example 2.2). The gain of electrons by an atom leads to the formation of a negative ion (anion). Ionic compounds consist of positive and negative ions held together by strong attractive forces. In such a compound, the sum of positive charges equals the sum of negative charges. Certain elements and compounds contain discrete structural units called molecules. Molecules consist of two or more atoms joined by strong chemical bonds.

A mole represents Avogadro's number (6.022×10^{23}) of items, which may be atoms (Example 2.4), molecules, ions, etc. The molar mass of a substance can be found from its formula (Example 2.5). It is numerically equal to the sum of the atomic masses of the element(s) present. Thus, we have

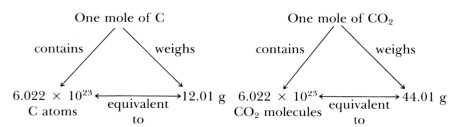

Knowing the molar mass of a substance, we can readily relate moles to grams (Examples 2.6 and 2.7).

The number of moles of a solute in solution depends upon the volume of the sample and the concentration of solute. The concentration unit, molarity, relates the number of moles of solute to the volume of solution in liters (Examples 2.8 and 2.9).

The basic concepts described above evolved from the atomic theory first proposed by John Dalton. Atomic theory leads directly to three of the basic laws of chemistry: (1) Law of Conservation of Mass; (2) Law of Constant Composition; and (3) Law of Multiple Proportions.

KEY WORDS

amu	ion	mole
anion	ionic compound	molecule
atomic mass	isotope	neutron
atomic number	Law of Conservation of Mass	nonstoichiometric compounds
atomic theory	Law of Constant Composition	
atoms	Law of Multiple Proportions	nuclear symbol
Avogadro's number	mass number	nucleus
carbon-12 scale	mass spectrometer	proton
cation	molar mass	solute
electron	molarity	solvent
formula mass		

CHAPTER 3

The formula of a compound shows its chemical composition. The subscripts in the formula give the number of moles of each element in one mole of the compound. From the formula, we can calculate the mass percents of the elements (Example 3.1) or the mass of an element in a weighed sample of the compound (Example 3.2). By the same token, if we know the mass percents, we can calculate the simplest formula (Example 3.3). Combustion analysis serves to determine the mass percents of carbon and hydrogen in an organic compound (Example 3.4). We can readily obtain the molecular formula from the simplest formula if we know the molar mass (Example 3.5).

We can predict the simplest formula of an ionic compound from the charges of the ions (Examples 3.6 and 3.7). Charges of monatomic ions are indicated in Figure 3.4, those of polyatomic ions are given in Table 3.1. In naming ionic compounds, as in writing their formulas, the cation comes first. If a metal forms more than one cation, a Roman numeral is used to indicate the charge of the ion (Example 3.8). In naming binary compounds of the nonmetals, Greek prefixes are used to indicate the number of atoms of each type in the formula (Example 3.9). Certain molecular compounds react with water to produce acidic solutions containing H^+ ions; the acids you are most likely to use in the laboratory are hydrochloric acid (HCl), nitric acid (HNO_3), and sulfuric acid (H_2SO_4).

In order to write a chemical equation for a reaction, the formulas and physical states of reactants and products must be known (Example 3.10). The coefficients in a balanced equation represent numbers of moles. We can use them directly to relate moles of different substances taking part in a reaction (Example 3.11). Knowing the molar masses, we can go a step further and relate masses in grams of reactants and products (Example 3.12).

Ordinarily, we do not mix reactants in exactly the ratio called for by the coefficients of the balanced equation. One reactant is in excess; the other reactant is limiting. The amount of product we obtain if all the limiting reactant is consumed is referred to as the theoretical yield (Example 3.13). The actual yield is ordinarily less than the theoretical yield; in other words, the percent yield is less than 100.

KEY WORDS

actual yield	limiting reactant	percent yield
binary molecule	metal	polyatomic ion
chemical equation	mole	product
coefficient	molecular formula	reactant
electrical neutrality	monatomic ion	simplest formula
empirical formula	nonmetal	theoretical yield
hydrate	percent composition	transition metal

CHAPTER 4

The variables in gas behavior are pressure, volume, amount (number of moles), and temperature. These are related by the Ideal Gas Law:

$$PV = nRT$$

where n is the number of moles, T is the temperature in K, and R is a constant [0.0821 L·atm/(mol·K)]. This law can be used for many different kinds of calculations involving gases. It can be applied to determine relationships between two or more variables (Example 4.3) or to calculate one variable knowing the other three (Example 4.4). By using the relation $n = g/MM$, the Ideal Gas Law can be extended to find the molar mass of a gas (Example 4.5) or its density (Example 4.6). Going one step further, we can use the Ideal Gas Law to relate the volumes of gases involved in reactions

(Examples 4.7 to 4.9). Dalton's Law enables us to obtain the partial pressures of gases in a mixture (Example 4.10) or their mole fractions (Example 4.11).

The kinetic theory of gases deals with the random, continuous motion of gas molecules from place to place (translational motion). One of its basic postulates is that the average translational energy is directly proportional to the absolute temperature; the proportionality constant is the same for all gases. This relation can be used to compare the rates of effusion of different gases (Example 4.12) or to calculate average molecular speeds (Example 4.13). The distribution of molecular speeds at different temperatures is given in Figure 4.12.

Real gases deviate at least slightly from the Ideal Gas Law because of the finite size of gas molecules and attractive forces between them. These deviations become more significant when the molecules are close together, at high pressures and low temperatures. The van der Waals equation, although more complex than the Ideal Gas Law, better represents the behavior of real gases.

IMPORTANT EQUATIONS

Ideal Gas Law $\qquad PV = nRT = gRT/MM$

Dalton's Law $\qquad P_{tot} = P_1 + P_2 + - - \; ; P_i = X_i P_{tot}$

Graham's Law $\qquad \dfrac{\text{rate}_2}{\text{rate}_1} = \dfrac{\text{time}_1}{\text{time}_2} = \left(\dfrac{MM_1}{MM_2}\right)^{1/2}$

Molecular velocity $\qquad u = (3RT/MM)^{1/2}; \; R = 8.314 \text{ kg·m}^2/(\text{s}^2\text{·mol·K})$

KEY WORDS

atmosphere	Law of Combining Volumes
average speed of a gas particle	manometer
Avogadro's Law	millimeters of Hg (mm Hg)
Boyle's Law	molar mass (MM)
Charles' and Gay-Lussac's Law	mole
Dalton's Law	mole fraction
density	partial pressure
diffusion	pounds per square inch (psi)
effusion	R—gas constant
elastic collisions	rate of effusion
Graham's Law	real gases
Ideal Gas Law	STP
Kelvin temperature scale	torr
kilopascal	translational energy
Kinetic Theory of Gases	Van der Waal's equation

CHAPTER 5

Atoms can have only certain quantized (discrete) energies; that is, their electrons can be located only in certain energy levels. When an electron moves from a higher to a lower energy level, energy is ordinarily given off in the form of light. The difference in energy between the two levels is inversely related to the wavelength of the light (Example 5.1) and directly related to its frequency (Example 5.2). The wavelengths of lines in the spectrum of the hydrogen atom can be quite accurately calculated from the Bohr model (Examples 5.3 and 5.4).

The quantum mechanical model of the atom describes electrons in terms of quantum numbers. There are four such numbers:

n designates the principal level; **n** = 1, 2, 3, . . .

ℓ designates the sublevel (s, p, d, or f); ℓ = 0, 1, . . . , (**n** − 1)

$\mathbf{m}_\ell$ designates the orbital; $\mathbf{m}_\ell$ = $+\ell$, . . . , +1, 0, −1, . . . , $-\ell$

$\mathbf{m_s}$ designates the electron spin; $\mathbf{m_s}$ = $+\frac{1}{2}$, $-\frac{1}{2}$

No two electrons in an atom can have the same set of four quantum numbers. This means, for example, that two electrons in the same orbital must have opposed spins (Tables 5.3 and 5.4).

These rules fix the capacities of principal levels, sublevels, and orbitals (Example 5.5). With one more piece of information, the order in which sublevels fill (Fig. 5.11 and Table 5.6), we can derive electron configurations of atoms (Example 5.6). Orbital diagrams can be written (Example 5.7) taking into account Hund's rule. Finally, we can write a set of four quantum numbers for each electron in an atom (Example 5.8).

Monatomic ions derived from main-group elements ordinarily have noble gas electronic structures. The configurations of transition metal ions are readily derived by taking into account the fact that they do not contain the outer s electrons found in the corresponding atoms (Example 5.9).

IMPORTANT EQUATIONS

Energy vs. wavelength

$$E_{\text{hi}} - E_{\text{lo}} = \frac{1.986 \times 10^{-25}}{\lambda} \frac{\text{J·m}}{\text{particle}}$$

$$= \frac{1.196 \times 10^5}{\lambda} \frac{\text{kJ·nm}}{\text{mol}}$$

Frequency vs. wavelength

$$\nu = \frac{2.998 \times 10^8 \text{ m/s}}{\lambda}$$

Bohr equation

$$E_n = \frac{-2.179 \times 10^{-18}}{n^2} \frac{\text{J}}{\text{particle}}$$

$$= \frac{-1312}{n^2} \text{ kJ/mol}$$

KEY WORDS

abbreviated electron
  configuration
amplitude
atomic spectra
Balmer series
Bohr model
deBroglie relation
diamagnetic
electron configuration
electron cloud
electron spin
electronic energy
excited state

frequency
ground state
hertz
Hund's Rule
ionization energy
Lyman series
$n + l$ rule
nanometer
noble gas structure
opposed spins
orbital
orbital diagram
paired electrons

parallel spins
paramagnetic
Pauli Exclusion Principle
photon
Planck's constant
principal energy level
quantum mechanics
quantum number
quantum theory
Schrödinger equation
sublevel
unpaired electron
wavelength

CHAPTER 6

The discovery of the Periodic Table is usually credited to Mendeleev; Meyer and Moseley also made important contributions. In the table, elements are arranged in order of increasing atomic number in horizontal periods and vertical groups. Elements in the same group have the same outer electron configuration, which explains why they have similar chemical properties. Main-group elements fill s sublevels (Groups 1 and 2) or p sublevels (Groups 3 to 8). Transition metals fill inner d sublevels (3d, 4d, 5d, 6d); lanthanides fill the 4f sublevel, actinides the 5f.

As one moves across from left to right in the Periodic Table, effective nuclear charge and ionization energy increase; atomic radius and metallic character decrease. As one moves down in the table, effective nuclear charge stays constant and ionization energy decreases; atomic radius and metallic character increase. Trends in ionic radius parallel those in atomic radius; in addition, negative ions are larger than the corresponding atoms, while positive ions are smaller. Electron affinity ordinarily decreases as metallic character increases.

Of 109 known elements, about 85 are metals, all of which are good conductors of heat and electricity. Beyond that, metals are ductile, malleable, and have a characteristic luster. Many of the properties of metals can be explained in terms of a simple bonding theory referred to as the electron-sea model (Figure 6.8).

The alkali and alkaline earth metals show trends in physical properties (Table 6.5) related to their positions in the Periodic Table. The atomic spectra of these elements yield characteristic flame colors. Chemically, the Group 1 and Group 2 metals are extremely reactive (Table 6.6). They react with hydrogen and the halogens to form binary compounds containing $+1$ cations (Group 1 metals) or $+2$ cations (Group 2 metals) and -1 anions (H^-, F^-, Cl^-, Br^-, I^-). The ionic compounds formed with oxygen include normal oxides (O^{2-} ion), peroxides (O_2^{2-} ion), and superoxides (O_2^- ion).

With water, hydrogen is evolved; a solution of the Group 1 or Group 2 hydroxide is formed at the same time. Aluminum resembles the alkali and alkaline earth metals in its chemical properties except that it is somewhat less reactive and forms a +3 cation. Aluminum is prepared commercially by the electrolysis of aluminum oxide; cryolite, Na_3AlF_6, is added to lower the temperature. Sodium metal is prepared by a similar method, starting with sodium chloride.

Relatively few new concepts have been introduced in this chapter; Examples 6.1 (outer electron configurations), 6.2 and 6.3 (atomic radius, ionic radius, ionization energy), 6.5 and 6.7 (reactions of metals) deal with these. The remaining examples review concepts introduced in previous chapters.

KEY WORDS

actinides
alkali metal
alkaline earth metal
atomic number
atomic radius
Coulomb's Law
cryolite
ductility
effective nuclear charge
electrical conductivity
electrode
electrolysis
electron affinity
electron-sea model
first ionization energy
flame test
group (Periodic Table)
Hall process
hydride
ionic radii
ionization energy

lanthanides
luster
main-group elements
malleability
Mendeleev
metal
metallic bonding
metallic character
metalloid
Newlands
nitride
noble gas configuration
nonmetal
outer electron configuration
oxide
period (Periodic Table)
peroxide
shielding effect
superoxide
thermal conductivity
transition metal

CHAPTER 7

The heat flow that accompanies a chemical reaction or a phase change has both direction and magnitude. In an exothermic process, the system gives off heat to the surroundings; in an endothermic process, heat flows from the surroundings to the system. The magnitude of the heat flow can be measured in a calorimeter. In a coffee-cup calorimeter, all the heat is absorbed by the water (Example 7.2). In a bomb calorimeter, part of the heat is absorbed by the metal bomb (Example 7.3).

We can relate the heat flow for a reaction carried out in an open container at constant pressure to the difference in enthalpy (H) between products and reactants:

exothermic reaction   *endothermic reaction*

$\Delta H < 0$ $\Delta H > 0$

$H_{\text{products}} < H_{\text{reactants}}$ $H_{\text{products}} > H_{\text{reactants}}$

A thermochemical equation differs from an ordinary chemical equation in that it includes the value of ΔH for the reaction. The magnitude of ΔH depends upon the amounts of substances reacting or formed (Examples 7.4 and 7.5). The ΔH for a reaction in one direction is equal in magnitude but opposite in sign to ΔH for the reverse reaction (Example 7.6). Because ΔH is independent of path, enthalpy changes for successive steps in a process can be added to obtain ΔH for the overall reaction (Hess's Law; Example 7.7).

The enthalpy change for a reaction can be calculated by using molar heats of formation of compounds or of ions in aqueous solution (Tables 7.3 and 7.4). The relation used is

$$\Delta H = \Sigma \, \Delta H_f \text{ products } - \Sigma \, \Delta H_f \text{ reactants}$$

Here it is understood that the heat of formation of an element in its stable state is zero; ΔH_f of $H^+(aq)$ is also taken to be zero (Examples 7.8 through 7.10).

The First Law of Thermodynamics relates the change in energy for a process (ΔE) to the heat flow (q) and the work term (w). It tells us that $\Delta E = q + w$ (Example 7.11). Energy, E, like enthalpy, is a state property. For a chemical reaction, ΔE can be determined (Example 7.12) by measuring the heat flow in a bomb calorimeter. Ordinarily, ΔH and ΔE for a reaction are nearly equal (Example 7.13).

IMPORTANT EQUATIONS

Heat flow

$$q = C \times \Delta t = (\text{S.H.}) \times m \times \Delta t$$

Calorimetry

$$q_{\text{reaction}} = - \, q_{\text{calorimeter}}$$
$$= -(\text{S.H.}) \times m \times \Delta t$$
$$\text{(coffee cup)}$$
$$= -C \times \Delta t \text{ (bomb)}$$

Laws of thermochemistry

$$\Delta H_{\text{forward}} = - \, \Delta H_{\text{reverse}}$$
$$\Delta H = \Delta H_1 + \Delta H_2 + \, - \, - \, -$$

Heats of Formation

$$\Delta H^0 = \Sigma \, \Delta H_f^0 \text{ products } - \Sigma \, \Delta H_f^0 \text{ reactants}$$

First Law of Thermodynamics

$$\Delta E = q + w$$

Energy vs. enthalpy

$$\Delta H = \Delta E + \Delta(PV) = \Delta E + \Delta n_{\text{gas}} RT$$

KEY WORDS

bomb calorimeter
calorimetry
coffee cup calorimeter
endothermic
enthalpy (H)
exothermic
First Law of Thermodynamics
fission
fossil fuels
fusion
heat flow
heat of formation
heat of fusion
heat of vaporization
Hess's Law
internal energy (E)
joule

kilojoule
Law of Conservation of Energy
Laws of Thermochemistry
mole
nuclear energy
sign of heat flow
solar energy
specific heat
standard enthalpy change
state property
surroundings
system
thermochemistry
work

CHAPTER 8

In a molecule or polyatomic ion, nonmetal atoms are joined by covalent (electron pair) bonds. Two atoms may share one, two, or three electron pairs to form a single, double, or triple bond, respectively. Covalent bonds are nonpolar if the two atoms joined are the same; otherwise, they are polar (Example 8.5). Polar bonds are shorter and stronger than one would expect if they were nonpolar. Multiple bonds are shorter and stronger than single bonds. The strength of a bond is measured by the bond energy; ΔH for a gas-phase reaction can be estimated from tables of bond energies (Example 8.6).

In molecules or ions, atoms of main-group elements ordinarily have octet structures; H atoms are surrounded by only one shared electron pair. Lewis structures are drawn using the octet rule to locate shared and unshared pairs. These structures are derived for molecules or polyatomic ions using the rules given on p. 254. The application of these rules is illustrated in Examples 8.1 and 8.2. For certain species, the concept of resonance is invoked when a single Lewis structure is inadequate (Example 8.3).

Certain molecules do not "obey" the octet rule. Most such species fall into one or the other of two categories:

a. Molecules where the central atom is surrounded by fewer than four pairs of valence electrons. The principal examples here are the elements beryllium (2 electron pairs in BeF_2, − −) and boron (3 electron pairs in BF_3, − −).
b. Molecules where the central atom has an "expanded octet," i.e., more than four pairs of valence electrons. The number of electron pairs around

the central atom may be 5 (P in PF_5) or 6 (S in SF_6). The Lewis structure of one such molecule is derived in Example 8.4.

KEY WORDS

atomic radius	free radical	partial ionic character
bond distance	Hess's Law	polar bond
bond energy	Lewis structure	resonance form
central atom	multiple bond	single bond
covalent bond	nonpolar bond	skeleton structure
double bond	nuclear stability	triple bond
electronegativity	octet rule	unshared electron pairs
exceptions to octet rule	paramagnetic	valence electron

CHAPTER 9

The geometry (shape) of a polyatomic ion or simple molecule can be predicted from its Lewis structure by considering repulsions among electron pairs about the central atom (Examples 9.1 through 9.3). The geometry depends upon the number of bonds and unshared pairs about the central atom (Table 9.1 and Fig. 9.6). Insofar as geometry is concerned, a multiple bond behaves as if it were a single bond.

A polar molecule has a net dipole. The polarity of each bond *and* the molecular geometry must be considered to predict whether a molecule is polar or nonpolar (Example 9.4).

Valence bond theory considers a covalent bond to consist of a pair of electrons of opposed spin occupying an atomic orbital. In most cases, the atomic orbitals used in bonding are hybridized. This way, atoms make maximum use of their valence electrons to form two bonds (Be), three bonds (B), four bonds (C), five bonds (P), or six bonds (S). The corresponding hybridizations are sp, sp^2, sp^3, sp^3d, and sp^3d^2, respectively (Example 9.5). In a multiple bond, the extra electron pairs are not in hybridized orbitals; instead they form pi bonds (Examples 9.6 and 9.7).

In the molecular orbital model, atomic orbitals are combined to form an equal number of molecular orbitals. A given MO may be bonding or antibonding (Figs. 9.14 and 9.15). The relative energies of different molecular orbitals govern the order in which they are filled (Table 9.4, Example 9.8).

KEY WORDS

absorption spectroscopy	bent molecule	bonding orbital
antibonding orbital	bond angle	central atom
atomic orbital	bond distance	delocalized orbital
atomic orbital model	bond order	dipole
axial position	bond polarity	dipole moment

electron diffraction

electron pair

electron pair repulsion

equatorial position

equilateral triangle

expanded octet

hybrid orbitals

hybridization

Lewis structure

linear molecule

molecular geometry

molecular orbital

multiple bond

nonpolar molecule

octahedron

pi bond

polar bond

polar molecule

pyramidal molecule

sigma bond

square pyramid

terminal atom

tetrahedron

triangular bipyramid

unshared electron pairs

valence bond model

valence electron

VSEPR model

CHAPTER 10

One way to describe the equilibrium between a liquid and its vapor is to cite the vapor pressure at a particular temperature. For an equilibrium system, vapor pressure is independent of container volume (Fig. 10.1 and Example 10.1) but increases exponentially with temperature (Fig. 10.2). There is a linear relation between the logarithm of the vapor pressure and the reciprocal of the absolute temperature (Example 10.2). The vapor pressure reaches 1 atm at the normal boiling point. It has its maximum value, called the critical pressure, at the critical temperature. Above that temperature, the liquid cannot exist.

The temperature/pressure conditions for phase changes are summarized in a phase diagram (Fig. 10.5 and Example 10.3). A solid sublimes below the triple point if the vapor pressure is reduced below the equilibrium value. In solid-liquid equilibria, a pressure increase results in formation of the more dense phase (Fig. 10.7).

A substance can be classified into one of four types (molecular, network covalent, ionic, or metallic), depending upon its basic structural units (Fig. 10.8). Properties depend upon structure (Example 10.7 and Table 10.6). Molecular substances are held together by relatively weak intermolecular forces whose magnitude determines their boiling point (Example 10.4). These forces include hydrogen bonds (Example 10.5), dipole forces, and dispersion forces (Example 10.6).

Crystals have a repeating pattern called a unit cell. Three types of unit cells are simple cubic, face-centered cubic, and body-centered cubic (Fig. 10.17, Examples 10.8 and 10.9). Ions in ionic solids also are arranged in repeating patterns of this type (Fig. 10.19).

IMPORTANT EQUATIONS

Vapor pressure vs. T

$$\ln \frac{P_2}{P_1} = \frac{\Delta H_{vap}}{R} \left(\frac{T_2 - T_1}{T_2 T_1} \right)$$

Unit cells

simple cubic: 1 atom/cell; $2r = s$

FCC: 4 atoms/cell; $4r = s\sqrt{2}$

BCC: 2 atoms/cell; $4r = s\sqrt{3}$

APPENDIX 5   CHAPTER SUMMARIES

KEY WORDS

allotrope

body-centered cubic cell (BCC)

boiling point

Clausius-Clapeyron equation

condensation

Coulomb's Law

covalent bond

critical pressure

critical temperature

crystal structure

dense phase

dipole force

dispersion force

dynamic equilibrium

evaporation

face-centered cubic cell (FCC)

fluid

freezing point

heat of vaporization

hydrogen bond

intermolecular force

intramolecular force

ionic bond

ionic compound

ionic crystal

ionic substance

lattice energy

liquid-vapor equilibrium

melting point

metal

molecular substance

natural logarithm

network covalent substance

nonconductor

nonpolar molecule

normal boiling point

phase diagram

radius-edge relation

simple cubic cell

sublimation

triple point

unit cell

vapor pressure

volatile

CHAPTER 11

A solution consists of solute(s) and solvent. In this chapter we have dealt mainly with water solutions. Solutes can be classified as electrolytes or non-electrolytes, depending upon their dissociation behavior in water (Table 11.2 and Example 11.1). Nonelectrolytes tend to be insoluble in water unless they are capable of forming hydrogen bonds.

Solubilities are affected by temperature. If the solution process is en-dothermic, as is most often the case with solids, the solute becomes more soluble as temperature increases. If the solution process is exothermic, as with gases dissolving in water, solubility decreases with increasing temper-ature. The solubility of gaseous solutes is increased by increasing the pres-sure of the gas over the solution (Example 11.9).

Solutes lower the vapor pressure of a solvent, decrease the freezing point, and increase the boiling point (Fig. 11.15 and Examples 11.10 and 11.11). These properties, along with the osmotic pressure (Figs. 11.11–11.13), are colligative. The extent to which vapor pressure is lowered depends primarily upon the concentration rather than the type of solute particle. The same is true of freezing point depression and boiling point elevation. Colligative properties, particularly freezing point lowering and osmotic pressure, can be used to determine the molar mass of a nonelectrolyte (Examples 11.12 and 11.13). Since electrolytes dissociate in water, they have a greater effect on colligative properties than do nonelectrolytes (Equation 11.17, and Example 11.14).

In this chapter, we have referred to four different concentration units: mass percent, mole fraction, molality, and molarity. To compare these units, consider a solution containing 15.0 g of sugar ($MM = 342$ g/mol) dissolved in 110 g of water. This solution is found to have a density of 1.05 g/cm³ and hence a volume of

$$125 \text{ g} \times \frac{1 \text{ cm}^3}{1.05 \text{ g}} = 119 \text{ cm}^3$$

Thus, we have

$$\text{molarity} = \frac{\text{no. moles solute}}{\text{no. liters solution}} = \frac{15.0/342 \text{ mol}}{0.119 \text{ L}} = 0.369 \text{ mol/L}$$

$$\text{molality} = \frac{\text{no. moles solute}}{\text{no. kilograms solvent}} = \frac{15.0/342 \text{ mol}}{0.110 \text{ kg H}_2\text{O}} = 0.399 \text{ mol/kg H}_2\text{O}$$

$$X_{\text{sugar}} = \frac{\text{no. moles sugar}}{\text{no. moles sugar} + \text{no. moles water}} = \frac{15.0/342}{15.0/342 + 110/18.0}$$

$$= 0.00713$$

$$X_{\text{water}} = 1 - 0.00713 = 0.99287$$

$$\text{mass percent sugar} = \frac{\text{mass sugar}}{\text{total mass solution}} \times 100$$

$$= \frac{15.0 \text{ g}}{15.0 \text{ g} \times 110 \text{ g}} \times 100 = 12.0\%$$

The use of these concentration units is further illustrated in Examples 11.2 and 11.3. Ways to prepare a solution to a desired molarity are considered in Example 11.4, where we start with a pure solid, and Example 11.5, where we start with a concentrated solution. The molarity of a solute is readily converted to molarities of the corresponding ions (Example 11.6). Conversions between different concentration units are considered in Examples 11.7 and 11.8; a general approach is shown in Figure 11.7.

IMPORTANT EQUATIONS

Henry's Law $\qquad C_g = k\, P_g$

Raoult's Law $\qquad P_1 = X_1 P_1^0$

Osmotic pressure $\qquad \pi = MRT$

Boiling point elevation $\qquad \Delta T_b = 0.52°\text{C} \times m$ (nonelectrolytes in water); multiply by i for electrolytes

Freezing point depression $\qquad \Delta T_f = 1.86°\text{C} \times m$ (nonelectrolytes in water); multiply by i for electrolytes

APPENDIX 5 CHAPTER SUMMARIES

KEY WORDS

boiling point elevation

colligative property

concentrated solution

density

dialysis

dilute solution

electrolyte

freezing point depression

Henry's Law

hydrogen bond

mass percent

molal boiling point constant

molal freezing point constant

molality

molarity

mole fraction

nonelectrolyte

osmosis

osmotic pressure

parts per billion (ppb)

parts per million (ppm)

Raoult's Law

reverse osmosis

saturated solution

semipermeable membrane

solubility

solute

solution

solvent

supersaturated solution

unsaturated solution

vapor pressure lowering

CHAPTER 12

In this chapter, we considered three different kinds of reactions taking place in aqueous solution.

1. *Precipitation,* in which a cation from one solution combines with an anion from a second solution to form an insoluble solid. Net ionic equations for such reactions are readily written (Examples 12.1 and 12.2), knowing the solubility rules (Table 12.1).

2. *Acid-base,* where the equation written (Example 12.3) depends upon whether the acid and base are strong or weak (Table 12.2). To decide upon the nature of the equation, it is essential to recognize the principal species present in solution (Table 12.3).

3. *Oxidation-reduction,* which involves an exchange of electrons from a reducing agent to an oxidizing agent. One species is oxidized (increased oxidation number), while the other is reduced (decreased oxidation number). Oxidation number is calculated using a set of arbitrary rules (Example 12.4). Redox equations can be balanced by a systematic approach illustrated within the text and in Example 12.5.

The mass-mole relations involved in solution reactions are basically no different from those discussed in Chapter 3 in connection with reactions between pure substances. The use of these relations is shown in Examples 12.6 through 12.8; see also Figures 12.6 and 12.7. The principles developed are readily applied to determine the concentration of a species by volumetric analysis (Examples 12.9 and 12.10).

KEY WORDS

acid	net ionic equation	solubility rules
amine	oxidation	spectator ions
base	oxidation number	strong acid
disproportionation	oxidizing agent	strong base
electronegative	precipitation	titrant
end point	quantitative analysis	titration
equivalence point	reacting species	volumetric analysis
half-equation	reducing agent	weak acid
indicator	reduction	weak base
limiting reactant		

CHAPTER 13

When a reaction system is confined, it reaches a dynamic equilibrium where forward and reverse reactions occur at the same rate. From that point on, there is no net change in the concentrations of products or reactants. The equilibrium system can be described in terms of a ratio known as the equilibrium constant, K_c. The expression for K_c follows directly from the chemical equation (Examples 13.1 and 13.2). It is important to understand that the expression for K_c and hence its numerical value depend upon the equation written to represent the equilibrium system. This is reflected in the Coefficient Rule (Equation 13.3), the Reciprocal Rule (Equation 13.4), and the Rule of Multiple Equilibria (Equation 13.5).

To decide the direction in which a reaction system will move to establish equilibrium, we compare the original concentration quotient to the equilibrium constant, K_c. If the original concentration quotient, Q, is less than K_c, reaction occurs in the forward direction (Example 13.5). If this quotient is greater than K_c, the reverse reaction must occur to establish equilibrium.

Several calculations involving equilibrium systems were considered in this chapter. Generally, they are of three types, depending upon what is known and what is to be calculated.

KNOWN	TO BE CALCULATED	EXAMPLE
1. Equilibrium concentrations of all species	K_c	13.3
2. Initial concentrations of all species, equilibrium concentration of one species	K_c	13.4
3. K_c and all but one equilibrium concentration	equilibrium concentration of a species	13.6
4. K_c and initial concentrations	equilibrium concentrations of all species	13.7, 13.8, 13.9

If a system at equilibrium is disturbed in some way, either the forward or the reverse reaction will occur to restore equilibrium. We can use Le

Châtelier's Principle to predict which way the equilibrium will shift under stress (Examples 13.10 and 13.11). In general, we find that

— a decrease in volume (compression) causes the reaction to occur that decreases the number of moles of gas (Tables 13.3 and 13.4).

— an increase in temperature causes the reaction that is endothermic to occur; in this case, the magnitude of K_c changes (Table 13.5).

The equilibrium constant K_c is related in a simple way to the equilibrium constant K_p (Equation 13.13 and Example 13.13). The expression for K_p involves partial pressures rather than molar concentrations (Example 13.12). Calculations involving K_p (Example 13.14) are very similar to those carried out with K_c.

IMPORTANT EQUATIONS

Coefficient rule	$K_c' = (K_c)^n$
Reciprocal rule	$K_c'' = 1/K_c$
Multiple equilibria	$K = K_1 \times K_2 \times --$
K_p vs. K_c	$K_p = K_c \times (RT)^{\Delta n}$ gas

KEY WORDS

Coefficient Rule
direction of a reaction
equilibrium concentration
equilibrium constant (K_c)
equilibrium constant (K_p)
equilibrium expression
extent of a reaction
heterogeneous equilibria
homogeneous equilibria

Le Chatelier's Principle
original concentration quotient (Q)
partial pressure
position of equilibrium
quadratic formula
Reciprocal Rule
Rule of Multiple Equilibria
square bracket

CHAPTER 14

The acidity or basicity of a water solution can be expressed in terms of [H$^+$], [OH$^-$], pH, or pOH. These quantities are related by the equations:

$$[H^+] \times [OH^-] = K_w = 1.0 \times 10^{-14} \text{ at } 25°C \quad \text{(Example 14.1)}$$

$$pH = -\log_{10}[H^+]; \ pOH = -\log_{10}[OH^-] \quad \text{(Examples 14.2 and 14.3)}$$

acidic solution: [H$^+$] > [OH$^-$]; [H$^+$] > $10^{-7}\,M$ \quad pH < 7

basic solution: [OH$^-$] > [H$^+$]; [OH$^-$] > $10^{-7}\,M$ \quad pOH < 7

neutral solution: [H$^+$] = [OH$^-$] = $10^{-7}\,M$ \quad pH = pOH = 7

Weak acids include molecules (HClO, $HC_2H_3O_2$), cations (NH_4^+, Zn^{2+}), and a few anions, notably HSO_4^-. Chemical equations for the dissociation of these species are considered in Example 14.4. The dissociation constant for a weak acid, K_a:

$$HB(aq) \rightleftharpoons H^+(aq) + B^-(aq); K_a = \frac{[H^+] \times [B^-]}{[HB]}$$

can be determined if the pH of a weak acid solution is known (Example 14.5). Conversely, knowing K_a, it is possible to calculate the pH of a solution of a weak acid (Examples 14.6 and 14.7; for a general approach, see Figure 14.6). With a polyprotic acid such as H_3PO_4, virtually all of the H^+ ions come from the first dissociation (Example 14.8).

Weak bases are either molecules (NH_3, CH_3NH_2) or anions (F^-, CO_3^{2-}); there are no basic cations (Example 14.9). K_b, the dissociation constant of a weak base:

$$B^-(aq) + H_2O \rightleftharpoons HB(aq) + OH^-(aq); K_b = \frac{[HB] \times [OH^-]}{[B^-]}$$

can be calculated using the relation:

$$K_b \times K_a = K_w$$

where K_a represents the dissociation constant of the conjugate weak acid (Example 14.11). It can be used to obtain $[OH^-]$ in a solution of a weak base (Example 14.10); the calculation is very similar to those involving K_a.

An aqueous solution of a salt can be either neutral, acidic, or basic (Table 14.7 and Example 14.12). A general approach to predicting the pH of salt solutions is shown in Figure 14.8.

The model of acid-base reactions upon which most of the discussion in this chapter is based is that of Arrhenius. Other more general models have been proposed (Table 14.9). One of the most useful of these is the Brönsted-Lowry model (Example 14.13).

IMPORTANT EQUATIONS

Dissociation of water $K_w = [H^+] \times [OH^-] = 1.0 \times 10^{-14}$

K_b vs. K_a $K_b \times K_a = K_w$

KEY WORDS

acid
acid dissociation constant (K_a)
acidic ions
acidic solution
Arrhenius acid, base
base
base dissociation constant (K_b)
basic ions

basic solution
Brönsted-Lowry acid, base
conjugate acid
conjugate base
diprotic acid
dissociation constant of water (K_w)
electron pair acceptor
electron pair donor

equilibrium concentrations

hydronium ion

indicator

ionizable hydrogen atom

Lewis acid, base

method of successive approximations

monoprotic acid

neutral ions

neutral solution

percent dissociation

polyprotic acid

proton

proton acceptor

proton donor

pH

pH meter

pK_a

pOH

quadratic equation

Rule of Multiple Equilibria

salt

strong acid

strong base

weak acid

weak base

CHAPTER 15

A buffer consists of a weak acid and its conjugate base, both present in solution at roughly equal concentrations. For a buffer (Example 15.1):

$$[H^+] = K_a \times \frac{[HB]}{[B^-]} = K_a \times \frac{\text{no. moles HB}}{\text{no. moles B}^-}$$

To choose a buffer system, we pick a weak acid (Example 15.2) for which:

$$pK_a \approx pH \text{ desired}$$

Addition of a strong acid to a buffer increases the ratio $[HB]/[B^-]$ and hence $[H^+]$; the pH drops at least slightly. Addition of base has the reverse effect; the ratio $[HB]/[B^-]$ decreases, lowering $[H^+]$ and increasing the pH (Example 15.3).

The way in which pH changes during an acid-base titration depends upon the nature of the acid and base. If both are strong, the pH is 7 at the equivalence point and changes very rapidly near that point (Example 15.4 and Fig. 15.8). Titration of a weak acid with a strong base produces a smaller pH change near the equivalence point, which occurs at a pH above 7 (Example 15.5 and Fig. 15.9). Conversely, titration of a weak base with a strong acid gives a pH below 7 at the equivalence point (Example 15.6 and Figure 15.10). Table 15.3 summarizes these relations; it can be used to calculate the pH of any mixture prepared from an acid and a base (Example 15.7).

The relation between the concentrations of ions in equilibrium with a slightly soluble solid is given by the solubility product expression, K_{sp} (Example 15.8):

$$M_xN_y(s) \rightleftharpoons xM^{y+}(aq) + yN^{x-}(aq); \qquad K_{sp} = [M^{y+}]^x \times [N^{x-}]^y$$

This equilibrium constant can be used to determine

— the concentration of one ion in equilibrium with a known concentration of the other ion (Example 15.9).

— whether or not a precipitate will form when two solutions are mixed (Examples 15.10 and 15.11).

— the solubility of an ionic solid in pure water (Examples 15.12, 15.13) or in a solution containing a common ion (Example 15.14).

KEY WORDS

acid dissociation constant (K_a)

acid-base indicator

acid-base titration

base dissociation constant (K_b)

buffer

buffer capacity

common ion effect

conjugate acid

conjugate base

equivalence point

five percent rule

Henderson-Hasselbalch equation

ion product (Q)

Le Chatelier's Principle

net ionic equation

neutralization

precipitation reaction

pH

pK_a

solubility product constant (K_{sp})

titration curve

water solubility

CHAPTER 16

A complex ion consists of a central metal cation surrounded by ligands that may be molecules or anions. A ligand may form one or more bonds with the metal ion; if it forms more than one bond, it is called a chelating agent. The charge of the complex is the sum of the charges of the metal ion and the ligands (Example 16.1).

The coordination number of the metal ion is the number of bonds that it forms (Example 16.2). Coordination numbers of 6, 4, and 2, in that order, are most common (Table 16.2). Complexes with a coordination number of 2 are linear. Those with a coordination number of 4 may be square planar or tetrahedral. A coordination number of 6 gives an octahedral complex (Fig. 16.5). Square planar and octahedral complexes can show *cis-trans* isomerism (Example 16.4).

Complex ions and coordination compounds are named in a systematic way (Example 16.3). The name identifies the number and type of ligands present as well as the nature and oxidation number of the central metal. It also indicates whether the complex ion is a cation or anion.

Two different models of bonding in complex ions were presented in this chapter. Both start with the electron configuration of the central cation (Table 16.3).

In the valence bond model, electron pairs from the ligands are fed into hybrid orbitals of the metal ion (sp, sp^3, dsp^2, sp^{3}d^2, or d^{2}sp^3). With this model, it is possible to derive orbital diagrams for complex ions (Example 16.5). In the crystal-field model, the only effect of the ligands is to change the relative energies of the d orbitals of the metal ion. In an octahedral complex, two of these orbitals are raised in energy as compared to the other three. This leads to the possibility of "high-spin" and low-spin complexes, differing in the number of unpaired electrons (Example 16.6).

The formation constant of a complex ion is a measure of its stability; the larger the value of K_f, the more stable the complex. Knowing K_f, we can readily calculate the ratio of free to complexed cation (Example 16.7).

KEY WORDS

bidentate ligand

central atom

chelating agent

complex ion

coordinate covalent bond

coordination compound

coordination number

crystal field model

crystal-field splitting
    energy

diamagnetic

formation constant
    (K_f)

geometric (*cis, trans*)
    isomerism

high spin complex

hybridization

Lewis acid, base

ligand

linear

low spin complex

monodentate ligand

octahedral

orbital diagram

paramagentic

polydentate ligand

spectrochemical series

square planar

tetrahedral

valence bond model

CHAPTER 17

The qualitative analysis of cations is carried out in the chemistry laboratory by first separating them into groups based on selective precipitation (Table 17.1 and Examples 17.1 and 17.2). The Group I ions precipitate as chlorides ($AgCl$, $PbCl_2$, Hg_2Cl_2). These solids can be separated by taking advantage of the fact that $AgCl$ dissolves in NH_3, whereas $PbCl_2$ dissolves in hot water (Example 17.3).

Anions are commonly identified by spot tests which take advantage of differences in chemical properties (Example 17.4). The test may involve the formation of a gas, a precipitate, or a colored species:

$$CO_3^{2-}(aq) + 2H^+(aq) \rightarrow CO_2(g) + H_2O$$

$$SO_4^{2-}(aq) + Ba^{2+}(aq) \rightarrow BaSO_4(s)$$

$$SCN^-(aq) + Fe^{3+}(aq) \rightarrow FeSCN^{2+}(aq); \text{ red}$$

Many different kinds of reactions are involved in qualitative analysis. Most of these were discussed in Chapters 15 and 16. They include the

— formation of complex ions containing NH_3 or OH^- as ligands (Table 17.2 and Example 17.5).

— decomposition of complex ions; those containing NH_3 or OH^- are unstable in strong acid (Example 17.6).

— formation of precipitates. Cations can be precipitated by adding anions directly (Example 17.7) or by adding a molecule, such as H_2S or NH_3, that dissociates to form the appropriate anion (Example 17.8).

— dissolving of precipitates. Solids containing a basic anion can be dissolved by adding a strong acid (Example 17.9). Ammonia and sodium hydroxide often bring solids into solution by forming a complex with the cation (Example 17.10).

Equilibrium constants for solution reactions are useful in determining whether or not a precipitate will form at a particular stage of analysis (Examples 17.11 and 17.12) or whether a precipitate will dissolve under a particular set of conditions (Examples 17.13 and 17.14). Often, to calculate the required equilibrium constant, it is necessary to use the Reciprocal Rule, or Rule of Multiple Equilibria, discussed in Chapter 13.

KEY WORDS

cation	net ionic equation
flow sheet	qualitative analysis
groups 1, 2, 3, 4	Reciprocal Rule
K_a	Rule of Multiple Equilibria
K_f	selective precipitation
K_{sp}	solubility equilibria

CHAPTER 18

The rate of a reaction indicates how the concentration of a reactant or product changes with time (Examples 18.1, 18.2). A rate expression gives the relation between concentration of reactant and rate of reaction. The rate expression includes a rate constant, k, and concentration terms. The exponent of a concentration term indicates the order of the reaction with respect to that species. Reaction order can be determined by observing how rate varies as the initial concentration of a species changes (Examples 18.3 and 18.5). Once the rate expression has been established, data relating rate to concentration can be used to calculate the rate constant k (Example 18.4).

As a reaction proceeds, the reactant concentration decreases with time. The relationships between concentration and time are summarized for zero-, first-, and second-order reactions in Table 18.3 and Figure 18.4. These relationships can be used to determine a concentration at a given time (Example 18.6) or to find the order of a reaction (Example 18.7).

A certain minimum energy, the activation energy, is necessary for a reaction to occur when two molecules collide (Fig. 18.7). In general, reaction rate is inversely related to activation energy; a fast reaction implies a low activation energy. A catalyst increases reaction rate by lowering the activation energy.

As temperature increases, a greater fraction of molecules have energies that exceed the activation energy (Fig. 18.8). Thus, we expect an increase in temperature to increase the rate of reaction (Table 18.4). Equation 18.23 relates temperatures, rate constants, and activation energy (Example 18.8).

A reaction mechanism shows the individual steps by which reaction occurs. Using certain rules, a rate expression may be derived from the mechanism (Example 18.9).

APPENDIX 5   CHAPTER SUMMARIES

IMPORTANT EQUATIONS

Zero-order reaction	$X = X_0 - kt$; $t_{1/2} = X_0/2k$
First-order reaction	$\ln X_0/X = kt$; $t_{1/2} = 0.693/k$
Second-order reaction	$1/X - 1/X_0 = kt$; $t_{1/2} = 1/kX_0$
Rate constant vs. T	$\ln \dfrac{k_2}{k_1} = \dfrac{\Delta E_a}{R} \left(\dfrac{T_2 - T_1}{T_2 T_1} \right)$

KEY WORDS

activated complex
activation energy
activation energy diagram
Arrhenius equation
average rate
catalyst
chain reaction
chemical kinetics

first-order reaction
half-life
heterogeneous catalysis
homogeneous catalysis
initial rate
instantaneous rate
order of a reaction
overall order

rate constant, k
rate expression
rate-determining step
reaction mechanism
reaction rate
second-order
  reaction
zero-order reaction

CHAPTER 19

This chapter describes selected topics in atmospheric chemistry, including

— the liquefaction of air and separation of its components
— the fixation of nitrogen to form ammonia (Haber process) and the conversion of ammonia to nitric acid (Ostwald process).
— the effect of water and carbon dioxide upon weather and climate (relative humidity (Example 19.3), cloud formation, greenhouse effect).
— photochemical reactions taking place in the upper atmosphere (Table 19.3 and Example 19.5).
— the sources, effects, and treatment of air pollution by sulfur oxides, sulfuric acid, carbon monoxide, and nitrogen oxides.

Several concepts from Chapters 13 and 18 were reviewed in this chapter. These include

— equilibrium calculations (Example 19.8).
— effect of changes in conditions upon the position of an equilibrium and the rate at which it is reached (Example 19.2).
— reaction mechanisms (Example 19.6).
— rate constant and its temperature dependence (Examples 19.7, 19.9).
— determination of reaction order (Example 19.10).

KEY WORDS

acid rain	infrared radiation	partial pressure
air pollution	liquifaction of air	photochemical smog
ammonia	mean free path	relative humidity
catalytic converter	nitric acid	roasting
cloud formation	nitrogen fixation	scrubber
cloud seeding	nuclear winter	stratosphere
dynamite	Ostwald process	troposphere
greenhouse effect	ozone	ultraviolet radiation
Haber process	ozone hole	

CHAPTER 20

Two new state functions, entropy (S) and free energy (G), were introduced in this chapter. We also made use of the enthalpy function (H), discussed in Chapter 7. Our interest is in how these quantities change in a reaction.

CHANGE	NATURE OF REACTION	EXAMPLES
$\Delta H = \Sigma \, \Delta H_{f \, products} - \Sigma \, \Delta H_{f \, reactants}$	$\Delta H < 0$; exothermic $\Delta H > 0$; endothermic	20.5–20.8
$\Delta S = \Sigma \, S_{products} - \Sigma \, S_{reactants}$	$\Delta S > 0$; randomness increases $\Delta S < 0$; order increases	20.1–20.3, 20.5–20.8
$\Delta G = \Delta H - T\Delta S$	$\Delta G < 0$; spontaneous $\Delta G > 0$; nonspontaneous $\Delta G = 0$; equilibrium	20.5–20.8

Both ΔS and ΔG depend upon the pressures or concentrations of species taking part in the reaction. We dealt only with the standard entropy change, ΔS^0, and the standard free energy change, ΔG^0 (1 atm, 1 M). Both ΔH^0 and ΔS^0 are taken to be independent of temperature; ΔG^0 is a linear function of T (Fig. 20.8). The fact that ΔG^0 becomes zero at equilibrium allows us to calculate the temperature at which a reaction is at equilibrium at standard pressures and concentrations (Example 20.9).

The Second Law of Thermodynamics deals with the spontaneity of physical and chemical changes. It tells us that a spontaneous change is one for which

$\Delta S_{sys} + \Delta S_{surr} > 0$ (Example 20.4)

$\Delta G = \Delta H - T\Delta S_{sys} < 0$ (constant P, T)

Free energy changes, like enthalpy changes, are additive (Example 20.10); in effect, Hess's Law applies to ΔG as well as to ΔH. The free energy change is related to the equilibrium constant by the equation: $\Delta G^0 = -RT \ln K$ (Examples 20.11, 20.12). For gas phase reactions, the proper equilibrium constant here is K_p. Finally, the free energy change is a measure of the maximum amount of work that can be obtained from a reaction.

APPENDIX 5 CHAPTER SUMMARIES

IMPORTANT EQUATIONS

Entropy change	$\Delta S^0 = \Sigma\, S^0_{products} - \Sigma\, S^0_{reactants}$
Second Law of Thermodynamics	$\Delta S_{sys} + \Delta S_{surr} > 0$
Gibbs-Helmholtz equation	$\Delta G^0 = \Delta H^0 - T\Delta S^0$
ΔG^0 vs. K	$\Delta G^0 = -RT \ln K$

KEY WORDS

coupled reactions
endothermic
enthalpy (H)
enthalpy change
entropy (S)
entropy change
equilibrium constant
exothermic
free energy
free energy change (G)
Gibbs-Helmholtz equation
heat of vaporization
high-probability state
low-probability state
nonspontaneous
order

osmosis
randomness
Second Law of Thermodynamics
spontaneous process
standard concentration
standard free energy change
standard free energy of formation
standard molar entropy
standard pressure
state property
surroundings
system
Third Law of Thermodynamics
universe
work

CHAPTER 21

There are two kinds of electrochemical cells:

1. A voltaic cell, in which a spontaneous redox reaction generates electrical energy. Such cells are often of the salt-bridge type (Figs. 21.2–21.4; Example 21.2). There are a variety of commercial voltaic cells, both primary and secondary (Figs. 21.10–21.13).

2. An electrolytic cell, in which electrical energy is used to bring about a nonspontaneous redox reaction (Fig. 21.14). The cell reaction may involve cations, anions, or water molecules (Example 21.11). The amount of product is directly proportional to the number of coulombs passed through the cell (Example 21.12). The energy in joules required for electrolysis can be found by multiplying the voltage by the number of coulombs (Example 21.13).

In both types of cells, oxidation occurs at the anode, reduction at the cathode. Anions move to the anode, cations to the cathode. The two electrodes used must be electrical conductors; platinum and graphite are often used as inert electrodes.

Voltaic cells can be used to measure standard voltages, E^0_{tot}. By arbitrarily setting the standard reduction voltage (E^0_{red}) of H^+ equal to zero, it is possible to obtain standard potentials for the reduction (E^0_{red}) of various species (Table 21.1). Standard voltages for oxidation (E^0_{ox}) can be obtained by changing the sign of the E^0_{red} value for the reverse half-reaction. Standard voltages for half-reactions can be used to do the following:

1. Compare the strengths of oxidizing agents and reducing agents (Example 21.3). A strong oxidizing agent has a large positive E^0_{red} value; a strong reducing agent has a large positive E^0_{ox} value.

2. Calculate the standard cell voltage (Example 21.4): $E^0_{tot} = E^0_{ox} + E^0_{red}$.

3. Determine whether or not a redox reaction is spontaneous (Example 21.5). The reaction is spontaneous if $E^0_{tot} > 0$.

Knowing the standard voltage corresponding to a given redox reaction, we can calculate either the standard free energy change (Example 21.6) or the equilibrium constant (Example 21.7). The appropriate relations are

$$\Delta G^0 \text{ (in kJ)} = -96.5 \, n \, E^0_{tot} \qquad \ln K = n \, E^0_{tot}/0.0257$$

The voltage of a cell depends upon the concentrations of reactants and products. The effect can be calculated from the Nernst equation (Example 21.8). The Nernst equation can also be used to obtain the concentration of a species in solution, if the cell voltage is known (Examples 21.9 and 21.10).

IMPORTANT EQUATIONS

ΔG^0 vs. E^0_{tot} $\Delta G^0 = -nF \, E^0_{tot}$ ($F = 96.485$ kJ/V)

E^0_{tot} vs. K $\ln K = nE^0_{tot}/0.0257$

Nernst equation
$$E = E^0_{tot} - \frac{0.0591}{n} \log_{10}Q$$
$$= E^0_{tot} - \frac{0.0257}{n} \ln Q$$

KEY WORDS

ampere	half-cell	reduction
anion	joule	salt bridge
anode	kilowatt-hour	secondary cell
cathode	lead storage battery	spontaneity
cation	LeClanché cell	standard cell conditions
cell voltage	mercury cell	standard cell voltage
coulomb	Nernst equation	standard free energy change
disproportionation	Nicad battery	standard oxidation voltage
dry cell	overvoltage	standard potential
electrolysis	oxidation	standard reduction voltage
electrolytic cell	oxidizing agent	storage cell
equilibrium constant	primary cell	volt
Faraday constant	redox reaction	voltaic cell
fuel cell	reducing agent	work

CHAPTER 22

In this chapter, we have considered the chemistry of the more common transition metals and the cations and oxyanions derived from them. Most of these reactions were of the oxidation-reduction type. For example, we discussed

— reduction of metal cations in sulfide and oxide ores (Section 22.1).

— reaction with nonmetals (Section 22.2).

— oxidation of metals by acids (Section 22.2).

— redox equilibria between two different cations of the same metal (Section 22.3).

— reduction of oxyanions of chromium and manganese (Section 22.4).

We have also looked briefly at other types of reactions involving transition metal cations, in particular precipitation (Example 22.5).

Many of the examples worked in this chapter involved redox reactions of the type discussed in Chapters 12 and 21. Principles reviewed include

— balancing redox equations (Examples 22.4, 22.9).

— quantitative relationships in electrolysis (Example 22.3).

— calculation and interpretation of E^0_{tot} values (Examples 22.6, 22.7).

— use of the Nernst equation (Example 22.8).

KEY WORDS

alloy	electrolysis	pig iron
amphoteric	ferromagnetism	roasting
aqua regia	flotation	slag
basic oxygen process	limestone	sulfide ore
blister copper	net ionic equation	thermal
coke	nonstoichiometric compound	conductivity
disproportionation	oxide ore	thermite process
electrical conductivity	oxyanion	transition metal

CHAPTER 23

The halogens are all relatively strong oxidizing agents; their oxidizing ability decreases in the order $F_2 > Cl_2 > Br_2 > I_2$. Of these, the most commonly used is Cl_2, which reacts with water to form hypochlorous acid (Example 23.1). Commercially, chlorine is made by electrolysis of a water solution of NaCl (Example 23.2); it is used to prepare Br_2 and I_2 by oxidation of the corresponding anions.

Of the hydrogen halides, all except HF are strong acids. They react readily with strong and weak bases, both in solution (Example 23.3) and in the solid state (Example 23.4). The hydrogen halides can be prepared by direct reaction between the elements (Example 23.5) or by heating a sodium halide with H_3PO_4 or H_2SO_4 (Example 23.6).

The halide ions (F^-, Cl^-, Br^-, I^-) can act as Lewis bases; only the F^- ion is an Arrhenius base. Most ionic halides are soluble in water, but there are important exceptions (Table 23.4, Example 23.7). Ease of oxidation increases in the order $F^- < Cl^- < Br^- < I^-$ (Example 23.8).

In the oxyacids of the halogens (e.g., $HClO_4$, $HClO_3$, $HClO_2$, $HClO$), the hydrogen atom is bonded to oxygen. Acid strength decreases as oxidation number decreases and as the atomic number of the halogen increases (Example 23.9). Oxyanions of the halogens commonly take part in redox reactions. Generally speaking, they are stronger oxidizing than reducing agents (species in the +7 state cannot act as reducing agents). Their oxidizing strength is greatest in acidic solution (Example 23.10). Certain oxyanions and oxyacids spontaneously disproportionate (Example 23.11) and hence are unstable. All of these species are named systematically, as described in Section 23.4.

KEY WORDS

disproportionation	Nernst equation	principal species
electronegativity	oxidation number	reducing agent
halogen molecule	oxidizing agent	standard cell voltage
hydrogen halide	oxyacid	standard oxidation voltage
hydrohalic acid	oxyanion	standard reduction voltage
Lewis acid, base		

CHAPTER 24

Of the four elements considered in this chapter, all except nitrogen show allotropy. Sulfur has allotropes in all three physical states; phosphorus has several solid allotropes (P_4, P_x; Example 24.1), while oxygen is allotropic in the gas phase (O_2, O_3). Nitrogen and oxygen are obtained commercially from liquid air. Sulfur is obtained largely from underground deposits of the element, using the Frasch process (Figure 24.5); phosphorus is made from calcium phosphate (Equation 24.4 and Example 24.2).

Two of the more important hydrides of these elements are ammonia, NH_3, and hydrogen sulfide, H_2S. Ammonia can act as a weak base, a complexing agent, a precipitating agent to form metal hydroxides, and a reducing agent. Hydrogen sulfide is a weak acid, a precipitating agent for sulfides, and a reducing agent. Hydrogen peroxide, H_2O_2, is a bent molecule (Example 24.3) which can act as either an oxidizing agent or a reducing agent (Example 24.4).

Nitrogen, phosphorus, and sulfur form several different oxides, many of which react with water to give an acidic solution. The structures of the oxides of these elements are shown in Figures 24.8 and 24.9; see also Example 24.6.

The oxyacids of these elements include HNO_3 and H_2SO_4, both of which are strong, and the weak acids HNO_2, H_2SO_3 (Example 24.9), and H_3PO_4 (Example 24.10). Nitric acid is a strong oxidizing agent (Example 24.7); sulfuric acid is a considerably weaker oxidizing agent. Nitrous acid and sulfurous acid can act as either oxidizing or reducing agents. Commercially, nitric acid is made from ammonia (Ch. 19). Sulfuric acid is made from sulfur by a three-step process; the critical step is the oxidation of SO_2 to SO_3 (Example 24.8). Phosphoric acid forms three different kinds of salts, containing the $H_2PO_4^-$ ion (acidic), the HPO_4^{2-} ion (slightly basic), or the PO_4^{3-} ion (more strongly basic; Example 24.10). The acid itself can be made from calcium phosphate with sulfuric acid.

KEY WORDS

acid anhydride	Frasch process	reducing agent
allotropy	Lewis acid, base	resonance form
bond angle	Lewis structure	standard cell voltage
Brönsted acid, base	oxidizing agent	standard oxidation voltage
catalyst	paramagnetic	standard reduction voltage
contact process	precipitating agent	triprotic acid
diprotic acid	principal species	valence electron
disproportionation		

CHAPTER 25

Nuclei with an even number of protons and/or neutrons tend to be more stable than those with an odd number of nuclear particles (Example 25.1). Certain numbers of neutrons or protons (2, 8, 20, 28, 50, 82, and 126) seem to be particularly stable. Unstable radioactive nuclei decay by emitting

— alpha particles ($^4_2$He nuclei). Emission of an alpha particle decreases the atomic number by two units and the mass number by four units.

— beta particles (electrons). Emission of a beta particle increases the atomic number by one unit, leaving the mass number unchanged.

— gamma radiation.

Nuclear decay can be represented by a nuclear equation (Example 25.2). Stable nuclei can be made radioactive by bombardment with neutrons or high-energy, positively charged particles. The biological effects of radiation are most often expressed in rems (Tables 25.1, 25.3).

Radioactive decay follows a first-order rate law (Example 25.4). Frequently the rate is expressed in curies; a curie corresponds to 3.700×10^{10} atomic disintegrations per second (Example 25.3). Measurements of decay rate can be used to estimate the age of rocks or dead organic matter (Example 25.5).

Changes in mass and energy associated with a nuclear reaction can be calculated using the Einstein equation (Example 25.6). The mass defect of a nucleus is the difference between its mass (g/mol) and that of the corresponding number of protons and neutrons. The corresponding energy (kJ/mol) is referred to as the binding energy (Example 25.7).

In nuclear fission, energy is released when a heavy nucleus splits into lighter ones. A critical mass of fissionable material is required for a self-sustaining chain reaction. Such reactions in nuclear reactors evolve heat to produce steam for generating electricity (Fig. 25.7). Nuclear fusion releases energy when light nuclei combine to form heavier nuclei (Example 25.8). Current research efforts to achieve controlled fusion use magnetic containment (Fig. 25.8) or laser bombardment (Fig. 25.9).

IMPORTANT EQUATIONS

Rate of decay $\ln X_0/X = kt$; $k = \text{rate}/N_t$

Mass-energy $\Delta E = 9.00 \times 10^{10} \dfrac{\text{kJ}}{\text{g}} \times \Delta m$

KEY WORDS

activity	first-order reaction	natural radioactivity
alpha radiation	fission	neutron activation analysis
belt of stability	fusion	nuclear binding energy
beta radiation	gamma radiation	nuclear stability
bombardment reaction	half-life	nucleon
carbon-14 dating	induced radioactivity	parent nucleus
curie	isotope	positron
daughter nucleus	K-electron capture	rad
decay constant	magic numbers	rate of decay
Einstein equation	mass defect	rem

CHAPTER 26

The simplest type of organic compound contains only carbon and hydrogen atoms. Among the different kinds of hydrocarbons are alkanes, alkenes, alkynes, and aromatics. These compounds are named in a systematic, consistent way (Examples 26.2, 26.3).

Organic compounds may contain halogen, oxygen, and nitrogen atoms in addition to carbon and hydrogen. In such compounds, an atom or small group of atoms serves as a functional group (Table 26.2). Functional groups discussed here include those found in alkyl halides (Example 26.4), alcohols, carboxylic acids and esters (Example 26.5), and amines (Example 26.6). Compounds containing a given functional group show similar chemical properties.

Isomerism is common among organic molecules. Isomers have the same molecular formula but differ in their structures and properties. Three types of isomerism were considered in this chapter:

TYPE	DIFFERENCE BETWEEN ISOMERS	EXAMPLE
Structural	bonding pattern of atoms	26.1, 26.7
Geometric	distance between groups	26.8
Optical	molecules not superimposable	26.9

Organic reactions can be classified into various types. Among these are

— *addition,* in which a small molecule adds across a multiple bond (Example 26.10).

— *elimination* of groups from adjacent carbon atoms to form a multiple bond.

— *condensation,* in which two molecules combine by eliminating a small molecule such as H_2O.

— *substitution* of one group in a molecule by another (Example 26.11).

KEY WORDS

addition reaction	chlorination	optical isomerism
alcohol	*cis* isomer	polar molecule
alkane	condensation reaction	primary amine
alkene	condensed ring structure	saturated hydrocarbon
alkyl halide	elimination reaction	secondary amine
alkyne	enantiomer	straight chain
amine	ester	structural formula
arene	functional group	structural isomer
aromatic	geometric isomerism	substitution reaction
hydrocarbons	hydrocarbon	tertiary amine
branched chain	isomerism	*trans* isomer
carboxylic acid	multiple bond	unsaturated hydrocarbon
chiral center	nitration	

CHAPTER 27

An organic polymer contains monomer units joined in an extended chain. Addition polymers are made by direct addition of monomer molecules containing double bonds (Table 27.1 and Example 27.1). A condensation polymer is made by splitting out a small molecule such as H_2O between two monomer molecules. Synthetic condensation polymers may be polyesters (Example 27.2) or polyamides (Example 27.3).

Natural polymers include

— *carbohydrates,* in which the monomer unit is a simple sugar such as α-glucose (Examples 27.4, 27.5).

— *proteins,* which can be broken down to, or synthesized from, α-amino acids (Examples 27.7, 27.8). Amino acids themselves commonly exist in water solution as zwitterions, cations, or anions, depending upon pH (Example 27.6).

— *nucleic acids* (DNA, RNA). These species give three products upon hydrolysis: phosphoric acid, a five-carbon sugar (deoxyribonucleic acid, ribonucleic acid), and an amine containing a 5- or 6-membered ring (Table 27.3 and Figure 27.13). The nucleic acids are responsible for the synthesis of proteins (Example 27.9).

KEY WORDS

acid dissociation constant

amino acid

axial

Brönsted acid, base

carbohydrate

cellulose

deoxyribose

dimer

dipeptide

disaccharide

DNA

enantiomer

equatorial

free radical

functional group

genetic code

glycogen

head-to-head, head-to-tail polymer

isoelectric point

monomer

monosaccharide

mRNA

nucleic acid

peptide linkage

polyamide

polyester

polyethylene

polymer

polypeptide

polysaccharide

polyvinyl chloride

primary structure

protein

purine

pyrimidine

random polymer

ribose

RNA

secondary structure

starch

sugar

synthetic addition polymer

tertiary structure

zwitterion

CREDITS

INDEX/GLOSSARY

Note: Italic page numbers indicate figures; t indicates tables; q indicates elements from problems within the question section at the end of each chapter.

TABLE OF ATOMIC MASSES* (Based on Carbon-12)

	Symbol	Atomic No.	Atomic Mass		Symbol	Atomic No.	Atomic Mass
Actinium	Ac	89	227.0278	Mercury	Hg	80	200.59
Aluminum	Al	13	26.981539	Molybdenum	Mo	42	95.94
Americium	Am	95	[243]†	Neodymium	Nd	60	144.24
Antimony	Sb	51	121.75	Neon	Ne	10	20.1797
Argon	Ar	18	39.948	Neptunium	Np	93	237.0482
Arsenic	As	33	74.92159	Nickel	Ni	28	58.69
Astatine	At	85	[210]	Niobium	Nb	41	92.90638
Barium	Ba	56	137.327	Nitrogen	N	7	14.00674
Berkelium	Bk	97	[247]	Nobelium	No	102	[259]
Beryllium	Be	4	9.012182	Osmium	Os	76	190.2
Bismuth	Bi	83	208.98037	Oxygen	O	8	15.9994
Boron	B	5	10.811	Palladium	Pd	46	106.42
Bromine	Br	35	79.904	Phosphorus	P	15	30.973762
Cadmium	Cd	48	112.411	Platinum	Pt	78	195.08
Calcium	Ca	20	40.078	Plutonium	Pu	94	[244]
Californium	Cf	98	[251]	Polonium	Po	84	[209]
Carbon	C	6	12.011	Potassium	K	19	39.0983
Cerium	Ce	58	140.115	Praseodymium	Pr	59	140.90765
Cesium	Cs	55	132.90543	Promethium	Pm	61	[145]
Chlorine	Cl	17	35.4527	Protactinium	Pa	91	231.03588
Chromium	Cr	24	51.9961	Radium	Ra	88	226.0254
Cobalt	Co	27	58.93320	Radon	Rn	86	[222]
Copper	Cu	29	64.546	Rhenium	Re	75	186.207
Curium	Cm	96	[247]	Rhodium	Rh	45	102.90550
Dysprosium	Dy	66	162.50	Rubidium	Rb	37	85.4678
Einsteinium	Es	99	[252]	Ruthenium	Ru	44	101.07
Erbium	Er	68	167.26	Samarium	Sm	62	150.36
Europium	Eu	63	151.965	Scandium	Sc	21	44.9559
Fermium	Fm	100	[257]	Selenium	Se	34	78.96
Fluorine	F	9	18.998403	Silicon	Si	14	28.0855
Francium	Fr	87	[223]	Silver	Ag	47	107.868
Gadolinum	Gd	64	157.25	Sodium	Na	11	22.989768
Gallium	Ga	31	69.723	Strontium	Sr	38	87.62
Germanium	Ge	32	72.61	Sulfur	S	16	32.066
Gold	Au	79	196.96654	Tantalum	Ta	73	180.9479
Hafnium	Hf	72	178.49	Technetium	Tc	43	[98]
Helium	He	2	4.002602	Tellurium	Te	52	127.60
Holmium	Ho	67	164.93032	Terbium	Tb	65	158.92534
Hydrogen	H	1	1.00794	Thallium	Tl	81	204.3833
Indium	In	49	114.82	Thorium	Th	90	232.0381
Iodine	I	53	126.90447	Thulium	Tm	69	168.9342
Iridium	Ir	77	192.22	Tin	Sn	50	118.710
Iron	Fe	26	55.847	Titanium	Ti	22	47.88
Krypton	Kr	36	83.80	Tungsten	W	74	185.85
Lanthanum	La	57	138.9055	Uranium	U	92	238.0289
Lawrencium	Lr	103	[260]	Vanadium	V	23	50.9415
Lead	Pb	82	207.2	Xenon	Xe	54	131.29
Lithium	Li	3	6.941	Ytterbium	Yb	270	173.04
Lutetium	Lu	71	174.967	Yttrium	Y	39	88.90585
Magnesium	Mg	12	24.3050	Zinc	Zn	30	65.39
Manganese	Mn	25	54.93805	Zirconium	Zr	40	91.224
Mendelevium	Md	101	[258]				

*Atomic masses given here are 1985 IUPAC values.

†A value given in brackets denotes the mass number of the longest-lived or best-known isotope.